The Revolutions of Wisdom

STUDIES IN THE CLAIMS AND PRACTICE OF ANCIENT GREEK SCIENCE

G. E. R. Lloyd

University of California Press

BERKELEY · LOS ANGELES · LONDON

University of California Press
Berkeley and Los Angeles, California

University of California Press, Ltd.
London, England

© 1987 by
The Regents of the University of California
First Paperback printing 1989

Library of Congress Cataloging-in-Publication Data

Lloyd, G. E. R. (Geoffrey Ernest Richard), 1933–
    The revolutions of wisdom.

    (Sather classical lectures ; v. 52)
    Bibliography: p.
    Includes index.
    1. Science—Greece—History.   2. Science, Ancient.
3. Greece—Civilization—To 146 B.C.   I. Title.   II. Series.
Q127.G7L595   1987   509'.38   86-16055
ISBN 0-520-06742-8

Printed in the United States of America

1   2   3   4   5   6   7   8   9

Contents

Preface   vii

Note on Editions and Abbreviations   xi

CHAPTER 1   The Displacements of Mythology   1

CHAPTER 2   Tradition and Innovation, Text and Context 50

CHAPTER 3   Dogmatism and Uncertainty   109

CHAPTER 4   Metaphor and the Language of Science   172

CHAPTER 5   Measurement and Mystification   215

CHAPTER 6   Idealisations and Elisions   285

Bibliography   337

General Index   421

Index Locorum   431

bly kept the supporting documentation and illustration to a minimum. To meet the demands of supplying such documentation involved considerable expansion of the text of the lectures, although the overall strategy of this book still corresponds closely to that of the lectures as delivered. Much of the documentation is confined to the extensive notes, printed here at the foot of the page so that, while the argument in the text can be read independently, the reader can see at a glance where there are supplementary points and questions to be pursued. As I have attempted to bring to bear ideas that derive from my own reading in many different fields—the philosophy and sociology of science, social anthropology, Oriental studies, as well as the scholarly literature on Greco-Roman antiquity—I have provided the work with a full, though still far from exhaustive, bibliography. I have done so not just from the obligation to acknowledge my sources, but also in the hope that those from different disciplines who may be interested in the problems raised here may have an introduction to some of the relevant literature from other cognate fields.

Many friends and colleagues have been kind enough to read and comment on drafts of this work. I owe special debts of thanks first to Anthony Bulloch and Linda Coleman, who gave me their detailed and most perceptive reactions to drafts of the lectures, and also to Giovanni Ferrari, whose constructive and critical reading of the typescript of the whole book has saved me from many mistakes and enabled me to make many improvements. I owe much to Andrew Barker for his advice on music theory, to Simon Goldhill and Mary Hesse on metaphor, to John Ray on Egyptological issues, to Andrew Stewart on art historical problems, and to Jack Goody, Caroline Humphrey, and Alan Macfarlane especially on anthropological questions. Many others too have helped with comments on particular points or on the arguments of whole sections: Myles Burnyeat, Richard Gordon, Peter Khoroche, Wilbur Knorr, Martha Nussbaum, Thomas Rosenmeyer, Malcolm Schofield, David Sedley, Richard Sorabji, Gregory Vlastos. On many different occasions I have had the benefit of questions and comments from audiences at lectures and seminars based on this material. My graduate seminar in ancient philosophy in Cambridge in 1983 proved one of my most consistently tough and creative audiences, and I learnt much also in Cambridge at the History and Philosophy of Science and Social Anthropology seminars, in London at University College and at Chelsea College, at the Queen's University at Belfast, and in North America at the University of British Columbia, at Stanford, and at St. Mary's College of California. Most notably the comments from the audiences at my lectures and graduate seminars at Berkeley itself throughout the spring semester of 1984 stimulated me to clarify, justify, or modify my positions.

The hospitality accorded to Sather lecturers is legendary, and in reality the kindness of Leslie Threatte, and of all his colleagues in the Department of Classics, to myself, my wife and my family was indeed overwhelming. We were entertained, guided, instructed, and amused, with generosity, warmth, tact, and imagination, introduced by turns to Californian Nature and to Californian Culture and enchanted by both. No expressions of gratitude can begin to be adequate: our thanks, nevertheless, to all our hosts, and especially to Bill and Deidre Anderson, Esperance and Jock Anderson, Alan Code, Alan and Carolyn Dundes, Crawford Greenwalt, Jr., Mark Griffith, Eric Gruen, John Heilbron, Sylvia Lark, Kay and Tony Long, Don Mastronarde, Charles Murgia, Michael Nagler, Lila and Tom Rosenmeyer, Allan and Annie Silverman, Connie and Ron Stroud, Leslie Threatte.

Finally I wish to express my thanks to the officers of the University of California Press, and especially to Doris Kretschmer and to Mary Lamprech, for their exemplary efficiency in overseeing all the stages of the production of this book.

G. E. R. L.
May 1986

Editions and Abbreviations Used

Except where otherwise stated, the fragments of the pre-Socratic philosophers are quoted according to the edition of Diels, revised by Kranz, *Die Fragmente der Vorsokratiker* (6th ed., 1951–52) (referred to as D.-K.); the works of Plato according to Burnet's Oxford text; the treatises of Aristotle according to Bekker's Berlin edition; and the fragments of Aristotle according to the numeration in W. D. Ross, *Fragmenta selecta* (Oxford, 1955). Greek medical texts are cited, for preference, according to the *Corpus medicorum Graecorum* (*CMG*) editions. For those Hippocratic treatises not edited in *CMG*, I use E. Littré, *Oeuvres complètes d'Hippocrate*, 10 vols. (Paris, 1839–61) (L), except that for *On Sevens* I use the edition of W. H. Roscher (Paderborn, 1913). For those works of Rufus not in *CMG*, I use C. V. Daremberg and C. E. Ruelle, *Oeuvres de Rufus d'Ephèse* (Paris, 1879). Galen is cited according to *CMG* and Teubner editions (where these exist), but the reference is also given to the edition of C. G. Kühn (Leipzig, 1821–33) (K), which is also used for works neither in *CMG* nor Teubner; the later books of *On Anatomical Procedures,* extant only in an Arabic version, are cited according to the translation of W. L. H. Duckworth (D) (ed. M. C. Lyons and B. Towers, Cambridge, 1962).

Euclid's *Elements* are cited according to the edition of Heiberg, revised by Stamatis, 4 vols. (Leipzig, 1969–73) (HS), and the works of Archimedes according to Heiberg, revised by Stamatis, 2 vols. (Leipzig, 1972) (HS), with the third volume, containing Eutocius' commen-

tary. Ptolemy's *Syntaxis* is cited according to the two-volume edition of
Heiberg (Leipzig, 1898, 1903) (cited as H 1 and H 2); his *Tetrabiblos*
according to F. Boll and A. Boer (Leipzig, 1942); his *Optics* according
to A. Lejeune (Louvain, 1956) (L); and his *Harmonics* according to
I. Düring (Göteborg, 1930) (D.). Porphyry's *Commentary on Ptolemy's
Harmonics* is cited according to I. Düring (Göteborg, 1932) (D.).

Other Greek authors are cited according to the editions named in
the *Greek-English Lexicon* of H. G. Liddell and R. Scott, revised by
H. S. Jones, with Supplement (1968) (LSJ), though, where relevant,
references are also provided to more recent editions, and Latin authors
are cited according to the editions named in the new *Oxford Latin
Dictionary* (*OLD*), supplemented, where necessary, from Lewis and
Short.

Abbreviations are those in LSJ and *OLD*, supplemented, where
necessary, from Lewis and Short and with the following abbreviations
of works of Galen: *AA* (*De anatomicis administrationibus*), *PHP* (*De
placitis Hippocratis et Platonis*).

Full details of modern works referred to will be found in the bibliography on pp. 337 ff. They are cited in my text and notes by author's
name and publication date or dates. A double date is used to distinguish, where this has seemed relevant, the original publication from
the revised or reprinted version used. Such works are listed in the bibliography by the first date but cited according to the second. Thus Kuhn
1961/1977 refers to the revised 1977 version of an article originally
published in 1961; Scholz 1930/1975 refers to the 1975 translation of
an article originally published in 1930.

The translations of Greek and Latin texts that I offer are in general
my own but I have made extensive use of existing translations and in
particular of the following: Chadwick and Mann 1978, W. H. S. Jones
1923–31, and Lonie 1981a for Hippocratic works; Dengler 1927,
Hort 1916, Ross and Fobes 1929 for Theophrastus; Macran 1902 for
Aristoxenus; Heath 1913 for Aristarchus; Spencer 1935–38 for Celsus; Temkin 1956 for Soranus; Toomer 1984 for Ptolemy; Duckworth
1962, May 1968, and Singer 1956 for Galen.

CHAPTER ONE

The Displacements of Mythology

Jane Harrison thrilled to the dark shapes she thought she could discern behind the bright splendours of the masterpieces of Greek literature.[1] E. R. Dodds, in his preeminently distinguished contribution to the Sather series, began from the puzzlement that the Greeks had been thought to lack something of "the awareness of mystery" and "the ability to penetrate to the deeper, less conscious levels of human experience."[2] The irrational then and subsequently has been much pursued—in classical studies, in social anthropology, in philosophy, and in psychology—but has proved, predictably, an elusive quarry, escaping clear characterisation, let alone elucidation.

I shall certainly not attempt, in this set of studies, to reopen the whole of this vast and ill-defined dossier. My aim is a more limited one, with a narrower focus, though it is still perhaps ambitious enough,

---

1. J. E. Harrison 1925, pp. 86f: "I mention these ritual dances, this ritual drama, this bridge between art and life, because it is things like these that I was all my life blindly seeking. A thing has little charm for me unless it has on it the patina of age. Great things in literature, Greek plays for example, I most enjoy when behind their bright splendours I see moving darker and older shapes. That must be my *apologia pro vita mea*."
2. On p. 1 of his 1951 Dodds wrote: "To a generation whose sensibilities have been trained on African and Aztec art, and on the work of such men as Modigliani and Henry Moore, the art of the Greeks, and Greek culture in general, is apt to appear lacking in the awareness of mystery and in the ability to penetrate to the deeper, less conscious levels of human experience."

since it concerns the invention of the category—the ancient Greek category—of the rational. Acknowledging, but leaving to one side, much of the material that Dodds and others collected to illustrate the irrational in Greek culture at every period, I wish to focus attention on one of the citadels of presumed Greek rationality (presumed by many of them, as well as by some of us), namely, what they called the "inquiry concerning nature." My plan, broadly, is to investigate where, or if, it may be said to break new ground in the understanding of the world, and where, on the contrary, what it shares with its antecedents is more impressive than the points at which it diverges from them. The character of the "science" on offer in the ancient world is one of our targets, then, though less with a view to matching their science against ours (to vindicate or to undermine the claim that they were *doing science*) than to explore the complexities of ancient disputes and confrontations. We shall try to make some sense of some highly perplexing and challenging phenomena, though the perplexities and challenges are ones that the anthropologists, used to dealing with problems concerning the nature of "primitive thought," probably appreciate more fully than the majority of classicists.

We may take heart for the assault on Greek science from the realisation that scientific thought as a whole and, especially, the nature of scientific inventiveness have latterly come increasingly to be recognised as less translucent, more complex, puzzling, and problematic, than many of Dodds' generation and before took them to be.[3] But while that realisation makes our inquiry easier in one respect, in that it releases us from one set of preconceptions concerning the purity of the scientific enterprise, in another it makes it harder, since the very criteria of science are now more highly contested than ever. My chief concern in

---

3. Among the fundamental contributions to this debate have been Popper 1935/1968, 1963, Quine 1953/1961, 1960, 1969, Kuhn 1962/1970, 1977, 1983, Feyerabend 1962, 1975, 1978, 1981a, 1981b, Habermas 1968/1978, 1971/1974, Hesse 1974, 1980, S. B. Barnes 1974, 1977, Putnam 1975a, 1975b, 1981, 1983, Lakatos 1976, 1978a, 1978b, Bloor 1976, Laudan 1977, Holton 1978, Van Fraassen 1980a, Newton-Smith 1981, Hollis and Lukes, edd., 1982, Hacking 1983.

what follows is not directly with those current controversies in the philosophy of science, though I shall have occasion to join battle where they impinge on the assessment of ancient investigations. Rather, my main problem is the characterisation of those ancient investigations themselves, particularly in relation to *their* background. For while those who engage in them often make extravagant claims on their behalf (as also do some modern commentators), just how far such claims can be sustained and just how far the principles and ideals they stated were implemented in practice will be among our major preoccupations. To put our problem in its most general terms: *Was* there a revolution of wisdom with regard to the understanding of nature? What *kind* of revolution was there?

In the chapters that follow I shall address some very general questions concerning the nature of Greek inquiry and speculation about the physical world, where I have chosen to concentrate not on such traditional topics as the experimental method but, rather, on certain characteristics that relate to, and reveal, the ancient investigators' own aims and ambitions, even their self-image, their theory of what they were doing and their actual practice and the matches and mismatches between the two. We shall consider the tension between tradition and authority on the one hand, and innovativeness on the other, broaching here issues in the wider social background to the intellectual changes with which we are concerned. We shall study the aggressions and bluff of dogmatism, but also—to set against that—the scrupulous avoidance of the dogmatic and the willingness to acknowledge failures and ignorance, and then again the turning of the anti-dogmatic into a conventional stance or even pose. We shall discuss the development, indeed the invention, of the category of the metaphorical, and again the tension between the desire to exclude this from, and its continued use in, the inquiry concerning nature. We shall examine the extent to which Greek science remained purely qualitative in character—where we shall discuss both the use of measurement and its abuse, that is, the mystifications involved in some appeals to it and to the quantitative. Finally we shall tackle the use of idealisations and simplifications, and again their abuse in the discounting or eliding of parts of what is there to be explained.

In this opening chapter I want to take certain concrete topics which will provide test-cases to illuminate the nature and the strength of the challenge, from the side of *logos,* to some traditional attitudes and patterns of belief. If we consider some phenomena that lie at or near the centre of most naive or sophisticated configurations of the irrational, we may be able to see to what extent the inquiry concerning nature offered an alternative to what had long been accepted. It is not that that inquiry was necessarily obliged to present any such alternative in relation to those phenomena; it may not even have been well advised to try to do so. Yet in the controversy between would-be science and the irrational, it is important to look at certain of the topics that are, on the face of it, among the *least* favourable to the rationalist takeover, not just at those areas where the triumphs of rationalism may seem predictable enough. It is important to do so to help to determine the character and the limitations of the wisdom that came to be offered from the side of *logos.*

Many of the phenomena discussed in *The Greeks and the Irrational* look promising from the point of view I have specified, but among those that seem particularly so—in that they appear to offer some of the greatest problems for, or the maximum resistance to, any scientific takeover—are death, disease, madness, dreams, divination, and fate. These were the province of myth, religion, and ritual long before science and natural philosophy, and long *after* their first hesitant appearance in Greek thought. It was mainly through myth, in belief, and through ritual, in practice, that the Greeks, like others, responded to the facts of death and disease, for example—and it remained so, even after the inquiry concerning nature was some kind of going concern. yet to say the Greeks "responded to" natural facts through myth is not quite accurate. For myth is not, and does not aim to be, explicitly systematic and coherent.[4] I am not denying, of course, the findings of

---

4. The point stands even though, to be sure, current theories on the interpretation of mythology can still hardly be said to provide a satisfactory framework for its understanding (see, for example, Lévi-Strauss 1958/1968, 1971/1981, Leach, ed., 1967, P. S. Cohen 1969, Smith and Sperber 1971). Thus despite, for instance, the claims of Van Riet 1960, p. 63, the sense in which sys-

structuralism, which has decoded remarkably coherent messages in groups of myths, even whole mythologies. But those messages, as structuralism itself insists, remain implicit, below the surface. On the surface, the intelligibility provided by myth is metaphorical, both in the sense that it is of the nature of metaphor and in the sense that it is a qualified intelligibility.[5] Myth does not, in any case, normally attempt to give the kind of direct answers to questions that ordinary practical experience is used to and demands. To be effective, myth must work below the surface, while on the surface the appearance is often of inconsistencies, of a lack of coherent unity. The encoded messages are vulnerable to question, to challenge, and like books in Plato, they cannot answer back.[6] Equally, ritual comforts, in part, because in the already given and socially sustained patterns of behaviour it is simply the right thing to do. But again the vulnerability to the question "why?" is evident—as is shown by the dismay registered by an earlier tradition of ethnography when that question, pressed in the field, led with some

---

tems of myth can be said to constitute some kind of protometaphysics is only a very attenuated one and this may obscure more than it illuminates: for the essential point of difference is that metaphysics is explicit, even if the point of similarity is that in the most general sense a "world-view" may be conveyed by myths or otherwise by implicit beliefs and attitudes as much as by self-conscious philosophical statements. The flexibility of myth, stressed by T. S. Turner 1977, 1980, for instance, is both its strength and its weakness. P. Smith's statement, 1973, p. 77, that "[myths] taken as a whole, aim not so much to define the real as to speculate upon its latent potentialities; not so much to think something through as to walk the boundaries of the thinkable," suggestive as it is, is made, rather, from outside the boundaries of myth itself. His equally suggestive remark, p. 86, that "when *logos* recognizes *mythos* as such, and so deprives it of its efficacity, at the same time it takes over its place and becomes a new working myth," and his tentative "it may be that if one is to do full justice to science, one must acknowledge the portion of myth it has in it" (cf. also Derrida 1967/1976, 1972/1981, 1972/1982) may be said to encapsulate the *problems* explored here, the sense in which what replaces myth, in ancient Greece, was or was not *just* more myth, and the difference that that recognition made.

5. See further below, Chap. 4, pp. 176ff. Chap. 6, pp. 285ff.

6. The point, in Plato, applies precisely to the written word as opposed to the spoken exchange; see *Protagoras (Prt.)* 329a, *Seventh Letter (Ep.* 7) 343a, as well as famous texts in the *Phaedrus (Phdr.),* 274bff., especially 275d–e,

inevitability to the—to *logos* unacceptable[7]—answers that "we have always done so," "this is laid down," "this is the way our forefathers did it."[8]

But if myth and ritual provide some imperfect means of responding, in various ways, to various manifestations of the apparently intractable or refractory in experience, what did the "response" of the new investigations into nature amount to? What business had they, in any event—to pick up my earlier question—with such phenomena as death, disease, and the others I listed, or how far did they abandon them; or should they have abandoned them, renouncing any claim to be able to provide alternative, and no doubt also imperfect, resources for a response? To be sure, that question, like my earlier questions, has to be unpacked even to begin to attempt an answer. What, in particular, were the problems to which solutions were required? What kinds of explanation were needed for what kinds of explananda? Are we dealing—to start with—with puzzles concerning the that (or what), the how, or the why?[9]

Death

The "that" of death (for instance), the fact that men die, cannot be treated as an unproblematic cultural universal.[10] We have only to reflect on beliefs in various modes of symbolic death to see that here, as so often elsewhere, there may be wide cultural divergences and substan-

---

277d–e (discussed from different points of view by, among others, Havelock 1963, cf. 1982, and Derrida 1967/1976, 1967/1978). I shall return, in Chap. 2, to some aspects of the issue of the labile or unstable nature of the oral tradition.

7. Cf. further below, Chap. 2, on the appeal to tradition as such as justification for a belief or practice.

8. For some comments on this theme, see Horton 1982, pp. 239ff.; Sperber 1985, pp. 59f.

9. Not that these questions are themselves unambiguous, a point already made by Aristotle in respect of the διὰ τί question, for example at *Physics* (*Ph.*) 194b17ff., and reiterated by many others for the "why?" question in English, as for example by Waismann 1965, p. 41; Bromberger 1966; Van Fraassen 1980a, pp. 126ff., 141ff.

10. See, for example, V. W. Turner 1964, p. 231.

tial difficulties in matching actor and observer categories.[11] *A fortiori* what counts as disease or illness and what as mental illness or madness vary strikingly between cultures. Yet so far as ancient Greek views of death go (the subject of another distinguished contribution to the Sather lectures),[12] a resolute acceptance *that* men die is strongly marked in Homer,[13] even if there is afterwards a shadowy existence in Hades, and even if some exceptional individuals escape that fate and achieve semi-divine status as heroes.[14] But acceptance of the brute fact of death gives no consolation for, indeed may even heighten, personal bereavement. That acceptance does not qualify, rather it lends resonance to, Achilles' anguished cry that he would rather be a bondsman on earth than rule among the dead.[15]

---

11. Furthermore, the point that death may be viewed very differently as it affects the young and old is emphasised, for example, by Cassin 1981, p. 321.

12. Vermeule 1979. Two recent collections of essays contain important discussions of Greek and Roman attitudes along with comparative studies of other cultures: Humphreys and King 1981, and Gnoli and Vernant 1982. Sourvinou-Inwood 1983 explores, in particular, changes in attitudes that take place at different periods in antiquity: cf. also Garzón Díaz 1981, Wankel 1983.

13. See, for example, *Iliad* (*Il.*) 12.322ff., 18.115ff., 21.106ff., 24.525ff., and from a god's perspective, 21.462ff. The point has often been brought out forcefully, as by Rohde 1925, chap. 4; Guthrie 1950, pp. 305f; Sourvinou-Inwood 1983, pp. 34f. The centrality of the topic of death in the Homeric poems has recently also been stressed by Segal 1978.

14. See, for example, *Odyssey* (*Od.*) 4.561ff. (Menelaus), 11.300ff. (Castor and Polydeuces), 11.601ff. (Heracles). Some individuals are, of course, subject to exemplary punishment: see especially *Od.* 11.576–600 (Tityus, Tantalus, Sisyphus).

15. *Od.* 11.488ff. How far, on this or any other of the issues germane to our discussion, Homer should be taken to represent common attitudes is, to be sure, highly problematic, but the influence and prestige of the Homeric poems in the fifth and fourth centuries insure at least their relevance to our understanding of the background to natural philosophical speculation. Aspects of the themes of the transience of human life, human helplessness, and the preponderance of evil, expressed, for example, at *Il.* 6.145ff., 21.464ff., 24.527ff., *Od.* 18.130ff., and in Hesiod, e.g., *Works and Days* (*Opera, Op.*) 101ff., are reiterated in early lyric and in tragedy, e.g., Solon 1.35ff. (Diehl), Mimnermus 1 and 2 (Diehl): the theme that it is better not to have been born at all, found, for example, in Theognis 425ff., Bacchylides 5.160ff. (Snell-Maehler) and Sophocles *Oedipus Coloneus* 1224ff., reappears in a particularly emphatic statement in Aristotle's lost dialogue the *Eudemus* fr. 6 (Ross).

On the how of death, Greek physics eventually had, as usual, a multitude of theories to offer. Yet they provided little understanding and no reassurance. There was Aristotle's suggestion, for example, that death is the extinction of the vital heat, which may take place, he believes, either from cold or from an excess of heat.[16] That theories that appeal just to the hot, the cold, and the like are quite inadequate had already been argued in the Hippocratic treatise *On Ancient Medicine*. There the writer criticises those who use such newfangled "hypotheses" in part on the grounds that to do so is to narrow down the causal principles of death and disease.[17] What is needed, he believes, is a more complex account, taking into consideration all the manifold powers in the body and their combinations.[18] Again even Plato had a suggestion to make on the subject in the *Timaeus,* namely, that the material cause of death is a deterioration in the structure of the atomic triangles that constitute the physical elements of which the body (and everything else) is made.[19]

To be sure, each of those, and many other, hardheaded naturalistic explanations entailed the denial of the *literal* truth of Hesiod's mythology of death as presented in the myth of the metals in the *Works and Days*, with its complex counterpoint on the way each race meets its end.[20] Those of the Golden Age are as if overcome by sleep; those of the Age of Silver, who remain children for a hundred years, are "hidden" by Zeus and become the blessed ones of the underworld; the Bronze Age race destroy themselves; some from the Age of Heroes go

---

16. See, for example, *De juventute* (*Juv.*) 469b18ff., 21ff., *De respiratione* (*Resp.*) 478b22ff., 479a32ff. Contrast *Ethica Nicomachea* (*EN*) 1115a26, where Aristotle recognises that death is the most fearful thing there is.

17. περὶ ἀρχαίης ἰητρικῆς (*De vetere medicina, VM*) 1, *Corpus medicorum graecorum* (*CMG*) 1.1.36.2ff., 4ff., ἐς βραχὺ ἄγοντες τὴν ἀρχὴν τῆς αἰτίης τοῖσιν ἀνθρώποισι νούσων τε καὶ θανάτου καὶ πᾶσι τὴν αὐτὴν ἐν ᾗ δύο ὑποθέμενοι. See further below, p. 15, and Chap. 2 at nn. 63f.

18. See, for example, *VM* 14, *CMG* 1.1.45.26ff., 15. *CMG* 1.1.46.27ff., 22, *CMG* 1.1.53.1ff.

19. *Timaeus* (*Ti.*) 81b–e.

20. *Op.* 109ff., on which see, for example, Kirk 1970, pp. 233ff., J.-P. Vernant 1983, pp. 3ff., 33ff.

to the Islands of the Blest; and Zeus will destroy the last Age of Iron when men are born grey-haired.[21] Again, theories of the physics of death were not compatible with a *literal* reading of Plato's own myth, in the *Politicus,* of the age of Cronos—the anti-cosmos when time flows in reverse[22]—while they had no comment to make on the values implicit in the ideology of the "beautiful death"—the death while young, in battle, securing lasting fame.[23] No prosaic naturalistic account of the how of death had, of course, anything to offer on the why, nor on how we as mortals should live with our mortality. They offered nothing to replace the lesson obliquely taught by Hesiod's myth: we must realise that, since we are born in the Age of Iron, there is an imperative upon us to accept death, along with toil and pain.

Such comfort as was on offer from the philosophers in the classical period, at least,[24] came principally from a very different quarter, from the essentially religious belief in the immortality of the soul found first in the Pythagorean tradition, then in Plato and others.[25] Yet that was certainly not *science* replacing earlier attitudes or patterns of belief.

---

21. See *Op.* 116, 137ff., 152ff., 170ff., 180f.

22. *Politicus* (*Plt.*) 268e ff., especially 270c–e referring to periodic destructions of the human race and the reversal of aging, with the old becoming young. Compare the discussions in Herter 1958, Rosen 1979–80, and especially Vidal-Naquet 1975/1986.

23. See especially Loraux 1981/1986, pp. 98ff., 1982, pp. 27ff.; J.-P. Vernant 1982, pp. 45ff.; cf. Dover 1974, p. 229; Sourvinou-Inwood 1983, p. 43.

24. There were, however, also philosophical arguments designed to establish that "death is nothing to us." That death is not to be feared is a belief already put into the mouth of Socrates by Plato: for example, *Apology* (*Ap.*) 28b ff., 39a, 40c ff., *Gorgias* (*Grg.*) 522e (contrast Aristotle *EN* 1115a26, cited above, n. 16). But it is in the Hellenistic period, especially, that attempts were made to demonstrate that death is "nothing to us": see, for example, Epicurus Κύριαι δόξαι (*Sententiae, Sent.*) 2 and 11, *Letter to Herodotus* (*Ep.Hdt.*) 10.63ff., *Letter to Menoeceus* (*Ep.Men.*) 10.124ff., Lucretius 3.417ff., 830ff. For the Stoics, see Diogenes Laertius (D.L.) 7.102, cf. 106; Stobaeus *Eclogae* (*Ecl.*) 2.7.5a (2.57.18ff. Wachsmuth), cf. Lanza 1980b, Furley 1986.

25. Of course it was not just from among the philosophical writers that comfort of this kind was on offer, but also, as early or earlier, from within the growing and altering religious traditions, notably with the development of mystery religions: see Burkert 1977/1985, chap. 6, pp. 276ff., cf. Nilsson 1957, for Hellenistic continuations.

To see this in the right perspective is more complicated than might appear. First it is as well to stress that not *all* physical ailments and disabilities were deemed by the ancient Greeks—or have been by anyone else—to be the products of divine or demonic forces. Medical anthropologists have, to be sure, only comparatively recently begun to insist that there is much more to the map of most societies' beliefs about physical ailments than the parts that have generally received most attention in the ethnographical accounts, that is, those that relate to the severest diseases and the most dramatic ones, such as epilepsy.[32] For many minor ailments, as it might be the common cold, minor stomach upsets, bruises, or bunions, many societies have no recourse to supernatural explanation. Homer has no occasion to talk about the common cold. But apart from the fact that there are many straightforward accounts of wounds and lesions caused by men in battle[33] there is an important contrast between the plague sent by Apollo in *Iliad* 1[34] and references to diseases that are not directly attributed to a god, such as, for example, the "long disease" contrasted with the arrows of Artemis at *Od.* 11.172.[35]

However, this is not to deny that notions that diseases are often sent by the gods or that diseases are themselves semi-divine creatures stalk-

---

des 2.47 and 53—though such beliefs and expectations persisted in many quarters. In the Hellenistic period a large part of Epicurean natural philosophy was to be devoted to excluding divine or supernatural agencies from the explanation of natural phenomena: cf. further below, n. 163, and Chap. 3, n. 239, in the context of Epicurean appeals to the principle of plural explanations of obscure phenomena.

32. See, for example, G. Lewis 1975, pp. 196ff., 248ff.

33. Those wounds are caused by men, even though the success or failure of a blow may often be ascribed in addition or in part to a god—as with other phenomena that have been discussed under the rubric of "double determination," where both a divine and a human explanation are invoked: see, for example, Dodds 1951, chap. 1.

34. *Il.* 1.43ff. This has, of course, a particularly important role in initiating the action of the epic. But when there is less at stake for the purposes of the narrative, the darts of Apollo or Artemis are still often invoked as causes of death or disease, for example at *Il.* 6.428, 19.59, 24.758f., *Od.* 3.279f., 5.123f., 11.324, 15.478f., 18.202ff., 20.61ff., and cf. also 9.411.

35. Cf. *Od.* 11.200f. See also *Il.* 13.666–70, *Od.* 15.407f.

the belly at the right time, or dysentery. In many cases the crisis was not reached with the appearance of just one of the signs described, but in most cases all were experienced successively and the patients seemed to be in great distress: but all who experienced them recovered. . . . I know of no woman who died in whom one of these signs had appeared properly. For the daughter of Philo, who had a violent nose-bleed, dined rather intemperately on the seventh day: she died.[51]

We are evidently far from the world of Apollo sending the plague or of Hesiod's diseases randomly roaming the earth.[52] Interestingly enough, however, a feature that provides both a link and a contrast with earlier patterns of thought is the residual moralising tone of some Hippocratic comments on the causes or predisposing factors to diseases. *Epidemics* 1.9 gives a list of the types of persons who died in a particular epidemic. They include: "boys, young men, men in the prime of life, those with smooth skins, those of a pallid complexion, those with straight hair, the black-haired, the black-eyed," and so on, but also "those who had been given to reckless and loose living."[53]

---

51. *Epid.* 1.9 (L) 2.656.7–658.6, 9–12: ἐν δὲ τῇ καταστάσει ταύτῃ ἐπὶ σημείων μάλιστα τεσσάρων διεσώζοντο· οἷσι γὰρ ἦν ἢ διὰ ῥινῶν αἱμορραγῆσαι, ἢ κατὰ κύστιν οὖρα πουλλά, καὶ πουλλὴν ὑπόστασιν καὶ καλὴν ἔχοντα ἔλθοι, ἢ κατὰ κοιλίην ταραχώδεα, χολώδεα, ἐπικαίρως, ἢ δυσεντερικοὶ γενοίατο· πολλοῖσι δὲ ξυνέπιπτε μὴ ἐφ᾽ ἑνὸς κρίνεσθαι τῶν ὑπογεγραμμένων σημείων, ἀλλὰ διεξιέναι διὰ πάντων τοῖσι πλείστοισι, καὶ δοκέειν μὲν ἔχειν ὀχληροτέρως· διεσώζοντο δὲ πάντες, οἷσι ταῦτα ξυνεμπίπτοι. . . . καὶ οὐδεμίην οἶδα ἀπολομένην, ᾗσι τουτέων τι καλῶς γένοιτο· Φίλωνος γὰρ τῇ θυγατρὶ ἐκ ῥινῶν λαῦρον ἐρρύη, ἑβδομαίη δὲ ἐοῦσα ἐδείπνησεν ἀκαιροτέρως, ἀπέθανεν.

52. *Op.* 102–4. To claim as Fränkel did (1975, p. 118) that "from Hesiod's realization that diseases and other pains afflict mankind according to their own impulse and nature, a straight line leads to the theory and empirical methods of a Hippocrates" is drastically to underestimate the remaining differences. Even if αὐτόματοι and σιγῇ are taken to imply that these diseases are not sent by the gods (see also R. M. Frazer 1972, but cf. Kudlien 1968b, pp. 315ff.), Hesiod still strongly suggests that they are both frightening and unpredictable. There is no hint here of the possibility of understanding and control through the art of medicine.

53. *Epid.* 1.9 (L) 2.656.2–5: ἐκ δὲ τῶν καμνόντων ἀπέθνησκον μάλιστα μειράκια, νέοι, ἀκμάζοντες, λεῖοι, ὑπολευκοχρῶτες, ἰθύτριχες, μελανότριχες, μελανόφθαλμοι, οἱ εἰκῇ καὶ ἐπὶ τὸ ῥᾴθυμον βεβιωκότες.

food was a starvation diet, for which there was even a technical term, λιμοκτονία.[62] Moreover, as happens so often in Greek medicine, a simple point was subjected to massive theoretical over-elaboration. One of our Hippocratic treatises is entirely devoted to the subject of its title, A Regimen for Health. Although the final chapter ends with the laudable sentiment that "an intelligent man ought to reckon that health is man's most valuable possession and learn how to gain help in illnesses by *his own* judgement,"[63] the work as a whole sets out some very elaborate recommendations about foods, exercises, emetics, and enemas that would have gladdened the heart of any ancient hypochondriac and that also implicitly laid claim to much esoteric learning on the subject.[64]

The topic of physical illnesses offered one of the clearest openings for the rationalist takeover. There are plenty of signs of the *hubris* of Greek rationalism in the Hippocratic treatises, as also of its tendency to run to excess. Yet one of the strengths of the new conceptual framework they present, and one of its originalities, lies in its absolute, un-

---

62. See, for example, περὶ διαίτης ὀξέων (νόθα) (De victus ratione in morbis acutis, spuria, Acut. (Sp.)) 24 (L) 2.508.8, Vict. 3.71 (L) 6.610.16f., cf. Galen (K) 10.264.6, 11.182.13f. Alternatively λιμαγχία is used, e.g., by Rufus, De renum et vesicae affectionibus (Ren. Ves.) 1, CMG 3.1.90.9, cf. Caelius Aurelianus De morbis chronicis (Morb.Chron.) 1.171, cf. λιμαγχονία in Galen (K) 11.182.15, 185.5ff. In the Hippocratic Corpus an excessively reduced diet is criticised, for example, at περὶ διαίτης ὀξέων (De victus ratione in morbis acutis, Acut.) 11 (L) 2.310.1ff., 316.9ff., cf. ἀφορισμοί (Aphorismi, Aph.) 1.5 (L) 4.462.3ff., but something similar is recommended not just in the passages from Acut. (Sp.) and Vict. 3 cited already, but also, for example, at Aph. 7.60 (L) 4.596.1f.; cf. περὶ χυμῶν (De humoribus, Hum.) 6 (L) 5.484.19ff. At Plato Prt. 354a, starvation is referred to among cures that are painful but beneficial.

63. περὶ διαίτης ὑγιεινῆς (De salubri victus ratione, Salubr.) 9 (Nat.Hom. 24), CMG 1.1.3.220.8–10: ἄνδρα χρή, ὅστις ἐστὶ συνετός, λογισάμενον ὅτι τοῖσιν ἀνθρώποισιν πλείστου ἄξιόν ἐστιν ἡ ὑγιείη, ἐπίστασθαι ἀπὸ τῆς ἑωυτοῦ γνώμης ἐν τῆσι νούσοισιν ὠφελεῖσθαι.

64. See, for example, Salubr. 5 (Nat.Hom. 20), CMG 1.1.3.212.20ff. ("After bathing in warm water, let the patient first drink a cotyle of neat wine: then let him eat food of all sorts and not drink either with the food or after it, but wait enough time to walk ten stades; then mix three wines, dry, sweet, and acidic, and give him these to drink, first rather neat and in small sips and at long intervals, then more diluted, more quickly, and in larger quantities.")

compromising character. The assumptions to be made (about the naturalness of diseases) and the way forward in research are confidently sketched out, even if the elements of promise are greater than those of fulfilment, for in practice delivery fell short both in the matter of understanding and in that of control—that is, the cures achieved.

Madness

Mental illness posed problems for the rationalists that were at points importantly different from those of physical illnesses.[65] While physical sickness was never exactly celebrated (though the case of Philoctetes illustrates that it could be viewed with awe),[66] there were what Dodds called, after Plato, the *blessings* of madness,[67] especially the gift of prophecy and the inspiration of poetry. There is no need to rehearse the rich variety of phenomena to which Dodds drew attention other than to recall that they included not just the star examples of the statement of the power of Dionysus in the *Bacchae* and the exceptional recognition of the positive manifestations of madness in Plato's *Phaedrus*, but much else besides in Greek religious belief and practice as well as in Greek literature.[68] The question I wish to address is, rather, the following: in the face of these proofs of the hold, so to speak, of madness

---

65. Two recent books, B. Simon 1978 and Pigeaud 1981, provide helpful general discussions of many aspects of Greek attitudes towards mental illness.

66. To be sure, Philoctetes is a case where physical condition, moral persona, and position in relationship both to the gods and to human society are inextricably interwoven.

67. Dodds 1951, chap. 3, begins from *Phdr.* 244a, τὰ μέγιστα τῶν ἀγαθῶν ἡμῖν γίγνεται διὰ μανίας, "our greatest blessings come to us by way of madness," and the account of the four types of madness under divine patronage, prophetic madness (whose patron is Apollo), telestic (Dionysus), poetic (the Muses), and erotic (Aphrodite and Eros): see *Phdr.* 244a6ff., 245b1f., 265b2ff., cf. Linforth 1946, Brisson 1974. Cf. also *Ti.* 71e–72b, where prophecy is a gift to human folly: no one who is in his senses—ἔννους—lays hold of godlike and true divination; it is achieved only when the power of intelligence is shackled by sleep or one is distraught from disease or ἐνθουσιασμός; but interpretation is the work of reason.

68. See Dodds 1951, chap. 3, which often builds, as he acknowledges, on Rohde 1925 especially.

on the Greek imagination, how did the would-be rationalists fare? Among those would-be rationalists the pre-Socratic natural philosophers were the first in the field,[69] but I shall concentrate once again on the fuller material available in the medical writers.

Their ambition to naturalise mental illness as well as physical, to treat it both conceptually and in practice no differently from physical, is clear, but we must ask with what success they did so, and at what price. First some of the material that is important to us, much of it less familiar now than the texts on which Dodds focused, should be set out, and we may begin with another of the case histories from the *Epidemics,* the account of a condition that was evidently taken to be at least in part psychological in origin:

> A woman at Thasos became morose as the result of a grief with a reason for it, and although she did not take to her bed, she suffered from insomnia, anorexia, thirst, and nausea. . . . Early on the night of the first day, she complained of fears and talked much; she showed despondency and a very slight fever. In the morning she had many convulsions; whenever the frequent convulsions intermitted, she talked at random and used foul language; many intense and continuous pains. On the second day, condition unchanged, no sleep, higher fever. Third day: the convulsions ceased but coma and lethargy supervened followed by renewed wakefulness, when she kept leaping up and losing control. There was much random talk and high fever. That night she sweated profusely all over with warm sweat. She lost her fever and slept, becoming quite lucid and reaching the crisis. About the third day the urine was dark and thin and contained suspended matter, for the most part round particles, which did not sediment. Near the crisis, copious menstruation.[70]

▪ As this and many other examples show, Hippocratic accounts of symptoms move in a continuous gradation from thirst and nausea,

---

69. Thus according to Caelius Aurelianus *Morb.Chron.* 1.145, the followers of Empedocles explained one kind of madness as a disorder of the mind arising from a bodily cause, though another arises from a purification of the soul.

70. *Epid.* 3 case 11 of the second series (L) 3.134.2–15: ἐν Θάσῳ γυνὴ δυσήνιος, ἐκ λύπης μετὰ προφάσιος ὀρθοστάδην ἐγένετο ἄγρυπνός τε καὶ ἄσιτος, καὶ διψώδης ἦν καὶ ἀσώδης. . . . τῇ πρώτῃ, ἀρχομένης νυκτός, φόβοι, λόγοι πουλλοί, δυσθυμίη, πυρέτιον λεπτόν· πρωῒ σπασμοὶ πολλοί·

through anorexia and insomnia, to despondency and depression, or from high fever, to the delirium that so often accompanies it, to the patients being out of their minds, or from twitching and convulsions, to agitation and anger, to hallucinations.[71] As we noted before, attempts are made to establish correlations. For example, the third constitution in *Epidemics* 1 chap. 9 states: "High fever attended the start of the illness along with slight shivering fits, insomnia, thirst, nausea, slight sweating about the forehead and over the clavicles (but in no case all over), much random talk, fears and despondency, while the extremities such as the toes, and especially the hands, were chilled."[72] The doctors were concerned to collect cases of cold toes along with those of fear and despondency: all formed part of a total homogeneous epidemiological picture.

The strength of the Hippocratic approach to madness lies, as before, in its naturalism.[73] There is no question of any of these writers

---

ὅτε δὲ διαλίποιεν οἱ σπασμοὶ οἱ πολλοί, παρέλεγεν, ᾐσχρομύθει · πολλοὶ πόνοι, μεγάλοι, ξυνεχέες. δευτέρῃ, διὰ τῶν αὐτῶν· οὐδὲν ἐκοιμᾶτο· πυρετὸς ὀξύτερος. τρίτῃ, οἱ μὲν σπασμοὶ ἀπέλιπον· κῶμα δὲ, καὶ καταφορή, καὶ πάλιν ἔγερσις· ἀνήϊσσε, κατέχειν οὐκ ἠδύνατο, παρέλεγε πολλά· πυρετὸς ὀξύς· ἐς νύκτα δὲ ταύτην ἴδρωσε πολλῷ θερμῷ δι᾽ ὅλου· ἄπυρος· ὕπνωσε, πάντα κατενόει, ἐκρίθη. περὶ δὲ τὴν τρίτην ἡμέρην, οὖρα μέλανα, λεπτά, ἐναιώρημα δὲ ἐπὶ πουλὺ στρογγύλον, οὐχ ἵδρυτο· περὶ δὲ κρίσιν γυναικεῖα πουλλὰ κατέβη.

71. Hippocratic concern to bring what we should call psychological complaints or conditions within the ambit of their art extends to discussion of such matters as how to treat anxiety or worry, φροντίς, described as a "difficult disease" at *Morb.* 2.72 (L) 7.108.25ff. For one notable case history involving fear and hallucinations, see *Epid.* 5.81 (L) 5.250.10ff. (≃ *Epid.* 7.86 [L] 5.444.13ff.) and cf. *Epid.* 5.82 (L) 5.250.14ff. (≃ *Epid.* 7.87 [L] 5.444.17ff.). See Heiberg 1927, Kudlien 1968b, pp. 326ff.; Laín Entralgo 1970, pp. 139ff.; and Pigeaud 1981, pp. 70ff. especially.

72. *Epid.* 1.9 (L) 2.650.14–652.4: αὐτίκα γὰρ ἀρχομένοισι πυρετὸς ὀξύς, ἐπερρίγεον σμικρά· ἄγρυπνοι, διψώδεες, ἀσώδεες· σμικρὰ ἐφίδρουν, περὶ μέτωπον καὶ κληῖδας, οὐδεὶς δι᾽ ὅλου· πουλλὰ παρέλεγον· φόβοι, δυσθυμίαι, ἄκρεα περίψυχρα, πόδες ἄκροι, μάλιστα δὲ τὰ περὶ χεῖρας.

73. A naturalistic attitude towards madness and physical explanations of its origin can be illustrated from non-medical literature in the fifth and fourth centuries in, for example, Xenophon *Memorabilia* (*Mem.*) 3.12.6. But passages in Herodotus, for instance, illustrate how such an attitude may still be combined with traditional beliefs about the possibility of divine intervention:

thinking of madness as the result or the manifestation of ἄτη, no question, here, of any concessions to the blessings of madness. Madness is mental illness, and mental illness, like any other, is investigable and treatable. Yet the assumptions that are made are considerable. There is no sign of any realisation of the particular difficulty of specifying *what* mental illness is, what it takes for a patient to be mad. Foul language and random talk (as the case cited shows), even (as other cases do) "much talking, laughter, and singing"[74] are signs of abnormality; so too is loss of memory (not specified further):[75] so too, on some occasions, is silence.[76] The doctor is confident that the patient was merely babbling, or was unnaturally silent. He is confident, too, that he can tell the difference between depression arising from a distinct external stimulus, μετὰ προφάσιος,[77] and straight depression.

While the resolute matter-of-factness robs of its purchase any attempt to glorify madness, there seems no recognition that some modification in approach when dealing with mental illness might be called for. Treatments are, in any case, not often discussed in the case histo-

---

in 3.33 the possible reasons for Cambyses going mad are either that he offended Apis or that it came about because he suffered from the sacred disease (evidently treated here as primarily a bodily condition), and cf. the alternative accounts reported on Cleomenes' madness, 6.75ff. and 84. Cf. G. E. R. Lloyd 1979, pp. 29ff.

74. *Epid.* 1 case 2 (L) 2.686.6–7: λόγοι πολλοί, γέλως, ᾠδή. Cf. *Aph.* 6.53 (L) 4.576.13f., which states that delirium accompanied by laughter is less serious than when not.

75. E.g. *Epid.* 3 case 13 of the second series (L) 3.140.7.

76. E.g. *Epid.* 3 case 15 of the second series (L) 3.142.8, 146.5. Elsewhere, at *Aph.* 2.6 (L) 4.470.17f., insensitivity to pain is taken as a sign of mental disorder; at *Aph.* 5.40 (L) 4.544.16f., it is said to be a sign of madness when blood congeals around a woman's nipples; at *Aph.* 5.65 (L) 4.558.7f., madness is said to follow when the swellings that accompany wounds in the front of the body suddenly disappear; at *Aph.* 6.21 (L) 4.568.7f., varicose veins and haemorrhoids are said to bring an end to madness; at *Aph.* 7.5 (L) 4.578.14, it is said to be a good sign when madness is followed by dysentery, dropsy, or an ecstatic state. In *Aër.* 7, *CMG* 1.1.2.36.12f., when the effects of drinking stagnant water are described, pneumonia and madness are said to attack young people in winter.

77. See *Epid.* 3 (L) 3.134.3, cited above n. 70.

ries in the *Epidemics,* but one theme is not reassuring. Several of the patients (as in the case cited) "lost control"—κατέχειν οὐκ ἠδύνατο— and some were clearly actively restrained.[78] We do not know what kinds of restraint were attempted by the Hippocratics, nor how severe, nor in what precise circumstances, but there is no need to agree with the more extreme themes developed in some modish modern psychology[79] to see that this has an ominous ring. We hear from later writers such as Celsus in the first century A.D. and Caelius Aurelianus in the fourth that some medical theorists advocated violent and gruesome "cures" for μανία. These included chaining the patients, drugging them, starving them, keeping them in the dark, making them drunk, and flogging them, and though Caelius is outspoken in his criticisms of most of these,[80] Celsus gives some of them

---

78. *Epid.* 3 (L) 3.134.10. This expression can, and sometimes clearly should, be taken intransitively to refer primarily to a loss of *self*-control: cf., e.g., *Epid.* 1 case 8 (L) 2.702.18. But in *Epid.* 7.11 (L) 5.382.13ff., 384.8ff., in the case of the wife of Hermoptolemos, κατέχειν is undoubtedly used of someone else restraining her: ἔργον κατέχειν ἦν . . . ὡς καταλαβών τις αὐτὴν κατάσχοι χρόνον ὀλίγον. ("It was a job to control her . . . until someone held her and restrained her for a time.")

79. For the notion of mental illness as a metaphorical disease used to facilitate social control, see Szasz 1962/1974, Illich 1976, cf. Laing 1960, Goffman 1961/1973. With regard to the ongoing disputes on the questions of the validity and significance of the use of the vocabulary of disease for what is seen as mental abnormality, it is worth remarking that νόσος, νόσημα, and πάθος have a very wide application in Greek in the classical period and can be used in referring to any unwelcome condition or misfortune: cf. Dover 1974, pp. 125ff., who instances, among many other examples, cases where the "disease" is being in love (Eubulus fr. 41.6), showing piety towards bad people (Sophocles *Antigone* 732), and opposing the gods (Sophocles *Trachiniae* 491f.). Dover goes on to link this with the "absence of distinction between insanity and wrongdoing," although such distinctions are, of course, drawn in the philosophical tradition especially.

80. *Morb.Chron.* 1.144ff., 171ff. In presenting what he calls the Methodist account, which includes, nevertheless, criticisms of some early Methodist doctors, Caelius Aurelianus is no doubt drawing mainly on Soranus—though here and elsewhere it would be rash to assume he is simply translating Soranus (cf. G. E. R. Lloyd 1983a, p. 186 n. 258). Other less violent remedies recommended by other physicians but criticised by Caelius include cooling substances (the idea that madness results from heat is ascribed to Aristotle and

A man with the knowledge of how to produce by means of a regimen dryness and moisture, cold and heat in the human body, could cure this disease too [that is, epilepsy, and that is in addition to other diseases, indeed every other disease, madness included],[87] provided that he could distinguish the right moment for the application of the remedies. He would not need to resort to purifications and magic and all that kind of charlatanism.[88]

As in the attack on the topic of physical diseases, some of the rationalists are loud in their claims both to superior knowledge and to superior therapeutic efficacy, but in the case of mental illness the bluffing is even more transparent. The establishing of a naturalistic basis for the understanding of madness, the ruling out of references to the divine or demonic, is a release from one mystification. But it was achieved at the cost of the substitution of another of a different kind, at least when the theorists' own proposed explanations were quite unsubstantiated and imaginary. Nor did the positive and constructive help on offer amount to very much. To be told that your madness was not sent by the gods might (if you were convinced) be reassuring. At the same time the convinced rationalists cut themselves off from such support as had been available from traditional social, let alone religious, resources. But otherwise, to have any great expectations of improvement from adopting the anti-bilious or anti-phlegmatic diet of cold, or alternatively warm, food and exercise, prescribed in *On the Sacred Disease,* was clearly, and equally, principally a matter of faith.[89]

---

87. In 18 (L) 6.394.19ff., the writer sets out therapeutic principles that he claims to be valid not just for epilepsy but for all other diseases: ἐν ταύτῃ τῇ νούσῳ καὶ ἐν τῇσιν ἄλλῃσιν ἁπάσῃσιν, 396.1.

88. *Morb.Sacr.* 18 (L) 6.396.5–9: ὅστις δ' ἐπίσταται ἐν ἀνθρώποισι ξηρὸν καὶ ὑγρὸν ποιεῖν καὶ ψυχρὸν καὶ θερμὸν ὑπὸ διαίτης, οὗτος καὶ ταύτην τὴν νοῦσον ἰῷτο ἄν, εἰ τοὺς καιροὺς διαγινώσκοι τῶν συμφερόντων, ἄνευ καθαρμῶν καὶ μαγίης καὶ πάσης τῆς τοιαύτης βαναυσίης. No doubt the writer's chief point, in this final chapter, is the negative one that purifications are not necessary, but it is precisely the extravagance of the positive claims that he makes that I am concerned to point out.

89. Similar points can be made about the attempts to give purely naturalistic, physical accounts both for passions and the emotions (both included under the πάθη by the Greeks) and for the development of character. Some writers favoured humoral theories (like the one invoked in the account of mad-

ness we have considered from *Morb.Sacr.*); others attempted physiognomical correlations between character and bodily signs; still others put forward different psychophysical correlations or proposed combinations of these ideas. The more elaborate and systematic versions of both humoral and physiognomical theories are post-classical developments. This is true, for example, of the classification of human characters into four main types corresponding to which of the four humours—bile, black bile, phlegm, blood—predominates in the body (on which see, for example, Müri 1953; Flashar 1966; Kudlien 1967, pp. 77ff.; Pearcy 1984; cf. Klibansky, Panofsky, Saxl 1964: compare Galen's treatise devoted to the general thesis that the character of the soul "follows" the physical constitution or temperament, κρᾶσις, of the body, ὅτι ταῖς τοῦ σώματος κράσεσιν αἱ τῆς ψυχῆς δυνάμεις ἕπονται). Yet associations of "melancholy" or black bile with depression and fear occur already in the Hippocratic Corpus, e.g., *Aph.* 6.23 (L) 4.568.11f., *Epid.* 3 case 2 of the second series (L) 3.112.11f., as also with madness, *Aph.* 3.20 (L) 4.494.16ff., 3.22 (L) 4.496.7f., 6.56 (L) 4.576.19ff. At the same time μελαγχολικά is frequently used elsewhere with no close connection with psychological symptoms, and some recognition of the range of application of the term may be indicated by the fact that in the text from *Epid.* 3 cited the writer adds περὶ τὴν γνώμην to specify what he has in mind. Humoral theory applied to character thus appears to be an extension of ideas that originally related primarily to physical and physiological experience and speculation.

The first extant complete physiognomical treatise is the spurious *Physiognomonica* in the Aristotelian Corpus, though this develops theories along lines already adumbrated by Aristotle in, for example, *Historia Animalium* (*HA*) 1.8ff., 491b9ff., and *Analytica Priora* (*APr.*) 2.27, 70b7ff. (cf. G. E. R. Lloyd 1983a, pp. 22ff.: ancient physiognomical writings are collected in Förster 1893). Moreover, Aristotle also attempts more generally to correlate the character and intelligence of different species of animals with the qualities (especially the temperature, purity, and "thinness") of their blood in the *De Partibus Animalium* (*PA*) 2.2.647b31ff., and 2.4.650b14ff., cf. *HA* 521a2ff.

Before that Plato provides us with one of the most notable texts attempting to explain sexual incontinence in physical terms, when, at *Ti.* 86c ff., this is ascribed to excessive production of semen. The passage goes on to generalise the point to apply to other cases of apparent lack of self-control in the matter of pleasures, and bad temper, rashness and cowardice, forgetfulness, and slowness in learning are in turn associated with the effects of various humours trapped in the body. Plato here uses these ideas to support the Socratic dictum that "no one does wrong willingly" (*Ti.* 86d–e), but it should be noted that it is not just a bad condition of the body but also faulty upbringing, ἀπαίδευτος τροφή, that makes people bad.

Earlier still we find other correlations proposed in the medical writers. In the Hippocratic *On Airs, Waters, Places*, differences in climatic conditions are held responsible, in part, both for physical constitutions (including the susceptibility to certain diseases) and for the psychological character of different races, though in the latter case the political institutions and government un-

only distinguishing the various types of dreams—predictive and non-predictive, allegorical and non-allegorical—but also setting out in detail how they should be interpreted. Many of the writers in question, such as Artemidorus,[97] are sophisticated, at points quite cautious, restrained, even self-deprecatory. Many topics of interest might be pursued here. One we may note in passing is the extent to which the importance of wish-fulfilment is recognised by Greek dream-theorists: Freud himself remarked, rather defensively, in the 1914 edition of *Die Traumdeutung,* that those who attach any importance to such anticipations can go back to classical antiquity, and he cited Herophilus in particular in this connection, while he still insisted that no one before him had held that *every* dream is a wish-fulfilment.[98] However, our chief concern here must be with the kinds of theories and explanations our early would-be rationalists offered to account for the phenomena.

Once again we have a whole Hippocratic treatise devoted to the subject, the fourth book of *On Regimen* (sometimes called *On Dreams*), as well as an important discussion in Aristotle which (whether or not he knew *On Regimen* 4) develops a similar theory.[99] Since Aristotle con-

---

97. Thus the *Onirocritica* repeatedly criticises the exaggerations of dream interpretation as it had been and continued to be practised (e.g., 4.22.255.13ff., 59.283.20ff., 63.286.13ff., cf. 1 pr. 2.1ff., 1.3.11.7ff., 2.69.195.10ff.). Artemidorus' own classification, dividing predictive dreams, ὄνειροι, from non-predictive ones, ἐνύπνια, e.g., 1.1.3.9ff., acknowledges that many dreams contain no indication for the future. Moreover, among the predictive ones only a subgroup require interpretation, for some are plain and direct: 1.2.4.22ff. He also insists both that all the circumstances of the dreamer's situation need to be taken into account (e.g., 1.8.17.11ff., 9.18.16ff., 4.59.283.4ff.) and, as Freud was also to emphasise, that all the details of the dream itself should be (e.g., 1.12.20.18f., 4.4.248.5ff., 28.263.14ff.), even though in practice Artemidorus often works with broad categories of dreamer, classifying them, for example, according to profession, and many of the dreams he reports are described and interpreted without any *great* attention to detail. See especially Behr 1968, R. J. White 1975, Winkler 1982, and for other late dream theorists, for example Kessels 1969.

98. Freud 1953, vol. 4, p. 132 n. 2.

99. Among pre-Socratic speculations about dreams and sleep may be mentioned Heraclitus' view (fr. 89) that while the world of those who are awake is "one and common," in sleep we each withdraw into a private world; Empedo-

fines himself largely to providing a general framework of explanation, it will be convenient to reverse the chronological order and take him first.[100]

Dreams correspond to movements in the body, notably in the sense organs themselves, these movements being transmitted to the soul.[101] During the day many of these movements go unnoticed in the welter of impressions the soul receives. But at night, when the soul is less preoccupied, traces of some of the daytime impressions may be registered in it, provided the soul is itself in a stable condition.[102] No dreams are sent by the gods, though Aristotle says that they are δαιμόνια, on the grounds that nature herself is δαιμονία.[103] That makes dreams natural, but serves to remind us that for Aristotle nature is something to be

---

cles' idea that thoughts in general, and according to some reports dreams in particular, vary with changes in the body (fr. 108, cf. fr. 106 and cf. reports of Pythagoreans' views in D.L. 8.32); and the atomist explanation of dreams as images thrown off objects when these images penetrate the dreamer's body through its pores (Democritus fr. 166, Plutarch *Quaestiones convivales* 8.10.2.735a–b). The idea that some part of the person becomes more active in sleep occurs already in Pindar fr. 116 Bowra, which puts it that the "image of life" (αἰῶνος εἴδωλον) sleeps when the limbs are active, but to men asleep reveals in dreams the pleasant or painful outcome of future events (cf. also Xenophon *Cyropaedia* 8.7.21), cf. Dodds 1951, pp. 135ff.

100. On the date of *Vict.* views have differed widely, and no precision is possible. Internal evidence shows that the author of *Vict.* 1.4–5 is familiar with the work of Heraclitus, Empedocles, and especially Anaxagoras, but the use made of their ideas is compatible with a fourth- as well as a late fifth-century date. W. D. Smith's recent attempt (1979) to establish that this is an authentic work of Hippocrates himself has not won wide agreement. The work shows no signs of Hellenistic influence, however, and a date before Aristotle's treatises *De somno et vigilia* (*Somn. Vig.*), *De insomniis* (*Insomn.*) and *De divinatione per somnia* (*Div.Somn.*) seems likely.

101. *Insomn.* 459a11ff., 24ff.

102. *Insomn.* 460b28ff., 461a3ff., cf. also fr. 12a (Ross) of the περὶ φιλοσοφίας.

103. At *Div.Somn.* 463b12ff. Aristotle first notes that some animals have dreams, and then remarks: θεόπεμπτα μὲν οὐκ ἂν εἴη τὰ ἐνύπνια, οὐδὲ γέγονε τούτου χάριν (δαιμόνια μέντοι· ἡ γὰρ φύσις δαιμονία, ἀλλ' οὐ θεία). Cf. also *Div.Somn.* 463b15ff., 464a19ff., where he notes that dreams come to anyone, whereas if god sent them they would happen also by day and come (especially) to the wise.

ory, only assertions, although some of the underlying assumptions, and his use of symbolism, are transparent, and traditional, enough.[113] Broadly and simplemindedly, to see good things in dreams is good, and bad things bad, and again it is good to see things that correspond to daytime thoughts and actions and that represent them as occurring in an orderly fashion.[114] More specifically we are told that it is a sign of health if, in the dream, when a star seems to fall out of its orbit, it appears pure and bright and moves eastwards.[115] Conversely, "when [a star] seems dark or dim or to move westwards, or towards the sea, or towards the earth, or upwards, these signify diseases. Upward movements indicate fluxes in the head; movements towards the sea, diseases of the bowels; and those towards the earth, usually tumors growing in the flesh."[116]

Moreover, confident recommendations for treatment match confident diagnoses:

> Should one of the stars seem to be injured, or to disappear, or to be obstructed in its orbit, if this happens because one sees it affected by mist or cloud, this is a weak sign, but it is a more severe one if by water or hail. It signifies that an excretion of moisture and phlegm has occurred in the body, and has fallen towards the outermost circuit.[117] In such cases it is beneficial for the patient to take long runs, well wrapped up: they should gradually be increased to cause as much sweating as possible. The exercise should be followed by long walks and the patient

---

113. As, for example, in the use of black-white symbolism in *Vict.* 4.91–92 (L) 6.658.8, 10, 13, 18.

114. As, for example, at *Vict.* 4.88 (L) 6.642.11ff., cf. 4.93 (L) 6.660.15f., which suggests that seeing customary things indicates a desire of the soul.

115. *Vict.* 4.89 (L) 6.650.4ff.

116. *Vict.* 4.89 (L) 6.650.9–14: ὅ τι δ' ἂν τούτων μέλαν καὶ ἀμυδρὸν καὶ πρὸς ἑσπέρην δοκῇ φέρεσθαι ἢ ἐς θάλασσαν ἢ ἐς τὴν γῆν ἢ ἄνω, ταῦτα σημαίνει τὰς νούσους· τὰ μὲν ἄνω φερόμενα κεφαλῆς ῥεύματα· ὅσα δὲ ἐς θάλασσαν, κοιλίης νοσήματα· ὅσα δὲ ἐς γῆν, φύματα μάλιστα σημαίνει τὰ ἐν τῇ σαρκὶ φυόμενα.

117. The writer has a theory of three main circuits or orbits of the heavenly bodies, the stars outermost, the sun in the middle, and the moon below that (*Vict.* 4.89 [L] 6.644.18ff.), and he assumes that the main parts of the human body are disposed in three corresponding circuits (cf. *Vict.* 1.10 [L] 6.486.3ff.).

should go without breakfast. Food should be cut by a third and the normal diet restored gradually over five days. If the disorder appears more severe, prescribe steam baths in addition.[118]

Analogies with more modern health faddists are not hard to suggest. The limitations of this rationalist takeover are twofold. First, the field of what is taken over is restricted. This writer is not concerned (though Aristotle was) with the whole range of predictive dreams, about some of which he is quite indifferent; he concentrates, rather, on what can be discovered about the state of health of patients from their dreams. He has an entirely naturalistic theory of the correlations between the two. No god sends these signs; they are the natural by-products of physical disturbances, a theory he elaborates with some persistence. But it is limited also in a second sense, in that, although superior knowledge is claimed, in practice the theory draws heavily on, and at points is merely a rationalisation of, popular beliefs. Yet the ambition to go one better than traditional views and even than specialist interpretations is evident from those claims to superior knowledge.[119] The specialists are said not to know what they are talking about—whereas the Hippocratic writer, armed with his naturalistic theory of physical-psychical correlations, can, if you believe him, put the whole "science" on a firm footing.

---

118. *Vict.* 4.89 (L) 6.644.19–646.7: ὅ τι μὲν οὖν δοκέοι τῶν ἄστρων βλάπτεσθαι ἢ ἀφανίζεσθαι ἢ ἐπίσχεσθαι τῆς περιόδου, ἢν μὲν ὑπ' ἠέρος ἢ νεφέλης, ἀσθενέστερον· εἰ δὲ καὶ ὑπὸ ὕδατος ἢ χαλάζης, ἰσχυρότερον· σημαίνει δὲ ἀπόκρισιν ἐν τῷ σώματι ὑγρὴν καὶ φλεγματώδεα γενομένην ἐς τὴν ἔξω περιφορὴν ἐσπεπτωκέναι. συμφέρει δὲ τούτῳ τοῖσί τε δρόμοισιν ἐν τοῖσιν ἱματίοισι χρῆσθαι πολλοῖσιν, ἐξ ὀλίγου προσάγοντα, ὅκως ἐξιδρώσῃ ὡς μάλιστα, καὶ τοῖσι περιπάτοισιν ἀπὸ τοῦ γυμνασίου πολλοῖσι, καὶ ἀνάριστον διάγειν· τῶν τε σίτων ἀφελόμενον τὸ τρίτον μέρος προσάγειν ἐς πέντε ἡμέρας· εἰ δὲ δοκέοι ἰσχυρότερον εἶναι καὶ πυρίῃσι χρῆσθαι.

119. The opening chapter of book 4 (*Vict.* 4.86 [L] 6.640.13f.) states that being able to make correct judgements in the matter of dreams is a "large part of wisdom," μέγα μέρος . . . σοφίης, and the last sentence (4.93 [L] 6.662.8f.), which probably relates to the treatise as a whole, claims that using this advice a man will be healthy and that the author has set out as much as it is possible for any man to know about regimen.

Divination and Fate

The limitations, and pretensions, of the inquiry concerning nature emerge clearly once again in relation to the final pair of topics I mentioned, divination and fate.[120] Here one might have expected the proto-scientists to have abstained from confrontation with the likes of Teiresias and Cassandra.[121] Even if some dreams might be scrutinised for the diagnostic signs they might yield about the dreamer's *current* state of health, the idea of trying to predict the future, especially the individual's future, was, one might have thought, a palpably unpromising area for any kind of research that purported to involve the inquiry concerning nature, even if it might be the concern of a moral philosophy, whether deterministic or anti-deterministic.[122]

---

120. Many aspects of divination in the ancient world are well analysed in the collection of essays in J.-P. Vernant et al. 1974: see especially Vernant's own contribution, pp. 9ff., which points out, for example, the solidarity, in many ancient societies, between divination and other forms of rationality, and cf. Bottéro's study of divination in Mesopotamia, pp. 70ff., especially 153ff., 168ff., 190ff., in which he argues, among other things, that curiosity in general may be stimulated by the ambition to foretell the future, and that in that sense divination may be seen as leading to science as well as being itself a science insofar as it has claims to be deductive, analytic, and systematic. Among recent anthropological discussions of divination should be mentioned those of Moore 1957, Park 1963, Jules-Rosette 1978, and Ahern 1981, particularly. Moore and Park in particular provide comparative material to illustrate how divination may sometimes serve to take the question of responsibility for particular decisions out of what is perceived as the human domain or to introduce a randomisation process into such decisions.

121. Xenophanes, indeed, is reported to have rejected all forms of divination (Cicero *De divinatione* (*Div.*) 1.3.5, Aetius 5.1.2) but he was clearly exceptional, if not unique, in the pre-Socratic period.

122. As is well known, the issue of determinism itself once it became, as it did in the post-Aristotelian period, a central topic of philosophical debate was discussed as a physical as well as a moral problem. Both the Stoics, who asserted determinism, and the Epicureans, who denied it, argued their case in the first instance by reference to natural causation, the Stoics maintaining that there is an inexorable nexus of physical causes and effects and the Epicureans postulating the swerve precisely to constitute an uncaused exception to that rule. It is clear that the Epicureans used this doctrine not just to explain cosmogony but also to insure free will, though quite how what is, *ex hypothesi,* an uncaused movement of soul atoms is to secure the latter remains controver-

Yet first, the discourse of prediction, which is prominent in our scientific vocabulary when we talk of the predictive value of a hypothesis, encompassed in the ancient world too a wide range of phenomena. In two areas, especially, it is legitimate to talk of ancient *scientific* predictions. Whereas modern medicine is concerned with diagnoses, the ancients often focused, rather, on "prognosis," [123] especially the outcome of the disease, and we need not doubt that, drawing on a wide experience and sometimes despite some simpleminded theories, many Greek doctors were often able to anticipate not just the recovery or death of their patients but also the general progress of their ailments.[124] Again,

---

sial (any such movement in the soul at the moment of choice would appear to tell against personal control of that choice or moral responsibility for the action: the issue is helpfully discussed by Furley 1967; cf. Long 1974, pp. 56ff.; and, most recently, Sedley 1983a and Don Fowler 1983). The notions of insisting on a separation first between physical and psychological determinism, and secondly between the question of the nature of physical laws and the issue of moral responsibility, which have figured prominently in many modern discussions of the problems (e.g., Pears, ed., 1963, Popper 1965/1972, Lucas 1970, Anscombe 1971, O'Connor 1971, Kenny 1975) run counter to the general tendency, in ancient debate, to run together the physical and the moral philosophical questions. Not even the Stoics, who held the doctrines both of moral responsibility and of physical determinism, argued that the latter is irrelevant to the former but, rather, that the former is qualified by the latter. Again, the Epicureans may have assumed that to secure free will it was necessary (if not also enough) to show or, rather, to assert exceptions to the nexus of physical causes and effects. The literature on the post-Aristotelian debate is immense: apart from the studies already mentioned, see especially Long 1971, 1977, Donini 1973, 1974–75, M. Frede 1974, 1980, Sharples 1975a, 1975b, 1981, 1983, Reesor 1978, Stough 1978, Sorabji 1980a, 1980b, Sedley 1980, D. Frede 1982, Annas 1986.

123. But "prognosis," for the ancients, concerned the past and the present as well as the future of the disease, as is clear, for example, from *Prog.* 1 (L) 2.110.2f., quoted below, n. 126.

124. Thus confident and often well-grounded pronouncements on the likely results of certain lesions are common in the surgical treatises: see, for example, *Art.* 63 (L) 4.270.3ff.; 69 (L) 4.288.3ff. We shall be returning later to the ancient debate on whether or how far medicine can be deemed to be an exact science, but many of those who insisted that it is not maintained nevertheless that it is a rational inquiry that can and does yield knowledge: see below Chap. 3, at nn. 88ff., Chap. 5, at nn. 134ff.

engage in such "divinations" but will set out reliable signs by which one can recognise which patients will recover and which die.[137] Other predictions should be made ἀνθρωπινωτέρως—in a more modest fashion, as befits mere human beings[138]—though he too will set out what you have to know if you want to succeed in this kind of competition.[139] As for the exactness sometimes claimed for medical forecasts, he says that he listens and he laughs.[140]

The doctors thus evidently used their ability to foretell the future and to retrodict the past as a means of impressing their patients and indeed of building up their practice. Yet there was a risk of the doctor being assimilated to the soothsayer, a risk some Hippocratic writers try to guard against, and which much later Galen repeatedly tells us he still had to contend with.[141] Some of the medical men actively sought a reputation for being able to predict the future,[142] even while they dis-

---

137. *Prorrh.* 2.1 (L) 9.8.2–4: ἐγὼ δὲ τοιαῦτα μὲν οὐ μαντεύσομαι, σημεῖα δὲ γράφω οἷσι χρὴ τεκμαίρεσθαι τούς τε ὑγιέας ἐσομένους τῶν ἀνθρώπων καὶ τοὺς ἀποθανουμένους, τούς τε ἐν ὀλίγῳ χρόνῳ ἢ ἐν πολλῷ ὑγιέας ἐσομένους ἢ ἀπολουμένους.

138. *Prorrh.* 2.2 (L) 9.8.11ff., cf. 2.3 (L) 9.10.23ff.

139. *Prorrh.* 2.2 (L) 9.10.4ff.: ἀλλὰ χρὴ προλέγειν καταμανθάνοντα πάντα ταῦτα, ὅστις τῶν τοιουτέων ἐπιθυμέει ἀγωνισμάτων. Cf. also διαγωνίζεσθαι at 2.2 (L) 9.8.15. The writer advises caution in the matter of forecasting since, although successes lead to a doctor being "wondered at" (θαυμασθείη), in failure he is hated and thought to be mad: 2.2 (L) 9.10.10ff.

140. *Prorrh.* 2.4 (L) 9.14.10f., cf. 20.11ff. The difficulty of making predictions before the disease has become established is mentioned at 2.3 (L) 9.12.20ff., 14.2ff.

141. Galen tells us that his use of the pulse in diagnosis was considered to be mere divination by his critics, e.g., *CMG* 5.8.1.106.21ff. ([K] 14.637.10ff.) and elsewhere reports that he was suspected of magic, charlatanry, and divination (see [K] 7.354.4ff., 11.299.10ff., 301.10ff., 12.263.6ff., *CMG* 5.8.1.84.5ff., 94.18f. [(K) 14.615.4ff., 625.16f.]), although he himself accuses others in similar vein (e.g., [K] 11.793.12ff., 795.16f., 796.7ff.). It is quite clear that on occasion he deliberately sought to amaze his audience (e.g., [K] 8.361.12ff., 365.9ff.), especially in the context of spectacular exhibition dissections (e.g., *AA* 8.4 [K] 2.669.7f., 15) while he criticises others for doing the same (e.g., [K] 11.797.10ff.), cf. Kollesch 1965, Vegetti 1981.

142. Indeed, after the classical period medical divination became systematically linked with astrology in the iatro-mathematical tradition: see, for example, Ptolemy *Tetrabiblos* (*Tetr.*) 1.3.16.7ff.; 3.13.147.9ff.; 4.9.200.12ff. Boll-Boer; cf. Bouché-Leclercq 1899, chap. 15, pp. 517–42.

sociated themselves from others who, in other contexts, had precisely the same ambition.[143]

The ambivalence of the relationship between—to use our terms—astronomy and astrology is more highly charged still, and many have simply dismissed the latter as an aberration. The fact that most prominent ancient astronomers, including Hipparchus and Ptolemy, also engaged in astrology is often taken to be irrelevant to Greek *science* and as evidence only of the failure of the Greeks to be *scientific*. Yet not to be guilty of gross anachronism, we must take as our explananda not just those parts of ancient mathematics and natural philosophy that we approve or consider fruitful, but the whole of the corpus of work of those who engaged in different branches of those complex and manifold traditions. To ignore astrology would be to miss the insights it can offer both about ancient controversies concerning what those traditions comprised and about the ambitions some theorists entertained concerning some areas that they were certainly eager to include.

That some parts of their work were better grounded than others goes without saying; it went without saying to the ancients themselves, even while they, like us, argued about the criteria of superiority. Ptolemy, for one, clearly distinguishes between the two types of prediction or prognostication to be made from the study of the heavenly bodies: on the one hand, predictions of their movements (astronomy in our sense); and on the other, prediction concerning events on earth.[144] Moreover, he explicitly emphasises the conjectural nature and the difficulty of the latter study,[145] criticising the excessive claims made

---

143. Among the early philosophers, too, there were those who assimilated themselves, or were assimilated by others, to prophets or diviners. The oracular pronouncements of Heraclitus invite comparison with those of Delphi, where Apollo—as he put it in fr. 93—"neither speaks nor hides but indicates" his meaning (cf., e.g., Hölscher 1952, Merlan 1953, Kahn 1979, pp. 123f.). Empedocles, who casts himself in the role of a wonder-worker (fr. 111), speaks of the people of Acragas thronging to ask him for oracles and for the "word of healing for every kind of disease" (where βάξις carries some of the associations of inspired utterance) (fr. 112). Cf. the comparison between Socrates' sign and μαντική at Xenophon *Mem.* 1.1.2ff., cf. Plato *Ap.* 40a, though elsewhere Plato undercuts such a claim; see below, Chap. 2 n. 134.

144. See, for example, Ptolemy *Tetr.* 1.1.2.16ff.

145. *Tetr.* 1.1.3.5ff., 2.8.1ff.; 3.2.109.1ff.

by some past and contemporary practitioners[146] and limiting his own discussion to generalisations based on the supposed beneficence or maleficence of various heavenly bodies or their configurations.[147] The validity of astrology as a whole was disputed,[148] but we should remember that there were similar foundational disputes in many other areas of the inquiry concerning nature, including about astronomical model-building itself.[149] Based on a belief in a connection, συμπάθεια,[150] between heaven and earth, which could be illustrated, in the first instance, by such uncontroversial examples as the seasons and the tides,[151] astrology was usually defended (like medicine) primarily by reference to what were claimed as its results,[152] and as in medicine again, there was considerable indeterminacy in evaluating these.

---

146. *Tetr.* 1.2.7.20ff., 22, 52.11ff., and especially 3.4.113.18ff., where Ptolemy dismisses aspects of genethlialogy as currently practised as superfluous nonsense and as lacking any plausibility.

147. *Tetr.* 1 and 2 deal with general astrology, 3 and 4 with genethlialogy.

148. Cicero's *De divinatione* provides the most comprehensive ancient account of the arguments used *pro* and *contra* the possibility of divination in general.

149. Thus although Proclus gives a detailed introductory account of astronomy in his *Hypotyposis astronomicarum positionum* (*Hyp.*), this is prefaced by a passage (pr. 1.2.1ff.) where he cites Plato for the view that the true study concerns the region "beyond" the heavens and is directed to the immutable Forms. Moreover, he repeatedly expresses his view, both in *Hyp.* and in his *Commentaries* to the *Republic* and *Timaeus,* that the epicycle-eccentric model is complicated, farfetched, and arbitrary (see, for example, *In Platonis Timaeum commentarii* (*In Ti.*) 3.56.28ff., 146.14ff., *Hyp.* 7.236.10–238.21); cf. Samburksy 1965. From the life sciences we shall be considering later the dispute over the validity of the practice of animal dissection: Chap. 3 at nn. 220ff.

150. The idea was used by the Stoics especially (see, e.g., Cicero, *ND* 2.7.19, *Div.* 2.14.33ff., Sextus *Adversus mathematicos* [*M.*] 9.75ff., 79ff., cf. 5.4ff., Cleomedes *De motu circulari corporum caelestium* 1.1.4.10ff., 8.15ff., Alexander *De mixtione* [*Mixt.*] 3.216.14ff., 11.226.30ff., 12.227.5ff.) but was not confined to them: see, for example, Philoponus *In Aristotelis libros de Generatione et Corruptione commentaria* (*In GC*) 41.25f., *In Aristotelis Physica commentaria* (*In Ph.*) 113.8f., and for its use in Soranus' gynaecology, for instance, see G. E. R. Lloyd 1983a, pp. 178ff.

151. See Ptolemy *Tetr.* 1.2.4.3ff. The way Ptolemy moves from these incontestable cases to far more dubious ones has been analysed especially by Long 1982.

152. Thus Ptolemy repeatedly refers to what he says the ancients had ob-

The case for it seemed untenable to many,[153] but others exploited what seemed to them a splendid opportunity to bring this area of divination too into the orbit of rigorous mathematical disciplines. Even though astrology began in Greece with vague ideas about the influences of the stars such as we find in Hesiod,[154] it came to have, in Hellenistic times, an elaborate theoretical framework most of which it owed to astronomy.[155] In this, astrology was importantly different both from divination by the consultation of books, for instance (as in the sortes Homericae or Virgilianae), which involved no study of natural objects at all, and from hepatoscopy, for there the marks on the surface of the liver that the diviner studies are, as Rufus gives us to understand, of no significance for the medical man.[156] The use of planetary tables and of spherical geometry is *common* to astronomy and astrology, both of which engage in the determinations of planetary positions. Initially just impressionistic, astrology came to have claims to be, in some respects at least, an exact study. Certain assumptions (and they were of course the crucial factor) had to be made about what were claimed as the natural effects of different heavenly bodies or at least about how they could be used as signs; and the application of general rules to individual cases was always a matter of the astrologer's own judgement. Yet planetary configurations could be worked out with impeccable mathematical precision and deductive rigour. In that respect

---

served to be the properties or qualities of the heavenly bodies, e.g., *Tetr.* 1.3.17.7ff., 9.22.21ff., 10.30.6f., cf. 12.32.23ff., though he is aware of competing traditions on some points, e.g., 1.21–22, 44.22–53.13. Analogies with medicine, navigation, and archery are invoked, e.g., at 1.2.10.2ff., 3.13.16ff.; 3.2.109.11ff., to distinguish, for example, between the errors of individual practitioners and the general soundness of the art.

153. At *Tetr.* 1.1.3.15ff., before Ptolemy sets out to defend astrology against the specious arguments brought against it, he remarks that what is hard to attain is easily attacked by "the many."

154. See, for example, *Op.* 417ff., 587f. (cf. 609ff.) on the effects brought about by, or associated with, the star Sirius.

155. As has been remarked, for example, by Neugebauer 1952/1957, p. 171; cf. 1975, vol. 2, pp. 607ff.

156. See Rufus περὶ ὀνομασίας (*Onom.*) 158.5ff., which lists some of the terms used concerning the external surface of the liver in divination but remarks that the nomenclature of such parts is not necessary for medical purposes.

the casting of horoscopes or genethlialogy could claim to be more exact than most areas of medicine or of natural philosophy. The symbiosis of the two studies of the heavenly bodies is remarkable, for on the one hand the aspirations of astrology helped to keep elementary astronomy alive, and on the other the prestige of astrology depended partly on its incorporating the same mathematical procedures used in astronomy. It was thanks, in part, to his mathematics that the "mathematician," as he was often known, won his reputation as a superior diviner.

Finally, the importance of astrology, for Ptolemy at least, was underpinned by its ultimately moral aim of helping us view the future with calm and steadiness.[157] But this moral concern is in no way exceptional in the inquiry concerning nature in the ancient world. On the contrary, as we shall see in due course,[158] it is a recurrent one. For now, we may simply note that, following Plato, Ptolemy held that astronomy too is good for the character,[159] and that following Aristotle, Galen claimed that the study of the parts of animals reveals the wonders, beauty, and goodness of creation and instils in the student true piety towards its wise and benevolent creator.[160]

---

157. See *Tetr.* 1.3.10.14ff. In the case of what is inevitable, foreknowledge enables one to view this μετ' εἰρήνης καὶ εὐσταθείας (12.4f.). But not everything is inevitable, and influences that might lead to one outcome can, if they are anticipated, be resisted and averted, so that on this score, too, astrology can be claimed to be useful and beneficial.

158. See below, Chap. 6, pp. 319 and 336.

159. At *Ti.* 47b–e Plato suggested that by beholding the revolutions of reason in the heavens we can stabilise the revolutions of our own thinking. Ptolemy claimed that astronomy can help to make men good: *Syntaxis* 1.1, Heiberg (H) 1.7.17ff., quoted below, Chap. 6 n. 144.

160. Aristotle stressed the value and the pleasures to be gained from the study of animals in his famous protreptic in *PA* 1.5.644b22ff., cf. *De anima* (*De An.*) 402a1ff., and emphasised particularly that that study was undertaken primarily for the sake of the revelation of form and the final cause; the admirable (θαυμαστόν) and the fine (καλόν) are present in all natural things: *PA* 645a16f., 22f. For Galen, too, the study of anatomy is motivated, in part, by its revealing that "nature does nothing without an aim," a motivation he explicitly states at *AA* 2.2 (K) 2.286.5ff., while the principle itself is cited on countless occasions. Again, the study of the parts of animals in general teaches us about the power, beauty, and justice of nature, e.g., *De usu partium* (*UP*)

Our later investigations will give us ample opportunity to consider aspects of Greek science where traditional beliefs and attitudes are less prominent in the background or even—as often with Archimedes, for instance—quite irrelevant. This first study has been deliberately directed at a set of highly problematic and difficult topics for science, where myth, religion, and ritual provided the usual resources for stabilising belief. Those who engaged in the inquiry into nature—those who invented that inquiry—exhibit some well-grounded confidence in their ability to provide an alternative, naturalistic, rationalist framework for understanding. At the same time they often, it may seem to us, fail to recognise the limitations of what they had achieved or of what they could hope to achieve, both where the questions they raised are simply not amenable to their approach (certainly not in their day, and in some cases not even now) and where the answers they proposed are vulnerable, if in different ways, to criticisms similar to those they themselves brought against earlier beliefs.

Yet that recurrent phenomenon may be understandable in part at least in terms of the problems the new investigation into nature faced in establishing itself alongside and in confrontation with other more traditional sources of wisdom, comfort, and understanding. Some of the investigators themselves claimed that theirs was the way not just to understand nature, but to gain a correct apprehension of the divine:[161]

3.10 (H) 1.172.15ff., (K) 3.235.6ff.; 7.14 (H) 1.418.24ff., (K) 3.576.8ff.; 7.15 (H) 1.422.24ff., (K) 3.582.1ff.; 11.2 (H) 2.116.10ff., (K) 3.846.13ff.; 17.1 (H) 2.448.9ff., (K) 4.361.7ff.; 17.2 (H) 2.449.15ff., (K) 4.362.16ff.

161. It would no doubt be an exaggeration to claim (as Jaeger 1947 did, cf. the more moderate thesis of Hussey 1972) that the chief motivation of pre-Socratic philosophy was the reform of theology, although both Xenophanes and Heraclitus—like others outside the philosophical traditions, notably Pindar, Herodotus, and Euripides—criticised aspects of common religious beliefs and practices (cf. further below, n. 163, and Chap. 4, pp. 176ff. on Xenophanes). But the evidence that a number of early philosophers suggested that the natural elements or their cosmological principles are divine is substantial, if mostly secondhand (cf. G. E. R. Lloyd 1979, p. 11), and we have seen that there are medical writers, too, who subscribe to the belief that all nature is divine (see above, n. 39). Plato, to be sure, saw those who appealed merely to nature and to chance—to the exclusion of reference to the good and to the

true piety consisted in the type of study in which they were themselves engaged.[162] Others did not so much transmute as undermine traditional systems of belief, attacking in particular some of the authority figures who sustained them—the prophets, diviners, purifiers, and the like.[163]

---

craftsmanlike force in the cosmos—as atheists (*Leges* [*Lg.*] 889a ff., cf. below n. 163). His own views on the proper study of nature are clear from the *Timaeus*, where the chief emphasis is on the investigation of just those factors the materialists omitted, the revelation of the workings of a divine craftsman described as free from envy. For Aristotle in turn the correct, teleologically oriented study of nature reveals the immanent craftsmanlike force at work in it, a view echoed, modified, and developed not just in the Lyceum (for Strato, for instance, see Cicero *ND* 1.13.35) but also by the Stoics. Much later, both Ptolemy and Galen suggest a direct connection between the study of nature and the true understanding of the divine. Thus although Ptolemy distinguishes mathematical astronomy from theology in *Syntaxis* 1.1 (H) 1.5.7ff., the former is said to be "the best science to help theology along its way, since it is the only one which can make a good guess at [the nature of] that activity which is unmoved and separated" (7.5–7: τό τε γὰρ θεολογικὸν εἶδος αὐτη μάλιστ᾽ ἂν προοδοποιήσειε μόνη γε δυναμένη καλῶς καταστοχάζεσθαι τῆς ἀκινήτου καὶ χωριστῆς ἐνεργείας). Mathematical astronomy studies bodies that are not merely "heavenly" (οὐράνια) but in a real sense "divine" (θεῖα, 6.23). Galen, more explicitly still, describes the study of animals in terms drawn from religion: see below, n. 162.

162. Thus in *UP* 3.10 (H) 1.174.6–13, (K) 3.237.10–17, Galen calls his work "a sacred account which I compose as a true hymn to him who created us: for I believe that true piety consists not in sacrificing many hecatombs of oxen to him and burning ten thousand talents of cassia, but in discovering first myself, and then showing to the rest of mankind, his wisdom, his power and his goodness." (ἱερὸν λόγον, ὃν ἐγὼ τοῦ δημιουργήσαντος ἡμᾶς ὕμνον ἀληθινὸν συντίθημι, καὶ νομίζω τοῦτ᾽ εἶναι τὴν ὄντως εὐσέβειαν, οὐκ εἰ ταύρων ἑκατόμβας αὐτῷ παμπόλλας καταθύσαιμι καὶ τάλαντα μυρία θυμιάσαιμι κασίας, ἀλλ᾽ εἰ γνοίην μὲν αὐτὸς πρῶτος, ἔπειτα δὲ καὶ τοῖς ἄλλοις ἐξηγησαίμην, οἷος μέν ἐστι τὴν σοφίαν, οἷος δὲ τὴν δύναμιν, ὁποῖος δὲ τὴν χρηστότητα.) The theme is picked up and elaborated elsewhere in *UP*: cf. below, Chap. 6 n. 161.

163. There were, of course, as we have already noted (n. 161), others besides those who were directly and principally engaged in the study of nature who criticised conventional religious beliefs and practices. The most radical attack came from philosophers and sophists who rationalised the origin of such beliefs. Thus Prodicus is reported to have suggested that men treated as gods things from which particular benefit was derived, such as bread and wine, as well as water, fire and the sun, moon and rivers (see, e.g., Sextus *M.*

But, implicitly or explicitly, the investigators into nature laid claim to a new kind of wisdom, a wisdom that purported to yield superior enlightenment, even superior practical effectiveness, and in part (though this can only be justified by the detailed studies that follow) they were surely right in their conviction of their own originality and their belief in the potentiality of the approach they adopted, even though their strengths had their corresponding weaknesses, especially in the excesses and exaggerations of many of those claims. They were wise men of a different kind, unlike the old seers in important respects, though again much closer to them in others than aspects of the self-image they projected would lead one to expect. They successfully demystified many a mythical, mystical, symbolic, or traditional assumption. For all that, the science they presented was, in some cases, no more than the myth of the elite that produced it. These are themes that will be explored more fully in the remainder of this book.

9.18 and 52, cf. Henrichs 1975). But a recurrent motif in such rationalisations is an association of the divine with terrifying natural phenomena. In the famous fragment from Critias' *Sisyphus* (fr. 25) that suggests that the gods were a deliberate human invention introduced to provide a sanction to insure good behaviour, the imagined human inventor locates the gods in the upper circuit from which lightning and thunder come to frighten mankind. Although Democritus did not dismiss notions of the gods entirely, he too is reported by Sextus (*M.* 9.24) to have argued that belief in the gods is in part a mistaken inference from terrifying natural phenomena such as thunder, lightning, and eclipses, and the atheists criticised by Plato in the *Laws* (*Lg.* 889e ff.) are represented as claiming that the gods exist "by art" and "by convention" rather than by nature. But whether or not it was the *intention* of any natural philosopher who attempted purely physical explanations of thunder, lightning, eclipses, and so on to substitute a naturalistic account for a religious or supernaturalistic one, the *effect* of the new search for aetiologies was to make available what could be seen as adequate alternative frameworks for understanding. "What Zeus?" as Socrates is made to ask in Aristophanes' *Clouds* (*Nubes, Nu.*) 367ff., "there is no Zeus." And to the question of Strepsiades, "who then rains?" Socrates replies that it never rains without clouds.

Moreover, the explicit *intention* to remove any involvement of the gods in such natural phenomena is clear in Epicurus, for whom the primary motivation of the study of nature is, precisely, to rid men of such superstitious beliefs: *Ep.Hdt.* 10.76ff., *Letter to Pythocles* (*Ep.Pyth.*) 10.85ff., 97, 113, *Sent.* 11, 12.

than to figures securely framed as what we should call historical personages—though we should note that the Greek "first discoverers"
also include plenty of the former as well as of the latter: Prometheus
and his like, as well as Archytas, Archimedes, and theirs.[7]

Even though, as I noted in Chapter 1, tradition is what is usually
appealed to both to justify certain ways of behaviour and to block certain types of question, what is included in concrete science or as common knowledge is, nevertheless, subject to adjustment—to the tinker-
ing characteristic of the *bricoleur*,[8] if not to planned or self-conscious
revision. Even the actors' own protestations of the sacrosanctity of the
tradition do not preclude the possibility of aspects of that tradition un-
dergoing modification and scrutiny, even if not necessarily formal or
institutionalised scrutiny. In two areas especially, recent anthropological work has thrown light on the degree to which innovations are possible, occur, and are even inevitable within what is still conceived as an
unaltered tradition. First, such studies of oral literature as Goody's on
the myth of the Bagre,[9] Phillips' on Sijobang,[10] and Finnegan's general
analysis[11] have demonstrated the tolerance of variation in what is still
thought of, and unreservedly claimed to be, precisely the *same* nar-

---

7. Thus Diogenes Laertius, who refers repeatedly to the topic of first discoverers, offers a very heterogeneous collection of examples, both in terms of
the techniques or ideas claimed as invented and in terms of the inventors he
names. They range from the attribution of the first predictions of solar eclipses
to Thales (1.23) and of the first map of the earth and sea to Anaximander (2.2)
to that of the first purifications of houses and fields to Epimenides (1.112). In
some cases what is represented by Diogenes as a Greek invention is, rather, a
matter of the introduction into Greece of ideas or devices found elsewhere,
as, for example, with the gnomon ascribed to Anaximander (2.1) which
Herodotus (2.109) says was brought to Greece from Babylonia. Diodorus
Siculus, who records Egyptian claims that certain inventions came from their
gods, 1.14.1, 15.8, 43.5f., refers explicitly to disputes between one race and
another as to who were the first discoverers of things useful for life, 1.9.3, cf.
2.38.1ff., 4.1.6f., 2.5.
    8. Cf Lévi-Strauss 1962/1966.
    9. See Goody 1972, 1977 chap. 3, and forthcoming.
    10. Phillips 1981.
    11. Finnegan 1977.

rative.[12] Secondly, using fieldwork in Thailand, Tambiah has recently investigated the outer limits within which rituals may be altered and adapted to new politico-religious ends and still be apprehended as unchanged.[13]

The inference to past innovation is even more compelling where its intellectual products are more obviously exceptional, as is the case with some of those of ancient societies, for example, the mathematics and astronomy of Babylonian civilisation or the mathematics and medicine of ancient Egypt. Here our evidence is written, though some of the extant didactic mathematical texts give glimpses of the oral situation of instruction within which they were presumably used. Thus in the Rhind mathematical papyrus from Egypt we read: "If the scribe says to thee '10 has become 2/3rds and 1/10th of what?' let him hear . . ." and there then follow the workings and the conclusion.[14]

Admittedly, we are usually in no position to chart the stages of the developments that occurred, let alone to identify individuals responsible for particular features of them. Nor should we Whiggishly assume that there was a single continuous development, a sustained onward-and-upward march, as opposed to periods of advance interspersed with others of retrenchment, stagnation, and regression. But even though we are in the dark about most of the circumstances surrounding the *growth* of mathematics, astronomy, and medicine in the ancient Near East, the documentary evidence we possess for the *end*

---

12. That in oral literature poems are composed not for, but in, performance, and that a given poem is thus recomposed, in a sense, at each performance, were points stressed already by A. B. Lord 1960, pp. 13ff., cf. Nagy 1983. Compare also Detienne 1977/1979, pp. 6, 16, and 1981/1986, pp. 22ff., 37ff., 124ff.

13. Tambiah 1977, cf. 1982. Compare also Goody 1977, pp. 29f., and forthcoming. In ancient Mesopotamia major myths evidently underwent certain transformations in part in response to changes in the dominant political power, the most notable example being, perhaps, the substitution of one supreme god—Enlil, Marduk, or Assur—for another in the creation myth Enuma Elish: see, for example, Jacobsen 1949, pp. 182ff., and 1976.

14. See Peet 1923, pp. 65ff., on number 30; cf. p. 87 on number 47, p. 104 on numbers 61b and 62, and p. 111 on number 68.

results is substantial, and those end results themselves certainly imply innovation at *some* stage. It is worth laying some stress on that feature of those great civilisations, since it tends to be brushed aside when attention is focused on their undoubted elements of conservativeness and of deference to traditional authority, both in the sense of deference to the customary authority figures and in that of deference to the past.

Those elements, too, are, to be sure, very prominent and they can be exemplified in many different aspects of Egyptian and Babylonian culture, ranging from the social and political sphere to medicine and astronomy themselves. True, our Greek sources typically exaggerate the contrast between, for instance, Greek and Egyptian attitudes. Herodotus,[15] Plato,[16] and Aristotle[17] are among those in the classical period who remark on the conservative tendencies in Egyptian culture, on which Plato, especially, often commented with approval.[18] Later, Dio-

---

15. Herodotus frequently refers to the care the Egyptians devoted to the preservation of the memory of the past, including their keeping of written records (e.g., 2.77, 100, 145), as well as to their adherence to traditional customs (2.79).

16. Thus in the *Laws*, 656d–657a, cf. also 798e–799b, Plato cites Egypt as the one country in which poetry, music, and artistic productions generally are controlled by the state and innovation is prohibited. The correct forms and tunes have been laid down long ago: they are posted up in the temples and "apart from them it was not permissible either for painters or for any other artists . . . to innovate, καινοτομεῖν, or to devise anything outside what is traditional, nor is it permissible now in these matters or in any other branch of culture. And if you look you will find there that the things painted or engraved ten thousand years ago—not in a manner of speaking, but literally ten thousand years ago—are in no respect better or worse than those done today, but accomplished according to the same art."

17. In his discussion of whether it is better to be governed by the best laws or by the best men, Aristotle cites Egyptian medicine, evidently as an example of the former (*Politics, Pol.*, 1286a7ff., 12ff.), though he notes that Egyptian doctors are allowed to alter their treatment after four days; if they do so before, it is at their own risk.

18. As in the text cited above, n. 16. Sparta and Crete are favourably contrasted with other Greek states for being less inclined to artistic innovation, and in that respect closer to the Egyptian model, at *Lg.* 660b. According to Plutarch *Instituta Laconica* 238c, the Spartans even attempted to forbid innovation.

dorus[19] reported that Egyptian doctors ran a risk of legal sanctions—the death penalty, no less—if they deviated from what the sacred medical books laid down. Moreover, in the case of medicine we do not, of course, have to rely simply on what the Greeks tell us: the extant Egyptian medical texts themselves exhibit the enormous strength and authority of tradition. We have a striking example of this in the Edwin Smith papyrus, a text which preserves medical lore handed down over an extended period, where the glosses contained in the version we have represent the attempts by the final redactors to explain difficult points in the diagnoses of the cases and the treatments prescribed.[20] Clearly the chief aim of later physicians was the conservation and faithful interpretation of the wisdom of their predecessors.[21]

---

19. Diodorus 1.82.3: ἐὰν δέ τι παρὰ τὰ γεγραμμένα ποιήσωσι, θανάτου κρίσιν ὑπομένουσιν, ἡγουμένου τοῦ νομοθέτου τῆς ἐκ πολλῶν χρόνων παρατετηρημένης θεραπείας καὶ συντεταγμένης ὑπὸ τῶν ἀρίστων τεχνιτῶν ὀλίγους ἂν γενέσθαι συνετωτέρους. ("If they do anything contrary to what is written down, they are liable to a capital charge, since the legislator thought that few would be likely to be cleverer than the treatment laid down over a long period and established by the best craftsmen.")

20. The Edwin Smith papyrus itself dates from around 1600 B.C., but the material it contains comes from a much earlier period. Each of the case-histories consists of five sections: Title, Examination, Diagnosis, Treatment, and Glosses. While the first four were composed, according to Breasted 1930, pp. 9f., sometime during the Old Kingdom, between 3000 and 2500 B.C., the glosses were added at the end of that period: "In the latter part of the Old Kingdom, probably not later than 2500 B.C., some 'modern' surgeon, as unknown to us as the original author, equipped the document with a commentary in the form of brief definitions and explanations, which we term glosses, appended to each case. For example when the original treatise directed the practitioner to 'moor him [the patient] at his mooring stakes,' the commentator knows that this curious idiom is no longer intelligible and appends the explanation, 'It means put him on his accustomed diet and do not administer to him any medicine.'" Cf. also Breasted 1930, pp. 61ff.

21. As Wildung, for example, notes in his discussion of Imhotep and Amenhotep (1977, p. 298), in Egypt wisdom was often represented as a matter of learning, the wise man being the man who had direct access to the sources of knowledge, namely, books. On occasion, however, some dissatisfaction appears to be expressed at the constraints imposed by the authority of written texts. This seems to be the gist of a passage in the "Complaints of Khakheperre-

Even so the point should not be *over*-emphasised. We may turn to another Greek text to suggest another side to the picture, a report in Herodotus which, despite its patent inaccuracy at certain points, nevertheless has a lesson to teach us. Herodotus says that, lacking doctors, the Babylonians take their sick down to the marketplace so that any passerby may comment and say how he or anyone else he knew recovered from a similar illness.[22] Now it is just not true that the Babylonians had no doctors—though those who acted as such may have looked to Herodotus less distinctive than Greek or Egyptian physicians.[23] Yet whatever the flaws in Herodotus' story as a historical report, it serves to remind us that even in generally conservative societies there may be, and often is, a certain open-mindedness about therapy, a readiness to canvass diagnostic opinion and to listen to suggestions concerning treatment.[24]

Greek Innovations and Egotism

I have been at some pains to suggest that a certain innovativeness may, indeed must, be supposed to have been at work in the ancient Near East—including in fields that are directly relevant to the understanding of what we call natural phenomena—*even though* we are usually in no position either to date the innovations or to identify their authors. But the significance of that point, in turn, should not be missed. Even though, obviously, innovativeness is no Greek prerogative,

---

Sonb"—to be dated, according to Gardiner 1909, pp. 96f., to the reign of Sesotris II (1897–1878 B.C.), though our extant text belongs to the mid-eighteenth dynasty—where we read: "Had I unknown phrases, / sayings that are strange, / novel, untried words, / free of repetition; / not transmitted sayings, / spoken by the ancestors! / . . . Ancestor's words are nothing to boast of, / They are found by those who come after." (Lichtheim 1973, pp. 145ff.)

22. Herodotus 1.197.

23. The extant Mesopotamian medical texts indicate that there were several different complementary, if not rival, approaches to diagnosis and treatment. The main features are summarised by Oppenheim 1962; cf. 1964, pp. 289ff.

24. The point can be illustrated extensively from medical anthropology: see, for example, G. Lewis 1975, pp. 248ff.

that does not mean that the manifestations of innovation are everywhere the same. The extant remains of Egyptian and Babylonian medicine, mathematics, and astronomy can be combed in vain for a single example of a text where an individual author explicitly distances himself from, and criticises, the received tradition in order to claim originality for himself;[25] whereas our Greek sources repeatedly do just that. Even where we can *infer* innovations in Egyptian or Babylonian texts, that is to say, it is not the *style* of their authors to publicise the fact or even to mention it.[26]

The contrast with Greek writers is striking, and we are dealing, of

---

25. Such culture heroes as Imhotep (see Wildung 1977), renowned for the benefits he conferred on mankind, testify to a general Egyptian appreciation of certain types of inventiveness. Moreover, my colleague John Ray draws my attention to two specific pieces of evidence that indicate recognition of particular innovations. The first is a Ramses II graffito that celebrates the achievement of Zoser, more than 1,300 years earlier, who had been the first king to make extensive use of stone in monumental building (see Yoyotte 1960, pp. 57f.). Secondly, the Amenemhet inscription provides an example of a claim for originality in technology, the construction of an improved clepsydra. Amenemhet was active in the reign of Amenhotep I (1545–1525 B.C.). In the inscription published by Borchardt 1920, Amenemhet proudly points out that his clepsydra tells the time accurately throughout the year—that is, that it can register differences between the hours from one season to another (cf. Neugebauer and Parker 1960, p. 119). Two points are, however, to be noted. First, he prefaces his account of the instrument by referring to his study of "all the books of the words of God" (as Borchardt observes, 1920, p. 62, it is not evident whether the knowledge that Amenemhet says he discovered was what he discovered by reading the literature or for himself). Secondly, while he claims that his instrument is better than all others, he expresses this point not in the form that no such instrument had ever been made before him, but in the form that no such instrument had been made "since antiquity." Nevertheless, this inscription is an important—if exceptional—example of a claim to do better than one's immediate contemporaries.

26. As Grapow has pointed out, 1954–73, 2, pp. 25f., the use of the first person singular in Egyptian medical writings is a characteristic of the magical texts (for example, where the god announces his presence or is invoked as aid, "I am Horus," "I am Theuth," "I have come from Heliopolis . . . assuredly, I have come from Sais with the mother of the gods," cf. Ebbell 1937, p. 29), rather than of the descriptive accounts of the surgical or medical case-histories incorporating observations of particular patients and setting out diagnoses and recipes, though "I" may be used in the verdict on whether the disease is to be treated, e.g., Ebbell 1937, pp. 120ff.

goddess's instruction to Parmenides takes a form that marks a far greater distancing from that tradition, for she tells him to "judge by *logos* the much-contested refutation pronounced by me."[37] It is, we should say, the strength of the deductive argument, not the appeal to divine authority, on which Parmenides depends to secure conviction in his hearers, even though he equates the latter with the former. In Ionian speculative natural philosophy, too, Diogenes of Apollonia begins his work by telling us about his method and again leaves us in no doubt that it is *his*: "it seems to me that every *logos* should begin from an incontestable starting-point."[38]

Egotism, to be sure, is not necessarily connected with innovativeness,[39] but the two often go together in early Greek philosophy, especially in claims to set forth the truth that had eluded everyone else. One after another, the major pre-Socratic philosophers from Xenophanes onwards state or imply that no one else had got the answers right, establishing their own presence in the text with copious criticisms of other writers, their predecessors or contemporaries, named or unnamed, at the limit by criticisms of what everybody else believed. Xenophanes attacks Homer and Hesiod by name for "ascribing to the gods everything that is shameful and a blame among men, thieving, adultery, and deceiving each other."[40] Heraclitus, in turn, hammers

---

see, for example, frr. 8.1, 9.5, 17.1 and 16, 35.1, 38, 111.2, 112.4, 113.2, 114.1ff.)

37. Parmenides fr. 7.5–6: κρῖναι δὲ λόγῳ πολύδηριν ἔλεγχον ἐξ ἐμέθεν ῥηθέντα. Whether ἔλεγχος here means "proof" or "refutation" is disputed: cf., most recently, Lesher 1984. But that does not affect the point that the persuasive force of the Way of Truth depends on the—largely original—methods of argument that Parmenides there deploys.

38. Diogenes of Apollonia fr. 1: cf. also the recurrence of "it seems to me" (ἐμοί or μοι δοκεῖ) in frr. 2, 5 and 8.

39. As many examples from the Greek orators could be used to illustrate: see, for instance, Antiphon 1.1, 5, 11f.; 6.15f., Lysias 1.5, 22; 3.4, 14; 12.3, 37, among many texts where an orator stresses a personal view with the first-person-singular pronoun or—in ways that are analogous to themes common in natural philosophy and medicine—sets out what he claims to be able to demonstrate.

40. Xenophanes fr. 11, cf. fr. 12 and the more general attack on anthropomorphic religious beliefs, frr. 14–16 (cf. below, Chap. 4, pp. 176ff.).

Xenophanes along with Hesiod, Pythagoras, and Hecataeus for "much learning," πολυμαθίη,[41] and in many richly abusive fragments expresses his contempt for the ignorance and folly of mankind in general.[42] For Parmenides, too, what ordinary men believe is mere illusion, a world of Seeming.[43]

These features of early Greek philosophical writing are well known and need no elaboration. The corresponding points in relation to early Greek medical texts are in some respects more complex, reflecting in part the heterogeneity of the extant treatises, but we can give examples of both egotism and innovativeness—and their conjunction—in works of very different types. We may take first such comparatively polished or pretentious exhibition pieces, ἐπιδείξεις, as *On Breaths* and *On the Art*. The author of *On Breaths* repeatedly introduces himself into his text with first-person statements, specifying, for example, what he claims to be able to show, ἐπιδείκνυμι.[44] The writer of *On the Art*, who refers to himself with the first-person-singular pronoun twice in the first three lines of the work,[45] repeatedly parades what he represents as his personal views. For example, chapter 2 gives the author's views, introduced by "I at any rate think," about the relationship between

---

41. Heraclitus fr. 40: πολυμαθίη νόον ἔχειν οὐ διδάσκει· Ἡσίοδον γὰρ ἂν ἐδίδαξε καὶ Πυθαγόρην αὖτίς τε Ξενοφάνεά τε καὶ Ἑκαταῖον. ("Much learning does not teach sense: for otherwise it would have taught Hesiod and Pythagoras and again Xenophanes and Hecataeus.") Cf. also frr. 42, 56, 57, 106.

42. See especially frr. 1, 2, 5, 17, 19, 29, 34, 78, 104, 108.

43. Parmenides frr. 1.30ff., 6.4ff., 8.51ff.

44. *Flat.* 5, CMG 1.1.94.6f., picked up at 15, CMG 1.1.101.17ff., cf. also 2, CMG 1.1.92.16f.; 6, CMG 1.1.94.8ff.; 10, CMG 1.1.97.12ff.; 14, CMG 1.1.99.20ff.

45. περὶ τέχνης (*De arte*) 1, CMG 1.1.9.2ff. The text begins: εἰσίν τινες, οἳ τέχνην πεποίηνται τὸ τὰς τέχνας αἰσχροποιεῖν, ὡς μὲν οἴονται, οὐ τοῦτο διαπρησσόμενοι, ὃ ἐγὼ λέγω, ἀλλ᾽ ἱστορίης οἰκείης ἐπίδειξιν ποιεύμενοι. ("There are those who make an art out of vilifying the arts, though they think, not that they are accomplishing what I say they do, but that they are making an exhibition of their own research.") Thus while itself an exhibition piece, this treatise attacks others who also claim to make an *epideixis*. The text continues with that quoted in n. 48 below, which opens with a further first-person pronoun.

language and reality, no less,[46] and the following chapter offers a definition of medicine with two more first-person-singular verbs: "first I
shall define what I take medicine to be."[47] And apart from his obses-
sive self-advertising, he lines himself up firmly on the side of innovation: "it seems to me that it is the aim and function of intelligence to
discover what was unknown before, wherever such a discovery confers
a benefit over ignorance."[48]

First-person-singular statements appear with great frequency also
in treatises that display a much greater knowledge of actual clinical
practice. *On Airs, Waters, Places,* for instance, gives us in the opening
chapter a summary of what the medical student should consider, the
effects of the seasons, the changes in the weather, the effects of warm
and cold winds, the properties of waters, the position of the city, and
so on, and then proceeds: "I shall explain clearly, ἐγὼ φράσω σαφέως,
the way in which each of these subjects should be investigated and
what tests are to be applied."[49]

The severely professional *Epidemics,* especially, presents a particularly intriguing case. Thus in book 1, chapter 4, of the second "constitution," we read: "I know of no case of καῦσος, ardent fever, which
was fatal on that occasion"[50] and "I have no instance to record where a
cough was either harmful or beneficial on that occasion."[51] In chapter
8 of the third constitution: "I know of no woman who died, if these
indications occurred properly, but so far as I know, ἃς καὶ ἐγὼ οἶδα,
all who fell ill while pregnant aborted."[52] Again in case 4 of the set of
case-histories in the same book: "I myself examined the urine which
was of the colour and thickness of that of cattle."[53] Yet we should be

---

46. *De arte* 2, CMG 1.1.10.10, cf. 10.2.
47. *De arte* 3, CMG 1.1.10.19, cf. below Chap. 3 at nn. 27ff.
48. *De arte* 1, CMG 1.1.9.4–6: ἐμοὶ δὲ τὸ μέν τι τῶν μὴ εὑρημένων
ἐξευρίσκειν, ὅ τι καὶ εὑρεθὲν κρέσσον ἢ ἀνεξεύρετον, ξυνέσιος δοκέει
ἐπιθύμημά τε καὶ ἔργον εἶναι.
49. *Aër.* 3, CMG 1.1.2.26.22f.
50. *Epid.* 1.4 (L) 2.620.5f.
51. *Epid.* 1.4 (L) 2.626.3f.
52. *Epid.* 1.8 (L) 2.648.4ff.
53. *Epid.* 1 case 4 (L) 2.692.13ff.

careful not to conclude, from those references by themselves, that the "I" is the same individual in each case. These books generally reflect several physicians' experience: they were built up by a process of accretion and have been subject not so much to interpolation (for there is no definite original text into which interpolations have been inserted) as to a series of additions by authors who were no doubt seeking to improve their usefulness. The process may not have been too dissimilar from the way in which modern textbooks of pathology or physiology are subject to successive (if more easily identifiable) stages of rewriting and reediting in the light of what is taken to be the latest knowledge. Evidently some of those Greek doctors who inserted their contributions sometimes chose to vouch for a particular piece of information by indicating that it came from their *personal* observations. The chief point, in any event, is this: unlike the Egyptian case-histories in the Edwin Smith papyrus and elsewhere, where the authors do not intrude to vouch for their opinions or observations personally, just such claims punctuate our extant Greek clinical records.

But as in philosophy, so too in medicine first-person statements may point up claims for originality. Three treatises of rather different types that exemplify this are *On Regimen in Acute Diseases, On Fleshes,* and *On Regimen.* Thus the author of *On Regimen in Acute Diseases,* who holds forth on the shortcomings of previous writers, especially the authors and revisers of the work called *Cnidian Sentences,* begins his own positive account of diet in acute diseases with the claim: "it seems to me to be worthwhile to write down such matters as are *not yet known* to doctors even though they are of great importance and bring great benefits and injuries."[54] *On Fleshes* draws a contrast between the opening chapter of the work, which sets out certain preliminary considerations for the study of the formation of the human body and where the author says he draws on "common opinions," and the sequel, expressing his personal views: "now I myself declare my own

---

54. *Acut.* 3 (L) 2.238.8−10: δοκέει δέ μοι ἄξια γραφῆς εἶναι ταῦτα μάλιστα, ὁκόσα τε ἀκαταμάθητά ἐστι τοῖσιν ἰητροῖσιν, ἐπίκαιρα ἐόντα εἰδέναι, καὶ ὁκόσα μεγάλας ὠφελείας φέρει ἢ μεγάλας βλάβας. Cf. also 2 (L) 2.230.1ff.

opinions."[55] *On Regimen,* especially, describes his theory of the balance between food and exercise as a discovery, ἐξεύρημα, that is a "fine thing for me the discoverer and useful for those who have learnt it: and none of my predecessors attempted to understand it though I judge it to be in every respect of great importance."[56]

The surgical treatises, too, are often much concerned with innovation in surgical practice. On the one hand we find passages praising inventiveness[57] and claiming or implying that the only correct treatment is that which the author himself sets out.[58] On the other, these authors also frequently criticise some of their own colleagues for *their misplaced* striving after effects.[59] The opening chapter of *On Fractures,*

---

55. περὶ σαρκῶν (*De carnibus, Carn.*) 1 (L) 8.584.1ff. and 8: νῦν δὲ ἀποφαίνομαι αὐτὸς ἐμεωυτοῦ γνώμας.

56. *Vict.* 3.69 (L) 6.606.3–5: τόδε δὲ τὸ ἐξεύρημα καλὸν μὲν ἐμοὶ τῷ εὑρόντι, ὠφέλιμον δὲ τοῖσι μαθοῦσιν, οὐδεὶς δέ πω τῶν πρότερον οὐδὲ ἐπεχείρησε συνεῖναι, πρὸς ἅπαντα δὲ τὰ ἄλλα κρίνω αὐτὸ εἶναι πολλοῦ ἄξιον. The topic of the author's discoveries recurs throughout the work, e.g., 1.1 (L) 6.466.16ff., 468.2ff.; 1.2 (L) 6.470.13ff., 472.4; 3.67 (L) 6.592.12ff.; 4.93 (L) 6.662.8f.

57. See especially *Art.* 42 (L) 4.182.13ff., which first criticises those who used succussion on a ladder (see below, n. 59) but then continues, 182.22ff.: "Yet the contrivance is an ancient one, and I at any rate praise heartily the man who first invented it—both this contrivance and any other that is in accordance with nature." Cf. 77 (L) 4.308.7ff., 10ff., which refers to the original intentions of the inventor of the method in a passage setting out an approved use of the wine-skin for reductions.

58. Thus *Fract.* 8 (L) 3.444.1ff. specifies the correct method of reducing fractures of the humerus in contradistinction to a number of faulty methods, where the specifications are sufficiently detailed and concrete to indicate the writer's own procedures. Again at *Art.* 11 (L) 4.104.16ff., the writer states that he knows of no practitioner (we are to understand, by implication, except himself) who treats dislocated shoulders correctly. Cf. also 46 (L) 4.198.5ff., 9ff., with its general condemnation of those who claimed to be able to treat inward dislocation of the vertebrae of the spine.

59. In addition to *Fract.* 1 mentioned immediately in my text, and *Art.* 35, discussed below (at n. 72), see especially *Art.* 42 (L) 4.182.13ff., which attacks those who attempted to reduce humpback by succussion on a ladder. Those doctors who do so are chiefly "those who desire to make the crowd gape: for to such it seems marvellous to see a man suspended or shaken or treated in such ways; and they always applaud these performances, never troubling themselves about the result of the operation, whether bad or good," (L)

for instance, first criticises practitioners who "get a name for wisdom"[60] for their over-sophisticated treatments of fractured arms, and then remarks on the general problem: "Many other parts of this art are judged thus: for men praise what seems outlandish before they know whether it is good, rather than the customary which they already know to be good; the bizarre, rather than the obvious."[61]

This is one of several treatises in which the *new* point that the authors present as *their* contribution is, precisely, the *dangers* of the *newfangled* in medicine: texts that are trebly suggestive, first because of the indirect evidence they supply concerning the positive value set— in some quarters—on originality in medical practice, secondly because of the way they exemplify the strong authorial presence that is such a distinctive feature of early Greek medical texts, and thirdly because they illustrate how the question of *whether* or *how far* to follow tradition was openly contested.

The treatise *On Ancient Medicine*, especially, takes as its chief theme the need to return to the tried and tested methods of earlier

---

4.182.15–20: χρέονται δὲ οἱ ἰητροὶ μάλιστα αὐτῇ οὗτοι οἱ ἐπιθυμέοντες ἐκχαυνοῦν τὸν πολὺν ὄχλον· τοῖσι γὰρ τοιούτοισι ταῦτα θαυμάσιά ἐστιν, ἢν ἢ κρεμάμενον ἴδωσιν, ἢ ῥιπτεόμενον, ἢ ὅσα τοῖσι τοιούτοισιν ἔοικε, καὶ ταῦτα κληίζουσιν αἰεί, καὶ οὐκέτι αὐτοῖσι μέλει, ὁκοῖόν τι ἀπέβη ἀπὸ τοῦ χειρίσματος, εἴτε κακόν, εἴτε ἀγαθόν. Again at the end of the same chapter, (L) 4.184.2ff., the writer says that he is himself ashamed to treat all such cases in this way, since such methods belong, rather, to charlatans, ἀπατεῶνες. In general, the writers of the surgical treatises tend to specify that the simplest treatment should be preferred, e.g., *Art.* 62 (L) 4.268.3ff., 78 (L) 4.312.1ff., cf. 34 (L) 4.156.5ff. (reading ἤ), *Fract.* 15 (L) 3.472.14ff. Yet on occasion they are prepared to recommend recourse to mechanical methods, e.g., *Fract.* 13 (L) 3.462.6ff., 15 (L) 3.472.16ff., 20 (L) 3.484.7ff., 30 (L) 3.516.14ff., 524.6ff., 31 (L) 3.528.16ff.; *Art.* 7 (L) 4.88.15ff., 43 (L) 4.184.5ff., 47 (L) 4.202.5ff., 67 (L) 4.278.10ff.; *Mochl.* 38 (L) 4.382.3ff.

60. *Fract.* 1 (L) 3.414.4–5: σοφοὺς δόξαντας εἶναι. Cf. 414.1: σοφιζόμενοι.

61. *Fract.* 1 (L) 3.414.6–9: ἀλλὰ γὰρ πολλὰ οὕτω ταύτης τῆς τέχνης κρίνεται· τὸ γὰρ ξενοπρεπὲς οὔπω ξυνιέντες εἰ χρηστόν, μᾶλλον ἐπαινέουσιν, ἢ τὸ ξύνηθες, ὃ ἤδη οἴδασιν ὅτι χρηστόν, καὶ τὸ ἀλλόκοτον, ἢ τὸ εὔδηλον. Strange new names for diseases and newfangled methods of treating them also attract Plato's criticisms at *R.* 404e ff., 405d5f.

medical practice, what the author himself calls *ancient* medicine.[62] His main endeavour is to refute those who try to base medicine on the *new*[63] method of postulates (or "hypotheses") and who thereby, in his view, drastically oversimplify the problems. Yet while allying himself with the past, he constantly uses expressions that signify his personal view. "I at any rate have not thought" (οὐκ ἠξίουν . . . ἔγωγε), he says in chapter 1, "that medicine needs a new hypothesis."[64] "It seems to me of the greatest importance," he says in the next chapter, "that anyone speaking about this art [medicine] should be intelligible to laymen."[65] Introducing his theory of the origin of medicine he begins: "I at any rate hold" (ἔγωγε ἀξιῶ) that it comes from dietetics.[66] Having explained his view of the balance of powers in the body, he proceeds:[67] "I think I have set forth this subject sufficiently. But certain doctors and sophists assert that it is impossible to know medicine if you do not know what a man is"—that is, his origin and elemental constitution. But that takes you into "philosophy," the kind of subject dealt with by Empedocles. "I, however, ἐγὼ δέ, think that what has been said or written by any sophist or doctor about nature has less to do with medicine than with painting."[68] In some nineteen-and-a-half

---

62. See *VM* 12, *CMG* 1.1.44.3, cf. 2, *CMG* 1.1.37.1–3: ἰητρικῇ δὲ πάλαι πάντα ὑπάρχει, καὶ ἀρχὴ καὶ ὁδὸς εὑρημένη, καθ' ἣν τὰ εὑρημένα πολλά τε καὶ καλῶς ἔχοντα εὕρηται ἐν πολλῷ χρόνῳ. ("Medicine has long had all its means to hand, a principle, and a method that has been discovered, according to which many fine discoveries have been made over an extended period of time.") Cf. also *Hebd.* 53.80.4ff. Roscher, (L) 9.466.8ff.

63. See *VM* 1, *CMG* 1.1.36.15ff., reading καινῆς with Heiberg at 16. In any case the description of the method of "hypothesis" as "new" is put beyond doubt at 13, *CMG* 1.1.44.8 ("the new method of those who investigate the art by means of hypothesis").

64. *VM* 1, *CMG* 1.1.36.15–16.

65. *VM* 2, *CMG* 1.1.37.9f., and see further below at nn. 142f. Cf. also Diogenes of Apollonia fr. 1, requiring, in cosmology, an exposition that is simple and solemn, ἀπλῆν καὶ σεμνήν.

66. *VM* 3, *CMG* 1.1.37.26ff.

67. *VM* 20, *CMG* 1.1.51.6–8: περὶ μὲν οὖν τούτων ἱκανῶς μοι ἡγεῦμαι ἐπιδεδεῖχθαι, λέγουσι δέ τινες ἰητροὶ καὶ σοφισταί, ὡς οὐκ εἴη δυνατὸν ἰητρικὴν εἰδέναι, ὅστις μὴ οἶδεν, ὅ τι ἐστὶν ἄνθρωπος. . . .

68. *VM* 20, *CMG* 1.1.51.12–14: ἐγὼ δὲ τοῦτο μέν, ὅσα τινὶ εἴρηται ἢ σοφιστῇ ἢ ἰητρῷ ἢ γέγραπται περὶ φύσιος, ἧσσον νομίζω τῇ ἰητρικῇ

pages in the Heiberg (*CMG*) text, this author uses the first-person-singular pronoun no fewer than thirty times, as well as first-person-singular verbs, φημί, οἷμαι, and the like, without the addition of ἐγώ or ἔγωγε, on a further twenty-two occasions. His notion of the past, clearly, is *his* construction. Moreover, for all his backward looking, he has an eye also to the future, for he says he believes that the *whole* of medicine will one day be discovered, using the traditional methods he recommends.[69] Traditional medicine, evidently, does not yet have all the answers, even if in this author's view it shows how they are eventually to be obtained.

Several Hippocratic authors thus engage in an active debate on innovation in medical theory and practice. Their public, or at least parts of it, were evidently also much taken by the latest fads or fancies in treatment. *On Regimen in Acute Diseases,* for instance, criticises laymen for not appreciating true excellence in treatment but, rather, being preoccupied with praising or blaming strange remedies,[70] and *Precepts* chides patients who ask for treatment that is outlandish or obscure, though this writer puts it that they should not be punished for this prejudice, merely disregarded.[71] The writer of *On Joints* refers to the wonder and delight registered by patients and their friends at specially intricate techniques of bandaging (a new fad, presumably) though he goes on to remark that after a time the patients become bored with wearing their complicated bandages.[72]

τέχνῃ προσήκειν ἢ τῇ γραφικῇ. On the meaning of γραφική here, see Festugière 1948, pp. 6of., Dihle 1963, pp. 146ff.

69. See especially *VM* 2, *CMG* 1.1.37.3–4: καὶ τὰ λοιπὰ εὑρεθήσεται, ἤν τις ἱκανός τε ὢν καὶ τὰ εὑρημένα εἰδὼς ἐκ τούτων ὁρμώμενος ζητέῃ. ("And the rest will be discovered, if the inquirer is capable and takes as his starting-point, in his inquiry, the knowledge of what has already been discovered.") Cf. also *VM* 8, *CMG* 1.1.41.8–9: ταῦτα δὴ πάντα τεκμήρια, ὅτι αὕτη ἡ τέχνη πᾶσα ἡ ἰητρικὴ τῇ αὐτῇ ὁδῷ ζητεομένη εὑρίσκοιτο ἄν. ("All this is evidence that the whole of this art of medicine could be discovered, using the same method of inquiry.")

70. *Acut.* 2 (L) 2.234.2ff.

71. παραγγελίαι (*Praeceptiones, Praec.*) 5, *CMG* 1.1.31.26–27: καίτοι ἔνιοι νοσέοντες ἀξιοῦσι τὸ ξενοπρεπὲς καὶ τὸ ἄδηλον προκρίνοντες, ἄξιοι μὲν ἀμελίης, οὐ μέντοι γε κολάσιος.

72. *Art.* 35 (L) 4.158.4ff.

Today we can understand a similar twentieth-century obsession with the latest fashion in treatment in part in terms of the effects of certain identifiable pressures, not least those from the sales forces of pharmaceutical companies. But in the ancient world there were no equivalent pressures, but just doctors trying to impress patients—or potential pupils—and patients in turn making demands on doctors.[73] In these circumstances the phenomenon we encounter in Greece is more surprising, particularly since to justify the use of new therapies the ancients had only insecure analogues to the argument that appeals to the full apparatus of "modern medical science."[74] Unlike in natural philosophy—where what was at stake was merely a matter of belief and not usually one of immediate practical consequence—the treatment a patient received from a Hippocratic or from any other type of healer might, at the limit, make the difference between death and recovery. In this context we might expect a reasonably deep-seated caution, if not conservatism, to prevail, and indeed some aspects of our Hippocratic evidence suggest just that.[75] Yet the degree of innovativeness tolerated, even favoured, is striking, nor is this just a matter of

---

73. Many doctors earned their living partly as teachers (as the *Oath* demonstrates) and so were, no doubt, like other teachers, sometimes subject to the pressures towards innovativeness that may arise from an ambition to collect pupils (see further below, at n. 163 and p. 100f.). Meanwhile the Hippocratic Corpus illustrates some of the complex interactions between doctors and patients. The surgical treatises, especially, provide several examples where a doctor appears to be responding to pressure from his patients to employ a popular or modish treatment against his better judgement. In addition to *Fract.* 1 (mentioned above, n. 61), where the criticism is directed at the newfangled in particular, at *Fract.* 16 (L) 3.474.16ff. the writer first expresses his doubts about what to advise concerning the use of hollow splints in the treatment of fractured legs. At 476.8ff. he proceeds: ἔστιν οὖν σὺν σωλῆνι καὶ ἄνευ σωλῆνος καὶ καλῶς καὶ αἰσχρῶς κατασκευάσασθαι· πιθανώτερον δὲ τοῖσι δημότῃσίν ἐστι, καὶ τὸν ἰητρὸν ἀναμαρτητότερον εἶναι, ἢν σωλὴν ὑποκέηται· καίτοι ἀτεχνέστερόν γέ ἐστιν. ("Thus it is possible both with and without the hollow splint to arrange things well or clumsily. But ordinary people have greater faith in it and the doctor will be more free from blame, if a splint is applied, even though it is rather bad practice.") Cf. also *Art.* 67 (L) 4.280.11ff., 77 (L) 4.308.12ff.
74. Cf., however, below, n. 175.
75. Thus the famous first Aphorism (*Aph.* 1.1 [L] 4.458.1ff.) specifies not just that "opportunity is elusive" but that "experience is treacherous" (πεῖρα

fancy bandaging, nor one where, the case being desperate, *any* remedy is worth trying. Among the surgical practices the Hippocratic doctors elaborated are some of the daunting, if not foolhardy, techniques I mentioned in Chapter 1, the forcible straightening on the Hippocratic bench,[76] and the succussion of the patient, often upside down, used not just in certain cases of reduction but even, amazingly, in some of difficult childbirth.[77] Some of those who practised succussion are accused of doing so just for the performance, "to make the crowd gape": that is the criticism made in chapter 42 of *On Joints*, although that treatise goes on itself to endorse and recommend a modified version of succussion in some cases.[78]

σφαλερή) and judgement difficult. (*Aph.* 1.20 [L] 4.468.8ff. and *Hum.* 6 [L] 5.484.13ff. warn the doctor against νεωτεροποιέειν, though that is just a matter of avoiding innovation during the course of the treatment of particular cases; cf. *Aph.* 2.52 [L] 4.484.13f., which recommends continuing with the same treatment if the diagnosis remains the same. Cf. further below, Chap. 3 at nn. 82f., on passages warning doctors about the dangers of certain treatments.) Much later, Galen, too, criticises certain practitioners for their overeagerness to innovate, e.g., (K) 11.252.10ff.

76. See above, Chap. 1 n. 60. Who invented this and various other complex surgical procedures described in our Hippocratic texts cannot usually be determined, though Celsus provides us with a list of names (including "Hippocrates") of those who invented surgical instruments (*Med.* 8.20.4, *CML* 1.407.7ff.). There is, however, good reason to believe that the writers represented in our Hippocratic texts were themselves responsible for certain developments in surgical procedures, as, for example, when they refer to their own adaptation or modification of existing methods: see *Art.* 38 (L) 4.168.9ff., 13ff.; 78 (L) 4.312.5ff.

77. Of the texts cited above, Chap. 1 n. 59, in which succussion is recommended, that at *Foet.Exsect.* 4 (L) 8.514.14ff. concerns difficulties at delivery, cf. also *Mul.* 1.68 (L) 8.142.20ff. At *Epid.* 5.103 (L) 5.258.9ff. (≈ *Epid.* 7.49 [L] 5.418.1ff.) one case of a woman who had been succussed in childbirth is reported which ends fatally, though success in the use of succussion in a case where a patient suffers from a liver complaint is claimed at *Epid.* 6.8.28 (L) 5.354.4f.

78. *Art.* 42 (L) 4.182.13ff., cited above, n. 59. Despite the criticisms the writer makes of those who used the method, he notes, nevertheless, that "I think it is not hopeless, if one has proper apparatus and does the succussion properly, that some cases may be straightened out," 42 (L) 4.184.1f., though that remark is then undercut by his statement that such methods belong, rather, to charlatans. However, the immediately following chapter describes what the writer calls the proper arrangement for succussion if one wants to

In the prominence of the authorial ego, the prizing of innovation both theoretical and practical, the possibility of engaging in explicit criticism of earlier authorities, even in the wholesale rejection (at times) of custom and tradition—in all these features there are marked contrasts of degree, if not also in kind, between *parts* of early Greek, and most ancient Near Eastern, speculative thought, for example, not just in styles of presentation but also in the substance of what could be presented and discussed. These are far from being the only contrasts that we might consider in relation to the study of nature, nor are they features that are confined to that general domain of inquiry. But they raise a set of problems that any evaluation of the early stages of science must confront, even if we must acknowledge that the issues stretch far beyond the range of our discussion here.

The Role of Literacy

One thesis already in the field—that aims to explain the growth of both critical and innovative attitudes, both in their ancient Near East-

---

use it, 43 (L) 4.184.5ff., 11ff. Here he is still dealing with cases of humpback and he distinguishes where succussion should be made with the patient attached to the ladder feet downwards (where the curvature is near the neck) and where with the patient head downwards (when the curvature is lower down). He ends this chapter by saying that it is distasteful to go on at length about these matters, but all the same succussion would best be carried out by the methods he describes (43 [L] 4.186.11f.). Cf. also 44 (L) 4.188.1ff., where he concludes his instructions at 188.13–16: ταῦτα μέντοι τοιουτοτρόπως ποιη- τέα, εἰ πάντως δέοι ἐν κλίμακι κατασεισθῆναι· αἰσχρὸν μέντοι καὶ ἐν πάσῃ τέχνῃ καὶ οὐχ ἥκιστα ἐν ἰητρικῇ πουλὺν ὄχλον, καὶ πολλὴν ὄψιν, καὶ πουλὺν λόγον παρασχόντα, ἔπειτα μηδὲν ὠφελῆσαι. ("This is the sort of thing that must be done if it is absolutely necessary to use succussion on a ladder: however, it is disgraceful in any art and especially in medicine to make parade of much trouble, display and talk, and then do no good")—as if the question of shamefulness were subordinate to the matter of the results achieved. Moreover, at *Art.* 70 (L) 4.288.11ff., suspension of the patient by his feet from a cross-beam to reduce dislocation of the thigh at the hip is recommended as a "good and correct method" and as "having something of the competitive about it" (καὶ δή τι καὶ ἀγωνιστικὸν ἔχουσα, 70 [L] 4.288.13) for someone who takes pleasure in ingenious devices—as if the writer there were not above exploiting the striking effects that certain treatments might make to further his reputation (cf. also 48 [L] 4.214.1f.).

ern manifestations and in their Greek ones—has it that the key factor is the development in techniques of communication. One influential statement of such a thesis is the article by Goody and Watt entitled "The Consequences of Literacy,"[79] and Goody has subsequently returned to the issues on more than one occasion, modifying his thesis especially on the nature of the contrast between Greek and Near Eastern achievements and focusing increasingly on the latter.[80] Thus Goody argues first that many of the pre-Greek achievements depend essentially on the existence of written texts, and secondly that so far as Greece itself is concerned, the *spread* of literacy made possible by the introduction of an *alphabetic* system of writing is all-important. Literacy by itself, of course, does not discriminate between those who composed Egyptian medical texts or Babylonian astronomical cuneiform tablets on the one hand, and Hippocratic authors or pre-Socratic natural philosophers on the other.[81]

Many questions raised by this thesis are controversial and beyond the scope of our discussion here, notably the historical problems surrounding the development of the alphabet itself, the contributions of various Semitic groups, and the range of possible intermediaries between them and the Greeks,[82] as well as the thorny issues in dispute between Havelock and his opponents on the timing and extent of the spread of literacy within Greece itself.[83] But three points can readily be

---

79. Goody and Watt 1968, cf., for example, Vansina 1971, superseding Vansina 1961/1965, and two recent collections of articles, Gentili and Paioni 1977, and Vegetti, ed., 1983.

80. Goody 1977 and forthcoming, cf. Street 1984 and Parry 1985. The strength of the claims made for the "literacy thesis" has varied in different formulations, while it remains clear that it is invoked in order to give a causal, not merely a descriptive, account of certain changes or developments. Goody and Watt 1968 spoke of "consequences"; cf. Goody 1977, p. 75, "implications." While Goody 1977, pp. 10, 46, 51, explicitly disavows monocausal or single-factor explanations, as does Goody forthcoming, the thrust of his argument nevertheless has been effectively to discount other considerations.

81. For one attempt to evaluate the extent and the spread of literacy in ancient Egypt, see Baines and Eyre 1983, and cf. Baines 1983.

82. See, for example, Jeffery 1961, 1982, Snodgrass 1971, pp. 348f., Driver 1976, Coldstream 1977, pp. 295ff., Isserlin 1982.

83. See Havelock 1963 and the essays collected in Havelock 1982, especially; cf. Kenyon 1951; E. G. Turner 1951; Davison 1962; Harvey 1966;

agreed. First, the existence of written texts obviously permits a different kind of critical inspection, more leisurely and more formalised than is possible with spoken discourse.[84] It *permits*, though it does not necessarily *dictate*, critical scrutiny, since the existence of written texts may and often does positively *inhibit* it,[85] a point that has been made often enough by anthropologists (for example, by Shirokogoroff in his study of the psychomental complex of the Tungus)[86] and by Orientalists (for example, by Oppenheim in relation to Mesopotamian

---

Knox 1968; Reynolds and Wilson 1968/1974; Pfeiffer 1968, pt. 1, chap. 2; Robb 1970; Finley 1964–65/1975a, 1977; Burns 1981; Bremmer 1982; Gentili 1983. The question of the extent to which, at any period, a male citizen at Athens could do more than merely write his name should, of course, be kept separate from the issue of the role of written texts in the culture of those who numbered themselves among the most literate sections of society (cf. Finley 1983, pp. 29ff.). In Plato's day references to learning how to read and write as part of primary education are commonplace: see, for example, *Prt.* 325e, 326d, *Charmides* (*Chrm.*) 159c. However when Aeschines 1.6–11 suggests that primary schooling had been compulsory since Solon, this is dubious as a piece of historical evidence (cf. Havelock 1982, p. 205 n. 4), though it is certainly a pointer to what some Athenians would have liked to believe about their past. Yet even though Aristophanes *Frogs* (*Ranae*) 52ff. *may* allude to silent reading, such allusions are very rare in the classical period (see Knox 1968, Woodbury 1976), and it is agreed on all sides that in the fifth and fourth centuries a written text, when read, was usually read out aloud. The opening exchanges in Plato's *Parmenides* (*Prm.*) are among the many texts that illustrate this. There Zeno has brought his book, and Socrates, keen to learn its contents, asks him, not to lend him the text, but to read it out (*Prm.* 127c, cf. 127d–e).

84. See, for example, Goody 1977, p. 149.

85. That is to put the point negatively. Goody has the further argument, however, that there is a positive side to the limitations to a field of inquiry presented by written texts. Once the medical texts (for instance) were there for consultation, this not only preserved certain medical lore but provided a focus of attention that gave medical practitioners guiding ideas about what to look for in individual cases. Moreover, more generally, the scrupulous making and transmission of records in the ancient Near East clearly gave new depth to a sense of the past and thus transformed the notion of tradition itself, lending greater authority to appeals to the precedents it afforded: cf. Baines 1983, pp. 587ff.

86. Shirokogoroff 1935, pp. 108, 340ff. Cf. Lévi-Strauss 1958/1968, Lotman and Pjatigorskij 1977.

medicine)[87] and the validity of which for late Greco-Roman antiquity is obvious enough.[88] Secondly, when recorded in writing, innovations have a greater chance of being recognised as such and of being cumulative.[89] Thirdly, certain types of writing that are taken to be characteristic of early literacy (tables, lists, formulas, recipes) may stimulate certain types of question—for example, problems of classification—and thus affect cognitive processes themselves.[90]

All these are positive contributions to our understanding of these complex questions. Yet reservations about how far Goody's thesis, and others like it, go to resolve our main problems in relation to early Greek speculative thought must also be expressed. One area where some of the points Goody made can be tested, but about which he has so far had comparatively little to say in his three main discussions, is mathematics, where, as we have already noted, we have extensive Egyptian and Babylonian texts as well as Greek ones—the last beginning much later than the other two, of course.[91] Each of these three cultures developed its own arithmetical notation or notations, the Greek, like the Egyptian, being a decimal, the main Babylonian a sexagesimal, system. But as Goody himself has remarked,[92] mathematics rests on universal logographic symbols, not on restricted phonetic

---

87. Oppenheim 1962, p. 104.

88. See below, pp. 104ff.

89. See Goody 1977, chap. 3.

90. See Goody 1977, pp. 99ff., where, commenting on Egyptian Onomastica, Goody writes (p. 102): "We can see here the dialectical effect of writing upon classification. On the one hand it sharpens the outlines of the categories; one has to make a decision as to whether rain or dew is of the heavens or of the earth; furthermore it encourages the hierarchisation of the classificatory system. At the same time, it leads to questions about the nature of the classes through the very fact of placing them together."

91. The evidence for Babylonian mathematics is collected in Neugebauer and Sachs 1945. For a first orientation in Egyptian mathematics, see Neugebauer 1934, chap. 4, 1952/1957, pp. 71ff.; van der Waerden 1954/1961, pp. 15ff.; among the principal primary sources are those edited by Peet 1923, Struve 1930, R. A. Parker 1972. Note the emphatic reminder in Neugebauer 1952/1957, pp. 53ff., about how much of the ancient Near Eastern, especially Mesopotamian, material remains unpublished.

92. Goody 1977, p. 122.

ones. In that respect all three notations are equivalent. In any case none has any *evident* superiority over the others on this score that could be compared to the superiority of an alphabetic system of writing over syllabaries, let alone over pictograms. Indeed, if there are advantages, these lie with the Babylonians, for their notation incorporated the place-value system and so revealed what the Greek use of letters for numbers conceals, namely, the equivalence of operation involved in the multiplication and division by the base and by its powers, by 10 and by 100, or, in the sexagesimal system, by 60 and 60^2.

The fact that in all three cultures the development of complex mathematical manipulations depended on the existence of *some written* notation can certainly be taken to confirm *that* element in Goody's thesis. The role of tables here is particularly clear. Like our multiplication tables, tables of reciprocals, for instance, such as we have from Babylonia,[93] encapsulate knowledge that is itself decontextualised and that can be put to a variety of uses in various concrete situations. Yet Goody's thesis by itself does nothing to help explain certain major differences in the mathematics developed in these three cultures; it can hardly explain them since, as noted, the technical processes of communication involved, the notations, are in the crucial respect *broadly* equivalent.

Take one important development that is confined, so far as we know, to Greece, namely, that of the explicit notion of, and demand for, demonstration or proof.[94] By *proof* I do not mean the confirmation or checking of a result that regularly occurs in Egyptian arithmetic, for example, and that is often translated into English, legitimately enough, as the "proof": the scribe works his way through to the solution of a

---

93. See, for example, Neugebauer and Sachs 1945, pp. 11ff. Cf. the rarer evidence for multiplication tables in Egypt, Peet 1923, pp. 103f., R. A. Parker 1972, pp. 72f.

94. The attempts by Seidenberg (e.g., 1960–62, 1977–78, cf. van der Waerden 1980a, 1980b) to trace proofs and proof-theoretical interests back to Vedic mathematics are unconvincing, foundering on the difficulty that inadequate attention is paid to the fundamental distinction between the ability to get results, and having clear and explicit concepts of aims and methods. Cf., e.g., Knorr 1981, pp. 147f.

problem and then checks that his result is correct.[95] What the Greeks eventually developed was the concept of proof in the more rigorous sense of demonstration by deductive argument from clearly identified premises: and the qualification "eventually" should be stressed, since this was a gradual and hard-won development and not a concept available to the Greeks in, say, the days of Thales or Pythagoras.[96] There is accordingly no call whatsoever in this respect (or indeed in any other) to speak of the Greeks as endowed with some special natural characteristic, some distinctive mental ability, as those who fantasised about the "Greek miracle" liked to do.

Moreover, in connection with the development of the concept, and practice of, mathematical proof, there is already by the late fifth century B.C. a concern with foundational problems, not the famous *Grundlagenkrisis* postulated by some historians who saw Greek mathematics as brought to an abrupt standstill when the incommensurability of the side and diagonal of the square was discovered,[97] but more simply an interest in the *elements,* that is, the fundamental principles

---

95. See, for example, in the Rhind papyrus, Peet 1923, pp. 64f. on number 29, pp. 68f. on number 33.

96. The sources that ascribe a variety of mathematical knowledge, and indeed attempts at mathematical proof, to Thales are generally untrustworthy, and attributions of sophisticated mathematical concepts to Pythagoras himself are often fantastical (see Burkert 1972). Our first extended extant Greek mathematical text is that of Hippocrates of Chios, preserved in Simplicius *In Ph.* 60.22–68.32. This already shows considerable mathematical knowledge and a clear grasp of deductive argument. But while the very fact that Hippocrates is reported to have composed a treatise of *Elements* by itself shows an interest in the starting-points from which mathematics can be built up, there is no evidence to suggest that he had a clear notion corresponding to that of axiom, nor that he had distinguished among different types of basic undemonstrated premises. It is striking that in the proof of the quadrature of lunes, at *In Ph.* 61.5ff., the "starting-point" or ἀρχή for that piece of reasoning is a proposition that is itself proved from a prior proposition.

97. Against Tannery 1887, p. 98, Hasse and Scholz 1928, von Fritz 1945/1970 (who mostly favour an early date—in the early fifth, if not in the sixth century—for the discovery of incommensurability), see especially Burkert 1972, pp. 455ff.; Knorr 1975, pp. 306ff. The arguments are set out briefly in G. E. R. Lloyd 1979, pp. 112ff.

which the rest of mathematics presupposes and on the basis of which
the rest of mathematics can be built up. From the very first—that is,
from the work of Hippocrates of Chios, sometime around 430 B.C.—
the attempt to systematise mathematics depended on decisions as to
what to take *as* the elements. By the fourth century we know that
Greek mathematicians were actively exploring the possibility of alternative starting-points to geometry,[98] though not (despite what has
sometimes been claimed)[99] the possibility of alternative geometries:
when Aristotle mentions the possibility of denying that the internal
angles of a triangle add up to two right angles, it is not in connection
with any proposal to construct an alternative geometry on the basis of
this denial.[100] After Euclid, too, the question of what should be included among the axioms continued to be disputed, as the particularly
well-documented and famous controversy over the status of the parallel postulate illustrates sufficiently conclusively.[101]

---

98. In his commentary on Euclid, *In primum Euclidis Elementorum li-
brum commentarius* (*In Euc.*) 66.14–68.6, Proclus refers to a sequence of
mathematicians after Hippocrates and Theodorus and before Euclid himself
who continued and extended work on the elements. They include Leodamas,
Archytas, Theaetetus, Neoclides, Leon, Eudoxus, Amyclas of Heraclea, Menaechmus, Dinostratus, Theudius, Athenaeus of Cyzicus, Hermotimus of Colo-
phon, and Philip of Mende. But they should not be imagined as being in fundamental agreement either about the types of starting-points necessary for the
construction of mathematics or on particular issues relating to specific definitions, postulates, or common opinions. On the contrary, it is clear that there
were disputes of a more than purely nominal kind on such questions. Thus
those concerning the conceptions of number, plurality, and unity are discussed
by Klein 1968, while the ambiguity of the term *elements* itself was remarked
by Menaechmus: see Proclus *In Euc.* 72.23f.

99. See, for example, Toth 1966–67, 1977, Bruins 1968, Hösle 1982, and
cf. Kayas 1976.

100. See, for example, *Analytica Posteriora* (*APo.*) 93a33ff., *Ph.* 200a16ff.,
29f., *Metaphysics* (*Metaph.*) 1052a6f.

101. See especially Proclus *In Euc.* 191.21ff. Elsewhere Proclus discusses
the alternative definition of parallel proposed by Posidonius (176.5ff.), and
Geminus' objection to the parallel postulate (183.14ff., 192.5ff.), as well as setting out Ptolemy's, and his own, attempted demonstrations (365.5ff., 371.10ff.,
23ff.). See the discussion of this and other ancient evidence, as well as of aspects of the later history of the issue, in Heath 1926, 1, pp. 202ff.

None of these developments (some of course quite late) can be paralleled in extant Egyptian or Babylonian mathematics, even though, let me repeat, we can infer that they too were highly innovative in many other respects. None depends simply on technical advances in notation and the like. They do, however, all have fairly obvious affinities with the features I exemplified in early Greek natural philosophy and medicine. In mathematics, too, there is not just *manifest* disagreement between rival views and the demand for validation, but also, we may infer, a fair degree of egotism. To be sure, in this case, with fewer early primary texts at our disposal, we cannot cite extensive passages to illustrate the use of the authorial ego—not before Archimedes, at least.[102] Yet even in the wreck of pre-Euclidean geometry we have enough reliable evidence for the individual contributions of named theorists—for example, on special problems such as squaring the circle or the duplication of the cube,[103] or in the discovery of particular theorems or

---

102. Archimedes often speaks in the first person, especially but not exclusively, in the prefaces to his treatises: see, for example, *Arenarius* (*Aren.*) 1.1 (HS) 2.216.15ff., 1.8 (HS) 2.220.10ff., 2.4 (HS) 2.236.8f., 3.1 (HS) 2.236.17, and cf. in the definitions and postulates in *De sphaera et cylindro* (*Sph.Cyl.*) (HS) 1.6.6, 15, 20; 8.2. And he frequently refers directly to his own discoveries and to those of other mathematicians: see, for example, *De conoidibus et sphaeroidibus* (*Con.Sph.*) proem (HS) 1.246.2ff., *De lineis spiralibus* (*Spir.*) proem (HS) 2.2.2ff., 13ff., 18ff., 4.1ff., *Quadratura parabolae* (*Quadr.*) proem (HS) 2.262.3ff., 264.4, *Methodus* proem (HS) 2.426.4ff., and cf. below, n. 104. The terminology of *sects* for rival groups of mathematicians can be illustrated in Nicomachus *Arithmetica introductio* (*Ar.*) 143.1ff. In the developing mathematical sciences, similarly, we find in the tradition of writers on harmonics, for example, some, such as Aristoxenus, who make emphatic claims for their originality: see *Harmonica* (*Harm.*) 1.1, 2–3, 4 ("No one has had any idea of these matters as yet, but in dealing with them all it has been necessary for us to make a new beginning, for we have had nothing handed down to us worthy of note"), 5–6 ("On these points no account, either with or without a demonstration, has yet been given. . . . No one has touched on this part of the subject at all as yet, except Eratocles, who attempted a partial enumeration without demonstration") 2.35–36, 37–38.

103. The most famous of the early discussions of quadratures, after that of Hippocrates of Chios, are those of Antiphon and Bryson (mentioned by Aristotle, e.g., *Topica, Top.* 171b12ff., 172a2ff.), Hippias (to whom the quadratrix is attributed by Proclus *In Euc.* 272.7ff., 356.11, though whether this is

their proofs[104]—to be fairly confident that, like early Greek philosophers and medical writers, Greek mathematicians were often ambitious innovators and proprietorial towards their own ideas.

The Argument from Politics

Elsewhere, following Vernant and others, I have argued that *in addition to* other factors that must be held to be relevant, including the
spread of literacy, the political dimension is crucial for our understanding of some of the distinctive characteristics of early Greek speculative thought.[105] Dealing with some Greek attacks on magic in particular, and more generally with the development of a certain openness
and dialectical acuteness in parts of Greek philosophy and science
(and I stressed then as I do again now that it is not the *whole* of Greek
philosophy and science that can be so characterised),[106] I argued that

---

the sophist Hippias of Elis is disputed), and Dinostratus (see, for example,
Pappus *Collectio* 4.30–32 [1.250.33–258.19 Hultsch]). In his Commentary
on Archimedes' *Sph.Cyl.* Eutocius refers to several early attacks on the problem of duplicating a cube, including those of Hippocrates of Chios, Archytas,
Eudoxus, and Menaechmus, (HS) 3.54.26ff., 78.13ff., 84.12ff., 88.17ff.,
90.4ff. See Heath 1921, 1, pp. 220ff., 244ff.; Knorr 1986.

104. Thus Archimedes scrupulously distinguishes between the discovery of
the *proof* of the theorems relating the volumes of a cone and a cylinder, and
those of the pyramid and prism, and the discovery of those theorems themselves. The latter, he tells us, is to be credited to Democritus, the former to
Eudoxus: Archimedes *Methodus* proem (HS) 2.430.1ff., 6ff., cf. *Sph.Cyl.*
proem (HS) 1.4.2ff.

105. See J.-P. Vernant 1957/1983, 1962/1982, 1983; Vidal-Naquet 1967/
1986, 1970/1986; Detienne 1967; G. E. R. Lloyd 1979, chap. 4.

106. It is certainly not the case that the pre-Socratic philosophers and Hippocratic writers unanimously adopted an attitude of openness towards a lay
public. Although the alleged secrecy of the Pythagorean sects has been exaggerated in some of our late sources, particularly in relation to the penalties imposed on Hippasus for the supposed divulgence of the incommensurability of
the side and the diagonal of the square, some teaching even before the end of
the fourth century B.C. seems to have been esoteric in character, taking the
form of apophthegms, σύμβολα, whose meaning was certainly not transparent to the layman: see Burkert 1972, pp. 178ff., 454ff. Meanwhile the restriction of medical instruction to certain specified individuals is a feature of
the Hippocratic *Oath* (ὅρκος: *Jusjurandum*), CMG 1.1.4.7ff.; indeed the *Law*

these reflect the very considerable experience that many Greek citizens acquired in the evaluation of evidence and arguments in the contexts of politics and the law. True, that experience is uneven, far greater in the democracies than in the oligarchies, but it extended also to the latter on a restricted scale, even though the proportion of the population engaged in decision-taking was smaller and the occasions when they did so were fewer.[107] It is well known, however, that in some of the democracies that experience could be very extensive indeed, on jury service, in the Assemblies, and in any one of a wide range of offices held by lot or by election.[108] Certain aspects of the Greek experience of the pro-

---

($νόμος$: *Lex*) speaks of initiates, 5, CMG 1.1.8.15ff. (see below, Chap. 6 n. 160), while *Decent.* 16, CMG 1.1.29.13ff. advises withholding certain information from patients (cf. also *Decent.* 18, CMG 1.1.29.32f.). It has, too, been suggested that the obscurity of some of the aphoristic works—especially $περὶ$ $τροφῆς$ (*De alimento*, *Alim.*), *Hum.*, parts of *Praec.*, and parts of *Decent.*—has been *deliberately* cultivated, to restrict access to medical lore to those who were taught personally by practitioners: cf. G. E. R. Lloyd 1979, pp. 227–29. Conversely, though obscurantism is a feature of some Greek astrological and alchemical writings, the first three books of Artemidorus' *Onirocritica* were addressed to a general public, and even the last two, conveying more specialist lore for his son's personal use (according to book 4 pr. 238.1ff.), are not markedly more esoteric in style or content.

107. We know appreciably less about the political institutions of Sparta and Crete, for example, than about those of Athens, but in the oligarchies, too, as Finley 1983, p. 62, has put it, "we know that at times there were sharp disagreements over policy that had to be resolved politically." Thus Thucydides reports one debate at Sparta at 1.79ff., commenting that the decisions in the Lacedaemonian assembly were reached by acclaim rather than by voting (1.87). Moreover, Aristotle distinguishes between different types of oligarchy precisely by reference to the way in which the class of those who shared in deliberation was defined, *Pol.* 1298a3ff., 34ff.; cf. 1297a17f. At *Pol.* 1272a10f., for instance, he reports that at Crete all the citizens shared in the Assembly, though this only ratified the decisions of the elders and the Cosmoi, and at 1270b18ff. he puts it that at Sparta the people—$δῆμος$—keep quiet since they have a share in the greatest magistracy, the Ephorate, the Ephors being elected from the whole *demos*. See Andrewes 1966; D. M. Lewis 1977, pp. 36ff.; de Ste. Croix 1972, pp. 27ff.

108. The issues have been fully discussed recently by Finley 1983, chap. 4. To mention just two of the most striking estimates he gives concerning Athens in the fifth and fourth centuries: "the best analysis of the evidence, some of it archaeological, suggests that attendance [at the Assembly] ran to 6000 in the fifth century, to substantially more in the fourth"; "in any decade, something

cess Goody called the domestication of the savage mind—a process never, of course, completed, transparently not so in Greece—depend less, I would claim, on technical improvements in communication than on developments in, broadly, the political domain; they owe more to the experience that many Greeks gained there in types of argument and scrutiny than to the spread of literacy among them.[109]

On the more general issues of the domestication of the savage mind, I shall not repeat my earlier arguments. But there is more to be said on the specific questions we are concerned with here, where the problem presented by the material we reviewed is not so much one of innovation *tout court*, for to some degree innovation is, we said, universal. Rather, it arises from the combination of the degree of contestability of tradition and of what we may call the pressures towards *overt* innovativeness, the fashion for, even the obsession with, the novel, familiar enough to us today, but scarcely to ancient civilisations.

Now there are very evidently political dimensions to this issue too. We shall need to come back to these in due course, but three of the most obvious points may be mentioned at once. Clearly, first, innovativeness is just as prominent, or even more prominent, an aspect of Greek political life as of Greek speculative thought, and in the former is no mere theoretical matter, as new ideas could be and were put into

---

between a fourth and a third of the total citizenry over thirty would have been Council members, serving daily (in principle) throughout the year and for a tenth of that year on full duty as so-called *prytaneis*" (Finley 1983, pp. 73f.). Finley is also careful to point out both that "even in Athens under what modern historians tend to call the 'radical democracy', the *demos* never produced spokesmen in the Assembly from their own ranks" and that "the evidence strongly suggests that even in Athens few exercised their right of *isegoria*" (1983, pp. 27, 139f.; and cf. the reservations about the running of the democracy in R. Osborne 1985). Yet whether or not they participated as speakers in the debate, all fundamental policy decisions rested with the citizens as voters in the Assembly, just as it was they who constituted the standing bodies of jurors with whom lay the decision in the courts: see A. R. W. Harrison 1971, pp. 43ff. Cf. Hansen 1974, 1976/1983, Lanza and Vegetti 1975.

109. See G. E. R. Lloyd 1979, pp. 246ff., especially 252ff., which discuss the parallelisms between the development of the notions of evidence and witnessing, testing, scrutiny, and accountability, in the contexts of law and politics and in speculative thought.

practice in the framing and reform of constitutions and in legislative measures of every kind—a trait often disapproved of, and feared, to be sure,[110] but also often greatly admired.[111] Secondly, the possibility of dissent from deep-seated traditional views presupposes a certain measure of political freedom of speech, though that measure should certainly not be *over*estimated.[112] Thirdly and most importantly, the revisability of political constitutions and of the laws of various Greek city-states not only parallels, but at points interacts with, the revisability accepted in such other areas as cosmology, religion, and moral philosophy.

Thus political and moral philosophical innovation are often intrinsically related, nor can we doubt, surely, that political revisability helped to release inhibitions about revisability in other domains of thought, even while there may also be feedback from those other domains of thought to political revisability. An extended text in Aristotle's *Politics* shows that he, for one, recognised the special importance of political innovativeness in relation to other manifestations of innovativeness in other fields, for he remarked on both the similarities and the differences in this respect between politics and such arts as medicine. So far as the similarities go, he notes the argument that the advances that have been made in both domains depend on innovative-

---

110. As is well known, καινοτομία and νεωτερίζειν were often used not just of political innovations but of revolutionary movements, *res novae*, e.g., Plato *Lg.* 950a, Thucydides 1.97. Cf. the accusations against Socrates for innovating περὶ τῶν θείων: *Euthyphro* (*Euthphr.*) 5a7f., cf. *Ap.* 24c1 (Plato's own views on the dangers of political changes are expressed very forcibly at, e.g., *Lg.* 758c, 797d ff., though he also shows he is well aware of the novelty of his own proposals in the *Laws,* 797a, 805b, 810d). On the workings of the γραφὴ παρανόμων, the fifth-century Athenian procedure whereby any citizen could prosecute another for having made an "illegal" proposal, even when the Assembly had approved it, see H. J. Wolff 1970, Hansen 1974, Finley 1983, pp. 53ff.

111. For example at Thucydides 1.71, a Corinthian speaker contrasts Spartan conservativeness unfavourably with Athenian adventurousness, πολυπειρία, and innovativeness, and draws a comparison with art to suggest that as in that case, so too in politics, "what supervenes necessarily wins out": ἀνάγκη δὲ ὥσπερ τέχνης αἰεὶ τὰ ἐπιγιγνόμενα κρατεῖν.

112. See below, at nn. 187–88.

ness, putting it, very much with the voice of Greek rationalism, that all
men seek the good rather than the traditional.[113] At the same time he
points to a contrast, in that politics deals with what is established not
just by rules but by custom, so that frequent changes in the laws have
the effect of weakening the capacity of the law. Thus in this regard
"much caution is necessary."[114] But the very contrast suggests that, like
many other Greek theorists, Aristotle saw politics as a master art that
controls the very framework within which the other arts are exercised.
The reason why particular caution is needed as regards innovation in
politics is, precisely, that it has such far-reaching repercussions.

To throw light on the problems presented by both the positive and
the negative aspects of the pressures towards overt innovativeness in
Greece we may investigate some of the *contexts* in which it is mani-

---

113. *Pol.* 1269a3–4: ζητοῦσι δ᾽ ὅλως οὐ τὸ πάτριον ἀλλὰ τἀγαθὸν
πάντες.

114. At *Pol.* 1267b22ff. Aristotle first outlines and then criticises the ideas
of Hippodamus on the best constitution: these included the proposal that
those who discovered something beneficial for the state should be rewarded
(1268a6ff., cf. Xenophon *Hiero* 9.9f.). When Aristotle comes to criticise this
view at 1268b22ff., 33ff., he raises the general question of the advisability of
changes to the νόμοι (here covering both "laws" and "customs") of states: "It
might seem better to make alterations; at least in other branches of knowledge
it has proved beneficial, for example when medicine has been altered from the
traditional system, or gymnastic training, and in general in all the arts and
faculties: so that since political skill is counted as one of these, it is clear that
the same thing necessarily holds in this case too." Aristotle then offers various
evidence from past political developments that could be taken as endorsing
that point of view, for example the abandonment of such "simple" and "bar-
barian" customs as carrying arms and buying wives. Yet at 1269a12ff., having
said that "it is clear from these considerations that even some of the laws are
sometimes to be altered," he proceeds: "but from another point of view it is
agreed that much caution is necessary." Small improvements in the law are not
worthwhile: the analogy with the arts is false, since it is not the same thing to
change an art and to change a law or custom, as the latter have no power to
compel obedience apart from habit (ἔθος) and that only becomes established
after a long time. Thus, lightly to change from the existing laws to new ones
will be to weaken the capacity of law (1269a14–15, 19–24). The aporetic
nature of Aristotle's discussion and its implications for understanding his conception of νόμος have recently been explored in an acute study by Brunschwig
1980b, cf. de Romilly 1971.

fested. In Chapter 1 I broached some aspects of the growth and transformation of rivalry in claims to wisdom, σοφία, and some points should now be elaborated further.

Σοφία and the Sophistic Debate

Both in the archaic and the classical periods the term σοφία had, as is well known, an enormous range.[115] It is often and foremost skill in poetry that is in question.[116] But in the classical period you can be called σοφός in any one of the arts, painting or sculpting or flute-playing, in athletic skills, wrestling, or throwing the javelin or horsemanship, and in any of the crafts, not just in piloting a ship or healing the sick or farming but, at the limit, in cobbling or carpentry or cooking: all those examples can be illustrated from the Platonic corpus.[117] From the seventh century onwards, many different kinds of leader gained a repu-

---

115. The use and development of the term σοφία and its relationship with both σοφιστής and φιλοσοφία have been discussed by, for example, Snell 1924; Nestle 1942; Gernet 1945/1981, pp. 355ff.; Detienne 1967; Gladigow 1967; Guthrie 1969, pp. 27ff.; Vegetti 1973; Lanza and Vegetti 1975; Svenbro 1976; Kerferd 1976; Classen, ed., 1976; Humphreys 1978, pp. 209ff.; Lanza 1979.

116. See below, n. 128.

117. For these and many other examples in the Platonic corpus, see, for instance, *La.* 194d–e, *Lysis* (*Ly.*) 210a, 214a, *Meno* (*Men.*) 93d, *Euthydemus* (*Euthd.*) 271d, 279e, 292c, 294e, *Prt.* 312c, *Hippias Minor* (*Hp.Mi.*) 368b, *Symposium* (*Smp.*) 175d–e, *Theages* (*Thg.*) 123b–126d, *Epinomis* (*Epin.*) 974e ff. In *Ap.* 21a–22e, it will be remembered, Socrates says how he set out to prove wrong the Delphic oracle's statement that no man was wiser than he, by going the rounds of those with a reputation for σοφία, where the three main groups he cross-examines are the statesmen, the poets, and the craftsmen. The question of who counts as σοφός is frequently problematised with respect to specific well-known figures, such as Pittacus (*Prt.* 339c), Simonides (*Prt.* 343b–c, *R.* 331e, cf. 335e), Themistocles and Pericles (*Men.* 93e ff., *Prt.* 320a, *Alcibiades* I [*Alc.* 1] 118c ff.), as well as the major "sophists" and natural philosophers (as repeatedly, in *Prt.*, e.g., 309d, *Grg.*, e.g., 487a ff., *Hippias Major* [*Hp.Ma.*], 281a ff., especially, cf. *Lg.* 888e, 890a) not to mention in connection with the Athenians as a whole, as at *Prt.* 319b ff. For σοφός in relation to craft skills, see also Aristotle περὶ φιλοσοφίας fr. 8, *EN* 1141a9ff.

tation for *sophia* in general. They included seers, holy men, wonder-
workers. Men such as Epimenides, Aristeas, Hermotimus,[118] were con-
sulted in crises or disasters, plagues or pollutions,[119] which shows how
wise men may be not just spokesmen of traditional lore but called in
where that knowledge faces an impasse—though, to be sure, in offering a way forward any wise man may represent himself as the true exponent of tradition as much as the mediator of knowledge that goes
beyond the common store. But already in the sixth century the variety
of σοφοί is considerable. Among those who appear in the lists of the
Seven Wise Men (starting with Plato's)[120] those who had a reputation
as statesmen figure prominently: they include Solon, Pittacus, and
Periander, and it is possible that Thales won his place in their number
as much for the political advice he gave his fellow Ionians[121] as for his
ideas about water as the origin of things—which, as is well known, are
formidably difficult to interpret with any confidence, being, indeed, a
matter of conjecture already for Aristotle.[122]

---

118. Epimenides, for example, is mentioned by Plato as an ἀνὴρ θεῖος
who sacrificed at Athens (*Lg.* 642d–e, cf. 677d–e) and the tradition that he
purified Athens is in Aristotle's *Constitution of Athens* (*Ath.*) 1, as well as in
Plutarch *Solon* 12, D.L. 1.110, and elsewhere. But much of the evidence for
these and other early wonder-workers is doubtful. For a balanced assessment,
see Burkert 1972, pp. 147ff., whose reservations (Preface and pp. 164f.)
should be noted concerning earlier speculative discussions (Meuli 1935; Dodds
1951, chap. 5, pp. 135ff.; Eliade 1964) ascribing knowledge of, or even participation in, shamanistic beliefs and practices to the ancient Greeks.

119. Consultations not just of oracles, but also of those deemed to be experts in religious matters, for example, when pollution has occurred or is suspected, continue to be referred to in fifth- and fourth-century texts, as is
clearly illustrated in Plato's *Euthphr.* 4b–c and 9a. See R. C. T. Parker 1983,
p. 141.

120. Plato *Prt.* 343a, cf. *R.* 335e f. D.L. 1.41f. records contrasting traditions concerning the membership of the Seven.

121. See, for example, Herodotus 1.170 (advice to set up a common centre
of government), D.L. 1.25, and compare the stories relating to his practical
skills, Herodotus 1.75 (diverting the river Halys), Aristotle *Pol.* 1259a6ff.
(forecasting a bumper crop of olives and cornering the presses). When at
Hp.Ma. 281c, Socrates says that most of the early wise men down to Anaxagoras kept clear of politics, this is more than a little ironical: he has just referred to Pittacus and Bias, as well as Thales.

122. It is well known that when at *Metaph.* 983b20ff. Aristotle discusses

The existence of more or less formalised competitions in "wisdom" of one kind or another from as early as the eighth century B.C. provides us with a clue concerning the *eventual opening* for the inquiry concerning nature.[123] I noted that Hesiod already tells us that he won a poetry competition and that there is a similar allusion to such a competition in the Homeric *Hymn to the Delian Apollo*.[124] Trials of skill at solving riddles, γρῖφοι, are referred to not only in the legend of Oedipus, but in our evidence for such admittedly shadowy figures as Mopsus and Calchas,[125] and riddles did not just remain a feature of oracular discourse,[126] for the ability to resolve them continued to be, in popular legend, a mark of the wise man.[127] Xenophanes provides pre-

---

the reasons that may have led Thales to make the ἀρχή water, he introduces his remarks with the words: λαβὼν ἴσως τὴν ὑπόληψιν ἐκ . . . ("perhaps he derived the idea from the following considerations . . .").

123. One striking illustration of this is the way in which, in the opening book of the *Metaphysics*, Aristotle relates his own philosophising to earlier traditions and indeed to general human capacities: for "all men naturally desire to know" (*Metaph.* 980a21), and philosophising springs from wonder (τὸ θαυμάζειν, 982b12ff.), so that in a sense even the lover of myth is a philosopher (b18ff., Ross's text). In the development of the arts, the discoverers were admired as wise, σοφός, as much as for producing what is useful (981b13ff.), and "everyone supposes that what is called wisdom is concerned with the first causes and principles" (981b28ff., 982a1–3). Moreover, when recording the views of the early philosophers, beginning with Thales (note φιλοσοφήσαντες at 983b1ff., 6ff., τοιαύτης φιλοσοφίας at 983b18ff.), Aristotle is ready enough to draw comparisons with the ideas of those whom he describes as the first to "theologise" (θεολογήσαντες, 983b27ff., cf. 984b23ff. and 1000a9ff. on Hesiod, cf. 1071b26ff., 1075b26f.), even though he does not, in so many words, suggest that the "philosophers" supplanted or superseded the "theologians"—nor would that be in tune with the view he expresses in *Metaphysics* Λ especially, that the highest form of philosophy is, precisely, the study of divine substance; cf. also 1026a18ff.

124. See above, n. 29. On the apocryphal story of the competition between Homer and Hesiod, see West 1967, N. J. Richardson 1981.

125. Hesiod fr. 278 Merkelbach and West: cf. the material collected in Ohlert 1912 and in Schultz 1914, and cf. Veyne 1983, pp. 41ff.

126. As is recognised directly in, for example, Herodotus (e.g., 1.53ff. and 71), and indirectly in the parodies of Aristophanes, e.g., *Peace* (*Pax*) 1070ff., *Birds* (*Aves*, *Av.*) 960ff.

127. Thus it is a recurrent theme in Diogenes Laertius that the "wise men" or "philosophers" whose lives he records show a particular ability both to formulate and to resolve riddles or puzzling questions of one type or another.

cious early evidence of rivalry (though not in a formal ἀγών) between different types of claimant to excellence when in fr. 2 he complains that the useless achievements of athletes are prized more than *his sophia*—where he speaks, no doubt, both as statesman and poet in general and as someone involved in the investigation into nature in particular.[128] We may recall, too, Heraclitus criticising others for "much learning," πολυμαθίη:[129] in one of the particular contexts in which he attacks Homer, Heraclitus specifically calls him "wiser than all the Greeks,"[130] and he expresses his contempt also for the "bards of the peoples" (δημῶν ἀοιδοί),[131] while reserving the title of "the wise" for his own teaching,[132] including his own teaching about the true Zeus.[133] Here, then, a space could be won for philosophy, including the kind we call science, in an area already associated with poetry and religion.[134]

This can be illustrated not just in his accounts of the proverbial wise men in book 1 (Thales, 35f.; Chilon, 68f.; Pittacus, 77f.; Bias, 86f.; Cleobulus, 89) but in later books as well (for example, Aristippus, 2.68ff.; Aristotle, 5.17ff.; Theophrastus, 5.39f.). That wisdom may be enigmatic is a theme in Plato *Chrm.* 161c, 162a–b, 164e ff., cf., e.g., *Alc.* 2.147c–d. It may not be too far-fetched to recall that some of the classic problems in Greek mathematics are posed in the form of riddles, for example, the Delian problem of the duplica-tion of a cube (see Theon 2.3ff., Eutocius [HS] 3.88.5ff.) and Archimedes' Cattle Problem in the work of that name, (HS) 2.528.5ff.

128. For the common use of σοφός, σοφία, in relation to poets and poetry in particular, see, for example, among many other texts, Solon 1.51f. Diehl, Pindar *O.* 1.9, Aristophanes *Nu.* 520, cf. Plato *Ly.* 214a, *Smp.* 175d–e, Aris-totle *Rhetorica* (*Rh.*) 1398b9ff.

129. Heraclitus fr. 40; see above, n. 41.

130. Heraclitus fr. 56: ὃς ἐγένετο τῶν Ἑλλήνων σοφώτερος πάντων. Cf. fr. 42 suggesting that Homer and Archilochus should be thrown out of the contests.

131. Heraclitus fr. 104: "For what mind or sense do they have? Not know-ing that 'the many are bad, the good few,' they believe the bards of the peoples and use the mob as teacher."

132. Heraclitus frr. 41, 50, 108.

133. Heraclitus fr. 32: "The one wise thing is not willing, and is willing, to be called by the name of Zeus."

134. Much of the traditional wisdom, and poetry, passed as enigmatic, but so too did some of the new learning. While some of the natural philosophers had some ambition to make knowledge open (for example, Diogenes of Apol-lonia in fr. 1, cited above, n. 65), Heraclitus implicitly compares himself with the Delphic oracle (fr. 93); cf. above, n. 106, on deliberate obscurity in some

The "wise man" thus afforded some of the early cosmologists a category within which to work, one that was flexible enough to *permit* innovation.[135] Thus far, extensive parallels for at least some of what we know about Greece can be found in other societies, for instance in the competitions in riddle-solving or in other aspects of wisdom reported from India, Sumeria, Babylonia, and many other parts of the Near East.[136] But one important eventual difference in degree, if not in kind—as, again, others before me have stressed[137]—is that *some* of the Greek competitions were a matter of *public* debate, adjudicated by lay audiences with (as we noted) considerable experience in evaluating arguments in such other contexts as the Assemblies and the law courts. In India, the contests reported in the Upaniṣads, at least, are essentially esoteric.[138] It was, in general, up to the wise men themselves to claim

---

Hippocratic works. Plato has Socrates exploit riddles or appear to his interlocutors to do so, but frequently undercuts references to Socrates as some kind of μάντις, e.g., *Smp.* 198a–b, cf. *Men.* 80a, *Phdr.* 241e, 242c—while he is heavily ironical with regard to the "initiations" of sophistry, e.g., *Euthd.* 277d–e.

135. One well-known passage referring to the claims of the inquiry concerning nature to be σοφία is Plato *Phd.* 96a6ff. There Socrates speaks of his youthful eagerness for "this wisdom which indeed they call the inquiry concerning nature," ταύτης τῆς σοφίας ἣν δὴ καλοῦσι περὶ φύσεως ἱστορίαν (cf. also *Ly.* 214b ff.)—though his hopes were later dashed (*Phd.* 98b). In the Hippocratic treatise *Decent.* 5, *CMG* 1.1.27.1ff.; cf. 1, *CMG* 1.1.25.1ff., medicine is assimilated to σοφίη in a general sense.

136. On the wisdom literature of the ancient Near East, see especially van Dijk 1953, Lambert 1960, Bottéro 1974, 1977 (for Mesopotamian versions of a tradition of Seven Sages, though not as historical personages, see Reiner 1961; Bottéro 1981b, pp. 110f.). On riddles in general see Huizinga 1944/1970, pp. 127ff.; Dundes 1975, pp. 95ff. On riddling and riddling games in ancient Indian sacred literature in particular, including the Ṛgveda, see especially Winternitz 1927, pp. 117ff., 183ff., 352ff.; Gonda 1975, pp. 132ff., and cf. pp. 379ff. on the brahmodyas.

137. See especially J.-P. Vernant 1957/1983, 1962/1982, 1983; Vidal-Naquet 1967/1986; Detienne 1967, chap. 5.

138. This appears to apply both to the Chāndogya Upaniṣad and to the Bṛhadāraṇyaka Upaniṣad, as following Ruben 1929 and 1954, chap. 8, I argued in G. E. R. Lloyd 1979, pp. 60f. That, in some sense, the debate between Yājñavalkya and other sages in the Bṛhadāraṇyaka Upaniṣad is one between new and old wisdom (as Ruben 1979, p. 150, suggests) does not substantially alter the esoteric, specialised nature of the discussion.

victory or to acknowledge defeat.[139] As for the ancient Sumerian wisdom debates, recently adduced by Frischer as parallels to the Greek material,[140] in them the judgement of the contest is represented as in the hands of a god—Šamaš or Enlil.[141] However we *interpret* what that verdict means, it is firmly assigned to non-human authority.

In the Greek context, the speakers often addressed, and had to be intelligible to, a far wider public. The author of *On Ancient Medicine,* as we noted, insisted that it is "of the greatest importance that anyone speaking about this art should be intelligible to laymen." [142] But that in turn *involved* the layman, in that case in making up his mind about medical theory, in others about physics or cosmology, in yet others

---

139. Thus in the debates in the Bṛhadāraṇyaka Upaniṣad, the sages yield to Yājñavalkya and acknowledge defeat by falling silent (e.g., Hume 1931, pp. 109–19). While various types of speaker are there involved, the audience as such plays no direct part: at one point, indeed, Yājñavalkya takes another sage aside and says, "We two only will know of this. This is not for us two (to speak of) in public" (Hume 1931, p. 110). Elsewhere, however, Gonda 1975, p. 380, notes an implicit or passive role for the audience when, for example, in the Brāhmaṇas the author "after stating two different opinions, lets his audience take their choice"; cf. Renou and Silburn 1949a, pp. 22ff. Cf. Keith 1928, p. 408, on discussions at the Sabhās held by rich kings or patrons: "any new doctrine which desired to establish itself was only able to do so if its supporter could come forward on such an occasion and by his advocacy secure the verdict of those assembled and the favour of the king or patron of the assembly."

140. Frischer 1982, pp. 14f., drawing on van Dijk 1953. Yet van Dijk, p. 39, specifies that the judgement is given by a god (cf. pp. 49f.) and while the story of Enkimdu and Dumuzi is an exception to this general rule, even there they both address Inanna Queen of Heaven.

141. See, for example, Lambert 1960, pp. 150f. Lambert notes that "there is no certainty that a judgement did take place in all the Babylonian texts," though there are traces of such scenes both in *Nisaba and Wheat* and in *The Fable of the Fox* (where Šamaš is involved, p. 201).

142. VM 2, CMG 1.1.37.9f., see above, n. 65. Vict. 3.68 (L) 6.594.3ff., cf. 598.4ff., explicitly addresses itself to "the many," as also does *Salubr.* 1 (*Nat. Hom.* 16), CMG 1.1.3.204.22ff. In the question-and-answer sessions envisaged in *Morb.* 1.1 (L) 6.140.1ff. (see below at nn. 177 and 180) laymen as well as doctors are involved. Again *De arte* 4, CMG 1.1.11.5f., perhaps echoing Diogenes of Apollonia fr. 1, demands that the starting-point should be agreed by all—where, presumably, "all" is not restricted to medical theorists.

about aspects of morality or even religion.[143] There are still rules about winning and criteria for success, but in principle, at least, those rules are entirely open. They are not made to depend on the authority of individuals, human or divine, let alone on the authority of some corporate notion of the past or of what is hallowed by tradition, even though what they do depend on chiefly, the appeal to *logos,* still cannot be totally dissociated from those who purported to be its representatives. In medicine that meant most of the Hippocratics, though other Greek healers, those in the temples of Asclepius or the itinerant purifiers,[144] no doubt refused to enter that kind of debate, to play the

---

143. The lay public were, in different contexts, called on to exercise their judgement on a wide variety of topics, not just as the jury in the dicasteries and as voters in the Assemblies (which Euthyphro describes himself as addressing on religious matters, Plato *Euthphr.* 3b–c). Judges chosen by lot, or the audience as a whole, decided many poetry, drama, and music competitions. At *Lg.* 659b–c Plato remarks on this as a current custom in music competitions in Sicily and Italy, and there and elsewhere does not conceal his contempt for the practice (*Lg.* 658a ff., 700c ff., *R.* 492b, cf. Aristotle *Poetica* [*Po.*] 1451b35ff., *Pol.* 1281b7ff.). Moreover, it was the bystanders who adjudicated the type of *epideixis* competition referred to in the Hippocratic *Nat.Hom.* (see below at nn. 156 ff.)—in that case, a debate on the elemental constitution of the human body. Again, Plato frequently makes play with the hubbub and applause that might greet a sophist's speech or eristic questioning, e.g., *Prt.* 339d–e, *Euthd.* 276b–d, and the caricatures that he offers of the styles of Prodicus and Hippias, for instance (*Prt.* 337a ff., c ff.), might be taken to suggest that once a sophist became known for a particular set of mannerisms he might well become the prisoner of his own reputation and find it hard not to give the audience what it had come to expect (cf. also the association between writers of manuals on rhetoric and the particular tropes they invented at *Phdr.* 266d ff.). Plato himself, of course, repeatedly emphasised that what mattered was not the verdict of the crowd, but the truth (e.g., *Grg.* 471e, 472b–c, 474a). The continued role of the crowd as judges in music and other competitions emerges from, for example, Lucian *Harmonides* (66) 2f., and Galen several times refers to competitive public anatomical dissections in front of an audience quick to ridicule failure, e.g., *AA* 7.10 (K) 2.619.16ff., 7.16 (K) 2.642.3ff., 645.7ff., cf. *CMG* 5.8.1.96.9ff., 98.9ff. ([K] 14.627.5ff., 629.1ff.), and see Vegetti 1981, pp. 54ff.

144. The evidence from the temple inscriptions at Epidaurus certainly suggests that those who set them up were keen to claim *efficacy* for the god's healing (cf. below, Chap. 3 at n. 112), and at one point the divergence between the

game by the rules those Hippocratic writers themselves laid down. The new wisdom did not, of course, drive out the old, in medicine especially, though it evidently proved its attractiveness *at least to a certain kind of audience* of those keen, in principle, to judge what was said by the case made out for it, rather than just by the standing of the speakers.[145] The double bind on the new-style wise men was that they sought to be not just admired (like athletes) but understood, even while they insisted that what they offered to be understood was no merely common understanding.

In the open debates that we know took place each participant, striving to win, would naturally try to justify his own position and undermine those of his opponents, and one way he might attempt to claim superiority for his own ideas was by stressing their novelty. Moreover, the occasions for display that occurred (both in connection with contests of wisdom and independently of them) did not just permit, but must sometimes positively have *favoured* open, indeed ostentatious, claims to originality. We know, for example, that the pan-Hellenic games provided one context not just for music, drama, and poetry

---

god's treatment and what ordinary mortals would prescribe is underlined. This is in case 48, Herzog 1931, p. 28, where the first instruction the god gives to the patient is that he should not follow the treatment (cauterisation) recommended by ordinary doctors. Similarly, later, Aelius Aristides refers often enough to the god overruling the diagnoses or therapies of ordinary physicians, e.g., *Or.* 47.61–64, 67–68, 49.7–9, where the god is always right. But while, implicitly or explicitly, differences between styles of treatment, and especially in their comparative success, are noticed, there is nothing to suggest that the healers at the temples of Asclepius or the itinerant purifiers attacked in *On the Sacred Disease* chose to *debate* with other doctors such questions as the causation of diseases, the constitution of the body, and the right type of treatment and its justification, along the lines of the discussions presupposed in such Hippocratic works as *Flat., De arte, Nat.Hom., Morb.* 1 and *VM*, let alone in the debates on the foundations of medical method in the Hellenistic sects.

145. This is not to deny that the standing of the speaker must often have played a role in the evaluation of his performance in an *epideixis* or a debate. Indeed, Aristotle was explicitly and repeatedly to emphasise the importance of the apparent character of the speaker as a factor in his carrying conviction with his audience: *Rh.* 1356a1ff., 1366a10ff., 23ff., 1377b24ff., 1378a6ff.

competitions, but for other intellectual exhibitions—part education, part entertainment—of various types,[146] including lectures not only on morally uplifting or cultural subjects but even on such topics as element theory or the fundamental constituents of the human body.[147] We hear, for instance, of the frequent attendance of Hippias at the Olympic games. He took along, according to the evidence in Plato's *Hippias Major* and *Minor*,[148] his own poems and prose works and was ready to speak on any subject on which he had prepared an exhibition piece or *epideixis*, and to answer questions afterwards.[149] Gorgias, too, we know, gave speeches at the Olympian and the Pythian games, and according to Plato was prepared to answer questions on any subject anyone cared to propose.[150]

Most of our specific evidence relates, to be sure, to well-known sophists[151] such as the two just mentioned, and accepting the sophists as making any serious contribution to developments relating to science still presents difficulties, since discussion still continues to be concentrated rather narrowly on the work of a small group of the most famous individuals, beginning with Protagoras, and to be preoccupied with the criticisms that Plato brought against them. These criticisms pose a major obstacle to our understanding, since he figures so promi-

---

146. This is clear, for example, from the evidence concerning Hippias; see below at n. 149, and cf. n. 166. The festivals continued to be the, or a, context in which the dream-interpreter might hope to attract a public, as we may infer from Artemidorus's references to frequenting them to collect dreams, 1 pr. 2.18, 5 pr. 301.10f.

147. See below at nn. 156ff. on *Nat.Hom.*

148. The continuing debates on the authenticity of these works do not materially affect their usefulness as evidence of the type of interests Hippias was believed to have.

149. See especially *Hp.Mi.* 363c–364a, 368b–369a; cf. *Hp.Ma.* 281a ff., 282d–e, Philostratus *Vitae Sophistarum* (*VS*) 1.11.7.

150. See, for example, Plato *Men.* 70c, *Grg.* 449b–c, Aristotle *Rh.* 1414b29ff., and cf. Cicero *De finibus* 2.1.1, Philostratus *VS* 1.9.4ff.

151. Among the more important recent discussions of the sophists have been those of Guthrie 1969; Stanton 1973; Welskopf 1974; Vlastos 1975b, pp. 155ff.; Martin 1976; Classen, ed., 1976; Moreau 1979; Kerferd 1981; and Kerferd, ed., 1981.

nently among our early sources of information.[152] Yet we must recognise that there was far more to what is called the sophistic movement than the work of the named individuals Plato attacks or even of the generality he abuses. The category of sophist, in Plato himself, as well as elsewhere,[153] is far from hard-edged, and there were important over-

---

152. It is a central part of Plato's enterprise to distinguish between the sophists and Socrates, though of course Socrates is treated as a sophist by Aristophanes in *his* attacks on the new learning. The acceptance of money for instruction is often used as the defining characteristic of a sophist, and what separates the later fifth-century teachers from earlier generations, going back from Anaxagoras and Damon, through Pythagoras, all the way to Thales and the Wise Men or even to Homer and Hesiod themselves. But first the well-known fact that σοφιστής as well as σοφός are used of Pythagoras, Parmenides, and Anaxagoras, as well as of the wise men, of physiologists as well as poets (see next note) must give us pause. Secondly, while we hear a good deal from Plato about the money earned by Protagoras, Hippias, and Prodicus, for example (see, for instance, *Prt.* 349a, *Hp.Ma.* 282b–e, *Hp.Mi.* 364d, *Thg.* 127e f.) and about Socrates *not* taking fees from pupils, that hardly provides a satisfactory criterion. Acceptance of money for instruction in such *technai* as medicine or sculpture was a well-established and uncontroversial practice, and at *Prt.* 311b ff. the possibility of paying Hippocrates, Polyclitus, or Pheidias for such instruction causes Hippocrates' namesake, the son of Apollodorus, no embarrassment. It is, rather, fee-taking for teaching such subjects as "virtue" or excellence, ἀρετή, for which Plato reserves his bitterest attacks. But although at *Prt.* 349a we are told that Protagoras was the first to have accepted money for such teaching, elsewhere the evidence from the Platonic *Alcibiades* 1.119a might be taken to suggest that perhaps even before him Zeno of Elea, not otherwise usually classed as a sophist, might rate the title. There we are told that Pythodorus and Callias became "wise," σοφός, and famous, through Zeno, by paying him 100 minae each, implying that both criteria are fulfilled—certainly the payment of a fee, and probably also, since σοφός seems to have a general sense, for instruction that included "excellence."

The attempt to see rhetoric as the principal interest of, and link between, the "sophists" (as in H. Gomperz 1912) also founders. It is clear from Plato that several of those he calls *sophists,* Hippias especially, taught such subjects as astronomy, mathematics, grammar, and philology (see below, n. 154). Conversely, the early writers of *Arts* of Rhetoric, mentioned in the *Phaedrus* (266d ff.), by Aristotle (e.g., *Rh.* 1354a11ff., 1402a17) and elsewhere, include some, such as Corax and Teisias, who are not otherwise spoken of as sharing the interests of a Protagoras or a Hippias. It has been argued that the sophist Antiphon and the orator of the same name are one and the same (see Morrison 1961, 1963, cf. Avery 1982), but this is not certain; more importantly, the ora-

laps not only between sophists and natural philosophers but also and more especially between sophists and medical writers or lecturers.[154] There is a permeability in those categories, as well as a permeability in the audiences the individuals concerned took as their targets.

Certainly the extant medical texts yield excellent evidence that

---

tor's *Tetralogies* serve as a reminder that training in the art of speaking might more often be by example than by theory, and that it was certainly not just in connection with speculative theories of one type or another that arguing on both sides of a question was practised.

The activities of Protagoras, Gorgias, Hippias, Prodicus, and the sophist Antiphon no doubt have certain features in common, but these men form no self-contained group, let alone one that constituted itself self-consciously as a movement or school. Rather, their work, and that of many others, should be set against a wider background of intellectual, social, and political developments in the late fifth century—developments that include (1) an increasing interest in rhetoric and dialectic (the origins of which Aristotle took back to Empedocles and Zeno, respectively, in a fragment of the lost work the *Sophist,* fr. 1 Ross), (2) some spread in the demand for more than merely elementary instruction in such subjects as mathematics, medicine, and natural philosophy, as well as (3) the developments of those subjects themselves, in addition to (4) a growing interest in political and moral questions, including the relativisation of moral judgements and other aspects of the complex set of issues often debated under the rubric of the controversy between nature and convention.

Finally, as regards other aspects of the activity of those whom Plato attacks besides their educational role, it is as well to recall that several of those concerned served as spokesmen or ambassadors for their home cities, as Hippias did for Elis and Gorgias for Leontini (see *Hp.Ma.* 281a–b, 282b).

153. The linguistic data are well known. Not only is the verb σοφίζεσθαι originally used non-pejoratively (e.g., Hesiod *Op.* 649, Theognis 19) but the same is true of the noun σοφιστής. It is used by Pindar (*Isthmians,* 5.28), of poets, by Herodotus of Solon and others (1.29), of the institutors of the Dionysiac cult (2.49) and of Pythagoras (4.95). Later, in Isocrates it is used of the Seven Wise Men (15.235, also in Aristotle's περὶ φιλοσοφίας, fr. 5 Ross) and of Alcmaeon, Parmenides, Melissus, Empedocles, and Ion, as well as Gorgias (15.268, cf. 313). In Plato, too, it is used in a non-derogatory sense of, for example, geometers (*Men.* 85b), nor should we forget that at *Prt.* 316c ff. Protagoras represents what he does as in line with, and the successor to, the work of a wide variety of earlier specialists, including poets, such as Homer, Hesiod, and Simonides; those who concerned themselves with initiations and prophecies, such as the followers of Orpheus and Musaeus; with "gymnastics," such as Iccus and Herodicus of Selymbria (where it may be that aspects of his medical practice are in view, cf. above, Chap. 1 n. 61); and with "music," such as Agathocles and Pythocleides of Ceos. To be sure, Plato makes Protagoras con-

some medical writers fought hard to differentiate themselves both from those *they* called "sophists" and from what they call "philosophy." The writer of *On Ancient Medicine,* as we saw, emphasises the point when he rejects speculative theorising about the nature of man, where he specifies that it is not just what he calls sophists but also doctors, ἰητροί, who are at fault.[155] We can examine one example of such theorising in the treatise *On the Nature of Man.* The author of the first eight chapters of that work may well have been a medical practitioner himself,[156] but he has more than a touch of the sophist about him too,[157] even though he is also concerned to distinguish the way he treats the subject of the nature of man from the way other lecturers do when they go beyond what is, in his view, relevant to medicine. Those lecturers can be seen to be ignoramuses, the writer says,[158] among

---

cede that they did not call themselves *sophists* "to escape the odium of that term"—an odium, however, that in Plato's day owed much to Plato himself. Conversely, the terms φιλοσοφία and φιλοσοφεῖν were far from confined to the activities that Plato himself would have recognised as philosophy: the verb is used in Herodotus 1.30 of Solon's travels, and by Pericles in Thucydides 2.40 of a characteristic of the Athenians as a whole.

154. Plato himself makes Hippias claim to teach astronomy, geometry, arithmetic, linguistics, and music (*Hp.Ma.* 285b ff., *Hp.Mi.* 366c ff., cf. *Prt.* 318e) and at *Prt.* 315c represents him as holding forth on nature and "meteorology" (cf. the μετεωροσοφισταί of Aristophanes *Nu.* 360, cf. 333). Aspects of the interaction of rhetoric and natural science are discussed in G. E. R. Lloyd 1979, pp. 86ff., where p. 87 n. 146 outlines the main evidence for ascribing an interest in physical questions to some of the major sophists whom Plato attacks.

155. See *VM* 20, *CMG* 1.1.51.6ff. and 12ff., quoted above, nn. 67–68.

156. *Nat.Hom.* is fairly clearly a composite work, and the fact that the blood-vascular theory of 11, *CMG* 1.1.3.192.15ff., corresponds to that ascribed to Polybus by Aristotle *HA* 512b12ff., and, further, that the report of Polybus' theories in Anon. Lond. 19.1ff. tallies broadly with the views set out in *Nat.Hom.* 1–8, does not necessarily mean that we should attribute the whole of the treatise as we have it to him, though that cannot be ruled out.

157. Whoever the author of *Nat.Hom.* 1–8 was, his debating style clearly owes a good deal to the model of a legal ἀγών: see G. E. R. Lloyd 1979, pp. 92ff.

158. *Nat.Hom.* 1, *CMG* 1.1.3.164.8ff., 166.2–7: δῆλον ὅτι οὐδὲν γινώσκουσι. γνοίη δ' ἂν τῷδέ τις μάλιστα παραγενόμενος αὐτοῖσιν ἀντιλέγουσιν· πρὸς γὰρ ἀλλήλους ἀντιλέγοντες οἱ αὐτοὶ ἄνδρες τῶν αὐτῶν ἐναντίον ἀκροατέων οὐδέποτε τρὶς ἐφεξῆς ὁ αὐτὸς περιγίνεται ἐν τῷ λόγῳ, ἀλλὰ

other reasons because the same man never wins the argument three times in succession, but whoever happens to have the glibbest tongue in front of the crowd: important, if well known, evidence for the existence of those competitive lectures on physiology judged by a lay audience.[159] It is clear that some of the medical texts that seek to distance themselves from those that offered general disquisitions on topics marginally relevant to medicine have more in common with the works in question than might seem likely from the way they set out to stress the distance.[160] Moreover, as we noted earlier, there are other texts that do not dissociate themselves from the sophistic *epideixis*, but *follow* that model and exemplify it. We have mentioned *On the Art* and *On Breaths* as the two most striking cases.[161] The reaction of an earlier generation of commentators was to suggest that maybe either Protagoras or Hippias himself wrote *On the Art*.[162] That is unlikely, in all conscience, but it was certainly not just foolish of Theodor Gomperz and others to entertain such a possibility.

On Breaths and *On the Art* may be taken to establish that the sophistic *epideixis* marks one extreme end of the spectrum that our extant Greek medical treatises represent. Those treatises do form a *spectrum*: there are important distinctions to be drawn between more, and less, popular works, between general lectures and practical manuals or

---

τοτὲ μὲν οὗτος ἐπικρατεῖ, τοτὲ δὲ οὗτος, τοτὲ δὲ ᾧ ἂν τύχῃ μάλιστα ἡ γλῶσσα ἐπιρρυεῖσα πρὸς τὸν ὄχλον. ("It is clear that they do not know what they are talking about. One can discover this most easily by being present at their debates. When the same men debate with each other in front of the same audience, the same speaker never wins three times in succession, but now one does, now another, now whoever happens to have the glibbest tongue in front of the crowd.")

159. *Nat.Hom.* 1 envisages those who attempt to prove that man consists of a single physical element, air, water, fire, or earth, but 2, *CMG* 1.1.3.166.12ff., attacks doctors who similarly argue that man is constituted by blood, bile, or phlegm alone.

160. Thus while *Nat.Hom.* 1–8 attacks monistic physiological and pathological theories, the pluralist doctrines the writer himself recommends are based on similar ideas concerning the elements in the body and the causes of diseases: see below, Chap. 3 at nn. 42ff.

161. See above at nn. 44f.

162. See T. Gomperz 1910, pp. 22ff.; W. H. S. Jones 1923–31, vol. 2, p. 187.

collections of notes (for instance), between authors with more, and less, clinical experience, or with none at all. But the extension of that spectrum *as far as* the sophistic *epideixis* can throw light on at least some aspects of our specific problem concerning what I called the pressures towards overt innovativeness. If we turn back to the sophistic *epideixis*, three features stand out. First and most obviously, in the context of an exhibition performance, whether at one of the great Games or on some other public occasion, caution and reserve are not likely to be the most highly prized qualities. On the contrary, every effort will be made to attract and hold an audience, to make the "sales pitch" as effective as possible. This was, after all, one of the chief ways in which teachers attracted fee-paying pupils.[163] We expect, and we duly find, in examples of the genre, both from Gorgias and in the Hippocratic Corpus, a striving after originality as well as after effects of every kind.[164]

Secondly, from the side of the hearers, we may imagine that most were aware of, and so must surely to some extent have discounted, the elements of exaggeration in this kind of performance. Although the analogy should certainly not be pressed too far—and maybe Huizinga did press it too far[165]—the occasion of the sophistic *epideixis* has some of the razzmatazz of the fairgound sideshow.[166] Most of the audience at Delphi and Olympia were away from home, and all must have been

163. Thus we hear from Plato that Prodicus had both a fifty-drachma and a one-drachma *epideixis* on the correctness of names (*Cratylus, Cra.,* 384b–c) and Aristotle at *Rh.* 1415b15ff. writes of his recommending slipping in sections from the fifty-drachma exposition when the audience showed signs of nodding off.

164. Thus when Diodorus 12.53.2–5 records the sensation that Gorgias created when he visited Athens on an embassy from Leontini (in 427 B.C.), this is explicitly attributed, in part, to the novelty of his rhetorical style. Again in Xenophon *Mem.* 4.4.6–7, Hippias expressly claims to attempt to say something "new" on each occasion, while later Isocrates also claimed originality for his own work: 9.8–11, cf. 4.7–10, 12.10ff. Similarly, Lysias' speech is explicitly praised by Phaedrus in Plato's *Phaedrus* (227c5ff.) for its subtlety and inventiveness.

165. Huizinga 1944/1970. Compare Bakhtin's discussion of the ancient origins of what he calls the carnivalisation of literature: Bakhtin 1973, pp. 87, 93, 100ff., and cf., e.g., Carrière 1979.

conscious of the contrast between festival and everyday experience. Certainly the element of playfulness is commented on directly by Gorgias at the end of his *Helen*.[167] In Thucydides too, when in the Mytilenean debate Cleon is made to chide the Athenian Assembly for their lack of seriousness, he puts it that they are behaving like an audience at a performance of sophists,[168] and Thucydides himself underlines the seriousness of his own historical enterprise by insisting on the contrast between it and the competition pieces of earlier writers of chronicles, λογογράφοι.[169]

The point extends to the inquiry concerning nature and to medicine. Exploiting what became a standard device to put down opponents, the author of *On Ancient Medicine* contrasts his own serious interests in the art with their speculations, which, in his memorable phrase, belong rather to painting than to medicine.[170] Plato too undercuts all attempts at accounts of the changing world of becoming—his own included—by categorizing them as a mere pastime, παιδιά, even though a moderate and intelligent one.[171] We cannot represent the

---

166. Diogenes Laertius records several stories, some no doubt apocryphal, of the way in which wise men of various kinds attracted attention to themselves. At 6.27, for instance, he says of Diogenes the Cynic that when, one day, he was discoursing seriously and no one paid attention, he started humming or babbling, τερετίζειν, whereupon a crowd gathered, whom he then reproached for coming in all seriousness to hear nonsense, but being slow to listen to a serious speech.

167. Gorgias *Helen* (fr. 11) 21; cf. Aristotle's report at *Rh.* 1419b3f. (fr. 12) that Gorgias recommended that one should destroy one's opponents' earnestness with laughter, and their laughter with earnestness.

168. Thucydides 3.37ff., especially 38.4 and 7.

169. Thucydides 1.21–22.

170. *VM* 20, *CMG* 1.1.51.12ff., cf. above, n. 68.

171. Plato *Ti.* 59d. Plato further undercuts the seriousness of writing as a whole, notably in passages in *Phdr.*, e.g., 274b ff., 276a–d, 277d–e, and *Ep.* 7.341b ff., much discussed by, for example, Gaiser 1963, pp. 3ff; Gadamer 1964/1980, 1968/1980; and Havelock 1963, 1982. On the other hand, the lack of seriousness is a charge repeatedly brought against eristic or antilogic, e.g., *Euthd.* 278b ff., 283b–c, 288b–c, *Grg.* 500b–c, *R.* 539b ff., *Sophist* (*Sph.*) 251b–c, *Philebus* (*Phlb.*) 15d–e. Aspects of the interactions of the serious and the playful in Plato have been discussed from different points of view by, for example, de Vries 1949; Friedländer 1958, vol. 1, chap. 5; Derrida 1972/1981, pp. 65ff.; Brisson 1982; Ferrari forthcoming.

whole of early Greek science just as fun—though that would suit Feyerabend as well as Huizinga.[172] At the same time we should not ignore what the signs of speculative playfulness in parts of it can tell us about the aims of authors and the expectations of audiences.

Thirdly and more generally, it is worth emphasising that some of our Greek material relates to contexts, such as inter-state games, where at least some of the constraints that existed within any given city-state were suspended—though not all were, and some additional ones were operative. Again, of the main groups of "intellectuals" concerned, nearly all of the most notable "sophists," many of the doctors, and indeed quite a number of the natural philosophers too had spheres of influence that were not confined to a single state. They could and did move freely from one city to another, both simply to earn their living and, if need be, to avoid political trouble. That is, after all, what Aristotle was to do when he withdrew from Athens in 323 B.C., and it was probably what most people expected of Socrates.[173] Here too the link with politics is clear, and while this aspect of political pluralism no doubt facilitated, and may even have been a necessary condition of, intellectual innovativeness, we should not underestimate the possible influence in the reverse direction, the effect that such intellectual innovativeness could and did have on the development of political pluralism.[174]

Far more than their counterparts in most other ancient civilisations, Greek doctors, philosophers, sophists, even mathematicians, were alike faced with an openly competitive situation of great intensity. While the modalities of their rivalries varied, in each the premium—to a greater or lesser degree—was on skills of self-justification and self-

---

172. See Feyerabend 1975 and 1978.

173. When news of Alexander's death reached Athens in 323 B.C., this sparked off a wave of anti-Macedonian feeling and Aristotle was charged with impiety, on the ground that he had composed a hymn to Hermias, according to D.L. 5.5f., whereupon he withdrew to Chalcis, where he died in the following year. The expectation that Socrates would accept the help of his friends to escape from prison before he was executed provides the dramatic setting of Plato's *Crito*, but was clearly no merely dramatic device.

174. See, for example, Finley 1973a, chap. 3, 1983, pp. 123ff.

advertisement, and this had far-reaching consequences for the way they practised their investigations as well as on how they presented their results. To be sure, to stress the novelty of your own ideas was not the only possible tactic to adopt in such a situation. Some medical writers, as we saw, took the opposite stance, criticising newfangled theories and treatments and siding with what they represent as traditional methods: yet we also found that when the author of *On Ancient Medicine* takes that line, his arguments—and it is to be noted that he does *argue*—are punctuated by expressions that underline his own authorial presence. The temptation to claim to introduce new ideas and practices was there and often not resisted—it was, indeed, yielded to with some abandon. As I remarked earlier, new medical treatments could not be justified with appeals to the authority of "modern medical science"—to the outcome of laboratory tests and the like—even though some ancient doctors were certainly not above attempts to mystify their clients with esoteric talk of the supposed humoral or elemental analysis of drugs and other therapies.[175] But since there were no legally recognised medical degrees or qualifications for them to cite as basic credentials, they had to start further back, as it were, and rely more on the force of direct argumentative persuasion to get remedies, new or old, accepted.

To win and hold an audience demanded a strong personality and the gift of the gab, whether in the surgery or in the public lecture or debate: while those contexts no doubt look very different *to us*, in Greece they were readily connected.[176] You could not, or at any rate you did not, if you were an exponent of Hippocratic rationalism, simply

---

175. The rationalist doctors prided themselves precisely on being able to explain the effects of drugs and were not content with merely empirical remedies in the modern medical sense of that term; but such theories as they produced usually took the form of highly speculative, if not quite arbitrary, appeals to opposites, elements or humours.

176. Thus *Praec.* 12, CMG 1.1.34.5ff., warns against turning a consultation into a public lecture or occasion for display. ("And if for the sake of the crowd, you wish to hold a lecture, your ambition is no laudable one; but at least avoid citations from poetry, for that betrays an incapacity for industry.") Cf. *Decent.* 2, CMG 1.1.25.15ff.

shelter behind, or assimilate yourself to, the "tradition"—at least not without justifying your so doing. We know from the treatise *On Diseases* book 1,[177] as well as from a famous text in Plato's *Laws*,[178]—that some doctors might expect to have to justify their diagnoses and treatments not just to other doctors (behind the scenes, as it were), but to and in front of their patients and their patients' friends and relatives, sometimes with other doctors present seen as rivals eager to take over the case if the opportunity arose.[179] Thus *On Diseases* 1 provides tips on how to deal with the veritable cross-examination you might have to withstand.[180] There was no deference to the professional in the white coat. Yet externals were not irrelevant. Another Hippocratic work, *Precepts,* warns the doctor against the use of luxurious headgear and exotic perfumes to impress: that clearly indicates where the temptations lay, even though this particular author says firmly that they should be resisted.[181]

The natural philosophers did not similarly have to amaze their audiences to get them to take their medicine. They were not competing for patients, though they were for pupils—and fame. Yet Empedocles at

---

177. See *Morb.* 1.1 (L) 6.140.1ff.; cf. also *Decent.* 3, CMG 1.1.25.20ff., 25ff.; 12, CMG 1.1.28.25f.; again at *Art.* 1 (L) 4.78.1ff., 9ff., the author refers to a discussion of the diagnosis and treatment of dislocated shoulder in which both doctors and laymen are involved.

178. Plato *Lg.* 720a ff., cf. 857c ff.; see Kudlien 1968c, R. Joly 1969–70, Hošek 1973.

179. In the Hippocratic Corpus, *Praec.* sees fit to recommend that the doctor should not be reluctant to call in other doctors, if need be, for consultation and warns that there should be no jealousy between doctors (8, CMG 1.1.33.5ff.; 12ff.; cf. *Praec.* 7, CMG 1.1.32.22ff.). Galen later cites several cases where he takes over from other doctors when their treatment (according to Galen) brought no results, and he even relates how he countermanded other doctors' orders without their knowing: see, for example, (K) 10.536.11ff., 538.12ff., cf. (K) 11.299.10ff., CMG 5.8.1.82.25ff. ([K] 14.614.9ff.); cf. Nutton 1979, p. 169.

180. *Morb.* 1.1 (L) 6.140.1ff., especially 142.7–12. Once again there is a parallel in the literature concerned with dream interpretation, for Artemidorus too offers advice about how to deal with the questions put to the interpreter, 4 pr. 237.25f., 4.84,299.15ff.

181. *Praec.* 10, CMG 1.1.33.32ff., reading Triller's θρύψις for Heiberg's τρῦψις; cf. *Decent.* 2, CMG 1.1.25.17ff.

least, who was the only prominent early cosmologist, so far as we know, to have some of his work delivered by a rhapsode at Olympia,[182] was certainly not easy to outdo in showmanship. "He liked," as Guthrie put it,[183] drawing on Diogenes Laertius, "to walk about with a grave expression, wearing a purple robe with a golden girdle, a Delphic wreath, shoes of bronze, and a luxuriant growth of hair, and attended by a train of boys."[184] But although the styles and contents of their speculations differ widely, what Anaxagoras, Democritus, and the rest have in common with Empedocles is that explicitly or implicitly they too claim to have found the solutions to physical and cosmological problems that had defeated everyone else. However much they differ in their other interests, they were rivals there, and were in business to argue that their own ideas were different from, and superior to, everyone else's.

Both in medicine and in natural philosophy the written text had an important and, as time went on, an increasing role as the object of critical reflection, though (as we noted) the texts, when read, were still usually read out.[185] Yet overt innovativeness in speculative thought and the corresponding self-distancing from tradition stem not only from the spread of literacy (by itself no guarantee that such attitudes will be adopted), but also from a complex, pluralistic social and cultural situation. What may be particularly important there is the development of new modes of rivalry and competition, calling for new styles of self-justification. In philosophy too, as in medicine, the individual

---

182. D.L. 8.63. At D.L. 2.10 a different type of success is ascribed to Anaxagoras, when he went to Olympia: he wrapped himself up in a leather cloak as if it were going to rain, which it duly did. Diogenes recounts other stories of the visits of other wise men at Olympia (e.g., Plato, 3.25; Diogenes the Cynic, 6.43) and while these may well be apocryphal, the idea of such visits is not implausible. Cf. also Lucian *Herodotus* (62) 1ff., who reports that Herodotus went there to have his history recited.

183. Guthrie 1965, p. 132.

184. D.L. 8.73, cf. 66. Aelian *Varia Historia* 12.32 refers to Hippias and Gorgias also wearing the purple robes that were associated particularly with rhapsodes.

185. See above, n. 83.

often thought of himself as participating in—and sometimes literally participated in—a debate in which the personal contribution of each participant was clearly marked as *his,* even when he did not go out of his way (as so many did) to stress his originality explicitly.[186] When we speak of Greek writers needing to win and hold an audience, *audience* is often the apposite term, and it may be to that interaction with audiences, and to the development of contexts for that interaction, that we have to look for the chief clues to the understanding of the particular positive and negative modalities of innovativeness in ancient Greece.

Conclusions

My theme has been that one of the striking and distinctive features of much of early Greek thought, particularly when we contrast it with what we know from some other ancient civilisations, relates to the degree of overtness of innovation and of the contestability of tradition. The actual measure of free speech that the political situation secured in different city-states, at different junctures, over different types of political, moral, religious, and cosmological subjects, poses problems of great intricacy that cannot be explored here.[187] Yet in the grossest terms, there is certainly a gulf between Athens in the fifth century, even the Athens that prosecuted Anaxagoras and was to put Socrates to death,[188] and the Babylonia of Darius or the Egypt of Amasis.[189]

---

186. I have taken my chief examples from early Greek medicine and pre-Socratic philosophy. But even though the authority of earlier writers plays an increasingly important role in parts of both science and philosophy from the late fourth century onwards, claims for originality are still often made. Aristotle himself offers a notable example at the end of the *Sophistici Elenchi* (*SE*) when he claims for his own studies in logic that they have initiated a new inquiry, 183a37ff., b34ff. Cf. also above, n. 102, with regard to Archimedes and Aristoxenus, and below, nn. 200, 206.

187. The issues have been discussed by Momigliano 1973, Dover 1975, Finley 1975b, 1977, Lanza 1979, especially. On trials for impiety—where some of the stories generated concerning other philosophers show the influence of the fame of the model of Socrates—see especially Derenne 1930.

188. The continuing difficulties that philosophers might encounter at Athens can be illustrated not just by the instance of Aristotle (see above, n.

Rivalry in claims to be wise starts almost as soon as we have any evidence to go on in Greece, and what counted as wisdom was an extraordinarily open-ended and negotiable question. Anyone could set himself up as a philosopher or as a sophist or, come to that, as a doctor. You depended not on legally recognised qualifications (there were none, we said, not even for doctors),[190] nor even simply on accreditation—though that was undeniably important.[191] What you had to rely on, largely, was your own wits and personality, and they were often judged by the verbal dexterity with which you presented your case, even when such verbal dexterity itself came to be suspect and so turned into a quality that had to be concealed to be fully effective.[192] Plato makes the sophist Gorgias say that he could take on and defeat any ordinary doctor in argument, whether in front of the Assembly (in

---

173) but also by the report in Diogenes Laertius 5.38 of the moves made against Theophrastus and others by Sophocles the son of Amphiclides, though in that case, according to the report, Sophocles was himself prosecuted, and after a year's exile Theophrastus returned.

189. We have, however, once again to discount some of the Greeks' own elaborations of the theme of the contrast between Greek freedom and Eastern tyranny in writers as diverse as the Hippocratic treatise *On Airs, Waters, Places*, e.g., 16, CMG 1.1.2.62.13ff., 23, CMG 1.1.2.78.3ff., and Plato, e.g., *Lg.* 694a–696b, even while this remains evidence of the Greeks' own attitudes.

190. The lack of legal sanctions in relation to the practice of medicine in Greece is criticised in the Hippocratic *Lex* 1, CMG 1.1.7.5ff.; see Amundsen 1977, cf. Preiser 1970. On the legal position of doctors in Hellenistic times and later, see Nutton 1970, Kollesch 1974, Kudlien 1979.

191. For example, although the doctrinal coherence of the doctors on Cos should not be exaggerated, and the extent to which the island offered more formal medical education than was available elsewhere is an open question, the evidence analysed by Cohn-Haft 1956 suggests that Coan doctors were particularly successful in obtaining appointments as public physicians in the fourth century and later. See also Sherwin-White 1978, chap. 7.

192. It soon became a commonplace with public speakers of various kinds to disclaim any special skill in speaking themselves and to represent their opponents as dangerous and unscrupulous manipulators of argument: see, for example, Antiphon 5.1–7, Lysias 12.3, 86, 17.1, Isocrates 15.26, 42, and cf. Plato *Ap.* 17a–18a and the counterpoint in the exchanges between Socrates and Protagoras at *Prt.* 316d ff., 342a ff. See in general Dover 1974, pp. 25ff., de Romilly 1975, and cf. on Gorgias in particular, Segal 1962, cf. Rosenmeyer 1955, Verdenius 1981.

competition for a post as public physician) or, indeed, at the bedside[193] (where again we may remark the ease with which those two contexts are juxtaposed). No doubt Plato means his audience to see this as an exaggerated claim.[194] But in ancient Greece, where what passed for medical knowledge was both far less technical and more widely shared than now, the point was not an extravagant one; one of the elements of exaggeration, rather, we might say, is that, to judge from some of our Hippocratic texts, there were doctors who would have been well able to look after themselves in debate, even with a Gorgias. Even those who there appealed to what they represent as tradition, to the good old ways of medical practice, for example, argued to justify doing so. Tradition by itself, in many of the areas we are concerned with, at least,[195] carried little authority.[196] Pre-Socratic philosophers do not assert that earlier ideas should be accepted simply because of the prestige of those who had first proposed them; no more do Plato and Aristotle. Even those Hippocratic writers who saw the danger as one of an obsession with the newfangled do not base their case simply on appeals to authority figures.

In time, to be sure, the balance between these two, tradition and innovation, was to change very drastically, though I cannot here go into the stages, let alone discuss the possible underlying causes, of this

---

193. Plato *Grg.* 456b–c, cf. also 452e, 459a–c, 514d ff. For other evidence of doctors called upon to address the Assembly, see Xenophon *Mem.* 4.2.5. At Plato *Plt.* 297e ff, 298c (cf. *Prt.* 319b f.) too, it becomes clear that the Athenians called in experts of various kinds including on medicine, though there the context is merely to take advice from them. Aristotle at *Rh.* 1403b32ff., 1404a12, draws a general contrast between the rhetorical style of delivery used in drama and in the Assemblies with that appropriate for teaching such a subject as geometry.

194. In the continuation, *Grg.* 464d ff., Socrates draws a comparison with a competition between a cook and a doctor before an audience of boys.

195. Here a contrast may be suggested with the role of appeals to the "ancestral constitution" at certain junctures in Athenian political debates: see Finley 1975a, chap. 2, 1983, p. 25.

196. It is notable that according to *Vict.* 1.1 (L) 6.466.18–468.2, it takes the same intelligence to evaluate what has already been said correctly as to make original discoveries.

complex process. We may simply note that from the end of the fourth century, increasingly, a series of great names—some, like Hippocrates, very largely the constructs of the commentators [197]—came to be turned into just such authority figures, to whom appeals could be made as some kind of guarantee of the validity of the ideas associated with them. "Hippocrates," Plato, Aristotle, and later Ptolemy and Galen were transformed into such figures, and even though at an earlier period the written texts of Plato (for example) may well have helped Aristotle (for one) to develop and press home his *objections* to Plato's philosophy, the explicit aim of some of the late commentators was not to criticise those texts so much as to show how they contain the truth. Indeed, the sixth-century Aristotelian commentator Simplicius sought to show how Plato and Aristotle were in substantial agreement,[198] just as in the second century A.D. Galen already often aimed to reconcile Plato and Hippocrates.[199] One of the principal manifestations of that shift towards tradition [200] was, indeed, the turning of the written text into a vehicle for the transmission of authority rather than one for

---

197. See, for example, Edelstein 1931, 1935, 1939/1967a, and especially the recent analysis of the growth and influence of the Hippocratic tradition in W. D. Smith 1979, especially chap. 3, pp. 177ff.

198. See, for example, Blumenthal 1981; cf. Moraux 1984, pp. 441ff.

199. This is the chief theme in the *De placitis Hippocratis et Platonis* (*PHP*), *CMG* 5.4.1.2 (K) 5.181.1ff., but the topic recurs.

200. Yet innovativeness continued, of course, to be prized in many contexts, including in speculative thought. The very considerable power and originality of aspects of Hellenistic philosophy, for example, both dogmatic and anti-dogmatic, have only recently begun to be fully appreciated, thanks to such studies as those collected in Brunschwig, ed., 1978; Schofield, Burnyeat, and Barnes, edd., 1980; Barnes, Brunschwig, Burnyeat, and Schofield, edd., 1982; Schofield and Striker, edd., 1985. Moreover, as an explicit topic the desire to claim to be innovative can be illustrated in the second century A.D. in, for example, Lucian, who in the *Gallus* (22) 18 implies that it was from such a desire that the Pythagoreans introduced many of their more arcane rules and doctrines, while in that century or the next the obsession of Diogenes Laertius with the theme of first discoverers—along with the heterogeneity of the examples he cites—have already been remarked (n. 6 above). For further innovations both in the substantive theories, and the methodologies, of the exact and life sciences, see, for example, below, Chap. 3 pp. 163ff., Chap. 4, pp. 206ff., Chap. 5, pp. 230ff., 249ff.

challenging it—thereby producing the opposite effect to the one Goody claimed for increasing literacy at an earlier period.[201]

Yet we should be careful not to suppose that the tendency to appeal to the authority of the past was uniform and all-pervasive in natural scientific inquiry, even in late antiquity. While when Galen cites Hippocrates it is almost always to agree with him,[202] the reverse is true of Galen's slightly older contemporary, Soranus. On nearly all the occasions when he cites Hippocrates or his followers in the *Gynaecology* it is to criticise them and to expose their mistakes.[203] Ptolemy, too, dissents from Hipparchus often enough, greatly though he admires him.[204] Nor should we underestimate the originality of Galen and Ptolemy themselves, for all their repeatedly expressed deference to the past. To say, as was once fashionable,[205] that they are just eclectic, is nonsense, though some of their own rhetoric tends to mislead in that direction. Their own contributions to their subjects, both as observers and as theorists, are of the highest order,[206] even when they present these as the elabora-

---

201. Goody 1977, p. 37, does, however, allow that literacy also eventually encouraged what he calls the orthodoxy of the book.

202. Galen does, however, on occasion refer to his own modest additions to what Hippocrates taught, especially in the domains of anatomy, e.g., *UP* 2.3 (H) 1.70.10ff., (K) 3.96.8ff., and of therapeutics, e.g., (K) 10.420.9ff., 425.6ff., 632.1ff., and cf. also with regard to pulse theory, *CMG* 5.8.1.134.1ff. ([K] 14.665.2ff.).

203. See Soranus *Gynaecia* (*Gyn.*) 1.45, *CMG* 4.31.26ff., 3.29, *CMG* 4.112.14ff.; 4.13, *CMG* 4.144.2ff., 4.14–15, *CMG* 4.144.21ff., 145.14ff. In Celsus, too, Hippocrates is sometimes criticised for mistakes, e.g., *Med.* 3.4.12, *CML* 1.107.1ff., 6.6.1e, *CML* 1.260.3ff.

204. See, for example, *Syntaxis* 3.1 (H) 1.194.3ff., 4.11 (H) 1.338.5ff., 5.19 (H) 1.450.11ff., 6.9 (H) 1.525.14ff.

205. See, for example, Dampier-Whetham 1930, p. 53, on Ptolemy; Wightman 1950, pp. 330f., on Galen.

206. Yet Ptolemy explicitly claims his theory of the moon's second anomaly and his solutions to the models of the five planets as his own: see *Syntaxis* 5.2 (H) 1.354.20ff., 9.2 (H) 2.210.8ff. Cf. also 4.9 (H) 1.327.16ff., 328.3ff., where he claims that the use of new methods enabled him to improve on his predecessors' work. The appeal to past authority has a more dominant role in key contexts in the *Tetrabiblos* (cf. above, Chap. 1 n. 152), which also provides a notable example of the prestige sometimes accorded to ancient manuscripts believed to contain esoteric learning; see *Tetr.* 1.21.49.14ff. That the

tion of the work of those of their predecessors of whom they most approve.

Down to the sixth century A.D. and even beyond, a Kuhnian tension is still a feature of parts of ancient speculative thought, even though the balance had shifted after the classical period from innovation towards tradition—to innovation mainly within, and represented as faithful to, the tradition, or rather to *one or another* of the plurality of rival traditions that still continued in most fields of investigation. What in some areas of thought was to alter the balance irrevocably—indeed by the sixth century A.D. had already done so in those areas—was the appeal to a particular text, the Bible, as revealed truth. The shift from reference to the "divine Hippocrates," the "divine Plato," and so on, to reference to the word of God may seem not so great in verbal terms, but it reflects fundamental differences not least in the underlying institutional realities: the creation of a church, the constitution of Christianity as the official religion of the empire, and the availability of a new battery of sanctions that could be deployed against the deviant. But those topics, too, are beyond the scope of our discussion here. What this study has attempted, rather, is to sketch out some of the problems presented by the balance of the tension at the very earliest stages of the Greek inquiry into nature. There in the classical period one crucial development was the opening up of the possibility, precisely, of development—if the oxymoron can be excused, the initiation of a tradition of, precisely, the contestability of tradition.

To conclude that the bias towards innovativeness characteristic of parts of early Greek speculative thought just confirms a Kuhnian verdict[207] that what we have here is, after all, not proper science—not "normal" science working within a dogmatic tradition or set of paradigms—is tempting and has an element of truth, but is one-sided and premature. On the matter of its one-sidedness, what that ver-

---

commentary form itself could provide the framework for a claim to originality can be illustrated, for example, from Porphyry, who justifies his undertaking the exegesis of Ptolemy's *Harmonics* in part on the grounds that no one had done this before him: *In Ptolemaei Harmonica* (*In Harm.*) 3.16ff. (Düring).

207. See, most recently, Kuhn 1983, p. 567.

dict importantly leaves out of account is the stages through which proto-science itself passed. Fifth-century Greek speculative thought was no merely aberrant—rhetorical—interlude, intervening between tradition-oriented Egyptian and Babylonian medicine, mathematics, and astronomy, and again tradition-oriented Hellenistic science. The ancient Near Eastern evidence suggests some of the weaknesses, as well as some of the strengths, of the opposite bias towards conservatism—the negative effects, the constraints, of monolithic authority. By contrast, the early Greek material we have reviewed illustrates not just the *excesses* to which egotism often led (though it does that) but also some of the positive aspects of aggressive innovativeness, in the canvassing of alternatives and the development of criticism through competition, as debate is opened up between rival theories and attention is focused on their grounds and articulation, indeed, on the question of the nature and foundations of science, medicine, and mathematics themselves. While too much attention paid to such second-order issues may detract from the business of getting on with the inquiries themselves, to pay no attention at all runs the risk of leaving the inquiries blind. A certain self-consciousness in the investigations and an awareness of alternatives, at least of rivals, were tolerably durable legacies bequeathed by early Greek to Hellenistic science, part of what then became, for some, revered tradition. For those early developments themselves to occur, however, what was needed was not just written texts, texts in which the figure of the author may not be visible against the background of the tradition, but (among other things) texts that through a strong authorial presence implied a personal accountability for the claims they contained.

And as to the matter of the prematurity of that judgement, our exploration of the Greek experience in the following chapters will provide the basis for the expression of certain other reservations and qualifications.

CHAPTER THREE

Dogmatism and Uncertainty

On several occasions already I have drawn attention to the elements of bluff and dogmatism in parts of early Greek science. Yet anti-dogmatic opinions are also prominent in other—sometimes even in the very same—works. A readiness to admit to not knowing the answers and to grant that you have been mistaken is still often thought part of the scientific, indeed a general intellectual, ideal. Examples where the ideal is put into practice can be given from modern science, although so too can cases where it has been ignored, and some writers would want to recommend that it should be ignored at least in certain circumstances.[1] We find what look like anticipations of those principles in some early Greek texts. The general question that this raises is, then, the interplay, or tension, between the dogmatic and the anti-dogmatic strains in Greek investigations into nature. In particular at what point, under what circumstances, with what motives and intentions did ancient scientists begin to acknowledge the possibility of their own mistakes?

As before, it is useful to establish a benchmark by the use of broad cross-cultural comparisons. First, scepticism about certain claims or claimants to special knowledge can be attested in many contexts in many peoples. Shirokogoroff pointed this out in his classic study of the

---

1. On the function of dogma in research see, for example, Kuhn 1963; cf. more generally in Kuhn 1962/1970 and the elaboration and modification of his position in Kuhn 1977.

Tungusi.[2] Evans-Pritchard stressed that the Azande often suspected particular witch-doctors of being frauds.[3] In his study of Ifa divination Bascom similarly noted that the honesty or knowledge of individual diviners may be questioned,[4] and Turner pointed out how attempts may be made to trip up individual Ndembu diviners.[5] The case of the Kwakiutl Quesalid, reported by Boas and popularised by Lévi-Strauss, is a poignant one.[6] Quesalid himself ended up as a shaman, but he had begun with the intention of showing that the ways of the local shamans were fraudulent, that their techniques were a set of tricks. What happened was that he tried other tricks that he learnt from other shamans from neighbouring groups and discovered that they worked: the sick reported remarkable recoveries, and Quesalid found himself, willy-nilly, a shaman. Again, in some mundane contexts, the recognition that there are limits to what any human being knows and can know is widespread and needs no illustration. It is a wise man who knows his own father, or, as Telemachus puts it, no one does.[7]

Our evidence from the ancient Near East is, once again, of exceptional value. Medicine, well represented in our extant texts, provides a particularly promising field of inquiry, since whether a disease has

---

2. Shirokogoroff 1935, e.g., pp. 332ff., 389ff.

3. See Evans-Pritchard 1937, p. 183: "Many people say that the great majority of witch-doctors are liars whose sole concern is to acquire wealth. I found that it was quite a normal belief among Azande that many of the practitioners are charlatans who make up any reply which they think will please their questioner, and whose sole inspiration is love of gain." But Evans-Pritchard went on to deny, p. 185, disbelief in witch-doctorhood in general.

4. See Bascom 1969, p. 11: "The honesty or knowledge of individual babalawo may be questioned," though he went on: "but most are highly esteemed, and the system itself is rarely doubted." Cf. p. 70, where he notes that the blame for failures is shifted "from the system of divination to other causes, such as the ignorance or dishonesty of the diviner." Cf. Lienhardt 1961, p. 73, and more generally, and in connection with the ancient world, Jacques Vernant 1948 and the papers collected in J.-P. Vernant et al. 1974.

5. V. W. Turner 1964, p. 242. Herodotus 1.46ff. (cf. 2.174), for example, provides Greek evidence for the testing of oracles.

6. F. Boas 1930, pp. 1–41; cf. Lévi-Strauss 1958/1968, pp. 175ff.

7. *Od.* 1.214ff. On various other occasions in Homer attention is drawn to certain limitations to human knowledge, e.g., *Il.* 2.484ff., *Od.* 10.190ff.

been diagnosed correctly and whether the treatment adopted is the right one are questions of more than merely theoretical interest. Although, as we noted before,[8] the authors of Egyptian medical documents do not, as a general rule, intrude to vouch for their personal observations, reference is quite often made in general terms to experience. The Papyrus Ebers, for instance, on several occasions ends its account of a charm or remedy with the comment: "really excellent, [proved] many times."[9] Elsewhere the issue of the effectiveness of treatments is implicit. The relationship between the healer and the disease is frequently represented as a conflict, a hard-fought battle between them. In both Egyptian and Mesopotamian medicine, what causes the disease—the peccant material or force—is often apostrophised, commanded or cajoled to leave the patient, that departure being construed as a matter of negotiation.[10] Again, Egyptian, like later Greek, medicine explicitly recognised a category of complaints "where there is no treatment"[11] (though in practice in some such cases treatment is nevertheless attempted).

All of this goes to show that ancient Egyptian doctors, especially, were often aware of the limitations of their art and conscious of its difficulties. When claims for the effectiveness of remedies are made, they can, in principle, be controverted. Yet so far as our extant evidence goes, that mostly remained just a theoretical possibility. Neither Egyptian nor Mesopotamian medicine developed a tradition of the

---

8. See above, Chap. 2, pp. 6f. and 63.

9. See Ebbell 1937, pp. 29, 30, 42, 73.

10. See, for example, Ebbell 1937, p. 105; Breasted 1930, p. 477; R. Campbell Thompson 1923–24, p. 31, 1925–26, p. 59. The general point remains, even though there are, to be sure, important differences within the diverse medical traditions in both Egypt and Mesopotamia.

11. See, for example, Breasted 1930, cases 7, 8, 17, and 20; Ebbell 1937, pp. 127f. The recognition of a category of cases that are hopeless and that cannot be treated can also be illustrated from the ethnographic reports: see, e.g., Shirokogoroff 1935, p. 334: "some shamans may refuse to attend cases which are known to be absolutely hopeless." Shirokogoroff further remarks, p. 385, on a case of a shaman who admitted to him and to a Manchu friend that he did not understand a situation, but that, from the report, appears to have been a private, not a public, admission of ignorance.

criticism of current practice, any more than they did criticism of past custom and tradition themselves. In general, if doubts were felt about the efficacy of treatments or on the correctness of diagnoses, these were not usually expressed. Even when a case was deemed untreatable, this was generally asserted dogmatically.[12] Above all, there are no detailed records of particular failures of diagnosis or of cure (as opposed to mere expressions of despair), no debate between alternative treatments, let alone between rival schools of medicine with competing theories of disease.[13]

Dogmatism in Early Greek Natural Philosophy

One of the first things that strikes a student turning to the beginnings of Greek speculative thought, and first to pre-Socratic natural philosophy, is its dogmatism.[14] The wildest generalisations are offered with no suspicion that they may require qualification. True, this impression is partly one created by the doxographical sources on whom we often have to rely. They are concerned to record a sequence of positive theories ascribable to Thales, Anaximander, and the rest, uncomplicated by reservations or provisos.[15] Yet this impression is often confirmed when, as for several of the later pre-Socratics, we have more substantial evidence, in the form of original quotations.[16]

---

12. As in the cases from Breasted 1930 cited in the previous note. Cf. J. A. Wilson 1952, p. 77.

13. Cf., however, Bottéro 1974 who, in his study of divination in ancient Mesopotamia, notes (pp. 133f.) certain expressions of the difficulties encountered in particular problems in divination, and further draws attention (pp. 183ff.) to evidence that points to the development of different "schools" of omen interpretation, though without suggesting explicit debate between them.

14. I discussed some aspects of this in G. E. R. Lloyd 1979, pp. 139ff. On other features of the issue of pre-Socratic dogmatism, compare Cornford 1952, chap. 3, with Matson 1954–55 and Vlastos 1955/1970 and 1975a, e.g., p. 87.

15. This follows from the organisation of the material topic by topic in the doxographic tradition: see Diels 1879, cf. McDiarmid 1953/1970.

16. These quotations themselves, however, have always to be related to the

It is not as if there is much divergence, on this score, between otherwise radically divergent figures, such as Empedocles and Anaxagoras. Empedocles, for instance, announces categorically that bone consists of a certain definite proportion of the four "roots" or elements, earth, water, air, and fire.[17] Anaxagoras, who represents what is in many ways a quite different, Ionian, tradition of research, is sometimes just as positive in his assertions, for example, on the original state of the cosmos, when "all things were together" and "air and aether held all things,"[18] or on the production of earth from water and of stones from earth under the influence of cold.[19] Even those who were much later hailed as the forerunners of scepticism, such as Xenophanes and Democritus, were, on occasion, categorical enough.[20] Xenophanes certainly states that "there never was a man, nor will there ever be, who knows the certain truth about the gods and all the other things about which I speak" and that "seeming is wrought over all things."[21] But elsewhere he is prepared to speak of earth stretching down indefinitely below our feet, of the ocean as the begetter of the winds, and of our all being born from earth and water.[22] Democritus, too, though quoted as saying that we understand nothing exactly, ἀτρεκές, and know nothing truly, ἐτεῇ, about anything,[23] is also cited as confidently asserting nevertheless that atoms and the void alone are true or real, ἐτεῇ.[24]

---

contexts and concerns of those who report them, as has recently been emphasised by C. Osborne in her study of Hippolytus: C. Osborne forthcoming.

17. Fr. 96, cf. fr. 98, Aetius 5.22.1, Aristotle *De An.* 410a1ff., cf. 408a18ff., *PA* 642a18ff.

18. Fr. 1, often quoted by Aristotle, e.g., *Ph.* 203a25, and Simplicius, e.g., *In Ph.* 155.23ff.

19. Fr. 16.

20. Sextus Empiricus is, indeed, often our source for earlier epistemological views that can be given a sceptical interpretation.

21. Fr. 34 (quoted by many ancient writers; see Guthrie 1962, p. 395 n. 1). The difficulty of gaining knowledge of the gods is a topos that recurs, for example, in Protagoras fr. 4.

22. Frr. 28, 30, 33 (with fr. 29).

23. Frr. 6–10 and 117, on which see Sextus *M.* 7.135ff. especially. The most recent discussion of Democritus as a sceptic is that of Wardy forthcoming.

24. Fr. 125; cf. fr. 9 and fr. 11 on the contrast between "legitimate," γνησίη, and "bastard," σκοτίη, cognition.

The Hippocratic Medical Writers

For more sustained expressions of doubt and uncertainty we have to turn to our other and more extensive main early source, the medical writers—not that they do not also provide examples of dogmatism to equal or surpass anything we find in pre-Socratic natural philosophy. On this, as on so many other topics, the positions adopted in our extant fifth- and fourth-century B.C. medical texts vary widely—and initially rather puzzlingly—from extreme dogmatism on the one hand to a self-conscious anti-dogmatism on the other.[25] How far, we may ask, are these apparently strongly contrasting attitudes to be correlated with different types of treatise, types of writer, types of audience, or a combination of some or all of these? In what respects are the attitudes in question indeed alternative and conflicting, or how far can we suggest a framework of explanation to cover both apparently opposed tendencies?

Dogmatism in the Hippocratic Corpus

We must begin with a fairly detailed review of the modalities and manifestations of dogmatism in the medical writers, since it is against that background that what I have called anti-dogmatism must be evaluated. The treatise *On the Art,* which we have considered before as an example of authorial egotism,[26] shows to what lengths some writers went to protect themselves and the medical profession against any possible charge of incompetence or even of fallibility. Chapter 3 sets out what the author hopes to demonstrate, the word used being *apodeixis.* Medicine is first defined in terms of its aims, which include "the complete removal of the sufferings of the sick" and the "alleviation of the violences of diseases," and the writer claims that medicine achieves

---

25. Some aspects of this problem have been discussed by Di Benedetto 1966, and by R. Joly 1966, pp. 240ff., 1980, pp. 287f.
26. See above, Chap. 2 at nn. 45ff.

these ends and "is ever capable of achieving them."[27] Against those who demolish the art of medicine by citing the misfortunes of those who die from their illnesses, he counters with a passage that is worth quoting at length:

> As if it is possible for doctors to give the wrong instructions but not possible for the sick to disobey their orders. And yet it is far more probable that the sick are not able to carry out the orders than that the doctors give wrong instructions. For the doctors come to a case healthy in both mind and body; they assess the present circumstances as well as past cases that were similarly disposed, so they are able to say how treatment led to cures then. But the patients receive their orders not knowing what they are suffering from, nor why they are suffering from it, nor what will succeed their present state, nor what usually happens in similar cases. . . . Which is then more likely? That people in such a condition will carry out the doctors' orders, or do something quite different from what they are told—or that the doctors, whose very different condition has been indicated, give the wrong orders? Is it not far more likely that the doctors give proper orders, but the patients probably are unable to obey and, by not obeying, incur their deaths—for which those who do not reason correctly ascribe the blame to the innocent while letting the guilty go free?[28]

---

27. *De arte* 3, CMG 1.1.10.20f.: τὸ δὴ πάμπαν ἀπαλλάσσειν τῶν νοσεόντων τοὺς καμάτους καὶ τῶν νοσημάτων τὰς σφοδρότητας ἀμβλύνειν. The writer also adds the refusal to tackle cases where the disease has already won the mastery: see further below, n. 105.

28. *De arte* 7, CMG 1.1.13.10–19, 22–29: ὡς τοῖσι μὲν ἰητροῖς ἔνεστι τὰ μὴ δέοντα ἐπιτάξαι, τοῖσι δὲ νοσέουσιν οὐκ ἔστι τὰ προσταχθέντα παραβῆναι. καὶ μὴν πολύ γε εὐλογώτερον τοῖσι κάμνουσιν ἀδυνατέειν τὰ προστασσόμενα ὑπουργέειν ἢ τοῖς ἰητροῖσι τὰ μὴ δέοντα ἐπιτάσσειν· οἱ μὲν γὰρ ὑγιαινούσῃ γνώμῃ μεθ' ὑγιαίνοντος σώματος ἐγχειρέουσι λογισάμενοι τά τε παρεόντα τῶν τε παροιχομένων τὰ ὁμοίως διατεθέντα τοῖσι παρεοῦσιν, ὥστε ποτὲ θεραπευθέντα εἰπεῖν, ὡς ἀπήλλαξαν, οἱ δ' οὔτε ἃ κάμνουσιν, οὔτε δι' ἃ κάμνουσιν, οὔθ' ὅ τι ἐκ τῶν παρεόντων ἔσται, οὔθ' ὅ τι ἐκ τῶν τούτοισιν ὁμοίων γίνεται, εἰδότες ἐπιτάσσονται . . . οὕτω δὲ διακειμένους πότερον εἰκὸς τούτους τὰ ὑπὸ τῶν ἰητρῶν ἐπιτασσόμενα ποιέειν ἢ ἄλλα ποιέειν, ἢ ἃ ἐπετάχθησαν, ἢ τοὺς ἰητροὺς τοὺς ἐκείνως διακειμένους, ὡς ὁ πρόσθεν λόγος ἡρμήνευσεν, ἐπιτάσσειν τὰ μὴ δέοντα;

Chapter 9 proceeds to distinguish between two main classes of diseases, a small group in which the signs are easily seen—where the disease is manifest to sight or to touch, for instance—and a larger one where they are not so clear. In the former group "in all cases the cures should be infallible, not because they are easy, but because they have been discovered."[29] So far as the second group goes, "the art should not be at a loss in the case of the unclear diseases too."[30] The difficulty in achieving cures stems largely from delays in diagnosis, but this is more often due to the nature of the disease and to the patient than to the physician. The patients' own descriptions of their complaints are unreliable, for they have opinion rather than knowledge.[31] "For if they had understood [their diseases], they would not have incurred them. For it belongs to the same skill to know the causes of diseases and to understand how to treat them with all the treatments that prevent diseases from growing worse."[32] Again the writer's naive optimism comes out: the nature of our bodies is such that where a sickness admits of being seen, it admits of being healed.[33]

The breathtaking self-confidence of this treatise is far from unique. Drastically oversimplified pathological, therapeutic, and physiological doctrines—stated with apparently total self-assurance despite the manifest controversiality of the subjects in question—figure not just in

---

ἆρ᾽ οὐ πολὺ μᾶλλον τοὺς μὲν δεόντως ἐπιτάσσειν, τοὺς δὲ εἰκότως ἀδυνατέειν πείθεσθαι, μὴ πειθομένους δὲ περιπίπτειν τοῖσι θανάτοισιν, ὧν οἱ μὴ ὀρθῶς λογιζόμενοι τὰς αἰτίας τοῖς οὐδὲν αἰτίοις ἀνατιθέασι τοὺς αἰτίους ἐλευθεροῦντες.

29. *De arte* 9, CMG 1.1.15.11–13: τῶν μὲν οὖν τοιούτων πάντων ἐν ἅπασι τὰς ἀκεσίας ἀναμαρτήτους δεῖ εἶναι, οὐχ ὡς ῥηιδίας, ἀλλ᾽ ὅτι ἐξεύρηνται. Cf. *Praec.* 2, CMG 1.1.31.3ff.

30. *De arte* 10, CMG 1.1.15.17: δεῖ γε μὴν αὐτὴν οὐδὲ πρὸς τὰ ἧσσον φανερὰ ἀπορέειν.

31. *De arte* 11, CMG 1.1.16.23f.

32. *De arte* 11, CMG 1.1.16.24–27: εἰ γὰρ ἠπίσταντο, οὐκ ἂν περιέπιπτον αὐτοῖς· τῆς γὰρ αὐτῆς ξυνέσιός ἐστιν, ἥσπερ τὸ εἰδέναι τῶν νούσων τὰ αἴτια, καὶ τὸ θεραπεύειν αὐτὰς ἐπίστασθαι πάσῃσι τῇσι θεραπείῃσιν, αἳ κωλύουσι τὰ νοσήματα μεγαλύνεσθαι.

33. *De arte* 11, CMG 1.1.17.5f.

34. *Flat.* 2, CMG 1.1.92.13–17: τῶν δὲ δὴ νούσων ἁπασέων ὁ μὲν

other exhibition pieces, such as *On Breaths*,[34] but also, for example, in *On Affections*,[35] *On Diseases* 1,[36] *On the Sacred Disease*,[37] *On Fleshes*,[38] *On Regimens* 1,[39] and so on. *On the Places in Man*, for instance, is a work chiefly devoted to a quite detailed account first of certain anatomical topics and then of a range of morbid conditions and their treatments. Towards the end of the treatise as we have it[40] we find a chapter that announces: "The whole of medicine, thus constituted,

---

τρόπος ωὑτός, ὁ δὲ τόπος διαφέρει. δοκέει μὲν οὖν οὐδὲν ἐοικέναι τὰ νοσήματα ἀλλήλοισιν διὰ τὴν ἀλλοιότητα τῶν τόπων, ἔστι δὲ μία πασέων νούσων καὶ ἰδέη καὶ αἰτίη· ταύτην δέ, ἥτις ἐστίν, διὰ τοῦ μέλλοντος λόγου φράσαι πειρήσομαι. ("Of all diseases the manner is the same, but the place varies. Diseases are thought not to resemble one another at all, because of the difference in their locations, but in fact the kind and cause of all diseases is one and the same; I shall try to declare its character in the coming discourse.")

35. *Aff.* 1 (L) 6.208.7ff.: "in men, all diseases are caused by bile and phlegm. Bile and phlegm give rise to diseases when they become too dry or too wet or too hot or too cold in the body."

36. *Morb.* 1.2 (L) 6.142.13ff.: "all diseases come to be, as regards things inside the body, from bile and phlegm, and as regards external things, from exercise and wounds, from the hot being too hot, the cold too cold, the dry too dry, the wet too wet."

37. *Morb.Sacr.* 18 (L) 6.394.14–16: φύσιν δὲ ἕκαστον ἔχει καὶ δύναμιν ἐφ᾽ ἑωυτοῦ, καὶ οὐδὲν ἄπορόν ἐστιν οὐδ᾽ ἀμήχανον. ἀκεστά τε τὰ πλεῖστά ἐστι τοῖς αὐτοῖσι τούτοισιν ἀφ᾽ ὧν καὶ γίνεται. ("Each [disease] has its own nature and power and there is nothing that is unmanageable or beyond expedient. The majority may be cured by the very same things from which they arise"), and cf. 18 (L) 6.396.5ff., quoted above, Chap. 1 at n. 88.

38. The writer of *Carn.* sets out his version of a four-element theory in the opening two chapters as his own opinion, e.g., "it *seems to me* that what we call hot is immortal" (2 [L] 8.584.9), "the ancients *seem to me* to have called this aither" ([L] 8.584.12, and cf. 5 [L] 8.590.5). Yet in the sequel there are few signs of tentativeness as he develops some highly speculative physiological and embryological theories about, for example, the interaction of the two principles he calls the glutinous and the fatty in the formation of the main viscera: see, e.g., 3 (L) 8.584.18ff., 4 (L) 8.588.14ff., and the claims to demonstrate in 9 (L) 8.596.9 and 16. Cf. also 1 (L) 8.584.5.

39. *Vict.* 1.3 (L) 6.472.12ff., for example, states: "All the other animals and man are composed of two things, different in power, but complementary in their use, I mean fire and water."

40. Cf., however, below at nn. 98f. on the recognition of the complexities of medicine in περὶ τόπων τῶν κατ᾽ ἄνθρωπον (*De locis in homine, Loc.*

seems to me to have been discovered already. . . . He who understands medicine thus, waits for chance least of all, but would be successful with or without chance. The whole of medicine is well established and the finest of the theories it comprises appear to stand least in need of chance."[41]

On the Nature of Man, in particular, makes repeated claims to be able to demonstrate the theories it proposes.[42] While his opponents add to their speeches "evidences and proofs that amount to nothing,"[43] the author says that he will "produce evidences and declare the necessities through which each thing is increased or decreased in the body."[44] Yet his own positive evidences turn out to be very much of the same general type as theirs, even though their monistic conclusions are more extreme than his. He suggests that what influenced the monistic theorists he attacks was the observation that a certain substance may

---

Hom.) chap. 41 (L) 6.330.20ff., and of the difficulty of seizing the right moment, καιρός, in medicine, in 44 (L) 6.338.6ff.

41. *Loc.Hom.* 46 (L) 6.342.4 and 5–9: ἰητρικὴ δή μοι δοκέει ἤδη ἀνευρῆσθαι ὅλη, ἥτις οὕτως ἔχει. . . . ὃς γὰρ οὕτως ἰητρικὴν ἐπίσταται, ἐλάχιστα τὴν τύχην ἐπιμένει, ἀλλὰ καὶ ἄνευ τύχης καὶ ξὺν τύχῃ εὐποιηθείη ἄν. βέβηκε γὰρ ἰητρικὴ πᾶσα, καὶ φαίνεται τῶν σοφισμάτων τὰ κάλλιστα ἐν αὐτῇ συγκείμενα ἐλάχιστα τύχης δεῖσθαι.

42. See especially ἀποδείξω at *Nat.Hom.* 2, CMG 1.1.3.170.3, and ἀποδείκνυμι in 5, CMG 1.1.3.178.9, and cf. ἀποφανεῖν at 2, CMG 1.1.3.170.6 and 5, CMG 1.1.3.174.11.

43. *Nat.Hom.* 1, CMG 1.1.3.164.14. His opponents in chap. 1 are monists who discourse about the nature of man beyond what is relevant to medicine and who claim that man is composed of air or water or fire or earth. In *Nat.Hom.* 2, CMG 1.1.3.166.12ff., he turns to attack monistic doctors who take blood, bile, or phlegm as the sole element of man. He has a general argument, against these, that if man were a unity he would never feel pain, since there would be nothing by which, being a unity, it could be hurt (*Nat.Hom.* 2, CMG 1.1.3.168.4f., with which compare Melissus fr. 7, para. 4). But against those who asserted that man consists of blood alone, for example, he demands that they should be able to show that there is a time of year or of human life when blood is obviously the sole constituent in the body (*Nat.Hom.* 2, CMG 1.1.3.168.9ff.).

44. *Nat.Hom.* 2, CMG 1.1.3.170.6f.

be purged from the body when a man dies. In some cases where a patient dies from an overdose of a purgative drug he vomits bile, in others maybe phlegm, and the monists, seeing this, then concluded that the human body consists of this one thing.[45] But while destructively the author sets about demolishing monism with powerful dialectical arguments, constructively when he seeks to establish that the body consists of the four humours, blood, phlegm, yellow bile, and black bile, his own chief argument too depends on the simple observation that all four are found in the excreta. This shows, to be sure, that all four are present in the body, but spectacularly fails to demonstrate that they are the elements of which it is composed.[46]

Alongside the frequent use of the vocabulary of evidence and proof, one of the key terms this author employs is necessity, ἀνάγκη, and its cognates, and the deployment of this word in this and other treatises offers an insight into their dogmatic character.[47] From the rich collection of uses in *On the Nature of Man* itself, the following may be cited. In chapter 3 he writes: "first, necessarily generation does not arise from a single thing: for how could one thing generate another unless it united with something?"[48] Later on he says that it is not likely that generation could take place from one thing, when it does not even occur from many unless those many are combined in the right proportions. He proceeds: "necessarily, then, since such is the nature of man and of everything else, man is not a single thing."[49] Further on in the

45. *Nat.Hom.* 6, CMG 1.1.3.178.11–14.
46. *Nat.Hom.* 5, CMG 1.1.3.176.10ff.; 6, CMG 1.1.3.180.2ff.; 7, CMG 1.1.3.182.12ff. In 5, CMG 1.1.3.178.5ff., he claims that the humours are congenital, on the grounds that they are present at every age and in both parents. Yet even if that were conceded, it would still not show that they are the chief, let alone that they are the sole elemental, constituents of the body.
47. In what follows I concentrate on the use of the term ἀνάγκη in this and other treatises (for the general history of this term, cf. Schreckenberg 1964): the use of δεῖ and χρή could, in many cases, also be cited to corroborate similar points: cf. Redard 1953, pp. 47ff.; Benardete 1965.
48. *Nat.Hom.* 3, CMG 1.1.3.170.8–9; cf. also 2, CMG 1.1.3.168.6.
49. *Nat.Hom.* 3, CMG 1.1.3.172.2–3.

same chapter we find: "necessarily, each thing returns again to its own nature when the body of the man dies, the wet to the wet, the dry to the dry, the hot to the hot, the cold to the cold."[50] Chapter 4 argues that when the humours in the body are well mixed and in the right proportion, the body is healthy, but that pain occurs when one of them is in excess or defect or is separated off from the others. "Necessarily, when one of them is separated and stands by itself, not only the place from which it has come becomes diseased, but also that where it collects and streams together causes pain and distress."[51] Again in chapter 5, having suggested that blood, bile, and phlegm differ to sight, to touch, in temperature, and in humidity, he goes on: "necessarily, then, since they are so different from one another in appearance and power, they cannot be one, if fire and water are not one."[52]

Clearly, logical and physical, conceptual and causal, necessity are not here differentiated. Many instances represent a conflation of one or more ideas that we might distinguish. Often the underlying idea seems merely to be the claim that something is always or usually the case. At the limit, the addition of the term *necessarily* appears to reflect little more than the writer's desire to assert his point with emphasis.

Similar uses of the term ἀνάγκη are common elsewhere in the Hippocratic Corpus, not only in the types of treatise that have provided most of our examples so far[53] but also in other major works,[54]

---

50. *Nat.Hom.* 3, CMG 1.1.3.172.5–8.

51. *Nat.Hom.* 4, CMG 1.1.3.174.3–6, cf. also 174.9f.

52. *Nat.Hom.* 5, CMG 1.1.3.176.8–9. Cf. also 7, CMG 1.1.3.186.3; 8, CMG 1.1.3.186.17ff., and from after the main physiological section of the treatise (chaps. 1–8), e.g., *Nat.Hom.* 10, CMG 1.1.3.192.10; 12, CMG 1.1.3.198.5, 200.3 and 8.

53. See, e.g., *De arte* 5, CMG 1.1.12.2 and 6; *Flat.* 7, CMG 1.1.95.7; 10, CMG 1.1.98.16; *Aff.* 37 (L) 6.246.20; *Morb.* 1.3 (L) 6.144.4, 17, 4 (L) 6.146.6, 9, 12, 13, 8 (L) 6.156.2, 4, 22 (L) 6.184.4, 186.10, 24 (L) 6.190.1, 7, 25 (L) 6.192.2; *Morb.Sacr.* 8 (L) 6.376.6, 13 (L) 6.386.7, 14 (L) 6.388.6ff., 17 (L) 6.392.19; *Carn.* 19 (L) 8.614.16; *Vict.* 1.4 (L) 6.474.15, 1.7 (L) 6.480.11, 1.9 (L) 6.484.4, 1.30 (L) 6.504.19, 1.36 (L) 6.524.7, 2.37 (L) 6.528.4; 2.38 (L) 6.530.14, 532.7, 2.40 (L) 6.538.4ff., 3.68 (L) 6.598.8, 3.71 (L) 6.610.9.

54. Further examples could be given from other pathological treatises, e.g., *Morb.* 2.5 (L) 7.12.24f.; *Morb.* 3.16, CMG 1.2.3.90.1ff.; from the embryologi-

including some which, as we shall see later, are otherwise remarkable for their undogmatic or anti-dogmatic traits. Examples could be given from *Aphorisms*,[55] *On Ancient Medicine*,[56] *Wounds in the Head, On Joints,* and *On Fractures.*[57] The treatise *On Airs Waters Places,* too, frequently presents as matters of necessity the correlations it proposes

---

cal works, e.g., περὶ γονῆς (*De Genitura*) 8 (L) 7.480.9f.; περὶ φύσιος παιδίου (*De natura pueri, Nat.Puer.*) 12 (L) 7.488.8f., 18 (L) 7.504.21, 26f.; Morb. 4.34 (L) 7.548.7ff.; and from the gynaecological treatises, e.g., *Mul.* 1.25 (L) 8.64.13ff., 1.34 (L) 8.78.11ff., 2.133 (L) 8.280.12ff., 2.138 (L) 8.312.2ff., and περὶ ἀφόρων (*De Mulieribus sterilibus, Steril.*) 222 (L) 8.428.15ff., 223 (L) 8.432.4f., 244 (L) 8.458.5.

55. From the *Aphorisms* we may cite 5.54 (L) 4.552.4f. (when the mouth of the womb is hard, it necessarily closes), 6.20 (L) 4.568.5f. (haemorrhage into the abdominal cavity is necessarily followed by suppuration), 6.45 (L) 4.574.8f. (when ulcers last a year or longer, the bone necessarily exfoliates and the scars become hollow) and 6.50 (L) 4.576.4f. (laceration of the brain is necessarily followed by fever and bilious vomiting, with which compare Κωακαὶ προγνώσιες [*Praenotiones coacae, Coac.*] 4.490 [L] 5.696.5ff., dealing with wounding of the brain, where the consequence is not said to be a matter of necessity, but of what is the case "for the most part"), and cf., e.g., *Aph.* 6.58 (L) 4.578.3, 7.58 (L) 4.594.10f., 7.85 (L) 4.606.10ff.

56. See, for example, *VM* 22, *CMG* 1.1.54.6–10 ("as for what produces flatulence and colic, it belongs to the hollow and broad parts, such as the stomach and chest, to produce noise and rumbling. For when a part is not completely full so as to be at rest, but instead undergoes changes and movements, necessarily these produce noise and clear signs of movement"), and cf., e.g., *VM* 19, *CMG* 1.1.50.7ff., in the writer's general statement about causation (cf. below, Chap. 6 n. 14).

57. In the surgical treatises, among the types of consequences and connections that are sometimes presented as matters of necessity are (1) the real or assumed consequences of lesions, (2) real or assumed anatomical facts and their consequences, and (3) the consequences of treatments, especially of faulty treatments. As examples of (1) we may cite *VC* 4 (L) 3.196.1f. (if the bone in the head is fractured when wounded, then necessarily contusion occurs), and *Art.* 63 (L) 4.272.14ff. (the doctor must bear in mind that in certain severe dislocations of the bones of the leg, when they project right through the ankle joint, the patient will necessarily be deformed and lame), and cf., e.g., *VC* 7 (L) 3.204.8f., 11 (L) 3.220.7f., 15 (L) 3.244.1ff.; *Art.* 13 (L) 4.116.23ff., 38 (L) 4.168.9f. As examples of (2): *Fract.* 3 (L) 3.424.10ff. (bending of a fractured arm necessarily causes a change in the position of the muscles and bones) and *Art.* 47 (L) 4.200.15ff. (in curvature of the spine one of the vertebrae necessarily appears to stand out more prominently than the rest) and cf.,

between the aspect of a city and the character of its water, or between both of those and the constitutions and endemic diseases of the inhabitants, or even between the political constitution and the character of the people. We may again illustrate very selectively from the rich fund of examples.

Thus we are told that in a city sheltered from the northerly winds but exposed to warm prevailing southerly ones, the water is "necessarily plentiful, brackish, surface water, warm in the summer and cold in the winter," [58] while in a city that faces the risings of the sun, the water is "necessarily clear, sweet-smelling, soft, and pleasant," [59] As for the effects of waters of different types, the writer states, for instance, that "stagnant, standing, marshy water is in summer necessarily warm, thick, and of an unpleasant smell, because it does not flow. But by continually being fed by the rains and evaporated by the sun it is necessarily discoloured, harmful, and productive of biliousness." [60] Dealing with physical constitutions and endemic diseases, the writer claims, for instance, that in northerly-facing cities that generally have hard, cold water, the inhabitants are "necessarily vigorous and lean." [61] Pleurisies and acute diseases are common, "for this is necessarily the case when bellies are hard." [62] Correlating the character and changes of the seasons with the diseases to be expected in them, the writer says:

---

e.g., *Fract.* 23 (L) 3.492.7ff. As an example of (3) we may cite *Fract.* 25 (L) 3.498.8ff., criticising bandaging that leaves the wound exposed ("the treatment, too, is itself evidence: for in a patient so bandaged the swelling necessarily arises in the wound itself, since if even healthy tissue were bandaged on this side and that, and a vacancy left in the middle, it would be especially at the vacant part that swelling and discoloration would occur. How then could a wound fail to be affected in this way? For it necessarily follows that the wound is discoloured with everted edges, and has a watery discharge devoid of pus"), and cf., e.g., *Art.* 14 (L) 4.122.16ff., *Fract.* 7 (L) 3.442.7ff., 16 (L) 3.476.11ff., 34 (L) 3.536.9ff.

58. *Aër.* 3, CMG 1.1.2.26.23ff., 28.2f.
59. *Aër.* 5, CMG 1.1.2.32.10ff., 13ff.
60. *Aër.* 7, CMG 1.1.2.34.19–23, cf. also 36.25, 38.7f.; and 9, CMG 1.1.2.44.15f., 20f.
61. *Aër.* 4, CMG 1.1.2.30.4.
62. *Aër.* 4, CMG 1.1.2.30.8f., cf. 12f.; and 6, CMG 1.1.2.34.1f.

"If the winter be dry, with northerly winds prevailing and the spring wet, with southerly winds, the summer will necessarily be feverish and productive of ophthalmia." [63] Finally, correlating political constitutions and characters, the second half of the treatise suggests, for example, that "where men are ruled by kings, there necessarily they are most cowardly. . . . For their souls are enslaved and they are unwilling to run risks heedlessly for the sake of another's power." [64]

Even though other generalisations in this treatise are quite often explicitly qualified as holding only "for the most part" or just as being "likely," εἰκός, [65] the variety of connections claimed as being matters of "necessity" is, as these and many other examples demonstrate, considerable. Sometimes the grounds for the necessity are specified in a succeeding γάρ or because clause. [66] The point is important since it indicates at least an occasional recognition of the need, in principle, to support with evidence or argument the conclusions that are asserted with such emphasis; in that respect the dogmatists in the Hippocratic Corpus may be distinguished from even more extreme cases where no such recognition surfaces in the text at all. Yet it must also be remarked, first, that often no such grounds are adduced, and, secondly, that even when they are, they are often little more than cosmetic, and they generally fall far short of justifying the claims made as to the *necessity* of the conclusions.

---

63. *Aër.* 10, CMG 1.1.2.46.22–24: ἦν δὲ ὁ μὲν χειμὼν αὐχμηρὸς καὶ βόρειος γένηται, τὸ δὲ ἦρ ἔπομβρον καὶ νότιον, ἀνάγκη τὸ θέρος πυρετῶδες γίνεσθαι καὶ ὀφθαλμίας ἐμποιεῖν (compare *Aph.* 3.11 [L] 4.490.2ff.), and cf. also *Aër.* 10, CMG 1.1.2.46.24ff., 50.18ff.

64. *Aër.* 23, CMG 1.1.2.78.3–5; cf. 16, CMG 1.1.2.62.20ff. Physiological and pathological correlations claimed as necessary also occur in the second part of the treatise, e.g., *Aër.* 19, CMG 1.1.2.68.15ff., and 24, CMG 1.1.2.80.3ff.

65. See, for example, from the first part of the treatise, *Aër.* 3, CMG 1.1.2.28.5f.; 4, CMG 1.1.2.30.3, 7, 18, and from the second, *Aër.* 14, CMG 1.1.2.58.23; 24, CMG 1.1.2.78.15.

66. As, for example, in the texts from *Aër.* 7, CMG 1.1.2.34.19ff.; 4, CMG 1.1.2.30.12f; and 10, CMG 1.1.2.46.22ff., quoted at notes 60, 62, and 63 above.

Uncertainty

In diagnosis and therapeutics, in pathology, anatomy, and physiology, the overwhelming impression created by a very considerable body of texts in a wide variety of Hippocratic works is one of their authors overstating their cases, representing as incontestable assertions for which their ground were—and must even have seemed to many of their own contemporaries to have been—tenuous or nonexistent. Yet that is only one side of the picture. Alongside the dogmatic tendencies I have illustrated—sometimes, indeed in the very same treatises—there are signs of tentativeness and caution, a readiness to admit to doubts and to mistakes, a recognition of the rashness of unsupported claims, explicit qualifications concerning how far a general rule applies or about the limits of the writer's own firsthand knowledge, and statements insisting on the inexactness of the whole of medical practice.[67] In some cases, where, for example, the healer deliberately records his own errors, we are dealing with what appears to be—to judge from the extant remains of ancient medical literature, non-Greek as well as Greek—a quite unprecedented phenomenon.

We have noted before that criticisms of current medical practice are common in certain works,[68] but a critical attitude towards the mistakes of colleagues is of course quite compatible with and often accompanies overconfidence about the correctness of one's own ideas and procedures. In some Hippocratic texts, however, the author explicitly acknowledges that he was himself mistaken. Thus in *Epidemics* 5.27, which describes the case of one Autonomus who suffered from a wound in the head, the writer remarks: "It escaped my notice that he needed trepanning. The sutures which bore on themselves the lesion made by the weapon deceived my judgement, for afterwards it became

---

67. Some, though not all, of these points find parallels in that other major domain of ἱστορίη, the writing of history. Both Herodotus and Thucydides frequently express their uncertainty concerning facts or their explanations or their doubts about the reliability of the evidence available to them.

68. See above, Chap. 2 at nn. 59ff. and n. 78.

apparent." [69] The following chapter refers to the case of a young girl who was also wounded in the head, where trepanning was recognised to be indicated and was in fact carried out, but in this case, the writer says, not enough of the bone was removed. [70] The next two chapters describe two further cases where cauterisation was undertaken too late—in one case, we are told, thirty days too late—and both patients died. [71]

The author (or authors) of the surgical treatise *On Joints* not only describes some of his own mistakes but specifically notes that one such report is included so that others may learn from his own experience. Chapter 47 remarks on the difficulty of reducing humpback. "For my part . . . I know of no better or more correct modes of reduction than these. For straight-line extension on the spine itself, from below, at the so-called sacrum, gets no grip; from above, at the neck and head, it gets a grip indeed, but extension made here looks unseemly, and would also cause harm if carried to excess." [72] He then proceeds:

> I once tried to make extension with the patient on his back, and after
> putting an uninflated wineskin under the hump, then tried to blow air
> into the skin with a smith's bellows. But my attempt was not a success,
> for when I got the man well stretched, the skin collapsed, and air could

---

69. *Epid.* 5.27 (L) 5.226.10–12: τοῦτο παρέλαθέ με δεόμενον πρι-
σθῆναι· ἔκλεψαν δέ μου τὴν γνώμην αἱ ῥαφαὶ ἔχουσαι ἐν σφίσιν ἑωυτῇσι
τοῦ βέλεος τὸ σῖνος· ὕστερον γὰρ καταφανὲς γίνεται. In his *De Medicina*
Celsus was to express his admiration for Hippocrates' honesty and love of
truth in recording his errors in this and similar cases: *Med.* 8.4.3, *CML*
1.378.3ff. The problem of being deceived by the sutures in assessing a lesion to
the skull is mentioned in general terms at *VC* 12 (L) 3.222.6ff., 228.4ff., and
19 (L) 3.252.3ff. comments on failures to save patients through omitting to
trepan or use the rasp.

70. *Epid.* 5.28 (L) 5.226.17ff.

71. *Epid.* 5.29 (L) 5.228.5ff., and 30 (L) 5.228.10ff.

72. *Art.* 47 (L) 4.210.3–9: οὔκουν ἐγὼ ἔχω τουτέων ἀνάγκας καλλίους,
οὐδὲ δικαιοτέρας· ἡ γὰρ κατ' αὐτὴν τὴν ἄκανθαν ἰθυωρίη τῆς κατατάσιος
κάτωθέν τε καὶ κατὰ τὸ ἱερὸν ὀστέον καλεόμενον οὐκ ἔχει ἐπιλαβὴν οὐδε-
μίην· ἄνωθεν δὲ κατὰ τὸν αὐχένα καὶ κατὰ τὴν κεφαλήν, ἐπιλαβὴν μὲν
ἔχει, ἀλλ' ἐσιδέειν γε ἀπρεπὴς ταύτῃ τοι γινομένη ἡ κατάτασις, καὶ
ἄλλας βλάβας ἂν προσπαρέχοι πλεονασθεῖσα.

not be forced into it; it also kept slipping round at any attempt to bring the patient's hump and the convexity of the blown-up skin forcibly together; while when I made no great extension of the patient, but got the skin well blown up, the man's back was hollowed as a whole rather than where it should have been. I relate this on purpose: for those things also give good instruction which after trial show themselves failures, and show why they failed.[73]

That the author and his colleagues were at a loss as to how to cure or even help a patient is often admitted in both the surgical works and the *Epidemics*. *Epidemics* 3 case 9 of the first series ends an account of a woman who suffered from an attack of ileus with the grim note: "it was impossible to do anything to help her; she died."[74] Case 5 of the second series remarks of a man who suffered from a sudden pain in the right thigh that "no treatment that he received did him any good."[75] Chapter 8 in the Constitution in this book comments more generally that there was little response to treatment and that purgatives did more harm than good,[76] and elsewhere writers in the *Epidemics* note that if

---

73. *Art.* 47 (L) 4.210.9–212.5: ἐπειρήθην δὲ δή ποτε, ὕπτιον τὸν ἄνθρωπον κατατείνας, ἀσκὸν ἀφύσητον ὑποθεῖναι ὑπὸ τὸ ὕβωμα, κἄπειτα αὐλῷ ἐκ χαλκείου ἐς τὸν ἀσκὸν τὸν ὑποκείμενον ἐνιέναι φῦσαν. ἀλλά μοι οὐκ εὐπορεῖτο · ὅτε μὲν γὰρ εὖ κατατείνοιμι τὸν ἄνθρωπον, ἡσσᾶτο ὁ ἀσκός, καὶ οὐκ ἡδύνατο ἡ φῦσα ἐσαναγκάζεσθαι · καὶ ἄλλως ἕτοιμον περιολισθά-νειν ἦν, ἅτε ἐς τὸ αὐτὸ ἀναγκαζόμενον, τό τε τοῦ ἀνθρώπου ὕβωμα, καὶ τὸ τοῦ ἀσκοῦ πληρουμένου κύρτωμα. ὅτε δ' αὖ μὴ κάρτα κατατείνοιμι τὸν ἄνθρωπον, ὁ μὲν ἀσκὸς ὑπὸ τῆς φύσης ἐκυρτοῦτο, ὁ δὲ ἄνθρωπος πάντη μᾶλλον ἐλορδαίνετο ἢ ᾗ ξυνέφερεν. ἔγραψα δὲ ἐπίτηδες τοῦτο · καλὰ γὰρ καὶ ταῦτα τὰ μαθήματά ἐστιν, ἅ, πειρηθέντα, ἀπορηθέντα ἐφάνη, καὶ δι' ἅσσα ἠπορήθη.
74. *Epid.* 3 case 9 of the first series (L) 3.58.7f. Reporting of failures in clinical case-histories continues after the Hippocratic *Epidemics*. That Erasistratus' accounts of individual cases contained instances where the patient died is clear from the reports in Galen, e.g., (K) 11.200.1ff., 205.6ff., 206.5ff., 209.14ff., who exploits these failures for his own polemical purposes.
75. *Epid.* 3 case 5 of the second series (L) 3.118.8.
76. *Epid.* 3.8 (L) 3.88.2ff., cf. *Epid.* 5.18 (L) 5.218.8, 5.31 (L) 5.228.14f., 5.42 (L) 5.232.9f., 5.43 (L) 5.232.17f., 5.76 (L) 5.248.9ff. (≈ *Epid.* 7.38 [L] 5.406.5ff.).

the treatment had been different, a patient might have recovered or survived longer.[77]

On Joints, too, often refers to surgical cases where no remedy is possible[78] and repeatedly warns that the attempt to reduce certain intractable lesions does more harm than good.[79] Elsewhere the surgical writers explicitly say they do not know what to advise,[80] or withhold judgement.[81] The difficulties and dangers of treatment are mentioned also in other treatises, either in general terms, as in the famous first *Aphorism* ("life is short, art long, opportunity elusive, experience dangerous, judgement difficult"),[82] or in relation to particular remedies, as, for example, the administration of hellebore or the practice of cautery or that of venesection.[83]

Many works draw attention to the incurability of certain diseases, though the advice they offer differs. Some suggest that the doctor should at least do what he can to help,[84] but others warn or instruct

---

77. See, for example, *Epid.* 5.7 (L) 5.208.9ff., 5.15 (L) 5.214.18f., 5.18 (L) 5.218.2f., 12f., 5.31 (L) 5.228.20f., 5.33 (L) 5.230.4f., 5.95 (L) 5.254.19ff. (≃ *Epid.* 7.121 [L] 5.466.14ff.); on some occasions, however, the problem is located, in whole or in part, with the patient—with a deliberate or involuntary failure on the patient's part to take treatment, e.g., 3 case 14 of the second series (L) 3.140.18, *Epid.* 5.18 (L) 5.218.6f.

78. *Art.* 48 (L) 4.212.17ff., 63 (L) 4.270.7ff.

79. *Art.* 63 (L) 4.268.12ff., 64 (L) 4.274.8ff., 65 (L) 4.274.20ff., 66 (L) 4.276.12ff., 67 (L) 4.278.5ff.; cf. *Fract.* 35 (L) 3.536.13ff.; *Mochl.* 33 (L) 4.374.16f., 376.2f.; *Aph.* 6.38 (L) 4.572.5ff.

80. E.g., *Fract.* 16 (L) 3.474.17.

81. E.g., *Art.* 1 (L) 4.78.2 ff., 80.13f., 53 (L) 4.232.12ff.; and frequently in *Epid.,* e.g., 1.4 (L) 2.626.3ff. Cf. *Praec.* 8, *CMG* 1.1.33.5ff., which advises the doctor, when in difficulties, not to hesitate to consult others.

82. *Aph.* 1.1 (L) 4.458.1f.

83. See, for example, *Aph.* 4.16 (L) 4.506.9f., 5.31 (L) 4.542.12f., 6.27 (L) 4.570.3f., 7.45 (L) 4.590.4ff.; *Acut.* 11 (L) 2.306.9ff., 308.7ff., 316.6ff.; *Art.* 40 (L) 4.172.5ff., 69 (L) 4.284.8ff.; *Fract.* 25 (L) 3.496.15ff., 30 (L) 3.518.4ff., 31 (L) 3.524.19ff.; *VC* 21 (L) 3.256.11ff. (on the hazards of trepanning), and cf. *Morb.* 1.6 (L) 6.150.6ff., which sets out a whole list of errors in judgement or practice the doctor should avoid.

84. See, for example, *Art.* 58 (L) 4.252.8ff.

him not to undertake such cases.[85] *On Fractures* 36, dealing with fractures of the femur and humerus, illustrates the dilemma the doctor
sometimes faced. "One should especially avoid such cases if one has a
respectable excuse, for the favourable chances are few and the risks
many. Besides, if a man does not reduce the fracture he will be thought
unskillful, while if he does reduce it he will bring the patient nearer to
death than to recovery."[86] Yet if some of the Hippocratic writers register their unease on this topic, it is important to note that none recommends that those patients whom they cannot or will not treat should
have recourse to other modes of healing: none suggests that the sick
should turn to the cult of Asclepius,[87] let alone try their luck with the
itinerant sellers of charms and purifications.

The theme of the inexactness of the medical art is a prominent one
in several treatises and of particular interest for our inquiry. We shall
be returning later to aspects of this in connection with the use of measurement.[88] Here we may simply note the recurrence of the motif in a
variety of treatises. *On Ancient Medicine,* especially, develops the topic
at some length. Exactness (ἀκριβείη, τὸ ἀκριβές, or τὸ ἀτρεκές) in
the control of diet is difficult to achieve and small errors are bound to
occur.[89] "I would heartily praise the physician who makes only small
mistakes: exactness is rarely to be seen."[90] Up to a certain point the

---

85. See *De arte* 3, CMG 1.1.10.21ff.; *Mul.* 1.71 (L) 8.150.12ff. (≈ *Steril.*
233 [L] 8.446.20ff.); *Prorrh.* 2.12 (L) 9.34.15ff.
86. *Fract.* 36 (L) 3.540.9–12: μάλιστα δὲ χρὴ τὰ τοιαῦτα διαφυγεῖν,
ἅμα ἤν τις καλὴν ἔχῃ τὴν ἀποφυγήν· αἵ τε γὰρ ἐλπίδες ὀλίγαι, καὶ οἱ κίν-
δυνοι πολλοί · καὶ μὴ ἐμβάλλων ἄτεχνος ἂν δοκέοι εἶναι, καὶ ἐμβάλλων
ἐγγυτέρω ἂν τοῦ θανάτου ἀγάγοι, ἢ τῆς σωτηρίης.
87. That *Morb.Sacr.* 1 (L) 6.362.10ff. is not to be taken in that sense (despite Herzog 1931, p. 149) is, I believe, clear: see G. E. R. Lloyd 1979, p. 48 n.
209. On occasion, however, *Vict.* 4 recommends prayer to the gods: see above,
Chap. 1 n. 111; cf. n. 112.
88. See below, Chap. 5 at nn. 134ff.; cf. also at n. 187.
89. *VM* 9, CMG 1.1.41.18ff.: the correct diet cannot be determined by
reference to some number or weight: the only criterion is bodily feeling. Cf.
below, Chap. 5 n. 136.
90. *VM* 9, CMG 1.1.41.23–24: κἂν ἐγὼ τοῦτον τὸν ἰητρὸν ἰσχυρῶς
ἐπαινέοιμι τὸν σμικρὰ ἁμαρτάνοντα, τὸ δὲ ἀτρεκὲς ὀλιγάκις ἔστι κατιδεῖν.

subject can be, and has been, made exact, but perfect exactness (τὸ ἀτρεκέστατον) is unattainable. "But I assert that the ancient art of medicine should not be rejected as nonexistent or not well investigated because it has not attained exactness in every item. Much rather, since, as I think, it has been able to come close to perfect exactness by means of reasoning where before there was great ignorance, its discoveries should be a matter of admiration, as well and truly the result of discovery and not of chance."[91]

Other treatises, too, develop similar themes. *On Diseases* I, which presents a highly dogmatic general theory of diseases based on bile and phlegm,[92] states nevertheless that there is, in medicine, no ἀρχὴ ἀποδεδειγμένη, no demonstrated beginning or principle,[93] which is correct for the whole of the art of healing.[94] Discussing the καιροί, the turning-points of diseases which present the doctor with opportunities for intervention, the writer observes how much they differ from one disease to another and, after sketching out some of their variety, notes: "they have no exactness, ἀκριβείη, other than this."[95] Elsewhere too he stresses the differences between one body and another, one age and another, one illness and another, and repeats that "it is not possible to have exact knowledge, τὸ ἀκριβὲς εἰδέναι, nor to indicate at what

---

91. VM 12, CMG 1.1.44.2–7: οὔ φημι δὲ δεῖν διὰ τοῦτο τὴν τέχνην ὡς οὐκ ἐοῦσαν οὐδὲ καλῶς ζητεομένην τὴν ἀρχαίην ἀποβαλέσθαι, εἰ μὴ ἔχει περὶ πάντα ἀκριβείην, ἀλλὰ πολὺ μᾶλλον διὰ τὸ ἐγγὺς οἶμαι τοῦ ἀτρεκεστάτου δύνασθαι ἥκειν λογισμῷ ἐκ πολλῆς ἀγνωσίης θαυμάζειν τὰ ἐξευρημένα, ὡς καλῶς καὶ ὀρθῶς ἐξεύρηται καὶ οὐκ ἀπὸ τύχης. Cf. 20, CMG 1.1.51.6ff., where in his attack on those who argued that medicine should be based on natural philosophy the writer reverses the dependence and claims that such exact knowledge as can be attained about the nature of man and the causes of our coming-to-be is to be gained through medicine alone.

92. See above, n. 36, on *Morb.* 1.2 (L) 6.142.13ff.

93. The term ἀρχή is ambiguous, covering both beginning and principle. That the former is part of the meaning here is clear from the denial that medicine has a second point, a middle and an end: *Morb.* 1.9 (L) 6.156.14ff.

94. *Morb.* 1.9 (L) 6.156.14ff.; with which compare *De arte* 4, CMG 1.1.11.5f.; *Carn.* 1 (L) 8.584.2ff.; and Diogenes of Apollonia fr. 1.

95. *Morb.* 1.5 (L) 6.146.15ff., 148.15f.

The importance of taking full account of the different audiences envisaged by the various types of writing extant in the Hippocratic Corpus needs no underlining. To elaborate some points from our discussion in Chapter 2: many suggestions that a medical man might make to his colleagues—and many ways in which he might wish to make them—would be totally inappropriate for a lay audience. This remains true, even though, as I stressed before, the lay/professional distinction was much less firm in ancient medicine than it is today, and there is ample evidence, from the fifth and fourth centuries, of an extended interest in medical topics—not just as a potential audience, but also as speakers and writers—among people who had no intention of actually engaging in medical practice. Plato would be one obvious example.[109]

Yet whatever features of this hypothesis we may wish eventually to retain, as stated it clearly will not do, for two main reasons. First we have seen that there are treatises (including some that are reasonably well-defined unities, not multi-author concoctions) that combine a certain dogmatism at some points with an apparent tentativeness at others. By itself this would not be surprising, for it might simply reflect the varying degrees of difficulty of the topics dealt with and the varying degrees of confidence of the authors in dealing with them.[110] Yet to that, in turn, it must be said that in several of the cases we have considered, principally from *On Diseases* 1, *On the Places in Man*, *On Regimen*, and *On the Art*, it could not be claimed that dogmatism is confined to elementary or straightforward topics on which the authors might, with some justification, feel on safe ground. We have only to recall the claim in *On the Art* that for diseases with visible signs, "in

---

109. The theory of diseases presented at *Ti.* 81e ff. was taken sufficiently seriously to be excerpted at length in the history of medicine in Anonymus Londinensis 14.11ff., 17.11ff. Other theorists there reported on might also be used to illustrate the point, for example, Philolaus 18.8ff., and Philistion 20.25ff., and we have mentioned before the non-specialist, general interest in medicine shown by some sophists.

110. This may well be the more likely explanation in the case of some of the material from the surgical treatises that we considered.

all cases the cures should be infallible" because they have been discovered. The *combination* of dogmatism and hesitancy in this and other works suggests a difficulty for any theory based on a clear-cut contrast between dogmatic treatises addressed to a general public and more cautious ones aimed at professional medical colleagues.

A second, more general objection to the hypothesis is that it is in danger of ignoring what most of the treatises we have considered have in common. Admittedly there are clearly identifiable differences between the two ends of what I referred to before as the *spectrum* represented by our extant texts—on the one hand the *epideixis* designed for public consumption, and on the other the almost exclusively technical notebooks. Yet there is a case for saying that, in their different ways and to different degrees, *both* types of production are exercises in persuasion.[111] That is obvious enough in the case of the sophistic *epideixis*. But even those writers who mainly had their fellow-practitioners in mind were also concerned to win their confidence, or at least to make sure that their own credentials were going to be recognised.

There is no reason to doubt the good faith of the author of the chapter in *On Joints* that sets out his own mistakes so that other practitioners may learn from them. At the same time we should not rule out the possibility that deliberate self-criticism may occasionally be motivated by a desire to suggest a mature experience in the art. Admittedly it seems paradoxical that confessions of failure should be used in order to inspire confidence. Yet for a medical writer to demonstrate that he is well aware of the dangers of overconfidence would have a salutary effect. It would reassure prospective clients that they were dealing with a man who would not rashly undertake risky treatments nor raise hopes of cure unjustifiably. And it would help to persuade professional colleagues that the author was a man of experience conscious of the complexity and limitations of the art.

There is an important contrast here, not just between the more ten-

---

111. Some aspects of the relationship between rhetoric and Hippocratic medicine are discussed in G. E. R. Lloyd 1979, pp. 88ff.; cf. Kudlien 1974.

tative and the more dogmatic Hippocratic texts, but between the former and the claims for unqualified success that are characteristic of temple medicine. In the inscriptions set up in the shrine of Asclepius at Epidaurus it is 100 percent success that is recorded.[112] Some Hippocratic writers might well have wanted to dissociate themselves from the implicit claims to infallibility made in religious healing, even while other medical authors represented in the Corpus adopt a tone that rivals temple medicine in self-assurance.

The idea that self-criticism was sometimes deliberately deployed with such an intention cannot be confirmed directly. But it is perhaps suggestive that the main context in which an apparent tentativeness is expressed in certain treatises is in general remarks concerning the inexactness or variability of medicine, as in *On Regimen* 3 and *On Diseases* 1. It looks as if the explanation of these apparently mixed cases is neither that the authors are simply expressing a variety of attitudes on different topics, nor that they are merely inconsistent, nor yet that we are dealing with divergent material in composite works. Rather, it may be that even in otherwise dogmatic works, the inclusion of some indication of the inexactness of medicine had become, or was becoming, something of a convention or a commonplace.

If so, we should accept the apparent paradox. Dogmatism is clearly a stance frequently adopted to impress people, especially a lay audience, and especially on such questions as the origins of diseases in general or the constituents of the human body. Yet professions of uncertainty may also have a certain persuasive role, and while detailed accounts of failure in individual cases are confined to the more technical works that record actual clinical practice,[113] even more theoretical or philosophically oriented treatises occasionally include among their otherwise doctrinaire assertions a note to the effect that medi-

---

112. See Herzog 1931.

113. This is not just a matter of surgical practice, although surgery provides most of the more striking cases (cf. R. Joly 1980, pp. 287f.): some of the examples of recorded mistakes or faulty treatments in the *Epidemics* relate to general medicine, as for instance those at *Epid.* 5.18 (L) 5.218.2ff., and 5.31 (L) 5.228.14f., cited above in nn. 76f.

cine is not certain. With some authors it becomes part of the definition of medicine, and of its claim to be the art that it is, that it is inexact. The recognition that it cannot do *everything* is sometimes used as a genuine warning, but it is also sometimes used to bolster claims (and they might be extravagant claims) that it *could* do a very great deal. That certain diseases are incurable is sometimes *not* taken as a sign of the inadequacy of the art in its current state but is turned into part of the medical man's knowledge,[114] part of what the medical man can be said to know.

Dogmatism and Uncertainty in the Fourth Century and Later

The continuing interactions of dogmatism and uncertainty have far-reaching repercussions in many areas of Greek science long after the fifth century B.C. This is not just a matter of tone or style but relates to a deep-seated epistemological conflict where what are at stake are the answers to fundamental questions concerning the status of scientific theories and the possibility of science itself. With a wealth of material to draw on from philosophers of science, mathematicians, natural philosophers, and medical writers, our discussion must be even more drastically selective than ever.

Plato and Aristotle

We may begin with two central issues in the philosophies of science of Plato and Aristotle. When Plato comes to discuss the generation of the physical world, in the *Timaeus,* he refers to this repeatedly as a "likely story," εἰκὼς μῦθος, but quite how we are to interpret this expression or evaluate the account we are given has been and continues to be much disputed.[115] Some suggestions that have been canvassed

---

114. Cf. *De arte* 3, CMG 1.1.10.19ff. (accepting Heiberg's text).

115. Among more recent discussions of the *Timaeus* in particular should be noted those of Witte 1964, Schulz 1966, Gadamer 1974/1980, Zeyl 1975, Vlastos 1975a, Scheffel 1976. On the general issue of the imperfection of per-

need not detain us long. The alternative expression, εἰκὼς λόγος, immediately shows that μῦθος here need not carry the connotation of fiction, over and above that of narrative account.[116] On the other hand, Taylor's claim that Plato was offering merely a provisional account[117] falls foul of the objection that an account of the physical world can, in Plato's view, under no circumstances be converted from a merely probable into a certain one.[118] Again, although Friedländer suggested that Heisenberg's uncertainty principle was in a sense anticipated by Plato,[119] it is as well to recognise where it differs from anything for which Plato's authority could be claimed. Two points are fundamental: first, the uncertainty principle is precise, in that it specifies that it is impossible to determine both the momentum and the location of a fundamental particle; secondly, it is grounded on reflections on the circumstances of experimental observation and intervention.

Both the nature of the reservations Plato expresses and their scope need to be considered carefully. The fundamental ontological distinction that dictates the status of any account of the physical world is, of course, that between being and becoming. What comes to be, insofar as it comes to be, cannot be the object of certain knowledge. That is stressed at *Ti.* 27d5ff. and repeatedly in what follows. Yet in respect of

---

ceptible phenomena, the contributions of Cooper 1970, Nehamas 1972–73, 1975, 1982–83, and Burnyeat 1976 are fundamental. Cf. Irwin 1977. For what follows see also G. E. R. Lloyd 1968a and 1983b.

116. The two phrases are used indifferently, and at *Ti.* 59c–d, for instance, μῦθος and λόγος are clearly equivalent. This is not to deny that there are many figurative, as well as narrative, elements in the account for which the term μῦθος is suitable enough. These figurative elements include, for example, the relationship between the Demiurge and the lesser divine craftsmen, if not also aspects of the description of the former himself, about whom Timaeus remarks (28c) that "to discover the father and maker of the whole is a difficult task, and once one has found him to declare him to all men impossible."

117. A. E. Taylor 1928, e.g., pp. 59ff.; criticised by Cornford 1937, pp. 29f.

118. Comparisons with the hypotheses of modern science are, then, liable to be misleading, at least insofar as they had better not be, in principle, beyond the reach of empirical support or refutation.

119. Friedländer 1958–69, vol. 1, p. 251. Heisenberg himself occasionally referred in admiring terms to Plato's atomic theory, e.g., 1945/1952, p. 57, 1955/1958, pp. 59f.; cf. also Feyerabend 1981b, p. 84.

being itself no such reservation applies; on the contrary, concerning what is stable Timaeus makes the considerable demand that the accounts should "so far as possible" be irrefutable and unchangeable (or invincible) ones.[120] Whenever the cosmologist or the natural philosopher has to do with the intelligible model—the Forms—after which the visible cosmos is constructed, there should, in principle, be no falling short.[121]

Moreover, the claim in respect to the particular cosmological account set out in the *Timaeus* is that it is "inferior to none in likelihood."[122] The visible cosmos is not of course identical with the intelligible model. In the work of creation the Craftsman has to bring order into what is already in chaotic motion.[123] He has to contend with the factor Plato calls necessity or the wandering cause.[124] Yet he made the cosmos as like the model as he could. Four points are worth emphasising. First, the model the Craftsman uses is itself eternal and unchanging; the importance of this is spelled out at *Ti.* 28a ff., where the inferiority of any production based on a created model is stressed. Secondly, the product of his workmanship is *good.* The theme is a recurrent one and is given a triumphant climax in the final sentence of the *Timaeus,* where the likeness of the intelligible model is described as a perceptible god, greatest and best and fairest and most perfect.[125]

---

120. *Ti.* 29b–d, reading either ἀκινήτοις or ἀνικήτοις at 29b8.

121. As notably in relation to the intelligible, ungenerated Forms that correspond to fire and the other three simple bodies, at *Ti.* 51b ff. In practice, however, the account Timaeus offers of the varieties of the simple bodies and of their interactions concerns the geometrical shapes that represent modifications in the Receptacle, *Ti.* 53c ff.—shapes that themselves are subject to the *transformations* he describes, e.g., 54c–d. Although at 56b the pyramid is claimed to be the element of fire according to the correct account, κατὰ τὸν ὀρθὸν λόγον, as well as according to the likely one, Timaeus enters reservations about the selection of the primary triangles; see below at nn. 129–131.

122. μηδενὸς ἧττον εἰκότας, *Ti.* 29c4, echoed at 48d; cf. μάλιστα εἰκός at 44c–d and 67d.

123. *Ti.* 30a, 52d ff.

124. *Ti.* 47e ff. Broadly, reason "persuades" necessity in the sense that the best ends are secured within the framework of the possibilities set by the inherent properties and characteristics of the material available.

125. *Ti.* 92c: μέγιστος καὶ ἄριστος κάλλιστός τε καὶ τελεώτατος.

major premise is universal, the conclusion is again not "for the most part" if that is taken to exclude "universally": "all B's are A" and "most C's are B" together do not rule out "*all* C's are A."

In the light of the difficulties in Aristotle's opaque and elliptical discussions [140] it has been suggested that "for the most part" is not purely statistical but is used, rather, as a temporal operator (i.e., "not always") or as a quasi-modal operator ("not necessarily") or corresponds to some admittedly unanalysed notion of what holds "by nature." [141] Yet Aristotle himself, it must be said, nowhere elucidates the concept, nor does he explain how syllogisms incorporating propositions true "for the most part" meet the requirements laid down for understanding in the opening chapters of the *Posterior Analytics,* notably the requirement that it is of what cannot be otherwise than it is.

Some alleviation of the general problem is possible. The *Posterior Analytics,* it has been argued, [142] has primarily a pedagogic aim: it presents certain recommendations about how a mature science is to be taught, or at least about how to set out a body of theorems in good deductive order so that their connections are revealed and the explanations they incorporate are grasped as the explanations they are. Manifestly, Aristotle has very little to say, in this work, on the problems of discovery, about how scientific understanding is acquired in the first

---

140. The chief texts in the *Organon* are in *APr.* 1.27, *APo.* 1.30 and 2.12. In *APo.* 1.30.87b19ff., Aristotle remarks that when, in syllogisms, the propositions are necessary, the conclusion is also necessary; when for the most part, the conclusion is also likewise—where "for the most part" is clearly contrasted with "necessary," 87b22–23—but the main aim of the chapter is to refute the notion that there is demonstrative knowledge of what happens by chance. At *APo.* 2.12.96a8–19, he stipulates that for the conclusion to be true "for the most part," as opposed to universally, the middle term must also hold "for the most part"—where this is contrasted with what holds universally, for all and always (96a15–16). However, the greatest difficulty for the statistical view is in *APr.* 1.27.43b33ff. There when the "problems" are "for the most part," the syllogisms consist of propositions that are—either all or some of them—"for the most part," and this appears to envisage the possibility of syllogisms with *both* premises true "for the most part."

141. Apart from the perceptive remarks in J. Barnes' commentary on *APo.,* 1975 ad loc., see the full discussion in the elegant paper devoted to the topic by Mignucci 1981.

142. Most forcefully, in recent times, by J. Barnes 1969/1975.

place.[143] At the same time, the examples he gives show that his discussion is not restricted to the already well-established disciplines such as mathematics and the exact sciences. Although most of his illustrations are drawn from such fields, a fair number, particularly in the second book, relate to zoological or botanical questions.[144] Presumably he has in mind an ideal that these studies can *eventually* attain, for certainly they had not done so in his day.[145] Yet for that ideal to be realised, *either* we have to imagine that the studies as set out will deal solely with universal and necessary propositions, *or* the difficulties in extending the schema to cover propositions true only "for the most part" have to be resolved—with corresponding modifications, no doubt, to the ideal itself.[146]

The value of the model in the *Posterior Analytics* as a *model* of demonstration, however, remains. If we recall the complex and confused uses of the terms for necessity and demonstration in the Hippocratic writers, we can see the advances made.[147] Aristotle stipulates precisely

---

143. The chief exception is the last chapter of *APo.*, 2.19.99b15ff., where Aristotle discusses briefly the faculty (νοῦς) and the procedures (ἐπαγωγή, "induction") by which we gain knowledge of the immediate starting-points. But the protracted controversies over the interpretation of this chapter are some indication of its problematic nature: see especially von Fritz 1964/1971, Kosman 1973, Lesher 1973, Raphael 1974, Hamlyn 1976, Engberg-Pedersen 1979, Hintikka 1980, McKirahan 1983.

144. See, for example, *APo.* 98a35ff., 99a23ff., b4ff.; cf. J. Barnes 1969/1975, pp. 70ff.

145. This is not to deny that connections can be found between the recommendations of the *Posterior Analytics* and the actual practice of the zoological treatises. Lennox 1987, for instance, has recently drawn attention to the concern, in the latter, to establish the widest class of which a character is true (cf. *APo.* 1.4.73b26ff. and 5.74a4ff.): cf. also Pellegrin 1986. But neither of these studies tackles the problems raised by physics dealing with what is true "for the most part."

146. At *APo.* 94a36ff., in his discussion of the different types of causes that may serve as middle terms, Aristotle even gives an example of a historical explanation to illustrate the efficient cause. Moreover, this is one that involves reference to a singular term (the Athenians' raid on Sardis, cited as provoking the Persian war) and so falls outside the scope of the theory of the syllogism set out in the *Prior Analytics*.

147. See above at nn. 47–66. Aristotle himself notes at *PA* 639b21ff., cf. *Metaph.* 1015a20ff., both that many of his predecessors reduced their expla-

what conditions have to be met to justify the claim that conclusions have been demonstrated. True premises and valid inference are not enough: the premises must be prior to and explanatory of the conclusions. In a sequence of demonstrations the ultimate starting-points (they comprise definitions, axioms, and hypotheses) must themselves be indemonstrable (on pain of an infinite regress) but known to be true.[148] Whatever other obscurities remain, necessity as logical consequence is now deployed with confidence, and we have a whole subtle discussion of necessity as a modal operator, even though, again, the precise interpretation of many points in Aristotle's treatment remains controversial.[149]

But the clarity of the model has been bought at a price in terms of the range of its applicability. In mathematics and the exact sciences there is little difficulty in fulfilling Aristotle's criteria: a body of theorems can be presented in systematic order and their derivation from a set of axioms and definitions made clear. Yet the situation is very different in the natural sciences, and not just for the reason already mentioned, that these deal with propositions some of which are true only "for the most part." For the model to be applicable here we have also to be able to answer the thorny question of the nature of the indemonstrable starting-points. Over and above the general regulative principles that govern all discourse—the laws of contradiction and of excluded middle—what will count as axioms in zoology and botany, in meteorology or geology?[150] Can we envisage the definitions in such fields having the status of such starting-points?

---

nations to the necessary (by which he means that they took no account of the final cause) and that they failed to distinguish the senses of necessity. This is not to deny, of course, that certain distinctions continue to be ignored by Aristotle himself, as is clearly shown by Sorabji, 1980a, and cf. Waterlow 1982b.

148. See *APo.* 72a5ff., cf. b18ff., 76a31ff.

149. See especially Sorabji 1980a, and cf. Lear 1980, chap. 1.

150. The dictum that "nature does nothing in vain" is often appealed to, in the zoological treatises especially, as the grounds for particular explanations, and it may be said to act as some kind of general regulative principle governing the zoologist's inquiry, one which must be accepted for that inquiry to be fruitful and one that is chiefly to be justified by the results obtained by its use. On the other hand, it is unlike both the laws of excluded middle and contradic-

Definitions and demonstrations are, as Aristotle points out in his acute if often problematic discussions of their interrelations in the *Posterior Analytics*,[151] crucially interdependent. Take first one of his astronomical examples. Lunar eclipse is not just any loss of light that the moon suffers (a cloud obscuring it will not count), but loss of light due to the interposition of the earth. But if you ask for the explanation, you will receive the information packed into the full definition. Why does it suffer eclipse? Because the earth intervenes.[152] Similarly, in one of the botanical examples alluded to:[153] deciduousness is not just any

---

tion, and the particular mathematical axioms that Aristotle mentions (such as the equality axiom that if equals are taken from equals, equals remain: e.g., *APo.* 76a41). No attempt is or can be made to prove the latter, while the former are to be supported by what he calls an "elenctic demonstration" (*Metaph.* 1006a11ff., 15ff.), which proceeds by pressing any opponent who would deny them to signify something, to himself or to another (cf. Lear 1980, pp. 98ff.). Clearly, opposition to the dictum that nature does nothing in vain cannot be dealt with in *that* way. Rather, we have several serious attempts to discuss the consequences of its denial, notably in *Ph.* 2.8.198b10ff. and *PA* 1.1, especially 640a18ff., even if in the body of the physical treatises it is thereafter generally assumed—as Aristotle may hold it has to be, for progress to be made in scientific inquiry.

151. See *APo.* 75b30ff., 2.8, 93a14ff., 10, 93b29ff., 94a11ff. Aristotle recognises that before we are in a position to give a definitive definition we sometimes have some grasp of the subject inquired into (93a21ff., 29ff.), as well as some understanding of the meaning of the term (93b29ff.), though these points do not receive much elaboration in his discussion. See R. Bolton 1976, Ackrill 1981.

152. See *APo.* 93a29ff., 98b15ff., 23ff. Syllogisms of the fact (ὅτι) are distinguished from syllogisms of the reasoned fact (διότι), for example at *APo.* 1.13.78a22ff. Aristotle insists that while reciprocal deductions are sometimes possible (one can show that planets are near from the fact that they twinkle, though, strictly speaking, their being near is the cause of their twinkling), the cause-effect relationship is asymmetrical and only demonstration through the cause is explanatory (e.g., *APo.* 2.16.98a35ff.).

153. See *APo.* 2.17.99a23ff. In the previous chapter, 98a35ff., b5ff., one of his examples of a demonstration through the cause had been that vines are deciduous because they are broad-leaved, but he then gives coagulation as an explanation of deciduousness (98b36ff.). Moreover, in *APo.* 2.17 itself, 99a27ff., his hesitation about what the actual explanation of deciduousness is emerges from the addition of "or something like that," ἤ τι ἄλλο τοιοῦτον. Coagulation is, evidently, merely an illustration of the *type* of explanation that might be invoked.

Nevertheless, one strand of a dogmatic tradition thereby attained a measure of philosophical respectability in the wake of the development of the theory of demonstration and of its practice in the mathematical sciences, and one clear benefit from this was a greater awareness of the questions of the formal analysis and validity of arguments—though Stoic logic takes as much of the credit for this as Aristotle's.[159] Yet over against that tradition, the recognition of the dangers of dogmatism, and a certain tentativeness and open-mindedness, can also be amply exemplified in some of Aristotle's successors, as they can in Aristotle himself. We may turn first to Theophrastus and to two works in the Aristotelian Corpus that are in the main the products of the Lyceum— the *Problemata* and the *Mechanics*—for excellent illustrations of the continuing tension between the dogmatic and the tentative.

Theophrastus

In a wide variety of contexts Theophrastus engages in a far-reaching examination of many of the fundamental assumptions on which natural scientific inquiry had been based, including in particular many Aristotelian positions, though in his criticism of these Theophrastus often elaborates points to which Aristotle himself had drawn attention. The short treatise *Metaphysics,* for instance, mainly consists of a review of difficulties—and certainly not just in Aristotle. Thus although Theophrastus accepts Aristotle's notion that the ultimate source of movement in the universe must itself be an *un*moved mover that acts as an object of desire, the nature of the impulse it imparts requires, he says, more discussion. The heavenly bodies so moved are a plurality, and their motions are complex and opposed to one another.

> For if that which imparts movement is one, it is strange that it does not move all the bodies with the same motion; and if [alternatively] that

---

159. See M. Frede 1974. The fact that Galen, who is in general no friend of the Stoics, uses Stoic propositional logic freely is good evidence of its widespread influence: see, for instance, the examples commented on by Furley and Wilkie 1984, pp. 53, 258f., 265f.

which imparts movement is different for each moving body and the sources of movement are more than one, then their harmony as they move in the direction of the best desire is by no means obvious. And the matter of the number of the spheres demands a fuller discussion of the reason for it; for the astronomers' account is not adequate. It is hard to see, too, how it can be that, though the heavenly bodies have a natural desire, they pursue not rest but motion.[160]

Developing points that were in most cases anticipated by Aristotle himself,[161] Theophrastus later raises questions concerning the limits of teleological explanation. "With regard to the view that all things are for the sake of an end and nothing is in vain," he says, "the assignation of ends is in general not easy . . . , and in particular some things are difficult because they do not seem to be for the sake of an end but to occur, some of them, by coincidence, and others, by some necessity, as in the case both of celestial and of most terrestrial things."[162] What purpose, he asks, do changes in sea level serve, or breasts in male animals? Some things—his example is outsize horns in deer—are even

---

160. Theophrastus *Metaphysics* (*Metaph.*) 5a17–25: εἴτε γὰρ ἓν τὸ κινοῦν, ἄτοπον τὸ μὴ πάντα τὴν αὐτήν· εἴτε καθ᾿ ἕκαστον ἕτερον αἵ τ᾿ ἀρχαὶ πλείους, ὥστε τὸ σύμφωνον αὐτῶν εἰς ὄρεξιν ἰόντων τὴν ἀρίστην οὐθαμῶς φανερόν. τὸ δὲ κατὰ τὸ πλῆθος τῶν σφαιρῶν τῆς αἰτίας μείζονα ζητεῖ λόγον· οὐ γὰρ ἀρκεῖ ὅ γε τῶν ἀστρολόγων. ἄπορον δὲ καὶ πῶς ποτε φυσικὴν ὄρεξιν ἐχόντων οὐ τὴν ἠρεμίαν διώκουσιν ἀλλὰ τὴν κίνησιν.

161. Lennox 1985 and Vallance forthcoming now provide careful studies of Theophrastus' critique of earlier views on the issue of teleology in *Metaph.* chap. 9. In the final analysis, as Vallance argues, Theophrastus' own position has more in common with that of Aristotle than with extreme positions on either side of him, that is, with either the out-and-out anti-teleologists, on the one hand, or, on the other, those who failed to recognise any limits at all to teleological explanation (which Aristotle certainly did). (Those explicitly named in this chapter include Speusippus, Plato, and the Pythagoreans, 11a23, 27.) At the same time Theophrastus focuses critically on some examples found in Aristotle (such as the position of the windpipe, 11a9ff.; cf. Aristotle *PA* 665a9–26) and is concerned to spell out more explicitly than Aristotle had done that teleological explanation is not applicable in many cases.

162. *Metaph.* 10a22–23, 25–28: ὑπὲρ δὲ τοῦ πάνθ᾿ ἕνεκά του καὶ μηδὲν μάτην, ἄλλως θ᾿ ὁ ἀφορισμὸς οὐ ῥᾴδιος. . . . καὶ δὴ ἔνια τῷ μὴ δοκεῖν ἔχειν οὕτως ἀλλὰ τὰ μὲν συμπτωματικῶς τὰ δ᾿ ἀνάγκῃ τινί, καθά περ ἔν τε τοῖς οὐρανίοις καὶ ἐν τοῖς περὶ τὴν γῆν πλείοσιν.

harmful to the animals that possess them.[163] There is even a certain
plausibility in the view that many things come about spontaneously
and "by the rotation of the universe."[164] "If they have no purpose, we
must set certain limits to the final cause and to the tendency towards
what is best, and not assume it absolutely in every case. . . . For even if
this is the desire of nature, it is clear that there is much that does not
obey nor receive the good."[165] He is confident in rejecting the view
that good is rare and that evil predominates in the universe, but he
ends his catalogue of problems with: "but at any rate these are the
questions we must inquire into."[166]

A similar searchingly aporetic tone characterises his discussion not
just of high-level metaphysical and methodological issues, but also of
several particular physical problems. Take, for example, his treatment
of the nature of fire. In the treatise devoted to that question he raises a
series of difficulties connected with the idea that fire is a simple body,
like earth, water, or air. "Of the simple bodies," he begins, "the nature
of fire has the most special powers."[167] None of the other simple bodies
can generate itself, but fire can do so. Most of the ways it comes to be,
whether natural or artificial, appear to involve force. Even if that is not
the case (he corrects himself) yet "at least this much is clear: fire has
many modes of coming-to-be, none of which belong to the other
simple bodies."[168] The most important difference, he proceeds, is that
the other simple bodies are self-subsistent and do not require a substratum, whereas fire does, "at least so far as is clear to our percep-

---

163. *Metaph.* 10a28ff., b11ff. That the reference to "incursions" and "re-
fluxes" of the sea at 10a28ff. is more probably one to changes in the general
level of the sea, rather than to tides, has recently been argued by Vallance
forthcoming.

164. *Metaph.* 10b26ff.

165. *Metaph.* 11a1–3, 13–15: εἰ δὲ μή, τοῦ θ' ἕνεκά του καὶ εἰς τὸ
ἄριστον ληπτέον τινὰς ὅρους καὶ οὐκ ἐπὶ πάντων ἁπλῶς θετέον. . . . εἰ
γὰρ καὶ ἡ ὄρεξις οὕτως, ἀλλ' ἐκεῖνό γ' ἐμφαίνει διότι πολὺ τὸ οὐχ ὑπακοῦον
οὐδὲ δεχόμενον τὸ εὖ.

166. *Metaph.* 11b24.

167. Theophrastus, *De igne* (*Ign.*) 1.3.1ff.: ἡ τοῦ πυρὸς φύσις ἰδιαιτάτας
ἔχει δυνάμεις τῶν ἁπλῶν. Cf. also 9.9.3ff.

168. *Ign.* 1.3.1ff., 2.3.12ff.

tion." [169] "In sum, everything that burns is always as it were in a process of coming-to-be, like movement,[170] and so it perishes, in a way, as it comes to be and as soon as what is combustible is lacking it too itself perishes." [171] "Hence it seems absurd to call this a primary [substance] and as it were a principle, if it cannot exist without matter"—that is, the fuel.[172]

By the end of the treatise he has exposed many of the weaknesses in common Greek assumptions about fire and has questioned the too easy assimilation of fire to the other so-called simple bodies. Yet he has clearly not abandoned that notion entirely. His dilemma is evident: he recognises many of the fundamental difficulties; he realises that many issues require further investigation and his parting remark, at the end of the work,[173] is to promise a more exact discussion of some topics on another occasion. Yet he has no *new* constructive element theory to propose, nor does he answer the question of the nature of fire that he set himself, beyond stressing the diversity of its forms and examining some of these.

A second instructive example that illustrates both his acute perception of weaknesses in widespread assumptions and also some of the difficulties he experienced in pressing home his critique comes from his botany, from his discussion of spontaneous generation. This is mentioned in the *Inquiry concerning Plants* as the first of the ways in which plants and trees may come to be.[174] In the *Causes* he begins his

---

169. *Ign.* 3.5.1ff. (ὅ γε τῇ περὶ ἡμᾶς αἰσθήσει φανερόν). Problems to do with the relationship between fire and its substratum are raised already by Aristotle: *PA* 649a20ff., *GA* 761b19f.

170. With Coutant's text at 3.5.9, καθάπερ ἡ κίνησις, compare Wimmer, καὶ τὸ πῦρ ἐν κινήσεως εἴδει.

171. *Ign.* 3.5.8ff.: ἁπλῶς δ' ἀεὶ καὶ πᾶν καιόμενόν τι καὶ ὥσπερ ἐν γενέσει καθάπερ ἡ κίνησις, διὸ καὶ γινόμενον φθείρεταί πως καὶ ἅμα τῷ ὑπολείπειν τὸ καυστὸν καὶ αὐτὸ συναπόλλυται.

172. *Ign.* 4.5.13–15: διὸ καὶ ἄτοπον φαίνεται πρῶτον αὐτὸ λέγειν καὶ οἷον ἀρχήν, εἰ μὴ οἷόν τ' εἶναι χωρὶς ὕλης.

173. *Ign.* 76.51.3f.

174. At *HP* 2.1.1, for example, he writes: "The ways in which trees and plants in general originate are these: spontaneous growth, growth from seed, from a root, from a piece torn off, from a branch or twig, from the trunk it-

discussion: "Spontaneous generation, broadly speaking, takes place in smaller plants, especially in those that are annuals and herbaceous. But still it occasionally occurs too in larger plants whenever there is rainy weather or some peculiar condition of air or soil. . . . Many believe that animals also come into being in the same way." [175] Yet having thus apparently endorsed the common view, he goes on to introduce reservations:

> But if, in truth, the air also supplies seeds, picking them up and carrying them about, as Anaxagoras says, then this fact is much more likely to be the explanation. . . . Moreover, rivers and the gathering together and breaking forth of waters purvey seed from everywhere. . . . Such growths would not appear spontaneous, but, rather, as sown or planted. Of the sterile sorts, one might, rather, expect them to be spontaneous, as they are neither planted nor grown from seed, and if they come to be in neither way, they must necessarily be spontaneous. But this may possibly not be true, at least for the larger plants; it may be, rather, that all the stages of development of their seeds escape our observation, just as was said in the *Inquiry* about willow and elm. Indeed, the development of seed escapes observation also in many of the smaller herbaceous plants, as we said about thyme and others, whose seeds are not evident to the eye, but evident in their effect, since the plant is produced by sowing the flowers. Further, in trees too some seeds are hard to see and small in size, as in the cypress. For here the seed is not the entire ball-shaped fruit, but

---

self. . . . Of these methods spontaneous growth comes first, one may say, but growth from seed or root would seem most natural; indeed, these methods too may be called spontaneous, wherefore they are found even in wild kinds, while the remaining methods depend on human skill or at least on human choice." As this passage shows, the term "spontaneous," αὐτόματον, is used in two different, but related, ways: (1) to distinguish between wild and cultivated— where what happens "of its own accord" is opposed to what happens by human agency; and (2) to distinguish reproduction without seed from reproduction by seed, root, or some other natural method.

175. *De causis plantarum* (CP) 1.5.1f.: αἱ δ' αὐτόματοι γίνονται μὲν ὡς ἁπλῶς εἰπεῖν τῶν ἐλαττόνων καὶ μάλιστα τῶν ἐπετείων καὶ ποιωδῶν· οὐ μὴν ἀλλὰ καὶ τῶν μειζόνων ἔστιν ὅτε συμβαίνουσιν ὅταν ἢ ἐπομβρίαι κατάσχωσιν ἢ ἄλλη τις ἰδιότης γένηται περὶ τὸν ἀέρα καὶ τὴν γῆν. . . . ὥσπερ καὶ τὴν τῶν ζώων γένεσιν οἱ πολλοὶ ποιοῦσιν.

the thin and unsubstantial bran-like flake produced within it. It is these that flutter away when the balls split open. This is why an experienced person is needed to gather it, who has the ability to observe the proper season and recognize the seed itself.[176]

In many cases, therefore, propagation comes from unnoticed seed. The succession of trees in wild forests and in the mountains could not easily be maintained by spontaneous generation. "Instead there are two alternatives: to come from a root or from seed." [177] He notes that woodcutters report that among trees of the same kind some individual specimens are sterile. There is still a possibility that their seed passes unnoticed; alternatively, the trees become sterile because all their nourishment is used up on other parts. But if this can happen in individuals or kinds that can and do bear fruit, it may not be impossible for the same thing to happen in whole kinds. He concludes: "Let this be given merely as our opinion; more accurate investigation must be made of the subject and the matter of spontaneous generation must be thoroughly inquired into. To sum the matter up generally: this phenomenon necessarily occurs when the earth is thoroughly warmed and when the

---

176. *CP* 1.5.2–4: εἰ δὲ δὴ καὶ ὁ ἀὴρ σπέρματα δίδωσι συγκαταφέρων ὥσπερ φησὶν Ἀναξαγόρας καὶ πολλῷ μᾶλλον. . . . ἔτι δ᾽ οἱ ποταμοὶ καὶ αἱ συρρόαι καὶ ἐκρήγματα τῶν ὑδάτων πολλαχόθεν ἐπάγουσι σπέρματα. . . . ἀλλ᾽ αὗται μὲν οὐκ αὐτόματοι δόξαιεν ἄν, ἀλλ᾽ ὥσπερ σπειρόμεναί τινες ἢ φυτευόμεναι. τὸ δὲ τῶν ἀκάρπων οἰηθείη τις ἂν μᾶλλον αὐτομάτους εἶναι, μήτε φυτευομένων μήτε ἀπὸ σπέρματος γινομένων, ὅπερ ἀναγκαῖον εἰ μηδ᾽ ἕτερον τούτων. ἀλλὰ μή ποτ᾽ οὐκ ᾖ τοῦτ᾽ ἀληθὲς ἐπί γε τῶν μειζόνων, ἀλλὰ μᾶλλον λανθάνουσιν αἱ πᾶσαι τῶν σπερμάτων φύσεις· ὅπερ καὶ ἐν ταῖς ἱστορίαις ἐλέχθη περί τε τῆς ἰτέας καὶ τῆς πτελέας. ἐπεὶ καὶ τῶν ἐλαττόνων πολλαὶ διαλανθάνουσι τῶν ποιωδῶν, ὥσπερ καὶ περὶ τοῦ θύμου καὶ ἑτέρων ὧν εἴπομεν ὡς κατὰ μὲν τὴν ὄψιν οὐ φανερά, κατὰ δὲ τὴν δύναμιν φανερά· σπειρομένων γὰρ τῶν ἀνθῶν γεννᾶται. καὶ δυσόρατα καὶ μικρὰ καὶ τῶν δένδρων ἔνια σπέρματα τυγχάνει καθάπερ καὶ τῆς κυπαρίττου. ταύτης γὰρ οὐχ ὅλος ὁ καρπὸς ὁ σφαιροειδής ἐστιν, ἀλλὰ τὸ ἐγγινόμενον ἐν τούτῳ λεπτὸν καὶ ὥσπερ πιτυρῶδες καὶ ἀμενηνὸν ὅπερ ἐκπέτεται διασχασκόντων τῶν σφαιρίων, διὸ καὶ ἐμπείρου τινός ἐστι συλλέξαι τήν θ᾽ ὥραν παρατηρεῖν αὐτό τε τὸ σπέρμα γνωρίζειν δυναμένου.

177. *CP* 1.5.4: ἀλλὰ δυοῖν θάτερον, ἢ ἀπὸ ῥίζης ἢ ἀπὸ σπέρματος βλαστάνει.

eventually come to a standstill? Do they stop when the force which
started them fails? Or because of being drawn in a contrary direction?
Or is it due to the downward tendency, which is stronger than the force
which threw them? Or is it absurd to discuss such questions, while the
principle escapes us?"[199]

Schools of Medical Thought: Dogmatism, Empiricism, Methodism

The texts we have considered illustrate some of the tensions between
the dogmatic and the tentative, the speculative and the self-restrained,
in post-Aristotelian natural philosophy. But in Hellenistic medicine,
varieties of dogmatism and scepticism or anti-dogmatism are elevated
into self-conscious methodologies. The so-called Dogmatic medical
school (δογματικοί or λογικοί)[200] takes its origin from the objections
of its opponents. Those labelled Dogmatists in our sources (they include Herophilus and Erasistratus and, often, Hippocrates himself)
would not have recognised themselves as forming a distinct sect with
shared principles and practices. But first the Empiricists—beginning
perhaps with Philinus of Cos around the middle of the third century B.C.—and then also the Methodists—followers of Themison (first

---

199. *Mech.* 32.858a13–16: διὰ τί παύεται φερόμενα τὰ ῥιφέντα; πότερον ὅταν λήγῃ ἡ ἰσχὺς ἡ ἀφεῖσα, ἢ διὰ τὸ ἀντισπᾶσθαι, ἢ διὰ τὴν ῥοπήν,
ἐὰν κρείττων ᾖ τῆς ἰσχύος τῆς ῥιψάσης; ἢ ἄτοπον τὸ ταῦτ᾽ ἀπορεῖν, ἀφέντα
τὴν ἀρχήν;
200. The question of the relationship between the medical Dogmatists and
the dogmatic philosophical schools, especially the Stoics, is highly problematic. The unqualified term δογματικοί can refer to either. Thus in the *Outlines
of Pyrrhonism* (Πυρρώνειοι ὑποτυπώσεις, *P.*) 1.65, Sextus specifies that he
has the Stoics particularly in mind (cf. *M.* 8.156, where he attacks both the
δογματικοί philosophers and the λογικοί doctors), while Galen often has
medical dogmatists in view, for example, at περὶ αἱρέσεων τοῖς εἰσαγο-
μένοις (*De Sectis ad eos qui introducuntur, Sect.Intr.*) 4; *Scripta Minora*
(*Scr.Min.*) 3.7.1ff. Helmreich (1.72.4ff. Kühn). More important, the extent
and the direction of influence between philosophy and medicine here are controversial and obscure: see, for example, M. Frede 1982; Sedley 1982b, e.g.,
p. 241; J. Barnes 1982a, 1983.

century B.C.) and of Thessalus (first century A.D.)[201]—set themselves apart from those of their predecessors and contemporaries whom they represented as having certain methodological principles in common.

The evidence we have to rely on is in many cases indirect and much of it comes from critical or hostile sources. Empiricism, especially, is poorly represented by original texts,[202] and so too is Methodism until we come to Soranus in the second century A.D. Neither Celsus nor, more obviously, Galen is an impartial witness, and aspects of their reports are suspect as historical accounts.[203] On the other hand, both are, obviously, evidence for the *currency* of certain ideas at the time they wrote,[204] and we can analyse their interpretations of the debate even if we have to bear in mind that they are *their* interpretations and even if the evidence to confirm or refute what they attribute to some of the contending parties is often not available.

Celsus presents a particularly full picture of the alternatives as he saw them in the proem to the first book of his *De medicina*.[205] The chief issues, as he reports them, relate to the aims, limits, and methods of the medical art. Those grouped together as Dogmatists are represented as holding that medicine should investigate not only (1) the so-called evident causes (such as heat and cold considered as causes), but also (2) hidden or obscure ones, as well as (3) natural actions (such as breathing and digestion, in other words, physiology) and, finally, (4) internal anatomy.[206]

Of these four inquiries the Empiricists are said to accept only the first, that into evident causes, alone. The other three are not just superfluous but impossible, since "nature cannot be comprehended"; the

---

201. Whether Themison and Thessalus are to be considered the founders or the forerunners of Methodism is disputed: see, for example, Edelstein 1935/1967a.

202. The evidence is collected in Deichgräber 1930/1965.

203. The point is given particular emphasis in Rubinstein 1985.

204. Celsus wrote in the first century A.D., Galen in the second.

205. *Med.* 1 pr. 12ff., *CML* 1.19.4ff. Celsus sets out his own position in the dispute at 1 pr. 45ff., *CML* 1.24.24ff.; see Mudry 1982.

206. *Med.* 1 pr. 13, *CML* 1.19.11ff.

doctor's task is to treat individual cases and for this purpose he must be guided by the manifest symptoms of the patient alone. Against the Dogmatists, the Empiricists rejected "reasoning" and accepted "experience" alone as the criterion. It is this that has suggested cures; it is from experience that medicine has been built up and on which it must continue to rely. It is not a discovery made following reasoning; rather, the discovery came first and the reason for it was sought afterwards. Moreover, where reasoning teaches the same as experience, it is unnecessary, and where different, it is opposed to experience and should be rejected.[207]

As Celsus makes the Empiricist argue:

> It does not matter what produces the disease, but what relieves it. Nor does it matter how digestion takes place, but what is best digested—whether concoction comes about from this cause or that, and whether the process is concoction or merely digestion.[208] We have no need to inquire in what way we breathe, but what relieves laboured breathing; nor what may move the blood-vessels, but what the various kinds of movements signify. All this is to be learnt through experiences. In all theorising over a subject it is possible to argue on either side, and so cleverness and fluency may get the best of it. However it is not by eloquence, but by remedies, that diseases are treated. A man of few words who learns by practice to discern well would make an altogether better practitioner than he who, unpractised, overcultivates his tongue.[209]

---

207. *Med.* 1 pr. 27f., 36, *CML* 1.22.1ff., 23.4ff. According to Celsus, the Empiricist response to the possibility, entertained by the Dogmatists, that new diseases may arise was still to insist that the practitioner should not attempt to theorise about causes, but see to which existing disease the condition is similar and try out remedies that had proved successful in such other similar cases.

208. This appears to allude to the long-standing debate on the nature of digestion, where Herophilus, following Aristotle, argues that it involves "concoction," while Erasistratus and the Erasistrateans explain the process in purely mechanical terms, as the result of the trituration or pounding that the food is subjected to in the stomach before being absorbed, as chyle, into the blood-vessels communicating with the liver.

209. *Med.* 1 pr. 38–39, *CML* 1.23.16–27: *sed has latentium rerum coniecturas ad rem non pertinere, quia non intersit, quid morbum faciat, sed quid tollat; neque ad rem pertineat, quomodo, sed quid optime digeratur, siue hac*

Again, "even students of philosophy would have become the greatest medical practitioners, if reasoning could have made them so. But as it is, they have words in plenty, but no knowledge of healing at all."[210]

The third main medical group, the Methodists, had their own subtle and often rather maligned ideas about treatment,[211] but on the essential topic we are concerned with here they are represented by both Celsus and Galen as agreeing with many of the criticisms that the Empiricists brought against the Dogmatists, for example, about their theorising about hidden causes.[212] While Celsus reports the Empiricists as *asserting* that nature cannot be comprehended,[213] Sextus makes it

---

de causa concoctio incidat siue illa, et siue concoctio sit illa siue tantum digestio. Neque quaerendum esse quomodo spiremus, sed quid grauem et tardum spiritum expediat; neque quid uenas moueat, sed quid quaeque motus genera significent. Haec autem cognosci experimentis. Et in omnibus eiusmodi cogitationibus utramque partem disseri posse; itaque ingenium et facundiam uincere, morbos autem non eloquentia sed remediis curari. Quae si quis elinguis usu discreta bene norit, hunc aliquanto maiorem medicum futurum, quam si sine usu linguam suam excoluerit.

210. *Med.* 1 pr. 29, *CML* 1.22.11–13: *Etiam sapientiae studiosos maximos medicos esse, si ratiocinatio hoc faceret: nunc illis uerba superesse, deesse medendi scientiam.* The rejection of the idea that medicine can be learnt from books has a long history. The Hippocratic surgical treatise *Art.* 33 (L) 4.148.13ff. refers to the difficulties of explaining surgical procedures, in particular, in writing, and cf. Plato *Phdr.* 268c, Aristotle *EN* 1181b2ff.

211. The Methodist idea of the three common conditions, the constricted, the lax, and the mixed, came under particular attack. To the chagrin of Galen, especially, (e.g., *Sect.Intr.* 6, *Scr.Min.* [H] 3.15.2ff., [K] 1.83.1ff.), the Methodists were reputed to have claimed that medicine could be learnt in six months: cf. M. Frede 1982. The three common conditions were neither themselves disease entities nor causes of diseases, but generalisations about the state of the body that guided the practitioner in deciding upon treatment (seen as a matter of counteracting the lax with the constricted and vice versa). As we can see from Soranus (see G. E. R. Lloyd 1983a, pp. 182ff.) and from Caelius Aurelianus, not only in principle but also in practice Methodist pathology and therapeutics stayed a good deal closer to what was directly observable than rival theories and were a good deal simpler than they were.

212. See Celsus *Med.* 1 pr. 57, *CML* 1.26.27f., cf. Galen *Sect.Intr.* 6, *Scr.Min.* (H) 3.13.21ff., 7.17.3ff., 18.1ff., (K) 1.81.6ff., 85.14ff., 86.17ff.

213. *Med.* 1 pr. 27, *CML* 1.22.4, where Celsus makes this the grounds, for the Empiricists, of the claim that such an inquiry is superfluous (*superuacuam*); cf. Sextus *P.* 1.236.

"only a small number of criminals."[222] But the Empiricists and Methodists are reported as rejecting human post mortem dissection as well, partly on the grounds that it is, if not cruel, at least nasty (*foedus*),[223] but partly also on the basis of the argument that what is observed in the dead is not relevant to the living, since on death the body is changed.[224]

The obscure not just in the sense of the theoretical or the speculative, but in the sense of what is literally hidden, cannot or need not be inquired into. So far as anatomy went, Celsus has this to add about the Empiricist position:

> If, however, there be anything to be observed while a man is still breathing, chance often presents it to the view of those treating him. For sometimes a gladiator in the arena or a soldier in battle or a traveller who has been set upon by robbers is so wounded that some or other interior part is exposed in one man or another. Thus, they say, an observant practitioner learns to recognise site, position, arrangement, shape, and suchlike, not when slaughtering, but while striving for health.[225]

Moreover, to judge from Soranus, the Methodists too showed a certain ambivalence on the question. Dissection is useless, Soranus says in the *Gynaecology*, but it is studied for the sake of "profound learning," χρηστομάθεια.[226] So he says he will teach what has been discovered by

---

222. *Med.* 1 pr. 26, *CML* 1.21.29–32: *Neque esse crudele, sicut plerique proponunt, hominum nocentium et horum quoque paucorum suppliciis remedia populis innocentibus saeculorum omnium quaeri.* ("Nor is it cruel, as most people state, to seek remedies for multitudes of innocent men of all future ages by means of the sacrifice of only a small number of criminals.")

223. *Med.* 1 pr. 44, *CML* 1.24.21f.

224. On the ancient disputes over dissection, see further Manuli and Vegetti 1977, Vegetti 1979.

225. *Med.* 1 pr. 43, *CML* 1.24.14–19: *Si quid tamen sit, quod adhuc spirante homine conspectu subiciatur, id saepe casum offerre curantibus. Interdum enim gladiatorem in harena uel militem in acie uel uiatorem a latronibus exceptum sic uulnerari, ut eius interior aliqua pars aperiatur, et in alio alia; ita sedem, positum, ordinem, figuram, similiaque alia cognoscere prudentem medicum, non caedem sed sanitatem molientem.*

226. *Gyn.* 1.5, *CMG* 4.6.6–8: ἥτις εἰ καὶ ἄχρηστός ἐστιν, ὅμως ἐπεὶ παραλαμβάνεται χρηστομαθείας ἕνεκα, διδάξομεν καὶ τὰ ἐκ ταύτης ἐπι-

it. "For we shall easily be believed when we say that dissection is useless, if we are first found to be acquainted with it, and we shall not arouse suspicion that we reject through ignorance something which is accepted as useful."[227]

Both Empiricists and Methodists thus went some way towards accommodating the findings of dissection. But both probably stopped well short of advocating the continued practice of the method. Here the rejection of dogmatism and speculation was also a rejection of new research. It was left to such a writer as Galen (who, even if he would himself have resisted the label, would certainly have been classed as a Dogmatist by his opponents)[228] to recommend the method. This he does in texts whose very eloquence and passion testify not just to Galen's personal commitment to the method but also to his sense of the need to come to its support against its detractors. In *On Anatomical Procedures* he sets out no fewer than four kinds of reasons for studying anatomy:

> Anatomical study has one use for the natural scientist who loves knowledge for its own sake, another for him who values it not for its own sake but, rather, to demonstrate that nature does nothing without an aim, a third for one who provides himself from anatomy with data for inves-

---

γνωσθέντα. ("Since [dissection], although useless, is nevertheless employed for the sake of profound learning, we shall also teach what has been discovered by it.") Cf. 1.2, CMG 4.4.6f., where Soranus remarks that the theoretical part of the subject is useless, although it "enhances profound learning" (χρηστομάθεια).

227. *Gyn.* 1.5, CMG 4.6.8–11: ῥᾳδίως τε γὰρ πιστευθησόμεθα λέγοντες ἄχρηστον τὴν ἀνατομήν, εἰ πρότερον αὐτὴν εἰδότες εὑρεθείημεν, καὶ οὐ παρέξομεν ὑπόνοιαν τοῦ δι᾽ ἄγνοιαν παραιτεῖσθαί τι τῶν ὑπειλημμένων εὐχρήστων.

228. Cf. M. Frede 1981. It may, however, be noted that while Galen does not often admit making mistakes, he does sometimes do so. Thus in *AA* 14.7.214 Duckworth, he does so with regard to operations attempting to reveal the courses of certain nerves. There and elsewhere, when he acknowledges that he was at first unsuccessful in a surgical or anatomical operation, it is often to emphasise the need for practice and experience: cf. *AA* 7.10 (K) 2.621.12ff. (cf. also 3.2 [K] 2.348.14ff., 8.4 [K] 2.674.6ff.). On occasion, too, he admits to some hesitation on points of detail, for instance concerning the nerves of the brachial plexus at *AA* 15.6.254 (D.).

as ran riot not just in medicine but in many other areas of the investigation of nature, led in time to a reaction, the rejection of theorising of any kind that went beyond the "appearances."[236] Where some Hippocratic writers had already rejected excessive claims for exactness and the use of arbitrary postulates, the Hellenistic medical schools evidently developed clearer and more powerful epistemologies that drew on the traditions of sceptical philosophy. Yet though the sceptic was an inquirer,[237] his insistence on the need to withhold judgement and on the idea that it is either impossible or useless to seek to comprehend the hidden causes of nature could and did inhibit, even stop dead, a certain kind of research. The sceptic raised questions and saw that much—in fact he thought *just* as much[238]—could be said on either side of disputed issues, but idle curiosity was pointless, and much that had been investigated, in an admittedly often over-sanguine way, had to be rejected as idle curiosity.

On the side of dogmatism, where the dogmatic Hellenistic philosophical sects met the sceptical challenge by upholding one or another positive view of the criterion of knowledge,[239] most of the dogmatic

---

236. The "appearances" often included the common opinions and beliefs, as well as what was perceived, as already in Aristotle. See Owen 1961/1986, cf. Burnyeat 1977, 1979, 1982b, Nussbaum 1986, pp. 240ff.

237. There is evidently some tension between (1) the sceptic continuing to inquire, pursuing an investigation that reveals that there is as much to be said on one side of each dispute about the obscure as on the other, and at the same time (2) his securing freedom from anxiety, ἀταραξία, as the end result of this process, thanks to his suspending judgement. This is part of the more general problem of whether or not, or in what sense, the sceptic can live his scepticism: see M. Frede 1979, Burnyeat 1980a, Striker 1980, 1983b, J. Barnes 1982b, Sedley 1983b, Stough 1984.

238. The doctrine known as ἰσοσθένεια.

239. On the one hand, the Stoics claimed that the sage is infallible and they developed a positive epistemology based on the notion of φαντασία καταληπτική. On the other, although the Epicureans had no doctrine of the infallible sage, they too held that there are positive criteria on which claims to knowledge can be based. However, over a range of problems in the study of nature concerning τὰ ἄδηλα or what is obscure, they insisted that various explanations are possible, and where this is the case, to opt for a single explanation is to dogmatise (see, e.g., *Ep.Hdt.* 10.79f., referring to "meteorology," the study of things in the heavens and things under the earth, in particular). In

practising scientists took for granted an affirmative answer to the question of whether knowledge is possible. But inordinately speculative theories and excessive claims for their correctness, even their necessity, can be illustrated in every branch of the inquiry into nature. Many of those who engaged in that inquiry, as we said, pursued the goal of certainty in part under the influence of the models provided by axiomatised mathematics. In the process, much of the complexity of their subject

---

Ep.Pyth. 10.114, Epicurus even accused those who want a single cause of trying to amaze, τερατεύεσθαι, the multitude, and at 93, for instance, referred contemptuously to the "slavish artifices of the astronomers," τὰς ἀνδραποδώδεις ἀστρολόγων τεχνιτείας. In such contexts as the explanation of lightning and thunder there was, we might say, a good deal of sense in invoking the possibility of plural causes, and occasionally Epicurus shows a certain independence of judgement in suggesting, for example, that the sun may not move with a constant speed (10.98; cf. Lucretius 5.696ff.). Yet on many astronomical problems the same principle led the Epicureans not just to insist that the subject is not capable of yielding certainty (as at Lucretius 5.509ff., 526ff.), but also to entertain some unpromising suggestions: for example, that the "turnings" of the sun and moon may be due to pressure from the air or to the lack of the appropriate fuel (*Ep.Pyth.* 10.93; cf. Lucretius 5.614ff.); that the phases of the moon may be due to the rotation of its body, or to the configuration of the air, or to the interposition of other bodies (*Ep.Pyth.* 10.94, where the possibility of its being explained by the relative positions of moon, sun, and earth is *not* clearly mentioned, though cf. Lucretius 5.705ff.); or that the non-setting of the circumpolar stars may be due to eddies in the air or to a dearth of fuel as well as to the turning of the heaven on its axis (*Ep.Pyth.* 10.112); or that eclipses may be due to the extinction of the heavenly body, sun or moon, as well as to the interposition of the earth—or of some invisible body (*Ep.Pyth.* 10.96). In all such cases the Epicureans insisted on being satisfied when *some* explanation had been given but thought it misguided to try to decide between alternatives. In principle they entertained only alternatives that were not ruled out by the appearances (οὐκ ἀντιμαρτύρησις) but in practice they were tenacious of suggestions that many contemporary practising scientists had shown good grounds to dismiss. Moreover, the essential point in assessing the tone of Epicurean natural philosophy is that although the principle of plural explanation applied to many detailed physical problems, on other, more basic issues they allowed no such doubt—on, for instance, the fundamental doctrine that atoms and the void alone exist (*Ep.Pyth.* 10.86; cf. Lucretius 1.675f.), on matters concerning human morality, on the blessed nature of the gods, and on the rejection of teleology (*Ep.Hdt.* 10.76–80, *Ep.Men.* 10.123f.). On Epicurean philosophy of science see especially Wasserstein 1978; cf. J.-P. Dumont 1982, Asmis 1984.

matter was sometimes ignored, finessed, elided; we shall return to that topic in Chapter 6.

At the same time the example of dissection, especially, shows how it was those who could be criticised for Dogmatism who upheld empirical research. Where the sceptical tradition could degenerate into defeatism [240] (even if a defeatism that is readily understandable in terms of the impasse reached in many areas of physical and biological study), it was the more dogmatic and speculative theorists who offered more justification and incentive for further inquiry. It should, however, be stressed that they did so against the background of that challenge from scepticism. The dogmatism in question was, in this respect, still very different from the monolithic traditions exemplified from the ancient Near East in Chapter 2.

Some of the Greek work was, to be sure, undertaken within a framework of regulative principles approximating to what we might call a research programme, and so may be deemed to lend support to the claims of Kuhn and others concerning the role of such in normal science. At the same time we should acknowledge that much ancient speculation had always been and continued to be both more individualistic and more opportunistic than the title *research programme* would suggest or allow. In an ancient perspective, we have seen that whatever inhibiting effects tentativeness and anti-dogmatism came to have, they were also, especially initially, characterised by a notable

---

240. Pessimism about reaching satisfactory solutions to the major problems in dispute in physical theory goes back to the pre-Socratic period (see above at nn. 21 and 23) and is thereafter a recurrent theme. One may, however, distinguish between doubts or reservations expressed on particular topics within or after a physical investigation, and a quite general dismissal of the possibility of the study of nature (as is reported for the Cyrenaics, for example, by Diogenes Laertius 2.92, cf. 7.160 on Ariston; and cf. Eusebius *Praeparatio evangelica* 15.62 paras. 7ff., 854c4ff., [2.494 Gifford, 2.423.26ff. Mras], on which see Ioppolo 1980, pp. 78ff.). In late antiquity the failure of science, particularly of astronomy, to secure agreed and consistent results was used by Proclus, for example, to support what he represents as the Platonic thesis, that the only proper objects that can be said to be known are transcendent Forms: see, e.g., *Hyp. pr.* 1ff., 2.1ff., 4.5ff., together with 7.238.9ff., and cf. *In Ti.* 3.56.28ff.; cf. Sambursky 1965.

boldness and originality. In Hippocratic medicine, expressions of uncertainty, statements of the difficulties encountered and of the failures that could have been avoided, at least sometimes reflect a remarkably open and direct response to day-to-day clinical experience and a new commitment to the principle of recording mistakes so that others may learn from them—even if some of these attitudes were themselves in turn conventionalised and became part of the fund of rhetorical commonplaces used by authors who were otherwise unrestrained in their pretensions to knowledge.

CHAPTER FOUR

Metaphor and the Language of Science

Metaphor, like mythology, had to be invented—that is to say, the explicit category had to be—and we can trace the steps in which it was made explicit in the fourth century B.C. in Greece. Moreover, our Greek evidence makes it clear that even if there was not quite the scandal that Detienne has recently suggested surrounded the development of the category of myth as fiction,[1] the invention of the category of the metaphorical took place against a background of overt polemic. Yet one outcome of the intense debates concerning theories of metaphor in the past few decades has been that increasingly sophisticated challenges have been mounted calling the literal/metaphorical dichotomy itself into fundamental question,[2] and this issue has repeatedly been at the centre of the most radical controversies in the philosophy of language, the philosophy of science, and literary critical theory. That in some sense all language is metaphorical has been argued with some force both by literary critics and by philosophers.[3] Where theories of mean-

---

1. Detienne 1981/1986. For an analysis of Plato's use that diverges in certain respects from Detienne's, see Brisson 1982, and cf. Moors 1982 with Ferrari 1983.

2. See, for example, Derrida 1972/1981 and 1972/1982. In addition to such classic studies as Black 1962 and Ricoeur 1975/1977, there have been three recent collections of articles in Ortony 1979, Sacks 1979, and Johnson 1981. Shibles 1971 presents an annotated bibliography of work on metaphor to that date.

3. For one sophisticated statement of such a thesis, see Hesse 1982. Some of the antecedents of such a claim go back, in the English-speaking tradition,

ing in the tradition that stems, precisely, from the Greeks represent the literal and the univocal as the norm, the metaphorical as the deviant, a case as strong or stronger can be made for the reverse reduction.[4] The univocal, at least, it can be argued, is the exception; certainly it is not overwhelmingly usual in most natural languages. What proportion of entries in Webster's or Collins' are single entries?

But the more we take note of this recent challenge, the more puzzling the original introduction and invention of that dichotomy are bound to appear. Those who were primarily responsible, Aristotle especially, were in part motivated by the aim of excluding the metaphorical from certain types of discourse. We shall be trying to come to

---

to I. A. Richards, who already inveighed against what he dubbed the One and Only One True Meaning Superstition (1936, p. 39), and who saw metaphor as "the omnipresent principle of language" (pp. 92ff.). But radical attacks on the question of metaphor are equally a feature of Continental scholarship, some of which takes its inspiration from Nietzsche, invoked explicitly by Derrida, for instance at 1972/1982, pp. 216f.

4. Standardly, the univocity/equivocity contrast, like that between synonymy and homonymy, is used in the characterisation of terms, the literal/ metaphorical contrast of their use, and it can be agreed readily enough that equivocity and homonymy are distinct from, and do not necessarily entail, ambiguity or vagueness in use. But whether it is correct to hold (as Searle has recently argued: 1979, chaps. 4 and 5) that metaphorical meaning is not a property of sentences, but always one of speaker's utterances, is controversial and leads to the heart of the question of the status and validity of the concept of "literal meaning" itself. Thus objections have been brought against Searle by, for example, Hesse (1982, p. 42 nn. 1 and 5), who develops a theory of metaphor based on Wittgenstein's family-resemblance theory of universals, a theory of meaning-as-use that does not recognise as fundamental the distinction between "sentence-meaning" and "utterance-meaning" (for the threefold distinction between utterance meaning, sentence meaning, and word meaning, see, for example, Grice 1968). Hesse accordingly rejects the idea of the literal meaning of a sentence as entirely determined by the meanings of its words and its syntactic rules: "If it is a *matter of fact* that the "literal meaning" of a sentence changes *frequently* in utterance because of metaphoric shifts in the meaning of words in different contexts, then the category of literal meaning becomes applicable only in the same kinds of local or limiting cases in which the category of natural kind is applicable" (Hesse 1982, p. 42 n. 5; original emphasis). For a variety of views on the different types of indeterminacy in terms, sentences, and speech acts, see, for example, Black 1937; Hempel 1939; Waismann 1945/1951, 1953; Lyons 1977, vol. 1, pp. 169f., 261ff., vol. 2, pp. 396ff., 550ff.; Dammann 1977–78; and see further below, n. 7.

Quite *how* their authors saw them, and how far and how explicitly they recognised problems to do with the meanings of certain terms, are indeed crucial, if delicate and at points not ultimately decidable, questions.

Some direct problematising of language can, however, be illustrated already in Heraclitus, for whom "the one wise thing is not willing, and is willing, to be called by the name of Zeus,"[25] and for whom the name of the bow is life (βίος, one name for the bow being βιός), but its work is death.[26] In Parmenides the attack on certain terms takes the form of the charge that they are vacuous, with no purchase on reality. "Coming-to-be and perishing, being and not being, change of place and alteration of bright colour" are names laid down by men confident that they are true, but the only thing there is to be named is what is,[27] and for Empedocles and Anaxagoras too coming-to-be and perishing are empty terms, merely conventional expressions.[28]

Of all the pre-Socratic philosophers, Empedocles, perhaps, comes closest to an explicit recognition of the extension involved in his use of a term for a cosmological principle, for he says of Philia, Love, that while she is acknowledged as inborn in the limbs of mortals and is called by the names of Joy and Aphrodite, yet no man is aware of her as she goes to work on the elements.[29] Yet however imperfect our ideas

---

25. Heraclitus fr. 32.   26. Heraclitus fr. 48.
27. Parmenides fr. 8.38–41:

   τῷ πάντ᾽ ὀνόμασται
   ὅσσα βροτοὶ κατέθεντο πεποιθότες εἶναι ἀληθῆ,
   γίγνεσθαί τε καὶ ὅλλυσθαι, εἶναί τε καὶ οὐχί,
   καὶ τόπον ἀλλάσσειν διά τε χρόα φανὸν ἀμείβειν.

Both the reading at the end of line 38 (ὄνομ᾽ ἔσται or ὀνόμασται) and the construction are disputed (see, for example, Woodbury 1958; Guthrie 1965, pp. 39ff.; Mourelatos 1970, pp. 180–85; Burnyeat 1982b, p. 19 n. 22), but it is clear that Parmenides rejects ordinary mortals' categories of change; cf. also fr. 19 from the Way of Seeming.

28. Empedocles frr. 8 and 9, Anaxagoras fr. 17.
29. Empedocles fr. 17.21–26:

   τὴν σὺ νόῳ δέρκευ, μηδ᾽ ὄμμασιν ἧσο τεθηπώς·
   ἥτις καὶ θνητοῖσι νομίζεται ἔμφυτος ἄρθροις,
   τῇ τε φίλα φρονέουσι καὶ ἄρθμια ἔργα τελοῦσι,
   Γηθοσύνην καλέοντες ἐπώνυμον ἠδ᾽ Ἀφροδίτην·

may be, it is that cosmic principle that is at work in us. Cosmic Love would have to be said to be no mere metaphor, then, if we chose to press that question, but that just points to the difficulty of the stretch involved in the application to cosmology of any such term, and that in turn says something about what it is to do cosmology.

Plato

By the time we reach Plato not only is there an extraordinary proliferation of images and analogies (now often recognised as such) deployed in cosmology, psychology, politics, and ethics, but their use, or some of their uses, become the subject of explicit comment.[30] Though the terms *muthos* and *logos* are not always contrasted, of course, they can be used, and were (notoriously) by Plato, to indicate a difference in the statuses of accounts.[31] A *logos* can be, and in certain cases should be, incontrovertible, a matter of demonstration or at least of verification and argument: a *muthos* may be believed to be true and yet be incapable of proof (though many *muthoi* are presented as mere fictions).

Though *muthoi* have their uses, one refrain from the Socratic dialogues onwards is a demand for definition, for clarity, for the giving of

---

τὴν οὔ τις μετὰ τοῖσιν ἑλισσομένην δεδάηκε
θνητὸς ἀνήρ.
("Contemplate her [Love] with your mind and do not sit gazing with your eyes; for she is acknowledged as inborn in the limbs of mortals, and by her they have a gentle disposition and achieve works of peace, calling her by the names of Joy and Aphrodite. No mortal man has learnt of her as she circles among these [the roots or elements].") In view of Xenophanes' critique of ideas about the gods it is striking that Empedocles recuperates the traditional names of the gods—Zeus, Hera and so on—as names of his elements (fr. 6).

30. Among the important studies of various aspects of this topic in Plato, see especially Goldschmidt 1947a; R. Robinson 1953, pp. 202ff.; Bambrough 1956/1967 and 1962/1967; J.-P. Vernant 1979, pp. 105ff.; Detienne 1981/1986; Brisson 1982; Ferrari forthcoming. Some issues are discussed in my 1966, pp. 389ff.

31. Some examples of this were given above, Chap. 1 at nn. 26–30, and cf. Chap. 3 at nn. 115–16.

account. In texts in the *Phaedo*, *Phaedrus*, *Theaetetus*, and *Sophist*, especially, aspects of the use of images, εἰκόνες, likenesses, ὁμοιότητες, and the plausible and specious, πιθανολογία, are discussed critically, with warnings as to the possible deceptiveness of all of these and to their inadequacy as a method of proof.[32] Here, then, are certain general statements concerning the validity of certain types of argumentative device. Even so none of these texts offers an explicit *definition* of the arguments in question, let alone a formal analysis of the type Aristotle was to undertake in connection with his theory of the syllogism. Although in the *Sophist*, especially, Plato begins the analysis of otherness, difference, contrariety, similarity, and identity, he undertakes no systematic classification of those relationships, nor does he directly investigate the relationships between the various modes of reasoning that we may say are based on implicit or explicit comparisons. Moreover, when he says that accounts that use images are charlatans, ἀλαζόνες,[33] we ignore at our peril that he uses a likeness to tell us that likenesses mislead. Or, again, when in the *Sophist* we are told that likenesses are a "most slippery tribe," ὀλισθηρότατον γένος, we might ask how slippery that characterisation is.[34]

Plato's ambivalence on this whole topic emerges not just from his

---

32. See especially *Phd.* 92c–d, *Phdr.* 262a–c, *Theaetetus* (*Tht.*) 162e–163a, *Sph.* 231a–b, 236a–b, 240a ff.

33. Simmias' attunement theory of the soul is characterised, by implication, as dependent on an image, εἰκών, at *Phd.* 87b3. At *Phd.* 92c11ff. Simmias himself says: ὅδε μὲν γάρ μοι γέγονεν ἄνευ ἀποδείξεως μετὰ εἰκότος τινὸς καὶ εὐπρεπείας, ὅθεν καὶ τοῖς πολλοῖς δοκεῖ ἀνθρώποις· ἐγὼ δὲ τοῖς διὰ τῶν εἰκότων τὰς ἀποδείξεις ποιουμένοις λόγοις σύνοιδα οὖσιν ἀλαζόσιν, καὶ ἄν τις αὐτοὺς μὴ φυλάττηται, εὖ μάλα ἐξαπατῶσι, καὶ ἐν γεωμετρίᾳ καὶ ἐν τοῖς ἄλλοις ἅπασιν. ("I put forward this theory without proof, but with a certain probability and speciousness, which is why most men accept it. But I am well aware that theories which base their proofs on what is probable are charlatans: unless one is on one's guard against them, they deceive one very badly, in both geometry and everything else.")

34. *Sph.* 231a8. A further striking example to which Dr. Ferrari has drawn my attention comes at *R.* 601b, where to describe what poetry looks like when stripped of its "colouring," Plato uses the image of a youth who has lost his bloom, ἄνθος, "flower"; here, then, doubly an image.

own very extensive *use* of similes, metaphors, and analogies, but *within* those explicit comments on likenesses, for while some texts issue warnings about their deceitfulness, others recognise their usefulness. Paradigms, especially, are allotted a positive role,[35] both for didactic purposes, to bring a student to an understanding of a difficult problem by considering first a simpler case or one analogous to it,[36] and for heuristic ones, where the dialectician himself is supposed to use a similar method to discover the truth.[37]

Aristotle's Critique of Metaphor

In Aristotle, the shift in emphasis towards a more negative evaluation—at least in certain contexts—is marked. First, he frequently censures the metaphors and images used by his predecessors. Thus Empedocles' notion of the salt sea as the sweat of the earth is "adequate, perhaps, for poetic purposes" but "inadequate for understand-

---

35. See especially Goldschmidt 1947a.

36. When the use of paradigms is itself illustrated and explained by a paradigm in the *Politicus*, 277d ff., the Eleatic Stranger takes the case of children learning to read. Once they have learnt to recognise letters in short and easy syllables, they can be taught to recognise them also in more complex combinations, by juxtaposing the known and the unknown and pointing to the same likeness and nature in both cases. In both the *Sophist* and the *Politicus*, paradigms serve to provide practice in method, when the method of division, to be used on the sophist and on the kingly art, is first exemplified with the easier cases of angling (*Sph.* 218e ff.; cf. 218b–d) and of weaving (*Plt.* 279a ff.; cf. 286a–b).

37. In the illustration of children being taught to read, the instructor himself already knows the letters. But in the problems investigated in the *Sophist* and *Politicus*, the hunt for the sophist and for the definition of the kingly art are represented as *searches*, where neither the leader, the Eleatic Stranger, nor his interlocutors have the answers when they set out. Moreover, in both dialogues the paradigm that is chosen as an illustration is particularly relevant to the substantive subject under investigation. In the *Sophist*, when the activity of dividing angling illustrates the method, angling turns out to be like sophistry (*Sph.* 221d8ff.), and in the *Politicus* weaving is chosen at the outset (*Plt.* 279a7ff.) for its similarity to politics (cf. *Plt.* 308d ff.). In these examples the activity in the case of the paradigm is an instance, not merely a likeness, of the activity also exemplified in the larger case.

ing the nature of the thing."[38] Other images of Empedocles and other
pre-Socratic philosophers are criticised on the grounds that they are
based on superficial similarities—or on none, that the illustrations are
obscure, or crude, or in need of qualification.[39] Thus milk, he insists at
one point, is formed by a process of concoction, not putrefaction, so
Empedocles was wrong, or he used a bad metaphor, when he spoke of
it as "whitish pus."[40] Similarly, Plato's own theory of Forms as a whole
is dismissed on the grounds that to say that the Forms are "models and
that other things share in them is to speak nonsense and to use poetic
metaphors"[41]—where again we may remark that *poetic* is used as a
term of censure.

Aristotle is especially uncompromising in his criticisms of the use of
μεταφορά in the context of his formal logic and theory of demonstration, μεταφορά, for him, being defined as the transfer of a term appropriate to one domain to another.[42] In the *Posterior Analytics* he
condemns them as a whole, especially their use in definitions. "If one

---

38. *Meteorologica* (*Mete.*) 357a24ff., 26–28: πρὸς ποίησιν μὲν γὰρ
οὕτως εἰπὼν ἴσως εἴρηκεν ἱκανῶς (ἡ γὰρ μεταφορὰ ποιητικόν), πρὸς δὲ τὸ
γνῶναι τὴν φύσιν οὐχ ἱκανῶς. Empedocles' business, in Aristotle's view, is
with φυσιολογία, natural philosophy. His verse is not really poetry; rather,
"there is nothing in common between Homer and Empedocles except the
metre: so it is right to call the first a poet, but the second a natural philosopher
rather than a poet" (*Po.* 1447b17ff.). So Aristotle does not allow Empedocles
a let-out from his criticisms that would appeal to the character of the medium
he used. On the contrary, when at *Rh.* 1407a32ff. he discusses those who deliberately cultivate ambiguity, Aristotle remarks that such people often write in
verse—as Empedocles does, for instance: he bamboozles (φενακίζει) his hearers with his circumlocutions and has a similar effect on them as diviners,
whose ambiguous remarks are greeted with nods of acquiescence.

39. See, for example, *Top.* 127a17ff., *GA* 747a34ff., 752b25ff., and cf. *De
sensu* (*Sens.*) 437b9ff., *PA* 652b7ff. Cf. Bremer 1980.

40. *GA* 777a7ff., quoting Empedocles fr. 68, where no doubt Empedocles
was exploiting a play on words, the similarity between πύον, short υ, meaning
pus, and πυός, long υ, a word for the first milk or beestings.

41. *Metaph.* 991a20ff., 1079b24ff. Other comparisons used by Plato are
criticised at *Pol.* 1264b4ff., 1265b18ff., for example.

42. *Po.* 1457b6ff. (μεταφορὰ δ' ἐστὶν ὀνόματος ἀλλοτρίου ἐπιφορά . . .).
On this definition and the illustrations that Aristotle offers of the four species
of μεταφορά, see, for example, Tamba-Mecz and Veyne 1979.

should not argue in metaphors, it is clear that one should not use metaphors or metaphorical expressions in giving definitions." [43] In the *Topics*, too, he repeats the criticism of definitions that contain metaphors on the grounds that "every metaphorical expression is obscure." [44]

There is, to be sure, another side to the picture. Elsewhere when he discusses style, especially, [45] he approves of certain types of metaphor, particularly those that express a proportion, for these, he says, are vivid, witty, and clear [46] (by which he does not mean to deny that from another point of view they are still "obscure"). He praises in the poet the ability to deploy metaphor and to discern resemblances; the latter is a skill that the philosopher too will need to exhibit. [47] In the *Topics*, moreover, the "investigation of likeness" is said to be a useful means by which to become well-supplied with arguments and even also, in certain contexts, for rendering definitions, that is, in securing the genera for them. [48] In the *Sophistici Elenchi* he is not above recommending

---

43. *APo.* 97b37–38: εἰ δὲ μὴ διαλέγεσθαι δεῖ μεταφοραῖς, δῆλον ὅτι οὐδ᾽ ὁρίζεσθαι οὔτε μεταφοραῖς οὔτε ὅσα λέγεται μεταφοραῖς. Alternatively the last clause may be taken as the object of ὁρίζεσθαι and the whole then translated: "if one should not argue in metaphors, it is clear that one should not use metaphors in giving definitions, nor define metaphorical expressions."

44. *Top.* 139b32ff., 34–35: πᾶν γὰρ ἀσαφὲς τὸ κατὰ μεταφορὰν λεγόμενον. In this chapter of the *Topics*, 6.2, which deals with various commonplaces regarding the unclear, Aristotle clearly distinguishes homonymy as a different mode of unclarity from metaphor: see 139b19ff. At 140a6ff. metaphor is contrasted with what is even worse than metaphor, expressions that are quite unclear (because not based on any similarity at all), where he points out that "in some sense metaphor does make its meaning intelligible, because of the similarity [on which it is based]." Further criticisms of metaphor used in giving the genus or of definitions otherwise incorporating metaphor are made at *Top.* 123a33ff., 158b8ff.

45. See *Rh.* 1405a8ff., 1407a14ff., 1410b36–1411b23.

46. For example at *Rh.* 1405a8ff., 1410b13ff.

47. For the importance of metaphor in poetry, see *Po.* 1459a5ff. (the ability to use metaphor well is a sign of natural genius and cannot be learnt from another) and cf. *Rh.* 1405a4ff. At *Rh.* 1412a11ff. Aristotle says that in philosophy too it is a sign of the εὔστοχος, the man who hits the mark, that he can discern similarity in things that are far apart, and cf., e.g., *APo.* 97b7ff. (Cf. above, Chap. 1 at n. 106, in the context of dream interpretation.)

48. *Top.* 105a21ff., 108a7ff., b7ff.

the work of the Craftsman in the *Timaeus*. Aristotle too compares the blood-vascular system to a system of irrigation channels,[57] and he too compares the crisscrossing of the blood-vessels to a wickerwork structure.[58] More simply, nature is repeatedly described as creating, ποιεῖν, δημιουργεῖν, devising, μηχανᾶσθαι, and adorning, ἐπικοσμεῖν, living creatures or their parts,[59] and most frequently of all, of course, her purposeful activity is expressed in the phrase "nature does nothing in vain," ἡ φύσις οὐδὲν μάτην ποιεῖ.[60]

For a philosopher who condemned all metaphor as obscure, Aristotle is, one might think, extraordinarily free with implicit and explicit comparisons of every kind between the role of φύσις and the τέχναι. But the first-stage defence he would offer is not far to seek. It is above all in relation to the workings of the final cause that these comparisons are developed. *Both* domains, Aristotle would insist, exemplify finality, though its modality in each is different: he points out, for instance, that nature does not deliberate, just as he also recognises that there are exceptions to finality, failures to secure the good, in both artistic and natural productions.[61] But in many of the comparisons he draws he would claim that there is no question of transferring conclusions from one particular instance *to* another *directly* (thereby encountering the difficulty he mentioned in his analysis of analogical argument). Rather, both particulars fall under a general rule for which he believes he has ample grounds. Art can be used to illustrate nature because both domains manifest certain general principles concerning, for example, the adaptation of form to function, the hierarchisation of ends, and the relationship between the end to be attained and the character of the matter necessary to attain it. To quote just one prominent example:

---

57. *PA* 668a13ff.; cf. Plato *Ti.* 77c ff.
58. *PA* 668b21ff.; at *Ti.* 77d–e Plato used the image of cleaving, σχίσαντες, and weaving together, πλέξαντες, the veins round the head to serve as a binding, δεσμός.
59. E.g., *GA* 731a24, *PA* 652a31, 658a32.
60. Bonitz's index, which is not exhaustive, cites twenty-three instances from the Corpus, 836b28ff. Cf. the discussions of Ulmer 1953; Solmsen 1960, pp. 102ff.; Bartels 1966; Fiedler 1978.
61. E.g., *Ph.* 199a33ff., b26ff.

just as an axe, to be used for chopping, must be made of a hard material such as iron or bronze, so each of the parts of the body must be of a material suitable for the function it is to perform.[62]

But if, in general, we can see why art may be invoked as an analogue to nature, this does nothing to explain why in any given case a particular technological analogy should be used, let alone guarantee that it will not mislead. The crisscrossing of the blood-vessels may suggest wickerwork, but it does not show that they do indeed have the function of binding the front and the back of the body.[63] Moreover, in this instance there is a fairly obvious negative analogy (or difference) that might have given Aristotle pause, in that the texture of the blood-vessels, the veins especially, might be thought ill suited to serve a binding function.

An even more disastrous example is Aristotle's theories concerning the role of the testicles, which he several times compares to the weights on looms.[64] He believes their function to be, not to produce the semen but, rather, simply to keep the seminal vessels taut. It is true that he believes he has independent evidence that even after castration bulls can fertilise cows successfully, a supposed fact that he took to suggest that the testes do not produce seed.[65] The tension of the seminal vessels, on the other hand, would—he thought—be released only gradually after the excision of the testes. The loom-weight idea offered the basis of an alternative theory, though the more immediately visible similarity it appealed to was—*we* should say—superficial.

Again, the general doctrine of the adaptation of the parts of living creatures to ends is expressed by Aristotle with the help both of particular comparisons with ὄργανα, tools or instruments, and of the term ὀργανικόν, instrumental, applied to such non-uniform parts as the hand. When he speaks of the organs of the body, the technological model plays an active heuristic role. A single text will serve to illustrate

---

62. *PA* 642a9ff.; cf., e.g., 639b23ff., 646a24ff., b3ff.
63. *PA* 668b21ff. Aristotle does, however, also hold that the blood-vessels serve to convey nourishment to the parts of the body, e.g., 668a12–21.
64. E.g., *GA* 717a34ff., 787b19ff., 788a3ff.
65. *GA* 717b3f.; cf. *HA* 510b3f.

the doctrine: "since every instrument (ὄργανον) is for the sake of something, and *each of the parts of the body is for the sake of something,* that is to say, some action, it is clear that the body as a whole arose for the sake of some complex action. *Just as* the saw came to be for the sake of sawing, and not sawing for the sake of the saw . . . *so* the body exists for the sake of the soul in a way and the parts of the body for the sake of the functions that each of them naturally fulfils."[66]

Definition of Terms

In many of the cases so far considered, Aristotle would justify the implicit or explicit comparisons he himself uses by referring to the general rule, of which both items compared can be seen as instances, a rule which can, or should in principle, be supported independently. But the broader questions that Aristotle's theory of meaning and his demand for precision and the literal raise concern also his reaction to and criticism of many of the complex and problematic theoretical terms that his predecessors and contemporaries used in their natural philosophical speculation, whether or not Aristotle saw these as, or as involving, metaphor. In some instances he proceeds in the way we might expect from his criticisms of the obscurity of metaphor and the like and from his general statements requiring the strict use of terms: that is, he goes all out to purge the terms of ambiguity and vagueness and to establish a single clear-cut definition, even though the strain

---

66. *PA* 645b14–21, reading πολυμεροῦς at b17. Aristotle speaks frequently of, for example, the hand (e.g., *PA* 687a10, 18ff.) and the tongue (*PA* 683a19ff.) as or as like organs, and of the organs of locomotion, e.g., *GA* 732b26ff., *PA* 683b5ff., *De incessu animalium* (*IA*) 713a3ff., of reproduction, e.g., *GA* 717a12f., 721a26, 766a3ff., 22f., of digestion, e.g., *GA* 788b20ff., and of respiration/refrigeration, *Resp.* 480a16ff., and he puts it that nature uses the seed (*GA* 730b19ff.), or pneuma (*GA* 789b7ff.) or heat and cold (*GA* 740b31f.) as or as like ὄργανα (cf. also *Mete.* 4.381a10f., where he says that it is indifferent whether the concoction that takes place by boiling is effected in artificial or in natural ὄργανα). Plants too are said to have ὄργανα, though only of the simplest kind: *De An.* 412b1ff., *PA* 656a1f. Cf. the discussions in Bartels 1966, pp. 8off.; Byl 1971, 1980, pp. 161ff.; Fiedler 1978, p. 285.

that this imposes on some parts of his scientific enterprise are, at times, as we shall see, considerable. In other cases, however, he allows that a term may be "said in many ways," πολλαχῶς λεγόμενον, but argues that these ways have a systematic relationship to a single central, "focal" meaning, a principle particularly important, as Owen showed,[67] in relation to many high-level metaphysical concepts such as essence (τὸ τί ἦν εἶναι), being, and substance themselves. The question that this raises is the extent to which this type of analysis implicitly modifies the ideals set out in the *Organon*. We may consider first two pairs of examples from his physics, heavy/light and hot/cold, to illustrate the former type of move and to analyse its strengths and weaknesses.

The pair heavy/light had been used in ordinary Greek primarily of what is difficult or easy to carry, though in both cases with a fair range of other meanings or applications as well, including difficult, and easy, more generally.[68] But signs of the strain under which the naive conception was coming are already visible in pre-Socratic philosophy, where various correlations are proposed with other pairs of opposites (such as dense/rare) or with the elements as well as with movements,[69] and

---

67. See especially Owen 1957/1986, 1960/1986, 1965a/1986.

68. βαρύς is already used in Homer as an epithet of ἄτη, ἔρις, and κακότης (*Il.* 2.111, 20.55, 10.71), and once in the *Odyssey*, 9.257, of the quality of a sound (as it was later to be used, with the antonym ὀξύς). κοῦφος appears in Homer only in the adverbial neuter plural, κοῦφα ποσὶ προβιβάς, stepping nimbly upon (*Il.* 13.158), and appears in Pindar and the tragedians in a sense LSJ take to be "metaph.," "easy," e.g., Pi. O. 13.83, Aeschylus *Septem contra Thebas* 260.

69. See, for example, Parmenides fr. 8.56ff., Empedocles fr. 21, together with the admittedly often tendentious reports and criticisms in Aristotle (e.g., *De generatione et corruptione* [*GC*] 314b20ff., 315a3ff., 10f.), and in Theophrastus (*De sensu, Sens.*, e.g., 59ff.). Whether or not weight is a primary property of the atoms for Leucippus and Democritus is much disputed (see below, Chap. 5 n. 41), but it appears from passages in both Aristotle (*GC* 326a9f.) and Theophrastus (*Sens.* 61) that heavy and light were sometimes referred to the size of atoms. Compounds, however, could also be distinguished by the proportion of atoms to void or the amount of void they contain. Theophrastus also reports (*Sens.* 62, 68) that light and heavy were correlated or associated with rare and dense (as also still in Aristotle *Ph.* 217b11ff.).

where Aristotle complains with some justice that the capacities in question were generally left undefined.[70]

Plato in the *Timaeus* first follows up the popular association with below and above and emphatically rejects the idea that this second pair relates to two distinct regions in the universe.[71] The universe is spherical, so it makes no sense to talk of one part of the sphere being above or below another. Imagining—boldly—a thought experiment in which someone stands in the heavens at the interface of fire and air and forces a larger, and a smaller, quantity of fire towards the air (i.e., towards the centre), he says that it is obvious that the smaller quantity will be moved more easily.[72] It then will be "lighter" and tend "upwards," the larger will be "heavier" and tend "downwards"—though "downwards" in this case is *to* the periphery, "upwards" away from it, whereas the ordinary Greek assumption was that, on earth at least, more fire is "lighter."[73] What is light in one region, Plato is prepared to say,[74] is the opposite of what is light in the other. Several aspects of the interpretation of this text remain highly disputed,[75] but it is beyond doubt that Plato has radically redefined heavy and light: they do not just depend on the quantities of the material concerned but, like up and down, are relativised to where in the universe you are or to which element is in question.

Aristotle, in turn, is no less emphatic that certain conventional views are mistaken. Modifying Plato's idea of the importance of the element in which the real or imagined weighing takes place, he distinguishes between the two simple bodies that are heavy (or light) absolutely (that is, earth and fire) and the two that are so only relatively

---

70. *De caelo* (*Cael.*) 308a3f.; cf. Theophrastus *Sens.* 59f., who makes partial exceptions of Democritus and Plato.

71. *Ti.* 62c ff.

72. *Ti.* 63b–c.

73. This is stated to be obviously true and to hold universally, by Aristotle at *Cael.* 308b3ff., 13ff., 18ff., in the course of his quite exceptionally polemical criticisms of the theories in the *Timaeus*.

74. *Ti.* 63d–e.

75. See, for example, Solmsen 1960, pp. 275ff.; Hahm 1976, pp. 59ff., 70; O'Brien 1984.

(water and air: that is, relative to other elements).[76] He is confident that air is light in comparison with earth and with water, but he raises as a puzzle the question of whether air has weight in air, deciding the issue positively by invoking a purported trial which, he claimed, showed that in air an inflated bladder weighs more than an uninflated one.[77] Evidently here the possibility of carrying out a measurement was appreciated, though its difficulty and delicacy are reflected in the fact that when Aristotle's conclusion was challenged by later commentators, first by Ptolemy and then by Simplicius, they obtained quite different results from the same test.[78]

Aristotle further diverges from Plato in insisting that heavy (and down) are always to be defined in relation to movement *to*—and light (and up) in relation to movement *away from*—the centre of the universe, deemed to coincide with the centre of the earth.[79] Plato too had held that the universe and the earth are spherical, but Aristotle now demonstrates the latter thesis with a battery of arguments.[80] Some of these, it is true, are not independent of the issue concerning the nature of heavy and light, for they attempt to show the earth's sphericity as a *consequence* of the doctrine that the natural movements of the simple bodies are to, or from, the centre of the universe, where Aristotle *assumes* that heavy bodies do not move downwards in parallel lines.[81]

---

76. See, for example, *Cael.* 4.1.308a7ff., 4.4.311a15ff., and cf. Seeck 1964, pp. 108ff.; Hahm 1976, p. 62.

77. *Cael.* 311b9ff.

78. Simplicius reports Ptolemy's result (that the bladder weighs less) and then proceeds to describe his own attempt to verify the facts at *In Aristotelis De caelo commentaria* (*In Cael.*) 710.24ff. The weight of a bladder inflated with air is also discussed in the *Problemata*, 25.13.939a33ff., and in Anon. Lond. 31.34ff., 32.22ff.

79. E.g. *Cael.* 308a14ff.

80. *Cael.* 297a8–298a20; cf. *Ti.* 62e, 63a. The importance of this point will be the greater if Furley 1976, pp. 97f., is right to argue that the rival atomist account of the natural motion of atoms (in Epicurus, certainly, and in Furley's view also in Democritus), according to which they move perpendicularly "downwards" in space, depends crucially on the doctrine that the earth is flat.

81. *Cael.* 296b6ff., 18ff., 297b17ff.

But some offer good independent grounds for his thesis, notably arguments that appeal to astronomical data, first, to changes in the visibility of the stars at different latitudes, and especially in the circumpolar stars that never set,[82] and, second, to the shape of the earth's shadow in eclipses of the moon.[83]

The example of heavy and light vividly illustrates the meaning shifts that occur as theory develops, shifts that are similar in kind to those that have been explored from later science, where one example often cited is that between the notions of mass in Newton and in Einstein.[84] While in assessing just how radical those ancient meaning shifts were it is fair to recognise that the theoretical framework within which heavy and light were entrenched in ancient debate was a good deal less sophisticated than many more modern examples, we should not, on the other side, underestimate just how much of Aristotle's account of both the sublunary and the superlunary region was at stake—a point not lost on some of his ancient critics such as Philoponus.[85] Meanwhile Aristotle's own view of the matter was that he was providing heavy and light with clear, univocal definitions, and ones that incorporated the adjustments to popular notions necessary to take into account the doctrine of the spherical earth.

My second example was the pair hot/cold. Once again Aristotle

---

82. *Cael.* 297b30ff.; cf. further below, Chap. 5 at n. 57.

83. *Cael.* 297b24ff.

84. See, for example, Kuhn 1964/1977, p. 259 n. 30. Aspects of the problems of meaning invariance have been discussed, taking as illustrations the differences between Aristotle and his predecessors on the question of up and down, falling and rising, by Feyerabend 1962, p. 85 (1981a, pp. 85ff.), and by Hesse 1974, pp. 33ff.

85. In his *De aeternitate mundi contra Aristotelem,* for which our chief source is the extensive quotations in Simplicius, Philoponus explores, among other things, the difficulties that Aristotle's theory encounters in squaring the doctrine of the four simple bodies with that of the two directions of natural sublunary movement, and he mounts a sustained attack on the Aristotelian doctrine of the fifth element, aether, lacking the primary qualities hot, cold, wet, and dry. See especially Wildberg forthcoming and cf. also the *De Aeternitate mundi contra Proclum* 13.6 and 13–17, 492.5ff., 512.17–531.21; cf. M. Wolff 1978, p. 156.

complains about the ambiguities of common usage—and about the disagreements among earlier theorists.[86] Sometimes touch is invoked as the criterion, sometimes various effects (melting, burning, and the like) that the substance claimed to be hot, or cold, has on other things, and the conflicts between these criteria are discussed. Thus, boiling water imparts heat better than flame, but flame can burn; again, boiling water, he says, is hotter to the touch than olive oil, but cools and solidifies more quickly.[87] The consequences of unclarity on this, and on the nature of the dry and the wet, are particularly drastic since, as he puts it, "it seems evident that [these four primary opposites] are practically the causes of death and of life, as also of sleep and waking, of maturity and old age, and of disease and health."[88] More even than that, they provide the basis of Aristotle's own essentially qualitative element theory.

In the *De generatione et corruptione* he presents not only a very full discussion of issues connected with element theory and of rival views to his own, but also a set of definitions of the four primary opposites, to which he believes other qualitative differences (hard and soft, rough and smooth, viscous and brittle, and so on) can be reduced.[89] "Hot," he says, is "that which combines things of the same kind" (τὸ συγκρῖνον τὰ ὁμογενῆ), "cold," "that which brings together and combines homogeneous and heterogeneous things alike" (τὸ συνάγον καὶ συγκρῖνον ὁμοίως τά τε συγγενῆ καὶ τὰ μὴ ὁμόφυλα). Again, "wet," ὑγρόν (though "fluid" is often a better translation) is "that which, being readily delimited [i.e., by something else], is not determined by its own boundary," and "dry" (or solid) is "that which, not being readily delimited [i.e., by something else], is determined by its own boundary."[90] Aris-

---

86. See, for example, *PA* 648a21ff., 36ff., and cf. also 649b9ff., *GC* 330a12ff., on dry and wet.

87. *PA* 648b12ff., 17ff., 26ff., 30ff. Aristotle further exploits the distinction between what is hot *per se* and what hot *per accidens,* e.g., *PA* 649a5ff., and between what is hot potentially and what hot actually, e.g., *PA* 649b3ff.

88. *PA* 648b4ff.

89. *GC* 2.2.329b7ff.

90. *GC* 329b26–32.

totle does not proceed *per genus et differentiam*, but he evidently aims
to give clear and distinct characterisations of the four primary opposites. The somewhat abstract nature of his account is, however,
striking. Moreover, as soon as we look at the range of types of case
where he uses the four opposites, we encounter instances where his initial characterisations seem inappropriate and hard to apply.

This is particularly true when he is discussing the role of vital heat,
one of the chief foundations of his whole biology. It is important, from
his point of view, that it is *heat* in question, since this gives him his link
with his general physical theory of the elements. But it is a quality that
sometimes seems remote both from anything that might be suggested
by the definition "that which combines things of the same kind" and
from what might be thought to have some justification either in terms
of popular usage or, indeed, of appeals to subjective impressions. Thus
in one of his several discussions of the main groups of animals[91] he arranges them in a hierarchy according to their methods of reproduction, which are themselves correlated with the four primary opposites.
The most perfect animals, the Vivipara, are "hotter and wetter and
less earthy by nature"; next come the ovoviviparous animals, the cartilaginous fishes (sharks and rays), which are cold and wet; the third
and fourth groups are Ovipara that lay perfect, and those that lay imperfect, eggs, and these are hot and dry, and cold and dry, respectively;
and the fifth and final group, the larvae-producing animals such as the
insects, are "coldest of all."

We can see why he claims that the Vivipara, which include humans,
are the most perfect creatures, and also why they are warmer than, for
instance, fish. Yet it is a puzzle why he should claim that the oviparous

---

91. GC 732b28–733b16. I shall not here reenter the controversial issue
of whether or how far Aristotle's concern with differentiae in the zoological
treatises is to be seen as connected with an interest in classification (see Balme
1961/1975 and, most recently, Pellegrin 1986). For my present purposes it is
enough to note that Aristotle himself tells us that the zoologist is chiefly concerned with the formal and the final causes (e.g., *PA* 639b14ff., 640b22ff.,
28ff., 645a30ff.) and, indeed, with οὐσίαι, and with the λόγος of the essence
(*PA* 639b14ff., 640a18f., 641a27, b32)—and so ultimately with definition
(see *PA* 642a25f. and especially *GA* 715a5, 8ff.).

fish are cold and *dry*—though we may notice that he has already used the combination cold and wet for the (superior) ovoviviparous fish.[92] The whole represents a schema that appears to owe more to Aristotle's preconceptions of the hierarchy of the animal kingdom, and especially to his views on the distance of the different groups from humans at the top of that hierarchy, than it does *either* to empirical considerations, or even to considerations derived from the general definitions of hot, cold, wet, and dry set out in the *De generatione et corruptione*.

Nor is it only in connection with animal taxonomy and vital heat that questions of this kind arise. At the very heart of the physical theory, some of the correlations proposed between simple bodies and pairs of primary opposites pose problems. Earth, Aristotle suggests, is cold and dry, air hot and wet, water cold and wet, fire hot and dry.[93] The "wetness" of both air and water corresponds, of course, both to the range of the term ὑγρόν in normal Greek usage and to Aristotle's definition as "that which, being readily delimited [by something else], is not determined by its own boundary." Yet conversely, while, for the sake of the schema, fire has to be hot and *dry,* and "dry" may seem unproblematic enough at first glance, when we reflect on his definition of that quality it becomes much harder to see its appropriateness as a characterisation of *fire:* for just as ὑγρόν corresponds rather to fluidity than to wetness, so ξηρόν as "that which, not being readily delimited [by something else], is determined by its own boundary," often corresponds to *solidity,* and so from that point of view does not look very suitable for fire.[94]

---

92. Compare, however, the account given of the dietetic qualities of different sea-animals at *Vict.* 2.48 (L) 6.548.9ff., where fish in general, and also shellfish, are said to be "dry," though the cartilaginous fish "moisten" ([L] 6.550.7f.). But in the discussion in the medical writer many other qualities are also taken into account, and differences are suggested between the flesh of different types of fish.

93. *GC* 330b3ff.

94. Cf. *GA* 761b18ff. and the discussion of the shape of flame in Theophrastus *Ign.* 52ff., 35.6ff. While Theophrastus represents flame as generally having a pyramidal shape, he recognises also not just that fire is a kind of movement (see above, Chap. 3 at n. 170), but also that flame is constantly moving and flowing (*Ign.* 54, 37.3ff.).

Focal Meaning, Proportional Analogy, and Homonymy in Aristotle's Science

Heavy and hot are, then, two terms of great theoretical importance where, diagnosing confusions in their use, Aristotle aims to establish and adhere consistently to a single univocal definition but in practice encounters difficulties in following through this programme. Elsewhere, however, as we said, he uses the concept of focal meaning, which preserves the centrality of a primary significance but allows a cluster of others to be related to it. We cannot do justice here, clearly, to the intricacies of this important concept, and of the related but distinct notion of proportional analogy, but one of Aristotle's canonical examples will serve as the briefest of introductions.[95] "Healthy," we are told,[96] is said primarily in relation to health itself, but also derivatively of signs of health (as when a blooming complexion is said to be healthy) or of what promotes or preserves health (as when regular exercise and a kind of climate are said to be healthy). "Healthy" is not to be understood and explicated in the same way when said of a climate or of exercise as when said of a patient who has recovered from illness, but the term is not merely ambiguous or homonymous, since all the other uses are to be connected with a primary one in relation to however we define health itself. This allows for what I have been calling semantic stretch, while it still privileges a primary application.

Here, then, is a device of great power and scope which Aristotle in fact uses repeatedly and to particular effect, as we noted, in connection with some high-level metaphysical principles such as essence and being. In his *Physics* and elsewhere such concepts as place, τόπος, or what it is to be "in" something, and contact, ἅπτεσθαι, are elucidated

---

95. Apart from Owen's own discussions, noted above, n. 67, see also J. Barnes 1971, Hamlyn 1977–78, Tarán 1978, Ferejohn 1980, Irwin 1980–81, Fine 1982.

96. *Metaph.* 1003a34ff., cf. *Top.* 106b33ff., *Metaph.* 1060b37ff. See, for example, Owen 1960/1986, pp. 192ff., 198ff., 1965a/1986, pp. 259ff., and cf., e.g., Mackinnon 1965.

in such a way.[97] It is characteristic of his discussion to move outwards from a central, familiar, unpuzzling usage, gradually widening the range of what is to be included under the original rubric.

He proceeds in a similar way when clarifying the concepts of matter and form, and potentiality and actuality, for example, via the notion of proportional analogy, though this is not a *kind* of focal meaning so much as an alternative to it.[98] Thus matter—where the term he coined, ὕλη, originally meant just wood, of course—is used first of the stuff physical substances are made of, but also of the substratum of change more generally.[99] That there *is* something that underlies and survives change is illustrated by such straightforward cases as a man becoming pale or educated, but the idea is then applied not just to the bronze the statue is made from, but also more problematically, in embryology, to the matter—the menses—that Aristotle holds to be supplied by the mother to the embryo.[100] Again, matter is said to individuate members of the same species, which are the same in form but numerically distinct,[101] and he feels entitled also to speak of intelligible matter in, for example, mathematics.[102] Two identical triangles used in a geometric proof are differentiated by their intelligible matter: not the triangles I draw on the blackboard, but the triangles we have specified and are reasoning about. As matter is what is characterised by form, and the genus receives determination from the species (also εἶδος), the genus too can be called matter.[103] But as that last example particularly illus-

---

97. See especially *Ph.* 210a14ff., 226b18ff., *GC* 322b29ff. and cf. *De interpretatione (Int.)* 23a7ff., *Metaph.* 1047b35ff., 1048a13ff. on δυνατόν; cf. *Metaph.* 1071a3ff.

98. As becomes clear at *EN* 1096b26ff., for example, when he discusses "good."

99. As, for example, in *Ph.* 1.6ff., 189a11ff., especially 7, 191a7ff., 9, 192a3ff.

100. See, for example, *GA* 727b31ff., 729a10ff., 28ff.; cf. G. E. R. Lloyd 1983a, p. 97.

101. *Metaph.* 1034a7f.

102. See, for example, *Metaph.* 1036a9f., 1037a4ff., 1045a33ff.; cf. 1059b14ff., 1061a28ff., and cf. Lear 1982, p. 181.

103. See especially *Metaph.* 1024b4ff. (but cf. b9ff.), 1045a34f., 1058a21ff., cf. 1016a24ff., 1023b2, 1038a5ff., 1071a36ff., *Ph.* 200b7f., *GC* 324b6ff.

trates—where whether the genus *is* matter, or is just *like* matter, is disputed[104]—the point at which Aristotle is using the term in an "as if" way (that is, in a way *he* has to recognise as such) may be quite unclear and controversial.

Thus in a variety of contexts, dealing especially with the fundamental notions that underpin the whole of his philosophy and science, Aristotle offers a kind or kinds of analysis that while certainly not in direct contradiction to anything in his logic, nevertheless represent a certain relaxation of the requirements of univocity and universal, *per se*, predication laid down in his accounts of definition and demonstration.[105] The balance between these two points is delicate—the more so as we do not have an extended formal discussion of focal meaning and so have to rely on the scattered comments that occur in texts that deploy the notion. But clearly, first, there is no question of Aristotelian *metaphora* being involved, in the sense of the transference of a term from one field to another. Nor, secondly, is focal meaning a matter of a comparison to be justified by reference to an (in principle) independently verifiable general rule exemplified in the particular cases compared. Thirdly, while the extent to which focal meaning is proposed by Aristotle as a *tertium quid* between what he calls synonymy (i.e., univocity) and homonymy is disputed,[106] at the very least it is marked out

---

104. See, for example, Balme 1962a; A. C. Lloyd 1962, 1970; Wieland 1960–61/1975, pp. 136f.; Rorty 1973, 1974; Grene 1974; M. J. White 1975; Lear 1982, p. 181; I. Mueller 1987.

105. The *Posterior Analytics* lays down, among other things, first that demonstrations must proceed from premises that are true, primary, immediate, better known than, and prior to the conclusions (1.2.71b20ff.; cf. above, Chap. 3, n. 138) and further (1.4.73a25ff.) that in such premises the terms must be predicated *per se* (here subdivided into four types: 73a34ff.) and universally, both κατὰ παντός and καθόλου, in a special sense that he elucidates at 73b26ff. as true not just of all the subject, but belonging to it *per se* and qua itself, the so-called commensurate universal, on which see Inwood 1979, but cf. J. Barnes 1975, pp. 247f.

106. Aristotle himself expresses the relationship between πρὸς ἕν predication and καθ᾿ ἕν or ὡσαύτως on the one side, and what is said ὁμωνύμως on the other, in rather different terms in different texts (though one feature of his discussion is that he generally focuses not on words but on the things to which they are applied). Thus at *Metaph.* 1030a32–b3, when he discusses τὸ ἰατρικόν,

from other cases of homonymy in that a systematic relationship can be exhibited between primary and peripheral significances. Nevertheless, fourthly, despite his evident dislike of some modes of reasoning based on likenesses, and despite the demand for the strictest univocity in all terms used in demonstrative reasoning, focal meaning and proportional analogy tacitly mark a departure from that ideal in many key concepts. This is not the reintroduction of imagery, but it is a loosening of the straightjacket of univocity, an implicit recognition (maybe) that the requirements specified for definition and predication in the *Posterior Analytics* are an *ideal*.[107]

---

the medical, this is clearly stated to be said neither καθ' ἕν nor ὁμωνύμως, but πρὸς ἕν; cf. *Magna Moralia* 1209a23ff., 29ff., and what is said πρὸς ἕν is again contrasted with what is said homonymously, at *Metaph.* 1003a33ff.; cf. *GC* 322b29ff. Elsewhere, however, Aristotle uses a weaker formulation, where what is said πρὸς ἕν is contrasted with "chance" homonymy (*EN* 1096b26ff.) or with "total" homonymy (πάμπαν, *Ethica Eudemia* 1236a15ff.). Owen himself (1960/1986 pp. 184ff., 192ff.) called focal meaning a tertium quid and a convincing extension of synonymy, but remarked that some texts distinguish focal meaning from homonymy for what he called "political reasons" (1965a/1986, p. 262 n. 5), and that there is a tertium quid has been denied, for example by Irwin 1980–81 and Fine 1982. That focal meaning is something other than paronymy, represented in the *Categories*, 1a12ff., as distinct from both synonymy and homonymy, is clear in that paronymy considers only cases where there is linguistic inflection. On the general point, that what is said in many ways is itself said in many ways, see also Owens 1963, pp. 118ff.; Hintikka 1973, chap. 1.

107. Thus in the demonstrations envisaged in *APo.* 2.16 and 17, e.g., 98a36ff., b5ff., 33ff., 99a23ff., involving such terms as "deciduous," "broad-leaved," "coagulation," and "sap" (ὀπός: but at 98b37 he had just said "fluid", ὑγρόν) (cf. above, Chap. 3, at n. 153), these predicates should presumably meet the requirements laid down in *APo.* 1.4, in respect of universality and belonging per se (cf. above, n. 105). Yet in practice, in the zoological treatises Aristotle acknowledges, for example, that the character of being biped differs in birds and in humans (*PA* 643a3f., 693b2ff.). Again, when he considers different kinds of blooded and bloodless animals, he not only uses comparisons between blood and what is analogous to it but remarks that the blood of different animals has very different qualities—qualities which, indeed, he invokes, as we have noted, to explain the different characters and intelligence of different species (see above, Chap. 1 n. 89, on *PA* 647b31ff., 650b14ff., and cf. G. E. R. Lloyd 1983a, pp. 32ff., on some of the difficulties in this idea); at *PA* 643a4f. he puts it that blood differs in different animals, or else it cannot be

The tension here mirrors and indeed exemplifies a further deep-seated tension within Aristotle's divergent statements on the relationship between philosophy and dialectic. Often that relationship is expressed in terms of a series of contrasts: the philosopher works—or can work—on his own, the dialectician in conjunction with a partner;[108] the philosopher deals with truth, the dialectician with opinion—for dialectical syllogisms reason from generally accepted views, demonstrative ones from premises that are true and primary.[109] Yet on other, admittedly rarer occasions, Aristotle recognises a fundamental role for dialectic, in the sense of the critical scrutiny of received opinions, as a means of securing the primary principles of each science.[110] But the snag is that the primary principles used in demonstrations, including definitions, are required to be better known than and prior to the conclusions.[111] To get round the difficulty Aristotle would no doubt invoke his distinction between what is "better known to us" and what is "better known simpliciter,"[112] but quite how the move from the first to the second is to be made, or how we are to recognise we have accomplished it when we have accomplished it, can be problematic[113]—

---

included in the οὐσία, and cf. *Top.* 148a29ff. on the term ζωή. Moreover, in both *PA* and *GC* he explicitly points out that "wet," ὑγρόν, and "dry," like hot and cold, are said in many ways (see above, n. 86).

108. See, for example, *Top.* 155b7ff.

109. See, for example, *Top.* 100a18ff., 105b30f., *Metaph.* 1004b17ff.; cf. *Rh.* 1355a33ff.

110. *Top.* 1.2.101a36–b4. Elsewhere, when discussing the way in which the primary premises of each science are secured, Aristotle invokes ἐμπειρία, experience (*APr.* 46a17ff.) or νοῦς, reason, and ἐπαγωγή, "induction" (*APo.* 2.19.99b20ff., 100b3ff.; see above, Chap. 3 n. 143; cf. *EN* 1098b2ff., 1139b29ff.). The issues have been extensively debated: in addition to the studies mentioned above, Chap. 3 n. 143, see especially Le Blond 1939, pp. 117ff.; Weil 1951/1975, pp. 88ff.; Wieland 1960–61/1975; Owen 1961/1986, 1965b/1986; J. D. G. Evans 1977, pp. 31ff.

111. As at *APo.* 1.2.71b20ff., referred to above, n. 105.

112. As at *APo.* 1.2.71b33ff.; cf. also *Top.* 142a6ff., *Ph.* 184a16ff., *Metaph.* 1029b3ff., *EN* 1095b2ff.

113. See, for example, Wieland 1962/1970, pp. 69ff.; Mignucci 1975, pp. 30f.; S. Mansion 1979.

particularly when we are dealing with cases that may involve focal meaning or proportional analogy. In any event the chief point that remains is that what is presented in the *Posterior Analytics* has to be seen as an ideal to which no more than approximations are to be expected in some key areas of inquiry.

Metaphor and the Development of Technical Terminology

I have so far focused largely on the evidence in Aristotle, since this offers by far the best opportunity to assess the match and mismatch between theoretical analysis and actual practice. But we should now extend the scope of our discussion to consider (once again, very selectively) some other aspects of the development and use of theoretical terms in Greek science more generally. Initially much of the inquiry into nature is almost entirely devoid of established technical terms: its discourse just *is* ordinary language and reflects—for better and for worse—its vaguenesses and unclarities. This is especially true of early Greek medicine, which takes over, more or less without modification, many popular terms for diseases.[114] The usual generic word for fever, πυρετός, which may simply mean fiery heat or fire, like πῦρ, in Homer,[115] is an example where the original associations of the term stayed with it. Whatever theory a given Hippocratic author might adopt on the causes of diseases, πυρετός remained semantically wedded to the idea of heat. So too did καῦσος, from καίω, burn, though this was rather more a term of art: it was often connected with a combination of symptoms, even though these were neither distinct enough nor sufficiently

---

114. Cf. Lonie 1983, pp. 152ff., who remarks on the comparative exiguousness of the Greek lexicon of disease names, with the exception of names for skin diseases and some other products of specialized medical knowledge; cf. also Kudlien 1967.

115. *Il.* 22.31, which speaks of the effects of the Dog Star, is given as a separate entry from *fever* in LSJ. Conversely, πῦρ is often used of fevers in the *Epidemics*, e.g., *Epid.* 1 case 6 (L) 2.698.7, *Epid.* 3 case 6 of the first series (L) 3.50.2.

widely agreed upon to justify the gloss in Liddell Scott Jones as "bilious remittent fever.[116]

The referential opacity of many popular, *and* newly coined, terms is particularly evident in words for diseases based on particular organs or parts of the body: ὀφθαλμίη, πλευρῖτις, νεφρῖτις, ὑστερικά.[117] The general sense was given by the root, in each case, though by itself this was not necessarily very informative. What counted as *the* disease of the pleura, or kidneys, let alone the uterus, was often a matter of dispute and depended on the writer's views on both the symptoms and the causes at work, though in some cases there were discernible limits to the sense and reference of the term and general agreement, for instance, that ὀφθαλμίη was an inflammation accompanied by discharges from the eyes (though these might be "dry").[118]

Similar points apply to many common terms in physiology. Both the advantages and the disadvantages of considerable semantic stretch can be illustrated in such a term as πέψις, usually translated "concoction." Originally used of the ripening of fruit (one of the root meanings of

---

116. Thus one of the fullest descriptions of καῦσος, at *Epid.* 3.6 (L) 3.80.5–82.17, makes no mention of bile or bilious symptoms, and in one context in *Aër.* 10, *CMG* 1.1.2.50.19ff., the writer specifies that καῦσος attacks the phlegmatic and the old, in contrast to the bilious, in certain climatic conditions. When at *Aff.* 11 (L) 6.218.13ff., 21ff., we are told that καῦσος comes from bile, this must be interpreted in the light of that treatise's general theory that all diseases come from bile or phlegm (1 [L] 6.208.7ff.). Compare the comments of Chadwick and Mann, 1978, p. 64, that while "in many cases, enteric fever is clearly described," the "condition, *causus,* cannot be generally identified as a single disease."

117. For examples of these, see *Aër.* 3, *CMG* 1.1.2.28.14, 9, *CMG* 1.1.2.44.4; cf. also νεφριτικά, e.g., *Aph.* 6.6 (L) 4.564.6f. The term ὑστερικός occurs in Aristotle, *GA* 776a10, and in the Hippocratic *Prorrh.* 1.119 (L) 5.550.7f., *Coac.* 343 and 543 (L) 5.658.2, 708.5, but already in the classical period elaborate theories were developed concerning the wanderings of the womb around the inside of the female body: see especially *Mul.* 2.123–31 (L) 8.266.11–280.3; cf. Plato *Ti.* 91c and many other texts cited from *Mul.*, *Nat.Mul.* and *Loc.Hom.* in G. E. R. Lloyd 1983a, p. 84 n. 100. These theories were later to be strongly criticised by Soranus at *Gyn.* 1.8, *CMG* 4.7.18ff.; 3.29, *CMG* 4.112.10ff., 113.3ff.; 4.36, *CMG* 4.149.21ff.

118. As, for example, at *Aër.* 10, *CMG* 1.1.2.48.19.

the verb),[119] then of cooking and digestion,[120] it came to be applied to a wide range of physiological processes (including the production of semen and its action on the menses, the hatching of eggs, the development of the embryo, and the formation of blood, fat, suet, milk, and residues such as urine),[121] as well as to various pathological changes, the formation of pus, catarrh, phlegm, and other humours[122] (compare, in English, talk of the "ripening" of boils). Finally πέψις was even used of the production of metals and stones from earth,[123] and of snow, hoarfrost, and hail from rain.[124]

---

119. As, for example, at *Od.* 7.119; cf. Theophrastus *CP* 6.16.1ff. πεπαίνω, πέπανσις and πέπων also share some of the same range as πέψις: see, for example, of the "ripening" of discharges, e.g., *VM* 18, *CMG* 1.1.49.14; 19, *CMG* 1.1.50.5, or of urine, *Epid.* 1.3 (L) 2.610.8f.; and of the "ripening" of a disease as a whole, *Prog.* 12 (L) 2.140.13; or of its ἀκμή, *Acut.* 11 (L) 2.304.5; cf. also *Epid.* 6.2.16 (L) 5.284.13ff.

120. The verb is so used, for example, at *VM* 11, *CMG* 1.1.43.15; cf. *Flat.* 7, *CMG* 1.1.95.6; cf. the verb in Aristotle at *PA* 677b31, *GA* 718b21, and the noun at *Resp.* 474a26, *PA* 650a4; cf. *Pr.* 1.15.861a6; *VM* 18, *CMG* 1.1.49.19. In *Mete.* 4.2–3.379b10ff., 380a11ff., ripening (πέπανσις) boiling (ἕψησις) and roasting/baking (ὄπτησις) are said to be types of πέψις; indeed, the claim is that there are natural counterparts to the last two, artificial, processes, even though these have not been given distinct names (381b3ff.).

121. For some representative texts in Aristotle, see *PA* 652a9f., *GA* 719a32ff., b2, 727a34ff., 744b1ff., 753a18ff., 756b28f., 775a17f., 776a20ff., b33ff., 780b6ff., *Mete.* 4.2.380a1ff., and cf. many passages that distinguish male and female by the capacity/incapacity to concoct semen, e.g., *GA* 728a18ff., 738a13, 34ff., 765b10ff., 766a30ff.

122. See especially *VM* 18, *CMG* 1.1.49.20f.; 19, *CMG* 1.1.49.26f., 50.1ff., 23ff., 51.5, and cf., e.g., of φύματα, *De medico* 10, *CMG* 1.1.23.25ff., Aristotle *Mete.* 4.2.379b29ff., and cf., of changes in the colour of the hair, *GA* 784a34ff., and in the pseudo-Aristotelian *De coloribus* (*Col.*) 795b22ff., 796b15ff., 799b12ff. *Mete.* 4.3 wrestles with the question of the "strict" or "proper" use of this and related terms and recognises, for example, that "boiling" said of gold or wood involves μεταφορά (*Mete.* 380b13ff., 28ff.).

123. As, it would appear, in the pseudo-Aristotelian *De plantis* 822a25ff.; for ancient beliefs concerning the *growth* of minerals see, for example, Halleux 1974, pp. 67 and 152, citing especially Proclus *In Ti.* 1.43.1ff.

124. *Pr.* 26.3.940b8ff., 12ff.; cf. *GA* 784b3ff. It is particularly striking that in his discussion of the claims of the natural elements and of the parts of animals to be substances, in *Metaph.* Z 16.1040b5ff., Aristotle rejects these on

This is a particularly clear example illustrating the difficulties of erecting *definite* boundaries between primary and derivative uses: there is no question of saying precisely where the term begins to be applied "metaphorically." As a portmanteau concept, it *both* enables a variety of different processes to be related and brought under the scope of a single theory, *and* it pays a price for this in the indeterminacy of the theory and a corresponding lack, at many points, of predictive or explanatory power.

But while large areas of Greek science, at every period, manifest a certain conceptual vagueness, there are important exceptions to this, cases where technical terms are coined and given clear working definitions. Anatomy, zoology,[125] harmonics,[126] and astronomy all provide examples, but we may concentrate on some from the first and the last of these. In anatomy, for instance, once some Greek investigators had begun to use dissection extensively,[127] many structures were discovered

---

the grounds that a substance must be a unity that results from πέψις (πρὶν ἤ πεφθῇ καὶ γένηταί τι ἐξ αὐτῶν ἕν, 1040b9f.). He is led to conclude, in the following chapter, 1041b28ff., that substances are constituted "according to nature" and "naturally": ὅσαι οὐσίαι κατὰ φύσιν καὶ φύσει συνεστήκασι.

125. Thus Aristotle was led to coin terms for some of the main groups of animals that he recognised, such as μαλακόστρακα, for the crustacea (though this term may also have been used by Speusippus: see fr. 8 Lang), μαλάκια for the cephalopods, and ὀστρακόδερμα for the testacea; this and other aspects of his zoological nomenclature are discussed in G. E. R. Lloyd 1983a, pp. 16ff.

126. Thus in his *Harmonica* Aristoxenus devotes a good deal of space to the careful explanation and definition of technical musicological terminology, as he does, for example, of ἄνεσις (tension), ἐπίτασις (relaxation), βαρύτης (depth), ὀξύτης (height), τάσις (pitch), at 1.3 and 10–13; of διάστημα (interval) and σύστημα (scale) at 1.4–6 and 15–16; of πυκνόν (the "close" or "compressed," used of the sum of the two small intervals of the tetrachord when that sum is less than the remainder of the fourth: Macran 1902, p. 247) at 1.24 and 2.48; as well as of the main genera (γένη) of melody, the enharmonic, chromatic, and diatonic, at 1.19, 21ff., and of the various keys, τόνοι, at 2.37ff.

127. The debate within the Hellenistic medical sects, relating in part to human dissection and vivisection, is discussed above, Chap. 3 at nn. 220–235. But before that Aristotle had both advocated and practised animal dissection (see, e.g., *PA* 645a26ff.), though the extent of dissection before him is controversial: see G. E. R. Lloyd 1975a, Manuli and Vegetti 1977, Vegetti 1979.

that had no popular names. Often the new coinages were constructed on the basis of analogies of one kind or another. What we call the retina, for example, was dubbed the net-like membrane (ἀμφιβληστροειδής in Greek: the origin of our own term, via the Latin intermediary, *rete*, net) though it was also called the spider's-web-like one (ἀραχνοει-δής).[128] This illustrates one recurrent problem, namely, the standard-isation of anatomical terms or, rather, the lack of it.[129] But whichever term was used, the associations inherent in the original comparison were unlikely to prove problematic, at least when the reference was clearly fixed. Even where the assumptions implicit in the new coinage bore a theoretical load, provided that the reference was definite, the term sometimes continued in use even after the theory changed. Thus the carotid arteries (καρωτίδες, stupefiers) were so called because originally they were believed to cause unconsciousness when pressed; but even after it was established that this was an effect of compressing the nerves in the neck rather than the arteries, the term was still used of the arteries in question.[130]

Finally, astronomy, especially, developed a wealth of technical termi-nology, clearly defined words for zenith, meridian, apogee, perigee, parallax, colure, station, retrogradation and many others,[131] let alone geometrical terms such as homocentric, epicycle, eccentric. Where im-agery continued to play a more prominent and indeterminate part is, rather, in the sister discipline of astrology, where conclusions were drawn about the influences of planets and constellations from their

---

128. Both ἀμφιβληστροειδής and ἀραχνοειδής may be coinages by Her-ophilus: see Rufus *Onom.* 154.9f., Celsus *Med.* 7.7.13b, CML 1.319.20–22. Rufus further mentions that this structure was called ὑαλοειδής from the vitreous humour it contained, as well as also simply the "third" membrane: *Onom.* 154.7ff.; cf. ἀνατομή (*Anat.*) 171.9ff.

129. Aspects of this are discussed in G. E. R. Lloyd 1983a, pp. 158ff.

130. See, for example, Rufus *Onom.* 163.9ff.; cf. Galen *PHP* 1.7, CMG 5.4.1.2.86.24ff., (K) 5.195.4ff.; *UP* 16.12 (H) 2.427.15ff., (K) 4.332.9ff.

131. In such introductory works as those of Geminus, Cleomedes, and Theon of Smyrna, such terms are often defined or explained (see, for example, Geminus 1.20.10.8ff. Manitius, on στηριγμός, "station," Theon 131.4ff. on ὁρίζων, "horizon," 132.2ff. on κόλουρος, "colure"), but in Ptolemy's *Syntaxis* knowledge of their meaning is more generally presupposed.

supposed masculine and feminine qualities, their commanding or obedient character, their being diurnal or nocturnal, beneficent or maleficent.[132]

The Category of Metaphor and the Criteria for Truth

A comprehensive study of "metaphor" in Greek science would be a comprehensive study of Greek science. But perhaps enough examples have been passed very rapidly under review to allow a certain perspective on the range of uses and to permit us to take stock of aspects of the polemic that some Greek philosophers directed against metaphor, myth, and other modes of reasoning involving likenesses or the non-univocal.

As we should expect, the Greek inquiry into nature is heavily dependent, in all periods, on every kind of more or less evidently stretched terms. Even those who might not accept that all language is in some sense metaphorical, will agree that it is often, even normally, through the generation of new, and the elaboration of old, models that science grows and acquires new ideas—and this is true, for sure, of ancient inquiries. The conceptions of concoction, of organ, and of matter itself provide examples, at different theoretical levels, of ingenious and surprisingly durable cases of creative semantic stretch from Greek science, even if we must grant that elsewhere such uses also permitted much merely wayward speculation.

---

132. The use of these terms can be extensively illustrated from Ptolemy's *Tetrabiblos* (which deals with masculine and feminine planets and signs at 1.6.20.8ff. and 13.34.9ff., with diurnal and nocturnal planets at 7.21.2ff., commanding and obeying signs at 15.37.3ff., and beneficent and maleficent planets at 5.19.17ff., developing the characterisations of 4.17.13ff.), as also can elaborate symbolism drawing on the animal or human forms and characters of the signs of the zodiac and other constellations (especially in 2.8.81.5ff., 82.15ff.). At the same time, astrology also possessed a technical vocabulary for certain geometrical relations (Ptolemy explains "trine"—triangular—"quartile," and "sextile" in 1.14.35.20ff.), as well as sharing with astronomy the terminology of zenith, meridian, ascension, syzygy, and so on.

Strong reservations are expressed first by Plato and then by Aristotle on the use of images, likenesses, myths, and metaphors, and the aggressiveness of their tone is a pointer to that underlying polemic. At the very earliest stages of the inquiry into nature, what it was replacing or attempting to replace was a view of the world put together (but not consciously) from elements relating to straightforward, concrete experience, and applied (again, not consciously) to the understanding of the otherwise inexplicable. The unexpected, the imaginary, the frightening, the occult, can only be comprehended within a coherent network of beliefs by some such extrapolation from the domain of the known, the familiar, the unproblematic—though that way of putting it runs the risk of representing the spiritual world as less well known, or at least less vividly apprehended in belief, than the world of tables and chairs, and that may well not be the case at all.[133]

The question of the status of ideas applied across a variety of contexts (as it might be, to the gods as well as to humans) was not an explicit issue until the philosophers made it such, until they made problematic the whole question of what *counts* as a *variety* of contexts. The effect of having an explicit category of the metaphorical was that it enabled issues of meaning and commitment to be brought into the open and, indeed, to be pressed—and the nature of the challenge to which the whole corpus of traditional beliefs might be subject was thereby transformed. It is particularly striking, in view of the way in which poetry was later viewed by some philosophy, that the fifth-century poets themselves, Pindar and the tragedians especially, can be seen as already frequently raising—more or less directly—problems concerning naming, meaning, understanding, and deception.[134] Yet it was not their concern, of course, to develop *explicit* theories to do with the relationship between language and reality. The special *sophia* they often claimed was—naturally—heavily dependent on what Aristotle would

---

133. The vividness with which the spirit world may be apprehended is well conveyed in such studies as Lambek's of the Mayotte: Lambek 1981.

134. These are especially prominent themes in the *Oresteia*, as has recently been shown in detail by Goldhill 1984.

sist, the constraints on language in the two areas are not *identical*. In particular, the natural scientist is bound to have to explore, and at points to delimit, the implications of his "metaphors" with one eye on the need for his theories to come eventually (and no doubt not one by one)[139] to a distinctive type of empirical or pragmatic test. The contrast with the way in which poetry works or is effective is obvious. So too, to take another field to make an analogous point, in law the lawyer will need to define and clarify his terms, though, again, without hoping for *complete* precision: Aristotle could be used to illustrate the recognition of the need,[140] Lysias the resistance to a demand for precision.[141]

So far as the sphere of the understanding of nature goes, the positive features of the conceptual moves we have been studying (and it is important to recognise that there *are* some positive features) may be said to lie in the favouring of the explicit over the merely allusive, of *comparative* determinacy over interdeterminacy, obscurity, even fudge. Suggestive though it may be to view sleep as poured over you, or to speak of the "channels" of communication as just that, πόροι, within the body, such ideas cannot be scrutinised or made the subject of further investigation—at least not until the limits of the commitment of the implied conceptions are made clearer. To be fair, however, even when the term νεῦρον had been (partially) disambiguated and the sensory and the motor nerves distinguished and made the topic of detailed research—as by Herophilus and Galen—the stretch of that term still permitted continuing indeterminacies,[142] and theories about the *mecha-*

---

139. See, for example, Hesse 1974, chap. 2, 1980, chap. 3.

140. In *Rh.* 1.13.1373b38ff., Aristotle discusses cases where the characterisation of what was done is in dispute, where the issues turn on the distinctions between, for example, "theft" and "sacrilege," "intercourse" and "adultery," "dealing with the enemy" and "treason" (1374a3ff., 6ff.). Cf. also Plato *Lg.* 943e–944c.

141. That the law does not attempt to specify all the terms that fall under a general rubric or have the same general sense is asserted by Lysias at 10.6ff., though elsewhere often—and unsurprisingly—enough his argument depends on insisting on distinctions between types of case assimilated or confused by his adversaries, e.g., 3.41ff., 4.9.

142. Following Herophilus, whose work he often praises, Galen draws careful distinctions both between the nerves and other structures (ligaments,

nisms of the transmission of movement and sensation (as opposed to descriptions of the courses of the nerves) [143] remained extremely vague. There was a long way to go.

But the—more obvious—negative features of this polemic include especially the rejection of, or at least lack of sympathy for, much that was heuristically fruitful in imagery, analogy, myth. Nor were these three by any means always characterised by—and they certainly had no monopoly of—obscurity, fudge, and the allusive. Moreover, the expectations of transparency and univocity that the philosophers generated were not ones they were usually or even often in a position to fulfil, even though they were, up to a point at least, useful expectations to raise, if only to focus on the ideal or the limiting case. The vocabulary of *muthos* and *logos* and of metaphor enabled a distinct type of challenge to be pressed, that of specifying the limits of the commitment to a theory, even though, maybe, in the strictest versions of the tests, with the emphasis on the *precision* of the limits, they were ones that science itself (then and now) was bound finally to fail—at least if it continues to use natural language or, rather, because it has to do so.

Some of the most outspoken condemnations of metaphor come in the context of Aristotle's discussion of the conditions for *episteme*, understanding, in the *Posterior Analytics*. Only universal, per se, predications using univocal terms will do for the purposes of demon-

---

tendons) often confused with them, and between the sensory and motor nerves, but apart from pointing out differences between the "ancient" and his own use of the term νεῦρον, he has many occasions to criticise the continuing errors of his own contemporaries: see, for example, *AA* 7.8 (K) 2.612.2ff., 15ff., 613.1ff., *UP* 1.17 (H) 1.33.26ff., (K) 3.47.1ff., 4.369.1ff., *CMG* 5.10.2.2.9.10ff., (K) 17A.803.14ff. For Herophilus' work in this area, see Rufus *Anat.* 184.15ff., Galen (K) 8.212.13ff., and cf. 7.605.7ff.; cf. Solmsen 1961, von Staden forthcoming.

143. While Galen's accounts of the courses of the nerves in *AA* 3.3f., 9.13, 14, 15, and *UP* 9 and 16 are, on the whole, detailed and meticulous, his view of the way in which movement and sensation are transmitted depends largely on his speculative theories concerning the action of the πνεῦμα ψυχικόν: see, for example, *PHP* 7.3, *CMG* 5.4.1.2.444.12ff., (K) 5.606.16ff., and cf. L. G. Wilson 1959.

stration, and accordingly there is no way to make room for *metaphorai* here. Yet to have confined the scientist to what would meet the requirements for demonstration according to the official programme in the *Analytics* would have been massively restrictive. Nor is there any question of Aristotle in practice actually so limiting the domain of physics. In fact, however, when, under the influence of that programme or not, he attempts precise definitions of complex concepts, the result is sometimes a certain arbitrariness (as we saw with his definitions of the four primary opposites). More often he is more tentative as he moves towards the delineation of fundamental concepts, especially when he uses the notion of focal meaning in their explication, even though that notion remains (we said) in a somewhat anomalous position in relation to the official programme.

The programme is a part, in a way the culmination, of that aggressive attack mounted by the new wisdom, at this point, against the old (and against some rivals from among the new) in a bid to supplant them. Yet once again, in practice, the new had, and continued to have, more in common with the old than the form that the attack took might lead one to expect. Certainly the effect of the forging of an explicit terminology to distinguish different types of discourse and different claims for truth was to raise the most radical questions with the most far-reaching repercussions, reverberating to the present day. Yet although Aristotle's desideratum for the well-ordered presentation of a mature science is that it should limit itself to strictly univocal terms, aspects of the actual science he does are evidently a good deal less rigorous—to their advantage. While there is no way in which he will allow *poetry* back into *science*,[144] in acknowledging the role of dialectic he recognises that the language of the scientist will often remain some distance from the ideal formalisations of the *Analytics*.

---

144. This remains true even though, in the *Poetics*, reacting perhaps in part to Plato's criticisms and his exclusion of most poetry from the ideal state in the *Republic*, Aristotle restores to poetry in general and to tragedy in particular an honourable place both in its educative role and as a source of pleasure: see especially *Po.* 4.1448b4ff.

CHAPTER FIVE

Measurement and Mystification

O ne radical criticism that has been levelled at ancient Greek science is that it was essentially qualitative in character. In both the physical and the life sciences, so it has been said, theories were neither given exact quantitative expression—as in modern chemical formulae—nor supported by exact quantitative data. One influential proponent of this thesis, who saw this as the distinguishing characteristic of ancient science, indeed, of all science up to Galileo—and a characteristic that seriously diminished its claims to *be* science—was Alexandre Koyré;[1] his view has been endorsed, and the thesis further elaborated independently, by several other prominent historians of science, including Temkin, Kuhn, and Joly.[2] In his paper entitled "From the World of the Approximate to the Universe of Precision," Koyré argued that the Greeks had no real technology, no real physics in our sense. No attempt was made to mathematise terrestrial physics; indeed, ancient science "never attempted to use on earth a measuring instrument or even to measure exactly anything except distances."[3] Referring to the whole of science

---

1. Koyré 1948/1961, pp. 311ff., and 1968, especially pp. 89ff.
2. Temkin 1961; Kuhn 1961/1977; R. Joly 1966, pp. 108ff. Cf. also Aaboe and Price 1964.
3. Koyré 1948/1961, p. 313: "Elle [i.e., Greek science] n'a jamais essayé . . . d'employer sur la terre un instrument de mesure et même de mesurer exactement quoi que ce soit en dehors des distances." In a footnote Koyré remarks that Vitruvius' description of a theodolite for measuring horizontal and vertical angles constitutes an exception and that precious metals were weighed

before Galileo, he put it that "no one had the idea of counting, of weighing and of measuring. Or, more exactly, no one ever sought to get beyond the practical uses of number, weight, measure in the imprecision of everyday life."[4]

Koyré's studies had the great merit of focusing on a fundamental issue that goes to the heart of the question of the quality of much ancient scientific work, and the thesis he propounds is bold and simple. Among the few qualifications he enters are the success of Greek celestial physics: the lack of physics is a lack of sublunary physics. Moreover, even so far as sublunary physics goes, the "superhuman" "divus" Archimedes, as he calls him, receives honorific mention, although the limitations of his achievement (a statics, but no dynamics) are stressed.[5] As those exceptions indicate, Koyré no doubt realised that the problem is more complex than some of his generalisations might seem to allow. In what has inevitably to be a highly selective discussion here I shall try to do three things: first, to see how far Koyré was right about a systematic failure to use measurement in ancient science; second, to illustrate some of the negative as well as the positive features of the ancients' search for exactness; and, third, to review briefly some of the underlying epistemological presuppositions at work. That is, in what contexts and in what forms did the ancients seek or demand exactness or even believe it to be possible? How far did they have a concept of measurement? As in our earlier studies, we shall try to evaluate not just the principles the ancients adopted but also the match between principles and practice, and we must be prepared to recognise once again that the complexity and ambivalence of ancient presuppositions and practice are such as to make generalisation hazardous.

---

precisely. The thesis that the Greeks did not and could not develop a real technology and lacked physics in the modern sense is propounded at pp. 311ff.; cf. also 1968, pp. 22ff., 34ff.

4. Koyré 1948/1961, p. 318: "Personne ne s'est avisé de compter, de peser et de mesurer. Ou, plus exactement, personne n'a jamais cherché à dépasser l'usage pratique du nombre, du poids, de la mesure dans l'imprécision de la vie quotidienne."

5. See Koyré 1968, pp. 14, 22, 32, and especially 38.

We may consider first two familiar fields that appear to provide strong evidence for Koyré's thesis: dynamics (his star example) and element theory.

Dynamics

In the field we call dynamics, the study of moving bodies, it is well known that up until Aristotle the Greeks had not progressed much beyond general statements of such vague principles as that "like" seeks "like," a principle applied by the fifth-century atomists, for example, to animate as well as to inanimate phenomena, to birds as well as to pebbles on a beach.[6] Notoriously, Aristotle himself attempts no *theory* of the factors influencing the speed of moving bodies, whether in "natural" or in "forced" motion, but merely introduces a number of general statements on that topic in the course of his discussion of other problems, such as the existence of the void or that of an infinite body, in the *Physics* and *De caelo*.[7] Many of these statements are quite indeterminate. For example, several suggest merely that in natural motion the more there is of a heavy body the faster it moves downwards, and, similarly, that the more there is of a light body, such as fire, the faster it moves upwards. Thus at *Cael.* 277b3ff., for instance, in the course of an argument denying that the natural movement of the elements is due to an external force, he says simply that "it is, on the contrary, always

---

6. Democritus fr. 164, and cf. Aetius 4.19.3. Furley 1976, p. 85, suggests, however, that the example of the birds may be meant purely "to illustrate that there are instances of natural sorting without the action of a discriminating mind." Cf. C. W. Müller 1965a, pp. 76ff. On early atomistic accounts of motion in general, see Furley 1967 and 1976, O'Brien 1977 and 1981.

7. There is an incisive discussion of this topic in Owen 1986a, pp. 315ff., following Owen 1970/1986, with powerful criticisms of such earlier studies as Drabkin 1938. Cf. also Carteron 1923, Cornford 1931, W. D. Ross 1936. I would now want to qualify some of the views I expressed in G. E. R. Lloyd 1964 and cf. 1968b, pp. 175ff. Among other important recent discussions, apart from Owen's, see de Gandt 1982; Hussey 1983, additional note B, pp. 185ff.; and Wardy 1985.

points, required. Thus in *Physics* book 7 chapter 5, especially, he considers the effects of "powers" of different strengths on the speed of the objects they move, where he has in mind, among other things, such familiar cases as a gang of men hauling a ship.[17] At *Ph.* 249b30ff. he says: "if A, which causes the motion, moves B a distance C in a time D, then in the same time the same power A will move half B twice the distance C, and in half the time D it will move it [half B] the distance C. For thus it will be proportional. And if the same power move the same object a certain distance in a certain time and half the distance in half the time,[18] then half the strength will move half the object an equal distance in an equal time; for example, let E be half the power A, and F half B, then the proportion of the strength to the weight will be the same, and so they will move an equal distance in an equal time."[19]

Thus far it might appear that we have here the makings of exact general laws of forced motion. Yet Aristotle immediately proceeds, at *Ph.* 250a9ff., with a statement that qualifies the field of application of the principles he has just sketched out. "And if E moves F a distance C in a time D, it is not necessary that E moves twice F half the distance C in the same time. If indeed A moves B a distance C in a time D, half A—that is, E—will not, in a time D or in any fraction of it, move B a part of the distance C which is in the same proportion to it as A is

---

17. See *Ph.* 250a17ff., cited below, and cf. *Ph.* 253b18.

18. I adopt the punctuation in W. D. Ross 1936. If a comma is read after κινεῖ and not after ἡμίσει, in a5, movement of half the distance in half the time will then be part of the apodosis, a conclusion, not a second premise. See de Gandt 1982, p. 106; Owen 1986a, p. 329.

19. *Ph.* 249b30–250a9: εἰ δὴ τὸ μὲν Α τὸ κινοῦν, τὸ δὲ Β τὸ κινούμενον, ὅσον δὲ κεκίνηται μῆκος τὸ Γ, ἐν ὅσῳ δέ, ὁ χρόνος, ἐφ' οὗ τὸ Δ, ἐν δὴ τῷ ἴσῳ χρόνῳ ἡ ἴση δύναμις ἡ ἐφ' οὗ τὸ Α τὸ ἥμισυ τοῦ Β διπλασίαν τῆς Γ κινήσει, τὴν δὲ τὸ Γ ἐν τῷ ἡμίσει τοῦ Δ· οὕτω γὰρ ἀνάλογον ἔσται. καὶ εἰ ἡ αὐτὴ δύναμις τὸ αὐτὸ ἐν τῳδὶ τῷ χρόνῳ τοσήνδε κινεῖ καὶ τὴν ἡμίσειαν ἐν τῷ ἡμίσει, καὶ ἡ ἡμίσεια ἰσχὺς τὸ ἥμισυ κινήσει ἐν τῷ ἴσῳ χρόνῳ τὸ ἴσον. οἷον τῆς Α δυνάμεως ἔστω ἡμίσεια ἡ τὸ Ε καὶ τοῦ Β τὸ Ζ ἥμισυ· ὁμοίως δὴ ἔχουσι καὶ ἀνάλογον ἡ ἰσχὺς πρὸς τὸ βάρος, ὥστε ἴσον ἐν ἴσῳ χρόνῳ κινήσουσιν. Note the switch from talk of δύναμις, here translated "power," to talk of ἰσχύς, here translated "strength."

to E. For it may be that it will not move it at all."[20] Otherwise, as he goes on to say, "one man might move a ship, since both the strength of the haulers and the distance they all cause the ship to move are divisible by the number [of the men]."[21]

This exception strongly suggests that the proportionalities adumbrated in this chapter are not intended to apply *strictly* as *universal rules*.[22] This becomes clearer still when we consider the range of phenomena that Aristotle believes to exemplify such proportionalities.[23] *Ph.* 250a28ff. states that growth or increase, and even qualitative changes, are generally subject to similar rules, for example, that "in

---

20. *Ph.* 250a9–16: καὶ εἰ τὸ Ε τὸ Ζ κινεῖ ἐν τῷ Δ τὴν Γ, οὐκ ἀνάγκη ἐν τῷ ἴσῳ χρόνῳ τὸ ἐφ᾽ οὗ Ε τὸ διπλάσιον τοῦ Ζ κινεῖν τὴν ἡμίσειαν τῆς Γ· εἰ δὴ τὸ Α τὴν τὸ Β κινεῖ ἐν τῷ Δ ὅσην ἡ τὸ Γ, τὸ ἥμισυ τοῦ Α τὸ ἐφ᾽ ᾧ Ε τὴν τὸ Β οὐ κινήσει ἐν τῷ χρόνῳ ἐφ᾽ ᾧ τὸ Δ οὐδ᾽ ἕν τινι τοῦ Δ τι τῆς Γ ἀνάλογον πρὸς τὴν ὅλην τὴν Γ ὡς τὸ Α πρὸς τὸ Ε· ὅλως γὰρ εἰ ἔτυχεν οὐ κινήσει οὐδέν.

21. *Ph.* 250a17–19: εἰς γὰρ ἂν κινοίη τὸ πλοῖον, εἴπερ ἤ τε τῶν νεωλκῶν τέμνεται ἰσχὺς εἰς τὸν ἀριθμὸν καὶ τὸ μῆκος ὃ πάντες ἐκίνησαν.

22. On attempts to reconcile the proportionalities of *Ph.* 7.5 with Newtonian dynamics, see especially Owen 1970/1986, pp. 156f. (compare Hussey 1983, pp. 196ff.). This book has the quite general task of arguing for a first *moved* mover by showing that an infinite sequence of movements is impossible. On the assumption (which he would not share) that celestial movements may be deemed to be such a sequence of *artificial* movements (though for Aristotle, of course, they are natural), chapter 5 would provide an argument to show that infinite artificial motion would have to be powered by an infinite mover (cf. the arguments in *Cael.* 273b30ff., considered above, and *Ph.* 8.10.266a10ff., which rule out an infinite body and an infinite magnitude). It must be acknowledged, however, that there is no explicit direct application of the analysis in chapter 5 to the general concerns of the book, and the question of its place in the overall strategy of that book and of the *Physics* as a whole remains problematic (cf. Manuwald 1971). Nevertheless, as Wardy (1985) has recently argued with some force, the ambition to find *general* laws of motion in this chapter founders not only (1) on the absence of any explicit distinction between natural and artificial motion, but also (2) on the inapplicability of the proportions to cases of artificial motion where the speed is not constant (e.g., the motion of projectiles), and (3) on the failure to specify clearly the limiting case marking out the exceptions that Aristotle allows.

23. The point was again made emphatically by Owen 1970/1986, pp. 156f.

twice the time, twice as much alteration will take place."[24] Again, in the *De caelo*, 274b34ff., when he proposes a similar analysis for a variety of modes of change, he begins by specifying "heating" and "pushing" and then generalises his point to apply to any affection or movement whatsoever. If we bear in mind Aristotle's essentially qualitative conception of the hot/cold spectrum, there is clearly no question of the proportions or ratios in such a case being expressible in exact quantities.

For anyone on the lookout for the first signs of an ambition to arrive at strict quantitative laws of motion, Aristotle's statements about how objects move, in natural or forced motion, are a disappointment. But then Aristotle clearly had no such ambition. Certain proportionalities are stated that might, at first sight, be taken to be part of such a general theory, set out in the form of exact equations governing the speed of moving bodies. Yet these statements are generally made *en passant* in the course of his discussion of other topics, where he has dialectical, often destructive, ends in view and is certainly not concerned with the positive development of any exact general theory of motion—and where the statements in question are intended to apply only loosely or subject to exceptions which are themselves not specified precisely.[25]

Aristotle's views were influential, but he was, of course, far from being the only theorist to discuss aspects of the problem of motion. Several later writers implicitly or explicitly contested his statements, and some certainly made some attempt to broaden the empirical base of the discussion. A passage in Simplicius shows that the third head of the Lyceum, Strato, tried to adduce evidence for the phenomenon of acceleration (as we call it) in natural movement. Yet as reported by Simplicius, at least, Strato's observations are typically imprecise: "If one observes water pouring down from a roof and falling *from a con-*

---

24. *Ph.* 250a28ff., b2. Again, the exception is noted, at b4ff., that it is not necessarily the case that if what causes the alteration is halved, half the alteration is brought about. Cf. *Ph.* 253b13ff., where discontinuities in increase and diminution and in other modes of change are accepted and indeed insisted upon.

25. Cf. above, n. 22, on *Ph.* 7.5.

siderable height, the flow at the top is seen to be continuous, but the water at the bottom falls to the ground in discontinuous parts. This would never happen to it unless it traversed each successive space more swiftly."[26] Again: "if one drops a stone or any other weight *from a height above the earth of about a finger's breadth,* the blow made on the ground will not be perceptible, but if one drops the object *from a height of a hundred feet or more,* the blow it will make will be a powerful one."[27]

Much later, in the sixth century A.D., Philoponus, the most devastating ancient critic of Aristotle's views on dynamics, certainly sought to refute those views by appeal to what he represents as empirical as well as to logical considerations.[28] He takes Aristotle's doctrine of natural motion to imply that in motion through the same medium, the times required for the movement will be inversely proportional to the

---

26. Simplicius *In Ph.* 916.15–19: τό τε γὰρ ἀπὸ τῶν κεράμων καταρρέον ὕδωρ, ἐάν τις ἀφ᾽ ὑψηλοῦ τόπου φερόμενον αὐτὸ θεωρῇ, ἄνωθεν μὲν συνεχὲς φαίνεται ῥέον, ἐν δὲ τῷ κάτω διεσπασμένον πίπτει ἐπὶ τὸ ἔδαφος. εἰ οὖν μὴ ἀεὶ τὸν ὕστερον τόπον θᾶττον ἐφέρετο, οὐκ ἄν ποτε συνέβαινεν αὐτῷ τοῦτο.

27. Simplicius *In Ph.* 916.21–24: ἐάν τις λίθον ἢ ἄλλο βάρος ἔχον ἀφῇ ἀποσχὼν τῆς γῆς ὅσον δακτυλιαῖον ὕψος, οὐ πάνυ ποιήσεται ἔνδηλον πληγὴν ἐν τῷ ἐδάφει, ἐὰν δὲ πλέθρον ἢ ἔτι πλέον ἀποσχὼν ἀφῇ ἄνωθεν, ἰσχυρὰν πληγὴν ποιήσει. Elsewhere Simplicius reports other discussions of acceleration in natural motion, particularly those of Hipparchus and of Alexander at *In Cael.* 264.25ff., 265.6ff. Thus Hipparchus argued, against Aristotle, that bodies are heavier the further they are from their natural place and that the slower movement at the beginning of free fall is due to the continuing influence of the power that kept them in position, an idea that has been compared to that of potential forces. Simplicius himself ends his discussion, 266.35ff., by saying that the explanation of acceleration should be investigated further, and he suggests that one might test whether a body is heavier when weighed on the surface of the earth or when weighed "in the air" "from a high tower or tree or sheer precipice" (though someone might reject the results of such a test on the grounds that the difference in weight is imperceptible). Here too, however, at no point are any actual tests, let alone any exact measurements, recorded by Simplicius.

28. That seems clear for instance from the appeal to ἐνάργεια in the text cited below, n. 30. That the actual form of Philoponus' arguments relies more on abstract considerations than empirical evidence has been maintained recently by M. Wolff 1971, pp. 23ff.; cf. 1978, pp. 75ff.

weights or impulses of the moving bodies.[29] To this he then comments: "This is completely false, and this can be established by what is manifestly evident more powerfully than by any sort of demonstration by arguments."[30] He envisages the possibility of a test to refute Aristotle's theory, and yet the imprecision of this, as he goes on to describe it, is remarkable. "For if you let fall at the same time from the same height *two weights that differ by a very large measure,* you will see that the proportion of the times of the motions does not correspond to the proportion of the weights, but that the difference in the times *is a very small one.* So if the weights were not to differ by a very large measure, but the one, for example, were to be double the other, there will be no difference in the times of the movements, or if there is one, it will be imperceptible, although the difference in the weights is by no means such, but the one has the ratio of double the other."[31]

Although elsewhere in his discussion Philoponus occasionally refers to some specific weights, distances, and times for purely illustrative purposes,[32] no attempt is made to report *precise* results of actual tests.

---

29. Philoponus *In Aristotelis Physica commentaria* 683.9ff. It should be noted that Philoponus presents this as an inference, introduced by the words εὔλογον δήπου. Moreover, as Wolff has rightly stressed (1971, pp. 23ff.), here as elsewhere in the *Corollarium de Inani* (*In Ph.* 675.12ff.) Philoponus' *strategic* aim is to refute Aristotle's denial of the possibility of motion on the assumption of a void. The attack on the assumed proportionalities of the natural movements of *different* objects in the *same* medium becomes the basis of an argument against Aristotle's views on the proportionalities of the natural movements of the *same* object in *different* media (*In Ph.* 683.29ff.).

30. Philoponus *In Ph.* 683.16–18: τοῦτο δὲ παντελῶς ἐστι ψεῦδος. καὶ τοῦτο ἔστι πιστώσασθαι κρεῖττον πάσης διὰ λόγων ἀποδείξεως ἐξ αὐτῆς τῆς ἐναργείας. Cf. also 682.29f.

31. *In Ph.* 683.18–25: πολλῷ γὰρ πάνυ μέτρῳ διαφέροντα ἀλλήλων δύο βάρη ἅμα ἀφεὶς ἐκ τοῦ αὐτοῦ ὕψους ὄψει ὅτι οὐχ ἕπεται τῇ ἀναλογίᾳ τῶν βαρῶν ἡ ἀναλογία τοῦ χρόνου τῶν κινήσεων, ἀλλὰ πάνυ ἐλαχίστη τις ἡ διαφορὰ κατὰ τοὺς χρόνους γίνεται, ὡς εἰ μὴ πολλῷ πάνυ μέτρῳ διαφέροιεν ἀλλήλων τὰ βάρη, ἀλλ' οἷον τὸ μὲν διπλάσιον εἴη τὸ δὲ ἥμισυ, οὐδὲ διαφοράν τινα σχήσουσιν οἱ χρόνοι τῶν κινήσεων, ἤ, εἰ καὶ σχήσουσιν, οὐκ αἰσθητὴν ἕξουσι, καίτοι τῶν βαρῶν οὐ τοιαύτην ἐχόντων τὴν διαφοράν, ἀλλὰ διπλασίῳ λόγῳ ἔχοντος τοῦ ἑτέρου πρὸς τὸ ἕτερον.

32. Thus he does so at *In Ph.* 683.12ff. to illustrate the implications of the

Even in these texts, which offer probably the most sustained ancient discussion of dynamical problems, Philoponus is content to show quite generally that the proportions he takes to be implied in Aristotle's statements are wide of the mark—without recording the exact measurements obtained in a series of particular trials (if, indeed, he carried these out).[33]

Yet this should not be said to be just a matter of a conceptual block. A technical factor is certainly at work, though the importance we attach to it may be a matter of some disagreement. The association of movement and time with *number* is already found in Aristotle, who calls time "the number of motion in respect of before and after."[34] Yet in practice neither Aristotle nor anyone else in the ancient world had any means of exactly measuring short intervals of time.[35] The day was divided into hours of variable length, an hour being a proportion of daylight or darkness. Shorter periods were measured by such devices as the water clock or sundial. But even after Ctesibius had introduced an improved constant-flow water clock in the third century B.C., accuracy in measuring short periods to within an interval corresponding

---

principle he attributed to Aristotle. Cf. in other contexts also at *In Ph.* 646.22ff., 677.20ff., 681.17, 30ff.

33. Cf. also his discussion of Aristotle's view of the role of the density of the medium in natural motion at *In Ph.* 647.12ff. At 647.18ff. (cf. 682.30ff.) he represents Aristotle as asserting that the times of the movements will have the same ratio as the density of the media (though Aristotle had put the point hypothetically: see above, n. 10). Philoponus observes that this principle is difficult to refute, because of the difficulty of evaluating the difference in the density of the media (*In Ph.* 683.1ff.).

34. *Ph.* 219b1f. (number here in the sense of what is counted, not that by which we count: cf. 219b5ff., 220b8ff.). Aristotle also points out that we not only measure movement by time, but also time by movement (*Ph.* 220b14ff., 23ff., 223b15ff.).

35. Contrast the situation regarding the evaluation of long intervals of time such as astronomical periodicities. Estimates of the solar year and lunar month began, in Greece, in the mid-fifth century B.C. (cf. below at n. 72) and eventually Ptolemy, drawing on the work not just of earlier Greek astronomers but also of Babylonian ones going back, in some cases, to the eighth century B.C., gave figures for the two main periodicities of each of the planets that are accurate to within 0.002 percent of the modern computed value in every case.

show that in some contexts, at least, it was recognised that gases expand with heat. In Philo, a sphere containing air and hermetically sealed is connected by a bent tube to a vessel containing water. When the sphere becomes hot "on being left in the sun" (though Philo notes that the same effect is obtained as well when the sphere is heated in other ways),[45] the air bubbles out of the vessel, and when the sphere cools, water is drawn back up the bent tube towards and into the sphere. Such an instrument *might* have been adapted to give rough measurements of temperature. Yet neither Philo nor Hero gives any hint that they appreciated this possible application. Their devices (like so many others described in their works) serve merely to illustrate a striking effect.[46] Again when in later writers, such as Galen, we encounter talk of *grades* or *degrees* of hot and cold, wet and dry,[47] this is no more than a *theoretical* elaboration; the grades are not thought to be measurable.[48]

---

matica 2.8 (224.2ff. Schmidt). In both cases Schmidt, in his edition, labelled the devices *thermoscopes*. Cf. A. G. Drachmann 1948, pp. 119ff.

45. Philo *Spir.* 7.474.27ff.

46. Hero himself describes the production of marvellous effects as one of his aims in the *Pneumatica*, in the Introduction to book 1, 2.19f., though he also claims there to set out devices that supply the most necessary wants of human life (2.18ff.: αἱ μὲν ἀναγκαιοτάτας τῷ βίῳ τούτῳ χρείας παρέχου-σαι, αἱ δὲ ἐκπληκτικόν τινα θαυμασμὸν ἐπιδεικνύμεναι). Cf. also Pappus *Collectio* 8.1–2 (3.1022.3ff., 1024.12ff. Hultsch); Proclus *In Euc.* 41.8ff.

47. The theory of the different standards by which hot and cold, hotter and colder, are to be judged is developed in the *De temperamentis* (= *Mixt.*), for example 1.6.21.20ff. Helmreich, 1.542.13ff. Kuhn; 1.8 (H) 29.3ff., (K) 1.554.12ff.; 1.9 (H) 32.24ff., (K) 1.560.13ff., and the terminology of degrees is common in the pharmacological treatises, for example, (K) 11.561.3ff., 571.9ff., 15ff., 739.12ff., 786.11ff., 12.104.18ff., 126.9ff., 16f., 129.15ff., 132.3ff. See especially Harig 1974.

48. Galen refers repeatedly to the ambiguities of hot, cold, wet, and dry (for example, [K] 1.476.8ff.; *Mixt.* 1.6 [H] 19.10ff., [K] 1.538.11ff.), to the difficulties of determining these qualities whether by touch or "by reason" and of discriminating between what merely appears hot, for instance, and what is really hot, whether potentially or in actuality (for example, [K] 1.381.12ff.; *Mixt.* 1.9 [H] 32.5ff., [K] 1.559.10ff., [H] 33.21ff., [K] 1.562.4ff.; 2.2 [H] 51.18ff., [K] 1.590.9ff., [H] 53.14ff., [K] 1.593.7ff.; 2.3 [H] 56.12ff., [K] 1.598.7ff.), and to the lack of any means of measuring them or of determining

Aristotle's own view of the explananda, and his explanations, are both resolutely qualitative. The problems that are to be resolved[49] concern the qualities of perceptible substances—such properties as whether they are or are not capable of solidification, or of being melted, or of being broken—and he is satisfied with an account of compounds that specifies merely which of the simple bodies *predominates* in them.[50] He does not attempt to state the precise proportions of the elements in various compounds,⁻ despite the fact that in an earlier four-element theory, that of Empedocles, some admittedly hesitant steps were taken in that direction.[51] The fourth book of the *Meteorologica* refers often enough not only to a wide variety of compounds but also to such phenomena as evaporation and combustion.[52] Yet there is not a *single*

---

them precisely (for example, *Mixt.* 2.4 [H] 62.25ff., [K] 1.608.13ff., [H] 63.12ff., [K] 1.609.9ff., 10.183.3f., 650.14ff., 11.285.12ff., 544.8ff., 552.13ff., 555.17ff.), although he sometimes writes as if it were possible to attach simple numerical values to deviations from the norm and to the corrections to be made to restore it (for example, [K] 1.383.14ff.). For another ancient theorist who attempted to attach numerical values to qualitative differences, see Philoponus *In Aristotelis libros De generatione et corruptione commentaria*, 170.13ff.; cf. 148.26ff.

49. See, for example, *GC* 2.2.329b7ff., *Mete.* 4.8.384b24ff., 385a10ff. While the authenticity of the fourth book of the *Meteorologica* has often been doubted (see, most recently, Furley 1983a, Strohm 1983), it may still be taken to reflect Aristotle's views on the questions that concern our discussion here.

50. While at *Mete.* 4.7.384a3ff. compounds of water and earth are said to be classifiable according to the amount, πλῆθος, of each, in practice the homoeomerous bodies are grouped into three broad categories; those that are watery; those in which earth predominates; and those that are composed of earth, water, and air (*Mete.* 4.10.389a7ff., 11ff., 19f.).

51. See Empedocles frr. 96 and 98. Interestingly, Aristotle sometimes applauds Empedocles for his anticipations of the notion that the varieties of natural compounds are to be differentiated according to the λόγος of their mixture (for example, *De An.* 410a1ff., *PA* 642a18ff.); yet what Aristotle himself required as an account of the form of the various homoeomerous substances focused less on any statement of the ratio of the component elements than on a (nonquantitative) analysis of their qualitative differentiae.

52. See, for example, *Mete.* 4.9.387a17ff., b10f., 18ff.: among the substances whose combustibility and inflammability are discussed are wood, wool, bone, fat, oil, pitch, wax, wine, stone, ice, and carbuncle (ἄνθραξ). At *Mete.* 2.3.358b16ff., Aristotle asserts that he has discovered by test (πεπειραμένοι)

based his calculation on observations of the shadow cast by a gnomon[60] at noon on the day of the summer solstice at two points on the earth's circumference, namely, Alexandria and Syene, which he assumed to be on the same meridian. At Syene there was said to be no shadow,[61] while at Alexandria there was one of a fiftieth of a circle, i.e., seven and one-fifth degrees. Taking the distance between the two places to be 5,000 stades,[62] Eratosthenes arrived by simple geometry at a figure of 250,000 stades for the circumference of the earth.[63] Then in the first century B.C., Posidonius is reported to have suggested a method based on comparing observations of the star Canopus above the horizon at Rhodes and at Alexandria. Taking these two locations to be 5,000 stades apart on the same meridian[64] and the difference in altitude of Canopus to be "a

---

60. Cleomedes refers to the pointers (γνώμονες) in the hemispherical sundial called the σκάφη, 1.10.98.3ff., 10ff., 22ff., 100.15ff.; cf., e.g., Vitruvius 9.8.1.

61. It was, however, recognised that the lack of a noon shadow at the solstice applied over a distance of 300 stades (Cleomedes 98.4f.), a point that I. Fischer 1975, p. 154, takes to be tantamount to an uncertainty statement about whether Syene lies precisely on the summer tropic: contrast Newton 1980a, p. 383, for whom this is a case of the transformation of a vague observation into a precise statement.

62. Cleomedes does not report how this figure was obtained and, indeed, includes it among Eratosthenes' "suppositions" (ὑποκείσθω, 96.2ff.). Modern conjectures have varied from references to the estimates of professional "bematists," or pacers, used in Egypt (I. Fischer 1975, p. 153), to the use of an already existing map (cf. Rawlins 1982b, pp. 214ff.). It is, however, clear that both this figure and that for the value of the angle of the sun's shadow at Alexandria are—as Neugebauer 1975, vol. 2, p. 653, has remarked—no more than "crude estimates, expressed in convenient round numbers."

63. The figure of 252,000 stades ascribed to Eratosthenes in other sources, e.g., Pliny HN 2.247, Strabo 2.5.7, is generally interpreted as an adjustment motivated by the wish to give a round number for each sixtieth division of the circumference of the circle (see Dicks 1960, p. 146; cf. Heath 1913, p. 339); it also yields a round number for the value of a degree (700 stades), but whether Eratosthenes already used the division of the circle into 360°—as Hipparchus later did—is not certain; see below, n. 76.

64. Again Cleomedes does not say how this figure was obtained. According to Strabo 2.5.24, Eratosthenes distrusted the estimates given for this distance (5,000 or 4,000 stades) on the basis of reports of sailing times and got a figure of 3,750 stades from sundial observations (which, if true, would involve pre-

quarter of a sign" (of the zodiac, i.e., seven and one-half degrees), he obtained a figure of 240,000 stades for the circumference.[65]

Apart from other difficulties relating to the interpretation of these reports, the accuracy of the various recorded estimates of the size of the earth has been the subject of a protracted debate. Yet this has inevitably been quite inconclusive, among other reasons because, although our sources give the figures in stades, we have no certain indication of which of the several different stades used in antiquity is in question on each occasion.[66] The very fact that the stade was not standardised is, of course, significant. Nor is it clear that later estimates always represent an improvement in accuracy over earlier ones, despite the assumption of steady progress that has often been made, on this and other topics, by modern commentators.

The methods used by Eratosthenes and Posidonius are certainly sound enough in principle. But in practice inaccuracies could and did

---

supposing the value of the circumference of the earth and reversing the procedure used in the Syene-Alexandria case): cf. Pliny *HN* 5.132; Neugebauer 1975, vol. 2, p. 653.

65. Cleomedes, 94.22, adds the rider that if the distance between Rhodes and Alexandria is not 5,000 stades, the figure for the circumference will be different but in the same ratio to that distance (see Taisbak 1973–74, who stresses the hypothetical nature of the argument and suggests that Posidonius was more concerned to describe a method than to reach a result; cf. I. Fischer 1975, pp. 161f.). Again, Strabo 2.2.2 ascribes a different figure for the circumference to Posidonius, namely, 180,000 stades, which some have taken as equivalent to 240,000 stades, using a different value for the stade (see below, n. 66), while others have seen it as a revised figure (Heath 1913, pp. 345f.), though Taisbak has recently argued that Strabo's reports are internally inconsistent. Ptolemy in turn adopted 180,000 stades for the circumference and at *Geographia* 7.5.12 claimed that this is based on "the more accurate measurements," but, once again, the value of the stade may not have been the same for Posidonius as for Ptolemy.

66. Following Hultsch 1882, pp. 42ff., Lehmann-Haupt 1929 distinguishes seven different values of the stade; cf. also Dicks 1960, pp. 42ff. At *HN* 2.247 Pliny converts Eratosthenes' stades at eight to the Roman mile, though his further report, at 12.53, that Eratosthenes took the schoenus to be forty stades has been taken to suggest a figure of ten stades to the mile, assuming the schoenus is equivalent to the parasang and so to four miles. Pliny himself, however, while commenting on the different values given to the schoenus, translates forty stades as five Roman miles.

our evidence we continue to have to rely on such sources as Ptolemy, writing much later, in the second century A.D. Ptolemy himself not only reports his predecessors' and contemporaries' observations on many occasions but also provides the first extant comprehensive star catalogue. This is particularly valuable evidence, as the observations it is based on are not subject to interference from planetary models.[77] Books 7 and 8 of the *Syntaxis* give the longitudes and latitudes of over 1,000 stars in degrees and fractions of a degree, using seven simple fractions corresponding to 10', 15', 20', 30', 40', 45', and 50'.[78] Ptolemy tells us that he used the armillary astrolabe for these and other observations, often providing a certain amount of circumstantial detail on this.

Now, whether Ptolemy actually carried out the careful observations he says he made has become, once again, in recent years, the subject of heated controversy;[79] and the suggestion has been revived that his star catalogue in particular was plagiarised from Hipparchus.[80] The view I have argued for elsewhere is that this is an oversimplification, to say the least. Though he has taken Hipparchus' figures as his starting-point[81] (*not* to have done so would have been foolish), he has added

---

See more generally Neugebauer 1975, vol. 1, pp. 274ff., who remarks, p. 277, that in Hipparchus' time a definite system of spherical coordinates for stellar positions did not yet exist; on the question of whether it was Hipparchus or Eratosthenes who first introduced the division of the circle into 360 degrees, see Neugebauer 1975, vol. 1, p. 305 n. 27, vol. 2, p. 590.

77. The theory of the sun is, however, implicated when star positions are determined with reference to it or to the moon.

78. Estimates are also given of the stars' magnitudes, though these are, of course, not based on measurement.

79. See Newton 1973, 1974a, 1974b, 1977, 1980b, Hartner 1977, 1980, Moesgaard 1980b, Gingerich 1980, 1981. References to earlier literature will be found in G. E. R. Lloyd 1979, p. 184 n. 308.

80. The idea that Ptolemy plagiarised an earlier Greek astronomer, Menelaus, was already suggested by Arabic astronomers: see Björnbo 1901; Dreyer 1916–17, pp. 533ff.; Vogt 1925, pp. 37f.

81. Perhaps Newton's most telling argument is based on an analysis of the pattern of error in Ptolemy's catalogue: Newton 1977, pp. 237ff., 1979, pp. 383ff.

stars that were not included by Hipparchus, and where comparisons are possible, these suggest that he has done more than just take over Hipparchus' results and adjust these for precession.[82] However, the ramifications of this controversy need not detain us further at this point, for the simple reason that whoever was chiefly responsible, whether Hipparchus or Ptolemy, the catalogue as we have it is excellent evidence of sustained observations. It reveals both the degree of *precision* aimed at (of the order of 10') and the *accuracy* obtained (the mean error in longitude is of the order of a degree; in latitude, of half a degree).[83]

When we turn to the observations carried out in connection with the determinations of the parameters of astronomical models, the picture is complicated, in Ptolemy's case especially, by that controversy over the issue of the match—or mismatch—between his protestations of a concern for accuracy and his actual practice. Yet, to begin with the protestations, the evidence that both Ptolemy and, before him, Hipparchus were at pains to draw attention to the problems posed by the reliability of the data they had to work with is impressive. Ptolemy often expresses his qualms about the accuracy of some of the observations conducted by earlier astronomers, criticising their rough-and-ready character, and he indicates that Hipparchus already had similar doubts or reservations.[84] They were also alert to the differences in reliability of different kinds of data. Those derived from eclipses or occultations were recognised as more trustworthy than those involving estimates of wide angular distances or of absolute positions. Thus, Hipparchus used lunar eclipse data for his theory of the moon, even though these presupposed, of course, his model for the sun.[85]

Furthermore, both Hipparchus and Ptolemy drew attention to particular sources of inaccuracy in both naked eye and instrumentally

---

82. G. E. R. Lloyd 1979, p. 184; cf. Gingerich 1981, pp. 42f.
83. Cf. Toomer 1984, p. 328 n. 51.
84. See *Syntaxis* 3.1 (H) 1.203.7ff., 14f., 205.15ff.; 7.1 (H) 2.2.22ff., 3.4f.; 9.2 (H) 2.209.5ff.
85. See *Syntaxis* 4.5 (H) 1.294.21ff.; cf. 4.1 (H) 1.265.18ff.

to be quite gratuitous unless they are a response to what *he* perceived to be mismatches between the simple model and *some* empirical data, however and by whomsoever these were obtained.[96] Many of his procedures would be considered sharp practice, as well as slapdash, today— in some cases also, maybe, in his own day. At the same time, there are many contexts in which his practice can be taken to bear out, at least to some extent, his expressed concern over securing a comprehensive and reliable data base.

However hesitant its beginnings, Greek astronomy eventually achieved outstanding successes in developing detailed, quantitative models to account for complex natural phenomena. The mathematical models themselves were rigorous exercises in deductive geometry. But they were evaluated not just as geometry but on how well they matched the data—an essential point we shall return to in Chapter 6.[97] Greek astronomers were certainly neither as active nor as systematic as they might have been in confronting—or in recording the confrontations between—predicted theoretical positions and actual sightings. Yet from Hipparchus onwards, and I should say including Ptolemy, the quality of the data obtainable was a major preoccupation, not just in principle but also in practice. The rigour and exactness of the inquiry were its pride. But the point was not—or was not so much—that astronomy deals with the unchanging heavens, as, more simply, that it is based on mathematics.[98] In particular, the realisation that the exact-

---

96. Cf., e.g., Gingerich 1980, pp. 261f.: "Ptolemy must surely have put credence in some specific observations here, or he would not have ended up with such an unnecessarily complicated mechanism for Mercury."

97. See below, Chap. 6, pp. 304ff., 312ff.

98. In the opening chapter of the first book of the *Syntaxis* (and cf. also *Tetrabiblos* 1.1.3.5ff.) Ptolemy does indeed draw an Aristotelian distinction between the "material and ever-moving quality" that belongs to perishable things below the sphere of the moon, the subject-matter of "physics" (*Syntaxis* 1.1 (H) 1.5.19ff.; cf. 6.14ff.) and the unchangeable form and "aetherial" nature of the eternal heavenly bodies (6.9–11). But it is as mathematics that astronomy is claimed to be superior to "physics": it alone achieves sure and unshakable knowledge because its methods of proof—arithmetic and geometry— are incontrovertible (6.17–21). While it is absolutely clear that Ptolemy has the subject matter of the *Syntaxis* principally in mind, "mathematics" is not

ness and reliability of the data *vary* in different contexts is important, since it shows that there is nothing *automatic* about the accuracy of the data and that the *degree* of accuracy was a matter that had to be evaluated in the given circumstances of each part of the inquiry.

Harmonics

Two other areas of investigation, neither of which is tied to super-lunary phenomena and both of which are regularly hailed as mathematical, will enable us to test the points I have just made. In harmonics there is a long-drawn-out dispute over the status of the perceptible phenomena, where the positions adopted range from an extreme empiricism all the way to the bid to reduce harmonics to pure number theory.[99] How far a particular investigator was committed to a search for exact quantitative data would depend on his position in that overall epistemological controversy. But, as is well known, Plato already knew a tradition in which the *measurement* of the phenomena was fundamental. In the seventh book of the *Republic,* 530dff., Socrates first agrees with a view he ascribes to the Pythagoreans, that harmonics and astronomy are sister sciences, but then he goes on to criticise as "useless labour" the business of *measuring* (ἀναμετροῦντες, 531a2) audible sounds and concords against one another. This contrasts, rather—at least at first sight—with the approval of measurement to deal with certain optical effects expressed later, in *Republic* book 10.[100]

---

here used simply as an alternative name for "astronomy": "mathematics," in this opening chapter, is defined quite generally as the study that is concerned with forms and locomotion, it investigates "shape, quantity, size and again place, time and suchlike" (5.25–6.3) and it belongs to *both* mortal and immortal things alike (6.6ff.). Again at 7.4 and 7.10ff. he explicitly notes how "mathematics" can contribute to, or cooperate with, συνεργεῖν, physics.

99. We shall consider this below, Chap. 6 at nn. 41ff.

100. Unlike *R.* 7.531a–b, where the ultra-empiricists use measurement beyond a point where it is appropriate, the positive aspects of the correct use of measurement are brought out at *R.* 10.602c ff. There Socrates first refers to the effect of distance on apparent size, then to refraction in water, and then to the illusion of depth created by colour contrasts. σκιαγραφία (here, probably, the creation of the illusion of distance within a painting; see further be-

support the doctrine that "all things are numbers."[108] Yet the contrast between the Pythagoreans and the ultra-empiricists shows that Plato had others in mind as well. Here, then, in an admittedly simple case, we can say that empirical inquiries involving measurement were undertaken before Plato—and we can follow their fortunes (not always auspicious fortunes, to be sure) in a long line of writers on harmonics from Aristoxenus down to Ptolemy, Porphyry, and beyond, though—to repeat—the importance attached to such investigations and the status accorded to the information obtained vary from one writer to another.[109] Harmonics is, however, certainly one of the first examples of the successful quantitative explanation of certain qualitative phenomena.

Optics

The evidence we have for the early stages of the development of optics relates mainly to certain purely geometrical aspects of the study.[110] Euclid's own optical treatise first sets out certain assumptions about

---

108. We shall be returning to discuss this doctrine below, pp. 275ff.

109. Thus in Ptolemy the inaccuracies of sense perception are sometimes related to its association with matter, ὕλη: for example, *Harm.* 1.1.3.14ff. D.; cf. Porphyry *In Harm.* 18.1ff. D. But as in astronomy, so also in harmonics, reason can reveal form and order. Ptolemy often, in fact, compares these two sciences, pointing out that both use mathematical methods (e.g., *Harm.* 3.3.94.9–20 D.), describing the aims of the two inquiries in similar terms, as being to "save" certain phenomena or certain hypotheses (*Harm.* 1.2.5.13ff. D., *Syntaxis* 13.2 (H) 2.532.22ff.; cf. below, Chap. 6 n. 35), and claiming that sight and hearing are the highest and most wonderful of the senses (*Harm.* 3.3.93.11ff. D.; cf. 1.2.5.23f. D. λογικώτεραι). In both disciplines, both reason and perception have a role. But that one deals with the heavenly bodies and the other with sounds does not affect the point that what reason reveals, in both cases, are form and order.

110. Aristotle already includes optics, along with harmonics and astronomy, among the "more physical of the mathematical inquiries" (*Ph.* 194a7f.), but the direct evidence for pre-Euclidean work is very limited: cf. Lejeune 1948, Mugler 1957, 1958.

FIGURE I.   After Cohen and Drabkin, edd., 1958, p. 269 n. 1.

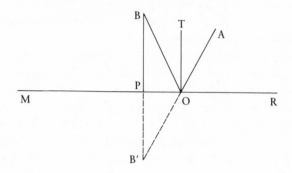

light rays and then proceeds deductively, *more geometrico*.[111] However, empirical tests confirming the laws of reflection are described, for instance, in Ptolemy's *Optics*, although we cannot pinpoint the date of their discovery.

He first sets out the three elementary laws (3.3.88.9ff. Lejeune): (1) objects that are seen in mirrors are seen in the direction of the visual ray that falls on them when reflected by the mirror; (2) things that are seen in mirrors are seen on the perpendicular that falls from the object to the surface of the mirror and is produced; and (3) the position of the reflected ray, from the eye to the mirror and from the mirror to the object, is such that each of its two parts contains the point of reflection and makes equal angles with the perpendicular to the mirror at that point. With reference to Figure 1, where MR is the mirror, A the eye, B the object, B' the image, O the point at which the visual ray

---

111. As with the *Elements*, the textual tradition of Euclid's *Optics* has been subject to much reworking. One of the two extant versions is the result of editing by Theon of Alexandria in the late fourth century A.D., and we cannot assume that the other has escaped similar revision and correction. Both versions are, however, strongly characterised by the deductive geometrical treatment of the problems, even though, as Suppes 1981 has recently stressed, the axiomatisation is, by modern standards, quite incomplete: cf. also Lear 1982, p. 189 n. 36.

strikes the mirror, and TO and BP perpendiculars to the mirror, these three laws state: (1)B′ lies on AO produced; (2) B′lies on BP produced; (3) ∠TOA = ∠TOB. He then provides experimental confirmation of these (3.4–13.89.4ff. Lejeune). [112]

For our purposes the evidence for investigations of refraction is particularly important since, to judge again from Ptolemy, they included not just general discussions of the phenomenon but also measurements of its amount for different pairs of media carried out with apparatus that he is at pains to describe (*Optics* 5.8 [227.5ff. Lejeune]). The tables in *Optics* 5.11, 18, and 21 (229.1ff., 233.10ff., 236.4ff. L) setting out the amount of refraction for angles of incidence at 10° intervals from 10° to 80° first for air to water, then for air to glass, and then for water to glass, are remarkable from several points of view. They provide one of the clearest cases of an ancient scientist doctoring his results. Ptolemy has evidently adjusted these to fit his general law, even though that law itself is not stated. This takes the form $r = ai - bi^2$, where r is the angle of refraction, i the angle of incidence (the incident ray being that from the eye to the refracting surface), and a and b constants for the media concerned. [113] Nevertheless, the complexities of that general law are *quite unmotivated* unless Ptolemy has made *some* (and we may believe quite extensive) [114] empirical investigations involv-

---

112. For discussion of these laws and of their pre-Ptolemaic background, see Boyer 1945–46; Lejeune 1946, 1957, pp. 47ff.

113. The equation stated is one formulated by Govi 1885, p. xxxiii, but as Lejeune 1940–46, pp. 97ff., noted, it does not depend on the use of an algebraic expression, for Ptolemy could easily have set out the relationship in words; moreover, the use of first and second differences had been standard in other contexts, such as astronomy, since Babylonian times. A. M. Smith 1982, p. 237, has, however, recently argued against the view that Ptolemy had any such general theory, and indeed, that he was unable to find the law of second differences for both mathematical and methodological reasons. Yet in view of the fact that every one of the results in all three tables tallies exactly with this law and that they do so even where there are notable discrepancies between them and what the application of the sine law would give, this must be thought to make this highly unlikely.

114. See Lejeune 1940–46, p. 94. A. M. Smith 1982, p. 234, also assumes that Ptolemy had observational data with a high degree of accuracy.

ing the measurement of angles of incidence and of refraction for these media—even if those investigations did not yield quite the results that were claimed.[115]

Weighing

So far I have concentrated exclusively on the exact sciences. But one simple measuring technique used in a wide variety of contexts was *weighing*,[116] and this will now take us further afield, including into what we call the life sciences. Heavy and light were often cited as, or among, the differentiae of natural substances in both physics and physiology, but we must be careful, since they are sometimes understood in purely qualitative terms, on a par with wet and dry, or sweet, salty, and bitter.[117] Thus when in certain contexts in his mineralogical work *On Stones* Theophrastus differentiates varieties of pumice or of

---

115. Lejeune 1940–46, p. 97, suggests that the second difference was applied to the middle range in the tables and that Ptolemy extrapolated from these to the (generally less accurate) results claimed for the extreme cases of angles of incidence for 10° and (especially) 80°. On the other hand, provided we assume, as in the cases Ptolemy discusses, that the incident ray from the eye passes from the less dense to the denser medium, the generalisations he sets out at 5.34.245.1ff. L, are unobjectionable, namely, that where i' is greater than i, (1) $i':i > r':r$, (2) $i':r' > i:r$, and (3) $(i'-i):i > (r'-r):r$.

116. Written evidence for standardised weights in Greece goes back to the Mycenaean period: see Chadwick 1973, pp. 54–58. Moreover, the archaeological record provides evidence for the standardisation of weights in the ancient Near East and the Indus valley at a much earlier date: see Hemmy 1931; F. G. Skinner 1954, pp. 779ff.

117. Thus at *GC* 329b18ff. (cf., e.g., *GC* 326a7f., 329a10ff.), Aristotle lists heavy and light along with hot and cold, dry and wet, hard and soft, viscous and brittle, rough and smooth, dense and rare, among the tangible contrarieties. Moreover, when taken as definable in terms of a natural tendency to move in a certain direction (up/down), heavy and light run counter to the stipulation, in the *Categories* 5b11ff., that quantities have no contraries. However, at *Metaph.* 1052b18–31 Aristotle includes weight with length, breadth, depth, and speed among examples of what can be measured, where it meets the criterion that there must be units or standards of measurement by which weight can be determined.

metal-bearing ore by "heaviness,"[118] no actual quantities are mentioned. "Pumices," he says, "differ from one another in colour, density,
and heaviness. They differ in colour inasmuch as the pumice from the
Sicilian lava-flow is black, while in density and heaviness it is quite like
a millstone. For pumice of this kind does indeed exist, heavy and dense
and more valuable in use than the other kind. This pumice from the
lava-flow is a better abrasive than the kind which is light [in weight]
and white in colour, although that which comes actually from the sea
is the best abrasive of all."[119]

Elsewhere, however, direct reference is made to *weighing* to distinguish heavier and lighter kinds of the same substance. The Hippocratic
treatise *On Airs Waters Places* is much preoccupied with the differences in the waters that occur in different places, distinguishing those
that are "hot" and "cold," "hard" and "soft," stagnant and free-
running, turbid and pure and bright, as well as—frequently—those
that are "heavy" and those that are "light."[120] The opening chapter

---

118. The usual Greek term, often translated "weight" in English, is, as here
in Theophrastus, βάρος, which, like βαρύτης, is cognate with the adjective
βαρύς, heavy.

119. Theophrastus *Lap.* 22. Cf. *Lap.* 39: "There are also many kinds
of stones extracted from mines. Of these some contain gold and silver,
though only the silver is clearly perceptible: they are rather heavy and strong-
smelling. . . . There is also another stone like charcoal in colour, but heavy."
At *Lap.* 46 the quantities of metals in gold alloys are said to be determinable
by the use of the touchstone.

120. See especially *Aër.* 7, *CMG* 1.1.2.34.16–40.6, and 8, *CMG*
1.1.2.40.7–44.2, with references to "light" waters, for example, at 38.8, 22,
40.8. In *Aër* 8, *CMG* 1.1.2.42.15ff., the writer refers to measuring quantities
in a test that he proposes to show that freezing causes the "lightest" part of
fluids to disappear: "Waters from snow and ice are all harmful because, once
they have frozen, they never return to their original nature. The bright, light,
and sweet part is frozen out and disappears, and the muddiest and heaviest
(σταθμωδέστατον) part is left. You may recognise this as follows, if you wish,
by pouring a measured amount of water into a jar and leaving it in the open, in
winter, where it may freeze best. Then the next day bring it into the warmth,
where the ice may melt most easily, and when it has dissolved, measure it (ἀνα-
μετρέων) again: you will find the quantity considerably less. This is evidence
that the lightest and finest part of the water disappears and is dried up in the
freezing, not the heaviest and thickest part, for it could not be so affected. For

suggests that here we are dealing not just with vague general impressions, but with something measurable, for there we are told that waters "vary both in taste and 'on the balance.'"[121]

Measurement is also clearly involved in Archimedes' famous hydrostatical investigations. The story of how he detected the adulteration of

---

this reason I consider these waters—from snow, ice, and related waters—to be most harmful for all purposes."

The interpretation of this passage occasioned a notable exchange between Cornford and Vlastos. Cornford 1952, pp. 6f., used this text, among others, to argue for a major distinction between the methods used by the doctors and those of the pre-Socratic natural philosophers. He contrasted Anaximenes, who had suggested a correlation between differences in temperature and differences in density (see fr. 1), but who might—Cornford claimed—have been led to revise his theory if he had "set a jar full of water outside his door on a frosty night and found it split in the morning." "It is significant," Cornford wrote after citing *Aër.* 8, "that the experiment should have been recorded by a doctor, and entirely neglected by the natural philosopher." To this Vlastos, for one, protested, not only on the grounds that Anaximenes' theory was not such as to be directly refutable by appeal to obvious facts, but also drawing attention to some of the "oddities" of the "experiment" of *Aër.* 8 (Vlastos 1955/1970, pp. 43ff. and 52). To establish his theory that freezing causes the "heavy" parts of the water to be separated out, what the Hippocratic writer needed to do—Vlastos suggested—was to measure the bulk of the frozen water, not that of the water after it was melted. Moreover, more damagingly, the observation that the quantity of water, on thawing, was less did nothing to confirm the theory the writer was advancing concerning the disappearance of the lightest and finest part of the water and "such loss of water as he observed . . . would be obviously due to evaporation."

As a testing procedure, the investigation suggested is evidently quite inconclusive: as so often elsewhere in ancient writers, the description of the results presupposes the theory it purports to be testing. Furthermore, although the reference to measuring shows some recognition of the possibility of verifying a general impression by a quantitative investigation, the writer does not, in fact, record any actual quantities, here or elsewhere.

121. *Aër.* 1, *CMG* 1.1.2.24.9: ὥσπερ γὰρ ἐν τῷ στόματι διαφέρουσι καὶ ἐν τῷ σταθμῷ. Contrast *Aph.* 5.26 (L) 4.542.1f. (cf. *Epid.* 2.2.11 [L] 5.88.15ff.), for example, which suggests much more vaguely that water that is heated quickly and cools quickly is the "lightest." Pliny, at *HN* 31.32, is one who remarks on the difficulties of determining which type of water is "lightest," noting that the differences in weight between one kind of water and another are very small (*nullo paene momento ponderis*). Yet in one context, at least, the difference in buoyancy between salt water and fresh was well known.

a gold crown by observing that it displaced more water than the equivalent weight of pure gold may well be inaccurate in the form in which we have it from Vitruvius.[122] But the extant treatise *On Floating Bodies* shows that he had a clear working conception of—even if he does not explicitly formulate—what we call specific gravity.[123] In book 1, chapters 3 ff., he distinguishes between solids that are "equal in weight" (ἰσοβαρέοντα) with a given fluid, those that are "heavier" and those that are "lighter" than it, where he clearly has in mind not absolute weight but weight in relation to a given volume,[124] and in chapter 7 he enunciates the principle since named after him: "solids heavier than the fluid will, if placed in the fluid, be carried down to the bottom of the fluid, and they will be lighter in the fluid by the weight of the amount of fluid that has the same volume as the solid."[125]

Further evidence from the medical writers shows that they referred readily enough to weighing and measuring in particular contexts. For instance, in their pharmacology, the proportions of the ingredients in compound drugs, and the dose to be used, are often—though certainly far from invariably—specified by weight or otherwise by exact quantity, that is, by dry or liquid measure.[126] Thus *On the Diseases of*

Aristotle notes the effect of this on the seaworthiness of laden cargo ships, and though he puts this down primarily to the density of salt water he also remarks that it weighs more than fresh (*Mete.* 2.359a5–11; at 11–14 he proposes a test to illustrate buoyancy, namely, adding salt to water to make eggs float in it). Cf. also *Problemata* 23.3.931b9ff., 20, 933b21ff. and 38, 935b17ff.

122. Vitruvius 9 praef. 9ff.

123. The Arabic writer Al Khazini ascribes to Archimedes a device that could be used to determine relative specific gravities of different metals when weighed in water, in the *Book of the Balance of Wisdom* 4.1, on which see, for example, Knorr 1982b.

124. *De corporibus fluitantibus* (*Fluit.*) 1.3ff. (HS) 2.320.32ff.

125. *Fluit.* 1.7 (HS) 2.332.21ff.: τὰ βαρύτερα τοῦ ὑγροῦ ἀφεθέντα εἰς τὸ ὑγρὸν οἰσεῖται κάτω, ἔστ' ἂν καταβάντι, καὶ ἐσσοῦνται κουφότερα ἐν τῷ ὑγρῷ τοσοῦτον, ὅσον ἔχει τὸ βάρος τοῦ ὑγροῦ τοῦ ταλικούτον ὄγκον ἔχοντος, ἁλίκος ἐστὶν ὁ τοῦ στερεοῦ μεγέθεος ὄγκος.

126. Already much earlier in Egyptian pharmacology, quantities are sometimes specified (as, for example, in para. 2 of the Papyrus Ebers: "to expel diseases in the belly: Another [remedy] for the belly, when it is ill: cumin ½ ro,

Women book 1 gives this prescription to promote parturition: "one obol of dittany, one obol of myrrh, two obols of anis, one obol of nitre: pound these till they are smooth, pour on them a cyathus of sweet wine and two cyathi of hot water; give to the patient to drink and wash her in warm water." [127] Many similar examples could be given—though

---

goosefat 4 ro, milk 20 ro, are boiled, strained and taken. Another: figs 4 ro, sebesten 4 ro, sweet beer 20 ro, likewise" [Ebbell 1937, p. 30]), though this is not invariably the case (cf. para. 3 of the Papyrus Ebers: "another: wine, honey, colocynth, are strained and taken in one day" [Ebbell 1937, p. 31]). F. L. Griffith 1898, pp. 5ff., commenting on the prescriptions in the Petrie papyri, noted that the "quantities to be used are often left to the discretion of the practitioner to determine; but where necessary the amount is specified, though in round terms, by measure and not by weight," and he went on to argue that "a great advance was made when weight was substituted for measure, as in the Greek medical works." As we shall see, however, there is still plenty of indeterminacy in Hippocratic prescriptions too, as well as in those of later periods. On the measures used in the Ebers Papyrus, see Ebers 1890; for a comparison between Greek and ancient Near Eastern pharmacological recipes, see Goltz 1974, Harig 1975, 1977, 1980, Harig and Kollesch 1977. On the possibility of the deliberate withholding of information concerning quantities for reasons of secrecy, see, for example, Goody 1977, pp. 137f.

127. *Mul.* 1.77 (L) 8.170.14ff. Cf., among many other texts where weights or liquid measures are specified, *Mul.* 1.74 (L) 8.156.15ff., 1.75 (L) 8.164.7ff., 168.7ff., 1.77 (L) 8.170.9f., 1.78 (L) 8.176.3ff., 18ff., *Mul.* 2.119 (L) 8.258.23ff., 260.2ff., 2.172 (L) 8.352.19ff. From outside the gynaecological works similar prescriptions that specify weights or measures appear, for example, at *Morb.* 3.17, CMG 1.2.3.96.19ff., 22ff., 27ff., 98.2f., 9ff.; *Salubr.* 5 (*Nat.Hom.* 20), CMG 1.1.3.212.16ff.; *Int.* 23 (L) 7.226.13ff., 26 (L) 7.234.15ff., 31 (L) 7.248.9ff.; cf. 20 (L) 7.216.22f., which specifies that there should be a measure (μέτρον) for each of the ingredients in a particular prescription for a clyster, though the text had just referred to taking an amount of one ingredient "as big as a sheep's knucklebone."

Later pharmacologists too exhibit a similar general pattern, of the common, but far from invariable, specification of weights and measures. They are indicated far more often than not in, for example, Celsus *Med.* 5.18–25, CML 1.194.31–215.3, though less systematically in, for instance, Dioscorides *De materia medica* (but see 4.69 [2.228.2ff. Wellmann], 72 [2.231.3ff. W.], 73 [2.232.12ff. W.], 75 [2.235.10ff. W.]). Galen sometimes demands that pharmacology should be mastered "exactly," ἀκριβῶς, for example at (K) 10.180.9ff., though he often acknowledges the element of conjecture, e.g., (K) 10.209.4ff., CMG 5.9.1.197.6ff., (K) 15.585.6ff., putting it at (K) 11.285.10off. (cf. 293.13ff., 294.12f.) that nothing shows so clearly that medicine is in prac-

so too can others where the quantity of one or more of the ingredients, or the dose, is *not* specified exactly,[128] and after the Hippocratics, references to the problems of the standardisation of weights and measures and of correlating those used in different parts of the Greco-Roman world appear in the pharmacological sections of such writers as Celsus, Scribonius Largus, and Galen,[129] while tables of weights and measures begin to become common in specialist metrological writings.[130]

---

tice a matter of guesswork as the question of the quantity of each remedy. Sextus too, in his account of the seventh sceptical mode, on quantities, *P.* 1.129ff., suggests at *P.* 1.133 that the mixture of compound drugs must be exact, πρὸς ἀκρίβειαν, and that a slight oversight in weighing in their preparation can make them harmful and poisonous; elsewhere, at *M.* 7.35ff., in his attack on the notion of the "criterion," he uses weighing to illustrate the threefold distinction among (1) the agent, (2) the means, and (3) the application of the means.

128. Thus in the prescription to promote conception at *Mul.* 1.75 (L) 8.164.17ff., five ingredients are named but no quantities specified: they are simply to be ground together and applied as a pessary. Elsewhere prescriptions specify an "equal amount" but do not indicate whether this is by weight or by volume, e.g., *Mul.* 1.74 (L) 8.154.15f.; *Morb.* 2.54 (L) 7.82.21ff.; or they offer such rough-and-ready directions as to quantity as "the size of a bean," ὅσον κύαμον, e.g., *Mul.* 1.74 (L) 8.156.9, 12f., "the size of a knucklebone," ὅσον ἀστράγαλον, e.g., *Mul.* 1.74 (L) 8.154.19, sometimes specified as a sheep's knucklebone, or a deer's, e.g., *Int.* 20 (L) 7.216.20; 23 (L) 7.226.14f., 31 (L) 7.248.10; or a "handful," χεῖρα πλέην, *Morb.* 3.17, CMG 1.2.3.98.7, cf. *Mul.* 1.75 (L) 8.162.8, or they speak merely of a "little" of some ingredient, e.g, *Mul.* 1.75 (L) 8.170.4. On other occasions the quantities of ingredients are allowed to vary ("the whites of three or four eggs," *Morb.* 3.17, CMG 1.2.3.98.12), or their method of preparation is ("boil twice or three times in a chous of water," *Morb.* 3.17, CMG 1.2.3.98.15f.), and very often the dose is not specified, nor whether nor how often it is to be repeated: the references to "for five days" at *Mul.* 1.78 (L) 8.174.19f., and "twice a day" at *Mul.* 2.192 (L) 8.372.4, are the exception rather than the rule in those chapters and in general, while in the instruction to take parts of an ingredient "by threes" at *Mul.* 1.75 (L) 8.162.11, for instance, the influence of symbolic factors cannot be ruled out (cf. Aristotle *Metaph.* 1092b28ff.).

129. See, for example, Celsus *Med.* 5.17.1c, CML 1.194.5ff.; Scribonius Largus praef. 15.5.23ff.; Galen (K) 13.435.1ff., 616.1ff., 789.2ff., 893.4ff. Cf. also Pliny *HN* 21.185 (though at 22.117–18 Pliny says that it is not possible to weigh out the powers of drugs "scruple by scruple," and at 29.24f. remarks that Mithridates' antidote that contains fifty-four ingredients no two of which have the same weight is clearly the product of ostentatious boasting).

130. The remains of Greek and Roman metrological writings have been collected by Hultsch 1864 and 1866. The treatise devoted to weights and mea-

A twofold contrast suggests itself. On the one hand, the simpler notion, found already in Empedocles' element theory,[131] that a compound consists of certain *proportions* of the constituent substances may be contrasted with the more precise idea that the quantities of the constituents are to be determined *by weight*.[132] Yet on the other, despite the progress made towards exact quantitative specification, that progress was still very incomplete. Moreover, quantitative specification when we find it—even when all the relevant quantities are stated—was often no more than window-dressing.

In interpreting this evidence we have to bear in mind, first, that the ingredients used are not chemically pure substances, and, secondly, that ancient doctors are frequently urged to modify the drug and the dosages *in relation to particular patients*.[133] Thirdly, as we noted in Chapter 3, some early medical writers insist that medicine, though a genuine *techne*, art or skill, cannot be made an *exact* study,[134] and

---

sures in the Galenic corpus, (K) 19.748.1–781.3, is spurious, as is some of the corresponding material in the works of Hero: see Hultsch 1882, pp. 7ff.

131. Empedocles frr. 96 and 98.

132. Apart from in the pharmacological contexts we have considered, the specification of the weights and measures of ingredients is common also in the extant Greek chemical and alchemical texts. See, for example, from the Leyden Papyrus X, pagina 1a.21ff. and 25ff. (Leemans 1885, p. 205); pag. 8a.28ff. (p. 225); pag. 11a.8ff. (p. 233); 24ff. (p. 235); Halleux 1981, nos. 4, 5, 56, 81, and 83; and Berthelot and Ruelle 1888, part 1.13.10ff., 2.31.7ff., part 4.19.1ff., 2.285.6ff. Cf. also Preisendanz 1973–74, P. 12.193ff., 2 p. 71. Although the reactions of various natural substances to fire were often remarked on, for example by Aristotle (cf. above, n. 52) and by Theophrastus, especially *Lap.* 9–17, no ancient scientist thought to make systematic observations of the weights of substances before and after combustion. Vitruvius 2.5.3, however, does note that in the manufacture of quicklime "about a third" of the weight of the stone is lost.

133. See, for example, *Vict.* 1.2 (L) 6.470.7, 14ff.; *Mul.* 2.192 (L) 8.372.7ff.; cf., e.g., Pliny *HN* 25.150. Alternatively the dose is to be modified in accordance with the strength of the disease, as, e.g., at *Mul.* 1.78 (L) 8.184.17. It may also be noted that the problem of the identification of the active ingredients in compound drugs is further complicated when beliefs about their interactions, including their "sympathies" and "antipathies," have to be taken into account: cf. Pliny *HN* 22.106. Cf. Müri 1950, p. 189; Harig 1974, pp. 64ff., 83ff., 133ff., and 1980.

134. Cf. above, Chap. 3 at nn. 89–103, on texts in *VM, Morb.* 1, *Loc.Hom.* and *Vict.* 3, especially.

some object specifically to appeals to such a procedure as weighing. When the writer of *On Ancient Medicine* protests that exactness in the control of diet is difficult to achieve, he says that "one should aim at some measure," [135] but he then goes on: "but as a measure you will find *neither number nor weight* by referring to which you will know what is exact, and no other measure than the feeling of the body." [136] The treatise *On Sterile Women,* too, writes that treatment should be adapted to the particular patient, having regard to her condition and strength, which are *not* a matter of *weighing, τούτων γὰρ οὐδεὶς σταθμός ἐστιν.* [137] The question of when it is appropriate to have re-course to weighing was, in fact, a matter of dispute, for some writers were for making medicine exact, or at least for representing it as such, [138] while others were suspicious of attempts to do so and critical of what I have just called window-dressing. Nevertheless, *some* refer-ence to weighing and measuring in pharmacological contexts is com-mon enough, even if often the concern is not so much with exact for-mulae as with the proportionalities between the "strength" of the drug and that of the patient.

To these pharmacological cases we can add an admittedly limited number of other examples from medical writers at different periods where quantitative reasoning is in play in various physiological or pathological contexts. In the general description of the climatic and

---

135. *VM* 9, CMG 1.1.41.19f.: δεῖ γὰρ μέτρου τινὸς στοχάσασθαι. It is relevant to recall that μέτρον may signify not just measure, but the due mea-sure or mean.

136. *VM* 9, CMG 1.1.41.20–22: μέτρον δὲ οὔτε ἀριθμὸν οὔτε σταθμὸν ἄλλον, πρὸς ὃ ἀναφέρων εἴσῃ τὸ ἀκριβές, οὐκ ἂν εὔροις ἀλλ᾽ ἢ τοῦ σώματος τὴν αἴσθησιν. Yet later, *VM* 19, CMG 1.1.50.25f., this treatise still makes use of the idea of numerable periods and crises in diseases. The notion that there is no measure, μέτρον, nor balance, σταθμός, by which health may be deter-mined recurs, for example, in Galen *Mixt.* 2.4 (H) 62.25ff., (K) 1.608.13ff.; (H) 63.12ff., (K) 1.609.9ff.; cf. (K) 11.171.4ff., 151.17ff., in the context of venesection; while the idea of medicine as a stochastic art is widely used by many others besides medical writers, cf. above, Chap. 3 n. 217.

137. *Steril.* 230 (L) 8.444.1f.

138. Cf. above, Chap. 3 at nn. 26ff., on the dogmatic claims to certain knowledge in such treatises as *De arte,* and below at nn. 150ff. on Hippocratic numerology.

epidemiological conditions encountered that is set out in the Constitution in *Epidemics* book 3, it is remarked, at one point, that the urine discharged was out of proportion to the fluid drunk, though here no specific quantities are mentioned.[139] In one of the case-histories in *Epidemics* book 7, however, we are told that a patient discharged more than a chous[140] of fresh blood in his stool and then, after a short while, a further third of a chous of coagulated globlets.[141] Specifications of the quantities of the lochial discharge or of the menses are also sometimes given in the gynaecological and the embryological treatises—though in several cases the quantities reported appear fanciful.[142]

Then Erasistratus, in a remarkable experiment recorded in Anonymus Londinensis,[143] tried to prove that animals emit invisible effluvia, by keeping a bird in a closed vessel without food for a period and then weighing the bird and its visible excreta. Comparing this with the original weight, he found, we are told, that there had been a "great loss of weight"—another case where, in our source at any rate, an observed *difference* in weight is remarked without any *actual* weights being reported.[144]

---

139. *Epid.* 3.10 (L) 3.90.7f., cf., e.g., *Morb.* 4.42 (L) 7.564.4ff.

140. A *chous* is estimated as between 2.52 and 3.96 litres in OCD².

141. *Epid.* 7.10 (L) 5.380.20ff.; cf., e.g., 7.3 (L) 5.370.23ff., 372.1ff., where the exceptional quantities of milk consumed by a particular patient are specified; *Epid.* 5.14 (L) 5.214.1ff., 5.18 (L) 5.218.10; 5.50 (L) 5.236.16.

142. See *Mul.* 1.6 (L) 8.30.8ff.: menses of two Attic cotylae "or a little more or less," i.e., c. 0.45 litres (cf. Aristotle, who claimed generally that female humans produce more menses than any other animal, e.g., *HA* 521a26f., and estimated the discharge of a cow in heat as "about half a cotyle or a little less," *HA* 573a5ff.; and contrast Soranus *Gyn.* 1.20, *CMG* 4.14.4, who gives a maximum figure for menstruation as two cotylae but who then devotes two chapters to pointing out how the quantity and duration may vary, 1.21–22, *CMG* 4.14.6ff., 15.1ff.). *Mul.* 1.72 (L) 8.152.3ff., *Nat.Puer.* 18 (L) 7.502.3ff.: the lochial discharge is one and a half Attic cotylae "at first" "or a little more" (*Nat.Puer.* adds "or a little less"). For discussion of these figures, see Bourgey 1953, p. 178 and n. 2; R. Joly 1970, p. 62 n. 2; Lonie 1981a, pp. 190ff.

143. *Anon. Lond.* 33.43ff.; see von Staden 1975, pp. 179ff., and forthcoming. Further tests involving the weighing of fresh and "high" meat, and of a bladder empty and full of air, are reported in other contexts in *Anon. Lond.* at 31.10ff., 34ff. (purporting to present an Empiricist view), 32.22ff.

144. It appears from a report in Galen *UP* 7.8 (H) 1.392.25ff., (K)

Galen, especially, uses quantitative arguments on several occasions. In *On the Use of Parts* he remarks generally on the proportionalities between the fluids and solids taken into the body and those discharged or lost,[145] and elsewhere he specifies actual amounts of, for example, pus expectorated.[146] In *On the Natural Faculties* the difference in size between, on the one hand, the vena cava (together with the right auricle) and, on the other, the pulmonary artery is cited among the arguments to support the conclusion that some blood must pass directly from the right ventricle to the left through invisible pores in the septum, though—unlike Harvey—Galen does not attempt to measure the quantities or flow of blood exactly or even approximately.[147] Most notably of all, perhaps, a quantitative argument is adduced in the refutation of Lycus' view that urine is the residue from the nourishment of the kidneys.[148] That cannot be the case, Galen claims, if one considers the amounts discharged, which in exceptional cases may be as much as three or four choes.[149] If that is produced from nourishing the

---

3.540.8ff., that Erasistratus attempted to distinguish between different types of "air" by their "thinness" and "thickness," claiming that the air from burning coals is "thinner" than "pure" air, but Galen records no measurement in this connection.

145. See *UP* 4.13 (H) 1.223.10ff., (K) 3.304.7ff. (where the quantity of drink consumed is proportional to the urine discharged), and *UP* 16.14 (H) 2.433.4ff., (K) 4.340.2ff. (where the nourishment taken in is equal to the material lost from the body).

146. E.g., (K) 8.321.15ff. Cf. also (K) 11.227.9ff., blood expectorated up to two cotylae.

147. *De naturalibus facultatibus* (*Nat.Fac.*) 3.15 (H) 3.252.13ff., (K) 2.208.11ff. Cf. *UP* 6.17 (H) 1.362.7ff., (K) 3.497.9ff., where Galen reverses the explanation, putting it that there is good reason for the vena cava to be larger than the pulmonary artery, since blood is taken over from the right ventricle to the left through the interventricular pores.

148. *Nat.Fac.* 1.17 (H) 3.152.17ff., (K) 2.70.10ff., on which see Temkin 1961, and cf. Temkin 1973, pp. 153f.

149. *Nat.Fac.* 1.17 (H) 3.153.23ff., (K) 2.72.4ff.; cf. also (H) 3.153.13ff., (K) 2.71.12ff. Altman and Dittmer 1972–74, vol. 3, p. 1496, give a normal figure, for a 70 kg body, of 1.4 litres, with upper and lower limits of 2.94 and 0.49 litres. Galen's "three choes" is clearly more than five times the normal figure and more than twice the upper limit.

kidneys, one would expect even greater amounts of residue from the nourishment of the other principal viscera, where there is no sign of this.

Counting

Exactness in the medical writers is sometimes a matter not of weighing or measuring, but of *counting*.[150] Great importance is attached by many Hippocratic authors to the study of numerical relationships in connection with the determination of periodicities, notably in two types of context: (1) pregnancy and childbirth; and (2) the phases of diseases, especially their "crises," the points at which exacerbations or remissions are to be expected. In both contexts some of the ideas expressed have a solid basis. The normal time of gestation in humans is fixed to within fairly well-defined limits.[151] Before the advent of antibiotics, studies were carried out that went to show that certain acute conditions such as certain pneumonias and malaria manifest quite marked periodicities.[152] In both fields, however, the proposals about periods and relations made in some Hippocratic texts go far beyond the range of what could be justified fairly straightforwardly by appeals to readily accessible evidence. Here the search for exactness led not to Koyré's "universe of precision" but to spurious quantification and ad hoc numerological elaboration.[153]

---

150. The relationship between measuring and counting is discussed by Aristotle at, for example, *Metaph.* 1052b18ff., 1088a4–11, *Ph.* 220b18ff.: normally, counting is deemed a kind of measuring, but at *Metaph.* 1020a8ff. the two are contrasted where he distinguishes numbering quantities constituted by discontinuous parts and measuring magnitudes that are continua.

151. Altman and Dittmer 1972–74, vol. 1, pp. 137f., specify a range of 253 to 303 days for humans and give corresponding figures for various other species of animals. Apart from in the medical writers, an interest in the topic is shown by Aristotle, who represents humans as exceptional among animals in the variation shown in the times of gestation of viable infants, e.g., *HA* 584a33ff., *GA* 772b7ff., and cf. *Problemata* 10.41.895a24ff.

152. See, for example, Musser and Norris, cited by Osler 1947, pp. 49f., on pneumococcus lobar pneumonia, and Osler 1947, p. 491, on malaria.

153. Aspects of this question have been discussed by Lichtenthaeler 1963,

Number lore in Greek medicine must be interpreted in part against a background of Pythagorean beliefs, not just the general doctrine that "all things are numbers" but also more particular ideas concerning, for example, the importance of odd and even numbers and the correlation of that pair with other pairs of opposites. Odd is associated with right, male, and good, and even with left, female, and bad in the Table of Opposites reported by Aristotle, and we have other evidence that suggests that above/below, front/back, and other pairs were also sometimes incorporated into similar schemata.[154]

Yet the patterns of beliefs to which the medical theories we are interested in can be related include much besides Pythagoreanism. Many of the ideas attributed to the Pythagoreans are, in any case, widespread in popular belief. The positive and negative associations of some of the pairs of opposites included in the συστοιχία certainly antedate Pythagoras.[155] The classification of numbers into odd and even is general throughout Greek arithmetic. The idea that the days of the month may be good or evil can be traced back to Hesiod.[156] Among other aspects of number lore, the idea of the special significance of the number seven occurs in sources before Pythagoras, notably in a famous poem of Solon's, not to mention the more controversial question of possible non-Greek influences dating from earlier still.[157]

---

pp. 109ff.; R. Joly 1966, pp. 108ff.; Heinimann 1975; and Kudlien 1980, especially.

154. Aristotle reports the Pythagorean συστοιχία at *Metaph.* 986a22ff., and although at *Cael.* 285a10ff. he says that the Pythagoreans neglected the pairs up/down and front/back, Simplicius quotes from Aristotle's own lost work on the Pythagoreans a passage that suggests that they correlated those pairs with right/left and good/bad: *In Cael.* 386.20ff., cf. 392.16ff.; cf. Burkert 1972, pp. 51f., 294f.

155. Cf G. E. R. Lloyd 1966, part 1, especially pp. 41ff.

156. Hesiod *Op.* 765ff., 822ff.

157. Solon 19 Diehl, and cf. below at n. 208, on Aristotle's criticism of farfetched theories correlating sevens. On the provenance of the ideas set out in the Hippocratic treatise *Hebd.* and on the date of that work, see Mansfeld 1971 with references to earlier literature. On the question of Near Eastern parallels to, and possible influence on, Greek ideas here, see Roscher 1904, pp. 85ff., 1906, 1911, p. 10 n. 9; Götze 1923; Reitzenstein and Schaeder 1926; Kranz 1938b; Mansfeld 1971, pp. 21ff. and 65; Burkert 1972, pp. 468ff.

We must recognise at the outset, therefore, that the pattern of beliefs against which Hippocratic numerological ideas are to be judged is complex. Moreover, those ideas themselves are extraordinarily heterogeneous. We may begin with some of those connected with pregnancy and childbirth. It would, of course, be futile to attempt to determine at what stage the Greeks were aware of the approximate time of gestation of the human embryo. When we reach the classical period, the view that babies born in the seventh, ninth, or tenth month are viable, whereas those born in the eighth month generally die, is widespread.[158] But many other beliefs in periods and relations are also found. Thus the idea that the male embryo moves first in the third month, the female in the fourth, appears in the gynaecological treatises.[159] *On the Nature of the Child,* to which we shall be returning, states that the male foetus takes thirty days at most to form, the female forty-two days, and also maintains that the lochial discharge lasts thirty days for a boy and forty-two days for a girl, a view also expressed in *On the Diseases of Women* book 1.[160] *On Sevens,* a treatise of admittedly doubtful date, claims that the human seed is "set" in seven days,[161] and *On Fleshes* states that it takes seven days for the embryo to acquire all its parts and elsewhere develops other theories of periodicities based on the number seven.[162] *On Regimen* puts forward an obscure theory

---

158. Apart from the theories set out in the works περὶ ὀκταμήνων (*De octimestri partu, Oct.*) and περὶ ἑπταμήνου (*De septimestri partu, Septim.*), discussed below, see also *Epid.* 2.6.4 (L) 5.134.2ff., *Carn.* 19 (L) 8.612.3f., and cf. Aristotle *HA* 583b31ff., 584a36ff., b2ff. and 6ff. (where it is suggested that few eighth-month babies survive in Greece, but in Egypt they live, even if many of them are deformed).

159. *Mul.* 1.71 (L) 8.150.9ff. ≃ *Steril.* 233 (L) 8.446.17f.; *Nat.Puer.* 21 (L) 7.510.19ff.; contrast Aristotle *HA* 583b2ff.: males move first on the right side in forty days, females on the left in ninety, though "there is nothing exact about this."

160. *Nat.Puer.* 18 (L) 7.498.27ff., 500.4ff., 504.2ff., *Mul.* 1.72 (L) 8.152.8ff. Another obscure set of suggestions concerning the periods of formation of the embryo where no distinction is drawn between male and female is to be found in περὶ τροφῆς (*De alimento*) 42, CMG 1.1.83.7−10.

161. *Hebd.* 1.1.8ff. Roscher, (L) 9.433.3f.

162. *Carn.* 19 (L) 8.608.22ff., 612.1ff., 5ff.

about the concords or harmonies to which the movements of the developing foetus must correspond.[163] *Epidemics* book 2 section 3 chapter 17 suggests that the pains in pregnancy occur every third day when there is movement after seventy days, and, further, that they occur on the third day after the fiftieth, and on the sixth after the one-hundredth, and in the second and fourth months.[164]

It is not the case that suggestions about such topics as when a male or a female embryo begins to move in the womb invariably take the form of a proposal of a definite number. *Epidemics* book 6 section 2 chapter 25, for instance, probably suggests merely that males move earlier, and develop more slowly after they are born.[165] But references to particular numbers of days are very common, even though there is considerable disagreement about which are the significant ones. In some cases we may assume that the proposals are intended to be interpreted flexibly, merely as approximate suggestions of what may, in general, be expected.[166] But in others the theories are stated without qualification. Often the role of symbolic schemata is obvious enough, though, again, in other cases we can do no more than guess on what basis certain numerical relations were proposed. We may, for instance, compare the suggestion that male embryos move in the third month, females in the fourth, with the correlation of male with odd and female with even in the Pythagorean Table of Opposites. Again, it has been suggested that a figure of thirty days for males in *On the Nature of the Child* corresponds to a musical interval of a fourth (two-and-a-half tones, with each tone as twelve days), while one of forty-two days for females is equivalent to a fifth (three-and-a-half tones).[167] That is

---

163. *Vict.* 1.8 (L) 6.480.21ff., 482.5ff.; cf. 1.26 (L) 6.498.17ff.

164. *Epid.* 2.3.17 (L) 5.116.12f., 16ff. Cf. *Epid.* 6.8.6 (L) 5.344.10ff., 15ff.

165. *Epid.* 6.2.25 (L) 5.290.7ff.; cf. *Epid.* 2.3.17 (L) 5.116.15f. ≈ *Epid.* 6.8.6, 344.13ff. Cf. also Aristotle *HA* 583b23ff., 584a26f.

166. Thus *Nat.Puer.* 18 (L) 7.498.27ff. is concerned, in the first instance, to establish the upper limit to the periods considered, and at (L) 7.500.2ff. states that the rule applies generally and with some variation.

167. Cf. Lonie 1981a, pp. 192ff.; cf. Delatte 1930.

conjectural, but more transparently that treatise maintains that the basis for the difference between the sexes here is that the female seed is weaker.[168]

Some insight into these theories can be gained from passages where the Hippocratic authors themselves are more tentative or reflective about their proposals. The writer in *On the Eighth Month Child* raises the question of whether women report their experiences in pregnancy correctly. "One should not disbelieve what women say about childbirth," we are told in one context.[169] Yet on the difficulties experienced in the eighth month the writer says: "Women neither state nor recognise the days uniformly. For they are misled because it does not always happen in the same way; for sometimes more days are added from the seventh month, sometimes from the ninth, to arrive at the forty days. . . . But the eighth month is undisputed."[170]

The writer's own view is that the principal phases of pregnancy consist of periods of forty days, and he is at pains to calculate the beginning of the seventh month with some precision: it begins after 182 days and a fraction, that is, half a solar year.[171] He endorses, in the main, the general view of the difficulties of the eighth month but at the same time claims superior, more exact, knowledge of how to calculate it. It is notable that he does not seek to contradict, so much as to make

---

168. *Nat.Puer.* 18 (L) 7.504.24ff.

169. *Oct.* 7 Grensemann (= *Septim.* 4 Littré), CMG 1.2.1.92.15, (L) 7.440.13. χρὴ δὲ οὐκ ἀπιστεῖν τῇσι γυναιξὶν ἀμφὶ τῶν τόκων. I have discussed this text and the question of male doctors' evaluations of the information they received from their female patients in G. E. R. Lloyd 1983a, pp. 76ff.

170. *Oct.* 6 Grensemann (= *Septim.* 3 Littré), CMG 1.2.1.92.7ff. (cf. [L] 7.440.4ff.): ἀλλὰ καὶ ἡμέραι πρόσεισιν ἀπό τε τοῦ ἑβδόμου μηνὸς καὶ ἀπὸ τοῦ ἐνάτου. ἀλλὰ τὰς ἡμέρας οὐχ ὁμοίως οὔτε λέγουσιν οὔτε γινώσκουσιν αἱ γυναῖκες. πλανῶνται γὰρ διὰ τὸ μὴ κατὰ τὸ αὐτὸ γίνεσθαι, ἀλλὰ τὸ μὲν ἀπὸ τοῦ ἑβδόμου μηνὸς πλείονας ἡμέρας προσγενέσθαι ἐς τὰς τεσσαράκοντα, τὸ δὲ ἀπὸ τοῦ ἐνάτου. . . . ὁ δὲ δὴ ὄγδοος ἀναμφισβήτητός ἐστι.

171. *Oct.* 4, CMG 1.2.1.88.17ff. (*Septim.* 1 [L] 7.436.1ff.). The writer's view that the main phases of pregnancy consist of forty-day periods is set out, for example, at *Oct.* 1f., 5f., and 8, CMG 1.2.1.78.6, 80.13ff., 82.21ff., 90.9ff., 22ff., 94.1–14 ([L] *Septim.* 9, 446.15f., 448.21ff., *Oct.* 10, 452.13ff., *Septim.* 2, 436.15ff., 3, 438.14ff., 4, 442.7–22).

more precise, the traditional conception, including that of the danger to any child born in the eighth month, and indeed he continues to talk of the "eighth month child" even when his own theory is that, strictly speaking, this is inexact.[172]

On the Nature of the Child is another treatise that is critical of what women say about their pregnancies, flatly denying that they can be right when they assert that a pregnancy can last longer than ten months.[173] When he proposes his theory about the periods required for the formation of the male and female embryo the writer first argues on the basis of the analogy of the equivalent periods taken for the lochial discharge,[174] but when he recapitulates "for the sake of clarity" he cites what he calls a piece of research, ἱστόριον, to support his view. The first consideration he mentions is that on the receipt of the seed the flow of blood into the womb is least, though it subsequently increases (while the reverse is true concerning the lochial discharge) where direct observation of such changes in the flow of blood is clearly out of the question.[175] But he goes on to refer to what might have been the far more impressive evidence of miscarriages. "Again, many women have miscarried with a male child a little earlier than thirty days, and the embryo has been observed to be without limbs; whereas those that were miscarried at a later time, or on the thirtieth day, were clearly articulated. So too in the case of female embryos which are miscarried, the corresponding period being forty-two days, articulation of the limbs is observed. Hence both the earlier and the later miscarriages

---

172. *Oct.* 2, CMG 1.2.1.82.19 and 21; 5, CMG 1.2.1.90.18; 10, CMG 1.2.1.96.12 ([L] *Oct.* 10, 7.452.10 and 12, *Septim.* 2, 438.10, 8, 446.7).

173. *Nat.Puer.* 30 (L) 7.532.14ff.: "But those women who imagine that they have been pregnant for more than ten months—a thing I have often heard them say—are quite mistaken" (cf. Aristotle *HA* 584b18ff., 21ff.). The Hippocratic author goes on to identify the source of their error, (L) 7.532.16ff.: "it can happen that the womb becomes inflated and swells as the result of flatulence from the stomach, and the women of course then think that they are pregnant," and it may be too that the menses are interrupted; cf. also (L) 7.534.10ff.

174. *Nat.Puer.* 18 (L) 7.500.4ff.

175. *Nat.Puer.* 18 (L) 7.504.2ff., 8ff.

show both by reasoning and by necessity, that the period of articulation is, for a girl, forty-two days, and for a boy, thirty."[176]

What is so striking about this passage is the disparity between the impeccable statement of method, and what the writer provides by way of the results of its purported application. He recognises very clearly that miscarriages would, provided the time of the miscarriage is known, yield telling evidence about the various stages in the development of the human embryo, male or female. Yet what he claims as his result is simply the complete and total endorsement of his theory. His statement of what miscarriages reveal is suspiciously vague and general, and although it may be too much to say that he has no actual evidence at his disposal at all, at least he does not here provide detailed documentation of any single case.[177]

Finally, the continuation of the text already quoted from *Epidemics* 2.3.17 shows that, within limits, questions could be raised about some of the periodicities that were proposed. After advancing his theory about pains on every third day when there is movement after seventy days, the writer proceeds: "Should the nine months be numbered from the [last] menstruation or from conception? Do the Greek months amount to 270 days, or is there an addition to these? Does the same apply for males as for females, or the opposite?"[178] Yet it is significant

---

176. *Nat.Puer.* 18 (L) 7.504.16–23: πολλαὶ δὲ γυναῖκες ἤδη διέφθειραν κοῦρον ὀλίγῳ πρόσθεν τριήκοντα ἡμερέων, καὶ ἄναρθρον ἐφαίνετο· ὁκόσα δὲ ὕστερον ἢ ἅμα τῆσι τριήκοντα ἡμέρῃσι, διηρθρωμένα ἐφαίνετο ἐόντα · καὶ ἐπὶ τῇ κούρῃ κατὰ λόγον τῶν τεσσαράκοντα καὶ δύο ἡμερέων, ὁκόταν διαφθαρῇ, φαίνεται ἡ διάρθρωσις τῶν μελέων · ἤν τε πρόσθεν φθαρῇ τὸ παιδίον ἤν τε ὕστερον, ὧδε φαίνεται καὶ λόγῳ καὶ ἀνάγκῃ ἡ διάρθρωσις ἐοῦσα, ἐπὶ μὲν τῇ κούρῃ ἐν τεσσαράκοντα καὶ δύο ἡμέρῃσιν, ἐπὶ δὲ τῷ κούρῳ ἐν τριήκοντα.

177. Contrast *Nat.Puer.* 13 (L) 7.488.22ff., which provides some circumstantial detail concerning the writer's observations of what he takes to be an aborted six-day-old embryo discharged by a prostitute owned by a kinswoman. Compare also the examination of the aborted embryo at *Carn.* 19 (L) 8.610.3ff., 5ff.

178. *Epid.* 2.3.17 (L) 5.118.1–5: εἰ ἀπὸ τῶν γυναικείων ἀριθμητέοι οἱ ἐννέα μῆνες, ἢ ἀπὸ τῆς ξυλλήψιος, καὶ εἰ ἑβδομήκοντα καὶ διακοσίῃσιν οἱ ἑλληνικοὶ μῆνες γίνονται, καὶ εἴ τι προσέτι τούτοισι, καὶ εἴ τι τοῖς ἄρσεσιν ἢ καὶ τῇσι θηλείῃσι ταὐτὰ ποιέεται ἢ τἀναντία.

that even when, as here, certain questions are raised about accepted beliefs, those questions are formulated within the framework of those very beliefs. The writer clearly assumes that pregnancy generally lasts "nine months"; *that* is not in doubt. What is in question, rather, is how the nine months are to be calculated, that is, to put it bluntly, how the presumption of the nine-month period is to be validated.

There is thus a fair degree of disagreement both about what the significant periods in pregnancy and childbirth are and about how they are to be calculated. But that *some* calculation of days for some relations is correct is common ground to many authors. Theories about the periods at which the child born is or is not likely to survive are, in the main, based on popular beliefs which we may suppose to have originated in many cases long before the earliest Hippocratic treatises. The Hippocratic writers, for their part, are often critical of such beliefs, and sometimes they support their criticisms with appeals to what is claimed to be direct evidence. The importance of such empirical support is, we may say, certainly appreciated in principle. Yet in practice, in this context, what the Hippocratic writers offer is often little more than a more or less elaborate rationalisation of popular beliefs. In many cases the criticism is not that some popular assumption is too dogmatic and too precise, but that it is too imprecise—where the Hippocratic writer claims more accurate knowledge of the periodicities in question.

The second main area in which the medical writers develop complex theories of numerical relations concerns the periodicities of diseases, especially of "acute" diseases, that is, those accompanied by high fever. Here less is owed to popular assumptions, or at least there is no good evidence that the development of the classification of fevers into tertians, quartans, and so on antedates the period in which the Hippocratic writers themselves worked, although such a notion is not, of course, confined to them.

As already noted, certain diseases do in fact exhibit marked periodicities, and it is not too difficult to see this as one important and continuing stimulus to the elaboration of Hippocratic theories on the subject. Naturally enough, many writers share the general classification of acute diseases according to their periodicities: there were not just tertians, quartans, quintans, septans, and nonans, but also semitertians, and as fevers that did not fall into any other category could be termed "irregular," πλάνητες, the classification could be made exhaustive. But in addition a wide variety of specific proposals are made concerning complex periodicities, especially doctrines associating groups of even, or of odd, days together. Thus *Epidemics* book 1 chapter 12 states:

> Where paroxysms are on even days, the crisis too is on even days. Where the paroxysms are on odd days, the crisis is on odd days. The first period in those with crises on even days is 4, then 6, 8, 10, 14, 20, 30, 40, 60, 80, or 120 days. In those with crises on odd days the first period is 3, then 5, 7, 9, 11, 17, 21, 27, or 31 days. Further, one must know that if the crisis is on other days than those mentioned, there will be relapses and also it may prove a fatal sign.[179]

Offering a theory about the days on which sweating is beneficial in fevers, one of the *Aphorisms* repeats the same sequence of odd days, though adds to these the fourteenth and the thirty-fourth day.[180] The treatise *On Humours* recommends that if the paroxysms occur on odd days, the patient should be evacuated upwards on odd days, and that if the paroxysms are on even days, the evacuation should be downwards on even days—although if the periods of the paroxysms are different,

---

179. *Epid.* 1.12 (L) 2.678.5–680.6: τὰ δὲ παροξυνόμενα ἐν ἀρτίῃσι, κρίνεται ἐν ἀρτίῃσιν· ὧν δὲ οἱ παροξυσμοὶ ἐν περισσῇσι, κρίνεται ἐν περισσῇσιν. ἔστι δὲ πρώτη περίοδος τῶν ἐν τῇσιν ἀρτίῃσι κρινόντων, τετάρτη, ἕκτη, ὀγδόη, δεκάτη, τεσσαρεσκαιδεκάτη, εἰκοστή, τριακοστή, τεσσαρακοστή, ἑξηκοστή, ὀγδοηκοστή, ἑκατοστή· τῶν δὲ ἐν τῇσι περισσῇσι κρινόντων περίοδος πρώτη, τρίτη, πέμπτη, ἑβδόμη, ἐνάτη, ἑνδεκάτη, ἑπτακαιδεκάτη, εἰκοστὴ πρώτη, εἰκοστὴ ἑβδόμη, τριακοστὴ πρώτη. εἰδέναι δὲ χρὴ, ὅτι, ἢν ἄλλως κριθῇ ἔξω τῶν ὑπογεγραμμένων, ἐσομένας ὑποστροφὰς σημαίνοιτο, γένοιτο δ' ἂν καὶ ὀλέθρια.
180. *Aph.* 4.36 (L) 4.514.8ff. Other texts where the emphasis is on odd days are *Aph.* 4.61 (L) 4.524.3f.; *Morb.* 2.41 (L) 7.58.9ff., *Morb.* 3.3, *CMG* 1.2.3.72.14f., *Morb.* 4.46 (L) 7.572.1ff.; and cf. also *Acut.* 4 (L) 2.250.11ff.; *Aph.* 4.64 (L) 4.524.10ff.; *Coac.* 79 (L) 5.600.15f., 142 (L) 5.614.3ff.; *Epid.* 2.5.12 (L) 5.130.14f., 5.15 (L) 5.130.17f., 6.8 (L) 5.134.13ff., 6.10 (L) 5.134.16ff. See Kudlien 1980.

though counting the days is the usual method of measuring the periods, they are sometimes told not to assume that periodicities will consist of multiples of whole days.[189]

Yet all this excellent advice is given on the basis of the assumption that the periodicities are there to find. They may be hard to identify: many fevers may simply be "irregular." But the presumption is that the periodicities will usually be determinable, and even that complex cycles of exacerbations and remissions will be. The more care and attention the doctors devoted to establishing the times of the crises, the more confident they could feel in their conclusions, not just in particular cases but in general. The grounds themselves of the general theory, however, were not examined critically, or not critically enough, and reflections on the *causes* at work generally presupposed that theory.[190] It was enough for the more cautious doctors that periodicities could sometimes be spotted. Meanwhile the more speculative theorists had no compunction in making the most extravagant proposals concerning complex numerological relationships.[191]

---

189. This point is picked up and elaborated by Galen, for example, (K) 9.870.13ff., 933.12ff., 937.3ff., CMG 5.10.1.123.12ff., (K) 17A.246.4ff.

190. Typical in this area are such suggestions as that quartans are produced by or associated with black bile, tertians and quotidians with other kinds of bile: *Nat.Hom.* 15, *CMG* 1.1.3.202.10ff., 204.8ff., 11ff.; cf. *Morb.* 2.40–43 (L) 7.56.3–60.24; Caelius Aurelianus *De Morbis acutis* 1.108 on Asclepiades.

191. Later writers sometimes criticised the periodicities proposed by "Hippocrates," as Celsus, for example, did partly on the grounds of the inconsistencies detected between one Hippocratic text and another: see *Med.* 3.4.11ff., *CML* 1.106.25ff., and compare Galen's comments on this issue at (K) 9.868.11ff. and *CMG* 5.10.1.123.12ff., (K) 17A.246.4ff.; at *Med.* 3.4.12, *CML* 1.107.2ff., Celsus quotes the view of Asclepiades that no day was more dangerous to a patient for being even or odd, and at *Med.* 3.4.15, *CML* 1.107.23ff., Pythagorean numerology is singled out for criticism and said to have misled ancient doctors. Later still Caelius Aurelianus, for instance, notes that the periods in epilepsy, for example, are not regular and recommends that treatment should not depend on the number of the days but on changes in the disease, but he nevertheless takes the three-day periods as the starting-point for his discussion and offers advice as to how these are to be recognised: *Morb.Chron.* 1.105, 126.

The Underlying Epistemological Factors

The evidence we have reviewed is enough to show that no simple hypothesis to the effect that the ancients totally failed to make use of measurement will do. But we must now raise the question of the underlying epistemological factors at work. There was, of course, no orthodoxy on the question of the foundations of knowledge in antiquity, whether in the investigation of nature or elsewhere—no one standard set of views shared by all who engaged in that investigation, any more than among those who were more purely philosophical in their interests.[192] But how far can we go towards identifying the factors that militated for and against the appeal to measurement?

For Koyré and no doubt many others, the key factor would be the influence of Platonism. To be sure, the dichotomy between reason and perception and the preference for reason over perception—even for reason to the exclusion of perception—have strong roots already in the pre-Socratic period.[193] But the theme of the untrustworthiness of perceptible phenomena is associated particularly with prominent statements in Plato, especially the Plato of the middle dialogues,[194] where the doctrine takes various forms. The emphasis is sometimes on the simple fact that such phenomena are subject to change,[195] but more often also on the further point that particulars bear the predicates they bear in a qualified or relative fashion: what is beautiful in one respect may be said to be ugly in another, appear beautiful to some people but not to others, at one time but not at others, and so on.[196]

---

192. I have discussed aspects of what follows at greater length in G. E. R. Lloyd 1982, pp. 128ff.

193. See, for example, Heraclitus fr. 107 ("eyes and ears are bad witnesses for men if they have souls that do not understand the language"), Parmenides fr. 7, Melissus fr. 8, Empedocles frr. 2, 3, Anaxagoras fr. 21, Democritus frr. 9, 11, 125.

194. See, for example, *Phd.* 65b–c, 79a–c, *R.* 529b–c (in the context of the astronomical programme of the Guardians), *R.* 532a, cf. *Ti.* 28b, 52a–b, *Phlb.* 59a–b. Yet at *Phd.* 74b, 75a–b, perception stimulates the soul to recollect the Forms.

195. For example, *Smp.* 210e6–211a2.

196. The classic statement is at *Smp.* 211a2ff., cf. *Phd.* 74b8–9, *R.*

ignore nor dismiss the empirical phenomena entirely. But it might still be argued that the search for exact data was inhibited by a general expectation that any data gained from observation will fall far short of the true reality. The problem can be stated simply, but it is of the very greatest complexity, and it would be foolish to try to generalise about the expectations of ancient investigators even within a single discipline. Obviously, those expectations will vary, depending on, among other things, the individual's view of the difficulties encountered in conducting observations, whether with or without instrumental aids, and especially on the confidence he had in his theories.[202] Yet—to take our best-documented example again—although Ptolemy often acknowledges inexactnesses in the astronomical data he uses, it is not that he is *indifferent* to their magnitude. It is not that he has a metaphysical principle that allows him to *disregard* such inexactnesses. On the contrary, his concern is always to insist that the inexactnesses he tolerates are minor and fall within the bounds appropriate to the particular problem in question.[203]

Paradoxically, perhaps, the very fact that he engaged in some selection and adjustment of his data in the light of his theories reveals *his* expectation that the fit between them will, in general, be a good one. This is true in certain contexts in the *Syntaxis,* but the evidence we considered from the *Optics* is even more striking in this regard. There, in the tables of refraction for the three pairs of media studied, the results are given as correct to within half a degree.[204] But they all tally exactly with the underlying general law. Yet this very feature of his account—which shows that Ptolemy has adjusted his results—*also* reveals that his assumption is that a *perfect* fit between the observed data and the theory is possible, not just a perfect fit between the generalisations derived from the observations and the theory but one between

---

202. Cf. further below, Chap. 6, pp. 315ff.
203. Cf. Ptolemy's frequent appeal to the notion of "negligible difference," both in his astronomy, e.g., *Syntaxis* 3.1 (H) 1.194.10ff., 196.21ff., cf. *Syntaxis* 5.10 (H) 1.394.6, 400.11f., and cf. below, p. 305. on *Syntaxis* 9.2 (H) 2.212.9f., and in his harmonics, *Harm.* 1.4 (9.23ff. D.), 1.16 (39.20 D.), cf. 1.14 (32.20f. D.).
204. See above, p. 246: note *ad prope* at *Optics* 5.11.229.5, cf. 18.234.2, 21.236.9 L.

what he represents as the observed results themselves and that theory. Here there are no signs of inhibitions stemming from a belief that the data are bound to prove intractable. On the contrary, this example shows very clearly that the error is, at least on occasion, on the side of expecting, or assuming the possibility of, *too close* a fit between theory and data rather than on the side of the opposite assumption.

Thus far I have concentrated on aspects of the epistemological background that might be thought, or have been thought, to work *against* a realisation of the importance of quantitatively precise data. But one other influential idea that tells, rather, in the opposite direction is the Pythagorean doctrine that "all things are numbers." This was admittedly a highly obscure, at points perhaps even obscurantist, principle. The relationship between "numbers" and "things" is expressed in different, even incompatible, forms in our sources for early Pythagoreanism, for sometimes things are said to *be* numbers, sometimes merely *like* them.[205] More important still, the examples cited to illustrate and support the principle include many that have nothing to do with natural philosophical inquiry, as when justice is associated with the number four, or marriage with the number five (the sum of three and two, identified with male and female, respectively).[206] Again, we noted that other symbolic associations (not confined to those made by the Pythagoreans) appear to underlie many of the complex numerological relationships found in Greek medicine.[207] Moreover, Aristotle reports and criticises overenthusiasm for the number seven: to the reflection that there are seven vowels in Greek, seven notes to the scale, seven

---

205. See, for example, Aristotle *Metaph.* 985b27ff., 32ff., 986a2ff., 987a19, b11f., 27f., 1080b16ff., 1083b11ff. On the interpretation of these reports, see, for example, Guthrie 1962, pp. 229ff., Burkert 1972, chap. 6. As has been emphasised recently by Huffman (in an unpublished paper on "Philolaus and Early Greek mathematics" presented to a conference on Greek mathematics held at Cambridge, England, in May 1984) the attribution of the doctrine that things are numbers is more often an inference from what Aristotle believes the Pythagoreans are committed to, than a direct report.

206. See, for example, Aristotle *Metaph.* 985b29ff., 1078b22f. Other ancient testimonies and examples are collected and discussed by Burkert 1972, pp. 466ff.

207. See above, pp. 258ff.

Pleiades, at seven years children lose their first teeth—or at least some do—and that there were seven who fought against Thebes, Aristotle's reaction is to say that such theorisers are like the Homeric scholars who see small resemblances but neglect important ones.[208]

While in many contexts the interests shown by Pythagoreans and others in the classification of numbers [209] and in proportions, concords, and harmonies [210] reflect ethical, symbolic, or aesthetic considerations, in others the theory that "all things are numbers" could and did act as a stimulus to find those numbers, by measurement, in the phenomena. The Pythagoreans, we said, had no monopoly of interest in the numerical relationships investigated in the study of harmonics. But the expression of the concords of octave, fifth, and fourth in terms of the

---

208. Aristotle *Metaph.* 1093a13ff., 26ff.

209. The association of odd and even with other pairs of opposites, including good and evil, in the Pythagorean συστοιχία reported by Aristotle has already been noted, and the role of the "perfect" number ten in their cosmology attracts his criticism at *Metaph.* 986a8ff. We have extensive evidence for the development of the classification of numbers in Theon, Nicomachus, and Iamblichus especially, who discuss, for example, the notions of "perfect" and of "friendly" numbers: see Theon 18.3ff., 25.5ff., 45.9ff., Nicomachus *Arithmetica introductio* 14.13ff., 19.9ff., 39.14ff., Iamblichus *In Nicomachi arithmeticam introductionem* (*In Nic.*) 20.7ff., 32.20ff., 34.26ff.

210. Many of the terms used to express mathematical relationships have important ethical and political connotations. ἰσότης, equality, like ἰσονομία, is frequently used in the context of a just or equitable distribution (e.g., Plato *Lg.* 757b–c) and in a famous passage in Plato's *Gorgias* (508a) geometrical equality is elevated to a principle of cosmic order (see, for example, Vlastos 1947/1970, 1953, Mau and Schmidt, edd., 1964, Heinimann 1975). Again συμφωνία (concord) and ἁρμονία (scale, attunement) were extensively used outside music, notably in a variety of cosmological contexts including the astronomical theory of the "harmony of the spheres" (an idea already criticised by Aristotle in *Cael.* 2.9.290b12ff., and later elaborated in considerable detail in Ptolemy *Harmonica* 3.8–16 (100.18ff. D.), and cf. also Nicomachus *Harmonicum enchiridium* (*Harm.*) 3.241.3ff., and Aristides Quintilianus *De musica* 3.20ff., 119.21ff.). Although at *Metaph.* 1092b26ff. Aristotle questions the idea that a mixture is better if its constituents are in a particular ratio (λόγος), he holds nevertheless that the most pleasant colours are those in which the elements stand in a simple ratio to one another, comparing these explicitly with concordant sounds: *Sens.* 439b25ff., 30ff., cf. Kucharski 1954, Sorabji 1972a, Barker 1978c, 1981a.

simple ratios 1:2, 2:3, 3:4 ranked for them, we may be sure, as a paradigm of the application of numbers to things. The exactness here is a matter of the simplicity of the mathematical relationships: the ratios are either multiplicate or superparticular. Yet those ratios had broadly to be confirmed, if not discovered, by reference to measurable data,[211] and various investigations involving measurement were attempted, not just on the monochord but also, for example, measuring lengths of pipe or the quantities of water in jars that gave different notes when struck, and even weighing hammers that did so.[212]

As is well known, the stories that report some of these inquiries contain many elements of pure fantasy, especially concerning the results that were supposed to have been obtained.[213] Yet that does not affect their value to us as evidence for the aims and methods of such investigations. Sometimes the inquiry involves working back from the results expected: thus in the story where predetermined quantities of water are poured into jars, the quantities are *chosen* to *yield* the harmonies. Sometimes the investigation proceeds from what is already given: thus in the story about the hammers, they were clearly not *made* by anyone to give the notes they were supposed to have done. But what the two types of inquiry have in common is the attempt, or

---

211. This is true of the principal concords of the octave, fifth, and fourth, even though in the mathematical development of musical theory there could be no question of empirical verification of such ratios as that of the lemma (256:243)—corresponding to a fifth less three tones or to a fourth less two— or of the various theoretical subdivisions of the semitone; see Burkert 1972, p. 385. It is noteworthy that the numerical ratios for the principal concords were common ground not just to Pythagoreans but also to other musical theorists, for example, those working in the Peripatetic tradition: see, e.g., pseudo-Aristotle *Problemata* 19.35.920a27ff., 41.921b1ff.

212. The chief ancient testimonies are listed at G. E. R. Lloyd 1979, p. 144 n. 95. Raasted 1979 has recently suggested, on the basis of a study of the Hagiopolites MSS published in 1847 by J. H. Vincent, that there may be a confusion between the word σφυρῶν used in some sources for hammer (cf. ῥαιστήρων) and the word σφαῖρα, σφαιρῶν, with which may be connected the less implausible set of stories of the investigation of the sounds made by disks, as in the report in a scholium to Plato *Phd.* 108d, attributing such an inquiry to Hippasus.

213. Cf. further below, Chap. 6 at nn. 37ff.

cients' approach lies. Far from being inadequately quantitative, some areas of ancient inquiry were excessively so,[218] in part under the influence of the very successes obtained in such fields as harmonics and astronomy. An important recurrent phenomenon in Greek speculations about nature is a premature or insecurely grounded quantification or mathematisation. The excessive elaborations of numerical relations in theories concerning the periodicities of diseases and in embryology are examples of this. Another instance is Galen's attempt to distinguish four different grades of hot, cold, wet, and dry. In various versions of atomism, too, atomic shapes are manipulated in a way that is interesting geometrically, but almost wholly arbitrary. Numbers and geometrical relations could be the key to the understanding of the phenomena, but they were often merely the focus of symbolic attention—as on many occasions, notoriously, in Plato.[219] The mathematical rigour of an entire inquiry—as in the casting of horoscopes, to revert to an earlier example[220]—could be impeccable, but the inquiry remains with too little purchase, with too little grip, on the phenomena. The appeal to the mathematical often gave a spurious air of certainty, the precise

---

218. Learned attempts to engage in measuring are already the butt of Aristophanes' humour in the famous passage in *Nu.* 143ff., where he represents Socrates attempting to measure the length of a flea's jump. In a variety of contexts the desire for ἀκρίβεια, exactness or nicety, was considered illiberal or ungentlemanly: see, for example, Aristotle *Metaph.* 995a6ff., cf. *Pol.* 1258b33ff.; differing expressions of the point are found in Plato, e.g., *Hippias Major* 284e1f., 295a5–6 (presumably ironical), *Cratylus* 414e f., cf. *Amatores* 135c–d, 136a, Isocrates 15.261–65, and Demosthenes 23.148. Contrast Herodotus' delight in recording purported measurements, especially in connection with the marvels he reports from foreign lands: see Hartog 1980, pp. 246ff., 346ff.

219. As, for example, in the discussion of the "nuptial number" in *R.* 546b–d, in the element theory and account of the construction of the world-soul in the *Timaeus*, 31b ff., 35b ff., and cf. *Epinomis* 990e f.

220. Extravagant numerological speculation is easy to exemplify in astrology, as, for example, in the correlations proposed by Ptolemy, *Tetrabiblos* 3.11.129.2–142.15, between the number of years of life and the number of degrees, despite his critical remarks about some traditional methods of calculating these, and cf. his correlations between the seven ages of man and the seven planets, 4.10.204.6ff.

being confused with the accurate.[221] Yet this very feature of some ancient work, the pursuit of exactness where it is *in*appropriate, is itself the subject of critical comment by other ancient authors, as for example, by the medical writers who protest against some bids to turn medicine into an exact science.[222]

Yet although the characterisation of ancient science as essentially qualitative stands in need of drastic modification, it has a certain limited validity. Appeal to measurement is rare in dynamics and in element theory, and even where it occurs, *actual* measurements are generally not recorded. In some fields the way the ancients usually formulated the problems directed attention to qualitative, rather than to quantitative, aspects. The instruments available to carry out exact measurements are of varying accuracy (a symptom as well as a cause of the problem), adequate enough for weighing and measuring medium-sized lengths and volumes but not, for instance, for measuring short intervals of time. The example of astronomy shows that when there was sufficient motivation, the ancient Greeks could develop some quite sophisticated instruments,[223] but in general the improvements made in measuring instruments were modest. Outside astronomy, the weighing and measuring of the ingredients of drugs was the chief context in which ancient investigators were repeatedly engaged, as a matter of

---

221. As I pointed out in G. E. R. Lloyd 1982, p. 130 n. 3, Greek ἀκρίβεια sometimes connotes exactness or precision, sometimes accuracy. The distinction we draw between precision in the sense of "the degree of refinement with which an operation is performed or a measurement stated" (Webster) and accuracy in the sense of "degree of conformity to some recognised standard value" or "value accepted as true" would have to be made in Greek by contrasting ἀκρίβεια with ὀρθότης, correctness, or ἀλήθεια, truth (cf., e.g., Ptolemy *Syntaxis* 3.1 [H] 1.200.15f.). As Webster also observes, however, exactness, precision, and accuracy are often used loosely and interchangeably.

222. Cf. above at nn. 134ff. and n. 215, Chap. 3 at nn. 89ff. Compare also Aristotle's insistence that ethics, unlike mathematics, is not an exact study, *EN* 1094b23ff., 25ff.

223. In astronomy, however, the ancient Greeks produced nothing to compare with the imposing bronze armillaries made by Chinese technologists, the development of which is described by J. Needham 1954– , vol. 3, pp. 339ff.; cf. Needham, Ling, Price 1960.

undeterred, invented a special kind of water clock that could be calibrated according to the age of the patient, although we do not know what degree of accuracy he obtained or expected from this. Secondly, there is the evident *ambition* to make the inquiry an exact one, to construct pulse theory on the model of music, the successful mathematisation of harmonics. If the main concords are expressible in terms of simple numerical relationships, why not also the main ratios between the dilations and contractions of the arteries? But, thirdly, it is clear that we have yet another example of premature or insecurely grounded quantification. As in the Hippocratic study of the periodicities of diseases, there is, to be sure, *some* basis for the investigation. But that basis could not sustain the elaborate theoretical superstructure erected upon it. The attempt to reduce the data concerning the pulse to mathematically expressible relations like those of music theory shows how the ancients sometimes exercised considerable ingenuity and persistence in exploring such possibilities, but it also illustrates how in practice that ambition could turn out to be, in part, misplaced.

ὥστε κλεψύδραν κατασκευάσαι χωρητικὴν ἀριθμοῦ ῥητοῦ τῶν κατὰ φύσιν σφυγμῶν ἑκάστης ἡλικίας εἰσιόντα τε πρὸς τὸν ἄρρωστον καὶ τιθέντα τὴν κλεψύδραν ἅπτεσθαι τοῦ πυρέσσοντος· ὅσῳ δ' ἂν πλείονες παρέλθοιεν κινήσεις τῶν σφυγμῶν παρὰ τὸ κατὰ φύσιν εἰς τὴν ἐκπλήρωσιν τῆς κλεψύδρας, τοσούτῳ καὶ τὸν σφυγμὸν πυκνότερον ἀποφαίνειν, τουτέστι πυρέσσειν ἢ μᾶλλον ἢ ἧττον. ("There is a story that Herophilus was so confident in the frequency of the pulse, using it as a reliable sign, that he constructed a water clock able to contain a specified amount for the natural pulses of each age. Going in to the patient, he would set up the water clock and feel the pulse of the person suffering from fever. By as much as the movements of the pulse exceeded what is natural for the filling of the water clock, by that much he declared the pulse too frequent, that is, that there was greater or less fever.") See the discussion of this report in von Staden forthcoming, whose reading τῶν σφυγμῶν for Schöne's τῷ σφυγμῷ at line 265 is adopted here. Aspects of the problems encountered in timing pulse rates in antiquity and later are discussed by Kümmel 1974, cf. 1977.

CHAPTER SIX

Idealisations and Elisions

I n the last chapter I discussed both the positive and the negative features of the use of measurement and the search for quantitative exactness in ancient science. This final chapter will be devoted to further aspects of the problem of the match expected between data and theory, between explanandum and explanation. An element of simplification and idealisation is present in all science: it is only by ignoring certain features of what is given that the underlying relationships governing the phenomena can be revealed. Again, a theory is not held to be refuted when what is predicted on its basis is found to disagree, within certain limits, with observed results. The questions we may pose are: What kinds of simplification did ancient scientists allow themselves? What phenomena did they permit themselves to discount and what constraints did they recognise on that discounting? According to a still highly influential view, that of Duhem,[1] the ancient slogan of "saving the phenomena" involved, in astronomy, precisely the production of mathematical theories from which the positions of the heavenly bodies could be predicted independently of any physical considerations. The theories were purely calculating devices; they had nothing to do with any underlying physical realities. Whatever the constraints on the theory from the side

---

1. Duhem 1908, 1954b. Aspects of Duhem's interpretation of the ancient evidence are criticised in G. E. R. Lloyd 1978b. See further below at nn. 93ff.

of the match between predicted and observed positions, there were no constraints at least from the side of the physics of the question, since that was of no concern.

The Development of the Notion of Explanation

The validity of that interpretation of areas of ancient inquiry will be examined in due time, but certain preliminary remarks should first be made on the development of the notion of explanation itself. In a sense myths too, as I noted in Chapter 1, provide explanations—of a sort—of the subjects they deal with, but only in a sense, and only attenuated explanations. The myths in question range from major quasi-cosmological statements about the origin of the universe or of man's place in it, through particular aetiological accounts, down to "just-so" stories about "how the leopard got his spots." In interpreting these the first important point is a sociological one, namely, the context of delivery. Anthropologists themselves took some time fully to come to terms with the problems of context and intentionality in their material. Many stories do not record what adults seriously believe, only what adults habitually say in response to—and maybe in the hope of blocking—a certain kind of inquisitive questioning, from children, for example, or even from anthropologists. It is not as if we seriously believe that babies are brought by storks or found under gooseberry bushes. It is not as if many of us believe that the century plant does actually bloom only once in a hundred years.[2]

But apart from the question of whether such stories are claimed to be *true*, the extent to which they are or contain *explanations* is problematic. They may contain the equivalent of explanations, that is, the answers to "how," "why," or "what" questions. But in two ways especially, they are likely to be defective. First, the problem may not be made explicit, and, secondly, the proposed solution may consist of a set

---

2. Cf. in the 1970 edition of the *Encyclopaedia Britannica, s.v.* "century plant": "The century plant is a name given to *Agave americana* from the erroneous supposition that it flowers only when 100 years old."

of arbitrary assertions, the range of applicability of which is left quite indeterminate. This applies not only to tales told to children (under what circumstances storks bring babies is not discussed) but also to quasi-cosmological myths, about the origin of the world, or of humans or animals, or about how fire came to be used or skills discovered. Under what circumstances Marduk split the primeval water goddess Tiamat to make the sky with its celestial waters on the one side and Apsu, the deep, and Esharra, the "great abode," on the other, just does not occur *as a question:* [3] no more does why the stones thrown by Deucalion and Pyrrha became men and women.[4] That this happened is simply asserted, and it is understood that this was an exceptional occasion with a special outcome. But why thrown stones do not usually metamorphose is not an issue, though it is known that they do not. That very way of querying the story presupposes a framework of natural causation that became self-conscious and explicit only with difficulty and with time—even though that realisation could and did build on what was, in a sense, already common knowledge, or at least commonly assumed.[5]

The emergence of what can begin to be called fully fledged explanations of classes of natural phenomena is an important new development, though a hesitant one, in early Greek philosophy, with the practice of such explanations preceding the theory. The sequence of ideas that Aristotle reports in *De caelo* 2.13 about the shape and position of

---

3. *Enuma Elis* Tablet 4.135ff., Pritchard 1969, p. 67. Compare the translation in Lambert 1975, p. 55: "Bel [i.e., Marduk] rested, surveying the corpse, / To divide the lump by a clever scheme. / He split her into two like a dried fish, / One half of her he set up and stretched out as the heavens. / He stretched a skin and appointed a watch, / With the instruction not to let her waters escape. / He crossed over the heavens, surveyed the celestial parts, / And adjusted them to match the Apsû, Nudimmud's [i.e., Ea's] abode. / Bel measured the shape of the Apsû, / And set up Esharra, a replica of the Eshgalla. / In Eshgalla, Esharra which he had built, and the heavens, / He settled in their shrines Anu, Enlil, and Ea."

4. Pindar *Olympian* 9.41ff., Apollodorus *Bibliotheca* 1.7.2ff., Hyginus *Fabulae* 153, Ovid *Metamorphoses* 1.395ff. (where it is worth recalling the parenthesis, v. 400: *quis hoc credat, nisi sit pro teste vetustas?*).

5. Cf. G. E. R. Lloyd 1979, pp. 49ff.

the earth and on the question of why it does not move illustrates both the advances and the limitations of pre-Socratic natural philosophical accounts, even if it is evidence that must be used with caution.[6] For one thing, it was Aristotle who chose to present these ideas as a *sequence* of replies to the same set of questions, and that may well distort the original context in which they were proposed. Even so, although it would be quite wrong to represent later theories as progressively more sophisticated (let alone truer or in some sense more correct) than earlier ones, some of the *constraints* on what counts as an answer appear to be grasped more fully as time goes on.

Three features are worth remarking very briefly. First, there is the phenomenon of the *regression of the explanandum*. A common suggestion was that the earth does not move because it is supported on something, such as water (according to Thales) or air (as in Anaximenes).[7] That resolved one difficulty by raising another: what keeps the water, or the air, itself in place—a point that was evidently appreciated by Aristotle and may already have been by Anaximander.[8]

Secondly, in Anaximander's suggestion—that the earth does not move because it is equally balanced on all sides and there is no reason, then, for it to move in one direction rather than in any other[9]—we

---

6. Aristotle *Cael.* 2.13.293a15–296a23.

7. Aristotle *Cael.* 294a28ff. (Thales), 294b13ff. (on Anaximenes, Anaxagoras, and Democritus, who are said to appeal to the flatness of the earth as the explanation of its being at rest: it does not cleave the air beneath it, but settles on it like a lid).

8. See Aristotle *Cael.* 294a32ff. It seems likely from the evidence in Aristotle *Ph.* 204b24ff., together with Simplicius *In Ph.* 479.32ff., that Anaximander arrived at his cosmological principle, the Boundless, in part by reflecting on the difficulties presented by any view—such as Thales' doctrine of water—that started from a single determinate substance. If so it is possible—though of course far from certain—that a similar line of reasoning led Anaximander to his radically different solution to the problem of the earth being at rest, in which he appealed not to any underlying support but to its being "equally balanced" on all sides (see next note). The proponents of the view that the earth does not move because it rests on air like a lid appear to have argued that the air itself cannot move for lack of room, *Cael.* 294b19ff., 25ff.

9. *Cael.* 295b10ff., cf. Hippolytus *Refutatio Omnium Haeresium (Haer.)* 1.6.3.

have an example of *suspending* some of the commonly assumed data. A clod of earth, as Aristotle was prosaically to insist, moves in a certain direction, "downwards." Aristotle, with a spherical earth, defined that as towards the centre of the universe, deemed to coincide with the centre of the earth.[10] That answer was not available to Anaximander, who thought the earth flat.[11] But then in his case that truth about pieces of earth has to be assumed *not* to apply to the earth as a whole, for his suggestion to be an answer to the problem of why the earth does not move.

Thirdly, we find in the same chapter an example of the *denial of the data* that are supposed to generate the problem. The full motivation of the suggestion that Aristotle ascribes to certain Pythagoreans, namely, that the earth is like a planet,[12] is unclear and controversial, but the effect of the suggestion is to make the earth move in space. The question "Why does the earth not move?" thus gets answered by denying the assumed fact: "But it *does* move"—though we evidently have another case of the regression of the explanandum, since *how* it moves and how, on the hypothesis of its movement, other phenomena are to be accounted for involve a series of other problems a stage further back.

These first attempts to resolve questions concerning the position of the earth may look indistinguishable from myths, or at least subject to criticisms that are similar in kind to those I made of the types of

---

10. See above, Chap. 4 at n. 79.

11. See pseudo-Plutarch *Stromateis* 2, Hippolytus *Haer.* 1.6.3, Aetius 3.10.2.

12. *Cael.* 293a21ff.: the earth is one of the "stars" and makes night and day as it travels round the centre in a circle. At the centre itself is the Central Fire, and Aristotle reports that the Pythagoreans held that fire is more honourable than earth and so occupies this honourable position (*Cael.* 293a30ff.). The Pythagoreans further postulated an invisible "counter-earth" and held that this and maybe other invisible bodies accounted for eclipses of the moon being more frequent than those of the sun (that is, presumably, as seen at any given position on earth) (*Cael.* 293b23ff.). Aristotle himself criticises them for introducing the counter-earth merely to bring the number of the heavenly bodies up to the perfect number ten (*Metaph.* 986a8ff.) and for forcing the appearances to fit their own preconceived opinions (*Cael.* 293a25ff.).

explanations that are offered in myths. But apart from the well-known point that the philosophers' accounts are naturalistic ones,[13] they are in principle subject to open challenge. A new suggestion on an old topic implicitly claims superiority to others in the field and has, accordingly, to give an account of itself. It is in that crucible of debate on contested issues that clearer working notions of what will count as an explanation, and of what an explanation should be, come to be elaborated.

For the first more explicit discussions of that topic we have to wait until Plato, though several of the Hippocratic writers made, rather more incidentally, important contributions to the understanding of such distinctions as that between causal and merely coincidental factors.[14] Two of the key ideas for which Plato himself appears to have been responsible are, first, the explicit distinction between necessary condition and cause or explanation, and, secondly, the more general contrast between essence and accident. The first distinction is made in the *Phaedo,* where reference to what is true merely of the material conditions of a situation (without which, to be sure, it would not be the situation it is) is contrasted with reference to the αἴτιον, which must specify some good.[15] The further point here, that explanation must be

---

13. Cf. Farrington's much-quoted dictum 1944–49/1961, p. 37: "What Thales did was to leave Marduk out."

14. See, for example, *Vict.* 3.70 (L) 6.606.20ff. ("the sufferer always lays blame, αἰτιῆται, on the thing he may happen to do at the time of the illness, even though this is not responsible, οὐκ αἴτιον ἐόν"); *VM* 21, *CMG* 1.1.52.17ff. ("if the patient has done something unusual near the day of the disease, such as taking a bath, or going for a walk, or eating something different . . . I know that most doctors, like laymen, assign the cause, αἰτίη, [of the disease] to one of these things, and in their ignorance of the responsible factor, αἴτιον, they stop what may have been most advantageous") and *VM* 19, *CMG* 1.1.50.7ff ("we must therefore consider the causes, αἴτια, of each condition to be those things which are such that, when they are present, the condition necessarily occurs, but when they change to another combination, it ceases"). On these and other Hippocratic texts, cf. G. E. R. Lloyd 1979, pp. 53ff.

15. Plato *Phd.* 99a–b, where Socrates denies that the "that without which," ἐκεῖνο ἄνευ οὗ, can truly be said to be an αἴτιον, for the αἴτιον of an event must state why it occurs in terms of the good aimed at. The literature on the notions of αἴτιον and αἰτία in Plato is extensive: see especially Vlastos 1969/1973, C. C. W. Taylor 1969, M. Frede 1980, Fine 1987.

in terms of what a thing is for or the good it serves—that is, that explanation must be teleological—was fraught with significance for the future, and we shall be returning to it later.[16]

The second, more general distinction between essence and accident is crucial for the theory of Forms but is present already in the Socratic search for definitions. The *Euthyphro* puts it that definition is directed at the οὐσία, what the thing really is, rather than at the πάθος, that is, some attribute that it may happen to possess.[17] The frequent insistence in the Socratic dialogues on the *equivalence of extension* of definition and definiendum provides one of the clearest early contexts for a demand for an exact match between a *logos* and that of which it is the *logos*. Even though in practice, in the natural sciences, the distinction between essence (or the lawlike) and the accidental will often be problematic and hard to apply, once some such distinction is available it can be appealed to in attempts to determine what, in the phenomena under review, can and should be discounted.

These points are all very familiar. My aim in recalling them is simply to stress the moral they convey, that an explanation—in science or anywhere else—must focus on *certain* aspects of the phenomena in preference to others (causes, not preconditions) and to the exclusion of yet others (that is, must focus on the essential, not the accidental).

Mathematics and Physics in Plato and Aristotle

Further pressure positively to discard certain features of the phenomena comes from the side of the model of mathematical knowledge—it, too, prominent in Plato. The nature of mathematical truths and of the objects that mathematics studies had become, already by Aristotle's time, topics of intense controversy.[18] Where Platonism construes mathematics as to do with separate intelligible objects and

---

16. See below, pp. 319ff.
17. Plato *Euthphr.* 11a.
18. In *Metaphysics* M and N especially, Aristotle engages in sustained debate with Plato, with Pythagorean positions, and with those of his own contemporaries Xenocrates and Speusippus. See Annas 1976 and the papers collected in Graeser, ed., 1987, especially.

accepts and insists that the truths of geometry, for instance, are never unqualifiedly instantiated in *physical* objects (the diagram on the blackboard, for example), Aristotle argued that mathematics had no *separate entities* as its objects: mathematics studies certain features *of physical* objects taken in abstraction from certain others, namely, the features that make them the physical objects they are.[19] Mathematical truths are, then, truths about the mathematical properties of physical objects. Indeed, it has recently been argued, with some force and sophistication,[20] that Aristotle does not merely *not deny,* he even requires that there are physical straight lines that fully and perfectly instantiate the geometrical truths about straight lines. It is true that the line drawn in chalk on the blackboard will not do as an example, nor even its outer edge, but, then, it would be wise to say that they are not straight lines. The truths about straight lines will nevertheless be instantiated in *any* of a number of straight lines that are present in any physical object.[21] That interpretation is disputed, but at least it can be agreed that there is no need for Aristotle to say that in principle it is impossible for physical objects to instantiate mathematical truths; they certainly will not fail to instantiate truths of arithmetic,[22] and he certainly has some perfect spheres—in the heavens.

Whatever the disagreements between Plato and Aristotle in the philosophy of mathematics, both held that mathematics is exact, and that point is fundamental, even though it requires as a gloss that pure mathematics also has to admit approximations in certain contexts (for the values of surds, for example).[23] But in a bid for exactness, applied

---

19. See especially *Physics* 2.2.193b22ff., 24ff., 194a9ff.

20. See Lear 1982: contrast I. Mueller 1970/1979 and 1982a, pp. 70ff.; Annas 1976, pp. 29ff.; and cf. Annas 1987; Hussey 1983, pp. 176ff.

21. Cf. Lear 1982, pp. 175ff., 180f. The possibility that some early Greek *definitions* of a straight line are based on or derived from the physical law of the rectilinear propagation of a ray of light is discussed by Mugler 1957 and 1958.

22. On Aristotle's philosophy of arithmetic, see Lear 1982, pp. 183f.; J. Barnes 1985; Mignucci 1987. Aristotle himself does not, however, draw attention to differences between arithmetic and geometry in the relevant connection.

23. Cf. Archimedes' extraction of a value for π in his *Dimensio circuli*. In

mathematics too will discard, even for the Aristotelian, some, at least, of the physical aspects of the phenomena.[24] Although the precise conditions under which the procedure called abstraction, ἀφαίρεσις, can be carried out are controversial, *some* discarding under *some*, more or less rigorous, conditions is clearly involved.[25] It will not matter if the line in the diagram is not straight or is not a foot long, for the geometer will say: take the line as straight. And if, in fact, it is not so, nevertheless, as Aristotle put it, the falsehood does not lie in the premises.[26]

"Saving the Phenomena"

After these rather cursory preliminaries concerning the philosophical background, we may now turn to our principal concern, the kinds of idealisations found in the ancient inquiry into nature. We are told by Simplicius that Plato set as a problem the saving of the apparent wanderings of the planets, by means of regular, orderly—we are to understand, circular—motions.[27] But the question of the conditions un-

---

Ptolemy's *Syntaxis,* for instance, approximations are certainly just as prominent in the purely mathematical parts of his calculations as they are in the adjustments made to observational data: cf. G. E. R. Lloyd 1982, pp. 153–59.

24. Thus at *Metaph.* 1078a14ff., harmonics and optics study their objects not *as* sight or *as* sound, but as lines and numbers, and, he adds, "similarly with mechanics," though at *Ph.* 194a11ff., optics, for instance, is contrasted in turn with geometry as concerned with a mathematical line but not *as* mathematical but *as* physical. The more that is discarded, the more exact the study is: see *APo.* 87a31ff., *Metaph.* 982a25ff.

25. On Aristotelian abstraction see, for example, Philippe 1948; I. Mueller 1970/1979, pp. 98ff.; Lear 1982 (who gives a sophisticated formal analysis of the *qua* operator as a predicate filter); and cf. Cleary 1985; Annas 1987; Mignucci 1987.

26. See *APr.* 49b34ff., *APo.* 76b39ff., *Metaph.* 1078a17ff., cf. 1089a21ff.; see Lear 1982, pp. 171ff.

27. Simplicius *In Cael.* 488.18ff., 492.31ff. (the latter passage specifies circular motions explicitly). In the first text Simplicius gives Sosigenes as his authority and he has just referred to Sosigenes' use of Eudemus' *History of Astronomy.* But who first formulated the problem for astronomy in exactly these terms is unclear: neither Plato nor Aristotle speaks of "saving" the "phenomena" as such, though Aristotle, for instance, refers to saving a "hypothesis" at *Cael.* 306a29f.

to exchange hammers, but that does not make any difference to the sounds the hammers make. So it is not the strength of the smith that counts. He then weighs the hammers and—according to our sources— obtains his result—even though this is impossible: the note will vary with the anvil, not the hammer.[40]

Those who actually engaged in the study of harmonics (as opposed to merely fantasising about the discoveries of Pythagoras) disagreed about how much of the phenomena to discount, and the epistemological debate, extensively reported in Porphyry especially,[41] is sometimes conducted in rather simplistic terms, as if it were a matter merely of deciding whether reason or perception is the ultimate criterion.[42] At

---

Gaudentius *Harmonica introductio* 11.340.4ff., Macrobius *Somnium Scipionis* 2.1.9ff., 2.96.23ff., Boethius *Mus.* 1.10.197.3ff.

40. The story of Hippasus constructing bronze disks to yield the harmonies is closer to what is possible: see schol. *Phd.* 108d, and cf. Raasted's discussion, 1979, mentioned above, Chap. 5 n. 212, relating this to the story of the hammers.

41. *In Harm.* 25.3ff. D.; cf. Ptolemy *Harm.* 1.8 (19.16ff. D.).

42. See, e.g., Ptolemais of Cyrene, quoted by Porphyry *In Harm.* 25.9ff. D., where three groups are distinguished: (1) those who "preferred" reason; (2) those "instrumentalists" who preferred perception; and (3) those who used both. Some of the "Pythagoreans" are here categorised as belonging to the first group, though Pythagoras and his "successors" are in group 3 (cf. Didymus in Porphyry *In Harm.* 26.15ff. D., who suggests that the Pythagoreans preferred reason but used perception with regard to the starting-point of the inquiry only, and cf. Porphyry's own opinion concerning the Pythagoreans at *In Harm.* 9.1ff., 15ff., 33.5ff. D.). As Barker has pointed out, e.g., 1981b, pp. 10f. and 16, many of those who were said to value reason more highly than perception did not totally reject the latter (cf. Aristoxenus at *Harm.* 2.33 and Didymus in Porphyry *In Harm.* 28.12ff. D., who point out that while the geometer can take what is not straight as straight, the musician cannot take what is not a fourth as a fourth). The point was often, rather, that perception cannot make fine discriminations. Nor can it settle the dispute as to whether or not the octave, fifth, and fourth are exactly six tones, three-and-a-half tones, and two-and-a-half tones, respectively, nor the question of whether or not either the tone or the semitone can be divided exactly into two equal intervals.

In one of the opposing traditions, that of Aristoxenus, the claim was that the unit of measurement must be something identifiable by perception. Thus Aristoxenus *defines* the tone as the difference between the fifth and the fourth, both concords immediately apprehensible to perception—whereas to the

one extreme there were those who sought to reduce the subject to number theory: some, we are told, maintained that since 8:3 is neither a multiplicate ratio (like 2:1 or 4:1) nor superparticular (like 3:2 and 4:3),[43] the interval of an octave plus a fourth cannot *be* a concord, even if it sounds like one.[44] Yet to that Theophrastus pertinently remarked

---

Pythagoreans the tone may, rather, be defined mathematically as the difference between sounds whose "speeds" stand in a ratio of 9:8. Whilst for the Pythagoreans musical relations are to be expressed as ratios between numbers, for Aristoxenus musical intervals are construed on the model of line segments and their interrelations investigated quasi-geometrically: cf. Barker 1978b, p. 4, 1981b, p. 3.

Each of these two major traditions faced its own corresponding difficulties or anomalies. Aristoxenus found it impossible, in practice, to carry through his programme of founding music theory on the basis solely of an appeal to what can be heard, on the principle that "what [the voice] cannot produce and [the ear] cannot discriminate must be excluded" from the sphere of useful and practically possible musical intervals (*Harm.* 1.14). In particular the principle of concordance, on which his theory relies, will not allow the construction of intervals smaller than the semitone (though he attempts to discuss these, e.g., at *Harm.* 1.21) (cf. Barker 1978a, pp. 15f.). Conversely, as I point out in my text, one problem for the Pythagoreans was that since 8:3 is neither multiplicate nor superparticular, they were led to ignore or deny that the interval of the octave plus a fourth is a concord—and similarly with the double octave plus a fourth.

For Ptolemy and Porphyry, who distance themselves from both the earlier Pythagoreans and from the Aristoxeneans, the chief problems connected with perception are identified as: (1) that different observers obtain different results (e.g., Porphyry *In Harm.* 18.12ff. D.); (2) that instruments may be unreliable (e.g., Ptolemy *Harm.* 1.8 [16.32ff. D.], Porphyry *In Harm.* 119.13ff. D., Ptolemy *Harm.* 2.12 [66.6ff. D.]); and (3) that perception cannot give the exact measure of very small intervals (e.g., Ptolemy *Harm.* 1.1 [4.13ff. D.], 1.10 [21.25ff. D., 24.20ff. D.], Porphyry *In Harm.* 20.12ff., 129.18ff. D., and cf. 75.25ff. D., quoting the pseudo-Aristotelian *De audibilibus*, and 80.22ff. D., reporting Aristoxenus). More generally the "rough-and-ready" character of perception is often referred to, e.g., Porphyry *In Harm.* 16.13ff., 17.6ff., 18.9ff., 19.2ff., 21ff. D.

43. A superparticular, or epimoric, ratio is defined as $n + 1 : n$, where n is a positive integer greater than 1.

44. See, for example, the discussion in Ptolemy *Harm.* 1.5f. (11.5ff., 13.1ff. D.), and in Porphyry *In Harm.* 95.25ff., 105.12ff., 124.4ff. D., and cf. Barker 1981b, pp. 9ff.

ἐνέργεια, or a tension, τάσις.[54] Divergent positions were also maintained on the further fundamental question of whether visual—or
light—rays form a continuum (as, for example, Ptolemy insisted)[55] or
are discontinuous (as appears to be assumed in Euclid's *Optics*),[56] and
this in turn affected beliefs concerning how far the programme of geo-
metrising optics could be carried through and on the constraints on
such a programme. Since it is quite clear that in the *Elements* Euclid
assumes that geometrical magnitudes are infinitely divisible,[57] it was
presumably not for purely *geometrical* reasons that he would have de-
parted from that assumption for optical phenomena in his *Optics*,
but, rather, for reasons to do with problems connected with the visibility of objects at a distance.[58] Yet some geometrisation of optics is
common ground to most investigations of perspective, reflection, and
refraction—including those of both Euclid and Ptolemy—to the extent at least that it was assumed, first, that visual/light rays can be
treated as straight lines,[59] and, secondly, that for some purposes the eye

---

54. Aristotle's own theory was that light is an actuality, ἐνέργεια, but
he records and criticises Empedocles' view that it is a movement, *De An.*
418b9ff., 20ff., *Sens.* 446a26ff., b27ff. In the Hellenistic debate, the Epicureans held that vision takes place through the reception of images (Epicurus
Ep. Hdt. 10.46, 49ff.), while for the Stoics it involved the tension of the medium between the eye and the object seen (see, e.g., Diogenes Laertius 7.157,
Aetius 4.15.3, Alexander *De anima libri mantissa* 130.14ff. Bruns).
55. At *Optics* 2.50ff., 37.4ff. Lejeune, Ptolemy insists, against Euclid, that
the visible flux is a continuum: cf. also *Optics* 2.48, 35.18ff., 36.5ff. L.
56. Euclid *Optics* Definition 3, 2.7ff., Propositions 1, 3, and 9, 2.22ff.,
4.26ff., 16.7ff., and cf. Theon's *Opticorum recensio* (*Opt.Rec.*) introduction
146.18ff. The interpretation of *Optics* Proposition 3 especially is, however,
disputed: see Brownson 1981, p. 174.
57. There were, however, those who denied this assumption and held that
geometrical magnitudes are made up of indivisibles, notably, in the Hellenistic
period, the Epicureans, and cf. also the pseudo-Aristotelian treatise *On Indi-
visible Lines*: see, most recently, Sedley 1976b.
58. This is the issue in Euclid *Optics* Proposition 3, 4.26ff., especially.
59. The rectilinear propagation of light, or the visual ray, was, however,
often justified, or argued for, on quasi-physical grounds, namely, that it will
take the shortest or quickest path: see Hero *Catoptrics* 2–4, Nix Schmidt, 2,
320.15ff., 24ff., 322.18ff., 324.21ff.; Olympiodorus *In Aristotelis Meteora
commentaria* 212.5ff. Stüve; Damianus *Opt.* 14.20.12ff. (quoting Hero's

can be considered as a point, the vertex of the visual cone.[60] Moreover, we have good grounds for supposing that the second of these assumptions was clearly recognised by some *as an idealisation*—since on certain occasions the fact that vision takes place not from a point but from a certain area was acknowledged. Archimedes, in particular, provides a sophisticated discussion of the allowance that has to be made for this in the context of his determination of the angular diameter of the sun in the *Sand-Reckoner*.[61]

Statics and Hydrostatics

As a third example we may take statics. Although here Archimedes does not state all his assumptions, he evidently discounts, for the purposes of his investigation of the lever, such factors as the possible variation in the material constitution of an actual metal bar and, more

---

Catoptrics); cf. also Euclid's use of the assumption that visibility depends on the *movement* of the visual rays along the object seen (*Optics* Proposition 1, 4.6ff., cf. Theon *Opt.Rec.* 148.17ff.). As noted above, n. 21, Mugler, 1957, 1958, has suggested that the *geometrical* idea of the straight line was derived from, or at least influenced by, optical phenomena, as when the straight line is defined as that in which the middle points are "in front of" the extremes: see [Euclid] *Catoptrics* Definition 1, 286.1f., and cf. Plato *Parmenides* 137e3f. (where, however, Mugler emends ἐπίπροσθεν to ἐπιπροσθέον), cf. Aristotle *Top.* 148b27f. Other cases where optics draws on physical analogies are when reflection of sight is compared with that of sound (see Aristotle *APo.* 98a25ff., *De An.* 419b28ff., cf. *Pr.* 11.45.904a30ff.), or when sight is compared with moving bodies (see Ptolemy *Optics* 3.19 [98.13ff. L.], cf. *Pr.* 16.13.915b18ff., 30ff.), though disanalogies are also noted, e.g., *Pr.* 11.58.905a35ff.

60. As in Euclid *Optics* Definition 2, 2.4ff., cf. Theon *Opt.Rec.* 154.8: where Euclid had had the vertex of the cone "in," ἐν, the eye, Theon substitutes "at," πρός, cf., e.g., Diogenes Laertius 7.157 reporting the Stoic view, and contrast the reservation expressed by Damianus *Opt.* 11, 12.12ff. In Aristotle, already, the source of light, and the eye, are occasionally treated as points, notably in his geometrical analyses of the halo and the rainbow, *Mete.* 3.3 and 5.373a4ff., 16ff., 375b19ff.; cf. Boyer 1945–46; Lennox forthcoming; Owen 1986a, pp. 317f.

61. Archimedes *Aren.* 1.10 (HS) 2.222.3ff., on which see, for example, Lejeune 1947–48, Shapiro 1975, pp. 82f. Cf. also the discussion of binocular vision in Ptolemy *Optics* 2.27ff. (26.18ff. L.), and 3.25ff. (102.13ff. L.) with Lejeune 1948, pp. 130ff., 145ff.

importantly, that the movement of a bar about a fulcrum will be accompanied by friction.[62] Similarly, in hydrostatics he stipulates explicitly that the fluid be perfectly homogeneous and totally inelastic.[63] Moreover, in his investigation, in the second book of *On Floating Bodies,* of the conditions of stability of segments of paraboloids of revolution of varying shapes and of varying specific gravities in a fluid, he assumes that he may talk, ideally, of the *centres of gravity* of plane segments of geometrical figures, as well as of the paraboloids themselves.[64]

These are, as I noted, on the whole comparatively straightforward cases, and they are, of course, among the most commonly cited examples of the successes of Greek science. There has, to be sure, been much, rather laboured, discussion of the possible circularity of the argument in Archimedes' statics—of the relationship between the first postulate, which states that "equal weights at equal distances are in equilibrium," [65] and the law of the lever subsequently demonstrated, on its basis, in propositions 6 and 7 of book 1 of *On the Equilibrium of Planes.*[66] Yet to the charge of circularity it might be countered that of course the law of the lever is in *some* sense presupposed at the beginning, but that is no objection: there is no vicious circularity but, rather, a quite unproblematic, indeed unavoidable, mutual entailment here between postulates and subsequent propositions.[67] That point aside, the type of idealisation involved in the studies we have consid-

---

62. Archimedes' assumptions are set out in the first book of *De planorum aequilibriis (Aequil.)* (HS) 2.124.3ff.

63. Archimedes *Fluit.* 1 Postulate (HS) 2.318.2ff.

64. Archimedes *Fluit.*, e.g., 2.2 (HS) 2.350.11ff., 27ff., cf. 1 Postulate 2 (HS) 2.336.14ff., 1.8 (HS) 2.338.26ff., 1.9 (HS) 2.342.15ff., *Aequil.* 1 Postulates 4f. (HS) 2.124.13ff., 16ff. As to whether the postulates in *Aequil.* and *Fluit.* can be thought of as defining implicitly the notion of centre of gravity, see Stein 1931, but contrast Suppes 1981, pp. 207ff., cf. also Schmidt 1975.

65. Archimedes *Aequil.* 1 Postulate 1 (HS) 2.124.3f.

66. *Aequil.* 1.6 (HS) 2.132.14ff. (for commensurable magnitudes), 1.7 (HS) 2.136.18ff. (for incommensurable ones), on which see Mach 1893, pp. 13f., cf. 1960, pp. 19f.; Duhem 1905–6, vol. 1, pp. 9ff.; contrast Knorr 1978b, p. 185; Suppes 1981, p. 212 n. 3.

67. Cf. the discussions of Knorr and Suppes cited in the last note.

ered so far is uncontroversial, indeed, not just uncontroversial but the fundamental factor on which the advances in understanding that were made depended. It is only by thinking away some of the features of the phenomenal situation that the underlying, mathematically expressible relations can be revealed.

Dynamics

We can go further: it was partly because of a failure to think away *sufficient* of the factors in the phenomenal situation that the ancients, Aristotelians and anti-Aristotelians alike, failed to arrive at satisfactory resolutions to the problems in the field we identify as dynamics. Aristotle himself, for instance, *argues,* as we saw in Chapter 5, that motion *must* be through a medium; in some of the texts setting out the proportionalities of natural motion he has the express purpose of *disproving* the void.[68] Philoponus attacked the Aristotelian position on the role of the medium and maintained that it acts purely as a resistance to the moving object.[69] Yet Philoponus, like Aristotle, assumed that weight is *one* of the factors that determine the speed of a freely falling object and, moreover, not only that it does so in a plenum but also that it would do so in a void.[70] But if you take as your explanandum, or as one of your explananda, motion *through a medium,* this is bound to prove a major stumbling block to analysis, if only because of the difficulty, indeed the impossibility, of *quantifying* the factor that corresponds to the density of the medium—a problem that is expressly remarked on by Philoponus.[71]

---

68. Aristotle *Ph.* 4.8.215a31ff., b12ff.; see above, Chap. 5, pp. 217ff.

69. Philoponus *In Ph.* 647.9ff., 681.10ff., 682.29ff., especially, and cf. above, Chap. 5 at n. 29.

70. Philoponus denies that an actual continuous void exists in nature: see Sedley 1982a and forthcoming, Furley 1982. We may contrast the position of the Epicureans, who both asserted the void and maintained that in the void heavy and light atoms move with equal speed "as quick as thought," Epicurus *Ep. Hdt.* 10.61, cf. 48, and cf. Furley 1967, pp. 121ff.

71. Philoponus *In Ph.* 683.1ff., see above, Chap. 5 n. 33.

Aristarchus' heliocentric theory, as reported by Archimedes, namely, that not just the earth but the circle in which the earth moves around the sun is as a point in relation to the sphere of the fixed stars (see Figure 2).[82] Archimedes' own comment is that that is, strictly speaking, impossible, since a point has zero magnitude and the fixed stars would then be at infinite distance[83] (a similar point applies, of course, to the first type of parallax case as well). What Aristarchus needs is not that the stars be infinitely, only that they be indefinitely, far away.

The interesting feature is that he evidently incorporated this *into his assumptions,* in part in order to meet a possible objection to heliocentricity. If the earth moves in a circle around the sun (rather than the sun around the earth) there should be, one might think, observable differences in the shapes of the constellations as viewed from different points in the earth's orbit—from the points representing the spring and autumn equinoxes, for instance, at opposite ends of the same diameter of the orbit. Yet no such variation was observed; indeed, stellar parallax was not confirmed until well into the nineteenth century, with the work of Bessel and others around 1835. Aristarchus seems to have appreciated that this otherwise very damaging objection to heliocentricity was no objection at all *provided* that the stars are sufficiently far away. If the diameter of the earth's orbit around the sun is negligible in comparison to the diameter of the sphere of the fixed stars, then you would not expect *observable* variations in the relative positions of the stars, certainly not within the limits of ancient techniques of observation. Unlike Ptolemy's discussion of the size of the earth in *Syntaxis* 1.6, the assumption in the form adopted by Aristarchus was not itself justified by reference to independently observable phenomena; there was no way in which it could be. Rather, this reveals precisely what has to be accepted *among the assumptions* in order for an apparent objection from the side of the phenomena *not* to be the objection it seems. No doubt Aristarchus could have argued that the

82. Archimedes *Aren.* 1.4 (HS) 2.218.7ff.
83. *Aren.* 1.6 (HS) 2.218.18ff.

FIGURE 2.   Three cases of discounted parallax. In each case circle B is treated as a point.

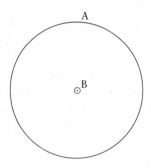

Case 1: Ptolemy *Syntaxis* 1.6: circle A = fixed stars;
circle B = the earth.

Case 2: Aristarchus' heliocentric theory: circle A = fixed stars;
circle B = circle in which the earth orbits the sun.

Case 3: Aristarchus' *On the Sizes and Distances:* circle A = moon's orbit round earth;
circle B = the earth.

inability to confirm an assumption directly does not make it untrue—and Copernicus would have said the same.[84]

The third type of case again comes from Aristarchus, this time from the extant treatise *On the Sizes and Distances of the Sun and Moon.* The second hypothesis set out there is that the earth is as a point, not in relation to the sphere of the fixed stars, but in relation to the *moon's orbit.*[85] In this form the assumption involves discounting *lunar* paral-

---

84. Cf. Copernicus *De revolutionibus orbium coelestium (Rev.)* 1.6. Already those Pythagoreans who treated the earth as one of the planets (see above at n. 12) had argued, according to Aristotle, that there is no difficulty in supposing that the phenomena would be the same as they would be if the earth were at the centre—even though they denied that. "For there is nothing to show, even on the current view, that we are distant half the earth's diameter [viz. from the centre]," Aristotle *Cael.* 293b25ff., 29f.

85. Aristarchus Hypothesis 2 (352.5f. Heath).

fifth takes it that the breadth of the earth's shadow, viz., at the moon, is two moons.[91] Moreover, the second hypothesis itself not only discounts lunar parallax but takes the moon to move in a simple circle with the earth as centre—and no serious Greek astronomer had thought *that* since before Eudoxus.

Such a set of hypotheses would, of course, be an unmitigated disaster in any attempt to arrive at concrete determinations of the actual sizes and distances of the moon and sun. What Aristarchus is doing, rather, is exploring the geometry of the problems. Given certain assumptions—and it will not matter, for the sake of the geometry, that some of the values are a little, and others wildly, inaccurate—what follows? The study is certainly *relevant to* astronomy, in particular because it shows *how* one *could* obtain *actual* solutions to the astronomical parameters, that is, it offers one set of answers to the question of the premises, or data, needed in order to arrive at such solutions. Yet it remains itself essentially a study of the geometry of the problems.[92]

The Aims and Assumptions of Greek Astronomers

As this last example shows, certain types of simplifying assumption involve not so much discounting a value that can—with greater or less justification—be deemed to be negligible, as a veritable *mutation* of the problem. Once certain of the known empirical data are suspended, the study becomes one of pure geometry and does not offer to solve, though it remains relevant to, the astronomical problems themselves. Now, it is just such a shift that Duhem and his followers saw as typical of the dominant strand in Greek astronomy: a lack of concern with the physics of the problems in favour of a preoccupation with the mathematics, the construction of models that are purely calculating devices with nothing to do with any underlying physical realities.[93] It is

---

91. Hypothesis 5 (352.13 Heath).

92. Cf. Neugebauer 1975, vol. 2, p. 643, quoted above, n. 88.

93. In his 1908, e.g., pp. 120, 281, 284 especially, Duhem drew a fundamental contrast between two views on astronomical hypotheses. On the one hand, they might be treated as "simple mathematical fictions," "pure concep-

indeed undeniable that there are instances (Aristarchus' treatise is one) where the problems are treated, at least for the time, as problems of geometry. There is certainly a tradition of the investigation of the mathematics relevant and useful to astronomy that exists side by side with astronomy itself—a tradition that goes back to Autolycus' work *On the Moving Sphere*.[94] But that cannot be said to vindicate the line of interpretation that Duhem advocated. What that line of interpretation itself discounts, or at least seriously underestimates, is an equally undeniable concern with more than just the mathematics of the problems in much—indeed, in my opinion, most—ancient Greek astronomy.

In what are admittedly complex issues,[95] the chief objection to Duhem can most easily be illustrated in relation to Ptolemy himself.

---

tions," where there is no question of their being "true" or even "probable," where "true" is glossed as "in conformity with the nature of things." On the other, they might also be held to describe "concrete bodies" and "movements that are actually accomplished." Duhem is in no doubt that the former represents the position of the major Greek astronomers and commentators: cf. Wasserstein 1962, p. 54, and contrast G. E. R. Lloyd 1978b.

94. Autolycus is represented as active around 330 B.C.: see Mogenet 1950, pp. 5ff.; Neugebauer 1975, vol. 2, pp. 573, 750f.; Aujac 1979, pp. 8ff. On Autolycus and the tradition he represents, see also I. Mueller 1980, pp. 106ff.

95. The major difficulty we face is that for several important astronomers, such as Apollonius, we have no direct evidence, and for others, such as Hipparchus, only secondary reports of doubtful reliability. Thus Theon 188.15ff. does not—pace Duhem 1908, pp. 119f.; cf. G. E. R. Lloyd 1978b, pp. 217ff.—characterise Hipparchus as indifferent to physics. On the contrary, he represents him as adopting the epicyclic hypothesis in preference to the eccentric one not for purely mathematical reasons, but on general and cosmological grounds: "it is more plausible that all the heavenly bodies should lie symmetrically with regard to the centre of the universe and be joined together similarly." (Moreover, Theon is not only a naive realist himself, but represents Greek astronomy as a whole as founded on the study of nature, contrasting it with Babylonian, Chaldaean, and Egyptian astronomy in just this regard, 177.20ff.). While the positions of Eudoxus and Callippus on this issue are a matter of conjecture (see above, n. 73), it is clear that Aristotle demanded a physically unified system in *Metaph.* Λ 8.1073b38ff. As for our single most important and most comprehensive source, Ptolemy, both the *Planetary Hypotheses* (*Plan.Hyp.*) and the *Syntaxis* make it abundantly clear, in my view, that his strategic intention was to provide not just a mathematical, but also a physical account of the phenomena.

FIGURE 3.   Ptolemy's model to explain the moon's second anomaly.

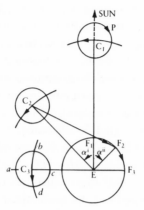

The moon's epicycle, centre C, moves round a centre (F) which itself describes a circle round the earth (E). F moves round E with the same angular velocity as, but in the contrary sense to, the movement of C round E. At position (1) (syzygy) the model is equivalent to a simple epicycle model. At position (3) (quadrature), when the moon is at apogee or perigee on the epicycle (that is, at *a* or at *c*) the model is again equivalent to a simple epicycle model so far as the moon's angular distance from the sun in concerned. But at position (3), when the moon is at *b* or *d*, midway between apogee and perigee on the epicycle, the effect of the new model is to increase the apparent diameter of the epicycle. Derived from G. E. R. Lloyd, *Greek Science after Aristotle*, fig. 27.

minimum distance (see Figure 3).[101] The angular diameter of the moon should, in turn, vary roughly by a similar amount, a factor of 2. Yet it patently does not. Nor does the evidence in the *Syntaxis* suggest that Ptolemy thought that it did. It is true that he gives specific values only for the maximum and minimum apparent diameter at syzygy (not for the maximum at quadrature),[102] but those he offers are of the right

---

101. Taking the distance from the earth (the centre of the ecliptic) and the centre of the moon's epicycle (R = EC, in the diagram) as usual as 60 parts, Ptolemy estimated the radius of the epicycle (r = C_1P in the diagram) as 5; 15p expressed sexagesimally (i.e., 5 15/60ths) and the eccentricity (e = EF in the diagram) as 10; 19p, *Syntaxis* 5.4 (H) 1.366.15ff. Maximum geocentric distance, at syzygy, i.e., R + r, will then be 65; 15p. Minimum geocentric distance, at quadrature, i.e., R − 2e − r, will be 34; 7p, only slightly over half the value for the maximum. Cf. Neugebauer 1952/1957, pp. 194ff., 1975, vol. 1, pp. 86ff., cf. Pedersen 1974, pp. 192ff.

102. The modification that Ptolemy introduced to Hipparchus' lunar model was to suppose that the centre of the deferent circle (F in the diagram) itself moves in a circle round the earth (E) at the same angular velocity as, but in the

order of magnitude, a minimum of 31′ 20″ and a maximum at perigree at syzygy of 35′ 20″.[103] Moreover, and to complicate matters further, the *Planetary Hypotheses* tackles the problem of how the spheres of the heavenly bodies nest into one another, where it is assumed that the maximum distance of one body corresponds to the minimum distance of the one above it, and there Ptolemy clearly accepts the geometrical consequences of his epicycle-eccentric model in the *Syntaxis* as correct.[104] So far from being embarrassed by those consequences, he takes them as the basis of his calculations of the absolute distances of the moon and other bodies.

Here, then, we have a major discrepancy between the theory and the data of observation, and Ptolemy's lack of embarrassment just increases ours, since it looks as if he has simply discarded part of the phenomena quite arbitrarily. That he *has* discarded part of his data is clear. That it is quite arbitrary is more debatable. We have to recall that the theory worked extremely well as a theory of the longitudinal positions of the moon, where it represented a quite marked improvement on Hipparchus' lunar model, which itself gave tolerably serviceable results.[105] As regards the lack of appreciable variation in the apparent size of the moon, how—without recourse to desperate expedients— Ptolemy thought he could get round the difficulty we do not know.[106] But he may not have despaired—presumably he did not despair—of

---

contrary sense to, the movement of the centre of the epicycle (C) round the earth: angle a^i = angle a^{ii} in the diagram. The effect of this modification on the geocentric distance of the moon, and so on its apparent diameter, is at a maximum at quadrature, but decreases to zero at syzygy (conjunction or opposition): in the latter case the difference between apogee and perigee corresponds simply to the epicycle's diameter.

103. See *Syntaxis* 5.14 (H) 1.421.3ff., 6.5 (H) 1.479.14ff. In his comments on the hypotheses in Aristarchus' *On the Sizes and Distances*, Pappus attributes the same figures to Ptolemy, apparently as if they were his definitive values: *Collectio* 6.37.71 (2.556.17ff. Hultsch).

104. See *Plan.Hyp.* 1 part 2 chap. 3, Goldstein 1967, pp. 7f., and cf. *Plan.Hyp.* 2.16 (138.14ff. H.).

105. On the superiority of Ptolemy's final lunar model, in *Syntaxis* 5.5 (H) 1.367.3ff., to its predecessors, see Petersen 1969 and Gingerich 1980, pp. 256ff.

106. Thus there is no suggestion that Ptolemy wavered in his belief in the constancy of the radius of the moon's epicycle, though this possibility has been canvassed in the case of Hipparchus; see Neugebauer 1975, vol. 1, pp. 315f.

some explanation being possible. In his view, we may assume, the difficulty presented by this phenomenon was not great enough to justify his abandoning the model as a whole.

As we have noted before, until such time as a superior model is available, any scientist would be justified in continuing to maintain, in the face of *prima facie* counter-evidence, a theory that had shown its ability to account for at least part of the phenomena, though every scientist *should* in principle, to be sure, be *especially* self-critical on the question of when the strength of the counter-evidence is such as to make a new model imperative.[107] Yet the price Ptolemy has paid in this case is clear: the elision of part of the data is here no mere temporary

---

107. This raises, of course, one of the most hotly contested topics in recent philosophy of science, namely, the rationality of different responses to anomalies or counter-examples in scientific research programmes. The issues have been debated in connection especially with Kuhn's ideas on the generation of "crisis" in normal, puzzle-solving, science, and with those of Lakatos on the relation between the "hard core" of a scientific research programme ("irrefutable" by the methodological decision of its proponents) and its "protective belt" of auxiliary hypotheses (where anomalies may abound and the programme still remain "progressive"—defined in terms of increasing content) and where Lakatos distinguishes between *different* types of ad hoc procedure, including the strategies he dubbed "monster-barring," "exception-barring," and "monster-adjustment" (this last defined as turning a counter-example, in the light of a new theory, into an example). See most notably Kuhn 1962/1970, and 1977; Lakatos 1976, 1978a, e.g., p. 48, p. 63 n. 3; Laudan 1977, 1981a; Newton-Smith 1981; and Hacking 1983. Ptolemy's astronomy as a whole, which has been much discussed in relation to Copernicus especially (Kuhn 1957; Lakatos 1978a, chap. 4, pp. 168ff.), represents, from an ancient perspective, the most elaborate and fully developed research programme in ancient science—and it was one that was to prove extremely durable. The example of the variation in the angular diameter of the moon shows that, on occasion, Ptolemy, confronted with an anomaly, offered no adjustment but simply passed it by in silence. When the general problem of the apparent complexity of his devices is explicitly raised *as a problem,* in *Syntaxis* 13.2 (H) 2.532.12ff., his response, as we have seen, n. 78, is to *claim* that he has attempted to use, so far as possible, the simplest hypotheses, but then also to argue that "simplicity" in regard to heavenly things should not be judged by our ordinary criteria of simplicity. At this, the point where the question of the viability of the system as a whole comes, perhaps, closest to the surface, there is an uncomfortable tension between simplicity invoked as a *criterion* to judge between theories, and the *axiomatic* assertion that the movements of the heavenly bodies are simple, however complicated they may appear to us.

simplification for the sake of argument, but represents a limitation on the viability of the model itself.

Conclusions for the Exact Sciences

Our survey, to date, of some of the main types of simplification and idealisation in Greek science has been drastically selective, but some points have, I hope, emerged sufficiently clearly. The move to discount some of the phenomena in question is associated with some of the most notable successes of Greek science. On occasion, to one looking back with the benefits of hindsight, it seems as if the problem in ancient science is too little abstraction from the complexities of the phenomenal situation, rather than too much, though there was—there is—no way of telling in advance when this may be the case. Some attention, at least, is paid by Ptolemy, for one, to the conditions under which simplification is possible, though he provides no exact rules, only rough-and-ready practical guidelines, with his appeal to a vague, certainly unquantified, notion of "negligible difference." That already indicates some awareness of the problem of discounting parts of the phenomena to leave in play only what is readily explicable. While this was often a sensible policy, it could also prove all too facile a manoeuvre, when recalcitrant data that are central to a problem are simply ignored and when there is no question of their being reintroducible, in principle, at a later stage, with the theory remaining intact.

The Life Sciences: Teleology

Thus far we have taken our examples from the exact sciences, but comparable moves can be documented also in other areas of ancient speculative thought with results that must provoke further reflection on the aims and methods of ancient investigators. Teleology offers a general rubric under which we can discuss some especially striking examples from various areas of the life sciences.[108] From the time of

---

108. On the pre-Platonic background see especially Theiler 1924, Solmsen 1963, Balme 1965; cf. Sorabji 1980a.

Plato, at least, the notion that the world as a whole is well-ordered, the product of divine design, is one of the most powerful motive forces in ancient science, though it remained a far from unanimous, indeed a much contested, view.[109] The question we may raise in this regard is whether or how far teleological accounts were secured at a cost of discarding part of the phenomena in a way that is broadly comparable with the elisions we have studied from the exact sciences.

In the first developed statement of a teleological cosmology, in Plato's *Timaeus*—and often thereafter—it is explicitly recognised that other factors besides the good have to be taken into account. Plato's Demiurge and the workings of reason are confronted with the factor of necessity, exemplified by the concomitance of material properties, as when the hardness of bone necessary to protect the head is inseparable from a heaviness that weighs it down and makes for insensitivity.[110] The human race would have been longer-lived but less intelligent if the head had been covered by a thicker layer of bone: as it is, unable to secure *both* long life *and* intelligence, god sacrifices the former, lesser end to insure for humans, as far as possible, a noble and intelligent, if shorter, life. In Aristotle, too, the final cause is often contrasted with what he too calls necessity—that is, simple, not conditional, necessity[111]—again associated with the material properties of things. Some

---

109. Teleological explanation was, of course, denied, both before and after Aristotle, by the atomists especially.

110. Plato *Ti.* 74e ff. On the role of necessity and of what Plato calls the "wandering cause" (*Ti.* 48a7) in the *Timaeus,* see especially A. E. Taylor 1928, pp. 299f.; Vlastos 1939/1965, 1975a; Morrow 1950/1965; Alt 1978.

111. As we have noted before, Chap. 3 at n. 149, Aristotle distinguishes between several different types or modes of necessity in passages in the Organon, the *Metaphysics,* and the physical and biological works, and the interpretation of his distinctions and the question of their consistency have been the subject of protracted recent debate. Apart from (1) necessity in the sense of the compulsory or the violent, and (2) absolute or unqualified necessity relating to "eternal" things, including both (2a) the eternal heavenly bodies, and (2b) the (timeless) truths of mathematics, two further types of necessity are of particular concern to the natural philosopher. (3) "Conditional" or "hypothetical" necessity is explained in terms of the material conditions necessary for an end to be realised; thus for a house to be built, matter of a particular kind must be available; for an axe to be able to cut, it must be of suitable material. But this is

things are so because they have to be so, or happen to be—the colour of the eyes, for instance—not for the sake of some end.[112]

Since teleology does not apply without exception in Plato or Aristotle,[113] there is no need for failures of the good to be *denied*—though both philosophers will insist that the Demiurge or nature has secured

---

in turn contrasted with (4) the necessary consequences of the natures of things or of their being as they are. While (3), conditional necessity, works *with* the final cause, as the necessary condition for the realisation of some good, (4) is sometimes contrasted with the good. The distinction is drawn at *PA* 642a31ff. in connection with the example of respiration, where (4) is illustrated by the necessary behaviour of the hot substance moving in and out and of the air flowing in. Again, at *PA* 677a11ff., discussing bile, Aristotle contrasts those residues that nature is able to turn to some advantage with those where this is not the case. At *PA* 662b23ff., 663b22ff., for instance, he had pointed out that in the case of horns, surplus earthy matter is made to serve the purposes of self-defence and attack, but bile itself is not "for the sake of something" but arises merely as the *consequence* of other things that *are* for some good (*PA* 677a16ff.). Again at *GA* 5.1.778b10ff., 16ff., when he contrasts what arises merely in the process of generation or formation with what is contained in the essence of an animal and is for the sake of some good, he distinguishes the necessity for an animal to have an eye (where that characteristic belongs to the essence of the animal in question) and the necessity for it to have an eye of a particular kind (the result merely of the natural processes of formation). Cf., for example, *APo.* 94b27ff., *Ph.* 2.8.198b10ff., *GC* 337a34ff., *PA* 1.1.639b21ff., 642a1ff., *Metaph.* 1015a20ff. The issues are controversial, but among the most important recent studies are Balme 1965, 1987; Kullmann 1974a, 1979; Preus 1975; Gotthelf 1976–77; Nussbaum 1978, pp. 59ff.; Sorabji 1980a; Waterlow 1982a; Lennox 1981, 1985; Leszl 1982; Cooper 1982, 1987; Furley 1985.

112. Aristotle states emphatically in such passages as *PA* 677a14ff. and *GA* 5.1.778a32ff. that some things are not for the sake of some good, specifying some residues in the former text, and a range of conditions such as the colour of the eyes, the pitch of the voice, and differences in colour of hair or of feathers in the latter chapter, 778a16ff. It is, however, striking that while he notes that the colours of the eyes in humans vary and initially appears to treat this as a matter of indifference, he comes to express the view that colour correlates with the amount of fluid in the eye, that there is an advantage in this being neither in excess nor deficient, and that the best eyes will correspondingly be intermediate in colour, neither "black" nor "grey-blue" (γλαυκός), the latter in particular being characterised as a weakness and due to a lack of concoction: see *GA* 779a26ff., b12ff., 34ff., 780b6ff.

113. The limits of teleological explanation are also discussed in Theophrastus' *Metaphysics:* see above, Chap. 3 at nn. 161ff.

the best possible results, within the constraints of necessity.[114] Failures need not simply be elided, since they can be laid at the door of necessity or the recalcitrance of matter, even though for Plato, and to some extent for Aristotle, that risks amounting to a concession that they are, in that respect at least, *beyond* explanation, since a *proper* explanation, by definition, will be in terms of the good.[115]

Moreover, the *normative* role of the concept of nature in Aristotle, in the zoological treatises especially, deserves remarking. His official and explicit statement, many times repeated, is that nature corresponds to what happens always or for the most part.[116] In practice, however, "natural" is sometimes reserved not for what happens usually, but for what is quite exceptional.[117] This is the case, for instance, where we are told that in humans alone the natural parts are fully "according to nature": in humans alone the "upper" part is directed to

---

114. See, e.g., Plato *Ti.* 30a, Aristotle *Ph.* 259a10ff., 260b21f., *GC* 336b27f., among a very large number of other texts. Similar themes often recur, of course, in both later philosophy and science, as, most notably, in the Stoic doctrine of providence and in Proclus, both in his commentaries on Plato (e.g., *In Ti.* 1.370.13ff.) and in *Hyp.* (e.g., 1.4.20ff.). Cf. also on Galen, below at nn. 126ff.

115. As we have seen (above at nn. 15f.), in the *Phaedo* 99a–b necessary conditions are contrasted with the αἴτιον, though in the *Timaeus* 48a7 (above, n. 110) the "wandering cause" so-called is an αἰτία of a kind (cf. the contrast between αἴτια and συναίτια at *Ti.* 46d, cf. 68e). In Aristotle, the material cause certainly answers to a legitimate question, but he makes clear his view that the natural philosopher is less concerned with it than with the formal and final causes (especially *PA* 1.1 and 5.640b22ff., 641a10ff., 642a13ff., 645a30ff.).

116. See especially *Ph.* 2.8.198b10ff., 34ff., 199b23ff., on which see most recently Furley 1985.

117. Aspects of the normative role of the concept of nature in Aristotle are discussed in my 1983a, especially part 1, chap. 3, pp. 26ff., on the use of man as model in his zoology. From one point of view, where the nature of a living creature is identified with its peak, ἀκμή, or τέλος in the sense of goal, old age and decay, like any loss of power, are "against nature," παρὰ φύσιν, *Cael.* 288b14ff. But from another he recognises that they are regular and indeed distinguishes cases where old age, decay, and death happen "naturally" or "according to nature," κατὰ φύσιν, from those where an external factor, violence or constraint, is involved: see *Mete.* 4.1.379a3ff., *Resp.* 478b24ff.

the "upper" part of the universe, and the right side is *most* right-sided.[118] Here what Aristotle describes as "natural" is what he deems to be best, where he uses the human species as the norm by which the rest of the animal kingdom is to be judged. By that standard all other animals fall short. Yet the ideal they fall short of is still referred to as what is "according to nature" or "natural," despite the fact that this picks out not what happens always nor even for the most part, but what is true of humans alone.

Many of the parts and functions of the lower animals are thus evaluated from the point of view of those of higher species, especially of the supreme species, humans. The heuristic value of this idea is clear. It enables Aristotle to recognise, for example, that in the so-called blood-less animals there is an analogous fluid that performs the same functions as blood,[119] or, again, that there are analogues to the heart, in his view the control centre of the vital functions.[120] Yet, equally clearly, he is led to make some very dubious value judgements. He speaks repeatedly of the parts of certain animals, or of whole species, as *deformed* or *maimed,* using the very same terms,[121] such as ἀνάπηρος, πεπηρωμένος, κολοβός, that are used of deformed individual specimens (as it might be an octopus with a tentacle missing). We can understand the use of such terms in relation to the mole's eyes, for example, which he believes not to function as eyes.[122] Yet he also calls the whole genus of

---

118. See especially *PA* 656a10ff., *IA* 706a16ff., 20ff., b9f., *HA* 494a26ff., 33ff.

119. See especially *HA* 489a20ff., *PA* 645b8ff., 648a1ff., 19ff., 678a8f., *GA* 728a20f.

120. What is analogous to the heart exists in lower groups of animals not only as the receptacle for what is analogous to blood (e.g., *PA* 665b11ff.), but also as the centre of perception, imagination, and locomotion in those animals that have them and indeed of life in general (e.g., *Juv.* 469b3ff., *PA* 647a30f., 678b1ff., *De Motu Animalium* 703a14ff., *GA* 735a22ff., 738b16f., 741b15ff., 742b35ff., 781a20ff.).

121. See, for example, *PA* 684a32ff. (on lobsters), *IA* 714a6ff. (on flatfish), *PA* 695b2ff. (on fish in general), and other texts discussed in G. E. R. Lloyd 1983a, pp. 40f.

122. *HA* 533a2ff., but cf. *HA* 491b27ff. and *De An.* 425a10f. Compare also *PA* 657a22ff. on the seal's "ears."

the ὀστρακόδερμα, testacea (including, for example, the snails, mussels, and oysters), "maimed, as it were," in that the way they move is "contrary to nature."[123] Again, compared with humans, *all* other creatures are said to be "dwarf-like"—in that they have their "upper" parts, or those near the head, larger than the "lower" ones.[124] It is not that Aristotle's teleology leads him straightforwardly to *deny* the phenomena, but he certainly denies that some are (fully) natural, and this tends to downgrade them as the subject matter for the inquiries of the physicist.[125]

Similar tendencies are particularly prominent in Galen. The whole of his treatise *On the Use of Parts* and many other extended passages in other works are devoted to establishing and illustrating the thesis that nature does nothing in vain, which he often, indeed generally, construes as not just a general but an exceptionless rule: *every* part has a purpose, and nature is perfect.[126] But the tension this thesis sets up can be seen in his anatomy, his physiology, and his pathology.

---

123. *IA* 714b8ff., 10ff., 14ff. Cf. also *PA* 683b18ff.: the testacea, having their head downwards, are said to be "upside down," as also are the plants, in that they take in food through their roots (cf. *PA* 686b31ff., *IA* 705b2ff., 706b5ff.).

124. "Upper" is defined functionally in relation to the intake and distribution of food. See *PA* 686b2ff., 20ff., 689b25ff., 695a8ff. At *IA* 710b12ff. infants are said to be dwarf-like in comparison with adults.

125. As we have noted before (Chap. 1 n. 160), Aristotle explicitly states at *PA* 645a15ff., 21ff., that all natural things, and especially all animals, have some share of the admirable and the beautiful—and this insures that all are accordingly and to that extent worthy of investigation. Yet the shares are evidently unequal, and while in principle there need be no reason why downgrading in regard to a creature higher in the hierarchy should entail downgrading as a subject of study, it follows from Aristotle's insistence on the physicist's concern with the final cause (and the good) that the extent to which they are manifested in any given subject matter has direct repercussions on how worthwhile the investigation of that subject matter is, at least under that heading. Moreover, in practice, in the zoological treatises, although there are exceptions, the attention Aristotle devotes to the various kinds of animals broadly reflects, even if it is not to be precisely correlated with, his view of their position in the overall hierarchy.

126. The strong thesis—that there is no possible improvement to the work of nature—appears, for example, in *UP* 1.5 (H) 1.6.18ff., (K) 3.9.4ff., 3.10 (H) 1.177.20ff., (K) 3.242.5ff., 3.16 (H) 1.190.10ff., (K) 3.259.3ff., and 5.5

Thus, developing a common Greek idea,[127] Galen represents apes as caricatures of human beings.[128] Yet, as is well known, he often uses apes—as others had done before him—as the basis of his anatomical descriptions of humans; his account of the muscles, for example, explicity derives from his dissections of apes.[129] From the point of view of transferring conclusions to human anatomy, the ape had better be as close to us as possible.[130] Yet so far from some of Galen's great admiration for humans extending to the ape, he calls the ape ridiculous: it has a ridiculous soul and so also a ridiculous body.[131] Yet why, if nature does nothing in vain, it plays such jokes, is not explained—and similar points apply also, regrettably, to Galen's account of the human female.[132]

(H) 1.267.12ff., (K) 3.364.17ff. (where the Erasistrateans are rebuked for failing to demonstrate how nature is worthy of praise in detail—by considering each organ in turn). Elsewhere, however, Galen recognises that nature cannot achieve perfection (see below [at n. 139]) and must weigh the balance of advantage (e.g., *UP* 5.4 (H) 1.260.1ff., (K) 3.354.17ff.).

127. See, for example, Aristotle *Top.* 117b17ff., cf. Archilochus 81, 83, Semonides 7.71ff., Plato *Hp.Ma.* 289a–b, cf. Vegetti 1983, pp. 59ff.

128. See, for example, *UP* 1.22 (H) 1.58.18ff., (K) 3.79.18ff., 3.8 (H) 1.152.21ff., (K) 3.208.15ff., 3.16 (H) 1.194.11ff., (K) 3.264.9ff., 11.2 (H) 2.117.14ff., (K) 3.848.8ff., 15.8 (H) 2.367.15ff., (K) 4.252.5ff., *AA* 4.1 (K) 2.416.3ff.

129. See, for example, *AA* 4.2 (K) 2.423.5ff., 5.9 (K) 2.526.4ff. and cf. 1.2 (K) 2.222.2ff., 3.5 (K) 2.384.12ff., 6.1 (K) 2.532.5ff., 6.3 (K) 2.548.2ff.

130. Galen recognises, indeed, that different species of apes resemble humans to different degrees, and he recommends using those that are most like man, with short jaws and small canines: *AA* 1.2 (K) 2.222.5ff., 223.9ff., 6.1 (K) 2.532.5–535.15 (specifying apes with an upright gait, a thumb in the hand, and a temporal muscle like that in humans), *UP* 11.2 (H) 2.114.17ff., (K) 3.844.7ff.

131. See *UP* 13.11 (H) 2.273.8ff., 23ff., (K) 4.126.1ff., 15ff., in addition to the passages cited above in n. 128.

132. Galen follows Aristotle in maintaining that the female is colder and wetter than the male, and that the woman is less perfect than the man. He holds that while she produces seed, this is imperfect, and he even repeats Aristotle's view that the female is like a deformed creature, οἷον ἀνάπηρον. See *De semine* 1.7 (K) 4.536.16ff., 1.10 (K) 4.548.6ff., *UP* 11.14 (H) 2.154.20ff., (K) 3.900.10ff., 14.5 (H) 2.295.27ff., (K) 4.157.12ff., 14.6 (H) 2.296.8ff., 299.3ff., 19ff., (K) 4.158.3ff., 161.12ff., 162.10ff. In the last text especially he remarks that "you should not think that our creator would willingly make half

Elsewhere—in, for example, his account of the blood-vascular system—we can see the heuristic value of the principle that each part
serves a purpose, yet his identification of purposes is selective. Thus
he infers that blood is transferred from the right to the left side of
the heart, in the adult, through pores in the septum separating the ven-
tricles. He does not claim that these pores can be seen, though pits in
the septum suggest their beginnings.[133] But they are necessary to account for blood in the arterial system.[134] However, he knows very well
that in the *embryo* there is a direct route for the blood between the
two atria, namely, through the foramen ovale, and he also knows that
this closes after birth.[135] *Why* it should close should be a problem,
since if it had remained open nature would not have needed the interventricular pores. Yet Galen quite fails to discuss this, merely remarking how marvellous it is that the foramen closes after birth and *asserting* that it would be of no use in the adult.[136] It is enormously to his
credit that he realises that the communications in the embryo heart
differ from those in the child once born, and it is to his credit too that

---

the whole race imperfect and as it were maimed unless there were going to
be some great use for this mutilation." μὴ γὰρ δὴ νομίσῃς, ὡς ἑκὼν ἄν ποτε
τὸ ἥμισυ μέρος ὅλου τοῦ γένους ἡμῶν ὁ δημιουργὸς ἀτελὲς ἀπειργάσατο
καὶ οἷον ἀνάπηρον, εἰ μή τις καὶ τούτου τοῦ πηρώματος ἔμελλεν ἔσεσθαι
χρεία μεγάλη.

133. *Nat.Fac.* 3.15 (H) 3.251.27ff., (K) 2.207.17ff.

134. Galen refuted the view, held by Erasistratus among others, that the
arteries normally contain air (though Erasistratus appreciated that blood flows
from a severed artery, arguing that in lesions blood passes from the veins to the
arteries through invisible pores, συναναστομώσεις): see especially the treatise
An in arteriis natura sanguis contineatur (K) 4.703ff., Furley and Wilkie 1984,
pp. 144ff. But holding that the blood is produced in the liver, from which it
enters the venous system, Galen was faced with the difficulty of explaining
how the blood passes into the arteries.

135. See especially *UP* 6.21 (H) 1.371.4ff., (K) 3.510.2ff., 15.6 (H)
2.360.19ff., 361.12ff., (K) 4.242.18ff., 243.18ff., *AA* 12.5 (pp. 118ff. D.).
Galen also clearly knows of the ductus arteriosus, the communication, in the
foetal heart, between the pulmonary artery and the aorta, e.g., *AA* 12.5
(pp. 118ff. D.), 13.10 (p. 179 D.), cf. (K) 2.828.10ff.

136. See *UP* 6.21 (H) 1.374.4ff., (K) 3.514.2ff., and 15.6 (H) 2.362.1ff.,
(K) 4.244.14ff.

he does not fudge the function of the foramen ovale, of which he provides the first extant description.[137] Thus there is no question of his attempting to *deny* that the foramen exists or that it acts *as* a foramen. Nevertheless, it is striking that his thinking is sufficiently compartmentalised for the difficulty I have mentioned *not* to have occurred to him.

Finally, a far more massive elision is involved in his treatment of diseases, where we may recall some of the problems raised in Chapter 1. How, we may ask, does an out-and-out teleologist account for diseases? Here, if anywhere, there is evidence of a failure of the good.[138] One argument that was available, and that Galen duly uses, is that nature can only achieve as good results as the material she has to work with will allow.[139] Left to herself, Galen says, nature would have made us immortal.[140] He also argues that residues, for example, are formed as the necessary by-products of other physiological processes that are essential to secure some good, and, again, that potentially damaging bile is needed to counteract the potentially damaging phlegmatic residues:[141] one thing leads to another. He asserts that when the animal is healthy, there is no danger, but adds that nature foresaw that it would be easy for excessive residues to be purged from the stomach by vomiting—a remedy long used by Greek doctors.[142] Nature evidently needs a helping hand; but Galen still fails to confront the question of the break-down of the system in ill-health. What *good* does that do?

---

137. Whether Galen himself was the discoverer is not clear and has been doubted principally on the inconclusive grounds that he makes no claims in that direction (see May 1968, vol. 1, p. 331 n. 102, with references to earlier literature).

138. The Stoics, however, argued that disease, like health, is indifferent, though among things that are indifferent health, like wealth, is "preferred," προηγμένον: see Stobaeus *Ecl.* 2.7.5a (2.57.18ff., 58.2ff. Wachsmuth), 2.7.7b (81.4ff. W.), Diogenes Laertius 7.102–7.

139. See, for example, *UP* 3.10 (H) 1.175.3ff., (K) 3.238.11ff., 5.4 (H) 1.260.5ff., (K) 3.355.4ff., 17.1 (H) 2.446.11ff., 19ff., (K) 4.358.14ff., 359.6ff.

140. *UP* 14.2 (H) 2.285.7ff., (K) 4.143.5ff., cf. also 5.4 (H) 1.260.7ff., 13ff., (K) 3.355.5ff., 11ff.

141. See *UP* 5.3 (H) 1.255.6ff., 257.4ff., (K) 3.348.4ff., 350.16ff., and 5.4 (H) 1.258.26ff., 259.6ff., 263.20ff., (K) 3.353.7ff., 15ff., 359.17ff., especially.

142. *UP* 5.4 (H) 1.259.11ff., 262.17ff., 263.1ff., (K) 3.354.3ff., 358.9ff., 18ff.

None, obviously; it is simply the necessary consequence of the materials we are made of. But the problem is that that factor had been appealed to before, to account for how we come to have residues in health. Why those potentially damaging residues should get out of control, why there is a disruption of the *status quo* in the body, is left unexplained, at least unexplained in teleological terms.

The thesis that nature *always* acts for the good can only be sustained by resolutely focusing on some parts of the phenomena to the exclusion of others that had also to be reckoned as part of common knowledge. The failure of the animal kingdom to be humans is not allowed to count as counter-evidence even to the weaker, qualified Aristotelian version of the thesis, for many animals, as degenerate or deformed, are not fully "natural." Nor are evident inconveniences in the anatomy or the physiological processes of humans allowed to tell against the stronger, Galenic position, and no more are diseases. It is not that teleology as such is mistaken in principle. On the contrary, in many areas and on many questions it proved itself in the ancient world—as it was also to do later—a marvellously powerful heuristic tool. Yet the negative features of its use, as a device for exclusion, for foreclosure, are manifest.

To try to understand this dominant—though, it should be repeated, far from universal—trend in ancient science, it may be helpful to recall our earlier discussions of the extent to which Greek scientists offered accounts that did not merely differ from but directly rivalled and aimed to supersede traditional mythical and religious beliefs and attitudes.[143] While many ancient scientists had no intention of incorporating a moral message in their work, many others had and did. Even where that was not the primary motivation of their inquiry, it was sometimes an adjunct to it. Ptolemy not only draws personal comfort from the order revealed by astronomy, he claims (following Plato) that it improves men's characters,[144] and that the same is true also of the

---

143. Cf. above, Chap. 1, pp. 46ff.
144. *Syntaxis* 1.1 (H) 1.7.17–24: πρός γε μὴν τὴν κατὰ τὰς πράξεις καὶ τὸ ἦθος καλοκαγαθίαν πάντων ἂν αὕτη μάλιστα διορατικοὺς κατασκευά-σειεν ἀπὸ τῆς περὶ τὰ θεῖα θεωρουμένης ὁμοιότητος καὶ εὐταξίας καὶ συμ-

study of harmonics.[145] But the revelation of order is not, of course, by itself bad science. The nub of the question is what kind of order, and how—with what scruples—secured. The double bind on the teleologists was that the greater the potential strength of the moral message concerning the beauty and goodness of nature, the more had to be set aside and either ignored completely or set down to necessity and to the recalcitrance of matter. The most striking examples of the difficulty arise in connection with claims made in the life sciences, but the exact sciences too exemplify the point,[146] and even where the good is not at issue, there were hesitations and waverings on the limits of permissible, and those of necessary, idealisations.

μετρίας καὶ ἀτυφίας ἐραστὰς μὲν ποιοῦσα τοὺς παρακολουθοῦντας τοῦ θείου τούτου κάλλους, ἐνεθίζουσα δὲ καὶ ὥσπερ φυσιοῦσα πρὸς τὴν ὁμοίαν τῆς ψυχῆς κατάστασιν. ("Of all studies this one especially would prepare men to be perceptive of nobility both of action and of character: when the sameness, good order, proportion, and freedom from arrogance of divine things are being contemplated, this study makes those who follow it lovers of this divine beauty and instils, and as it were makes natural, the same condition in their soul.") Cf. also above, Chap. 1 at n. 157, on *Tetrabiblos* 1.3.10.14ff.

145. See especially *Harm.* 3.4ff. (94.24ff. D.), on which see E. A. Lippmann 1964, chap. 2. The more general idea that harmonics reveals and demonstrates the orderliness and rationality of the works of nature (e.g., *Harm.* 1.2 [5.19ff. D.]) is echoed also by Porphyry *In Harm.* 24.22ff. D. Again, the notion of the moral value of music has the authority of Plato, e.g., *Ti.* 47c–d, cf. 80a–b, *R.* 522a. At *Harm.* 2.31–32, Aristoxenus mentions the view that the study of harmonics can make you a better person, but there expresses his own reservations on the topic.

146. Thus in astronomy the belief that the heavenly bodies are divine is invoked, on occasion, to rule out any apparent irregularities as *merely* apparent: see, for example, Proclus *Hyp.* 1.4.15–6.11, and cf. 7.236.12–15, where he criticises current astronomical theories precisely because they tend to show that the substance of the heavenly bodies is irregular and full of modifications. But we find recurrent, often passionate, expression in Greek cosmology and science of the motifs of the positive evaluation of the orderly, the regular, the limited as (1) what is good (as in Plato's *Philebus* 25 e ff., and *Timaeus* 30a, 87c, "all the good is fair, and the fair is not disproportionate," and cf. above, Chap. 3, pp. 137ff. and n. 126); (2) what is natural (as in Aristotle *Ph.* 252a11ff., "there is nothing disorderly in the things that are natural and according to nature: for nature is in everything the cause of order"; cf. *Ph.* 259a10ff., *PA* 641b18ff.); and (3) what is knowable (as in Aristotle *Ph.* 187b7ff., the ἄπειρον, indefinite or infinite, is as such unknowable).

The Value and Effect of Science in the Ancient World

These case-studies have raised a number of extremely general issues in the interpretation of ancient Greek science, and we may, in conclusion—and at the risk of still further sweeping generalisations—broach as a final topic the question of some of its values and effects. From the point of view of the ancient world it is worth asking what difference science made. The question relates to *their* point of view, since from *ours* parts of the answer, concerning what difference their science made, must be clear. From the Renaissance on, the myths and realities of Greek science have been enormously influential: myths, because the ancients' ideas have often been distorted when invoked on either side of later disputes, whether to be idealised or to be reviled; realities, because not everything that Greek science has been taken to stand for is mere fantasy, in particular not certain key methodological notions, including those of the value of empirical research, of the application of mathematics to the understanding of the physical world, and of an axiomatic deductive system. The repercussions both of those myths and of those realities have been immense, even though it goes without saying that not every idea influential in the rise of modern science has an ancient antecedent, real or mythical—in particular not our intense preoccupation with the possibility that science, by being applied, may provide the key to material progress and prosperity (an idea only modestly represented in the ancient world).[147] Moreover, when translated into modern terms and given an institutional framework as a result partly of that preoccupation, what were mere aspirations towards understanding and control in the ancient world have certainly been transformed in the process of their very actualisation.

---

147. While the idea of the *past* advance of civilisation from a state of primitivism can be exemplified readily enough in poetry, in history, and in philosophy (the evidence is collected and discussed by Edelstein 1967b and by Dodds 1973 especially), the notion of *future* progress tends to find expression primarily in the context of the spiritual, not the material. Aristotle, for instance, occasionally states that nearly all possible discoveries and knowledge have been secured already (see *Politics* 1264a1ff., *Metaph.* 981b20ff.).

The blunter and in some ways harder question I proposed concerned the ancients themselves. There the inquiry into nature was generally an activity confined to a tiny elite and intelligible to not many more. Natural science, and even mathematics, established only very limited bridgeheads in what passed for moderately general education. Long after the correct explanation of eclipses was available, ordinary soldiers—and some of their generals—were still capable of being frightened by them, as the debacle of the Athenian retreat from Syracuse illustrates.[148] Moreover, in that instance, those ordinary soldiers were nevertheless able, at least according to the story in Plutarch, to win their freedom, in some cases, by reciting passages from Euripides.[149]

Even among the literate elite themselves, the gap between those who were capable of independent research and those who merely knew something about it was very great, as the immensely learned, but at points quite uncritical and confused, Pliny illustrates.[150] Introductory or elementary textbooks came to be produced in mathematics, astronomy, and medicine, but in late antiquity this increasingly had the negative effect of defining the outer limits of what there was to know, as much as the positive one of increasing the chance of what was within those limits being preserved.[151] Medicine, to be sure, was al-

---

148. Thucydides 7.50; cf. Plutarch *Nicias* 23, which claims that while many understood that *solar* eclipses were caused by the moon, they had no explanation for lunar eclipses. Among well-known earlier texts that express some consternation at solar eclipses are Archilochus 74 and Pindar Paean 9.

149. *Nicias* 29.

150. Aspects of this in connection with the botanical and pharmacological sections in *HN* are discussed in G. E. R. Lloyd 1983a, part 3, chap. 3, pp. 135ff. Cf. Green 1954; Vegetti 1983, pp. 91ff.

151. Thus in astronomy the phenomenon of the precession of the equinoxes, which had been discovered by Hipparchus and was the subject of a careful discussion in Ptolemy, was either ignored completely or flatly denied by later writers: see, for example, Proclus *In Ti.* 3.125.15ff., *Hyp.* 5.136.4ff., 7.234.7ff., Philoponus *De opificio mundi* 3.4.117.12ff. While Galen prides himself on his use of proof *more geometrico*, he expresses some diffidence in offering a geometrical demonstration of some optical phenomena in *UP* 10.12ff. (H) 2.93.5ff., (K) 3.812.14ff., putting it in 10.14 (H) 2.109.8ff., (K) 3.835.17ff., that most people would rather suffer anything than have to do geometry, and claiming that he has omitted many proofs that require astronomy,

ways of exceptional general interest.[152] The other main area that involves the inquiry concerning nature where knowledge and interest extended beyond a small minority was astrology, which must be granted
an important positive role in keeping some scientific knowledge alive,
since, as we remarked before, some of the same framework of theory
underpinned it as underpinned the study of planetary motions in
themselves.[153]

Much of the otherwise reasonably well-educated or well-read public
remained very largely ignorant of advanced natural science. There were
particular discoveries—such as that of the vast size of the universe in
comparison with the earth—that might have had important repercussions on common assumptions, but they did not, or did not to any
great extent, even when they were not totally ignored.[154] The day when
science could shake the whole foundation of the belief in the privileged
place of man in the universe was not yet. The ancients themselves often
maintained the belief in some form, and they tended rather to be for-
tified in it by their scientific studies.[155]

---

geometry, music, and so on from his works so that they will not be utterly
hated by doctors, (H) 2.110.9ff., (K) 3.837.7ff.

152. Yet the shrinking of medical knowledge can be illustrated in connection with the production of medical encyclopedias in late antiquity. In the mid-
fourth century, Oribasius, encouraged as he tells us by the emperor Julian,
collected "all that is most important from all the best doctors" in a comprehensive *Medical Collection* in seventy books (see *Coll.Med.Reliq.* 1 praef. 2,
CMG 6.1.1.4.7f.). But Paul of Aegina, in the seventh century, referring to
Oribasius' work in the proem to his own treatise, says that it is too bulky and
offers his own shorter compendium: *CMG* 9.1 praef. 4.6ff.

153. See above, Chap. 1, pp. 45f.

154. On ancient views of the dimensions of the universe, see Préaux 1968,
1973, pp. 202ff. especially. There were, to be sure, those who denied that the
universe is finite, but when, for example, infinite worlds were located beyond
our heavens, that idea was not necessarily, and not even often, combined with
any definite notions of the dimensions of our world: cf. Furley 1981a. One
writer who did, however, draw out some of the implications of the minuteness
of the earth in comparison with the sphere of the fixed stars is Seneca, for example at *Quaestiones Naturales* 1 praef. 11–13.

155. This is especially true of zoological and anatomical studies such as
those of Aristotle and Galen which we have considered, in which man is
treated as the supreme animal.

One general moral was, however, quite widely learned. Natural scientific explanations appeared to have enough success to justify the general claim that natural phenomena have naturalistic explanations. Yet even here bad examples (such as the quite speculative and largely imaginary theories of thunder and lightning) were cited as often as good ones (such as eclipses), and it is notable that those who used this argument in late antiquity were more often philosophers—such as the Epicureans[156]—than those who actually engaged extensively in advanced scientific research. Meanwhile, more sinisterly, those successes of science, especially the demonstration of the orderliness of heavenly motions, were also appealed to, as early as Plato, in order to support a particular view of the moral governance of the cosmos, itself invoked—in Plato's case—as justification for drastic measures of social control directed against atheists and dissidents of every kind.[157] More generally, whenever the order revealed in nature could be represented as hierarchical, this provided grist to authoritarian mills.[158]

As seen by the average theatre-goer, those who studied nature were figures of fun, in Aristophanes' day, in Plautus'—and in Molière's.[159] Natural science was thus assimilated to any other kind of mumbo-jumbo or wonder-work, including to some more traditional modes of

---

156. Yet, for the Epicureans, inquiry ceased when *some* explanation was available: see above, Chap. 3 n. 239.

157. In *Laws* 10, the various types of atheists, who include those who attempted materialist cosmologies based on the denial that soul is prior to body, are made subject to penalties that are set out at 907d ff. The mildest of these—for anyone who does not heed warnings to reform and is convicted of impiety—is imprisonment, but death is prescribed in many cases.

158. As Aristotle puts it at *Metaph.* 1075a16ff., "everything is ordered together in a way, but not in the same way," and he uses the example of a household and the distinction between slave and free to illustrate how different places are occupied by different kinds of being. Cf. *Politics* 1256b15ff., where plants are said to be for the sake of animals, and the other animals for humans, and the extended argument in *Politics* 1.1–5, especially 1253b14ff., 1254a17ff., maintaining that the distinction between the function of ruling and of being ruled—and so also that between master and slave—is natural.

159. Aristophanes' targets range over a wide spectrum, including not just the new learning (represented in different ways by Socrates and by Euripides) but also more traditional modes: prophecy and divination are often his butt (e.g., *Pax* 1045ff., *Av.* 961ff.) and temple-medicine is at *Plutus* 665ff.

special wisdom such as prophecy or divination. Furthermore, although some ancient natural scientists were keen to differentiate themselves from rivals, whether from divination, from philosophy, or from within science itself, others, as we have seen, sought, rather, to associate their activity to moral philosophy (astronomy is good for the soul) or even to religion,[160] as when Galen talks about the study of the parts of animals in terms drawn from the mystery religions and speaks of his own book on that subject as a *hymn* to nature and, indeed, superior to ordinary hymns.[161] In part this simply reflects the modalities of the expression of the theoretical motivation of scientific research (which was not the only possible motivation; there were other, practical ones as well, especially in medicine).[162] But while to assimilate science to moral

---

160. The Hippocratic treatise *Law* already borrows the language of the mystery religions in speaking of the secrets of medicine: 5, *CMG* 1.1.8.15ff.: τὰ δὲ ἱερὰ ἐόντα πρήγματα ἱεροῖσιν ἀνθρώποισι δείκνυται, βεβήλοισι δὲ οὐ θέμις, πρὶν ἢ τελεσθῶσιν ὀργίοισιν ἐπιστήμης. ("Holy things are revealed only to holy men. Such things must not be made known to the profane until they are initiated into the mysteries of knowledge.") Cf. also *Decent.* 6, *CMG* 1.1.27.13ff., and *Decent.* 18, *CMG* 1.1.29.32f. The idea that the aim not just of philosophy as a whole, but of the study of the mathematical sciences in particular, is to achieve some kind of purification of the soul, is adumbrated by Plato (*R.* 527d–e, cf. *Phd.* 67a–b) and then elaborated by such writers as Theon and Iamblichus: see, for example, Theon 1.1ff. and the use of the language of initiation at 14.18ff., Iamblichus *De Communi Mathematica Scientia* 15.54.25ff., 30.90.28ff.

161. *UP* 3.10 (H) 1.174.6ff., (K) 3.237.10ff., has already been mentioned above, Chap. 1 n. 162. Other passages where Galen speaks of that work as a hymn include *UP* 7.9 (H) 1.396.23, (K) 3.545.13, 12.4 (H) 2.190.19ff., (K) 4.13.3ff., cf. 17.3 (H) 2.451.21ff., (K) 4.365.15ff., and at 7.15 (H) 1.423.12ff., (K) 3.582.15ff., he goes further and suggests it is greater than (ordinary) hymns. Elsewhere he speaks of anatomical research as an initiation, *UP* 7.14 (H) 1.418.19ff., (K) 3.576.3ff., 12.6 (H) 2.196.5ff., (K) 4.20.16ff., and indeed as the starting-point, or principle, of exact theology, superior to the mysteries of Eleusis and Samothrace, 17.1 (H) 2.447.22ff., (K) 4.360.12ff., for "while they exhibit what they teach only dimly, the works of nature are manifest in all animals," (H) 2.448.7ff., (K) 4.361.5ff.: ἀμυδρὰ μὲν γὰρ ἐκεῖνα (viz., the rites, ὄργια, of Eleusis and Samothrace) πρὸς ἔνδειξιν ὧν σπεύδει διδάσκειν· ἐναργῆ δὲ τὰ τῆς φύσεώς ἐστι κατὰ πάντα τὰ ζῷα.

162. The possible practical applications of the knowledge gained from anatomical dissection, for instance, are emphasised by Galen in *AA* 2.2–3 in texts discussed above, Chap. 3 at nn. 229ff.

philosophy, to the pious worship of nature, even to prophecy, might make it more prestigious, even more comprehensible to a certain audience, this also masked certain differences between science and other modes of wisdom. Anatomical research might be described by Galen as an initiation into the mysteries of nature, but in many respects it was unlike any other initiation: it was harder work, and the results obtained were subject to a different kind of scrutiny and verification. Ptolemy too might hope that studying examples of heavenly order might make you more orderly in everyday life. But, again, that heavenly order was to be revealed only by distinct and rather rigorous methods.

Much of the ancient inquiry concerning nature was formalised common knowledge, and much was fantastic speculation. But some of it was neither, as we can see from such examples as the proofs of the sphericity of the earth, or of Archimedes' principle, or of the role of the valves of the heart, or by such discoveries as that of the precession of the equinoxes, or the nervous system, or the diagnostic value of the pulse. To express an allegiance to the *principles* of engaging in research and of securing a comprehensive and reliable data base, to the need to put theories to the test, to expose and root out unexamined assumptions, to withhold judgement where the evidence was insufficient, to acknowledge your own mistakes and uncertainty—all this was often no more than a matter of paying lip-service to high-sounding ideals. But if this was to bluff (and as we have seen, it often was), it was a bluff that could be called, and we have also seen how, on occasion, it was called, and how the ideals were at least sometimes lived up to and the promises they implied fulfilled.

For all the more or less ill-informed, at the limit actually malicious, confusion of science with some other kinds of wisdom—a confusion to some extent fed by the scientists themselves—it was, as those very scientists were, to judge from their practice, well aware, a wisdom with a difference. It was a wisdom committed to different procedures of discovery and of the justification of belief—even if the full force of those differences was hardly *generally* appreciated, and even if the full demonstration of its potential had to wait until modern times. It was more vulnerable than other modes of wisdom, since in principle it incorporated within itself an invitation to challenge its results (it gave

hostages to its opponents); at the same time it was more secure, insofar as parts of those challenges could be withstood successfully. It
played for higher stakes, and sometimes won, even if not as often as it
claimed. The distinction between science and myth, between the new
wisdom and the old, was often a fine one, and the failures of ancient
science to practise what it preached are frequent; yet what it preached
was different from myth, and not *just* more of the same, more myth.
The rhetoric of rationality was powerful and cunning rhetoric, yet it
was exceptional rhetoric, not so much in that it claimed not to be
rhetoric at all (for any rhetoric may aim to conceal itself), as in supplying the wherewithal for its own unmasking—even if some of its expo-
nents did not notice that the mask was still in place. If many of the new
wise men were short on delivery, they were long on aspiration, and the
aspirations were of a kind that were, in time, to produce extraordinary
delivery.

Meanwhile, however, the fact that in its beginnings, science was
often explicitly concerned, if sometimes rather naively, with the moral
dimension of the activity of science itself reminds us, if we need re-
minding, that it originated in no merely intellectualist debate. Indeed,
its offering an alternative world view, in the widest sense an alternative
morality, was central to some of its confrontations with traditional
wisdom, though the fact that science may and to some extent must
incorporate such values was rather to be lost sight of in the aftermath
of the scientific revolution and has only gradually come to be recognised once more in recent times. There is a moral for us today, too, in
the point that, again from its very beginnings, we can detect some tension in disciplines that professed that they must give a public account
of themselves but that, to a greater or lesser degree, were bound to remain specialised, if not exclusive, studies. We have had many occasions
to point to the mystifications of the ancient inquiry into nature, but of
course that is a feature that is still with us today, and one whose threat
has increased immeasurably with the increasing remoteness and spe-
cialisations of science—as one might say of the massive superstruc-
tures that have been erected on or, rather, built over the foundations
laid by some ancient visionaries.

Bibliography

The bibliography provides details of all the books and articles cited in my text, together with a very selective list of other studies that, though not mentioned in my discussion, bear directly on the issues raised.

Aaboe, A. 1974. "Scientific Astronomy in Antiquity." In *The Place of Astronomy in the Ancient World*, ed. D. G. Kendal et al., pp. 21–42. Oxford.

———. 1980. "Observation and Theory in Babylonian Astronomy." *Centaurus* 24:13–35.

Aaboe, A., and D. J. de S. Price. 1964. "Qualitative Measurement in Antiquity." In *L'Aventure de la science*, Mélanges A. Koyré, vol. 1, pp. 1–20. Paris.

Ackerknect, E. H. 1971. *Medicine and Ethnology*. Baltimore.

Ackrill, J. L. 1981. "Aristotle's Theory of Definition: Some Questions on *Posterior Analytics* II 8–10." In Berti, ed., 1981, pp. 359–84.

Adkins, A. W. H. 1960. *Merit and Responsibility*. Oxford.

———. 1970. *From the Many to the One*. London.

———. 1972. *Moral Values and Political Behaviour in Ancient Greece*. London.

———. 1973. "ἀρετή, τέχνη, Democracy, and Sophists: *Protagoras* 316b–328d." *Journal of Hellenic Studies* 93:3–12.

Ahern, E. M. 1981. *Chinese Ritual and Politics*. Cambridge.

Aiton, E. J. 1981. "Celestial Spheres and Circles." *History of Science* 19:75–114.

Allan, D. J. 1965. "Causality, Ancient and Modern." *Proceedings of the Aristotelian Society* Suppl. 39, pp. 1–18.

———. 1970. *The Philosophy of Aristotle*. 2nd ed. (1st ed. 1952). Oxford.

Allen, R E., ed. 1965. *Studies in Plato's Metaphysics*. London.

Allen, R. E., and D. J. Furley, edd. 1975. *Studies in Presocratic Philosophy*, vol. 2. London.

Alt, K. 1978. "Die Überredung der Ananke zur Erklärung der sichtbaren Welt in Platons Timaios." *Hermes* 106:426–66.

Altman, P. L., and D. S. Dittmer. 1972–74. *Biology Data Book*. 2nd ed., 3 vols. Bethesda, Maryland.

Amory, A. 1966. "The Gates of Horn and Ivory." *Yale Classical Studies* 20:
3–57.

Amundsen, D. W. 1973. "The Liability of the Physician in Roman Law." In
*International Symposium on Society, Medicine and Law, Jerusalem, March
1972*, ed. H. Karplus, pp. 17–30. Amsterdam.

———. 1977. "The Liability of the Physician in Classical Greek Legal Theory
and Practice." *Journal of the History of Medicine and Allied Sciences* 32:
172–203.

Andrewes, A. 1954. *Probouleusis: Sparta's Contribution to the Technique of
Government*. Oxford.

———. 1956. *The Greek Tyrants*. London.

———. 1966. "The Government of Classical Sparta." In *Ancient Society and
Institutions: Studies presented to V. Ehrenberg*, ed. E. Badian, pp. 1–20.
Oxford.

Annas, J. 1975. "Aristotle, Number and Time." *Philosophical Quarterly* 25:
97–113.

———. 1976. *Aristotle's Metaphysics, Books M and N*. Oxford.

———. 1980. "Truth and Knowledge." In Schofield, Burnyeat, and Barnes,
edd., 1980, pp. 84–104.

———. 1986. "Doing without Objective Values: Ancient and Modern Strategies." In Schofield and Striker, edd., 1986, pp. 3–29.

———. 1987. "Die Gegenstände der Mathematik bei Aristoteles." In Graeser,
ed., 1987, pp. 131–47.

Anscombe, G. E. M. 1971. *Causality and Determination*. Cambridge.

Anton, J. P., ed. 1980. *Science and the Sciences in Plato*. New York.

Anton, J. P., and G. L. Kustas, edd. 1971. *Essays in Ancient Greek Philosophy*.
Albany, New York.

Anton, J. P., and A. Preus, edd. 1983. *Essays in Ancient Greek Philosophy*, vol.
2. Albany, New York.

Apostle, H. G. 1952. *Aristotle's Philosophy of Mathematics*. Chicago.

Arnim, H. von. 1905–24. *Stoicorum veterum fragmenta*, 4 vols. Leipzig.

Artelt, W. 1937. *Studien zur Geschichte der Begriffe "Heilmittel" und "Gift."*
Leipzig.

Asmis, E. 1984. *Epicurus' Scientific Method*. Ithaca, New York.

Aubenque, P. 1966. *Le Problème de l'être chez Aristote*. 2nd ed. (1st ed. 1962).
Paris.

———. 1979. "La Pensée du simple dans la *Métaphysique* (Z 17 et Θ 10)." In
Aubenque, ed., 1979, pp. 69–80.

Aubenque, P., ed. 1979. *Etudes sur la métaphysique d'Aristote*. Actes du VIᵉ
Symposium Aristotelicum. Paris.

———. 1980. *Concepts et catégories dans la pensée antique*. Paris.

Aujac, G. 1979. *Autolycos de Pitane: La Sphère en mouvement, levers et cou-
chers héliaques*. Paris.

Avery, H. C. 1982. "One Antiphon or Two?" *Hermes* 110:145–58.

Bachelard, G. 1972. *La Formation de l'esprit scientifique*. 8th ed. (1st ed.
1947). Paris.

Baines, J. 1983. "Literacy and Ancient Egyptian Society." *Man* n.s. 18: 572–99.

Baines, J., and C. J. Eyre. 1983. "Four Notes on Literacy." *Göttinger Miszellen* 61:65–96.

Bakhtin, M. M. 1973. *Problems of Dostoevsky's Poetics,* trans. R. W. Rotsel. Ann Arbor, Michigan.

———. 1981. *The Dialogic Imagination,* ed. M. Holquist, trans. C. Emerson and M. Holquist. Austin, Texas.

Balme, D. M. 1939. "Greek Science and Mechanism. I. Aristotle on Nature and Chance." *Classical Quarterly* 33:129–38.

———. 1941. "Greek Science and Mechanism. II. The Atomists." *Classical Quarterly* 35:23–28.

———. 1961/1975. "Aristotle's Use of Differentiae in Zoology." From S. Mansion, ed., 1961, pp. 195–212. In Barnes, Schofield, and Sorabji, edd., 1975, pp. 183–93.

———. 1962a. "γένος and εἶδος in Aristotle's Biology." *Classical Quarterly* n.s. 12:81–98.

———. 1962b. "Development of Biology in Aristotle and Theophrastus: Theory of Spontaneous Generation." *Phronesis* 7:91–104.

———. 1965. "Aristotle's Use of the Teleological Explanation." Inaugural Lecture, Queen Mary College, London, 1 June 1965.

———. 1970. "Aristotle and the Beginnings of Zoology." *Journal of the Society for the Bibliography of Natural History* 5 (1968–71):272–85.

———. 1980. "Aristotle's Biology Was Not Essentialist." *Archiv für Geschichte der Philosophie* 62:1–12.

———. 1987. "Teleology and Necessity." In Gotthelf and Lennox, edd., 1987, pp. 275–85.

Bambrough, J. R. 1956/1967. "Plato's Political Analogies." From *Philosophy, Politics and Society,* ed. P. Laslett (Oxford, 1956), pp. 98–115. In Bambrough, ed., 1967, pp. 152–69.

———. 1962/1967. "Plato's Modern Friends and Enemies." From *Philosophy* 37:97–113. In Bambrough, ed., 1967, pp. 3–19.

Bambrough, [J.] R., ed. 1965. *New Essays on Plato and Aristotle.* London.

———. 1967. *Plato, Popper and Politics.* Cambridge.

Barker, A. D. 1977. "Music and Mathematics: Theophrastus against the Number-Theorists." *Proceedings of the Cambridge Philological Society* n.s. 23:1–15.

———. 1978a. "Music and Perception: A Study in Aristoxenus." *Journal of Hellenic Studies* 98:9–16.

———. 1978b. "οἱ καλούμενοι ἁρμονικοί: The Predecessors of Aristoxenus." *Proceedings of the Cambridge Philological Society* n.s. 24:1–21.

———. 1978c. "σύμφωνοι ἀριθμοί: A Note on *Republic* 531C1–4." *Classical Philology* 73:337–42.

———. 1981a. "Aristotle on Perception and Ratios." *Phronesis* 26:248–66.

———. 1981b. "Methods and Aims in the Euclidean *Sectio Canonis.*" *Journal of Hellenic Studies* 101:1–16.

———. 1982a. "Aristides Quintilianus and Constructions in Early Music Theory." *Classical Quarterly* n.s. 32:184–97.

———. 1982b. "The Innovations of Lysander the Kitharist." *Classical Quarterly* n.s. 32:266–69.

———. 1984. *Greek Musical Writings.* Vol. 1: *The Musician and His Art.* Cambridge.

Barnes, J. 1969/1975. "Aristotle's Theory of Demonstration." From *Phronesis* 14:123–52. In Barnes, Schofield, and Sorabji, edd., 1975, pp. 65–87.

———. 1971. "Homonymy in Aristotle and Speusippus." *Classical Quarterly* n.s. 21:65–80.

———. 1975. *Aristotle's Posterior Analytics.* Oxford.

———. 1976. "Aristotle, Menaechmus, and Circular Proof." *Classical Quarterly* n.s. 26:278–92.

———. 1979. *The Presocratic Philosophers,* 2 vols. London.

———. 1980a. "Socrates and the Jury: Paradoxes in Plato's Distinction between Knowledge and True Belief." *Proceedings of the Aristotelian Society* Suppl. 54:193–206.

———. 1980b. "Proof Destroyed." In Schofield, Burnyeat, and Barnes, edd., 1980, pp. 161–81.

———. 1981. "Proof and the Syllogism." In Berti, ed., 1981, pp. 1–59.

———. 1982a. "Medicine, Experience and Logic." In Barnes, Brunschwig, Burnyeat, and Schofield, edd., 1982, pp. 24–68.

———. 1982b. "The Beliefs of a Pyrrhonist." *Proceedings of the Cambridge Philological Society* n.s. 28:1–29.

———. 1983. "Ancient Skepticism and Causation." In Burnyeat, ed., 1983, pp. 149–203.

———. 1985. "Aristotle's Arithmetic." *Revue de Philosophie Ancienne* 3:97–133.

———. Forthcoming. "Galen on Logic and Therapy." In *Proceedings of the Second International Galen Conference,* Kiel 1982, ed. F. Kudlien, forthcoming.

Barnes, J., J. Brunschwig, M. Burnyeat, and M. Schofield, edd. 1982. *Science and Speculation.* Cambridge.

Barnes, J., M. Schofield, and R. Sorabji, edd. 1975. *Articles on Aristotle.* I: *Science.* London.

———. 1979. *Articles on Aristotle.* III: *Metaphysics.* London.

Barnes, S. B. 1974. *Scientific Knowledge and Sociological Theory.* London.

———. 1977. *Interests and the Growth of Knowledge.* London.

Barnett, M. K. 1956. "The Development of Thermometry and the Temperature Concept." *Osiris* 12:269–341.

Bartels, K. 1965. "Der Begriff Techne bei Aristoteles." In Flashar and Gaiser, edd., 1965, pp. 275–87.

———. 1966. *Das Techne-Modell in der Biologie des Aristoteles.* Tübingen.

Barthes, R. 1953/1967. *Writing Degree Zero,* trans. A. Lavers and C. Smith of *Le Degré zéro de l'écriture* (Paris, 1953). London.

———. 1964/1967. *Elements of Semiology,* trans. A. Lavers and C. Smith of *Eléments de sémiologie* (Paris, 1964). London.

Bascom, W. 1969. *Ifa Divination: Communication between Gods and Men in West Africa.* Bloomington, Indiana.

Becker, O. 1957. *Das mathematische Denken der Antike.* Göttingen.

Behr, C. A. 1968. *Aelius Aristides and the Sacred Tales.* Amsterdam.

Belfiore, E. 1980. "*Elenchus, Epode,* and Magic: Socrates as Silenus." *Phoenix* 34:128–37.

Below, K.-H. 1953. *Der Arzt im römischen Recht.* Münchener Beiträge zur Papyrusforschung und antiken Rechtsgeschichte 37. Munich.

Benacerraf, P., and H. Putnam, edd. 1983. *Philosophy of Mathematics, Selected Readings.* 2nd ed. Cambridge.

Benardete, S. 1965. "χρή and δεῖ in Plato and Others." *Glotta* 43:285–98.

———. 1971. "On Plato's *Timaeus* and Timaeus' Science Fiction." *Interpretation* 2:21–63.

Ben-David, J. 1984. *The Scientist's Role in Society.* 2nd ed. (1st ed. 1971). Chicago.

Ben-David, J., and T. N. Clark, edd. 1977. *Culture and Its Creators: Essays in Honor of E. Shils.* Chicago.

Benveniste, E. 1945. "La Doctrine médicale des Indo-Européens." *Revue de l'histoire des religions* 130:5–12.

———. 1971. *Problems in General Linguistics,* trans. M. E. Meek of *Problèmes de linguistique générale* (2 vols., Paris, 1966). Coral Gables, Florida.

———. 1973. *Indo-European Language and Society,* trans. E. Palmer of *Le Vocabulaire des institutions indo-européennes* (2 vols., Paris, 1969). London.

Berger, H. 1903. *Geschichte der wissenschaftlichen Erdkunde der Griechen.* 2nd ed. Leipzig.

Berger, P. L., and T. Luckmann. 1967. *The Social Construction of Reality.* London.

Berggren, D. 1962–63. "The Use and Abuse of Metaphor." *Review of Metaphysics* 16:237–58, 450–72.

Berka, K. 1963. "Aristoteles und die axiomatische Methode." *Das Altertum* 9:200–205.

———. 1983. *Measurement,* trans. A. Riska. Dordrecht.

Bernstein, B. 1964. "Elaborated and Restricted Codes: Their Social Origins and Some Consequences." *American Anthropologist* special publications 66, 6, part 2:55–69.

Berthelot, M. 1885. *Les Origines de l'alchimie.* Paris.

Berthelot, M., and C.-E. Ruelle. 1888. *Collection des anciens alchimistes grecs.* 3 vols. Paris.

Berti, E. 1977. *Aristotele: Dalla dialettica alla filosofia prima.* Padua.

———. 1978. "The Intellection of 'Indivisibles' according to Aristotle, *De Anima* III 6." In Lloyd and Owen, edd., 1978, pp. 141–63.

———. 1980. "La critica di Aristotele alla teoria atomistica del vuoto." In Romano, ed., 1980, pp. 135–59.

Berti, E., ed. 1981. *Aristotle on Science: The Posterior Analytics.* Proceedings of the Eighth Symposium Aristotelicum. Padua.

Björnbo, A. A. 1901. "Hat Menelaos aus Alexandria einen Fixsternkatalog verfasst?" *Bibliotheca Mathematica* Dritte Folge 2:196–212.

Black, M. 1937. "Vagueness: An Exercise in Logical Analysis." *Philosophy of Science* 4:427–55.

———. 1962. *Models and Metaphors.* Ithaca, New York.

Bloch, M., ed. 1975. *Political Language and Oratory in Traditional Society.* London.

Block, I. 1960–61. "Aristotle and the Physical Object." *Philosophy and Phenomenological Research* 21:93–101.

———. 1961. "Truth and Error in Aristotle's Theory of Sense Perception." *Philosophical Quarterly* 11:1–9.

Bloor, D. 1971. "The Dialectics of Metaphor." *Inquiry* 14:430–44.

———. 1976. *Knowledge and Social Imagery.* London.

———. 1978. "Polyhedra and the Abominations of Leviticus." *British Journal for the History of Science* 11:245–72.

Blumenthal, H. J. 1981. "Some Platonist Readings of Aristotle." *Proceedings of the Cambridge Philological Society* n.s. 27:1–16.

Boas, F. 1930. *The Religion of the Kwakiutl Indians,* Part 2. Columbia University Contributions to Anthropology 10. New York.

Boas, G. 1959. "Some Assumptions of Aristotle." *Transactions of the American Philosophical Society* n.s. 49, Part 6.

———. 1961. *Rationalism in Greek Philosophy.* Baltimore.

Bochner, S. 1966. *The Role of Mathematics in the Rise of Science.* Princeton.

Bodson, L. 1978. ΊΕΡΑ ΖΩΙΑ. Académie Royale de Belgique, Mémoires de la Classe des Lettres, 2nd ser. 63, 2. Brussels.

Boeder, H. 1959. "Der frühgriechische Wortgebrauch von Logos und Aletheia." *Archiv für Begriffsgeschichte* 4:82–112.

———. 1968. "Der Ursprung der 'Dialektik' in der Theorie des 'Seienden.' Parmenides und Zenon." *Studium Generale* 21:184–202.

Böhme, G. 1976. "Platons Theorie der exakten Wissenschaften." *Antike und Abendland* 22:40–53.

Bogaard, P. A. 1979. "Heaps or Wholes: Aristotle's Explanation of Compound Bodies." *Isis* 70:11–29.

Bolkestein, H. 1929. *Theophrastos' Charakter der Deisidaimonia als religionsgeschichtliche Urkunde.* Religionsgeschichtliche Versuche und Vorarbeiten 21.2. Giessen.

Boll, F. 1894. "Studien über Claudius Ptolemäus. Ein Beitrag zur Geschichte der griechischen Philosophie und Astrologie." *Jahrbücher für classische Philologie* Suppl. Bd. 21:49–244.

———. 1899. "Beiträge zur Ueberlieferungsgeschichte der griechischen Astrologie und Astronomie." *Sitzungsberichte der philosophisch-philologischen und der historischen Classe der k. b. Akademie der Wissenschaften zu München,* pp. 77–140.

———. 1901. "Die Sternkataloge des Hipparch und des Ptolemaios." *Bibliotheca Mathematica* Dritte Folge 2:185–95.

Boll, F., and C. Bezold. 1917/1931. *Sternglaube und Sterndeutung. Die Geschichte und das Wesen der Astrologie.* 4th ed., ed. W. Gundel (1st ed. 1917). Leipzig.

Bollack, J. 1965–69. *Empédocle.* 3 vols. in 4. Paris.

———. 1971. "Mythische Deutung und Deutung des Mythos." In *Terror und Spiel,* Poetik und Hermeneutik 4, ed. M. Fuhrmann, pp. 67–119. Munich.

Bolton, J. D. P. 1962. *Aristeas of Proconnesus.* Oxford.

Bolton, R. 1976. "Essentialism and Semantic Theory in Aristotle: *Posterior Analytics* II, 7–10." *Philosophical Review* 85:514–44.

Boncompagni, R. 1970. "Empirismo e osservazione diretta nel περὶ διαίτης del *Corpus Hippocraticum.*" *Physis* 12:109–32.

Borchardt, L. 1920. *Die altägyptische Zeitmessung.* Band 1, B of *Die Geschichte der Zeitmessung und der Uhren,* ed. E. von Bassermann-Jordan. Berlin.

Bottéro, J. 1974. "Symptômes, signes, écritures en Mésopotamie ancienne." In Vernant et al. 1974, pp. 70–197.

———. 1977. "Les Noms de Marduk, l'écriture et la 'logique' en Mésopotamie ancienne." In *Essays on the Ancient Near East,* ed. M. de Jong Ellis, Memoirs of the Connecticut Academy of Arts and Sciences 19, pp. 5–28. Hamden, Connecticut.

———. 1981a. 'L'Ordalie en Mésopotamie ancienne." *Annali della scuola normale superiore di Pisa* 11:1005–67.

———. 1981b. "Mésopotamie: L'Intelligence et la fonction technique du pouvoir: Enki/Ea." In *Dictionnaire des Mythologies,* ed. Y. Bonnefoy, vol. 2, pp. 102–11. Paris.

Bouché-Leclercq, A. 1879–82. *Histoire de la divination dans l'antiquité,* 4 vols. Paris.

———. 1899. *L'Astrologie grecque.* Paris.

Bourdieu, P. 1977. *Outline of a Theory of Practice,* trans. R. Nice of *Esquisse d'une théorie de la pratique* (Geneva, 1972). Cambridge.

Bourgey, L. 1953. *Observation et expérience chez les médecins de la collection hippocratique.* Paris.

———. 1955. *Observation et expérience chez Aristote.* Paris.

Bourgey, L., and J. Jouanna, edd. 1975. *La Collection hippocratique et son rôle dans l'histoire de la médecine.* Leiden.

Bouteiller, M. 1950. *Chamanisme et guérison magique.* Paris.

Boyancé, P. 1937. *Le Culte des Muses chez les philosophes grecs.* Bibliothèque des écoles françaises d'Athènes et de Rome 141. Paris.

Boyer, C. B. 1945–46. "Aristotelian References to the Law of Reflection." *Isis* 36:92–95.

Boylan, M. 1983. *Method and Practice in Aristotle's Biology.* Washington, D.C.

Brannigan, A. 1981. *The Social Basis of Scientific Discoveries.* Cambridge.

Brătescu, G. 1975. "Eléments archaïques dans la médecine hippocratique." In Bourgey and Jouanna, edd., 1975, pp. 41–49.

Breasted, J. H. 1930. *The Edwin Smith Surgical Papyrus,* vol. 1. Chicago.
Bremer, D. 1980. "Aristoteles, Empedokles und die Erkenntnisleistung der Metapher." *Poetica Zeitschrift für Sprach- und Literaturwissenschaft* 12: 350–76.
Bremmer, J. N. 1982. "Literacy and the Origins and Limitations of Greek Atheism." In *Actus: Studies in Honour of H. L. W. Nelson,* ed. J. den Boeft and A. H. M. Kessels, pp. 43–55. Utrecht.
———. 1983. *The Early Greek Concept of the Soul.* Princeton.
Brisson, L. 1974. "Du bon usage du dérèglement." In J.-P. Vernant et al. 1974, pp. 220–48.
———. 1976. *Le Mythe de Tirésias.* Leiden.
———. 1982. *Platon, les mots et les mythes.* Paris.
Britton, J. P. 1969. "Ptolemy's Determination of the Obliquity of the Ecliptic." *Centaurus* 14:29–41.
Brody, B. A. 1972. "Towards an Aristotelean Theory of Scientific Explanation." *Philosophy of Science* 39:20–31.
Bröcker, W. 1958. "Gorgias contra Parmenides." *Hermes* 86:425–40.
Bromberger, S. 1966. "Why-Questions." In *Mind and Cosmos,* ed. R. G. Colodny, University of Pittsburgh Studies in the Philosophy of Science 3, pp. 86–111. Pittsburgh.
Brown, A. L. 1984. "Three and Scene-Painting Sophocles." *Proceedings of the Cambridge Philological Society* n.s. 30:1–17.
Brown, P. 1971. *The World of Late Antiquity.* London.
———. 1978. *The Making of Late Antiquity.* Cambridge, Massachusetts.
Brownson, C. D. 1981. "Euclid's Optics and Its Compatibility with Linear Perspective." *Archive for History of Exact Sciences* 24:165–94.
Bruin, F., and M. Bruin. 1976. "The Equator Ring, Equinoxes and Atmospheric Refraction." *Centaurus* 20:89–111.
Bruins, E. M. 1968. *La Géométrie non euclidienne dans l'antiquité.* Université de Paris, Conférences du Palais de la découverte, Sér. D, 121. Paris.
Brunschvicq, L. 1949. *L'Expérience humaine et la causalité physique.* 3rd ed. (1st ed. 1922). Paris.
Brunschwig, J. 1963. "Aristote et les pirates tyrrhéniens." *Revue Philosophique de la France et de l'Etranger* 153:171–90.
———. 1967. *Aristote, Topiques,* vol. 1. Paris.
———. 1973. "Sur quelques emplois d' ΟΨΙΣ." In *Zetesis: Festschrift de Strycker,* pp. 24–39. Antwerp.
———. 1980a. "Proof Defined." In Schofield, Burnyeat, and Barnes, edd., 1980, pp. 125–60.
———. 1980b. "Du mouvement et de l'immobilité de la loi." *Revue Internationale de Philosophie* 133–34: 512–40.
———. 1981. 'L'Objet et la structure des *Seconds Analytiques* d'après Aristote." In Berti, ed., 1981, pp. 61–96.
Brunschwig, J., ed. 1978. *Les Stoïciens et leur logique.* Actes du Colloque de Chantilly, 18–22 septembre 1976. Paris.
Burkert, W. 1959. "ΣΤΟΙΧΕΙΟΝ. Eine semasiologische Studie." *Philologus* 103:167–97.

————. 1962. "ΓΟΗΣ. Zum griechischen 'Schamanismus.'" *Rheinisches Museum* N.F. 105:36–55.

————. 1963. "Iranisches bei Anaximandros." *Rheinisches Museum* N.F. 106:97–134.

————. 1968. "Orpheus und die Vorsokratiker. Bemerkungen zum Derveni-Papyrus und zur pythagoreischen Zahlenlehre." *Antike und Abendland* 14:93–114.

————. 1970. "La genèse des choses et des mots." *Les Etudes Philosophiques:* 443–55.

————. 1972. *Lore and Science in Ancient Pythagoreanism*, revised trans. E. L. Minar of *Weisheit und Wissenschaft* (Nürnberg, 1962). Cambridge, Massachusetts.

————. 1977/1985. *Greek Religion: Archaic and Classical*, trans. J. Raffan of *Griechische Religion der archaischen und klassischen Epoche* (Stuttgart, 1977). Oxford.

————. 1979. *Structure and History in Greek Mythology and Ritual.* Berkeley.

————. 1983. *Homo Necans*, trans. P. Bing of 1972 German ed. Berkeley.

Burnet, J. 1948. *Early Greek Philosophy.* 4th ed. (1st ed. 1892). London.

Burns, A. 1981. "Athenian Literacy in the Fifth Century B.C." *Journal of the History of Ideas* 42:371–87.

Burnyeat, M. F. 1976. "Plato on the Grammar of Perceiving." *Classical Quarterly* n.s. 26:29–51.

————. 1977. "Examples in Epistemology: Socrates, Theaetetus and G. E. Moore." *Philosophy* 52:381–98.

————. 1978. "The Philosophical Sense of Theaetetus' Mathematics." *Isis* 69:489–513.

————. 1979. "Conflicting Appearances." *Proceedings of the British Academy* 65:69–111.

————. 1980a. "Can the Sceptic Live His Scepticism? In Schofield, Burnyeat, and Barnes, edd., 1980, pp. 20–53. (Reprinted in Burnyeat, ed., 1983, pp. 117–48.)

————. 1980b. "Socrates and the Jury: Paradoxes in Plato's Distinction be-pp. 97–139.

————. 1982a. "The Origins of Non-Deductive Inference." In Barnes, Brunschwig, Burnyeat, and Schofield, edd., 1982, pp. 193–238.

————. 1982b. "Idealism and Greek Philosophy: What Descartes Saw and Berkeley Missed." *Philosophical Review* 91:3–40.

————. 1984. "The Sceptic in His Place and Time." In *Philosophy in History*, ed. R. Rorty, J. B. Schneewind, and Q. Skinner, pp. 225–54. Cambridge.

————. 1987. "Platonism and Mathematics: A Prelude to Discussion." In Graeser, ed., 1987, pp. 213–40.

Burnyeat, M. [F.] ed. 1983. *The Skeptical Tradition.* Berkeley.

Burton, H. E. 1945. "The Optics of Euclid." *Journal of the Optical Society of America* 35:357–72.

Cohen, M. R., and I. E. Drabkin, edd. 1958. *A Source Book in Greek Science.* 2nd ed. (1st ed. 1948). Cambridge, Massachusetts.

Cohen, P. S. 1969. "Theories of Myth." *Man* n.s. 4:337–53.

Cohen, S. Marc. 1977–78. "Essentialism in Aristotle." *Review of Metaphysics* 31:387–405.

Cohen, Sheldon. 1984. "Aristotle's Doctrine of the Material Substrate." *Philosophical Review* 93:171–94.

Cohn-Haft, L. 1956. *The Public Physicians of Ancient Greece.* Smith College Studies in History 42. Northampton, Massachusetts.

Coldstream, J. N. 1977. *Geometric Greece.* London.

Cole, T. 1967. *Democritus and the Sources of Greek Anthropology.* American Philological Association, Philological Monographs 25.

Cook, R. M. 1972. *Greek Painted Pottery.* 2nd ed. (1st ed. 1960). London.

Cooper, J. M. 1970. "Plato on Sense Perception and Knowledge: *Theaetetus* 184–186." *Phronesis* 15:123–46.

———. 1975. *Reason and Human Good in Aristotle.* Cambridge, Massachusetts.

———. 1982. "Aristotle on Natural Teleology." In Schofield and Nussbaum, edd., 1982, pp. 197–222.

———. 1987. "Hypothetical Necessity and Natural Teleology." In Gotthelf and Lennox, edd., 1987, pp. 243–74.

Corcoran, J. 1973. "A Mathematical Model of Aristotle's Syllogistic." *Archiv für Geschichte der Philosophie* 55:191–219.

Cornford, F. M. 1912. *From Religion to Philosophy.* London.

———. 1922. "Mysticism and Science in the Pythagorean Tradition. I." *Classical Quarterly* 16:137–50.

———. 1923. "Mysticism and Science in the Pythagorean Tradition. II." *Classical Quarterly* 17:1–12.

———. 1931. *The Laws of Motion in Ancient Thought.* Cambridge.

———. 1932/1965. "Mathematics and Dialectic in the *Republic* VI–VII." From *Mind* n.s. 41:37–52, 173–90. In Allen, ed., 1965, pp. 61–95.

———. 1937. *Plato's Cosmology.* London.

———. 1938. "Greek Natural Philosophy and Modern Science." In *Background to Modern Science,* ed. J. Needham and W. Pagel, pp. 3–22. Cambridge.

———. 1950. *The Unwritten Philosophy and Other Essays.* Cambridge.

———. 1952. *Principium Sapientiae.* Cambridge.

Couissin, P. 1929. "L'Origine et l'évolution de l'ἐποχή." *Revue des Etudes Grecs* 42:373–97.

———. 1929/1983. "The Stoicism of the New Academy." Originally "Le Stoïcisme de la Nouvelle Académie." *Revue d'Histoire de la Philosophie* 3:241–76. In Burnyeat, ed., 1983, pp. 31–63.

Couloubaritsis, L. 1978. "Sophia et Philosophia chez Aristote." *Annales de l'Institut de Philosophie de l'Université Libre de Bruxelles* 7–38.

———. 1980a. *L'Avènement de la science physique.* Brussels.

———. 1980b. "Y a-t-il une intuition des principes chez Aristote?" *Revue*

Internationale de Philosophie 133–34: 440–71.

Croissant, J. 1932. *Aristote et les mystères.* Bibliothèque de la Faculté de Philosophie et Lettres de l'Université de Liège 51. Paris.

Crombie, A. C., ed. 1963. *Scientific Change.* London.

Crombie, I. M. 1962. *An Examination of Plato's Doctrines.* I: *Plato on Man and Society.* London.

———. 1963. *An Examination of Plato's Doctrines.* II: *Plato on Knowledge and Reality.* London.

Cumont, F. 1912. *Astrology and Religion among the Greeks and Romans.* New York.

Curi, U. 1972. "Historia e polymathia. A proposito del frammento 35 di Eraclito." *Giornale di Metafisica* 27:569–74.

Czwalina, A. 1956–58. "Über einige Beobachtungsfehler des Ptolemäus und die Deutung ihrer Ursachen." *Centaurus* 5:283–306.

———. 1959. "Ptolemaeus: Die Bahnen der Planeten Venus und Merkur." *Centaurus* 6:1–35.

Dambska, I. 1961. "Le Problème des songes dans la philosophie des anciens Grecs." *Revue Philosophique de la France et de l'Etranger* 151:11–24.

Dammann, R. M. J. 1977–78. "Metaphors and Other Things." *Proceedings of the Aristotelian Society* n.s. 78:125–40.

Dampier-Whetham, W. C. D. 1930. *A History of Science.* 2nd ed. (1st ed. 1929). Cambridge.

Daremberg, C. V. 1879. *Oeuvres de Rufus d'Éphèse,* completed by C. E. Ruelle. Paris.

Davidson, D. 1980. *Essays on Actions and Events.* Oxford.

———. 1984. *Inquiries into Truth and Interpretation.* Oxford.

Davison, J. A. 1962. "Literature and Literacy in Ancient Greece." *Phoenix* 16:141–56, 219–33.

Decharme, P. 1904. *La Critique des traditions religieuses chez les grecs.* Paris.

De Fidio, P. 1969. "ΑΛΗΘΕΙΑ: dal mito alla ragione." *La Parola del Passato* 24:308–20.

Deichgräber, K. 1930/1965. *Die griechische Empirikerschule: Sammlung der Fragmente und Darstellung der Lehre.* (Berlin, 1930.) Repr. Berlin.

———. 1933a. *Die Epidemien und das Corpus Hippocraticum.* Abhandlungen der preussischen Akademie der Wissenschaften, Jahrgang 1933, 3, phil.-hist. Kl. Berlin.

———. 1933b. "Die ärztliche Standesethik des hippokratischen Eides." *Quellen und Studien zur Geschichte der Naturwissenschaften und der Medizin* 3, 2:79–99.

———. 1933c. "ΠΡΟΦΑΣΙΣ: Eine terminologische Studie." *Quellen und Studien zur Geschichte der Naturwissenschaften und der Medizin* 3, 4: 209–25.

———. 1935. *Hippokrates, Über Entstehung und Aufbau des menschlichen Körpers (περὶ σαρκῶν).* Leipzig.

———. 1939. "Die Stellung des griechischen Arztes zur Natur." *Die Antike* 15:116–38.

————. 1957. "Galen als Erforscher des menschlichen Pulses." *Sitzungsberichte der deutschen Akademie der Wissenschaften zu Berlin,* Klasse für Sprachen, Literatur und Kunst, Jahrgang 1956, 3. Berlin.

De Jong, H. W. M. 1959. "Medical Prognostication in Babylon." *Janus* 48: 252–57.

De Lacy, P. H. 1958. "οὐ μᾶλλον and the Antecedents of Ancient Scepticism." *Phronesis* 3:59–71.

————. 1972. "Galen's Platonism." *American Journal of Philology* 93:27–39.

De Lacy, P. H., and E. A. De Lacy. 1978. *Philodemus On Methods of Inference.* Revised ed. Naples.

Delambre, J. B. J. 1817. *Histoire de l'astronomie ancienne.* 2 vols. Paris.

Delatte, A. 1915. *Etudes sur la littérature Pythagoricienne.* Bibliothèque de l'Ecole des Hautes Etudes 217. Paris.

————. 1922. *La Vie de Pythagore de Diogène Laërce.* Académie Royale de Belgique, Mémoires de la Classe des Lettres, 2nd ser. 17. Brussels.

————. 1930. "Les Harmonies dans l'embryologie hippocratique." In *Mélanges Paul Thomas,* pp. 160–71. Bruges.

————. 1961. *Herbarius: Recherches sur le cérémonial usité chez les anciens pour la cueillette des simples et des plantes magiques.* Académie Royale de Belgique, Mémoires de la Classe des Lettres, 2nd ser. 54, 4. 3rd ed. Brussels.

Del Corno, D. 1962. "Ricerche sull' onirocritica greca." In *Rendiconti dell' Istituto Lombardo,* Classe di Lettere e Scienze morali e storiche 96, pp. 334–66.

Demuth, G. 1972. *Ps.-Galeni De dignotione ex insomniis.* Göttingen.

Dengler, R. E. 1927. *Theophrastus De causis plantarum, Book One.* Philadelphia.

Denyer, N. 1985. "The Case against Divination: An Examination of Cicero's *De Divinatione." Proceedings of the Cambridge Philological Society* n.s. 31:1–10.

Derenne, E. 1930. *Les Procès d'impiété intentés aux philosophes à Athènes au Vme et au IVme siècles avant J.-C.* Bibliothèque de la Faculté de Philosophie et Lettres de l'Université de Liège 45. Liège.

Derrida, J. 1967/1976. *Of Grammatology,* trans. G. C. Spivak of *De la grammatologie* (Paris, 1967). Baltimore.

————. 1967/1978. *Writing and Difference,* trans. A. Bass of *L'Ecriture et la différence* (Paris, 1967). London.

————. 1972/1981. *Dissemination,* trans. B. Johnson of *La Dissémination* (Paris, 1972). London.

————. 1972/1982. *Margins of Philosophy,* trans. A. Bass of *Marges de la philosophie* (Paris, 1972). Brighton, Sussex.

De Ste. Croix, G. E. M. 1972. *The Origins of the Peloponnesian War.* London.

————. 1981. *The Class Struggle in the Ancient Greek World.* London.

Detel, W. 1975. "αἴσθησις und λογισμός. Zwei Probleme der epikureischen Methodologie." *Archiv für Geschichte der Philosophie* 57:21–35.

Detienne, M. 1960. "La Notion mythique d'ΑΛΗΘΕΙΑ." *Revue des Études Grecques* 73:27–35.

————. 1963. *De la pensée religieuse à la pensée philosophique: La Notion de Daïmôn dans le Pythagorisme ancien.* Bibliothèque de la Faculté de Philosophie et Lettres de l'Université de Liège 165. Paris.

————. 1967. *Les Maîtres de vérité dans la grèce archaïque.* Paris.

————. 1972/1977. *The Gardens of Adonis,* trans. J. Lloyd of *Les Jardins d'Adonis* (Paris, 1972). Hassocks, Sussex.

————. 1977/1979. *Dionysos Slain,* trans. M. and L. Muellner of *Dionysos mis à mort* (Paris, 1977). Baltimore.

————. 1981/1986. *The Creation of Mythology,* trans. M. Cook of *L'Invention de la mythologie* (Paris, 1981). Chicago.

Detienne, M. and J.-P. Vernant. 1974/1978. *Cunning Intelligence in Greek Culture and Society,* trans. J. Lloyd of *Les Ruses de l'intelligence: La Mètis des grecs* (Paris, 1974). Hassocks, Sussex.

Deubner, L. 1900. *De Incubatione.* Leipzig.

Di Benedetto, V. 1966. "Tendenza e probabilità nell' antica medicina greca." *Critica Storica* 5:315–68.

Dicks, D. R. 1953–54. "Ancient Astronomical Instruments." *Journal of the British Astronomical Association* 64:77–85.

————. 1959. "Thales." *Classical Quarterly* n.s. 9:294–309.

————. 1960. *The Geographical Fragments of Hipparchus.* London.

————. 1966. "Solstices, Equinoxes and the Presocratics." *Journal of Hellenic Studies* 86:26–40.

————. 1970. *Early Greek Astronomy to Aristotle.* London.

Diels, H. 1879. *Doxographi Graeci.* Berlin.

————. 1884. "Gorgias und Empedokles." *Sitzungsberichte der königlich preussischen Akademie der Wissenschaften zu Berlin.* Jahrgang 1884, pp. 343–68. Berlin.

————. 1893a. "Über die Excerpte von Menons Iatrika in dem Londoner Papyrus 137." *Hermes* 28:407–34.

————. 1893b. *Anonymi Londinensis ex Aristotelis Iatricis Menoniis et aliis medicis Eclogae.* Supplementum Aristotelicum 3, 1. Berlin.

————. 1924. *Antike Technik.* 3rd ed. (1st ed. 1914). Leipzig.

Dierauer, U. 1977. *Tier und Mensch im Denken der Antike.* Amsterdam.

Dihle, A. 1963. "Kritisch-exegetische Bemerkungen zur Schrift Über die Alte Heilkunst." *Museum Helveticum* 20:135–50.

Dijk, J. J. A. van. 1953. *La Sagesse suméro-accadienne: Recherches sur les genres littéraires des textes sapientiaux.* Leiden.

Dijksterhuis, E. J. 1956. *Archimedes.* Acta Historica Scientiarum Naturalium et Medicinalium 12. Copenhagen.

Diller, A. 1949. "The Ancient Measurements of the Earth." *Isis* 40:6–9.

Diller, H. 1932. "ὄψις ἀδήλων τὰ φαινόμενα." *Hermes* 67:14–42. (Reprinted in *Kleine Schriften zur antiken Literatur* [Munich, 1971], pp. 119–43.)

————. 1934. *Wanderarzt und Aitiologe.* Philologus Suppl. Bd. 26, 3. Leipzig.

————. 1952. "Hippokratische Medizin und attische Philosophie." *Hermes* 80:385–409. (Reprinted in Diller 1973, pp. 46–70.)

————. 1964. "Ausdrucksformen des methodischen Bewusstseins in den hippokratischen Epidemien." *Archiv für Begriffsgeschichte* 9:133–50. (Reprinted in Diller 1973, pp. 106–23.)

———. 1971. "Der griechische Naturbegriff." In *Kleine Schriften zur antiken Literatur* (Munich), pp. 144–61.

———. 1973. *Kleine Schriften zur antiken Medizin*. Berlin.

———. 1974. "Empirie und Logos: Galen's Stellung zu Hippokrates und Platon." In Döring and Kullmann, edd., 1974, pp. 227–38.

———. 1975. "Das Selbstverständnis der griechischen Medizin in der Zeit des Hippokrates." In Bourgey and Jouanna, edd., 1975, pp. 77–93.

Dirlmeier, F. 1937. *Die Oikeiosis-Lehre Theophrasts*. Philologus Suppl. Bd. 30, 1. Leipzig.

Dodds, E. R. 1951. *The Greeks and the Irrational*. Berkeley.

———. 1968. *Pagan and Christian in an Age of Anxiety* (1st ed. 1965). Repr. Cambridge.

———. 1973. *The Ancient Concept of Progress and Other Essays on Greek Literature and Belief*. Oxford.

Döring, K., and W. Kullmann, edd. 1974. *Studia Platonica: Festschrift H. Gundert*. Amsterdam.

Dörrie, H. 1981. "Mysterien (in Kult und Religion) und Philosophie." In *Die orientalischen Religionen im Römerreich*, ed. M. J. Vermaseren, pp. 341–62. Leiden.

Dolby, R. G. A. 1971. "Sociology of Knowledge in Natural Science." *Science Studies* 1:3–21.

Donini, P. L. 1973. "Crisippo e la nozione del possibile." *Rivista di filologia e di istruzione classica* 3rd ser., 101:333–51.

———. 1974–75. "Fato e volontà umana in Crisippo." *Atti dell' Accademia delle Scienze di Torino*, Classe di Scienze morali, storiche e filologiche 109, pp. 187–230.

———. 1980. "Motivi filosofici in Galeno." *La Parola del Passato* 35:333–70.

Douglas, M. 1966. *Purity and Danger*. London.

———. 1970. *Natural Symbols*. London.

———. 1975. *Implicit Meanings*. London.

———. 1978. *Cultural Bias*. Royal Anthropological Institute of Great Britain and Ireland, Occasional Paper 35. London.

Douglas, M., ed. 1970. *Witchcraft Confessions and Accusations*. London.

———. 1973. *Rules and Meanings*. London.

Dover, K. J. 1968. *Lysias and the Corpus Lysiacum*. Berkeley.

———. 1974. *Greek Popular Morality in the Time of Plato and Aristotle*. Oxford.

———. 1975. "The Freedom of the Intellectual in Greek Society." *Talanta* (Proceedings of the Dutch Archaeological and Historical Society) 7:24–54.

———. 1983. "The Originality of the First Greek Historians." *Humanities* 17:1–10.

Drabkin, I. E. 1938. "Notes on the Laws of Motion in Aristotle." *American Journal of Philology* 59:60–84.

———. 1942–43. "Posidonius and the Circumference of the Earth." *Isis* 34:509–12.

————. 1944. "On Medical Education in Greece and Rome." *Bulletin of the History of Medicine* 15:333–51.

————. 1951. "Soranus and His System of Medicine." *Bulletin of the History of Medicine* 25:503–18.

————. 1955. "Remarks on Ancient Psychopathology." *Isis* 46:223–34.

————. 1957. "Medical Education in Ancient Greece and Rome." *Journal of Medical Education* 32:286–95.

Drachmann, A. B. 1922. *Atheism in Pagan Antiquity*. London.

Drachmann, A. G. 1948. *Ktesibios, Philon and Heron: A Study in Ancient Pneumatics*. Copenhagen.

————. 1963. *The Mechanical Technology of Greek and Roman Antiquity*. Copenhagen.

————. 1967–68. "Archimedes and the Science of Physics." *Centaurus* 12: 1–11.

Drake, S. 1978. "Ptolemy, Galileo, and Scientific Method." *Studies in History and Philosophy of Science* 9:99–115.

Dreyer, J. L. E. 1906. *History of the Planetary Systems from Thales to Kepler*. Cambridge.

————. 1916–17. "On the Origin of Ptolemy's Catalogue of Stars." *Monthly Notices of the Royal Astronomical Society* 77:528–39.

————. 1917–18. "On the Origin of Ptolemy's Catalogue of Stars." *Monthly Notices of the Royal Astronomical Society* 78: 343–49.

Driver, G. R. 1976. *Semitic Writing*. Revised ed. (1st ed. 1948). London.

Ducatillon, J. 1977. *Polémiques dans la Collection Hippocratique*. Lille.

Duchemin, J. 1955. *Pindare, poète et prophète*. Paris.

————. 1968. *L'ΑΓΩΝ dans la tragédie grecque*. 2nd ed. (1st ed. 1945). Paris.

Duchesne-Guillemin, J. 1953. *Ormazd et Ahriman, l'aventure dualiste dans l'antiquité*. Paris.

————. 1956. "Persische Weisheit in griechischem Gewande?" *Harvard Theological Review* 49:115–22.

————. 1962. *La Religion de l'Iran ancien*. Paris.

Duckworth, W. L. H. 1962. *Galen On Anatomical Procedures, the Later Books*, trans. W. L. H. Duckworth, ed. M. C. Lyons and B. Towers. Cambridge.

Düring, I. 1944. *Aristotle's Chemical Treatise, Meteorologica Book IV*. Göteborgs Högskolas Årsskrift 50. Göteborg.

————. 1961. "Aristotle's Method in Biology." In S. Mansion, ed., 1961, pp. 213–21.

————. 1966. *Aristoteles: Darstellung und Interpretation seines Denkens*. Heidelberg.

Düring, I., ed. 1969. *Naturphilosophie bei Aristoteles und Theophrast*. Verhandlungen des 4. Symposium Aristotelicum. Heidelberg.

Duhem, P. 1905–6. *Les Origines de la statique*. 2 vols. Paris.

————. 1908. "ΣΩΖΕΙΝ ΤΑ ΦΑΙΝΟΜΕΝΑ." *Annales de Philosophie Chrétienne* 6:113–39, 277–302, 352–77, 482–514, 561–92.

————. 1954a. *The Aim and Structure of Physical Theory*, trans. P. P. Wiener of 2nd ed. of *La Théorie physique: Son Objet, sa structure* (Paris, 1914). Princeton.

―――. 1954b. *Le Système du monde: Histoire des doctrines cosmologiques de Platon à Copernic,* vols. 1 and 2. 2nd ed. Paris.

Dumont, J.-P. 1971. *Le Scepticisme et le phénomène.* Paris.

―――. 1982. "Confirmation et disconfirmation." In Barnes, Brunschwig, Burnyeat, and Schofield, edd., 1982, pp. 273–303.

Dumont, L. 1970. *Homo Hierarchicus,* trans. M. Sainsbury of French ed. (Paris, 1966). London.

Dundes, A. 1975. *Analytic Essays in Folklore.* The Hague.

―――. 1980. *Interpreting Folklore.* Bloomington, Indiana.

Durkheim, E. 1912/1976. *The Elementary Forms of the Religious Life,* trans. J. W. Swain of *Les Formes élémentaires de la vie religieuse* (Paris, 1912). 2nd ed. London.

Durkheim, E., and M. Mauss. 1901–2/1963. *Primitive Classification,* trans. R. Needham of "De quelques formes primitives de classification," *L'Année Sociologique* 6 : 1–72. London.

Ebbell, B. 1937. *The Papyrus Ebers.* Copenhagen.

Ebbinghaus, K. 1964. *Ein formales Modell der Syllogistik des Aristoteles.* Hypomnemata 9. Göttingen.

Ebers, G. 1890. *Papyrus Ebers: Die Maasse und das Kapitel über die Augenkrankheiten.* I: *Die Gewichte und Hohlmaasse des Papyrus Ebers.* Abhandlungen der philologisch-historischen Classe der königlich sächsischen Gesellschaft der Wissenschaften, 11, 2. Leipzig.

Ebert, T. 1974. *Meinung und Wissen in der Philosophie Platons.* Berlin.

Edelstein, E. J., and L. Edelstein. 1945. *Asclepius.* 2 vols. Baltimore.

Edelstein, L. 1931. ΠΕΡΙ ΑΕΡΩΝ *und die Sammlung der hippokratischen Schriften.* Problemata 4. Berlin.

―――. 1932–33/1967a. "The History of Anatomy in Antiquity." Originally "Die Geschichte der Sektion in der Antike." *Quellen und Studien zur Geschichte der Naturwissenschaften und der Medizin* 3, 2: 100–156. In Edelstein 1967a, pp. 247–301.

―――. 1935. "Hippokrates." *Pauly-Wissowa Real-Encyclopädie der classischen Altertumswissenschaft,* Suppl. Bd. 6, cols. 1290–1345.

―――. 1935/1967a. "The Methodists." Originally "Methodiker," *Pauly-Wissowa Real-Encyclopädie der classischen Altertumswissenschaft,* Suppl. Bd. 6, cols. 358–73. In Edelstein 1967a, pp. 173–91.

―――. 1937/1967a. "Greek Medicine in Its Relation to Religion and Magic." From *Bulletin of the Institute of the History of Medicine* 5 : 201–46. In Edelstein 1967a, pp. 205–46.

―――. 1939/1967a. "The Genuine Works of Hippocrates." From *Bulletin of the History of Medicine* 7 : 236–48. In Edelstein 1967a, pp. 133–44.

―――. 1943/1967a. *The Hippocratic Oath.* From Supplements to the Bulletin of the History of Medicine 1 (Baltimore, 1943). In Edelstein 1967a, pp. 3–63.

―――. 1952a/1967a. "The Relation of Ancient Philosophy to Medicine."

From *Bulletin of the History of Medicine* 26:299–316. In Edelstein 1967a, pp. 349–66.

———. 1952b/1967a. "Recent Trends in the Interpretation of Ancient Science." From *Journal of the History of Ideas* 13:573–604. In Edelstein 1967a, pp. 401–39.

———. 1967a. *Ancient Medicine*, ed. O. Temkin and C. L. Temkin. Baltimore.

———. 1967b. *The Idea of Progress in Classical Antiquity*. Baltimore.

Edlow, R. B. 1977. *Galen On Language and Ambiguity*. Philosophia Antiqua 31. Leiden.

Ehrenberg, V. 1935. *Ost und West*. Schriften der philosophischen Fakultät der deutschen Universität in Prag 15. Brno.

———. 1940. "Isonomia." *Pauly-Wissowa Real-Encyclopädie der classischen Altertumswissenschaft*, Suppl. Bd. 7, cols. 293–301.

———. 1946. *Aspects of the Ancient World*. Oxford.

———. 1973. *From Solon to Socrates*. 2nd ed. (1st ed. 1968). London.

Einarson, B. 1936. "On Certain Mathematical Terms in Aristotle's Logic." *American Journal of Philology* 57:33–54, 151–72.

Eisenstein, E. L. 1979. *The Printing Press as an Agent of Change*. 2 vols. Cambridge.

Eliade, M. 1946. "Le Problème du chamanisme." *Revue de l'Histoire des Religions* 131:5–52.

———. 1963. *Myth and Reality*, trans. W. R. Trask of *Aspects du mythe* (Paris, 1963). New York.

———. 1964. *Shamanism: Ancient Techniques of Ecstasy*, trans. W. R. Trask of *Le Chamanisme et les techniques archaïques de l'extase* (Paris, 1951). London.

Engberg-Pedersen, T. 1979. "More on Aristotelian Epagoge." *Phronesis* 24:301–19.

Evans, E. C. 1941. "The Study of Physiognomy in the Second Century A.D." *Transactions and Proceedings of the American Philological Association* 72:96–108.

Evans, James. 1984. "Fonction et origine probable du point équant de Ptolémée." *Revue d'Histoire des Sciences* 37:193–213.

Evans, J. D. G. 1975. "The Codification of False Refutations in Aristotle's *De Sophisticis Elenchis*." *Proceedings of the Cambridge Philological Society* n.s. 21:42–52.

———. 1977. *Aristotle's Concept of Dialectic*. Cambridge.

Evans, M. G. 1958–59. "Causality and Explanation in the Logic of Aristotle." *Philosophy and Phenomenological Research* 19:466–85.

Evans-Pritchard, E. E. 1937. *Witchcraft, Oracles and Magic among the Azande*. Oxford.

———. 1956. *Nuer Religion*. Oxford.

Fahr, W. 1969. ΘΕΟΥΣ ΝΟΜΙΖΕΙΝ. *Zum Problem der Anfänge des Atheismus bei den Griechen*. Spudasmata 26. Hildesheim.

Farrington, B. 1939. *Science and Politics in the Ancient World*. London.

———. 1944–49/1961. *Greek Science*. Revised ed. (1st ed., 2 vols., 1944–49). Harmondsworth.

————. 1953–54. "The Rise of Abstract Science among the Greeks." *Centaurus* 3:32–39.

Ferejohn, M. T. 1980. "Aristotle on Focal Meaning and the Unity of Science." *Phronesis* 25:117–28.

————. 1982–83. "Definition and the Two Stages of Aristotelian Demonstration." *Review of Metaphysics* 36:375–95.

Fernandez, J. W. 1977. "The Performance of Ritual Metaphors." In Sapir and Crocker, edd., 1977, pp. 100–131.

————. 1982. *Bwiti: An Ethnography of the Religious Imagination in Africa.* Princeton.

Ferrari, G. R. F. 1983. Review of Moors 1982, *Ancient Philosophy* 3:219–25.

————. Forthcoming. *Listening to the Cicadas.* Cambridge.

Festugière, A. J. 1944–54. *La Révélation d'Hermès trismégiste.* 4 vols. Paris.

————. 1948. *Hippocrate, l'ancienne médecine.* Etudes et Commentaires 4. Paris.

Feyerabend, P. K. 1961. *Knowledge without Foundations.* Oberlin.

————. 1962. "Explanation, Reduction, and Empiricism." In *Scientific Explanation, Space and Time,* Minnesota Studies in the Philosophy of Science 3, ed. H. Feigl and G. Maxwell, pp. 28–97. Minneapolis.

————. 1975. *Against Method.* London.

————. 1978. *Science in a Free Society.* London.

————. 1981a. *Realism, Rationalism and Scientific Method. Philosophical Papers,* vol. 1. Cambridge.

————. 1981b. *Problems of Empiricism. Philosophical Papers,* vol. 2. Cambridge.

Fiedler, W. 1978. *Analogiemodelle bei Aristoteles.* Studien zur antiken Philosophie 9. Amsterdam.

Filliozat, J. 1943. *Magie et médecine.* Paris.

————. 1949/1964. *The Classical Doctrine of Indian Medicine: Its Origins and Its Greek Parallels,* trans. D. R. Chanana of *La Doctrine classique de la médecine indienne: Ses Origines et ses parallèles grecs* (Paris, 1949). Delhi.

Findlay, J. N. 1978. "The Myths of Plato." *Dionysius* 2:19–34.

Fine, G. 1982. "Aristotle and the More Accurate Arguments." In Schofield and Nussbaum, edd., 1982, pp. 155–77.

————. 1984. "Separation." *Oxford Studies in Ancient Philosophy* 2:31–87.

————. 1987. "Forms as Causes." In Graeser, ed., 1987, pp. 69–112.

Finley, M. I. 1954/1977. *The World of Odysseus.* 2nd ed. (1st ed. 1954). New York.

————. 1964–65/1975a. "Myth, Memory, and History." From *History and Theory* 4:281–302. In Finley 1975a, pp. 11–33.

————. 1965. "Technical Innovation and Economic Progress in the Ancient World." *Economic History Review* 2nd ser., 18:29–45.

————. 1970. *Early Greece: The Bronze and Archaic Ages.* London.

————. 1973a. *Democracy Ancient and Modern.* London.

————. 1973b. *The Ancient Economy.* London.

————. 1974. "Athenian Demagogues." From *Past and Present* 21 (1962): 3–24. In *Studies in Ancient Society,* ed. M. I. Finley, pp. 1–25. London.

————. 1975a. *The Use and Abuse of History.* London.

————. 1975b. "The Freedom of the Citizen in the Greek World." *Talanta* (Proceedings of the Dutch Archaeological and Historical Society) 7:1–23.

————. 1977. Censura nell'antichità classica." *Belfragor* 32:605–22.

————. 1980. *Ancient Slavery and Modern Ideology.* London.

————. 1983. *Politics in the Ancient World.* Cambridge.

Finnegan, R. 1977. *Oral Poetry.* Cambridge.

Fischer, I. 1975. "Another Look at Eratosthenes' and Posidonius' Determinations of the Earth's Circumference." *Quarterly Journal of the Royal Astronomical Society* 16:152–67.

Fischer, N. 1982. "Die Ursprungsphilosophie in Platons 'Timaios'." *Philosophisches Jahrbuch* 89:247–68.

Flashar, H. 1962. *Aristoteles, Problemata Physica.* Aristoteles Werke in deutscher Übersetzung 19. Darmstadt.

————. 1966. *Melancholie und Melancholiker in den medizinischen Theorien der Antike.* Berlin.

Flashar, H., and K. Gaiser, edd. 1965. *Synusia: Festgabe W. Schadewaldt.* Pfullingen.

Fleming, D. 1955. "Galen on the Motions of the Blood in the Heart and Lungs." *Isis* 46:14–21.

Förster, R. 1893. *Scriptores Physiognomonici Graeci et Latini.* 2 vols. Leipzig.

Fontenrose, J. 1959. *Python.* Berkeley.

————. 1971. *The Ritual Theory of Myth.* Berkeley.

————. 1974. "Work, Justice, and Hesiod's Five Ages." *Classical Philology* 69:1–16.

Fotheringham, J. K. 1915. "The Probable Error of a Water-Clock." *Classical Review* 29:236–38.

————. 1923. "The Probable Error of a Water-Clock." *Classical Review* 37:166–67.

————. 1928. "The Indebtedness of Greek to Chaldaean Astronomy." *The Observatory* 51:301–15. (Also in "Quellen und Studien zur Geschichte der Mathematik, Astronomie und Physik" B 2, 1 [1933]: 28–44. Berlin.)

Foucault, M. 1963/1973. *The Birth of the Clinic,* trans. A. M. Sheridan Smith of *Naissance de la clinique* (Paris, 1963). London.

————. 1966/1970. *The Order of Things,* trans. of *Les Mots et les choses* (Paris, 1966). London.

————. 1969/1972. *The Archaeology of Knowledge,* trans. A. M. Sheridan Smith of *L'Archéologie du savoir* (Paris, 1969). London.

————. 1976/1978. *The History of Sexuality,* trans. R. Hurley of *La Volonté du savoir,* vol. 1 of *Histoire de la sexualité* (Paris, 1976). New York.

————. 1984/1985. *The Use of Pleasure,* trans. R. Hurley of *L'Usage des plaisirs,* vol. 2 of *Histoire de la sexualité* (Paris, 1984). New York.

————. 1984/forthcoming trans. of *Le Souci de soi,* vol. 3 of *Histoire de la sexualité* (Paris, 1984). New York.

Fowler, D. H. 1979. "Ratio in Early Greek Mathematics." *Bulletin of the American Mathematical Society* n.s. 1:807–46.

Fowler, Don. 1983. "Lucretius on the Clinamen and 'Free Will' (II 251–93)."

In ΣΥΖΗΤΗΣΙΣ: *Studi sull' epicureismo greco e romano offerti a M. Gigante*, pp. 329–52. Naples.

Fränkel, H. 1921. *Die homerischen Gleichnisse*. Göttingen.

———. 1960. *Wege und Formen frühgriechischen Denkens*. 2nd ed. (1st ed. 1955). Munich.

———. 1975. *Early Greek Poetry and Philosophy*, trans. M. Hadas and J. Willis of 2nd ed. of *Dichtung und Philosophie des frühen Griechentums* (Munich, 1962). Oxford.

Franciosi, F. 1976. "Die Entdeckung der mathematischen Irrationalität." *Acta Antiqua Academiae Scientiarum Hungaricae* 24:183–203.

Frankfort, H. 1948. *Kingship and the Gods*. Chicago.

Frankfort, H., ed. 1949. *Before Philosophy*. 2nd ed. 1st ed., *The Intellectual Adventure of Ancient Man* (Chicago, 1946). London.

Fraser, P. M. 1969. "The Career of Erasistratus of Ceos." In *Rendiconti dell'Istituto Lombardo*, Classe di Lettere e Scienze Morali e Storiche 103, pp. 518–37.

———. 1972a. *Ptolemaic Alexandria*, 3 vols. Oxford.

———. 1972b. "Eratosthenes of Cyrene." *Proceedings of the British Academy* 56:175–207.

Frazer, J. G. 1911–15. *The Golden Bough*, 12 vols. 3rd ed. London.

Frazer, R. M. 1972. "Pandora's Diseases, *Erga* 102–04." *Greek, Roman and Byzantine Studies* 13:235–38.

Frede, D. 1970. *Aristoteles und die Seeschlacht*. Hypomnemata 27. Göttingen.

———. 1982. "The Dramatization of Determinism: Alexander of Aphrodisias' *De Fato*." *Phronesis* 27:276–98.

Frede, M. 1974. *Die stoische Logik*. Göttingen.

———. 1979. "Des Skeptikers Meinungen." *Neue Hefte für Philosophie* 15–16:102–29.

———. 1980. "The Original Notion of Cause." In Schofield, Burnyeat, and Barnes, edd., 1980, pp. 217–49.

———. 1981. "On Galen's Epistemology." In Nutton, ed., 1981, pp. 65–86.

———. 1982. "The Method of the So-Called Methodical School of Medicine." In Barnes, Brunschwig, Burnyeat, and Schofield, edd., 1982, pp. 1–23.

———. 1983. "Stoics and Skeptics on Clear and Distinct Impressions." In Burnyeat, ed., 1983, pp. 65–93.

Freud, S. 1953. *The Interpretation of Dreams*. Originally *Die Traumdeutung* (Leipzig, 1900). In vols. 4 and 5 of *The Standard Edition of the Complete Psychological Works*, trans. under general editorship of J. Strachey, 24 vols. (London, 1953–74). London.

Freudenthal, H. 1966. "Y avait-il une crise des fondements des mathématiques dans l'antiquité?" *Bulletin de la Société Mathématique de Belgique* 18:43–55.

Friedländer, P. 1958–69. *Plato*, trans. H. Meyerhoff of 2nd ed. of *Platon* (Berlin, 1954–60). 3 vols. London.

Frischer, B. 1982. *The Sculpted Word: Epicureanism and Philosophical Recruitment in Ancient Greece*. Berkeley.

Frisk, H. 1935. *"Wahrheit" und "Lüge" in den indogermanischen Sprachen.* Göteborgs Högskolas Årsskrift 41, 3. Göteborg.

Fritz, H. von. 1945/1970. "The Discovery of Incommensurability by Hippasus of Metapontum." From *Annals of Mathematics* 2nd ser., 46: 242–64. In Furley and Allen, edd., 1970, pp. 382–412.

———. 1955/1971. "Die APXAI in der griechischen Mathematik." From *Archiv für Begriffsgeschichte* 1: 13–103. In Fritz 1971, pp. 335–429.

———. 1959/1971. "Gleichheit, Kongruenz und Ähnlichkeit in der antiken Mathematik bis auf Euklid." From *Archiv für Begriffsgeschichte* 4: 7–81. In Fritz 1971, pp. 430–508.

———. 1960. "Mathematiker und Akusmatiker bei den alten Pythagoreern." *Sitzungsberichte der bayerischen Akademie der Wissenschaften, phil.-hist. Kl.,* 1960, 11. Munich.

———. 1964/1971. "Die ΕΠΑΓΩΓΗ bei Aristoteles." *Sitzungsberichte der bayerischen Akademie der Wissenschaften, phil.-hist. Kl.,* 1964, 3 (Munich, 1964). In Fritz 1971, pp. 623–76.

———. 1971. *Grundprobleme der Geschichte der antiken Wissenschaft.* Berlin.

———. 1978. *Schriften zur griechischen Logik.* 2 vols. Stuttgart.

Fujisawa, N. 1974. "Ἔχειν, Μετέχειν and Idioms of 'Paradeigmatism' in Plato's Theory of Forms." *Phronesis* 19: 30–58.

Furley, D. J. 1967. *Two Studies in the Greek Atomists.* Princeton.

———. 1969. "Aristotle and the Atomists on Infinity." In Düring, ed., 1969, pp. 85–96.

———. 1976. "Aristotle and the Atomists on Motion in a Void." In Machamer and Turnbull, edd., 1976, pp. 83–100.

———. 1978. "Self-Movers." In Lloyd and Owen, edd., 1978, pp. 165–79.

———. 1981a. "The Greek Theory of the Infinite Universe." *Journal of the History of Ideas* 42: 571–85.

———. 1981b. "Antiphon's Case against Justice." In Kerferd, ed., 1981, pp. 81–91.

———. 1982. "The Greek Commentators' Treatment of Aristotle's Theory of the Continuous." In Kretzmann, ed., 1982, pp. 17–36.

———. 1983a. "The Mechanics of *Meteorologica* IV: A Prolegomenon to Biology." In Moraux and Wiesner, edd., 1983, pp. 73–93.

———. 1983b. "Weight and Motion in Democritus' Theory." *Oxford Studies in Ancient Philosophy* 1: 193–209.

———. 1985. "The Rainfall Example in *Physics* II.8." In Gotthelf, ed., pp. 177–82.

———. 1986. "Nothing to Us?" In Schofield and Striker, edd., 1986, pp. 75–91.

Furley, D. J., and R. E. Allen, edd. 1970. *Studies in Presocratic Philosophy,* vol. 1. London.

Furley, D. J., and J. S. Wilkie. 1984. *Galen On Respiration and the Arteries.* Princeton.

Gadamer, H.-G. 1964/1980. "Dialectic and Sophism in Plato's Seventh Letter." Originally "Dialektik und Sophistik im siebenten platonischen Brief,"

Sitzungsberichte der Heidelberger Akademie der Wissenschaften, phil.-hist. Kl., 2 (Heidelberg, 1964): In Gadamer 1980, pp. 93–123.

————. 1968/1980. "Plato's Unwritten Dialectic." Originally "Platons ungeschriebene Dialektik," in Gadamer, Gaiser, Gundert, Krämer, and Kuhn 1968, pp. 9–30. In Gadamer 1980, pp. 124–55.

————. 1974/1980. "Idea and Reality in Plato's *Timaeus*." Originally "Idee und Wirklichkeit in Platos *Timaios*," *Sitzungsberichte der Heidelberger Akademie der Wissenschaften, phil.-hist. Kl.*, 2 (Heidelberg, 1974). In Gadamer 1980, pp. 156–93.

————. 1980. *Dialogue and Dialectic: Eight Hermeneutical Studies on Plato*, trans. P. Christopher Smith. New Haven.

Gadamer, H.-G., K. Gaiser, H. Gundert, J. Krämer, and H. Kuhn. 1968. *Idee und Zahl*. Abhandlungen der Heidelberger Akademie der Wissenschaften, phil.-hist. Kl., Jahrgang 1968, 2. Heidelberg.

Gaiser, K. 1963. *Platons ungeschriebene Lehre*. Stuttgart.

————. 1969. "Das zweifache Telos bei Aristoteles." In Düring, ed., 1969, pp. 97–113.

Gandt, F. de. 1975. "La *Mathésis* d'Aristote: Introduction aux *Analytiques seconds*." *Revue des Sciences Philosophiques et Théologiques* 59: 564–600.

————. 1976. "La *Mathésis* d'Aristote: Introduction aux *Analytiques seconds*." *Revue des Sciences Philosophiques et Théologiques* 60: 37–84.

————. 1982. "Force et science des machines." In Barnes, Brunschwig, Burnyeat, and Schofield, edd., 1982, pp. 96–127.

Gardiner, A. H. 1909. *The Admonitions of an Egyptian Sage*. Leipzig.

————. 1938. "The House of Life." *Journal of Egyptian Archaeology* 24: 157–79.

Garland, R. S. J. 1985. *The Greek Way of Death*. London.

Garzón Díaz, J. 1981. "La muerte: consciencia de la muerte desde Homero a Pindaro." *Helmantica* 32: 353–89.

Gatzemeier, M. 1970. *Die Naturphilosophie des Straton von Lampsakos*. Meisenheim.

Gaukroger, S. 1980. "Aristotle on Intelligible Matter." *Phronesis* 25: 187–97.

Geertz, C. 1973. *The Interpretation of Cultures*. New York.

Geertz, H. 1975–76. "An Anthropology of Religion and Magic." *Journal of Interdisciplinary History* 6: 71–89.

Gellner, E. 1962/1970. "Concepts and Society." From *The Transactions of the Fifth World Congress of Sociology* 1: 153–83. In B. R. Wilson, ed., 1970, pp. 18–49.

————. 1973. "The Savage and the Modern Mind." In Horton and Finnegan, edd., 1973, pp. 162–81.

————. 1974. *Legitimation of Belief*. Cambridge.

————. 1985. *Relativism and the Social Sciences*. Cambridge.

Gentili, B. 1983. "Oralità e scrittura in Grecia." In Vegetti, ed., 1983, pp. 30–46.

Gentili, B., and G. Paioni, edd. 1977. *Il mito greco*. Rome.

Gernet, L. 1917. *Recherches sur le développement de la pensée juridique et morale en Grèce.* Paris.

———. 1945/1981. "The Origins of Greek Philosophy." Originally "Les Origines de la philosophie." *Bulletin de l'Enseignement Public du Maroc* 183: 1–12. In Gernet 1968/1981, pp. 352–64.

———. 1955. *Droit et société dans la Grèce ancienne.* Paris.

———. 1968/1981. *The Anthropology of Ancient Greece,* trans. J. Hamilton and B. Nagy of *Anthropologie de la Grèce antique* (Paris, 1968). Baltimore.

———. 1983. *Les Grecs sans miracle.* Paris.

Giannantoni, G., ed. 1981. *Lo Scetticismo antico.* Naples.

Gigante, M. 1981. *Scetticismo e Epicureismo.* Naples.

Gigon, O. 1936. "Gorgias 'Über das Nichtsein.'" *Hermes* 71:186–213.

———. 1946. "Die naturphilosophischen Voraussetzungen der antiken Biologie." *Gesnerus* 3:35–58.

———. 1968. *Der Ursprung der griechischen Philosophie.* 2nd ed. (1st ed. 1945). Basel.

———. 1973. "Der Begriff der Freiheit in der Antike." *Gymnasium* 80:8–56.

Gilbert, O. 1907. *Die meteorologischen Theorien des griechischen Altertums.* Leipzig.

———. 1910. "Spekulation und Volksglaube in der ionischen Philosophie." *Archiv für Religionswissenschaft* 13:306–32.

Gill, C. 1985. "Ancient Psychotherapy." *Journal of the History of Ideas* 46: 307–25.

Gingerich, O. 1980. "Was Ptolemy a Fraud?" *Quarterly Journal of the Royal Astronomical Society* 21:253–66.

———. 1981. "Ptolemy Revisited." *Quarterly Journal of the Royal Astronomical Society* 22:40–44.

Ginzburg, C. 1979. "Spie. Radici di un paradigma indiziario." In A. Gargani, ed., *Crisi della ragione,* pp. 57–106. Turin.

Gladigow, B. 1965. *Sophia und Kosmos.* Spudasmata 1. Hildesheim.

———. 1967. "Zum Makarismos des Weisen." *Hermes* 95:404–33.

Glidden, D. K. 1979. "Epicurus on Self-Perception." *American Philosophical Quarterly* 16:297–306.

———. 1983. "Epicurean Semantics." In ΣΥΖΗΤΗΣΙΣ: *Studi sull epicureismo greco e romano offerti a M. Gigante,* pp. 185–226. Naples.

Gnoli, G., and J.-P. Vernant, edd. 1982. *La Mort, les morts dans les sociétés anciennes.* Cambridge.

Götze, A. 1923. "Persische Weisheit in griechischem Gewande: Ein Beitrag zur Geschichte der Mikrokosmos-Idee." *Zeitschrift für Indologie und Iranistik* 2:60–98, 167–77.

Goffman, E. 1961/1973. *Asylums: Essays on the Social Situation of Mental Patients and Other Inmates.* Revised ed. (1st ed. 1961). Chicago.

Gohlke, P. 1924. "Die Entstehungsgeschichte der naturwissenschaftlichen Schriften des Aristoteles." *Hermes* 59:274–306.

———. 1936. *Die Entstehung der aristotelischen Logik.* Berlin.

Goldhill, S. D. 1984. *Language, Sexuality, Narrative: The Oresteia.* Cambridge.

Goldschmidt, V. 1947a. *Le Paradigme dans la dialectique platonicienne.* Paris.
———. 1947b. *Les Dialogues de Platon: Structure et méthode dialectique.* Paris.
———. 1970. *Questions platoniciennes.* Paris.
Goldstein, B. R. 1967. "The Arabic Version of Ptolemy's *Planetary Hypotheses.*" *Transactions of the American Philosophical Society* 57, 4.
———. 1980. "The Status of Models in Ancient and Medieval Astronomy." *Centaurus* 24:132–47.
Goldstein, B. R., and A. C. Bowen. 1983. "A New View of Early Greek Astronomy." *Isis* 74:330–40.
Goltz, D. 1972. *Studien zur Geschichte der Mineralnamen in Pharmazie, Chemie und Medizin von den Anfängen bis Paracelsus.* Sudhoffs Archiv Beiheft 14. Wiesbaden.
———. 1974. *Studien zur altorientalischen und griechischen Heilkunde, Therapie, Arzneibereitung, Rezeptstruktur.* Sudhoffs Archiv Beiheft 16. Wiesbaden.
Gómez-Lobo, A. 1976–77. "Aristotle's Hypotheses and the Euclidean Postulates." *Review of Metaphysics* 30:430–39.
Gomperz, H. 1912. *Sophistik und Rhetorik.* Leipzig.
———. 1943. "Problems and Methods of Early Greek Science." *Journal of the History of Ideas* 4:161–76.
Gomperz, T. 1910. *Die Apologie der Heilkunst.* 2nd ed. Leipzig.
Gonda, J. 1975. *Vedic Literature: Saṃhitas and Brāhmaṇas.* A History of Indian Literature 1, 1. Wiesbaden.
———. 1977. *The Ritual Sūtras.* A History of Indian Literature 1, 2. Wiesbaden.
Goodman, N. 1978. *Ways of Worldmaking.* Hassocks, Sussex.
Goody, J. 1972. *The Myth of the Bagre.* Oxford.
———. 1977. *The Domestication of the Savage Mind.* Cambridge.
———. Forthcoming. *The Interface between the Written and the Oral.* Cambridge.
Goody, J., ed. 1968. *Literacy in Traditional Societies.* Cambridge.
Goody, J., and I. P. Watt. 1968. "The Consequences of Literacy." Originally published in *Comparative Studies in Society and History* 5 (1962–63): 304–45. In Goody, ed., 1968, pp. 27–68.
Gordon, R. L., ed. 1981. *Myth, Religion and Society: Structuralist Essays by M. Detienne, L. Gernet, J.-P. Vernant and P. Vidal-Naquet.* Cambridge.
Gotthelf, A. 1976–77. "Aristotle's Conception of Final Causality." *Review of Metaphysics* 30:226–54.
Gotthelf, A., ed. 1985. *Aristotle on Nature and Living Things.* Pittsburgh.
Gotthelf, A., and J. G. Lennox, edd. 1987. *Philosophical Issues in Aristotle's Biology.* Cambridge.
Gottschalk, H. B. 1961. "The Authorship of *Meteorologica,* Book IV." *Classical Quarterly* n.s. 11:67–79.
———. 1965. *Strato of Lampsacus: Some Texts.* Proceedings of the Leeds Philosophical and Literary Society, Literary and Historical Section 11 (1964–66), Part 6. Leeds.

Gould, J. B. 1970. *The Philosophy of Chrysippus.* Albany, New York.
Gouldner, A. W. 1967. *Enter Plato.* London.
Gourevitch, D. 1969. "Déontologie médicale: Quelques Problèmes, I." *Mélanges d'Archéologie et d'Histoire* 81:519–36.
———. 1970. "Déontologie médicale: Quelques Problèmes, II." *Mélanges d'Archéologie et d'Histoire* 82:737–52.
Govi, G. 1885. *L'Ottica di Claudio Tolomeo.* Turin.
Gracia, D. 1978. "The Structure of Medical Knowledge in Aristotle's Philosophy." *Sudhoffs Archiv* 62:1–36.
Graeser, A., ed. 1987. *Mathematics and Metaphysics in Aristotle.* Proceedings of the 10th Symposium Aristotelicum. Bern.
Granet, M. 1934. *La Pensée chinoise.* Paris.
Granger, G. G. 1976. *La Théorie aristotélicienne de la science.* Paris.
Granger, H. 1980. "Aristotle and the Genus-Species Relation." *Southern Journal of Philosophy* 18:37–50.
Grapow, H. 1954–73. *Grundriss der Medizin der alten Ägypter,* 9 vols. Berlin.
Green, P. M. 1954. "Prolegomena to the Study of Magic and Superstition in the *Natural History* of Pliny the Elder, with Special Reference to Book XXX and Its Sources." Ph.D. diss. Cambridge.
Grene, M. 1974. "Is Genus to Species as Matter to Form? Aristotle and Taxonomy." *Synthese* 28:51–69.
Grensemann, H. 1968a. *Die hippokratische Schrift "Über die heilige Krankheit."* Ars Medica, Abt. II, Bd. 1. Berlin.
———. 1968b. *Hippocratis, De Octimestri Partu, De Septimestri Partu.* Corpus Medicorum Graecorum 1, 2, 1. Berlin.
———. 1975. *Knidische Medizin.* I: *Die Testimonien zur ältesten knidischen Lehre und Analysen knidischer Schriften im Corpus Hippocraticum.* Ars Medica Abt. II, Bd. 4, 1. Berlin.
Grice, H. P. 1957. "Meaning." *Philosophical Review* 66:377–88.
———. 1968. "Utterer's Meaning, Sentence-Meaning, and Word-Meaning." *Foundations of Language* 4:225–42. (Reprinted in Searle, ed., 1971, pp. 54–70.)
———. 1969. "Utterer's Meaning and Intentions." *Philosophical Review* 78:147–77.
Griffith, F. L. 1898. *The Petrie Papyri: Hieratic Papyri from Kahun and Gurob.* London.
Griffith, G. T. 1966. "Isegoria in the Assembly at Athens." In *Ancient Society and Institutions: Studies Presented to V. Ehrenberg,* ed. E. Badian, pp. 115–38. Oxford.
Grmek, M. D., ed. 1980. *Hippocratica.* Actes du Colloque Hippocratique de Paris. Paris.
Groningen, B. A. van. 1953. *In the Grip of the Past: Essay on an Aspect of Greek Thought.* Philosophia Antiqua 6. Leiden.
———. 1960. *La Composition littéraire archaïque grecque.* 2nd ed. (1st ed. 1958). Amsterdam.

Güterbock, H. G. 1962. "Hittite Medicine." *Bulletin of the History of Medicine* 36:109–13.

Guidorizzi, G. 1973. "L'opuscolo di Galeno *De dignotione ex insomniis.*" *Bollettino del comitato per la preparazione dell'edizione nazionale dei classici greci e latini* n.s. 21:81–105.

Gundel, H. G. 1968. *Weltbild und Astrologie in den griechischen Zauberpapyri.* Munich.

Gundel, W. 1922. *Sterne und Sternbilder im Glauben des Altertums und der Neuzeit.* Leipzig.

———. 1936. *Neue astrologische Texte des Hermes Trismegistos.* Abhandlungen der bayerischen Akademie der Wissenschaften, phil.-hist. Abt., N.F. 12, 1935. Munich.

Gundel, W., and H. G. Gundel. 1966. *Astrologumena: Die astrologische Literatur in der Antike und ihre Geschichte.* Sudhoffs Archiv Beiheft 6. Wiesbaden.

Gundert, H. 1965. "Zum Spiel bei Platon." In *Beispiele: Festschrift E. Fink,* ed. L. Landgrebe, pp. 188–221. The Hague.

———. 1971. *Dialog und Dialektik: Zur Struktur des platonischen Dialogs.* Amsterdam.

———. 1973. "'Perspektivische Täuschung' bei Platon und die Prinzipienlehre." In *Zetesis: Festschrift de Strycker,* pp. 80–97. Antwerp.

Guthrie, W. K. C. 1950. *The Greeks and Their Gods.* London.

———. 1962. *A History of Greek Philosophy.* Vol. 1: *The Earlier Presocratics and the Pythagoreans.* Cambridge.

———. 1965. *A History of Greek Philosophy.* Vol. 2: *The Presocratic Tradition from Parmenides to Democritus.* Cambridge.

———. 1969. *A History of Greek Philosophy.* Vol. 3: *The Fifth-Century Enlightenment.* Cambridge.

———. 1975. *A History of Greek Philosophy.* Vol. 4: *Plato the Man and His Dialogues: Earlier Period.* Cambridge.

———. 1978. *A History of Greek Philosophy.* Vol. 5: *The Later Plato and the Academy.* Cambridge.

———. 1981. *A History of Greek Philosophy.* Vol. 6: *Aristotle, an Encounter.* Cambridge.

Haas, A. E. 1907. "Antike Lichttheorien." *Archiv für Geschichte der Philosophie* N.F. 13:345–86.

———. 1908–9. "Die Grundlagen der antiken Dynamik." *Archiv für die Geschichte der Naturwissenschaften und der Technik* 1:19–47.

Habermas, J. 1968/1978. *Knowledge and Human Interests,* trans. J. J. Shapiro of *Erkenntnis und Interesse* (Frankfurt, 1968). New ed. London.

———. 1971/1974. *Theory and Practice,* trans. J. Viertel of *Theorie und Praxis* (Frankfurt, 1971). London.

Hacking, I. 1975. *The Emergence of Probability.* Cambridge.

———. 1979. Review of T. S. Kuhn 1977, *History and Theory* 18:223–36.

———. 1983. *Representing and Intervening.* Cambridge.

Hacking, I., ed. 1981. *Scientific Revolutions.* Oxford.

Hahm, D. E. 1972. "Chrysippus' Solution to the Democritean Dilemma of the Cone." *Isis* 63:205–20.

————. 1976. "Weight and Lightness in Aristotle and His Predecessors." In Machamer and Turnbull, edd., 1976, pp. 56–82.

————. 1977. *The Origins of Stoic Cosmology.* Columbus, Ohio.

Halleux, R. 1974. *Le Problème des métaux dans la science antique.* Bibliothèque de la Faculté de Philosophie et Lettres de l'Université de Liège 209. Paris.

————. 1981. *Les Alchimistes grecs,* vol. 1. Paris.

Halliday, W. R. 1913. *Greek Divination.* London.

Hamelin, O. 1907. *Essai sur les éléments principaux de la représentation.* Paris.

————. 1931. *Le Système d'Aristote.* 2nd ed. (1st ed. 1920). Paris.

Hamilton, M. 1906. *Incubation, or the Cure of Disease in Pagan Temples and Christian Churches.* London.

Hamlyn, D. W. 1959. "Aristotle's Account of Aesthesis in the *De Anima.*" *Classical Quarterly* n.s. 9: 6–16.

————. 1961. *Sensation and Perception.* London.

————. 1968. *Aristotle's De Anima, Books II and III.* Oxford.

————. 1976. "Aristotelian Epagoge." *Phronesis* 21: 167–84.

————. 1977–78. "Focal Meaning." *Proceedings of the Aristotelian Society* n.s. 78: 1–18.

Hammer-Jensen, I. 1915. "Das sogenannte IV. Buch der *Meteorologie* des Aristoteles." *Hermes* 50: 113–36.

Hand, W. D. 1980. *Magical Medicine.* Berkeley.

Hansen, M. H. 1974. *The Sovereignty of the People's Court in Athens in the 4th Century B.C.* Odense.

————. 1976/1983. "How Many Athenians Attended the Ecclesia?" *Greek, Roman and Byzantine Studies* 17: 115–34. In Hansen 1983, pp. 1–20, 21–23.

————. 1983. *The Athenian Ecclesia.* Copenhagen.

Hanson, N. R. 1958. *Patterns of Discovery.* Cambridge.

Happ, H. 1965. "Der chemische Traktat des Aristoteles." In Flashar and Gaiser, edd., 1965, pp. 289–322.

————. 1969. "Die *Scala Naturae* und die Schichtung des Seelischen bei Aristoteles." In *Beiträge zur Alten Geschichte und deren Nachleben: Festschrift F. Altheim,* ed. R. Stiehl and H. E. Stier, vol. 1, pp. 220–44. Berlin.

————. 1971. *Hyle. Studien zum aristotelischen Materiebegriff.* Berlin.

Hare, R. M. 1965. "Plato and the Mathematicians." In Bambrough, ed., 1965, pp. 21–38.

Harig, G. 1974. *Bestimmung der Intensität im medizinischen System Galens.* Berlin.

————. 1975. Review of Goltz 1974, *Deutsche Literaturzeitung* 96, cols. 654–58.

————. 1976. "Der Begriff der lauen Wärme in der theoretischen Pharmakologie Galens." *NTM: Schriftenreihe für Geschichte der Naturwissenschaften, Technik und Medizin* 13, 2: 70–76.

————. 1977. "Bemerkungen zum Verhältnis der griechischen zur altorientalischen Medizin." In Joly, ed., 1977, pp. 77–94.

————. 1980. "Anfänge der theoretischen Pharmakologie in Corpus Hippocraticum." In Grmek, ed., 1980, pp. 223–45.

————. 1983. "Die philosophischen Grundlagen des medizinischen System des Asklepiades von Bithynien." *Philologus* 127:43–60.

Harig, G., and J. Kollesch. 1973–74. "Arzt, Kranker und Krankenpflege in der griechisch-römischen Antike und im byzantinischen Mittelalter." *Helikon* 13–14: 256–92.

————. 1974. "Diokles von Karystos und die zoologische Systematik." *NTM: Schriftenreihe für Geschichte der Naturwissenschaften, Technik und Medizin* 11, 1:24–31.

————. 1975. "Galen und Hippokrates." In Bourgey and Jouanna, edd., 1975, pp. 257–74.

————. 1977. "Neue Tendenzen in der Forschung zur Geschichte der antiken Medizin und Wissenschaft." *Philologus* 121:114–36.

————. 1978. "Der hippokratische Eid. Zur Entstehung der antiken Deontologie." *Philologus* 122:157–76.

Harris, C. R. S. 1973. *The Heart and the Vascular System in Ancient Greek Medicine from Alcmaeon to Galen.* Oxford.

Harrison, A. R. W. 1968. *The Laws of Athens.* Vol. 1: *The Family and Property.* Oxford.

————. 1971. *The Laws of Athens.* Vol. 2: *Procedure.* Oxford.

Harrison, J. E. 1903/1922. *Prolegomena to the Study of Greek Religion.* 3rd ed. (1st ed. 1903). Cambridge.

————. 1912/1927. *Themis.* 2nd ed. (1st ed. 1912). Cambridge.

————. 1925. *Reminiscences of a Student's Life.* London.

Hartman, E. 1977. *Substance, Body and Soul: Aristotelian Investigations.* Princeton.

Hartner, W. 1968. *Oriens-Occidens.* Hildesheim.

————. 1977. "The Role of Observations in Ancient and Medieval Astronomy." *Journal of the History of Astronomy* 8:1–11.

————. 1980. "Ptolemy and Ibn Yūnus on Solar Parallax." *Archives Internationales d'Histoire des Sciences* 30:5–26.

Hartog, F. 1980. *Le Miroir d'Hérodote.* Paris.

Harvey, F. D. 1966. "Literacy in the Athenian Democracy." *Revue des Etudes Grecques* 79:585–635.

Hasse, H., and H. Scholz. 1928. *Die Grundlagenkrisis der griechischen Mathematik.* Berlin. (Also in *Kant-Studien* 33 [1928]: 4–34.)

Havelock, E. A. 1963. *Preface to Plato.* Oxford.

————. 1982. *The Literate Revolution in Greece and Its Cultural Consequences.* Princeton.

Havelock, E. A., and J. P. Hershbell, edd. 1978. *Communication Arts in the Ancient World.* New York.

Heath, T. E. 1913. *Aristarchus of Samos.* Oxford.

————. 1921. *A History of Greek Mathematics,* 2 vols. Oxford.

————. 1926. *The Thirteen Books of Euclid's Elements,* 3 vols. 2nd ed. (1st ed. 1908). Cambridge.

———. 1949. *Mathematics in Aristotle.* Oxford.

Heiberg, J. L. 1925. *Geschichte der Mathematik und Naturwissenschaften im Altertum.* Munich.

———. 1927. "Geisteskrankheiten im klassischen Altertum." *Allgemeine Zeitschrift für Psychiatrie* 86:1–44.

Heidel, W. A. 1909–10. "περὶ φύσεως: A Study of the Conception of Nature among the Pre-Socratics." *Proceedings of the American Academy of Arts and Sciences* 45:77–133.

———. 1933. *The Heroic Age of Science.* Baltimore.

———. 1940/1970. "The Pythagoreans and Greek Mathematics." From *American Journal of Philology* 61:1–33. In Furley and Allen, edd., 1970, pp. 350–81.

———. 1941. *Hippocratic Medicine: Its Spirit and Method.* New York.

Heinimann, F. 1945. *Nomos und Physis.* Schweizerische Beiträge zur Altertumswissenschaft 1. Basel.

———. 1961. "Eine vorplatonische Theorie der τέχνη." *Museum Helveticum* 18:105–30.

———. 1975. "Mass-Gewicht-Zahl." *Museum Helveticum* 32:183–96.

Heisenberg, W. 1945/1952. *Philosophic Problems of Nuclear Science,* trans. F. C. Hayes of *Wandlungen in den Grundlagen der Naturwissenschaft* (6th ed. Leipzig, [1945]). London.

———. 1955/1958. *The Physicist's Conception of Nature,* trans. A. J. Pomerans of *Das Naturbild der heutigen Physik* (Hamburg, 1955). London.

Heitsch, E. 1962. "Die nicht-philosophische ΑΛΗΘΕΙΑ." *Hermes* 90:24–33.

———. 1963. "Wahrheit als Erinnerung." *Hermes* 91:36–52.

———. 1970. *Gegenwart und Evidenz bei Parmenides.* Akademie der Wissenschaften und der Literatur, Mainz, Abhandlungen der geistes- und sozialwissenschaftlichen Kl., Jahrgang 1970, 4. Wiesbaden.

———. 1974. "Evidenz und Wahrscheinlichkeitsaussagen bei Parmenides." *Hermes* 102:411–19.

Held, K. 1980. *Heraklit, Parmenides und der Anfang von Philosophie und Wissenschaft. Eine phänomenologische Besinnung.* Berlin.

Hemmy, A. S. 1931. "System of Weights at Mohenjo-daro." In *Mohenjo-daro and the Indus Civilization,* ed. J. Marshall, vol. 2, pp. 589–98. London.

Hempel, C. G. 1939. "Vagueness and Logic." *Philosophy of Science* 6:163–80.

———. 1958. "The Theoretician's Dilemma: A Study in the Logic of Theory Construction." In *Concepts, Theories and the Mind-Body Problem,* Minnesota Studies in the Philosophy of Science 2, ed. H. Feigl, M. Scriven, and G. Maxwell, pp. 37–98. Minneapolis.

———. 1973. "The Meaning of Theoretical Terms: A Critique of the Standard Empiricist Construal." In *Logic, Methodology and Philosophy of Science* 4, ed. P. Suppes et al., pp. 367–78. Amsterdam.

Hempel, C. G., and P. Oppenheim. 1948. "Studies in the Logic of Explanation." *Philosophy of Science* 15:135–75, 350–52.

Henle, P. 1958. "Metaphor." In *Language, Thought and Culture*, ed. P. Henle, pp. 173–95. Ann Arbor.

Henrichs, A. 1975. "Two Doxographical Notes: Democritus and Prodicus on Religion." *Harvard Studies in Classical Philology* 79:93–123.

Herter, H. 1957. "Bewegung der Materie bei Platon." *Rheinisches Museum* N.F. 100:327–47.

———. 1958. "Gott und die Welt bei Platon." *Bonner Jahrbücher* 158: 106–17.

———. 1963a. "Die kulturhistorische Theorie der hippokratischen Schrift von der alten Medizin." *Maia* 15:464–83.

———. 1963b. "Die Treffkunst des Arztes in hippokratischer und platonischer Sicht." *Sudhoffs Archiv für Geschichte der Medizin und der Naturwissenschaften* 47:247–90.

Herzog, R. 1931. *Die Wunderheilungen von Epidauros*. Philologus Suppl. Bd. 22, 3. Leipzig.

Hess, W. 1970. "Erfahrung und Intuition bei Aristoteles." *Phronesis* 15: 48–82.

Hesse, M. 1961. *Forces and Fields: The Concept of Action at a Distance in the History of Physics*. London.

———. 1963. *Models and Analogies in Science*. London.

———. 1974. *The Structure of Scientific Inference*. London.

———. 1980. *Revolutions and Reconstructions in the Philosophy of Science*. Brighton, Sussex.

———. 1982. "The Cognitive Claims of Metaphor." In *Metaphor and Religion*, ed. J. P. von Noppen, pp. 27–45. Brussels.

Heusch, L. de. 1981. "The Madness of the Gods and the Reason of Men." In *Why Marry Her?* trans. J. Lloyd of *Pourquoi l'épouser?* (Paris, 1974), pp. 165–95. Cambridge.

Hintikka, J. 1972. "On the Ingredients of an Aristotelian Science." *Nous* 6:55–69.

———. 1973. *Time and Necessity*. Oxford.

———. 1974. *Knowledge and the Known*. Dordrecht and Boston.

———. 1980. "Aristotelian Induction." *Revue Internationale de Philosophie* 133–34: 422–39.

———. 1981. "Aristotelian Axiomatics and Geometrical Axiomatics." In Hintikka, Gruender, and Agazzi, edd., 1981, pp. 133–44.

Hintikka, J., D. Gruender, and E. Agazzi, edd. 1981. *Theory Change, Ancient Axiomatics and Galileo's Methodology*. Proceedings of the 1978 Pisa Conference on the History and Philosophy of Science, vol. 1. Dordrecht.

Hintikka, J., and U. Remes. 1974. *The Method of Analysis*. Boston Studies in the Philosophy of Science 25. Dordrecht.

Hirschberger, J. 1959. "Paronymie und Analogie bei Aristoteles." *Philosophisches Jahrbuch* 68:191–203.

Hirzel, R. 1903. Ἄγραφος νόμος. Abhandlungen der phil.-hist. Classe der königlich sächsischen Gesellschaft der Wissenschaften 20, 1, 1900. Leipzig.

———. 1907. *Themis, Dike und Verwandtes*. Leipzig.

Hocutt, M. 1974. "Aristotle's Four Becauses." *Philosophy* 49:385–99.
Höffe, O. 1976. "Grundaussagen über den Menschen bei Aristoteles." *Zeitschrift für philosophische Forschung* 30:227–45.
Hölscher, U. 1952. "Der Logos bei Heraklit." In *Varia Variorum: Festgabe K. Reinhardt*, ed. F. Klingner, pp. 69–81. Münster.
———. 1953/1970. "Anaximander and the Beginnings of Greek Philosophy." Originally "Anaximander und die Anfänge der Philosophie." *Hermes* 81: 257–77, 385–418. In Furley and Allen, edd., 1970, pp. 281–322.
———. 1968. *Anfängliches Fragen*. Göttingen.
———. 1976. "Der Sinn von Sein in der älteren griechischen Philosophie." *Sitzungsberichte der Heidelberger Akademie der Wissenschaften*, Jahrgang 1976, 3. Heidelberg.
Hösle, V. 1982. "Platons Grundlegung der Euklidizität der Geometrie." *Philologus* 126:184–97.
Hoffmann, E. 1925. *Die Sprache und die archaische Logik*. Tübingen.
Hollis, M., and S. Lukes, edd. 1982. *Rationality and Relativism*. Oxford.
Holton, G. 1973. *Thematic Origins of Scientific Thought: Kepler to Einstein*. Cambridge, Massachusetts.
———. 1978. *The Scientific Imagination*. Cambridge.
Holwerda, D. 1955. ΦΥΣΙΣ. Groningen.
Hookway, C., and P. Pettit, edd. 1978. *Action and Interpretation*. Cambridge.
Hooykaas, R. 1972. *Religion and the Rise of Modern Science*. Edinburgh.
Hopfner, T. 1925. *Orient und griechische Philosophie*. Leipzig.
———. 1928. "Mageia." *Pauly-Wissowa Real-Encyclopädie der classischen Altertumswissenschaft* 27 Halbband, 14, 1, cols. 301–93.
———. 1937. "Traumdeutung." *Pauly-Wissowa Real-Encyclopädie der classischen Altertumswissenschaft* 2nd ser., 12 Halbband, 6, 2, cols. 2233–45.
Hort, A. 1916. *Theophrastus, Enquiry into Plants*, Loeb ed. 2 vols. London.
Horton, R. 1960. "A Definition of Religion and Its Uses." *Journal of the Royal Anthropological Institute* 90:201–26.
———. 1967. "African Traditional Thought and Western Science." *Africa* 37:50–71, 155–87. (Abbreviated version reprinted in B. R. Wilson, ed., 1970, pp. 131–71.)
———. 1973. "Lévy-Bruhl, Durkheim and the Scientific Revolution." In Horton and Finnegan, edd., 1973, pp. 249–305.
———. 1982. "Tradition and Modernity Revisited." In Hollis and Lukes, edd., 1982, pp. 201–60.
Horton, R., and R. Finnegan, edd. 1973. *Modes of Thought*. London.
Hošek, R. 1973. Review of Kudlien 1968c, in *Eirene* 11:177–79.
Howald, E. 1922. "ΕΙΚΩΣ ΛΟΓΟΣ." *Hermes* 58:63–79.
Hubert, H. 1904. "Magia." In *Dictionnaire des antiquités grecques et romaines*, ed. C. Daremberg, E. Saglio, and E. Pottier, vol. 3, pp. 1494–521. Paris.
Huby, P. M. 1979. "The Paranormal in the Works of Aristotle and His Circle." *Apeiron* 13, 1:53–62.
Hübner, W. 1980. "Die geometrische Theologie des Philolaos." *Philologus* 124:18–32.

Huizinga, J. 1944/1970. *Homo Ludens.* Trans. R. F. C. Hull of 1944 German edition (original Dutch 1938). 2nd ed. London.

Hultsch, F. 1864. *Metrologicorum Scriptorum Reliquiae,* vol. 1. Leipzig. (Reprinted Stuttgart 1971.)

———. 1866. *Metrologicorum Scriptorum Reliquiae,* vol. 2. Leipzig. (Reprinted Stuttgart 1971.)

———. 1882. *Griechischen und römischen Metrologie.* 2nd ed. Berlin.

Hume, R. E. 1931. *The Thirteen Principal Upanishads.* 2nd ed. (1st ed. 1921). Oxford.

Humphreys, S. C. 1978. *Anthropology and the Greeks.* London.

———. 1983. *The Family, Women and Death.* London.

Humphreys, S. C., and H. King, edd. 1981. *Mortality and Immortality: The Anthropology and Archaeology of Death.* London.

Hussey, E. 1972. *The Presocratics.* London.

———. 1983. *Aristotle's Physics, Books III and IV.* Oxford.

Ilberg, J. 1889. "Über die Schriftstellerei des Klaudios Galenos, I." *Rheinisches Museum* N.F. 44:207–39.

———. 1982. "Über die Schriftstellerei des Klaudios Galenos, II." *Rheinisches Museum* N.F. 47:489–514.

———. 1896. "Über die Schriftstellerei des Klaudios Galenos, III." *Rheinisches Museum* N.F. 51:165–96.

———. 1897. "Über die Schriftstellerei des Klaudios Galenos, IV." *Rheinisches Museum* N.F. 52:591–623.

———. 1931. *Rufus von Ephesos. Ein griechischer Arzt in trajanischer Zeit.* Abhandlungen der phil.-hist. Klasse der sächsischen Akademie der Wissenschaften 41, 1, 1930. Leipzig.

Illich, I. 1976. *Limits to Medicine. Medical Nemesis: The Expropriation of Health.* London.

Ingenkamp, H. G. 1981. "Erkenntniserwerb durch στοχάζεσθαι bei Aristoteles." *Hermes* 109:172–78.

Inwood, B. 1979. "A Note on Commensurate Universals in the *Posterior Analytics.*" *Phronesis* 24:320–29.

Ioppolo, A. M. 1980. *Aristone di Chio e lo stoicismo antico.* Naples.

Irigoin, J. 1980. "La Formation du vocabulaire de l'anatomie en grec: Du mycénien aux principaux traités de la Collection hippocratique." In Grmek, ed., 1980, pp. 247–56.

Irmer, D. 1980. "Die Bezeichnung der Knochen in *Fract.* und *Art.*" In Grmek, ed., 1980, pp. 265–83.

Irmscher, J., and R. Müller, edd. 1983. *Aristoteles als Wissenschaftstheoretiker.* Schriften zur Geschichte und Kultur der Antike 22. Berlin.

Irwin, T. H. 1977. "Plato's Heracliteanism." *Philosophical Quarterly* 27:1–13.

———. 1977–78. "Aristotle's Discovery of Metaphysics." *Review of Metaphysics* 31:210–29.

———. 1980–81. "Homonymy in Aristotle." *Review of Metaphysics* 34:523–44.

———. 1982. "Aristotle's Concept of Signification." In Schofield and Nussbaum, edd., pp. 241–66.

Isnardi Parente, M. 1961. "Techne." *La Parola del Passato* 16:257–96.

———. 1966. *Techne: Momenti del pensiero greco da Platone ad Epicuro.* Florence.

Isserlin, B. S. J. 1982. "The Earliest Alphabetic Writing." In *Cambridge Ancient History,* 2nd ed., vol. 3, ed. J. Boardman et al., pp. 794–818. Cambridge.

Iversen, E. 1939. *Papyrus Carlsberg No. VIII.* Det kgl. Danske Videnskabernes Selskab. Historisk-filologiske Meddelelser 26, 5. Copenhagen.

Izard, M., and P. Smith, edd. 1979. *La Fonction symbolique.* Paris.

Jacobsen, T. 1949. "Mesopotamia." In Frankfort, ed., 1949, pp. 137–234.

———. 1976. *The Treasures of Darkness.* New Haven.

Jacoby, F. 1923–58. *Die Fragmente der griechischen Historiker.* Berlin (1923–30), Leiden (1940–58).

Jaeger, W. 1938. *Diokles von Karystos. Die griechische Medizin und die Schule des Aristoteles.* 2nd ed. Berlin.

———. 1939–45. *Paideia: The Ideals of Greek Culture,* trans. G. Highet, 3 vols. Oxford.

———. 1947. *The Theology of the Early Greek Philosophers* (Gifford Lectures 1936), trans. E. S. Robinson. Oxford.

———. 1948. *Aristotle: Fundamentals of the History of His Development,* trans. R. Robinson. 2nd ed. (1st ed. 1934). Oxford.

Jakobson, R., and M. Halle. 1956. *Fundamentals of Language.* The Hague.

Jammer, M. 1961. *Concepts of Mass in Classical and Modern Physics.* Cambridge, Massachusetts.

Jardine, N. 1979. "The Forging of Modern Realism: Clavius and Kepler against the Sceptics." *Studies in History and Philosophy of Science* 10: 141–73.

———. 1984. *The Birth of History and Philosophy of Science.* Cambridge.

Jarvie, I. C. 1972. *Concepts and Society.* London.

———. 1976. "On the Limits of Symbolic Interpretation in Anthropology." *Current Anthropology* 17:687–91.

Jarvie, I. C., and J. Agassi. 1967/1970. "The Problem of the Rationality of Magic." From *British Journal of Sociology* 18:55–74. In B. R. Wilson, ed., 1970, pp. 172–93.

Jeffery, L. H. 1961. *The Local Scripts of Archaic Greece.* Oxford.

———. 1982. "Greek Alphabetic Writing." In *Cambridge Ancient History,* 2nd ed., vol. 3, ed. J. Boardman et al., pp. 819–33. Cambridge.

Joachim, H. 1904. "Aristotle's Conception of Chemical Combination." *Journal of Philology* 29:72–86.

———. 1922. *Aristotle, On Coming-to-Be and Passing-Away.* Oxford.

Johnson, M., ed. 1981. *Philosophical Perspectives on Metaphor.* Minneapolis.

Joly, H. 1974. *Le Renversement platonicien: Logos, Episteme, Polis.* Paris.

Joly, R. 1960. *Recherches sur le traité pseudo-hippocratique du Régime.* Bibliothèque de la Faculté de Philosophie et Lettres de l'Université de Liège 156. Paris.

———. 1966. *Le Niveau de la science hippocratique.* Paris.

———. 1968. "La Biologie d'Aristote." *Revue Philosophique de la France et de l'Etranger* 158:219–53.

———. 1969–70. "Esclaves et médecins dans la Grèce antique." *Sudhoffs Archiv* 53:1–14.

———. 1970. *Hippocrate, De la génération, De la nature de l'enfant, Des maladies IV, Du foetus de huit mois*. Paris.

———. 1980. "Un Peu d'épistémologie historique pour hippocratisants." In Grmek, ed., 1980, pp. 285–97.

Joly, R., ed. 1977. *Corpus Hippocraticum*. Editions Universitaires de Mons, Série Sciences Humaines 4. Mons.

Jones, J. F. 1983. "Intelligible Matter and Geometry in Aristotle." *Apeiron* 17, 2:94–102.

Jones, J. W. 1956. *The Law and Legal Theory of the Greeks*. Oxford.

Jones, W. H. S. 1923–31. *Hippocrates*, Loeb ed., 4 vols. London.

———. 1946. *Philosophy and Medicine in Ancient Greece*. Suppl. to the Bulletin of the History of Medicine 8. Baltimore.

———. 1947. *The Medical Writings of Anonymus Londinensis*. Cambridge.

Jope, J. 1972. "Subordinate Demonstrative Science in the Sixth Book of Aristotle's *Physics*." *Classical Quarterly* n.s. 22:279–92.

Jouanna, J. 1961. "Présence d'Empédocle dans la Collection hippocratique." *Bulletin de l'Association Guillaume Budé:* 452–63.

———. 1966. "La Théorie de l'intelligence et de l'âme dans le traité hippocratique 'Du régime': Ses Rapports avec Empédocle et le 'Timée' de Platon." *Revue des Etudes Grecques* 79:xv–xviii.

———. 1974. *Hippocrate: Pour une archéologie de l'école de Cnide*. Paris.

Jürss, F. 1967. "Über die Grundlagen der Astrologie." *Helikon* 7:63–80.

Jules-Rosette, B. 1978. "The Veil of Objectivity: Prophecy, Divination, and Social Inquiry." *American Anthropologist* 80:549–70.

Junge, G. 1958. "Von Hippasus bis Philolaus. Das Irrationale und die geometrischen Grundbegriffe." *Classica et Mediaevalia* 19:41–72.

Justesen, P. T. 1928. *Les Principes psychologiques d'Homère*. Copenhagen.

Kahn, C. H. 1960a. *Anaximander and the Origins of Greek Cosmology*. New York.

———. 1960b. "Religion and Natural Philosophy in Empedocles' Doctrine of the Soul." *Archiv für Geschichte der Philosophie* 42:3–35. (Reprinted in Anton and Kustas, edd., 1971, pp. 3–38.)

———. 1966. "Sensation and Consciousness in Aristotle's Psychology." *Archiv für Geschichte der Philosophie* 48:43–81.

———. 1970. "On Early Greek Astronomy." *Journal of Hellenic Studies* 90:99–116.

———. 1973. *The Verb "Be" in Ancient Greek*. Foundations of Language Suppl. 16. Dordrecht.

———. 1979. *The Art and Thought of Heraclitus*. Cambridge.

Kapp, E. 1931/1975. "Syllogistic." Originally "Syllogistik," in *Pauly-Wissowa Real-Encyclopädie der classischen Altertumswissenschaft*, 2nd ser., 7 Halbband, 4, 1, cols. 1046–67. In Barnes, Schofield, and Sorabji, edd., 1975, pp. 35–49.

———. 1942. *Greek Foundations of Traditional Logic*. New York.

Kattsoff, L. O. 1947–48. "Ptolemy and Scientific Method." *Isis* 38:18–22.

Kayas, G. J. 1976. "Aristote et les géométries non-euclidiennes avant et après Euclide." *Revue des Questions Scientifiques* 147:175–94, 281–301, 457–65.

Keith, A. B. 1925. *The Religion and Philosophy of the Veda and Upanishads*, 2 vols. Cambridge, Massachusetts.

———. 1928. *A History of Sanskrit Literature*. Oxford.

Kennedy, G. 1963. *The Art of Persuasion in Greece*. London.

Kenny, A. 1967. "The Argument from Illusion in Aristotle's *Metaphysics* (Γ, 1009–10)." *Mind* n.s. 76:184–97.

———. 1975. *Will, Freedom and Power*. Oxford.

Kenyon, F. G. 1951. *Books and Readers in Ancient Greece and Rome*. 2nd ed. (1st ed. 1932). Oxford.

Kerferd, G. B. 1955. "Gorgias on Nature or That Which Is Not." *Phronesis* 1:3–25.

———. 1956–57. "The Moral and Political Doctrines of Antiphon the Sophist: A Reconsideration." *Proceedings of the Cambridge Philological Society* n.s. 4:26–32.

———. 1976. "The Image of the Wise Man in Greece in the Period before Plato." In *Images of Man in Ancient and Medieval Thought: Studia Gerardo Verbeke . . . dicata*, ed. F. Bossier et al., pp. 17–28. Louvain.

———. 1981. *The Sophistic Movement*. Cambridge.

Kerferd, G. B., ed. 1981. *The Sophists and Their Legacy*. Proceedings of the Fourth International Colloquium of Ancient Greek Philosophy at Bad Homburg 1979, Hermes Einzelschriften 44. Wiesbaden.

Kerschensteiner, J. 1945. *Platon und der Orient*. Stuttgart.

———. 1962. *Kosmos: Quellenkritische Untersuchungen zu den Vorsokratikern*. Zetemata 30. Munich.

Kessels, A. H. M. 1969. "Ancient Systems of Dream-Classification." *Mnemosyne*, 4th ser., 22:389–424.

———. 1978. *Studies on the Dream in Greek Literature*. Utrecht.

Keuls, E. C. 1978. *Plato and Greek Painting*. Leiden.

Keyt, D. 1971. "The Mad Craftsman of the *Timaeus*." *Philosophical Review* 80:230–35.

Kirk, G. S. 1954. *Heraclitus: The Cosmic Fragments*. Cambridge.

———. 1960/1970. "Popper on Science and the Presocratics." From *Mind* n.s. 69:318–39. In Furley and Allen, edd., 1970, pp. 154–77.

———. 1961. "Sense and Common-Sense in the Development of Greek Philosophy." *Journal of Hellenic Studies* 81:105–17.

———. 1970. *Myth, Its Meaning and Functions in Ancient and Other Cultures*. Berkeley.

———. 1974. *The Nature of Greek Myths*. Harmondsworth.

Kirk, G. S., J. E. Raven, and M. Schofield. 1983. *The Presocratic Philosophers*. 2nd ed. (1st ed. 1957). Cambridge.

Klein, J. 1968. *Greek Mathematical Thought and the Origin of Algebra*, trans. E. Brann of *Die griechische Logistik und die Entstehung der Algebra*, Quel-

len und Studien zur Geschichte der Mathematik, Astronomie und Physik B
3, 1 (1934), pp. 18–105, and B 3, 2 (1936), pp. 122–235. Cambridge,
Massachusetts.

Kleingünther, A. 1933. ΠΡΩΤΟΣ ΕΥΡΕΤΗΣ *Untersuchungen zur Geschichte einer Fragestellung.* Philologus Suppl. Bd. 26, 1. Leipzig.

Klibansky, R., E. Panofsky, and F. Saxl. 1964. *Saturn and Melancholy.* London.

Klowski, J. 1966a. "Das Entstehen der Begriffe Substanz und Materie." *Archiv für Geschichte der Philosophie* 48:2–42.

———. 1966b. "Der historische Ursprung des Kausalprinzips." *Archiv für Geschichte der Philosophie* 48:225–66.

———. 1970. "Zum Entstehen der logischen Argumentation." *Rheinisches Museum* N.F. 113:111–41.

Kneale, W., and M. Kneale. 1962. *The Development of Logic.* Oxford.

Knorr, W. R. 1975. *The Evolution of the Euclidean Elements.* Dordrecht.

———. 1975–76. "Archimedes and the Measurement of the Circle: A New Interpretation." *Archive for History of Exact Sciences* 15:115–40.

———. 1978a. "Archimedes and the *Elements:* Proposal for a Revised Chronological Ordering of the Archimedean Corpus." *Archive for History of Exact Sciences* 19:211–90.

———. 1978b. "Archimedes and the Pre-Euclidean Proportion Theory." *Archives Internationales d'Histoire des Sciences* 28:183–244.

———. 1981. "On the Early History of Axiomatics: The Interaction of Mathematics and Philosophy in Greek Antiquity." In Hintikka, Gruender, and Agazzi, edd., 1981, pp. 145–86.

———. 1982a. "Infinity and Continuity: The Interaction of Mathematics and Philosophy in Antiquity." In Kretzmann, ed., 1982, pp. 112–45.

———. 1982b. "Ancient and Medieval Balances." *Annali dell' Instituto e Museo di Storia della Scienza* Supplement 1982, 2:121–35.

———. 1983. "Construction as Existence Proof in Ancient Geometry." *Ancient Philosophy* 3:125–48.

———. 1986. *The Ancient Tradition of Geometric Problems.* Boston.

Knox, B. M. W. 1968. "Silent Reading in Antiquity." *Greek, Roman and Byzantine Studies* 9:421–35.

Knutzen, G. H. 1964. *Technologie in den hippokratischen Schriften περὶ δι-αίτης ὀξέων, περὶ ἀγμῶν, περὶ ἄρθρων ἐμβολῆς.* Akademie der Wissenschaften und der Literatur, Mainz, Abhandlungen der geistes- und sozialwissenschaftlichen Kl., Jahrgang 1963, 14. Wiesbaden.

Köcher, F. 1963–71. *Die babylonische-assyrische Medizin in Texten und Untersuchungen,* 4 vols. Berlin.

Koelbing, H. M. 1975. "Der hippokratische Arzt ohne Nimbus." *Praxis (Schweizerische Rundschau für Medizin)* 29:933–39.

———. 1977. *Arzt und Patient in der antiken Welt.* Zürich.

König, E. 1970. "Aristoteles' erste Philosophie als universale Wissenschaft von der ΑΡΧΑΙ." *Archiv für Geschichte der Philosophie* 52:225–46.

Koller, H. 1959–60. "Das Modell der griechischen Logik." *Glotta* 38:61–74.

Kollesch, J. 1965. "Galen und seine ärztlichen Kollegen." *Das Altertum* 11:47–53.

————. 1974. "Die Medizin und ihre sozialen Aufgaben zur Zeit der Polis-krise." In Welskopf, ed., 1974, vol. 4, pp. 1850–71.

————. 1976. "Vorstellungen vom Menschen in der hippokratischen Medi-zin." In R. Müller, ed., 1976, pp. 269–82.

————. 1979. "Ärztliche Ausbildung in der Antike." *Klio* 61: 507–13.

————. 1981. "Galen und die zweite Sophistik." In Nutton, ed., 1981, pp. 1–11.

Konstan, D. 1972. "Epicurus on 'Up' and 'Down.'" *Phronesis* 17: 269–78.

————. 1979. "Problems in Epicurean Physics." *Isis* 70: 394–418.

Kosman, L. A. 1973. "Understanding, Explanation and Insight in the *Posterior Analytics*." In Lee, Mourelatos, and Rorty, edd., 1973, pp. 374–92.

Koyré, A. 1948/1961. "Du monde de l'à peu près à l'univers de la précision." From *Critique* 4, 28: 806–23. In *Etudes d'histoire de la pensée philosophique*, pp. 311–29. Paris.

————. 1957. *From the Closed World to the Infinite Universe*. Baltimore.

————. 1968. *Metaphysics and Measurement*. London.

Krafft, F. 1965. "Der Mathematikos und der Physikos. Bemerkungen zu der angeblichen platonischen Aufgabe, die Phänomene zu retten." *Beiträge zur Geschichte der Wissenschaft und der Technik* 5: 5–24.

————. 1970. *Dynamische und statische Betrachtungsweise in der antiken Mechanik*. Wiesbaden.

————. 1982. "Zielgerichtetheit und Zielsetzung in Wissenschaft und Natur." *Berichte zur Wissenschaftsgeschichte* 5: 53–74.

Kranz, W. 1938a. "Gleichnis und Vergleich in der frühgriechischen Philoso-phie." *Hermes* 73: 99–122.

————. 1938b. "Kosmos und Mensch in der Vorstellung frühen Griechen-tums." *Nachrichten von der Gesellschaft der Wissenschaften zu Göttingen*, phil.-hist. Kl., N.F. 2, 1938, pp. 121–61.

————. 1938–39. "Kosmos als philosophischer Begriff frühgriechischer Zeit." *Philologus* 93: 430–48.

————. 1957. "Zwei kosmologische Fragen." *Rheinisches Museum* N.F. 100: 114–29.

————. 1961. "SPHRAGIS. Ichform und Namensiegel als Eingangs- und Schlussmotiv antiker Dichtung." *Rheinisches Museum* 104: 3–46, 97–124.

Kretzmann, N., ed. 1982. *Infinity and Continuity in Ancient and Medieval Thought*. Ithaca, New York.

Kripke, S. 1980. *Naming and Necessity*. Originally in *Semantics of Natural Language*, ed. G. Harman and D. Davidson (Dordrecht 1972), pp. 253–355. Oxford.

Krips, H. 1980. "Aristotle on the Infallibility of Normal Observation." *Studies in History and Philosophy of Science* 11: 79–86.

Kube, J. 1969. TEXNH *und* APETH. Berlin.

Kucharski, P. 1949. *Les Chemins du savoir dans les derniers dialogues de Platon*. Paris.

————. 1954. "Sur la théorie des couleurs et des saveurs dans le 'De Sensu' aristotélicien." *Revue des Etudes Grecques* 67: 355–90.

————. 1963. "Sur la notion pythagoricienne du καιρός." *Revue Philosophique de la France et de l'Etranger* 153: 141–69.

————. 1965. "Sur l'évolution des méthodes du savoir dans la philosophie de Platon." *Revue Philosophique de la France et de l'Etranger* 155:427–40.

Kudlien, F. 1962. "Poseidonios und die Ärzteschule der Pneumatiker." *Hermes* 90:419–29.

————. 1963. "Probleme um Diokles von Karystos." *Sudhoffs Archiv für Geschichte der Medizin und der Naturwissenschaften* 47:456–64.

————. 1964. "Herophilos und der Beginn der medizinischen Skepsis." *Gesnerus* 21:1–13.

————. 1967. *Der Beginn des medizinischen Denkens bei den Griechen.* Zürich.

————. 1968a. "Der Arzt des Körpers und der Arzt der Seele." *Clio Medica* 3:1–20.

————. 1968b. "Early Greek Primitive Medicine." *Clio Medica* 3:305–36.

————. 1968c. *Die Sklaven in der griechschen Medizin der klassischen und hellenistischen Zeit.* Forschungen zur antiken Sklaverei 2. Wiesbaden.

————. 1969. "Antike Anatomie und menschlicher Leichnam." *Hermes* 97: 78–94.

————. 1970a. "Medical Education in Classical Antiquity." In *The History of Medical Education,* ed. C. D. O'Malley, pp. 3–37. Berkeley.

————. 1970b. "Medical Ethics and Popular Ethics in Greece and Rome." *Clio Medica* 7:91–121.

————. 1973. "'Schwärzliche' Organe im frühgriechischen Denken." *Medizinhistorisches Journal* 8:53–58.

————. 1974. "Dialektik und Medizin in der Antike." *Medizinhistorisches Journal* 9:187–200.

————. 1976. "Medicine as a 'Liberal Art' and the Question of the Physician's Income." *Journal of the History of Medicine and Allied Sciences* 31: 448–59.

————. 1977. "Das Göttliche und die Natur im hippokratischen *Prognostikon.*" *Hermes* 105:268–74.

————. 1979. *Der griechische Arzt im Zeitalter des Hellenismus.* Akademie der Wissenschaften und der Literatur, Mainz, Abhandlungen der geistes- und sozialwissenschaftlichen Kl., Jahrgang 1979, 6. Wiesbaden.

————. 1980. "Die Bedeutung des Ungeraden in der hippokratischen Krisenarithmetik." *Hermes* 108:200–205.

Kühn, J.-H. 1956. *System- und Methodenprobleme im Corpus Hippocraticum.* Hermes Einzelschriften 11. Wiesbaden.

Kümmel, W. F. 1974. "Der Puls und das Problem der Zeitmessung in der Geschichte der Medizin." *Medizinhistorisches Journal* 9:1–22.

————. 1977. *Musik und Medizin.* Freiburger Beiträge zur Wissenschafts- und Universitätsgeschichte 2. Freiburg.

Kugler, F. X. 1907–35. *Sternkunde und Sterndienst in Babel,* 4 vols. with supplements, ed. J. Schaumberger. Münster.

Kuhn, T. S. 1957. *The Copernican Revolution.* Cambridge, Massachusetts.

————. 1959/1977. "The Essential Tension: Tradition and Innovation in Scientific Research." Originally in *The Third (1959) University of Utah Re-*

search Conference on the Identification of Scientific Talent, ed. C. W. Taylor (Salt Lake City, 1959), pp. 162–74. In Kuhn 1977, pp. 225–39.

———. 1961/1977. "The Function of Measurement in Modern Physical Science." From *Isis* 52:161–93. In Kuhn 1977, pp. 178–224.

———. 1962/1970. *The Structure of Scientific Revolutions*. 2nd ed. (1st ed. 1962). Chicago.

———. 1963. "The Function of Dogma in Scientific Research." In A. C. Crombie, ed., 1963, pp. 347–69.

———. 1964/1977. "A Function for Thought Experiments." From *L'Aventure de la science*, Mélanges A. Koyré (Paris, 1964), vol. 2, pp. 307–34. In Kuhn 1977, p. 240–65.

———. 1970/1977. "Logic of Discovery or Psychology of Research?" From Lakatos and Musgrave, edd., 1970, pp. 1–23. In Kuhn 1977, pp. 266–92.

———. 1974/1977. "Second Thoughts on Paradigms." From *The Structure of Scientific Theories*, ed. F. Suppe (Urbana, 1974), pp. 459–82. In Kuhn 1977, pp. 293–319.

———. 1977. *The Essential Tension*. Chicago.

———. 1979. "Metaphor in Science." In Ortony, ed., 1979, pp. 409–19.

———. 1983. "Rationality and Theory Choice." *The Journal of Philosophy* 80:563–70.

Kullmann, W. 1965. "Zur wissenschaftlichen Methode des Aristoteles." In Flashar and Gaiser, edd., 1965, pp. 247–74.

———. 1974a. *Wissenschaft und Methode: Interpretationen zur aristotelischen Theorie der Naturwissenschaft*. Berlin.

———. 1974b. "Der platonische Timaios und die Methode der aristotelischen Biologie." In Döring and Kullmann, edd., 1974, pp. 139–63.

———. 1979. *Die Teleologie in der aristotelischen Biologie*. Sitzungsberichte der Heidelberger Akademie der Wissenschaften, phil.-hist. Kl., Jahrgang 1979, 2. Heidelberg.

———. 1980. "Der Mensch als politisches Lebewesen bei Aristoteles." *Hermes* 108:419–43.

———. 1982. "Aristoteles' Grundgedanken zu Aufbau und Funktion der Körpergewebe." *Sudhoffs Archiv* 66:209–38.

Kung, J. 1977. "Aristotle on Essence and Explanation." *Philosophical Studies* 31:361–83.

Kurz, D. 1970. AKPIBEIA: *Das Ideal der Exaktheit bei den Griechen bis Aristoteles*. Göppingen.

Kutsch, F. 1913. *Attische Heilgötter und Heilheroen*. Religionsgeschichtliche Versuche und Vorarbeiten 12, 3 (1912–13). Giessen.

Labat, R. 1951. *Traité Akkadien de diagnostics et pronostics médicaux*. Paris.

Lämmli, F. 1962. *Vom Chaos zum Kosmos*. Schweizerische Beiträge zur Altertumswissenschaft 10. Basel.

Lafrance, Y. 1980. "Platon et la Géométrie: La Méthode dialectique en République 509d–511e." *Dialogue* 19:46–93.

Laín Entralgo, P. 1970. *The Therapy of the Word in Classical Antiquity*, trans. L. J. Rather and J. M. Sharp of *La curación por la palabra en la Antigüedad clásica* (Madrid, 1958). New Haven.

Laing, R. D. 1960. *The Divided Self: A Study of Sanity and Madness.* London.
Laing, R. D., and A. Esterson. 1970. *Sanity, Madness and the Family.* 2nd ed. (1st ed. 1964). London.
Lakatos, I. 1970. "Falsification and the Methodology of Scientific Research Programmes." In Lakatos and Musgrave, edd., 1970, pp. 91–195. (Reprinted in Lakatos 1978a, pp. 8–101.)
——. 1976. *Proofs and Refutations,* ed. J. Worrall and E. G. Zapar. Revised version of *British Journal for the Philosophy of Science* 14 (1963–64): 1–25, 120–39, 221–45, 296–342. Cambridge.
——. 1978a. *The Methodology of Scientific Research Programmes, Philosophical Papers,* vol. 1, ed. J. Worrall and G. Currie. Cambridge.
——. 1978b. *Mathematics, Science and Epistemology, Philosophical Papers,* vol. 2, ed. J. Worrall and G. Currie. Cambridge.
Lakatos, I., and A. Musgrave, edd. 1970. *Criticism and the Growth of Knowledge.* Cambridge.
Lakoff, G., and M. Johnson. 1980. *Metaphors We Live By.* Chicago.
Laloy, L. 1973. *Aristoxène de Tarente, disciple d'Aristote, et la musique de l'antiquité.* (1st ed. Paris, 1904). Repr. Geneva.
Lambek, M. 1981. *Human Spirits: A Cultural Account of Trance in Mayotte.* Cambridge.
Lambert, W. G. 1960. *Babylonian Wisdom Literature.* Oxford.
——. 1975. "The Cosmology of Sumer and Babylon." In *Ancient Cosmologies,* ed. C. Blacker and M. Loewe, pp. 42–65. London.
Lanata, G. 1967. *Medicina Magica e Religione Popolare in Grecia.* Rome.
Landels, J. G. 1978. *Engineering in the Ancient World.* London.
——. 1979. "Water-Clocks and Time Measurement in Classical Antiquity." *Endeavour* n.s. 3, 1:32–37.
Lanza, D. 1972. "'Scientificità' della lingua e lingua della scienza in Grecia." *Belfragor* 27:392–429.
——. 1979. *Lingua e discorso nell' Atene delle professioni.* Naples.
——. 1980a. "La morte esclusa." *Quaderni di storia* 6, 11:157–72.
——. 1980b. "La massima epicurea 'Nulla è per noi la morte.'" In Romano, ed., 1980, pp. 357–65.
Lanza, D., and M. Vegetti. 1975. "L'ideologia della città." *Quaderni di storia* 1, 2:1–37.
Lasserre, F. 1954. *Plutarque De la musique.* Olten.
——. 1964. *The Birth of Mathematics in the Age of Plato,* trans. H. Mortimer. London.
——. 1966. *Die Fragmente des Eudoxos von Knidos.* Texte und Kommentare 4. Berlin.
Lasserre, F., and P. Mudry, edd. 1983. *Formes de pensée dans la Collection hippocratique.* Actes du IVᵉ Colloque International Hippocratique, Lausanne, 21–26 September 1981. Geneva.
Laudan, L. 1976. "Two Dogmas of Methodology." *Philosophy of Science* 43:585–97.
——. 1977. *Progress and Its Problems.* Berkeley.

————. 1981a. "A Problem-Solving Approach to Scientific Progress." In Hacking, ed., 1981, pp. 144–55.
————. 1981b. "A Confutation of Convergent Realism." *Philosophy of Science* 48:19–49.
Leach, E. R. 1961. *Rethinking Anthropology.* London.
————. 1969. *Genesis as Myth and Other Essays.* London.
Leach, E. R., ed. 1967. *The Structural Study of Myth and Totemism.* London.
————. 1968. *Dialectic in Practical Religion.* Cambridge.
Leach, E. R., and D. Alan Aycock. 1983. *Structuralist Interpretations of Biblical Myth.* Cambridge.
Lear, J. 1980. *Aristotle and Logical Theory.* Cambridge.
————. 1982. "Aristotle's Philosophy of Mathematics." *Philosophical Review* 91:161–92.
Le Blond, J. M. 1938. *Eulogos et l'argument de convenance chez Aristote.* Paris.
————. 1939. *Logique et méthode chez Aristote.* Paris.
Lee, E. N. 1967. "On Plato's *Timaeus*, 49D4–E7." *American Journal of Philology* 88:1–28.
————. 1971. "On the 'Gold-Example' in Plato's *Timaeus* (50A5–B5)." In Anton and Kustas, edd., 1971, pp. 219–35.
————. 1978. "The Sense of an Object: Epicurus on Seeing and Hearing." In Machamer and Turnbull, edd., 1978, pp. 27–59.
Lee, E. N., A. P. D. Mourelatos, and R. M. Rorty, edd. 1973. *Exegesis and Argument.* Assen.
Lee, H. D. P. 1935. "Geometrical Method and Aristotle's Account of First Principles." *Classical Quarterly* 29:113–24.
————. 1962. *Aristotle, Meteorologica,* Loeb ed. 2nd ed. (1st ed. 1952). London.
Leemans, C. 1843. *Papyri Graeci Musei Antiquarii Publici Lugduni-Batavi,* vol. 1. Leyden.
————. 1885. *Papyri Graeci Musei Antiquarii Publici Lugduni-Batavi,* vol. 2. Leyden.
Lefebure, G. 1956. *Essai sur la médecine Egyptienne de l'époque pharaonique.* Paris.
Lefkowitz, M. R. 1963. "ΤΩ ΚΑΙ ΕΓΩ: The First Person in Pindar." *Harvard Studies in Classical Philology* 67:177–253.
————. 1981. *Lives of the Greek Poets.* London.
Lehmann-Haupt, C. F. 1929. "Stadion." *Pauly-Wissowa Real-Encyclopädie der classischen Altertumswissenschaft* 2nd ser., 6 Halbband, 3, 2, cols. 1930–63.
Leibbrand, W., and A. Wettley. 1961. *Der Wahnsinn: Geschichte der abendländischen Psychopathologie.* Freiburg.
Lejeune, A. 1940–46. "Les Tables de réfraction de Ptolémée." *Annales de la Société Scientifique de Bruxelles.* Série 1: *Sciences Mathématiques, Physiques et Astronomiques* 60: 93–101.
————. 1946. "Les Lois de la réflexion dans l'*Optique* de Ptolémée." *L'Antiquité Classique* 15:241–56.

————. 1947–48a. "Archimède et la loi de la réflexion." *Isis* 38 : 51–53.

————. 1947–48b. "La Dioptre d'Archimède." *Annales de la Société Scientifique de Bruxelles*. Série 1: *Sciences Mathématiques, Astronomiques et Physiques* 61 : 27–47.

————. 1948. *Euclide et Ptolémée: Deux stades de l'optique géométrique grecque*. Université de Louvain Recueil de Travaux d'Histoire et de Philologie, 3ᵉ série, 31. Louvain.

————. 1956. *L'Optique de Claude Ptolémée*. Université de Louvain Recueil de Travaux d'Histoire et de Philologie, 4ᵉ série, 8. Louvain.

————. 1957. *Recherches sur la Catoptrique grecque*. Académie Royale de Belgique, Mémoires de la Classe des Lettres, Série 2, 52, 2, 1954. Brussels.

Lennox, J. G. 1980. "Aristotle on Genera, Species, and 'The More and the Less.'" *Journal of the History of Biology* 13 : 321–46.

————. 1981. "Teleology, Chance, and Aristotle's Theory of Spontaneous Generation." *Journal of the History of Philosophy* 20 : 219–38.

————. 1984. "Aristotle on Chance." *Archiv für Geschichte der Philosophie* 66 : 52–60.

————. 1985. "Theophrastus on the Limits of Teleology." In *Theophrastus of Eresus: On His Life and Work*, ed. W. W. Fortenbaugh, P. M. Huby and A. A. Long, Rutgers University Studies in Classical Humanities 2, pp. 143–63. New Brunswick.

————. 1987. "Divide and Explain: The *Posterior Analytics* in Practice." In Gotthelf and Lennox, edd., 1987, pp. 90–119.

————. Forthcoming. "Aristotle, Galileo and 'Mixed Sciences.'" In *Studies in Galileo*, ed. W. A. Wallace.

Lerner, M.-P. 1969. *Recherches sur la notion de finalité chez Aristote*. Paris.

Lesher, J. H. 1973. "The Meaning of NOYΣ in the *Posterior Analytics*." *Phronesis* 18 : 44–68.

————. 1983. "Heraclitus' Epistemological Vocabulary." *Hermes* 111 : 155–70.

————. 1984. "Parmenides' Critique of Thinking: The Poludēris Elenchos of Fragment 7." *Oxford Studies in Ancient Philosophy* 2 : 1–30.

Lesky, E. 1951. *Die Zeugungs- und Vererbungslehren der Antike und ihr Nachwirken*. Akademie der Wissenschaften und der Literatur, Mainz, Abhandlungen der geistes- und sozialwissenschaftlichen Kl., Jahrgang 1950, 19. Wiesbaden.

Leszl, W. 1970. *Logic and Metaphysics in Aristotle*. Padua.

————. 1972–73. "Knowledge of the Universal and Knowledge of the Particular in Aristotle." *Review of Metaphysics* 26 : 278–313.

————. 1980. "Unity and Diversity of the Sciences: The Methodology of the Mathematical and of the Physical Sciences and the Role of Nominal Definition." *Revue Internationale de Philosophie* 133–34: 384–421.

————. 1981. "Mathematics, Axiomatization and the Hypotheses." In Berti, ed., 1981, pp. 271–328.

————. 1982. "Principi, cause e spiegazione teologica in Aristotele." *Rivista critica di storia della filosofia* 37 : 123–68.

Lévêque, P., and P. Vidal-Naquet. 1964. *Clisthène l'Athénien*. Annales Lit-téraires de l'Université de Besançon 65. Paris.

Levin, F. R. 1980. "πληγή and τάσις in the *Harmonika* of Klaudios Ptole-maios." *Hermes* 108:205–29.

Lévi-Strauss, C. 1958/1968. *Structural Anthropology*, trans. C. Jacobson and B. G. Schoepf of *Anthropologie structurale* (Paris, 1958). London.

———. 1962/1966. *The Savage Mind*, trans. of *La Pensée sauvage* (Paris, 1962). London.

———. 1962/1969. *Totemism*. Revised trans. R. Needham of *Le Totémisme aujourd'hui* (Paris, 1962). London.

———. 1964/1969. *The Raw and the Cooked*, trans. J. and D. Weightman of *Le Cru et le cuit* (Paris, 1964). New York.

———. 1967/1973. *From Honey to Ashes*, trans. J. and D. Weightman of *Du Miel aux cendres* (Paris, 1967). New York.

———. 1968/1978. *The Origin of Table Manners*, trans. J. and D. Weightman of *L'Origine des manières de table* (Paris, 1968). London.

———. 1971/1981. *The Naked Man*, trans. J. and D. Weightman of *L'Homme nu* (Paris, 1971). London.

———. 1973/1976. *Structural Anthropology 2*, trans. M. Layton of *Anthro-pologie structurale Deux* (Paris, 1973). New York.

Lévy-Bruhl, L. 1923. *Primitive Mentality*, trans. L. A. Clare of *La Mentalité primitive* (Paris, 1922). London.

———. 1926. *How Natives Think*, trans. L. A. Clare of *Les Fonctions men-tales dans les sociétés inférieures* (Paris, 1910). London.

———. 1936. *Primitives and the Supernatural*, trans. L. A. Clare of *Le Sur-naturel et la nature dans la mentalité primitive* (Paris, 1931). London.

———. 1975. *The Notebooks on Primitive Mentality*, trans. P. Rivière of *Car-nets* (Paris, 1949). London.

Lewis, D. K. 1969. *Convention*. Cambridge, Massachusetts.

———. 1973. *Counterfactuals*. Oxford.

———. 1974. "Radical Interpretation." *Synthèse* 27:331–44.

Lewis, D. M. 1977. *Sparta and Persia*. Leiden.

Lewis, G. 1975. *Knowledge of Illness in a Sepik Society*. London.

———. 1980. *Day of Shining Red*. Cambridge.

Lewis, J. D. 1971. "Isegoria at Athens: When Did It Begin?" *Historia* 20:129–40.

Lichtenthaeler, C. 1948. *La Médecine hippocratique*. I. *Méthode expéri-mentale et méthode hippocratique*. Lausanne.

———. 1957. "De l'origine sociale de certains concepts scientifiques et phi-losophiques grecs." In *La Médecine hippocratique II–V*, pp. 91–114. Neuchâtel.

———. 1963. "Le Logos mathématique de la première clinique hippocra-tique." In *Quatrième série d'études hippocratiques VII–X*, pp. 109–35. Geneva.

Lichtheim, M. 1973. *Ancient Egyptian Literature*, vol. 1. Berkeley.

Lienhardt, G. 1961. *Divinity and Experience: The Religion of the Dinka*. Oxford.

Lieshout, R. G. A. van. 1980. *The Greeks on Dreams.* Utrecht.

Linforth, I. M. 1941. *The Arts of Orpheus.* Berkeley.

————. 1946. "Telestic Madness in Plato, *Phaedrus* 244de." *University of California Publications in Classical Philology* 13:163–72.

Lippmann, E. A. 1964. *Musical Thought in Ancient Greece.* New York.

Lippmann, E. O. von. 1910. "Chemisches und Alchemisches aus Aristoteles." *Archiv für die Geschichte der Naturwissenschaften und der Technik* 2:233–300.

————. 1919. *Entstehung und Ausbreitung der Alchemie.* Berlin.

Littré, E. 1839–61. *Oeuvres complètes d'Hippocrate,* 10 vols. Paris.

Lloyd, A. C. 1962. "Genus, Species and Ordered Series in Aristotle." *Phronesis* 7:67–90.

————. 1970. "Aristotle's Principle of Individuation." *Mind* n.s. 79:519–29.

Lloyd, G. E. R. 1964. "Experiment in Early Greek Philosophy and Medicine." *Proceedings of the Cambridge Philological Society* n.s. 10:50–72.

————. 1966. *Polarity and Analogy.* Cambridge.

————. 1967. "Popper versus Kirk: A Controversy in the Interpretation of Greek Science." *British Journal for the Philosophy of Science* 18:21–38.

————. 1968a. "Plato as a Natural Scientist." *Journal of Hellenic Studies* 88:78–92.

————. 1968b. *Aristotle, the Growth and Structure of His Thought.* Cambridge.

————. 1975a. "Alcmaeon and the Early History of Dissection." *Sudhoffs Archiv* 59:113–47.

————. 1975b. "The Hippocratic Question." *Classical Quarterly* n.s. 25:171–92.

————. 1978a. "The Empirical Basis of the Physiology of the *Parva Naturalia.*" In Lloyd and Owen, edd., 1978, pp. 215–39.

————. 1978b. "Saving the Appearances." *Classical Quarterly* n.s. 28:202–22.

————. 1979. *Magic, Reason and Experience.* Cambridge.

————. 1982. "Observational Error in Later Greek Science." In Barnes, Brunschwig, Burnyeat, and Schofield, edd., 1982, pp. 128–64.

————. 1983a. *Science, Folklore and Ideology.* Cambridge.

————. 1983b. "Plato on Mathematics and Nature, Myth and Science." *Humanities* 17:11–30.

Lloyd, G. E. R., ed. 1978. *Hippocratic Writings.* Harmondsworth.

Lloyd, G. E. R., and G. E. L. Owen, edd. 1978. *Aristotle on Mind and the Senses.*

Lohne, J. 1963. "Zur Geschichte des Brechungsgesetzes." *Sudhoffs Archiv für Geschichte der Medizin und der Naturwissenschaften* 47:152–72.

Long, A. A. 1966. "Thinking and Sense-Perception in Empedocles: Mysticism or Materialism?" *Classical Quarterly* n.s. 16:256–76.

————. 1971. "Freedom and Determinism in the Stoic Theory of Human Action." In Long, ed., 1971, pp. 173–99.

————. 1974. *Hellenistic Philosophy.* London.

————. 1976. "The Early Stoic Concept of Moral Choice." In *Images of Man in Ancient and Medieval Thought,* Studia Gerardo Verbeke . . . dicata, ed. F. Bossier et al., pp. 77–92. Louvain.

———. 1977. "Chance and Natural Law in Epicureanism." *Phronesis* 22: 63–88.

———. 1981. "Aristotle and the History of Greek Scepticism." In *Studies in Aristotle,* ed. D. J. O'Meara, Studies in Philosophy and the History of Philosophy 9:79–106. Washington, D.C.

———. 1982. "Astrology: Arguments Pro and Contra." In Barnes, Brunschwig, Burnyeat, and Schofield, edd., 1982, pp. 165–92.

———. Forthcoming. "Ptolemy *On the Criterion:* An Epistemology for the Practising Scientist." In *Truth in Greek Philosophy,* ed. P. Huby and G. Neale. Liverpool.

Long, A. A., ed. 1971. *Problems in Stoicism.* London.

Longrigg, J. 1963. "Philosophy and Medicine, Some Early Interactions." *Harvard Studies in Classical Philology* 67:147–75.

———. 1975. "Elementary Physics in the Lyceum and Stoa." *Isis* 66:211–29.

———. 1980. "The Great Plague of Athens." *History of Science* 18:209–25.

———. 1981. "Superlative Achievement and Comparative Neglect: Alexandrian Medical Science and Modern Historical Research." *History of Science* 19:155–200.

Lonie, I. M. 1964. "Erasistratus, the Erasistrateans, and Aristotle." *Bulletin of the History of Medicine* 38:426–43.

———. 1965. "Medical Theory in Heraclides of Pontus." *Mnemosyne,* 4th series, 18:126–43.

———. 1973. "The Paradoxical Text 'On the Heart.'" *Medical History* 17: 1–15, 136–53.

———. 1977a. "*De natura pueri,* Ch. 13." In Joly, ed., 1977, 123–35.

———. 1977b. "A Structural Pattern in Greek Dietetics and the Early History of Greek Medicine." *Medical History* 21:235–60.

———. 1978. "Cos versus Cnidus and the Historians." *History of Science* 16:42–75, 77–92.

———. 1981a. *The Hippocratic Treatises "On Generation," "On the Nature of the Child," "Diseases IV."* Ars Medica Abt. 2, Bd. 7. Berlin.

———. 1981b. "Hippocrates the Iatromechanist." *Medical History* 25: 113–50.

———. 1983. "Literacy and the Development of Hippocratic Medicine." In Lasserre and Mudry, edd., 1983, pp. 145–61.

Loraux, N. 1981/1986. *The Invention of Athens: The Funeral Oration in the Classical City,* trans. A. Sheridan of *L'Invention d'Athènes: Histoire de l'oraison funèbre dans la "cité classique"* (Paris, 1981). Cambridge, Mass.

———. 1981. *Les Enfants d'Athéna: Idées athéniennes sur la citoyenneté et la division des sexes.* Paris.

———. 1982. "Mourir devant Troie, tomber pour Athènes: De la gloire du héros à l'idée de la cité." In Gnoli and Vernant, edd., 1982, pp. 27–43.

Lord, A. B. 1960. *The Singer of Tales.* Cambridge.

Lord, C. 1978. "Politics and Philosophy in Aristotle's *Politics.*" *Hermes* 106:336–57.

Lorenzen, P. 1960. *Die Entstehung der exakten Wissenschaften.* Berlin.

————. 1975. "L'Etablissement constructif des fondements des sciences exactes." *Bulletin de l'Association Guillaume Budé:* 467–77.

Lotman, J. M., and A. M. Pjatigorskij. 1977. "Text and Function." In *Soviet Semiotics,* ed. D. P. Lucid, pp. 125–35. Baltimore.

Louis, P. 1945. *Les Métaphores de Platon.* Paris.

————. 1955a. "Remarques sur la classification des animaux chez Aristote." In *Autour d'Aristote: Receuil d'études . . . offert à M. Mansion,* pp. 297–304. Louvain.

————. 1955b. "Le Mot Ἱστορία chez Aristote." *Revue de Philologie* 29: 39–44.

————. 1975. "Monstres et monstruosités dans la biologie d'Aristote." In *Le Monde Grec: Hommages à Claire Preaux,* ed. J. Bingen, G. Cambier, and G. Nachtergael, pp. 277–84. Brussels.

Lovejoy, A. O. 1909. "The meaning of φύσις in the Greek Physiologers." *Philosophical Review* 18: 369–83.

————. 1936. *The Great Chain of Being.* Cambridge, Massachusetts.

Lovejoy, A. O., and G. Boas. 1935. *Primitivism and Related Ideas in Antiquity.* Baltimore.

Lucas, J. R. 1970. *The Freedom of the Will.* Oxford.

Lucchetta, G. A. 1978. *Una fisica senza matematica: Democrito, Aristotele, Filopono.* Trento.

Lukasiewicz, J. 1957. *Aristotle's Syllogistic.* 2nd ed. (1st ed. 1951). Oxford.

Lukes, S. 1977. *Essays in Social Theory.* London.

————. 1982. "Relativism in Its Place." In Hollis and Lukes, edd., 1982, pp. 261–305.

Luria, S. 1927. "Studien zur Geschichte der antiken Traumdeutung." *Bulletin de l'Académie des Sciences de l'URSS,* 6th ser., 21:441–66, 1041–72.

————. 1933. *Die Infinitesimaltheorie der antiken Atomisten.* Quellen und Studien zur Geschichte der Mathematik, Astronomie und Physik B, 2, 2 (1932), pp. 106–85. Berlin.

————. 1963. *Anfänge griechischen Denkens.* Berlin.

Luther, W. 1935. *"Wahrheit" und "Lüge" im ältesten Griechentum.* Göttingen.

————. 1958. "Der frühgriechische Wahrheitsgedanke im Lichte der Sprache." *Gymnasium* 65:75–107.

Lycos, K. 1964. "Aristotle and Plato on 'Appearing.'" *Mind* n.s. 73:496–514.

Lynch, J. P. 1972. *Aristotle's School.* Berkeley.

Lyons, J. 1963. *Structural Semantics: An Analysis of Part of the Vocabulary of Plato.* Publications of the Philological Society 20. Oxford.

————. 1977. *Semantics,* 2 vols. Cambridge.

Lyotard, J.-F. 1976. "Sur la force des faibles." *L'Arc* (Aix-en-Provence) 64: 4–12.

McCall, M. H. 1969. *Ancient Rhetorical Theories of Simile and Comparison.* Cambridge, Massachusetts.

MacDermot, V. 1971. *The Cult of the Seer in the Ancient Middle East.* London.

McDermott, W. C. 1938. *The Ape in Antiquity.* Baltimore.

McDiarmid, J. B. 1953/1970. "Theophrastus on the Presocratic Causes." From *Harvard Studies in Classical Philology* 61:85–156. In Furley and Allen, edd., 1970, pp. 178–238.

Mach, E. 1893. *The Science of Mechanics*, trans. T. J. McCormack of 2nd German ed. (1888). Chicago.

———. 1960. *The Science of Mechanics*, trans. T. J. McCormack of 9th German ed. (1933). Chicago.

Machamer, P. K. 1978. "Aristotle on Natural Place and Natural Motion." *Isis* 69:377–87.

Machamer, P. K., and R. G. Turnbull, edd. 1976. *Motion and Time, Space and Matter*. Columbus, Ohio.

———. 1978. *Studies in Perception*. Columbus, Ohio.

MacIntyre, A. 1967/1970. "The Idea of a Social Science." From *Proceedings of the Aristotelian Society*, Suppl. 41. In B. R. Wilson, ed., 1970, pp. 112–30.

McKeon, R. 1947. "Aristotle's Conception of the Development and the Nature of Scientific Method." *Journal of the History of Ideas* 8:3–44.

———. 1964–65. "The Flight from Certainty and the Quest for Precision." *Review of Metaphysics* 18:234–53.

MacKinney, L. 1964. "The Concept of Isonomia in Greek Medicine." In Mau and Schmidt, edd., 1964, pp. 79–88.

MacKinnon, D. M. 1965. "Aristotle's Conception of Substance." In Bambrough, ed., 1965, pp. 97–119.

McKirahan, R. D. 1978. "Aristotle's Subordinate Sciences." *British Journal for the History of Science* 11:197–220.

———. 1983. "Aristotelian Epagoge in *Prior Analytics* 2.21 and *Posterior Analytics* 1.1." *Journal of the History of Philosophy* 21:1–13.

McMullin, E., ed. 1963. *The Concept of Matter in Greek and Medieval Philosophy*. Notre Dame, Indiana.

Macran, H. S. 1902. *The Harmonics of Aristoxenus*. Oxford.

Madarász-Zsigmond, A. 1978. "Die Anfänge der griechischen Logik." *Acta Antiqua Academiae Scientiarum Hungaricae* 26:291–345.

Madden, E. H. 1957. "Aristotle's Treatment of Probability and Signs." *Philosophy of Science* 24:167–72.

Maeyama, Y. 1984. "Ancient Stellar Observations: Timocharis, Aristyllus, Hipparchus, Ptolemy—The Dates and Accuracies." *Centaurus* 27:280–310.

Mahoney, M. S. 1968–69. "Another Look at Greek Geometrical Analysis." *Archive for History of Exact Sciences* 5:318–48.

Majno, G. 1975. *The Healing Hand*. Cambridge, Massachusetts.

Manetti, D. 1973. "Valore semantico e risonanze culturali della parola ΦΥΣΙΣ." *La Parola del Passato* 28:426–44.

Mansfeld, J. 1964. *Die Offenbarung des Parmenides und die menschliche Welt*. Assen.

———. 1971. *The Pseudo-Hippocratic Tract* ΠΕΡΙ ῾ΕΒΔΟΜΑΔΩΝ *Ch. 1–11 and Greek Philosophy*. Assen.

————. 1975. "Almaeon: 'Physikos' or Physician?" In *Kephalaion: Studies . . .
offered to C. J. de Vogel*, ed. J. Mansfeld and L. M. de Rijk, pp. 26–38.
Assen.

————. 1980a. "Theoretical and Empirical Attitudes in Early Greek Scientific
Medicine." In Grmek, ed., 1980, pp. 371–90.

————. 1980b. "Plato and the Method of Hippocrates." *Greek, Roman and
Byzantine Studies* 21:341–62.

————. 1981. "Bad World and Demiurge: A 'Gnostic' Motif from Parmenides
and Empedocles to Lucretius and Philo." In *Studies in Gnosticism and
Hellenistic Religions*, ed. R. Van den Broek and M. J. Vermaseren, pp. 261–
314. Leiden.

Mansion, A. 1946. *Introduction à la physique aristotélicienne*. 2nd ed. (1st ed.
1913). Louvain.

————. 1956. "L'objet de la science philosophique suprême d'après Aristote,
Métaphysique E, 1." In *Mélanges de philosophie grecque offerts à Mgr.
Diès*, pp. 151–68. Paris.

Mansion, P. 1899. "Note sur le caractère géométrique de l'ancienne astronomie." *Abhandlungen zur Geschichte der Mathematik* 9:275–92.

Mansion, S. 1969. "L'Objet des mathématiques et l'objet de la dialectique selon Platon." *Revue Philosophique de Louvain* 67:365–88.

————. 1976. *Le Jugement d'existence chez Aristote*. 2nd ed. (1st ed. 1946).
Louvain.

————. 1979. "'Plus Connu en soi' 'plus connu pour nous': Une Distinction
épistémologique importante chez Aristote." *Pensamiento* 35:161–70.

Mansion, S., ed. 1961. *Aristote et les problèmes de méthode*. Louvain.

Manuli, P. 1980. "Fisiologia e patologia del femminile negli scritti ippocratici
dell' antica ginecologia greca." In Grmek, ed., 1980, pp. 393–408.

————. 1981. "Claudio Tolomeo: Il criterio e il principio." *Rivista critica di
storia della filosofia* 36:64–88.

Manuli, P., and M. Vegetti. 1977. *Cuore, sangue e cervello*. Milan.

Manuwald, B. 1971. *Das Buch H der aristotelischen "Physik."* Beiträge zur
klassischen Philologie 36. Meisenheim.

Maracchia, S. 1979–80. "Aristotele e l'incommensurabilità." *Archive for History of Exact Sciences* 21:201–28.

Marrou, H. I. 1956. *A History of Education in Antiquity*, trans. G. Lamb.
London.

Marsden, E. W. 1969. *Greek and Roman Artillery: Historical Development*.
Oxford.

————. 1971. *Greek and Roman Artillery: Technical Treatises*. Oxford.

Martin, J. 1976. "Zur Entstehung der Sophistik." *Saeculum* 27:143–64.

Martin, R., and H. Metzger. 1976. *La Religion grecque*. Vendôme.

Masson-Oursel, P. 1916. "La Sophistique: Etude de philosophie comparée."
Revue de Métaphysique et de Morale 23:343–62.

————. 1917a. "Etudes de logique comparée I: Evolution de la logique indienne." *Revue Philosophique de la France et de l'Etranger* 83:453–69.

————. 1917b. "Etudes de logique comparée II: Evolution de la logique
chinoise." *Revue Philosophique de la France et de l'Etranger* 84:59–76.

———. 1918. "Etudes de logique comparée, III: Confrontations et analyse comparative." *Revue Philosophique de la France et de l'Etranger* 85: 148–66.

Mates, B. 1961. *Stoic Logic*. Berkeley.

Matson, W. I. 1952–53. "The Naturalism of Anaximander." *Review of Metaphysics* 6: 387–95.

———. 1954–55. Review of Cornford 1952, *Review of Metaphysics* 8: 443–54.

Mau, J. 1954. *Zum Problem des Infinitesimalen bei den antiken Atomisten*. Berlin.

———. 1969. "Zur Methode der aristotelischen Ableitung der Elementar-Körper." In Düring, ed., 1969, pp. 133–46.

Mau, J., and E. G. Schmidt, edd. 1964. *Isonomia*. Deutsche Akademie der Wissenschaften zu Berlin, Arbeitsgruppe für hellenistisch-römische Philosophie, Veröffentlichung 9. Berlin.

Maula, E. 1974. *Studies in Eudoxus' Homocentric Spheres*. Helsinki.

Mauss, M. 1950/1972. *A General Theory of Magic*, trans. R. Brain, from *Sociologie et Anthropologie* (Paris, 1950). London. (Originally "Esquisse d'une théorie générale de la magie," with H. Hubert, in *L'Année Sociologique* 7 [1902–3] 1904: 1–146.)

May, M. T. 1968. *Galen On the Usefulness of the Parts of the Body*, 2 vols. Ithaca, New York.

Mead, H. L. 1975. "The Methodology of Ptolemaic Astronomy: An Aristotelian View." *Laval Théologique et Philosophique* 31: 55–74.

Méautis, G. 1922. *Recherches sur le Pythagorisme*. Neuchâtel.

Mellor, D. H. 1965. "Experimental Error and Deducibility." *Philosophy of Science* 32: 105–22.

———. 1966. "Inexactness and Explanation." *Philosophy of Science* 33: 345–59.

———. 1967. "Imprecision and Explanation." *Philosophy of Science* 34: 1–9.

———. 1969. "Physics and Furniture." In *Studies in the Philosophy of Science*, ed. N. Rescher, pp. 171–87. Oxford.

Merlan, P. 1953. "Ambiguity in Heraclitus." *Proceedings of the 11th International Congress of Philosophy, Brussels, 1953*, vol. 12, pp. 56–60. Louvain.

———. 1963. *Monopsychism, Mysticism, Metaconsciousness*. Archives Internationales d'Histoire des Idées 2. The Hague.

———. 1968a. *From Platonism to NeoPlatonism*. 3rd ed. (1st ed. 1953). The Hague.

———. 1968b. "On the Terms 'Metaphysics' and 'Being-qua-Being.'" *The Monist* 52: 174–94.

Meuli, K. 1935. "Scythica." *Hermes* 70: 121–76.

Meyer-Steineg, T. 1912. *Chirurgische Instrumente des Altertums. Ein Beitrag zur antiken Akiurgie*. Jenaer medizin-historische Beiträge 1. Jena.

———. 1916. *Das medizinische System der Methodiker*. Jenaer medizin-historische Beiträge 7–8. Jena.

Michel, J.-H. 1981. "La Folie avant Foucault: *Furor* et *ferocia*." *L'Antiquité Classique* 50:517–25.

Michler, M. 1962a. 'Das Problem der westgriechischen Heilkunde.' *Sudhoffs Archiv für Geschichte der Medizin und der Naturwissenschaften* 46: 137–52.

———. 1962b. "Die praktische Bedeutung des normativen Physis-Begriffes in der hippokratischen Schrift De fracturis—De Articulis." *Hermes* 60: 385–401.

Mignucci, M. 1965. *La teoria aristotelica della scienza*. Florence.

———. 1975. *L'argomentazione dimostrativa in Aristotele*. Padua.

———. 1981. "'Ὡς ἐπὶ τὸ πολύ et nécessaire dans la conception aristotélicienne de la science." In Berti, ed., 1981, pp. 173–203.

———. 1987. "Aristotle's Arithmetics." In Graeser, ed., 1987, pp. 175–211.

Milhaud, G. 1903. "Aristote et les mathématiques." *Archiv für Geschichte der Philosophie* N.F 9:367–92.

Miller, H. W. 1952. "*Dynamis* and *Physis* in *On Ancient Medicine*." *Transactions and Proceedings of the American Philological Association* 83.: 184–97.

———. 1953. "The Concept of the Divine in De Morbo Sacro." *Transactions and Proceedings of the American Philological Association* 84:1–15.

Mittelstrass, J. 1962. *Die Rettung der Phänomene*. Berlin.

———. 1979. "*Phaenomena bene fundata*: From 'Saving the Appearances' to the Mechanisation of the World-Picture." In *Classical Influences on Western Thought A.D. 1650–1870*, ed. R. R. Bolgar, pp. 39–59. Cambridge.

Moesgaard, K. P. 1980a. "The Full Moon Serpent: A Foundation Stone of Ancient Astronomy?" *Centaurus* 24:51–96.

———. 1980b. Review of Newton 1977. *Journal of the History of Astronomy* 11:133–35.

Mogenet, J., ed. 1950. *Autolycus de Pitane*. Université de Louvain, Recueil de Travaux d'Histoire et de Philologie 37. Louvain.

Mohr, R. D. 1985. *The Platonic Cosmology*. Philosophia Antiqua 42. Leiden.

Momigliano, A. 1973. "Freedom of Speech in Antiquity." In *Dictionary of the History of Ideas*, ed. P. P. Wiener, vol. 2, pp. 252–62. New York.

———. 1975. *Alien Wisdom: The Limits of Hellenization*. Cambridge.

———. 1978. "The Historians of the Classical World and Their Audiences: Some Suggestions." *Annali della Scuola Normale Superiore di Pisa* 8:59–75.

———. 1980. *Sesto contributo alla storia degli studi classici e del mondo antico*. 2 vols. Rome.

Moore, O. K. 1957. "Divination—A New Perspective." *American Anthropologist* 59:69–74.

Moors, K. F. 1982. *Platonic Myth: An Introductory Study*. Washington, D.C.

Moraux, P. 1968. "La Joute dialectique d'après le huitième livre des *Topiques*." In Owen, ed., 1968, pp. 277–311.

———. 1973. *Der Aristotelismus bei den Griechen*, vol. 1. Berlin.

———. 1984. *Der Aristotelismus bei den Griechen*, vol. 2. Berlin.

Moraux, P., and J. Wiesner, edd. 1983. *Zweifelhaftes im Corpus Aristotelicum.* Akten des 9. Symposium Aristotelicum, Peripatoi 14. Berlin.

Moravcsik, J. M. E. 1974. "Aristotle on Adequate Explanations." *Synthese* 28:3–17.

Moravcsik, J. M. E., ed. 1967. *Aristotle.* New York.

Moreau, J. 1959. "L'Eloge de la biologie chez Aristote." *Revue des Etudes Anciennes* 61:57–64.

———. 1968. "Aristote et la dialectique platonicienne." In Owen, ed., 1968, pp. 80–90.

———. 1979. "Qu'est-ce qu'un sophiste?" *Les Etudes Philosophiques* 1979: 325–35.

Morrison, J. S. 1941. "The Place of Protagoras in Athenian Public Life (460–415 B.C." *Classical Quarterly* 35:1–16.

———. 1961. "Antiphon." *Proceedings of the Cambridge Philological Society* n.s. 7:49–58.

———. 1963. "The *Truth* of Antiphon." *Phronesis* 8:35–49.

Morrow, G. R. 1950/1965. "Necessity and Persuasion in Plato's *Timaeus*." From *Philosophical Review* 59:147–63. In Allen, ed., 1965, pp. 421–37.

———. 1960. *Plato's Cretan City.* Princeton.

———. 1969. "Qualitative Change in Aristotle's *Physics*." In Düring, ed., 1969, pp. 154–67.

———. 1970a. "Plato and the Mathematicians: An Interpretation of Socrates' Dream in the *Theaetetus* (201e–206c)." *Philosophical Review* 79:309–33.

———. 1970b. *Proclus: A Commentary on the First Book of Euclid's Elements.* Princeton.

Mosshammer, A. A. 1981. "Thales' Eclipse." *Transactions of the American Philological Association* 111:145–55.

Motte, A. 1981. "Persuasion et violence chez Platon." *L'Antiquité Classique* 50:562–77.

Moulinier, L. 1952. *Le Pur et l'impur dans la pensée des Grecs d'Homère à Aristote.* Etudes et Commentaires 12. Paris.

Mourelatos, A. P. D. 1967. "Aristotle's 'Powers' and Modern Empiricism." *Ratio* 9:97–104.

———. 1970. *The Route of Parmenides.* New Haven.

———. 1980. "Plato's 'Real Astronomy,' *Republic* 527D–531D." In Anton, ed., 1980, pp. 33–73.

———. 1981. "Astronomy and Kinematics in Plato's Project of Rationalist Explanation." *Studies in History and Philosophy of Science* 12:1–32.

Mudry, P. 1982. *La Préface du De medicina de Celse.* Bibliotheca Helvetica Romana 19. Lausanne.

Müller, C. W. 1965a. *Gleiches zu Gleichem. Ein Prinzip frühgriechischen Denkens.* Wiesbaden.

———. 1965b. "Die Heilung 'durch das Gleiche' in den hippokratischen Schriften *De morbo sacro* und *De locis in homine*." *Sudhoffs Archiv für Geschichte der Medizin und der Naturwissenschaften* 49:225–49.

———. 1967. "Protagoras über die Götter." *Hermes* 95:140–59.

Mueller, I. 1969. "Euclid's *Elements* and the Axiomatic Method." *British Journal for the Philosophy of Science* 20:289–309.

———. 1970/1979. "Aristotle on Geometrical Objects." From *Archiv für Geschichte der Philosophie* 52:156–71. In Barnes, Schofield, and Sorabji, edd., 1979, pp. 96–107.

———. 1974. "Greek Mathematics and Greek Logic." In *Ancient Logic and Its Modern Interpretations*, ed. J. Corcoran, pp. 35–70. Dordrecht.

———. 1980. "Ascending to Problems: Astronomy and Harmonics in *Republic* VII." In Anton, ed., 1980, pp. 103–21.

———. 1981. *Philosophy of Mathematics and Deductive Structure in Euclid's Elements.* Cambridge, Massachusetts.

———. 1982a. "Geometry and Scepticism." In Barnes, Brunschwig, Burnyeat, and Schofield, edd., 1982, pp. 69–95.

———. 1982b. "Aristotle and the Quadrature of the Circle." In Kretzmann, ed., 1982, pp. 146–64.

———. 1987. "Aristotle's Approach to the Problem of Principles in *Metaphysics* M and N." In Graeser, ed., 1987, pp. 241–59.

Müller, I. von. 1897. "Ueber Galens Werk vom wissenschaftlichen Beweis." *Abhandlungen der philosophisch-philologischen Classe der königlich bayerischen Akademie der Wissenschaften* 20, 2 (1894–95), pp. 403–78. Munich.

Müller, R., ed. 1976. *Der Mensch als Mass der Dinge.* Berlin.

Müri, W. 1947/1976. "Bemerkungen zur hippokratischen Psychologie." From *Festschrift für Edouard Tièche,* Schriften der literarischen Gesellschaft Bern 6 (1947), pp. 71–85. In Müri 1976, pp. 100–114.

———. 1950. "Der Massgedanke bei griechischen Ärzten." *Gymnasium* 57:182–201. (Reprinted in Müri 1976, pp. 115–38.)

———. 1953. "Melancholie und schwarze Galle." *Museum Helveticum* 10:21–38. (Reprinted in Müri 1976, pp. 139–64).

———. 1976. *Griechische Studien,* ed. E. Vischer. Schweizerische Beiträge zur Altertumswissenschaft 14. Basel.

Mugler, C. 1948. *Platon et la recherche mathématique de son époque.* Strasbourg.

———. 1957. "Sur l'histoire de quelques définitions de la géométrie grecque et les rapports entre la géométrie et l'optique, I." *L'Antiquité Classique* 26:331–45.

———. 1958. "Sur l'histoire de quelques définitions de la géométrie grecque et les rapports entre la géométrie et l'optique, II." *L'Antiquité Classique* 27:76–91.

———. 1973. "Sur quelques points de contact entre la magie et les sciences appliquées des anciens." *Revue de Philologie* 47:31–37.

Mulkay, M. J. 1972. *The Social Process of Innovation.* London.

———. 1979. *Science and the Sociology of Knowledge.* London.

Murphy, J. G. 1976. "Rationality and the Fear of Death." *The Monist* 59:187–203.

Nagler, M. N. 1974. *Spontaneity and Tradition: A Study in the Oral Art of Homer.* Berkeley.

Nagy, G. 1979. *The Best of the Achaeans*. Baltimore.

———. 1982. "Theognis of Megara: The Poet as Seer, Pilot, and Revenant." *Arethusa* 15:109–28.

———. 1983. "Poet and Tyrant: *Theognidea* 39–52, 1081–1082b." *Classical Antiquity* 2:82–91.

Nakamura, H. 1960. *The Ways of Thinking of Eastern Peoples*. Tokyo.

Narcy, M. 1978. "Aristote et la géométrie." *Les Etudes Philosophiques* 1978: 13–24.

Needham, J. 1954–. *Science and Civilisation in China*. Vol. 1: *Introductory Orientations*, 1954; vol. 2: *History of Scientific Thought*, 1956; vol. 3: *Mathematics and the Sciences of the Heavens and the Earth*, 1959; vol. 5: *Chemistry and Chemical Technology*, part 2, 1974. Cambridge.

Needham, J., W. Ling, and D. J. de S. Price. 1960. *Heavenly Clockwork*. Cambridge.

Needham, R. 1972. *Belief, Language, and Experience*. Oxford.

———. 1978. *Essential Perplexities*. Oxford.

———. 1980. *Reconnaissances*. Toronto.

Needham, R., ed. 1973. *Right and Left: Essays on Dual Symbolic Classification*. Chicago.

Nehamas, A. 1972–73. "Predication and Forms of Opposites in the *Phaedo*." *Review of Metaphysics* 26:461–91.

———. 1975. "Plato on the Imperfection of the Sensible World." *American Philosophical Quarterly* 12:105–17.

———. 1975–76. "Confusing Universals and Particulars in Plato's Early Dialogues." *Review of Metaphysics* 29:287–306.

———. 1979. "Self-Predication and Plato's Theory of Forms." *American Philosophical Quarterly* 16:93–103.

———. 1982–83. "Participation and Predication in Plato's Later Thought." *Review of Metaphysics* 36:343–74.

Neitzel, H. 1980. "Hesiod und die lügenden Musen. Zur Interpretation von *Theogonie* 27f." *Hermes* 108:387–401.

Nelson, A. 1909. *Die hippokratische Schrift περὶ φυσῶν: Text und Studien*. Uppsala.

Nenci, G. 1963. "Il sigillo di Teognide." *Rivista di filologia e di istruzione classica* 3rd ser. 91:30–37.

Nestle, W. 1903. "Kritias." *Neue Jahrbücher für das klassische Altertum, Geschichte und deutsche Literatur* 11:81–107, 178–99. (Reprinted in Nestle 1948, pp. 253–320.)

———. 1922. "Die Schrift des Gorgias 'über die Natur oder über das Nichtseiende.'" *Hermes* 57:551–62. (Reprinted in Nestle 1948, pp. 240–52.)

———. 1938. "Hippocratica." *Hermes* 73:1–38. (Reprinted in Nestle 1948, pp. 517–66.)

———. 1942. *Vom Mythos zum Logos*. 2nd ed. Stuttgart.

———. 1948. *Griechische Studien*. Stuttgart.

Neuburger, M. 1932. *The Doctrine of the Healing Power of Nature throughout the Course of Time*, trans. L. J. Boyd of *Die Lehre von der Heilkraft der Natur im Wandel der Zeiten* (Stuttgart, 1926). New York.

Neugebauer, O. 1928. "Zur Geschichte des pythagoräischen Lehrsatzes." *Nachrichten von der Gesellschaft der Wissenschaften zu Göttingen*, math.-phys. Kl., pp. 45–48. Berlin.

———. 1929. *Über vorgriechische Mathematik.* Hamburger mathematische Einzelschriften 8. Leipzig.

———. 1931. *Zur Geschichte der babylonischen Mathematik.* Quellen und Studien zur Geschichte der Mathematik, Astronomie und Physik B, 1, 1 (1929), pp. 67–80. Berlin.

———. 1934. *Vorlesungen über Geschichte der antiken mathematischen Wissenschaften*, Bd. 1. Vorgriechische Mathematik. Berlin.

———. 1938–39. *Über eine Methode zur Distanzbestimmung Alexandria-Rom bei Heron.* Det kgl. Danske Videnskabernes Selskab. Historisk-filologiske Meddelelser 26, 2 and 26, 7. Copenhagen.

———. 1945/1983. "The History of Ancient Astronomy: Problems and Methods." From *Journal of Near Eastern Studies* 4:1–38. In Neugebauer 1983, pp. 33–98.

———. 1947/1983. "The Water-Clock in Babylonian Astronomy." From *Isis* 37:37–43. In Neugebauer 1983, pp. 239–45.

———. 1952/1957. *The Exact Sciences in Antiquity.* 2nd ed. (1st ed. 1952). Providence, Rhode Island.

———. 1955. *Astronomical Cuneiform Texts*, 3 vols. Princeton.

———. 1956/1983. "Notes on Hipparchus." From *The Aegean and the Near East: Studies Presented to H. Goldman*, ed. S. S. Weinberg (New York, 1956), pp. 292–96. In Neugebauer 1983, pp. 320–24.

———. 1972/1983. "On Some Aspects of Early Greek Astronomy." From *Proceedings of the American Philosophical Society* 116:243–51. In Neugebauer 1983, pp. 361–69.

———. 1975. *A History of Ancient Mathematical Astronomy*, 3 vols. Berlin.

———. 1983. *Astronomy and History: Selected Essays.* New York.

Neugebauer, O., and H. B. van Hoesen. 1959. *Greek Horoscopes.* American Philosophical Society Memoirs 48. Philadelphia.

Neugebauer, O., and R. A. Parker. 1960. *Egyptian Astronomical Texts.* Providence, Rhode Island.

Neugebauer, O., and A. Sachs. 1945. *Mathematical Cuneiform Texts.* American Oriental Series 29. New Haven.

Newton, R. R. 1973. "The Authenticity of Ptolemy's Parallax Data, Part I." *Quarterly Journal of the Royal Astronomical Society* 14:367–88.

———. 1974a. "The Authenticity of Ptolemy's Parallax Data, Part II." *Quarterly Journal of the Royal Astronomical Society* 15:7–27.

———. 1974b. "The Authenticity of Ptolemy's Eclipse and Star Data." *Quarterly Journal of the Royal Astronomical Society* 15:107–21.

———. 1977. *The Crime of Claudius Ptolemy.* Baltimore.

———. 1979. "On the Fractions of Degrees in an Ancient Star Catalogue." *Quarterly Journal of the Royal Astronomical Society* 20:383–94.

———. 1980a. "The Sources of Eratosthenes' Measurement of the Earth." *Quarterly Journal of the Royal Astronomical Society* 21:379–87.

———. 1980b. "Comments on 'Was Ptolemy a Fraud?' by Owen Gingerich." *Quarterly Journal of the Royal Astronomical Society* 21:388–99.

Newton-Smith, W. H. 1981. *The Rationality of Science.* London.

Nickel, D. 1972. "Ärztliche Ethik und Schwangerschaftsunterbrechung bei den Hippokratikern." *NTM: Schriftenreihe für Geschichte der Naturwissenschaften, Technik und Medizin* 9, 1:73–80.

Niebyl, P. H. 1971. 'Old Age, Fever and the Lamp Metaphor." *Journal of the History of Medicine and Allied Sciences* 26:351–68.

Nilsson, M. P. 1907. *Die Kausalsätze im griechischen bis Aristoteles.* Würzburg.

———. 1940. *Greek Popular Religion.* New York.

———. 1955–61. *Geschichte der griechischen Religion.* 2nd ed., 2 vols. Munich.

———. 1957. *The Dionysiac Mysteries of the Hellenistic and Roman Age.* Lund.

Nissen, H. 1903. "Die Erdmessung des Eratosthenes." *Rheinisches Museum* N.F. 58:231–45.

Nock, A. D. 1972. *Essays on Religion and the Ancient World,* 2 vols. Oxford.

Nock, A. D., and A. J. Festugière. 1945–54. *Corpus Hermeticum,* 4 vols. Paris.

Nörenberg, H. W. 1968. *Das Göttliche und die Natur in der Schrift über die heilige Krankheit.* Bonn.

Nougayrol, J. 1966. "Trente Ans de recherches sur la divination babylonienne (1935–1965)." In *La Divination en mésopotamie ancienne,* XIVᵉ Rencontre Assyriologique Internationale, Strasbourg, 2–6 July 1965, pp. 5–19. Paris.

Novak, J. A. 1978. "A Geometrical Syllogism: *Posterior Analytics,* II, 11." *Apeiron* 12, 2:26–33.

Nussbaum, M. C. 1978. *Aristotle's De motu animalium.* Princeton.

———. 1979. "Eleatic Conventionalism and Philolaus on the Conditions of Thought." *Harvard Studies in Classical Philology* 83:63–108.

———. 1982. "Saving Aristotle's Appearances." In Schofield and Nussbaum, edd., 1982, pp. 267–93.

———. 1986. *The Fragility of Goodness.* Cambridge.

Nutton, V. 1970. "The Medical Profession in the Roman Empire from Augustus to Justinian." Ph.D. diss. Cambridge.

———. 1972. "Galen and Medical Autobiography." *Proceedings of the Cambridge Philological Society* n.s. 18:50–62.

———. 1979. *Galen on Prognosis. Corpus Medicorum Graecorum* 5, 8, 1. Berlin.

———. 1983. "The Seeds of Disease: An Explanation of Contagion and Infection from the Greeks to the Renaissance." *Medical History* 27:1–34.

———. 1984. "Galen in the Eyes of His Contemporaries." *Bulletin of the History of Medicine* 58:315–24.

Nutton, V., ed. 1981. *Galen: Problems and Prospects.* London.

Oberhelman, S. M. 1983. "Galen, On Diagnosis from Dreams." *Journal of the History of Medicine and Allied Sciences* 38:36–47.

Obeyesekere, G. 1981. *Medusa's Hair*. Chicago.

O'Brien, D. 1967. "Anaximander's Measurements." *Classical Quarterly* n.s. 17:423–32.

———. 1969. *Empedocles' Cosmic Cycle*. Cambridge.

———. 1977. "Heavy and Light in Democritus and Aristotle: Two Conceptions of Change and Identity." *Journal of Hellenic Studies* 97:64–74.

———. 1978. "Aristote et la catégorie de quantité: Divisions de la quantité." *Les Etudes Philosophiques* 1978:25–40.

———. 1981. *Theories of Weight in the Ancient World*. Vol. 1: *Democritus, Weight and Size*. Paris.

———. 1984. *Theories of Weight in the Ancient World*. Vol. 2: *Plato, Weight and Sensation*. Paris.

O'Connor, D. J. 1971. *Free Will*. Garden City, New York.

Ogle, W. 1882. *Aristotle on the Parts of Animals*. London.

———. 1897. *Aristotle on Youth and Old Age, Life and Death and Respiration*. London.

Ohlert, K. 1912. *Rätsel und Rätselspiele der alten Griechen*. 2nd ed. (1st ed. 1886). Berlin.

Olson, R. 1978. "Science, Scientism and Anti-Science in Hellenic Athens: A New Whig Interpretation." *History of Science* 16:179–99.

Onians, R. B. 1951. *The Origins of European Thought*. Cambridge.

Oppenheim, A. Leo. 1962. "Mesopotamian Medicine." *Bulletin of the History of Medicine* 36:97–108.

———. 1964. *Ancient Mesopotamia: Portrait of a Dead Civilization*. Chicago.

Ortony, A., ed. 1979. *Metaphor and Thought*. Cambridge.

Osborne, C. Forthcoming. *Rethinking Early Greek Philosophy*. London.

Osborne, R. 1985. *Demos: The Discovery of Classical Attika*. Cambridge.

Osler, W. 1947. *The Principles and Practice of Medicine*. 16th ed., ed. H. A. Christian. New York.

Ostwald, M. 1969. *Nomos and the Beginnings of the Athenian Democracy*. Oxford.

Owen, G. E. L. 1957/1986. "A Proof in the περὶ ἰδεῶν." From *Journal of Hellenic Studies* 77:103–11. In Owen, 1986b, pp. 165–79.

———. 1960/1986. "Logic and Metaphysics in Some Earlier Works of Aristotle." From *Aristotle and Plato in the Mid-Fourth Century*, ed. I. Düring and G. E. L. Owen (Göteborg, 1960), pp. 163–90. In Owen, 1986b, pp. 180–99.

———. 1961/1986. "Tithenai ta phainomena." From *Aristote et les problèmes de méthode*, ed. S. Mansion (Louvain, 1961), pp. 83–103. In Owen, 1986b, pp. 239–51.

———. 1965a/1986. "Aristotle on the Snares of Ontology." From Bambrough, ed., 1965, pp. 69–95. In Owen, 1986b, pp. 259–78.

———. 1965b/1986. "The Platonism of Aristotle." From *Proceedings of the British Academy* 51:125–50. In Owen, 1986b, pp. 200–220.

———. 1968/1986. "Dialectic and Eristic in the Treatment of the Forms." From Owen, ed., 1968, pp. 103–25. In Owen, 1986b, pp. 221–38.

————. 1970/1986. "Aristotle: Method, Physics and Cosmology." From *Dictionary of Scientific Biography*, ed. C. C. Gillispie (New York, 1970), vol. 1, pp. 250–58. In Owen, 1986b, pp. 151–64.

————. 1986a. "Aristotelian Mechanics." In Owen, 1986b, pp. 315–33.

————. 1986b. *Logic, Science and Dialectic*. London.

Owen, G. E. L., ed. 1968. *Aristotle on Dialectic*. Oxford.

Owens, J. 1963. *The Doctrine of Being in the Aristotelian Metaphysics*. Revised ed. (1st ed. 1951). Toronto.

————. 1968. "Teleology of Nature in Aristotle." *The Monist* 52:159–73.

Pack, R. A. 1941. "Artemidorus and the Physiognomists." *Transactions and Proceedings of the American Philological Association* 72:321–34.

Padel, R. 1981. "Madness in Fifth-Century B.C. Athenian Tragedy." In *Indigenous Psychologies*, ed. P. Heelas and A. Lock, pp. 105–31. London.

————. 1983. "Women: Model for Possession by Greek Daemons." In *Images of Women in Antiquity*, ed. A. Cameron and A. Kuhrt, pp. 3–19. London.

Palm, A. 1933. *Studien zur hippokratischen Schrift* ΠΕΡΙ ΔΙΑΙΤΗΣ. Tübingen.

Palter, R. 1970–71. "An Approach to the History of Early Astronomy." *Studies in History and Philosophy of Science* 1:93–133.

Pannekoek, A. 1955. "Ptolemy's Precession." In *Vistas in Astronomy*, ed. A. Beer, vol. 1, pp. 60–66. London.

Park, G. K. 1963. "Divination and Its Social Contexts." *Journal of the Royal Anthropological Institute* 93:195–209.

Parker, R. A. 1972. *Demotic Mathematical Papyri*. Providence, Rhode Island.

Parker, R. C. T. 1983. *Miasma*. Oxford.

Parry, J. P. 1985. "The Brahmanical Tradition and the Technology of the Intellect." In *Reason and Morality*, ed. J. Overing, pp. 200–225. London.

Patzig, G. 1960–61/1979. "Theology and Ontology in Aristotle's *Metaphysics*." Originally "Theologie und Ontologie in der *Metaphysik* des Aristoteles." *Kant-Studien* 52:185–205. In Barnes, Schofield, and Sorabji, edd., 1979, pp. 33–49.

————. 1968. *Aristotle's Theory of the Syllogism*, trans. J. Barnes of *Die aristotelische Syllogistik*, 2nd ed. (Göttingen, 1963). Dordrecht.

Pearcy, L. T. 1984. "Melancholy Rhetoricians and Melancholy Rhetoric: 'Black Bile' as a Rhetorical and Medical Term in the Second Century A.D." *Journal of the History of Medicine and Allied Sciences* 39:446–56.

Pears, D. F., ed. 1963. *Freedom and the Will*. London.

Pedersen, O. 1974. *A Survey of the Almagest*. Odense.

Peet, T. E. 1923. *The Rhind Mathematical Papyrus*. London.

Pellegrin, P. 1986. *Aristotle's Classification of Animals*, trans. A. Preus of *La Classification des animaux chez Aristote* (Paris, 1982). Berkeley.

Pembroke, S. G. 1971. "Oikeiōsis." In Long, ed., pp. 114–49.

Pépin, J. 1958. *Mythe et allégorie*. Paris.

Pera, M. 1980. "Le teorie come metafore e l'induzione." *Physis* 22:433–61.

Peradotto, J. 1979. "Originality and Intentionality." In *Arktouros: Festschrift B. M. W. Knox*, ed. G. W. Bowersock, W. Burkert, and M. C. J. Putnam, pp. 3–11. Berlin.

Perelman, C. 1970. *Le Champ de l'argumentation*. Brussels.

Perelman, C., and L. Olbrechts-Tyteca. 1969. *The New Rhetoric: A Treatise on Argumentation*, trans. J. Wilkinson and P. Weaver of *La Nouvelle Rhétorique* (Paris, 1958). Notre Dame.

Peters, C. H. F. 1877. "Ueber die Fehler des Ptolemäischen Sternverzeichnisses." *Vierteljahrsschrift der astronomischen Gesellschaft*, 12 Jahrgang: 296–99.

Peters, C. H. F., and E. B. Knobel. 1915. *Ptolemy's Catalogue of Stars: A Revision of the Almagest*. Washington, D.C.

Petersen, V. M. 1969. "The Three Lunar Models of Ptolemy." *Centaurus* 14:142–71.

Petersen, V. M., and O. Schmidt. 1967–68. "The Determination of the Longitude of the Apogee of the Orbit of the Sun according to Hipparchus and Ptolemy." *Centaurus* 12:73–96.

Petrie, Flinders. 1938. "The Present Position of the Metrology of Egyptian Weights." *Journal of Egyptian Archaeology* 24:180–81.

Pfeffer, F. 1976. *Studien zur Mantik in der Philosophie der Antike*. Beiträge zur klassischen Philologie 64. Meisenheim.

Pfeiffer, R. 1968. *History of Classical Scholarship from the Beginnings to the End of the Hellenistic Age*. Oxford.

———. 1976. *History of Classical Scholarship from 1300 to 1850*. Oxford.

Pfister, F. 1935. "Katharsis." *Pauly-Wissowa Real-Encyclopädie der classischen Altertumswissenschaft*, Suppl. Bd. 6, cols. 146–62.

Philip, J. A. 1966. *Pythagoras and Early Pythagoreanism*. Phoenix Suppl. 7. Toronto.

Philippe, M.-D. 1948. "'Αφαίρεσις, πρόσθεσις, χωρίζειν dans la philosophie d'Aristote." *Revue Thomiste* 56, 48:461–79.

Phillips, N. 1981. *Sijobang*. Cambridge.

Pieri, S. N. 1978. *Carneade*. Padua.

Pigeaud, J. M. 1978a. "Une Physiologie de l'inspiration poétique." *Les Etudes Classiques* 46:23–31.

———. 1978b. "Du rhythme dans le corps: Quelques Notes sur l'interprétation du pouls par le médecin Hérophile." *Bulletin de l'Association Guillaume Budé* 1978, 3:258–67.

———. 1980. "La Physiologie de Lucrèce." *Revue des Etudes Latines* 58: 176–200.

———. 1981. *La Maladie de l'âme*. Paris.

Pingree, D. 1974. "Concentric with Equant." *Archives Internationales d'Histoire des Sciences* 24:26–28.

———. 1976. "The Recovery of Early Greek Astronomy from India." *Journal for the History of Astronomy* 7:109–23.

Plamböck, G. 1964. *Dynamis im Corpus Hippocraticum*. Akademie der Wissenschaften und der Literatur, Mainz, Abhandlungen der geistes- und sozialwissenschaftlichen Kl., Jahrgang 1964, 2. Wiesbaden.

Pohle, W. 1971. "The Mathematical Foundations of Plato's Atomic Physics." *Isis* 62:36–46.

Pohlenz, M. 1937. "Hippokratesstudien." *Nachrichten von der Gesellschaft der Wissenschaften zu Göttingen*, phil.-hist. Kl., N.F. 2, 4, pp. 67–101.

Göttingen. (Reprinted in *Kleine Schriften*, vol. 2, ed. H. Dörrie [Hildesheim, 1965], pp. 175–209.)
———. 1938. *Hippokrates und die Begründung der wissenschaftlichen Medizin.* Berlin.
———. 1964. *Die Stoa.* 3rd ed. Göttingen.
Polanyi, M. 1958. *Personal Knowledge.* London.
———. 1967. *The Tacit Dimension.* London.
Pollak, K. 1968–69. *Wissen und Weisheit der alten Ärzte,* 2 vols. Düsseldorf.
Pollner, M. 1974. "Mundane Reasoning." *Philosophy of the Social Sciences* 4:35–54.
Popper, K. R. 1935/1968. *The Logic of Scientific Discovery.* Trans. K. R. Popper with J. and L. Freed of *Logik der Forschung* (1935). 2nd ed. London.
———. 1958–59/1963. "Back to the Presocratics." From *Proceedings of the Aristotelian Society* n.s. 59:1–24. In Popper 1963, pp. 136–65. (Also reprinted in Furley and Allen, edd., 1970, pp. 130–53.)
———. 1962. *The Open Society and its Enemies,* 2 vols. 4th ed. (1st ed. 1945). London.
———. 1963. *Conjectures and Refutations.* London.
———. 1965/1972. "Of Clouds and Clocks" (A. H. Compton Memorial Lecture, 21 April 1965, Washington University, St Louis, Missouri). In Popper 1972, pp. 206–55.
———. 1970. "Normal Science and Its Dangers." In Lakatos and Musgrave, edd., 1970, pp. 51–58.
———. 1972. *Objective Knowledge: An Evolutionary Approach.* Oxford.
Porkert, M. 1974. *The Theoretical Foundations of Chinese Medicine.* Cambridge, Massachusetts.
Porzig, W. 1934. "Wesenhafte Bedeutungsbeziehungen." *Beiträge zur Geschichte der deutschen Sprache und Literatur* 58:70–97.
Potter, P. 1976. "Herophilus of Chalcedon: An Assessment of His Place in the History of Anatomy." *Bulletin of the History of Medicine* 50:45–60.
Pra, M. dal. 1975. *Lo scetticismo greco.* Revised ed., 2 vols. Rome.
Préaux, C. 1966. "Sur la stagnation de la pensée scientifique à l'époque hellénistique." *American Studies in Papyrology* 1:235–50.
———. 1968. 'L'Elargissement de l'espace et du temps dans la pensée grecque." *Bulletin de la Classe des Lettres et des Sciences Morales et Politiques, Académie Royale de Belgique,* 5th ser. 54:208–67.
———. 1973. *La Lune dans la pensée grecque.* Académie Royale de Belgique, Mémoires de la Classe des Lettres, 2nd ser. 61, 4. Brussels.
Preisendanz, K. 1973–74. *Papyri Graecae Magicae.* 2nd ed., ed. A. Henrichs. Stuttgart.
Preiser, G. 1970. "Über die Sorgfaltspflicht der Ärzte von Kos." *Medizinhistorisches Journal* 5:1–9.
———. 1976. *Allgemeine Krankheitsbezeichnungen im Corpus Hippocraticum.* Ars Medica Abt. II, Bd. 5. Berlin.
Preus, A. 1975. *Science and Philosophy in Aristotle's Biological Works.* Hildesheim.

Price, D. J. de S. 1957. "Precision Instruments: To 1500." In *A History of Technology*, vol. 3, ed. C. Singer et al., pp. 582–619. Oxford.

———. 1964. "Automata and the Origins of Mechanism and Mechanistic Philosophy." *Technology and Culture* 5:9–23.

———. 1964–65. "The Babylonian 'Pythagorean Triangle' Tablet." *Centaurus* 10:1–13.

———. 1974. "Gears from the Greeks: The Antikythera Mechanism—A Calendar Computer from *ca.* 80 B.C." *Transactions of the American Philosophical Society* n.s. 64:7.

Prier, R. A. 1976. *Archaic Logic*. The Hague.

Pritchard, J. B. 1969. *Ancient Near Eastern Texts*. 3rd ed. (1st ed. 1955). Princeton.

Pritzl, K. 1983. "Aristotle and Happiness after Death: *Nicomachean Ethics* I 10–11." *Classical Philology* 78:101–11.

Pucci, P. 1971. "Lévi-Strauss and Classical Culture." *Arethusa* 4:103–17.

———. 1977. *Hesiod and the Language of Poetry*. Baltimore.

Putnam, H. 1975a. *Mathematics, Matter and Method: Philosophical Papers*, vol. 1. Cambridge.

———. 1975b. *Mind, Language and Reality: Philosophical Papers*, vol. 2. Cambridge.

———. 1981. *Reason, Truth and History*. Cambridge.

———. 1983. *Realism and Reason: Philosophical Papers*, vol. 3. Cambridge.

Quine, W. Van O. 1953/1961. *From a Logical Point of View*. 2nd ed. (1st ed. 1953). Cambridge, Massachusetts.

———. 1960. *Word and Object*. Cambridge, Massachusetts.

———. 1969. *Ontological Relativity and Other Essays*. New York.

———. 1970. "On the Reasons for Indeterminacy of Translation." *Journal of Philosophy* 67:178–83.

———. 1974. *The Roots of Reference*. La Salle, Illinois.

Raasted, J. 1979. "A Neglected Version of the Anecdote about Pythagoras's Hammer Experiment." *Cahiers de l'Institut du Moyen-Age Grec et Latin* 31a:1–9.

Rabel, R. J. 1981. "Diseases of Soul in Stoic Psychology." *Greek, Roman and Byzantine Studies* 22:385–93.

Radermacher, L. 1951. *Artium Scriptores*. Österreichische Akademie der Wissenschaften, phil.-hist. Kl., Sitzungsberichte 227, 3. Abhandlung, Vienna.

Radin, M. 1927. "Freedom of Speech in Ancient Athens." *American Journal of Philology* 48:215–30.

Rambaux, C. 1980. "La Logique de l'argumentation dans le *De Rerum Natura*, III, 830–1094." *Revue des Etudes Latines* 58:201–19.

Ramnoux, C. 1970. *Etudes présocratiques*. Paris.

Randall, J. H. 1960. *Aristotle*. New York.

Ranulf, S. 1924. *Der eleatische Satz vom Widerspruch*. Copenhagen.

Raphael, S. 1974. "Rhetoric, Dialectic and Syllogistic Argument: Aristotle's Position in 'Rhetoric' I–II." *Phronesis* 19:153–67.

Raven, J. E. 1948. *Pythagoreans and Eleatics*. Cambridge.

Rawlings, H. R. 1975. *A Semantic Study of PROPHASIS to 400 B.C.* Hermes Einzelschriften 33. Wiesbaden.

Rawlins, D. 1982a. "Eratosthenes' Geodesy Unraveled: Was There a High-Accuracy Hellenistic Astronomy?" *Isis* 73:259–65.

———. 1982b. "The Eratosthenes-Strabo Nile Map: Is It the Earliest Surviving Instance of Spherical Cartography? Did It Supply the 5000 Stades Arc for Eratosthenes' Experiment?" *Archive for History of Exact Sciences* 26:211–19.

Reale, G. 1967. *Il Concetto di filosofia prima e l'unità della metafisica di Aristotele*. 3rd ed. (1st ed. 1961). Milan.

Redard, G. 1953. *Recherches sur XPH, XPHΣΘAI: Etude sémantique*. Bibliothèque de l'Ecole des Hautes Etudes, Sciences Historiques et Philologiques 303. Paris.

Redfield, J. M. 1975. *Nature and Culture in the "Iliad."* Chicago.

Reesor, M. E. 1978. "Necessity and Fate in Stoic Philosophy." In Rist, ed., 1978, pp. 187–202.

Regenbogen, O. 1931. *Eine Forschungsmethode antiker Naturwissenschaft.* Quellen und Studien zur Geschichte der Mathematik, Astronomie und Physik B, 1, 2 (1930), pp. 131–82. Berlin. (Reprinted in *Kleine Schriften* [Munich, 1961], pp. 141–94.)

———. 1937. "Eine Polemik Theophrasts gegen Aristoteles." *Hermes* 72: 469–75. (Reprinted in *Kleine Schriften* [Munich, 1961], pp. 276–85.)

———. 1940. "Theophrastos." *Pauly-Wissowa Real-Encyclopädie der classischen Altertumswissenschaft*, Suppl. Bd. 7, cols. 1354–562.

Rehm, A. 1899. "Zu Hipparch und Eratosthenes." *Hermes* 34:251–79.

Rehm, A., and K. Vogel. 1933. *Exakte Wissenschaften*. 4th ed. Leipzig.

Reiche, H. A. T. 1971. "Myth and Magic in Cosmological Polemics: Plato, Aristotle, Lucretius." *Rheinisches Museum* N.F. 114:296–329.

Reidemeister, K. 1949. *Das exakte Denken der Griechen*. Hamburg.

Reiner, E. 1961. "The Etiological Myth of the 'Seven Sages.'" *Orientalia* n.s. 30:1–11.

Reinhardt, K. 1916. *Parmenides und die Geschichte der griechischen Philosophie*. Bonn.

———. 1926. *Kosmos und Sympathie*. Munich.

Reitzenstein, R., and H. H. Schaeder. 1926. *Studien zum antiken Synkretismus aus Iran und Griechenland*. Studien der Bibliothek Warburg 7. Berlin.

Renehan, R. 1981. "The Greek Anthropocentric View of Man." *Harvard Studies in Classical Philology* 85:239–59.

Renou, L., and L. Silburn. 1949a. "Sur la notion de *Bráhman*." *Journal Asiatique* 237:7–46.

———. 1949b. "Un Hymne à énigmes du Ṛgveda." *Journal de Psychologie Normale et Pathologique* 42:266–73.

Reymond, A. 1927. *History of the Sciences in Greco-Roman Antiquity*, trans. R. G. de Bray of 1st ed. of *Histoire des sciences exactes et naturelles dans l'antiquité gréco-romaine* (Paris, 1924). London.

Reynolds, L. D., and N. G. Wilson. 1968/1974. *Scribes and Scholars*. 2nd ed. (1st ed. 1968). Oxford.

Rhodes, P. J. 1972. *The Athenian Boule*. Oxford.

Richards, I. A. 1936. *The Philosophy of Rhetoric*. London.

Richardson, N. J. 1975. "Homeric Professors in the Age of the Sophists." *Proceedings of the Cambridge Philological Society* n.s. 21:65–81.

———. 1981. "The Contest of Homer and Hesiod and Alcidamas' *Mouseion*." *Classical Quarterly* n.s. 31:1–10.

Richardson, W. F. 1979. "Celsus on Medicine." *Prudentia* 11:69–93.

Ricoeur, P. 1959. "Le Symbole donne à penser." *Esprit* 27, 275, n.s. 7–8: 60–76.

———. 1975/1977. *The Rule of Metaphor*, trans. R. Czerny with K. McLaughlin and J. Costello of *La Métaphore vive* (Paris, 1975). Toronto.

Riddell, R. C. 1979. "Eudoxan Mathematics and the Eudoxan Spheres." *Archive for History of Exact Sciences* 20:1–19.

Riddle, J. M. 1971. "Dioscorides." In *Dictionary of Scientific Biography*, ed. C. C. Gillispie, vol. 4, pp. 119–23. New York.

———. 1985. *Dioscorides on Pharmacy and Medicine*. Austin, Texas.

Riondato, G. 1954. "ἱστορία ed ἐμπειρία nel pensiero aristotelico." *Giornale di Metafisica* 9:303–35.

Rist, J. M. 1965. "Some Aspects of Aristotelian Teleology." *Transactions and Proceedings of the American Philological Association* 96:337–49.

———. 1969. *Stoic Philosophy*. Cambridge.

———. 1978. "The Stoic Concept of Detachment." In Rist, ed., 1978, pp. 259–72.

Rist, J. M., ed. 1978. *The Stoics*. Berkeley.

Robb, K. 1970. "Greek Oral Memory and the Origins of Philosophy." *The Personalist* 51:5–45.

Robert, F. 1982. "Hippocrate, Platon, Aristote et les notions de genre et d'espèce." *History and Philosophy of the Life Sciences* 4:173–201.

Robin, L. 1928. *Greek Thought*, trans. M. R. Dobie. London.

Robinson, J. M. 1971. "Anaximander and the Problem of the Earth's Immobility." In Anton and Kustas, edd., 1971, pp. 111–18.

———. 1973. "On Gorgias." In Lee, Mourelatos, and Rorty, edd., 1973, pp. 49–60.

Robinson, R. 1936/1969. "Analysis in Greek Geometry." From *Mind* n.s. 45:464–73. In *Essays in Greek Philosophy*, pp. 1–15. Oxford.

———. 1953. *Plato's Earlier Dialectic*. 2nd ed. (1st ed. 1941). Oxford.

Roccasalvo, J. F. 1980. "Greek and Buddhist Wisdom: An Encounter between East and West." *International Philosophical Quarterly* 20:73–85.

Rodis-Lewis, G. 1975. "L'Articulation des thèmes du 'Phèdre.'" *Revue Philosophique de la France et de l'Etranger* 165:3–34.

Rösler, W. 1983. "Der Anfang der 'Katharmoi' des Empedokles." *Hermes* 111:170–79.

Rohde, E. 1925. *Psyche*, trans. W. B. Hillis. London.

Romano, F., ed. 1980. *Democrito e l'atomismo antico*. Atti del convegno internazionale, Catania, 18–21 April 1979, Siculorum Gymnasium 33, 1. Catania.

Romilly, J. de. 1956. *Histoire et raison chez Thucydide.* Paris.

———. 1971. *La Loi dans la pensée grecque des origines à Aristote.* Paris.

———. 1973. "Gorgias et le pouvoir de la poésie." *Journal of Hellenic Studies* 93:155–62.

———. 1975. *Magic and Rhetoric in Ancient Greece.* Cambridge, Massachusetts.

Rorty, R. 1973. "Genus as Matter." In Lee, Mourelatos, and Rorty, edd., 1973, pp. 393–420.

———. 1974. "Matter as Goo: Comments on Grene's Paper." *Synthese* 28: 71–77.

Rosaldo, M. Z. 1972. "Metaphors and Folk Classification." *Southwestern Journal of Anthropology* 28:83–99.

———. 1980. *Knowledge and Passion: Ilongot Notions of Self and Social Life.* Cambridge.

Roscher, W. H. 1904. *Die Sieben- und Neunzahl im Kultus und Mythus der Griechen.* Abhandlungen der philologisch-historischen Klasse der königlich sächsischen Gesellschaft der Wissenschaften, 24, 1. Leipzig.

———. 1906. *Die Hebdomadenlehren der griechischen Philosophen und Ärzte.* Abhandlungen der philologisch-historischen Klasse der königlich sächsischen Gesellschaft der Wissenschaften, 24, 6. Leipzig.

———. 1911. *Über Alter, Ursprung und Bedeutung der hippokratischen Schrift von der Siebenzahl.* Abhandlungen der philologisch-historischen Klasse der königlich sächsischen Gesellschaft der Wissenschaften 28, 5. Leipzig.

———. 1913. *Die hippokratische Schrift von der Siebenzahl in ihrer vierfachen Überlieferung.* Studien zur Geschichte und Kultur des Altertums 6. Paderborn.

Rosen, S. 1979–80. "Plato's Myth of the Reversed Cosmos." *Review of Metaphysics* 33:59–85.

Rosenmeyer, T. G. 1955. "Gorgias, Aeschylus and *Apate.*" *American Journal of Philology* 76:225–60.

———. 1960. "Plato's Hypothesis and the Upward Path." *American Journal of Philology* 81:393–407. (Reprinted in Anton and Kustas, edd., 1971, pp. 354–66.)

Ross, J. F. 1981. *Portraying Analogy.* Cambridge.

Ross, W. D. 1924/1953. *Aristotle, Metaphysics,* 2 vols. Revised ed. (1st ed. 1924). Oxford.

———. 1936. *Aristotle, Physics.* Oxford.

———. 1953. *Plato's Theory of Ideas.* 2nd ed. (1st ed. 1951). Oxford.

Ross, W. D., and F. H. Fobes. 1929. *Theophrastus Metaphysics.* Oxford.

Rossitto, C. 1977–78. "La possibilità di un' indagine scientifica sugli oggetti della dialettica nella *Metafisica* di Aristotele." *Atti dell' Istituto Veneto di Scienze, Lettere ed Arti, Classe di scienze morali, lettere ed arti* 136: 363–89.

Schmidt, O. 1975. "A System of Axioms for the Archimedean Theory of Equilibrium and Centre of Gravity." *Centaurus* 19:1–35.

Schöne, H. 1907. "Markellinos' Pulslehre. Ein griechisches Anekdoton." In *Festschrift zur 49 Versammlung deutscher Philologen und Schulmänner in Basel im Jahre 1907,* pp. 448–72. Basel.

Schöne, R., ed. 1897. *Damianos Schrift über Optik.* Berlin.

Schöner, E. 1964. *Das Viererschema in der antiken Humoralpathologie.* Sudhoffs Archiv Beihefte 4. Wiesbaden.

Schofield, M. 1978. "Aristotle on the Imagination." In Lloyd and Owen, edd., 1978, pp. 99–140.

———. 1980a. *An Essay on Anaxagoras.* Cambridge.

———. 1980b. "Preconception, Argument, and God." In Schofield, Burnyeat, and Barnes, edd., 1980, pp. 283–308.

Schofield, M., M. Burnyeat, and J. Barnes, edd. 1980. *Doubt and Dogmatism: Studies in Hellenistic Epistemology.* Oxford.

Schofield, M., and M. C. Nussbaum, edd. 1982. *Language and Logos: Studies . . . Presented to G. E. L. Owen.* Cambridge.

Schofield, M., and G. Striker, edd. 1986. *The Norms of Nature: Studies in Hellenistic Ethics.* Cambridge.

Scholz, H. 1928. "Warum haben die Griechen die Irrationalzahlen nicht aufgebaut?" *Kant-Studien* 33:35–72.

———. 1930/1975. "The Ancient Axiomatic Theory." Originally "Die Axiomatik der Alten," *Blätter für deutsche Philosophie* 4:259–78. In Barnes, Schofield, and Sorabji, edd., 1975, pp. 50–64.

Schon, D. A. 1963. *Displacement of Concepts.* London.

Schramm, M. 1962. *Die Bedeutung der Bewegungslehre des Aristoteles für seine beiden Lösungen der zenonischen paradoxie.* Philosophische Abhandlungen 19. Frankfurt.

Schreckenberg, H. 1964. *Ananke. Untersuchungen zur Geschichte des Wortgebrauchs.* Zetemata 36. Munich.

Schrijvers, P. H. 1980. "Die Traumtheorie des Lukrez." *Mnemosyne* 4th series, 33:128–51.

Schüssler, I. 1982. *Aristoteles Philosophie und Wissenschaft.* Frankfurt.

Schuhl, P. M. 1949. *Essai sur la formation de la pensée grecque.* 2nd ed. (1st ed. 1934). Paris.

Schultz, W. 1914. "Rätsel." *Pauly-Wissowa Real-Encyclopädie der classischen Altertumswissenschaft,* 2nd ser., 1 Halbband, 1, 1, cols 62–125.

Schulz, D. J. 1966. *Das Problem der Materie in Platons "Timaios."* Abhandlungen zur Philosophie, Psychologie und Pädagogik 31. Bonn.

Schumacher, J. 1963. *Antike Medizin.* 2nd ed. Berlin.

Searle, J. R. 1965. "What Is a Speech Act?" In *Philosophy in America,* ed. M. Black, pp. 221–39. Ithaca, New York. (Reprinted in Searle, ed., 1971, pp. 39–53.)

———. 1969. *Speech Acts.* Cambridge.

———. 1979. *Expression and Meaning.* Cambridge.

Searle, J. R., ed. 1971. *The Philosophy of Language.* Oxford.

Sedley, D. N. 1974. "The Structure of Epicurus' On Nature." *Cronache Ercolanesi* 4:89–92.

———. 1976a. "Epicurus and His Professional Rivals." *Cahiers de Philologie* 1:121–59.

———. 1976b. "Epicurus and the Mathematicians of Cyzicus." *Cronache Ercolanesi* 6:23–54.

———. 1977. "Diodorus Cronus and Hellenistic Philosophy." *Proceedings of the Cambridge Philological Society* n.s. 23:74–120.

———. 1980. "The Protagonists." In Schofield, Burnyeat, and Barnes, edd., 1980, pp. 1–20.

———. 1981. "The End of the Academy." *Phronesis* 26:67–75.

———. 1982a. "Two Conceptions of Vacuum." *Phronesis* 27:175–93.

———. 1982b. "On Signs." In Barnes, Brunschwig, Burnyeat, and Schofield, edd., 1982, pp. 239–72.

———. 1983a. "Epicurus' Refutation of Determinism." In ΣΥΖΗΤΗΣΙΣ: *Studi sull' epicureismo greco e romano offerti a M. Gigante*, pp. 11–51. Naples.

———. 1983b. "The Motivation of Greek Skepticism." In Burnyeat, ed., 1983, pp. 9–29.

———. Forthcoming. "Philoponus' Conception of Space." In *John Philoponus and the Rejection of Aristotelian Science*, ed. R. Sorabji. London.

Seeck, G. A. 1964. *Über die Elemente in der Kosmologie des Aristoteles*. Zetemata 34. Munich.

———. 1965. "*Nachträge" im achten Buch der Physik des Aristoteles*. Akademie der Wissenschaften und der Literatur, Mainz, Abhandlungen der geistes- und sozialwissenschaftlichen Kl., Jahrgang 1965, 3. Wiesbaden.

———. 1969. "Leicht-schwer und der Unbewegte Beweger. (*DC* IV 3 und *Phys*. VIII 4)." In Düring, ed., 1969, pp. 210–16. (Reprinted in Seeck, ed., 1975, pp. 391–99.)

Seeck, G. A., ed. 1975. *Die Naturphilosophie des Aristoteles*. Darmstadt.

Seeskin, K. R. 1976. "Platonism, Mysticism, and Madness." *The Monist* 59: 574–86.

Segal, C. P. 1962. "Gorgias and the Psychology of the Logos." *Harvard Studies in Classical Philology* 66:99–155.

———. 1978. "'The Myth was Saved': Reflections on Homer and the Mythology of Plato's *Republic*." *Hermes* 106:315–36.

Seide, R. 1981a. "Kontinuum und geometrischer Atomismus bei Demokrit." *Sudhoffs Archiv* 65:105–16.

———. 1981b. "Zum Problem des geometrischen Atomismus bei Demokrit." *Hermes* 109:265–80.

Seidenberg, A. 1960–62. "The Ritual Origin of Geometry." *Archive for History of Exact Sciences* 1:488–527.

———. 1974–75. "Did Euclid's *Elements*, Book I, Develop Geometry Axiomatically?" *Archive for History of Exact Sciences* 14:263–95.

———. 1977–78. "The Origin of Mathematics." *Archive for History of Exact Sciences* 18:301–42.

Striker, G. 1974. Κριτήριον τῆς ἀληθείας. *Nachrichten der Akademie der Wissenschaften in Göttingen,* phil.-hist. Kl. 1974, 2, pp. 47–110. Göttingen.

———. 1977. "Epicurus on the Truth of Sense-Impressions." *Archiv für Geschichte der Philosophie* 59:125–42.

———. 1980. "Sceptical Strategies." In Schofield, Burnyeat, and Barnes, edd., 1980, pp. 54–83.

———. 1981. "Über den Unterschied zwischen den Pyrrhoneern und den Akademikern." *Phronesis* 26:153–71.

———. 1983a. "The Role of *Oikeiosis* in Stoic Ethics." *Oxford Studies in Ancient Philosophy* 1:145–67.

———. 1983b. "The Ten Tropes of Aenesidemus." In Burnyeat, ed., 1983, pp. 95–115.

Strömberg, R. 1937. *Theophrastea. Studien zur botanischen Begriffsbildung.* Göteborg.

Strohm, H. 1983. "Beobachtungen zum vierten Buch der aristotelischen Meteorologie." In Moraux and Wiesner, edd., 1983, pp. 94–115.

Struve, W. W. 1930. *Mathematischer Papyrus des staatlichen Museums der schönen Künste in Moskau.* Quellen und Studien zur Geschichte der Mathematik, A, 1. Berlin.

Stückelberger, A. 1974. "Empirische Ansätze in der antiken Atomphysik." *Archiv für Kulturgeschichte* 56:124–40.

———. 1979. *Antike Atomphysik.* Munich.

Suppes, P. 1974. "Aristotle's Concept of Matter and Its Relation to Modern Concepts of Matter." *Synthese* 28:27–50.

———. 1981. "Limitations of the Axiomatic Method in Ancient Greek Mathematical Sciences." In Hintikka, Gruender, and Agazzi, edd., 1981, pp. 197–213.

Svenbro, J. 1976. *La Parole et le marbre.* Lund.

Swerdlow, N. 1969. "Hipparchus on the Distance of the Sun." *Centaurus* 14:287–305.

———. 1979–80. "Hipparchus' Determination of the Length of the Tropical Year and the Rate of Precession." *Archive for History of Exact Sciences* 21:291–309.

Szabó, Á. 1960–62. "Anfänge des euklidischen Axiomensystems." *Archive for History of Exact Sciences* 1:37–106.

———. 1964–66. "The Transformation of Mathematics into Deductive Science and the Beginnings of Its Foundation on Definitions and Axioms." *Scripta Mathematica* 27:27–48a, 113–39.

———. 1969/1978. *The Beginnings of Greek Mathematics.* Trans. A. M. Ungar of *Anfänge der griechischen Mathematik* (Vienna, 1969). Budapest.

Szasz, T. S. 1962/1974. *The Myth of Mental Illness.* Revised ed. (1st ed. 1962). London.

Taisbak, C. M. 1973–74. "Posidonius Vindicated at All Costs? Modern Scholarship versus the Stoic Earth Measurer." *Centaurus* 18:253–69.

———. 1984. "Eleven Eighty-Thirds: Ptolemy's Reference to Eratosthenes in *Almagest* I 12." *Centaurus* 27:165–67.

Tamba-Mecz, I., and P. Veyne. 1979. "*Metaphora* et comparaison selon Aristote." *Revue des Etudes Grecques* 92:77–98.
Tambiah, S. J. 1968. "The Magical Power of Words." *Man* n.s. 3:175–208.
———. 1973. "Form and Meaning of Magical Acts: A Point of View." In Horton and Finnegan, edd., 1973, pp. 199–229.
———. 1977. "The Galactic Polity." *Annals of the New York Academy of Sciences* 293:69–97.
———. 1982. "Famous Buddha Images and the Legitimation of Kings." *Res* 4:5–19.
———. 1984. *The Buddhist Saints of the Forest and the Cult of Amulets.* Cambridge.
Tannery, P. 1887. *La Géométrie grecque.* Paris.
———. 1893. *Recherches sur l'histoire de l'astronomie ancienne.* Paris.
———. 1912–43. *Mémoires scientifiques,* 16 vols. Paris.
———. 1930. *Pour l'histoire de la science Hellène,* 2nd ed. Paris.
Tarán, L. 1965. *Parmenides.* Princeton.
———. 1978. "Speusippus and Aristotle on Homonymy and Synonymy." *Hermes* 106:73–99.
Tasch, P. 1947–48. "Quantitative Measurements and the Greek Atomists." *Isis* 38:185–89.
Taylor, A. E. 1928. *A Commentary on Plato's Timaeus.* Oxford.
Taylor, C. C. W. 1967. "Plato and the Mathematicians: An Examination of Professor Hare's Views." *Philosophical Quarterly* 17:193–203.
———. 1969. "Forms as Causes in the *Phaedo.*" *Mind* 78:45–59.
———. 1980. "All Perceptions are True." In Schofield, Burnyeat, and Barnes, edd., 1980, pp. 105–24.
Taylor, F. K. 1979. *The Concepts of Illness, Disease and Morbus.* Cambridge.
Temkin, O. 1935. "Celsus' 'On Medicine' and the Ancient Medical Sects." *Bulletin of the Institute of the History of Medicine* 3:249–64.
———. 1955. "Medicine and Greco-Arabic Alchemy." *Bulletin of the History of Medicine* 29:134–53.
———. 1956. *Soranus' Gynecology.* Baltimore.
———. 1961. "A Galenic Model for Quantitative Physiological Reasoning?" *Bulletin of the History of Medicine* 35:470–75.
———. 1973. *Galenism: The Rise and Decline of a Medical Philosophy.* Ithaca, New York.
Thayer, H. S. 1979. "Aristotle on the Meaning of Science." *Philosophical Inquiry* 1:87–104.
Theiler, W. 1924. *Zur Geschichte der teleologischen Naturbetrachtung bis auf Aristoteles.* Zurich.
———. 1967. "Historie und Weisheit." In *Festgabe H. von Greyerz,* ed. E. Walder et al., pp. 69–81. Bern. (Reprinted in Theiler 1970, pp. 447–59.)
———. 1970. *Untersuchungen zur antiken Literatur.* Berlin.
Thivel, A. 1975. "Le 'Divin' dans la *collection hippocratique.*" In Bourgey and Jouanna, edd., 1975, pp. 57–76.
———. 1981. *Cnide et Cos.* Paris.

Thomas, K. 1971. *Religion and the Decline of Magic.* London.
———. 1975–76. "An Anthropology of Religion and Magic, II." *Journal of Interdisciplinary History* 6:91–109.
Thompson, D'A. W. 1913. *On Aristotle as a Biologist.* Oxford.
———. 1936. *A Glossary of Greek Birds.* 2nd ed. (1st ed. 1895). Oxford.
———. 1940. "Aristotle the Naturalist." In *Science and the Classics,* p. 37–78. London.
———. 1946. *A Glossary of Greek Fishes.* London.
Thompson, R. Campbell. 1923–24. "Assyrian Medical Texts." *Proceedings of the Royal Society of Medicine* 17, Section of the History of Medicine:1–34.
———. 1925–26. "Assyrian Medical Texts." *Proceedings of the Royal Society of Medicine* 19, Section of the History of Medicine: 29–78.
Thomson, G. 1946. *Aeschylus and Athens.* 2nd ed. (1st ed. 1941). London.
———. 1954. *Studies in Ancient Greek Society.* I: *The Prehistoric Aegean.* 2nd ed. (1st ed. 1949). London.
———. 1955. *Studies in Ancient Greek Society.* II: *The First Philosophers.* London.
Thorndike, L. 1923–58. *A History of Magic and Experimental Science,* 8 vols. New York.
Tigner, S. 1974. "Empedocles' Twirled Ladle and the Vortex-Supported Earth." *Isis* 65:433–47.
Todd, R. B. 1980. "Some Concepts in Physical Theory in John Philoponus' Aristotelian Commentaries." *Archiv für Begriffsgeschichte* 24:151–70.
Todorov, T. 1970. "Synecdoques." *Communications* 16:26–35.
———. 1973. "Le Discours de la magie." *L'Homme* 13, 4:38–65.
———. 1973/1983. *Symbolism and Interpretation,* trans. C. Porter of *Symbolisme et interprétation* (Paris, 1973). London.
———. 1977/1982. *Theories of the Symbol,* trans. C. Porter of *Théories du symbole* (Paris, 1977). Oxford.
———. 1978. *Les Genres du discours.* Paris.
Toeplitz, O. 1931. *Das Verhältnis von Mathematik und Ideenlehre bei Plato.* Quellen und Studien zur Geschichte der Mathematik, Astronomie und Physik B, 1, 1 (1929), pp. 3–33. Berlin.
Toomer, G. J. 1967–68. "The Size of the Lunar Epicycle according to Hipparchus." *Centaurus* 12:145–50.
———. 1973–74. "The Chord Table of Hipparchus and the Early History of Greek Trigonometry." *Centaurus* 18:6–28.
———. 1974–75. "Hipparchus on the Distances of the Sun and Moon." *Archive for History of Exact Sciences* 14:126–42.
———. 1975. "Ptolemy." In *Dictionary of Scientific Biography,* ed. C. C. Gillispie, vol. 11, pp. 186–206. New York.
———. 1976. *Diocles on Burning Mirrors.* Berlin.
———. 1980. "Hipparchus' Empirical Basis for His Lunar Mean Motions." *Centaurus* 24:97–109.
———. 1984. *Ptolemy's Almagest.* London.

Tóth, I. 1966–67. "Das Parallelenproblem im Corpus Aristotelicum." *Archive for History of Exact Sciences* 3:249–422.

———. 1977. "Geometria More Ethico." In ΠΡΙΣΜΑΤΑ: *Festschrift W. Hartner*, ed. Y. Maeyama and W. G. Saltzer, pp. 395–415. Wiesbaden.

Tracy, T. 1969. *Physiological Theory and the Doctrine of the Mean in Plato and Aristotle.* Studies in Philosophy 17. The Hague.

Trapp, H. 1967. *Die hippokratische Schrift DE NATURA MULIEBRI. Ausgabe und textkritischer Kommentar.* Hamburg.

Turner, E. G. 1951. "Athenian Books in the Fifth and Fourth Centuries B.C." Inaugural Lecture, University College, London.

Turner, T. S. 1977. "Narrative Structure and Mythopoesis: A Critique and Reformulation of Structuralist Concepts of Myth, Narrative and Poetics." *Arethusa* 10:103–63.

———. 1980. "Le Dénicheur d'oiseaux en contexte." *Anthropologies et sociétés* 4:85–115.

Turner, V. W. 1964. "An Ndembu Doctor in Practice." In *Magic, Faith, and Healing*, ed. A. Kiev, pp. 230–63. London.

———. 1970. *The Forest of Symbols.* Ithaca, New York.

———. 1974. *Dramas, Fields and Metaphors.* Ithaca, New York.

Turrini, G. 1977. "Contributo all'analisi del termine ΕΙΚΟΣ. I: L'età arcaica." *Acme* 30:541–58.

———. 1979. "Contributo all'analisi del termine ΕΙΚΟΣ. II: Linguaggio, verosimiglianza e immagine in Platone." *Acme* 32:299–323.

Ullmann, S. 1962. *Semantics: An Introduction to the Science of Meaning.* Oxford.

Ulmer, K. 1953. *Wahrheit, Kunst und Natur bei Aristoteles.* Tübingen.

Unguru, S. 1975–76. "On the Need to Rewrite the History of Greek Mathematics." *Archive for History of Exact Sciences* 15:67–114.

Usener, H. 1896. *Götternamen.* Bonn.

Vallance, J. T. 1986. "The Physiology and Pathology of Asclepiades of Bithynia." Ph.D. diss. Cambridge.

———. Forthcoming. "Theophrastus and the Study of the Intractable." In *Theophrastus On Science*, ed. W. W. Fortenbaugh. New Brunswick.

Van Fraassen, B. C. 1980a. *The Scientific Image.* Oxford.

———. 1980b. "A Re-Examination of Aristotle's Philosophy of Science." *Dialogue* 19:20–45.

Van Riet, G. 1960. "Mythe et vérité." *Revue Philosophique de Louvain* 58: 15–87.

Vansina, J. 1961/1965. *Oral Tradition.* Trans. H. M. Wright of *De la tradition orale* (Tervuren, 1961). London.

———. 1971. "Once upon a Time: Oral Traditions as History in Africa." *Daedalus* 100:442–68.

Vegetti, M. 1973. "Nascita dello scienzato." *Belfragor* 28:641–63.

———. 1979. *Il Coltello e lo Stilo.* Milan.

———. 1981. "Modelli di medicina in Galeno." In Nutton, ed., 1981, pp. 47–63. (Reprinted in Vegetti 1983, pp. 113–37.)

Vogel, C. J. de. 1966. *Pythagoras and Early Pythagoreanism.* Assen.

Vogt, H. 1925. "Versuch einer Wiederherstellung von Hipparchs Fixstern-verzeichnis." *Astronomische Nachrichten* 224:17–54.

Vries, G. J. de. 1949. *Spel bij Plato.* Amsterdam.

Vygotsky, L. S. 1962. *Thought and Language,* trans. E. Hanfmann and G. Vakar. Cambridge, Massachusetts.

Wächter, T. 1910. *Reinheitsvorschriften im griechischen Kult.* Religionsge-schichtliche Versuche und Vorarbeiten 9, 1. Giessen.

Waerden, B. L. van der. 1940–41. "Zenon und die Grundlagenkrise der griech-ischen Mathematik." *Mathematische Annalen* 117:141–61.

———. 1954/1961. *Science Awakening,* trans. A. Dresden of *Ontwakende Wetenschap.* 2nd ed. (1st ed. 1954). New York.

———. 1974. *Science Awakening II: The Birth of Astronomy.* Leiden.

———. 1977–78. "Die Postulate und Konstruktionen in der frühgriechischen Geometrie." *Archive for History of Exact Sciences* 18:343–57.

———. 1980a. "On Pre-Babylonian Mathematics I." *Archive for History of Exact Sciences* 23:1–25.

———. 1980b. "On Pre-Babylonian Mathematics II." *Archive for History of Exact Sciences* 23:27–46.

Waismann, F. 1945/1951. "Verifiability." From *Proceedings of the Aristotelian Society* Suppl. 19:119–50. In *Logic and Language* (first series), ed. A. Flew, pp. 117–44. Oxford.

———. 1953. "Language Strata." In *Logic and Language* (second series), ed. A. Flew, pp. 11–31. Oxford.

———. 1965. *The Principles of Linguistic Philosophy,* ed. R. Harré. London.

Wankel, H. 1983. "'Alle Menschen müssen sterben': Variationen eines Topos der griechischen Literatur." *Hermes* 111:129–54.

Wardy, R. B. B. 1985. "A Study of Physics VII." Ph.D. diss. Cambridge.

———. Forthcoming. "Eleatic Pluralism." *Archiv für Geschichte der Philosophie.*

Wartofsky, M. W. 1968. *Conceptual Foundations of Scientific Thought.* London.

Waschkies, H.-J. 1970–71. "Eine neue Hypothese zur Entdeckung der inkom-mensurablen Grössen durch die Griechen." *Archive for History of Exact Sciences* 7:325–53.

———. 1977. *Von Eudoxos zu Aristoteles.* Studien zur antiken Philosophie 8. Amsterdam.

Wasserstein, A. 1962. "Greek Scientific Thought." *Proceedings of the Cam-bridge Philological Society* n.s. 8:51–63.

———. 1972. "Le Rôle des hypothèses dans la médecine grecque." *Revue Phi-losophique de la France et de l'Etranger* 162:3–14.

———. 1978. "Epicurean Science." *Hermes* 106:484–94.

Waterlow, S. 1982a. *Nature, Change, and Agency in Aristotle's Physics.* Oxford.

———. 1982b. *Passage and Possibility: A Study of Aristotle's Modal Con-cepts.* Oxford.

Wear, A. 1981. "Galen in the Renaissance." In Nutton, ed., 1981, pp. 229–62.

Wedberg, A. 1955. *Plato's Philosophy of Mathematics.* Stockholm.

Wehrli, F. 1967–78. *Die Schule des Aristoteles,* 2nd ed., 10 vols., 2 suppl. vols. Basel.

———. 1982. "Die aristotelische Anthropologie zwischen Platonismus und Sophistik." *Museum Helveticum* 39:179–205.

Weidlich, T. 1894. *Die Sympathie in der antiken Literatur.* Stuttgart.

Weil, E. 1951/1975. "The Place of Logic in Aristotle's Thought." Originally "La Place de la logique dans la pensée aristotélicienne." *Revue de métaphysique et de morale* 56:283–315. In Barnes, Schofield, and Sorabji, edd., 1975, pp. 88–112.

Weinreich, O. 1909. *Antike Heilungswunder.* Religionsgeschichtliche Versuche und Vorarbeiten 8, 1. Giessen.

Weiss, H. 1942. *Kausalität und Zufall in der Philosophie des Aristoteles.* Basel.

Wellmann, M. 1895. *Die pneumatische Schule.* Philologische Untersuchungen 14. Berlin.

———. 1901. *Die Fragmente der sikelischen Ärzte Akron, Philistion und des Diokles von Karystos.* Berlin.

Welskopf, E. C. 1974. "Sophisten." In Welskopf, ed., 1974, vol. 4, pp. 1927–84.

Welskopf, E. C., ed. 1974. *Hellenische Poleis,* 4 vols. Berlin.

West, M. L. 1967. "The Contest of Homer and Hesiod." *Classical Quarterly* n.s. 17:433–50.

———. 1971. *Early Greek Philosophy and the Orient.* Oxford.

Westermann, A. 1839. ΠΑΡΑΔΟΞΟΓΡΑΦΟΙ *Scriptores rerum mirabilium Graeci.* Braunschweig.

Wheelwright, P. 1962. *Metaphor and Reality.* Bloomington, Indiana.

White, F. C. 1975. "Plato on Geometry." *Apeiron* 9, 2:5–14.

White, M. J. 1975. "Genus as Matter in Aristotle?" *Studi Internazionali di Filosofia* 7:41–56.

White, R. J. 1975. *The Interpretation of Dreams, Oneirocritica by Artemidorus.* Park Ridge, New Jersey.

Wieland, W. 1960–61/1975. "Aristotle's Physics and the Problem of Inquiry into Principles." Originally "Das Problem der Prinzipienforschung und die aristotelische Physik," *Kant-Studien* 52:206–19. In Barnes, Schofield, and Sorabji, edd., 1975, pp. 127–40.

———. 1962/1970. *Die aristotelische Physik.* 2nd ed. (1st ed. 1962). Göttingen.

———. 1962/1975. "The Problem of Teleology." Originally "Zum Teleologieproblem," chap. 16 of Wieland 1962/1970. In Barnes, Schofield, and Sorabji, edd., 1975, pp. 141–60.

———. 1972. "Zeitliche Kausalstrukturen in der aristotelischen Logik." *Archiv für Geschichte der Philosophie* 54:229–37.

———. 1982. *Platon und die Formen des Wissens.* Göttingen.

Wiesner, J. 1978. "The Unity of the Treatise *De Somno* and the Physiological Explanation of Sleep in Aristotle." In Lloyd and Owen, edd., 1978, pp. 241–80.

General Index

abstraction, 243, 293, 304, 319
acceleration, 219n16, 222–23
Aeschines, 72n83
aether, 113, 117n38, 194n85, 240n98, 314n98
air, 16, 95n159, 113, 118n43, 152, 192–93, 197, 218n10, 226–27, 256n144, 314
alchemy, 79n106, 253n132
Alexander of Aphrodisias, 223n27
alphabet, 71, 74
ambiguity, 173n4, 184n38, 190, 195, 198
Amenemhet, 57n25
analogy, 179, 181–82, 186–89, 207, 213, 227, 262, 283, 301n59, 323; proportional, 198–99, 201, 203
anatomy, 46n160, 106n202, 117, 121n57, 124, 159, 164–67, 206–7, 324–27, 334–35
Anaxagoras, 92n152, 101–2, 113, 152, 179–80, 271n193, 299n52
Anaximander, 52n7, 288–89
Anaximenes, 249n120, 288
anthropomorphism, 176–79, 209
Antiphon (orator), 60n39, 92–93n152
Antiphon (sophist), 31n96, 77n103, 92–93n152
ape, 325
Apollo, 12, 17–18, 43n143
Apollonius, 40n125, 304n73, 313n95
appearances, 163, 168, 169n239, 242n100. *See also* phenomena, saving the
approximations, 239, 292, 315
Archigenes, 283n229

Archilochus, 58, 86n130, 331n148
Archimedes, 47, 52, 77, 78n104, 147n156, 216, 238nn.86, 89, 249–50, 292n23, 301–3, 308, 311, 335
Archytas, 52, 76n98, 78n103, 243n107, 294n35
Aristarchus, 308–13
Aristides, Aelius, 90n144
Ariston, 170n240
Aristophanes, 49n163, 58n30, 92n152, 280n218, 333
Aristotle, 46, 54, 81–82, 85n123, 90n145, 98, 104–5, 154–55, 160n208, 168n236, 205nn.120–24, 206n127, 227, 230n53, 231, 244n110, 250n121, 257nn.150–51, 272, 275–76, 287–89, 299n53, 300n54, 313n95, 320n111; on death, 8; on demonstration, 141, 143–47, 184, 200, 210, 213–14; dialectic, 202, 210, 214; on dreams, 32–34, 37; element theory, 151n169, 194–97, 229–30, 247n117; logic, 102n186, 140–48, 182, 184, 190–92, 198–203, 210–14; mathematics, 291–93; on metaphor, 173–74, 183–87, 209–14; movement, theory of, 157n198, 191–94, 217–25, 289, 303; on nature, 187–90; physics, 140, 143–47, 191–97, 291–93; teleology, 48n161, 188–89, 321–24, 328; zoology, 46, 143, 196–97, 206n125, 322–25
Aristoxenus, 77n102, 206n126, 243n102, 244, 273, 296–98, 329n145
Aristyllus, 235

Euclid, 76, 146–47, 244–45, 298, 300–301, 307
Eudemus, 293n27
Eudoxus, 40n125, 76n98, 78nn.103–4, 235, 304, 312, 313n95
Euripides, 47n161, 58n30, 331, 333n159
evidence, 80n109, 118–19, 123, 263–64, 335
exactness, 41–42, 45–46, 210, 215–84. *See also* inexactness
experience, 68–69, 111, 127, 160, 163, 202n110
experiment, thought, 192, 246
explanation, 28, 32–33, 138–39n126, 142–46, 149, 156, 168–69n239, 178, 285–91, 322

fate, 4, 38, 175n9, 176
fees, 92–93n152, 96
Feyerabend, P., 98
Finnegan, R., 52
fire, 16, 95n159, 113, 117n39, 118n43, 137n121, 150–51, 192–93, 197, 203, 217–18, 226–30, 289n12
forms: Platonic, 44n149, 137, 139n126, 170n240, 184, 271n194, 291; in Aristotle, 46n160, 188, 196n91, 199, 322n115
free speech, 81, 102
Freud, S., 32

Galen, 29n89, 30–31, 42, 46, 48nn.161–62, 69n75, 89n143, 100n179, 105–6, 131n107, 147, 148n159, 159, 161, 165–67, 212, 213n143, 228, 251n127, 252, 254n136, 256–57, 270nn.189, 191, 280, 282–84, 324–27, 331n151, 332n155, 334–35
Galileo, 215–16, 226
games, Pan-Hellenic, 90–91, 96–98
Geminus, 76n101, 207n131
genethlialogy, 44, 46. *See also* horoscopes
geocentricity, 193–94, 288–89, 304–5, 308
geophysics, 231–34, 273
gods, 8–9, 12, 17–18, 23n73, 27–28, 30–31, 33–35, 37, 48–49, 51, 57n26, 58n27, 60–61, 88, 90n144, 107, 113, 137, 140, 169n239, 176–79, 181n29, 209, 287, 299n52, 320

Gomperz, T., 95
good, 46, 137, 140, 150, 188, 199n98, 242nn.100–101, 258, 276, 290–91, 320–22, 327–29
Goody, J., 52, 71–74, 106
Gorgias, 91, 93nn.152, 153, 96–97, 101n184, 103–4
gravity, specific, 250

harmonics, 77n102, 206n126, 241–44, 273, 274n203, 276–77, 280, 293n24, 294n35, 295–98, 329
Harrison, J., 1
Harvey, W., 256
Havelock, E. A., 71
heart, 166, 256, 323, 326–27, 335
heavy/light, 191–94, 198, 217–19, 224, 247–50
Hecataeus, 59n32, 61
Heisenberg, W., 136
heliocentricity, 308, 314
Hempel, C. G., 227
Heraclides, 304n73
Heraclitus, 32n99, 43n143, 47n161, 59–61, 86, 179–80, 271n193
Herodicus, 19, 93n153
Herodotus, 23–24n73, 47n161, 54, 56, 59n32, 101n182, 124n67, 280n218
heroes, 8, 35, 51
Hero of Alexandria, 227–28, 234n68, 253n130, 299n52
Herophilus, 31–32, 158, 160n208, 162n218, 163, 207n128, 212, 282–84
Hesiod, 8–9, 45, 58, 60–61, 85, 92n152, 93n153, 258
Hipparchus, 40n125, 43, 106, 223n27, 232n63, 235–40, 273, 310n86, 317, 331n151
Hippasus, 78n106, 277n212, 296n40
Hippias, 77–78n103, 89n143, 91–96, 101n184
Hippocrates of Chios, 75n96, 76, 77–78n103
Hippocrates of Cos, 33n100, 92n152, 105–7, 158, 270n191
Hippocratic writers, 8, 13–32, 34–37, 39–42, 61–69, 71, 78n106, 88–90, 93–100, 104, 114–35, 143, 168, 203–5, 248–49, 250–55, 257, 259–70, 290
Hippodamus, 82n114

history, 59n32, 97, 124n67, 330n147
hodometer, 234n68
Homer, 7, 12, 18n56, 40, 58, 60,
    85n124, 86, 92n152, 93n153, 110n7,
    177–78, 184n38
homonymy, 173n4, 185n44, 198–201
horoscopes, 46, 280
hot/cold, 8, 14–15, 27–28, 99n175,
    117n36, 120, 138n126, 159, 162,
    190n66, 194–98, 202n107, 203, 214,
    221–22, 226–30, 247–48, 280
Huizinga, J., 96, 98
humans, 196–97, 320, 322–23, 325,
    328, 332
humours, 14, 26–30, 95n159, 99, 117–
    22, 129, 204–5, 270n190, 327
hydrostatics, 249–50, 301–3
hymn, 48n162, 98n173, 334
hypotheses, 8, 15, 39, 66, 136n118, 144,
    168, 244n109, 293n27, 294n35,
    309–12

Iamblichus, 276n209, 334n160
images, 179, 181–84, 187, 201, 209,
    213
Imhotep, 55n21, 57n25
impiety, 98n173, 102n187, 140,
    333n157
incommensurability, 75, 78n106,
    302n66
incontrovertibility, 146, 240n98
incubation, 30
indemonstrables, 144, 146–47
India, 87–88
induction, 143n143, 186, 202n110,
    272n198
inexactness, 41–42, 124, 128–31, 134–
    35, 162, 168, 253–54, 269n187, 274,
    281. *See also* exactness
infallibility, 134, 168n239
instruments, 215, 225–26, 228, 274,
    281, 299n52; astronomical, 231–32,
    234n68, 236–39, 281, 299n52,
    311n89; surgical, 69n76
inventions, 51–52, 57n25, 64, 69n76
Ion, 93n153
irrational, 1–2, 4

Joly, R., 215

kairos, 28, 118n40, 129–30
Koyré, A., 215–16, 226, 230, 234–35,

257, 271
Kuhn, T. S., 50, 107–8, 170, 215
language and reality, problem of, 62, 179,
    209
law, 79–82, 87, 94n157, 179, 212
laypersons, 66, 78n106, 87–88, 95,
    100n177, 131–32, 134, 290n14
Leucippus, 191n69, 226n40
Lévi-Strauss, C., 51, 110
light, ray of, 292n21, 299
lightning, 49n163, 169n239, 333
likeness, 182–86, 201, 208–9
literacy, 70–78, 80, 101, 105–6
lochia, 259, 262
logic, 102n186, 146, 148, 184, 210
logos, 3–6, 10–11, 59–60, 89, 181,
    196n91, 213, 229n51, 276n210, 291
love, 179–81
Lucian, 105n200
Lucretius, 169n239, 227n40
Lyceum, 48n161, 148, 155, 157n195, 222
Lysias, 60n39, 96n164, 212

madness, 4, 7, 11, 21–28
magic, 28, 42n141, 78
male/female, 205n121, 258–63, 275,
    325
Marduk, 287, 290n13
mathematics, 43, 45–46, 48n161, 53–
    54, 57, 73–78, 86n127, 92–93n152,
    93n153, 94n154, 98, 108, 143–48,
    155, 169, 182n33, 199, 210, 215,
    227, 231–33, 240–47, 258, 276–77,
    279–81, 284–85, 291–93, 295–98,
    300–303, 311–15, 320n111, 330–
    31, 334n160
meaning, 172–73, 179–80, 190–94,
    209; focal, 198–201, 203, 214
measurement, 138n126, 193, 215–284
mechanical devices, 19, 64–65, 68–70
mechanics, 157, 293n24
Melissus, 93n153, 118n43, 271n193
Menaechmus, 76n98, 78n103
Menelaus, 236n80
Mesopotamia, 111, 177. *See also*
    Babylonians
metaphora, 174n6, 176, 184, 186–87,
    200, 214
meteorology, 15n46, 94n154, 144, 147,
    168n239
Methodist medicine, 25n80, 158–59,
    161–65

mind, 178–79
miscarriages, 262–63
Mithridates, 252n129
models, 187, 189, 208, 318–19; astronomical, 40, 44, 106n206, 236–37, 239–40, 279n217, 285–86, 295n36, 304–6, 312–19; in Plato, 137–40, 184
moon, 36n117, 106n206, 169n239, 236n77, 237, 239, 309–12, 315–18
morality, 17–18, 38, 46, 81, 89, 93n152, 155, 169n239, 177, 230n53, 276n210, 281n222, 328, 336
movement, 151, 191, 193, 194n85, 213, 217–25, 247n117, 303; of light, 299–301
Muses, 58–59
music, 54n16, 58–59n31, 77n102, 89n143, 90, 93n153, 94n154, 138n126, 155, 206, 242n101, 260, 276–78, 283–84, 329n145. *See also* harmonics
mystery religions, 9n25, 334
mystification, 3, 13, 28, 99, 215–84, 336
myth, 4–6, 8–11, 47, 51, 53n13, 85n123, 135–36, 172, 181, 208–10, 213, 286–87, 289–90, 328, 336

naturalistic accounts, 8–9, 11–18, 21–30, 33–37, 47–49, 287, 290, 333
nature, concept of, 13–14, 46–47, 187–90, 322–24
necessity, 118–23, 137, 142–44, 153, 154, 156, 169, 263, 320–22, 329
nerves, 165n228, 166–67, 207, 212–13, 335
Newton, I., 194, 221n22
Nicomachus, 276n209
number, 128n89, 138n126, 216, 225, 241–43, 254, 257–68, 275–78, 289n12, 298. *See also* counting

odd/even, 258, 260, 265–67, 269n187, 270n191, 276
old age, 9, 195, 322n117
oligarchy, 79
Olympia, 91, 96, 101
Oppenheim, A. L., 72–73
optics, 241, 244–47, 273–75, 293n24, 299–301
oracles, 43n143, 83–86, 110n5
orality, 52–53, 101

order, 137–38, 140, 244n109, 276n210, 328–29, 333, 335
organs, 189–90, 208
Oribasius, 332n152
Owen, G. E. L., 191, 219

painting, 54n16, 66, 97, 241n100, 299n52
papyri: Ebers, 111, 250–51n126; Leyden, 253n132; Petrie, 251n126; Rhind, 53, 75n95; Edwin Smith, 55, 63
paradigm, 107, 277; in Aristotle, 186; in Plato, 183. *See also* models
paradox, 179, 305–6, 315
parallax, 307–12
Parmenides, 59–61, 92n152, 93n153, 179–80, 191n69, 271n193
pathology, 14–16, 116–24, 130, 147, 162, 205, 324, 327–28
Paul of Aegina, 332n152
perception, 150–53, 230n53, 241, 244n109, 271–73, 296
Periander, 84
Pericles, 83n117, 94n153
periods, 254n136, 257–70, 280
perspective, 299n52, 300
Petron, 15n45
pharmacology, 228n47, 250–54, 331n150. *See also* drugs
phenomena, saving the, 244n109, 271–72, 285, 293–319
Philinus, 158
Philistion, 15n45, 132n109
Philolaus, 15n45, 132n109
Philo of Byzantium, 227–28
Philoponus, 194, 223–25, 229n48, 303
phlegm, 14, 26–28, 28–29n89, 95n159, 99, 117nn.35–36, 118n43, 119–20, 129, 204–5, 327
physicians, public, 103n191, 104
physics, 8–9, 88, 101, 136, 140–41, 143, 147, 150–51, 154–56, 168–69n239, 170, 191–200, 210, 214–15, 240–41n98, 247, 279n215, 285–86, 291–93, 299, 304n73, 312–15
physiognomy, 29n89
physiology, 15, 28–30n89, 95, 116–19, 122–24, 130, 134, 147, 159, 204–5, 247, 254–57, 324–27
Pindar, 33n99, 47n161, 58n30, 209–210,

331n148
Pittacus, 83n117, 84, 86n127
planets, 45, 139, 145n152, 207–8,
   225n35, 236–37, 289, 293, 304–6,
   314, 332
plants, 151–54, 190n66, 324n123
Plato, 5, 8–11, 21, 46–49, 54, 65n59,
   83n117, 89n143, 91–94, 97, 103–5,
   107, 132, 135–41, 149n161,
   154n179, 181–84, 187–88, 192–93,
   209–11, 214n144, 227nn.40, 42,
   230n53, 241–44, 271–73, 279–80,
   290–94, 320–22, 328–29, 333
Pliny, 252n129, 331
Plutarch, 331
pneuma, 162, 190n66, 213n143
poetry, 21, 54n16, 58, 83, 86, 89–93,
   99n176, 175–76, 183–85, 209–12,
   214, 330n147
Polemarchus of Cyzicus, 304n73, 305
politics, 30n89, 53–54, 78–83, 84n121,
   93n152, 98, 122–23, 276n210
Polybus, 15n45, 94n156
Porphyry, 107n206, 244, 273, 296, 298
Posidonius, 76n101, 232–34, 273
potentiality/actuality, 195n87, 199,
   228n48, 299
Praxagoras, 282
prayer, 35, 128n87
precession, 237, 239n94, 331n151, 335
prediction, 32, 37–44, 206, 240, 285–
   86, 315
Proclus, 44n149, 76nn.98, 101,
   170n240, 295n36, 322n114,
   329n146
Prodicus, 48n163, 89n143, 92–93n152,
   96n163
prognosis, 39–41, 43–46, 48
proof. *See* demonstration
prophecy, 21, 34, 40–41, 43n143, 48–
   49, 334–35. *See also* divination
proportions, 119–20, 130, 138n126,
   139, 185–86, 229, 250–56, 276, 303,
   311
Protagoras, 91–95, 113n21, 230n53
psychology, 22, 25, 28–30n89, 181. *See
   also* soul
Ptolemais of Cyrene, 296n42
Ptolemy, 40n125, 42–44, 46, 48n161,
   76n101, 193, 207n131, 225n35,
   233n65, 234n68, 236–41, 244–47,
   272–74, 279n217, 293n23, 298–

300, 305–8, 310, 313–19, 328,
   331n151, 335
pulse, 42n141, 106n202, 282–84
purge, 119, 126, 266
purifications, 22n69, 27–28, 52n7,
   84n118, 89, 128, 334n160
Pyrrho, 162
Pythagoras, 61, 75, 92–93, 258, 295–96
Pythagoreans, 9, 33n99, 78n106,
   149n161, 241–44, 258, 260,
   270n191, 275–78, 289, 291n18,
   295–97, 304n73, 309n84

Quesalid, 110

reason, 129, 137n124, 160–63,
   202n110, 228n48, 244n109, 263,
   271–73, 296, 320
Receptacle, 137n121, 138n126
reflection, 245, 300
refraction, 234, 238, 241n100, 246–47,
   274–75, 300
regimen, 36–37, 64, 130. *See also* diet
religion, 4, 9, 11, 21, 27–28, 47, 49, 81,
   86, 89, 176–79, 210–11, 328, 334
research, 16, 154, 165, 168–70, 212,
   262, 330, 334–35
rhetoric, 89n143, 92–94, 104n193,
   108, 133n111, 171, 186n51, 219, 336
riddles, 85–87, 210
ritual, 4–5, 35, 47, 53
Rufus, 45, 207n128, 282–83

Sappho, 58
scepticism, 109–10, 113, 158, 162, 168,
   170, 252n127, 272n199, 273
Scribonius Largus, 252
seed, 152–54, 189, 205, 259, 262,
   325n132
seers, 41, 49, 84
semantic stretch, 174–75, 177, 179, 198,
   208
Seneca, 332n154
sensation, 213. *See also* perception
Sextus Empiricus, 113n20, 161–62,
   252n127
shaman, 84n118, 110, 111n11
Shirokogoroff, S. M., 72, 109–11
similes, 182, 186n50. *See also* comparison; likeness
Simonides, 83n117, 93n153

simplicity, 278, 285, 307n78. *See also*
 simplification
Simplicius, 14n43, 105, 193, 222, 293,
 304
simplification, 285, 312, 315, 319
slaves, 333n158
sleep, 32–33, 176, 195, 212
Snell, B., 58
Socrates, 43n143, 81n110, 83n117,
 87n134, 92n152, 98, 102, 280n218,
 291, 333n159
Solon, 72n83, 84, 93–94, 258
sophists, 48–49n163, 66, 83n117,
 87n134, 89n143, 91–98, 103, 130,
 132n109, 133
Soranus, 25n80, 44n150, 106, 159,
 161n211, 164–65, 204n117
Sosigenes, 293n27, 304n72
soul, 9–11, 29n89, 33–34, 280n219,
 325. *See also* psychology
Sparta, 54n18, 79n107, 81n111
Speusippus, 149n161, 206n125, 291n18
sphericity: of earth, 193–94, 231, 289,
 307n79, 335; of universe, 192–93,
 314n98
spontaneous generation, 150–54
star, 36, 45, 169n239, 194, 207–8,
 231–35, 307–8; catalogues, 236–37
statics, 216, 249–50, 301–3
Stoics, 9n24, 38–39n122, 44n150,
 48n161, 148, 158n200, 168n239,
 272, 300n54, 301n60, 322n114,
 327n138
Strato, 48n161, 157n195, 222–23
sublunary/superlunary, 194, 216, 235,
 239–41
Sumeria, 87–88
sun, 36n117, 169n239, 236–37, 308
syllogism, 141–42, 143n146, 145n152,
 146, 182, 186, 202
symbolism, 36, 49, 208n132, 252n128,
 260, 275–76, 280
sympatheia, 44, 155, 253n133

tables, 45, 73–74, 252, 258, 260
Tambiah, S., 53
Taylor, A. E., 136
technical terms, 203–8
Teisias, 92n152
teleology, 48n161, 140, 149–50,
 169n239, 290–91, 319–29

temperature, measurement of, 227–28.
 See also hot/cold
temple medicine, 30, 89–90, 134,
 333n159
tentativeness, 124, 131–34, 148, 154,
 156, 158, 162, 170, 214
tests, 99, 140n133, 193, 212–13,
 223n27, 224–25, 229–30n52,
 248–50, 255, 295–96
Thales, 52n7, 75, 84, 85n123, 86n127,
 92n152, 112, 288, 290n13
Theaetetus, 76n98
Themison, 158, 159n201
Themistocles, 83n117
Theodorus, 76n98
theology, 46–49, 85n123, 179, 334n161
Theon of Alexandria, 245n111
Theon of Smyrna, 207n131, 276n209,
 295n36, 313n95, 334n160
Theophrastus, 103n188, 148–55,
 197n94, 247–49, 253n132, 297–98,
 321n113
Thessalus, 159
thought experiment. *See* experiment,
 thought
Thucydides, 11n31, 59n32, 81n111, 97,
 124n67
thunder, 49n163, 169n239, 178, 333
time, measurement of, 225–26, 234n68,
 281, 283–84
Timocharis, 235
Timotheus, 59n31
tradition, 28, 36–37, 47–48, 50–108,
 170, 239
trepanning, 19, 124–25, 127n83
Turner, V. W., 110

univocity, 173–74, 194, 198, 200–201,
 208, 213–14
Upanishads, 87–88

venesection, 127, 254n136
Vernant, J.-P., 78
Vitruvius, 215n3, 250, 253n132
vivisection, 163–64, 206n127
void, 191n69, 217, 219, 303

water, 16, 24n76, 62, 84, 95n159, 113,
 117n39, 118n43, 122, 137n121, 150,
 192–93, 197, 217–18, 226–30,
 248–49

water clock, 225–26, 284. *See also*
    clepsydra
weighing, 192, 216, 230, 242n100,
    247–57, 273n200, 277, 281, 295–96
wet/dry, 8, 14–15, 27–28, 99n175,
    117n36, 120, 159, 162, 190n66,
    194–98, 202n107, 214, 226–30,
    247–48, 280
wisdom, 37n119, 47–49, 83–87, 92–
    93, 97n166, 103, 168n239, 209–10,
    214, 242n100, 333–35
wonder-workers, 84

Xenocrates, 291n18
Xenophanes, 38n121, 47n161, 60–61,
    85–86, 113, 176–79, 181n29
Xenophon, 23n73

Zeno of Elea, 92–93n152
Zeus, 8–9, 86, 176, 178, 180, 181n29
zoology, 46n160, 143–44, 146, 149–
    50, 196, 201n107, 206, 322–24,
    332n155

Index Locorum

AELIAN
  Varia Historia
    12.32   101n184
AESCHINES
    1.6–11   72n83
AESCHYLUS
  Agamemnon
    249   30n91
  Choephori
    523–34   30n91
  Septem contra Thebas
    260   191n68
AETIUS
    3.10.2   289n11
    4.15.3   300n54
    4.19.3   217n6
    5.1.2   38n121
    5.2.3   31n94
    5.22.1   113n17
ALEXANDER
  De Anima libri Mantissa
    130.14ff.   300n54
  De Mixtione
    3.216.14ff.   44n150
    11.226.30ff.   44n150
    12.227.5ff.   44n150
ANAXAGORAS
    fr. 1   113n18
    fr. 12   179n23
    fr. 13   179n23
    fr. 16   113n19
    fr. 17   180n28
    fr. 21   271n193
ANAXIMENES
    fr. 1   249n120
ANONYMUS LONDINENSIS (Anon. Lond.)
    14.11ff.   132n109

    17.11ff.   132n109
    18.8ff.   15n45
                             132n109
    19.1ff.   15n45
                             94n156
    20.1ff.   15n45
    20.25ff.   15n45
                             132n109
    31.10ff.   255n143
    31.34ff.   193n78
                             255n143
    32.22ff.   193n78
                             255n143
    33.43ff.   255n143
ANTIPHON
    1.1   60n39
    1.5   60n39
    1.11ff.   60n39
    5.1–7   103n192
    6.15f.   60n39
APOLLODORUS
  Bibliotheca
    1.7.2ff.   287n4
ARCHILOCHUS
    74   331n148
    81   325n127
    83   325n127
ARCHIMEDES (edd. Heiberg Stamatis)
  (Aequil.) De Planorum
  Aequilibriis
    1 Postulates 1ff. HS
      2.124.3ff.   302nn62,
                             65
    1 Postulates 4f. HS
      2.124.13ff.   302n64
    1.6 HS 2.132.14ff.   302n66
    1.7 HS 2.136.18ff.   302n66

ARISTOTLE (*continued*)

  (*Resp.*) De Respiratione

474a26	205n120
476a5ff.	141n139
478b22ff.	8n16
478b24ff.	322n117
479a32ff.	8n16
480a16ff.	190n66

  (*HA*) Historia Animalium

489a20ff.	323n119
1.8ff.491b9ff.	29n89
491b27ff.	323n122
494a26ff.	323n118
494a33ff.	323n118
510b3f.	189n65
512b12ff.	94n156
515a34ff.	187n53
521a2ff.	29n89
521a26f.	255n142
533a2ff.	323n122
573a5ff.	255n142
583b2ff.	259n159
583b23ff.	260n165
583b31ff.	259n158
584a26f.	260n165
584a33ff.	257n151
584a36ff.	259n158
584b2ff.	259n158
584b6ff.	259n158
584b18ff.	262n173
584b21ff.	262n173

  (*PA*) De Partibus Animalium

639b14ff.	196n91
1.1.639b21ff.	143n147
	321n111
639b23ff.	189n92
640a18ff.	145n150
	196n91
640b22ff.	196n91
	322n115
640b28ff.	196n91
641a10ff.	322n115
641a27	196n91
641b18ff.	329n146
641b32	196n91
642a1ff.	321n111
642a9ff.	189n92
642a13ff.	322n115
642a18ff.	113n17
	229n51
642a25f.	196n91
642a31ff.	321n111

643a3ff.	201n107
1.5.644b22ff.	46n160
645a15ff.	324n125
645a16–23	34n104
645a16f.	46n160
645a21ff.	324n125
645a22f.	46n160
645a26ff.	206n127
645a30ff.	196n91
	322n115
645b8ff.	323n119
645b14–21	190n66
646a24ff.	189n62
646b3ff.	189n62
647a30f.	323n120
2.2.647b31ff.	29n89
	201n107
648a1ff.	323n119
648a19ff.	323n119
648a21ff.	195n86
648a36ff.	195n86
648b4ff.	195n88
648b12ff.	195n87
648b17ff.	195n87
648b26ff.	195n87
648b30ff.	195n87
649a5ff.	195n87
649a20ff.	151n169
649b3ff.	195n87
649b9ff.	195n86
650a4	205n120
2.4.650b14ff.	29n89
	201n107
652a9f.	205n121
652a31	188n59
652b7ff.	184n39
654b29ff.	187n53
656a1f.	190n66
656a10ff.	323n118
657a22ff.	323n122
658a32	188n59
662b23ff.	321n111
663b22ff.	321n111
665a9–26	149n161
665b11ff.	323n120
668a12–31	189n63
668a13ff.	188n57
668a16ff.	187n55
668b21ff.	188n58
	189n63
675b20ff.	187n56
677a11ff.	321n111

677a14ff.	321n112
677a16ff.	321n111
677b31	205n120
678a8f.	323n119
678b1ff.	323n120
683a19ff.	190n66
683b5ff.	190n66
683b18ff.	324n123
684a32ff.	323n121
686b2ff.	324n124
686b20ff.	324n124
686b31ff.	324n123
687a10	190n66
687a18ff.	190n66
689b25ff.	324n124
693b2ff.	201n107
695a8ff.	324n124
695b2ff.	323n121

De Motu Animalium

703a14ff.	323n120

(IA) De Incessu Animalium

705b2ff.	324n123
706a16ff.	323n118
706a20ff.	323n118
706b5ff.	324n123
706b9f.	323n118
710b12ff.	324n124
713a3ff.	190n66
714a6ff.	323n121
714b8ff.	324n123
714b10ff.	324n123
714b14ff.	324n123

(GA) De Generatione Animalium

715a5	196n91
715a8ff.	196n91
717a12f.	190n66
717a34ff.	189n64
717b3f.	189n65
718b21	205n120
719a32ff.	205n121
719b2	205n121
721a1f.	141n139
721a14ff.	141n139
721a26	190n66
727a34ff.	205n121
727b31ff.	199n100
728a18ff.	205n121
728a20f.	323n119
729a10ff.	199n100
729a28ff.	199n100
730b19ff.	190n66
730b27ff.	187n53

731a24	188n59
732b26ff.	190n66
732b28–733b16	196n91
735a22ff.	323n120
738a13	205n121
738a34ff.	205n121
738b16f.	323n120
740b31f.	190n66
741a34ff.	141n139
741b15ff.	323n120
742b35ff.	323n120
743a1ff.	187n53
743b20ff.	187n54
744b1ff.	205n121
744b16ff.	187n56
746b4ff.	141n139
747a34ff.	184n39
752b25ff.	184n39
753a18ff.	205n121
756b28f.	205n121
757b22f.	141n139
760b27ff.	141n139
761b18ff.	197n94
761b19f.	151n169
762a33ff.	141n139
764b30f.	187n54
765b10ff.	205n121
766a3ff.	190n66
766a22f.	190n66
766a30ff.	205n121
772b7ff.	257n151
775a17f.	205n121
776a10	204n117
776a20ff.	205n121
776b33ff.	205n121
777a7ff.	184n40
5.1.778a16ff.	321n112
778a32ff.	321n112
778b10ff.	321n111
778b16ff.	321n111
779a26ff.	321n112
779b12ff.	321n112
779b34ff.	321n112
780b6ff.	205n121
	321n112
781a20ff.	323n120
784a34ff.	205n122
784b3ff.	205n124
787b19ff.	189n64
788a3ff.	189n64
788b20ff.	190n66
789b7ff.	190n66

ARISTOTLE *Rhetoric* (*continued*)

1406b20ff.	186n50
1407a10ff.	186n50
1407a14ff.	185n45
1407a32ff.	184n38
1410b13ff.	185n46
1410b17f.	186n50
1410b36–1411b23	185n45
1412a11ff.	185n47
1413a4ff.	186n50
1414b29ff.	91n150
1415b15ff.	96n163
1419b3f.	97n167

(Po.) *Poetics*

1447b17ff.	184n38
	210n136
4.1448b4ff.	214n144
1449a18	299n52
1451b35ff.	89n143
1457b6ff.	184n42
1459a5ff.	185n47

Constitution of Athens

1	84n118

Eudemus (ed. Ross)

fr. 6	7n15

Περὶ φιλοσοφίας (ed. Ross)

fr. 5	93n153
fr. 8	83n117
fr. 12a	33n102

Sophist (ed. Ross)

fr. 1	93n152

ARISTOXENUS

(*Harm.*) *Harmonica*

1.1	77n102
	298n47
1.2–3	77n102
1.3	206n126
1.4	77n102
1.4–6	206n126
1.5–6	77n102
1.10–13	206n126
1.14	297n42
1.15–16	206n126
1.19	206n126
1.21ff.	206n126
	297n42
1.24	206n126
	298n50
1.28	243n102
2.31–2	329n145
2.33	296n42
2.35–6	77n102

2.37ff.	77n102
	206n126
2.48	206n126

ARTEMIDORUS

Onirocritica

1.Pr. 2.1ff.	32n97
1.Pr. 2.18	91n146
1.1. 3.9ff.	32n97
1.2. 4.22ff.	32n97
1.3. 11.7ff.	32n97
1.6. 15.19–16.9	31n94
1.8. 17.11ff.	32n97
1.9 18.16ff.	32n97
1.12. 20.18f.	32n97
2.69. 195.10ff.	32n97
4.Pr. 237.25ff.	100n180
4.Pr. 238.1ff.	79n106
4.4. 248.5ff.	32n97
4.22. 255.13ff.	32n97
4.28. 263.14ff.	32n97
4.59. 283.4ff.	32n97
4.59. 283.20ff.	32n97
4.63. 286.13ff.	32n97
4.84. 299.15ff.	100n180
5.Pr. 301.10ff.	91n146

ATHENAEUS

122c–d	59n31

BACCHYLIDES (edd. Snell Maehler)

5.160ff.	7n15

BOETHIUS

(*Mus.*) *De Institutione Musica*
(ed. Friedlein)

1.10. 197.3ff.	296n39
3.11. 285.9ff.	243n107

CAELIUS AURELIANUS

De Morbis Acutis

1.108	270n190

(*Morb. Chron.*) *De Morbis
Chronicis*

1.105	270n191
1.126	270n191
1.144ff.	25n80
1.145	22n69
1.155ff.	26n80
1.157	26n80
1.158–61	26n80
1.171	20n62
1.171ff.	25n80
1.173	26n80

CELSUS

(*Med.*) *De Medicina* (CML 1)

1 Proem 12ff. 19.4ff.	159n205

1 Proem 13. 19.11ff.   159n206
1 Proem 23–4.
  21.15–21   163n220
1 Proem 26. 21.29f.   163n221
1 Proem 26.
  21.29–32   164n222
1 Proem 27f. 22.1ff.   160n207
1 Proem 27. 22.4   161n213
1 Proem 29.
  22.11–13   161n210
1 Proem 36. 23.4ff.   160n207
1 Proem 38–39.
  23.16–27   160n209
1 Proem 43.
  24.14–19   164n225
1 Proem 44. 24.21f.   164n223
1 Proem 45ff.
  24.24ff.   159n205
1 Proem 57. 26.26ff.   162n215
1 Proem 57. 26.27f.   161n212
1 Proem 74f.
  29.17–22   163n221
3.4.11ff. 106.25ff.   270n191
3.4.12. 107.1ff.   106n203
3.4.12. 107.2ff.   270n191
3.4.15. 107.23ff.   270n191
3.18. 122.14–
  127.15   26n81
3.18.21. 126.27ff.   26n81
5.17.1c. 194.5ff.   251n129
5.18–25. 194.31–
  215.3   251n127
6.6.1e. 260.3ff.   106n203
7.7.13b. 319.20–2   207n128
8.4.3. 378.3ff.   125n69
8.20.4. 407.7ff.   69n76

CICERO
  (*Div.*) *De Divinatione*
  1.3.5   38n121
  2.14.33ff.   44n150
  De Finibus
  2.1.1   91n150
  (*N.D.*) *De Natura Deorum*
  1.13.35   48n161
  1.18.46ff.   31n94
  2.7.19   44n150
CLEOMEDES
  De Motu Circulari Corporum
  Caelestium
  1.1. 4.10ff.   44n150
  8.15ff.   44n150
  1.10. 90.20ff.   231n59

94.22   233n65
96.2ff.   232n62
98.3ff.   232n60
98.4f.   232n61
98.10ff.   232n60
98.22ff.   232n60
100.15ff.   232n60
1.11. 102.23ff.   307n80
106.9ff.   307n80
2.6. 222.28ff.   238n87
224.11ff.   238n87
CRITIAS
  fr. 25   49n163
DAMIANUS
  (*Opt.*) *Optica* (ed. Schöne)
  11. 12.12ff.   301n60
  14. 20.12ff.   300n59
  24.7ff.   299n53
  24.16ff.   299n53
  28–30   299n52
  30.10–11   299n52
DEMOCRITUS
  frr. 6–10   113n23
  fr. 9   113n24
     271n193
  fr. 11   113n24
     271n193
  fr. 117   113n23
  fr. 119   18n56
  fr. 125   113n24
     271n193
  fr. 164   217n6
  fr. 166   33n99
  fr. 175   18n56
  fr. 234   18n56
     35n110
DEMOSTHENES
  23.148   280n218
DIOCLES
  On Burning Mirrors
  (ed. Toomer)
  par. 18ff.   307n80
DIODORUS SICULUS
  1.9.3   52n7
  1.14.1   52n7
  1.15.8   52n7
  1.43.5f.   52n7
  1.82.3   55n19
  2.38.1ff.   52n7
  4.1.6f.   52n7
  4.2.5   52n7
  12.53.2–5   96n164

GALEN (*continued*)
(*PHP*) *De Placitis Hippocratis
et Platonis* (ed. de Lacy) (*Corpus
Medicorum Graecorum* 5.4.1.2)
  1.7. CMG 5.4.1.2
    86.24ff.   207n130
    7.3. 444.12ff.   213n143
(*Sect. Intr.*) *De Sectis ad Introdu-
cendos* (ed. Helmreich, *Scripta
Minora* 3)
    4. H 3.7.1ff.   158n200
    6f. H 3.13.21ff.   161n212
    6. H 3.14.14ff.   162n215
    6. H 3.15.2ff.   161n211
    7. H 3.17.3ff.   161n212
    7. H 3.18.1ff.   161n212
Subfiguratio Empirica (ed.
Deichgräber)
    78.26ff.   31n95
(*Mixt.*) *De Temperamentis* (ed.
Helmreich)
    1.6. 19.10ff.   228n48
    1.6. 21.20ff.   228n47
    1.8. 29.3ff.   228n47
    1.9. 32.5ff.   228n48
    1.9. 32.24ff.   228n47
    1.9. 33.21ff.   228n48
    2.2. 51.18ff.   228n48
    2.2. 53.14ff.   228n48
    2.3. 56.12ff.   228n48
    2.4. 62.25ff.   229n48
                        254n136
    2.4. 63.12ff.   229n48
                        254n136
Thrasybulus (ed. Helmreich,
Scripta Minora 3)
    26. H 3.66.18ff.   14n43
(*UP*) *De Usu Partium* (ed.
Helmreich)
    1.5. H 1.6.18ff.   324n126
    1.17. H 1.33.26ff.   213n142
    1.22. H 1.58.18ff.   325n128
    2.3. H 1.70.10ff.   106n202
    3.8. H 1.152.21ff.   325n128
    3.10. H 1.172.15ff.   47n160
    3.10. H 1.174.6–13   48n162
                        334n161
    3.10. H 1.175.3ff.   327n139
    3.10. H 1.177.20ff.   324n126
    3.16. H 1.190.10ff.   324n126
    3.16. H 1.194.11ff.   325n128
    4.13. H 1.223.10ff.   256n145
    5.3. H 1.255.6ff.   327n141

    5.3. H 1.257.4ff.   327n141
    5.4 H 1.258.26ff.   327n141
    5.4. H 1.259.6ff.   327n141
    5.4. H 1.259.11ff.   327n142
    5.4. H 1.260.1ff.   325n126
    5.4. H 1.260.5ff.   327n139
    5.4. H 1.260.7ff.   327n140
    5.4. H 1.262.17ff.   327n142
    5.4. H 1.263.1ff.   327n142
    5.4. H 1.263.20ff.   327n141
    5.5. H 1.267.12ff.   325n126
    6.17. H 1.362.7ff.   256n147
    6.21. H 1.371.4ff.   326n135
    6.21. H 1.374.4ff.   326n136
    7.8. H 1.392.25ff.   255n144
    7.9. H 1.396.23   334n161
    7.14. H 1.418.19ff.   334n161
    7.14. H 1.418.24ff.   47n160
    7.15. H 1.422.24ff.   47n160
    7.15. H 1.423.12ff.   334n161
    10.12ff. H 2.93.5ff.   331n151
    10.14. H 2.109.8ff.   331n151
    10.14. H 2.110.9ff.   332n151
    11.2. H 2.114.17ff.   325n130
    11.2. H 2.116.10ff.   47n160
    11.2. H 2.117.14ff.   325n128
    11.14. H 2.154.20ff.   325n132
    12.4. H 2.190.19ff.   334n161
    12.6. H 2.196.5ff.   334n161
    13.11. H 2.273.8ff.   325n131
    13.11. H 2.273.23ff.   325n131
    14.2. H 2.285.7ff.   327n140
    14.5. H 2.295.27ff.   325n132
    14.6. H 2.296.8ff.   325n132
    14.6. H 2.299.3ff.   325n132
    14.6. H 2.299.19ff.   325n132
    15.6. H 2.360.19ff.   326n135
    15.6. H 2.361.12ff.   326n135
    15.6. H 2.362.1ff.   326n136
    15.8. H 2.367.15ff.   325n128
    16.12. H 2.427.15ff.   207n130
    16.14. H 2.433.4ff.   256n145
    17.1. H 2.446.11ff.   327n139
    17.1. H 2.446.19ff.   327n139
    17.1. H 2.447.22ff.   334n161
    17.1. H 2.448.7ff.   334n161
    17.1. H 2.448.9ff.   47n160
    17.2. H 2.449.15ff.   47n160
    17.3. H 2.451.21ff.   334n161
Corpus Medicorum Graecorum
  5.4.1.2. 64.6ff.   105n199
    86.24ff.   207n130
    444.12ff.   213n143

5.8.1.	76.29ff.	39n95
	82.25ff.	100n179
	84.5ff.	42n141
	94.18f.	42n141
	96.9ff.	89n143
	98.9ff.	89n143
	106.21ff.	42n141
	134.1ff.	106n202
5.9.1.	197.6ff.	251n127
5.10.1.	108.1ff.	39n95
	123.12ff.	270nn189, 191
5.10.2.2.9.10ff.		213n142
	19.5ff.	131n107
	69.19f.	131n107
	75.25ff.	131n107
	79.8	131n107
	80.16ff.	131n107
	227.27ff.	131n107
	253.4–259.6	14n43
(ed. Kühn)		
1	72.4ff.	158n200
	81.6ff.	161n212
	82.6ff.	162n215
	83.1ff.	161n211
	85.14ff.	161n212
	86.17ff.	161n212
	381.12ff.	228n48
	383.14ff.	229n48
	476.8ff.	228n48
	538.11ff.	228n48
	542.13ff.	228n47
	554.12ff.	228n47
	559.10ff.	228n48
	560.13ff.	228n47
	562.4ff.	228n48
	590.9ff.	228n48
	593.7ff.	228n48
	598.7ff.	228n48
	608.13ff.	229n48
		254n136
	609.9ff.	229n48
		254n136
2	70.10ff.	256n148
	71.12ff.	256n149
	72.4ff.	256n149
	207.17ff.	326n133
	208.11ff.	256n147
	222.2ff.	325n129
	222.5ff.	325n130
	223.9ff.	325n130
	283.12–17	167n235
	284.8–11	167n235

	286.3–12	166n229
	286.5ff.	46n160
	287.4–6	166n230
	288.3–13, 14–15	166n231
	288.15ff.	167n232
	289.3–9	167n233
	289.17ff.	167n234
	342.4ff.	40n129
	348.14ff.	165n228
	384.12ff.	325n129
	396.18ff.	40n129
	416.3ff.	325n128
	423.5ff.	325n129
	526.4ff.	325n129
	532.5–535.15	325nn129, 130
	548.2ff.	325n129
	612.2ff.	213n142
	612.15ff.	213n142
	613.1ff.	213n142
	619.16ff.	89n143
	621.12ff.	165n228
	642.3ff.	89n143
	645.7ff.	89n143
	669.7f.	42n141
	669.15	42n141
	674.6ff.	165n228
	828.10ff.	326n135
3	9.4ff.	324n126
	47.1ff.	213n142
	79.18ff.	325n128
	96.8ff.	106n202
	208.15ff.	325n128
	235.6ff.	47n160
	237.10–17	48n162
		334n161
	238.11ff.	327n139
	242.5ff.	324n126
	259.3ff.	324n126
	264.9ff.	325n128
	304.7ff.	256n145
	348.4ff.	327n141
	350.16ff.	327n141
	353.7ff.	327n141
	353.15ff.	327n141
	354.3ff.	327n142
	354.17ff.	325n126
	355.4ff.	327n139
	355.5ff., 11ff.	327n140
	358.9ff.	327n142
	358.18ff.	327n142
	359.17ff.	327n141
	364.17ff.	325n126

HERO (*continued*)
Dioptra (ed. Schöne)
    34. 292.16ff.   234n68
    35. 302.3ff.   234n68
Pneumatica (ed. Schmidt)
    1.Pr. 2.18ff.   228n46
    2.8. 224.2ff.   227n44
HERODOTUS
    1. 29   93n153
       30   94n153
       46ff.   110n5
       53ff.   85n126
       71   85n126
       75   84n121
       170   84n121
       197   56n22
    2. 49   93n153
       77   54n15
       79   54n15
       100   54n15
       109   52n7
       145   54n15
       174   110n5
    3. 33   24n73
    4. 95   93n153
    6. 75ff.   24n73
       84   24n73
HESIOD
(*Op.*) Opera
    11ff.   58n29
    26   58n29
    101ff.   7n15
    102ff.   13n36
                                       17n52
    109ff.   8n20
    116   9n21
    137ff.   9n21
    152ff.   9n21
    170ff.   9n21
    180f.   9n21
    242ff.   13n36
    417ff.   45n154
    587f.   45n154
    609ff.   45n154
    649   93n153
    654ff.   58n29
    765ff.   258n156
    822ff.   258n156
(*Th.*) Theogonia
    22ff.   58n28
    32   40n128
Fragments (edd. Merkelbach West)
    278   85n125

HIPPARCHUS
(*In Arat.*) In Arati et Eudoxi Phae-
nomena (ed. Manitius)
    1.2.11. 14.13ff.   235n73
    1.5.19. 52.1ff.   235n73
    1.6.2. 54.23ff.   235n74
    1.6.2. 56.2ff.   235n74
HIPPOCRATIC CORPUS
(*Acut.*) De Victus Ratione in Mor-
bis Acutis (ed. Littré)
    2. L 2.230.1ff.   63n54
    2. 234.2ff.   67n70
    3. 238.8−10   63n54
    3. 242.3ff.   41n135
    4. 250.11ff.   265n180
    11. 304.5   205n119
    11. 306.9ff.   127n83
    11. 308.7ff.   127n83
    11. 310.1ff.   20n62
    11. 316.6ff.   127n83
    11. 316.9ff.   20n62
(*Acut. Sp.*) De Victus Ratione in
Morbis Acutis Spuria (ed. Littré)
    24. L 2.508.8   20n62
(*Aër.*) De Aëre, Aquis, Locis
(ed. Diller, Corpus Medicorum
Graecorum 1.1.2)
    1. CMG 1 1.2. 24.9   249n121
    3. 26.22f.   62n49
    3. 26.23ff.   122n58
    3. 28.2f.   122n58
    3. 28.5f.   123n65
    3. 28.14   204n117
    4. 30.3   123n65
    4. 30.4   122n61
    4. 30.7   123n65
    4. 30.8f.   122n62
    4. 30.12f.   122n62
                                     123n66
    4. 30.18   123n65
    5. 32.10ff.   122n59
    5. 32.13ff.   122n59
    6. 34.1f.   122n62
    7. 34.16−40.6   248n120
    7. 34.19−23   122n60
                                     123n66
    7. 36.12f.   24n76
    7. 36.25   122n60
    7. 38.7f.   122n60
    7. 38.8   248n120
    7. 38.22   248n120
    8. 40.7−44.2   248n120
    8. 40.8   248n120

8. 42.15ff.	248n120
9. 44.4	204n117
9. 44.15f.	122n60
9. 44.20f.	122n60
10. 46.22ff.	123nn63, 66
	156n188
10. 46.24ff.	123n63
	156n190
10. 48.13ff.	157n193
10. 48.19	204n118
10. 50.18ff.	123n63
10. 50.19ff.	204n116
10. 50.21ff.	157n193
10. 52.2ff.	157n193
10. 52.4ff.	157n193
14. 58.23	123n65
16. 62.2ff.	30n89
16. 62.13ff.	103n189
16. 62.20ff.	123n64
19. 68.15ff.	123n64
22. 72.14–17	13n39
23. 76.20ff.	30n89
23. 78.3ff.	103n189
	123n64
24. 78.9ff.	30n89
24. 78.15	123n65
24. 80.3ff.	123n64

(Aff.) De Affectionibus (ed. Littré)

1. L 6.208.7ff.	14n44
	117n35
	204n116
1. 208.9f.	15n45
11. 218.13ff.	204n116
11. 218.21ff.	204n116
37. 246.20	120n53

(Alim.) De Alimento (ed. Heiberg, *Corpus Medicorum Graecorum* 1.1)

42. CMG 1.1. 83.7–10	259n160

(Aph.) Aphorismi (ed. Littré)

1.1. L 4.458.1ff.	68n75
	127n82
1.5. 462.3ff.	20n62
1.20. 468.8ff.	69n75
2.6. 470.17f.	24n76
2.19. 474.12f.	41n135
2.24. 476.11ff.	267n184
2.52. 484.13f.	69n75
3.11. 490.2ff.	123n63
	156n188
3.20. 494.16ff.	29n89

3.22. 496.7f.	29n89
4.16. 506.9f.	127n83
4.36. 514.8ff.	265n180
4.61. 524.3ff.	265n180
4.64. 524.10ff.	265n180
5.26. 542.1f.	249n121
5.31. 542.12f.	127n83
5.40. 544.16f.	24n76
5.54. 552.4f.	121n55
5.65. 558.7f.	24n76
6.6. 564.4f.	204n117
6.20. 568.5f.	121n55
6.21. 568.7f.	24n76
6.23. 568.11f.	29n89
6.27. 570.3f.	127n83
6.38. 572.5ff.	127n79
6.45. 574.8f.	121n55
6.50. 576.4f.	121n55
6.53. 576.13f.	24n74
6.56. 576.19ff.	29n89
6.58. 578.3	121n55
7.5. 578.14	24n76
7.45. 590.4ff.	127n83
7.58. 594.10f.	121n55
7.60. 596.1f.	20n62
7.85. 606.10ff.	121n55

(Art.) De Articulis (ed. Littré)

1. L 4.78.1ff.	100n177
1. 78.2ff.	127n81
1. 78.9ff.	100n177
1. 80.13f.	127n81
7. 88.15ff.	65n59
9. 100.3f.	41n134
11. 104.16ff.	64n58
13. 116.20f.	41n134
13. 116.23ff.	121n57
14. 122.16ff.	121n57
14. 128.1f.	269n187
33. 148.13ff.	161n210
34. 156.5ff.	65n59
35. 158.4ff.	67n72
38. 168.9ff.	69n76
	121n57
38. 168.13ff.	69n76
40. 172.5ff.	127n83
41. 182.11f.	41n134
42–44. 182.13ff.	19n59
42. 182.13ff.	64nn57, 59
	69n78
42. 182.15–20	65n59
42. 182.22ff.	64n57
42. 184.1ff.	69n78

HIPPOCRATIC CORPUS (*continued*)
38. 406.5ff.	126n76
49. 418.1ff.	69n77
86. 444.13ff.	23n71
87. 444.17ff.	23n71
121. 466.14ff.	127n77

(*Fist.*) *Fistulae* (ed. Littré)
7. L 6.454.23	19n58

(*Flat.*) *De Flatibus* (ed. Heiberg,
Corpus Medicorum Graecorum
1.1)
  2. CMG 1.1.
92.13–17	16n48
	117n34
2. 92.16f.	61n44
5. 94.6f.	61n44
6. 94.8ff.	61n44
7. 95.6	205n120
7. 95.7	120n53
10. 97.12ff.	61n44
10. 98.16	120n53
14. 99.20ff.	61n44
15. 101.17ff.	61n44

(*Foet. Exsect.*) *De Foetus Exsectione* (ed. Littré)
4. L 8.514.14ff.	19n59
	69n77

(*Fract.*) *De Fracturis* (ed. Littré)
1. L 3.412.1ff.	14n43
1. 414.1	65n60
1. 414.4–5	65n60
1. 414.6–9	65n61
3. 424.10ff.	121n57
5. 432.8ff.	269n187
6. 436.11ff.	269n187
7. 440.2ff.	269n187
7. 442.7ff.	122n57
8. 444.1ff.	64n58
9. 450.5ff.	269n187
13. 462.6ff.	65n59
13. 462.7ff.	19n60
13. 464.12ff.	19n60
13. 466.3ff.	19n60
15. 472.14ff.	65n59
15. 472.16ff.	65n59
16. 474.16ff.	68n73
16. 474.17	127n80
16. 476.8ff.	68n73
16. 476.11ff.	122n57
16. 478.8ff.	269n187
20. 484.7ff.	65n59

23. 492.7ff.	122n57
25. 496.15ff.	127n83
25. 498.8ff.	122n57
30. 516.14ff.	65n59
30. 518.4ff.	127n83
30. 524.6ff.	65n59
31. 524.19ff.	127n83
31. 528.16ff.	65n59
33. 532.17f.	269n187
33. 532.21ff.	269n187
34. 536.9ff.	122n57
35. 536.13ff.	127n79
35. 538.5f.	41n133
36. 540.9–12	128n86

(*Genit.*) *De Genitura* (ed. Littré)
8. L 7.480.9f.	121n54

(*Hebd.*) *De Hebdomadibus* (ed. Roscher)
1.1.8ff. (L 9.433.3f.)	259n161
45.66f.	
(L 9.460.17ff.)	31n93
53.80.4ff.	
(L 9.466.8ff.)	66n62

(*Hum.*) *De Humoribus* (ed. Littré)
4. L 5.480.17	31n93
6. 484.13ff.	69n75
6. 484.19ff.	20n62
6. 486.4ff.	266n181

(*Int.*) *De Affectionibus internis* (ed. Littré)
20. L 7.216.20	252n128
20. 216.22f.	251n127
23. 226.13ff.	251n127
23. 226.14f.	252n128
26. 234.15ff.	251n127
27. 238.3ff.	19n58
31. 248.9ff.	251n127
31. 248.10	252n128

(*Jusj.*) *Jusjurandum* (ed. Heiberg,
Corpus Medicorum Graecorum
1.1)
CMG 1.1. 4.7ff.	78n106

Lex (ed. Heiberg, *Corpus Medicorum Graecorum* 1.1)
1. CMG 1.1. 7.5ff.	103n190
5. 8.15ff.	79n106
	334n160

(*Loc. Hom.*) *De Locis in Homine* (ed. Littré)
41. L 6.330.20ff.	118n40
	130n98

44. 338.6ff.	118n40
	130n99
46. 342.4ff.	118n41
	130n97

(*Medic.*) *De Medico* (ed. Heiberg, *Corpus Medicorum Graecorum* 1.1)

10. CMG 1.1.	
23.25ff.	205n122

(*Mochl.*) *Mochlicon* (ed. Littré)

33. L 4.374.16f.	127n79
33. 376.2f.	127n79
33. 376.3ff.	41n133
38. 382.3ff.	65n59
38. 384.15ff.	19n60

(*Morb.* 1) *De Morbis* 1 (ed. Littré)

1. L 6.140.1ff.	88n142
	100nn177, 180
1. 142.7−12	100n180
2. 142.13ff.	14n44
	117n36
	129n92
2. 142.15−20	15n45
3. 144.4	120n53
3. 144.17	120n53
4. 146.6	120n53
4. 146.9	120n53
4. 146.12	120n53
4. 146.13	120n53
5. 146.15ff.	129n95
	269n187
5. 148.15f.	129n95
	269n187
6. 150.6ff.	127n83
8. 156.2	120n53
8. 156.4	120n53
9. 156.14ff.	129nn93, 94
16. 168.23ff.	130n96
	269n187
16. 170.2−4	130n96
	269n187
22. 184.4	120n53
22. 186.10	120n53
24. 190.1	120n53
24. 190.7	120n53
25. 192.2	120n53

(*Morb.* 2) *De Morbis* 2 (ed. Littré)

5. L 7.12.24f.	120n54
40−43. 56.3−60.24	270n190
41. 58.9ff.	265n180

54. 82.21ff.	252n128
61. 96.5f.	267n184
72. 108.25ff.	23n71

(*Morb.* 3) *De Morbis* 3 (ed. Potter, *Corpus Medicorum Graecorum* 1.2.3)

1. CMG 1.2.3.	
70.15	19n58
3. 72.14f.	265n180
16. 90.1ff.	120n54
17. 96.19ff.	251n127
17. 96.22ff.	251n127
17. 96.27ff.	251n127
17. 98.2f.	251n127
17. 98.7	252n128
17. 98.9ff.	251n127
17. 98.12	252n128
17. 98.15f.	252n128

(*Morb.* 4) *De Morbis* 4 (ed. Littré)

34. L 7.548.7ff.	121n54
42. 564.4ff.	255n139
46. 572.1ff.	265n180
47. 574.13ff.	266n182

(*Morb. Sacr.*) *De Morbo Sacro* (ed. Littré)

1. L 6.352.1ff.	13n40
1. 354.12ff.	27n83
1. 360.13−362.6	27n85
1. 362.10ff.	128n87
2. 364.9ff.	13n40
7. 372.4ff.	27n84
8. 376.6	120n53
13. 386.7	120n53
14. 388.6ff.	120n53
15. 388.12−24	26n82
17. 392.19	120n53
18. 394.9ff.	13n40
18. 394.12−15	13n39
18. 394.14−16	27n86
	117n37
18. 394.19ff.	28n87
18. 396.1	28n87
18.396.5−9	28n88
	117n37

(*Mul.* 1) *De Mulierum Morbis* 1 (ed. Littré)

6. L 8.30.8ff.	255n142
25. 64.13ff.	121n54
34. 78.11ff.	121n54
68. 142.20ff.	69n77
71. 150.9ff.	259n159
71. 150.12ff.	128n85

THEOPHRASTUS *De Lapidibus*
(*continued*)
22	248n119
39	248n119
46	248n119
51–52	19n58

(*Metaph.*) *Metaphysics*
5a17–25	149n160
10a22–8	149n162
10a28ff.	150n163
10b11ff.	150n163
10b26ff.	150n164
11a1–3	150n165
11a9ff.	149n161
11a13–15	150n165
11a23	149n161
11a27	149n161
11b24	150n166

(*Sens.*) *De Sensu*
49ff.	227n42
59ff.	191n69
	192n70
61	191n69
61ff.	227n42
62	191n69
68	191n69

THUCYDIDES
1.21–22	97n169
1.71	81n111
1.79ff.	79n107
1.87	79n107
1.97	81n110
2.40	94n153
2.47	12n31
2.53	12n31
3.37ff.	97n168
3.38	97n168
7.50	331n148

VITRUVIUS
2.5.3	253n132
7 Praef. 11	299n52
9 Praef. 9ff.	250n122
9.8.1	232n60
10.9.1–4	234n68
10.9.5ff.	234n68

XENOPHANES
fr. 2	86
fr. 11	60n40
	177n11
fr. 12	60n40
	177n11
fr. 14	177n12
frr. 14–16	60n40
fr. 15	177n13
fr. 23	179n18
fr. 24	178n17
fr. 25	178n17
fr. 28	113n22
fr. 29	113n22
fr. 30	113n22
fr. 33	113n22
fr. 34	59
	113n21

XENOPHON
Cyropaedia
8.7.21	33n99

Hiero
9.9f.	82n114

(*Mem.*) *Memorabilia*
1.1.2ff.	43n143
1.1.9	242n100
3.12.6	23n73
4.2.5	104n193
4.4.6–7	96n164

Financial Accounting and Reporting

Financial Accounting and Reporting

SEVENTEENTH EDITION

Barry Elliott and Jamie Elliott

Harlow, England • London • New York • Boston • San Francisco • Toronto • Sydney • Auckland • Singapore • Hong Kong
Tokyo • Seoul • Taipei • New Delhi • Cape Town • São Paulo • Mexico City • Madrid • Amsterdam • Munich • Paris • Milan

Pearson Education Limited
Edinburgh Gate
Harlow CM20 2JE
United Kingdom
Tel: +44 (0)1279 623623
Web: www.pearson.com/uk

First published 1993 (print)
Second edition 1996 (print)
Third edition 1999 (print)
Fourth edition 2000 (print)
Fifth edition 2001 (print)
Sixth edition 2002 (print)
Seventh edition 2003 (print)
Eighth edition 2004 (print)
Ninth edition 2005 (print)
Tenth edition 2006 (print)
Eleventh edition 2007 (print)
Twelfth edition 2008 (print)
Thirteenth edition 2009 (print)
Fourteenth edition 2011 (print)
Fifteenth edition 2012 (print and electronic)
Sixteenth edition 2013 (print and electronic)
Seventeenth edition 2015 (print and electronic)

ISBN: 978-1-292-08050-5 (print)
      978-1-292-08057-4 (PDF)
      978-1-292-08058-1 (eText)

British Library Cataloguing-in-Publication Data
A catalogue record for the print edition is available from the British Library

Library of Congress Cataloging-in-Publication Data
Elliott, Barry.
    Financial accounting and reporting / Barry Elliott and Jamie Elliott. – Seventeenth edition.
        pages cm
    ISBN 978-1-292-08050-5 (print) – ISBN 978-1-292-08057-4 (PDF) – ISBN 978-1-292-08058-1 (eText)
    1. Accounting.   2. Financial statements.   I. Elliott, Jamie.   II. Title.
    HF5636.E47 2015
    657–dc23

                                            2015001769

10 9 8 7 6 5 4 3 2 1
18 17 16 15

Front cover image: Getty Images

Print edition typeset in 10/12pt Ehrhardt MT Std by 35
Printed in Malaysia (CTP-VVP)

NOTE THAT ANY PAGE CROSS REFERENCES REFER TO THE PRINT EDITION

Brief contents

Preface   xxi
Publisher's acknowledgements   xxvii

Part 1
PREPARATION OF FINANCIAL STATEMENTS   1

 1 Accounting and reporting on a cash flow basis   3
 2 Accounting and reporting on an accrual accounting basis   21
 3 Preparation of financial statements of comprehensive income,
   changes in equity and financial position   32
 4 Annual Report: additional financial disclosures   67
 5 Statements of cash flows   99

Part 2
INCOME AND ASSET VALUE MEASUREMENT SYSTEMS   127

 6 Income and asset value measurement: an economist's approach   129
 7 Accounting for price-level changes   148
 8 Revenue recognition   184

Part 3
REGULATORY FRAMEWORK – AN ATTEMPT TO ACHIEVE
UNIFORMITY   209

 9 Financial reporting – evolution of global standards   211
10 Concepts – evolution of an international conceptual framework   234
11 Ethical behaviour and implications for accountants   248

Part 4
STATEMENT OF FINANCIAL POSITION – EQUITY,
LIABILITY AND ASSET MEASUREMENT AND DISCLOSURE   273

12 Share capital, distributable profits and reduction of capital   275
13 Liabilities   298
14 Financial instruments   321
15 Employee benefits   360

16 Taxation in company accounts   388
17 Property, plant and equipment (PPE)   412
18 Leasing   449
19 Intangible assets   468
20 Inventories   497
21 Construction contracts   523

Part 5
CONSOLIDATED ACCOUNTS   547

22 Accounting for groups at the date of acquisition   549
23 Preparation of consolidated statements of financial position after
   the date of acquisition   568
24 Preparation of consolidated statements of income, changes in equity
   and cash flows   582
25 Accounting for associates and joint arrangements   605
26 Introduction to accounting for exchange differences   632

Part 6
INTERPRETATION   653

27 Earnings per share   655
28 Review of financial statements for management purposes   680
29 Analysis of published financial statements   713
30 An introduction to financial reporting on the Internet   755

Part 7
ACCOUNTABILITY   773

31 Corporate governance   775
32 Sustainability – environmental and social reporting   808

Index   837

Contents

Preface   xxi
Publisher's acknowledgements   xxvii

Part I
PREPARATION OF FINANCIAL STATEMENTS   I

I Accounting and reporting on a cash flow basis   3
    I.I   Introduction   3
    I.2   Shareholders   3
    I.3   What skills does an accountant require in respect of external reports?   4
    I.4   Managers   4
    I.5   What skills does an accountant require in respect of internal reports?   5
    I.6   Procedural steps when reporting to internal users   5
    I.7   Agency costs   8
    I.8   Illustration of periodic financial statements prepared under the cash flow concept to disclose realised operating cash flows   8
    I.9   Illustration of preparation of statement of financial position   12
    I.10   Treatment of non-current assets in the cash flow model   14
    I.11   What are the characteristics of these data that make them reliable?   15
    I.12   Reports to external users   16
    Summary   17
    Review questions   18
    Exercises   18
    References   20

2 Accounting and reporting on an accrual accounting basis   21
    2.1   Introduction   21
    2.2   Historical cost convention   22
    2.3   Accrual basis of accounting   22
    2.4   Mechanics of accrual accounting – adjusting cash receipts and payments   23
    2.5   Reformatting the statement of financial position   24
    2.6   Accounting for the sacrifice of non-current assets   24
    2.7   Published statement of cash flows   27
    Summary   28
    Review questions   29
    Exercises   29
    References   31

3 Preparation of financial statements of comprehensive income, changes in equity and financial position   **32**
3.1   Introduction   32
3.2   Preparing an internal statement of income from a trial balance   32
3.3   Reorganising the income and expenses into one of the formats required for publication   35
3.4   Format 1: classification of operating expenses and other income by function   36
3.5   Format 2: classification of operating expenses according to their nature   39
3.6   Other comprehensive income   39
3.7   How non-recurring or exceptional items can affect operating income   40
3.8   How decision-useful is the statement of comprehensive income?   42
3.9   Statement of changes in equity   42
3.10   The statement of financial position   43
3.11   The explanatory notes that are part of the financial statements   44
3.12   Has prescribing the formats meant that identical transactions are reported identically?   47
3.13   Fair presentation   50
3.14   What does an investor need in addition to the primary financial statements to make decisions?   51
Summary   55
Review questions   56
Exercises   57
References   66

4 Annual Report: additional financial disclosures   **67**
4.1   Introduction   67
4.2   IAS 10 *Events after the Reporting Period*   67
4.3   IAS 8 *Accounting Policies, Changes in Accounting Estimates and Errors*   69
4.4   What do segment reports provide?   71
4.5   IFRS 8 *Operating Segments*   72
4.6   Benefits and continuing concerns following the issue of IFRS 8   76
4.7   Discontinued operations – IFRS 5 *Non-current Assets Held for Sale and Discontinued Operations*   79
4.8   Held for sale – IFRS 5 *Non-current Assets Held for Sale and Discontinued Operations*   80
4.9   IAS 24 *Related Party Disclosures*   82
Summary   87
Review questions   87
Exercises   88
References   98

5 Statements of cash flows   **99**
5.1   Introduction   99
5.2   Development of statements of cash flows   99
5.3   Applying IAS 7 (revised) *Statements of Cash Flows*   100
5.4   Step approach to preparation of a statement of cash flows – indirect method   103
5.5   Additional notes required by IAS 7   106
5.6   Analysing statements of cash flows   107

5.7   Approach to answering questions with time constraints   112
5.8   Preparing a statement of cash flows when no statement of income is
        available   115
5.9   Critique of cash flow accounting   117
Summary   117
Review questions   117
Exercises   118
References   125

Part 2
INCOME AND ASSET VALUE MEASUREMENT SYSTEMS   127

6 Income and asset value measurement:
    an economist's approach   129
6.1   Introduction   129
6.2   Role and objective of income measurement   129
6.3   Accountant's view of income, capital and value   132
6.4   Critical comment on the accountant's measure   135
6.5   Economist's view of income, capital and value   136
6.6   Critical comment on the economist's measure   142
6.7   Income, capital and changing price levels   142
Summary   144
Review questions   144
Exercises   145
References   147
Bibliography   147

7 Accounting for price-level changes   148
7.1   Introduction   148
7.2   Review of the problems of historical cost accounting (HCA)   148
7.3   Inflation accounting   149
7.4   The concepts in principle   149
7.5   The four models illustrated for a company with cash purchases
        and sales   150
7.6   Critique of each model   154
7.7   Operating capital maintenance – a comprehensive example   157
7.8   Critique of CCA statements   168
7.9   The ASB approach   169
7.10   The IASC/IASB approach   171
7.11   Future developments   172
Summary   175
Review questions   175
Exercises   176
References   183
Bibliography   183

8 Revenue recognition   184
8.1   Introduction   184
8.2   IAS 18 Revenue   185
8.3   The issues involved in developing a new standard   186

8.4	The challenges under both IAS 18 and IFRS 15	187
8.5	IFRS 15 *Revenue from Contracts with Customers*	188
8.6	Five-step process to identify the amount and timing of revenue	189
8.7	Disclosures	200
	Summary	201
	Review questions	201
	Exercises	203
	References	207

Part 3
**REGULATORY FRAMEWORK – AN ATTEMPT TO
ACHIEVE UNIFORMITY** — 209

9	**Financial reporting – evolution of global standards**	**211**
9.1	Introduction	211
9.2	Why do we need financial reporting standards?	211
9.3	Why do we need standards to be mandatory?	212
9.4	Arguments in support of standards	214
9.5	Arguments against standards	214
9.6	Standard setting and enforcement by the Financial Reporting Council (FRC) in the UK	215
9.7	The International Accounting Standards Board	217
9.8	Standard setting and enforcement in the European Union	219
9.9	Standard setting and enforcement in the US	221
9.10	Advantages and disadvantages of global standards for publicly accountable entities	223
9.11	How do reporting requirements differ for non-publicly accountable entities?	224
9.12	IFRS for SMEs	225
9.13	Why have there been differences in financial reporting?	226
9.14	Move towards a conceptual framework	230
	Summary	230
	Review questions	231
	Exercises	231
	References	232

10	**Concepts – evolution of an international conceptual framework**	**234**
10.1	Introduction	234
10.2	Different countries meant different financial statements	234
10.3	Historical overview of the evolution of financial accounting theory	235
10.4	IASC *Framework for the Presentation and Preparation of Financial Statements*	237
10.5	*Conceptual Framework for Financial Reporting* 2010	238
10.6	Chapter 4 content	242
10.7	A Review of the *Conceptual Framework for Financial Reporting*	243
	Summary	245
	Review questions	245
	Exercises	246
	References	247

11 Ethical behaviour and implications for accountants | 248

11.1	Introduction	248
11.2	The meaning of ethical behaviour	248
11.3	The accounting standard-setting process and ethics	249
11.4	The IFAC *Code of Ethics for Professional Accountants*	250
11.5	Implications of ethical values for the principles versus rules-based approaches to accounting standards	253
11.6	Ethics in the accountant's work environment – a research report	256
11.7	Implications of unethical behaviour for stakeholders using the financial reports	258
11.8	The increasing role of whistle-blowing	263
11.9	Legal requirement to report – national and international regulation	265
11.10	Why should students learn ethics?	266
	Summary	267
	Review questions	268
	Exercises	270
	References	272

Part 4
STATEMENT OF FINANCIAL POSITION – EQUITY, LIABILITY AND ASSET MEASUREMENT AND DISCLOSURE | 273

12 Share capital, distributable profits and reduction of capital | 275

12.1	Introduction	275
12.2	Common themes	275
12.3	Total owners' equity: an overview	276
12.4	Total shareholders' funds: more detailed explanation	277
12.5	Accounting entries on issue of shares	279
12.6	Creditor protection: capital maintenance concept	280
12.7	Creditor protection: why capital maintenance rules are necessary	280
12.8	Creditor protection: how to quantify the amounts available to meet creditors' claims	281
12.9	Issued share capital: minimum share capital	282
12.10	Distributable profits: general considerations	282
12.11	Distributable profits: how to arrive at the amount using relevant accounts	284
12.12	When may capital be reduced?	284
12.13	Writing off part of capital which has already been lost and is not represented by assets	284
12.14	Repayment of part of paid-in capital to shareholders or cancellation of unpaid share capital	289
12.15	Purchase of own shares	290
	Summary	292
	Review questions	292
	Exercises	292
	References	297

13 Liabilities | 298

| 13.1 | Introduction | 298 |
| 13.2 | Provisions – a decision tree approach to their impact on the statement of financial position | 299 |

13.3	Treatment of provisions	300
13.4	The general principles that IAS 37 applies to the recognition of a provision	300
13.5	Management approach to measuring the amount of a provision	301
13.6	Application of criteria illustrated	303
13.7	Provisions for specific purposes	303
13.8	Contingent liabilities	306
13.9	Contingent assets	306
13.10	ED IAS 37 *Non-financial Liabilities*	307
13.11	ED/2010/1 *Measurement of Liabilities in IAS 37*	314
	Summary	314
	Review questions	315
	Exercises	315
	References	320

14 Financial instruments — **321**

14.1	Introduction	321
14.2	Financial instruments – the IASB's problem child	321
14.3	IAS 32 *Financial Instruments: Disclosure and Presentation*	324
14.4	IAS 39 *Financial Instruments: Recognition and Measurement*	329
14.5	IFRS 7 *Financial Instruments: Disclosure*	342
14.6	Financial instruments developments	346
	Summary	350
	Review questions	351
	Exercises	352
	References	359

15 Employee benefits — **360**

15.1	Introduction	360
15.2	Greater employee interest in pensions	360
15.3	Financial reporting implications	361
15.4	Types of scheme	361
15.5	Defined contribution pension schemes	364
15.6	Defined benefit pension schemes	364
15.7	IAS 19 (revised 2011) *Employee Benefits*	365
15.8	The asset or liability for pension and other post-retirement costs	365
15.9	Changes in the pension asset or liability position	366
15.10	Comprehensive illustration	369
15.11	Multi-employer plans	370
15.12	Disclosures	371
15.13	Other long-service benefits	371
15.14	Short-term benefits	371
15.15	Termination benefits	372
15.16	IFRS 2 *Share-based Payment*	373
15.17	Scope of IFRS 2	374
15.18	Recognition and measurement	374
15.19	Equity-settled share-based payments	374
15.20	Cash-settled share-based payments	377
15.21	Transactions which may be settled in cash or shares	378
15.22	IAS 26 *Accounting and Reporting by Retirement Benefit Plans*	379

	Summary	381
	Review questions	382
	Exercises	382
	References	387

16 Taxation in company accounts — **388**

16.1	Introduction	388
16.2	Corporation tax	388
16.3	Corporation tax systems – the theoretical background	389
16.4	Corporation tax and dividends	390
16.5	Corporation tax systems – avoidance and evasion	391
16.6	IAS 12 – accounting for current taxation	395
16.7	Deferred tax	396
16.8	A critique of deferred taxation	404
16.9	Value added tax (VAT)	406
	Summary	407
	Review questions	407
	Exercises	408
	References	411

17 Property, plant and equipment (PPE) — **412**

17.1	Introduction	412
17.2	PPE – concepts and the relevant IASs and IFRSs	412
17.3	What is PPE?	413
17.4	How is the cost of PPE determined?	414
17.5	What is depreciation?	416
17.6	What are the constituents in the depreciation formula?	418
17.7	Calculation of depreciation	419
17.8	Measurement subsequent to initial recognition	423
17.9	IAS 36 *Impairment of Assets*	425
17.10	IFRS 5 *Non-current Assets Held for Sale and Discontinued Operations*	431
17.11	Disclosure requirements	432
17.12	Government grants towards the cost of PPE	433
17.13	Investment properties	435
17.14	Effect of accounting policy for PPE on the interpretation of the financial statements	436
	Summary	438
	Review questions	439
	Exercises	439
	References	448

18 Leasing — **449**

18.1	Introduction	449
18.2	Background to leasing	499
18.3	Why was the IAS 17 approach so controversial?	451
18.4	IAS 17 – classification of a lease	452
18.5	Accounting requirements for operating leases	453
18.6	Accounting requirements for finance leases	454
18.7	Example allocating the finance charge using the sum of the digits method	455

18.8   Example allocating the finance charge using the actuarial method   457
18.9   Disclosure requirements for finance leases   458
18.10   Accounting for the lease of land and buildings   459
18.11   Leasing – a form of off-balance-sheet financing   460
18.12   Accounting for leases – a new approach   460
Summary   463
Review questions   463
Exercises   464
References   467

19 Intangible assets   468
19.1   Introduction   468
19.2   Intangible assets defined   468
19.3   Accounting treatment for research and development   471
19.4   Why is research expenditure not capitalised?   472
19.5   Capitalising development costs   473
19.6   Disclosure of R&D   474
19.7   IFRS for SMEs' treatment of intangible assets   474
19.8   Internally generated and purchased goodwill   475
19.9   The accounting treatment of goodwill   475
19.10   Critical comment on the various methods that have been used
        to account for goodwill   477
19.11   Negative goodwill/Badwill   480
19.12   Brands   480
19.13   Accounting for acquired brands   482
19.14   Emissions trading   483
19.15   Intellectual capital disclosures (ICDs) in the annual report   484
19.16   Review of implementation of IFRS 3   485
19.17   Review of implementation of identified intangibles under IAS 38   486
Summary   488
Review questions   488
Exercises   489
References   495

20 Inventories   497
20.1   Introduction   497
20.2   Inventory defined   497
20.3   The impact of inventory valuation on profits   498
20.4   IAS 2 Inventories   499
20.5   Inventory valuation   500
20.6   Work in progress   506
20.7   Inventory control   508
20.8   Creative accounting   509
20.9   Audit of the year-end physical inventory count   511
20.10   Published accounts   512
20.11   Agricultural activity   513
Summary   516
Review questions   517
Exercises   517
References   522

21 Construction contracts **523**
21.1   Introduction   523
21.2   The need to replace IAS 11 *Construction Contracts*   523
21.3   Identification of contract revenue under IAS 11   525
21.4   Identification of contract costs under IAS 11   525
21.5   Public–private partnerships (PPPs)   528
21.6   IFRS 15 treatment of construction contracts   532
21.7   An approach when a contract can be separated into components   534
21.8   Accounting for a contract – an example   535
21.9   Illustration – loss-making contract using the step approach   538
        Summary   539
        Review questions   540
        Exercises   540
        References   546

Part 5
CONSOLIDATED ACCOUNTS   **547**

22 Accounting for groups at the date of acquisition   **549**
22.1   Introduction   549
22.2   Preparing consolidated accounts for a wholly owned subsidiary   549
22.3   IFRS 10 *Consolidated Financial Statements*   549
22.4   Fair values   551
22.5   Ilustration where there is a wholly owned subsidiary   551
22.6   Preparing consolidated accounts when there is a partly
        owned subsidiary   553
22.7   The treatment of differences between a subsidiary's fair value
        and book value   556
22.8   The parent issues shares to acquire shares in a subsidiary   557
22.9   IFRS 3 *Business Combinations* treatment of goodwill at the date
        of acquisition   558
22.10   When may a parent company not be required to prepare
        consolidated accounts?   558
22.11   When may a parent company exclude or not exclude a subsidiary
        from a consolidation?   558
22.12   IFRS 13 *Fair Value Measurement*   559
22.13   What advantages are there for stakeholders from requiring groups
        to prepare consolidated accounts?   560
        Summary   561
        Review questions   561
        Exercises   562
        References   567

**23 Preparation of consolidated statements of financial position
after the date of acquisition**   **568**
23.1   Introduction   568
23.2   Uniform accounting policies and reporting dates   568
23.3   Pre- and post-acquisition profits/losses   568
23.4   The Bend Group – assuming there have been no inter-group
        transactions   569

23.5	Inter-company transactions	571
23.6	The Prose Group – assuming there have been inter-group transactions	572
Summary		574
Review questions		575
Exercises		575
References		581

24 Preparation of consolidated statements of income, changes in equity and cash flows   582

24.1	Introduction	582
24.2	Eliminate inter-company transactions	582
24.3	Preparation of a consolidated statement of income – the Ante Group	583
24.4	The statement of changes in equity (SOCE)	585
24.5	Other consolidation adjustments	585
24.6	A subsidiary acquired part-way through the year	587
24.7	Published format statement of income	589
24.8	Consolidated statements of cash flows	590
Summary		592
Review questions		592
Exercises		593
References		604

25 Accounting for associates and joint arrangements   605

25.1	Introduction	605
25.2	Definitions of associates and of significant influence	605
25.3	The treatment of associated companies in consolidated accounts	606
25.4	The Brill Group – group accounts with a profit-making associate	606
25.5	The Brill Group – group accounts with a loss-making associate	609
25.6	The acquisition of an associate part-way through the year	611
25.7	Joint arrangements	612
25.8	Disclosure in the financial statements	616
25.9	Parent company use of the equity method in its separate financial statements	617
Summary		619
Review questions		619
Exercises		620
References		631

26 Introduction to accounting for exchange differences   632

26.1	Introduction	632
26.2	How to record foreign currency transactions in a company's own books	633
26.3	Boil plc – a more detailed illustration	635
26.4	IAS 21 *Concept of Functional and Presentation Currencies*	636
26.5	Translating the functional currency into the presentation currency	638
26.6	Preparation of consolidated accounts	639
26.7	How to reduce the risk of translation differences	642
26.8	Critique of use of presentational currency	643
26.9	IAS 29 *Financial Reporting in Hyperinflationary Economies*	643

Summary   645
Review questions   645
Exercises   646
References   652

Part 6
INTERPRETATION   653

27 Earnings per share   **655**
27.1   Introduction   655
27.2   Why is the earnings per share figure important?   655
27.3   How is the EPS figure calculated?   656
27.4   The use to shareholders of the EPS   657
27.5   Illustration of the basic EPS calculation   658
27.6   Adjusting the number of shares used in the basic EPS calculation 658
27.7   Rights issues   661
27.8   Adjusting the earnings and number of shares used in the diluted
       EPS calculation   666
27.9   Procedure where there are several potential dilutions   668
27.10 Exercise of conversion rights during the financial year   670
27.11 Disclosure requirements of IAS 33   670
27.12 The Improvement Project   672
27.13 The Convergence Project   672
Summary   673
Review questions   673
Exercises   674
References   679

28 Review of financial statements for management purposes   **680**
28.1   Introduction   680
28.2   Overview of techniques for the analysis of financial data   681
28.3   Ratio analysis – a case study   682
28.4   Introductory review   683
28.5   Financial statement analysis, part 1 – financial performance   686
28.6   Financial statement analysis, part 2 – liquidity   692
28.7   Financial statement analysis, part 3 – financing   695
28.8   Peer comparison   697
28.9   Report based on the analysis   698
28.10 Caution when using ratios for prediction   699
Summary   701
Review questions   702
Exercises   703
Reference   712

29 Analysis of published financial statements   **713**
29.1   Introduction   713
29.2   Improvement of information for shareholders   714
29.3   Published financial statements – their limitations for interpretation
       purposes   715

29.4	Published financial statements – additional entity-wide cash-based performance measures	716
29.5	Ratio thresholds to satisfy shariah compliance	719
29.6	Use of ratios in restrictive loan covenants	721
29.7	Investor-specific ratios	723
29.8	Determining value	726
29.9	Predicting corporate failure	731
29.10	Professional risk assessors	735
29.11	Valuing shares of an unquoted company – quantitative process	736
29.12	Valuing shares of an unquoted company – qualitative process	739
Summary		741
Review questions		741
Exercises		743
References		753

30 An introduction to financial reporting on the Internet — **755**

30.1	Introduction	755
30.2	The objectives of financial reporting	755
30.3	Reports and the flow of information pre-XBRL	757
30.4	What are HTML, XML and XBRL?	758
30.5	Reports and the flow of information post-XBRL	759
30.6	Why are companies adopting XBRL?	760
30.7	What are the processes followed to adopt XBRL for outputting information?	761
30.8	What is needed when receiving XBRL output information?	764
30.9	Progress of XBRL development for internal accounting	766
30.10	Real-time reporting	767
30.11	Further developments	768
Summary		768
Review questions		769
Exercises		769
References		771
Bibliography		771

Part 7
ACCOUNTABILITY — **773**

31 Corporate governance — **775**

31.1	Introduction	775
31.2	A systems perspective	775
31.3	Different jurisdictions have different governance priorities	777
31.4	Pressures on good governance behaviour vary over time	778
31.5	Types of past unethical behaviour	779
31.6	The effect on capital markets of good corporate governance	780
31.7	Risk management	781
31.8	The role of internal control and internal audit in corporate governance	782
31.9	External audits in corporate governance	784
31.10	Executive remuneration in the UK	789
31.11	Corporate governance, legislation and codes	792
31.12	Corporate governance – the UK experience	793

Summary   802
Review questions   802
Exercises   805
References   806

32 Sustainability – environmental and social reporting   808
32.1   Introduction   808
32.2   An overview – stakeholders' growing interest in corporate social
       responsibility (CSR)   810
32.3   An overview – business's growing interest in corporate social
       responsibility   811
32.4   Companies' voluntary adoption of guidelines and certification   814
32.5   The accountant's role in a capitalist industrial society   816
32.6   The nature of the accountant's involvement   817
32.7   Summary on environmental reporting   819
32.8   Concept of social accounting   820
32.9   Background to social accounting   821
32.10 Corporate social responsibility reporting   824
32.11 Need for comparative data   824
32.12 Investors   825
32.13 The accountant's changing role   826
Summary   828
Review questions   829
Exercises   830
References   833
Bibliography   835

Index   837

Preface

Our objective is to provide a balanced and comprehensive framework to enable students to acquire the requisite knowledge and skills to appraise current practice critically and to evaluate proposed changes from a theoretical base. To this end, the text contains:

- extracts from current IASs and IFRSs;
- illustrations from published accounts;
- a range of review questions;
- exercises of varying difficulty;
- extensive references.

Solutions to selected exercises can be found on MyAccountingLab.

We have assumed that readers will have an understanding of financial accounting to a foundation or first-year level, although the text and exercises have been designed on the basis that a brief revision is still helpful. For the preparation of financial statements in Part 1 and Part 5 we have structured the chapters to assist readers who may have no accounting knowledge.

Lecturers are using the text selectively to support a range of teaching programmes for second-year and final-year undergraduate and postgraduate programmes. We have therefore attempted to provide subject coverage of sufficient breadth and depth to assist selective use.

The text has been adopted for financial accounting, reporting and analysis modules on:

- second-year undergraduate courses for Accounting, Business Studies and Combined Studies;
- final-year undergraduate courses for Accounting, Business Studies and Combined Studies;
- MBA courses;
- specialist MSc courses; and
- professional courses preparing students for professional accountancy examinations.

Changes to the seventeenth edition

Our emphasis has been on keeping the text current and responsive to constructive comments from reviewers and lecturers.

National accounting standards and the IASB

Since 2005 UK listed companies have followed international standards EU-IFRS for their consolidated accounts.

From 2013 large and medium-sized private companies in the UK will follow FRS 102 *The Financial Reporting Standard*. This standard is based (with UK modifications) on *IFRS for SMEs* which was issued by the IASB in 2009 – the indications are that companies might adopt FRS 102 instead of the IFRS in the future.

Smaller entities will continue to follow FRSSE.

Accounting standards – seventeenth edition updates

Chapters covering the following International Standards have been revised. They are as follows:

Chapter 3	Preparation of financial statements	IAS 1
Chapter 4	Preparation of additional financial statements	IAS 10, IAS 24, IFRS 5 and IFRS 8
Chapter 5	Statements of cash flows	IAS 7
Chapter 7	Accounting for price-level changes	IAS 29
Chapter 8	Revenue recognition	IFRS 15
Chapter 13	Liabilities	IAS 37/ED/2010/1
Chapter 14	Financial instruments	IAS 32, IFRS 7 and IFRS 9
Chapter 15	Employee benefits	IAS 19 (revised 2011), IAS 26 and IFRS 2
Chapter 16	Taxation in company accounts	IAS 12
Chapter 17	Property, plant and equipment (PPE)	IAS 16, IAS 20, IAS 23, IAS 36, IAS 40 and IFRS 5
Chapter 18	Leasing	IAS 17 and ED/2013/6
Chapter 19	Intangible assets	IAS 38 and IFRS 3
Chapter 20	Inventories	IAS 2
Chapter 21	Construction contracts	IAS 11 and IFRS 15
Chapters 22–26	Consolidation	IAS 21, IAS 28, IFRS 3, 10, 11, 12 and 13
Chapter 27	Earnings per share	IAS 33

Part 1 Preparation of financial statements

Chapters 1 and 2 continue to cover accounting and reporting on a cash flow and accrual basis. Chapters 3 to 5 have been revised. They cover the preparation of statements of income, changes in equity, financial position and cash flows.

Part 2 Income and asset value measurement systems

Chapters 6 and 7 covering the economic income approach and accounting for price-level changes have been retained. Chapter 8 discusses the application of IFRS 15.

Part 3 Regulatory framework – an attempt to achieve uniformity

Chapters 9 and 10 have been revised.

Part 4 Statement of financial position

Chapters 12– 21 are core chapters which have been retained and updated as appropriate.

Part 5 Consolidated accounts

Chapters 22– 26 have been updated and revised to improve accessibility with explanations from first principles.

Part 6 Interpretation

Chapters 28 and 29 are retained, aiming at encouraging good report writing based on the pyramid approach to ratios and an introduction to other tools and techniques for specific assignments. Chapter 30 has been revised to discuss an overview of financial reporting on the internet.

Part 7 Accountability

Chapters 31 and 32 have been updated and continue to focus on the accountant's role in corporate governance and in the development of sustainability and integrated reporting.

Recent developments

In addition to the issue by the IASB of IFRS 15 and various EDs, the issue of the new EU Accounting Directive and the issue of statutory regulations on directors' remuneration in the UK in response to the apparent lack of any relationship between increases in directors' remuneration and company performance, there has been:

- guidance by the FRC in the UK on the implementation of the Strategic Report,
- the issue by the International Integrated Reporting Council of the Integrated Reporting Framework which aims to explain how the company interacts with the external environment and the capitals such as the Financial and Manufactured capitals to create value over the short term, medium term and long term, and
- the launch by the Collective Engagement Working Group of the 'Investors Forum' to encourage investors and companies to develop a shared sense of partnership to promote long-term strategies that can generate sustainable wealth creation for all stakeholders.

The content of financial reports continues to be subjected to discussion with tension between preparers, stakeholders, auditors, academics and standard setters; this is mirrored in the tension that exists between theory and practice.

- Preparers favour reporting transactions on a historical cost basis, which is reliable but does not provide shareholders with relevant information to appraise past performance or to predict future earnings.
- Shareholders favour forward-looking reports relevant in estimating future dividend and capital growth and in understanding environmental and social impacts.

- Stakeholders favour quantified and narrative disclosure of environmental and social impacts and the steps taken to reduce negative impacts.
- Auditors favour reports that are verifiable so that the figures can be substantiated to avoid them being proved wrong at a later date.
- Academic accountants favour reports that reflect economic reality and are relevant in appraising management performance and in assessing the capacity of the company to adapt.
- Standard setters lean towards the academic view and favour reporting according to the commercial substance of a transaction.

In order to understand the tensions that exist, students need:

- the skill to prepare financial statements in accordance with the historical cost and current cost conventions, both of which appear in annual financial reports;
- an understanding of the main thrust of mandatory and voluntary standards;
- an understanding of the degree of flexibility available to the preparers and the impact of this on reported earnings and the figures in the statement of financial position;
- an understanding of the limitations of financial reports in portraying economic reality; and
- an exposure to source material and other published material in so far as time permits.

Acknowledgements

Financial reporting is a dynamic area and we see it as extremely important that the text should reflect this and be kept current. Assistance has been generously given by colleagues and many others in the preparation and review of the text and assessment material. This seventeenth edition continues to be very much a result of the authors, colleagues, reviewers and Pearson editorial and production staff working as a team and we are grateful to all concerned for their assistance in achieving this.

We owe particular thanks to Charles Batchelor, formerly of FTC Kaplan, for 'Financial instruments' (Chapter 14); Ozer Erman of Kingston University for 'Share capital, distributable profits and reduction of capital' (Chapter 12); Paul Robins of the Financial Training Company for Leasing (Chapter 18); Professor Garry Tibbits of the University of Western Sydney for 'Revenue recognition' (Chapter 8); Hendrika Tibbits of the University of Western Sydney for 'An introduction to financial reporting on the Internet' (Chapter 30); and David Towers, formerly of Keele University, for 'Taxation in company accounts' (Chapter 16).

The authors are grateful for the constructive comments received over various editions from the following reviewers who have assisted us in making improvements: Pik Liew of Essex University; Anitha Majeed of Coventry University; Allison Wylde of London Metropolitan University; Terry Morris of Queen Mary University of London; Ajjay Mandal of London South Bank University; and Michael Jeffrey of Manchester Metropolitan University. We would also like to thank the reviewers of this new edition of the book.

Thanks are owed to Keith Brown, formerly of De Montfort University; Kenneth N. Field of the University of Leeds; Sue McDermott of London Metropolitan Business School; David Murphy of Manchester Business School; Bahadur Najak of the University of Durham; Graham Sara of the University of Warwick; and Laura Spira of Oxford Brookes University.

Thanks are also due to the following organisations: the Financial Reporting Council, the International Accounting Standards Board, the Association of Chartered Certified

Accountants, the Association of International Accountants, the Chartered Institute of Management Accountants, the Institute of Certified Public Accountants (CPA) in Ireland, the Institute of Chartered Accountants of England and Wales, the Institute of Chartered Accountants of Scotland and the Institute of Chartered Secretaries and Administrators.

We would also like to thank the authors of some of the end-of-chapter exercises. Some of these exercises have been inherited from a variety of institutions with which we have been associated, and we have unfortunately lost the identities of the originators of such material with the passage of time. We are sorry that we cannot acknowledge them by name and hope that they will excuse us for using their material.

We are indebted to Lucy Winder and the editorial team at Pearson Education for active support in keeping us largely to schedule and the attractively produced and presented text.

Finally we thank our wives, Di and Jacklin, for their continued good-humoured support during the period of writing and revisions, and Giles Elliott for his critical comment from the commencement of the project. We alone remain responsible for any errors and for the thoughts and views that are expressed.

Barry and Jamie Elliott

Publisher's acknowledgements

We are grateful to the following for permission to reproduce copyright material:

Figures

Figure on page 48 from AstraZeneca PLC (private limited company), Annual Report, Inventories, http://www.astrazeneca-annualreports.com/2007/financial_statements/accounting_policies_group.asp; Figure on page 730 from Thomas Reuters. Granted by permission of Marks & Spencer, Marks and Spencer plc (company); Figure 30.1 from www.xbrl.org.au/training/NSWWorkshop.pdf, https://www.xbrl.org/the-consortium/about/legal/copyright-information/. Used by permission; Figure 30.2 from http://xbrl.org.au/training/NSWWorkshop.pdf, https://www.xbrl.org/the-consortium/about/legal/copyright-information/. Used by permission; Figure 30.3 from http://xbrl.org.au/training/NSWWorkshop.pdf, https://www.xbrl.org/the-consortium/about/legal/copyright-information/. Used by permission; Figure 32.1 adapted from http://www.imperial.co.za/CMSFiles/File/Documents/2011AnnualResults/ImperialIntegratedReport2011.pdf

Text

Exercise 3.7 from The Association of International Accounts; Quote on page 85 from IAS 24 *Related Party Disclosures*, IASB, revised 2009, Copyright © IFRS Foundation. All rights reserved. Reproduced by Pearson Education Limited with the permission of the IFRS Foundation ®. No permission granted to third parties to reproduce or distribute.; Exercises 4.8, 4.10, 4.12, 5.5, 7.7, 13.10, 17.7, 17.8, 18.8, 19.8, 21.7, 23.4, 25.3, 25.8, 26.4, 26.5, 27.8, 27.9, 27.9, 28.5, 29.7 from The Association of International Accountants, © 2012 AIA. All rights reserved; Exercise 4.11 from Association of Chartered Certified Accountants, We are grateful to the Association of Chartered Certified Accountants (ACCA) for permission to reproduce past examination questions. The suggested solutions in the exam answer bank have been prepared by us, unless otherwise stated.; Exercises 4.8, 4.10, 4.12, 5.5, 7.7, 13.10, 17.7, 17.8, 18.8, 19.8, 21.7, 23.4, 25.3, 25.8, 26.4, 26.5, 27.8, 27.9, 27.9, 28.5, 29.7 from The Association of International Accountants; Exercise on page 97 from Institute of Certified Public Accountants in Ireland (CPA Ireland), Professional I Stage Corporate Reporting Examination, August 2010, http://www.cpaireland.ie/docs/default-source/Students/Study-Support/P1-Corporate-Reporting/august-2011.pdf?sfvrsn=0, used by permission; Quote on page 185 from The accounting standard IAS18, © Copyright IFRS Foundation. Used by permission; Quote on page 186 from The accounting standard IAS18. IAS © 2014,

© Copyright IFRS Foundation. Used by permission; Quote on page 186 from The accounting standard IAS18 IAS © 2014, © Copyright IFRS Foundation. Used by permission; Quote on page 188 from International Financial Reporting Standards http://www.ifrs.org/Pages/Terms-and-Conditions.aspx#IP, © Copyright IFRS Foundation. Used by permission; Extract on page 190 from Hewlett-Packard Company and Subsidiaries, Annual Report 2013, Note 1, pp. 86–8; Exercise on page 201 from http://accountingonion.com/2014/05/the-wild-west-of-nonauthoritative-gaap.html, Copyright 2014 © Thomas I. Selling. All Rights Reserved. Used by permission; Quote on page 217 from The International Accounting Standards Board (IASB). The IASB website, www.iasb.org.uk, © Copyright IFRS Foundation. Used by permission; Exercise 10.1 from SFAS 109, Accounting for Income Taxes, FASB, 1992, paragraphs 77 and 78, Copyright © 2010 by Financial Accounting Foundation. Reproduced with permission; Exercise 10.2 from 'Comments of Leonard Spacek', in R.T. Sprouse and M. Moonitz, *A Tentative Set of Broad Accounting Principles for Business Enterprises*, Accounting Research Study no. 3, AICPA, New York, 1962, pp. 77–9 American Institute of Certified Public Accountants; Exercise 10.3 from *Accountancy Age*, 25 January 200, Used by permission; Exercise 10.4 from *Louder than Words – Principles and Actions for Making Corporate Reports Less Complex and More Relevant included a call for action FRC 2009*. Financial Reporting Council; Quote on page 250 from The IFAC Code of Ethics for Professional Accountants – The IFAC Fundamental Principles, This text is an extract from *Handbook of the Code of Ethics for Professional Accountants*, 2014 edition of the International Ethics Standards Board (IESBA), Copyright © July 2014 by the International Federation of Accountants (IFAC). All rights reserved. Used with permission of IFAC; Quote on page 252 from Extract from the KPMG Code of Conduct, KPMG's Code of Conduct, *Our Promise of Professionalism*, p. 41 © 2010 KPMG LLP. Used by permission; Quote on page 253 from IFAC, Code of Ethics for Professional Accountants, This text is an extract from *Handbook of the Code of Ethics for Professional Accountants*, 2014 edition of the International Ethics Standards Board (IESBA), © July 2014 by the International Federation of Accountants (IFAC). All rights reserved. Used with permission of IFAC; Quotes on page 308, page 310, page 311, page 312, page 312 from ED IAS 37, © Copyright IFRS Foundation. Used by permission; Quote on page 342 from IFRS 7, © Copyright IFRS Foundation. Used by permission; Exercise 15.4 from Dip IFR December 2006 http://www.accaglobal.com/content/dam/acca/global/PDF-students/EC/2012-d/DipIFR_2006_dec_q.pdf, We are grateful to the Association of Chartered Certified Accountants (ACCA) for permission to reproduce past examination questions. The suggested solutions in the exam answer bank have been prepared by us, unless otherwise stated; Extract on page 400 from FASB Discussion Memorandum, Conceptual Framework for Financial Accounting and Reporting: Elements of Financial Statements and Their Measurement (Scope and Implications of the Conceptual Framework Project booklet, page 5), Copyright © 2010 by Financial Accounting Foundation. Used with permission; Quote on page 405 from Framework for the Preparation and Presentation of Financial Statements, IASB, 2001, para. 22, © Copyright IFRS Foundation. Used by permission; Quote on page 405 from Framework for the Preparation and Presentation of Financial Statements, IASB, 2001, para. 53, © Copyright IFRS Foundation. Used by permission; Quote on page 405 from Framework for the Preparation and Presentation of Financial Statements, IASB, 2001, para. 35, © Copyright IFRS Foundation. Used by permission; Quote on page 406 from IAS 18 Revenue, IASB, 2001, para. 8, © Copyright IFRS Foundation. Used by permission; Exercise 17.4 from (ACCA); Exercise 20.8 from DipIFR 2004; Quote on page 526 from Balfour Beatty 2013 annual accounts, © 2014 Balfour Beatty plc. Used by permission; Exercise 24.11 from from Institute of Certified Public Accountants in Ireland (CPA Ireland), Professional I Stage Corporate Reporting Examination, August

2010), used by permission; Quote on page 619 from IAS 28, © Copyright IFRS Foundation. Used by permission; Quotes on page 633, page 638 from IAS 21, © Copyright IFRS Foundation. Used by permission; Exercise 29.6 from ACCA, We are grateful to the Association of Chartered Certified Accountants (ACCA) for permission to reproduce past examination questions. The suggested solutions in the exam answer bank have been prepared by us, unless otherwise stated; Exercises 4.8, 4.10, 4.12, 5.5, 7.7, 13.10, 17.7, 17.8, 18.8, 19.8, 21.7, 23.4, 25.3, 25.8, 26.4, 26.5, 27.8, 27.9, 27.9, 28.5, 29.7 from The Association of International Accountants, © 2012 AIA. All rights reserved.

In some instances we have been unable to trace the owners of copyright material, and we would appreciate any information that would enable us to do so.

PART 1

Preparation of financial statements

CHAPTER 1

Accounting and reporting on a cash flow basis

1.1 Introduction

Accountants are communicators. Accountancy is the art of communicating financial information about a business entity to users such as shareholders and managers. The communication is generally in the form of financial statements that show in money terms the economic resources under the control of the management. The art lies in selecting the information that is relevant to the user and is reliable.

Shareholders require periodic information that the managers are accounting properly for the resources under their control. This information helps the shareholders to evaluate the performance of the managers. The performance measured by the accountant shows the extent to which the economic resources of the business have grown or diminished during the year.

The shareholders also require information to **predict future performance**. At present companies are not required to publish forecast financial statements on a regular basis and the shareholders use the report of past performance when making their predictions.

Managers require information in order to control the business and make investment decisions.

Objectives

By the end of this chapter, you should be able to:

- explain the extent to which cash flow accounting satisfies the information needs of shareholders and managers;
- prepare a cash budget and operating statement of cash flows;
- explain the characteristics that make cash flow data a reliable and fair representation;
- critically discuss the use of cash flow accounting for predicting future dividends.

1.2 Shareholders

Shareholders are external users. As such, they are unable to obtain access to the same amount of detailed historical information as the managers, e.g. total administration costs are disclosed in the published profit and loss account, but not an analysis to show how the figure is made up. Shareholders are also unable to obtain associated information, e.g. budgeted sales and costs. Even though the shareholders own a company, their entitlement to information is restricted.

The information to which shareholders are entitled is restricted to that specified by statute, e.g. the Companies Acts, or by professional regulation, e.g. Financial Reporting Standards, or by market regulations, e.g. listing requirements. This means that there may be a tension between the **amount** of information that a shareholder would like to receive and the amount that the directors are prepared to provide. For example, shareholders might consider that forecasts of future cash flows would be helpful in predicting future dividends, but the directors might be concerned that such forecasts could help competitors or make directors open to criticism if forecasts are not met. As a result, this information is not disclosed.

There may also be a tension between the **quality** of information that shareholders would like to receive and that which directors are prepared to provide. For example, the share-holders might consider that judgements made by the directors in the valuation of long-term contracts should be fully explained, whereas the directors might prefer not to reveal this information given the high risk of error that often attaches to such estimates. In practice, companies tend to compromise: they do not reveal the judgements to the shareholders, but maintain confidence by relying on the auditor to give a clean audit report.

The financial reports presented to the shareholders are also used by other parties such as lenders and trade creditors, and they have come to be regarded as general-purpose reports. However, it may be difficult or impossible to satisfy the needs of all users. For example, users may have different timescales – shareholders may be interested in the long-term trend of earnings over three years, whereas creditors may be interested in the likelihood of receiving cash within the next three months.

The information needs of the shareholders are regarded as the primary concern. The government perceives shareholders to be important because they provide companies with their economic resources. It is shareholders' needs that take priority in deciding on the nature and detailed content of the general-purpose reports.[1]

1.3 What skills does an accountant require in respect of external reports?

For external reporting purposes the accountant has a twofold obligation:

- an obligation to ensure that the financial statements comply with statutory, professional and Listing requirements; this requires the accountant to possess **technical expertise**;
- an obligation to ensure that the financial statements present the substance of the commercial transactions the company has entered into; this requires the accountant to have **commercial awareness**.

1.4 Managers

Managers are internal users. As such, they have access to detailed financial statements showing the current results, the extent to which these vary from the budgeted results and the future budgeted results. Other examples of internal users are sole traders, partners and, in a company context, directors and managers.

There is no statutory restriction on the amount of information that an internal user may receive; the only restriction would be that imposed by the company's own policy. Frequently, companies operate a 'need to know' policy and only the directors see all the financial statements; employees, for example, would be most unlikely to receive information that would assist them in claiming a salary increase – unless, of course, it happened to be a time of recession, when information would be more freely provided by management as a means of containing claims for an increase.

1.5 What skills does an accountant require in respect of internal reports?

For the internal user, the accountant is able to tailor his or her reports. The accountant is required to produce financial statements that are specifically relevant to the user requesting them.

The accountant needs to be skilled in identifying the information that is needed and conveying its implication and meaning to the user. The user needs to be confident that the accountant understands the user's information needs and will satisfy them in a language that is understandable. The accountant must be a skilled communicator who is able to instil confidence in the user that the information is:

- relevant to the user's needs;
- measured objectively;
- presented within a timescale that permits decisions to be made with appropriate information;
- verifiable, in that it can be confirmed that the report represents the transactions that have taken place;
- reliable, in that it is as free from bias as is possible;
- a complete picture of material items;
- a fair representation of the business transactions and events that have occurred or are being planned.

The accountant is a trained reporter of financial information. Just as for external reporting, the accountant needs commercial awareness. It is important, therefore, that he or she should not operate in isolation.

1.5.1 Accountants' reporting role

The accountant's role is to ensure that the information provided is useful for making decisions. For external users, the accountant achieves this by providing a general-purpose financial statement that complies with statute and is reliable. For internal users, this is done by interfacing with the user and establishing exactly what financial information is relevant to the decision that is to be made.

We now consider the steps required to provide relevant information for internal users.

1.6 Procedural steps when reporting to internal users

A number of user steps and accounting action steps can be identified within a financial decision model. These are shown in Figure 1.1.

Note that, although we refer to an accountant/user interface, this is not a single occurrence because the user and accountant interface at each of the user decision steps.

At **step 1**, the accountant attempts to ensure that the decision is based on the appropriate appraisal methodology. However, the accountant is providing a service to a user and, while the accountant may give guidance, the final decision about methodology rests with the user.

Figure 1.1 General financial decision model to illustrate the user/accountant interface

USER		ACCOUNTANT
User step 1		Identify the material information
Identify decision and		needed by the user
how it is to be made		Measure the relevant
User step 2		information
Establish with the	**USER/ACCOUNTANT**	Prepare report for user to
accountant the	**INTERFACE**	allow user to make decision
information necessary		Provide an understandable
for decision making		report to the user
User step 3		
Seek relevant data		
from the accountant		

At **step 2**, the accountant needs to establish the information necessary to support the decision that is to be made.

At **step 3**, the accountant needs to ensure that the user **understands** the full impact and financial implications of the accountant's report, taking into account the user's level of understanding and prior knowledge. This may be overlooked by the accountant, who feels that the task has been completed when the written report has been typed.

It is important to remember in following the model that the accountant is attempting to satisfy the information needs of the individual user rather than those of a 'user group'. It is tempting to divide users into groups with apparently common information needs, without recognising that a group contains individual users with different information needs. We return to this later in the chapter, but for the moment we continue by studying a situation where the directors of a company are considering a proposed capital investment project.

Let us assume that there are three companies in the retail industry: Retail A Ltd, Retail B Ltd and Retail C Ltd. The directors of each company are considering the purchase of a warehouse. We could assume initially that, because the companies are operating in the same industry and are faced with the same investment decision, they have identical information needs. However, enquiry might establish that the directors of each company have a completely different attitude to, or perception of, the primary business objective.

For example, it might be established that Retail A Ltd is a large company and under the Fisher–Hirshleifer separation theory the directors seek to maximise profits for the benefit of the equity investors; Retail B Ltd is a medium-sized company in which the directors seek to obtain a satisfactory return for the equity shareholders; and Retail C Ltd is a smaller company in which the directors seek to achieve a satisfactory return for a wider range of stakeholders, including, perhaps, the employees as well as the equity shareholders.

The accountant needs to be aware that these differences may have a significant effect on the information required. Let us consider this diagrammatically in the situation where a capital investment decision is to be made, referring particularly to user step 2: 'Establish with the accountant the information necessary for decision making'.

Figure 1.2 Impact of different user attitudes on the information needed in relation to a capital investment proposal

	USER A Directors of Retail A Ltd	USER B Directors of Retail B Ltd	USER C Directors of Retail C Ltd
User attitude	PROFIT MAXIMISER for SHAREHOLDERS	PROFIT SATISFICER for SHAREHOLDERS	PROFIT SATISFICER for SHAREHOLDERS/ STAFF
Relevant data to measure	CASH FLOWS	CASH FLOWS	CASH FLOWS
Appraisal method (decided on by user)	IRR	NPV	NPV
Appraisal criterion (decided on by user)	HIGHEST IRR	NPV but only if positive	NPV possibly even if negative

We can see from Figure 1.2 that the accountant has identified that:

● the relevant financial data are the same for each of the users, i.e. cash flows; but
● the appraisal methods selected, i.e. internal rate of return (IRR) and net present value (NPV), are different; and
● the appraisal criteria employed by each user, i.e. higher IRR and NPV, are different.

In practice, the user is likely to use more than one appraisal method, as each has advantages and disadvantages. However, we can see that, even when dealing with a single group of apparently homogeneous users, the accountant has first to identify the information needs of the particular user. Only then is the accountant able to identify the relevant financial data and the appropriate report. It is the user's needs that are predominant.

If the accountant's view of the appropriate appraisal method or criterion differs from the user's view, the accountant might decide to report from both views. This approach affords the opportunity to improve the user's understanding and encourages good practice.

The diagrams can be combined (Figure 1.3) to illustrate the complete process. The user is assumed to be Retail A Ltd, a company that has directors who are profit maximisers.

The accountant is reactive when reporting to an internal user. We observe this characteristic in the Norman example set out in Section 1.8. Because the cash flows are identified as relevant to the user, it is these flows that the accountant will record, measure and appraise.

The accountant can also be proactive, by giving the user advice and guidance in areas where the accountant has specific expertise, such as the appraisal method that is most appropriate to the circumstances.

Figure 1.3 User/accountant interface where the user is a profit maximiser

General model USER	Specific application for Retail A Ltd A PROFIT MAXIMISER		General model ACCOUNTANT	Specific application for Retail A Ltd ACCOUNTANT
Step 1 Decision to be made	Appraise which project warrants capital investment		Identify information needed by the user	User decision criterion is IRR
Step 2 Information needed		USER/ ACCOUNTANT INTERFACE	Measure	Measure the project cash flows
Step 3 Seek relevant data	Project with the highest IRR		Prepare report	Prepare report of highest IRR
	Report of IRR project		Provide report	Submit report of project with highest IRR per £ invested

1.7 Agency costs[2]

The information in Figure 1.2 assumes that the directors have made their investment decision based on the assumed preferences of the shareholders. However, in real life, the directors might also be influenced by how the decision impinges on their own position. If, for example, their remuneration is a fixed salary, they might select not the investment with the highest IRR, but the one that maintains their security of employment. The result might be suboptimal investment and financing decisions based on risk aversion and over-retention. To the extent that the potential cash flows have been reduced, there will be an agency cost to the shareholders. This agency cost is an opportunity cost – the amount that was forgone because the decision making was suboptimal – and, as such, it will not be recorded in the books of account and will not appear in the financial statements.

1.8 Illustration of periodic financial statements prepared under the cash flow concept to disclose realised operating cash flows

In the above example of Retail A, B and C, the investment decision for the acquisition of a warehouse was based on an appraisal of cash flows. This raises the question: 'Why not continue with the cash flow concept and report the financial changes that occur after the investment has been undertaken using that same concept?'

To do this, the company will record the consequent cash flows through a number of subsequent accounting periods; report the cash flows that occur in each financial period; and produce a balance sheet at the end of each of the financial periods. For illustration we follow this procedure in Sections 1.8.1 and 1.8.2 for transactions entered into by Mr S. Norman.

1.8.1 Appraisal of the initial investment decision

Mr Norman is considering whether to start up a retail business by acquiring the lease of a shop for five years at a cost of £80,000.

Our first task has been set out in Figure 1.1 above. It is to establish the information that Mr Norman needs, so that we can decide what data need to be collected and measured. Let us assume that, as a result of a discussion with Mr Norman, it has been ascertained that he is a profit satisficer who is looking to achieve at least a 10% return, which represents the time value of money. This indicates that, as illustrated in Figure 1.2:

- the relevant data to be measured are **cash flows**, represented by the outflow of cash invested in the lease and the inflow of cash represented by the realised operating cash flows;
- the appropriate appraisal method is **NPV**; and
- the appraisal criterion is a **positive NPV** using the discount rate of 10%.

Let us further assume that the cash to be invested in the lease is £80,000 and that the realised operating cash flows over the life of the investment in the shop are as shown in Figure 1.4. This shows that there is a forecast of £30,000 annually for five years and a final receipt of £29,000 in 20X6 when he proposes to cease trading.

We already know that Mr Norman's investment criterion is a positive NPV using a discount factor of 10%. A calculation (Figure 1.5) shows that the investment easily satisfies that criterion.

Figure 1.4 Forecast of realised operating cash flows

	Annually years 20X1–20X5	Cash in year 20X6 after shop closure
	£	£
Receipts from		
Customers	400,000	55,000
Payments to		
Suppliers	(342,150)	(20,000)
Expense creditors	(21,600)	(3,000)
Rent	(6,250)	(3,000)
Total payments	(370,000)	(26,000)
Realised operating cash flows	30,000	29,000

Figure 1.5 NPV calculation using discount tables

	£	£
Cost of lease		(80,000)
£30,000 annually for 5 years (30,000 × 3.79)	113,700	
£29,000 received in year 6 (29,000 × 0.564)	16,356	
		130,056
Positive net present value		50,056

1.8.2 Preparation of periodic financial statements under the cash flow concept

Having **predicted** the realised operating cash flows for the purpose of making the invest-ment decision, we can assume that the owner of the business will wish to obtain **feedback** to evaluate the correctness of the investment decision. He does this by reviewing the actual results on a regular **timely** basis and **comparing** these with the predicted forecast. Actual results should be reported quarterly, half-yearly or annually in the same format as used when making the decision in Figure 1.4. The actual results provide management with the feedback information required to audit the initial decision; it is a technique for achieving accountability. However, frequently, companies do not provide a report of actual cash flows to compare with the forecast cash flows, and fail to carry out an audit review.

In some cases, the transactions relating to the investment cannot be readily separated from other transactions, and the information necessary for the audit review of the investment cannot be made available. In other cases, the routine accounting procedures fail to collect such cash flow information because the reporting systems have not been designed to provide financial reports on a cash flow basis; rather, they have been designed to produce reports prepared on an accrual basis.

What would financial reports look like if they were prepared on a cash flow basis?

To illustrate cash flow period accounts, we will prepare half-yearly accounts for Mr Norman. To facilitate a comparison with the forecast that underpinned the investment decision, we will redraft the forecast annual statement on a half-yearly basis. The data for the first year given in Figure 1.4 have therefore been redrafted to provide a forecast for the half-year to 30 June, as shown in Figure 1.6.

We assume that, having applied the net present value appraisal technique to the cash flows and ascertained that the NPV was positive, Mr Norman proceeded to set up the business on 1 January 20X1. He introduced capital of £50,000, acquired a five-year lease for £80,000 and paid £6,250 in advance as rent to occupy the property to 31 December 20X1. He has decided to prepare financial statements at half-yearly intervals. The information given in Figure 1.7 concerns his trading for the half-year to 30 June 20X1.

Mr Norman was naturally eager to determine whether the business was achieving its forecast cash flows for the first six months of trading, so he produced the statement of

Figure 1.6 Forecast of realised operating cash flows

	Half-year to 30 June 20X1
	£
Receipts from	
Customers	165,000
Payments to	
Suppliers	(124,000)
Expense creditors	(18,000)
Rent	(6,250)
Total payments	(148,250)
Realised operating cash flows	16,750

Figure 1.7 Monthly sales, purchases and expenses for six months ended 30 June 20X1

Month	Sales invoiced £	Cash received £	Purchases invoiced £	Cash paid £	Expenses invoiced £	Cash paid £
January	15,000	7,500	16,000		3,400	3,100
February	20,000	17,500	19,000	16,000	3,500	3,400
March	35,000	27,500	29,000	19,000	3,800	3,500
April	40,000	37,500	32,000	29,000	3,900	3,800
May	40,000	40,000	33,000	32,000	3,900	3,900
June	45,000	42,500	37,000	33,000	4,000	3,900
TOTAL	195,000	172,500	166,000	129,000	22,500	21,600

Note: The following items were included under the Expenses invoiced heading:
Expense creditors – amount
Wages – £3,100 per month paid in the month
Commission – 2% of sales invoiced payable one month in arrears

Figure 1.8 Monthly realised operating cash flows

	Jan £	Feb £	Mar £	Apr £	May £	Jun £	Total £
Receipts							
Customers	7,500	17,500	27,500	37,500	40,000	42,500	172,500
Less payments							
Suppliers		16,000	19,000	29,000	32,000	33,000	129,000
Expense creditors	3,100	3,400	3,500	3,800	3,900	3,900	21,600
Rent	6,250						6,250
Realised	(1,850)	(1,900)	5,000	4,700	4,100	5,600	15,650

realised operating cash flows (Figure 1.8) from the information provided in Figure 1.7. From this statement we can see that the business generated positive cash flows after the end of February. These are, of course, only the cash flows relating to the trading transactions.

The information in the 'Total' row of Figure 1.7 can be extracted to provide the financial statement for the six months ended 30 June 20X1, as shown in Figure 1.9.

The figure of £15,650 needs to be compared with the forecast cash flows used in the investment appraisal. This is a form of auditing. It allows the assumptions made on the initial investment decision to be confirmed. The forecast/actual comparison (based on the information in Figures 1.6 and 1.9) is set out in Figure 1.10.

What are the characteristics of these data that make them relevant?

● The data are **objective**. There is no judgement involved in deciding the values to include in the financial statement, as each value or amount represents a verifiable cash transaction with a third party.

Figure 1.9 Realised operating cash flows for the six months ended 30 June 20X1

		£
Receipts from		
Customers		172,500
Payments to		
Suppliers	(129,000)	
Expense creditors	(21,600)	
Rent	(6,250)	
		156,850
Realised operating cash flow		15,650

Figure 1.10 Forecast/actual comparison

		Actual £	Forecast £
Receipts from			
Customers		172,500	165,000
Payment to			
Suppliers	(129,000)		(124,000)
Expense creditors	(21,600)		(18,000)
Rent	(6,250)		(6,250)
Total payments		(156,850)	(148,250)
Realised operating cash flow		15,650	16,750

- The data are **consistent**. The statement incorporates the same cash flows within the periodic financial report of trading as the cash flows that were incorporated within the initial capital investment report. This permits a logical comparison and confirmation that the decision was realistic.
- The results have a **confirmatory** value by helping users confirm or correct their past assessments.
- The results have a **predictive** value, in that they provide a basis for revising the initial forecasts if necessary.
- There is **no requirement for accounting standards** or disclosure of accounting policies that are necessary to regulate accrual accounting practices, e.g. depreciation methods.

1.9 Illustration of preparation of statement of financial position

Although the information set out in Figure 1.10 permits us to compare and evaluate the initial decision, it does not provide a sufficiently sound basis for the following:

- assessing the stewardship over the total cash funds that have been employed within the business;
- signalling to management whether its working capital policies are appropriate.

1.9.1 Stewardship

To assess the stewardship over the total cash funds we need to:

(a) evaluate the effectiveness of the accounting system to make certain that all transactions are recorded;

(b) extend the cash flow statement to take account of the capital cash flows; and

(c) prepare a statement of financial position or balance sheet as at 30 June 20X1.

The additional information for (b) and (c) above is set out in Figures 1.11 and 1.12 respectively.

The cash flow statement and statement of financial position, taken together, are a means of assessing stewardship. They identify the movement of **all** cash and derive a **net** balance figure. These statements are a normal feature of a sound system of internal control, but they have not been made available to external users.

1.9.2 Working capital policies

By 'working capital' we mean the current assets and current liabilities of the business. In addition to providing a means of making management accountable, cash flows are the raw data required by financial managers when making decisions on the management of working capital. One of the decisions would be to set the appropriate terms for credit policy. For example, Figure 1.11 shows that the business will have a £14,350 overdraft at 30 June 20X1. If this is not acceptable, management will review its working capital by reconsidering the

Figure 1.11 Cash flow statement to calculate the net cash balance

	Jan	Feb	Mar	Apr	May	Jun	Total
	£	£	£	£	£	£	£
Operating cash	(1,850)	(1,900)	5,000	4,700	4,100	5,600	15,650
New capital	50,000						50,000
Lease payment	(80,000)						(80,000)
Cash balance	(31,850)	(33,750)	(28,750)	(24,050)	(19,950)	(14,350)	(14,350)

Figure 1.12 Statement of financial position

	Opening 1 Jan 20X1 £	Closing 30 Jun 20X1 £
Capital introduced	50,000	50,000
Net operating cash flow		15,650
	50,000	65,650
Lease		80,000
Net cash balance	50,000	−14,350
	50,000	65,650

credit given to customers, the credit taken from suppliers, stock-holding levels and the timing of capital cash inflows and outflows.

If, in the example, it were possible to obtain 45 days' credit from suppliers, then the creditors at 30 June would rise from £37,000 to a new total of £53,500. This increase in trade credit of £16,500 means that half of the May purchases (£33,000/2) would not be paid for until July, which would convert the overdraft of £14,350 into a positive balance of £2,150. As a new business it might not be possible to obtain credit from all of the suppliers. In that case, other steps would be considered, such as phasing the payment for the lease of the warehouse or introducing more capital.

An interesting research report[3] identified that for small firms survival and stability were the main objectives rather than profit maximisation. This, in turn, meant that cash flow indicators and managing cash flow were seen as crucial to survival. In addition, cash flow information was perceived as important to external bodies such as banks in evaluating performance.

1.10 Treatment of non-current assets in the cash flow model

The statement of financial position in Figure 1.12 does not take into account any **unrealised** cash flows. Such flows are deemed to occur as a result of any rise or fall in the realisable value of the lease. This could rise if, for example, the annual rent payable under the lease were to be substantially lower than the rate payable under a new lease entered into on 30 June 20X1. It could also fall with the passing of time, with six months having expired by 30 June 20X1. We need to consider this further and examine the possible treatment of non-current assets in the cash flow model.

Using the cash flow approach, we require an independent verification of the realisable value of the lease at 30 June 20X1. If the lease has fallen in value, the difference between the original outlay and the net realisable figure could be treated as a negative unrealised operating cash flow.

For example, if the independent estimate was that the realisable value was £74,000, then the statement of financial position would be prepared as in Figure 1.13. The fall of £6,000 in realisable value is an unrealised cash flow and, while it does not affect the calculation of the net cash balance, it does affect the statement of financial position.

Figure 1.13 Statement of financial position as at 30 June 20X1 (assuming that there were unrealised operating cash flows)

	£
Capital introduced	50,000
Net operating flow: **realised**	15,650
: **unrealised**	(6,000)
	59,650
Lease: **net realisable value**	74,000
Net cash balance	−14,350
	59,650

The same approach would be taken to all non-current assets and could result in there being an unrealised cash flow where there is limited resale market for an asset, even though it might be productive and have value in use by the firm that owns it.

The additional benefit of the statement of financial position, as revised, is that the owner is able clearly to identify the following:

- the operating cash inflows of £15,650 that have been realised from the business operations;
- the operating cash outflow of £6,000 that has not been realised, but has arisen as a result of investing in the lease;
- the net cash balance of −£14,350;
- the statement provides a **stewardship-oriented** report: that is, it is a means of making the management accountable for the cash within its control.

1.11 What are the characteristics of these data that make them reliable?

We have already discussed some characteristics of cash flow reporting which indicate that the data in the financial statements are **relevant**, e.g. their predictive and confirmatory roles. We now introduce five more characteristics of cash flow statements which indicate that the information is also **reliable**, i.e. free from bias. These are prudence, neutrality, completeness, faithful representation and substance over form.

1.11.1 Prudence characteristic

Revenue and profits are included in the cash flow statement only when they are realised. Realisation is deemed to occur when cash is received. In our Norman example, the £172,500 cash received from debtors represents the revenue for the half-year ended 30 June 20X1. This policy is described as prudent because it **does not anticipate** cash flows: cash flows are recorded only when they actually occur and not when they are reasonably certain to occur. This is one of the factors that distinguishes cash flow from accrual accounting.

1.11.2 Neutrality characteristic

Financial statements are not neutral if, by their selection or presentation of information, they influence the making of a decision in order to achieve a predetermined result or outcome. With cash flow accounting, the information is not subject to management selection criteria.

Cash flow accounting avoids the tension that can arise between prudence and neutrality because, whilst neutrality involves freedom from deliberate or systematic bias, prudence is a potentially biased concept that seeks to ensure that, under conditions of uncertainty, gains and assets are not overstated and losses and liabilities are not understated.

1.11.3 Completeness characteristic

The cash flows can be verified for completeness provided there are adequate internal control procedures in operation. In small and medium-sized enterprises there can be a weakness if one person, typically the owner, has control over the accounting system and is able to under-record cash receipts.

1.11.4 Faithful representation characteristic

Cash flows can be depended upon by users to represent faithfully what they purport to represent provided, of course, that the completeness characteristic has been satisfied.

1.11.5 Substance over form

Cash flow accounting does not necessarily possess this characteristic which requires that transactions should be accounted for and presented in accordance with their substance and economic reality and not merely their legal form.

1.12 Reports to external users

1.12.1 Stewardship orientation

Cash flow accounting provides objective, consistent and prudent financial information about a business's transactions. It is stewardship-oriented and offers a means of achieving account-ability over cash resources and investment decisions.

1.12.2 Prediction orientation

External users are also interested in the ability of a company to pay dividends. It might be thought that the past and current cash flows are the best indicators of future cash flows and dividends. However, the cash flow might be misleading, in that a declining company might sell non-current assets and have a better **net cash position** than a growing company that buys non-current assets for future use. There is also no matching of cash inflows and out-flows, in the sense that a benefit is matched with the sacrifice made to achieve it.

Consequently, it has been accepted accounting practice to view the income statement prepared on the accrual accounting concept as a better predictor of future cash flows to an investor than the cash flow statements that we have illustrated in this chapter.

However, the operating cash flows arising from trading and the cash flows arising from the introduction of capital and the acquisition of non-current assets can become significant to investors, e.g. they may threaten the company's ability to survive or may indicate growth.

In the next chapter, we revise the preparation of the same three statements using the **accrual accounting** model.

1.12.3 Going concern

The Financial Reporting Council suggests in its Consultation Paper *Going Concern and Financial Reporting*[4] that directors in assessing whether a company is a going concern may prepare monthly cash flow forecasts and monthly budgets covering, as a minimum, the period up to the next statement of financial position date. The forecasts would also be sup-ported by a detailed list of assumptions which underlie them.

1.12.4 Tax authorities[5]

In the UK accounts prepared on a cash flow basis will, from April 2013, be acceptable to the tax authorities for small unincorporated micro-businesses. This is seen as cutting costs and reducing the need to engage accountants. Companies are still required to determine income on an accrual accounting basis on the grounds that this better reflects economic substance in that most incorporated businesses would have inventory or work in progress, or have creditors for the supply of materials.

Summary

To review our understanding of this chapter, we should ask ourselves the following questions.

How useful is cash flow accounting for internal decision making?

Forecast cash flows are relevant for the appraisal of proposals for capital investment. Actual cash flows are relevant for the confirmation of the decision for capital investment.

Cash flows are relevant for the management of working capital. Financial managers might have a variety of mathematical models for the efficient use of working capital, but cash flows are the raw data upon which they work.

How useful is cash flow accounting for making management accountable?

The cash flow statement is useful for confirming decisions and, together with the statement of financial position, provides a stewardship report. Lee states that 'Cash flow accounting appears to satisfy the need to supply owners and others with stewardship-orientated information as well as with decision-orientated information.'[6]

Lee further states that:

> By reducing judgements in this type of financial report, management can report factually on its stewardship function, whilst at the same time disclosing data of use in the decision-making process. In other words, cash flow reporting eliminates the somewhat artificial segregation of stewardship and decision-making information.

This is exactly what we saw in our Norman example – the same realised operating cash flow information was used for both the investment decision and financial reporting. However, for stewardship purposes it was necessary to extend the cash flow to include **all** cash movements and to extend the statement of financial position to include the **unrealised** cash flows.

How useful is cash flow accounting for reporting to external users?

Cash flow information is relevant:

- as a basis for making internal management decisions in relation to both non-current assets and working capital;
- for stewardship and accountability; and
- for assessing whether a business is a going concern.

Cash flow information is reliable and a fair representation, being:

- objective; consistent; prudent; and neutral.

However, professional accounting practice requires reports to external users to be on an accrual accounting basis. This is because the accrual accounting profit figure is a better predictor for investors of the future cash flows likely to arise from the dividends paid to them by the business, and of any capital gain on disposal of their investment. It could also be argued that cash flows may not be a fair representation of the commercial substance of transactions, e.g. if a business allowed a year's credit to all its customers there would be no income recorded.

REVIEW QUESTIONS

1 Explain why it is the user who should determine the information that the accountant collects, measures and reports, rather than the accountant who is the expert in financial information.

2 'Yuji Ijiri rejects decision usefulness as the main purpose of accounting and puts in its place account-ability. Ijiri sees the accounting relationship as a tripartite one, involving the accountor, the accountee, and the accountant . . . the decision useful approach is heavily biased in favour of the accountee . . . with little concern for the accountor . . . in the central position Ijiri would put fairness.'[7] Discuss Ijiri's view in the context of cash flow accounting.

3 Explain the effect on the statement of financial position in Figure 1.13 if the non-current asset consisted of expenditure on industry-specific machine tools rather than a lease.

4 'It is essential that the information in financial statements has a prudent characteristic if the financial statements are to be objective.' Discuss.

5 Explain why realised cash flow might not be appropriate for investors looking to predict future dividends.

6 Discuss why it might not be sufficient for a small business person who is carrying on business as a sole trader to prepare accounts on a cash flow basis.

7 'Unrealised operating cash flows are only of use for internal management purposes and are irrelevant to investors.' Discuss.

8 'While accountants may be free from bias in the measurement of economic information, they cannot be unbiased in identifying the economic information that they consider to be relevant.' Discuss.

EXERCISES

* Question I

Sasha Parker is going to set up a new business on 1 January 20X1. She estimates that her first six months in business will be as follows:

(i)   She will put £150,000 into a bank account for the firm on 1 January 20X1.

(ii) On 1 January 20X1 she will buy machinery £30,000, motor vehicles £24,000 and premises £75,000, paying for them immediately.

(iii) All purchases will be effected on credit. She will buy £30,000 goods on 1 January and will pay for these in February. Other purchases will be: rest of January £48,000; February, March, April, May and June £60,000 each month. Other than the £30,000 worth bought in January, all other purchases will be paid for two months after purchase.

(iv) Sales (all on credit) will be £60,000 for January and £75,000 for each month after. Customers will pay for the goods in the fourth month after purchase, i.e. £60,000 is received in May.

(v)   She will make drawings of £1,200 per month.

(vi) Wages and salaries will be £2,250 per month and will be paid on the last day of each month.

(vii) General expenses will be £750 per month, payable in the month following that in which they are incurred.

(viii) Rates will be paid as follows: for the three months to 31 March 20X1 by cheque on 28 February 20X1; for the 12 months ended 31 March 20X2 by cheque on 31 July 20X1. Rates are £4,800 per annum.

(ix) She will introduce new capital of £82,500 on 1 April 20X1.

(x) Insurance covering the 12 months of 20X1 of £2,100 will be paid for by cheque on 30 June 20X1.

(xi) All receipts and payments will be by cheque.

(xii) Inventory on 30 June 20X1 will be £30,000.

(xiii) The net realisable value of the vehicles is £19,200, machinery £27,000 and premises £75,000.

Required: Cash flow accounting

(i) Draft a cash budget (includes bank) month by month for the period January to June, showing clearly the amount of bank balance or overdraft at the end of each month.

(ii) Draft an operating cash flow statement for the six-month period.

(iii) Assuming that Sasha Parker sought your advice as to whether she should actually set up in business, state what further information you would require.

* Question 2

Mr Norman set up a new business on 1 January 20X8. He invested €50,000 in the new business on that date. The following information is available.

1 Gross profit was 20% of sales. Monthly sales were as follows:

Month	Sales €
January	15,000
February	20,000
March	35,000
April	40,000
May	40,000
June	45,000
July	50,000

2 50% of sales were for cash. Credit customers (50% of sales) pay in month following sale.

3 The supplier allowed one month's credit.

4 Monthly payments were made for rent and rates €2,200 and wages €600.

5 On 1 January 20X8 the following payments were made: €80,000 for a five-year lease of business premises and €3,500 for insurances on the premises for the year. The realisable value of the lease was estimated to be €76,000 on 30 June 20X8 and €70,000 on 31 December 20X8.

6 Staff sales commission of 2% of sales was paid in the month following the sale.

Required:

(a) A purchases budget for each of the first six months.

(b) A cash flow statement for the first six months.

(c) A statement of operating cash flows and financial position as at 30 June 20X8.

(d) Write a brief letter to the bank supporting a request for an overdraft.

References

1 *Framework for the Preparation and Presentation of Financial Statements*, IASC, 1989, para. 10.
2 G. Whittred and I. Zimmer, *Financial Accounting: Incentive Effects and Economic Consequences*, Holt, Rinehart & Winston, 1992, p. 27.
3 R. Jarvis, J. Kitching, J. Curran and G. Lightfoot, *The Financial Management of Small Firms: An Alternative Perspective*, ACCA Research Report No. 49, 1996.
4 *Going Concern and Financial Reporting – Proposals V. Revise the Guidance for Directors of Listed Companies*, FRC, 2008, para. 29.
5 https://www.gov.uk/simpler-income-tax-cash-basis
6 T.A. Lee, *Income and Value Measurement: Theory and Practice* (3rd edition), Van Nostrand Reinhold (UK), 1985, p. 173.
7 D. Solomons, *Making Accounting Policy*, Oxford University Press, 1986, p. 79.

CHAPTER 2

Accounting and reporting on an accrual accounting basis

2.1 Introduction

The main purpose of this chapter is to extend cash flow accounting by adjusting for the effect of transactions that have not been completed by the end of an accounting period.

Objectives

By the end of this chapter, you should be able to:

- explain the historical cost convention and accrual concept;
- adjust cash receipts and payments in accordance with IAS 18;
- account for the amount of non-current assets used during the accounting period;
- prepare a statement of income and a statement of financial position;
- reconcile cash flow accounting and accrual accounting data.

2.1.1 Objective of financial statements

The *Conceptual Framework for Financial Reporting*[1] states that the objective of general purpose financial statements is to provide information about the financial position, performance and cash flows of an enterprise that is useful to existing and potential investors, lenders and creditors in making economic decisions about providing resources.

Common information needs for decision making

The IASB recognises that all the information needs of all users cannot be met by financial statements, but it takes the view that some needs are common to all users: in particular, they have some interest in the financial position, performance and adaptability of the enterprise as a whole. This leaves open the question of which user is the primary target; the IASB states that, as investors are providers of risk capital, financial statements that meet their needs would also meet most of the needs of other users.

Stewardship role of financial statements

In addition to assisting in making economic decisions, financial statements also show the results of the stewardship of management: that is, the accountability of management for the resources entrusted to it. The IASB view is that users who assess the stewardship do so in order to make economic decisions, e.g. whether to hold or sell shares in a particular company or change the management.

2.1.2 Statements making up the financial statements published for external users

In 2007 the IASB stated[2] that a complete set of financial statements should comprise:

- a statement of financial position as at the end of the period;
- a statement of comprehensive income for the period;
- a statement of changes in equity for the period;
- a statement of cash flows for the period;
- notes comprising a summary of significant accounting policies and other explanatory information.

In this chapter we consider two of the conventions under which the statement of comprehensive income and statement of financial position are prepared: the historical cost convention and the accrual accounting concept. In Chapters 3–5 we consider each of the above statements.

2.2 Historical cost convention

The historical cost convention results in an appropriate measure of the economic resource that has been withdrawn or replaced.

Under it, transactions are reported at the £ amount recorded at the date the transaction occurred. Financial statements produced under this convention provide a basis for determining the outcome of agency agreements with reasonable certainty and predictability because the data are relatively objective.[3]

By this we mean that various parties who deal with the enterprise, such as lenders, will know that the figures produced in any financial statement are objective and not manipulated by subjective judgements made by the directors. For example, being confident that the revenue and expenses in the statement of income are stated at the £ amount that appears on the invoices. This means that the amount is objective and can be independently verified.

Because of this, the historical cost convention has strengths for stewardship purposes, i.e. providing an objective record of the resources that managers have had under their control. However, the price-level-adjusted figures which we discuss in Chapter 7 may well be more appropriate for decision making.

2.3 Accrual basis of accounting

The accrual basis dictates when transactions with third parties should be recognised and, in particular, determines the accounting periods in which they should be incorporated into the financial statements. Under this concept the cash receipts from customers and payments to creditors are replaced by revenue and expenses respectively.

Revenue and expenses are derived by adjusting the realised operating cash flows to take account of business trading activity that has occurred during the accounting period, but has not been converted into cash receipts or payments by the end of the period.

2.3.1 Accrual accounting is a better indicator than cash flow accounting of ability to generate cash

The IASB view is that financial statements prepared on an accrual basis inform users not only of past transactions involving the payment and receipt of cash, but also of obligations to pay cash in the future and of resources that represent cash to be received in the future, and that they provide the type of information about past transactions and other events that is most useful in making economic decisions about the future.[4]

Having briefly considered why accrual accounting may be more useful than cash flow accounting, we will briefly revise the preparation of financial statements under the accrual accounting convention.

2.4 Mechanics of accrual accounting – adjusting cash receipts and payments

Let us assume that the cash flows for the six months were as illustrated in Chapter 1:

Receipts £172,500, Materials purchased £129,000, Services paid for £21,600 and Rent paid £6,250.

Additional information that now has to be taken into account relating to the half year's transactions:

Invoices issued to customers but unpaid totalled £22,500; Invoices received from suppliers but unpaid totalled £21,600; Invoices for services received but unpaid totalled £900 and half of the £6,250 payment for rent relates to the following half year.

As these all relate to the current six month's business activity, the cash figures need to be adjusted in both the statement of income and statement of financial position as in Figure 2.1 and Figure 2.2.

Figure 2.1 Statement of income for the six months ended 30 June 20X1

	Operating cash flow	ADJUST cash flow	Business activity
	£	£	£
Revenue from business activity	172,500	22,500	195,000
Less: Matching expenses			
Transactions for materials	129,000	37,000	166,000
Transactions for services	21,600	900	22,500
Transaction with landlord	6,250	(3,125)	3,125
OPERATING CASH FLOW from business activity	15,650		
Transactions NOT converted to cash or relating to a subsequent period		(12,275)	
PROFIT from business activity			3,375

Figure 2.2 Statement of financial position adjusted to an accrual basis

	£
Capital	50,000
Net operating cash flow: **realised**	15,650
Net operating cash flow: **to be realised next period**	(12,275)
	53,375
Lease	80,000
Net cash balance (refer to Figure 1.11)	(14,350)
Net amount of activities not converted to cash or relating to subsequent periods	(12,275)
	53,375

2.5 Reformatting the statement of financial position

The item 'net amount of activities not converted to cash or relating to subsequent periods' is the net trade receivable/trade payable balance. If we wished, the statement of financial position could be reframed into the customary statement of financial position format, where items are classified as assets or liabilities. The IASB defines assets and liabilities in its *Conceptual Framework*:

- An asset is a resource:
  - controlled by the enterprise;
  - as a result of past events;
  - from which future economic benefits are expected to flow.
- A liability is a present obligation:
  - arising from past events
  - the settlement of which is expected to result in an outflow of resources.

The reframed statement set out in Figure 2.3 is in accordance with these definitions.

2.6 Accounting for the sacrifice of non-current assets

The statement of income and statement of financial position have both been prepared using verifiable data that have arisen from transactions with third parties outside the business. However, in order to determine the full sacrifice of economic resources that a business has made to achieve its revenue, it is necessary also to take account of the use made of the non-current assets during the period in which the revenue arose.

2.6.1 Going concern assumption

For all non-current assets a decision has to be made as to the amount to be charged against the current period's profits for the use of the asset. This is where the going concern assumption

Figure 2.3 Reframed statement as at 30 June

	Reframed	
	£	£
CAPITAL	50,000	50,000
Net operating cash flow: **realised**	15,650	
Net operating cash flow: **to be realised**	(12,275)	
NET INCOME		3,375
	53,375	53,375
NON-CURRENT ASSETS	80,000	80,000
NET CURRENT ASSETS		
Net amount of activities not converted to cash	(12,275)	
CURRENT ASSETS		
Trade receivables		22,500
Other receivables: prepaid rent		3,125
CURRENT LIABILITIES		
Trade payables		(37,000)
Other payables: service suppliers		(900)
Net cash balance	(14,350)	(14,350)
	53,375	53,375

comes into effect by assuming that the business enterprise will continue in operational existence for the foreseeable future and not be disposed of at the end of the current period. It is for this reason that an estimate is made of the amount to be charged and we do not use the amount for which the asset could actually be sold.

Treatment of a lease

For example, assuming that the lease could be transferred to a third party at the end of the current period for £50,000 there would be a loss of £30,000 to be charged against the current period's profit IF this action was taken. However, as the intention is to retain and continue to use the asset, the approach is to estimate how much of the cost has been used up in the current period. In our example this would be £8,000 for the half-year. The effect of this charge (referred to as amortisation) on the statements of income and position are shown in Figures 2.4 and 2.5.

Treatment of a tangible non-current asset

If the business was using a tangible non-current asset such as a vehicle then there would be a similar approach to calculate the amount to charge (referred to as depreciation) in the statement of income. The approach is first to calculate the depreciable amount which is the cost of the vehicle less its residual value on disposal, then to estimate its useful economic life and finally to decide how much of the depreciable amount to charge to each of the years that it is in use. Depreciation is discussed in greater detail in Chapter 17.

Figure 2.4 Statement of income for the six months ending 30 June

	Operating cash flow CURRENT period £	Adjust cash flow £	Business activity CURRENT period £
Revenue from business activity	172,500	22,500	195,000
Less			
Expenditure to support this activity:			
Transactions with suppliers	129,000	37,000	166,000
Transactions with service providers	21,600	900	22,500
Transaction with landlord	6,250	(3,125)	3,125
OPERATING CASH FLOW from activity	15,650		
TRANSACTIONS NOT CONVERTED TO CASH		(12,275)	
INCOME from business activity			3,375
Allocation of non-current asset cost to this period			8,000
INCOME			(4,625)

Figure 2.5 Statement of financial position as at 30 June

	Transaction cash flows £	Notional flows £	Reported £
CAPITAL	50,000		50,000
Net operating cash flow: **realised**	15,650		
Net operating cash flow: **to be realised**	(12,275)		
Net income before depreciation		3,375	
AMORTISATION		(8,000)	
Net income after amortisation			(4,625)
	53,375		45,375
NON-CURRENT ASSETS	80,000	80,000	
Less amortisation		(8,000)	
Net book value			72,000
NET CURRENT ASSETS			
Net amount not converted to cash	(12,275)		
CURRENT ASSETS			
Trade receivables			22,500
Other receivables – prepaid rent			3,125
CURRENT LIABILITIES			
Trade payables			(37,000)
Other payables: service suppliers			(900)
Net cash balance	(14,350)		(14,350)
	53,375		45,375

2.6.2 Financial capital maintenance concept

The financial capital maintenance concept recognises a profit only after the original monetary investment has been maintained. This means that, as long as the cost of the assets representing the initial monetary investment is recovered against the profit, by way of a depreciation charge, the initial monetary investment is maintained.

The concept has been described in the IASC *Framework for the Preparation and Presentation of Financial Statements*:

> a profit is earned only if the financial or money amount of the net assets at the end of the period exceeds the financial or money amount of the net assets at the beginning of the period, after excluding any distributions to, and contributions from, owners during the period. Financial capital maintenance can be measured in either nominal monetary units [as we are doing in this chapter] or in units of constant purchasing power [as we will be doing in Chapter 7].

2.6.3 Summary of views on accrual accounting

Standard setters

The profit (loss) is considered to be a guide when assessing the amount, timing and uncertainty of prospective cash flows as represented by future income amounts. The IASB, the FASB in the USA and the ASB in the UK clearly stated that the accrual accounting concept was more useful in predicting future cash flows than cash flow accounting.

Academic researchers

Academic research provides conflicting views. In 1986, research carried out in the USA indicated that the FASB view was inconsistent with its findings and that cash flow information was a better predictor of future operating cash flows;[5] research carried out in the UK, however, indicated that accrual accounting using the historical cost convention was 'a more relevant basis for decision making than cash flow measures'.[6]

2.7 Published statement of cash flows

In Figure 2.3 we reframed the statement of financial position to illustrate how it would be reported when published. In Figure 2.6 we set out the statement of cash flows under the standard headings required by IAS 7 *Statement of Cash Flows*[7] to arrive at the change in cash. The required headings are cash flows from:

- Operating activities
- Investing activities
- Financing activities.

In addition to explaining how the cash has increased or decreased, a reconciliation statement is prepared to explain why there is a positive operating cash flow of £15,650 but a loss in the statement of income of £4,625.

We discuss statements of cash flow further in Chapter 5.

Figure 2.6 Statement of cash flows in accordance with IAS 7 *Statement of Cash Flows*

	£
Net cash inflow from operating activities	15,650
Investing activities	
Payment to acquire lease	(80,000)
Net cash outflow before financing	(64,350)
Financing activities	
Issue of capital	50,000
Decrease in cash	(14,350)
Reconciliation of operating loss to net cash	
inflow from operating activities	£
Operating profit/loss	(4,625)
Amortisation charges	8,000
Increase in trade receivables	(22,500)
Increase in prepayments	(3,125)
Increase in trade payables	37,000
Increase in accruals	900
	15,650

Summary

Accrual accounting replaces cash receipts and payments with revenue and expenses by adjusting the cash figures to take account of trading activity which has not been converted into cash.

Accrual accounting is preferred to cash accounting by the standard setters on the assumption that accrual-based financial statements give investors a better means of predicting future cash flows.

The financial statements are transaction-based, applying the historical cost accounting concept which attempts to minimise the need for personal judgements and estimates in arriving at the figures in the statements.

Under accrual-based accounting the expenses incurred are matched with the revenue earned. In the case of non-current assets, a further accounting concept has been adopted, the going concern concept, which allows an entity to allocate the cost of non-current assets over their estimated useful life.

REVIEW QUESTIONS

1 'Cash flow accounting and accrual accounting information are both required by a potential shareholder.' Discuss.

2 'The asset measurement basis applied in accrual accounting can lead to financial difficulties when assets are due for replacement.' Discuss.

3 'Accrual accounting is preferable to cash flow accounting because the information is more relevant to all users of financial statements.' Discuss.

4 'Information contained in a statement of income and a statement of financial position prepared under accrual accounting concepts is factual and objective.' Discuss.

5 The *Conceptual Framework for Financial Reporting* identifies seven user groups: investors, employees, lenders, suppliers and other trade creditors, customers, government and the public.

Discuss which of the financial statements illustrated in Chapters 1 and 2 would be most useful to each of these seven groups if they could only receive one statement.

6 The annual financial statements of companies are used by various parties for a wide variety of purposes. Discuss which of the three statements of income, financial position and cash flows would be of most interest to (a) a loan creditor and (b) a trade creditor.

7 Discuss how amounts are reported in a statement of financial position if the accounts are not prepared on a going concern basis.

EXERCISES

* Question 1

Sasha Parker is going to set up a new business in Bruges on 1 January 20X1. She estimates that her first six months in business will be as follows:

(i) She will put €150,000 into the firm on 1 January 20X1.

(ii) On 1 January 20X1 she will buy machinery €30,000, motor vehicles €24,000 and premises €75,000, paying for them immediately.

(iii) All purchases will be effected on credit. She will buy €30,000 goods on 1 January and she will pay for these in February. Other purchases will be: rest of January €48,000; February, March, April, May and June €60,000 each month. Other than the €30,000 worth bought in January, all other purchases will be paid for two months after purchase, i.e. €48,000 in March.

(iv) Sales (all on credit) will be €60,000 for January and €75,000 for each month after that. Customers will pay for goods in the third month after purchase, i.e. €60,000 in April.

(v) Inventory on 30 June 20X1 will be €30,000.

(vi) Wages and salaries will be €2,250 per month and will be paid on the last day of each month.

(vii) General expenses will be €750 per month, payable in the month following that in which they are incurred.

(viii) She will introduce new capital of €75,000 on 1 June 20X1. This will be paid into the business bank account immediately.

(ix) Insurance covering the 12 months of 20X1 of €26,400 will be paid for by cheque on 30 June 20X1.

(x) Local taxes will be paid as follows: for the three months to 31 March 20X1 by cheque on 28 February 20X2, delay due to an oversight by Parker; for the 12 months ended 31 March 20X2 by cheque on 31 July 20X1. Local taxes are €8,000 per annum.

(xi) She will make drawings of €1,500 per month by cheque.

(xii) All receipts and payments are by cheque.

(xiii) Depreciate motor vehicles by 20% per annum and machinery by 10% per annum, using the straight-line depreciation method.

(xiv) She has been informed by her bank manager that he is prepared to offer an overdraft facility of €30,000 for the first year.

Required:

(a) Draft a cash budget (for the firm) month by month for the period January to June, showing clearly the amount of bank balance at the end of each month.

(b) Draft the projected statement of comprehensive income for the first six months' trading, and a statement of financial position as at 30 June 20X1.

(c) Advise Sasha on the alternative courses of action that could be taken to cover any cash deficiency that exceeds the agreed overdraft limit.

* Question 2

Mr Norman is going to set up a new business in Singapore on 1 January 20X8. He will invest $150,000 in the business on that date and has made the following estimates and policy decisions:

1 Forecast sales (in units) made at a selling price of $50 per unit are:

Month	Sales units
January	1,650
February	2,200
March	3,850
April	4,400
May	4,400
June	4,950
July	5,500

2 50% of sales are for cash. Credit terms are payment in the month following sale.

3 The units cost $40 each and the supplier is allowed one month's credit.

4 It is intended to hold inventory at the end of each month sufficient to cover 25% of the following month's sales.

5 Administration $8,000 and wages $17,000 are paid monthly as they arise.

6 On 1 January 20X8, the following payments will be made: $80,000 for a five-year lease of the business premises and $350 for insurance for the year.

7 Staff sales commission of 2% of sales will be paid in the month following sale.

Required:

(a) A purchases budget for each of the first six months.

(b) A cash flow forecast for the first six months.

(c) A budgeted statement of comprehensive income for the first six months' trading and a budgeted statement of financial position as at 30 June 20X8.

(d) Advise Mr Norman on the investment of any excess cash.

References

1 *Conceptual Framework for Financial Reporting*, IASB, 2010.
2 IAS 1 *Presentation of Financial Statements*, IASB, revised 2007, para. 10.
3 M. Page, *British Accounting Review*, vol. 24(1), 1992, p. 80.
4 *Framework for the Preparation and Presentation of Financial Statements*, IASC, 1989, para. 20.
5 R.M. Bowen, D. Burgstahller and L.A. Daley, 'Evidence on the relationship between earnings and various measures of cash flow', *Accounting Review*, October 1986, pp. 713–25.
6 J.I.G. Board and J.F.S. Day, 'The information content of cash flow figures', *Accounting and Business Research*, Winter 1989, pp. 3–11.
7 IAS 7 *Statement of Cash Flows*, IASB, revised 2007.

Preparation of financial statements of comprehensive income, changes in equity and financial position

3.1 Introduction

Annual Reports consist of primary financial statements, additional disclosures and narrative.

The primary financial statements should be presented using standardised formats as prescribed by International Financial Reporting Standards:

- a statement of income;
- a statement of other comprehensive income;
- a statement of changes in equity;
- a statement of financial position;
- a statement of cash flows (covered in Chapter 5);
- explanatory notes to the accounts.

Objectives

By the end of this chapter, you should be able to:

- understand the structure and content of published financial statements;
- prepare statements of comprehensive income, changes in equity and financial position;
- explain the nature of and reasons for notes to the accounts.

3.2 Preparing an internal statement of income from a trial balance

In this section we revise the steps taken to prepare an internal statement of income from a trial balance. These are to:

- identify year-end adjustments;
- calculate these adjustments;
- prepare an internal statement of income taking adjustments into account.

3.2.1 The trial balance of Wiggins SA

Accounts will be prepared for Wiggins SA from the trial balance set out in Figure 3.1.

Figure 3.1 The trial balance for Wiggins SA as at 31 December 20X3

	€000	€000
Issued share capital (€1)		16,500
Share premium		750
Retained earnings		57,500
10% long-term loan (20X9)		63,250
Bank overdraft		6,325
Trade payables		30,650
Depreciation – buildings		2,300
– equipment		3,450
– vehicles		9,200
Freehold land	57,500	
Freehold buildings	57,500	
Equipment	14,950	
Motor vehicles	20,700	
Inventory at 1 January 20X3	43,125	
Trade receivables	28,750	
Cash in hand	4,600	
Purchases	258,750	
Bank interest	1,150	
Dividends	1,725	
Interest on loan	6,325	
Insurance	5,290	
Salaries and wages	20,355	
Motor expenses	9,200	
Taxation that was under provided	750	
Light, power, miscellaneous	4,255	
Sales		345,000
	534,925	534,925

3.2.2 Identify the year-end adjustments

During the year cash and credit transactions are recorded by posting to the individual ledger accounts as cash is paid or received and invoices received or issued. It is only when financial statements are being prepared that adjustments are made to ensure that the statement of income includes only income and expenses related to the current financial period.

The following information relating to accruals and prepayments has not yet been taken into account in the amounts shown in the trial balance:

● Inventory valued at cost at 31 December 20X3 was €25,875,000.

● Depreciation is to be provided as follows:

  ● 2% on freehold buildings using the straight-line method;

  ● 10% on equipment using the reducing balance method;

  ● 25% on motor vehicles using the reducing balance method.

● €2,300,000 was prepaid for light, power and miscellaneous expenses and €5,175,000 has accrued for wages.

● Freehold land was revalued on 31 December 20X3 at €77,500,000, resulting in a gain of €20,000,000.

● Assume income tax at 20% of pre-tax profit.

● 1,500 €1 shares had been issued on 1 January 20X3 at a premium of 50c each.

3.2.3 Calculate the year-end adjustments

In this example they relate to accrued and prepaid expenses and depreciation.

W1 Salaries and wages:

  €20,355,000 + accrued €5,175,000 = €25,530,000

W2 Depreciation:

Buildings	2% of €57,500,000	€1,150,000
Equipment	10% of (€14,950,000 − €3,450,000)	€1,150,000
Vehicles	25% of (€20,700,000 − €9,200,000)	€2,875,000
Total		€5,175,000

W3 Light, power and miscellaneous

  €4,255,000 − prepaid €2,300,000 = €1,955,000

3.2.4 Prepare an internal statement of income after making the year-end adjustments

By way of revision, we have set out a statement of income prepared for internal purposes in Figure 3.2. We have arranged the expenses in descending monetary value. The method for doing this is not prescribed and companies are free to organise the expenses in other ways, for example in alphabetical order.

Figure 3.2 Statement of income of Wiggins SA for the year ended 31 December 20X3

		€000	€000
Sales			345,000
Less:			
Opening inventory		43,125	
Purchases		258,750	
		301,875	
Closing inventory		25,875	
Cost of sales			276,000
Gross profit			69,000
Less expenses:			
Salaries and wages	W1	25,530	
Motor expenses		9,200	
Loan interest		6,325	
Depreciation	W2	5,175	
Insurance		5,290	
Bank interest		1,150	
Light, power and miscellaneous	W3	1,955	
			54,625
Profit before tax			14,375
Income taxation (includes under-provision)			3,625
Profit after tax			10,750
Dividends (are disclosed in Statement of Changes in Equity in published format)			1,725
Retained earnings			9,025

3.3 Reorganising the income and expenses into one of the formats required for publication

Public companies are required to present their statement of income in a prescribed format to assist users making inter-company comparisons. IAS 1 allows a company two choices in the way in which it analyses the expenses, and the formats[1] are as follows:

- Format 1: Vertical with costs analysed according to function, e.g. cost of sales, distribution costs and administration expenses; or
- Format 2: Vertical with costs analysed according to nature, e.g. raw materials, employee benefits expenses, operating expenses and depreciation.

Many companies use Format 1 (unless there is an industry preference or possible national requirement to use Format 2) with the costs analysed according to function. If this format is used the information regarding the nature of expenditure (e.g. raw materials, wages and depreciation) must be disclosed in a note to the accounts. The analysis of expenses classified either by the nature of the expenses or by their function within the entity is decided by whichever provides information that is reliable and more relevant.

Figure 3.3 Assumptions made in analysing the costs

	Total €000	Cost of sales €000	Distribution costs €000	Administration expenses €000
Allocation of salaries and wages				
Factory staff	12,650	12,650		
Sales and warehouse	10,580		10,580	
Administration and accounts staff	2,300			2,300
Subtotal	25,530	12,650	10,580	2,300
An analysis of depreciation				
Freehold buildings	1,150	575	287.5	287.5
Equipment	1,150	575	287.5	287.5
Motor vehicles (allocated)	2,875		2,875	
Subtotal	5,175	1,150	3,450	575
Motor expenses (allocated)	9,200		9,200	
An apportionment of operating expenses				
on the basis of space occupied				
Insurance	5,290	2,645	1,322.5	1,322.5
Light, power and miscellaneous	1,955	977.5	488.75	488.75
Subtotal	16,445	3,622.5	11,011.25	1,811.25
TOTAL EXPENSES	47,150	17,422.5	25,041.25	4,686.25
Add material consumed	276,000	276,000		
TOTALS for statement of income	323,150	293,422.5	25,041.25	4,686.25

3.4.8 The statement of income using Format 1

Format 1 is favoured by capital markets and provides a multi-stage presentation reporting four profit measures for gross, operating, pre-tax and post-tax profit, as in Figure 3.4.

Figure 3.4 Statement of income of Wiggins SA for the year ended 31 December 20X3

	€000
Revenue	345,000.00
Cost of sales	293,422.50
Gross profit	51,577.50
Distribution costs	25,041.25
Administrative expenses	4,686.25
Operating profit	21,850.00
Finance costs	7,475.00
Profit on ordinary activities before tax	14,375.00
Income tax (2,875 + 750)	3,625.00
Profit for the year	10,750.00

Figure 3.5 Wiggins SA statement of income for the year ended 31 December 20X3

	€000	€000
Revenue		345,000
Decrease in inventory	(17,250)	
Raw materials	(258,750)	(276,000)
Employee benefit expense		
Salaries		(25,530)
Depreciation		(5,175)
Other operating expenses		
Motor expenses	(9,200)	
Insurance	(5,290)	
Light, power and miscellaneous	(1,955)	(16,445)
Operating profit		21,850

3.5 Format 2: classification of operating expenses according to their nature

Note that if Format 2 is used the expenses are classified as change in inventory, raw materials, employee benefits expense, other expenses and depreciation as in Figure 3.5. The operating profit is unchanged from that appearing in Figure 3.4 using Format 1. If this format is used, the cost of sales has to be disclosed.

This method differs in that classification by nature does not require the allocation of expenses to functions. It is a format that is seen to be appropriate to particular industries such as the airline industry where it is adopted by Air China and EasyJet.

3.6 Other comprehensive income

When IAS 1 was revised in 2008 the profit and loss account or 'income statement' was replaced by the statement of comprehensive income, and a new section of 'Other comprehensive income' was added to the previous statement of income.

Other comprehensive income includes **unrealised** gains and losses resulting from changes in fair values of assets/liabilities such as changes in the fair value of intangible assets and property, plant and equipment; actuarial gains and losses on defined benefit plans and gains and losses from translating the financial statements of a foreign operation.

The statement was then retitled as 'Statement of Comprehensive Income'.

3.6.1 What is meant by comprehensive income?

Comprehensive income recognises the gains and losses, both realised **and unrealised**, that have increased or decreased the owners' equity in the business. Such gains and losses arise, for example, from the revaluation of non-current assets and from other items that are discussed later in Chapters 14 (Financial instruments) and 15 (Employee benefits). These are referred to as *Other comprehensive income*.

Figure 3.6 Statement of comprehensive income of Wiggins SA for the year ended 31.12.20X3

	€000
Revenue	345,000.00
Cost of sales	293,422.50
Gross profit	51,577.50
Distribution costs	25,041.25
Administrative expenses	4,686.25
Operating profit	21,850.00
Finance costs	7,475.00
Profit on ordinary activities before tax	14,375.00
Income tax	3,625.00
Profit for the year	10,750.00
Other comprehensive income:	
Gains on property revaluation	20,000.00
Comprehensive income for the year	30,750.00

3.6.2 How to report other comprehensive income

IAS 1 allows a choice. It can be presented as a separate statement or as an extension of the statement of income. In our example we have presented it as an extension of the statement of income.

In this example, there is a revaluation gain on the freehold land which needs to be added to the profit on ordinary activities for the year in order to arrive at the comprehensive income. This is shown in Figure 3.6.

3.6.3 Analysing other comprehensive income

Analysts consider the implication for future profits and growth. For example, it prompts questions such as:

- If there is a gain on non-current asset revaluation: what will be the cash impact on plans for replacing or increasing operating capital? What do the notes to the accounts indicate about capital commitments?
- If there is a gain or loss on foreign exchange: does it indicate a weakening or strengthening of the domestic currency? Will the translation of future overseas sales and profits result in higher or lower reported earnings per share?

3.7 How non-recurring or exceptional items can affect operating income

Operating income is one of the measures used by investors when attempting to predict future income. Management are, therefore, keen to highlight if the current year's operating income has been adversely affected by events that are unlikely to occur in future periods – these are referred to as 'exceptional items'. Such items are within the normal operating activities of the business but require to be separately disclosed because they are significant due to their non-recurring nature and materiality in both size and nature. A company's **quality of earnings** is important as seen in the following extract from the 2012 ITV plc Annual Report:

> Exceptional items are material and non-recurring items excluded from management's assessment of profit because by their nature they could distort the Group's underlying **quality of earnings**. These are excluded to reflect performance in a consistent manner and are in line with how the business is managed and measured on a day-to-day basis.

There could be a number of exceptional reasons that result in a lower profit, for example costs incurred in restructuring the business or unusually high allowances for bad debts or material write-downs of inventories to net realisable value or non-current assets to recoverable amounts.

Exceptional items are not, however, always adverse – there might, for example, have been significant gains arising from the disposal of non-current assets. The following is an extract from the 2012 Aer Lingus Annual Report:

> At the pre-tax level, our 2012 profit of €40.6 million was €43.8 million lower than 2011. The decrease can be specifically attributed to the fact that we incurred exceptional costs of €26.5 million in 2012 while in 2011 we benefited from exceptional gains of €37.2 million.

They are a problem, however, when used to manipulate the figure for maintainable earnings, which led to the UK's FRC issuing a note in 2013 *FRC seeks consistency in the reporting of exceptional items* which aims to discourage companies from smoothing profits by creating exceptional charges – which are then fed back in a later period as part of earnings.

3.7.1 Notes to the accounts

It is important to refer to information in the Notes because these items can have a material impact as seen in the Carrefour 2011 Annual Report where an operating profit is turned into an operating loss:

	2011	2010	% change
Total revenue	82,764	81,840	1.1%
Cost of sales	(64,912)	(63,969)	1.5%
Gross margin from recurring operations	17,852	17,871	(0.1)%
Sales, general and administrative expenses	(13,969)	(13,494)	3.5%
Depreciation, amortisation and provisions	(1,701)	(1,675)	1.6%
Recurring operating income	2,182	2,701	(19.2)%
Non-recurring income and expenses, net	(2,662)	(999)	–
Operating profit/(loss)	(481)	1,703	(128.2)%

Non-recurring income and expenses consist mainly of gains and losses on disposal of property and equipment or intangible assets, impairment losses on property and equipment or intangible assets (including goodwill), restructuring costs and provisions for claims and litigation that are material at Group level. They are presented separately in the income statement to 'help users of the financial statements to better understand the Group's underlying operating performance and provide them with useful information to assess the earnings outlook'.

The maintainable figure to concentrate on is the Recurring operating income – in 2012 and 2013 the company reports this as remaining relatively stable at €2,124m and €2,238m respectively.

3.7.2 Need for consistency in presentation

In the UK the Financial Reporting Council (FRC) issued a reminder in 2013 to Boards on the need to improve the reporting of additional and exceptional items by companies and ensure consistency in their presentation to comply with the Corporate Governance Code principle that the annual report and accounts as a whole should be fair, balanced and understandable. The Financial Reporting Review Panel (FRRP) made the point that it is important that investors should be able to identify the trend in underlying, i.e. maintainable profits.

This means, for example, that where the same category of material items recurs each year and in similar amounts (for example, restructuring costs), companies should consider whether such amounts should be included as part of underlying profit. Also where significant items of expense may be subsequently reversed, they should be treated as exceptional items in the subsequent period.

3.7.3 Columnar format

Whilst the information could be disclosed as a note or a separate line item on the face of the statement of income, some companies emphasise the impact by preparing a three-column statement of income which is an alternative method for disclosing permitted by IAS 1.

3.8 How decision-useful is the statement of comprehensive income?

A key question we should ask whenever there is a proposal to present additional financial information is 'How will this be useful to users of the accounts?' There is no definitive answer, because some commentators[2] argue that there is no decision-usefulness in providing the comprehensive net income figure for investors, whereas others[3] take the opposite view. Intuitively, one might take a view that investors are interested in the total movement in equity regardless of the cause, which would lead to support for the comprehensive income figure. However, given that there is this difference of opinion and research findings, this would seem to be an area open to further empirical research to further test the decision-usefulness of each measure to analysts.

Interesting research[4] has since been carried out which supports the view that net income and comprehensive income are both decision-useful. The findings suggested that comprehensive income was more decision-relevant for assessing share returns and traditional net income more decision-relevant for setting executive bonus incentives.

3.9 Statement of changes in equity

This statement is designed to show the following:

● *Prior period adjustments.* The effect of any prior period adjustments is shown by adjusting the retained earnings figure brought forward (we will cover this in Chapter 4).

Figure 3.7 Statement of changes in equity for the year ended 31 December 20X3

	Share capital	Share premium	Retained earnings	Revaluation surplus	Total
Balance as at 1 January 20X3	15,000	—	57,500		74,750
Changes in equity for 20X3					
New shares issued	1,500	750			
Dividends			(1,725)		(1,725)
Total comprehensive income for the year			10,750	20,000	30,750
Balance as at 31 December 20X3	16,500	750	66,525	20,000	103,775

● *Capital transactions with the owners.* This includes dividends and a reconciliation between the opening and closing equity capital, reporting any change such as increases from bonus, rights or new cash issues and decreases from any buyback of shares.

● *Transfers from revaluation reserves.* When a revalued asset is disposed of, any revaluation surplus may be transferred directly to retained earnings, or it may be left in equity under the heading 'revaluation surplus'.

● *Comprehensive income.* The comprehensive income for the period is disclosed.

The statement for Wiggins SA is shown in Figure 3.7.

Note that the statement of changes in equity is a primary statement and is required to be presented with the same prominence as the other primary statements.

3.10 The statement of financial position

IAS 1 specifies which items are to be included on the face of the statement of financial position. These are referred to as alpha headings (a) to (r) – for example (a) Property, plant and equipment, (b) Investment property, . . . , (g) Inventories, . . . , (k) Trade and other payables.

It does not prescribe the order and presentation that are to be followed. It would be acceptable to present the statement as assets less liabilities equalling equity, or total assets equalling total equity and liabilities.

3.10.1 Current/non-current classification

● The standard does not absolutely prescribe that enterprises need to split assets and liabilities into current and non-current. However, it does state that this split would need to be done if the nature of the business indicates that it is appropriate.

● If it is more relevant, a presentation could be based on liquidity and, if so, all assets and liabilities would be presented broadly in order of liquidity. However, in almost all cases it would be appropriate to split items into current and non-current and the statement in Figure 3.8 follows the headings prescribed in IAS 1.

Figure 3.10 Disclosure note: Property, plant and equipment movements

	Freehold land €000	Freehold buildings €000	Equipment €000	Motor vehicles €000	Total €000
Cost/valuation					
As at 1.1.20X3	57,500	57,500	14,950	20,700	150,650
Revaluation	20,000				20,000
Additions					
Disposals					
As at 31.12.20X3	77,500	57,500	14,950	20,700	170,650
Accumulated depreciation					
As at 1.1.20X3		2,300	3,450	9,200	14,950
Charge for the year		1,150	1,150	2,875	5,175
As at 31.12.20X3		3,450	4,600	12,075	20,125
Net book value					
As at 31.12.20X3	**77,500**	**54,050**	**10,350**	**8,625**	**150,525**
As at 31.12.20X2	57,500	55,200	11,500	11,500	135,700

(c) **Notes giving additional information to assist prediction of future cash flows**

These are notes intended to assist in predicting future cash flows. They give information on matters such as:

● capital commitments that have been contracted for but not provided in the accounts;

● capital commitments that have been authorised but not contracted for;

● future commitments, e.g. share options that have been granted; and

● contingent liabilities, e.g. guarantees given by the company in respect of overdraft facilities arranged by subsidiary companies or customers.

In deciding upon disclosures, management have an obligation to consider whether the omission of the information is material and could influence users who base their decisions on the financial statements. The management decision would be influenced by the size or nature of the item and the characteristics of the users. They are entitled to assume that the users have a reasonable knowledge of business and accounting and a willingness to study the information with reasonable diligence.

(d) **Notes giving information that is of interest to other stakeholders**

An example is information relating to staff. It is common for enterprises to provide a disclosure of the average number of employees in the period or the number of employees at the end of the period. IAS 1 does not require this information but it is likely that many businesses would provide and categorise the information, possibly following functions such as production, sales, and administration as in the following extract from the 2012 Annual Report of Wienerberger:

	Total (2012)	Total (2011)
Production	8,673	8,408
Administration	1,142	980
Sales	3,245	2,865
Total	13,060	12,253
Apprentices	95	59

This shows a significant increase in staff numbers with reasons given within the report. However, the annual report is not the only source of information – there might be separate employee reports and information obtained during labour negotiations such as the ratio of short-term and long-term assets to employee numbers, the capital–labour ratios and the average revenue and net profits per employee in the company with inter-period and inter-firm comparisons. For example, in 2012 Asda reported revenue per employee as £117,605 compared to Tesco which reported £207,931 per employee – the questions should then be raised as to whether the higher rate means that there is less customer satisfaction and the possibility of a fall with lower revenue or more employees.

3.12 Has prescribing the formats meant that identical transactions are reported identically?

That is the intention, but there are various reasons why there may still be differences. For example, let us consider some of the reasons for differences in calculating the cost of sales: (a) how inventory is valued, (b) the choice of depreciation policy, (c) management attitudes, and (d) the capability of the accounting system.

(a) Differences arising from the choice of the inventory valuation method

Different companies may assume different physical flows when calculating the cost of direct materials used in production. This will affect the inventory valuation. One company may assume a first-in-first-out (FIFO) flow, where the cost of sales is charged for raw materials used in production as if the first items purchased were the first items used in production. Another company may use an average basis. This is illustrated in Figure 3.11 for a company that started trading on 1 January 20X1 without any opening inventory and sold 40,000 items on 31 March 20X1 for £4 per item.

Figure 3.11 Effect on sales of using FIFO and weighted average

Physical flow assumption	Items	£	FIFO £	Average £
Raw materials purchased				
On 1 Jan 20X1 at £1 per item	20,000	20,000		
On 1 Feb 20X1 at £2 item	20,000	40,000		
On 1 Mar 20X1 at £3 per item	20,000	60,000		
On 1 Mar 20X1 in inventory	60,000	120,000	120,000	120,000
On 31 Mar 20X1 in inventory	20,000		60,000	40,000
Cost of sales	40,000		60,000	80,000

Figure 3.12 Effect of physical inventory flow assumptions on the percentage gross profit

	Items	FIFO £	Average £	% difference in gross profit
Sales	40,000	160,000	160,000	
Cost of sales	40,000	60,000	80,000	
		100,000	80,000	
Gross profit %		62.5%	50%	25%

Inventory valued on a FIFO basis is £60,000 with the 20,000 items in inventory valued at £3 per item, on the assumption that the purchases made on 1 January 20X1 and 1 February 20X1 were sold first. Inventory valued on an average basis is £40,000 with the 20,000 items in inventory valued at £2 per item on the assumption that sales made in March cannot be matched with a specific item.

The effect on the gross profit percentage would be as shown in Figure 3.12. This demonstrates that, even from a single difference in accounting treatment, the gross profit for the same transaction could be materially different in both absolute and percentage terms.

How can an investor determine the effect of different assumptions?

Although companies are required to disclose their inventory valuation policy, the level of detail provided varies and we are not able to quantify the effect of different inventory valuation policies.

For example, a clear description of an accounting policy is provided by AstraZeneca. Even so, it does not allow the user to know how net realisable value was determined. Was it, for example, primarily based upon forecasted short-term demand for the product?

AstraZeneca inventory policy (2011) Annual Report

Inventories

Inventories are stated at the lower of cost or net realisable value.

The first in, first out or an average method of valuation is used.

For finished goods and work in progress, cost includes directly attributable costs and certain overhead expenses (including depreciation).

Selling expenses and certain other overhead expenses (principally central administration costs) are excluded.

Net realisable value is determined as estimated selling price less all estimated costs of completion and costs to be incurred in selling and distribution.

Write-downs of inventory occur in the general course of business and are included in cost of sales in the income statement. However, if the write-off is regarded as material it would be reported as an exceptional item.

The following illustration is an extract from the 2011 Annual Report of R & R Ice Cream plc:

	Before exceptional items *(€000)*	*Exceptional items* *(€000)*	*After exceptional items* *(€000)*
Revenue	501,028	—	501,028
Cost of sales (Note 2)	(394,932)	(1,150)	(396,082)
Gross profit	106,096	(1,150)	104,946

Note 2. Exceptional items:
Recognised in arriving at results from operating activities (€000):

	2011
Inventory write-off	(873)
Restructuring and redundancies	(277)
Total cost of sales exceptional items	(1,150)

Inventory write-off:
In the year, we incurred two exceptional inventory write-offs, one in respect of poor-quality bought-in products which the third-party supplier refused to refund in full, and one in respect of flash-priced packaging which became redundant as a result of the price increases we were forced to implement to recover rising commodity prices.

(b) Differences arising from the choice of depreciation method and estimates

Companies may make different choices:

- the accounting base to use, e.g. historical cost or revaluation; and
- the method that is used to calculate the charge, e.g. straight-line or reducing balance.

Companies make estimates that might differ:

- assumptions as to an asset's productive use, e.g. different estimates made as to the economic life of an asset; and
- assumptions as to the total cost to be expensed, e.g. different estimates of the residual value.

(c) Differences arising from management attitudes

Losses might be anticipated and measured at a different rate. For example, when assessing the likelihood of the net realisable value of inventory falling below the cost figure, the management decision will be influenced by the optimism with which it views the future of the economy, the industry and the company. There could also be other influences. For example, if bonuses are based on net income, there is an incentive to overestimate the net realisable value; whereas, if management is preparing a company for a management buy-out, there is an incentive to underestimate the net realisable value in order to minimise the net profit for the period.

(d) Differences arising from the capability of the accounting system to provide data

Accounting systems within companies differ and costs that are collected by one company may well not be collected by another company. For example, the apportionment of costs might be more detailed with different proportions being allocated or apportioned.

3.13 Fair presentation

IAS 1 *Presentation of Financial Statements* requires financial statements to give a **fair presentation** of the financial position, financial performance and cash flows of an enterprise. In paragraph 17 it states that:

> In virtually all circumstances, a fair presentation is achieved by compliance with applicable IFRSs. A fair presentation also requires an entity to:
>
> **(a)** select and apply accounting policies in accordance with IAS 8 *Accounting Policies, Changes in Accounting Estimates and Errors* [this is dealt with in Chapter 4];
>
> **(b)** present information, including accounting policies, in a manner that provides relevant, reliable, comparable and understandable information;
>
> **(c)** provide additional disclosures when compliance with the specific requirements in IFRSs is insufficient to enable users to understand the impact of particular transactions, other events and conditions on the entity's financial position and financial performance.

3.13.1 Legal opinions

In the UK we require financial statements to give a **true and fair view**. True and fair is a legal concept and can be authoritatively decided only by a court. However, the courts have never attempted to define 'true and fair'. In the UK the Accounting Standards Committee (ASC) obtained a legal opinion which included the following statement:

> Accounts will not be true and fair unless the information they contain is sufficient in quantity and quality to satisfy the reasonable expectations of the readers to whom they are addressed.

Accounting standards are an authoritative source of accounting practice.

However, an Opinion obtained by the FRC in May 2008 advised that true and fair still has to be taken into consideration by preparers and auditors of financial statements whether prepared under UK company law or IFRSs. Directors have to consider whether the statements are appropriate and auditors have to exercise professional judgement when giving an audit opinion – it is not sufficient for either directors or auditors to reach a conclusion solely because the financial statements were prepared in accordance with applicable accounting standards.

3.13.2 Fair override

Standards are not intended to be a straitjacket and IAS 1 recognises that there may be occasions when application of an IAS/IFRS might be misleading and departure from the IAS/IFRS treatment is permitted. This is referred to as the **fair override** provision. If a company makes use of the override it is required to explain why compliance with IASs/IFRSs would be misleading and also give sufficient information to enable the user to calculate the adjustments required to comply with the standard.

Although IAS 1 does not refer to true and fair, the International Accounting Standards Regulation 1606/2002 (paragraph 9) states that: 'To adopt an international accounting standard for application in the Community, it is necessary . . . that its application results in a true and fair view of the financial position and performance of an enterprise'. However, overrides under IFRS are likely to be less common, as IAS 1 states that departures from IAS should be 'extremely rare' (paragraph 17) and should only happen where compliance with the standard together with additional disclosure would not result in a fair presentation. Examples of the use of the IAS 1 override among European companies are very rare.

When do companies use the fair override?

Fair override can occur for a number[5] of reasons with the most frequent being the situation where the Accounting Standards may prescribe one method, which contradicts company law and thus requires an override, for example, providing no depreciation on investment properties. Next would be where Accounting Standards may offer a choice between accounting procedures, at least one of which contradicts company law. If that particular choice is adopted, the override should be invoked, for example grants not being shown as deferred income.

Fair override can be challenged

If a company in the UK relies on the fair override provision, it may be challenged by the Financial Reporting Review Panel and the company's decision overturned. For example, although Eurovestech had adopted an accounting policy in its 2005 and 2006 accounts not to consolidate two of its subsidiaries because its directors considered that to do so would not give a true and fair view, the FRRP decision was that this was unacceptable because the company was unable to demonstrate special circumstances warranting this treatment.

3.14 What does an investor need in addition to the primary financial statements to make decisions?

Investors attempt to estimate future cash flows when making an investment decision. As regards future cash flows, these are normally perceived to be influenced by past profits as reported in the statements of income and the asset base as shown by the statement of financial position.

In order to assist shareholders to predict future cash flows with an understanding of the risks involved, more information has been required by the IASB. For example:

- More quantitative information (discussed in Chapter 4):
  - financial statements are required to take account of events and information becoming available after the period-end;
  - segment reports are required;
  - disclosure is required of the impact of changes on the operation, e.g. a breakdown of turnover, costs and profits for both new and discontinued operations.
- More qualitative narrative information, including:
  - mandatory disclosures;
  - IFRS practice statement – management commentary.
- UK requirements:
  - Chairman's Statement;
  - Directors' Report;
  - Disclosure: Operating and Financial Review;
  - Strategic Report.

We will comment briefly on the qualitative narrative disclosures.

3.14.1 IFRS mandatory disclosures

When making future predictions, investors need to be able to identify that part of the net income that is likely to be maintained in the future. IAS 1 provides assistance to users in this by requiring that certain items are separately disclosed. These are items within the ordinary activities of the enterprise which are of such size, nature or incidence that their separate disclosure is required in the financial statements in order for the financial statements to show a fair view.

These items are not extraordinary and must, therefore, be presented above the tax line. It is usual to disclose the nature and amount of these items in a note to the financial statements, with no separate mention on the face of the statement of comprehensive income; however, if sufficiently material, they can be disclosed on the face of the statement.

Examples of the type of item that may give rise to separate disclosures are:

- the write-down of assets to realisable value or recoverable amount;
- the restructuring of activities of the enterprise; discontinued operations; disposals of items of property, plant and equipment and long-term investments; litigation settlements.

3.14.2 Subjective nature of items classified as exceptional

The items reported as exceptional require the exercise of judgement and so need to be approached with a certain amount of scepticism.

Exercise of judgement

Judgement is required in determining the best manner in which information is presented. IAS 1 is not prescriptive and companies may choose to present exceptional items as a line item on the face of the accounts, as a disclosure note or in columnar format.

Judgement is also required when classifying items as operating or exceptional. IAS 1 states that it would be misleading and would impair comparability if items of an operating nature were excluded from the results of operating activities. This is aimed at preventing companies from classifying operating costs as exceptional in order to improve the headline figure that is frequently used to calculate key performance indicators such as return on equity and earnings per share.

Exercise of scepticism

Reporting an item as exceptional may cause operating profits to be boosted, as illustrated in a 2014 S&P study *'Why Inconsistent Reporting of Exceptional Items Can Cloud Underlying Profitability'* of non-financial FTSE 100 companies which reported that 89% adjusted profits and in 73% of cases this boosted operating profits.

A similar review[6] in 2013 by the Irish Auditing and Advisory Authority found that the costs presented as exceptional exceeded the income presented as exceptional by a factor of 5:1.

Investors need to exercise scepticism and carefully scrutinise underlying earnings and exceptional items before reaching their own view of a company's performance.

3.14.3 IFRS management commentary

In December 2010 the IASB issued an IFRS Practice Statement *Management Commentary*. Management commentary is defined in the statement as:

> A narrative report that relates to financial statements that have been prepared in accordance with IFRSs. Management commentary provides users with historical explanations of the amounts presented in the financial statements, specifically the entity's financial position, financial performance and cash flows. It also provides commentary on an entity's prospects and other information not presented in the financial statements. Management commentary also serves as a basis for understanding management's objectives and its strategies for achieving those objectives.

The commentary should give management's view not only about what has happened, including both positive and negative circumstances, but also why it has happened and what the implications are for the entity's future.

Following the Practice Statement is not mandatory and the financial statements and annual report of a business can still be compliant with IFRS even if the requirements are not followed. However, it is the first document to be issued by the IASB that solely covers information that is provided by companies outside the financial statements.

The guidance does not attempt to dictate exactly how management commentary should be prepared so as to avoid the tick-box approach to compliance. Instead it indicates the information that should be included within the commentary:

(a) the nature of the business;

(b) management's objectives and its strategies for meeting those objectives;

(c) the entity's most significant resources, risks and relationships;

(d) the results of operations and prospects; and

(e) the critical performance measures and indicators that management uses to evaluate the entity's performance against stated objectives.

It will be interesting to observe the effect that this has on future annual reports.

Who presents and approves a management commentary may depend on jurisdictional requirements.

3.14.4 Strategic Report

In the UK the 2006 Companies Act[7] provides that from 2013 quoted companies should issue a Strategic Report to help members to assess how the directors have performed their duty of promoting the success of the company.

The report must contain a description of the company's strategy, its business model, the principal risks and uncertainties facing the company and a balanced and comprehensive analysis of the development and performance of the company's business during the financial year and its position at the end of that year.

The review must also, where appropriate, include an analysis using financial and other key performance indicators including information relating to environmental, social, employee and human rights matters.

Potentially useful key performance indicators

The intention is that there should not be a list of all performance measures but only those key indicators considered by the board and used in management reporting.

They might include:

● Economic measures of ability to create value (with the terms defined):

  ● Return on capital employed;

  ● Economic profit-type measures, i.e. post-tax profits less cost of capital;

● Market position;

● Market share;

● Development, performance and position:

  ● traditional financial measures such as asset turnover rates;

  ● industry-specific measures such as sales per square metre;

- Customers, employees and suppliers: how they rate the company;
- Social, environmental and community issues.

Guidance on Strategic Report

An Exposure Draft[8] was issued by the FRC in 2013. Its objective is to ensure that relevant information that meets the needs of shareholders is presented in the strategic report and to encourage companies to experiment and be innovative in the drafting of their annual reports.

3.14.5 Chairman's Statement

This often tends to be a brief upbeat comment on the current year. For example, the following is a brief extract from Findel plc's 2012 Annual Report to illustrate the type of information provided:

> Second-half group sales were ahead by 1.6%. This reflects the actions that we are taking and highlights the momentum building within the business. Profit before tax for the year increased by 53% to £10.7m (FY2011: £7.0m) as a result of substantially reduced finance costs following the refinancing. The group incurred exceptional restructuring and finance costs, primarily relating to various internal restructurings and management changes undertaken during the year, of £14.5m. . . . We continue to have sufficient headroom in our banking facilities and ample liquidity to execute our Full Potential plan.

3.14.6 Directors' Report

The paragraph headings from Findel's 2010 Annual Report illustrate the type of information that is published. The report headings were Activities; Review of the year and future prospects; Dividends; Capital structure; Suppliers' payment policy; Directors; Employees' Donations; Substantial holdings and Auditors.

There is a brief comment under each heading, for example:

Activities

The principal activities of the group are home shopping and educational supplies through mail order catalogues and the provision of outsourced healthcare services.

Review of the year and future prospects

The key performance indicators which management consider important are:

- operating margins;
- average order-value.

3.14.7 Survey of meeting narrative reporting needs for the future

In 2010 the ACCA issued the results of an international survey[9] of CFOs' views on narrative reporting, '*Hitting the notes, but what's the tune?*', based on a joint survey with Deloitte of some 230 chief financial officers and other preparers in listed companies in nine countries (Australia, China, Kenya, Malaysia, Singapore, Switzerland, the UAE, the UK and the US). The major findings were that:

- the principal audiences for narrative information were shareholders and regulators;
- the most important disclosures for shareholders were the explanation of financial results and financial position, identifying the most important risks and how they were managed, an outline of future plans and prospects, a description of the business model and a description of key performance indicators (KPIs);
- the interviewees supported a reporting environment with more discretion and less regulation.

As with all approaches to standard setting, there is the need to balance discretion and regulation.

The need for a mature user response

The following is an extract from an ACCA paper 'Writing the narrative: the triumphs and tribulations'[10] by Afra Sajjad:

> We believe that the future of narrative reporting lies in reconciliation of competing information needs and expectations of primary users of annual reports, i.e. regulators and shareholders. This should be accompanied by nurturing of a culture of corporate reporting where integrity, probity and transparency are fundamental to reporting. Regulators also need to facilitate change in the culture of reporting by giving preparers the flexibility to use discretion and facilitate market led best practices. Shareholders . . . need also to be mature enough to encourage real transparency. If they respond with panic to disappointing news, it will inhibit the preparer's disclosure process.

Summary

In this chapter we have revised the preparation of internal financial statements making accrual adjustments to trial balance figures.

In order to assess stewardship and management performance, there have been mandatory requirements for standardised presentation, using the two formats prescribed by International Financial Reporting Standards. The required disclosures were explained for both formats.

The importance of referring to Notes to the accounts was illustrated with discussion of exceptional items and their impact on predicting operating income.

The disclosure of accounting policies which allow shareholders to make comparisons between years by requiring companies to be consistent in the application of accounting policies or requiring disclosure if there has been a change was discussed.

The need for explanatory notes was explained and described.

The need for financial statements to give a true and fair view of the income and net assets was explained with recognition that this requires the exercise of professional judgement. Having recorded the transactions and made the normal adjustments for accruals, do the resulting financial statements give a fair presentation?

The evolving practices for narrative reporting under IASB and UK were discussed with the IFRS Practice Statement Management Commentary and the UK Strategic Report.

REVIEW QUESTIONS

1 Explain the effect on income and financial position if (a) the amount of accrued expense were to be underestimated and (b) the inventory at the year-end omitted inventory held in a customs warehouse awaiting clearance.

2 Explain why two companies carrying out identical trading transactions could produce different gross profit figures.

3 Classify the following items into cost of sales, distribution costs, administrative expenses, other operating income or item to be disclosed after trading profit:

(a) Personnel department costs

(b) Computer department costs

(c) Cost accounting department costs

(d) Financial accounting department costs

(e) Bad debts

(f) Provisions for warranty claims

(g) Interest on funds borrowed to finance an increase in working capital

(h) Interest on funds borrowed to finance an increase in property, plant and equipment.

4 'We analyze a sample of UK public companies that invoked a True and Fair View (TFV) override during 1998–2000 to assess whether overrides are used opportunistically. We find overrides increase income and equity significantly, and firms with weaker performance and higher levels of debt employ overrides that are more costly . . . financial statements are not less informative than control sample.'[11]

Discuss the enquiries and action that you think an auditor should take to ensure that the financial statements give a more true and fair view than from applying standards.

5 When preparing accounts under Format 1, how would a bad debt that was materially larger than normal be disclosed?

6 'Annual accounts have been put into such a straitjacket of overemphasis on uniform disclosure that there will be a growing pressure by national bodies to introduce changes unilaterally which will again lead to diversity in the quality of disclosure. This is both healthy and necessary.' Discuss.

7 Explain the relevance to the user of accounts if expenses are classified as 'administrative expenses' rather than as 'cost of sales'.

8 IAS 1 *Presentation of Financial Statements* requires 'other comprehensive income' items to be included in the statement of comprehensive income and it also requires a statement of changes in equity. Explain the need for publishing this information, and identify the items you would include in them.

9 Discuss the major benefit to an investor from the UK Strategic Report.

10 The following are three KPIs for the retail sector:[12] capital expenditure, expected return on new stores and customer satisfaction. Discuss two further KPIs that might be significant.

EXERCISES

* Question 1

The following trial balance was extracted from the books of Old NV on 31 December 20X1.

	€000	€000
Sales		12,050
Returns outwards		313
Provision for depreciation		
Plant		738
Vehicles		375
Rent receivable		100
Trade payables		738
Debentures		250
Issued share capital – ordinary €1 shares		3,125
Issued share capital – preference shares (treated as equity)		625
Share premium		350
Retained earnings		875
Inventory	825	
Purchases	6,263	
Returns inwards	350	
Carriage inwards	13	
Carriage outwards	125	
Salesmen's salaries	800	
Administrative wages and salaries	738	
Land	100	
Plant (includes €362,000 acquired in 20X1)	1,562	
Motor vehicles	1,125	
Goodwill	1,062	
Distribution costs	290	
Administrative expenses	286	
Directors' remuneration	375	
Trade receivables	3,875	
Cash at bank and in hand	1,750	
	19,539	19,539

Note of information not taken into the trial balance data:

(a) Provide for:

    (i)   An audit fee of €38,000.

    (ii) Depreciation of plant at 20% straight-line.

    (iii) Depreciation of vehicles at 25% reducing balance.

    (iv) The goodwill suffered an impairment in the year of €177,000.

    (v) Income tax of €562,000.

    (vi) Debenture interest of €25,000.

(b) Closing inventory was valued at €1,125,000 at the lower of cost and net realisable value.

(c) Administrative expenses were prepaid by €12,000.

(d) Land was to be revalued by €50,000.

Required:

(a) **Prepare a statement of income for internal use for the year ended 31 December 20X1.**

(b) **Prepare a statement of comprehensive income for the year ended 31 December 20X1 and a statement of financial position as at that date in Format 1 style of presentation.**

* Question 2

Formatone plc produced the following trial balance as at 30 June 20X6:

	£000	£000
Land at cost	2,160.0	—
Buildings at cost	1,080.0	—
Plant and Equipment at cost	1,728.0	—
Intangible assets	810.0	—
Accum. depreciation – 30.6.20X5		
Buildings	—	432.0
Plant and equipment	—	504.0
Interim dividend paid	108.0	
Receivables and payables	585.0	532.8
Cash and bank balance	41.4	—
Inventory as at 30.6.20X6	586.8	—
Taxation	—	14.4
Deferred tax	—	37.8
Distribution cost	529.2	—
Administrative expenses	946.8	—
Retained earnings b/f	—	891.0
Sales revenue	—	9,480.6
Cost of sales	5,909.4	—
Ordinary shares of 50p each	—	2,160.0
Share premium account	—	432.0
	14,484.6	14,484.6

The following information is available:

(i) A revaluation of the Land and Buildings on 1 July 20X5 resulted in an increase of £3,240,000 in the Land and £972,000 in the Buildings. This has not yet been recorded in the books.

(ii) Depreciation:
Plant and Equipment are depreciated at 10% using the reducing balance method.
Intangible assets are to be written down by £540,000.
Buildings have an estimated life of 30 years from date of the revaluation.

(iii) Taxation
The current tax is estimated at £169,200.
There had been an overprovision in the previous year.
Deferred tax is to be increased by £27,000.

(iv) Capital
150,000 shares were issued and recorded on 1 July 20X5 for 80p each.
A further dividend of 5p per share has been declared on 30 June 20X6.

Required:

Prepare for the year ended 30 June 20X6 the statement of comprehensive income, statement of changes in equity and statement of financial position.

* Question 3

Basalt plc is a wholesaler. The following is its trial balance as at 31 December 20X0.

	Dr £000	Cr £000
Ordinary share capital: £1 shares		300
Share premium		20
General reserve		16
Retained earnings as at 1 January 20X0		55
Inventory as at 1 January 20X0	66	
Sales		962
Purchases	500	
Administrative costs	10	
Distribution costs	6	
Plant and machinery – cost	220	
Plant and machinery – provision for depreciation		49
Returns outwards		25
Returns inwards	27	
Carriage inwards	9	
Warehouse wages	101	
Salesmen's salaries	64	
Administrative wages and salaries	60	
Hire of motor vehicles	19	
Directors' remuneration	30	
Rent receivable		7
Trade receivables	326	
Cash at bank	62	
Trade payables		66
	1,500	1,500

The following additional information is supplied:

(i) Depreciate plant and machinery 20% on straight-line basis.

(ii) Inventory at 31 December 20X0 is £90,000.

(iii) Accrue auditors' remuneration £2,000.

(iv) Income tax for the year will be £58,000 payable October 20X1.

(v) It is estimated that 7/11 of the plant and machinery is used in connection with distribution, with the remainder for administration. The motor vehicle costs should be allocated to distribution.

Required:
Prepare a statement of income and statement of financial position in a form that complies with IAS 1. No notes to the accounts are required.

The following information is available:

1 Freehold premises acquired for £1.8 million were revalued in 20X4, recognising a gain of £600,000. These include a warehouse, which cost £120,000, was revalued at £150,000 and was sold in June 20X7 for £225,000. Phoenix does not depreciate freehold premises.

2 Phoenix wishes to report plant and machinery at open market value which is estimated to be £1,960,000 on 1 July 20X6.

3 Company policy is to depreciate its assets on the straight-line method at annual rates as follows:

Plant and machinery   10%
Furniture and fittings   5%

4 Until this year the company's policy has been to capitalise development costs, to the extent permitted by relevant accounting standards. The company must now write off the development costs, including £124,000 incurred in the year, as the project no longer meets the capitalisation criteria.

5 During the year the company has issued one million shares of £1 at £1.20 each.

6 Included within administrative expenses are the following:

Staff salary (including £125,000 to directors)	£468,000
Directors' fees	£96,000
Audit fees and expenses	£86,000

7 Income tax for the year is estimated at £122,000.

8 Directors propose a final dividend of 4p per share declared and an obligation, but not paid at the year-end.

Required:
In respect of the year ended 30 June 20X7:
(a) The statement of comprehensive income.
(b) The statement of financial position as at 30 June 20X7.
(c) The statement of movement of property, plant and equipment.

Question 6

Olive A/S, incorporated with an authorised capital consisting of one million ordinary shares of €1 each, employs 64 persons, of whom 42 work at the factory and the rest at the head office. The trial balance extracted from its books as at 30 September 20X4 is as follows:

	€000	€000
Land and buildings (cost €600,000)	520	—
Plant and machinery (cost €840,000)	680	—
Proceeds on disposal of plant and machinery	—	180
Fixtures and equipment (cost €120,000)	94	—
Sales	—	3,460
Carriage inwards	162	—
Share premium account	—	150
Advertising	112	—
Inventory on 1 Oct 20X3	211	—
Heating and lighting	80	—
Prepayments	115	—
Salaries	820	—
Trade investments at cost	248	—
Dividend received (net) on 9 Sept 20X4	—	45
Directors' emoluments	180	—
Pension cost	100	—
Audit fees and expense	65	—
Retained earnings b/f	—	601
Sales commission	92	—
Stationery	28	—
Development cost	425	—
Formation expenses	120	—
Receivables and payables	584	296
Interim dividend paid on 4 Mar 20X4	60	—
12% debentures issued on 1 Apr 20X4	—	500
Debenture interest paid on 1 Jul 20X4	15	—
Purchases	925	—
Income tax on year to 30 Sept 20X3	—	128
Other administration expenses	128	—
Bad debts	158	—
Cash and bank balance	38	—
Ordinary shares of €1 fully called	—	600
	5,960	5,960

You are informed as follows:

(a) As at 1 October 20X3 land and buildings were revalued at €900,000. A third of the cost as well as all the valuation is regarded as attributable to the land. Directors have decided to report this asset at valuation.

(b) New fixtures were acquired on 1 January 20X4 for €40,000; a machine acquired on 1 October 20X1 for €240,000 was disposed of on 1 July 20X4 for €180,000, being replaced on the same date by another acquired for €320,000.

(c) Depreciation for the year is to be calculated on the straight-line basis as follows:

Buildings: 2% p.a.
Plant and machinery: 10% p.a.
Fixtures and equipment: 10% p.a.

(d) Inventory, including raw materials and work in progress on 30 September 20X4, has been valued at cost at €364,000.

(e) Prepayments are made up as follows:

	€000
Amount paid in advance for a machine	60
Amount paid in advance for purchasing raw materials	40
Prepaid rent	15
	€115

(f) In March 20X3 a customer had filed legal action claiming damages at €240,000. When accounts for the year ended 30 September 20X3 were finalised, a provision of €90,000 was made in respect of this claim. This claim was settled out of court in April 20X4 at €150,000 and the amount of the underprovision adjusted against the profit balance brought forward from previous years.

(g) The following allocations have been agreed upon:

	Factory	Administration
Depreciation of buildings	60%	40%
Salaries other than to directors	55%	45%
Heating and lighting	80%	20%

(h) Pension cost of the company is calculated at 10% of the emoluments and salaries.

(i) Income tax on 20X3 profit has been agreed at €140,000 and that for 20X4 estimated at €185,000.

(j) Directors wish to write off the formation expenses as far as possible without reducing the amount of profits available for distribution.

Required:
Prepare for publication:

(a) The statement of comprehensive income of the company for the year ended 30 September 20X4,

(b) the statement of financial position as at that date along with as many notes (other than the one on accounting policy) as can be provided on the basis of the information made available, and

(c) the statement of changes in equity.

Question 7

The following is an extract from the trial balance of Imecet at 31 October 2005:

	$000	$000
Property valuation	8,000	
Factory at cost	2,700	
Administration building at cost	1,200	
Delivery vehicles at cost	500	
Sales		10,300
Inventory at 1 November 2004	1,100	
Purchases	6,350	
Factory wages	575	
Administration expenses	140	
Distribution costs	370	
Interest paid (6 months to 30 April 2005)	100	
Accumulated profit at 1 November 2004		3,701
10% loan stock		2,000
$1 ordinary shares (incl. issue on 1 May 2005)		4,000
Share premium (after issue on 1 May 2005)		1,500
Dividends (paid 1 June 2005)	400	
Revaluation reserve		2,500
Deferred tax		650

Other relevant information:

(i)   One million $1 ordinary shares were issued 1 May 2005 at the market price of $1.75 per ordinary share.

(ii) The inventory at 31 October 2005 has been valued at $1,150,000.

(iii) A current tax provision for $350,000 is required for the period ended 31 October 2005 and the deferred tax liability at that date has been calculated to be $725,000.

(iv) The property has been further revalued at 31 October 2005 at the market price of $9,200,000.

(v)   No depreciation charges have yet been recognised for the year ended 31 October 2005.

(vi) The depreciation rates are:

Factory – 5% straight-line.
Administration building – 3% straight-line.
Delivery vehicles – 25% reducing balance. The accumulated depreciation at 31 October 2004 was $10,000. No new vehicles were acquired in the year to 31 October 2005.

Required:
(a) Prepare the income statement for Imecet for the year ended 31 October 2005.
(b) Prepare the statement of changes in equity for Imecet for the year ended 31 October 2005.

(The Association of International Accountants)

Question 8

Scott Ross, CFO of Ryan Industries PLC, is discussing the publication of the annual report with his managing director Nathan Davison. Graydon says: 'The law requires us to comply with accounting standards and at the same time to provide a true and fair view of the results and financial position. As half of the business consists of the crockery and brickmaking business which your great-great-grandmother Sasha started, and the other half is the insurance company which your father started, I am not sure that the consolidated accounts are very meaningful. It is hard to make sense of any of the ratios as you don't know what industry to compare them with. What say we also give them the comprehensive income statements and balance sheets of the two subsidiary companies as additional information, and then no one can complain that they didn't get a true and fair view?'

Nathan says: 'I don't think we should do that. The more information they have the more questions they will ask. Also they might realise we have been smoothing income by changing our level of pessimism in relation to the provisions for outstanding insurance claims. Anyway I don't want them to interfere with my business. Can't we just include a footnote, preferably a vague one, that stresses we are not comparable to insurance companies or brickmakers or crockery manufacturers because of the unique mix of our businesses? Don't raise the matter with the auditors because it will put ideas into their heads. But if it does come up we may have to charge head office costs to the two subsidiaries. You need to think up some reason why most of the charges should be passed on to the crockery operations. We don't want to show everyone how profitable that area is. I trust you will give that some thought so you will have a good answer ready.'

Required:
Discuss the professional, legal and ethical implications for Ross.

References

1 IAS 1 *Presentation of Financial Statements*, IASB, December 2008.
2 D. Dhaliwal, K. Subramnayam and R. Trezevant, 'Is comprehensive income superior to net income as a measure of firm performance?', *Journal of Accounting and Economics*, vol. 26(1), 1999, pp. 43–67.
3 D. Hirst and P. Hopkins, 'Comprehensive income reporting and analysts' valuation judgments', *Journal of Accounting Research*, vol. 36 (Supplement), 1998, pp. 47–74.
4 G.C. Biddles and J.-H. Choi, 'Is comprehensive income irrelevant?', 12 June 2002. Available at SSRN: http://ssrn.com/abstract=316703.
5 G. Livne and M. McNichols, *An Empirical Investigation of the True and Fair Override*, LBS Accounting Subject Area Working Paper No. 031 (http://www.bm.ust.hk/acct/acsymp2004/Papers/Livne.pdf).
6 https://www.iaasa.ie/publications/IAS1/ebook/IAS1_Commentary.pdf
7 Companies Act 2006 (Strategic Report and Directors Report) Regulations 2013.
8 Exposure Draft: Guidance for Strategic Report, FRC, August 2013.
9 http://www.accaglobal.com/gb/en/technical-activities/technical-resources-search/2010/september/hitting-the-notes.html
10 http://www.accaglobal.com/content/dam/acca/global/PDF-technical/narrative-reporting/writing_the_narrative.pdf
11 G. Livne and M.F. McNichols, 'An empirical investigation of the true and fair override', *Journal of Business, Finance and Accounting*, pp. 1–30, January/March 2009.
12 http://www.pwc.com/gx/en/corporate-reporting/assets/pdfs/UK_KPI_guide.pdf

CHAPTER 4

Annual Report: additional financial disclosures

4.1 Introduction

The main purpose of this chapter is to explain the additional content in an Annual Report that assists users to make informed assessments of stewardship and informed estimates of future financial performance. Investors need to be able to assess the effect on the published accounts of (a) transactions occurring after the year-end and (b) transactions occurring during the year that might not have been at arm's length. In looking at the future, investors need information on (a) the profitability of different product lines and markets and (b) the potential financial impact if any part of the business has been discontinued.

> ### Objectives
>
> By the end of this chapter, you should be able to:
>
> - make appropriate entries in the financial statements and/or disclosure in the notes to the accounts in accordance with IAS 10 *Events after the Reporting Period*;
> - make appropriate entries in the financial statements in accordance with IAS 8 *Accounting Policies, Changes in Accounting Estimates and Errors*;
> - identify reportable segments in accordance with IFRS 8 *Operating Segments*;
> - critically discuss the benefits and continuing concerns of segmental reporting;
> - explain the meaning of the term and account for 'discontinued operations' in accordance with IFRS 5 *Non-current Assets Held for Sale and Discontinued Operations*;
> - prepare financial statements applying IFRS 5;
> - discuss the impact of such operations on the statement of comprehensive income;
> - explain the criteria laid out in IFRS 5 that need to be satisfied before an asset (or disposal group) is classified as 'held for sale';
> - explain how to identify key personnel for the purposes of IAS 24 *Related Party Disclosures* and why this is considered to be important.

4.2 IAS 10 *Events after the Reporting Period*[1]

We have seen that transactions listed in the trial balance need to be adjusted for accruals and prepayments. They may also need to be adjusted as a result of further information becoming available after the year-end. This is covered in IAS 10.

IAS 10 requires preparers of financial statements to review events that occur after the reporting date but before the financial statements have been authorised for issue by the directors to decide whether an **adjustment** is required to be made to the financial statements or explanatory information is required to be **disclosed** by way of a note.

4.2.1 Adjusting events

These are events after the reporting period that provide additional evidence of conditions that existed at the period-end. Examples of such events include, but are not limited to:

- *Inventory*: After-date sales of inventory that provide additional evidence that the net realisable value of the inventory at the reporting date was lower than cost.
- *Liabilities*: Evidence received after the year-end that provides additional evidence of the appropriate measurement of a liability that existed at the reporting date, such as the settlement of a contingent liability or the calculation of bonuses for which an obligation existed at the end of the reporting period.
- *Non-current assets*: The revaluation of an asset such as a property that indicates the likelihood of impairment at the reporting date.
- The discovery of fraud or errors that show that the financial statements are incorrect.

Under IAS 10 such information becoming available after the period-end means that the financial statements themselves have to be adjusted **provided** the information becomes available before the accounts have been approved. It is important to consider the date of the period-end, the date when the financial statements are approved and the date when transactions/events occurred.

For example, consider the following scenario. Financial statements are being prepared for the year ended 31 March 20X4 and are expected to be approved in the Annual General Meeting announced for 25 June 20X4. Reviewing the audit file on 15 May, it was noted that the audit staff had identified on 29 April that stores staff had misappropriated a material amount of stock before the year-end and concealed it by reporting it as damaged. The police have been informed and are investigating. Should the financial statements be adjusted?

Solution: As the fraud involves a material amount occurring during the reporting period but which is only discovered after the period-end, it is classified as an adjusting event and the financial statements would require amendment.

Consideration would then be required of the accounting implications. For example, what is the impact on the cost of sales and the gross profit if closing inventory has been overstated? Is it necessary to disclose the loss as an exceptional item? What is the likelihood of recovering recompense from the staff themselves or the company's insurers? Is any recovery an asset or contingent asset?

4.2.2 Non-adjusting events

These are events occurring after the reporting period that concern conditions that did not exist at the statement of financial position date. Examples would include:

- Dividends proposed. These must be disclosed in the notes [IAS 1.137]: 'the amount of dividends proposed or declared before the financial statements were authorised for issue but not recognised as a distribution to owners during the period'. The concept of a 'dividend liability' for equity shares has effectively disappeared.

- Interim dividends are not non-adjusting events, because they will have been paid during the reporting period, whereas final dividends are at the discretion of the reporting entity until approved by shareholders at a general meeting.

- An issue or redemption of shares after the reporting date, as in the following extract from the 2010 Annual Report of Wolters Kluwer:

  Events after the reporting period
  The company intends to execute a €100 million share buy-back plan in 2011.

- Potential restructuring implications, as in the following extract from the 2013 Annual Report of Mothercare plc:

  As part of the Transformation and Growth plan an in-depth organisational review was conducted to streamline the group's structure and processes. As a result of the potential restructuring a number of employees in the head office in the UK and the overseas sourcing offices are in consultation. There are likely to be additional exceptional costs of approximately £5 million in respect of the implementation of this review and these will be charged in the next financial year.

- An announcement after the reporting date of a plan to discontinue an operation or entering into binding agreements to sell.

- The loss or other decline in value of assets due to events occurring after the reporting date.

- Entering into significant contracts, as in the following extract from the Deutz 2011 Annual Report:

  On 12 January 2012, DEUTZ AG signed an agreement with the Chinese construction and agricultural equipment manufacturer Shandong Changlin Machinery Group to establish a company for the production of engines. . . . Over the medium term, the new plant will have a production capacity of around 65,000 engines. . . . At the moment, China represents the greatest area of potential growth for DEUTZ within the Asia region as a whole.

4.2.3 Going concern issues

Deterioration in the operating results or other major losses that occur after the period-end are basically non-adjusting events. However, if they are of such significance as to affect the going concern basis of preparation of the financial statements, then this impacts on the numbers in the financial statements, because the going concern assumption would no longer be appropriate. In this limited set of circumstances if the going concern assumption is no longer appropriate, IAS 10 requires the financial statements to be produced on a liquidation rather than going concern basis.

4.3 IAS 8 Accounting Policies, Changes in Accounting Estimates and Errors[2]

IAS 8 gives guidance when deciding whether to make a retrospective or prospective change to financial statements. A retrospective change means that the financial statements of the current and previous years will be affected. A prospective change means that accounting treatments in future years will be affected.

Let us now consider how to treat accounting policy changes, prior period adjustments and changes in accounting estimates in accordance with IAS 8.

4.5 IFRS 8 *Operating Segments*[3]

IFRS 8 applies to both separate and consolidated financial statements of entities

- whose debt or equity instruments are traded in a public market; or
- that file financial statements with a securities commission or other regulatory organisation for the purpose of issuing any class of instruments in the public market.

We will comment briefly on the following four key areas:

- identification of segments;
- identification of reportable segments;
- measurement of segment information; and
- disclosures.

4.5.1 Identification of segments

IFRS 8 requires the identification of operating segments on the basis of internal reports that are regularly reviewed by the entity's chief operating decision maker (CODM) in order to allocate resources to the segment and assess its performance. A segment that sells exclusively or mainly to other operating segments of the group meets the definition of an operating segment if the business is managed in that way.

Criteria for identifying a segment

An operating segment is a component of an entity:

(a) that engages in business activities from which it may earn revenues and incur expenses;

(b) whose operating results are regularly reviewed by the entity's chief operating decision maker, to make decisions about resources to be allocated to the segment and to assess its performance; and

(c) for which discrete financial information is available.

Not every part of the entity will necessarily be an operating segment. For example, a corporate headquarters may not earn revenues.

Criteria for identifying the chief operating decision maker

The chief operating decision maker (CODM) may be an individual or a group of directors or others. The key identifying factors will be those of performance assessment and resource allocation. Some organisations may have overlapping sets of components for which managers are responsible, e.g. some managers may be responsible for specific geographic areas and others for products worldwide. If the CODM reviews the operating results of both sets of components, the entity determines which constitutes the operating segments using the core principles (a)–(c) above.

4.5.2 Identifying reportable segments

Once an operating segment has been identified, a decision has to be made as to whether it has to be reported. The segment information is required to be reported for any operating segment that meets any of the following criteria:

(a) its reported revenue, from internal and external customers, is 10% or more of the combined revenue (internal and external) of all operating segments; or

(b) the absolute measure of its reported profit or loss is 10% or more of the greater in absolute amount of (i) the combined profit of all operating segments that did not report a loss and (ii) the combined reported loss of all operating segments that reported a loss; or

(c) its assets are 10% or more of the combined assets of all operating segments.

Failure to meet any of the criteria does not, however, preclude a company from reporting a segment's results. Operating segments that do not meet any of the criteria may be disclosed voluntarily, if management think the information would be useful to users of the financial statements.

The 75% test

If the total external revenue of the reportable operating segments is less than 75% of the entity's revenue, additional operating segments need to be identified as reportable segments (even if they don't meet the criteria in (a)–(c) above) until 75% of the entity's revenue is included.

Combining segments

IFRS 8 includes detailed guidance on which operating segments may be combined to create a reportable segment, e.g. if they have mainly similar products, processes, customers, distribution methods and regulatory environments. Where there is an aggregation of operating segments an entity is required to disclose the judgements made by management in applying the aggregation criteria to operating segments.

Although IFRS 8 does not specify a maximum number of segments, it suggests that if the reportable segments exceed 10, the entity should consider whether a practical limit had been reached, as the disclosures may become too detailed.

EXAMPLE ● Varia plc is a large training and media entity with an important international component. It operates a state-of-the-art management information system which provides its directors with the information they require to plan and control the various businesses. The directors' reporting requirements are quite detailed and information is collected about the following divisions: Exam-based Training, E-Learning, Corporate Training, Print Media, Online Publishing and Cable Television. The following information is available for the year ended 31 December 2009:

Division	Total revenue £m	Profit £m	Assets £m
Exam-based Training	360	21	176
E-Learning	60	3	13
Corporate Training	125	5	84
Print Media	232	27	102
Online Publishing	124	2	31
Cable TV	73	5	39
	974	63	445

Question

Which of Varia plc's divisions are reportable segments in accordance with IFRS 8 *Operating segments*?

Solution

- The revenues of Exam-based Training, Corporate Training, Print Media and Online Publishing are clearly more than 10% of total revenues and so these segments are reportable.

- All three numbers for E-Learning and Cable TV are under 10% of entity totals for revenue, profit and assets and so, unless these segments can validly be combined with others for reporting purposes, they are not reportable separately, although Varia could choose to provide separate information.

As a final check we need to establish that the combined revenues of reportable sements we have identified (£360 million + £125 million + £232 million + £124 million = £841 million) is at least 75% of the total revenues of Varia of £974 million. £841 million is 86% of £974 million so this condition is satisfied. Therefore no other segments need to be added.

4.5.3 Measuring segment information

IFRS 8 specifies that the amount reported for each segment should be the measures reported to the chief operating decision maker for the purposes of allocating resources and assessing performance. It does not define segment revenue, segment expense, segment result, segment assets, and segment liabilities but rather requires an explanation of how segment profit or loss and segment assets and segment liabilities are measured for each reportable segment.

Allocations and adjustments to revenues and profit should only be included in segment disclosures if they are reviewed by the CODM.

4.5.4 Disclosure requirements for reportable segments

The principle in IFRS 8 is that an entity should disclose 'information to enable users to evaluate the nature and financial effect of the business activities in which it engages and the economic environment in which it operates'.

IFRS 8 requires disclosure of the following segment information:

(i)   Factors used to identify the entity's operating segments such as whether management organises the entity around products and services, geographical areas, regulatory environments, or a combination of factors, and whether segments have been aggregated.

(ii) Types of products and services from which each reportable segment derives its revenues.

(iii) A measure of profit or loss for each reportable segment.

(iv) A measure of liabilities for each reportable segment if it is regularly provided to the chief operating decision maker.

(v)   The following items if they are disclosed in the performance statement reviewed by the chief operating decision maker:

   - revenues from external customers and from transactions with other operating segments

   - interest revenue and interest expense

   - depreciation and amortisation

   - 'exceptional' items

   - income tax income or expense

   - other material non-cash items.

(vi) Total assets; total amounts for additions to non–current assets if they are regularly provided to the chief operating decision maker.

(vii) Reconciliations of profit or loss to the group totals for the entity.

(viii) Reliance on major customers. If revenues from a single external customer are 10% or more of the entity's total revenue, it must disclose that fact and the segment reporting the revenue. It need not disclose the identity of the major customer or the amount of the revenue.

4.5.5 Sample disclosures under IFRS 8

We consider (1) the format for disclosure of segment profits or loss, assets and liabilities, (2) the reconciliations of reportable segment revenues and assets, and (3) information about major customers.

(1) Format for disclosure of segment profits or loss, assets and liabilities

	Hotels	Software	Finance	Other	Entity totals
	£m	£m	£m	£m	£m
Revenue from external customers	800	2,150	500	100[a]	3,550
Intersegment revenue	—	450	—	—	450
Interest revenue	125	250	—	—	375
Interest expense	95	180	—	—	275
Net interest revenue[b]	—	—	100	—	100
Depreciation and amortisation	30	155	110	—	295
Reportable segment profit	27	320	50	10	407
Other material non-cash items – impairment of assets	20	—	—	—	20
Reportable segment assets	700	1,500	5,700	200	8,100
Expenditure for reportable segment non-current assets	100	130	60	—	290
Reportable segment liabilities	405	980	3,000	—	4,385

Reconciliations to group totals are in *bold italics*. Notes:

(a) Revenue from segments below the quantitative thresholds are attributed to four operating divisions. Those segments include a small electronics company, a warehouse leasing company, a retailer and an undertakers. None of these segments has ever met any of the quantitative thresholds for determining reportable segments.

(b) The finance segment derives most of its revenue from interest. Management primarily relies on net interest revenue, not the gross revenue and expense amounts, in managing that segment. Therefore, as permitted by paragraph 23, only net interest is disclosed.

(2) Reconciliations of reportable segment revenues and assets

Reconciliations are required for every material item disclosed. The following are just sample reconciliations.

Revenues	£m
Total revenues for reportable segments	3,900
Other revenues	100
Elimination of intersegment revenues	(450)
Entity's revenue	3,550

Profit or loss

	£m
Total profit or loss for reportable segments	397
Other profit or loss	10
Entity profit	407

Assets

	£m
Total assets for reportable segments	7,900
Other assets	200
Entity assets	8,100

(3) Information about major customers

A sample disclosure might be:

Revenues from one customer of the software and hotels segments represent approximately £400 million of the entity's total revenue.

(Note that disclosure is not required of the customer's name or of the revenue for each operating segment.)

4.6 Benefits and continuing concerns following the issue of IFRS 8

4.6.1 The benefits of segment reporting

The majority of listed and other large entities derive their revenues and profits from a number of sources (or segments). This has implications for the investment strategy of the entity, as different segments require different amounts of investment to support their activities. Conventionally produced statements of financial position and statements of comprehensive income capture financial position and financial performance in a single column of figures.

The following is an extract from the Tesco 2011/12 Annual Report reporting on five segments within the group:

	Trading profit	Trading margin %	Growth %	Sales	Growth %
Group results	3,761m	5.8%	1.3%	72,035m	7.4%
UK	2,480m	5.8%	(1.0)%	47,355m	6.2%
Asia	737m	6.8%	21.5%	11,627m	10.4%
Europe	529m	5.3%	(0.4)%	11,371m	7.8%
US	(153m)	(24.2)%	17.7%	638m	31.5%
Tesco Bank	168m	16.1%	(36.4)%	1,044m	13.6%

We can see that the group is showing a trading profit growth of 1.3%. Within that, segments vary from negative growth of 36.4% to positive growth of 21.5%. Individual segments are also interesting, with sales in the US increasing by 31.5% whilst there is a trading loss.

To put the segment results into a group context we can see their relative importance to the group expressed as a percentage of the group totals as follows:

	Trading profit %	Sales %
UK	66	66
Asia	20	16
Europe	14	16
US	(4)	1
Tesco Bank	4	1

The losses are a red flag to investors and the problem is addressed in the Annual Report by the Chairman, who writes as follows:

> Elsewhere, we have continued the substantial reorientation of the US business to give it the best possible opportunity to secure its future with all the potential for longer-term growth that would bring. We have announced our intention to exit from Japan. We are willing to invest for the long term but where we cannot see a profitable, scalable business earning good returns within an acceptable timescale, we prefer to pursue better opportunities. And we have slowed down the development of Tesco Bank to increase its focus on quality, service and risk management.

4.6.2 Concerns following the issue of IFRS 8

Despite the existence of IFRS 8, there are many concerns about the extent of segmental disclosure and its limitations must be recognised. A great deal of discretion is given to the directors concerning the **definition of each segment**. However, 'the factors which provide guidance in determining an industry segment are often the factors which lead a company's management to organise its enterprise into divisions, branches or subsidiaries'.

There is discretion concerning the **allocation of common costs** to segments on a reasonable basis. There is flexibility in the **definition of some of the items** to be disclosed (particularly net assets). These concerns have been recognised at government level and will be held under review by the European Parliament.

European Parliament reservations

In November 2007 the European Parliament accepted the Commission's proposal to endorse IFRS 8, incorporating US Statement of Financial Accounting Standard No. 131 into EU law, which will require EU companies listed in the European Union to disclose segmental information in accordance with the 'through-the-eyes-of-management' approach.

However, it regretted[4] that the impact assessment carried out by the Commission did not sufficiently take into account the interests of users as well as the needs of small and medium-sized companies located in more than one member state and companies operating only locally. Its view was that such impact assessments must incorporate quantitative information and reflect a balancing of interests among stakeholders.

It did not accept that the convergence of accounting rules was a one-sided process where one party (the IASB) simply copies the financial reporting standards of the other party (the FASB). In particular it expressed reservations that disclosure of geographical information on the basis of IFRS 8 would be comparable to that disclosed under IAS 14.

A post-implementation review was carried out by EFRAG in 2012 (www.efrag.org).

UK reservations

The FRRP reviewed a sample of 2009 interim accounts and 2008 annual accounts. On the basis of this review, the FRRP has highlighted situations where companies were asked to provide additional information:

- Only one operating segment is reported, but the group appears to be diverse with different businesses or with significant operations in different countries.
- The operating analysis set out in the narrative report differs from the operating segments in the financial report.

- The titles and responsibilities of the directors or executive management team imply an organisational structure which is not reflected in the operating segments.
- The commentary in the narrative report focuses on non-IFRS measures, whereas the segmental disclosures are based on IFRS amounts.

It also suggested a number of questions that directors should ask themselves when preparing segmental reports, such as:

- What are the key operating decisions made in running the business?
- Who makes the key operating decisions?
- Who are the segment managers and who do they report to?
- How are the group's activities reported in the information used by management?
- Have the reported segment amounts been reconciled to the IFRS aggregate amounts?
- Do the reported segments appear consistent with their internal reporting?

4.6.3 Post-implementation Review: IFRS 8 *Operating Segments*[5]

IFRS 8 was the first standard to be subjected to a post-implementation review by the IASB. The review identified different opinions among the stakeholders.

Preparers

Standard setters, accounting firms and auditors generally supported the standard subject to suggestions for improvement.

Investors

Views were mixed. Some were concerned that operating segments are aggregated inappropriately. Also there were concerns that, as the segmentation process is based on the management perspective, there is a risk that commercially sensitive information might be concealed or segments reported to conceal loss-making activities within individual segments.

Others welcome the fact that the report is audited and discloses information about how management views the business. They see added value if the segments agree with the management commentary and analyst presentations.

Suggestions for improvement

Replace the term 'chief operating decision maker CODM' with a more common term such as key management personnel or governing body. Also, many entities present different definitions of 'operating result' or 'operating cash flow', making comparison difficult between entities. Investors would like defined line items so that they could calculate their own subtotals for operating result or cash flow.

4.6.4 Constraints on comparison between entities

Segment reporting is intrinsically subjective. This means that there are likely to be major differences in the way segments are determined and because costs, for instance, may be allocated differently by entities in the same industry, it is difficult to make inter-entity comparisons at the segment level and the user still has to take a great deal of responsibility for the interpretation of that information.

4.7 Discontinued operations – IFRS 5 *Non-current Assets Held for Sale and Discontinued Operations*[6]

IFRS 5 deals, as its name suggests, with two separate but related issues. We will first discuss the treatment of discontinued operations.

4.7.1 Criteria

We need to be familiar with the IFRS 5 definition of a discontinued operation. It is a component of an entity that, during the reporting period, either:

- has been disposed of (whether by sale or abandonment); or
- has been classified as held for sale, and *also*
  - represents a separate major line of business or geographical area of operations; or
  - is part of a single coordinated plan to dispose of a separate major line of business or geographical area of operations; or
  - is a subsidiary acquired exclusively with a view to resale (possibly as part of the acquisition of an existing group with a subsidiary that does not fit into the long-term plans of the acquirer).

Defining a component

The IFRS defines a component as a part of an entity which comprises operations and cash flows that can be clearly distinguished, operationally and for financial reporting purposes, from the rest of the entity. This definition is somewhat subjective and the IASB is considering amending this definition to align it with that of an operating segment in IFRS 8 and has issued an exposure draft to this effect.

4.7.2 Disclosure in the statement of income

The results of discontinued operations should be separately disclosed from those of other, continuing, operations in the income statement. As a minimum, on the face of the statement, entities should show, as a single amount, the total of:

- the post-tax profit or loss of discontinued operations; and
- the post-tax gain or loss recognised on the measurement to fair value less cost to sell or on the disposal of the assets or disposal group(s) constituting the discontinued operation.

Further analysis of this amount required, either on the face of the statement of comprehensive income or in the notes:

- the revenue, expenses and pre-tax profit or loss of discontinued operations;
- the related income tax expense as required by IAS 12;
- the gain or loss recognised on the measurement to fair value less costs to sell or on the disposal of the assets or disposal group(s) constituting the discontinued operation; and
- the related income tax expense as required by IAS 12.

The following is an extract from Premier Foods' 2011 consolidated income statement:

	2011	2010
Continuing operations	*£m*	*£m*
Gross profit	**563.4**	711.7
Operating (loss)/profit	**(176.3)**	219.9
Before impairment and loss on disposal of operations	116.9	219.9
Impairment of goodwill and intangible assets	(282.0)	—
Loss on disposal of operations	(11.2)	—
Finance expense	(126.9)	(160.1)
Finance income	7.2	12.0
Net movement on fair valuation of interest rate financial instruments	36.9	(43.3)
(Loss)/profit before taxation from continuing operations	**(259.1)**	28.5
Taxation credit/(charge)	29.1	(24.4)
(Loss)/profit after taxation from continuing operations	**(230.0)**	4.1
Loss from discontinued operations	(109.0)	(103.4)
Loss for the year attributable to equity shareholders of		
the Parent Company	**(339.0)**	(99.3)

Note 11 in the Annual Report giving details of make-up of the post-tax loss of £109m is as follows:

	£m
Revenue	218.6
Operating expenses	(325.1)
Operating loss before loss on disposal	**(106.5)**
Interest payable	(0.1)
Interest receivable	—
Loss before taxation	**(106.6)**
Taxation credit	12.2
Loss after taxation on discontinued operations for the year	**(94.4)**
Loss on disposal before taxation	(14.6)
Tax credit on loss on disposal	—
Loss on disposal after taxation	**(14.6)**
Total loss arising from discontinued operations	**(109.0)**

4.8 Held for sale – IFRS 5 *Non-current Assets Held for Sale and Discontinued Operations*

Let us now discuss the treatment of assets which have been classified as held for sale.

IFRS 5 deals with the appropriate reporting of an asset (or group of assets – referred to in IFRS 5 as a 'disposal group') that management has decided to dispose of. It states that an asset (or disposal group) is classified as 'held for sale' if its carrying amount will be recovered principally through a sale transaction rather than through continuing use.

It further provides that:

● the asset or disposal group must be **available for immediate sale** in its present condition; and

● its sale must be **highly probable**.

The criteria for the sale to be highly probable are:

- The appropriate level of management must be committed to a plan to sell the asset or disposal group.
- An active programme to locate a buyer and complete the plan must have been initiated.
- The asset or disposal group must be actively marketed for sale at a price that is reasonable in relation to its current fair value.
- The sale should be expected to qualify for recognition as a completed sale within one year from the date of classification.
- Actions required to complete the plan should indicate that it is unlikely that significant changes to the plan will be made or that the plan will be withdrawn.

There is a pragmatic recognition that there may be events outside the control of the enterprise which prevent completion within one year. In such a case the 'held for sale' classification is retained, provided there is sufficient evidence that the entity remains committed to its plan to sell the asset or disposal group and has taken all reasonable steps to resolve the delay.

It is important to note that IFRS 5 specifies that this classification is appropriate for assets (or disposal groups) that are to be **sold** or distributed. The classification does not apply to assets or disposal groups that are to be **abandoned**.

4.8.1 IFRS 5 – implications of classification as held for sale

Assets, or disposal groups, that are classified as held for sale should be removed from their previous position in the statement of financial position and shown under a single 'held for sale' caption – usually as part of **current** assets. Any liabilities directly associated with disposal groups that are classified as held for sale should be separately presented within liabilities.

As far as disposal groups are concerned, it is acceptable to present totals on the face of the statement of financial position, with a more detailed breakdown in the notes. The following is a disclosure note from the published financial statements of Unilever for the year ended 31 December 2012:

Assets and liabilities held for sale	2012 £m	2011 £m
Groups held for sale		
Goodwill and intangibles	114	9
Property, plant and equipment	28	—
Inventories	26	—
Trade and other receivables	11	—
	179	9
Non-current assets held for sale		
Property, plant and equipment	13	12
Liabilities associated with assets held for sale	1	

Depreciable assets that are classified as 'held for sale' should not be depreciated from classification date, as the classification implies that the intention of management is primarily to recover value from such assets through sale, rather than through continued use.

When assets (or disposal groups) are classified as 'held for sale', their carrying value(s) at the date of classification should be compared with the 'fair value less costs to sell' of the asset (or disposal group). If the fair value less costs to sell exceeds the carrying value, the carrying value is the amount reported as the current asset. If the transfer occurs during an accounting period, the position is reassessed at the end of the period. For example, assume that a non-current asset acquired on 1 April 2010 at a cost of £100,000 and depreciated at 10% per annum on a straight-line basis is classified as held for sale on 1 October 2012 when fair value less cost to sell was £85,000 – reassessed on 31 March 2013 as £62,000.

The fair value of £85,000 exceeded the carrying value of £75,000 (£100,000 less 2.5 years depreciation) which means that it is recorded as a current asset at its carrying value of £75,000. If the carrying value exceeds fair value less costs to sell then the excess should be treated as an impairment loss. As the fair value at 31 March 2013 was £62,000 there is an impairment of £13,000 recognised in profit or loss.

In the case of a disposal group, the impairment loss should be allocated to the specific assets in the order specified in IAS 36 *Impairment of Assets*. The treatment of impairment losses is discussed in detail in Chapter 17.

4.9 IAS 24 *Related Party Disclosures*[7]

In the previous chapter we saw that after the financial statements have been drafted it is necessary to form a judgement as to whether or not they give a fair presentation of the entity's activities.

One of the considerations is whether there are any indications that transactions have not been carried out at arm's length. This can occur when one of the parties to the transaction is able to influence the management to enter into transactions which are not primarily in the best interest of the company. Where such a possibility exists, the person (or business) able to exert this influence is referred to in accounting terms as a 'related party'.

The users of financial statements would normally assume that the transactions of an entity have been carried out at arm's length and under terms which are in the best interests of the entity. The existence of related party relationships may mean that this assumption is not appropriate and IAS 24 therefore requires disclosure of such existence.

4.9.1 How to determine what is 'arm's length'

The Board of a company should consider a number of surrounding factors when determining whether a transaction has been at arm's length. These include considering:

- how the terms of the overall transaction compare with those of any comparable transactions between parties dealing on an arm's length basis in similar circumstances;
- the level of risk – how the transaction impacts on the company's financial position and performance, its ability to follow its business plan and the expected rate of return on the assets given the level of risk;
- other options – what other options were available to the company and whether any expert advice was obtained by the company.

4.9.2 IAS 24 disclosures required

The purpose of IAS 24 is to define the meaning of the term 'related party' and prescribe the disclosures that are appropriate for transactions with related parties (and in some cases for their mere existence). From the outset it is worth remembering that the term 'party' could refer to an individual (referred to as a person) or to another entity. IAS 24 breaks the definition down into two main sections relating to (a) persons and (b) entities. We will consider both below.

4.9.3 Definition of 'related party' when the party is a person

A person, or a close member of that person's family, whom we will refer to as P, is a related party to the reporting entity (RE) if:

- P has control or joint control over RE;
- P has significant influence over RE; or
- P is a member of the key management personnel of RE.

Close members of the family of P are those family members who may be expected to influence, or be influenced by, P in their dealings with RE and include:

- P's children and spouse or domestic partner; and
- children of the spouse or domestic partner; and
- dependants of P or P's spouse or domestic partner.

Key management personnel of RE are those persons having authority and responsibility for planning, directing and controlling the activities of RE, directly or indirectly, including any director (whether executive or otherwise) of RE.

Example: Individual as investor

Let us assume that Arthur has 60% of the shares in, and so controls, Garden Supplies Ltd and:

(a) he also has a 45% significant interest in Plant Growers Ltd. This means that in Garden Supplies Ltd's financial statements Plant Growers Ltd are a related party, and in Plant Growers Ltd's financial statements Garden Supplies Ltd are a related party; or

(b) a close member of his family (in this case his domestic partner) owns a 45% interest in Plant Growers Ltd. This means that a similar treatment would be required and the two companies are related; or

(c) Arthur still has the 60% interest in Garden Supplies Ltd but instead of having an investment in Plant Growers Ltd he is a member of Plant Growers Ltd's key management personnel. This means that a similar treatment would be required and the two companies are related.

4.9.4 Definition of 'related party' when the party is another entity

We have discussed the position where the related party relationship arises from an individual's relationship with two businesses. It also arises when companies are involved.

For example, let us now assume that Arthur, Garden Supplies and Plant Growers are all limited companies. We classify each company as follows:

- Arthur Ltd holds 60% of the shares and so is a parent of Garden Supplies Ltd.
- Plant Growers Ltd is an associate of Arthur Ltd because Arthur Ltd can exercise significant influence over Plant Growers Ltd.

This means that when any of the companies prepares its financial statements:

- Arthur Ltd is related to both Garden Supplies Ltd and Plant Growers Ltd.
- Garden Supplies Ltd is related to Plant Growers Ltd.
- Plant Growers Ltd is related to Garden Supplies Ltd.

4.9.5 Identifying related parties is not always clear

In the above examples we have clear knowledge of the relationship. However, there could be an intention to conceal the relationship, which requires ingenuity from any auditor. Steps might need to be taken such as discussions with lawyers and searching company records, referring to daily newspapers, trade magazines and phone books and, of course, using the Internet and social network sites.

The IASB has this area under review and, in its 2013 Annual Improvement Initiative, requires an entity that provides key management personnel services to be treated as a related party.

4.9.6 Parties deemed not to be related parties

IAS 24 emphasises that it is necessary to consider carefully the substance of each relationship to see whether or not a related party relationship exists. However, the standard highlights a number of relationships that would not normally lead to related party status:

● two entities simply because they have a director or other member of the key management personnel in common or because a member of the key management personnel of one entity has significant influence over the other entity;

● two venturers simply because they share control over a joint venture;

● providers of finance, trade unions, public utilities or government departments in the course of their normal dealings with the entity;

● a single customer, supplier, franchisor, distributor or general agent with whom an entity transacts a significant volume of business merely by virtue of the resulting economic dependence.

4.9.7 Disclosure of controlling relationships

IAS 24 requires that relationships between a parent and its subsidiaries be disclosed irrespective of whether there have been transactions between them. Where the entity is controlled, it should disclose:

● the name of its parent;

● the name of its ultimate controlling party (which could be an individual or another entity);

● if neither the parent nor the ultimate controlling party produces consolidated financial statements available for public use, the name of the next most senior parent that does produce such statements.

4.9.8 Exemption from disclosures re government-related entities

A reporting entity is exempt from the detailed disclosures referred to in Section 4.9.10 below in relation to related party transactions and outstanding balances with:

● a government that has control, joint control or significant influence over the reporting entity; and

● another entity that is a related party because the same government has control, joint control or significant influence over both parties.

If this exemption is applied, the reporting entity is nevertheless required to make the following disclosures about transactions with government-related entities:

- the name of the government and the nature of its relationship with the reporting entity;
- the following information in sufficient detail to enable users of the financial statements to understand the effect of related party transactions:
  - the nature and amount of each individually significant transaction; and
  - for other transactions that are collectively, but not individually, significant, a qualitative or quantitative indication of their extent.

The reason for the exemption is essentially pragmatic. In some jurisdictions where government control is pervasive it can be difficult to identify other government-related entities. In some circumstances the directors of the reporting entity may be genuinely unaware of the related party relationship. Therefore, the basis of conclusions to IAS 24 (BC 43) states that, in the context of the disclosures that are needed in these circumstances:

> The objective of IAS 24 is to provide disclosures necessary to draw attention to the possibility that the financial position and profit or loss may have been affected by the existence of related parties and by transactions and outstanding balances, including commitments, with such parties. To meet that objective, IAS 24 requires some disclosure when the exemption applies. Those disclosures are intended to put users on notice that related party transactions have occurred and to give an indication of their extent. The Board did not intend to require the reporting entity to identify **every** government-related entity, or to quantify in detail **every** transaction with such entities, because such a requirement would negate the exemption.

4.9.9 Disclosure of compensation of key management personnel

Compensation can be influenced by a person in this position. Consequently IAS 24 requires the disclosure of short-term employee benefits, post-employment benefits, other long-term benefits (e.g. accrued sabbatical leave), termination benefits, and share-based payment.

4.9.10 Disclosure of related party transactions

A related party transaction is a transfer of resources or obligations between a reporting entity and a related party, regardless of whether a price is charged. Where such transactions have occurred, the entity should disclose the nature of the related party relationship as well as information about those transactions and outstanding balances to enable a user to understand the potential effect of the relationship on the financial statements. As a minimum, the disclosures should include:

- the amount of the transactions;
- the amount of the outstanding balances and:
  - their terms and conditions, including whether they are secured, and the nature of the consideration to be provided in settlement; and
  - details of any guarantees given or received;
- provisions for doubtful debts related to the amount of outstanding balances; and
- the expense recognised during the period in respect of bad or doubtful debts due from related parties.

The following extract from the Unilever 2013 Annual Report is an example of the required disclosures:

30 Related party transactions
A related party is a person or entity that is related to the Group. These include both people and entities that have, or are subject to the influence or control of the Group.

The following related party balances existed with associate or joint venture businesses at 31 December:

Related party balances

	2013 €million	2012 €million
Trading and other balances due from joint ventures	130	116

4.9.11 Possible impact of transactions with related parties

It is possible that there could be both beneficial and prejudicial impacts.

Beneficial transactions with related parties

It could be that the related party is actually offering support to the business. For example, the business might have received benefits in a variety of ways ranging from financial support on favourable terms such as guarantees or low or no interest loans to the provision of goods or services at less than market rates.

Prejudicial transactions with related parties

These can arise when the business enters into transactions on terms that would not be offered to an unrelated party. There are numerous ways that this could be arranged, such as:

- **Loans**:
  - borrowing at above market rates;
  - lending at below market rates;
  - lending with no agreement as to date for repayment;
  - lending with little prospect of being repaid;
  - lending with the intention of writing off;
  - guaranteeing debts where there is no commercial advantage to the business.
- **Assets**:
  - selling non-current assets at below market value;
  - selling goods at less than normal trade price;
  - providing services at less than normal rates;
  - transfer of know-how, or research and development transfers.
- **Trading**:
  - sales made where there is secret agreement to repurchase to inflate current period revenue;
  - sales to inflate revenue with funds advanced to the debtor to allow the debt to be paid;
  - paying for services which have not been provided.

Summary

The published accounts of a listed company are intended to provide a report to enable shareholders to assess current-year stewardship and management performance and to predict future cash flows. Financial statements prepared from a trial balance and adjusted for accruals might require further adjustments. These arise from:

(a) events after the reporting period that provide additional evidence of conditions that existed at the period-end which might require the financial statements to be adjusted; or

(b) prior period errors that may require retrospective changes to the opening balances in the statement of financial position that could affect assets, liabilities and retained earnings.

In addition to these adjustments, in order to assist shareholders to predict future cash flows with an understanding of the risks involved, more information has been required by the IASB. This has taken two forms:

(a) more quantitative information in the accounts, e.g. segmental analysis, and the impact of changes on the operation, e.g. a breakdown of turnover, costs and profits for both new and discontinued operations; and

(b) more qualitative information, e.g. related party disclosures and events occurring after the reporting period.

REVIEW QUESTIONS

I Explain why non-adjusting items are not reported in the financial statements if they are of sufficient materiality to be disclosed.

2 Explain the criteria that have to be satisfied when identifying an operating segment.

3 Explain the criteria that have to be satisfied to identify a reportable segment.

4 Explain why it is necessary to identify a chief operating decision maker and describe the key identifying factors.

5 Discuss the review findings of the European Securities and Markets Authority (ESMA) in relation to the role of the chief operating decision maker.

6 A research report[8] found that users were worried about the lack of comparability among segmental disclosures of different companies following the issue of IFRS 8. Discuss:

(a) why it should have resulted in a lack of comparability;

(b) whether it is more relevant because its format and content are not closely defined;

(c) whether any of the other financial statements would be more relevant to users if they were free to format as they wished;

(d) whether inter-firm comparability is more important than inter-period comparability.

7 Explain the conditions set out in IFRS 5 for determining whether operations have been discontinued and the problems that might arise in applying them.

8 Explain the conditions that must be satisfied if a non-current asset is to be reported in the statement of financial position as held for sale.

9 Explain why it is important to an investor to be informed about assets held for sale.

10 Discuss how transactions with related parties can have

(a) a beneficial impact

(b) a prejudicial impact

on (i) the reported income and (ii) the financial position.

EXERCISES

Question 1

IAS 10 deals with events after the reporting period.

Required:
(a) Define the period covered by IAS 10.
(b) Explain when the financial statements should be adjusted.
(c) Why should non-adjusting events be disclosed?
(d) A customer made a claim for £50,000 for losses suffered by the late delivery of goods. The main part (£40,000) of the claim referred to goods due to be delivered before the year-end. Explain how this would be dealt with under IAS 10.
(e) After the year-end a substantial quantity of inventory was destroyed in a fire. The loss was not adequately covered by insurance. This event is likely to threaten the ability of the business to continue as a going concern. Discuss the matters you would consider in making a decision under IAS 10.
(f) The business entered into a favourable contract after the year-end that would see its profits increase by 15% over the next three years. Explain how this would be dealt with under IAS 10.

Question 2

Epsilon is a listed entity. You are the financial controller of the entity and its consolidated financial statements for the year ended 30 September 2008 are being prepared. Your assistant, who has prepared the first draft of the statements, is unsure about the correct treatment of a transaction and has asked for your advice. Details of the transaction are given below.

On 31 August 2008 the directors decided to close down a business segment which did not fit into its future strategy. The closure commenced on 5 October 2008 and was due to be completed on 31 December 2008. On 6 September 2008 letters were sent to relevant employees offering voluntary redundancy or redeployment in other sectors of the business. On 13 September 2008 negotiations commenced with relevant parties with a view to terminating existing contracts of the business segment and arranging sales of its assets. Latest estimates of the financial implications of the closure are as follows:

(i) Redundancy costs will total $30 million, excluding the payment referred to in (ii) below.

(ii) The cost of redeploying and retraining staff who do not accept redundancy will total $6 million.

(iii) Plant having a net book value of $11 million at 30 September 2008 will be sold for $2 million.

(iv) The operating losses of the business segment for October, November and December 2008 are estimated at $10 million.

Your assistant is unsure of the extent to which the above transactions create liabilities that should be recognised as a closure provision in the financial statements. He is also unsure as to whether or not the results of the business segment that is being closed need to be shown separately.

Required:
Explain how the decision to close down the business segment should be reported in the financial statements of Epsilon for the year ended 30 September 2008.

* Question 3

Epsilon is a listed entity. You are the financial controller of the entity and its consolidated financial statements for the year ended 31 March 2009 are being prepared. The board of directors is responsible for all key financial and operating decisions, including the allocation of resources.

Your assistant is preparing the first draft of the statements. He has a reasonable general accounting knowledge but is not familiar with the detailed requirements of all relevant financial reporting standards. He requires your advice and he has sent you a note as shown below.

We intend to apply IFRS 8 *Operating Segments* in this year's financial statements. I am aware that this standard has attracted a reasonable amount of critical comment since it was issued in November 2006.

The board of directors receives a monthly report on the activities of the five significant operational areas of our business. Relevant financial information relating to the five operations for the year to 31 March 2009, and in respect of our head office, is as follows:

Operational area	Revenue for year to 31 March 2009 $000	Profit/(loss) for year to 31 March 2009 $000	Assets at 31 March 2009 $000
A	23,000	3,000	8,000
B	18,000	2,000	6,000
C	4,000	(3,000)	5,000
D	1,000	150	500
E	3,000	450	400
Sub-total	49,000	2,600	19,900
Head office	Nil	Nil	6,000
Entity total	49,000	2,600	25,900

I am unsure of the following matters regarding the reporting of operating segments:

● How do we decide what our operating segments should be?

● Should we report segment information relating to head office?

● Which of our operational areas should report separate information? Operational areas A, B and C exhibit very distinct economic characteristics but the economic characteristics of operational areas D and E are very similar.

● Why has IFRS 8 attracted such critical comment?

Required:
Draft a reply to the questions raised by your assistant.

* Question 4

Filios Products plc owns a chain of hotels through which it provides three basic services: restaurant facilities, accommodation, and leisure facilities. The latest financial statements contain the following information:

Statement of financial position of Filios Products

	£m
ASSETS	
Non-current assets at book value	1,663
Current assets	
Inventories and receivables	381
Bank balance	128
	509
Total Assets	2,172
EQUITY AND LIABILITIES	
Equity	
Share capital	800
Retained earnings	1,039
	1,839
Non-current liabilities:	
Long-term borrowings	140
Current liabilities	193
Total Equity and liabilities	2,172

Statement of comprehensive income of Filios Products

	£m	£m
Revenue		1,028
Less: Cost of sales	684	
Administration expenses	110	
Distribution costs	101	
Interest charged	14	(909)
Net profit		119

The following breakdown is provided of the company's results into three divisions and head office:

	Restaurants	Hotels	Leisure	Head office
	£m	£m	£m	£m
Revenue	508	152	368	—
Cost of sales	316	81	287	—
Administration expenses	43	14	38	15
Distribution costs	64	12	25	—
Interest charged	10	—	—	4
Non-current assets at book value	890	332	364	77
Inventories and receivables	230	84	67	—
Bank balance	73	15	28	12
Payables	66	40	56	31
Long-term borrowings	100	—	—	40

Required:

(a) Outline the nature of segmental reports and explain the reason for presenting such information in the published accounts.

(b) Prepare a segmental statement for Filios Products plc complying, so far as the information permits, with the provisions of IFRS 8 *Operating Segments* so as to show for each segment and the business as a whole:

(i) revenue;

(ii) profit;

(iii) net assets.

(c) Examine the relative performance of the operating divisions of Filios Products. The examination should be based on the following accounting ratios:

(i) operating profit percentage;

(ii) net asset turnover;

(iii) return on net assets.

Question 5

The following is the draft trading and income statement of Parnell Ltd for the year ending 31 December 2003:

	$m	$m
Revenue		563
Cost of sales		310
		253
Distribution costs	45	
Administrative expenses	78	
		123
Profit on ordinary activities before tax		130
Tax on profit on ordinary activities		45
Profit on ordinary activities after taxation – all retained		85
Profit brought forward at 1 January 2003		101
Profit carried forward at 31 December 2003		186

You are given the following additional information, which is reflected in the above statement of comprehensive income only to the extent stated:

1 Distribution costs include a bad debt of $15 million which arose on the insolvency of a major customer. There is no prospect of recovering any of this debt. Bad debts have never been material in the past.

2 The company has traditionally consisted of a manufacturing division and a distribution division. On 31 December 2003, the entire distribution division was sold for $50 million; its book value at the time of sale was $40 million. The profit on disposal was credited to administrative expenses. (Ignore any related income tax.)

3 During 2003, the distribution division made sales of $100 million and had a cost of sales of $30 million. There will be no reduction in stated distribution costs or administration expenses as a result of this disposal.

4 The company owns offices which it purchased on 1 January 2001 for $500 million, comprising $200 million for land and $300 million for buildings. No depreciation was charged in 2001 or 2002, but the company now considers that such a charge should be introduced. The buildings were

expected to have a life of 50 years at the date of purchase, and the company uses the straight-line basis for calculating depreciation, assuming a zero residual value. No taxation consequences result from this change.

5 During 2003, part of the manufacturing division was restructured at a cost of $20 million to take advantage of modern production techniques. The restructuring was not fundamental and will not have a material effect on the nature and focus of the company's operations. This cost is included under administration expenses in the statement of comprehensive income.

Required:

(a) State how each of the items 1–5 above must be accounted for in order to comply with the requirements of international accounting standards.

(b) Redraft the income statement of Parnell Ltd for 2003, taking into account the additional information so as to comply, as far as possible, with relevant standard accounting practice. Show clearly any adjustments you make. Notes to the accounts are not required. Where an IAS recommends information to be on the face of the income statement it could be recorded on the face of the statement.

* Question 6

Springtime Ltd is a UK trading company buying and selling as wholesalers fashionable summer clothes. The following balances have been extracted from the books as at 31 March 20X4:

	£000
Auditor's remuneration	30
Income tax based on the accounting profit:	
For the year to 31 March 20X4	3,200
Overprovision for the year to 31 March 20X3	200
Delivery expenses (including £300,000 overseas)	1,200
Dividends: final (proposed – to be paid 1 August 20X4)	200
interim (paid on 1 October 20X3)	100
Non-current assets at cost:	
Delivery vans	200
Office cars	40
Stores equipment	5,000
Dividend income (amount received from listed companies)	1,200
Office expenses	800
Overseas operations: closure costs of entire operations	350
Purchases	24,000
Sales (net of sales tax)	35,000
Inventory at cost:	
At 1 April 20X3	5,000
At 31 March 20X4	6,000
Storeroom costs	1,000
Wages and salaries:	
Delivery staff	700
Directors' emoluments	400
Office staff	100
Storeroom staff	400

Notes:

1 Depreciation is provided at the following annual rates on a straight-line basis: delivery vans 20%; office cars 25%; stores 1%.

2 The following taxation rates may be assumed: corporate income tax 35%; personal income tax 25%.

3 The dividend income arises from investments held in non-current investments.

4 It has been decided to transfer an amount of £150,000 to the deferred taxation account.

5 The overseas operations consisted of exports. In 20X3/X4 these amounted to £5,000,000 (sales) with purchases of £4,000,000. Related costs included £100,000 in storeroom staff and £15,000 for office staff.

6 Directors' emoluments include:

Chairperson	100,000	
Managing director	125,000	
Finance director	75,000	
Sales director	75,000	
Export director	25,000	(resigned 31 December 20X3)
	£400,000	

Required:

(a) Produce a statement of comprehensive income suitable for publication and complying as far as possible with generally accepted accounting practice.

(b) Comment on how IFRS 5 has improved the quality of information available to users of accounts.

Question 7

Omega prepares financial statements under International Financial Reporting Standards. In the year ended 31 March 2007 the following transaction occurred:

Omega follows the revaluation model when measuring its property, plant and equipment. One of its properties was carried in the balance sheet at 31 March 2006 at its market value at that date of $5 million. The depreciable amount of this property was estimated at $3.2 million at 31 March 2006 and the estimated future economic life of the property at 31 March 2006 was 20 years.

On 1 January 2007 Omega decided to dispose of the property as it was surplus to requirements and began to actively seek a buyer. On 1 January 2007 Omega estimated that the market value of the property was $5.1 million and that the costs of selling the property would be $80,000. These estimates remained appropriate at 31 March 2007.

The property was sold on 10 June 2007 for net proceeds of $5.15 million.

Required:
Explain, with relevant calculations, how the property would be treated in the financial statements of Omega for the year ended 31 March 2007 and the year ending 31 March 2008.

* Question 8

The following trial balance has been extracted from the books of Hoodurz as at 31 March 2006:

	$000	$000
Administration expenses	210	
Ordinary share capital, $1 per share		600
Trade receivables	470	
Bank overdraft		80
Provision for warranty claims		205
Distribution costs	420	
Non-current asset investments	560	
Investment income		75
Interest paid	10	
Property, at cost	200	
Plant and equipment, at cost	550	
Plant and equipment, accumulated depreciation (at 31.3.2006)		220
Accumulated profits (at 31.3.2005)		80
Loans (repayable 31.12.2010)		100
Purchases	960	
Inventories (at 31.3.2005)	150	
Trade payables		260
Sales		2,010
2004/2005 final dividend paid	65	
2005/2006 interim dividend paid	35	
	3,630	3,630

The following information is relevant:

(i) The trial balance figures include the following amounts for a disposal group that has been classified as 'held for sale' under IFRS 5 *Non-current Assets Held for Sale and Discontinued Operations*:

	$000
Plant and equipment, at cost	150
Plant and equipment, accumulated depreciation	15
Trade receivables	70
Bank overdraft	10
Trade payables	60
Sales	370
Inventories (at 31.12.2005)	25
Purchases	200
Administration expenses	55
Distribution costs	60

The disposal group had no inventories at the date classified as 'held for sale'.

(ii) Inventories (excluding the disposal group) at 31.3.2006 were valued at $160,000.

(iii) The depreciation charges for the year have already been accrued.

(iv) The income tax for the year ended 31.3.2006 is estimated to be $74,000. This includes $14,000 in relation to the disposal group.

(v)   The provision for warranty claims is to be increased by $16,000. This is classified as administration expense.

(vi) Staff bonuses totalling $20,000 for administration and $20,000 for distribution are to be accrued.

(vii) The property was acquired during February 2006, therefore, depreciation for the year ended 31.3.2006 is immaterial. The directors have chosen to use the fair value model for such an asset. The fair value of the property at 31.3.2006 is $280,000.

Required:
Prepare for Hoodurz:
(a) an income statement for the year ended 31 March 2006; and
(b) a balance sheet as at 31 March 2006.
      Both statements should comply as far as possible with relevant International Financial Reporting Standards. No notes to the financial statements are required nor is a statement of changes in equity, but all workings should be clearly shown.

(The Association of International Accountants)

Question 9

Omega prepares financial statements under International Financial Reporting Standards. In the year ended 31 March 2007 the following transaction occurred. On 31 December 2006 the directors decided to dispose of a property that was surplus to requirements. They instructed selling agents to procure a suitable purchaser and advertised the property at a commercially realistic price.

The property was being measured under the revaluation model and had been revalued at $15 million on 31 March 2006. The depreciable element of the property was estimated as $8 million at 31 March 2006 and the useful economic life of the depreciable element was estimated as 25 years from that date. Omega depreciates its non-current assets on a monthly basis.

On 31 December 2006 the directors estimated that the market value of the property was $16 million, and that the costs incurred in selling the property would be $500,000. The property was sold on 30 April 2007 for $15.55 million, being the agreed selling price of $16.1 million less selling costs of $550,000. The actual selling price and costs to sell were consistent with estimated amounts as at 31 March 2007.

The financial statements for the year ended 31 March 2007 were authorised for issue on 15 May 2007.

Required:
Show the impact of the decision to sell the property on the income statement of Omega for the year ended 31 March 2007, and on its balance sheet as at 31 March 2007. You should state where in the income statement and the balance sheet relevant balances will be shown. You should make appropriate references to international financial reporting standards.

(IFRS)

Question 10

(a) In 20X3 Arthur is a large loan creditor of X Ltd and receives interest at 20% p.a. on this loan. He also has a 24% shareholding in X Ltd. Until 20X1 he was a director of the company and left after a disagreement. The remaining 76% of the shares are held by the remaining directors.

(b) Brenda joined Y Ltd, an insurance broking company, on 1 January 20X0 on a low salary but high commission basis. She brought clients with her that generated 30% of the company's 20X0 revenue.

(c) Carrie is a director and major shareholder of Z Ltd. Her husband, Donald, is employed in the company on administrative duties for which he is paid a salary of £25,000 p.a. Her daughter, Emma, is a business consultant running her own business. In 20X0 Emma carried out various consultancy exercises for the company for which she was paid £85,000.

(d) Fred is a director of V Ltd. V Ltd is a major customer of W Ltd. In 20X0 Fred also became a director of W Ltd.

Required:
Discuss whether parties are related in the above situations.

* Question 11

Maxpool plc, a listed company, owned 60% of the shares in Ching Ltd. Bay plc, a listed company, owned the remaining 40% of the £1 ordinary shares in Ching Ltd. The holdings of shares were acquired on 1 January 20X0.

On 30 November 20X0 Ching Ltd sold a factory outlet site to Bay plc at a price determined by an independent surveyor.

On 1 March 20X1 Maxpool plc purchased a further 30% of the £1 ordinary shares of Ching Ltd from Bay plc and purchased 25% of the ordinary shares of Bay plc.

On 30 June 20X1 Ching Ltd sold the whole of its fleet of vehicles to Bay plc at a price determined by a vehicle auctioneer.

Required:
Explain the implications of the above transactions for the determination of related party relationships and disclosure of such transactions in the financial statements of (a) Maxpool Group plc, (b) Ching Ltd and (c) Bay plc for the years ending 31 December 20X0 and 31 December 20X1.

(ACCA)

Question 12

Gamma is a company that manufactures power tools. Gamma was established by Mr Lee, who owns all of Gamma's shares. Mrs Lee, Mr Lee's wife, owns a controlling interest in Delta, a distributor of power tools. Delta is one of Gamma's biggest customers, accounting for 70% of Gamma's sales. Delta buys exclusively from Gamma.

Gamma's official price list is based on the policy of selling goods at cost plus 50%; however, sales to Delta are priced at normal selling price less a discount of 30% to reflect the scale of the business transacted.

Gamma's terms of sale require payment within one month, but Delta is permitted three months to pay.

Mrs Lee has decided to sell her shares in Delta and has provided a potential buyer with financial information including the following:

Sales revenue for the year ended 30 September 2011	$12.0m
Cost of sales	$8.0m
Gross profit %	33%
Current assets (including bank $0.3m)	$4.0m
Trade payables	$3.0m
Other current liabilities	$0.8m
Current ratio	1.1:1 (in line with the ratios reported in each of the past three years)

The buyer conducted a due diligence investigation and discovered the relationship between Gamma and Delta. She has decided to restate the figures provided in the table above to reflect a 'worst case' scenario before arriving at a final decision concerning the purchase.

Required:
(a) **Discuss the manner in which IAS 24 *Related Party Disclosures* should have alerted the potential buyer in this case.**
(b) **Recalculate the table of figures provided by Mrs Lee on the basis that Delta will not receive favourable terms from Gamma if Mrs Lee sells her shares, and discuss the resulting changes.**

(The Association of International Accountants)

* Question 13

IAS 8 *Accounting Policies, Changes in Accounting Estimates and Errors* lays down criteria for selection of accounting policies and prescribes circumstances in which an entity may change an accounting policy. The standard also deals with accounting treatment of changes in accounting policies, changes in accounting estimates and correction of prior errors.

You are the financial controller of Lifewest Ltd. The company began trading on 1 January 2007 and is currently involved in the preparation of financial statements for the year ended 31 December 2010. You have recently attended a one-day seminar on the application of International Financial Reporting Standards, organised by the Institute of Certified Public Accountants in Ireland (CPA). On 1 January 2010, the company had 50 million €1 ordinary shares in issue. On 30 June 2010, Lifewest Ltd issued 10 million 10% €1 irredeemable preference shares at par. There have been no other changes in share capital in the last five years. The appropriate dividend in respect of these shares was paid on 31 December 2010. A property revaluation at the year end gave a surplus of €250,000.

During the year ended 31 December 2010, Lifewest Ltd changed its accounting policy for depreciation in relation to the depreciation of property, plant and equipment. The depreciation charges calculated using the previous accounting policy and shown in the company's financial statements for three years ending 31 December 2009 were as follows:

	€000
Year to 31 December 2007	690
Year to 31 December 2008	810
Year to 31 December 2009	870

Assuming the new accounting policy had been applied in previous years, depreciation charges would have been:

	€000
Year to 31 December 2007	1,170
Year to 31 December 2008	930
Year to 31 December 2009	690

An extract from Lifewest Ltd's Statement of Comprehensive Income for the year to 31 December 2010 (before making any adjustments to reflect the change in accounting policy for the year 31 December 2009) shows the following:

	2010	2009
	€000	€000
Profit before depreciation	7,530	7,350
Depreciation of property, plant and equipment	570	870
Profit before taxation	6,960	6,480
Taxation	2,088	1,944
Profit after taxation	4,872	4,536
Other Comprehensive Income:		
Gains on property revaluation	250	

Lifewest Ltd's retained earnings were reported as €8,829,000 at 31 December 2008. No dividends have been paid in any year. The company pays tax at 30% on the profit.

Required:

(a) **Distinguish between accounting policies, accounting estimates and prior period errors.**

(b) **Present the extract from the Statement of Comprehensive Income so as to reflect the change in accounting policy, in accordance with IAS 8.**

(c) **Compute Lifewest Ltd's retained earnings at 31 December 2010 and the restated retained earnings as at 31 December 2008 and 2009.**

(d) **Prepare a statement of changes in equity for the year ended 31 December 2010.**

(Institute of Certified Public Accountants (CPA), Professional Stage 1 Corporate Reporting Examination, August 2011)

References

1 IAS 10 *Events after the Reporting Period*, IASB, revised 2003.
2 IAS 8 *Accounting Policies, Changes in Accounting Estimates and Errors*.
3 IFRS 8 *Operating Segments*, IASB, 2006.
4 www.europarl.europa.eu/sides/getDoc.do?Type=TA&Reference=P6-TA-2007-0526&language=EN
5 Post-implementation Review, IFRS 8 *Operating Segments*, IFRS Foundation, July 2013.
6 IFRS 5 *Non-current Assets Held for Sale and Discontinued Operations*, IASB, revised 2009.
7 IAS 24 *Related Party Disclosures*, IASB, revised 2009.
8 L. Crawford, H. Extance and C. Helliar, *Operating Segments: The Usefulness of IFRS 8*, The Institute of Chartered Accountants of Scotland, 2012.

CHAPTER 5

Statements of cash flows

5.1 Introduction

The main purpose of this chapter is to explain the reasons for preparing a statement of cash flows and how to prepare a statement applying IAS 7.

Objectives

By the end of this chapter, you should be able to:

- prepare a statement of cash flows in accordance with IAS 7;
- analyse a statement of cash flows;
- critically discuss their strengths and weaknesses.

5.2 Development of statements of cash flows

We saw in Chapter 3 that, at the end of an accounting period, a statement of income is prepared which explains the change in the retained earnings at the beginning and end of an accounting period. In this chapter we prepare a statement of cash flows in accordance with IAS 7 *Statements of Cash Flows*.

IAS 7 explains the changes that have occurred in the amount of liquid assets easily accessible – these are defined as cash + cash equivalents.

5.2.1 Statements of cash flows – their benefits

As far back as 1991 Professor John Arnold wrote in a report by the ICAEW Research Board and ICAS Research Advisory Committee *The Future Shape of Financial Reports*:[1]

> little attention is paid to the reporting entity's cash or liquidity position. Cash is the lifeblood of every business entity. The report . . . advocates that companies should provide a cash flow statement . . . preferably using the direct method.

Statements of cash flows are now primary financial statements and as important as statements of comprehensive income:

> The emphasis on cash flows, and the emergence of the statement of cash flows as an important financial report, does not mean that operating cash flows are a substitute for, or are more important than, net income. In order to analyse financial statements correctly we need to consider *both* operating cash flows and net income.[2]

They are now primary financial statements because the financial viability and survival prospects of any organisation rest on the ability to generate positive operating cash flows. These are necessary in order to be able to pay the interest on loans and repay the loans, finance capital expenditure to maintain or expand operating capacity, and reward the investors with an acceptable dividend policy. If there is still a positive cash flow after this, it will help to reduce the need for additional external loan or equity funding.

The message is that, independent of reported profits, if an organisation is unable to generate sufficient cash, it will eventually become insolvent and fail.

The following extract from Heath and Rosenfield's article on solvency[3] is a useful conclusion to our analysis of the benefits of cash flow statements, emphasising that they also provide a basis for predicting future performance:

> Solvency is a money or cash phenomenon. A solvent company is one with adequate cash to pay its debts; an insolvent company is one with inadequate cash . . . Any information that provides insight into the amounts, timings and certainty of a company's future cash receipts and payments is useful in evaluating solvency. Statements of past cash receipts and payments are useful for the same basic reason that statements of comprehensive income are useful in evaluating profitability: both provide a basis for predicting future performance.

5.3 Applying IAS 7 (revised) *Statements of Cash Flows*

5.3.1 IAS 7 format

The cash flows are analysed under three standard headings to explain the net increase/decrease in cash and cash equivalents and the effect on the opening amount of cash and cash equivalents. The headings are:

- Net cash generated by operating activities;
- Cash flows from investing activities;
- Cash flows from financing activities.

5.3.2 The two methods of presenting cash flows from operating activities

In the quote from *The Future Shape of Financial Reports* above, reference was made to the direct method. This preference was expressed because there are two methods, both of which are permitted by IAS 7. These are the direct method and the indirect method.

- The *direct* method reports cash inflows and outflows directly, starting with the major categories of gross cash receipts and payments. This means that cash flows such as receipts from customers and payments to suppliers are stated separately within the operating activities.
- The *indirect* method starts with the profit before tax and then adjusts this figure for non-cash items such as depreciation and changes in working capital.

5.3.3 Statement of cash flows illustrated using the direct method

The following shows the statement of cash flows for Tyro Bruce for the period ended 31.3.20X4.

	£000	£000
Cash flows from operating activities		
Cash received from customers (note (a))	11,740	
Cash paid to suppliers and employees (note (b))	(11,431)	
Cash generated from operations	309	
Interest paid (expense + (closing accrual – opening accrual))	(20)	
Income taxes paid (expense + (closing accrual – opening accrual))	(220)	
Net cash (used in) generated by operating activities		69
Cash flows from investing activities		
Purchase of property, plant and equipment	(560)	
Proceeds from sale of equipment	241	
Net cash used in investing activities		(319)
Cash flows from financing activities		
Proceeds from issue of shares at a premium	300	
Redemption of loan	(50)	
Dividends paid	(120)	
Net cash from financing activities		130
Net increase in cash and cash equivalents		(120)
Cash and cash equivalents at beginning of period		72
Cash and cash equivalents at end of period		(48)

Notes:

(a) Cash received from customers

	£000
Sales	12,000
Receivables increase	(260)
	11,740

(b) Cash paid to suppliers and employees

	£000
Cost of sales	10,000
Payables decreased	140
Inventory increased	900
Depreciation	(102)
Profit on sale	13
Distribution costs	300
Administration expenses	180
	11,431

5.3.4 Statement of cash flows illustrated using the indirect method

The two methods provide different types of information to the users. The indirect method applies changes in working capital to net income. In our illustration, for example, the cash generated from operations would be calculated as follows:

	£000
Cash flows from operating activities	
Profit before tax	1,500
Adjustments for non–cash items:	
Depreciation	102
Profit on sale of plant	(13)
Adjustments for changes in working capital:	
Increase in trade receivables	(260)
Increase in inventories	(900)
Decrease in trade payables	(140)
Interest expense (added back)	20
Cash generated from operations	309

5.3.5 Appraising the use of the direct method

The direct method demonstrates more of the qualities of a true cash flow statement because it provides more information about the sources and uses of cash. This information is not available elsewhere and helps in the estimation of future cash flows.

The principal advantage of the direct method is that it shows operating cash receipts and payments. Knowledge of the specific sources of cash receipts and the purposes for which cash payments were made in past periods may be useful in assessing future cash flows. Disclosure of *cash from customers* could provide additional information about an entity's ability to convert revenues to cash.

When is the direct method beneficial?

One such time is when the user is attempting to predict bankruptcy or future liquidation of the company. A research study looking at the cash flow differences between failed and non-failed companies[4] established that seven cash flow variables and suggested ratios captured statistically significant differences between failed and non-failed firms as much as five years prior to failure. The study further showed that the research findings supported the use of a direct cash flow statement, and the authors commented:

> An indirect cash flow statement will not provide a number of the cash flow variables for which we found significant differences between bankrupt and non-bankrupt companies. Thus, using an indirect cash flow statement could lead to ignoring important information about creditworthiness.

The direct method is the method preferred by the standard but preparers have a choice. In the UK the indirect method is often used; in other regions (e.g. Australia) the direct method is more common. It has been proposed in a review of IAS 7 that the direct method should be mandated and the alternative removed and this is the likely requirement in a new standard to eventually replace IAS 7.

5.3.6 Appraising the use of the indirect method

The principal advantage of the indirect method is that it highlights the differences between operating profit and net cash flow from operating activities to provide a measure of the quality of income. Many users of financial statements believe that such reconciliation is essential to give an indication of the quality of the reporting entity's earnings. Some investors and creditors assess future cash flows by estimating future income and then allowing for accruals adjustments; thus information about past accruals adjustments may be useful to help estimate future adjustments.

Preparer and user response

The IASB indicates that the responses to the discussion paper were mixed with the preparers tending to prefer the indirect method and the users having a mixed response. There was a view that the direct method would be improved if the movements on working capital were disclosed as supplementary information, and the indirect method would be improved if the cash from customers and payments to suppliers was disclosed as supplementary information; i.e. both are found useful.

5.3.7 Cash equivalents

IAS 7 recognised that companies' cash management practices vary in the amount of cash and range of short- to medium-term deposits that are held. The standard standardised the treatment of near-cash items by applying the following definition when determining whether items should be aggregated with cash in the statement of cash flows:

> Cash equivalents are short-term, highly liquid investments which are readily convertible into known amounts of cash and which are subject to an insignificant risk of changes in value.

Near-cash items are normally those that are within three months of maturity at the date of acquisition. Investments falling outside this definition are reported under the heading of 'investing activities'. In view of the variety of cash management practices and banking arrangements around the world and in order to comply with IAS 1 *Presentation of Financial Statements*, an entity discloses the policy which it adopts in determining the composition of cash and cash equivalents.

5.4 Step approach to preparation of a statement of cash flows – indirect method

We will now explain how to prepare a statement of cash flows for Tyro Bruce (Section 5.3.3) taking a step approach. We have shown our workings on the face of the statements of financial position and income.

Step 1: Calculate the differences in the statements of financial position and decide whether to report under operating, investing or financing activities or as a cash equivalent.

Statements of financial position of Tyro Bruce as at 31.3.20X3 and 31.3.20X4

	20X3		20X4		Calculate the differences	Decide which activities to report under
	£000	£000	£000	£000		
Non-current assets at cost	2,520		2,760		See PPE note	Investing/financing
Accumulated depreciation	452	2,068	462	2,298	for acquisitions or disposals	
Current assets						
Inventory	800		1,700		900	Operating
Trade receivables	640		900		260	Operating
Securities maturing less than 3 months at acquisition	—		20		20	Cash equivalent
Cash	80		10		70	Cash equivalent
	1,520		2,630			
Current liabilities						
Trade payables	540		400		140	Operating
Taxation	190		170		20	Operating
Overdraft	8		78		70	Cash equivalent
	738		648			
Net current assets		782		1,982		
		2,850		4,280		
Share capital	1,300		1,400		100	Financing
Share premium a/c	200		400		200	Financing
Retained earnings	1,150	2,650	1,150	2,950		
Profit for year		—		1,180		
10% loan 20×7		200	150		50	Financing
		2,850		4,280		

Step 2: Identify any items in the statement of income for the year ended 31.3.20X4 after profit before interest and tax (PBIT) to be entered under operating, investing or financing activities.

	£000	£000	
Sales		12,000	
Cost of sales		10,000	
Gross profit		2,000	
Distribution costs	300		
Administrative expenses	180	480	
PBIT		1,520	
Interest expense		(20)	Operating
Profit before tax		1,500	Operating
Income tax expense		(200)	Operating
Profit after tax		1,300	
Dividend paid		(120)	Financing
Retained earnings for year		1,180	

Step 3: Refer to the PPE schedule to identify any acquisitions, disposals and depreciation charges that affect the cash flows. The Tyro Bruce schedule showed:

Cost	£000
As at 31 March 20X3	2,520
(i) Additions	560
(iii) Disposal	(320)
As at 31.3.20X4	2,760
Accumulated depreciation	
As at 31.3.20X3	452
(ii) Charge for year	102
(iii) Disposal	(92)
As at 31.3.20X4	462
NBV as at 31.3.20X4	2,298
NBV as at 31.3.20X3	2,068

Note: Disposal proceeds were £241,000.

From Step 3 we can see that there are four impacts:

(i)   Additions: The cash of £560,000 paid out on additions will appear under Investing.

(ii) The depreciation charge: This is a non-cash item and the £102,000 will be added back as a non-cash item to the profit before tax in the operating activities section.

(iii) Disposal proceeds: The cash received of £241,000 from the disposal was given in the Note and will appear under Investing activities. *If the Note had provided you with the profit instead of the proceeds, then you would need to calculate the proceeds by taking the NBV and adjusting for any profit or loss. In this case it would be calculated as NBV of £228,000 (320,000 – 92,000) + the profit figure of £13,000 = £241,000.*

(iv) Profit on disposal: As the full proceeds of £241,000 are included under Investing activities there would be double counting to leave the profit of £13,000 within the profit before tax figure. It is therefore deducted as a non-cash item from PBT in the Operating activities section.

5.4.1 The statement of cash flows

The cash flow items can then be entered into the statement of cash flows in accordance with IAS 7.

		£000
Cash flows from operating activities		
Profit before tax		1,500
Adjustments for non-cash items:		
Depreciation	From Step 3 (ii)	102
Profit on sale of plant	From Step 3 (iv)	(13)
Adjustments for changes in working capital:		
Increase in trade receivables		(260)
Increase in inventories		(900)
Decrease in trade payables		(140)
Interest expense		20
Cash generated from operations		309
Interest paid (there are no closing		(20)
or opening accruals)		
Income taxes paid (expense +	200 + (190 − 170)	(220)
(opening accrual − closing accrual))		
Net cash (used in)/generated by operating activities		69
Cash flows from investing activities		
Purchase of property, plant and equipment	From Step 3 (i)	(560)
Proceeds from sale of equipment	From Step 3 (iii)	241
Net cash used in investing activities		(319)
Cash flows from financing activities		
Proceeds from issue of shares at a premium		300
Redemption of loan		(50)
Dividends paid		(120)
Net cash from financing activities		130
Net increase in cash and cash equivalents		(120)
Cash and cash equivalents at beginning of period	80 − 8	72
Cash and cash equivalents at end of period	(10 + 20) − 78	(48)

Note that interest paid and interest and dividends received could be classified either as operating cash flows or as financing (for interest paid) and investing cash flows (for receipts). Dividends paid could be presented either as financing cash flows or as operating cash flows. However, it is a requirement that whichever presentation is adopted by an enterprise should be consistently applied from year to year.

5.5 Additional notes required by IAS 7

As well as the presentation on the face of the cash flow statement, IAS 7 requires notes to the cash flow statement to help the user understand the information. The notes that are required are as follows.

Major non-cash transactions

If the entity has entered into major non-cash transactions that are therefore not represented on the face of the statement of cash flows, sufficient further information to understand the transactions should be provided in a note to the financial statements. Examples of major non-cash transactions might be:

- the acquisition of assets by way of finance leases;
- the conversion of debt to equity.

Components of cash and cash equivalents

An enterprise must disclose the components of cash and cash equivalents and reconcile these into the totals in the statement of financial position. An example of a suitable disclosure in the case of Tyro Bruce is:

	20X4	20X3
Cash	10	80
Securities	20	
Overdraft	(78)	(8)
Cash and cash equivalents	(48)	72

Disclosure must also be given on restrictions on the use by the group of any cash and cash equivalents held by the enterprise. These restrictions might apply if, for example, cash was held in foreign countries and could not be remitted back to the parent company.

Segmental information

IAS 7 encourages enterprises to disclose information about operating, investing and financing cash flows for each business and geographical segment. This disclosure may be relevant. IFRS 8 does not require a cash flow by segment.

5.6 Analysing statements of cash flows

Arranging cash flows into specific classes provides users with relevant and decision-useful information by classifying cash flows as cash generated from operations, net cash from operating activities, net cash flows from investing activities, and net cash flows from financing activities.

Lack of a clear definition

However, this does not mean that companies will necessarily report the same transaction in the same way. Although IAS 7 requires cash flows to be reported under these headings, it does not define operating activities except to say that it includes all transactions and other events that are not defined as investing or financing activities.

Alternative treatments

Alternative treatments for interest and dividends paid could be presented as either operating or financing cash flows. Whilst most companies choose to report the dividends as financing cash flows, when making inter-firm comparisons we need to see which alternative has been chosen. The choice can have a significant impact. If, for example, in the Tyro Bruce illustration the dividends of £120,000 were reported as an operating cash flow, then the net cash (used in)/ generated by operating activities would change from an inflow of £69,000 to an outflow of £51,000.

The classifications assist users in making informed predictions about future cash flows or raising questions for further enquiry which would be difficult to make using traditional accrual-based techniques.[5]

We will briefly comment on the implication of each classification.

5.6.1 Cash generated from operations

Cash flow generated by operations is one of the most significant numbers calculated after taking account of any investment in working capital. It shows the cash available from ongoing operations to service loans, pay tax, reinvest in the business, repay loans and pay a dividend to shareholders.

Lenders look to the cash generated from operations to pay interest and the revenue authorities to satisfy the company's tax liability. Both of these are unavoidable – it is an indication of the safety margin, i.e. how long a business could continue to pay unavoidable costs.

There are a number of examples where the failure to meet their tax liability has led to organisations being forced into administration or liquidation. Examples include football clubs: with Portsmouth FC going into administration; and Glasgow Rangers being put into liquidation by HMRC, owing £21.6m.

In the Tyro Bruce example (Section 5.3.3) we can see that there has been a significant increase in working capital of £1,300,000 (£260,000 + £900,000 + £140,000).

The effect is to reduce the profit before tax from £1,500,000 to the £309,000 reported as cash flow from operations.

Lenders in Tyro Bruce concerned with interest cover could see that there is sufficient cash available to meet their interest charges in the current year even though there has been a significant impact from the investment in working capital.

Interest cover

Interest cover is normally defined as the number of times the profit before interest and tax covers the interest charge: in the Tyro Bruce example this is 76 times (1,520/20). The position as disclosed in the statement of cash flows is a little weaker although, even so, the interest is still covered more than 15 times (309,000/20,000).

Current cash debt coverage ratio

This is a liquidity ratio which shows a company's ability to meet its current debt obligations. The ratio is the result of dividing *the net cash generated by operating activities* by *the average current liabilities.*

In the Tyro example the net cash generated by operations is £69,000 and the average current liabilities are £693,000 [(738,000 + 648,000)/2] giving a ratio of 0.1:1.

Cash debt coverage ratio

In addition to interest cover, lenders also have a longer-term view and want to be satisfied that their loan will be repaid on maturity. Failure to do so could lead to a going-concern problem for the company. One measure used is to calculate the ratio of *cash flow generated by operating activities* to *total debt* and, of more immediate interest, to *loans that are about to mature.*

The ratio can be adjusted to reflect the company's current position. For example, if there is a significant cash balance, it might be appropriate to add this on the basis that it would be available to meet the loan repayment.

In the Tyro example, the ratio is £69,000/(£693,000 + £150,000) giving a ratio of 0.08:1, which is low due to the heavy investment in working capital and payment of a dividend. If the company continued to achieve profits of £1,500,000 without a further significant investment in working capital, then the ratio is in excess of 1.5:1.

Cash dividend coverage ratio

The ratio of cash flow from operating activities less interest paid to dividends paid indicates the ability to meet the current dividend. If the dividend rate shows a rising trend, dividends declared might be used rather than the cash flow dividend paid figure. This would give a better indication of the coverage ratio for future dividends. In our example coverage is again reduced by the heavy working capital investment.

5.6.2 Future cash flows from operations

We need to consider trends, the discretionary costs and the investment in working capital.

Trends

We need to look at previous periods to identify the trend. Trends are important with investors naturally hoping to invest in a company with a rising trend. If there is a loss or a downward trend, this is a cause for concern and investors should make further enquiries to identify any proposed steps to improve the position.

This is where narrative may be helpful, such as that proposed in the IFRS Practice Statement *Management Commentary*, in the Strategic Review in the UK and in a Chairman's Statement. Reading these may give some indication as to how the company will be addressing the situation. For example, is the company planning a cost reduction programme or disposing of loss-making activities? If it is not possible to improve the trend or reverse the negative cash flow, then there could be future liquidity difficulties.

Discretionary costs

The implication for future cash flow is that such difficulties could have an impact on future discretionary costs, e.g. the curtailment of research, marketing or advertising expenditure; on investment decisions, e.g. postponing capital expenditure; and on financing decisions, e.g. the need to raise additional equity or loan capital.

Working capital

We can see the cash implication but would need to make further enquiries to establish the reasons for the change and the likelihood of similar cash outflow movements recurring in future years. If, for example, the increased investment in inventory resulted from an increase in turnover, then a similar increase could recur if the forecast turnover continued to increase. If, on the other hand, the increase was due to poor inventory control, then it is less likely that the increase will recur once management addresses the problem.

The cash flow statement indicates the cash **extent** of the change; additional ratios (see Chapter 28) and enquiries are required to allow us to **evaluate** the change.

5.6.3 Evaluating the investing activities cash flows

These arise from the acquisition and disposal of non-current assets and investments.

It is useful to consider how much of the expenditure is to replace existing non-current assets and how much is to increase capacity. One way is to relate the cash expenditure to the depreciation charge; this indicates that the cash expenditure is more than five times greater than the depreciation charge, calculated as £540,000/£102,000. This seems to indicate a possible increase in productive capacity. However, the cash flow statement does not itemise the expenditure, as the extract from the non-current asset schedule does not reveal how much was spent on plant – this information would be available in practice.

How to inform investors how much of the capital expenditure relates to replacing existing non-current assets

There has been a criticism that it is not possible to assess how much of the investing activities cash outflow related to simply maintaining operations by replacing non-current assets that were worn out rather than to increasing existing capacity with a potential for an increase in turnover and profits. The solution proposed was that investment that is merely maintained should be shown as an operating cash flow and that the investing cash flow should be restricted to increasing capacity. The IASB doubted the reliability of such a distinction but there is a view that such an analysis provides additional information, provided the breakdown between the two types of expenditure can be reliably ascertained.

Capital expenditure ratio

This is a ratio where the numerator is *net cash flow generated by operating activities* and the denominator is *capital expenditures*. This ratio measures the capital available for internal reinvestment and for meeting existing debt. We look for a ratio that exceeds 1.0, showing that the company has funds to maintain its operational capability and has cash towards meeting its debt repayments and dividends.

It is important to remember that this ratio is industry-specific and any comparator should be with another company that has similar capital expenditure ratio (CAPEX) requirements. The ratio would be expected to be lower for companies in growth industries as opposed to those in mature industries and more variable in cyclical industries, such as housing.

It should be recognised, however, that there is a risk if a company has significant free cash flow that its managers may be too optimistic about future performance. When they are not reliant on satisfying external funders there could be less constraint on their investment decisions. If there is negative free cash flow then the opposite applies and the business would require external finance which then means that it would be subject to any conditions imposed by the new source of finance.

5.6.4 Free cash flow (FCF)

This is a performance measure showing how much cash a company has for further investment after deducting from net cash generated by operating capital the amount spent on capital expenditure to maintain or expand its asset base. Many companies refer to it in their annual report with possible slight variations in definition.

For example, Colt SA in its 2011 Annual Report states:

> Free cash flow is net cash generated from operating activities less net cash used to purchase non-current assets *and net interest paid*.

Reasons for reporting FCF

It is emphasised by companies for different reasons – some emphasising its use as the way that the company manages its capital. For example, the following is an extract from the Kingfisher Group's 2013 Annual Report:

> The Group manages its capital by:
> Continued focus on free cash flow generation; Setting the level of capital expenditure and dividend in the context of current year and *forecast free cash flow* generation; Rigorous review of capital investments and post investment reviews to drive better returns; and Monitoring the level of the Group's financial and leasehold debt in the context of Group performance and its credit rating.

The company recognises the importance of free cash flow in maintaining its credit rating:

> The Group will maintain a high focus on free cash flow generation going forward to maintain its solid investment grade balance sheet, fund investment where economic returns are attractive and pay healthy dividends to shareholders.

Other companies might place their emphasis on liquidity. For example, the following is an extract from the Merck Group 2011 Annual Report:

> Free cash flow and underlying free cash flow are indicators that we use internally to measure the contribution of our divisions to liquidity.

Also there might be an emphasis on operational control as illustrated in this extract from the Merck Group 2013 Annual Report:

> Business free cash flow (BFCF)
> Apart from EBITDA pre and sales, business free cash flow (BFCF) is the third important Group and division KPI and therefore also used for internal target agreements and individual incentive plans. It comprises the major cash-relevant items that the individual businesses can influence. . . . The introduction of business free cash flow has led to considerable improvements in cash awareness as well as reduced working capital requirements.

The amount of free cash flow will be normally positive for a mature company and negative for a younger company. It will be impacted by the investment in working capital and capital expenditure and will depend on the industry. For example, free cash flow might be high in the tobacco industry and its products industry where there is low investment in either working capital or CAPEX and low in an industry such as petroleum and gas where, although the investment in working capital is low, CAPEX is high.

Ratios based on FCF

These include the cash conversion ratio (CCR) and cash dividend coverage ratio (CDCR).

The cash conversion ratio (CCR) is calculated as free cash flow divided by earnings before interest, tax, depreciation and amortisation (EBITDA). It indicates the rate at which profits are being turned into cash. From the point of view of the shareholders it indicates how much of the profit could be distributed as dividends without causing liquidity or cash flow problems for the company.

The cash dividend coverage ratio (CDCR) is calculated as free cash flow divided by dividends. It indicates that the company is able to generate earnings beyond maintaining its current operational capacity.

5.6.5 Evaluating the financing cash flows

Additional capital of £300,000 has been raised. After repaying a loan of £50,000 and payment of a dividend of £120,000, only £130,000 was left towards a negative free cash flow with a net outflow of £250,000 (£319,000 − £69,000).

This does not allow us to assess the financing policy of the company, e.g. whether the capital was raised the optimum way. Nor does it allow us to assess whether the company would have done better to provide finance by improved control over its assets, e.g. working capital reduction.[6]

The indications are healthy in that the company is relying on earnings and equity capital to finance growth. It is low-geared and further funds could be sought, possibly from the bank or private equity, particularly if it is required for capacity building purposes.

5.6.6 Reconciliation of net cash flows to net debt

A net debt reconciliation is useful in that it allows investors to see how business financing has changed over the year, identifying, for example, if a significant increase in cash has been achieved only by taking on increased debt.

It is not required by IFRS but is often sought by investors. The following illustrates the notes that would be prepared for Tyro Bruce (see Section 5.4 above) if the company decided to publish a reconciliation:

		20X4		20X3
1 Borrowings		(150)		(200)
Overdraft	(78)		(8)	
Securities	20			
Cash	10		80	
		(48)		72
		(198)		(128)

2 Reconcile net cash flow to movement in net debt

	20X4	20X3
Decrease in cash	(48 + 72)	(120)
Change in net debt resulting from cash	(200 – 150)	50
Movement in net debt	(198 – 128)	(70)
Net debt at beginning of period		(128)
Net debt at end of period		(198)

3 Analysis of net debt

	20X3	Cash flow	20X4
Cash at bank	80	(120)	10
Government securities			20
Overdraft	(8)		(78)
Debt outstanding	(200)	50	(150)
Net debt	(128)	(70)	(198)

5.6.7 Voluntary disclosures

IAS 7 (paragraphs 50–52) lists additional information, supported by a management commentary that may be relevant to understanding:

- liquidity, e.g. the amount of undrawn borrowing facilities;
- future profitability, e.g. cash flow representing increases in operating capacity separate from cash flow maintaining operating capacity; and
- risk, e.g. cash flows for each reportable segment, to better understand the relationship between the entity's cash flows and each segment's cash flows.

5.7 Approach to answering questions with time constraints

We have explained the step approach with the explanatory detail on the statements of financial position and income. In an examination it is preferable to show the workings on the statement of cash flows itself as shown in the examination question for Riddle worked below.

The following are the statements of financial position and income for Riddle plc.

Statements of financial position as at 31 March

	20X8 $000	$000	20X9 $000	$000
Non-current assets:				
Property, plant and equipment, at cost	540		720	
Less accumulated depreciation	(145)		(190)	
		395		530
Investments		115		140
Current assets:				
Inventory	315		418	
Trade receivables	412		438	
Bank	48	775	51	907
Total assets		1,285		1,577
Capital and reserves:				
Ordinary shares	600		800	
Share premium	40		55	
Retained earnings	217	857	311	1,166
Non-current liabilities:				
12% debentures		250		200
Current liabilities:				
Trade payables	139		166	
Taxation	39	178	45	211
Total equity and liabilities		1,285		1,577

Statement of income for the year ended 31 March 20X9

	$000	$000
Revenue		2,460
Cost of sales		1,780
Gross profit		680
Distribution costs	(124)	
Administration expenses	(300)	(424)
Operating profit		256
Interest on debentures		(24)
Profit before tax		232
Tax		(48)
Profit after tax		184

Note: The statement of changes in equity disclosed a dividend of $90,000.

Teaching note: Take an initial look at the statement of financial position and notes to check whether or not there has been any disposal of non-current assets which would give rise to a profit or loss adjustment as a non-cash adjustment to the profit after tax figure in the statement of income. In the case of Riddle there have only been acquisitions.

Required
(a) Prepare the statement of cash flows for Riddle plc for the year ended 31 March 20X9 and show the operating cash flows using the 'indirect method'.
(b) Calculate the cash generated from operations using the 'direct method'.

Solution

(a) Using indirect method

Statement of cash flows for the year ended 31 March 20X9

		$000	$000
Cash from operating activities			
Profit before tax	Income statement		232
Adjustments for:			
Depreciation	190 – 145	45	
Interest expense		24	69
Operating profit before working capital changes			301
Increase in inventory	418 – 315	(103)	
Increase in trade receivables	438 – 412	(26)	
Increase in trade payables	166 – 139	27	(102)
Cash generated from operations			199
Interest paid		(24)	
Tax paid	39 + 48 – 45	(42)	(66)
Net cash used in operating activities			133
Cash flows from investing activities:			
Purchase of PPE	720 – 540	(180)	
Disposal proceeds of PPE	None in question		
Investments	140 – 115	(25)	(205)
Cash flows from financing activities:			
Share capital	800 – 600	200	
Share premium	55 – 40	15	
Debentures	200 – 250	(50)	
Dividends paid	Given in note	(90)	75
Net increase in cash and cash equivalents			3
Cash and cash equivalents at beginning of year			48
Cash and cash equivalents at end of year			51

(b) Cash generated from operations using the direct method

	$000	$000
(i) Received from customers		2,434
(ii) Paid to suppliers	1,856	
(iii) Paid expenses (124 + 300 – depreciation 45)	379	2,235
Cash generated from operations		199

	$000
(i) Received from customers	
Trade receivables at beginning of year	412
Sales	2,460
	2,872
Less: Trade receivables at end of year	438
Cash received from customers	2,434

(ii) Paid to suppliers

	$000	$000
Trade payables at beginning of year		139
Cost of sales	1,780	
Closing inventory	418	
	2,198	
Less: Opening inventory	315	1,883
		2,022
Less: Trade payables at end of year		166
Cash paid to trade payables		1,856

Teaching note: Interest on the debentures is added back when preparing the statement using the indirect method. When using the direct method there is no need to include it within the payables calculation.

5.8 Preparing a statement of cash flows when no statement of income is available

Questions might be met where the statements of financial position are provided and figures have to be derived.

5.8.1 Flow Ltd – an example

As an example, the following statements of financial position have been provided for Flow Ltd for the years ended 31 December 20X5 and 20X6:

	20X5		20X6	
	€	€	€	€
Non-current assets				
Tangible assets				
PPE at cost	1,743,750		1,983,750	
Accumulated depreciation	551,250	1,192,500	619,125	1,364,625
Current assets				
Inventory		101,250		85,500
Trade receivables		252,000		274,500
		1,545,750		1,724,625
Capital and reserves				
Common shares of €1 each		900,000		1,350,000
Share premium				30,000
Retained earnings		387,000		176,625
Current liabilities				
Trade payables		183,750		159,750
Bank overdraft		75,000		8,250
		1,545,750		1,724,625

Notes stated that during the year ended 31 December 20X6:

1 Equipment that had cost €25,500 and with a net book value of €9,375 was sold for €6,225.

2 The company paid a dividend of €45,000.

3 A bonus issue was made at the beginning of the year of one bonus share for every three shares.

4 A new issue of 150,000 shares was made on 1 July 20X6 at a price of €1.20 for each share.

5 A dividend of €60,000 was declared but no entries had been made in the books of the company.

The requirement is to prepare a statement of cash flows for the year ended 31 December 20X6 that complies with IAS 7.

5.8.2 Solution to Flow Ltd

Step 1. Calculate the profit by working back from the end-of-period retained earnings

Retained earnings	176,625
Less opening retained earnings	387,000
	(210,375)
Add back the dividend already paid	45,000
Add back amount transferred to Capital on issue of bonus shares	300,000
	134,625

Step 2. Calculate the cash flow from operating activities

Profit	134,625
Depreciation	
619,125 – 551,250 = 67,875	
25,500 – 9,375 = 15,625	83,500
Loss on sale of plant	3,150
9,375 – 6,225	
Decrease in inventory	15,750
101,250 – 85,500	
Increase in receivables	(22,500)
252,000 – 274,500	
Decrease in payables	(24,000)
183,750 – 159,750	
Cash flow from operating activities	**190,525**

Step 3. Statement of cash flows for the year ended 31 December 20X6 for Flow Ltd

Net cash inflow from operating activities		190,025
Cash flows from investing activities		
Purchase of non-current assets	(265,500)	
(1,983,750 + 25,500 – 1,743,750)		
Receipts from sale of non-current assets	6,225	
Net cash paid on investing activities		(259,275)
Cash flows from financing activities		
Proceeds from issue of common shares	180,000	
Dividends paid	(45,000)	
(could be shown as operating cash flow)		
Net cash inflow from financing activities		135,000
Net increase in cash and cash equivalents		66,250
(75,000 – 8,250)		

5.9 Critique of cash flow accounting

IAS 7 (revised) applies uniform requirements to the format and presentation of cash flow statements. It still, however, allows companies to choose between the direct and the indirect methods, and the presentation of interest and dividend cash flows. It can be argued, therefore, that it has failed to rectify the problem of a lack of comparability between statements.

An important point is that, in its search for improved comparability, IAS 7 (revised) reduced the scope for innovation. It might be argued that standard setters should not be reducing innovation, but that there should be concerted effort to increase innovation and improve the information available to user groups. The acceptability of innovation is a fundamental issue in a climate that is becoming increasingly prescriptive.

Summary

IAS 7 (revised) defines the format and treatment of individual items within the cash flow statement. This leads to uniformity and greater comparability between companies. However, there is still some criticism of the current IAS 7:

- There are options within IAS 7 for presentation, since either the direct or the indirect method can be used; and there are choices about the presentation of dividends and interest.

- The cash flow statement does not distinguish between discretionary and non-discretionary cash flows, which would be valuable information to users.

- There is no separate disclosure of cash flows for expansion from cash flows to maintain current capital levels. This distinction would be useful when assessing the position and performance of companies, and is not always easy to identify in the current presentation.

- The definition of cash and cash equivalents can cause problems in that companies may interpret which investments are cash equivalents differently, leading to a lack of comparability. Statements of cash flows could be improved by removing cash equivalents and concentrating solely on the movement in cash, which is the current UK practice.

REVIEW QUESTIONS

1 Explain the entries in the statement of cash flows when a non-current asset is sold (a) at a loss and (b) at a profit.

2 Explain the two ways in which dividends received might be classified and discuss which provides the more relevant information.

3 Discuss if long-term debts are ever included with cash equivalents.

4 Discuss three ways in which free cash flow might be improved.

5 Discuss the significance of a ratio relating free cash flow to EBITDA.

6 Explain why depreciation appears in a statement of cash flows prepared applying the indirect method but not in that applying the direct method.

7 Explain the information that a user can obtain from a statement of cash flows that cannot be obtained from the current or comparative statements of financial position.

8 It is suggested that a reconciliation of net cash flows to net debt should be required by IFRS. Discuss the relevance of such a reconciliation and the suggestion that it should be a mandatory requirement.

9 Discuss the limitations of a statement of cash flows when evaluating a company's control over its working capital.

10 Discuss why the financing section of a statement of cash flows does not allow a user to assess a company's financing policy.

11 Access http://scheller.gatech.edu/centers-initiatives/financial-analysis-lab/index.html and discuss what accounts for the difference in free cash flow between the top three and bottom three industries.

EXERCISES

* Question 1

Direct plc provided the following information from its records for the year ended 30 September 20X9:

	€000	
Sales	316,000	
Cost of goods sold	110,400	
Other expenses	72,000	
Rent expense	14,400	
Dividends	10,000	
Amortisation expense – PPE	8,000	
Advertising expense	4,800	
Gain on sale of equipment	2,520	
Interest expense	320	
	20X9	20X8
Accounts receivable	13,200	15,200
Unearned revenue	8,000	9,600
Inventory	18,400	19,200
Prepaid advertising	0	400
Accounts payable	11,200	8,800
Rent payable	0	1,200
Interest payable	40	0

Required:
Using the direct method of presentation, prepare the cash flows from the operating activities section of the statement of cash flows for the year ended 30 September 20X9.

* Question 2

Marwell plc reported a profit after tax of €14.04m for 20X2 as follows:

	€m	€m
Revenue		118.82
Materials	29.70	
Wages	30.80	
Depreciation	22.68	
Loss on disposal of plant	3.78	
Profit on sale of buildings	(6.48)	
		80.48
Operating profit		38.34
Interest payable		16.20
Profit before tax		22.14
Income tax expense		8.10
Profit after tax		14.04

The statements of financial position and changes in equity showed:

(i)   Inventories at the year end were €5.94m higher than the previous year.

(ii) Trade receivables were €10.26m higher.

(iii) Trade payables were €4.86m lower.

(iv) Tax payable had increased by €2.7m.

(v)   Dividends totalling €18.36m had been paid during the year.

Required:
(a) Calculate the net cash flow from operating activities.
(b) Explain why depreciation and a loss made on disposal of a non-current asset are both treated as a source of cash.

* Question 3

The statements of financial position of Radar plc at 30 September were as follows:

	20X8		20X9	
	$000	$000	$000	$000
Non-current assets:				
Property, plant and equipment, at cost	760		920	
Less accumulated depreciation	(288)		(318)	
		472		602
Investments		186		214
Current assets:				
Inventory	596		397	
Trade receivables	332		392	
Bank	5	933	—	789
Total assets		1,591		1,605
Capital and reserves:				
Ordinary shares	350		500	
Share premium	75		125	
Retained earnings	137	562	294	919
Non-current liabilities:				
12% debentures		400		100
Current liabilities:				
Trade payables	478		396	
Accrued expenses	64		72	
Taxation	87		96	
Overdraft	—		22	
		629		586
Total equity and liabilities		1,591		1,605

The following information is available:

(i) An impairment review of the investments disclosed that there had been an impairment of $20,000.

(ii) The depreciation charge made in the statement of comprehensive income was $64,000.

(iii) Equipment costing $72,000 was sold for $54,000 which gave a profit of $16,000.

(iv) The debentures redeemed in the year were redeemed at a premium of 25%.

(v) The premium paid on the debentures was written off to the share premium account.

(vi) The income tax expense was $92,000.

(vii) A dividend of $25,000 had been paid and dividends of $17,000 had been received.

Required:
Prepare a statement of cash flows for the year ended 30 September using the indirect method.

Question 4

Shown below are the summarised final accounts of Martel plc for the last two financial years:

Statements of financial position as at 31 December

	20X1		20X0	
	£000	£000	£000	£000
Non-current assets				
Tangible				
Land and buildings	1,464		1,098	
Plant and machinery	520		194	
Motor vehicles	140		62	
		2,124		1,354
Current assets				
Inventory	504		330	
Trade receivables	264		132	
Government securities	40		—	
Bank	—		22	
	808		484	
Current liabilities				
Trade payables	266		220	
Taxation	120		50	
Proposed dividend	72		40	
Bank overdraft	184		—	
	642		310	
Net current assets		166		174
Total assets less current liabilities		2,290		1,528
Non-current liabilities				
9% debentures		(432)		(350)
		1,858		1,178
Capital and reserves				
Ordinary shares of 50p each fully paid		900		800
Share premium account	120		70	
Revaluation reserve	360		—	
General reserve	100		50	
Retained earnings	378		258	
		958		378
		1,858		1,178

Summarised statement of comprehensive income for the year ending 31 December

	20X1	20X0
	£000	£000
Operating profit	479	215
Interest paid	52	30
Profit before taxation	427	185
Tax	149	65
Profit after taxation	278	120

Additional information:

1 The movement in non-current assets during the year ended 31 December 20X1 was as follows:

	Land and buildings £000	Plant, etc. £000	Motor vehicles £000
Cost at 1 January 20X1	3,309	470	231
Revaluation	360	—	—
Additions	81	470	163
Disposals	—	(60)	—
Cost at 31 December 20X1	3,750	880	394
Depreciation at 1 January 20X1	2,211	276	169
Disposals	—	(48)	—
Added for year	75	132	85
Depreciation at 31 December 20X1	2,286	360	254

The plant and machinery disposed of during the year was sold for £20,000.

2 During 20X1, a rights issue was made of one new ordinary share for every eight held at a price of £1.50.

3 A dividend of £36,000 (20X0 £30,000) was paid in 20X1. A dividend of £72,000 (20X0 £40,000) was proposed for 20X1. A transfer of £50,000 was made to the general reserve.

Required:

(a) **Prepare a statement of cash flows for the year ended 31 December 20X1, in accordance with IAS 7.**

(b) **Prepare a report on the liquidity position of Martel plc for a shareholder who is concerned about the lack of liquid resources in the company.**

* Question 5

The statements of financial position of Maytix as at 31 October 2005 and 31 October 2004 are as follows:

	2005		2004	
	$000	$000	$000	$000
Non-current assets:				
Property, at cost	4,000		3,000	
Plant and equipment, at cost	7,390		4,182	
Less accumulated depreciation	(1,450)		(1,452)	
		9,940		5,730
Current assets:				
Inventory	5,901		4,520	
Trade receivables	2,639		2,233	
Bank	—	8,540	1,007	7,760
		18,480		13,490
Capital and reserves:				
Ordinary shares	5,000		3,500	
Share premium	2,500		1,000	
Retained earnings	2,110	9,610	3,090	7,590
Non-current liabilities:				
10% loan stock		4,750		3,750
Current liabilities:				
Trade payables	1,237		1,700	
Taxation	550		450	
Bank overdraft	2,333	4,120	—	2,150
		18,480		13,490

The statement of comprehensive income of Maytix for the year ended 31 October 2005 is as follows:

	$000	$000
Credit sales		9,500
Cash sales		1,047
Cost of sales		(8,080)
Gross profit		2,467
Distribution costs	(501)	
Administration expenses	(369)	(870)
Operating profit		1,597
Interest on loan stock		(425)
Loss on disposal of non-current assets		(102)
Profit before tax		1,070
Tax		(550)
Profit after tax		520

Notes:
(i) The 'statement of changes in equity' disclosed a dividend paid figure of $1,500,000 during the year to 31 October 2005.
(ii) The non-current asset schedule revealed the following details:

Property: Additions cost $1,000,000.

Plant and equipment	Cost	Depreciation	NBV
	$000	$000	$000
Balance at 31.10.2004	4,182	(1,452)	2,730
Additions	6,278	—	6,278
Annual charge	—	(540)	(540)
	10,460	(1,992)	8,468
Disposal	(3,070)	542	(2,528)
Balance at 31.10.2005	7,390	(1,450)	5,940

Required:

(a) Prepare the cash flow statement of Maytix for the year ended 31 October 2005. Use the format required by IAS 7 *Cash Flow Statements* and show operating cash flows using the indirect method.

(b) Describe the additional information that would be included in a cash flow statement showing operating cash flows using the direct method and discuss the proposition that such disclosures be made compulsory under IAS 7.

(The Association of International Accountants)

Question 6

The financial statements of Saturn plc have been prepared as follows:

Statements of financial position as at 30 June	20X2		20X1	
	€000	€000	€000	€000
Non-current assets:				
Property, plant and equipment at cost	6,600		5,880	
Accumulated depreciation	(1,680)	4,920	(1,380)	4,500
Development costs		540		480
Investments		420		300
Current assets:				
Inventory	1,665		1,872	
Trade receivables	1,446		1,188	
Cash	9	3,120	42	3,102
		9,000		8,382
Equity and reserves				
Ordinary shares of €1 each	3,000		2,700	
Share premium account	600		270	
Retained earnings	3,084	6,684	2,622	5,592
Non-current liability:				
7% debentures	—			1,200
Current liabilities:				
Trade payables	1,632		1,104	
Taxation	507		396	
Dividend declared	60		90	
Bank overdraft	117	2,316	—	1,590
		9,000		8,382

Further information:

(a) Extract from statement of income

	€000
Operating profit	1,008
Dividend received	36
Premium on debentures	(120)
Interest paid	(144)
Profit before taxation	780
Income tax	(258)
Profit after tax	522

(b) Operating expenses written off in the year include the following:

	€000
Amortisation of development costs	102
Depreciation of property, plant and equipment	318

(c) Equipment which had cost €240,000 was sold in the year, incurring a loss of €156,000.

(d) The debentures were redeemed at a premium of 10%.

Required:
(a) Prepare a statement of cash flows for the year ended 30 June 20X2.
(b) Briefly explain ways in which statements of cash flows may be more useful than statements of income.

References

1 J. Arnold et al., *The Future Shape of Financial Reports*, ICAEW and ICAS, 1991.
2 G.H. Sorter, M.J. Ingberman and H.M. Maximon, *Financial Accounting: An Events and Cash Flow Approach*, McGraw-Hill, 1990.
3 L.J. Heath and P. Rosenfield, 'Solvency: the forgotten half of financial reporting', in R. Bloom and P.T. Elgers (eds), *Accounting Theory and Practice*, Harcourt Brace Jovanovich, 1987, p. 586.
4 J.M. Gahlon and R.L. Vigeland, 'Early warning signs of bankruptcy using cash flow analysis', *Journal of Commercial Lending*, December 1988, pp. 4–15.
5 J.W. Henderson and T.S. Maness, *The Financial Analyst's Deskbook*, Van Nostrand Reinhold, 1989, p. 72.
6 G. Holmes and A. Sugden, *Interpreting Company Reports and Accounts* (5th edition), Woodhead-Faulkner, 1995, p. 134.

Income and asset value measurement systems

PART 2

Income and asset value
measurement systems

Income and asset value measurement: an economist's approach

6.1 Introduction

The main purpose of this chapter is to explain the need for income measurement, to compare the methods of measurement adopted by the accountant with those adopted by the economist, and to consider how both are being applied within the international financial reporting framework.

Objectives

By the end of this chapter, you should be able to:

- explain the role and objective of income measurement;
- explain the accountant's view of income, capital and value;
- critically comment on the accountant's measure;
- explain the economist's view of income, capital and value;
- critically comment on the economist's measure;
- define various capital maintenance systems.

6.2 Role and objective of income measurement

Although accountancy has played a part in business reporting for centuries, it is only since the Companies Act 1929 that financial reporting has become income-orientated. Prior to that Act, a statement of income was of minor importance. It was the statement of financial position that mattered, providing a list of capital, assets and liabilities that revealed the financial soundness and solvency of the business.

According to some commentators,[1] this scenario may be attributed to the sources of capital funding. Until the late 1920s, as in present-day Germany, external capital finance in the UK was mainly in the hands of bankers, other lenders and trade creditors. As the main users of published financial statements, they focused on the company's ability to pay trade creditors and the interest on loans, and to meet the scheduled dates of loan repayment: they were interested in the short-term liquidity and longer-term solvency of the entity.

Thus the statement of financial position was the prime document of interest. Perhaps in recognition of this, in the UK the statement of financial position, until recent times, tended to show liabilities on the left-hand side, thus making them the first part of the statement of financial position read.

The gradual evolution of a sophisticated investment market, embracing a range of financial institutions, together with the growth in the number of individual investors, caused a reorientation of priorities. Investor protection and investor decision-making needs started to dominate the financial reporting scene, and the revenue statement replaced the statement of financial position as the sovereign reporting document.

Consequently, attention became fixed on the statement of comprehensive income and on concepts of accounting for profit. Moreover, investor protection assumed a new meaning. It changed from simply protecting the **capital** that had **been invested** to protecting the **income information** used by investors when making an investment decision.

However, the sight of major companies experiencing severe liquidity problems over the past decade has revived interest in the statement of financial position; while its light is perhaps not of the same intensity as that of the profit and loss account, it cannot be said to be totally subordinate to its accompanying statement of income.

The main objectives of income measurement are to provide:

- a means of control in a micro- and macroeconomic sense;
- a means of prediction;
- a basis for taxation.

We consider each of these below.

6.2.1 Income as a means of control

Assessment of stewardship performance

Managers are the stewards appointed by shareholders. Income, in the sense of net income or net profit, is the crystallisation of their accountability. Maximisation of income is seen as a major aim of the entrepreneurial entity, but the capacity of the business to pursue this aim may be subject to political and social constraints in the case of large public monopolies, and private semi-monopolies such as British Telecommunications plc.

Maximisation of net income is reflected in the earnings per share (EPS) figure, which is shown on the face of the published statement of income. The importance of this figure to the shareholders is evidenced by contracts that tie directors' remuneration to growth in EPS. A rising EPS may result in an increased salary or bonus for directors and upward movement in the market price of the underlying security. The effect on the market price is indicated by another extremely important statistic, which is influenced by the statement of comprehensive income: namely, the price/earnings (PE) ratio. The PE ratio reveals the numerical relationship between the share's current market price and the last reported EPS. EPS and PE ratios are discussed in Chapters 27 and 29.

Actual performance versus predicted performance

This comparison enables the management and the investing public to use the lessons of the past to improve future performance. The public, as shareholders, may initiate a change in the company directorate if circumstances necessitate it. This may be one reason why management is generally loath to give a clear, quantified estimate of projected results – such an estimate is a potential measure of efficiency. The comparison of actual with projected results identifies apparent underachievement.

The macroeconomic concept

Good government is, of necessity, involved in managing the macroeconomic scene and as such is a user of the income measure. State policies need to be formulated concerning the allocation of economic resources and the regulation of firms and industries, as illustrated by the measures taken by Oftel and Ofwat to regulate the size of earnings by British Telecom and the water companies.

6.2.2 Income as a means of prediction

Dividend and retention policy

The profit generated for the year influences the payment of a dividend, its scale and the residual income after such dividend has been paid. Other influences are also active, including the availability of cash resources within the entity, the opportunities for further internal investment and the dividend policies of capital-competing entities with comparable shares.

However, some question the soundness of using the profit generated for the year when making a decision to invest in an enterprise. Their view is that such a practice misunderstands the nature of income data, and that the appropriate information is the prospective cash flows. They regard the use of income figures from past periods as defective because, even if the future accrual accounting income could be forecast accurately, 'it is no more than an imperfect surrogate for future cash flows'.[2]

The counter-argument is that there is considerable resistance by both managers and accountants to the publication of future operating cash flows and dividend payments.[3] This means that, in the absence of relevant information, an investor needs to rely on a surrogate. The question then arises: which is the best surrogate?

In the short term, the best surrogate is the information that is currently available, i.e. income measured according to the accrual concept. In the longer term, management will be pressed by the shareholders to provide the actual forecast data on operating cash flows and dividend distribution, or to improve the surrogate information, by for example reporting the cash earnings per share.

More fundamentally, Revsine has suggested that ideal information for investors would indicate the economic value of the business (and its assets) based on expected future cash flows. However, the Revsine suggestion itself requires information on future cash flows that it is not possible to obtain at this time.[4] Instead, he considered the use of replacement cost as a surrogate for the economic value of the business, and we return to this later in the chapter.

Future performance

While history is not a faultless indicator of future events and their financial results, it does have a role to play in assessing the level of future income. In this context, historic income is of assistance to existing investors, prospective investors and management.

6.2.3 Basis for taxation

The contemporary taxation philosophy, in spite of criticism from some economists, uses income measurement to measure the taxable capacity of a business entity.

However, the determination of income by HM Revenue and Customs is necessarily influenced by socioeconomic fiscal factors, among others, and thus accounting profit is subject to adjustment in order to achieve taxable profit. As a tax base, it has been continually eroded as the difference between accounting income and taxable income has grown.[2]

Her Majesty's Revenue and Customs in the UK has tended to disallow expenses that are particularly susceptible to management judgement. For example, a uniform capital allowance is substituted for the subjective depreciation charge that is made by management, and certain provisions that appear as a charge in the statement of income are not accepted as an expense for tax purposes until the loss crystallises, e.g. a charge to increase the doubtful debts provision may not be allowed until the debt is recognised as bad.

6.3 Accountant's view of income, capital and value

Variations between accountants and economists in measuring income, capital and value are caused by their different views of these measures. In this section, we introduce the account-ant's view and, in Section 6.5, the economist's, in order to reconcile variations in methods of measurement.

6.3.1 The accountant's view

Income is an important part of accounting theory and practice, although until 1970, when a formal system of propagating standard accounting practice throughout the accountancy profession began, it received little attention in accountancy literature. The characteristics of measurement were basic and few, and tended to be of an intuitive, traditional nature, rather than being spelled out precisely and given mandatory status within the profession.

Accounting tradition of historical cost

The statement of income is based on the actual costs of business transactions, i.e. the costs incurred in the currency and at the price levels pertaining at the time of the transactions.

Accounting income is said to be historical income, i.e. it is an *ex post* measure because it takes place after the event. The traditional statement of income is historical in two senses: because it concerns a past period, and because it utilises historical cost, being the cost of the transactions on which it is based. It follows that the statement of financial position, being based on the residuals of transactions not yet dealt with in the statement of income, is also based on historical cost.

In practice, certain amendments may be made to historical cost in both the statement of comprehensive income and statement of financial position, but historical cost still pre-dominates in both statements. It is justified on a number of counts which, in principle, guard against the manipulation of data.

The main characteristics of historical cost accounting are as follows:

- **Objectivity**. It is a predominantly objective system, although it does exhibit aspects of subjectivity. Its nature is generally understood and it is invariably supported by inde-pendent documentary evidence, e.g. an invoice, statement, cheque, cheque counterfoil, receipt or voucher.

- **Factual**. As a basis of fact (with exceptions such as when amended in furtherance of revaluation), it is verifiable and to that extent is beyond dispute.

- **Profit or income concept**. Profit as a concept is generally well understood in a capital market economy, even if its precise measurement may be problematic. It constitutes the difference between revenue and expenditure or, in the economic sense, between opening and closing net assets.

Unfortunately, historical cost is not without its weaknesses. It is not always objective, owing to alternative definitions of revenue and costs and the need for estimates.

For example, although inventories are valued at the lower of cost or net realisable value, the cost will differ depending upon the definition adopted, e.g. first-in-first-out, last-in-first-out or standard cost.

Estimation is needed in the case of inventory valuation, assessing possible bad debts, accruing expenses, providing for depreciation and determining the profit attributable to long-term contracts. So, although it is transaction-based, there are aspects of historical cost reporting that do not result from an independently verifiable business transaction. This means that profit is not always a unique figure.

Assets are often subjected to revaluation. In an economy of changing price levels, the historical cost system has been compromised by a perceived need to restate the carrying value of those assets that comprise a large proportion of a company's capital employed, e.g. land and buildings. This practice is controversial, not least because it is said to imply that a statement of financial position is a list of assets at market valuation, rather than a statement of unamortised costs not yet charged against revenue.

However, despite conventional accountancy income being partly the result of subjectivity, it is largely the product of the historical cost concept. A typical accounting policy specified in the published accounts of companies now reads as follows:

> The financial statements are prepared under the historical cost conventions as modified by the revaluation of certain non-current assets.

Nature of accounting income as an *ex post* measure

Accounting income is defined in terms of the business entity. It is the excess of revenue from sales over direct and allocated indirect costs incurred in the achievement of such sales. Its measure results in a net figure. It is the numerical result of the matching and accruals concepts discussed in the preceding chapter.

Accounting income is transaction-based and therefore can be said to be factual, in as much as the revenue and costs have been realised and will be reflected in cash inflow and outflow, although not necessarily within the financial year.

Under accrual accounting, the sales for a financial period are offset by the expenses incurred in generating such sales. Objectivity is a prime characteristic of accrual accounting, but the information cannot be entirely objective because of the need to break up the ongoing performance of the business entity into calendar periods or financial years for purposes of accountability reporting. The allocation of expenses between periods requires a prudent estimate of some costs, e.g. the provision for depreciation and bad debts attributable to each period.

Accounting income is presented in the form of the conventional statement of income. This statement of income, in being based on actual transactions, is concerned with a past-defined period of time. Thus accounting profit is said to be historical income, i.e. an *ex post* measure because it is after the event.

Nature of accounting capital

The business enterprise requires the use of non-monetary assets, e.g. buildings, plant and machinery, office equipment, motor vehicles, stock of raw materials and work in progress. Such assets are not consumed in any one accounting period, but give service over a number of periods; therefore, the unconsumed portions of each asset are carried forward from period to period and appear in the statement of financial position. This document itemises the unused asset balances at the date of the financial year-end. In addition to listing unexpired costs of non-monetary assets, the statement of financial position also displays monetary assets such as trade receivables and cash balances, together with monetary liabilities, i.e. moneys owing to trade creditors, other creditors and lenders. Funds supplied by shareholders and retained income following the distribution of dividend are also shown. Retained profits are usually added to shareholders' capital, resulting in what are known as shareholders' funds. These represent the company's equity capital.

Statement of income as a linking statement

The net assets of the firm, i.e. that fund of unconsumed assets which exceeds moneys attributable to payables and lenders, constitutes the company's net capital, which is the same as its equity capital. Thus the statement of income of a financial period can be seen as a linking statement between that period's opening and closing statement of financial positions: in other words, income may be linked with opening and closing capital. This linking may be expressed by formula, as follows:

$$Y_{0-1} = NA_1 - NA_0 + D_{0-1}$$

where Y_{0-1} = income for the period of time t_0 to t_1; NA_0 = net assets of the entity at point of time t_0; NA_1 = net assets of the entity at point of time t_1; and D_{0-1} = dividends or distribution during period t_{0-1}.

Less formally: Y = income of financial year; NA_0 = net assets as shown in the statement of financial position at beginning of financial year; NA_1 = net assets as shown in the statement of financial position at end of financial year; and D_{0-1} = dividends paid and proposed for the financial year. We can illustrate this as follows:

Income Y_{0-1} for the financial year t_{0-1} as compiled by the accountant was £1,200

Dividend D_{0-1} for the financial year t_{0-1} was £450

Net assets NA_0 at the beginning of the financial year were £6,000

Net assets NA_1 at the end of the financial year were £6,750.

The income account can be linked with opening and closing statements of financial position, namely:

$$\begin{aligned} Y_{0-1} &= NA_1 - NA_0 + D_{0-1} \\ &= £6,750 - £6,000 + £450 \\ &= £1,200 = Y_{0-1} \end{aligned}$$

Thus Y has been computed by using the opening and closing capitals for the period where capital equals net assets.

In practice, however, the accountant would compute income Y by compiling a statement of income. So, of what use is this formula? For reasons to be discussed later, the economist finds use for the formula when it is amended to take account of what we call **present values**. Computed after the end of a financial year, it is the *ex post* measure of income.

Nature of traditional accounting value

As the values of assets still in service at the end of a financial period have been based on the unconsumed costs of such assets, they are the by-product of compiling the statement of income. These values have been fixed not by direct measurement, but simply by an assessment of costs consumed in the process of generating period turnover. We can say, then, that the statement of financial position figure of net assets is a residual valuation after measuring income.

However, it is not a value in the sense of worth or market value as a buying price or selling price; it is merely a **value of unconsumed costs of assets**. This is an important point that will be encountered again later.

6.4 Critical comment on the accountant's measure

6.4.1 Virtues of the accountant's measure

As with the economist's, the accountant's measure is not without its virtues. These are invariably aspects of the historical cost concept, such as objectivity, being transaction based and being generally understood.

6.4.2 Faults of the accountant's measure

Principles of historical cost and profit realisation

The historical cost and profit realisation concepts are firmly entrenched in the transaction basis of accountancy. However, in practice, the two concepts are not free of adjustments. Because of such adjustments, some commentators argue that the system produces a heterogeneous mix of values and realised income items.[5]

For example, in the case of asset values, certain assets such as land and buildings may have a carrying figure in the statement of financial position based on a revaluation to market value, while other assets such as motor vehicles may still be based on a balance of unallocated cost. The statement of financial position thus pretends on the one hand to be a list of resultant costs pending allocation over future periods and on the other hand to be a statement of current values.

Prudence concept

This concept introduces caution into the recognition of assets and income for financial reporting purposes. The cardinal rule is that income should not be recorded or recognised within the system until it is realised, but unrealised losses should be recognised immediately.

However, not all unrealised profits are excluded. For example, practice is that attributable profit on long-term contracts still in progress at the financial year-end may be taken into account. As with non-current assets, rules are not applied uniformly.

Unrealised capital profits

Capital profits have been ignored as income until they are realised, when, in the accounting period of sale, they are acknowledged by the reporting system. This has meant that all the profit has been recognised in one financial period when, in truth, the surplus was generated over successive periods by gradual growth, albeit unrealised until disposal of the asset. Thus a portion of what are now realised profits applies to prior periods. Not all of this profit should be attributed to the period of sale. The introduction of the statement of comprehensive income has addressed this by including revaluation gains.

Going concern

The going concern concept is fundamental to accountancy and operates on the assumption that the business entity has an indefinite life. It is used to justify basing the periodic reports of asset values on carrying forward figures that represent unallocated costs, i.e. to justify the non-recognition of the realisable or disposal values of non-monetary assets and, in so doing, the associated unrealised profits/losses. Although the life of an entity is deemed indefinite, there is uncertainty, and accountants are reluctant to predict the future. When they are matching costs with revenue for the current accounting period, they follow the prudence concept of reasonable certainty.

In the long term, economic income and accountancy income are reconciled. The unrealised profits of the economic measure are eventually realised and, at that point, they will be recognised by the accountant's measure. In the short term, however, they give different results for each period.

What if we cannot assume that a business will continue as a going concern?

There may be circumstances, as in the case of HMV which in 2012 warned that, following falling sales, there was a material uncertainty on being able to continue as a going concern. The uncertainty may be reduced by showing that active steps are being taken such as introducing new sales initiatives, restructuring, cost reduction and raising additional share capital which will ensure the survival of the business. If survival is not possible, the business will prepare its accounts using net realisable values, which are discussed in the next chapter.

The key considerations for shareholders are whether there will be sufficient profits to support dividend distributions and whether they will be able to continue to dispose of their shares in the open market. The key consideration for the directors is whether there will be sufficient cash to allow the business to trade profitably. We can see all these considerations being addressed in the following extract from the 2011 Annual Report of Grontmij N.V.

> **Going concern**
> the Group faced declines in its operating results during 2011 and was unable to meet its original debt covenant ratios . . . the Company obtained a waiver . . . met the covenant levels set by the waiver . . . a deferral was granted . . . apparent that a redesign of the capital structure of the Company is required to sustain the operations of the Company in the long term . . . after a financial review by the management, it was concluded that the capital structure of the Company should consist of a committed credit facility agreement and additional equity ('the rights issue') . . . the Company reached, in principle, agreement with its major shareholders and the banks . . . as a consequence of the above, the 2011 financial statements are prepared on a going concern basis. The Company does, however, draw attention to the fact that the ability to continue as a going concern is dependent on the continuing support of its shareholders and banks . . .

6.5 Economist's view of income, capital and value

Let us now consider the economist's tradition of present value and the nature of economic income.

6.5.1 Economist's tradition of present value

Present value is a technique used in valuing a future money flow, or in measuring the money value of an existing capital stock in terms of a predicted cash flow *ad infinitum*.

Present value (PV) constitutes the nature of economic capital and, indirectly, economic income. Given the choice of receiving £100 now or £100 in one year's time, the rational person will opt to receive £100 now. This behaviour exhibits an intuitive appreciation of the fact that £100 today is worth more than £100 one year hence. Thus the mind has **discounted** the value of the future sum: £100 today is worth £100; but compared with today, i.e. **compared with present value**, a similar sum receivable in twelve months' time is worth less than £100. How much less is a matter of subjective evaluation, but compensation for the time element may be found by reference to interest: a person forgoing the spending of £1 today and spending it one year later may earn interest of, say, 10% per annum in compensation for the sacrifice undergone by deferring consumption.

So £1 today invested at 10% p.a. will be worth £1.10 one year later, £1.21 two years later, £1.331 three years later, and so on. This is the concept of compound interest. It may be calculated by the formula $(1 + r)^n$, where 1 = the sum invested; r = the rate of interest; and n = the number of periods of investment (in our case years). So for £1 invested at 10% p.a. for four years:

$$(1 + r)^n = (1 + 0.10)^4$$
$$= (1.1)^4$$
$$= £1.4641$$

and for five years:

$$= (1.1)^5$$
$$= £1.6105, \text{ and so on.}$$

Notice how the **future value** increases because of the compound interest element – it **varies** over time – whereas the investment of £1 remains constant. So, conversely, the sum of £1.10 received at the end of year 1 has a PV of £1, as does £1.21 received at the end of year 2 and £1.331 at the end of year 3.

It has been found convenient to construct tables to ease the task of calculating present values. These show the cash flow, i.e. the future values, at a constant figure of £1 and allow the investment to vary. So:

$$PV = \frac{CF}{(1+r)^n}$$

where CF = anticipated cash flow; and r = the discount (i.e. interest) rate. So the PV of a cash flow of £1 receivable at the end of one year at 10% p.a. is:

$$\frac{£1}{(1+r)^1} = £0.9091$$

and of £1 at the end of two years:

$$\frac{£1}{(1+r)^2} = £0.8264$$

and so on over successive years. The appropriate present values for years 3, 4 and 5 would be £0.7513, £0.6830, £0.6209 respectively.

£0.9091 invested today at 10% p.a. will produce £1 at the end of one year. The PV of £1 receivable at the end of two years is £0.8264 and so on.

Tables presenting data in this way are called 'PV tables', while the earlier method compiles tables usually referred to as 'compound interest tables'. Both types of table are compound interest tables; only the presentation of the data has changed.

To illustrate the ease of computation using PV tables, we can compute the PV of £6,152 receivable at the end of year 5, given a discount rate of 10%, as being £6,152 × £0.6209 = £3,820. Thus £3,820 will total £6,152 in five years given an interest rate of 10% p.a. So the PV of that cash flow of £6,152 is £3,820, because £3,820 would generate interest of £2,332 (i.e. 6,152 – 3,820) as compensation for losing use of the principal sum for five years. Future flows must be discounted to take cognisance of the time element separating cash flows. Only then are we able to compare like with like by reducing all future flows to the comparable loss of present value.

Figure 6.1 Dissimilar cash flows

Cash flows		
Machine A	Machine B	Receivable end of year
£	£	
1,000	5,000	1
2,000	4,000	2
7,000	1,000	3
10,000	10,000	

This concept of PV has a variety of applications in accountancy and will be encountered in many different areas requiring financial measurement, comparison and decision. It originated as an economist's device within the context of economic income and economic capital models, but in accountancy it assists in the making of valid comparisons and decisions. For example, two machines may each generate an income of £10,000 over three years. However, timing of the cash flows may vary between the machines. This is illustrated in Figure 6.1.

If we simply compare the profit-generating capacity of the machines over the three-year span, each produces a total profit of £10,000. But if we pay regard to the time element of the money flows, the machines are not so equal.

However, the technique has its faults. Future money flows are invariably the subject of **estimation** and thus the actual flow experienced may show variations from forecast. Also, the element of **interest**, which is crucial to the calculation of present values, is **subjective**. It may, for instance, be taken as the average prevailing rate operating within the economy or a rate peculiar to the firm and the element of risk involved in the particular decision. In this chapter we are concerned only with PV as a tool of the economist in evaluating economic income and economic capital.

6.5.2 Nature of economic income

Economics is concerned with the economy in general, raising questions such as: how does it function? how is wealth created? how is income generated? why is income generated? The economy as a whole is activated by income generation. The individual is motivated to generate income because of a need to satisfy personal wants by consuming goods and services. Thus the economist becomes concerned with the individual consumer's psychological state of personal **enjoyment and satisfaction**. This creates a need to treat the economy as a **behavioural entity**.

The behavioural aspect forms a substantial part of micro- and macroeconomic thought, emanating particularly from the microeconomic. We can say that the economist's version of income measurement is microeconomics-orientated in contrast to the accountant's business entity orientation.

The origination of the economic measure of income commenced with Irving Fisher in 1930.[6] He saw income in terms of consumption, and consumption in terms of individual perception of personal enjoyment and satisfaction. His difficulty in formulating a standard measure of this personal psychological concept of income was overcome by equating this individual experience with the consumption of goods and services and assuming that the cost of such goods and services formed the measure.

Thus, he reasoned, consumption (C) equals income (Y); so $Y = C$. He excluded savings from income because savings were not consumed. There was no satisfaction derived from savings; enjoyment necessitated consumption, he argued. Money was worthless until spent; so growth of capital was ignored, but reductions in capital became part of income because such reductions had to be spent.

In Fisher's model, capital was a stock of wealth existing at a point in time, and as a stock it generated income. Eventually, he reconciled the value of capital with the value of income by employing the concept of present value. He assessed the PV of a future flow of income by **discounting** future flows using the discounted cash flow (DCF) technique. Fisher's model adopted the prevailing average market rate of interest as the discount factor.

Economists since Fisher have introduced savings as part of income. Sir John Hicks played a major role in this area.[7] He introduced the idea that income was the maximum consumption enjoyed by the individual without reducing the individual's capital stock, i.e. the amount a person could consume during a period of time that still left him or her with the same value of capital stock at the end of the period as at the beginning. Hicks also used the DCF technique in the valuation of capital.

If capital increases, the increase constitutes savings and grants the opportunity of consumption. The formula illustrating this was given in Section 6.3.1, i.e. $Y_{0-1} = NA_1 - NA_0 + D_{0-1}$.

However, in the Hicksian model, $NA_1 - NA_0$, given as £6,750 and £6,000 respectively in that section, would have been discounted to achieve present values.

The same formula may be expressed in different forms. The economist is likely to show it as $Y = C + (K_1 - K_0)$ where C = consumption, having been substituted for dividend, and K_1 and K_0 have been substituted for NA_1 and NA_0 respectively.

Hicks's income model

Hicks's income model is often spoken of as an *ex ante* model because it is usually used for the measurement of **expected** income in advance of the time period concerned. Of course, because it specifically introduces the present value concept, present values replace the statement of financial position values of net assets adopted by the accountant. Measuring income **before the event** enables the individual to estimate the level of consumption that may be achieved without depleting capital stock. Before-the-event computations of income necessitate predictions of future cash flows.

Suppose that an individual proprietor of a business anticipated that his investment in the enterprise would generate earnings over the next four years as specified in Figure 6.2. Furthermore, such earnings would be retained by the business for the financing of new equipment with a view to increasing potential output.

We will assume that the expected rate of interest on capital employed in the business is 8% p.a.

Figure 6.2 Business cash flows for four years

Years	Cash inflows £
1	26,000
2	29,000
3	35,000
4	41,000

Figure 6.3 Economic value at K_0

Year	(a) Cash flow	(b) $DCF = \dfrac{1}{(1+r)^n}$	(c) $PV = (a) \times (b)$
	£		£
K_1	26,000	$\dfrac{1}{(1.08)^1} = 0.9259$	24,073
K_2	29,000	$\dfrac{1}{(1.08)^2} = 0.8573$	24,862
K_3	35,000	$\dfrac{1}{(1.08)^3} = 0.7938$	27,783
K_4	41,000	$\dfrac{1}{(1.08)^4} = 0.7350$	30,135
	131,000		106,853

The economic value of the business at K_0 (i.e. at the beginning of year 1) will be based on the discounted cash flow of the future four years. Figure 6.3 shows that K_0 is £106,853, calculated as the present value of anticipated earnings of £131,000 spread over a four-year term.

The economic value of the business at K_1 (i.e. at the end of year 1, which is the same as saying the beginning of year 2) is calculated in Figure 6.4. This shows that K_1 is £115,403 calculated as the present value of anticipated earnings of £131,000 spread over a four-year term.

From this information we are able to calculate Y for the period Y_1, as in Figure 6.5. Note that C (consumption) is nil because, in this exercise, dividends representing consumption have not been payable for Y_1. In other words, income Y_1 is entirely in the form of projected capital growth, i.e. savings.

By year-end K_1, earnings of £26,000 will have been received; in projecting the capital at K_2 such earnings will have been reinvested and at the beginning of year K_2 will have a PV of £26,000. These earnings will no longer represent a **predicted** sum because they will have been **realised** and therefore will no longer be subjected to discounting.

Figure 6.4 Economic value at K_1

Year	(a) Cash flow	(b) $DCF = \dfrac{1}{(1+r)^n}$	(c) $\dfrac{PV}{(a) \times (b)}$
	£		£
K_1	26,000	1.0000	26,000
K_2	29,000	$\dfrac{1}{(1+r)^1} = 0.9259$	26,851
K_3	35,000	$\dfrac{1}{(1+r)^2} = 0.8573$	30,006
K_4	41,000	$\dfrac{1}{(1+r)^3} = 0.7938$	32,546
	131,000		115,403

Figure 6.5 Calculation of Y for the period Y_1

$$Y = C + (K_1 - K_0)$$
$$Y = 0 + (115,403 - 106,853)$$
$$= 0 + 8,550$$
$$= £8,550$$

The income of £8,550 represents an anticipated return of 8% p.a. on the economic capital at K_0 of £106,853 (8% of £106,853 is £8,548, the difference of £2 between this figure and the figure calculated above being caused by rounding).

As long as the expectations of future cash flows and the chosen interest rate do not change, then Y_1 will equal 8% of £106,853.

What will the anticipated income for the year Y_2 amount to?

Applying the principle explained above, the anticipated income for the year Y_2 will equal 8% of the capital at the end of K_1 amounting to £115,403 = £9,233. This is demonstrated in Figure 6.6, which shows that K_2 is £124,636 calculated as the present value of anticipated earnings of £131,000 spread over a four-year term.

From this information we are able to calculate Y for the period Y_2 as in Figure 6.7. Note that capital value attributable to the end of year K_2 is being assessed at the beginning of K_2. This means that the £26,000 due at the end of year K_1 will have been received and reinvested, earning interest of 8% p.a. Thus by the end of year K_2 it will be worth £28,080. The sum of £29,000 will be realised at the end of year K_2 so its present value at that time will be £29,000.

If the anticipated future cash flows change, the expected capital value at the successive points in time will also change. Accordingly, the actual value of capital may vary from that forecast by the *ex ante* model.

Figure 6.6 Economic value at K_2

Year	(a) Cash flow	(b) $DCF = \dfrac{1}{(1+r)^n}$	(c) $PV = (a) \times (b)$
	£	£	£
K_1	26,000	1.08	28,080
K_2	29,000	1.0000	29,000
K_3	35,000	0.9259	32,407
K_4	41,000	0.8573	35,149
	131,000		124,636

Figure 6.7 Calculation of Y for the period Y_2

$$Y = C + (K_2 - K_1)$$
$$Y = 0 + (124,636 - 115,403)$$
$$= 0 + 9,233$$
$$= £9,233$$

6.6 Critical comment on the economist's measure

While the income measure enables us to formulate theories regarding the behaviour of the economy, it has inherent shortcomings not only in the economic field but particularly in the accountancy sphere.

● The calculation of economic capital, hence economic income, is subjective in terms of the present value factor, often referred to as the DCF element. The factor may be based on any one of a number of factors, such as opportunity cost, the current return on the firm's existing capital employed, the contemporary interest payable on a short-term loan such as a bank overdraft, the average going rate of interest payable in the economy at large, or a rate considered justified on the basis of the risk attached to a particular investment.

● Investors are not of one mind or one outlook. For example, they possess different risk and time preferences and will therefore employ different discount factors.

● The model constitutes a compound of unrealised and realised flows, i.e. profits. Because of the unrealised element, it has not been used as a base for computing tax or for declaring a dividend.

● The projected income is dependent upon the success of a planned financial strategy. Investment plans may change, or fail to attain target.

● Windfall gains cannot be foreseen, so they cannot be accommodated in the *ex ante* model. Our prognostic cash flows may therefore vary from the actual flows generated, e.g. an unexpected price movement.

● It is difficult to construct a satisfactory, meaningful statement of financial position detailing the unused stock of net assets by determining the present values of individual assets. Income is invariably the consequence of deploying a group of assets working in unison.

6.7 Income, capital and changing price levels

A primary concern of income measurement to both economist and accountant is the maintenance of the capital stock, i.e. the maintenance of capital values. The assumption is that income can only arise **after** the capital stock has been maintained at the same amount as at the beginning of the accounting period.

However, this raises the question of how we should define the capital that we are attempting to maintain. There are a number of possible definitions:

● **Money capital**. Should we concern ourselves with maintaining the fund of capital resources initially injected by the entrepreneur into the new enterprise? This is indeed one of the aims of traditional, transaction-based accountancy.

● **Potential consumption capital**. Is it this that should be maintained, i.e. the economist's present value philosophy expressed via the discounted cash flow technique?

● **Operating capacity capital**. Should maintenance of productive capacity be the rule, i.e. capital measured in terms of tangible or physical assets? This measure would utilise the current cost accounting system.

Revsine attempted to construct an analytical bridge between replacement cost accounting that maintains the operating capacity, and the economic concepts of **income** and **value**, by demonstrating that the distributable operating flow component of economic income is equal to the current operating component of replacement cost income, and that the unexpected income component of economic income is equal to the unrealisable cost savings of replacement cost income.[4] This will become clearer when the replacement cost model is dealt with in the next chapter.

- **Financial capital**. Should capital be maintained in terms of a fund of general purchasing power (sometimes called 'real' capital)? In essence, this is the consumer purchasing power (or general purchasing power) approach, but not in a strict sense, as it can be measured in a variety of ways. The basic method uses a general price index. This concept is likely to satisfy the criteria of the proprietor/shareholders of the entity. The money capital and the financial capital concepts are variations of the same theme, the former being founded on the historical cost principle and the latter applying an adjustment mechanism to take account of changing price levels.

The money capital concept has remained the foundation stone of traditional accountancy reporting, but the operating and financial capital alternatives have played a controversial secondary role over the past 25 years.

Potential consumption capital is peculiar to economics in terms of measurement of the business entity's aggregate capital, although, as discussed in Section 6.5.2, it has a major role to play as a decision-making model in financial management.

6.7.1 Why are these varying methods of concern?

The problem tackled by these devices is that plague of the economy known as 'changing price levels', particularly the upward spiralling referred to as **inflation**. Throughout this chapter we have assumed that there is a stable monetary unit and that income, capital and value changes over time have been in response to operational activity and the interaction of supply and demand or changes in expectations.

Following the historical cost convention, capital maintenance has involved a comparison of opening and closing capital in each accounting period. It has been assumed that the purchasing power of money has remained constant over time.

If we take into account moving price levels, particularly the fall in the purchasing power of the monetary unit due to inflation, then our measure of **income** is affected if we insist upon **maintaining capital in real terms**.

6.7.2 Is it necessary to maintain capital in real terms?

Undoubtedly it is necessary if we wish to prevent an erosion of the operating capacity of the entity and thus its ability to maintain real levels of income. If we do not maintain the capacity of capital to generate the current level of profit, then the income measure, being the difference between opening and closing capitals, will be overstated or overvalued. This is because the capital measure is being understated or undervalued. In other words, there is a danger of dividends being paid out of real capital rather than out of real income. It follows that, if the need to retain profits is overlooked, the physical assets will be depleted.

In accountancy there is no theoretical difficulty in measuring the impact of changing price levels. There are, however, two practical difficulties:

- A number of methods, or mixes of methods, are available and it has proved impossible to obtain consensus support for one method or compound of methods.

- There is a high element of subjectivity, which detracts from the objectivity of the information.

In the next chapter we deal with inflation and analyse the methods formulated, together with the difficulties that they in turn introduce into the financial reporting system.

Summary

In measuring income, capital and value, the accountant's approach varies from the sister discipline of the economist, yet both are trying to achieve similar objectives.

The accountant uses a traditional transaction-based model of computing income, capital being the residual of this model.

The economist's viewpoint is anchored in a behavioural philosophy that measures capital and deduces income to be the difference between the capital at the beginning of a period and that at its end.

The objectives of income measurement are important because of the existence of a highly sophisticated capital market. These objectives involve the assessment of stewardship performance, dividend and retention policies, comparison of actual results with those predicted, assessment of future prospects, payment of taxation and disclosure of matched costs against revenue from sales.

The natures of income, capital and value must be appreciated if we are to understand and achieve measurement. The apparent conflict between the two measures can be seen as a consequence of the accountant's need for periodic reporting to shareholders. In the longer term, both methods tend to agree.

Present value as a concept is the foundation stone of the economist, while historical cost, adjusted for prudence, is that of the accountant. Present value demands a subjective discount rate and estimates that time may prove incorrect; historical cost ignores unrealised profits and in application is not always transaction based.

The economist's measure, of undoubted value in the world of micro- and macroeconomics, presents difficulty in the accountancy world of annual reports. The accountant's method, with its long track record of acceptance, ignores any generated profits, which caution and the concept of the going concern deem not to exist.

The economic trauma of changing price levels is a problem that both measures can embrace, but consensus support for a particular model of measurement has proved elusive.

REVIEW QUESTIONS

1 What is the purpose of measuring income?

2 Explain the nature of economic income.

3 The historical cost concept has withstood the test of time. Specify the reasons for this success, together with any aspects of historical cost that you consider are detrimental in the sphere of financial reporting.

4 What is meant by present value? Does it take account of inflation?

5 Explain what you understand by an *ex ante* model.

6 Explain the principal criticisms of the economist's measure of income.

7 To an accountant, net income is essentially a historical record of the past. To an economist, net income is essentially a speculation about the future. Examine the relative merits of these two approaches for financial reporting purposes.

8 Examine and contrast the concepts of profit that you consider to be relevant to:

(a) an economist;

(b) a speculator;

(c) a business executive;

(d) the managing director of a company;

(e) a shareholder in a private company;

(f) a shareholder in a large public company.

EXERCISES

* Question 1

(a) 'Measurement in financial statements', Chapter 6 of the ASB's *Statement of Principles*, was published in 1999. Among the theoretical valuation systems considered is value in use, more commonly known as economic value.

Required:

Describe the Hicksian economic model of income and value, and assess its usefulness for financial reporting.

(b) Jim Bowater purchased a parcel of 30,000 ordinary shares in New Technologies plc for £36,000 on 1 January 20X5. Jim, an Australian on a four-year contract in the UK, has it in mind to sell the shares at the end of 20X7, just before he leaves for Australia. Based on the company's forecast growth and dividend policy, his broker has advised him that his shares are likely to fetch only £35,000 then.

In its annual report for the year ended 31 December 20X4 the company had forecast annual dividend payouts as follows:

Year ended: 31 December 20X5, 25p per share
31 December 20X6, 20p per share
31 December 20X7, 20p per share

Required:

Using the economic model of income:

(i) Compute Jim's economic income for each of the three years ending on the dates indicated above.

(ii) Show that Jim's economic capital will be preserved at 1 January 20X5 level. Jim's cost of capital is 20%.

* Question 2

(a) Describe briefly the theory underlying Hicks's economic model of income and capital. What are its practical limitations?

(b) Spock purchased a Space Invader entertainment machine at the beginning of year 1 for £1,000. He expects to receive at annual intervals the following receipts: at the end of year 1 £400; at end of year 2 £500; at end of year 3 £600. At the end of year 3 he expects to sell the machine for £400.

Spock could receive a return of 10% in the next best investment.

The present value of £1 receivable at the end of a period discounted at 10% is as follows:

End of year 1   £0.909
End of year 2   £0.826
End of year 3   £0.751

Required:
Calculate the ideal economic income, ignoring taxation and working to the nearest £.

Your answer should show that Spock's capital is maintained throughout the period and that his income is constant.

Question 3

Jason commenced with £135,000 cash. He acquired an established shop on 1 January 20X1. He agreed to pay £130,000 for the fixed and current assets and the goodwill. The replacement cost of the shop premises was £100,000, stock £10,000 and debtors £4,000; the balance of the purchase price was for the goodwill. He paid legal costs of £5,000. No liabilities were taken over. Jason could have resold the business immediately for £135,000. Legal costs are to be expensed in 20X1.

Jason expected to draw £25,000 per year from the business for three years and to sell the shop at the end of 20X3 for £150,000.

At 31 December 20X1 the books showed the following tangible assets and liabilities:

Cost to the business before any drawings by Jason:		He estimated that the net realisable values were:	
	£		£
Shop premises	100,000		85,000
Stock	15,500		20,000
Debtors	5,200		5,200
Cash	40,000		40,000
Creditors	5,000		5,000

Based on his experience of the first year's trading, he revised his estimates and expected to draw £35,000 per year for three years and sell the shop for £175,000 on 31 December 20X3.

Jason's opportunity cost of capital was 20%.

Required:
(a) Calculate the following income figures for 20X1:
   (i)   accounting income;
   (ii) income based on net realisable values;
   (iii) economic income *ex ante*;
   (iv) economic income *ex post*.
State any assumptions made.
(b) Evaluate each of the four income figures as indicators of performance in 20X1 and as a guide to decisions about the future.

References

1 T.A. Lee, *Income and Value Measurement: Theory and Practice* (3rd edition), Van Nostrand Reinhold (UK), 1985, p. 20.
2 D. Solomons, *Making Accounting Policy*, Oxford University Press, 1986, p. 132.
3 R.W. Scapens, *Accounting in an Inflationary Environment* (2nd edition), Macmillan, 1981, p. 125.
4 Ibid., p. 127.
5 T.A. Lee, *op. cit.*, pp. 52–54.
6 I. Fisher, *The Theory of Interest*, Macmillan, 1930, pp. 171–81.
7 J.R. Hicks, *Value and Capital* (2nd edition), Clarendon Press, 1946.

Bibliography

American Institute of Certified Public Accountants, *Objectives of Financial Statements*, Report of the Study Group, 1973.

The Corporate Report, ASC, 1975, pp. 28–31.

N. Kaldor, 'The concept of income in economic theory', in R.H. Parker and G.C. Harcourt (eds), *Readings in the Concept and Measurement of Income*, Cambridge University Press, 1969.

T.A. Lee, 'The accounting entity concept, accounting standards and inflation accounting', *Accounting and Business Research*, Spring 1980, pp. 1–11.

J.R. Little, 'Income measurement: an introduction', *Student Newsletter*, June 1988.

D. Solomons, 'Economic and accounting concepts of income', in R.H. Parker and G.C. Harcourt (eds), *Readings in the Concept and Measurement of Income*, Cambridge University Press, 1969.

R.R. Sterling, *Theory of the Measurement of Enterprise Income*, University of Kansas Press, 1970.

CHAPTER 7

Accounting for price-level changes

7.1 Introduction

The main purpose of this chapter is to explain the impact of inflation on profit and capital measurement and the concepts that have been proposed to incorporate the effect into financial reports by adjusting the historical cost data. These concepts are periodically discussed but there is no general support for any specific concept among practitioners in the field.

Objectives

By the end of the chapter, you should be able to:

- describe the problems of historical cost accounting (HCA);
- explain the approach taken in each of the price-level changing models;
- prepare financial statements applying each model (HCA, CPP, CCA, NRVA);
- critically comment on each model (HCA, CPP, CCA, NRVA);
- describe the approach being taken by standard setters and future developments.

7.2 Review of the problems of historical cost accounting (HCA)

The transaction-based historical cost concept was unchallenged in the UK until price levels started to hedge upwards at an ever-increasing pace during the 1950s and reached an annual rate of increase of 20% in the mid-1970s. The historical cost base for financial reporting witnessed growing criticism. The inherent faults of the system were discussed in Chapter 6, but inflation exacerbates the problem in the following ways:

- Profit is overstated when inflationary changes in the value of assets are ignored.

- Comparability of business entities, which is so necessary in the assessment of performance and growth, becomes distorted when assets are acquired at different times.

- The decision-making process, the formulation of plans and the setting of targets may be suboptimal if financial base data are out of date.

- Financial reports become confusing at best, misleading at worst, because revenue is mismatched with differing historical cost levels as the monetary unit becomes unstable.

- Unrealised profits arising in individual accounting periods are increased as a result of inflation.

In order to combat these serious defects, current value accounting became the subject of research and controversy as to the most appropriate method to use for financial reporting.

7.3 Inflation accounting

A number of versions of current value accounting (CVA) were eventually identified, but the current value postulate was said to suffer from the following disadvantages:

- It destroys the factual nature of HCA, which is transaction-based: the factual characteristic is to all intents and purposes lost as transaction-based historic values are replaced by judgemental values.
- It is not as objective as HCA because it is less verifiable from auditable documentation.
- It entails recognition of unrealised profit, a practice that is anathema to the traditionalist.
- The claimed improvement in comparability between commercial entities is a myth because of the degree of subjectivity in measuring current value by each.
- The lack of a single accepted method of computing current values compounds the subjectivity aspect. One fault-laden system is being usurped by another that is also faulty.

In spite of these criticisms, the search for a system of financial reporting devoid of the defects of HCA and capable of coping with inflation has produced a number of CVA models.

7.4 The concepts in principle

Several current income and value models have been proposed to replace or operate in tandem with the historical cost convention. However, in terms of basic characteristics, they may be reduced to the following three models:

- current purchasing power (CPP) or general purchasing power (GPP);
- current entry cost or replacement cost (RC);
- current exit cost or net realisable value (NRV).

We discuss each of these models below.

7.4.1 Current purchasing power accounting (CPPA)

The CPP model measures income and value by adopting a price index system. Movements in price levels are gauged by reference to price changes in a group of goods and services in **general** use within the economy. The aggregate price value of this **basket** of commodities-cum-services is determined at a base point in time and indexed as 100. Subsequent changes in price are compared on a regular basis with this base period price and the change recorded. For example, the price level of our chosen range of goods and services may amount to £76 on 31 March 20X1, and show changes as follows:

£76	at 31 March 20X1
£79	at 30 April 20X1
£81	at 31 May 20X1
£84	at 30 June 20X1

and so on.

The change in price may be indexed with 31 March as the base:

20X1	Calculation	Index
31 March	i.e. £76	100
30 April	i.e. $\frac{79}{76} \times 100$	103.9
31 May	i.e. $\frac{80}{76} \times 100$	106.6
30 June	i.e. $\frac{84}{76} \times 100$	110.5

In the UK, index systems similar in construction to this are known as the Retail or Consumer Price Index (RPI). The index is a barometer of fluctuating price levels covering a miscellany of goods and services as used by the average household. Thus it is a **general** price index. It is amended from time to time to take account of new commodities entering the consumer's range of choice and needs. As a model, it is unique owing to the introduction of the concept of gains and losses in **purchasing power**.

7.4.2 Current entry or replacement cost accounting (RCA)

The replacement cost (RC) model assesses income and value by reference to entry costs or current replacement costs of materials and other assets utilised within the business entity. The valuation attempts to replace like with like and thus takes account of the quality and condition of the existing assets. A motor vehicle, for instance, may have been purchased brand new for £25,000 with an expected life of five years, an anticipated residual value of nil and a straight-line depreciation policy. Its HCA carrying value in the statement of financial position at the end of its first year would be £25,000 less £5,000 = £20,000. However, if a similar new replacement vehicle cost £30,000 at the end of year 1, then its gross RC would be £30,000; depreciation for one year based on this sum would be £6,000 and the net RC would be £24,000. The increase of £4,000 is a holding gain and the vehicle with an HCA carrying value of £20,000 would be revalued at £24,000.

7.4.3 Current exit cost or net realisable value accounting (NRVA)

The net realisable value (NRV) model is based on the economist's concept of opportunity cost. It is a model that has had strong academic support, most notably in Australia from Professor Ray Chambers who referred to this approach as Continuous Contemporary Accounting (CoCoA). If an asset cost £25,000 at the beginning of year 1 and at the end of that year it had an NRV of £21,000 after meeting selling expenses, it would be carried in the NRV statement of financial position at £21,000. This amount represents the cash forgone by holding the asset, i.e. the opportunity of possessing cash of £21,000 has been sacrificed in favour of the asset. There is effectively a holding loss for the year of £25,000 less £21,000 = £4,000.

7.5 The four models illustrated for a company with cash purchases and sales

We will illustrate the effect on the profit and net assets of Entrepreneur Ltd.

Entrepreneur Ltd commenced business on 1 January 20X1 with a capital of £3,000 to buy and sell second-hand computers. The company purchased six computers on 1 January 20X1 for £500 each and sold three of the computers on 15 January for £900 each.

The following data are available for January 20X1:

	Retail Price Index	Replacement cost per computer £	Net realisable value £
1 January	100		
15 January	112	610	
31 January	130	700	900

The statements of income and financial position are set out in Figure 7.1 with the detailed workings in Figure 7.2.

7.5.1 Financial capital maintenance concept

HCA and CPP are both transaction-based models that apply the financial capital maintenance concept. This means that profit is the difference between the opening and closing net assets (expressed in HC £) or the opening and closing net assets (expressed in HC £ indexed for RPI changes) adjusted for any capital introduced or withdrawn during the month.

Figure 7.1 Trading account for the month ended 31 January 20X1

Statements of income for the month ended 31 January 20X1	HCA £		CPP CPP£		RCA £		NRVA £	
Sales	2,700	W1	3,134	W5	2,700	W1	2,700	W1
Opening inventory	—		—		—		—	
Purchases	3,000	W2	3,900	W6	3,000	W2	3,000	W2
Closing inventory	(1,500)	W3	(1,950)	W7	(1,500)	W3	(1,500)	W3
COSA	na		na		330	W10	na	
Cost of sales	1,500		1,950		1,830		1,500	
Holding gain	na		na		na		1,200	W15
Profit	**1,200**		**1,184**		**870**		**2,400**	

Statement of financial position as at 31 January 20X1	£		PCP£		£		£	
Current assets								
Inventory	1,500	W3	1,950	W7	2,100	W11	2,700	W14
Cash	2,700	W4	2,700		2,700		2,700	
Capital employed	**4,200**		**4,650**		**4,800**		**5,400**	
Capital	3,000		3,900	W8	3,000		3,000	
Holding gains								
On inventory consumed	na		na		330	W12		
On inventory in hand	na		na		600	W13		
Profit	1,200		1,184		870		2,400	
Loss on monetary items	na		(434)	W9	na		na	
	4,200		**4,650**		**4,800**		**5,400**	

na = not applicable

Figure 7.2 Workings (W)

HCA

W1 Sales	3 × £900 = £2,700	
W2 Purchases	6 × £500 = £3,000	
W3 Closing inventory	3 × £500 = £1,500	
W4 Cash	1 January 20X1 Capital	3,000
	1 January 20X1 Purchases	(3,000)
	1 January 20X1 Balance	nil
	15 January 20X1 Sales	
	3 × £900 =	£2,700
	31 January 20X1 Balance	£2,700

CPP

		CPP£
W5 Sales	£2,700 × 130/112 =	3,134
W6 Purchases	£3,000 × 130/100 =	3,900
W7 Closing inventory	£1,500 × 130/100 =	1,950
W8 Capital	£3,000 × 130/100 =	3,900

W9 Balance of cash was nil until 15 January when sales generated £2,700. This sum was held until 31 January during which period cash, a monetary item, lost purchasing power. The loss of purchasing power is measured by applying the general index to the cash held: £2,700 × 130/112 − £2,700 = CPP £434.

RCA

W10 Additional replacement cost of inventory consumed as at the date of sale is measured as a cost of sales adjustment (COSA). COSA is calculated as follows:

	3 × £610 =	1,830
Less:	3 × £500 =	1,500
	COSA	£330

W11 Closing inventory: 3 × £700 = £2,100

W12 Holding gains on inventory consumed: as for W10 = £330

W13 Inventory at replacement cost	= 3 × £700 = 2,100
Less: inventory at cost	= 3 × £500 = 1,500
Holding gains on closing inventory	£600

NRVA

W14 Closing inventory at net realisable value = 900 × 3 = £2,700

W15 3 × £900 =	2,700
3 × £500 =	1,500
Holding gain	£1,200

CPP adjustments

- All historical cost values are adjusted to a common index level for the month. In theory this can be the index applicable to any day of the financial period concerned. However, in practice it has been deemed preferable to use the last day of the period; thus the financial statements show the latest price level appertaining to the period.

- The application of a general price index as an adjusting factor results in the creation of an **alien** currency of **purchasing power**, which is used in place of sterling. Note, particularly, the impact on the entity's sales and capital compared with the other models. **Actual** sales shown on **invoices** will still read £2,700.

- Note the application of the concept of gain or loss on holding monetary items. In this example there is a monetary loss of CPP £434 as shown in Working 9 in Figure 7.2.

7.5.2 Operating capital maintenance concept

Under this concept capital is only maintained if sufficient income is retained to maintain the business entity's physical operating capacity, i.e. its ability to produce the existing level of goods or services. Profit is, therefore, the residual after increasing the cost of sales to the cost applicable at the date of sale.

- Basically, only two adjustments are involved: the additional replacement cost of inventory consumed and holding gains on closing inventories. However, in a comprehensive exercise an adjustment will be necessary regarding non-current assets and you will also encounter a gearing adjustment.

- Notice the concept of holding gains. This model introduces, in effect, unrealised profits in respect of closing inventories. The holding gain concerning inventory consumed at the time of sale has been realised and deducted from what would have been a profit of £1,200. The statement discloses profits of £870.

7.5.3 Capacity to adapt concept under the NRVA model

The HCA, CPP and RCA models have assumed that the business will continue as a going concern and only distribute realised profits after retaining sufficient profits to maintain either the financial or operating capital.

The NRVA concept is that a business has the capacity to realise its net assets at the end of each financial period and reinvest the proceeds and that the NRV accounts provide management with this information.

- This produces the same initial profit as HCA, namely £1,200, but a peculiarity of this system is that this realised profit is supplemented by **unrealised** profit generated by holding stocks. Under RCA accounting, such gains are shown in a separate account and are not treated as part of real income.

- This simple exercise has ignored the possibility of investment in non-current assets, thus depreciation is not involved. A reduction in the NRV of non-current assets at the end of a period compared with the beginning would be treated in a similar fashion to depreciation by being charged to the revenue account, and consequently profits would be reduced. An increase in the NRV of such assets would be included as part of the profit.

7.5.4 The four models compared

Dividend distribution

We can see from Figure 7.1 that if the business were to distribute the profit reported under HCA, CPP or NRVA the physical operating capacity of the business would be reduced and it would be paying dividends out of capital:

	HCA	CPP	RCA	NRVA
Realised profit	1,200	1,184	870	1,200
Unrealised profit	—	—	—	1,200
Profit for month	1,200	1,184	870	2,400

Shareholder orientation

The CPP model is shareholder-oriented in that it shows whether shareholders' funds are keeping pace with inflation by maintaining their purchasing power. Only CPP changes the value of the share capital.

Management orientation

The RCA model is management-oriented in that it identifies holding gains which represent the amounts required to be retained in order to simply maintain the operating capital.

RCA measures the impact of inflation on the individual firm, in terms of the change in price levels of its **raw materials and assets**, i.e. inflation peculiar to the company, whereas CPP measures general inflation in the economy as a whole. CPP may be meaningless in the case of an individual company. Consider a firm that carries a constant volume of stock valued at £100 in HCA terms. Now suppose that price levels double when measured by a general price index (GPI), so that its inventory is restated to £200 in a CPP system. If, however, the cost of that **particular** inventory has risen by 500%, then under the RCA model the value of the stock should be £500.

In the mid-1970s, when the accountancy profession was debating the problem of changing price-level measurement, the general price level had climbed by some 23% over a period during which petroleum-based products had risen by 500%.

7.6 Critique of each model

A critique of the various models may be formulated in terms of their characteristics and peculiarities as virtues and defects in application.

7.6.1 HCA

This model's virtues and defects have been discussed in Chapter 6 and earlier in this chapter.

7.6.2 CPP

Virtues

- It is an **objective measure** since it is still transaction-based, as with HCA, and the possibility of subjectivity is constrained if a GPI is used that has been constructed by a central agency such as a government department. This applies in the UK, where the Retail Price Index is currently published by the Office for National Statistics.

- It is a **measure of shareholders' capital** and that capital's maintenance in terms of purchasing power units. Profit is the residual value after maintaining the money value of capital funds, taking account of changing price levels. Thus it is a measure readily understood by the shareholder/user of the accounts. It can prevent payment of a dividend out of real capital as measured by GPPA.

- It **introduces the concept of monetary items** as distinct from non-monetary items and the attendant concepts of gains and losses in holding net monetary liabilities compared with holding net monetary assets. Such gains and losses are experienced on a disturbing scale in times of inflation. They are **real** gains and losses. The **basic RCA and NRV** models do not recognise such 'surpluses' and 'deficits'.

Defects

- It is **HCA-based but adjusted** to reflect general price movements. Thus it possesses the characteristics of HCA, good and bad, but with its values updated in the light of an arithmetic measure of general price changes. The major defect of becoming out of date is mitigated to a degree, but the impact of inflation on the entity's income and capital may be at variance with the rate of inflation affecting the economy in general.

- It may be **wrongly assumed that the CPP statement of financial position is a current value statement**. It is not a current value document because of the defects discussed above; in particular, asset values may be subject to a different rate of inflation than that reflected by the GPI.

- It **creates an alien unit of measurement** still labelled by the £ sign. Thus we have the HCA £ and the CPP £. They are different pounds: one is the *bona fide* pound, the other is a synthetic unit. This may not be fully appreciated or understood by the user when faced with the financial accounts for the recent accounting period.

- Its **concept of profit is dangerous**. It pretends to cater for changing prices, but at the same time it fails to provide for the additional costs of replacing stocks sold or additional depreciation due to the escalating replacement cost of assets. The inflation encountered by the business entity will not be the same as that encountered by the whole economy. Thus the maintenance of the CPP of shareholders' capital via this concept of profit is not the maintenance of the entity's operating capital in physical terms, i.e. its capacity to produce the same volume of goods and services. The use of CPP profit as a basis for decision making without regard to RCA profit can have disastrous consequences.

7.6.3 RCA

Virtues

- Its **unit of measurement** is the monetary unit and consequently it is understood and accepted by the user of accountancy reports. In contrast, the CPP system employs an artificial unit based on arithmetic relationships, which is different and thus unfamiliar.

- It **identifies and isolates holding gains** from operating income. Thus it can prevent the inadvertent distribution of dividends in excess of operating profit. It satisfies the prudence criterion of the traditional accountant and **maintains the physical operating capacity** of the entity.

- It introduces **realistic current values** of assets in the statement of financial position, thus making the statement of financial position a 'value' statement and consequently more meaningful to the user. This contrasts sharply with the statement of financial position as a list of unallocated carrying costs in the HCA system.

Defects

- It is a **subjective measure**, in that replacement costs are often necessarily based on estimates or assessments. It does not possess the factual characteristics of HCA. It is open to manipulation within constraints. Often it is based on index numbers which themselves may be based on a compound of prices of a mixture of similar commodities used as raw material or operating assets. This subjectivity is exacerbated in circumstances where rapid technological advance and innovation are involved in the potential new replacement asset, e.g. computers and printers.

- It **assumes replacement of assets** by being based on their replacement cost. Difficulties arise if such assets are not to be replaced by similar assets. Presumably, it will then be assumed that a replacement of equivalent value to the original will be deployed, however differently, as capital within the firm.

7.6.4 NRVA

Virtues

- It is a concept readily understood by the user. The value of any item invariably has two measures – a buying price and a selling price – and the twain do not usually meet. However, when considering the value of an **existing** possession, the owner instinctively considers its 'value' to be that in potential sale, i.e. NRV.

- It **avoids the need to estimate depreciation** and, in consequence, the attendant problems of assessing lifespan and residual values. Depreciation is treated as the arithmetic difference between the NRV at the end of a financial period and the NRV at its beginning.

- It is **based on opportunity cost** and so can be said to be more meaningful. It is the **sacrificial** cost of possessing an asset, which, it can be argued, is more authentic in terms of being a true or real cost. If the asset were not possessed, its cash equivalent would exist instead and that cash would be deployed in other opportunities. Therefore, NRV = cash = opportunity = cost.

Defects

- It is a **subjective measure** and in this respect it possesses the same major fault as RCA. It can be said to be less prudent than RCA because NRV will tend to be higher in some cases than RCA. For example, when valuing finished inventories, a profit content will be involved.

- **It is not a realistic measure** as most assets, except finished goods, are possessed in order to be utilised, not sold. Therefore, NRV is irrelevant.

- **It is not always determinable**. The assets concerned may be highly specialist and there may be no ready market by which a value can be easily assessed. Consequently, any particular value may be fictitious or erroneous, containing too high a holding gain or, indeed, too low a holding loss.

- **It violates the concept of the going concern**, which demands that the accounts are drafted on the basis that there is no intention to liquidate the entity. Admittedly, this concept was formulated with HCA in view, but the acceptance of NRV implies the possibility of a cessation of trading.

- It is less reliable and verifiable than HC.

- The statement of comprehensive income will report a more volatile profit if changes in NRV are taken to the statement of comprehensive income each year.

- The profit arising from the changes in NRV may not have been realised.

7.7 Operating capital maintenance – a comprehensive example

In Figure 7.1 we considered the effect of inflation on a cash business without fixed assets, credit customers or credit suppliers. In the following example, Economica plc, we now consider the effect where there are non-current assets and credit transactions.

The HCA statements of financial position as at 31 December 20X4 and 20X5 are set out in Figure 7.3 and index numbers required to restate the non-current assets, inventory and monetary items in Figure 7.4.

7.7.1 Restating the opening statement of financial position to current cost

The non-current assets and inventory are restated to their current cost as at the date of the opening statement as shown in W1 and W2 below. The increase from HC to CC represents an unrealised holding gain which is debited to the asset account and credited to a reserve account called a current cost reserve, as in W3 below.

Figure 7.3 Economica plc HCA statement of financial position

Statements of financial position as at 31 December on the basis of HCA				
		20X5		20X4
	£000	£000	£000	£000
Non-current assets:				
Cost	85,000		85,000	
Depreciation	34,000		25,500	
		51,000		59,500
Current assets:				
Inventory	25,500		17,000	
Trade receivables	34,000		23,375	
Cash and bank	17,000		1,875	
	76,500		42,250	
Current liabilities:				
Trade payables	25,500		17,000	
Income tax	8,500		4,250	
Dividend declared	5,000		4,000	
	39,000		25,250	
Net current assets	37,500		17,000	
Less: 8% debentures	11,000		11,000	
		26,500		6,000
		77,500		65,500
Share capital and reserves:				
Authorised and issued £1 ordinary shares		50,000		50,000
Share premium		1,500		1,500
Retained earnings		26,000		14,000
		77,500		65,500

Figure 7.4 Index data relating to Economica plc

1	Index numbers as prepared by the Office for National Statistics for non-current assets:

1 January 20X2	100
1 January 20X5	165
1 January 20X6	185
Average for 20X4	147
Average for 20X5	167

2 All non-current assets were acquired on 1 January 20X2. There were no further acquisitions or disposals during the four years ended 31 December 20X5.

3 Indices as prepared by the Office for National Statistics for inventories and monetary working capital adjustments were:

1 October 20X4	115
31 December 20X4	125
15 November 20X4	120
1 October 20X5	140
31 December 20X5	150
15 November 20X5	145
Average for 20X5	137.5

4 Three months' inventory is carried.

5 Depreciation: historical cost based on 10% p.a. straight-line with residual value of nil:

	£ HCA
20X4	8,500,000
20X5	8,500,000

The calculations are as follows. First we shall convert the HCA statement of financial position in Figure 7.3, as at 31 December 20X4, to the CCA basis, using the index data in Figure 7.4.

The **non-monetary items**, comprising the non-current assets and inventory, are converted and the converted amounts are taken to the CC statement and the increases taken to the current cost reserve, as follows.

(W1) Property, plant and equipment

	HCA £000	Index		CCA £000	Increase £000
Cost	85,000	$\times$	$\dfrac{165}{100}$	= 140,250	55,250
Depreciation	25,500	$\times$	$\dfrac{165}{100}$	= 42,075	16,575
	59,500			98,175	38,675

The CCA valuation at 31 December 20X4 shows a net increase in terms of numbers of pounds sterling of £38,675,000. The £59,500,000 in the HCA statement of financial position will be replaced in the CCA statement by £98,175,000.

(W2) Inventories

HCA		Index		CCA		Increase
£000				£000		£000
17,000	×	$\dfrac{125}{120}$	=	17,708	=	708

Note that Figure 7.4 specifies that three months' inventories are held. Thus on average they will have been purchased on 15 November 20X4, on the assumption that they have been acquired and consumed evenly throughout the calendar period. Hence, the index at the time of purchase would have been 120. The £17,000,000 in the HCA statement of financial position will be replaced in the CCA statement of financial position by £17,708,000.

(W3) Current cost reserve

The total increase in CCA carrying values for non-monetary items is £39,383,000, which will be credited to CC reserves in the CC statement. It comprises £38,675,000 on the non-current assets and £708,000 on the inventory.

Note that monetary items do not change by virtue of inflation. Purchasing power will be lost or gained, but the carrying values in the CCA statement will be identical to those in its HCA counterpart. We can now compile the CCA statement as at 31 December 20X4 – this will show net assets of £104,883,000.

7.7.2 Adjustments that affect the profit for the year

The statement of comprehensive income for the year ended 31 December 20X5 set out in Figure 7.5 discloses a profit before interest and tax of £26,350,000. We need to deduct realised holding gains from this profit to avoid the distribution of dividends that would reduce the operating capital. These deductions are a cost of sales adjustment (COSA), a depreciation adjustment (DA) and a monetary working capital adjustment (MWCA). The accounting treatment is to debit the statement of comprehensive income and credit the current cost reserve.

The adjustments are calculated as follows.

(W4) Cost of sales adjustment (COSA) using the average method

We will compute the cost of sales adjustment by using the average method. The average purchase price index for 20X5 is 137.5. If price increases have moved at an even pace throughout the period, this implies that consumption occurred, on average, at 30 June, the mid-point of the financial year.

	HCA		Adjustment		CCA		Difference
	£000				£000		£000
Opening inventory	17,000	×	$\dfrac{137.5}{120}$	=	19,479	=	2,479
Purchases	—		—		—		—
	17,000				19,479		
Closing inventory	(25,500)	×	$\dfrac{137.5}{145}$	=	24,181	=	1,319
	(8,500)				(4,702)		3,798

Figure 7.5 Economica plc HCA statement of comprehensive income

Statement of income for the year ended 31 December 20X5, on the basis of HCA

		20X5		20X4
		£000		£000
Turnover		42,500		38,250
Less: Cost of sales		(12,070)		(23,025)
Gross profit		30,430		15,225
Less: Distribution costs	2,460		2,210	
Less: Administrative expenses	1,620		1,540	
		(4,080)		(3,750)
Profit before interest and tax		26,350		11,475
Interest		(880)		(880)
Profit before tax		25,470		10,595
Income tax expense		(8,470)		(4,250)
Profit after tax		17,000		6,345
Dividend		(5,000)		(4,000)
Retentions		12,000		2,345
Balance b/f		14,000		11,655
Balance c/f		26,000		14,000
EPS		34p		13p

The impact of price changes on the cost of sales would be an increase of £3,798,000, causing a profit decrease of like amount and a current cost reserve increase of like amount.

(W5) Depreciation adjustment: average method

As assets are consumed throughout the year, the CCA depreciation charge should be based on average current costs.

	HCA		*Adjustment*		*CCA*		*Difference*
	£000				*£000*		*£000*
Depreciation	8,500	×	$\dfrac{167}{100}$	=	14,195	=	5,695

(W6) Monetary working capital adjustment (MWCA)

The objective is to transfer from the statement of comprehensive income to CC reserve the amount by which the need for monetary working capital (MWC) has increased due to rising price levels. The change in MWC from one statement of financial position to the next will be the consequence of a combination of changes in volume and escalating price movements. Volume change may be segregated from the price change by using an average index.

	20X5	20X4		Change
	£000	£000		£000
Trade receivables	34,000	23,375		
Trade payables	25,500	17,000		
MWC =	8,500	6,375	Overall change =	2,125

The MWC is now adjusted by the average index for the year. This adjustment will reveal the change in volume.

$$\left(8,500 \times \frac{137.5}{150}\right) - \left(6,375 \times \frac{137.5}{125}\right)$$

=	7,792	−	7,012	= Volume change	780

So price change = 1,345

The profit before interest and tax will be reduced as follows:

	£000	£000
Profit before interest and tax		26,350
Less:		
COSA (from W4)	(3,798)	
DA (from W5)	(5,695)	
MWCA (from W6)	(1,345)	
Current cost operating adjustments		(10,838)
Current cost operating profit		15,512

The adjustments will be credited to the current cost reserve.

7.7.3 Unrealised holding gains on non-monetary assets as at 31 December 20X5

The holding gains as at 31 December 20X4 were calculated in Section 7.7.1 above for non-current assets and inventory. A similar calculation is required to restate these at 20X5 current costs for the closing statement of financial position. The calculations are as in Working 7 below.

(W7) Non-monetary assets

(i) Holding gain on non-current assets

	£000
Revaluation at year-end	
Non-current assets at 1 January 20X5 (as W1) at CCA revaluation	140,250
CCA value at 31 December 20X5 = $140,250 \times \dfrac{185}{165}$ =	157,250
Revaluation holding gain for 20X5 to CC reserve in W8	17,000

This holding gain of £17,000,000 is transferred to CC reserves.

(ii) Backlog depreciation on non-current assets

	£000
CCA aggregate depreciation at 31 December 20X5 for CC statement of financial position	
$= \text{HCA } £34,000,000 \times \dfrac{185}{100} \text{ in CC statement of financial position}$	**62,900**
Less: CCA aggregate depreciation at 1 January 20X5 (as per W1 and statement of financial position at 1 January 20X5)	42,075
Being CCA depreciation as revealed between opening and closing statements of financial position	20,825
But CCA depreciation charged in revenue accounts (i.e. £8,500,000 in £HCA plus additional depreciation of £5,695,000 per W5) =	14,195
So total backlog depreciation to CC reserve in W8	6,630

	£000
The CCA value of non-current assets at 31 December 20X5:	
Gross CCA value (above)	157,250
Depreciation (above)	62,900
Net CCA carrying value in the CC statement of financial position in W8	94,350

This £6,630,000 is backlog depreciation for 20X5. Total backlog depreciation is not expensed (i.e. charged to revenue account) as an adjustment of HCA profit, but is charged against CCA reserves. The net effect is that the CC reserve will increase by £10,370,000, i.e. £17,000,000 – £6,630,000.

(iii) Inventory valuation at year-end

	CCA £000
CCA valuation at 31 December 20X5	
HCA £000 *CCA £000*	
$= 25,500 \times 150/145 = 26,379 = $ increase of	879
CCA valuation at 1 January 20X5 (per W2)	
$= 17,000 \times 125/120 = 17,708 = $ increase of	708
Inventory holding gain occurring during 20X5 to W8	171

7.7.4 Current cost statement of financial position as at 31 December 20X5

The current cost statement as at 31 December 20X5 now discloses non-current assets and inventory adjusted by index to their current cost and the retained profits reduced by the current cost operating adjustments. It appears as in Working 8 below.

(W8) Economica plc: CCA statement of financial position as at 31 December 20X5

	£000		20X5 £000	£000	20X4 £000
Non-current assets					
Cost	157,250	(W7(i))		140,250 (W1)	
Depreciation	62,900	(W7(ii))		42,075 (W1)	
			94,350 (W7(ii))		98,175
Current assets					
Inventory	26,379	(W7(iii))		17,708 (W2)	
Trade receivables	34,000			23,375	
Cash	17,000			1,875	
	77,379			42,958	
Current liabilities					
Trade payables	25,500			17,000	
Income tax	8,500			4,250	
Dividend declared	5,000			4,000	
	39,000			25,250	
Net current assets	38,379			17,708	
Less: 8% debentures	11,000			11,000	
			27,379		6,708
			121,729		104,883
Financed by					
Share capital: authorised					
and issued £1 shares			50,000		50,000
Share premium			1,500		1,500
CC reserve (Note 1)			55,067		39,383
Retained profit (Note 2)			15,162		14,000
Shareholders' funds			121,729		104,883

Note 1: **CC reserve**	£000		£000
Opening balance			39,383 (W3)
Holding gains			
Non-current assets	17,000	(W7(i))	
Inventory	171	(W7(iii))	
			17,171
COSA	3,798	(W4)	
MWCA	1,345	(W6)	
Less: backlog depreciation	(6,630)	(W7(ii))	(1,487)
			55,067

Note 2: **Retained profit**			
Opening balance			14,000 (Figure 7.5)
HCA profit for 20X5	12,000		
COSA	(3,798)	(W4)	
Extra depreciation	(5,695)	(W5)	
MWCA	(1,345)	(W6)	
			1,162
CCA profit for 20X5			15,162

7.7.5 How to take the level of borrowings into account

We have assumed that the company will need to retain £10,838,000 from the current year's earnings in order to maintain the physical operating capacity of the company. However, if the business is part financed by borrowings then part of the amount required may be assumed to come from the lenders. One of the methods advocated is to make a gearing adjustment. The gearing adjustment that we illustrate here has the effect of reducing the impact of the adjustments on the profit after interest, i.e. it is based on the realised holding gains only.

The gearing adjustment will change the carrying figures of CC reserves and retained profit, but not the shareholders' funds, as the adjustment is compensating. The gearing adjustment cannot be computed before the determination of the shareholders' interest because that figure is necessary in order to complete the gearing calculation.

(W9) Gearing adjustment

The CC operating profit of the business is quantified after making such retentions from the historical profit as are required in order to maintain the physical operating capacity of the entity. However, from a shareholder standpoint, there is no need to maintain in real terms the portion of the entity financed by loans that are fixed in monetary values. Thus, in calculating profit attributable to shareholders, that part of the CC adjustments relating to the proportion of the business financed by loans can be deducted:

Gearing adjustment =

$$\frac{\text{Average net borrowings for year}}{\text{Average net borrowings for year} + \text{Average shareholders' funds for year}} \times \frac{\text{Aggregate}}{\text{adjustments}}$$

This formula is usually expressed as $\dfrac{L}{(L+S)} \times A$ where L = loans (i.e. net borrowings); S = shareholders' interest or funds; and A = adjustments (i.e. extra depreciation + COSA + MWCA). Note that $L/(L+S)$ is often expressed as a percentage of A (see example below where it is 6.31%).

Net borrowings

This is the sum of all liabilities less current assets, excluding items included in MWC or utilised in computing COSA. In this instance it is as follows.

Note: in some circumstances (e.g. new issue of debentures occurring during the year) a weighted average will be used.

	Closing balance £000	Opening balance £000
Debentures	11,000	11,000
Income tax	8,500	4,250
Cash	(17,000)	(1,875)
Total net borrowings, the average of which equals *L*	2,500	13,375

$$\text{Average net borrowings} = \frac{2,500,000 + 13,375,000}{2} = £7,937,500$$

Net borrowings plus shareholders' funds

Shareholders' funds in CC £ (inclusive of proposed dividends)	126,729	108,883
Add: net borrowings	2,500	13,375
	129,229	122,258

Or, alternatively:

	£000	£000
Non-current assets	94,350	98,175
Inventory	26,379	17,708
MWC	8,500	6,375
	129,229	122,258

$$\text{Average } L + S = \frac{129,229,000 + 122,258,000}{2} = 125,743,500$$

$$\text{So gearing } = \frac{L}{L+S} \times A = \frac{7,937,500}{125,743,500} \times \frac{(\text{COSA} + \text{MWCA} + \text{extra depreciation})}{(3,798,000 + 1,345,000 + 5,695,000)}$$

$$= 6.31\% \text{ of } £10,838,000 = £683,877, \text{ say } £684,000$$

Thus the CC adjustment of £10,838,000 charged against historical profit may be reduced by £684,000 due to a gain being derived from net borrowings during a period of inflation as shown in Figure 7.6. The £684,000 is shown as a deduction from interest payable.

Figure 7.6 Economica plc CCA statement of income

*Economica plc CCA statement of comprehensive income for year ended 31 December 20X5
(i.e. under the operating capital maintenance concept)*

		£000
Turnover		42,500
Cost of sales		(12,070)
Gross profit		30,430
Distribution costs		(2,460)
Administrative expenses		(1,620)
Historical cost operating profit		26,350
Current cost operating adjustments (from Section 7.7.2 above)		**(10,838)**
Current cost operating profit		15,512
Interest payable	(880)	
Gearing adjustment	**684**	(196)
Current profit on ordinary activities before taxation		15,316
Tax on profit on ordinary activities		(8,470)
Current cost profit for the financial year		6,846
Dividends declared		(5,000)
Current cost profit retained		1,846
EPS		13.7p

7.7.6 The closing current cost statement of financial position

The closing statement with the non-current assets and inventory restated at current cost and the retained profit adjusted for current cost operating adjustments as reduced by the gearing adjustment is set out in Figure 7.7.

Figure 7.7 Economica plc CCA statement of financial position

Economica plc CCA statement of financial position as at 31 December 20X5

20X4				20X5	
£000	£000	*Non-current assets*		£000	£000
140,250		Property, plant and equipment		157,250	
42,075		Depreciation		62,900	
	98,175				94,350
		Current assets			
17,708		Inventory		26,379	
23,375		Trade receivables		34,000	
1,875		Cash		17,000	
42,958				77,379	
		Current liabilities			
17,000		Trade payables		25,500	
		Other payables			
4,250		— income tax		8,500	
4,000		— dividend declared		5,000	
25,250				39,000	
	17,708	*Net current assets*			38,379
		Non-current liabilities			
(11,000)					(11,000)
	6,708				27,379
	104,883				121,729
	£000	*Capital and reserves*			£000
	50,000	Called-up share capital			50,000
	1,500	Share premium account			1,500
	53,383	Total of other reserves			70,229
	104,883				121,729

Analysis of 'Total of other reserves'

	£000				£000
	14,000	Statement of income			15,846
	39,383	Current cost reserve			54,383
	53,383				70,229

continued

Figure 7.7 continued

Movements on reserves

(a) Statement of income: *£000*

  Balance at 1 January 20X5   14,000   (from Figure 7.5)

  Current cost retained profit   1,846   (from Figure 7.6)

  Balance at 31 December 20X5   15,846

(b) Current cost reserve:

	Total	Non-current assets	Inventory	MWCA	Gearing
	£000	*£000*	*£000*	*£000*	*£000*
Balance as at 1 January 20X5	39,383	38,675	708		
Movements during the year:					
Unrealised holding gains in year	10,541	10,370	171		
Gearing adjustment	(684)				(684)
MWCA	1,345			1,345	
COSA	3,798		3,798		
Balance as at 31 December 20X5	54,383	49,045	4,677	1,345	(684)

7.7.7 Real terms system

The real terms system combines both CPP and current cost concepts. This requires a calculation of total unrealised holding gains and an inflation adjustment as calculated in Workings 10 and 11 below.

(W10) Total unrealised holding gains to be used in Figure 7.8

[Closing statement of financial position at CC − Closing statement of financial position at HC] − [Opening statement of financial position at CC − Opening statement of financial position at HC]
= (£121,729,000 − £77,500,000) − (£104,883,000 − £65,500,000) = £4,846,000
  (Working 8)   (Figure 7.3)   (Working 8)   (Figure 7.3)

(W11) General price index numbers to be used to calculate the inflation adjustment in Figure 7.8

General price index at 1 January 20X5 = 317.2
General price index at 31 December 20X5   = 333.2
Opening shareholders' funds at CC × percentage change in GPI during the year =
$$104,883,000 \times \frac{333.2 - 317.2}{317.2} = £5,290,435, \text{ say } £5,290,000$$

The GPP (or CPP) real terms financial capital

The real terms financial capital maintenance concept may be incorporated within the CCA system as in Figure 7.8 by calculating an inflation adjustment.

Figure 7.8 Economica plc real terms statement of comprehensive income

Economica plc CCA statement of income under the real terms system
for the year ended 31 December 20X5

	£000	£000
Historical cost profit after tax for the financial year		17,000
Add: Total unrealised holding gains arising during the year (see W10)	4,846	
Less: Realised holding gains previously recognised as unrealised	none	
	4,846	
Less: Inflation adjustment to CCA shareholders' funds (W11)	(5,290)	
Real holding gains		(444)
Total real gains		16,556
Deduct: dividends declared		5,000
Amount retained		11,556

Real terms system: analysis of reserves

20X4		20X5
£000		*£000*
53,383	*Statement of income*	64,939
—	Financial capital maintenance reserve	5,290
53,383		70,229

Movements on reserves

	Income statement	Financial capital maintenance reserve
	£000	*£000*
Balances at 1 January 20X5	53,383	—
Amount retained	11,556	—
Inflation adjustment for year		5,290
Balances as at 31 December 20X5	64,939	5,290

7.8 Critique of CCA statements

Considerable effort and expense are involved in compiling and publishing CCA statements. Does their usefulness justify the cost? CCA statements have the following uses:

1 The operating capital maintenance statement reveals CCA profit. Such profit has removed inflationary price increases in raw materials and other inventories, and thus is more realistic than the alternative HCA profit.

2 Significant increases in a company's buying and selling prices will give the HCA profit a holding gains content. That is, the reported HCA profit will include gains consequent upon holding inventories during a period when the cost of buying such inventories increases. Conversely, if specific inventory prices fall, HCA profit will be reduced as it takes account of losses sustained by holding inventory while its price drops. Holding gains and

losses are quite different from operating gains and losses. HCA profit does not distinguish between the two, whereas CCA profit does.

3 HCA profit might be adjusted to reflect the moving price-level syndrome:

    (a) by use of the operating capital maintenance approach, which regards only the CCA **operating** profit as the authentic result for the period and which treats any holding gain or loss as a movement on reserves;

    (b) by adoption of the real terms **financial** capital maintenance approach, which applies a general inflation measure via the RPI, combined with CCA information regarding holding gains.

Thus the statement can reveal information to satisfy the demands of the management of the entity itself – as distinct from the shareholder/proprietor, whose awareness of inflation may centre on the **RPI**. In this way the concern of operating management can be accommodated with the different interest of the shareholder. The HCA profit would fail on both these counts.

4 CC profit is important because:

    (a) it quantifies cost of sales and depreciation after allowing for changing price levels; hence trading results, free of inflationary elements, grant a clear picture of entity activities and management performance;

    (b) resources are maintained, as a result of having eliminated the possibility of paying dividend out of real capital;

    (c) yardsticks for management performance are more comparable as a time series within the one entity and between entities, the distortion caused by moving prices having been alleviated.

7.9 The ASB approach

The ASB has been wary of this topic. It is only too aware that standard setters in the past have been unsuccessful in obtaining a consensus on the price-level adjusting model to be used in financial statements. The chronology in Figure 7.9 illustrates the previous attempts to deal with the topic. Consequently, the ASB has clearly decided to follow a gradualist approach and to require uniformity in the treatment of specific assets and liabilities where it is current practice to move away from historical costs.

The ASB view was set out in a Discussion Paper, *The Role of Valuation in Financial Reporting*, issued in 1993.[1] The ASB had three options when considering the existing system of modified historic costs:

- to remove the right to modify cost in the statement of financial position;

- to introduce a coherent current value system immediately;

- to make *ad hoc* improvements to the present modified historic cost system.

7.9.1 Remove the right to modify cost in the statement of financial position

This would mean pruning the system back to one rigorously based on the principles of historical costs, with current values shown by way of note.

Figure 7.9 Standard setters' unsuccessful attempts to replace HCA

1974	Statement of Accounting Practice SSAP 7 *Accounting for Changes in the Purchasing Power of Money* advocating the CPP model.
1975	*Inflation Accounting*, Report of the Inflation Accounting Committee (The Sandilands Report) advocating current cost accounting (CCA) rather than the CPP, RCA or NRVA model. The CCA system recommended by Sandilands was based on the deprival value of an asset, i.e. the value based on the loss, direct or indirect, sustainable by an entity if it were to be deprived of the asset concerned.
1984	SSAP 16 *Current Cost Accounting* was issued by the ASC requiring listed companies to produce CCA accounts as their primary financial report. There was widespread non-compliance and a new exposure draft ED 35 was issued effectively retaining HCA accounts as the primary financial report with supplementary current cost information.
1985	SSAP 16 was withdrawn and the ASC issued *Accounting for the Effects of Changing Prices: A Handbook*. The Handbook was interesting in that it set out four valuation bases if the financial capital maintenance concept was applied and four valuation bases if the operating capital maintenance concept was applied. Its preferred options were CCA under the financial capital maintenance concept which it referred to as real terms accounting (RTA) and CCA under the operating capital maintenance concept.

This option has strong support from the profession not only in the UK, e.g. 'in our view ... the most significant advantage of historical cost over current value accounting ... is that it is based on the actual transactions which the company has undertaken and the cash flows that it has generated ... this is an advantage not just in terms of reliability, but also in terms of relevance',[2] but also in the USA, e.g. 'a study showed that users were opposed to replacing the current historic cost based accounting model ... because it provides them with a stable and consistent benchmark that they can rely on to establish historical trends'.[3]

Although this would have brought UK practice into line with that of the USA and some of the EU countries, it has been rejected. This is no doubt on the basis that the ASB wishes to see current values established in the UK in the longer term.

7.9.2 Introduce a coherent current value system immediately

This would mean developing the system into one more clearly founded on principles embracing current values. One such system, advocated by the ASB in Chapter 6 of its *Statement of Accounting Principles*, is based on **value to the business**. The value to the business measurement model is eclectic in that it draws on various current value systems. The approach to establishing the value to the business of a specific asset is quite logical:

- If an asset is worth replacing, then use replacement cost (RC).
- If it is not worth replacing, then use:
  - value in use (economic value) if it is worth keeping; or
  - net realisable value (NRV) if it is not worth keeping.

The reasoning is that the value to the business is represented by the action that would be taken by a business if it were to be deprived of an asset – this is also referred to as the **deprival value**.

For example, assume the following:

	£
Historical cost	200,000
Accumulated depreciation (6 years straight line)	120,000
Net book value	80,000
Replacement cost (gross)	300,000
Aggregate depreciation	180,000
Depreciated replacement cost	120,000
Net realisable value (NRV)	50,000
Value in use (discounted future income)	70,565

If the asset were destroyed then it would be irrational to replace it at its depreciated replacement cost of £120,000 considering that the asset has a value in use of only £70,565.

However, the ASB did not see it as feasible to implement this system at that time because 'there is much work to be done to determine whether or not it is possible to devise a system that would be of economic relevance and acceptable to users and preparers of financial statements in terms of sufficient reliability without prohibitive cost'.[4]

7.9.3 Make *ad hoc* improvements to the present modified historical cost system

In the UK the *Statement of Accounting Principles* envisages that a mixed measurement system will be used and it focuses on the mix of historical cost and current value to be adopted.[5]

It is influenced in choosing this option by the recognition that there are anxieties about the costs and benefits of moving to a full current value system, and by the belief that a considerable period of experimentation and learning would be needed before such a major change could be successfully introduced.[6]

The historical cost-based system and the current value-based system have far more to commend them than the *ad hoc* option chosen by the ASB. However, as a short-term measure, it leaves the way open for the implementation in the longer term of its preferred value to the business model.

7.10 The IASC/IASB approach

The IASB has struggled in the same way as the ASB in the UK in deciding how to respond to inflation rates that have varied so widely over time. Theoretically there is a case for inflation-adjusting financial statements whatever the rate of inflation, but standard setters need to carry the preparers and users of accounts with them – this means that there has to be a consensus that the traditional HCA financial statements are failing to give a true and fair view. Such a consensus is influenced by the current rate of inflation.

When the rates around the world were in double figures, there was pressure for a **mandatory** standard so that financial statements were comparable. This led to the issue in 1983 of IAS 15 *Information Reflecting the Effects of Changing Prices* which required companies to restate the HCA accounts using either a general price index or replacement costs with adjustments for depreciation, cost of sales and monetary items.

As the inflation rates fell below double figures, there was less willingness by companies to prepare inflation-adjusted accounts and so, in 1989, the mandatory requirement was relaxed and the application of IAS 15 became **optional**.

In recent years the inflation rates in developed countries have ranged between 1% and 4% and so in 2003, 20 years after it was first issued, IAS 15 was **withdrawn** as part of the IASB Improvement Project.

These low rates have not been universal outside the developed world and there has remained a need to prepare inflation-adjusted financial statements where there is hyperinflation and the rates are so high that HCA would be misleading.

7.10.1 The IASB position where there is hyperinflation

What do we mean by hyperinflation?

IAS 29 *Financial Reporting in Hyperinflationary Economies* states that hyperinflation occurs when money loses purchasing power at such a rate that comparison of amounts from transactions that have occurred at different times, even within the same accounting period, is misleading.

What rate indicates that hyperinflation exists?

IAS 29 does not specify an absolute rate – this is a matter of qualitative judgement – but it sets out certain pointers, such as people preferring to keep their wealth in non-monetary assets, people preferring prices to be stated in terms of an alternative stable currency rather than the domestic currency, wages and prices being linked to a price index, or the cumulative inflation rate over three years approaching 100%.

Countries where hyperinflation has been a risk include Iran, Sudan and Venezuela.

How are financial statements adjusted?

The current year financial statements, whether HCA or CCA, must be restated using the domestic measuring unit current at the statement of financial position date. The domestic statements may be adjusted using an index as in the following extract from the Diageo 2013 Annual Report:

> Since December 2009 Venezuela has been classified as a hyperinflationary economy. Hyperinflationary accounting requires the restatement of the subsidiary undertaking's income statement to current purchasing power. The index used to calculate the hyperinflationary adjustment was the Indice Nacional de Precios al Consumidor which changed from 285.5 to 398.6 in the year ended 30 June 2013.

7.11 Future developments

A mixed picture emerges when we try to foresee the future of changing price levels and financial reporting. The accounting profession has been reluctant to abandon the HC concept in favour of a 'valuation accounting' approach. In the UK and Australia many companies have stopped revaluing their non-current assets, with a large proportion opting instead to revert to the historical cost basis, with the two main factors influencing management's decision being cost-effectiveness and future reporting flexibility.[7]

The pragmatic approach is prevailing with each class of asset and liability being considered on an individual basis. For example, non-current assets may be reported at depreciated replacement cost if this is lower than the value in use we discussed in Chapter 6; financial assets are reported at market value (exit value in the NRV model); and current assets reported at the lower of HC and NRV. In each case the resulting changes, both realised and unrealised, in value now find their way into the financial performance statement(s).

7.11.1 Increasing use of fair values

A number of IFRSs now require or allow the use of fair values, e.g. IFRS 3 *Business Combinations* in which fair value is defined as 'the amount for which an asset could be exchanged or a liability settled between knowledgeable, willing parties in an arm's length transaction'. This is equivalent to the NRVA model discussed above. It is defined as an exit value rather than a cost value but like NRVA it does not imply a forced sale, i.e. it is the best value that could be obtained.

It is very possible that the number of international standards requiring or allowing fair values will increase over time and reflect the adoption on a piecemeal basis. In the meantime, efforts[8] are in hand for the FASB and IASB to arrive at a common definition of fair value which can be applied to value assets and liabilities where there is no market value available. Agreeing a definition, however, is only a part of the exercise. If analysts are to be able to compare corporate performance across borders, then it is essential that both the FASB and the IASB agree that all companies should adopt fair value accounting – this has been proving difficult.

7.11.2 The move to defining how to measure fair value

The IASB addressed this by issuing IFRS 13 *Fair Value Measurement* in 2011. This standard[9] does not state when fair values are to be used but applies when the decision has been made to measure at fair value so that there is uniformity in the measurement process.

IFRS 13 *Fair Value Measurement*

The standard (a) defines fair value, (b) sets out a framework for measuring it and (c) sets out the disclosures that are required.

Fair value definition

IFRS 13 defines fair value as the price that would be received to sell an asset or paid to transfer a liability in an orderly transaction between market participants at the measurement date. This means that it is a market-based measurement we would refer to as an exit price – it is not an entity-specific measurement so that the entity's intention to hold an asset or to settle a liability is not relevant when measuring fair value.

Fair value measurement

An entity has to identify the particular asset or liability being measured, the market in which an orderly transaction would take place and the appropriate valuation technique.

Fair value hierarchy

It is not always possible to obtain a directly comparable market value. The IFRS establishes, therefore, a fair value hierarchy that categorises the inputs to a valuation into three levels. It provides a framework to increase comparability but it does not remove the judgement that is required in arriving at a fair value.

Level 1 typically applies to financial investments when there are inputs such as quoted prices in an active market for identical assets or liabilities at the date the fair value is being measured.

Level 2 applies when there are not quoted prices as in Level 1 but there is observable data such as the price per square metre that had been achieved locally in an orderly market when valuing retail space.

Level 3 applies when there are no comparable observable inputs and reliance has to be on judgement using data such as discounted cash flows.

Judgement is required in arriving at a fair value

Judgement is required in selecting the level input appropriate to a particular asset. For example, consider Retail Properties plc:

> Retail Properties plc has a portfolio of investments linked to the retail property market which it had acquired 5 years earlier when property prices were buoyant. At the end of the current financial period it had received an offer from a private equity vulture fund of £2m to acquire the portfolio. The company has been advised that this fund had acquired similar portfolios from companies that had gone into administration – however, Retail Properties plc was solvent and under no liquidity pressure to accept this offer.
>
> The company obtained advice from Commercial Property Valuers that from their current experience with sales in this sector the portfolio could be sold for £3m in the current market and, with the expected upturn in the retail sector, could probably realise up to £5m in 2 to 3 years' time.

There are three valuations and in determining the fair value the company has to (a) bear in mind that the fair value has to be that obtainable at the current date and (b) measured applying the IFRS 13 three hierarchy levels approach. So, taking each in turn:

Level 1 does not apply because it requires an active market such as the availability of quoted prices on a stock exchange.

Level 2 would seem to apply as there is *observable* evidence provided by commercial property valuers of the results on the sale of *similar* assets at the *current* time.

Level 3 is based on an estimated improvement of market conditions in the *future*. It is *not observable* and it is *not current* – it is not appropriate on those grounds.

The best estimate of fair value based on this analysis is the figure of £3m arrived at applying the Level 2 input which is observable and based on an orderly market – unlike the forced sale conditions that applied to the vulture fund offer.

Note. If there were no observable direct or indirect comparators and the Level 3 valuation used discounted cash flows, improvements in cash flows arising from action taken by the company would be acceptable provided those actions would also have been taken by any party taking over the asset. The cash flows used should reflect only the cash flows that market participants would take into account when assessing fair value. This includes both the type of cash flows (e.g. future capital expenditure) and the estimated amount of cash flows.

How will financial statements be affected if fair values are adopted?

The financial statements will have the same virtues and defects as the NRVA model (Section 7.6.4 above). Some concerns have been raised that reported annual income will become more volatile and the profit that is reported may contain a mix of realised and unrealised profits. Supporters of the use of fair values see the statements of comprehensive income and financial position as more relevant for decision making whilst accepting that the figures might be less reliable and not as effective as a means of assessing the stewardship by the directors.

This means that in the future historical cost and realisation will be regarded as less relevant[10] and investors, analysts and management will need to come to terms with increased volatility in reported annual performance.

This is one of the reasons that narrative reports such as the Strategic Report and Management Commentary are increasingly important when investors make their predictions about future performance and position.

Summary

The traditional HCA system reveals disturbing inadequacies in times of changing price levels, calling into question the value of financial reports using this system. Considerable resources and energy have been expended in searching for a substitute model able to counter the distortion and confusion caused by an unstable monetary unit.

Three basic models have been developed: RCA, NRVA and CPP. Each has its merits and defects; each produces a different income value and a different capital value.

In the search for more relevant decision useful financial statements we will see the gradual replacement of historical cost figures.

The contemporary financial reporting scene continues to be dynamic.

We see value in use used as a criterion in measuring the impairment of non-current tangible and intangible assets (discussed further in Chapters 16 and 17); we see financial assets value at fair values (discussed further in Chapter 14); we see current assets valued at lower of cost and NRV; we see the addition of a Statement of comprehensive income required as a primary financial statement to report fair value adjustments.

REVIEW QUESTIONS

1 Explain why financial reports prepared under the historical cost convention are subject to the following major limitations:

- periodic comparisons are invalidated; the depreciation charge may be understated;
- gains and losses on net monetary assets are undisclosed.

2 Explain how each of the limitations in Question 1 could be overcome.

3 Compare the operating and financial capital maintenance concepts and discuss if they are mutually exclusive.

4 Explain how the CPP model differs from the CCA model as a basis for making dividend decisions.

5 '... the IASB's failure to decide on a capital maintenance concept is regrettable as users have no idea as to whether total gains represent income or capital and are therefore unable to identify a meaningful "bottom line".'[11] Discuss.

6 'To be relevant to investors, the profit for the year should include both realised and unrealised gains/losses.' Discuss.

7 Discuss why there are objections to financial statements being prepared using the NRVA model.

8 Explain the criteria for determining whether hyperinflation exists.

9 'Investors benefit when unrealised changes in assets arising from fair value measurement are incorporated in the financial report even if this means that there is greater volatility in income and balance sheet ratios.' Discuss.

10 Retail plc had a portfolio linked to retail properties. Discuss the information that would be required if Level 1 and Level 2 inputs were unavailable. Explain the judgements that would be required.

EXERCISES

* Question 1

Raiders plc prepares accounts annually to 31 March. The following figures, prepared on a conventional historical cost basis, are included in the company's accounts to 31 March 20X5.

1 In the income statement:

	£000	£000
(i) Cost of goods sold:		
Inventory at 1 April 20X4	9,600	
Purchases	39,200	
	48,800	
Inventory at 31 March 20X5	11,300	37,500
(ii) Depreciation of equipment		8,640

2 In the statement of financial position:

	£000	£000
(iii) Equipment at cost	57,600	
Less: Accumulated depreciation	16,440	41,160
(iv) Inventory		11,300

The inventory held on 31 March 20X4 and 31 March 20X5 was in each case purchased evenly during the last six months of the company's accounting year.

Equipment is depreciated at a rate of 15% per annum, using the straight-line method. Equipment owned on 31 March 20X5 was purchased as follows: on 1 April 20X2 at a cost of £16 million; on 1 April 20X3 at a cost of £20 million; and on 1 April 20X4 at a cost of £21.6 million.

	Current cost of inventory	Current cost of equipment	Retail Price Index
1 April 20X2	109	145	313
1 April 20X3	120	162	328
30 September 20X3	128	170	339
31 December 20X3	133	175	343
31 March/1 April 20X4	138	180	345
30 September 20X4	150	191	355
31 December 20X4	156	196	360
31 March 20X5	162	200	364

Required:

(a) Calculate the following current cost accounting figures:

   (i)   The cost of goods sold of Raiders plc for the year ended 31 March 20X5.

   (ii) The statement of financial position value of inventory at 31 March 20X5.

   (iii) The equipment depreciation charge for the year ended 31 March 20X5.

   (iv) The net statement of financial position value of equipment at 31 March 20X5.

(b) Discuss the extent to which the figures you have calculated in (a) above (together with figures calculated on a similar basis for earlier years) provide information over and above that provided by the conventional historical cost statement of comprehensive income and statement of financial position figures.

(c) Outline the main reasons why the standard setters have experienced so much difficulty in their attempts to develop an accounting standard on accounting for changing prices.

Question 2

The finance director of Toy plc has been asked by a shareholder to explain items that appear in the current cost statement of comprehensive income for the year ended 31.8.20X9 and the statement of financial position as at that date:

		£	£
Historical cost profit			143,000
Cost of sales adjustment	(1)	10,000	
Additional depreciation	(2)	6,000	
Monetary working capital adjustment	(3)	2,500	18,500
Current cost operating profit before tax			124,500
Gearing adjustment	(4)		2,600
CCA operating profit			127,100
Non-current assets at gross replacement cost		428,250	
Accumulated current cost depreciation	(5)	(95,650)	332,600
Net current assets			121,400
12% debentures			(58,000)
			396,000
Issued share capital			250,000
Current cost reserve	(6)		75,000
Retained earnings			71,000
			396,000

Required:

(a) Explain what each of the items numbered 1–6 represents and the purpose of each.

(b) What do you consider to be the benefits to users of providing current cost information?

* Question 3

The statements of financial position of Parkway plc for 20X7 and 20X8 are given below, together with the income statement for the year ended 30 June 20X8.

Statement of financial position

	20X8			20X7		
	£000	£000	£000	£000	£000	£000
Non-current assets	Cost	Depn	NBV	Cost	Depn	NBV
Freehold land	60,000	—	60,000	60,000	—	60,000
Buildings	40,000	8,000	32,000	40,000	7,200	32,800
Plant and machinery	30,000	16,000	14,000	30,000	10,000	20,000
Vehicles	40,000	20,000	20,000	40,000	12,000	28,000
	170,000	44,000	126,000	170,000	29,200	140,800
Current assets						
Inventory		80,000			70,000	
Trade receivables		60,000			40,000	
Short-term investments		50,000			—	
Cash at bank and in hand		5,000			5,000	
		195,000			115,000	
Current liabilities						
Trade payables		90,000			60,000	
Bank overdraft		50,000			45,000	
Taxation		28,000			15,000	
Dividends declared		15,000			10,000	
		183,000			130,000	
Net current assets			12,000			(15,000)
			138,000			125,800
Financed by						
ordinary share capital			80,000			80,000
Share premium			10,000			10,000
Retained profits			28,000			15,800
			118,000			105,800
Long-term loans			20,000			20,000
			138,000			125,800

Statement of income of Parkway plc for the year ended 30 June 20X8

	£000
Sales	738,000
Cost of sales	620,000
Gross profit	118,000

Notes

1 The freehold land and buildings were purchased on 1 July 20X0. The company policy is to depreciate buildings over 50 years and to provide no depreciation on land.

2 Depreciation on plant and machinery and motor vehicles is provided at the rate of 20% per annum on a straight-line basis.

3 Depreciation on buildings and plant and equipment has been included in administration expenses, while that on motor vehicles is included in distribution expenses.

4 The directors of Parkway plc have provided you with the following information relating to price rises:

	RPI	Inventory	Land	Buildings	Plant	Vehicles
I July 20X0	100	60	70	50	90	120
I July 20X7	170	140	290	145	135	180
30 June 20X8	190	180	310	175	165	175
Average for year ending 30 June 20X8	180	160	300	163	145	177

Required:
(a) Making and stating any assumptions that are necessary, and giving reasons for those assumptions, calculate the monetary working capital adjustment for Parkway plc.
(b) Critically evaluate the usefulness of the monetary working capital adjustment.

* Question 4

The historical cost accounts of Smith plc are as follows:

Smith plc Statement of income for the year ended 31 December 20X8

	£000	£000
Sales		2,000
Cost of sales:		
Opening inventory 1 January 20X8	320	
Purchases	1,680	
	2,000	
Closing inventory at 31 December 20X8	280	
		1,720
Gross profit		280
Depreciation	20	
Administration expenses	100	
		120
Net profit		160

Statement of financial position of Smith plc as at 31 December 20X8

	20X7		20X8	
Non-current assets	£000		£000	
Land and buildings at cost	1,360		1,360	
Less aggregate depreciation	(160)		(180)	
	1,200		1,180	
Current assets				
Inventory	320		280	
Trade receivables	80		160	
Cash at bank	40		120	
	440		560	
Trade payables	200		140	
		240		420
		1,440		1,600
Ordinary share capital		800		800
Retained profit		640		800
		1,440		1,600

Notes

1 Land and buildings were acquired in 20X0 with the buildings component costing £800,000 and depreciated over 40 years.
2 Share capital was issued in 20X0.
3 Closing inventories were acquired in the last quarter of the year.
4 RPI numbers were:

Average for 20X0	120
20X7 last quarter	216
At 31 December 20X7	220
20X8 last quarter	232
Average for 20X8	228
At 31 December 20X8	236

Required:

(i) Explain the basic concept of the CPP accounting system.

(ii) Prepare CPP accounts for Smith plc for the year ended 20X8.
The following steps will assist in preparing the CPP accounts:
(a) Restate the statement of comprehensive income for the current year in terms of £CPP at the year-end.
(b) Restate the closing statement of financial position in £CPP at year-end, but excluding monetary items, i.e. trade receivables, trade payables, cash at bank.
(c) Restate the opening statement of financial position in £CPP at year-end, but including monetary items, i.e. trade receivables, trade payables and cash at bank, and showing equity as the balancing figure.
(d) Compare the opening and closing equity figures derived in (b) and (c) above to arrive at the total profit/loss for the year in CPP terms. Compare this figure with the CPP profit calculated in (a) above to determine the monetary gain or monetary loss.
(e) Reconcile monetary gains/loss in (d) with the increase/decrease in net monetary items during the year expressed in £CPP compared with the increase/decrease expressed in £HC.

* Question 5

Shower Ltd was incorporated towards the end of 20X2, but it did not start trading until 20X3. Its historical cost statement of financial position at 1 January 20X3 was as follows:

	£
Share capital, £1 shares	2,000
Loan (interest free)	8,000
	£10,000
Non-current assets, at cost	6,000
Inventory, at cost (4,000 units)	4,000
	£10,000

A summary of Shower Limited's bank account for 20X3 is given below:

		£	£
1 Jan 20X3	Opening balance		nil
30 Jun 20X3	Sales (8,000 units)		20,000
Less			
29 Jun 20X3	Purchase (6,000 units)	9,000	
	Sundry expenses	5,000	14,000
31 Dec 20X3	Closing balance		£6,000

All the company's transactions are on a cash basis.

The non-current assets are expected to last for five years and the company intends to depreciate its non-current assets on a straight-line basis. The non-current assets had a resale value of £2,000 at 31 December 20X3.

Notes

1 The closing inventory is 2,000 units and the inventory is sold on a first-in-first-out basis.
2 All prices remained constant from the date of incorporation to 1 January 20X3, but thereafter, various relevant price indices moved as follows:

		Specific indices	
	General price level	Inventory	Non-current assets
1 January 20X3	100	100	100
30 June 20X3	120	150	140
31 December 20X3	240	255	200

Required:

Produce statements of financial position as at December 20X3 and statements of comprehensive income for the year ended on that date on the basis of:

(i)   historical cost;
(ii) current purchasing power (general price level);
(iii) replacement cost;
(iv) continuous contemporary accounting (NRVA).

Question 6

Aspirations Ltd commenced trading as wholesale suppliers of office equipment on 1 January 20X1, issuing ordinary shares of £1 each at par in exchange for cash. The shares were fully paid on issue, the number issued being 1,500,000.

The following financial statements, based on the historical cost concept, were compiled for 20X1.

Aspirations Ltd

Statement of income for the year ended 31 December 20X1

	£	£
Sales		868,425
Purchases	520,125	
Less: Inventory 31 December 20X1	24,250	
Cost of sales		495,875
Gross profit		372,550
Expenses	95,750	
Depreciation	25,250	
		121,000
Net profit		251,550

Statement of financial position as at 31 December 20X1

	Cost	Depreciation	
Non-current assets	£	£	£
Freehold property	650,000	6,500	643,500
Office equipment	375,000	18,750	356,250
	1,025,000	25,250	999,750
Current assets			
Inventories		24,250	
Trade receivables		253,500	
Cash		1,090,300	
		1,368,050	
Current liabilities		116,250	
		1,251,800	
Non-current liabilities		500,000	751,800
			1,751,550
Issued share capital			
1,500,000 £1 ordinary shares			1,500,000
Retained earnings			251,550
			1,751,550

The year 20X1 witnessed a surge of inflation and in consequence the directors became concerned about the validity of the revenue account and statement of financial position as income and capital statements.

Index numbers reflecting price changes were:

Specific index numbers reflecting replacement costs

	1 January 20X1	31 December 20X1	Average for 20X1
Inventory	115	150	130
Freehold property	110	165	127
Office equipment	125	155	145
General price index numbers	135	170	155

Regarding current exit costs

Inventory is anticipated to sell at a profit of 75% of cost.

Value of assets at 31 December 20X1

	£
Freehold property	640,000
Office equipment	350,000

Initial purchases of inventory were effected on 1 January 20X1 amounting to £34,375; the balance of purchases was evenly spread over the 12-month period. The non-current assets were acquired on 1 January 20X1 and, together with the initial inventory, were paid for in cash on that day.

Required:

Prepare the accounts adjusted for current values using each of the three proposed models of current value accounting: namely, the accounting methods known as replacement cost, general (or current) purchasing power and net realisable value.

References

1 *The Role of Valuation in Financial Reporting*, ASB, 1993.
2 Ernst & Young, *UK GAAP* (4th edition), 1994, p. 91.
3 *The Information Needs of Investors and Creditors*, AICPA Special Committee on Financial Reporting.
4 *The Role of Valuation in Financial Reporting*, ASB, 1993, para. 31(ii).
5 *Statement of Accounting Principles*, ASB, December 1999, para. 6.4.
6 *The Role of Valuation in Financial Reporting*, ASB, 1993, para. 33.
7 Ernst & Young, 'Revaluation of non-current assets', Accounting Standard, Ernst & Young, January 2002, www.ey.com/Global/gcr.nsf/Australia.
8 *SFAS 157 Fair Value Measurement*, FASB, 2006.
9 IFRS 13 *Fair Value Measurement*, IASSB, 2011.
10 A. Wilson, 'IAS: the challenge for measurement', *Accountancy*, December 2001, p. 90.
11 N. Fry and D. Bence, 'Capital or income?', *Accountancy*, April 2007, p. 81.

Bibliography

W.T. Baxter, *Depreciation*, Sweet and Maxwell, 1971.

W.T. Baxter, *Inflation Accounting*, Philip Alan, 1984.

W.T. Baxter, *The Case for Deprival Accounting*, ICAS, 2003.

E. O. Edwards and P.W. Bell, *The Theory and Measurement of Business Income*, University of California Press, 1961.

J.R. Hicks, *Value and Capital* (2nd edition), Oxford University Press, 1975.

T.A. Lee, *Income and Value Measurement: Theory and Practice* (3rd edition), Van Nostrand Reinhold (UK), 1985, Chapter 5.

D.R. Myddleton, *On a Cloth Untrue – Inflation Accounting: The Way Forward*, Woodhead-Faulkner, 1984.

R.H. Parker and G.C. Harcourt (eds), *Readings in the Concept and Measurement of Income*, Cambridge University Press, 1969.

D. Tweedie and G. Whittington, *Capital Maintenance Concepts*, ASC, 1985.

D. Tweedie and G. Whittington, *The Debate on Inflation in Accounting*, Cambridge University Press, 1985.

CHAPTER 8

Revenue recognition

8.1 Introduction

Revenue recognition is at the core of the accounting process. A critical part of this process is to accurately identify those earnings outcomes which have been achieved during the period. The trend of earnings is important to investors, as it affects the share price, and to management, as it is often the basis for determining their bonuses.

It is in seeking to maintain an upward trend that scandals involving the manipulation of earnings have arisen on a regular basis, frequently caused by the overstatement by some companies of their revenue. The extent of such manipulation and its adverse impacts is evidenced by the considerable research undertaken in the US which has provided us with reliable statistics.

Adverse effect on capital markets

The US Government Accountability Office[1] reports that in 2005 6.8% of listed companies had to restate earnings, and during the period July 2002 to September 2005, the restatements affected market values by $36 billion. Of those restatements, 20.1% of the restatements were in relation to revenue. Other research finds that 'investors and dealers react negatively to restatements and are more concerned with revenue recognition problems than with other financial reporting errors'.[2] In recent years the magnitudes of the restatements have typically been smaller, but this is probably just a cyclical trend.

In individual cases the differences can be critical. The US company HP took over the British company Autonomy and subsequently suffered losses as a result of the transaction. In the review of what happened, the accounting of Autonomy was subjected to scrutiny. 'In the original accounts [of Autonomy Systems Limited for 2010], audited by Deloitte, revenues were £175.6m. The restated turnover of £81.3m, signed off by HP's auditors Ernst & Young and filed on Monday at Companies House, is less than half as much.'[3] Thus correct revenue recognition is important for the effective operation of the capital markets.

Adverse effect on staff prospects

It is not only investors who suffer. Collins et al. (2009)[4] suggest that chief financial officers of restating companies have enhanced likelihood of losing their job and find it harder to get comparable jobs subsequently.

Harmonisation of accounting standards

In recent years the USA and IASB have attempted to harmonise their standards wherever possible. However, there has been reluctance on the part of many in the USA to give up their standard-setting authority and a belief that US standards are better. Fortunately, however, in the case of revenue recognition there has been acceptance of the need to harmonise and with it the need to agree on broad principles and to move away from their reliance on industry-specific guidelines.

In the past US accountants have been happier to have very specific guidelines, which could reflect that they operate in a more litigious environment. The harmonisation process has also been partly driven by the need to provide more guidance in relation to longer-term contracts.

This chapter will discuss the principles underlying revenue recognition and measurement, and the ethical issues arising out of attempts to circumvent the rules. The discussions will primarily be based on IFRS 15 *Revenue from Contracts with Customers*, issued jointly by the Financial Accounting Standards Board in the USA and the International Accounting Standards Board in May 2014.[5]

Objectives

By the end of this chapter you should be able to:

- apply the principles of revenue recognition and measurement to typical accounting situations;
- understand the complexities of developing universally applicable revenue recognition standards;
- understand the importance of complying with the spirit as well as the detail of the revenue accounting standard;
- identify the situations in which there are industry-specific revenue recognition rules covered by separate standards.

8.2 IAS 18 *Revenue*

The accounting standard IAS 18 is currently in operation in accounting for (i) revenue for goods, (ii) revenue from services and (iii) interest, royalties and dividends. Each of those three categories had separate recognition criteria.

Sale of goods
'Revenue from sale of goods shall be recognised when all of the following conditions have been satisfied:

(a) the entity has transferred to the buyer the significant risks and rewards of ownership of the goods;

(b) the entity retains neither continuing managerial involvement to the degree usually associated with ownership nor effective control over the goods sold;

(c) the amount of revenue can be measured reliably;

(d) it is probable that the economic benefits associated with the transaction will flow to the entity; and

(e) the costs incurred or to be incurred in respect of the transaction can be measured reliably.'

In practical terms this normally means most goods are invoiced as they are shipped and ownership passes and the amount of the transaction is known.

Rendering of services

'When the outcome of a transaction involving the rendering of services can be estimated reliably, revenue associated with the transaction shall be recognised by reference to the stage of completion of the transaction at the end of the reporting period. The outcome of a transaction can be estimated reliably when all the following conditions are satisfied:

(a) the amount of revenue can be measured reliably;

(b) it is probable that the economic benefits associated with the transaction will flow to the entity;

(c) the stage of completion of the transaction at the end of the reporting period can be measured reliably; and

(d) the costs incurred for the transaction and the costs to complete the transaction can be measured reliably.'

In other words there was an attempt to measure revenue in a period if the calculation of the profit on the services could be measured with a reasonable degree of accuracy.

Interest, royalties and dividends

These are recognised when:

'(a) it is probable that the economic benefits associated with the transaction will flow to the entity; and

(b) the amount of the revenue can be measured reliably.'

Thus there are three separate criteria for revenue recognition. Although there are similarities it would be neater if there was one set of criteria that were applicable to all areas.

8.3 The issues involved in developing a new standard

Revenue broadly defined is the gross benefit arising from provision of goods and services, in the normal course of business, to external parties for which remuneration is receivable. Thus sales of non-current assets are not part of revenue. If the provision of goods or services is immediately followed by the receipt of remuneration in cash or cash equivalents (e.g. entitlement to cash from a credit card provider) then there is little controversy.

The accounting difficulties arise when:

● there is a significant probability that the full amount invoiced will not be received in full;

● gains arise from unusual or infrequent transactions with a decision required as to whether they are to be treated in revenue or kept separate;

● transactions are spread over several accounting periods so that it is not clear when the services have been provided;

● a single contract involves the supply of multiple goods and services which may not have similar patterns of delivery;

● the value of the transaction is difficult to determine because it involves payment in kind, volume discounts or the possibility of contract variations during the course of the contract;

● the application is so difficult in a particular industry that there has to be guidance to clarify the application of the general principles in that industry. This includes the construction industry where contracts may take several years, for example, road, ship and aircraft building.

Note that:

- the timing of revenue recognition affects the timing of transfers from inventory into cost of goods sold;
- separate standards are required as for the recognition of changes in market values of financial instruments (see Chapter 14), leases (see Chapter 18), biological assets (see Chapter 20), and the insurance, real estate and investment industries;
- all other standards which are impacted have to be modified to bring them into agreement with the revised revenue standard; and
- construction contracts are discussed in Chapter 21 but with the adoption of the new revenue standard a separate construction standard will not be necessary.

Changes from the Exposure Draft 2010

The changes from the exposure draft in 2010 highlights some of the issues that had to be resolved in order to agree a revenue standard. In particular, IFRS 15 simplified the exposure draft by:

- allowing revenue to be recorded at the gross amount if it is probable the revenue will eventuate and not net of expected bad debts as was previously suggested; also
- simplifying the treatment of warranties by allowing implied interest to be ignored in contracts which are less than one year; further
- allowing as a practical expedient costs of acquiring new contracts to be written off as an expense if the contract is for less than one year; and
- making more explicit the revenue recognition in construction contracts.

These modifications highlight that the standard setters are under continued pressure from financial statement preparers to sacrifice some of the conceptual niceties in order to make the application of the standards easier.

Effect of the standard being a joint IASB/FASB project

The other complication in developing IFRS 15 was that it was intended to be applied in the USA which is used to having much more detailed guidance in general, with many more industry guidelines. As a very litigious society, there is a philosophy in the USA that they need detailed and precise rules which they can follow. Further, there is a perception on the part of some that, if it is not covered by rules, the company is free to choose what suits it rather than trying to gauge the intent of the standard and attempting to implement that intent.

8.4 The challenges under both IAS 18 and IFRS 15

A fair view

Accountants, in addition to checking that each transaction has been recorded in accordance with the accounting standards, still need to ask whether the resulting accounting statements give a fair view of the situation. In the final analysis it will be the courts that decide whether the clever ploys to get around standards, or the taking advantage of technical accounting rules, is legitimate. In *New York* v *Ernst & Young* (Part II) 451586/2010 New York State Supreme Court (Manhattan) the complainant said it was still necessary to ensure the accounts were fair. If considered unfair, they need to disclose sufficient information to rectify the situation.

Recognising economic substance

In the case of revenue recognition, standard setters have been faced with the problem as to how to define revenue in such a way that it precludes transactions being artificially structured. Such a problem arose when Lehman Brothers in its last year entered into agreements to sell securities to a third party who agreed to sell them back after the reporting date allowing Lehman Bros to treat the transfer of securities as sales, thus increasing its revenue and reducing its leverage. So the major issue was whether the sale and the repurchase agreements should be treated as two separate transactions or as a single transaction.

This gave rise to the grey area as to whether to report according to the substance or according to the technical form. One could argue that reporting according to the technical form (i.e. applying an existing technical guideline) was the legal requirement and as a result should be reflected in the accounts – another party could argue that it was misleading and could be construed as fraudulent manipulation. Of course, it is desirable that accounting standards reduce the likelihood that such situations will arise, and following this case the SEC and FASB have issued new rules relating to sale and repurchase agreements, hoping to achieve reporting that better reflects the economic substance.

8.5 IFRS 15 *Revenue from Contracts with Customers*

Effective date

The new standard will be effective for financial years commencing on or after 1 January 2017. However, if a company wishes to provide comparative figures it will need to record revenue figures for the 2016 year on a comparable basis – as well as according to the previous standard. Early adoption is allowable. Thus companies which are adopting international financial standards in the interim period may adopt the new standard immediately so as to avoid a double set of changes in a short period of time.

Definition of revenue

There is no modification in the definition of revenues which is still:

> **Revenue** *is income arising in the course of an entity's ordinary activities* whilst **Income** represents *increases in economic benefits during the accounting period in the form of inflows or enhancements of assets or decreases in liabilities that result in increases in equity, other than those relating to contributions from equity participants.*

8.5.1 Transactions falling outside the IFRS 15 definition

The definition of revenue restricts it to income in the ordinary course of business through the provision of goods and services to customers. It therefore excludes:

- gains arising on the revaluation of biological assets;
- taxes collected on behalf of the government such as value added and sales tax. Thus a sale of goods on credit for £1,000 plus £125 for VAT would be recorded as:

Dr	Accounts receivable	1,125	
Cr	Revenue		1,000
Cr	VAT liability		125

- the effect of financing such as interest revenue or expense. Para 65 provides that '*an entity shall present the effects of financing (interest revenue or interest expense) separately from revenue from contracts with customers in the statement of comprehensive income*';
- transactions dealt with under other standards such as Leasing and Insurance contracts.

However, one of the major debates in terms of developing the revenue recognition standard was how to account for uncertainty.

8.5.2 Revenue recognition when there is uncertainty

There are two major elements in relation to uncertainty. There may be uncertainty as to (a) the amount that will be eventually received in accordance with the terms of the contract and (b) the amount that will be received due to failure to recover the amount due.

(a) Uncertainty as to amount finally due under the contract

There is uncertainty if the amount is variable in that it depends on future events. The standard setters were keen to ensure that revenue reflects, as far as is practical, the level of progress whilst not being over-optimistic. Thus the criteria for determining the revenue figure where the final outcome includes an element of variability is only to report amounts which are *highly probable* of being achieved.

This means not recognising revenue in the current period that will need to be reversed in a subsequent period.

In assessing whether it is highly probable that a significant reversal in the amount of cumulative revenue recognised will not occur, the standard gives examples of factors that could increase the likelihood. The factors include:

- consideration being highly susceptible to factors outside the entity's influence such as volatility in a market, the judgement or actions of third parties, weather conditions and a high risk of obsolescence of the promised good or service;
- the uncertainty about the amount of consideration is not expected to be resolved for a long period of time;
- the entity's experience (or other evidence) with similar types of contracts is limited, or that experience (or other evidence) has limited predictive value; and
- the entity has a practice of either offering a broad range of price concessions or changing the payment terms and conditions of similar contracts in similar circumstances.

(b) Failure to recover the amount due

In this case the uncertainty is dealt with in the traditional manner. Revenue is recorded at the gross amount of the contract and shortfalls in collections are dealt with as bad and doubtful debt expenses.

8.6 Five-step process to identify the amount and timing of revenue

In order to identify the amount and timing of the recognition of revenue, the entity should go through a five-step process:

(a) identify the contract with a customer;

(b) identify the separate performance obligations in the contract;

(c) determine the transaction price;

(d) allocate the transaction price to the separate performance obligations in the contract; and

(e) recognise revenue when (or as) the entity satisfies a performance obligation.

8.6.1 Identify the contract with a customer (Step a)

The first step is to ascertain that there is a legal (enforceable) contract. Obviously legal rules vary across jurisdictions and industry-standard terms may form part of the agreement. Only when the full terms of the contract are ascertainable are we able to record the transaction.

Thus the terms must be specified in the contract or refer to outside references such as the price on a specified market at a specified time and date. The rights and obligations of all parties to the agreement to supply goods and/or services must be identifiable. One of those terms will be when the items are to be paid for.

The contract may relate to a single commodity or may cover a combination of goods and services. In the event of the contract involving a combination it may be necessary to untangle the elements which relate to each item so as to be able to record them as separate items.

For example, the Hewlett-Packard Company in their statement of accounting policies in their 2013 Annual Report[6] refer to agreements which cover both hardware and software and the need to identify how much of the agreed price relates to the payment for the hardware and how much refers to the provision of the software.

The second step relates to the rules for determining whether the contract has to be split into individual components.

Policy illustrated: Hewlett-Packard Company and Subsidiaries (HP)
We have discussed the various accounting policies set out in the following extracts from the Hewlett-Packard 2013 Annual Report. These highlight the application of the above:

Accounting policies
HP recognizes revenue when persuasive evidence of a sales arrangement exists, delivery has occurred or services are rendered, the sale price or fee is fixed or determinable, and collectability is reasonably assured.

Additionally, HP recognizes hardware revenue on sales to channel partners, including resellers . . . at the time of delivery when the channel partners have economic substance apart from HP, and HP has completed its obligations related to the sale.

HP generally recognizes revenue for its stand-alone software sales to channel partners upon receiving evidence that the software has been sold to a specific end user.

When a sales arrangement contains multiple elements, such as hardware and software products, licenses and/or services, HP allocates revenue to each element based on a selling price hierarchy. The selling price for a deliverable is based on its vendor specific objective evidence (VSOE) of selling price, if available or third party evidence ('TPE') if VSOE of selling price is not available or estimated selling price ('ESP') if neither VSOE of selling price nor TPE is available.

HP establishes VSOE of selling price using the price charged for a deliverable when sold separately and, in rare instances, using the price established by management having the relevant authority.

HP establishes TPE of selling price by evaluating largely similar and interchangeable competitor products or services in standalone sales to similarly situated customers . . .

In arrangements with multiple elements, HP determines allocation of transaction price at the inception of the arrangement based on the relative selling price of each unit of accounting. . . .

HP limits the amount of revenue recognized for delivered element to the amount that is not contingent on the future delivery of products or services, future performance obligations or subject to customer-specified return or refund privileges. . . .

HP reports revenue net of any required taxes collected from customers and remitted to government authorities, with the collected taxes recorded as current liabilities until remitted to the relevant government authority. . . ."[6]

The full accounting policy is much more extensive than shown in the extracts and is worthy of review.

8.6.2 Separate performance obligations within the same contract (Step b)

What is a performance obligation?

A performance obligation would include the supply of goods, the provision of services such as consulting or dry cleaning, and compensation for making available assets such as capital (interest), intellectual property (royalties), property, plant and equipment (lease payments) and software.

Identifying separate performance obligations

If goods or services are stand-alone products or services which can be used separately, **or** together with other resources reasonably available to customers **and** the supply company has not agreed to integrate the two products/services, then they are separate performance obligations.

A contract to supply equipment and to service it for several years in the future where the items could be bought separately from one or more companies could constitute two distinct performance obligations.

8.6.3 Pricing the transaction (Step c)

The third step involves the pricing of the performance obligations in the contract. In many contracts that process is relatively simple. The contract is for one item and the price is unambiguously specified in the contract. However, it is possible that the contract may be more complicated.

Where the pricing of a contract is based on events that are unknown at the time *the guiding principle is to record revenue in a manner that is unlikely to result in significant reductions in revenue in the future.* Thus the aim is to be as accurate as possible but to prohibit companies from anticipating revenue that is unlikely to be received.

Next, some common pricing arrangements such as annual refunds, performance bonuses and payments in arrears will be examined.

Annual refund

Dee Pharmaceutical sells to pharmacies whereby it charges the standard price and at the close of the year pays a 5% discount based on the total purchases for the year. During January Dee Pharmaceutical invoices customers €5,000,000 for goods supplied. The entry would be:

Dr	Accounts receivable	5,000,000	
Cr	Sales Revenue		4,750,000
Cr	Sales Rebate liability		250,000

(Being gross sales of €5,000,000 less the obligation to refund 5% at the end of the financial year)

Performance bonuses

Suppose Gee Chemicals supplies chemicals to Perfection Manufacturers on the basis that it will supply 108,000 tons during the year, with shipments of 9,000 tons per month with an annual bonus of 3% on each shipment when the quality exceeds 99% purity, providing there are no late deliveries. The contract rate is £1,100 per ton.

Based on past experience, four of the months will produce a bonus and all deliveries will be on time. There are two ways of calculating the amount to record each month – (a) the most likely outcome and (b) the expected value.

(a) The most likely outcome approach

If we take this approach and focus on the most likely outcome, then we would calculate 9,000 tons per month at £1,100 per ton, giving £9,900,000 per month.

Dr	Accounts receivable	9,900,000
Cr	Sales revenue	9,900,000

(*Monthly sales based on the contract rate without recording any bonus.*)

Dr	Accounts receivable	xxxx
Cr	Sales revenue	xxxx

(*Recording additional revenue when bonuses are confirmed. This entry is based on the assumption that purity levels refer to the state in which they are delivered and thus depend on the outcome of testing at the customer.*)

(b) The expected value approach

Adopting this alternative approach we would work out the expected value based on expectations. The bonus on a month has an expected value of $(4/12) \times 3\% + (8/12) \times 0\%$ = 1%.

The 4/12 above represents the 4 months out of 12 months when they will earn a bonus and the 8/12 represents the 8 out of 12 months in which no bonus is expected.

Under this approach the monthly entry would be:

Dr	Accounts receivable	9,900,000	
Cr	Sales revenue		9,999,000
Dr	Bonus receivable	99,000	

(*Recording expected monthly revenue including anticipated bonuses.*)

Dr/Cr	Accounts receivable	XXX	
Cr	Bonus receivable	Amounts previously recorded	
Dr/Cr	Revenue	Over/under-recording of the revenue previously	

(*Adjusting accounts receivable to the right amount when the size of the bonus is known, eliminating bonus receivable, and getting the revenue figure correct.*)

The choice between the two depends on what experience has shown to be the more accurate in predicting the actual outcomes and the degree to which overstatements have been associated with each approach. Adjustments downward are a problem if they are significant for the particular company.

Treatment if payment is deferred

In a contract in which payment is not immediate on the performance of the obligation but is deferred then there is an implied financing element and the time value of money may have to be considered.

When the interest element may be ignored

If the timing of the delivery of the service is not specified in the contract but is instead determined by the customer, then there is not a definite financing period, so it can be ignored. Also if the total contract covers less than one year, or the gap between payment and performance is less than one year, then the standard allows the interest element to be ignored as a practical expedient.

When the interest element must be accounted for

The determination of an interest element assumes there are (a) pre-established times for delivery of the service and (b) payment at an agreed price. If the time value of money is relevant then the issues to consider include:

(a) the difference between the cash price (i.e. the price that would have been charged if the payment coincided with the performance of the obligation or occurred in accordance with normal industry terms) and the contract price;

(b) the interest rate in the contract; and

(c) the current interest rate in the market.

If interest is recognised then it is shown separately from other revenue in the statement of comprehensive income.

Accounting treatment illustrated

Let us assume that Heavy Goods plc entered into a contract for the sale of 10 coal trucks, at a cost of £242,000 each, which provided for deferred payment two years later. If the terms were payment on delivery, the contract price would have been £200,000 for a similar vehicle.

This would imply that each truck sold included interest of £42,000 which is at a compound interest rate of 10% per annum. If the market rate is also 10% p.a. then the recording would be:

Year one

Dr	Accounts receivable	2,420,000	
Cr	Sales revenue		2,000,000
Cr	Interest revenue		200,000
Cr	Deferred interest		220,000

(Being the recording of the sale of 10 coal trucks and receiving 10% interest for one year)

Year two

Dr	Deferred interest	220,000	
Cr	Interest revenue		220,000

(Being interest on 2,200,000 at ten percent for year two)

Dr	Bank	2,420,000	
Cr	Accounts receivable		2,420,000

(Being settlement of the account)

However, if the market interest rate was more than 10%, that would imply that the sale price was cheaper than competitors' prices or the interest rate was discounted or both were discounted. Such cases will be discussed in the next section.

8.6.4 Allocate the transaction price to the separate performance obligations in the contract (Step d)

Where Step b has identified that there are two or more performance obligations under a contract it is necessary to allocate the total contract consideration (amount) across the individual performance obligations. The method that will be adopted depends on the information available.

The simplest case is where there are readily available market prices for each and every performance obligation. These prices can be used to allocate the total remuneration between the various performance obligations. The highest-quality information is external market prices for identical or very similar items sold to comparable customers in similar circumstances.

In the absence of such high-quality information, the company's own selling prices for items sold individually can be used. If the company can only get reliable or consistent information on some of the performance obligations then the residual method can be used. It must be stressed that this residual method is the last resort rather than a desirable approach.

Suppose a company sells a combination of three performance obligations X, Y and Z for £32. If the market prices of X and Y as independent items are £10 and £15 respectively then the residual of £7 is deemed to be the price of item Z. However, if the company regularly sells bundles of (X + Y) at a discount price of £24, then the residual price would be £8.

Readily available market prices illustrated with Consensus Supplies plc

Let us illustrate the highest ranked approach whereby market prices for all performance obligations are used to allocate the contract price.

Consensus Supplies plc – information on contract

Let us assume that Consensus Supplies plc sells 10 printers on credit at a price of €4,000 each when the manufactured cost was €2,000 each. Let us further assume that Consensus offers its customers a combined contract for €4,800 for each printer which includes the provision of maintenance cover for two years. The cost of manufacture remains the same at €2,000 each and the cost of supplying maintenance is €250 per machine per year.

Let us also assume that customers could purchase separate maintenance cover from other suppliers for 2 years at a cost of €1,000 per printer.

Consensus Supplies plc – accounting for contract (ignore the financing elements to keep this introductory example simple)

The contract is for two separate performance obligations (with different timing of the services) and so the revenue has to be apportioned between the contracts and then recognised as the individual services are provided. The normal selling prices are:

Supply of printers	€40,000
Supply of maintenance in Year 1	€5,000
Supply of maintenance in Year 2	€5,000
Total services provided	€50,000
Combined price	€48,000

This shows that Consensus is selling at 48,000/50,000 or 96% of the normal price, i.e. 4% below normal selling price. Each component part of the contract is reduced by 4% as follows:

Supply of printers	40,000 × 0.96	€38,400
Supply of maintenance in Year 1	5,000 × 0.96	€4,800
Supply of maintenance in Year 2	5,000 × 0.96	€4,800
Combined price		€48,000

(This process represents Step d of the requirements which is to allocate the transaction price to the separate performance obligations in the contract.)

The entries for a contract made in accordance with the Standard are:

Dr	Trade receivables	€48,000	
Cr	Sales revenue (equipment sales)		€38,400
Cr	Sales revenue (maintenance)		€4,800
Cr	Revenue in advance liability		€4,800

(Being the recording of a sale and maintenance package.)

Dr	Cost of goods sold (Equipment)	€20,000	
Dr	Cost of goods sold (Maintenance)	€2,500	
Cr	Inventory		€20,000
Cr	Bank		€2,500

(Recording costs of providing services and the outlays for wages and materials used for maintenance.)

Disclosure at end of Year 1

The Standard requires the company to disclose in its annual report the amount and timing of the future revenue secured by existing contracts.

A possible way of disclosing this could be as follows:

Contracts for the supply of maintenance

	Period two	*Period three*
Prepaid amounts	£4,800	XXX
Executory contracts	XXX	XXX

Note, however, that this does not satisfy all the disclosure requirements. See paragraph 110 which says:

> The objective of the disclosure requirements is for an entity to disclose sufficient information to enable users of financial statements to understand the nature, amount, timing and uncertainty of revenue and cash flows arising from contracts with customers. To achieve that objective, an entity shall disclose qualitative and quantitative information about all of the following:
>
> **(a)** its contracts with customers . . .
>
> **(b)** the significant judgements, and changes in judgements, made . . .
>
> **(c)** any assets recognised from the costs to obtain or fulfil a contract . . .

Accounting entries in Year 2

These would be:

Dr	Revenue in advance	€4,800	
Cr	Sales revenue (maintenance)		€4,800

(Transferring revenue in advance to current performance.)

Dr	Cost of goods sold (maintenance)	€2,500	
Cr	Bank		€2,500

(Payment for materials and wages.)

Note that in this chapter we discuss the most common and simplest approach to price allocations. In special circumstances, where specific bundles of performance obligations that represent some but not all of the obligations in the current contract, are regularly sold as a bundle at a discount, then another allocation method may be used. See paragraph 82 of the Standard if you want more details.

8.6.5 Recognise revenue when (or as) the entity satisfies a performance obligation (Step e)

As a performance obligation is transferred to a customer in accordance with the terms of the contract, the revenue is recognised in the books of account. The major principle is that revenue is recognised when control of an asset passes to the customer.

Sale of goods

When the goods are delivered to the customer, normally the supplier has a right to payment for the goods, assuming they meet the contract specifications and the conditions demanded under the law (e.g. fit for its purpose). The requirement of the standard that the *revenue be recognised when the asset or service is transferred to the customer* is satisfied in that transfer of the goods is assumed to occur when control passes to customers. In the case of goods this is easy to envisage as once the customer receives the goods they are free to use the goods as they want including consuming them, using them in further manufacturing, selling them or just holding them as assets.

Providing services

When services are transferred, the customer receives an asset such as knowledge or consumable goods, which are immediately consumed. So as soon as the customer consumes the service they clearly have taken control of the asset and thus the revenue shall be recognised.

For example, when an accounting firm gives a client advice, the client has consumed the information and the accounting firm can, therefore, record the revenue. However, if the firm is performing an audit at an agreed price and, at the firm's balance date, it has completed a third of the expected audit work, then it could not justify recording the revenue, as the client has not received the product (the audit report) and does not have control of the audit papers.

More complex situations

If there are separate performance obligations then the agreed price has to be split, otherwise revenue is recognised for the complete transaction. Decisions may have to be made in complex situations such as when warranties, leases, long-term contracts, reservation of title, financial securities, transfers of intellectual property, equally unperformed contracts and onerous contracts are involved. We will briefly discuss each of these.

Warranties

Let us consider a situation where goods have been supplied under a manufacturer's warranty. This means that the full performance obligation has not been completed in that there are residual obligations in the form of warranties covering faults identified in use.

We have to decide whether the warranty is part of the primary performance obligation to supply the goods in satisfactory conditions or is part of a separate performance obligation.

As the warranty is restricted to making the goods fit for their stated purpose and does not constitute an additional performance obligation, the question of splitting the agreed price into separate components does not arise. The customer clearly has control of the asset and thus the revenue is recognised but the warranty has to be accounted for in accordance with IAS 37 *Provisions, Contingent Liabilities and Contingent Assets*.

Example where there is a manufacturer's warranty

Let us assume that Makem Manufacturing plc has shipped 2 million units to customers during the financial year with selling prices averaging £300 per unit. Over the last ten years the warranties cost an average of £1 per unit with only minor fluctuations from year to year.

The accounting entries would be as follows:

Dr	Accounts receivable	600,000,000	
Cr	Sales revenue		600,000,000

(Revenue recorded at their gross sales value)

Dr	Warranty expenses	2,000,000	
Cr	Liability (Provision) for warranty expenses		2,000,000

(Covering expected warranty costs)

If the warranty covers more than is normally associated with the supply of goods and if it represents more than an insignificant amount then the possibility of the warranty giving rise to the existence of a separate performance obligation has to be considered. If it is decided that there are separate performance obligations the approach would be as taken in the Consensus example above.

Leases

Under a lease the customer obtains control of the asset for a specified period and the benefits of use are related to the amount/period of use which can be made of the asset. This will be discussed further in the chapter on leasing.

Long-term contracts

Longer-term contracts such as building a road may take several years to complete. In such cases if the contractor is working on the road and fails to complete, such as when the contractor goes bankrupt, then the client has possession of that which has been accomplished to date and can organise for another contractor to take over the work and to complete the project. In such cases, a contractor may have agreed with the client that it will be entitled to bill the client when major components are completed; then revenue can be recognised when those milestones are reached. Obviously there are a number of other possible scenarios depending on the detailed facts of the case. These issues will be discussed in more detail in Chapter 21.

Reservation of title

Another sales agreement which raises questions as to whether the transaction can be recorded in the normal manner is a sale contract with a reservation of title. To reduce the likelihood of losses arising on the bankruptcy of a customer, some companies deliver goods on terms that include a specification that title to the goods does not pass until the customer has paid for them. Then if the customer gets into financial difficulties the supplier may be able to take the goods back if that is consistent with the laws of the country and provided the goods are identifiable. In the meantime the customer can deal with the goods in the normal way, such as to sell or use the asset.

When to recognise revenue if there is reservation of title

The question is whether the initial transfer of goods to the customer should be treated as revenue in the supplier's accounts.

The relevant provision in the standard is that *'an entity shall recognise revenue when (or as) the entity satisfies a performance obligation by transferring a promised good or service (i.e. an asset) to a customer. An asset is transferred when (or as) the customer obtains control of that asset'* (para. 31).

Clearly the asset has been transferred and the customer now has control of the asset. Thus the conditions for recognising the revenue have been met. Paragraph 38 (b) says: *'If an entity retains legal title solely as protection against the customer's failure to pay, those rights of the entity would not preclude the customer from obtaining control of an asset.'*

Financial securities

Another type of contract or combination of contracts that has been controversial in recent years has been the situation where a company sells an item, such as financial securities, and also agrees to buy them back at a later stage – such as the day after the financial reporting date. The repurchase amount would reflect the price paid for the security plus compensation for the financing cost for the holding period and the risk that the original seller will not be able undertake the repurchase at the specified date.

IFRS 15 explicitly covers sale agreements combined with a repurchase agreement or a put option. It does not matter whether the arrangement was incorporated in a single contract or there were two contracts, namely, one for the sale and the other for the repurchase if one (the sale) was executed with the understanding that the other contract (the repurchase) will also be agreed to.

IFRS 15 clearly specifies that separate contracts should be dealt with in the books as a combined contract if they are negotiated with a single commercial objective or if the amount paid in one contract depends on the consideration paid in the other contract (paragraph 17).

It also requires that revenue contracts must have commercial substance (paragraph 6). Here the sale and repurchase had a single commercial objective, which is for the initial selling company to get the financial instruments temporarily off their balance sheet and replace them with cash so as to make their balance sheet look stronger. However, no risk is being transferred as the repurchase agreement means the company temporarily holding the financial instrument gets a fixed price irrespective of what happens in the marketplace during the holding period.

Transfers of intellectual property

Two distinct categories were identified under this heading. These categories were:

(a) Where the purchaser has the right to use the intellectual property and will pay based on the level of sales; and

(b) Where the seller transfers to the customer all or part of the ownership of the intellectual property at a point in time.

The first situation might be where a band records its own music and receives a royalty from television and radio stations as they play the records and videos. The entry would be:

Dr Accounts receivable/Cash
Cr   Royalty revenue

The second situation might be where an entrepreneurial company develops a new drug which it has tested on mice but does not have the financial and medical connections and managerial resources to satisfy the regulatory requirements in the USA and sells its intellectual capital to a major US company. The sale of its knowledge would be a single transaction. If there was an additional undertaking to do support work for the US company that would be a separate performance obligation.

Equally unperformed contracts

A vexed question is what to do when a company enters into a contract to be a longer-term supplier of goods. If goods have not been supplied under the contract and either of the parties can terminate the contract at any time without compensating the other party, then there is no reason to record the event. The performance obligation has not been performed, so no revenue needs to be recognised.

Onerous contracts

After a contract has been entered into and the expected revenue has been allocated across performance obligations it may become apparent that one or more performance obligation is going to incur a loss. In such circumstances it is necessary to keep in mind two propositions; these are (a) the need to record revenue only at an amount that you are confident you will receive and (b) the need to recognise losses as soon as they are anticipated.

To estimate the amount of the anticipated loss it is necessary to ask: how can the losses be minimised? Generally the company faces two possibilities – either to complete the contract and absorb the resulting losses, or it may be able to negotiate with the customer to be released from the contract in return for payment of penalties. If both options are available the company will presumably choose the cheaper option, taking into account both direct costs and intangible costs (the negative impact on goodwill). The amount of probable losses will be based on the option chosen.

At balance date, revenue must not be overstated and any future losses need to be provided for as an onerous contract in accordance with IAS 37 *Provisions, Contingent Liabilities and Contingent Assets*. Finished goods and work in progress may also need to be reviewed to see if these need to be written down to their net realisable amount.

8.6.6 Modification of the terms of contracts

Sometimes the parties to a contract may modify an ongoing contract during the performance of the contract. An example of this may be when two businesses have an ongoing relationship and have negotiated a contract to formalise those ongoing transactions. It may be that the world price for the raw material that the producer uses to manufacture the goods, subject to the contract, has changed dramatically up or down. If the parties think it only fair that the contract be renegotiated then they are free to do so.

In general terms there are three types of situations involving modifications to contracts:

(a) Where the modification refers only to distinct goods or services which have not yet been supplied. In that case the future component of the contract can be treated like a new separate contract. Then the steps (a) to (e) would be applied separately to the future component to the contract.

(b) The modification may introduce another distinct performance obligation to the contract at a commercial price and will thus be separate from the previously agreed supply. Once again the additional performance obligation will be treated as if it were a new and separate contract.

(c) Where the modification relates to both previously completed performance obligations as well as to future supply. In such cases the revenue is allocated as if the new facts had been known at the commencement of the contract. That calculation will identify the extent to which revenue has previous been over- or understated and that amount will result in an immediate adjustment to revenue. In terms of the future supply those items will be recognised at the new price as the performance obligations take place.

Illustration where the modification relates to both previously completed performance obligations as well as to future supply

Let us consider International Manufacturing which has entered into a two-year contract to make a weekly delivery of biscuits priced at £100 a box. It finds that shortly after the contract is signed and before any shipments have taken place the cost of sugar has jumped 50% because of floods in several countries which are major producers of sugar. The customer of the biscuit company, a major grocery chain, does not offer any extra remuneration because they do not know whether the sugar price increase is a short-term fluctuation or a more sustained change.

In the first four months of the new contract International Manufacturing recognised revenue of £500,000 based on the original contract price. At that point, being confident that the sugar price increase would hold for the next two years, the grocery chain agreed to a 5% increase in price from the start of the contract. International Manufacturing therefore immediately recorded an additional 5% of £500,000 or £25,000 as revenue. Future shipments were then charged at £105 per box as they were delivered.

8.7 Disclosures

IFRS 15 para 110 states that '*The objective of the disclosure requirements is to enable users of financial statements to understand the nature, amount, timing and uncertainty of revenue and cash flows arising from contracts with customers . . .*'. This might require the revenue and assets to be disaggregated and detailed disclosures provided.

Disaggregate revenue

In particular, the business may have a range of products and services, or items in relation to different regions of the world, which are affected by different terms and conditions, economic forces and political risks. It is therefore desirable that management disaggregate total revenue in a manner which helps analysts and investors understand current performance and to predict future performance.

Disaggregate assets

The company has to show details of assets which arise as the result of accounting for revenue and they must be disaggregated as different levels of risk are associated with the different categories. The assets which are relevant include:

(a) Assets representing recoverable outlays in relation to obtaining contracts which relate to performance obligations in future periods;

(b) Accounts receivable representing performance obligations that have already been performed but not yet paid for; and

(c) Work in progress.

Disclose terms and conditions

In addition, the typical terms and conditions for the various categories of revenue need to be disclosed and the associated revenue recognition policies.

Disclose judgements

Further, any major judgements which were made in arriving at revenue figures must be disclosed, as must any changes in those judgements. In that way it is easier for outsiders to review trends in revenue.

Summary

(a) The recognition of revenue does not occur unless there is a fully specified contract.

(b) Revenue is recognised when the performance obligation has been performed for an unrelated/independent customer.

(c) Revenue is only recognised when there is a high probability that it will be paid for.

(d) If the contract involves multiple elements or performance obligations they are separated out and accounted for individually.

(e) To divide (allocate) the contract revenue between performance obligations the company looks to the most directly relevant price information.

(f) The need to distinguish the revenue stream from amounts collected on behalf of government.

REVIEW QUESTIONS

1 *The Accounting Onion*, a very good US accounting blog,[7] critically reviews the revenue recognition policy of a for-profit university and provides an extract from the policy as follows:

> Students are billed on a course-by-course basis. They are billed on the first day of attendance, and a journal entry is made to debit Accounts receivable A/R and credit deferred revenue for the amount of the billing. The A/R is ultimately adjusted by an allowance for uncollectible accounts of around 30%, and deferred revenue is recognized pro rata over the duration of the course.

Assuming the courses (subjects) go from September to February whilst the financial year ends on 31 December, discuss whether the policy is appropriate.

2 A continuing problem in accounting is where companies use multiple contracts to circumvent the intentions of rules. Sometimes this is called 'the need for accounting to reflect substance over form'. Lehman Bros in the USA entered into a contract for sale of some securities which it agreed to buy back after balance date. What are the advantages and disadvantages of the 'substance over form' approach?

3 One of the problems faced by the standard setters was the untangling of contacts which delivered multiple performance obligations. Identify a business which supplies multiple services in one contract and identify the individual service components. Then discuss the following:

(a) whether the two service components have similar or dissimilar patterns of delivery;

(b) whether the components are sold by that business or other businesses as separate contracts or as 'stand-alone' services;

(c) the significance of the presence of other companies selling the services as individual services;

(d) the likely impact on the business if the two service components have to be recognised individually.

4 The bloggers Anthony H. Catanach Jr and J. Edward Ketz discussed in their blog Grumpy Old Accountants (http://www.grumpyoldaccountants.com/) on 27 August 2012 the accounting for Internet companies. The article is titled 'What is Zynga's "Real" Growth Rate?' They quote from the accounting policies of Zynga as follows: 'We recognize revenue from the sale of durable virtual goods ratably over the estimated average playing period of paying players for the applicable game'. Discuss whether the preceding policy would be an appropriate policy under IFRS 15 and whether Internet companies need their own revenue standard.

5 The ASB made a submission on the 2010 proposed revenue standard and in that submission indicated that the revenue allocation in relation to a contract which involved multiple performance obligations should reflect the normal margins of the various performance obligations. This is in contrast to the then proposed standard which allocated revenue in relation to the stand-alone prices of the various performance obligations. Discuss the merits of the two alternatives.

6 In the technology sector there has been a high proportion of problems centred on what accountants call improper 'revenue recognition' – the recording of revenue that does not exist. Discuss why the technology sector might be more likely to do this and how it would have been justified to the auditors.

7 Do the proposed/new recognition rules give primary importance to calculating income or fairly presenting the statement of financial position (balance sheet)? How do you support that conclusion?

8 Executory contracts are contracts where both sides have not yet performed their obligations. If your company has a long-term contract for the supply of raw materials to XYZ Ltd for 50,000 tons per year for five years at a selling price of €10,000 per ton and the market price has fallen to €8,000 per ton, should this be recorded as a €2,000 a ton revenue in the current period? Justify your answer.

9 During the dot-com boom two major companies with excess data transmission lines in different areas arranged a sale whereby company X transferred lines in city A to company Y which in return transferred its excess lines in city B to company X. No cash changed hands. The contract specified the agreed value of the assets transferred. Identify and explain the potential problems in accounting for such a transaction. Would it make any difference if cash had changed hands?

10 There is a company which facilitates barter exchanges. Thus the barter company may ask a restaurant to make available a number of free meals which are then effectively exchanged for other services such as a painter repainting the building in which the restaurant is located. The painter can use only some of the free meals so with the help of the barter company he exchanges the balance of the meals with a manufacturer/paint supplier for his paint needs on several projects. The paint manufacturer uses the meals in their staff training activities. The barter company which facilitates these exchanges also gets remunerated by the companies involved for its brokering activities so may also get a small quantity of meals. Discuss the recording issues for the restaurant and the barter company.

11 As soon as the authorities identify new rules, accountants for hire will attempt to find ways around the rules. Identify as many ways as you can in which the new rules can be circumvented.

12 The Australian Securities and Investment Commission (ASIC) required a company called Flight Centre to amend the way in which it recorded revenue. It had previously recorded revenue based on the gross value of the flights it had booked for customers, whereas the ASIC wanted it to record revenue based on the commissions it received. The change would have no impact on the reported profit of the company. Given the company disclosed its accounting policies and was consistent from year to year it appears no one was misled. Explain why the company would prefer the gross approach, and the regulator the net revenue approach, and whether the issue is worthy of such a debate. (Provide justification for your conclusions.)

13 Access the IASB website and critically review the various decisions made to arrive at the final IFRS on revenue recognition.

14 The traditional method for recognising revenue made no allowance for possible bad debts. The first (2010) exposure draft reduced the amount of revenue by the estimated amount of bad debts. In the second exposure draft the standard setters reverted to the traditional method of revenue recognition. Given the contribution of possible bad debts to the great financial crisis, was the reversion to the traditional method wise? Discuss.

15 A property developer sells a building to a company which has little cash but is willing to exchange vacant land which it owns at a sought-after location. An independent valuer assesses the vacant land at £5,000,000 and on the basis of that the government charges the recipient £50,000 as land transfer taxes. Legal fees of £5,000 are incurred by the property developer. How much should be recorded as revenue and why?

16 HP limits the amount of revenue recognised for a delivered element to the amount that is not contingent on the future delivery of products or services, future performance obligations or subject to customer-specified return or refund privileges. Discuss if this is the same as recognising revenue when it is highly probable of being achieved.

EXERCISES

* Question 1

Senford PLC entered into a contract to sell 3,000 telephones each with a two-year provider contract. The total cost of the contract was €120 per month payable at the end of the month. The phones were bought for cash from a supplier for €480 each and the cost of providing the telephone service is estimated at €30 a month for each phone. Senford PLC sells two-year service contracts (without supplying a phone) for €90 a month. The balance date for the company is three months after the date of the sales.

Required:
(a) Prepare all the journal entries for the current financial year that are possible from the data given. (Ignore financing costs.)
(b) Show the disclosures which will be necessary in the annual report in relation to these phone contracts.
Show your calculations.

* Question 2

Strayway PLC sells two planes to Elliott & Elliott Budget Airlines PLC for 5 million euros each payable in two years' time on presentation of an accepted bill of exchange to be presented through Lloyds Bank. The face value of the bill is €10,000,000. Further enquiry ascertains that government bonds with two years to maturity yield 4% p.a. and Strayway borrows from their bank at 9%, and the average yield on commercial bills of exchange payable in two years time is 8%.

Required:
Record the sale and associated transactions in the books of Strayway PLC.

* Question 3

Penrith European Car Sales plc sells a new car with 'free' 5,000 kilometre and 20,000 kilometre services for a combined price of €41,500. The cost of the car from the manufacturer is €30,000. The two services normally cost €400 and €600 to do and are charged to casual customers at the rate of €800 and €1,200 respectively.

Required:
Record all the transactions associated with the sale.

* Question 4

Facts:

Henry Falk subscribes to an online monthly gardening magazine and selects the option of a three-year subscription from the following options:

One issue	€12
Twelve issues	€120
Twenty-four issues	€200
Thirty-six issues	€300

The publisher, English Magazine Specialties plc, estimates that it will cost €60, €62 and €64 per annum respectively to supply the magazines for the respective years.

Henry duly pays by credit card as part of the subscribing process.

Show the entries in the books of the English Magazine Specialties plc.

Required:
(a) Show the entries to record the transactions associated with the sale and supply for years one, two and three. (Ignore interest for part a.)
(b) Assuming a 10% borrowing rate and a constant revenue stream show the accounting entries. Use 2% for credit card fees.
(c) Provide the disclosures which would need to be made in the annual report at the end of the first year of the contract.
(d) What other patterns of revenue recognition could be used?

* Question 5

Assume the same facts as in Question 4(b) but add the presence of a 7.5% value added tax on the sales and a 1% transaction cost paid to the supplier of credit card transactions.

Required:
(a) Show the entries to record the transactions associated with the sale and supply for years one, two and three. (Ignore imputed interest costs.)
(b) Provide the disclosures which would need to be made in the annual report at the end of the first year of the contract.

* Question 6

Five G Telephones enters into telephone contracts on the following terms and options:

Xyz mobile phones	€1,000
Basic Y phones	€200

Basic connection service options:

A	€40 per month
B	€15 plus 50 cents per call

Combined services:

Supply of Xyz and connection service €79 per month with a minimum contract period of 24 months.

Required:
Record the entries for the first month for each of the possible transactions assuming the following information:
(a) Xyz phones are purchased from a supplier for €500.
(b) Basic phones are purchased from a supplier for €120.
(c) The average user makes 100 calls a month at a variable cost of five cents per call.
(d) Ignore imputed interest.

Question 7

Assume the facts as per Question 6 with the following additional costs:

There is a five euro cost to add a customer to the phone system and those who buy the phone outright do not default on payments for the phone or the service contract, but those who have a combined service contract default in 10% of the cases.

Required:
(a) Show the entries to record the transactions associated with the sale and supply for years one, two and three.
(b) Provide the disclosures which would need to be made in the annual report at the end of the first year of the contract.

Question 8

Assume that in Question 3 you had also been told that cars without the inclusion of free services are typically sold by other sales outlets for €40,000.

Required:
Re-do the entries for the sale and the servicing. (Ignore interest as the timing of the servicing is not given and the amounts are modest.) Justify your answer.

Question 9

Complete Computer Services (CCS) sells computer packages which include supply of a computer which carries the normal warranty against faulty parts plus a two-year assistance package covering problems encountered using any software sold or supplied with the computer. This package is designed for the person who lacks confidence in the use of the computer and likes to feel they have help available if they want it. The costs which CCS incurs are €600 for the purchase of the computer, €20 for normal warranty costs and €30 a year to service the assistance service component. The sales revenue is €950 being assigned as €100 for the service assistance and €850 for the computer.

Required:
Show all relevant journal entries.

Question 10

Henry plc is a company which has established a reputation as a company which generates growth in sales from year to year and those growth prospects have been incorporated into share prices. However, the current year has been more difficult and the managing director does not want to disappoint the market. He has approached a friend with an idea he got from one of the auditors for pulling a rabbit out of the hat. The friend spends €3,000 to establish a company Dreams Come True Pty Ltd and contributes a further 2 million euros as the new company's paid-up capital. Henry plc then enters into a contract to lend 3 million euros to Dreams Come True Pty Ltd. Next, Henry plc sells a building to Dreams Come True Pty Ltd for 3 million euros for cash. A week after the end of the financial year Henry plc enters into a contract to repurchase the building from Dreams Come True for 3.5 million euros with an effective date six months into the new financial year. The managing director wants the accountant to record the building as sales revenue.

Required:
(a) Discuss the technical issues of the proposals.
(b) Discuss the ethical issues of the proposals.

Question 11

Exess Steel plc specialises in steelmaking and is located in the northwest of the country. Due to an unexpected downturn in demand for its steel products it has excess coking coal. South East Steel Products plc has also been caught by the unexpected economic downturn and has an excess of steel pellets. At the steel producers annual conference the two managing directors discuss how their different auditors want them to write down the value of the excess stock because of the economic circumstances. They agree to do an exchange of the two commodities with the contracts including selling prices at the cost values in their books and a small cash payment to cover the difference between the two valuations. The two items are recorded in the books at the agreed purchase and selling prices respectively.

Required:
Comment critically on this proposal.

Question 12

New Management plc is a pharmaceutical company selling to wholesalers and retail pharmacies. The new CEO was appointed at the start of the financial year and was full of enthusiasm. For the first six months her new ideas created a 10% increase in sales and then the economy crashed as the government cut spending and monetary policy was tightened. Sales dropped 20% as customers had slower sales and were required by their banks to reduce their overdrafts. A new strategy was adopted in the last two months of the year. Sales representatives were told to sell on the basis that customers would not have to pay for three months, by which time they would have sold the stock. They were also told that if sales for the month to that customer were not 5% higher than the sales for the corresponding month for the previous year, they could say to the customer, off the record of course, that they could return any unsold stock after four months. In the last two months of the financial year sales were up 10 and 11% on the respective previous corresponding periods. The first month of the new financial year recorded a 10% drop in sales.

Required:
Critically discuss from the point of view of (a) an investor, (b) the auditor and (c) the CEO.

Question 13

Renee Aluminum Products plc enters into an agreement to supply Skyline Window Installers plc with standard window frames at the retail prices at the time less 40%. Renee supplies 300 windows a month at £66 each. However, the agreement provides for price increases based on Renee's increases in the cost of aluminium.

Required:
(a) In July Renee advises that the price will increase by 4%. Record revenue for the month of July.
(b) If in July Renee advises Skyline Windows that the price increase of 4% applies retrospectively from 1 March, what entry would be made in Renee's accounts for revenue in July?

References

1 US Government Accountability Office, *Financial Restatements Update of Public Company Trends, Market Impacts and Regulatory Enforcement Activities*, GAO, 2007.
2 K.L. Anderson and T.L. Yohn, The Effect of 10K Restatements on Firm Value, Information Asymmetries and Investors' Reliance on Earnings, SSRN, Sept. 2002.
3 Christopher Williams, Autonomy profits slashed by HP accounts restatement, *The Telegraph*, 3 Feb 2014.
4 D. Collins, A. Masli, A.L. Reitenga and J.M. Sanchez, 'Earnings Restatements, the Sarbanes–Oxley Act and the Disciplining of Chief Financial Officers', *Journal of Accounting, Auditing and Finance*, vol. 24(1), pp. 1–34.
5 In November 2011 there was an Exposure Draft *Revenue from Contracts with Customers* and there was a staff paper *Effects of Joint IASB and FASB Redeliberations on the November 2011 Exposure Draft Revenue from Contracts with Customers* in February 2013. There were extensive consultations in arriving at the standard.
6 Hewlett-Packard Company and Subsidiaries, Annual Report 2013, Note 1, pp. 86–8.
7 http://accountingonion.com/2014/05/the-wild-west-of-nonauthoritative-gaap.html

PART 3

Regulatory framework – an attempt to achieve uniformity

In its 2013 Annual Report the directors commented:

> At the start of the year, the board set ambitious targets, ahead of the prevailing industry analysts' consensus, but as the year unfolded, short term performance fell below that determined when setting the budget. As a result, no executive director bonuses will be paid this year, even though underlying earnings per share fell by just 2%.

This would not, however, preclude companies from taking typical steps such as **deferring discretionary expenditure**, e.g. research, advertising or training expenditure; **deferring amortisation**, e.g. making optimistic sales projections in order to classify research as development expenditure which can be capitalised; and **reclassifying** deteriorating current assets as non-current assets to avoid the need to recognise a loss under the lower of cost and net realisable value rule applicable to current assets.

The introduction of a mandatory standard that changes management's ability to adopt such measures **affects wealth distribution** within the firm. For example, if managers are able to delay the amortisation of development expenditure, then bonuses related to profit will be higher and there will effectively have been a transfer of wealth to managers from shareholders.

9.3 Why do we need standards to be mandatory?

Mandatory standards are needed, therefore, to define the way in which accounting numbers are presented in financial statements, so that their measurement and presentation are less subjective. It had been thought that the accountancy profession could obtain uniformity of disclosure by persuasion but, in reality, the profession found it difficult to resist management pressures.

During the 1960s the financial sector of the UK economy lost confidence in the accountancy profession when internationally known UK-based companies were seen to have published financial data that were materially incorrect. Shareholders are normally unaware that this occurs and it tends to become public knowledge only in restricted circumstances, e.g. when a third party has a **vested interest** in revealing adverse facts following a takeover, or when a company falls into the hands of an administrator, inspector or liquidator, **whose duty it is to enquire and report** on shortcomings in the management of a company.

Two scandals which disturbed the public at the time, GEC/AEI and Pergamon Press, were both made public in the restricted circumstances referred to above, when financial reports prepared from the same basic information disclosed a materially different picture.

9.3.1 GEC takeover of AEI in 1967

The first calamity for the profession involved GEC Ltd in its takeover bid for AEI Ltd when the pre-takeover accounts prepared by the old AEI directors differed materially from the post-takeover accounts prepared by the new AEI directors.

Under the control of the directors of GEC the accounts of AEI were produced for 1967 showing a **loss of £4.5 million**. Unfortunately, this was from basic information that was largely the same as that used by AEI when producing its profit forecast of £10 million.

There can be two reasons for the difference between the figures produced. Either the facts have changed or the judgements made by the directors have changed. In this case, it seems there was a change in the facts to the extent of a post-acquisition closure of an AEI factory; this explained £5 million of the £14.5 million difference between the forecast profit and the actual loss. The remaining £9.5 million arose because of differences in judgement. For example, the new directors took a different view of the value of stock and work in progress.

9.3.2 Pergamon Press

Audited accounts were produced by Pergamon Press Ltd for 1968 showing a profit of approximately £2 million.

An independent investigation by Price Waterhouse suggested that this profit should be reduced by 75% because of a number of unacceptable valuations, e.g. there had been a failure to reduce certain stock to the lower of cost and net realisable value, and there had been a change in policy on the capitalisation of printing costs of back issues of scientific journals – they were treated as a cost of closing stock in 1968, but not as a cost of opening stock in 1968.

9.3.3 Public view of the accounting profession following such cases

It had long been recognised that accountancy is not an exact science, but it had not been appreciated just how much latitude there was for companies to produce vastly different results based on the same transactions. Given that the auditors were perfectly happy to sign that those accounts showing either a £10 million profit or a £4.5 million loss were true and fair, the public felt the need for action if investors were to have any trust in the figures that were being published.

The difficulty was that each firm of accountants tended to rely on precedents within its own firm in deciding what was true and fair. In fairness, there could be consistency within an audit firm's approach but not across all firms in the profession. The auditors were also under pressure to agree to practices that the directors wanted because there were no professional mandatory standards.

This was the scenario that galvanised the City press and the investing public. An embarrassed, disturbed profession announced in 1969, via the ICAEW, that there was a majority view supporting the introduction of Statements of Standard Accounting Practice to supplement the legislation.

9.3.4 Does the need for standards and effective enforcement still exist in the twenty-first century?

The scandals involving GEC and Pergamon Press occurred more than 45 years ago. However, the need for the ongoing enforcement of standards for financial reporting and auditing continues unabated. We only need to look at the unfortunate events with Enron and Ahold to arrive at an answer.

Enron

Enron was formed in the mid-1980s and became by the end of the 1990s the seventh-largest company in revenue terms in the USA. However, this concealed the fact that it had off-balance-sheet debts and that it had overstated its profits by more than $500 million – falling into bankruptcy (the largest in US corporate history) in 2001.

Ahold

In 2003 Ahold, the world's third-largest grocer, reported that its earnings for the past two years were overstated by more than $500 million as a result of local managers recording promotional allowances provided by suppliers to promote their goods at a figure greater than the cash received. This may reflect on the pressure to inflate profits when there are option schemes for managers.

9.4 Arguments in support of standards

The setting of standards has both supporters and opponents. Those who support standards have a view that they are important in giving investors confidence and encouraging informed investment. In this section we discuss credibility, discipline and comparability.

Credibility

The accountancy profession would lose all credibility if it permitted companies experiencing similar events to produce financial reports that disclosed markedly different results simply because they could select different accounting policies. Uniformity was seen as essential if financial reports were to disclose a true and fair view. However, it has been a continuing view in the UK and IASB that standards should be based on principles and not be seen as rigid rules – they were not to replace the exercise of informed judgement in determining what constituted a true and fair view in each circumstance. The US approach has been different – its approach has been to prescribe detailed rules.

Discipline

Directors are under pressure to maintain and improve the market valuation of their company's securities. There is a temptation, therefore, to influence any financial statistic that has an impact on the market valuation, such as the trend in the earnings per share (EPS) figure, the net asset backing for the shares or the gearing ratios which show the level of borrowing relative to the amount of equity capital put in by the shareholders.

This is an ever-present risk and the Financial Reporting Council showed awareness of the need to impose discipline when it stated in its Annual Review as far back as 1991 that the high level of company failures in the then recession, some of which were associated with **obscure financial reporting**, damaged confidence in the high standard of reporting by the majority of companies.

Comparability

In addition to financial statements allowing investors to evaluate the management's performance, i.e. their stewardship, they should also allow investors to make predictions of future cash flows and make comparisons with other companies.

In order to be able to make valid inter-company comparisons of performance and trends, investors need relevant and reliable data that have been standardised. If companies were to continue to apply different accounting policies to identical commercial activities, innocently or with the deliberate intention of disguising bad news, then investors could be misled in making their investment decisions.

9.5 Arguments against standards

We have so far discussed the arguments in support of standard setting. However, there are also arguments that have been made against, such as consensus-seeking and information overload with IFRSs themselves exceeding 3,000 pages.

Consensus-seeking

Consensus-seeking can lead to the issuing of standards that are over-influenced by those who fear that a new standard will adversely affect their statements of financial position. For example, we see retail companies, who lease many of their stores, oppose the proposal to put operating leases onto the statement of financial position rather than reporting simply the future commitment as a note to the accounts.

Overload

Standard overload is not a new charge. It has been put forward by those who consider that:

- There are too many standard setters with differing requirements, e.g. the FRC in the UK with FRSs; the FASB in the US with the Accounting Standards Codification; the IASB with IFRSs and IFRICs; the EU with separate endorsement of IFRS giving us EU-IFRS; and the EU with its Directives and national Stock Exchange listing requirements.

- Standards are too detailed if rule-based and not sufficiently detailed if principle-based, leading to the need for yet further guidance from the standard setters. For example, further guidance has to be issued by the International Financial Reporting Interpretations Committee (IFRC) when existing IFRSs do not provide the answer.

- There are too many notes to the accounts to satisfy regulatory requirements, for example disclosing charitable donations.

- There are too many notes to the accounts put in by companies themselves. Various surveys by professional accounting firms including one by Baker Tilly in 2012[2] showed that the majority of financial directors were keen to cut 'clutter' from financial disclosures, believing that the financial statements are too long, and that key messages are being lost as a result. It was felt that existing standards lead to a checklist mentality and boiler-plate disclosures that are not material and can obscure relevant information.

- There has been no definition of a note by the standard setters. This is being addressed by EFRAG with the issue of a Discussion Paper in 2012, *Towards a Disclosure Framework for the Notes*, which proposes how notes should be defined. For example, it is proposed that relevance, for instance, should only apply to disclosures that fulfil some *specific* users' needs.

- International standards have, until 2009 with the issue of *IFRS for SMEs*, focused on the large multinational companies and failed to recognise the different users and information needs between large and smaller entities.

9.6 Standard setting and enforcement by the Financial Reporting Council (FRC) in the UK

The Financial Reporting Council (FRC) was set up in 1990 as an independent regulator. Under the FRC the Accounting Standards Board (ASB) issued standards and the Financial Reporting Review Panel (FRRP) reviewed compliance to encourage high-quality financial reporting.

Due to its success in doing this, the government decided, following corporate disasters such as that of Enron in the USA, to give it a more **proactive** role from 2004 onwards in the areas of corporate governance, compliance with statutes and accounting and auditing standards.

Figure 9.1 FRC structure from 2012

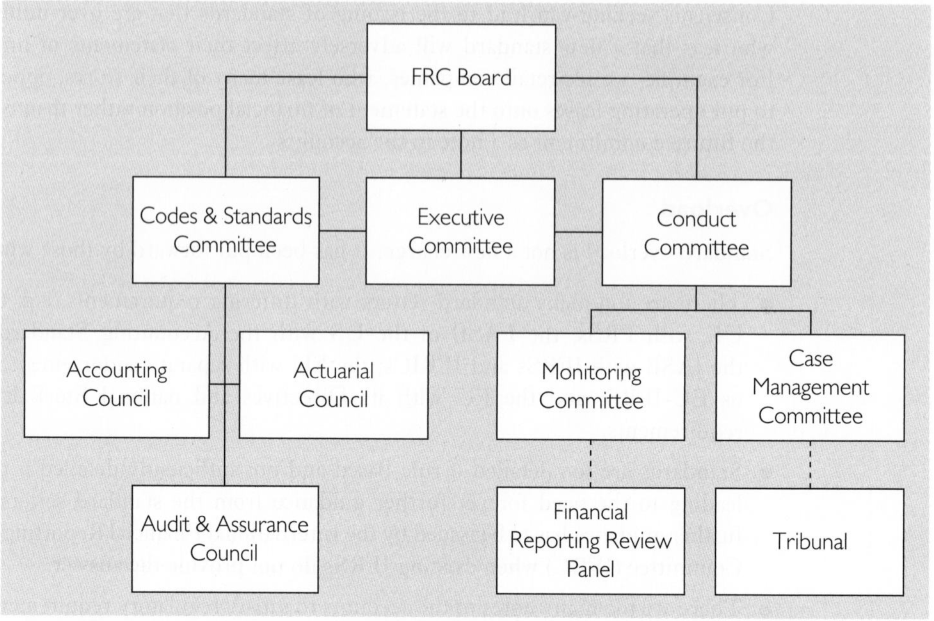

Countries experience alternating periods of favourable and unfavourable economic conditions – often described as 'boom and bust'. The FRC directs its reviews towards those sectors most likely to experience difficulties at the time. For example, it announced that its review activity in 2011/12 would focus on companies operating in niche markets, companies outside the FTSE 350, and companies providing support services with significant exposure to public spending cuts. It identified Commercial property, Insurance, Support services and Travel as priority sectors for review. It also pays particular attention to the reports and accounts of companies whose shareholders have raised concerns about governance or where there have been specific complaints.

9.6.1 The FRC structure

The FRC structure has evolved to meet changing needs. It was restructured in 2012 to operate as a unified regulatory body with enhanced independence. The new structure is shown in Figure 9.1.

9.6.2 The FRC Board

The FRC Board is supported by three committees: the Codes and Standards Committee, the Conduct Committee and the Executive Committee.

The Codes and Standards Committee

This will advise the FRC Board on matters relating to codes, standard setting and policy questions, through its Accounting, Actuarial and Audit & Assurance Councils. The *Accounting Council* replaces the Accounting Standards Board. It reports to the Codes and Standards Committee and is responsible for providing strategic input into the work-plan of the FRC as a whole and advising on draft national and international standards to ensure that high-quality, effective standards are produced.

The Conduct Committee

This will advise the FRC Board in matters relating to conduct to promote high-quality corporate reporting, including monitoring, oversight, investigative and disciplinary functions, through its Monitoring Committee and Case Management Committee. The *Monitoring Committee* will be concerned with the assessment and reviews of audit quality and decisions as to possible resulting sanctions and investigation leading to possible disciplinary action being taken.

The Executive Committee

This will support the Board by advising on strategic issues and providing day-to-day oversight of the work of the FRC.

9.6.3 The Financial Reporting Review Panel (FRRP)

Creative accounting

A research study[3] into companies that have been the subject of a public statement suggests that when a firm's performance comes under severe strain, even apparently well-governed firms can succumb to the pressure for creative accounting, and that good governance alone is not a sufficient condition for ensuring high-quality financial reporting. Enforcement is required.

Risk-based proactive approach to enforcement

The FRRP has a policing role with responsibility for overseeing some 2,500 companies. Its role is to review material departures from accounting standards and, where financial statements are defective, to require the company to take appropriate remedial action. It has the right to apply to the court to make companies comply, but it prefers to deal with defects by agreement and there has never been recourse to the court.

It selects companies and documents to be examined using a proactive risk-based approach or a mixed model where a risk-based approach is combined with a rotation and/or a sampling approach – a pure rotation approach or a pure reactive approach would not be acceptable.[4]

Cooperative approach to enforcement

The FRC published a report in 2013 showing that the FRRP had reviewed 264 sets of accounts selected from FTSE 100, FTSE 250, AIM and other listed and unlisted companies.[5] We continue to see good-quality corporate reporting by large public companies. The corporate reporting of FTSE 350 companies, in particular, remains at a good level.

There were, however, reservations expressed about the quality of reporting by some smaller listed and Alternative Investment Market (AIM) quoted companies that lacked the accounting expertise of their larger listed counterparts. In the EU this is being addressed by a new Accounting Directive which reduces the financial reporting applicable to smaller companies.

9.7 The International Accounting Standards Board

The International Accounting Standards Board (IASB) has responsibility for all technical matters including the preparation and implementation of standards. The IASB website (www.iasb.org.uk) explains that:

> The IASB is committed to developing, in the public interest, a single set of high quality, understandable and enforceable global accounting standards that require transparent and comparable information in general purpose financial statements.

In addition, the IASB co-operates with national accounting standard-setters to achieve convergence in accounting standards around the world.

The IASB adopted all current IASs and began issuing its own standards, International Financial Reporting Standards (IFRSs). The body of IASs and IFRSs is referred to collectively as 'IFRS'.

As a conceptual basis to assist when drafting Standards the IASC issued a *Framework for the Preparation and Presentation of Financial Statement*[6] in 1989 and adopted by the IASB in 2001.

9.7.1 The *Framework for the Preparation and Presentation of Financial Statements*

The position was that different social, economic and legal circumstances had led to countries producing financial statements using different criteria for defining elements, recognising and measuring items appearing in the profit and loss account and balance sheet. The IASB approach was to attempt to harmonise national regulations and move towards the adoption by countries of International Standards (IFRSs). The *Framework* is being gradually revised on the issue in 2010 of the Conceptual Framework. The objective of the Conceptual Framework project is to improve financial reporting by providing the IASB with a complete and updated set of concepts to use when it develops or revises standards. It is discussed further in Chapter 10.

Adoption by countries of IFRS

IFRS has been adopted in the EU for consolidated accounts since 2005. It is interesting to see how the other G20 countries are gradually moving towards the use of IFRS. Of these, the position with the BRICS countries is that Brazil has required individual companies to use IFRS since 2008, Russia already uses it, India is converging with IFRS, China has substantially converged national standards and South Africa has required IFRS for listed companies since 2005. Japan requires mandatory compliance by 2016.

What has been the impact of adopting IFRS

There have been numerous benefits claimed, for example:

● Multinationals see a reduction in the cost of capital and easier access to international equity markets.

● Investors see shares in companies adopting IFRS becoming more liquid and have greater confidence in earnings per share figures when making investment decisions.

● National standard setters see an advantage in the shared development of standards.

A detailed research report on the experience of converging IFRSs in China, *Does IFRS Convergence Affect Financial Reporting Quality in China?*, makes interesting reading.[7]

The report noted that there was a significant increase in the value relevance of reported earnings for the firms following mandatory adoption of IFRS-converged Chinese Accounting Standards (CAS). It also identified how access to external finance was an incentive to achieve improved quality of financial reporting, discussing the effect where companies are in the manufacturing sector, operating in less developed regions or operating under foreign ownership, in contrast to those under central government control or those that have financial problems and are tempted to manage earnings to avoid delisting.

Extant IASs and IFRSs are listed in Figure 9.2.

Figure 9.2 Extant international standards

IAS 1	Presentation of Financial Statements
IAS 2	Inventories
IAS 7	Statement of Cash Flows
IAS 8	Accounting Policies, Changes in Accounting Estimates and Errors
IAS 10	Events after the Reporting Period
IAS 11	Construction Contracts
IAS 12	Income Taxes
IAS 16	Property, Plant and Equipment
IAS 17	Leases
IAS 18	Revenue
IAS 19	Employee Benefits
IAS 20	Accounting for Government Grants and Disclosure of Government Assistance
IAS 21	The Effects of Changes in Foreign Exchange Rates
IAS 23	Borrowing Costs
IAS 24	Related Party Disclosures
IAS 26	Accounting and Reporting by Retirement Benefit Plans
IAS 27	Separate Financial Statements
IAS 28	Investments in Associates and Joint Ventures
IAS 29	Financial Reporting in Hyperinflationary Economies
IAS 32	Financial Instruments: Presentation
IAS 33	Earnings per Share
IAS 34	Interim Financial Reporting
IAS 36	Impairment of Assets
IAS 37	Provisions, Contingent Liabilities and Contingent Assets
IAS 38	Intangible Assets
IAS 39	Financial Instruments: Recognition and Measurement
IAS 40	Investment Properties
IAS 41	Agriculture
IFRS 1	First-time Adoption of International Financial Reporting Standards
IFRS 2	Share-based Payment
IFRS 3	(Revised) Business Combinations
IFRS 4	Insurance Contracts
IFRS 5	Non-current Assets Held for Sale and Discontinued Operations
IFRS 6	Exploration for and Evaluation of Mineral Resources
IFRS 7	Financial Instruments Disclosures
IFRS 8	Operating Segments
IFRS 9	Financial Instruments
IFRS 10	Consolidated Financial Statements
IFRS 11	Joint Arrangements
IFRS 12	Disclosure of Interests in Other Entities
IFRS 13	Fair Value Measurement
IFRS 14	Regulatory Deferral Accounts
IFRS 15	Revenue from Contracts with Customers

9.8 Standard setting and enforcement in the European Union (EU)[8]

A major aim of the EU has been to create a single financial market that requires access by investors to financial reports which have been prepared using common financial reporting standards. The initial steps were the issue of accounting directives – these were the Fourth Directive,[9] the Seventh Directive and the Eighth Directive which were required to be adopted by each EU country into their national laws – in the UK it is the Companies Act 2006. They were subsumed in 2013 into a new Accounting Directive.

9.8.1 The new Accounting Directive[10]

There has been ongoing pressure to reduce the regulatory burden on small and medium-sized enterprises. The European Commission responded to this pressure by issuing a new Accounting Directive.

It is not a conceptual rewrite of the Directives. It aimed to address two of the major problems with the existing Directives, which were the lack of comparability arising from use of Member State Options (MSOs) and the unnecessary regulatory burden placed on SMEs.

The lack of comparability

The new Directive has achieved a small reduction in the number of options available to member states. It formalises fundamental accounting principles (although still with some Member State Options) for recognition and presentation in the financial statements. These are:

(i) There is a going concern presumption.

(ii) Accounts are to be prepared on an accrual basis.

(iii) Accounting policies and measurement bases are to be applied consistently between accounting periods.

(iv) Recognition and measurement are to be on a prudent basis, and in particular:

    (a) Items are to be measured at price or production cost.

        There is an option that allows for the revaluation of non-current assets and the use of fair values for financial and non-financial assets.

    (b) Only profits made at the balance sheet date are to be recognised.

    (c) Individual assets and liabilities are to be valued separately and set-off is not permitted.

    (d) All liabilities arising in the course of a financial year are to be recognised even if identified after the year.

    (e) All items are to be accounted for and presented in accordance with the substance of the transaction.

    (f) All negative value adjustments are to be recognised whether the result for the financial year is a profit or a loss.

(v) Materiality applies to recognition, measurement, presentation, disclosure and consolidation.

There have also been some arbitrary changes such as the requirement to write off goodwill over a period of between 5 and 10 years in exceptional cases, where the useful life of goodwill and development costs cannot be reliably estimated, and continuing options such as the option for the related costs of borrowing to be added to the cost of fixed and current assets.

Regulatory burden on SMEs

In order to further simplify the requirements for SMEs and micro-undertakings the European Commission adopted a 'bottom-up' approach that started with the requirements for small undertakings and then added additional accounting and reporting requirements as undertakings passed the thresholds for medium and large undertakings. It has set new size thresholds for determining the category as follows:

Undertakings	Turnover (€)	Balance sheet total (€)	Average number of employees
Micro	<=0.7m	0.35m	10
Small	<=8.0m	<=4.0m	50
Medium-sized	<=40.0m	<=20.0m	250
Large	<=40.0m	<=20.0m	250

The undertaking must be within any two of the three thresholds for two successive accounting periods. The default thresholds for small undertakings' turnover and balance sheet total are €8 million and €4 million, respectively. However, member states have the option to increase either or both of these thresholds for small undertakings up to a maximum of €12 million and €6 million, respectively.

The need for small companies to be audited has been removed.

Progressive increase in requirements from the bottom up

The Directive starts by listing the reporting requirements applicable to a small company and then increases the disclosures required from a medium-sized and a large undertaking. A helpful summary of the requirements is provided by the Federation of European Accountants (FEE).[11]

9.8.2 Enforcement of standards in Europe

The European Enforcers Co-ordination Sessions (EECS) is a forum containing 37 European enforcers from 29 countries in the European Economic Area (EEA) which aims to promote a high level of consistency amongst enforcers in the decisions they take in respect of their reviews of financial statements.

It has as its main objective the coordination of the enforcement activities of member states in order to foster investor confidence.

A report[12] issued in 2013 provided an overview of the monitoring of compliance in 2012 with International Financial Reporting Standards (IFRS). It was not a rubber-stamping exercise, with the European enforcers performing full reviews of around 1,050 interim and annual accounts covering around 17% of listed entities' accounts in Europe. In addition, 1,200 accounts were subject to partial review, representing a coverage of 20% of the population of listed entities.

9.8.3 The importance of enforcement

There is research evidence[13] that the cost of capital falls following the mandatory adoption of IFRS and that there is an increase in foreign equity investment. However, in addition to the standards, effective enforcement has to be in place.[14]

Even following the mandatory adoption of IFRS and enforcement of reporting standards, investors continue to take into account national considerations such as the existence of good corporate governance, the degree of shareholder protection and the level of corruption.

9.9 Standard setting and enforcement in the US

Reporting standards are set by the Financial Accounting Standards Board (FASB) and enforced by the Securities Exchange Commission. Since 2002 it has also been necessary to satisfy the requirements of the Sarbanes–Oxley Act (normally referred to as SOX) which was passed following the Enron disaster.

9.9.1 Standard setting by the FASB and other bodies

The Financial Accounting Standards Board (FASB) is responsible for setting accounting standards in the USA. The FASB is financed by a compulsory levy on public companies, which should ensure its independence. (The previous system of voluntary contributions ran the risk of major donors trying to exert undue influence on the Board.) In 2009 the FASB launched the FASB *Accounting Standards Codification* as the single source of authoritative non-governmental US Generally Accepted Accounting Principles (GAAP), combining and replacing the jumbled mix of accounting standards that have evolved over the last half-century.

9.9.2 Enforcement by the SEC

The Securities and Exchange Commission (SEC) is responsible for requiring the publication of financial information for the benefit of shareholders. It has the power to dictate the form and content of these reports. The largest companies whose shares are listed must register with the SEC and comply with its regulations. The SEC monitors financial reports filed in great detail and makes useful information available to the public via its website.[15] However, it is important to note that the majority of companies fall outside the SEC's jurisdiction.

9.9.3 SOX (the Sarbanes–Oxley Act 2002)

SOX came as a response to the failures in Enron. It is different from the UK's Code of Corporate Governance in that, rather than the comply-or-explain approach, compliance is mandatory with significant potential sanctions for individual directors where there is non-compliance.

Prevention of fraud

The SOX objectives are to reduce the risk of fraud. It provides that

> Whoever knowingly alters, destroys, mutilates, conceals, covers up, falsifies, or makes a false entry in any record, document, or tangible object with the intent to impede, obstruct, or influence the investigation or proper administration of any matter within the jurisdiction of any department or agency of the United States . . . shall be fined under this title, imprisoned not more than 20 years, or both. (Section 802(a))

Following the Enron and other scandals, a number of weaknesses were identified which allowed the frauds to go undetected. Weaknesses included (a) the accounting profession where there was inadequate oversight and conflicts of interest, (b) company management that had poor internal controls and had been subject to weak corporate governance procedures, and (c) investors under-protected with stock analysts giving biased investment advice, the FASB which was responsible for inadequate disclosure rules, and an under-funded enforcement agency in the Securities and Exchange Commission (SEC).

Management

CEOs of publicly traded companies are now directly responsible for ensuring that financial reports are accurate. To protect themselves CEOs rely on a sound system of internal control and management is accountable for the quality of those controls. Under SOX, management is required to certify the company's financial reports and both management and an independent accountant are required to certify the organisation's internal controls.

Investors

SOX aimed to reduce fraud and improve investor confidence in financial reports and the capital market by seeking improvements in corporate accounting controls. In doing so it has created mandatory requirements that might have disadvantaged US companies operating in a global market where there is a comply-or-explain approach to compliance as in the UK and OECD countries.

9.9.4 Progress towards adoption by the USA of international standards

There has been progress since 2002 following the Norwalk Agreement on making the US standards and IFRS fully compatible and to coordinate future work programmes.

The FASB and IASB have worked together on joint projects such as Revenue Recognition and Leasing. However, there is a view[16] that the future of further convergence remains uncertain as the Boards shift attention to their own independent agendas.

It has to be recognised that there appears to be little enthusiasm for the adoption of IFRS in place of US GAAP. This is understandable when realising that the question as to whether moving to IFRS is actually in the best interests of the US securities markets generally and US investors individually is unresolved. Their decision is also influenced by their view that (a) the IASB is underfunded and too reliant on the major accountancy firms and (b) their assessment that there is neither a consistent application nor enforcement of IFRSs globally.

9.10 Advantages and disadvantages of global standards for publicly accountable entities

Publicly accountable entities are those whose debt or equity is publicly traded. Many are multinational and listed on a stock exchange in more than one country.

9.10.1 Advantages

The main advantages arising from the development of international standards are that it reduces the cost of reporting under different standards, makes it easier to raise cross-border finance, leads to a decrease in firms' costs of capital with a corresponding increase in share prices, and enables investors to compare performance. For developing countries there is also the incentive to improve accountants' technical training and expertise.

9.10.2 Disadvantages

Complexity

However, one survey[17] carried out in the UK indicated that finance directors and auditors surveyed felt that IFRSs undermined UK reporting integrity. In particular, there was little support for the further use of fair values as a basis for financial reporting, which was regarded as making the accounts less reliable with comments such as 'I think the use of fair values increases the subjective nature of the accounts and confuses unqualified users'.

There was further reference to this problem of understanding with a further comment: 'IFRS/US GAAP have generally gone too far – now nobody other than the Big 4 technical departments and the SEC know what they mean. The analyst community doesn't even bother trying to understand them – so who exactly do the IASB think they are satisfying?'

Impact on net profit and equity

IFRS 1 *First-time Adoption of International Financial Reporting Standards* requires companies to produce a reconciliation of their IFRS equity and profit/loss to their equity and profit/loss reported under national GAAP.

A research report[18] prepared for the Institute of Chartered Accountants in Scotland in 2008, *The Implementation of IFRS in the UK, Italy and Ireland*, analysed the impact on net profit and equity of selected standards which showed whether the standard had caused an increase or decrease in reported net profit due to the introduction of the standard.

Their analysis showed that adopting IFRS resulted in the net profit being increased in each of the countries with the net profit under national GAAP being 66% of the IFRS figure for UK companies, 89% for Italian and 89% for Irish companies. By contrast, the equity of the average company was less under IFRS, with the equity under national GAAP being 153% higher than that under IFRS.

These are average changes and the impact on an individual company might be very different. For example, there was a dramatic effect on the headline figures for Wassenan, a Dutch company, which reported an increase of over 400% in its net income figure when the Dutch GAAP accounts were restated under IFRS. In other cases, there may be some large adjustments to individual balances, but the net effect may be less obvious.

In the short term, these changes in reported figures can have important consequences for companies' contractual obligations (e.g. they may not be able to maintain the level of liquidity required by their loan agreements) and their ability to pay dividends. There may be motivational issues to consider where staff bonuses have traditionally been based on reported accounting profit. As a result, companies may find that they need to adjust their management accounting system to align it more closely with IFRS.

Volatility in the accounts

In most countries the use of IFRS will mean that earnings and statement of financial position values will be more volatile than in the past. This could be quite a culture shock for analysts and others used to examining trends that may have followed a fairly predictable straight line.

Lack of familiarity

While the change to IFRS was covered in the professional and the more general press, it was not clear whether users of financial statements fully appreciated the effect of the change in accounting regulations, although surveys by KPMG[19] and PricewaterhouseCoopers[20] indicated that most analysts and investors were confident that they understood the implications of the change. A survey following the issue of *IFRS for SMEs* in 2009 indicated that, as a new standard, there was naturally a fairly widespread lack of understanding of its provisions. This has been well addressed[21] by the IASB with supporting workshops and educational material.

9.11 How do reporting requirements differ for non-publicly accountable entities?

The EU, national governments and standard setters have realised that there are numerous small and medium-sized businesses that do not raise funds on the stock exchange and do not prepare general-purpose financial statements for external users. Countries adopting IFRS for publicly accountable entities have, therefore, been able to issue their own national standards for non-publicly accountable entities.

In the UK companies have a statutory obligation to submit accounts annually to the shareholders and file a copy with the Registrar of Companies. In recognition of the cost implications and need for different levels of privacy, there is provision for small and medium-sized companies to file abbreviated accounts and adopt *Financial Reporting Standard for Smaller Entities* (FRSSE).[22] This standard follows the top-down approach to standard setting by reducing some of the disclosures required by standards applicable to listed companies – such as disclosing an earnings per share figure.

As regards the long-term future of FRSSE, there is still uncertainty as there are currently consultations on accounting for micro-entities and EU plans for possible maximum harmonisation of reduced disclosures for small entities. Whereas the UK has the FRSSE, the IASB has issued the *IFRS for SMEs*.

9.12 IFRS for SMEs

The IASB issued *IFRS for SMEs* in July 2009. The approach follows that adopted by the ASB with FRSSE:

- some topics omitted, e.g. IAS 33 *Earnings per Share*, IFRS 8 *Operating Segments*, IAS 34 *Interim Financial Reporting*, IFRS 5 *Assets Held for Sale* and IFRS 4 *Insurance Contracts*;
- some additional requirements as companies adopting the IFRS will have to comply with its mandatory requirement to produce a statement of cash flows and more information as to related party transactions;
- simpler options allowed, e.g. expensing rather than capitalising borrowing cost;
- simpler recognition, e.g. allowing an amortisation (with a maximum life of 10 years) rather than an annual impairment review for goodwill;
- simpler measurement, e.g. using the historical cost-depreciation model for property, plant and equipment;
- SMEs are not prevented from adopting other options available under full IFRS and may elect to do this if they so decide.

However, in defining an SME it has moved away from the size tests towards a definition based on qualitative factors such as public accountability whereby an SME would be a business that does not have public accountability. Public accountability is implied if outside stakeholders have a high degree of investment, commercial or social interest and if the majority of stakeholders have no alternative to the external financial report for financial information.

It is intended to have a three-yearly review of the implementation of the standard and it is reasonable to expect that the IFRS will evolve based on review findings.

Longer-term future

This is in some doubt with the issue of the new Accounting Directive in 2013. This is designed to reduce unnecessary and disproportionate administrative costs on small companies by simplifying the preparation of financial statements and reducing the amount of information required by small companies in the notes to financial statements.

Under the Directive, small companies are only required to prepare a balance sheet, a profit and loss account and notes to meet regulatory requirements. When examining the various policy options available to replace the old Accounting Directives, the Commission examined and rejected the option to adopt the *IFRS for SMEs* at EU level as the Commission deemed that *IFRS for SMEs* did not meet the objective of reducing the administrative burden.

9.13 Why have there been differences in financial reporting?

Although there have been national standard-setting bodies, this has not resulted in uniform standards. A number of attempts have been made to identify reasons for differences in financial reporting.[23] The issue is far from clear but most writers agree that the following are among the main factors influencing the development of financial reporting:

- the character of the national legal system;
- the way in which industry is financed;
- the relationship of the tax and reporting systems;
- the influence and status of the accounting profession;
- the extent to which accounting theory is developed;
- accidents of history;
- language.

We will consider the effect of each of these.

9.13.1 The character of the national legal system

There are two major legal systems, that based on common law and that based on Roman law. It is important to recognise this because the legal systems influence the way in which behaviour in a country, including accounting and financial reporting, is regulated.

Countries with a legal system based on common law include the UK, Ireland, the USA, Australia, Canada and New Zealand. These countries rely on the application of equity to specific cases rather than a set of detailed rules to be applied in all cases. The effect in the UK, as far as financial reporting was concerned, was that there was limited legislation regulating the form and content of financial statements until the government was required to implement the EC Fourth Directive. The directive was implemented in the UK by the passing of the Companies Act 1981 and this can be seen as a watershed because it was the first time that the layout of company accounts had been prescribed by statute in the UK.

English common law heritage was accommodated within the legislation by the provision that the detailed regulations of the Act should not be applied if, in the judgement of the directors, strict adherence to the Act would result in financial statements that did not present a true and fair view.

Countries with a legal system based on Roman law include France, Germany and Japan. These countries rely on the codification of detailed rules, which are often included within their companies legislation. The result is that there is less flexibility in the preparation of financial reports in those countries. They are less inclined to look to fine distinctions to justify different reporting treatments, which is inherent in the common law approach. The existence of detailed rules or existing effective publication requirements also determines their approach to reporting standards as, for example, the reluctance in Germany to support the adoption of IFRS for SMEs.

However, it is not just that common law countries have fewer codified laws than Roman law countries. There is a fundamental difference in the way in which the reporting of commercial transactions is approached. In the common law countries there is an established practice of creative compliance. By this we mean that the spirit of the law is elusive[24] and management is more inclined to act with creative compliance in order to escape effective legal control. By creative compliance we mean that management complies with the form of the regulation but in a way that might be against its spirit, e.g. structuring leasing agreements in the most acceptable way for financial reporting purposes. This is addressed in an *ad hoc* manner with standards requiring the substance of a transaction to determine its treatment in the financial statements or revising individual standards to combat creative compliance.

9.13.2 The way in which industry is financed

Accountancy is the art of communicating relevant financial information about a business entity to users. One of the considerations to take into account when deciding what is relevant is the way in which the business has been financed, e.g. the information needs of equity investors will be different from those of loan creditors. This is one factor responsible for international financial reporting differences because the predominant provider of capital is different in different countries.[25] Figure 9.3 makes a simple comparison between domestic equity market capitalisation and Gross Domestic Product (GDP).[26] The higher the ratio, the greater the importance of the equity market compared with loan finance.

We see that in the USA companies have relied more heavily on individual investors to provide finance than in Europe or Japan. An active stock exchange has developed to allow shareholders to liquidate their investments. A system of financial reporting has evolved to satisfy a stewardship need where prudence and conservatism predominate, and to meet the capital market need for fair information[27] which allows interested parties to deal on an equal footing where the accruals concept and the doctrine of substance over form predominate. It is important to note that European statistics are *averages* that do not fully reflect the variation in sources of finance used between, say, the UK (where equity investment is very important) and Germany (where lending is more important). These could be important factors in the development of accounting.

Figure 9.3 Domestic equity market capitalisation/gross domestic product

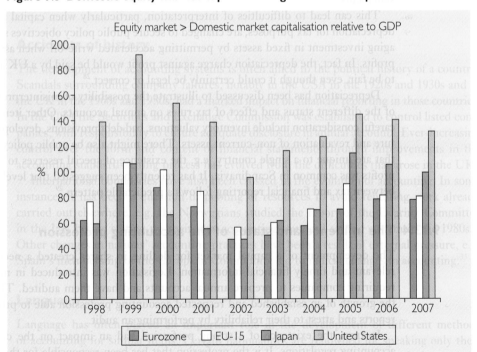

Typical economic decisions that are being made include:

- Assessing by stakeholders:
  - the stewardship or accountability of management;
  - when to buy, hold or sell an equity investment;
  - how much of the distributable profits to pay out as dividends;
  - the ability of the entity to pay and provide other benefits to its employees;
  - the security for amounts lent to the entity.
- Determining by regulatory authorities:
  - taxation policies;
  - how to regulate the activities of entities.

10.4.1 Revising the *Framework for the Preparation and Presentation of Financial Statements*

Just as with the convergence project referred to in Chapter 9, the IASB and FASB set up a joint project in 2004 to review the *Framework* in phases aiming to produce concepts relating to objectives and qualitative characteristics, elements and recognition, measurement and presentation and disclosure. There was also to be a phase aimed at reaching a converged IASB–FASB view on the secondary purpose of the framework.

The IASB and FASB worked jointly until 2010. After that date the FASB and the IASB suspended their work on the project to focus their resources on other projects. In 2012, following a public consultation that indicated that users regarded the revision of the Framework as a priority, the IASB revived the project as an *IASB-only comprehensive project*.

The position with regard to the proposed phases was at that time as follows:

Phase A: **Objectives and qualitative characteristics**

This is the only phase that has been completed and in 2010 the *Conceptual Framework for Financial Reporting* (2010) was issued.

Phase D: **Reporting entity**

An Exposure Draft ED/2010/2 *Conceptual Framework for Financial Reporting: The Reporting Entity* was published and is to be further considered as part of the IASB-only comprehensive project.

Phases which were to be considered further as part of the IASB-only comprehensive project:

Phase B: Elements and recognition

Phase C: Measurement

Phase E: Presentation and disclosure.

10.5 *Conceptual Framework for Financial Reporting 2010*

The *Framework* was to be produced in chapters as follows:

- Chapter 1 The objective of general purpose financial reporting
- Chapter 2 The reporting entity

- Chapter 3 Qualitative characteristics of useful financial information
- Chapter 4 The Framework (1989): this chapter contains the remaining text from the 1989 *Framework* and relevant paragraphs will be replaced by new chapters as the project progresses.

We comment briefly on each chapter.

10.5.1 Chapter 1 The objective of financial statements

This is a non-contentious chapter. In the USA, Australia, Canada, the UK and the IASB, the approach has been the same, i.e. commencing with a consideration of the objectives of financial statements, qualitative characteristics of financial information, definition of the elements, and when these are to be recognised in the financial statements. There is a general agreement on these areas.

The fundamental objective of general-purpose financial reporting is to provide financial information about the reporting entity that is useful to present and potential equity investors, lenders and other creditors when making investment and loan decisions. Information is needed to help them assess the prospects for future cash flows which, based to an extent on the review of past performance, will assist in assessing stewardship.

What information should be provided to satisfy the information needs?

The *Framework* proposes that information is required on resources controlled by the entity and claims against it and changes in those resources and claims.

The information on resources and claims is provided in the statement of financial position and on changes in resources and claims in the statements of income, cash flows and changes in equity.

Statement of financial position

This reports the economic resources and claims. It allows us to assess:

- the financial structure, i.e. capital gearing indicating how profits will be divided between the different sources of finance;
- the ability to repay or raise new financing;
- solvency and liquidity, i.e. current and liquid ratios;
- the possibility of obtaining cash from disposal of assets without disrupting continuing business, i.e. realise readily marketable securities that might have been built up as a liquid reserve.

Statement of income

This reports changes in economic resources on an accrual basis. It allows us to assess financial performance defined as the return an entity obtains from the resources it controls. It provides a means to assess past management performance, how effectively resources have been utilised and the capacity to generate cash flows.

Statement of cash flows

This reports changes in economic resources on a cash basis free from allocation and valuation issues. It allows us to identify net cash flows from operating, investment and financing activities. It provides a means to review cash flows that had been used when making CAPEX decisions and to assess the feasibility of free cash flows to meet future capital investment requirements and the possible impact on dividend policy.

Statement of changes in equity

This reports changes in economic resources – other than from financial performance. It identifies increases or decreases in issued capital and distributions to shareholders.

Do these financial statements satisfy all user needs?

The general-purpose financial reports do not and cannot provide all of the information that existing and potential investors, lenders and other creditors need. Users other than investors need to consider information from other sources, for example general economic conditions and expectations, political events and political climate, and industry and company outlooks.

10.5.2 Chapter 2 The reporting entity

This chapter on the reporting entity was to be inserted after re-deliberations by the IASB following its Exposure Draft ED/2010/2 issued in March 2010.

10.5.3 Chapter 3 Qualitative characteristics of useful financial information

There are two fundamental qualitative characteristics if information is to be decision-useful and not misleading. These are **relevance** and **faithful representation**. There are also characteristics which are referred to as **enhancing**. We will now discuss each of these.

(i) Relevance

Relevant financial information is capable of making a difference to the decisions made by users. Financial information is capable of making a difference to decisions if it has **predictive value, confirmatory value** or both.

Predictive value

Financial information has predictive value if it can be used as an input to processes employed by users to predict future outcomes. It is used by users in making their own predictions.

Confirmatory value

Financial information has confirmatory value if it provides feedback about (confirms or changes) previous evaluations.

Predictive and confirmatory values

Predictive and confirmatory values are interrelated. For example, revenue information for the current year, which can be used as the basis for predicting revenues in future years, can also be compared with revenue predictions for the current year that were made in past years.

Materiality in relation to relevance

It is defined in IAS 8 *Accounting Policies, Changes in Accounting Estimates and Errors* as being information whose omission or misstatement could, individually or collectively, influence the economic decisions of users.

It is a matter of judgement and decided on a case-by-case basis: what may be material for one company might not be for another – £1 million is a large amount but in relation to a potential misstatement of sales by a large multinational, it is likely to be immaterial. Conversely, if it relates to a disclosure required by legislation, then a comparatively small amount of £10,000 might be seen as material, even for a large multinational, if it relates to a benefit-in-kind which has been wrongly omitted from the disclosure of directors' remuneration.[1]

There are situations in which an item will be judged material and a lower-level value set as a benchmark because of the circumstances, such as if it results in non-compliance with a bank loan or turns a profit into a loss.

Materiality benchmarks

Materiality is entity-specific. It depends on the size of the item or error judged in the particular circumstances of its omission or misstatement. The need to exercise judgement means that the preparer needs to have a benchmark.

Many accountants and auditors have a rule of thumb,[2] assuming, for example, that if an item falls under a 5% threshold it is not material. This is to be regarded as a starting point only and exclusive reliance on this or any percentage or numerical threshold has no basis in the accounting literature or the law.

Progress on consideration of materiality in financial reports

ESMA issued a Consultation Paper[3] *Considerations of Materiality in Financial Reporting* in 2011. Following the feedback it received, it referred the topic to the IASB for their further consideration. In 2014 the IASB tentatively decided to undertake a project on materiality as part of its Disclosure Initiative. The project should also develop application guidance and educational material.

(ii) Faithful representation

Financial reports represent the effect of economic activities in words and numbers. A faithful representation would need to be complete, neutral and free from error.

Neutral

This means that the information has not been slanted, weighted, emphasised, de-emphasised or otherwise manipulated to increase the probability that financial information will be received favourably or unfavourably by users.

Freedom from error

Faithful representation does not mean the information is 100% accurate. 'Free from error' means there are no material errors or omissions in the description of the event or transaction and no errors in the process used to produce the reported information.

Taking the reporting of an estimate as an example, a representation of that estimate can be faithful if (a) the amount is described clearly and accurately as being an estimate and (b) the nature and limitations of the estimating process are explained and (c) no errors have been made in selecting and applying an appropriate process for developing the estimate.

(iii) Enhancing qualities

There are other characteristics that may make the information more useful. These are:

- comparability – allowing for both inter-company and inter-period comparisons;
- verifiability – means that different knowledgeable and independent observers could broadly agree that the report provides a faithful representation of transactions;
- timeliness – information is available to decision-makers in time to be capable of influencing their decisions;
- understandability – assumes that users have a reasonable knowledge of business and economic activities.

10.5.4 Chapter 4 The Framework (1989)

This contained the remaining content of the 1989 *Framework* which were to be eventually replaced after the various phases were completed.

10.6 Chapter 4 content

The content covers (a) underlying assumptions and (b) the definition, recognition and measurement of elements.

(a) Underlying assumptions

The underlying assumptions are that financial statements are prepared on the basis of the following.

Accruals
Transactions are recognised when they occur, rather than when cash or its equivalent is received or paid, and they are reported in the financial statements of the periods to which they relate. We have seen in Chapter 2 that, in addition to cash receipts and payments, obligations to pay cash in the future and resources that represent cash to be received in the future are also reported.

Going concern
The financial statements presume that an enterprise will continue in operation indefinitely or, if that presumption is not valid, disclosure and a different basis of reporting are required, such as preparing the statements using the net realisable value accounting model described in Chapter 7.

Consistency
In order to achieve comparability, the accounting policies are followed consistently from one period to another; any change in an accounting policy is made only to achieve a fairer representation or in response to a change in IFRS requirements.

Transactions reported in the financial statements are classified into groups according to their economic characteristics. These groups are referred to as elements. These elements are **defined** and there are rules as to how they are to be **recognised** and **measured** when reported in the financial statements.

(b) Defining elements

(i) Defining elements reported in the statement of financial position
The three elements that appear in the statement of financial position are Assets, Liabilities and Equity. They are defined as follows.

Assets
An asset is a resource *controlled* by the enterprise as a result of *past* events and from which *future* economic benefits are expected to flow to the enterprise.

Liabilities
A liability is a *present* obligation of the enterprise arising from *past* events, the settlement of which is expected to result in an *outflow* from the enterprise of resources embodying economic benefits.

Equity

Equity is the *residual* interest in the assets of the enterprise after deducting all its liabilities.

(ii) Defining elements reported in the statement of income

The elements that appear in the statement of income are income and expenses.

Income

Income is increases in economic benefits from revenue or gains during the accounting period that result in increases in equity, other than contributions from equity shareholders.

Expenses

Expenses are decreases in economic benefits from expenses incurred in the ordinary business and losses arising during the accounting period that result in decreases in equity, other than those relating to distributions to equity shareholders.

(c) Recognising elements

(i) Recognition in the statement of financial position

An **asset** is recognised in the statement of financial position when it is probable that the future economic benefits will flow to the enterprise and the asset has a cost or value that can be measured reliably.

A **liability** is recognised in the statement of financial position when it is probable that an outflow of resources embodying economic benefits will result from the settlement of a present obligation and the amount at which the settlement will take place can be measured reliably.

(ii) Recognition in the statement of income

Income is recognised in the statement of income when there has been an increase in an asset or a decrease of a liability has arisen that can be measured reliably.

Expenses are recognised when there has been a decrease in an asset or an increase of a liability has arisen that can be measured reliably.

(d) Measurement of all the elements of financial statements

Measurement requires elements to be reported in monetary amounts. The *Framework* recognises that elements are reported using a variety of bases which include those covered in Chapter 7, i.e. historical cost, current cost, net realisable value and present value. It does not give general guidance on the capital maintenance concept to apply but does specify a required base for particular elements on an individual IFRS basis – such as requiring inventory to be reported at the lower of cost and net realisable value.

10.7 A Review of the *Conceptual Framework for Financial Reporting*

The IASB view is that the existing *Conceptual Framework* had enabled it to develop high-quality IFRS. It will continue to adopt a principles-based rather than rule-based approach and will continue to revise the *Framework* following feedback to its 2013 Discussion Paper DP/2013/1 *A Review of the Conceptual Framework for Financial Reporting*.

(a) Principles versus rules

The *Conceptual Framework* will continue to adopt a principles-based approach. This is also supported by a report[4] from the Institute of Chartered Accountants of Scotland which concluded that the global convergence of accounting standards cannot be achieved by a 'tick-box' rules-driven approach but should rely on judgement-based principles.

A principles-based approach allows companies the flexibility to deal with new situations. A rules-based approach provides the auditor with protection against litigious claims because it can be shown that other auditors would have adopted the same accounting treatment. However, even in America the rules-based approach was heavily criticised, following the Enron disaster, where it was felt that a principles-based approach would have been more effective in preventing it.

A rules-based approach means that financial statements are more comparable. Recognising that a principles-based approach could lead to different professional judgements for the same commercial activity, it is important that there should be full disclosure and transparency.

(b) Proposals in the Discussion Paper

It proposed to adopt both Chapter 1, *The objective of financial statements*, and Chapter 3, *Qualitative characteristics of useful financial information*, which were issued as the *Conceptual Framework for Financial Reporting 2010*.

It invited comments on making significant changes to the existing *Conceptual Framework*. This related in part to the consideration of the remaining topics in the *Framework for the Preparation and Presentation of Financial Statements* that had not been revised on the issue of the 2010 Framework, such as stating its position on measurement.

Measurement

With regard to measurement, the proposal is that the current approach should continue where a judgement is made as to the measurement base relevant in the circumstances, i.e. a selective use of historic cost, current cost, net realisable value and value in use. This gives rise to the long-standing criticism of the statement of financial position as in the following extract from a 1988 report *Making Corporate Reports Valuable*:[5]

> The present statement of financial position almost defies comprehension. Assets are shown at depreciated historical cost, at amounts representing current valuations and at the results of revaluations of earlier periods (probably also depreciated); that is there is no consistency whatsoever in valuation practice. The sum total of the assets, therefore, is meaningless and combining it with the liabilities to show the entity's financial position does not in practice achieve anything worthwhile.

It also raised additional topics such as the items to be reported under other comprehensive income. Following consideration of the comments received, a resulting *Conceptual Framework* is expected in 2015. It is proposed that there will be significant additional guidance and educational material issued to accompany it.

The end of the piecemeal approach

The issue of the Discussion Paper is an encouraging move away from the piecemeal approach of the past. Ending the joint project with the FASB has allowed the IASB to move on to complete this comprehensive project which will have a continuing, significant impact on the development by the IASB of future IFRSs.

Summary

User needs have been accepted as paramount; qualitative characteristics of information have been specified; the elements of financial statements and the presentation of financial information are being revised in the IASB-only comprehensive project.

The intention remains to produce general-purpose financial statements that present a fair view. This is not achieved by detailed rules and regulations; the exercise of judgement will continue to be needed, based on a sound conceptual framework.

Judgement will be required at many levels. For example, in classifying and reporting an asset judgement is required as to (a) whether an economic resource exists, then (b) was it the result of a past event, then (c) is it controlled, then (d) what is the relevant measurement base, then (e) is it material in the context of the company or in the context of the company or group, then finally (f) should there be disclosure of the approach that had been taken when making estimates to arrive at the reported value.

Many of these judgements could be improved if there were to be more guidance from the IASB on the application of concepts and standards. It is the IASB's stated intention to provide this and additional educational material.

The question of the measurement base that should be used has yet to be settled. The IASB seems to favour choosing that which provides the most relevant information for investors making economic decisions.

The *Framework* sees the objective of financial statements as providing information about the financial position, performance and financial adaptability of an enterprise that is useful to a wide range of users for accountability and in making economic decisions. It recognises that they are limited because they largely show the financial effects of past events and do not necessarily show non-financial information.

This leaves the following point made by some critics unresolved. Accountability and the IASB's decision-usefulness are not compatible. Forward-looking decisions require forecasts of future cash flows, which in the economic model are what determines the values of assets. These values are too subjective to form the basis of accountability. The definition of assets proposed by the IASB and the recognition rules restrict assets to economic resources the enterprise controls as a result of past events. But economic decision-making requires examination of all sources of future cash flows, not just a restricted subset of them.

REVIEW QUESTIONS

1 'The replacement of accrual accounting with cash flow accounting would avoid the need for a conceptual framework.'[6] Discuss.

2 'The IASB is proposing to allow the selective choice of a measurement base.' Discuss why this may or may not be preferable to adopting a single measurement base for all elements.

3 Financial accounting theory has accumulated a vast literature. A cynic might be inclined to say that the vastness of the literature is in sharp contrast to its impact on practice.

(a) Describe the different approaches that have evolved in the development of accounting theory.

(b) Assess their varying impacts on standard setting.

4 'Rules-based accounting adds unnecessary complexity, encourages financial engineering and does not necessarily lead to a "true and fair view" or a "fair presentation".' Discuss.

5 'Tax avoidance would not occur if there was a principle- rather than rule-based approach.' Discuss.

6 Explain what you understand by a balance sheet approach to income determination and how this is demonstrated in the definition of elements.

7 Explain how a company assesses materiality when attempting to report a true and fair view of its income.

8 'The key qualitative characteristics in the *Conceptual Framework* are relevance and faithful representation. Preparers of financial statements may face a dilemma in satisfying both criteria at once.' Discuss situations where there might be a conflict.

9 'An asset is to be defined in the *Framework* as an economic resource which an entity controls as a result of past events.' Discuss whether property, plant and equipment automatically qualify as assets.

10 'The *Conceptual Framework* regards neutrality as necessary for financial statements to provide a faithful representation of transactions. However, the view of some of the respondents to the exposure draft was that neutrality is impossible to achieve because if it is accepted that information must be relevant as a tool to influence decision making then it could not be neutral.' Discuss.

EXERCISES

Question 1

The following extract is from *Conceptual Framework for Financial Accounting and Reporting: Elements of Financial Statements and Their Measurement*, FASB 3, December 1976:

> The benefits of achieving agreement on a conceptual framework for financial accounting and reporting manifest themselves in several ways. Among other things, a conceptual framework can (1) guide the body responsible for establishing accounting standards, (2) provide a frame of reference for resolving accounting questions in the absence of a specific promulgated standard, (3) determine bounds for judgement in preparing financial statements, (4) increase financial statement users' understanding of and confidence in financial statements, and (5) enhance comparability.

Required:
(a) Define a conceptual framework.
(b) Critically examine why the benefits provided in the above statements are likely to flow from the development of a conceptual framework for accounting.

Question 2

The following extract is from 'Comments of Leonard Spacek', in R.T. Sprouse and M. Moonitz, *A Tentative Set of Broad Accounting Principles for Business Enterprises*, Accounting Research Study No. 3, AICPA, New York, 1962, reproduced in A. Belkaoui, *Accounting Theory*, Harcourt Brace Jovanovich.

> A discussion of assets, liabilities, revenue and costs is premature and meaningless until the basic principles that will result in a fair presentation of the facts in the form of financial accounting and financial reporting are determined. This fairness of accounting and reporting must be for and to people, and these people represent the various segments of our society.

Required:
Discuss the extent to which the IASB conceptual framework satisfies the above definition of fairness.

Question 3

The following is an extract from *Accountancy Age*, 25 January 2001:

> A powerful and 'shadowy' group of senior partners from the seven largest firms has emerged to move closer to edging control of accounting standards from the world's accountancy regulators ... they form the Global Steering Committee ... The GSC has worked on plans to improve standards for the last two years after scathing criticism from investors that firms produced varying standards of audit in different countries.

Required:
Discuss the effect on standard setting if control were to be edged from the world's accountancy regulators and back in the hands of the profession.

Question 4

The FRC in its 2009 publication *Louder than Words – Principles and Actions for Making Corporate Reports Less Complex and More Relevant* included a call for action to 'Ensure disclosure requirements are relevant and proportionate to the risks', stating that 'We would like to see a project on disclosure which investigates the characteristics of useful disclosures and the main objectives of financial reporting disclosure. . . . Ideally, we believe another organisation could constructively kick off this work with a view to providing recommendations to the relevant regulators, including the International Accounting Standards Board (IASB).'

Required:
Critically discuss how a company could determine whether any disclosure is proportionate to the risks and whether this implies that there should be less mandatory disclosures which lead to ever more complexity.

References

1 Tech 03/08 Guidance on materiality in Financial Reporting by UK entities, ICAEW, 2008 http://www.icaew.com/~/media/Files/Technical/technical-releases/legal-and-regulatory/TECH-03-08-Guidance-on-Materiality-in-Financial-Reporting-by-UK-entities.pdf
2 https://www.sec.gov/interps/account/sab99.htm
3 http://www.esma.europa.eu/system/files/2013-218.pdf
4 ICAS, 'Principles not rules – a question of judgement', www.icas.org.uk/site/cms/contentViewArticle.asp?article=4597
5 ICAS, *Making Corporate Reports Valuable*, 1988, p. 35.
6 R. Skinner, *Accountancy*, January 1990, p. 25.

CHAPTER 11

Ethical behaviour and implications for accountants

11.1 Introduction

The main purpose of this chapter is for you to have an awareness of the need for ethical behaviour by accountants to complement the various accounting and audit standards issued by the International Accounting Standards Board (IASB), the International Auditing and Assurance Standards Board (IAASB) and professional accounting bodies.

Objectives

By the end of this chapter, you should be able to discuss:

- the meaning of ethical behaviour;
- the relationship of ethics to standard setting;
- the main provisions of the IFAC *Code of Ethics for Professional Accountants*;
- the implications of ethical values for the principles- versus rules-based approaches to accounting standards;
- the problem of defining principles and standards where there are cultural differences;
- the implications of unethical behaviour for stakeholders using the financial reports;
- the type of ethical issues raised for accountants in business;
- the role of whistle-blowing.

11.2 The meaning of ethical behaviour

Individuals in an organisation have their own ethical guidelines which may vary from person to person. These may perhaps be seen as social norms which can vary over time. For example, the relative importance of individual and societal responsibility varies over time.

11.2.1 Individual ethical guidelines

Individual ethical guidelines or personal ethics are the result of a varied set of influences or pressures. As an individual each of us 'enjoys' a series of ethical pressures or influences including the following:

- Parents – the first and, according to many authors, the most crucial influence on our ethical guidelines.

- Family – the *extended* family which is common in Eastern societies (aunts, uncles, grand-parents and so on) can have a significant impact on personal ethics; the *nuclear* family which is more common in Western societies (just parent(s) and siblings) can be equally as important but is more narrowly focused.

- Social group – the ethics of our 'class' (either actual or aspirational) can be a major influence.

- Peer group – the ethics of our 'equals' (again either actual or aspirational) can be another major influence.

- Religion – ethics based in religion are more important in some cultures, e.g. Islamic societies have some detailed ethics demanded of believers as well as major guidelines for business ethics. However, even in supposedly secular cultures, individuals are influenced by religious ethics.

- Culture – this is also a very effective formulator of an individual's ethics.

- Professional – when an individual becomes part of a professional body then they are subject to the ethics of the professional body.

Given the variety of influences it is natural that there will be a variety of views on what is acceptable ethical behaviour. For example, as an accounting student, how would you handle ethical issues? Would you personally condone cheating? Would you refrain from reporting cheating in exams and assignments by friends? Would you resent other students being selfish such as by hiding library books which are very helpful for an essay? Would you resent cheating in exams by others because you do not cheat and therefore are at a disadvantage? Would that resentment be strong enough to get you to report the fact that there is cheating to the authorities even if you did not name the individuals involved?

11.2.2 Professional ethical guidelines

A managing director of a well-known bank described his job as deciding contentious matters for which, after extensive investigation by senior staff, there was no obvious solution. The decision was referred to him because all proposed solutions presented significant downside risks for the bank. Ethical behaviour can be similarly classified. There are matters where there are clearly morally correct answers and there are dilemmas where there are conflicting moral issues.

Professional codes of conduct tend to provide solutions to common issues which the profession has addressed many times. However, the professional code of ethics is only the starting point in the sense that it can never cover all the ethical issues an accountant will face and does not absolve accountants from dealing with other ethical dilemmas.

11.3 The accounting standard-setting process and ethics

Standard setters seem to view the process as similar to physics in the sense of trying to set standards with a view to achieving an objective measure of reality. However, some academics suggest that such an approach is inappropriate because the concepts of profit and value are not physical attributes but 'man-made' dimensions. For instance, for profit we measure the progress of the business but the concept of progress is a very subjective attribute which has traditionally omitted public costs such as environmental and social costs. The criterion of fairness has been seen as satisfied by preparing income statements on principles such as going concern and accrual when measuring profit and neutrality when presenting the income statement.

What if fairness is defined differently? For example, the idea of basing accounting on the criterion of fairness to all stakeholders (financiers, workers, suppliers, customers and the community) was made by Leonard Spacek[1] before the formation of the FASB. However, this view was not appreciated by the profession at that time. We now see current developments in terms of environmental and social accounting which are moves in that direction but, even so, Corporate Social Responsibility (CSR) reports are not incorporated into the financial statements prepared under IFRSs and constitutes supplementary information that is not integrated into the accounting measures themselves.

Ethics and neutrality

The accounting profession sees ethical behaviour in standard setting as ensuring that accounting is neutral. Their opponents think that neutrality is impossible and that accounting has a wide impact on society and thus to be ethical the impact on all parties affected should be taken into consideration.

The ASB and EFRAG addressed this with the issue of a Discussion Paper: *Considering the Effects of Accounting Standards* in January 2011. The Paper aimed to obtain feedback on proposals to follow a systematic process for considering the effects of accounting standards when they are being developed. This has been in response to the endorsement process for the adoption of IFRSs within the European Union which requires an effects study to be carried out and the calls from users for standard setting proposals to be more evidence-based.

The accounting profession does not address ethics at the macro level other than in pursuing neutrality, but rather focuses its attention on actions after the standards and laws are in place. The profession seeks to provide ethical standards which will increase the probability of those standards being applied in an ethical fashion at the micro level where accountants apply their individual skills.

The accounting profession through its body the International Federation of Accountants (IFAC) has developed a *Code of Ethics for Professional Accountants*.[2] That code looks at fundamental principles as well as specific issues which are frequently encountered by accountants in public practice, followed by those commonly faced by accountants in business. The intention is that the professional bodies and accounting firms 'shall not apply less stringent standards than those stated in this code' (p. 4).

11.4 The IFAC *Code of Ethics for Professional Accountants*

The IFAC Fundamental Principles are:

(i) 'A distinguishing mark of the accountancy profession is its acceptance of the responsibility to act in the public interest . . .' (100.1).

(ii) 'A professional accountant shall comply with the following fundamental principles:

(a) *Integrity* – to be straightforward and honest in all professional and business relationships.

(b) *Objectivity* – to not allow bias, conflict of interest or undue influence of others to override professional or business judgements.

(c) *Professional Competence and Due Care* – to maintain professional knowledge and skill at the level required to ensure that a client or employer receives competent professional services and to act diligently in accordance with applicable technical and professional standards.

(d) *Confidentiality* – to respect the confidentiality of information acquired as a result of professional and business relationships and, therefore, not disclose any such information to third parties without proper and specific authority, unless there is a legal or professional right or duty to disclose, nor use the information for the personal advantage of the professional accountant or third parties.

(e) *Professional Behaviour* – to comply with relevant laws and regulations and avoid any action that discredits the profession' (100.5).

11.4.1 Acting in the public interest

The first underlying statement that accountants should act in the public interest is probably more difficult to achieve than is imagined. This requires accounting professionals to stand firm against accounting standards which are not in the public interest, even when politicians and company executives may be pressing for their acceptance. Owing to the fact that, in the conduct of an audit, the auditors have mainly dealings with the management it is easy to lose sight of who the clients actually are. For example, the expression 'audit clients' is commonly used in professional papers and academic books when they are referring to the management of the companies being audited. It immediately suggests a relationship which is biased towards management when, legally, the client may be either the shareholders as a group or specific stakeholders. Whilst it is a small but subtle distinction it could be the start of a misplaced orientation towards seeing the management as the client.

11.4.2 Fundamental principles

The five fundamental principles are probably uncontentious guides to professional conduct. It is the application of those guides in specific circumstances which provides the greatest challenges. The IFAC paper provides guidance in relation to public accountants covering appointments, conflicts of interest, second opinions, remuneration, marketing, acceptance of gratuities, custody of client assets, objectivity and independence. In regard to accountants in business they provide guidance in the areas of potential conflicts, preparation and reporting of information, acting with sufficient expertise, financial interests and inducements.

It is not intended to provide here all the guidance which the IFAC *Code of Ethics* provides, and if students want that detail they should consult the original document.[2] This chapter will provide a flavour of the coverage relating to accountants in public practice and accountants in business.

11.4.3 Problems arising for accountants in practice

Appointments

Before accepting appointments public accountants should consider the desirability of accepting the client given the business activities involved, particularly if there are questions of their legality. They also need to consider (a) whether the current accountant of the potential client has advised of any professional reasons for not becoming involved and (b) whether they have the competency required considering the industry and their own expertise. Nor should they become involved if they already provide other services which are incompatible with being the auditor or if the size of the fees would threaten their independence. (Whilst it is not stated in the code the implication is that it is better to avoid situations which are likely to lead to difficult ethical issues.)

Second opinions

When an accountant is asked to supply a second opinion on an accounting treatment it is likely that the opinion will be used to undermine an accountant who is trying to do the right thing. It is therefore important to ascertain that all relevant information has been provided before issuing a second opinion, and if in doubt decline the work.

Remuneration

Remuneration must be adequate to allow the work to be done in a professional manner.

Commissions received from other parties must not be such as to make it difficult to be objective when advising your client and in any event must at least be disclosed to clients. Some accountants have addressed that by passing the commissions on to their client and charging a flat fee for the consulting.

Marketing

Marketing should be professional and should not exaggerate or make negative comments about the work of other professionals.

Independence

The accountant and their close relatives should not accept gifts, other than trivial ones, from clients. IFAC provides that:

> A professional accountant in public practice who provides an assurance service shall be independent of the assurance client. Independence of mind and in appearance is necessary to enable the professional accountant in public practice to express a conclusion.

Professional firms have their own criterion level as to the value of gifts that can be accepted. For example, the following is an extract from the KPMG Code of Conduct:

> Qn: I manage a reproduction center at a large KPMG office. We subcontract a significant amount of work to a local business. The owner is very friendly and recently offered to give me two free movie passes. Can I accept the passes?

> Ans: Probably. Here, the movie passes are considered a gift because the vendor is not attending the movie with you. In circumstances where it would not create the appearance of impropriety, you may accept reasonable gifts from third parties such as our vendors, provided that the value of the gift is not more than $100 and that you do not accept gifts from the same vendor more than twice in the same year.

11.4.4 Problems arising for accountants in business

In relation to accountants in business the major problem identified by the code seems to be the financial pressures which arise from substantial financial interests in the form of shares, options, pension plans and dependence on employment income to support themselves and their dependants. When these depend on reporting favourable performance it is difficult to withstand the pressure.

Every company naturally wants to present its results in the most favourable way possible and investors expect this and it is part of an accountant's expertise to do this. However, the ethical standards require compliance with the law and accounting standards subject to the overriding requirement for financial statements to present a fair view. Misreporting and the omission of additional significant material which would change the assessment of the financial position of the company are unacceptable.

Accountants need to avail themselves of any internal steps to report pressure to act unethically and if that fails to produce results they need to be willing to resign.

11.4.5 Threats to compliance with the fundamental principles

The IFAC document has identified five types of threats to compliance with their fundamental principles and they will be outlined below. The objective of outlining these potential threats is to make you sensitive to the types of situations where your ethical judgements may be clouded and where you need to take extra steps to ensure you act ethically. The statements are deliberately broad to help you handle situations not covered specifically by the guidelines. IFAC para 100.12 provides the following classification:

Threats fall into one or more of the following categories:

(a) Self-interest threat – the threat that a financial or other interest will inappropriately influence the accountant's judgment or behavior;

(b) Self-review threat – the threat that a professional accountant will not appropriately evaluate the results of a previous judgment made or service performed by another individual within the professional accountant's firm or employing organization;

(c) Advocacy threat – the threat that a professional accountant will promote a client's or employer's position to the point that the professional accountant's objectivity is compromised;

(d) Familiarity threat – the threat that due to a long or close relationship with a client or employer, a professional accountant will be too sympathetic to their interests or too accepting of their work; and

(e) Intimidation threat – the threat that a professional accountant will be deterred from acting objectively because of actual or perceived pressures.

11.5 Implications of ethical values for the principles versus rules-based approaches to accounting standards

It is common in the literature for authors to quote Milton Friedman as indicating that the role of business is to focus on maximising profits, and also to cite Adam Smith as justification for not interfering in business affairs. In many cases those arguments are misinterpreting the authors.

Societal norms

Milton Friedman recognised that what business people should do was maximise profits *within the norms of society*. He knew that without laws to give greater certainty in regard to business activities, and the creation of trust, it was not possible to have a highly efficient economy. Thus he accepted laws which facilitated business transactions and norms in society which also helped to create a cooperative environment. Thus the norms in society set the minimum standards of ethical and social activity which businesses must engage in to be acceptable to those with whom they interact.

Equitable exchanges

Adam Smith (in *The Wealth of Nations*) did not say 'do not interfere with business'; rather, he assumed the existence of the **conditions necessary to facilitate fair and equitable exchanges**. He also suggested that government should interfere to prevent monopolies but should not interfere as a result of lobbying of business groups because their normal behaviour is designed to create monopolies. He also assumed that those who did not meet ethical standards might make initial gains but would be found out and shunned. His other major book (*The Theory of Moral Sentiments*) was on morality so there is no doubt that he thought ethics were a normal and essential part of society and business.

How does this relate to accounting standards?

The production of accounting standards is only the starting point in the application of accounting standards. We have seen that accountants can apply the standards to the letter of the law and still not achieve reporting that conveys the substance of the performance and financial state of the business. This is because businesses can structure transactions so as to avoid the application of a standard. For example, by taking liabilities off the balance sheet, such as when a company does not want to capitalise a lease, it arranges for a change in the lease terms so that it is reported as a note and not shown as a liability on the face of the accounts.

It is that type of gamesmanship which has worried accounting standard setters. The issue is whether such games are appropriate, and if they aren't, why haven't they been prevented by the ethical standards of the accountants?

How does the accounting profession attempt to ensure that financial reports reflect the substance of a transaction?

We have seen that standards have been set in many national jurisdictions and now internationally by the IASB, in order to make financial statements fair and comparable. The number of standards varies between countries and is described as rules-based or principles-based according to the number of standardised accounting treatments.

Rules-based

Where there are many detailed standards as in the US, the system is described as 'rules based' in that it attempts to specify the uniform treatment for many types of transactions. This is both a strength and a weakness in that the very use of precise standards as the only criterion leads to the types of games to get around the criteria that were mentioned earlier for lease accounting. One solution to combat this behaviour is to adopt a principles-based approach to support (or replace) the rules.

Principles-based

Where there are fewer standards as in the UK, the system is referred to as 'principles based'. In the principles-based system there is greater reliance on the application of the 'true and fair' override to (a) report unusual situations and (b) address the issue of whether the accounts prepared in accordance with existing standards provide a fair picture for the decisions to be made by the various users and provide additional information where necessary.

Whilst these are positive applications the override criteria can also be misused. For example, many companies during the 'dot-com' boom around the year 2000 produced statements of **normalised** earnings. The argument was that they were in the set-up phase and many of the costs they were incurring were one-offs. To get a better understanding of the business, readers were said to need to know what an ongoing result was likely to be. So they removed set-up costs and produced **normalised** or **sustainable** earnings which suggested the company was inherently profitable. Unfortunately, many of these companies failed because those one-off costs were not one-off and had to be maintained to keep a customer base.

Does a principles-based approach achieve true and fair reports?

The US regulators and the IASB have agreed that the principles-based approach should be adopted. However, this still leaves unanswered the question as to whether this approach can give a true and fair view to every stakeholder. Shareholders are recognised in all jurisdictions but the rights of other parties may vary according to the legal system. When, for example, do the rights of lenders become paramount? Should the accounts be tailored to suit employees when the legal system in some jurisdictions recognises that companies are not just there to support owners but have major responsibilities to recognise the preservation of employment wherever possible?

Can general-purpose accounts (whether rules-based or principles-based) ever be appropriate for the many purposes for which they are routinely used?

11.5.1 The problem of linking principles to accounting standards

The current conceptual framework assumes that we need to produce general-purpose financial accounts using understandability, relevance, reliability and comparability as guiding criteria. However, the individual standards do not demonstrate how those principles lead to the standards which have been produced.

11.5.2 The principles-based approach and ethics

The preceding discussion looked at the principle of true and fair or its equivalent from an accountant's perspective which is often seen as being achieved by following a reporting standard. However, even then there is an element of subjective judgement and it would be helpful to have an ethical concept of fairness expressed in everyday language.

Idea of superfairness

William Baumol, a celebrated economist, provides the interesting concept of superfairness[3] which would help with this type of ethical decision. He says if you didn't know what side of the transaction you were going to be on, what would you consider to be fair? If you didn't know whether you were going to be a company executive, or an auditor, or a buyer of shares, or a seller of shares, what do you think would be a fair representation of the company's performance and financial position?

Shift from shareholder orientation

It would, in order to avoid ambiguity, have to spell out 'fair to whom and for what purpose'. This is because at the present time society is in a process of reassessing the role of business relative to the demands by society to achieve high employment rates, to overcome environmental problems and to achieve fair treatment of all countries. Essentially this is suggesting that, given the changing orientation, consideration may have to be given to ethical criteria even if there is only a partial shift from a shareholder orientation to a balancing of competing claims in society.

Daniel Friedman (2008, p. 179)[4] says: 'The greatest challenge is to realign morals and markets so that they work together, rather than at cross purposes.' This will need a balancing act specific to the problem faced. In other words it would have to be principle driven.

11.5.3 The problem of defining principles and standards where there are cultural differences

Cultural differences may lead to different principles being formulated and applied. For example, the IASB has defined assets to be reported in the financial statements in such a way that human assets and social costs are not included. As regards application, an accountant in preparing accounts will always have a potential clash between what his or her employer and superior wants, what his or her profession requires and what is best from an ethical or community perspective.

This raises questions such as:

- 'What grounds are there for different accounting being applicable to different countries?'
- 'Should there be different principles if the purpose of accounting is not the same in all countries, with some countries placing, say, greater emphasis on the impact on employees or the community?'
- 'How do cultural norms and religion affect ethics in both the formulation and the interpretation of individual guidelines?'
- 'Is it correct to assume that shareholders in every country have identical information needs and apply identical ethical criteria in assessing a company's operations?'

11.5.4 Research into the impact of different cultural characteristics on behaviour

An interesting piece of research compared the attitudes of students in the USA and the UK to cheating and found the US students more likely to cheat.[5] The theoretical basis of the research was that different cultural characteristics, such as uncertainty avoidance or conversely the tolerance for ambiguity, lead to different attitudes to ethics.

Implication for multinationals

This means uniform ethical guidelines will not lead to uniform applications in multinational companies unless the corporate culture is much stronger than the country culture. This has implications for multinational businesses that want the accounts prepared in the different countries to be uniform in quality.

Implication for the profession

It is significant for audit firms that want their sister firms in other countries to apply the same standards to audit judgements. It is important to investment firms that are making investments throughout the world on the understanding that accounting and ethical standards mean the same things in all major security markets. It is one of the reasons that the US is reluctant to adopt IFRS as a replacement for its own US GAAP because it is not convinced that IFRSs are applied uniformly across the world.

Where there are differences in legal and cultural settings then potentially the correct accounting will also differ if a principles-based approach is adopted. Currently Western concepts dominate accounting but if the world power base shifts, either to several world centres of influence or to a new dominant world power, then principles of accounting may have to reflect that.

11.6 Ethics in the accountant's work environment – a research report

The Institute of Chartered Accountants in Scotland issued a discussion paper report[6] titled 'Taking ethics to heart' based on research into the application of ethics in practice. This section will discuss some of the findings of that report.

A student's perspective

From a student's perspective one of the interesting findings was that many accountants could not remember the work on ethics which they did as students and therefore had little to draw upon to guide them when problems arose. There was agreement that students need to get more experience in dealing with case studies so as to enhance their ethical-decision-making skills. This should be reinforced throughout their careers by continuing professional development. The training should sensitise accountants so that they can easily recognise ethical situations and develop skills in resolving the dilemmas.

A trainee's perspective

Exposure to ethical issues is usually low for junior positions, although even then there can be clear and grey issues. For example, padding an expense claim and overstating overtime are clear issues, whereas how to deal with information that has been heard in a private conversation between client staff is less clear. What if a conversation is overheard where one of the factory staff says that products which are known to be defective have been dispatched at the year-end? Would your response be different if you had been party to the conversation? Would your response be different if it had been suggested that there was a risk of injury due to the defect? Is it ethical to inform your manager or is it unethical not to inform?

A senior's perspective

Normally exposure to ethical issues increases substantially at the manager level and continues in senior management positions. However, the significance of ethical decision making has increased with the expansion of the size of both companies and accounting practices. The impact of decisions can be more widespread and profound. Further, there has been an increase in litigation, potentially exposing the accountant to more external review. Greater numbers of accounting and auditing standards can lead to a narrower focus, making it harder for individual accountants to envisage the wider ethical dimensions and to get people to consider more than the detailed rules.

Given the likelihood of internal or external review, the emphasis that many participants in the study placed on asking 'How would this decision look to others?' seems a sensible criterion. In light of that emphasis by participants in the research it is interesting to consider the 'Resolving conflicts' section of BT PLC's document called *The Way We Work*[7] which among other things says:

> How would you explain your decision to your colleagues in different countries?
> How would you explain your decision to your family or in public?
> Does it conflict with your own or BT's commitment to integrity?

This emphasis on asking how well ethical decisions would stand public scrutiny, including scrutiny in different countries, would be particularly relevant to accountants in businesses operating across national borders.

Ethical policies and advice

The role of the organisational setting in improving or worsening ethical decision making was given considerable attention in the ICAS report. A key starting point is having a set of ethical policies which are practical and are reinforced by the behaviour of senior management. Another support is the presence of a clearly defined process for referring difficult ethical decisions upward in the organisation.

For those in small organisations there needs to be an opportunity for those in difficult situations to seek advice about the ethical choice or the way to handle the outcomes of making an ethical stand. Most professional bodies either have senior mentors available or have organised referrals to bodies specialising in ethical issues.

The reality is that some who have taken ethical stands have lost their jobs, but some of those who haven't stood their ground have lost their reputations or their liberty.

11.7 Implications of unethical behaviour for stakeholders using the financial reports

One of the essential aspects of providing complete and reliable information which are taken seriously by the financial community is to have a set of rigorous internal controls. However, ultimately those controls are normally dependent on checks and balances within the system and the integrity of those with the greatest power within the system. In other words the checks and balances, such as requiring two authorisations to issue a cheque or transfer money, presume that at least one of those with authority will act diligently and will be alert to the possibility of dishonest or misguided behaviour by the other. Further, if necessary or desirable, they will take firm action to prevent any behaviour that appears suspicious. The internal control system depends on the integrity and diligence, in other words the ethical behaviour, of the majority of the staff in the organisation.

11.7.1 Increased cost of capital

The presence of unethical behaviour in an organisation will raise questions about the reliability of the accounts. If unethical behaviour is suspected by investors, they will probably raise the cost of capital for the individual business. If there are sufficient cases of unethical behaviour across all companies, the integrity of the whole market will be brought into question and the liquidity of the whole market is reduced. That would affect the cost of funds across the board and increase the volatility of share prices.

11.7.2 Hidden liabilities

A liability, particularly an environmental one, might not crystallise for a number of years, as with the James Hardie Group in Australia. The James Hardie Group was a producer of asbestos sheeting whose fibres can in the long term damage the lungs and lead to death. The challenge the company faced was the long gestation period between the exposure to the dust from the asbestos and the appearance of the symptoms of the disease. It can be up to 40 years before the victim finds out that they have a death sentence.

Liability transferred to a separate entity

The company reorganised so that there was a separate entity which was responsible for the liabilities and that entity was supposed to have sufficient funds to cover future liabilities as they came to light. When it was apparent that the funds set aside were grossly inadequate and that the assessment of adequacy had been based on old data rather than using the more recent data which showed an increasing rate of claims, there was widespread community outrage. As a result the James Hardie Group felt that irrespective of their legal position they had to negotiate with the state government and the unions to set aside a share of their cash flows from operations each year to help the victims.

Thus the unfair arrangements set in place came back to create the equivalent of liabilities and did considerable damage to the public image of the company. This also made some people reluctant to be associated with the company as customers or as employees. The current assessment of liability (as at 2009) is set out in a KPMG Actuarial Report.[8]

11.7.3 Effect of ethical collapse in an organisation

There is an increasing need to be wary of unethical behaviour by management leading to fraud.

Widespread awareness

Jennings (2006)[9] points out that while most of the major frauds that make the headlines tend to be attributed to a small number of individuals, there have to be many other participants who allow them to happen. For every CEO who bleeds the company through payment for major personal expenses, or through gross manipulation of accounts, or backdating of options, there have to be a considerable number of people who know what is happening but who choose not to bring it to the attention of the appropriate authorities. The appropriate authority could be the board of directors, or the auditors or regulatory authorities.

Signs of ethical collapse

Jennings attributes this to the culture of the organisation and suggests there are seven signs of ethical collapse in an organisation. They include pressure to maintain the numbers, suppression of dissent and bad news, iconic CEOs surrounding themselves with young executives whose careers are dependent on them, a weak board of directors, numerous conflicts of interest, innovation excess, and goodness in some areas being thought to atone for evil in others.

Others have suggested that companies with high levels of takeover activity and high leverage are often prime candidates for fraud because of the pressures to achieve the numbers. Also, if the attitude is that the sole purpose of the firm is to make money subject to compliance with the letter of the law, that is also a warning sign.

Pressure on accountants

The ICAS report 'Taking ethics to heart' noted that it appeared that the current business and commercial environment placed an enormous pressure on accountants, wherever they work, which may result in decisions and judgements that compromise ethical standards. It noted also that increased commercial pressures on accountants may be viewed by many within the profession as heralding a disquieting new era.

The accountant working within business has a different set of problems due to his or her dual position as an employee and a professional accountant. There is a potential clash of issues where the interests of the business could be at odds with professional standards.

11.7.4 Auditor reaction to risk of unethical behaviour

In addition to the above items, unethical behaviour should make auditors and investors scrutinise accounts more closely. Following the experiences with companies such as Enron, the auditing standards have placed greater emphasis on auditors being sceptical. This means that if they identify instances of unethical behaviour they should ask more searching questions. Depending on the responses they get, they may need to undertake more testing to satisfy themselves of the reliability of the accounts.

11.7.5 Action by professional accounting bodies to assist members

The various professional bodies provide members with support such as ICAEW's Ethics Advisory Service and ICMA's confidential advice.

The type of problem raised is a good indication of the ethical issues raised for accountants in business. They include:

- requests by employers to manipulate tax returns;
- requests to produce figures to mislead shareholders;
- requests to conceal information;
- requests to manipulate overhead absorption rates to extort more income from customers (an occurrence in the defence industries);
- requests to authorise and conceal bribes to buyers and agents, a common request in some exporting businesses;
- requests to produce misleading projected figures to obtain additional finance;
- requests to conceal improper expense claims put in by senior managers;
- requests to over- or undervalue assets to avoid breaching loan covenants;
- requests to misreport figures in respect of government grants;
- requests for information which could lead to charges of 'insider dealing';
- requests to redefine bad debts as 'good' or vice versa.

For accountants in industry the message is that if your employer has a culture which is not conducive to high ethical values then a good career move would be to look for employment elsewhere. For auditors the message is that the presence of symptoms suggested above is grounds for employing greater levels of scepticism in the audit.

11.7.6 Action taken by governments

The Sarbanes–Oxley Act (SOX) has had a major impact on company management and auditors to address what had been seen as an inadequate oversight of the accounting profession and conflicts of interest involving the auditors.

Management

Following the collapse of the auditors Arthur Andersen, the introduction of SOX placed personal responsibility on the CEO and the CFO for the accounts, with serious penalties for publishing misleading accounts. This led to these officers seeking reassurance that there were adequate systems and internal controls in place. This has in turn led to complaints that management effort is being directed away from growing the business and earnings.

Accountants and auditors

Management and audit committees since SOX are more focused on financial reporting. SOX gave rise to a major demand by business for internal auditors to undertake this work with the focus moving to assessing financial controls, as opposed to operational processes.

As for auditors, they have to confirm that companies have adequate systems and internal controls and are required to report to the audit committee rather than management.

11.7.7 Action by companies – company codes of ethics

Most companies now adopt codes of ethics. They may have alternative titles such as 'our values', codes of conduct, or codes of ethics. For example, BP has a code of conduct whose coverage, which is listed below, is what one would expect of a company involved in its industry and its activities covering a large number of countries. Its Code of Conduct includes the following major categories:

- Our commitment to integrity;
- Health, safety, security and the environment;
- Employees;
- Business partners;
- Governments and communities;
- Company assets and financial integrity.

However, the challenge is to make the code an integral part of the day-to-day behaviour of the company and to be perceived as doing so by outsiders. Obviously top management have to act in ways so as to reinforce the values of the code and to eliminate existing activities which are incompatible with the new values.

BP has been criticised for behaviour inconsistent with its values but such behaviour may relate to actions taken before the adoption of the code (see Beder[10]).

Thus it is important to ensure that the corporate behaviour is consistent with the code of conduct, and that staff are rewarded for ethical behaviour and suffer penalties for non-compliance. Breaches, irrespective of whether they are in the past, are difficult to erase from the memories of society.

Levels within codes

Stohl et al. (2009)[11] suggest that the content of codes of conduct can be divided into three levels:

- Level 1 is where there is an attempt to ensure that the company is in compliance with all the laws which impact on it in the various countries in which it operates.
- Level 2 focuses on ensuring fair and equitable relations with all parties with which the company has direct relations. In this category would be the well-publicised adverse publicity which Nike received when it was alleged that their subcontractors were exploiting child labour in countries where such treatment is legal. The adverse publicity and boycotts meant that many companies reviewed their operations and expanded their codes to cover such situations and thus moved into the second level of ethical awareness.
- Level 3 is where companies take a global perspective and recognise their responsibility to contribute to the likelihood of peace and favourable global environmental conditions. In most companies the level 1 concerns are more dominant than the level 2, and the level 2 more than the level 3. European firms are more likely than US firms to have a level 3 orientation.

11.7.8 Conflict between codes and targets

On the one hand we see companies developing codes of ethical conduct, whilst on the other hand we see some of these same companies developing management by objectives which set staff unachievable targets and create pressures that lead to unethical behaviour. Where this occurs there is the risk that an unhealthy corporate climate may develop, resulting in the manipulation of accounting figures and unethical behaviour.

There is a view[12] that there is a need to create an ethical climate that transcends a compliance approach to ethics and focuses instead on fostering socially harmonious relationships. An interesting article[13] proceeds to make the argument that the recent accounting scandals may be as much a reflection of a deficient corporate climate, with its concentration on setting unrealistic targets and promoting competition between the staff, as of individual moral failures of managers.

11.7.9 Multinationals face special problems

Modern multinational companies experience special problems in relation to ethics.

Firstly, the transactions are often extremely large, so that there are greater pressures to bend the rules so as to get the business.

Secondly, the ethical values as reflected in some of the countries may be quite different from those in the head office of the group. One company did business in a developing country where the wages paid to public officials were so low as to be insufficient to support a family even at the very modest living standards of that country. Many public officials had a second job so as to cope. Others saw it as appropriate to demand kickbacks in order for them to process any government approvals, as for them there was a strong ethical obligation to ensure their family was properly looked after, which in their opinion outweighs their obligation to the community.

Is it ethical for other nations to condemn such behaviour in the extreme cases? Should a different standard apply? What is the business to do if that is the norm in a country?

- **Decline business**
  Some may decline to do business in those countries.

- **Use intermediaries**
  Others may employ intermediaries. In the latter case, a company sells the goods to an intermediary company which then resells the goods in the problem country. The intermediary obviously has to pay fees and bribes to make the sale but that is not the concern of the multinational company! They deliberately do not ask the intermediary what they do. However, it could become a concern if a protest group identifies the questionable behaviour of the agent and decides to hold the multinational responsible.

- **Pay bribes**
  A third option is just to pay the fees and bribes. Unfortunately this reinforces the corrupt forces in the target country.

- **Risks in using intermediaries and/or paying bribes**
  The company may be held responsible by one of the countries in which they operate which has laws making it illegal to corrupt public officials in their country. The company may also be held liable by its own government.

For example, in the UK the Serious Fraud Office[14] and in the US the Department of Justice are actively investigating corrupt practices. The Serious Fraud Office made BAE Systems pay £30 million in relation to overpriced military radar sold to Tanzania whilst taking into account the implementation by the company of substantial ethical and compliance reforms. Part of the fine is being passed on to the people of Tanzania to compensate for the damage done. Ongoing investigations in 2013 include a criminal investigation into bribery and corruption at Rolls-Royce.

11.7.10 The support given by professional bodies in the designing of ethical codes

There are excellent support facilities available. For example, the Association of Chartered Certified Accountants' website (www.accaglobal.com) makes a toolkit available for accountants who might be involved with designing a code of ethics. The site also provides an overview which considers matters such as why ethics are important, links to other related sites, e.g. the Center for Ethics and Business at Loyola Marymount University in Los Angeles[15] with a quiz to establish one's ethical style as an ethic of justice or an ethic of care, and a toolkit from the Ethics Resource Center[16] to assist in the design of a code of ethics.

11.8 The increasing role of whistle-blowing

It is recognised that normally when the law or the ethical code is being broken by the company, a range of people inside and outside the company are aware of the illegal activities or have sufficient information to raise suspicions. To reduce the likelihood of illegal activity or to help identify its occurrence, a number of regulatory organisations have set up mechanisms for whistle-blowing to occur. Also a number of companies have set up their own units, often through a consulting firm, whereby employees can report illegal activities and breaches of the firm's code of ethics or any other activities which are likely to bring the company into disrepute.

11.8.1 Seeking advice

There are a variety of sources ranging from Public Concern at Work (PCaW) to the advice available from the professional accounting bodies themselves.

Public Concern at Work

There are whistle-blowing charities such as Public Concern at Work which provide confidential telephone advice, free of charge, to people who witness wrongdoing at work but are not sure whether or how to raise their concern. Since 1993, it has advised on over 10,000 actual and potential whistle-blowing concerns including tackling frauds.

It carried out a joint research project[17] in 2013 with the University of Greenwich, 'Whistle-blowing – the inside story', reviewing the story of 1,000 callers from a variety of professions including lawyers and accountants. One of the outcomes suggested that organisations seem to be better at correcting wrongdoing than at safeguarding the whistle-blower from harm.

Professional advice to members

All the professional bodies provide support to members. The ICAEW, for example, adopts the PCaW Guidelines and in addition has ethics advisers who can give ethics advice on the specific guidance that applies to its members. The ICAEW offers a Support Members Scheme throughout England, Wales and the Channel Islands.

11.8.2 Whistle-blowing – protection in the UK

In the UK the Public Interest Disclosure Act came into force in 1999 protecting whistle-blowers who raised genuine concerns about malpractice, from dismissal and victimisation in order to promote the public interest. The scope of malpractice is wide-ranging, including, e.g. the covering up of a suspected crime, a civil offence such as negligence, a miscarriage of justice, and health and safety or environmental risks.

11.8.3 Anonymous whistle-blowing

Whilst there is this statutory protection, and firms may well support whistle-blowers, they need to realise from the beginning that ultimately they may have to seek alternative employment.

Although the whistle-blowing policies might have been followed and the accountants protected by the provisions of the Public Interest Disclosure Act, whistle-blowing could result in a breakdown of trust making the whistle-blower's position untenable; this means that a whistle-blower might be well advised to have an alternative position in mind.

This is not to suggest they shouldn't blow the whistle. Rather it is to reflect the history of whistle-blowers.

Reasons for reporting anonymously

People will often be reporting on activities which they have been 'forced' to do or on activities of their superior or colleagues. Given that those colleagues will not take kindly to being reported on, and are capable of making life very difficult for the informant, it is important that reports can be made anonymously. Also, even those who are not directly affected will often view whistle-blowing as letting the side down. The whistle-blower, if identified, could well be ostracised.

However, action should be taken if it exposes you to criminal prosecution or, even if not at any personal risk, it might prevent the company from becoming involved further in inappropriate behavior. Take Enron as an example where the collapse of the company and Arthur Andersen might have been avoided with the loss of many jobs if there had been earlier response to staff concerns about dubious accounting practices.

11.8.4 Proportionate response

In spite of the above comments it is important to keep in mind that the steps taken should reflect the seriousness of the event and that the whistle-blowing should be the final strategy rather than the first. In other words, the normal actions should be to use the internal forums such as debating issues in staff meetings or raising the issue with an immediate superior or their boss when the superior is not approachable for some reason. Nor are disagreements over business issues a reason for reporting. The motivation should be to report breaches which represent legal, moral or public interest concerns and not matters purely relating to differences of opinion on operational issues, personality differences or jealousy.

11.8.5 Government support

There are legal protections against victimisation, but it would be more useful if the government provided positive support such as assistance with finding other employment or, perhaps, some form of financial reward such as is available in some countries to compensate for public-spirited actions that actually lead to professional or financial hardship for the whistle-blower.

11.8.6 Immunity to the first party to report

In many countries the regulatory authority responsible for pursuing price fixing has authority to give immunity or favourable treatment to the first party to report the occurrence of price fixing. It may be possible for the person's lawyer to ascertain whether the item has already been reported without disclosing the identity of their client. This arrangement is in place because of the difficulty of collecting information on such activities of sufficient quality and detail to prosecute successfully.

British Airways, for example, was fined about £270m after it admitted collusion in fixing the prices of fuel surcharges. The US Department of Justice fined it $300m (£148m) for colluding on how much extra to charge on passenger and cargo flights to cover fuel costs, and the UK's Office of Fair Trading fined it £121.5m after it held illegal talks with rival Virgin Atlantic. Virgin was given immunity after it reported the collusion and was not fined.

11.8.7 Breach of confidentiality

Auditors are protected from the risk of liability for breach of confidence provided that:

- disclosure is made in the public interest;
- disclosure is made to a proper authority;
- there is no malice motivating the disclosure.

11.9 Legal requirement to report – national and international regulation

It is likely that there will be an increase in formal regulation as the search for greater transparency and ethical business behaviour continues. We comment briefly on national and international regulation relating to money laundering and bribery.

Money laundering – overview

There are various estimates of the scale of money laundering ranging up to over 2% of global gross domestic product. Certain businesses are identified as being more prone to money laundering, e.g. import/export companies and cash businesses such as antiques and art dealers, auction houses, casinos and garages. However, the avenues are becoming more and more sophisticated with methods varying between countries, e.g. in the UK there is the increasing use of smaller non-bank institutions whereas in Spain it includes cross-border carrying of cash, money-changing at bureaux de change and investment in real estate.

Money laundering – implications for accountants

In 2010 the Auditing Practices Board (APB) in the UK issued a revised Practice Note 12 *Money Laundering* which required auditors to take the possibility of money laundering into account when carrying out their audit and to report to the appropriate authority if they become aware of suspected laundering.

Money laundering – the Financial Action Task Force (FATF)

The Financial Action Task Force (FATF) is an independent intergovernmental body that develops and promotes policies to protect the global financial system against money laundering and terrorist financing. Recommendations issued by the FATF define criminal justice and regulatory measures that should be implemented to counter this problem. These recommendations also include international cooperation and preventive measures to be taken by financial institutions and others such as casinos, real-estate dealers, lawyers and accountants. The recommendations are recognised as the global anti-money-laundering (AML) and counter-terrorist-financing (CTF) standard.

FATF issued a report[18] in 2009 entitled *Money Laundering through the Football Sector*. This report identified the vulnerabilities of the sector arising from transactions relating to the ownership of football clubs, the transfer market and ownership of players, betting activities and image rights, sponsorship and advertising arrangements. The report is an excellent introduction to the complex web that attracts money launderers.

11.10 Why should students learn ethics?

Survival of the profession

There is debate over whether the attempts to teach ethics are worthwhile. However, this chapter is designed to raise awareness of how important ethics are to the survival of the accounting profession. Accounting is part of the system to create trust in the financial information provided. The financial markets will not operate efficiently and effectively if there is not a substantial level of trust in the system. Such trust is a delicate matter and if the accounting profession is no longer trusted then there is no role for its members to play in the system. In that event the accounting profession will vanish. It may be thought that the loss of trust is so unlikely that it need not be contemplated. But who imagined that Arthur Andersen, one of the 'Big Five' as we knew it, would vanish from the scene so quickly? As soon as the public correctly or incorrectly decided they could no longer trust Arthur Andersen, the business crashed.

A future role for accountants in ethical assurance

The accountant within business could also be seeing a growth in the ethical policing role as internal auditors take on the role of assessing the performance of managers as to their adherence to the ethical code of the organisation. This is already partly happening as conflicts of interest are often highlighted by internal audits and comments raised on managerial practices. This is after all a traditional role for accountants, ensuring that the various codes of practice of the organisation are followed. The level of adherence to an ethical code is but another assessment for the accountant to undertake.

Implications for training

If, as is likely, the accountant has a role in the future as 'ethical guardian', additional training will be necessary. This should be done at a very early stage, as in the US, where accountants wishing to be Certified Public Accountants (CPAs) are required to pass formal exams on ethical practices and procedures before they are allowed the privilege of working in practice. Failure in these exams prevents the prospective accountant from practising in the business environment.

In the UK, for example, ethics is central to the ACCA qualification in recognition that values, ethics and governance are themes which organisations are now embedding into company business plans, and expertise in these areas is highly sought after in today's employment market. ACCA has adopted a holistic approach to a student's ethical development through the use of 'real-life' case studies and embedding ethical issues within the exam syllabi as in Paper 1 Governance, Risk and Ethics. In addition, as part of their ethical development, students will be required to complete a two-hour online training module, developed by ACCA. This will give students exposure to a range of real-life ethical case studies and will require them to reflect on their own ethical behaviour and values. Similar initiatives are being taken by the other professional accounting bodies, with case studies available from the Association of International Accountants and The Institute of Chartered Accountants of Scotland.

How will decisions be viewed?

Another aspect of ethical behaviour is that others will often be judging the morality of action using hindsight or whilst coming from another perspective. This is the 'how would it appear on the front page of the newspaper?' aspect. So being aware of what could happen is often part of ethical sensitivity. In other words, being able to anticipate possible outcomes or how other parties will view what you have done is a necessary part of identifying that ethical issues have to be addressed.

What if there are competing solutions?

Ethical behaviour involves making decisions which are as morally correct and fair as possible, recognising that sometimes there will have to be decisions in relation to two or more competing aspects of what is morally correct which are in unresolvable conflict. One has to be sure that any trade-offs are made for the good of society and that decisions are not blatantly or subtly influenced by self-interest. They must appear fair and reasonable when reviewed subsequently by an uninvolved outsider who is not an accountant. This is because the community places their trust in professionals because they have expertise that others do not, but at the same time it is necessary to retain that trust.

Summary

At the macro level the existence of the profession and the careers of all of us in it are dependent on the community's perception of the profession as being ethical. Students need to be very conscious of that as they will make up the profession of the future.

At a more micro level all accountants will face ethical issues during their careers, whether they recognise them or not. This chapter attempts to increase awareness of the existence of ethical questions. The simplest way to increase awareness is to ask the question:

● Who is directly or indirectly affected by this accounting decision?

Then the follow-up question is:

● If I was in their position how would I feel about the accounting decision in terms of its fairness?

By increasing awareness of the impact of decisions, including accounting decisions, on other parties, hopefully the dangers of decisions which are unfair will be recognised. By facing the implications head-on, the accountant is less likely to make wrong decisions. Also keep in mind those accountants who never set out to be unethical but by a series of small incremental decisions found themselves at the point of no return. The personal consequences of being found to be unethical can cover financial disasters, a long period of stress as civil or criminal cases wind their way through the courts, and at the extreme suicide or prison.

Another aspect of this chapter has been the attempt to highlight the vulnerability of companies to accusations of both direct and indirect unethical impacts and hence the need to be aware of trends to increasing levels of accountability.

Finally you need to be aware of the avenues for getting assistance if you find yourself under pressure to ignore ethics or to turn a blind eye to the inappropriate behaviour of others. You should be aware of built-in avenues for addressing such concerns within your own organisation. Further, you should make yourself familiar with the assistance which your professional body can provide, such as providing experienced practitioners to discuss your options and the likely advantages and disadvantages of those alternatives.

REVIEW QUESTIONS

1 Identify two ethical issues which university students experience and where they look for guidance. How useful is that guidance?

2 The following is an extract from a *European Accounting Review*[19] article:

> On the teaching front, there is a pressing need to challenge more robustly the tenets of modern day business, and specifically accounting, education which have elevated the principles of property rights and narrow self-interest above broader values of community and ethics.

Discuss how such a challenge might impact on accounting education.

3 The International Association for Accounting Education and Research states that: 'Professional ethics should pervade the teaching of accounting' (www.iaaer.org). Discuss how this can be achieved on an undergraduate accounting degree.

4 As a trainee auditor, what ethical issues are you most likely to encounter?

5 Explain what you think are four common types of ethical issues associated with (a) auditing, (b) public practice and (c) accounting in a corporate environment.

6 Lord Borrie QC has said[20] of the Public Interest Disclosure Bill that came into force in July 1999 that the new law would encourage people to recognise and identify with the wider public interest, not just their own private position, and it will reassure them that if they act reasonably to protect the legitimate interest of others, the law will not stand idly by should they be vilified or victimised. Confidentiality should only be breached, however, if there is a statutory obligation to do so. Discuss.

7 'Confidentiality means that an accountant in business has a loyalty to the business which employs him or her which is greater than any commitment to a professional code of ethics.' Discuss.

8 An interesting ethical case arose when an employee of a Swiss bank stole records of the accounts of international investors. The records were then offered for sale to the German government on the basis that many of them would represent unreported income and thus provide evidence of tax evasion. Should the government buy the records? Provide arguments for and against.

9 Refer to the Ernst & Young Code of Conduct and discuss the questions they suggest when putting their Global Code of Conduct into action.[21]

10 Should ethics be applicable at the standard-setting level? Express and justify your own views on this as distinct from repeating the material in the chapter.

11 Discuss the role of the accounting profession in the issue of ethics.

12 'The management of a listed company has a fiduciary duty to act in the best interest of the current shareholders and it would be unethical for them to act in the interest of other parties if this did not maximise the existing earnings per share.' Discuss.

13 How might a company develop a code of ethics for its own use?

14 Outline the advantages and disadvantages of a written code of ethics.

15 In relation to the following scenarios explain why it is a breach of ethics and what steps could have been taken to avoid the issue:

(a) The son of the accountant of a company is employed during the university holiday period to undertake work associated with preparation for a visit of the auditors.

(b) A senior executive is given a first-class seat to travel to Chicago to attend an industry fair where the company is launching a new product. The executive decides to cash in the ticket and to get two economy-class tickets so her boyfriend can go with her. The company picks up the hotel bill and she reimburses the difference between what it would have cost if she went alone and the final bill. The frequent flier points were credited to her personal frequent flier account. Would it make any difference if the company was not launching a new product at the fair?

(c) You pay a sizeable account for freight on the internal shipping of product deliveries in an underdeveloped country. At morning tea the gossip is that the company is paying bribes to a general in the underdeveloped country as protection money.

(d) The credit card statement for the managing director includes payments to a casino. The managing director says it is for the entertainment of important customers.

(e) You are processing a payment for materials which have been approved for repairs and maintenance when you realise the delivery is not to one of the business addresses of the company.

16 In each of the following scenarios, outline the ethical problem and suggest ways in which the organisation may solve the problem and prevent its recurrence:

(a) A director's wife uses his company car for shopping.

(b) Groceries bought for personal use are included on a director's company credit card.

(c) A director negotiates a contract for management consultancy services but it is later revealed that her husband is a director of the management consultancy company.

(d) The director of a company hires her son for some holiday work within the company but does not mention the fact to her fellow directors.

(e) You are the accountant to a small engineering company and you have been approached by the Chairman to authorise the payment of a fee to an overseas government employee in the hope that a large contract will be awarded.

(f) Your company has had some production problems which have resulted in some electrical goods being faulty (possibly dangerous) but all production is being dispatched to customers regardless of condition.

17 In each of the following scenarios, outline the ethical or potential ethical problem and suggest ways in which it could be resolved or avoided:

(a) Your company is about to sign a contract with a repressive regime in South America for equipment which **could** have a military use. Your own government has given you no advice on this matter.

(b) Your company is in financial difficulties and a large contract has just been gained in partnership with an overseas supplier who employs children as young as seven years old on their production line. The children are the only wage earners for their families and there is no welfare available in the country where they live.

(c) You are the accountant in a large manufacturing company and you have been approached by the manufacturing director to prepare a capital investment proposal for a new production line. After your calculations the project meets **none** of the criteria necessary to allow the project to proceed but the director instructs you to change the financial forecast figures to ensure the proposal is approved.

(d) Review the last week's newspapers and select **three** examples of failures of business ethics and justify your choice of examples.

(e) The company deducts from the monthly payroll employees' compulsory contribution to their superannuation accounts. The payment to the superannuation fund, which also includes the company's matching contribution, is being made only six-monthly because the cash flow of the company is tight following rapid expansion.

18 'It has been said that football clubs are seen by criminals as the perfect vehicles for money laundering.' Discuss the reason for this view.

19 Access www.saynotoolkit.net and discuss the definition and ethical risks that an accountant in business may encounter from bid rigging, facilitation payments, intermediaries, kick-back and market manipulation.

EXERCISES

Question 1

You have recently qualified and set up in public practice under the name Patris Zadan. You have been approached to provide accounting services for Joe Hardiman. Joe explains that he has had a lawyer set up six businesses and he asks you to do the books and to handle tax matters. The first thing you notice is that he is running a number of laundromats which are largely financed by relatives from overseas. As the year progresses you realise those businesses are extremely profitable, given industry averages.

Required:
Discuss: What do you do?

Question 2

Joe Withers is the chief financial officer for Withco plc responsible for negotiating bank loans. It has been the practice to obtain loans from a number of merchant banks. He has recently met Ben Billings who had been on the same undergraduate course some years earlier. They agree to meet for a game of squash and during the course of the evening Joe learns that Ben is the chief loans officer at The Swift Merchant Bank.

During the next five years Joe negotiates all of the company's loan requirements through Swift, and Ben arranges for Joe to receive substantial allocations in initial public offerings. Over that period Joe has done quite well out of taking up allocations and selling them within a few days on the market.

Required:
Discuss the ethical issues.

Question 3

Kim Lee is a branch accountant in a multinational company Green Cocoa plc responsible for purchasing supplies from a developing country. Kim Lee is authorised to enter into contracts up to $100,000 for any single transaction. Demand in the home market is growing and Head Office are pressing for an increase in supplies. A new government official in the developing country says that Kim needs an export permit from his department and that he needs a payment to be made to his brother-in-law for consulting services if the permit is to be granted. Kim quickly checks alternative sources and finds that the normal price combined with the extra 'facilitation fee' is still much cheaper than the alternative sources of supply. Kim faces two problems, namely, whether to pay the bribe and, if so, how to record it in the accounts so it is not obvious what it is.

Required:
Discuss the ethical issues.

Question 4

Jemma Burrett is a public practitioner. Four years earlier she had set up a family trust for a major client by the name of Simon Trent. The trust is for the benefit of Simon and his wife Marie. Marie is also a client of the practice and the practice prepares her tax returns. Subsequently Marie files for divorce. In her claim for a share of the assets she claims a third share of the business and half the other assets of the family which are listed. The assets of the family trust are not included in the list.

Required:
Discuss the ethical issues raised by the case and what action the accountant should take (if any).

Question 5

George Longfellow is a financial controller with a listed industrial firm which has a long period of sustained growth. This has necessitated substantial use of external borrowing.

During the great financial crisis it has become harder to roll over the loans as they mature. To make matters worse sales revenues have fallen 5% for the financial year, debtors have taken longer to pay, and margins have fallen. The managing director has said that he doesn't want to report a loss for the first time in the company's history as it might scare financiers.

The finance director (FD) has told George to make every effort to get the result to come out positively. He suggests that a number of expenses should be shifted to prepayments, provisions for doubtful debts should be lowered, and new assets should not be depreciated in the year of purchase but rather should only commence depreciation in the next financial year on the argument that new assets take a while to become fully operational.

In the previous year the company had moved into a new line of business where a small number of customers paid in advance. Because these were exceptional the auditors were persuaded to allow you to avoid the need to make the systems more sophisticated to decrease revenue and to recognise a liability. After all, it was immaterial in the overall group. Fortunately, that new line of business has grown substantially in the current financial year and it was suggested that the auditors be told that the revenue in advance should not be taken out of sales because a precedent had been set the year before.

George saw this as a little bit of creative accounting and was reluctant to do what he was instructed. When he tentatively made this comment to the FD he was assured that this was only temporary to ensure the company could refinance and that next year, when the economy recovered, all the discretionary adjustments would be reversed and everyone would be happy. After all, the employment of the 20,000 people who work for the group depends upon the refinancing and it was not as if the company was not going to be prosperous in the future. The FD emphasised that the few adjustments were, after all, a win–win situation for everyone and George was threatening the livelihood of all of his colleagues – many with children and with mortgage payments to meet.

Required:
Discuss who would or could benefit or lose from the Finance Director's proposals.

References

1 L. Spacek, 'The need for an accounting court', *The Accounting Review*, 1958, pp. 368–79.

2 IFAC, *Code of Ethics for Professional Accountants*, 2013.

3 W.J. Baumol, *Superfairness: Applications and Theory*, MIT Press, 1982.

4 D. Friedman, *Morals and Markets*, Palgrave Macmillan, 2008.

5 S.B. Salter, D.M. Guffey and J.J. McMillan, 'Truth, consequences and culture: a comparative examination of cheating and attitudes about cheating among US and UK students', *Journal of Business Ethics*, vol. 31(1), May 2001, pp. 37–50, Springer.

6 C. Helliar and J. Bebbington, 'Taking ethics to heart', ICAS, 2004, http://www.icas.org.uk

7 See http://www.btplc.com/TheWayWeWork/Businesspractice/twww_english.pdf

8 See http://www.ir.jameshardie.com.au/jh/asbestos_compensation.jsp

9 M.M. Jennings, *Seven Signs of Ethical Collapse: Understanding What Causes Moral Meltdowns in Organizations*, St Martin's Press, 2006.

10 S. Beder, *Beyond Petroleum*, http://www.uow.edu.au/~sharonb/bp.html

11 C. Stohl, M. Stohl and L. Popova, 'A new generation of codes of ethics', *Journal of Business Ethics*, vol. 90, 2009, pp. 607–22.

12 T. Morris, *If Aristotle Ran General Motors*, New York: Henry Holt and Company, 1997, pp. 118–145.

13 J.F. Castellano, K. Rosenweig and H.P. Roehm, 'How corporate culture impacts unethical distortion of financial numbers', *Management Accounting Quarterly*, vol. 5(4), Summer 2004.

14 See http://www.sfo.gov.uk

15 See http://www.lmu.edu/Page23070.aspx

16 See Ethics Resource Center, www.ethics.org

17 http:vwww.pcaw.org.uk/files/Whistleblowing%20-%20the%20inside%20story%20FINAL.pdf

18 See http://www.oecd.org

19 D. Owen, 'CSR after Enron: a role for the academic accounting profession?', *European Accounting Review*, vol. 14(2), 2005.

20 W. Raven, 'Social auditing', *Internal Auditor*, February 2000, p. 8.

21 See http://www.ey.com/Publication/vwLUAssets/Ernst-Young_Global_Code_of_Conduct/$FILE/EY_Code_of_Conduct.pdf

Statement of financial position – equity, liability and asset measurement and disclosure

- Equity itself is a residual figure in that the standard setters have taken the approach of defining assets and liabilities and leaving equity as the residual difference in the statement of financial position.
- Equity may consist of ordinary shares or equity elements of participating preference shares and compound instruments which include debt and equity, i.e. where there are conversion rights when there must be a split into their debt and equity elements, with each element being accounted for separately.
- Preference shareholders are not entitled (unless participating) to share in the residual income but may be entitled to a fixed or floating rate of interest on their investment.
- Distributable reserves equate to retained earnings when these have arisen from realised gains.
- Trade payables require protection to prevent an entity distributing assets to shareholders if creditors are not paid in full.
- Capital restructuring may be necessary when there are sound commercial reasons.

However, the rules are not static and there are periodic reviews in most jurisdictions, e.g. the proposal that an entity should make dividend decisions based on its ability to pay rather than on the fact that profits have been realised.

- The distributable reserves of entities are those that have arisen due to realised gains and losses (retained profits), as opposed to unrealised gains (such as revaluation reserves).
- There must be protection for trade payables to prevent an entity distributing assets to shareholders to the extent that the trade payables are not paid in full. An entity must retain net assets at least equal to its share capital and non-distributable reserves (a capital maintenance concept).
- The capital maintenance concept also applies with regard to reducing share capital, with most countries generally requiring a replacement of share capital with a non-distributable reserve if it is redeemed.

Because all countries have company legislation and these themes are common, the authors felt that, as the UK has relatively well-developed company legislation, it would be helpful to consider such legislation as illustrating a typical range of statutory provisions. We therefore now consider the constituents of total shareholders' funds (also known as total owners' equity) and the nature of distributable and non-distributable reserves. We then analyse the role of the capital maintenance concept in the protection of creditors, before discussing the effectiveness of the protection offered by the Companies Act 2006 in respect of both private and public companies.

12.3 Total owners' equity: an overview

Total owners' equity consists of the issued share capital stated at nominal (or par) value, non-distributable and distributable reserves. Here we comment briefly on the main constituents of total shareholders' funds. We go on to deal with them in greater detail in subsequent sections.

12.3.1 Right to issue shares

Companies incorporated[1] under the Companies Act 2006 are able to raise capital by the issue of shares and debentures. There are two main categories of company: private limited companies and public limited companies. Public limited companies are designated by the letters plc and have the right to issue shares and debentures to the public. Private limited companies are often family companies; they are not allowed to seek share capital by invitations to the public. The shareholders of both categories have the benefit of limited personal indemnity, i.e. their liability to creditors is limited to the amount they agreed to pay the company for the shares they bought.

12.3.2 Types of share

Broadly, there are two types of share: ordinary and preference.

Ordinary shares

Ordinary shares, often referred to as equity shares, carry the main risk and their bearers are entitled to the residual profit after the payment of any fixed interest or fixed dividend to investors who have invested on the basis of a fixed return. Distributions from the residual profit are made in the form of dividends, which are normally expressed as pence per share.

Preference shares

Preference shares usually have a fixed rate of dividend, which is expressed as a percentage of the nominal value of the share. The dividend is paid before any distribution to the ordinary shareholders. The specific rights attaching to a preference share can vary widely.

12.3.3 Non-distributable reserves

There are a number of types of **statutory** non-distributable reserve, e.g. when the paid-in capital exceeds the par value as a share premium. In addition to the statutory non-distributable reserves, a company might have restrictions on distribution within its memorandum and articles, stipulating that capital profits are non-distributable as dividends.

12.3.4 Distributable reserves

Distributable reserves are normally represented by the retained earnings that appear in the statement of financial position and belong to the ordinary shareholders. However, as we shall see, there may be circumstances where credits that have been made to the statement of comprehensive income are not actually distributable, usually because they do not satisfy the **realisation** concept.

Although the retained earnings in the statement of financial position contain the cumulative residual distributable profits, it is the earnings per share (EPS), based on the post-tax earnings for the year as disclosed in the profit and loss account, that influences the market valuation of the shares, applying the price/earnings ratio.

When deciding whether to issue or buy back shares, the directors will therefore probably consider the impact on the EPS figure. If the EPS increases, the share price can normally be expected also to increase.

12.4 Total shareholders' funds: more detailed explanation

12.4.1 Ordinary shares – risks and rewards

Ordinary shares (often referred to as equity shares) confer the right to:

- share proportionately in the rewards, i.e. the residual profit remaining after paying any loan interest or fixed dividends to investors who have invested on the basis of a fixed return;
- any dividends distributed from these residual profits;
- any net assets remaining after settling all creditors' claims in the event of the company ceasing to trade;

- share proportionately in the risks, i.e. lose a proportionate share of invested share capital if the company ceases to trade and there are insufficient funds to pay all the creditors and the shareholders in full.

12.4.2 Ordinary shares – powers

The owners of ordinary shares generally have one vote per share which can be exercised on a routine basis, e.g. at the Annual General Meeting to vote on the appointment of directors, and on an *ad hoc* basis, e.g. at an Extraordinary General Meeting to vote on a proposed capital reduction scheme.

However, there are some companies that have issued non-voting ordinary shares which may confer the right to a proportional share of the residual profits but not to vote.

Non-voting shareholders can attend and speak at the Annual General Meeting but, as they have no vote, are unable to have an influence on management if there are problems or poor performance – apart from selling their shares.

Non-voting shares are issued where typically the founders of a company wish to retain control as we can see with Facebook, LinkedIn and Google where 18% of the shares control 57% of the votes.

12.4.3 Methods and reasons for issuing shares

Methods of issuing shares

Some of the common methods of issuing shares are *offer for subscription*, where the shares are offered directly to the public; *placings*, where the shares are arranged (placed) to be bought by financial institutions; and *rights issues*, whereby the new shares are offered to the existing shareholders at a price below the market price of those shares. The rights issue might be priced significantly below the current market price but this may not mean that the shareholder is benefiting from cheap shares as the price of existing shares will be reduced, e.g. the British Telecommunications plc £5.9 billion rights issue announced in 2001 made UK corporate history in that no British company had attempted to raise so much cash from its shareholders. The offer was three BT shares for every 10 held and, to encourage take-up, the new shares were offered at a deeply discounted rate of £3 which was at a 47% discount to the share price on the day prior to the launch.

Reasons for issuing shares

- For future investment;
- As consideration on an acquisition, e.g. the Pfizer bid for AstraZeneca in 2014 was based on the consideration being partly funded by the issue of shares in Pfizer;
- To shareholders to avoid paying out cash from the company's funds;
- To directors and employees to avoid paying out cash in the form of salary from company funds, which is attractive to early-stage companies to preserve working capital;
- To shareholders to encourage reinvestment;
- To shareholders by way of a rights issue to shore up statements of financial position weakened in the credit crisis by reducing debt and to avoid breaching debt covenants;
- To loan creditors in exchange for debt;
- To obtain funds for future acquisitions;
- To reduce levels of debt to avoid credit rating agencies downgrading the company, which would make it difficult or more expensive to borrow;
- To overcome liquidity problems.

12.4.4 Types of preference shares

The following illustrate some of the ways in which specific rights can vary.

Cumulative preference shares
Dividends not paid in respect of any one year because of a lack of profits are accumulated for payment in some future year when distributable profits are sufficient.

Non-cumulative preference shares
Dividends not paid in any one year because of a lack of distributable profits are permanently forgone.

Participating preference shares
These shares carry the right to participate in a distribution of additional profits over and above the fixed rate of dividend after the ordinary shareholders have received an agreed percentage. The participation rights are based on a precise formula.

Redeemable preference shares
These shares may be redeemed by the company at an agreed future date and at an agreed price.

Convertible preference shares
These shares may be converted into ordinary shares at a future date on agreed terms. The conversion is usually at the preference shareholder's discretion.

There can be a mix of rights, e.g. Getronics entered into an agreement in 2005 with its cumulative preference shareholders whereby Getronics had the right in 2009 to repurchase (redeem) the shares and, if it did not redeem the shares, the cumulative preference share-holders had the right to convert into ordinary shares.

12.5 Accounting entries on issue of shares

12.5.1 Shares issued at nominal (par) value

If shares are issued at nominal value, the company simply debits the cash account with the amount received and credits the ordinary share capital or preference share capital, as appropriate, with the **nominal value** of the shares.

12.5.2 Shares issued at a premium

The market price of the shares of a company, which is based on the prospects of that company, is usually different from the par (nominal) value of those shares.

On receipt of consideration for the shares, the company again debits the cash account with the amount received and credits the ordinary share capital or preference share capital, as appropriate, with the **nominal value** of the shares.

Assuming that the market price exceeds the nominal value, a premium element will be credited to a share premium account. The share premium is classified as a **non-distributable reserve** to indicate that it is not repayable to the shareholders who have subscribed for their shares: it remains a part of the company's permanent capital.

The accounting treatment for recording the issue of shares is straightforward. For example, the journal entries to record the issue of 1,000 £1 ordinary shares at a market price of £2.50 per share payable in instalments of:

on application	on 1 January 20X1	25p
on issue	on 31 January 20X1	£1.75 including the premium
on first call	on 31 January 20X2	25p
on final call	on 31 January 20X4	25p

would be as follows:

1 Jan 20X1	Dr (£)	Cr (£)
Cash account	250	
Application account		250

31 Jan 20X1	Dr	Cr
Cash account	1,750	
Issue account		1,750

31 Jan 20X1	Dr	Cr
Application account	250	
Issue account	1,750	
Share capital account		500
Share premium in excess of par value		1,500

The first and final call would be debited to the cash account and credited to the share capital account on receipt of the date of the calls.

12.6 Creditor protection: capital maintenance concept

To protect creditors, there are often rules relating to the use of the total shareholders' funds which determine how much is distributable.

As a general rule, the paid-in share capital is not repayable to the shareholders and the reserves are classified into two categories: distributable and non-distributable. The directors have discretion as to the amount of the distributable profits that they recommend for distribution as a dividend to shareholders. However, they have no discretion as to the treatment of the non-distributable funds. There may be a statutory requirement for the company to retain within the company net assets equal to the non-distributable reserves. This requirement is to safeguard the interests of creditors and is known as **capital maintenance**.

12.7 Creditor protection: why capital maintenance rules are necessary

It is helpful at this point to review the position of unincorporated businesses in relation to capital maintenance.

12.7.1 Unincorporated businesses

An unincorporated business such as a sole trader or partnership is not required to maintain any specified amount of capital within the business to safeguard the interests of its creditors. The owners are free to decide whether to introduce or withdraw capital. However, they remain personally liable for the liabilities incurred by the business, and the creditors can have recourse to the personal assets of the owners if the business assets are inadequate to meet their claims in full.

When granting credit to an unincorporated business, the creditors may well be influenced by the personal wealth and apparent standing of the owners and not merely by the assets of the business as disclosed in its financial statements. This is why in an unincorporated business there is no external reason for the capital and the profits to be kept separate.

In partnerships, there are frequently internal agreements that require each partner to maintain his or her capital at an agreed level. Such agreements are strictly a matter of contract between the owners and do not prejudice the rights of the business creditors.

Sometimes owners attempt to influence creditors unfairly, by maintaining a lifestyle in excess of what they can afford, or try to frustrate the legal rights of creditors by putting their private assets beyond their reach, e.g. by transferring their property to relatives or trusts. These subterfuges become apparent only when the creditors seek to enforce their claim against the private assets. Banks are able to protect themselves by seeking adequate security, e.g. a charge on the owners' property.

12.7.2 Incorporated limited liability companies

Because of limited liability, the rights of creditors against the private assets of the owners, i.e. the shareholders of the company, are restricted to any amount unpaid on their shares. Once the shareholders have paid the company for their shares, they are not personally liable for the company's debts. Creditors are restricted to making claims against the assets of the company.

Hence, the legislature considered it necessary to ensure that the shareholders did not make distributions to themselves such that the assets needed to meet creditors' claims were put beyond creditors' reach. This may be achieved by setting out statutory rules.

12.8 Creditor protection: how to quantify the amounts available to meet creditors' claims

Creditors are exposed to two types of risk: the business risk that a company will operate unsuccessfully and will be unable to pay them; and the risk that a company will operate successfully, but will pay its shareholders rather than its creditors.

The legislature has never intended trade creditors to be protected against ordinary business risks, e.g. the risk of the debtor company incurring either trading losses or losses that might arise from a fall in the value of the assets following changes in market conditions.

In the UK, the Companies Act 2006 requires the amount available to meet creditors' claims to be calculated by reference to the company's annual financial statements. There are two possible approaches:

- The direct approach which requires the asset side of the statement of financial position to contain assets with a realisable value sufficient to cover all outstanding liabilities.

- The indirect approach which requires the liability side of the statement of financial position to classify reserves into distributable and non-distributable reserves (i.e. respectively, available and not available to the shareholders by way of dividend distributions).

The Act follows the indirect approach by specifying capital maintenance in terms of the total shareholders' funds. However, this has not stopped certain creditors taking steps to protect themselves by following the direct approach, e.g. it is bank practice to obtain a mortgage debenture over the assets of the company. The effect of this is to disadvantage the trade creditors. The statutory restrictions preventing shareholders from reducing capital accounts on the liability side are weakened when management grants certain parties priority rights against some or all of the company's assets.

We will now consider total shareholders' funds and capital maintenance in more detail, starting with share capital. Two aspects of share capital are relevant to creditor protection: minimum capital requirements and reduction of capital.

12.9 Issued share capital: minimum share capital

The creditors of public companies may be protected by the requirements that there should be a minimum share capital and that capital should be reduced only under controlled conditions.

In the UK, the minimum share capital requirement for a public company is currently set at £50,000 or its euro equivalent, although this can be increased by the Secretary of State for the Department for Business, Innovation and Skills.[2] A company is not permitted to commence trading unless it has issued this amount. However, given the size of many public companies, it is questionable whether this figure is adequate.

The minimum share capital requirement refers to the nominal value of the share capital. In the UK, the law requires each class of share to have a stated nominal value. This value is used for identification and also for capital maintenance. The law ensures that a company receives an amount that is at least equal to the nominal value of the shares issued, less a controlled level of commission, by prohibiting the issue of shares at a discount and by limiting any underwriting commissions on an issue. This is intended to avoid a material discount being granted in the guise of commission. However, the requirement is concerned more with safeguarding the relative rights of existing shareholders than with protecting creditors.

There is effectively no minimum capital requirement for private companies. We can see many instances of such companies having an issued and paid-up capital of only a few £1 shares, which cannot conceivably be regarded as adequate creditor protection. The lack of adequate protection for the creditors of private companies is considered again later in the chapter.

12.10 Distributable profits: general considerations

We have considered capital maintenance and non-distributable reserves. However, it is not sufficient to attempt to maintain the permanent capital accounts of companies unless there are clear rules on the amount that they can distribute to their shareholders as profit. Without such rules, they may make distributions to their shareholders out of capital. The question of what can legitimately be distributed as profit is an integral part of the concept of capital maintenance in company accounts. In the UK, there are currently statutory definitions of the amount that can be distributed by private, public and investment companies.

12.10.1 Distributable profits: general rule for private companies

The definition of distributable profits under the Companies Act 2006 is:

> Accumulated, realised profits, so far as not previously utilised by distribution or capitalisation, less its accumulated, realised losses, as far as not previously written off in a reduction or reorganisation of capital.

This means the following:

- Unrealised profits cannot be distributed.
- There is no difference between realised revenue and realised capital profits.
- All accumulated net realised profits (i.e. realised profits less realised losses) on the statement of financial position date must be considered.

On the key question of whether a profit is realised or not, the Companies Act (paragraph 853) simply says that realised profits or realised losses are:

> such profits or losses of the company as fall to be treated as realised in accordance with principles generally accepted, at the time when the accounts are prepared, with respect to the determination for accounting purposes of realised profits or losses.

Hence, the Act does not lay down detailed rules on what is and what is not a realised profit; indeed, it does not even refer specifically to 'accounting principles'. Nevertheless, it would seem reasonable for decisions on realisation to be based on generally **accepted accounting principles** at the time, subject to the court's decision in cases of dispute.

12.10.2 Distributable profits: general rule for public companies

According to the Companies Act, the undistributable reserves of a public company are its share capital, share premium, capital redemption reserve and also 'the excess of accumulated unrealised profits over accumulated unrealised losses at the time of the intended distribution and . . . any reserves not allowed to be distributed under the Act or by the company's own Memorandum or Articles of Association'.

This means that, when dealing with a public company, the distributable profits have to be reduced by any net unrealised loss.

12.10.3 Investment companies

The Companies Act 2006 allows for the special nature of some businesses in the calculation of distributable profits. There are additional rules for investment companies in calculating their distributable profits. For a company to be classified as an investment company, it must invest its funds mainly in securities with the aim of spreading investment risk and giving its members the benefit of the results of managing its funds.

Such a company has the option of applying one of two rules in calculating its distributable profits. These are either:

- the rules that apply to public companies in general, but excluding any realised capital profits, e.g. from the disposal of investments; or
- the company's accumulated realised revenue less its accumulated realised and unrealised revenue losses, provided that its assets are at least one and a half times its liabilities both before and after such a distribution.

The reasoning behind these special rules seems to be to allow investment companies to pass the dividends they receive to their shareholders, irrespective of any changes in the values of their investments, which are subject to market fluctuations. However, the asset cover ratio of liabilities can easily be manipulated by the company simply paying creditors, whereby the ratio is improved, or borrowing, whereby it is reduced.

12.11 Distributable profits: how to arrive at the amount using relevant accounts

In the UK, the Companies Act 2006 stipulates that the distributable profits of a company must be based on **relevant accounts**. Relevant accounts may be prepared under either UK GAAP or EU-adopted IFRS. On occasions a new IFRS might have the effect of making a previously realised item reclassified as unrealised, which would then become undistributable. For a more detailed description on the determination of realised profits for distribution refer to the ICAEW Technical Release 02/10. These would normally be the audited annual accounts, which have been prepared according to the requirements of the Act to give a true and fair view of the company's financial affairs.

12.11.1 Effect of fair value accounting on decision to distribute

In the context of fair value accounting, volatility is an aspect where directors will need to consider their fiduciary duties. The fair value of financial instruments may be volatile even though such fair value is properly determined in accordance with IAS 39 *Financial Instruments: Recognition and Measurement*. Directors should consider, as a result of their fiduciary duties, whether it is prudent to distribute profits arising from changes in the fair values of financial instruments considered to be volatile, even though they may otherwise be realised profits in accordance with the technical guidance.

12.12 When may capital be reduced?

Once the shares have been issued and paid up, the contributed capital together with any payments in excess of par value are normally regarded as permanent. However, there might be commercially sound reasons for a company to reduce its capital and we will consider three such reasons. These are:

- writing off part of capital which has already been lost and is not represented by assets;
- repayment of part of paid-up capital to shareholders or cancellation of unpaid share capital;
- purchase of own shares.

In the UK it has been necessary for both private and public companies to obtain a court order approving a reduction of capital. In line with the wish to reduce the regulatory burden on private companies, the government legislated[3] in 2008 for private companies to be able to reduce their capital by special resolution subject to the directors signing a solvency statement to the effect that the company would remain able to meet all of its liabilities for at least a year. At the same time a reserve arising from the reduction is treated as realised and may be distributed, although it need not be and could be used for other purposes, e.g. writing off accumulated trading losses.

12.13 Writing off part of capital which has already been lost and is not represented by assets

This situation normally occurs when a company has accumulated trading losses which prevent it from making dividend payments under the rules relating to distributable profits. The general approach is to eliminate the debit balance on retained earnings by setting it off against the share capital and non-distributable reserves.

12.13.1 Accounting treatment for a capital reduction to eliminate accumulated trading losses

The accounting treatment is straightforward. A capital reduction account is opened. It is debited with the accumulated losses and credited with the amount written off the share capital and reserves.

For example, assume that the capital and reserves of Hopeful Ltd were as follows at 31 December 20X1:

	£
200,000 ordinary shares of £1 each	200,000
Retained earnings	(180,000)

The directors estimate that the company will return to profitability in 20X2, achieving profits of £4,000 per annum thereafter. Without a capital reduction, the profits from 20X2 must be used to reduce the accumulated losses. This means that the company would be unable to pay a dividend for 45 years if it continued at that level of profitability and ignoring tax. Perhaps even more importantly, it would not be attractive for shareholders to put additional capital into the company because they would not be able to obtain any dividend for some years.

There might be statutory procedures such as the requirement for the directors to obtain a special resolution and court approval to reduce the £1 ordinary shares to ordinary shares of 10p each. Subject to satisfying such requirements, the accounting entries would be:

	Dr	Cr
	£	£
Capital reduction account	180,000	
Retained earnings:		180,000
Transfer of debit balance		
Share capital	180,000	
Capital reduction account:		180,000
Reduction of share capital		

Accounting treatment for a capital reduction to eliminate accumulated trading losses and loss of value on non-current assets – losses borne by equity shareholders

Companies often take the opportunity to revalue all of their assets at the same time as they eliminate the accumulated trading losses. Any loss on revaluation is then treated in the same way as the accumulated losses and transferred to the capital reduction account.

For example, assume that the capital and reserves and assets of Hopeful Ltd were as follows at 31 December 20X1:

	£	£
200,000 ordinary shares of £1 each		200,000
Retained earnings		(180,000)
		20,000
Non-current assets		
Plant and equipment		15,000
Current assets		
Cash	17,000	
Current liabilities		
Trade payables	12,000	
Net current assets		5,000
		20,000

The plant and equipment is revalued at £5,000 and it is resolved to reduce the share capital to ordinary shares of 5p each. The accounting entries would be:

	Dr £	Cr £
Capital reduction account	190,000	
Statement of income		180,000
Plant and machinery:		10,000
Transfer of accumulated losses and loss on revaluation		
Share capital	190,000	
Capital reduction account:		190,000
Reduction of share capital to 200,000 shares of 5p each		

The statement of financial position after the capital reduction shows that the share capital fairly reflects the underlying asset values:

	£	£
200,000 ordinary shares of 5p each		10,000
		10,000
Non-current assets		
Plant and equipment		5,000
Current assets		
Cash	17,000	
Current liabilities		
Trade payables	12,000	5,000
		10,000

Accounting treatment for a capital reduction to eliminate accumulated trading losses and loss of value on non-current assets – losses borne by equity and other stakeholders

In the Hopeful Ltd example above, the ordinary shareholders alone bore the losses. It might well be, however, that a reconstruction involves a compromise between shareholders and creditors, with an amendment of the rights of the latter. Such a reconstruction would be subject to any statutory requirements within the jurisdiction, e.g. the support, say, of 75% of each class of creditor whose rights are being compromised, 75% of each class of shareholder and the permission of the court. For such a reconstruction to succeed there needs to be reasonable evidence of commercial viability and that anticipated profits are sufficient to service the proposed new capital structure.

Assuming in the Hopeful Ltd example that the creditors agree to bear £5,000 of the losses, the accounting entries would be as follows:

	£	£
Share capital	185,000	
Creditors	5,000	
Capital reduction account:		190,000
Reduction of share capital to 200,000 shares of 7.5p each		

Reconstruction schemes can be complex, but the underlying evaluation by each party will be the same. Each will assess the scheme to see how it affects their individual position.

Trade payables

In their decision to accept £5,000 less than the book value of their debt, the trade payables of Hopeful Ltd would be influenced by their prospects of receiving payment if Hopeful were to cease trading immediately, the effect on their results without Hopeful as a continuing customer and the likelihood that they would continue to receive orders from Hopeful following reconstruction.

Loan creditors

Loan creditors would take into account the expected value of any security they possess and a comparison of the opportunities for investing any loan capital returned in the event of liquidation with the value of their capital and interest entitlement in the reconstructed company.

Preference shareholders

Preference shareholders would likewise compare prospects for capital and income following a liquidation of the company with prospects for income and capital from the company as a going concern following a reconstruction.

Relative effects of the scheme

In practice, the formulation of a scheme will involve more than just the accountant, except in the case of very small companies. An advising merchant bank, major shareholders and major debenture holders will undoubtedly be concerned. Each vested interest will be asked for its opinion on specific proposals: unfavourable reactions will necessitate a rethink by the accountant. The process will continue until a consensus begins to emerge.

Each stakeholder's position needs to be considered separately. For example, any attempt to reduce the nominal value of all classes of shares and debentures on a proportionate basis would be unfair and unacceptable. This is because a reduction in the nominal values of preference shares or debentures has a different effect from a reduction in the nominal value of ordinary shares. In the former cases, the dividends and interest receivable will be reduced; in the latter case, the reduction in nominal value of the ordinary shares will have no effect on dividends as holders of ordinary shares are entitled to the residue of profit, whatever the nominal value of their shares.

Total support may well be unachievable. The objective is to maintain the company as a going concern. In attempting to achieve this, each party will continually be comparing its advantages under the scheme with its prospects in a liquidation.

Illustration of a capital reconstruction

XYZ plc has been making trading losses, which have resulted in a substantial debit balance on the profit and loss account. The statement of financial position of XYZ plc as at 31 December 20X3 was as follows:

		£000
Ordinary share capital (£1 shares)		1,000
Less: Accumulated losses on retained earnings	Note 1	(800)
		200
10% debentures (£1)		600
Net assets at book value	Note 2	800

Notes:

1 The company is changing its product and markets and expects to make £150,000 profit before interest and tax every year from 1 January 20X4.

2 (a) The estimated break-up or liquidation value of the assets at 31 December 20X3 was £650,000.

   (b) The going concern value of assets at 31 December 20X3 was £700,000.

The directors are faced with a decision to liquidate or reconstruct. Having satisfied themselves that the company is returning to profitability, they propose the following reconstruction scheme:

- Write off losses and reduce asset values to £700,000.
- Cancel all existing ordinary shares and debentures.
- Issue 1,200,000 new ordinary shares of 25p each and 400,000 12.5% debentures of £1 each as follows:
  - the existing shareholders are to be issued with 800,000 ordinary 25p shares;
  - the existing debenture holders are to be issued with 400,000 ordinary 25p shares and the new debentures.

The stakeholders, i.e. the ordinary shareholders and debenture holders, have first to decide whether the company has a reasonable chance of achieving the estimated profit for 20X4. The company might carry out a sensitivity analysis to show the effect on dividends and interest over a range of profit levels.

Next, stakeholders must consider whether allowing the company to continue provides a better return than that available from the liquidation of the company. Assuming that it does, they assess the effect of allowing the company to continue without any reconstruction of capital and with a reconstruction of capital.

The accountant writes up the reconstruction accounts and produces a statement of financial position after the reconstruction has been effected.

The accountant will produce the following information:

Effect of liquidating

	£	Debenture holders £	Ordinary shareholders £
Assets realised	650,000		
Less: Prior claim	(600,000)	600,000	
Less: Ordinary shareholders	(50,000)		50,000
	—	600,000	50,000

This shows that the ordinary shareholders would lose almost all of their capital, whereas the debenture holders would be in a much stronger position. This is important because it might influence the amount of inducement that the debenture holders require to accept any variation of their rights.

Company continues without reconstruction

	£	Debenture holders £	Ordinary shareholders £
Expected annual income:			
Expected operating profit	150,000		
Debenture interest	(60,000)	60,000	
Less: Ordinary dividend	(90,000)		90,000
Annual income	—	60,000	90,000

However, as far as the ordinary shareholders are concerned, no dividend will be allowed to be paid until the debit balance of £800,000 has been eliminated, i.e. there will be no dividend for more than nine years (for simplicity the illustration ignores tax effects).

Company continues with a reconstruction

	£	Debenture holders £	Ordinary shareholders £
Expected annual income:			
Expected operating profit	150,000		
Less: Debenture interest	(50,000)	50,000	
(12.5% on £400,000)			
Less: Dividend on shares	(33,000)	33,000	
Less: Ordinary dividend	(67,000)		67,000
Annual income	—	83,000	67,000

How will debenture holders react to the scheme?

At first glance, debenture holders appear to be doing reasonably well: the £83,000 provides a return of almost 14% on the amount that they would have received in a liquidation (83,000/600,000 × 100), which exceeds the 10% currently available, and it is £23,000 more than the £60,000 currently received. However, their exposure to risk has increased because £33,000 is dependent upon the level of profits. They will consider their position in relation to the ordinary shareholders.

For the ordinary shareholders the return should be calculated on the amount that they would have received on liquidation, i.e. 134% (67,000/50,000 × 100). In addition to receiving a return of 134%, they would hold two-thirds of the share capital, which would give them control of the company.

A final consideration for the debenture holders would be their position if the company were to fail after a reconstruction. In such a case, the old debenture holders would be materially disadvantaged as their prior claim will have been reduced from £600,000 to £400,000.

Accounting for the reconstruction

The reconstruction account will record the changes in the book values as follows:

Reconstruction account

	£000		£000
Retained earnings	800	Share capital	1,000
Assets (losses written off)	100	Debentures	600
		(old debentures cancelled)	
Ordinary share capital (25p)	300		
12.5% debentures (new issue)	400		
	1,600		1,600

The post-reconstruction statement of financial position will be as follows:

Ordinary share capital (25p)	300,000
12.5% debentures of £1	400,000
	700,000

12.14 Repayment of part of paid-in capital to shareholders or cancellation of unpaid share capital

This can occur when a company wishes to return liquid funds that it considers to be more than it is able to profitably employ within the business – as a result the return on equity ratio is increased. The following is an extract from the Next plc 2013 Annual Report:

The Company has five core operational objectives . . . Underlying these operational goals is the ever present and overriding financial objective of delivering long term, sustainable growth in earnings per share.

One of these five objectives is to:

Focus on cash generation. Return funds that are not needed to develop the business to shareholders through share buybacks. This must be earnings-enhancing and in the interests of shareholders generally.

12.15 Purchase of own shares

This might take the form of the redemption of redeemable preference shares, the purchase of ordinary shares which are then cancelled and the purchase of ordinary shares which are not cancelled but held in treasury.

12.15.1 Redemption of preference shares

In the UK, when redeemable preference shares are redeemed, the company is required either to replace them with other shares or to make a transfer from distributable reserves to non-distributable reserves in order to maintain permanent capital. The accounting entries on redemption are to credit cash and debit the redeemable preference share account.

12.15.2 Buyback of own shares – intention to cancel

There are a number of reasons for companies buying back shares. These provide a benefit when taken as:

● a strategic measure, e.g. recognising that there is a lack of viable investment projects, i.e. expected returns being less than the company's weighted average cost of capital and so returning excess cash to shareholders to allow them to search out better growth investments;

● a defensive measure, e.g. an attempt to frustrate a hostile takeover or to reduce the power of dissident shareholders;

● a reactive measure, e.g. taking advantage of the fact that the share price is at a discount to its underlying intrinsic value or stabilising a falling share price;

● a proactive measure, e.g. creating shareholder value by reducing the number of shares in issue which increases the earnings per share, or making a distribution more tax-efficient than the payment of a cash dividend;

● a tax-efficient measure.

There is also a potential risk if the company has to borrow funds in order to make the buy-back, leaving itself liable to service the debt. Where it uses free cash rather than loans it is attractive to analysts and shareholders. For example, in the BP share buyback scheme (one of the UK's largest), the chief executive, Lord Browne, said that any free cash generated from BP's assets when the oil price was above $20 a barrel would be returned to investors over the following three years.

12.15.3 Buyback of own shares – treasury shares

The benefits to a company holding treasury shares are that it has greater flexibility to respond to investors' attitude to gearing, e.g. reissuing the shares if the gearing is perceived to be too high. It also has the capacity to satisfy loan conversions and employee share options without the need to issue new shares which would dilute the existing shareholdings.

National regimes where buyback is already permitted

In Europe and the USA it has been permissible to buy back shares, known as treasury shares, and hold them for reissue. In the UK this has been permissible since 2003. There are two common accounting treatments – the cost method and the par value method. The most common method is the cost method, which provides the following.

On purchase

The treasury shares are debited at gross cost to a Treasury Stock account – this is deducted as a one-line entry from equity, e.g. a statement of financial position might appear as follows:

Owners' equity section of statement of financial position

	£
Common stock, £1 par, 100,000 shares authorised, 30,000 shares issued	30,000
Paid-in capital in excess of par	60,000
Retained earnings	165,000
Treasury Stock (15,000 shares at cost)	(15,000)
Total owners' equity	240,000

On resale

- If on resale the sale price is higher than the cost price, the Treasury Stock account is credited at cost price and the excess is credited to Paid-in Capital (Treasury Stock).

- If on resale the sale price is lower than the cost price, the Treasury Stock account is credited with the proceeds and the balance is debited to Paid-in Capital (Treasury Stock). If the debit is greater than the credit balance on Paid-in Capital (Treasury Stock), the difference is deducted from retained earnings. Retained earnings may be decreased but never increased as a result of Treasury Stock transactions.

The UK experience

Treasury shares have been permitted in the UK since 2003. The regulations relating to Treasury shares are now contained in the Companies Act 2006.[4] These regulations permit companies with listed shares that purchase their own shares out of distributable profits to hold them 'in treasury' for sale at a later date or for transfer to an employees' share scheme.
There are certain restrictions whilst shares are held in treasury, namely:

- Their aggregate nominal value must not exceed 10% of the nominal value of issued share capital (if it exceeds 10% then the excess must be disposed of or cancelled).

- Rights attaching to the class of share – e.g. receiving dividends, and the right to vote – cannot be exercised by the company.

Treasury shares – cancellation

- Where shares are held as treasury shares, the company may at any time cancel some or all of the shares.

- If shares held as treasury shares cease to be qualifying shares, then the company must cancel the shares.

- On cancellation the amount of the company's share capital is reduced by the nominal amount of the shares cancelled.

Summary

Creditors of companies are not expected to be protected against ordinary business risks as these are taken care of by financial markets, e.g. through the rates of interest charged on different capital instruments of different companies. However, the creditors are entitled to depend on the non-erosion of the permanent capital unless their interests are considered and protected.

The chapter also discusses the question of capital reconstructions and the need to consider the effect of any proposed reconstruction on the rights of different parties.

REVIEW QUESTIONS

1 Discuss how the Companies Act 2006 defines distributable profits in the UK.

2 Why do companies reorganise their capital structure when they have accumulated losses?

3 What factors would a loan creditor take into account if asked to bear some of the accumulated loss?

4 Explain a debt/equity swap and the reasons for debt/equity swaps, and discuss the effect on existing shareholders and loan creditors.

5 The following relates to RWE AG:

> On April 22, 2009, the Annual General Meeting authorized us again to buy back shares. Hence, the Executive Board is entitled to acquire shares in the amount of up to 10% of the share capital until October 21, 2010.[5]

Explain why companies hold treasury shares.

EXERCISES

* Question 1

The draft statement of financial position of Telin plc at 30 September 20X5 was as follows:

	£000		£000
Ordinary shares of £1 each, fully paid	12,000	Product development costs	1,400
12% preference shares of £1 each, fully paid	8,000	Sundry assets	32,170
Share premium	4,000	Cash and bank	5,450
Retained (distributable) profits	4,600		
Payables	10,420		
	39,020		39,020

Preference shares of the company were originally issued at a premium of 2p per share. The directors of the company decided to redeem these shares at the end of October 20X5 at a premium of 5p per share. They also decided to write off the balances on development costs and discount on debentures.

All write-offs and other transactions are to be entered into the accounts according to the provisions of the Companies Acts and in a manner financially advantageous to the company and to its shareholders.

The following transactions took place during October 20X5:

(a) On 4 October the company issued for cash 2,400,000 10% debentures of £l each at a discount of 2¹/₂%.

(b) On 6 October the balances on development costs and discount of debentures were written off.

(c) On 12 October the company issued for cash 6,000,000 ordinary shares at a premium of 10p per share. This was a specific issue to help redeem preference shares.

(d) On 29 October the company redeemed the 12% preference shares at a premium of 5p per share and included in the payments to shareholders one month's dividend for October.

(e) On 30 October the company made a bonus issue, to all ordinary shareholders, of one fully paid ordinary share for every 20 shares held.

(f) During October the company made a net profit of £275,000 from its normal trading operations. This was reflected in the cash balance at the end of the month.

Required:
(a) Write up the ledger accounts of Telin plc to record the transactions for October 20X5.
(b) Prepare the company's statement of financial position as at 31 October 20X5.
(c) Briefly explain accounting entries which arise as a result of redemption of preference shares.

* Question 2

The following is the statement of financial position of Alpha Ltd as on 30 June 20X8:

	£000 Cost	£000 Accumulated depreciation	£000
Non-current assets			
Freehold property	46	5	41
Plant	85	6	79
	131	11	120
Investments			
Shares in subsidiary company		90	
Loans		40	130
Current assets			
Inventory		132	
Trade receivables		106	
		238	
Current liabilities			
Trade payables		282	
Bank overdraft		58	
		340	
Net current liabilities			(102)
Total assets less liabilities			148
Capital and reserves			
250,000 8¹/₂% cumulative redeemable preference shares of £1 each fully paid			250
100,000 ordinary shares of £1 each 75p paid			75
			325
Retained earnings			(177)
			148

The following information is relevant:

1 There are contingent liabilities in respect of (i) a guarantee given to bankers to cover a loan of £30,000 made to the subsidiary and (ii) uncalled capital of 10p per share on the holding of 100,000 shares of £1 each in the subsidiary.

2 The arrears of preference dividend amount to £106,250.

3 The following capital reconstruction scheme, to take effect as from 1 July 20X8, has been duly approved and authorised:

(i) the unpaid capital on the ordinary shares to be called up;

(ii) the ordinary shares thereupon to be reduced to shares of 25p each fully paid up by cancelling 75p per share and then each fully paid share of 25p to be subdivided into five shares of 5p each fully paid;

(iii) the holders to surrender three of such 5p shares out of every five held for reissue as set out below;

(iv) the $8^{1}/_{2}$% cumulative preference shares together with all arrears of dividend to be surrendered and cancelled on the basis that the holder of every 50 preference shares will pay to Alpha a sum of £30 in cash, and will be issued with:

   (a) one £40 convertible $7^{3}/_{4}$% note of £40 each, and

   (b) 60 fully paid ordinary shares of 5p each (being a redistribution of shares surrendered by the ordinary shareholders and referred to in (iii) above);

(v) the unpaid capital on the shares in the subsidiary to be called up and paid by the parent company whose guarantee to the bank should be cancelled;

(vi) the freehold property to be revalued at £55,000;

(vii) the adverse balance on retained earnings to be written off, £55,000 to be written off the shares in the subsidiary and the sums made available by the scheme to be used to write down the plant.

Required:

(a) **Prepare a capital reduction and reorganisation account.**

(b) **Prepare the statement of financial position of the company as it would appear immediately after completion of the scheme.**

Question 3

A summary of the statement of financial position of Doxin plc, as at 31 December 20X0, is given below:

	£		£
800,000 ordinary shares of £1 each	800,000	Assets other than bank (at book values)	1,500,000
300,000 6% preference shares of £1 each	300,000	Bank	200,000
General reserves	200,000		
Payables	400,000		
	1,700,000		1,700,000

During 20X1, the company:

(i) issued 200,000 ordinary shares of £1 each at a premium of 10p per share (a specific issue to redeem preference shares);

(ii) redeemed all preference shares at a premium of 5%. These were originally issued at 25% premium;

(iii) issued 4,000 7% debentures of £100 each at £90;

(iv) used share premium, if any, to issue fully paid bonus shares to members; and

(v) made a net loss of £500,000 by end of year which affected the bank account.

Required:

(a) Show the effect of each of the above items in the form of a moving statement of financial position (i.e. additions/deductions from original figures) and draft the statement of financial position of 31 December 20XI.

(b) Consider to what extent the interests of the creditors of the company are being protected.

Question 4

Discuss the advantages to a company of:

(a) purchasing and cancelling its own shares;

(b) purchasing and holding its own shares in treasury.

* Question 5

Speedster Ltd commenced trading in 1986 as a wholesaler of lightweight travel accessories. The company was efficient and traded successfully until 2000 when new competitors entered the market selling at lower prices which Speedster could not match. The company has gradually slipped into losses and the bank is no longer prepared to offer overdraft facilities. The directors are considering liquidating the company and have prepared the following statement of financial position and supporting information:

Statement of financial position	£000	£000
Non-current assets		
Freehold land at cost		1,500
Plant and equipment (NBV)		1,800
Current assets		
Inventories	600	
Trade receivables	1,200	
	1,800	
Current liabilities		
Payables	1,140	
Bank overdraft (secured on the plant and equipment)	1,320	
	2,460	
Net current assets		(660)
Non-current liabilities		
Secured loan (secured on the land)		(1,200)
		1,440
Financed by		
Ordinary shares of £1 each		3,000
Statement of comprehensive income		(1,560)
		1,440

Supporting information:

(i) The freehold land has a market value of £960,000 if it continues in use as a warehouse. There is a possibility that planning permission could be obtained for a change of use allowing the warehouse to be converted into apartments. If planning permission were to be obtained, the company has been advised that the land would have a market value of £2,500,000.

(ii) The net realisable values on liquidation of the other assets are:

Plant and equipment	£1,200,000
Inventory	£450,000
Trade receivables	£1,050,000

(iii) An analysis of the payables indicated that there would be £300,000 owing to preferential creditors for wages, salaries and taxes.

(iv) Liquidation costs were estimated at £200,000.

Required:
Prepare a statement showing the distribution on the basis that:
(a) planning permission was not obtained; and
(b) planning permission was obtained.

Question 6

Delta Ltd has been developing a lightweight automated wheelchair. The research costs written off have been far greater than originally estimated and the equity and preference capital has been eroded as seen on the statement of financial position.

The following is the statement of financial position of Delta Ltd as at 31.12.20X9:

	£000	£000
Intangible assets		
Development costs		300
Non-current assets		
Freehold property	800	
Plant, vehicles and equipment	650	1,450
		1,750
Current assets		
Inventory	480	
Trade receivables	590	
Investments	200	
	1,270	
Current liabilities		
Trade payables	(1,330)	
Bank overdraft	(490)	(550)
		1,200
10% debentures (secured on freehold premises)		(1,000)
Total assets less liabilities		200
Capital and reserves		
Ordinary shares of 50p each		800
7% cumulative preference shares of £1 each		500
Retained earnings (debit)		(1,100)
		200

The finance director has prepared the following information for consideration by the board.

1 Estimated current and liquidation values were estimated as follows:

	Current values £000	Liquidation values £000
Capitalised development costs	300	—
Freehold property	1,200	1,200
Plant and equipment	600	100
Inventory	480	300
Trade receivables	590	590
Investments	200	200
		2,390

2 If the company were to be liquidated there would be disposal costs of £100,000.

3 The preference dividend had not been paid for five years.

4 It is estimated that the company would make profits before interest over the next five years of £150,000 rising to £400,000 by the fifth year.

5 The directors have indicated that they would consider introducing further equity capital.

6 It was the finance director's opinion that for any scheme to succeed, it should satisfy the following conditions:

(a) The shareholders and creditors should have a better benefit in capital and income terms by reconstructing rather than liquidating the company.

(b) The scheme should have a reasonable possibility of ensuring the long-term survival of the company.

(c) There should be a reasonable assurance that there will be adequate working capital.

(d) Gearing should not be permitted to become excessive.

(e) If possible, the ordinary shareholders should retain control.

Required:
(a) advise the unsecured creditors of the minimum that they should accept if they were to agree to a reconstruction rather than proceed to press for the company to be liquidated.
(b) Propose a possible scheme for reconstruction.
(c) Prepare the statement of financial position of the company as it would appear immediately after completion of the scheme.

References

1 The Companies Act 2006.
2 Ibid., para 764.
3 Companies (Reduction of Share Capital) Order 2008.
4 The Companies Act 2006, paras 724–32.
5 http://www.rwe.com

CHAPTER 13

Liabilities

13.1 Introduction

In order for financial statements to show a true and fair view it is essential that reporting entities recognise all the liabilities that satisfy the *Framework* criteria, but **only** those liabilities that satisfy the criteria. Given that the recognition of a liability often involves a charge against profits, and the derecognition of a liability sometimes involves a credit to profits, there is the possibility that, unless this area of financial reporting is appropriately regulated, there is scope for manipulation of reporting profits when liabilities are recognised or derecognised inappropriately.

There are a number of financial reporting standards dealing with the recognition and measurement of specific liabilities that are dealt with elsewhere in this book:

- Financial liabilities (including, *inter alia*, trade payables and loans) are dealt with in IAS 39 *Financial Instruments: Recognition and Measurement* and, in the future, in IFRS 9 *Financial Instruments* (see Chapter 14).
- Pension liabilities are dealt with in IAS 19 *Employee Benefits* (see Chapter 15).
- Income tax liabilities are dealt with in IAS 12 *Income Taxes* (see Chapter 16).
- Lease liabilities are dealt with in IAS 17 *Leases* (see Chapter 18).

The above financial reporting standards deal with many types of liability but not with all liabilities. Examples of liabilities, or potential liabilities, not dealt with by the above financial reporting standards include:

- liabilities arising from legal disputes;
- liabilities arising due to corporate restructurings;
- environmental and decommissioning obligations;
- liabilities arising under contracts that have become onerous.

IAS 37 *Provisions, Contingent Liabilities and Contingent Assets* deals with the recognition, measurement and disclosure of these liabilities or potential liabilities.

Objectives

By the end of this chapter, you should be able to:

- account for provisions, contingent liabilities and contingent assets under IAS 37;
- explain the potential change the IASB is considering in relation to provisions.

13.2 Provisions – a decision tree approach to their impact on the statement of financial position

The IASC (now the IASB) approved IAS 37 *Provisions, Contingent Liabilities and Contingent Assets*[1] in July 1998. The key objective of IAS 37 is to ensure that appropriate recognition criteria and measurement bases are applied and that sufficient information is disclosed in the notes to enable users to understand their nature, timing and amount.

The IAS sets out a useful decision tree, shown in Figure 13.1, for determining whether an event requires the creation of a provision, the disclosure of a contingent liability or no action.

Figure 13.1 Decision tree

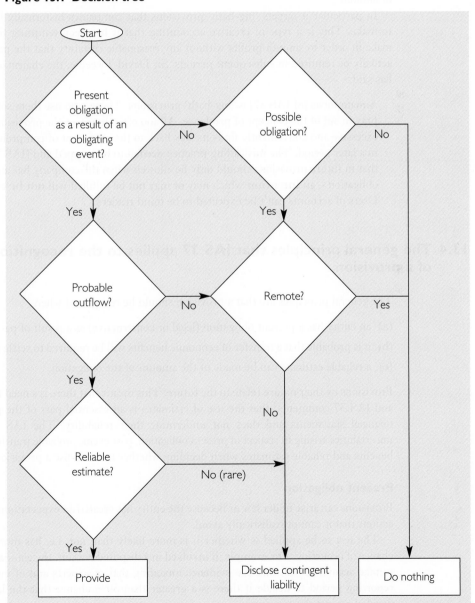

In June 2005 the IASB issued an exposure draft, IAS 37 *Non-financial Liabilities*, to revise IAS 37. A further exposure draft clarifying the proposed amendments was issued in January 2010. However, the current IASB timetable does not envisage a new accounting standard on liabilities very soon. We will now consider the current IAS 37 treatment of provisions, contingent liabilities and contingent assets.

13.3 Treatment of provisions

IAS 37 is mainly concerned with provisions and the distorting effect they can have on profit trends, income and capital gearing. It defines a provision as 'a liability of uncertain timing or amount'.

In particular it targets 'big-bath' provisions that companies historically have been able to make. This is a type of creative accounting that it has been tempting for directors to make in order to smooth profits without any reasonable certainty that the provision would actually be required in subsequent periods. Sir David Tweedie, the chairman of the IASB, has said:

> A main focus [of IAS 37] is 'big-bath' provisions. Those who use them sometimes pray in aid of the concept of prudence. All too often however the provision is wildly excessive and conveniently finds its way back to the statement of comprehensive income in a later period. The misleading practice needed to be stopped and [IAS 37] proposed that in future provisions should only be allowed when the company has an unavoidable obligation – an **intention** which may or may not be fulfilled will **not be enough**. Users of accounts can't be expected to be mind readers.

13.4 The general principles that IAS 37 applies to the recognition of a provision

The general principles are that a provision should be recognised when:[2]

(a) an entity has a present obligation (legal or constructive) as a result of past events;

(b) it is probable that a transfer of economic benefits will be required to settle the obligation;

(c) a reliable estimate can be made of the amount of the obligation.

Provisions by their nature relate to the future. This means that there is a need for estimation, and IAS 37 comments[3] that the use of estimates is an essential part of the preparation of financial statements and does not undermine their reliability. The IAS addresses the uncertainties arising in respect of present obligation, past event, probable transfer of economic benefits and reliable estimates when deciding whether to recognise a provision.

Present obligation

Provisions can arise under law or because the entity has created an expectation due to its past actions that it cannot realistically avoid.

The test to be applied is whether it is more likely than not, i.e. has more than a 50% chance of occurring. For example, if involved in a disputed lawsuit, the company is required to take account of all available evidence including that of experts and of events after the reporting period to decide if there is a greater than 50% chance that the lawsuit will be decided against the company.

Where it is more likely that no present obligation exists at the period-end date, the company discloses a contingent liability, unless the possibility of a transfer of economic resources is remote.

Past event[4]

A past event that leads to a present obligation is called an 'obligating event'. This is a new term with which to become familiar. It means that the company has no realistic alternative to settling the obligation. The IAS defines 'no alternative' as being only where the settlement of the obligation can be enforced by law or, in the case of a constructive obligation, where the event creates valid expectations in other parties that the company will discharge the obligation.

The IAS stresses that it is only those obligations arising from past events existing independently of a company's future actions that are recognised as provisions, e.g. clean-up costs for unlawful environmental damage that has occurred require a provision; environmental damage that is not unlawful but is likely to become so and involve clean-up costs will not be provided for until legislation is virtually certain to be enacted as drafted.

Probable transfer of economic benefits[5]

The IAS defines probable as meaning that the event is more likely than not to occur. Where it is not probable, the company discloses a contingent liability unless the possibility is remote.

13.5 Management approach to measuring the amount of a provision

IAS 37 states[6] that the amount recognised as a provision should be the *best estimate* of the expenditure required to settle the present obligation at the period-end date.

'Best estimate' is defined as the amount that a company would rationally pay to settle the obligation or to transfer it to a third party. The estimates of outcome and financial effect are determined by the judgement of management supplemented by experience of similar transactions and reports from independent experts. Management deal with the uncertainties as to the amount to be provided in a number of ways:

- A class obligation exists:
  - where the provision involves a large population of items as with a warranty provision – statistical analysis of expected values should be used to determine the amount of the provision.
- A single obligation exists but a number of outcomes may be possible:
  - where a single obligation is measured, the individual most likely outcome may be the best estimate;
  - more than one outcome exists or the outcome is anywhere within a range; or
  - expected values may be most appropriate.

For example, a company had been using unlicensed parts in the manufacture of its products and, at the year-end, no decision had been reached by the court. The plaintiff was seeking damages of $10 million.

In the draft accounts a provision had been made of $5.85 million using expected values. This had been based on the estimate by the entity's lawyers that there was a 20% chance that the plaintiff would be unsuccessful and a 25% chance that the entity would be required to pay $10 million and a 55% chance of $7 million becoming payable to the plaintiff. The provision had been calculated as 25% of $0 + 55% of $7 million + 20% of $10 million.

The finance director, however, disagreed with this on the grounds that it was a single obligation and more likely than not there would be an outflow of funds of $7 million, and required an additional $1.15 million to be provided.

Avoiding excessive provisions

Management must avoid creating excessive provisions based on a prudent view. Uncertainty does not justify the creation of excessive provisions.[7] If the projected costs of a particular adverse outcome are estimated on a prudent basis, that outcome should not then be deliberately treated as more probable than is realistically the case.

The measurement requirements of the current IAS 37 are somewhat imprecise and can be interpreted in more than one way. One of the objectives of the proposed amendment to IAS 37 is to remove the imprecision in the current standard. We will discuss the amendments proposed in this exposure draft in 13.10.

Approach when time value of money is material

The IAS states[8] that 'where the effect of the time value of money is material, the amount of a provision should be the present value of the expenditures expected to be required to settle the obligation'.

Present value is arrived at by discounting the future obligation at 'a pre-tax rate (or rates) that reflect(s) current market assessments of the time value of money and the risks specific to the liability. The discount rate(s) should not reflect risks for which future cash flow estimates have been adjusted.'

If provisions are recognised at present value, a company will have to account for the unwinding of the discounting. As a simple example, assume a company is making a provision at 31 December 2010 for an expected cash outflow of €1 million on 31 December 2012. The relevant discount factor is estimated at 10%. Assume the estimated cash flows do not change and the provision is still required at 31 December 2011.

	€000
Provision recognised at 31 December 2010 (€1m × 1/1.121)	826
Provision recognised at 31 December 2011 (€1m × 1/1.1)	909
Increase in the provision	83

This increase in the provision is purely due to discounting for one year in 2011 as opposed to two years in 2010. This increase in the provision must be recognised as an expense in profit or loss, usually as a finance cost, although IAS 37 does not make this mandatory.

The following is an extract from the Minefinders Corporation Ltd 2011 Annual Report:

> A provision for site closure and reclamation is recorded when the Company incurs liability for costs associated with the eventual retirement of tangible long-lived assets (for example, reclamation costs). The liability for such costs exists from the time the legal or constructive obligation first arises, not when the actual expenditures are made.
>
> Such obligations are based on estimated future cash flows discounted at a rate specific to the liability. . . . The amount added to the asset is amortized in the same manner as the asset.
>
> The liability is increased in each accounting period by the amount of the implied interest inherent in the use of discounted present value methodology. . . .

13.6 Application of criteria illustrated

Scenario 1

An offshore oil exploration company is required by its licence to remove the rig and restore the seabed. Management have estimated that 85% of the eventual cost will be incurred in removing the rig and 15% through the extraction of oil. The company's practice on similar projects has been to account for the decommissioning costs using the 'unit of production' method whereby the amount required for decommissioning was built up year by year, in line with production levels, to reach the amount of the expected costs by the time production ceased.

Decision process

1 **Is there a present obligation as a result of a past event?**
   The construction of the rig has created a legal obligation under the licence to remove the rig and restore the seabed.

2 **Is there a probable transfer of economic benefits?**
   This is probable.

3 **Can the amount of the outflow be reasonably estimated?**
   A best estimate can be made by management based on past experience and expert advice.

4 **Conclusion**
   A provision should be created of 85% of the eventual future costs of removal and restoration. This provision should be discounted if the effect of the time value of money is material. A provision for the 15% relating to restoration should be created when oil production commences.

The unit of production method is not acceptable in that the decommissioning costs relate to damage already done.

Scenario 2

A company has a private jet costing £24 million. Air regulations required it to be overhauled every four years. An overhaul costs £1.6 million. The company policy has been to create a provision for depreciation of £2 million on a straight-line basis over 12 years and an annual provision of £400,000 to meet the cost of the required overhaul every four years.

Decision process

1 **Is there a present obligation as a result of a past obligating event?**
   There is no present obligation. The company could avoid the cost of the overhaul by, for example, selling the aircraft.

2 **Conclusion**
   No provision for cost of overhaul can be recognised. Instead of a provision being recognised, the depreciation of the aircraft takes account of the future incidence of maintenance costs, i.e. an amount equivalent to the expected maintenance costs is depreciated over four years.

13.7 Provisions for specific purposes

Specific purposes could include considering the treatment of future operating losses, onerous contracts, restructuring and environmental liabilities. Let us consider each of these.

13.7.1 A provision for future operating losses

Such losses should not be recognised if there is no obligation at the reporting date on the basis that the entity could decide to discontinue that particular business activity. However, if it is contractually unable to discontinue then it classifies the contract as an onerous contract and makes provision.

13.7.2 Onerous contracts

A provision should be recognised if there is an onerous contract. An onerous contract is one entered into with another party under which the unavoidable costs of fulfilling the contract exceed the revenues to be received and where the entity would have to pay compensation to the other party if the contract was not fulfilled. A typical example in times of recession is the requirement to make a payment to secure the early termination of a lease where it has been impossible to sublet the premises. This situation could arise where there has been a downturn in business and an entity seeks to reduce its annual lease payments on premises that are no longer required.

The following is an extract from the 2011 Preliminary Results of the Spirit Pub Company plc:

Onerous lease provisions
The Group provides for its onerous obligations under operating leases where the property is closed or vacant and for properties where rental expense is in excess of income. The estimated timings and amounts of cash flows are determined using the experience of internal and external property experts; however, any changes to the estimated method of exiting from the property could lead to changes to the level of the provision recorded.

13.7.3 Restructuring provisions

- **A provision for restructuring** should only be recognised when there is a commitment supported by:
  - (a) a detailed formal plan for the restructuring identifying at least:
    - (i) the business or part of the business concerned;
    - (ii) the principal locations affected;
    - (iii) details of the approximate number of employees who will receive compensation payments;
    - (iv) the expenditure that will be undertaken; and
    - (v) when the plan will be implemented; and
  - (b) a valid expectation in those affected that the business will carry out the restructuring by implementing its restructuring plans or announcing its main features to those affected by it.
- **A provision for restructuring should not be created merely on the intention to restructure.** For example, a management or board decision to restructure taken before the reporting date does not give rise to a constructive obligation at the reporting date unless the company has, before the reporting date:
  - started to implement the restructuring plan, e.g. by dismantling plant or selling assets; or
  - announced the main features of the plan with sufficient detail to raise the valid expectation of those affected that the restructuring will actually take place.

- **A provision for restructuring** should only include the direct expenditures arising from the restructuring which are necessarily entailed and not associated with the ongoing activities of the company. For example, redundancy costs would be included, but note that the following costs which relate to the future conduct of the business are not included: retraining costs, relocation costs, marketing costs, and investment in new systems and distribution networks.

13.7.4 Environmental liabilities and decommissioning costs

- **A provision for environmental liabilities** should be recognised at the time and to the extent that the entity becomes obliged, legally or constructively, to rectify environmental damage or to perform restorative work on the environment. This means that a provision should be set up only for the entity's costs to meet its *legal* or *constructive* obligations. It could be argued that any provision for any additional expenditure on environmental issues is a public relations decision and should be written off.

- **A provision for decommissioning costs** should be recognised to the extent that decommissioning costs relate to damage already done or goods and services already received.

Provisions for decommissioning costs often relate to non-current assets, e.g. power stations. Where a liability for decommissioning exists at the date of construction, it is recognised, normally at the present value of the expected future outflow of cash, and added to the cost of the non-current asset.

EXAMPLE ● An entity constructs a nuclear power station at a cost of €20 million. The estimated useful life of the power station is 25 years. The entity has a legal obligation to decommission the power station at the end of its useful life and the estimated costs of this are €15 million in 25 years' time. A relevant annual discount factor is 5% and the present value of a payment of €15 million in 25 years' time is approximately €4.43 million.

In these circumstances a liability of €4.43 million is recognised at the completion of the construction of the facility. The debit side of this accounting entry is to property, plant and equipment, giving a total carrying amount for the power station of €24.43 million. This amount is then depreciated over 25 years which gives an annual charge (assuming straight-line depreciation with no residual value) of approximately €977,200.

The discounting of the liability is 'unwound' over the 25-year life of the power station, the annual unwinding being shown as a finance cost. The unwinding in the first year of operation is approximately €221,500 (€4.43 million × 5%).

13.7.5 Disclosures required by IAS 37 for provisions

Specific disclosures,[9] for each material class of provision, should be given as to the amount recognised at the year-end and about any movements in the year, e.g.:

- **Increases in provisions** – any new provisions; any increases to existing provisions; and, where provisions are carried at present value, any change in value arising from the passage of time or from any movement in the discount rate.

- **Reductions in provisions** – any amounts utilised during the period. Management are required to review provisions at each reporting date and adjust to reflect the current best estimates. If it is no longer probable that a transfer of economic benefits will be required to settle the obligation, the provision should be reversed. Note, however, that only expenditure that relates to the original provision may be set against that provision.

Disclosures need not be given in cases where to do so would be seriously prejudicial to the company's interests. For example, an extract from the Technotrans 2002 Annual Report states:

> A competitor filed patent proceedings in 2000, . . . the court found in favour of the plaintiff . . . paves the way for a claim for compensation which may have to be determined in further legal proceedings . . . the particulars pursuant to IAS 37.85 are not disclosed, in accordance with IAS 37.92, in order not to undermine the company's situation substantially in the ongoing legal dispute.

13.8 Contingent liabilities

IAS 37 deals with provisions and contingent liabilities within the same IAS because the IASB regarded all provisions as contingent as they are uncertain in timing and amount. For the purposes of the accounts, it distinguishes between provisions and contingent liabilities in that:

● Provisions are a present obligation requiring a probable transfer of economic benefits that can be reliably estimated – a provision can therefore be recognised as a liability.

● Contingent liabilities fail to satisfy these criteria, e.g. lack of a reliable estimate of the amount; not probable that there will be a transfer of economic benefits; yet to be confirmed that there is actually an obligation. A contingent liability cannot therefore be recognised in the accounts but may be disclosed by way of note to the accounts or not disclosed if an outflow of economic benefits is remote.

Where the occurrence of a contingent liability becomes sufficiently probable, it falls within the criteria for recognition as a provision as detailed above and should be accounted for accordingly and recognised as a liability in the accounts.

Where the likelihood of a contingent liability is possible but not probable and not remote, disclosure should be made, for each class of contingent liability, where practicable, of:

(a) an estimate of its financial effect, taking into account the inherent risks and uncertainties and, where material, the time value of money;

(b) an indication of the uncertainties relating to the amount or timing of any outflow; and

(c) the possibility of any reimbursement.

For example, an extract from the 2011 Annual Report of Nottingham Forest Football Club Ltd informs us that:

> Additional transfer fees amounting to £1,075,000 (2010: £1,470,000) in total will become payable to their previous clubs if certain players make an agreed number of international appearances, first team appearances or the club achieves defined feats. Signing on fees of £306,000 (2010: £451,000) will become due to certain players if they are still in the service of Nottingham Forest Football Club Limited on specific future dates.

13.9 Contingent assets

A contingent asset is a possible asset that arises from past events whose existence will be confirmed only by the occurrence of one or more uncertain future events not wholly within the entity's control.

Recognition as an asset is only allowed if the asset is *virtually certain*, and therefore by definition no longer contingent.

Disclosure by way of note is required if an inflow of economic benefits is *probable*. The disclosure would include a brief description of the nature of the contingent assets at the reporting date and, where practicable, an estimate of their financial effect taking into account the inherent risks and uncertainties and, where material, the time value of money.

No disclosure is required where the chance of occurrence is anything less than probable. For the purposes of IAS 37, probable is defined as more likely than not, i.e. with more than a 50% chance.

13.10 ED IAS 37 *Non-financial Liabilities*

In June 2005, the International Accounting Standards Board (IASB) proposed amendments to IAS 37 *Provisions, Contingent Liabilities and Contingent Assets*. These strip IAS 37 of the words 'Provisions', 'Contingent' and 'Assets' and add the term 'Non-financial' to create the new title IAS 37 *Non-financial Liabilities*.[10] It is interesting to see that the new standard has been developed around the *Framework*'s definitions of an asset and a liability.

It appears that the word 'non-financial' has been added to distinguish the subject from 'financial liabilities' which are covered by IAS 32 and IAS 39. It should be noted that whilst the exposure draft remains in issue a new standard based on these proposals is not in the current IASB work-plan.

13.10.1 The 'old' IAS 37 *Provisions, Contingent Liabilities and Contingent Assets*

To understand the 'new' approach in ED IAS 37 *Non-financial Liabilities*, it is necessary first to look at the 'old' IAS 37. The old treatment can be represented by the following table:

Probability	Contingent liabilities	Contingent assets
Virtually certain	Liability	Asset
Probable (p > 50%)	Provide	Disclose
Possible (p < 50%)	Disclose	No disclosure
Remote	No disclosure	No disclosure

Note that contingent liabilities are those items where the probability is less than 50% (p < 50%). Where, however, the liability is probable, i.e. the probability is p > 50%, the item is classified as a provision and not a contingent liability. Normally, such a provision will be reported as the product of the value of the potential liability and its probability.

Note that the approach to contingent assets is different in that the 'prudence' concept is used which means that only virtually certain assets are reported as assets. If the probability is probable, i.e. p > 50%, then contingent assets are disclosed by way of a note to the accounts, and if the probability is p < 50% then there is no disclosure.

Criticisms of the 'old' IAS 37

The criticisms included the following:

- The 'old' IAS 37 was not even-handed in its treatment of contingent assets and liabilities. In ED IAS 37 the treatment of contingent assets is similar to that of contingent liabilities, and provisions are merged into the treatment of contingent liabilities.

- The division between 'probable' and 'possible' was too strict or crude (at the $p = 50\%$ level) rather than being proportional. For instance, if a television manufacturer was considering the need to provide for guarantee claims (e.g. on televisions sold with a three-year warranty), then it is probable that each television sold would have a less than 50% chance of being subject to a warranty claim and so no provision would need to be made. However, if the company sold 10,000 televisions, it is almost certain that there would be some claims which would indicate that a provison should be made. A company could validly take either treatment, but the effect on the financial statements would be different.

- If there was a single possible legal claim, then the company could decide it was 'possible' and just disclose it in the financial statements. However, a more reasonable treatment would be to assess the claim as the product of the amount likely to be paid and its probability. This latter treatment is used in the new ED IAS 37.

13.10.2 Approach taken by ED IAS 37 *Non-financial Liabilities*

The new proposed standard uses the term 'non-financial liabilities' which it defines as 'a liability other than a financial liability as defined in IAS 32 *Financial Instruments: Presentation*'. In considering ED IAS 37, we will look at the proposed treatment of contingent liabilities/ provisions and contingent assets, starting from the *Framework*'s definitions of a liability and an asset.

The *Framework's* definition

The *Framework*, paragraph 91, requires a liability to be recognised as follows:

A liability is recognised in the statement of financial position when it is probable that an outflow of resources embodying economic benefits will result from the settlement of a present obligation and the amount at which the settlement will take place can be measured reliably.

The ED IAS 37 approach to provisions

Considering a provision first, the old IAS 37 (paragraph 10) defines it as follows:

A provision is distinguished from other liabilities because there is uncertainty about the timing or amount of the future expenditure required in settlement.

ED IAS 37 argues that a provision should be reported as a liability, as it satisfies the *Framework*'s definition of a liability. It makes the point that there is no reference in the *Framework* to 'uncertainty about the timing or amount of the future expenditure required in settlement'. It considers a provision to be just one form of liability which should be treated as a liability in the financial statements.

Will the item 'provision' no longer appear in financial statements?

One would expect that to be the result of the ED (exposure draft) classification. However, the proposed standard does not take the step of prohibiting the use of the term, as seen in the following extract (paragraph 9):

In some jurisdictions, some classes of liabilities are described as provisions, for example those liabilities that can be measured only by using a substantial degree of estimation. Although this [draft] Standard does not use the term 'provision', it does not prescribe how entities should describe their non-financial liabilities. Therefore, entities may describe some classes of non-financial liabilities as provisions in their financial statements.

The ED IAS 37 approach to contingent liabilities

Now considering contingent liabilities, the old IAS 37 (paragraph 10) defines these as:

(a) a possible obligation that arises from past events and whose existence will be confirmed only by the occurrence or non-occurrence of one or more uncertain future events not wholly within the control of the entity; or

(b) a present obligation that arises from past events, but is not recognised because:

    (i) it is not probable that an outflow of resources embodying economic benefits will be required to settle the obligation; or

    (ii) the amount of the obligation cannot be measured with sufficient reliability.

This definition means that the old IAS 37 has taken the strict approach of using the term 'possible' ($p < 50\%$) when it required no liability to be recognised.

ED IAS 37 is different in that it takes a two-stage approach in considering whether 'contingent liabilities' are 'liabilities'. To illustrate this, we will take the example of a restaurant where some customers have suffered food poisoning.

First determine whether there is a present obligation

The restaurant's year-end is 30 June 20X6. If the food poisoning took place after 30 June 20X6, then this is not a 'present obligation' at the year-end, so it is not a liability. If the food poisoning occurred up to 30 June, then it is a 'present obligation' at the year-end, as there are possible future costs arising from the food poisoning. This is the first stage in considering whether the liability exists.

Then determine whether a liability exists

The second stage is to consider whether a 'liability' exists. The *Framework*'s definition of a liability says it is a liability if 'it is probable that an outflow of resources will result from the settlement of the present obligation'. So, there is a need to consider whether any payments (or other expenses) will be incurred as a result of the food poisoning. This may involve settling legal claims, other compensation or giving 'free' meals. The estimated cost of these items will be the liability (and expense) included in the financial statements.

The rationale

ED IAS 37 explains this process as:

- the unconditional obligation (stage 1) establishes the liability; and
- the conditional obligation (stage 2) affects the amount that will be required to settle the liability.

The liability is the amount that the entity would rationally pay to settle the present obligation or to transfer it to a third party on the statement of financial position date. Often, the liability will be estimated as the product of the maximum liability and the probability of it occurring, or a decision tree will be used with a number of possible outcomes (costs) and their probability.

In many cases, the new ED IAS 37 will cover the 'possible' category for contingent liabilities and include the item as a liability (rather than as a note to the financial statements). This gives a more 'proportional' result than the previously strict line between 'probable' ($p > 50\%$) (when a liability is included in the financial statements) and 'possible' ($p < 50\%$) (when only a note is included in the financial statements and no charge is included for the liability).

What if they cannot be measured reliably?

For other 'possible' contingent liabilities, which have not been recognised because they cannot be measured reliably, the following disclosure should be made:

- a description of the nature of the obligation;
- an explanation of why it cannot be measured reliably;
- an indication of the uncertainties relating to the amount or timing of any outflow of economic benefits; and
- the existence of any rights to reimbursement.

What disclosure is required for maximum potential liability?

ED IAS 37 does not require disclosure of the maximum potential liability, e.g. the maximum damages if the entity loses the legal case.

13.10.3 Measured reliably

The *Framework*'s definition of a liability includes the condition 'and the amount at which the settlement will take place can be measured reliably'. This posed a problem when drafting ED IAS 37 because of the concern that an entity could argue that the amount of a contingent liability could not be measured reliably and that there was therefore no need to include it as a liability in the financial statements – i.e. to use this as a 'cop out' to give a 'rosier' picture in the financial statements. Whilst acknowledging that in many cases a non-financial liability cannot be measured exactly, it considered that it could (and should) be estimated. It then says that cases where the liability cannot be measured reliably are 'extremely rare'. We can see from this that the ED approach is that 'measured reliably' does not mean 'measured exactly' and that cases where the liability 'cannot be measured reliably' will be 'extremely rare'.

13.10.4 Contingent asset

The *Framework*, paragraph 89, requires recognition of an asset as follows:

> An Asset is recognised in the statement of financial position when it is probable that the future economic benefits will flow to the entity and the asset has a cost or value that can be measured reliably.

Note that under the old IAS 37, contingent assets included items where they were 'probable' (unlike liabilities, when this was called a 'provision'). However, probable contingent assets are not included as assets but only included in the notes to the financial statements.

The ED IAS 37 approach

ED IAS 37 takes a similar approach to 'contingent assets' as it does to 'provisions/contingent liabilities'. It abolishes the term 'contingent asset' and replaces it with the term 'contingency'. The term contingency refers to uncertainty about the amount of the future economic benefits embodied in an asset, rather than uncertainty about whether an asset exists.

Essentially, the treatment of contingent assets is the same as that of contingent liabilities. The first stage is to consider whether an asset exists and the second stage is concerned with valuing the asset (i.e. the product of the value of the asset and its probability). A major change is to move contingent assets to IAS 38 *Intangible Assets* (and not include them in IAS 37).

The treatment of 'contingent assets' under IAS 38 is now similar to that for 'contingent liabilities/provisions'. This seems more appropriate than the former 'prudent approach' used by the old IAS 37.

13.10.5 Reimbursements

Under the old IAS 37 an asset could be damaged or destroyed, when the expense would be included in profit or loss (and any future costs included as a provision). If the insurance claim relating to this loss was made after the year-end, it is likely that no asset could be included in the financial statements as compensation for the loss, as the insurance claim was 'not virtually certain'. In reality, this did not reflect the true situation when the insurance claim would compensate for the loss, and there would be little or no net cost.

With the new rules under ED IAS 37, the treatment of contingent assets and contingent liabilities is the same, so an asset would be included in the statement of financial position as the insurance claim, which would offset the loss on damage or destruction of the asset. The ED position is that an asset exists because there is an unconditional right to reimbursement – the only uncertainty is to the amount that will be received. But ED IAS 37 says the liability relating to the loss (e.g. the costs of repair) must be stated separately from the asset for the reimbursement (i.e. the insurance claim) – they cannot be netted off (although they will be in profit or loss).

13.10.6 Constructive and legal obligations

The term 'constructive obligation' is important in determining whether a liability exists. ED IAS 37 (paragraph 10) defines it as follows:

A constructive obligation is a present obligation that arises from an entity's past actions when:

(a) by an established pattern of past practice, published policies or a sufficiently specific current statement, the entity has indicated to other parties that it will accept particular responsibilities, and

(b) as a result, the entity has created a valid expectation in those parties, that they can reasonably rely on it to discharge those responsibilities.

It also defines a legal obligation as follows:

A legal obligation is a present obligation that arises from the following:

(a) a contract (through its explicit or implicit terms)

(b) legislation, or

(c) other operating law.

A contingent liability/provision is a liability only if it is either a constructive and/or a legal obligation. Thus, an entity would not normally make a provision (recognise a liability) for the potential costs of rectifying faulty products outside their guarantee period.

13.10.7 Present value

ED IAS 37 says that future cash flows relating to the liability should be discounted at the pre-tax discount rate. Unwinding of the discount would still need to be recognised as an interest cost.

Required:

(i) Explain why there was a need for detailed guidance on accounting for provisions.

(ii) Explain the circumstances under which a provision should be recognised in the financial statements according to IAS 37 *Provisions, Contingent Liabilities and Contingent Assets*.

(b) World Wide Nuclear Fuels, a public limited company, disclosed the following information in its financial statements for the year ending 30 November 20X9:

> The company purchased an oil company during the year. As part of the sale agreement, oil has to be supplied to the company's former holding company at an uneconomic rate for a period of five years. As a result, a provision for future operating losses has been set up of $135m, which relates solely to the uneconomic supply of oil. Additionally the oil company is exposed to environmental liabilities arising out of its past obligations, principally in respect of soil and ground water restoration costs, although currently there is no legal obligation to carry out the work. Liabilities for environmental costs are provided for when the group determines a formal plan of action on the closure of an inactive site. It has been decided to provide for $120m in respect of the environmental liability on the acquisition of the oil company. World Wide Nuclear Fuels has a reputation for ensuring the preservation of the environment in its business activities. The company is also facing a legal claim for $200 million from a competitor who claims they have breached a patent in one of their processes. World Wide Nuclear Fuels has obtained legal advice that the claim has little chance of success and the insurance advisers have indicated that to insure against losing the case would cost $20 million as a premium.

Required:
Discuss whether the provision has been accounted for correctly under IAS 37 *Provisions, Contingent Liabilities and Contingent Assets*, and whether any changes are likely to be needed under ED IAS 37.

* Question 2

On 20 December 20X6 one of Incident plc's lorries was involved in an accident with a car. The lorry driver was responsible for the accident and the company agreed to pay for the repair to the car. The company put in a claim to its insurers on 17 January 20X7 for the cost of the claim. The company expected the claim to be settled by the insurance company except for a £250 excess on the insurance policy. The insurance company may dispute the claim and not pay out; however, the company believes that the chance of this occurring is low. The cost of repairing the car was estimated as £5,000, all of which was incurred after the year-end.

Required:
Explain how this item should be treated in the financial statements for the year ended 31 December 20X6 according to both IAS 37 and ED IAS 37 *Non-financial Liabilities*.

Question 3

Plasma Ltd, a manufacturer of electrical goods, guarantees them for 12 months from the date of purchase by the customer. If a fault occurs after the guarantee period but is due to faulty manufacture or design of the product, the company repairs or replaces the product. However, the company does not make this practice widely known.

Required:
Explain how repairs after the guarantee period should be treated in the financial statements.

Question 4

In 20X6 Alpha AS made the decision to close a loss-making department in 20X7. The company proposed to make a provision for the future costs of termination in the 20X6 profit or loss. Its argument was that a liability existed in 20X6 which should be recognised in 20X6. The auditor objected to recognising a liability, but agreed to recognition if it could be shown that the management decision was irrevocable.

Required:
Discuss whether a liability exists and should be recognised in the 20X6 statement of financial position.

* Question 5

Easy View Ltd had started business publishing training resource material in ring binder format for use in primary schools. Later it diversified into the hiring out of videos and had opened a chain of video hire shops. With the growing popularity of a mail order video/DVD supplier the video hire shops had become loss-making.

The company's year-end was 31 March and in February the financial director (FD) was asked to prepare a report for the board on the implications of closing this segment of the business.

The position at the board meeting on 10 March was as follows:

1 It was agreed that the closure should take place from 1 April 2010 to be completed by 31 May 2010.

2 The premises were freehold except for one that was on a lease with six years to run. It was in an inner-city shopping complex where many properties were empty and there was little chance of sub-letting. The annual rent was £20,000 per annum. Early termination of the lease could be negotiated for a figure of £100,000. An appropriate discount rate is 8%.

3 The office equipment and vans had a book value of £125,000 and were expected to realise £90,000, a figure tentatively suggested by a dealer who indicated that he might be able to complete by the end of April.

4 The staff had been mainly part-time and casual employees. There were 45 managers, however, who had been with the company for a number of years. These were happy to retrain to work with the training resources operation. The cost of retraining to use publishing software was estimated at £225,000.

5 Losses of £300,000 were estimated for the current year and £75,000 for the period until the closure was complete.

A week before the meeting the managing director made it clear to the FD that he wanted the segment to be treated as a discontinued operation so that the continuing operations could reflect the profitable training segment's performance.

Required:
Draft the finance director's report to present to the MD before the meeting to clarify the financial reporting implications.

Question 6

Suktor is an entity that prepares financial statements to 30 June each year.

On 30 April 20X1 the directors decided to discontinue the business of one of Suktor's operating divisions. They decided to cease production on 31 July 20X1, with a view to disposing of the property, plant and equipment soon after 31 August 20X1.

On 15 May 20X1 the directors made a public announcement of their intentions and offered the employees affected by the closure termination payments or alternative employment opportunities elsewhere in the group. Relevant financial details are as follows:

(a) On 30 April 20X1 the directors estimated that termination payments to employees would total $12 million and the costs of retraining employees who would remain employed by other group companies would total $1.2 million. Actual termination costs paid out on 31 May 20X1 were $12.6 million and the latest estimate of total retraining costs is $960,000.

(b) Suktor was leasing a property under an operating lease that expires on 30 September 20Y0. On 30 June 20X1 the present value of the future lease rentals (using an appropriate discount rate) was $4.56 million. On 31 August 20X1 Suktor made a payment to the lessor of $4.56 million in return for early termination of the lease. There were no rental payments made in July or August 20X1.

(c) The loss after tax of Suktor for the year ended 30 June 20X1 was $14.4 million. Suktor made further operating losses totalling $6 million for the two-month period 1 July 20X1 to 31 August 20X1.

Required:

Compute the provision that is required in the financial statements of Suktor at 30 June 20X1 in respect of the decision to close.

* Question 7

On 1 April 20W9 Kroner began to lease an office block on a 20-year lease. The useful economic life of the office buildings was estimated at 40 years on 1 April 20W9. The supply of leasehold properties exceeded the demand on 1 April 2009 so as an incentive the lessor paid Kroner $1 million on 1 April 20W9 and allowed Kroner a rent-free period for the first two years of the lease, followed by 36 payments of $250,000, the first being due on 1 April 20X1.

Between 1 April 20W9 and 30 September 20W9 Kroner carried out alterations to the office block at a total cost of $3 million. The terms of the lease require Kroner to vacate the office block on 31 March 20Y9 and leave it in exactly the same condition as it was at the start of the lease. The directors of Kroner have consistently estimated that the cost of restoring the office block to its original condition on 31 March 20Y9 will be $2.5 million at 31 March 20Y9 prices.

An appropriately risk-adjusted discount rate for use in any discounting calculations is 6% per annum. The present value of $1 payable in $19^1/_2$ years at an annual discount rate of 6% is 32 cents.

Required:

Prepare extracts from the financial statements of Kroner that show the depreciation of leasehold improvements and unwinding of discount on the restoration liability in the statement of comprehensive income for *both* of the years ended 31 March 20X0 and 20X1.

Question 8

Epsilon is a listed entity. You are the financial controller of the entity and its consolidated financial statements for the year ended 31 March 2009 are being prepared. The board of directors is responsible for all key financial and operating decisions, including the allocation of resources.

Your assistant is preparing the first draft of the statements. He has a reasonable general accounting knowledge but is not familiar with the detailed requirements of all relevant financial reporting standards. There is one issue on which he requires your advice and he has sent you a note as shown below:

> I note that on 31 January 2009 the board of directors decided to discontinue the activities of a number of our subsidiaries. This decision was made, I believe, because these subsidiaries did not fit into the long-term plans of the group and the board did not consider it likely that the subsidiaries could be sold. This decision was communicated to the employees on 28 February 2009 and the activities of the subsidiaries affected were gradually curtailed starting on 1 May 2009, with an expected completion date of 30 September 2009. I have the following information regarding the closure programme:

> (a) All the employees in affected subsidiaries were offered redundancy packages and some of the employees were offered employment in other parts of the group. These offers had to be accepted or rejected by 30 April 2009. On 31 March 2009 the directors estimated that the cost of redundancies would be $20 million and the cost of relocation of employees who accepted alternative employment would be $10 million. Following 30 April 2009 these estimates were revised to $22 million and $9 million respectively.

> (b) Latest estimates are that the operating losses of the affected subsidiaries for the six months to 30 September 2009 will total $15 million.

> (c) A number of the subsidiaries are leasing properties under non-cancellable operating leases. I believe that at 31 March 2009 the present value of the future lease payments relating to these properties totalled $6 million. The cost of immediate termination of these lease obligations would be $5 million.

> (d) The carrying values of the freehold properties owned by the affected subsidiaries at 31 March 2008 totalled $25 million. The estimated net disposal proceeds of the properties are $29 million and all properties should realise a profit.

> (e) The carrying value of the plant and equipment owned by the affected subsidiaries at 31 March 2008 was $18 million. The estimated current disposal proceeds of this plant and equipment is $2 million and its estimated value in use (including the proceeds from ultimate disposal) is $8 million.

> I am unsure regarding a number of aspects of accounting for this decision by the board. Please tell me how the decision to curtail the activities of the three subsidiaries affects the financial statements.

Required:
Draft a reply to the questions raised by your assistant.

* Question 9

Epsilon is a listed entity. You are the financial controller of the entity and its consolidated financial statements for the year ended 30 September 2008 are being prepared. Your assistant, who has prepared the first draft of the statements, is unsure about the correct treatment of a transaction and has asked for your advice. Details of the transaction are given below.

The current projects of the IASB and FASB on financial instruments are part of the objective to achieve single high-quality global ccounting standards as supported by the G20.

14.3 IAS 32 *Financial Instruments: Disclosure and Presentation*[1]

The dynamic nature of the international financial markets has resulted in a great variety of financial instruments from traditional equity and debt instruments to derivative instruments such as futures or swaps. These instruments are a mixture of on- and off-balance-sheet instruments, and they can significantly contribute to the risks that an enterprise faces. IAS 32 was introduced to highlight to users of financial statements the range of financial instruments used by an enterprise and how they affect the financial position, performance and cash flows of the enterprise.

IAS 32 only considers the areas of presentation of financial instruments; recognition and measurement are considered in IAS 39.

14.3.1 Scope of the standards

IAS 32 (and IAS 39) should be applied by all enterprises and should consider all financial instruments with the exceptions[2] of

● share-based payments;
● interests in subsidiaries, associates and joint ventures;
● employers' rights and obligations under employee benefit plans; and
● rights and obligations arising under insurance contracts.

Further specific scope exemptions only for IAS 39 are highlighted later in this chapter.

14.3.2 Definition of terms[3]

The following are the principal definitions used in IAS 32 and also in IAS 39, which is to be considered later.

A **financial instrument** is any contract that gives rise to both a financial asset of one enterprise and a financial liability or equity instrument of another enterprise.

A **financial asset** is any asset that is:

(a) cash;
(b) a contractual right to receive cash or another financial asset from another entity;
(c) a contractual right to exchange financial instruments with another entity under conditions that are potentially favourable; or
(d) an equity instrument of another entity.

A **financial liability** is any liability that is a contractual obligation:

(a) to deliver cash or another financial asset to another entity; or
(b) to exchange financial instruments with another entity under conditions that are potentially unfavourable.

An **equity instrument** is any contract that evidences a residual interest in the assets of an entity after deducting all of its liabilities.

Following the introduction of IAS 39 extra clarification was introduced into IAS 32 in the application of the definitions. First, a commodity-based contract (such as a commodity future) is a financial instrument if either party can settle in cash or some other financial instrument. Commodity contracts would not be financial instruments if they were expected to be settled by delivery, and this was always intended. This is commonly referred to as the 'own-use' exemption and is commonly applied by businesses to commodity contracts, so avoiding the need to account for sale and purchase contracts at fair value.

The second clarification is for the situation where an enterprise has a financial liability that can be settled with either financial assets or the enterprise's own equity shares. If the number of equity shares to be issued is variable, typically so that the enterprise always has an obligation to give shares equal to the fair value of the obligation, they are treated as a financial liability.

14.3.3 Presentation of instruments in the financial statements

Two main issues are addressed in the standard regarding the presentation of financial instruments. These issues are whether instruments should be classified as liabilities or equity instruments, and how compound instruments should be presented.

Liabilities versus equity

IAS 32 follows a substance approach[4] to the classification of instruments as liabilities or equity. If an instrument has terms such that there is an obligation on the enterprise to transfer financial assets to redeem the obligation then it is a liability instrument regardless of its legal nature. Preference shares are the main instrument where legally they may be classified as equity but in substance they should be accounted as liabilities. The common conditions on the preference share that would indicate it is to be treated as a liability instrument are as follows:

- annual dividends are compulsory and not at the discretion of directors; or
- the share provides for mandatory redemption by the issuer at a fixed or determinable amount at a future fixed or determinable date; or
- the share gives the holder the option to redeem upon the occurrence of a future event that is highly likely to occur (e.g. after the passing of a future date).

If a preference share is treated as a liability instrument, it is presented as such in the statement of financial position. Any dividends paid or payable on that share are calculated in the same way as interest and presented as a finance cost in the statement of comprehensive income. The presentation on the statement of comprehensive income could be as a separate item from other interest costs, but this is not mandatory. Any gains or losses on the redemption of financial instruments classified as liabilities are also presented in profit or loss.

Impact on companies
The presentation of preference shares as liabilities does not alter the cash flows or risks that the instruments give, but there is a danger that the perception of a company may change. This presentational change has the impact of reducing net assets and increasing gearing. This could be very important, for example, if a company had debt covenants on other borrowings that required the maintenance of certain ratios such as gearing or interest cover. Moving preference shares to debt and dividends to interest costs could mean the covenants are breached and other loans become repayable.

In addition, the higher gearing and reduced net assets could mean the company is perceived as more risky, and therefore could result in the company being perceived to have a higher credit risk. This in turn might lead to a reduction in the company's credit rating, making obtaining future credit more difficult and expensive.

These very practical issues need to be managed by companies converting to IFRS from a local accounting regime that treats preference shares as equity or non-equity funds. Good communication with users is key to smoothing the transition.

Compound instruments[5]

Compound instruments are financial instruments that have the characteristics of both debt and equity. A convertible loan, which gives the holder the option to convert into equity shares at some future date, is the most common example of a compound instrument. The view of the IASB is that the proceeds received by a company for these instruments are made up of two parts: (i) a debt obligation and (ii) an equity option. Following the substance of the instruments, IAS 32 requires that the two parts be presented separately, a 'split accounting' approach.

The split is made by measuring the debt part and making the equity the residual of the proceeds. This approach is in line with the definitions of liabilities and equity, where equity is treated as a residual. The debt is calculated by discounting the cash flows on the debt at a market rate of interest for similar debt without the conversion option.

The following is an extract from the 2013 Balfour Beatty Annual Report relating to convertible preference shares:

> The Company's cumulative convertible redeemable preference shares and the Group's convertible bonds are compound instruments, comprising a liability component and an equity component. The fair value of the liability components were estimated using the prevailing market interest rates at the dates of issue for similar non-convertible instruments. The difference between the proceeds of issue of the preference shares and convertible bonds and the fair value assigned to the respective liability components, representing the embedded option to convert the liability components into the Company's ordinary shares, is included in equity. The interest expense on the liability components is calculated by applying applicable market interest rates for similar non-convertible debt prevailing at the dates of issue to the liability components of the instruments. The difference between this amount and the dividend/interest paid is added to the carrying amount of the liability component and is included in finance charges, together with the dividend/interest payable.

Illustration for compound instruments

Rohan plc issues 1,000 £100 5% convertible debentures at par on 1 January 2000. The debentures can be either converted into 50 ordinary shares per £100 of debentures, or redeemed at par at any date from 1 January 2005. Interest is paid annually in arrears on 31 December. The interest rate on similar debentures without the conversion option is 6%.

To split the proceeds the debt value must be calculated by discounting the future cash flows on the debt instrument. The value of debt is therefore:

Present value of redemption payment (discounted @ 6%)	£74,726
Present value of interest (5 years) (discounted @ 6%)	£21,062
Value of debt	£95,788
Value of the equity proceeds: (£100,000 − £95,788)	
(presented as part of equity)	£4,212

The following is an extract from the 2011 Annual Report of Aspen Pharmacare Holdings Ltd:

For accounting purposes the preference shares have been split into an equity and a liability component. Refer to the accounting policy for detail.

		R million
Preference shares – equity component		**162.0**
(per statement of changes in equity)		
Deferred tax effect		(8.7)
Net equity component		153.3
Preference shares – liability component		381.3
(per the statement of financial position)		
Amount expensed in 2011		(183.2)
Cumulative notional interest on liability component		
Opening balance	**20.1**	
For the year	5.3	25.4
		376.8

Perpetual debt

Following a substance approach, perpetual or irredeemable debt could be argued to be an equity instrument as opposed to a debt instrument. IAS 32, however, takes the view that it is a debt instrument because the interest must be paid (as compared to dividends which are only paid if profits are available for distribution and if directors declare a dividend approved by the shareholders). The present value of all the future obligations to pay interest will equal the proceeds of the debt if discounted at a market rate. The proceeds on issue of a perpetual debt instrument are therefore a liability obligation.

14.3.4 Calculation of finance costs on liability instruments

The finance costs will be charged to profit or loss. The finance cost of debt is the total payments to be incurred over the lifespan of that debt less the initial carrying value. Such costs should be allocated to profit or loss over the lifetime of the debt at a constant rate of interest based on the outstanding carrying value per period. If a debt is settled before maturity, any profit or loss should be reflected immediately in profit or loss – unless the substance of the settlement transaction fails to generate any change in liabilities and assets.

Illustration of the allocation of finance costs and the determination of carrying value

On 1 January 20X6 a company issued a debt instrument of £1,000,000 spanning a four-year term. It received from the lender £890,000, being the face value of the debt less a discount of £110,000. Interest was payable yearly in arrears at 8% per annum on the principal sum of £1,000,000. The principal sum was to be repaid on 31 December 20X9.

To determine the yearly finance costs and year-end carrying value it is necessary to compute:

- the aggregate finance cost;
- the implicit rate of interest carried by the instrument (referred to in IAS 39 as the effective yield);
- the finance charge per annum; and
- the carrying value at successive year-ends.

Figure 14.1 Allocation of finance costs and determination of carrying value

	(i)		(ii) Finance charge to statement of comprehensive income £000		(iii) Carrying value in statement of financial position £000
	Cash flows £000				
At 1 Jan 20X6	(890)	(1,000 – 110)	—		890
At 31 Dec 20X6	80	(8% × 1,000)	103.2	(11.59% × 890)	913.2
At 31 Dec 20X7	80	(8% × 1,000)	105.8	(11.59% × 913.2)	939.2
At 31 Dec 20X8	80	(8% × 1,000)	108.8	(11.59% × 939)	967.8
At 31 Dec 20X9	1,080	(1,000 + (8% × 1,000))	112.2	(11.59% × 967.8)	—
Net cash flow	430	= Cost	430		

Aggregate finance cost

This is the difference between the total future payments of interest plus principal, less the net proceeds received less costs of the issue, i.e. £430,000 in column (i) of Figure 14.1.

Implicit rate of interest carried by the instrument

This can be computed by using the net present value (NPV) formula:

$$\sum_{t=1}^{t=n} \frac{A_t}{(1+r)^t} - I = 0$$

where A is forecast net cash flow in year A, t time (in years), n the lifespan of the debt in years, r the company's annual rate of discount and I the initial net proceeds. Note that the application of this formula can be quite time-consuming. A reasonable method of assessment is by interpolation of the interest rate.

The aggregate formula given above may be disaggregated for calculation purposes:

$$\frac{A_1}{(1+r)} + \frac{A_2}{(1+r)^2} + \frac{A_3}{(1+r)^3} + \frac{A_4}{(1+r)^4} - I = 0$$

Using the data concerning the debt and assuming (allowing for discount and costs) an implicit constant rate of, say, 11%:

$$\sum = \frac{80,000}{(1.11)^1} + \frac{80,000}{(1.11)^2} + \frac{80,000}{(1.11)^3} + \frac{1,080,000}{(1.11)^4} - 890,000 = 0$$
$$= 72,072 + 64,930 + 58,495 + 711,429 - 890,000 = +16,926$$

The chosen implicit rate of 11% is too low. We now choose a higher rate, say 12%:

$$\sum = \frac{80,000}{(1.12)^1} + \frac{80,000}{(1.12)^2} + \frac{80,000}{(1.12)^3} + \frac{1,080,000}{(1.12)^4} - 890,000 = 0$$
$$= 71,429 + 63,776 + 56,942 + 686,360 - 890,000 = -11,493$$

This rate is too high, resulting in a negative net present value. Interpolation will enable us to arrive at an implicit rate:

$$11\% + \left[\frac{16,926}{16,926 + 11,493} \times (12\% - 11\%) \right] = 11\% + 0.59\% = 11.59\%$$

This is a trial and error method of determining the implicit interest rate. In this example the choice of rates, 11% and 12%, constituted a change of only 1%. It would be possible to choose, say, 11% and then 14%, generating a 3% gap within which to interpolate. This wider margin would result in a less accurate implicit rate and an aggregate interest charge at variance with the desired £430,000 of column (ii). The aim is to choose interest rates as close as possible to either side of the monetary zero, so that the exact implicit rate may be computed.

The object is to determine an NPV of zero monetary units, i.e. to identify the discount rate that will enable the aggregate future discounted net flows to equate to the initial net proceeds from the debt instrument. In the above illustration, a discount (interest) rate of 11.59% enables £430,000 to be charged to profit or loss after allowing for payment of all interest, costs and repayment of the face value of the instrument.

The finance charge per annum and the successive year-end carrying amounts

The charge to the statement of comprehensive income and the carrying values in the statement of financial position are shown in Figure 14.1.

14.3.5 Offsetting financial instruments[6]

Financial assets and liabilities can only be offset and presented net if the following conditions are met:

(a) the enterprise currently has a legally enforceable right to set off the recognised amounts; and

(b) the enterprise intends either to settle on a net basis, or to realise the asset and settle the liability simultaneously.

IAS 32 emphasises the importance of the intention to settle on a net basis as well as the legal right to do so. Offsetting should only occur when the cash flows and therefore the risks associated with the financial asset and liability are offset and therefore to present them net in the statement of financial position shows a true and fair view. An example of a situation where offsetting may be appropriate would be if a company has a receivable and a payable to the same counterparty, has a legal right to offset the two, and does offset the amounts in practice when settling the cash flows.

Situations where offsetting might be considered but which would not normally be appropriate are where:

- several different financial instruments are used to emulate the features of a single financial instrument;
- financial assets and financial liabilities arise from financial instruments having the same primary risk exposure but involve different counterparties;
- financial or other assets are pledged as collateral for non-recourse financial liabilities;
- financial assets are set aside in trust by a debtor for the purpose of discharging an obligation without those assets having been accepted by the creditor in settlement of the obligation;
- obligations incurred as a result of events giving rise to losses are expected to be recovered from a third party by virtue of a claim made under an insurance policy.

14.4 IAS 39 *Financial Instruments: Recognition and Measurement*

IAS 39 is the first comprehensive standard on the recognition and measurement of financial instruments and completes the guidance that was started with the introduction of IAS 32.

14.4.1 Scope of the standard

IAS 39 should be applied by all enterprises to all financial instruments except those excluded from the scope of IAS 32 (see Section 14.3.1) and the following additional instruments:

- rights and obligations under leases to which IAS 37 applies (except for embedded derivatives);
- equity instruments of the reporting entity including options, warrants and other financial instruments that are classified as shareholders' equity;
- contracts between an acquirer and a vendor in a business combination to buy or sell or acquire at a future date. The term of the forward contract should not exceed a reasonable period normally necessary to obtain any required approvals and to complete the transaction;
- rights to payments to reimburse the entity for expenditure it is required to make to settle a liability under ISA 37.

14.4.2 Definitions of the categories of financial instruments

The four categories of financial assets are (a) financial assets at fair value through profit or loss, (b) held-to-maturity investments, (c) loans and receivables, and (d) available-for-sale financial assets. The definition of each is as stated below.

Financial liabilities are less complicated to classify than financial assets. There are only two categories of financial liability, fair value through profit or loss, and other financial liabilities. Financial liabilities at fair value through profit or loss are defined consistently with those below. Other financial liabilities are measured at amortised cost.

(a) Financial assets or liabilities at fair value through profit or loss

Assets and liabilities under this category are reported in the financial statements at fair value. Changes in the fair value from period to period are reported as a component of net income. There are two types of investments that are accounted for under this heading, namely, *held-for-trading investments* and those *designated on initial recognition* under the fair value option.

Held-for-trading investments

These are financial instruments where (i) the investor's principal intention is to sell or repurchase a security in the near future and where there is normally active trading for profit-taking in the securities, or (ii) they are part of a portfolio of identified financial instruments that are managed together and for which there is evidence of a recent pattern of short-term profit-taking, or (iii) they are derivatives. This category includes commercial papers, certain government bonds and treasury bills.

A **derivative** is a financial instrument:

- whose value changes in response to the change in a specified interest rate, security price, commodity price, foreign exchange rate, index of prices or rates, a credit rating or credit index or similar variable (sometimes called the 'underlying');
- that requires no initial net investment or an initial net investment that is smaller than would be required for other types of contract that would be expected to have a similar response to changes in market factors; and
- that is settled at a future date.

Designated on initial recognition – the 'fair value option'

A company has the choice of designating as fair value through profit or loss on the initial recognition of an investment in the following situations:

- it eliminates or significantly reduces a measurement or recognition inconsistency (sometimes referred to as an 'accounting mismatch') that would otherwise arise from measuring assets or liabilities or recognising the gains and losses on them on different bases; or
- a group of financial assets, financial liabilities or both is managed and performance is evaluated on a fair value basis, in accordance with a documented risk management or investment strategy; or
- the financial asset or liability contains an embedded derivative that would otherwise require separation from the host.

The following is an extract from the Annual Consolidated Financial Statements of BNP Paribas Group for 2013:

Financial assets and liabilities designated at fair value through profit or loss (fair value option)

Financial assets or financial liabilities may be designated on initial recognition as at fair value through profit or loss, in the following cases:

- hybrid financial instruments containing one or more embedded derivatives which otherwise would have been separated and accounted for separately;
- where using the option enables the entity to eliminate or significantly reduce a mismatch in the measurement and accounting treatment of assets and liabilities that would arise if they were to be classified in separate categories;
- when a group of financial assets and/or financial liabilities is managed and measured on the basis of fair value, in accordance with a documented risk management and investment strategy.

Prior to October 2008 it was prohibited to transfer instruments either into or out of the fair value through profit or loss category after initial recognition of the instrument. Following significant pressure from the European Union that the international standards were more restrictive than US GAAP in this area, the IASB amended the standard to allow reclassification of financial instruments in rare circumstances. The financial crisis of 2008 was deemed to be a rare situation that would justify reclassification.

The reclassification requirements allow instruments to be transferred from the fair value through profit and loss to the loans and receivables category. They also allow reclassifications from the available-for-sale category (discussed below) to the loans and receivables category. The IASB allowed a short-term exemption from the general requirement that the transfer is at fair value, and permitted the transfers to be undertaken at the fair values of instruments on 1 July 2008, a date before significant reductions in fair value on debt instruments arose.

(b) Held-to-maturity investments

Held-to-maturity investments consist of instruments with fixed or determinable payments and fixed maturity for which the entity positively intends and has the ability to hold to maturity. For items to be classified as held-to-maturity an entity must justify that it will hold them to maturity. The tests that a company must pass to justify this classification are summarised in Figure 14.2.

The investments are initially measured at fair value (including transaction costs) and subsequently measured at amortised cost using the effective interest method, with the periodic

Figure 14.2 Tests for classification as held-to-maturity investment

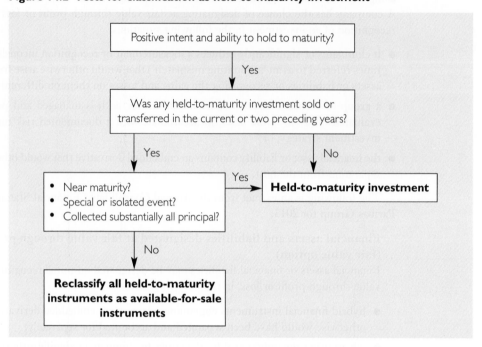

amortisation recorded in the statement of comprehensive income. As they are reported at amortised cost, temporary fluctuations in fair value are not reflected in the entity's financial statements.

Such investments include corporate and government bonds and redeemable preference shares which can be held to maturity. They do not include investments designated as at fair value through profit or loss on initial recognition, those designated as available for sale and those defined as loans and receivables. They also do not include ordinary shares in other entities because these do not have a maturity date.

(c) Loans and receivables

Loans and receivables include financial assets with fixed or determinable payments that are not quoted in an active market. They are initially measured at fair value (including transaction costs) and subsequently measured at amortised cost using the effective interest method, with the periodic amortisation in the statement of comprehensive income.

Amortised cost is normally the amount at which a financial asset or liability is measured at initial recognition minus principal repayments, minus the cumulative amortisation of any premium and minus any write-down for impairment.

This category includes trade receivables, accrued revenues for services and goods, loan receivables, bank deposits and cash at hand. It does not include financial assets held for trading, those designated on initial recognition as at fair value through profit or loss, those designated as available-for-sale and those for which the holder may not recover substantially all of its initial investment, other than because of credit deterioration.

(d) Available-for-sale financial assets

The available-for-sale category is a 'catch-all' that includes all financial assets that have not been classified as fair value through profit or loss, held-to-maturity or loans and receivables. Because of its nature as a catch-all it effectively gives entities a choice of classification for some instruments. For example, if an entity did not include an investment in a non-trading bond within loans and receivables or held-to-maturity it would, by default, be classified as available-for-sale.

A common financial asset that would be classified as available-for-sale is equity investments in another entity.

On initial recognition an asset is reported at fair value (including transaction costs) and at period-ends it is restated to fair value with changes in fair value reported under other comprehensive income. If the fair value falls below amortised cost and the fall is not determined to be temporary, it is reported in the investor's statement of comprehensive income.

Any revaluations to fair value of available-for-sale assets are recognised in other comprehensive income until the asset is derecognised. On derecognition any cumulative gain or loss recognised in other comprehensive income is transferred to profit or loss as part of the gain or loss on derecognition. Consider the example of Kathryn plc below.

Illustration of available-for-sale accounting

On 1 January 20X3 Kathryn plc acquires an equity investment for €100,000 which is classified as available-for-sale. On 31 December 20X3 the investment is valued at €110,000, and on 31 March 20X4 it is sold for €115,000. The accounting entries that would be reflected in the year ended 31 December 20X3 and 20X4 are as follows:

		€	€
20X3			
On acquisition of the investment:			
Dr	Available-for-sale asset	100,000	
Cr	Cash		100,000
On revaluation at the year-end:			
Dr	Available-for-sale asset	10,000	
Cr	Available-for-sale revaluation reserve		10,000
	(presented as a gain in other comprehensive income (OCI))		
20X4			
On disposal of the investment:		€	€
Dr	Cash	115,000	
Cr	Available-for-sale asset		110,000
Cr	Profit on sale		5,000
On disposal recycle the gain in OCI:			
Dr	Available-for-sale revaluation reserve	10,000	
Cr	Profit on sale		10,000

The impact of the above is that when in 20X4 the investment is sold, the profit recognised (€15,000) is the difference between the sale proceeds (€115,000) and cost (€100,000).

The fair value of publicly traded securities is normally based on quoted market prices at the year-end date. The fair value of securities that are not publicly traded is assessed using a variety of methods and assumptions based on market conditions existing at each year-end date referring to quoted market prices for similar or identical securities if available or employing other techniques such as option pricing models and estimated discounted values of future cash flows.

Available-for-sale does not include debt and equity securities classified as held for trading or held-to-maturity.

Example of accounting for an available-for-sale financial asset: the acquisition by Brighton plc of shares in Hove plc

On 1 September 20X9 Brighton purchased 15 million of the 100 million shares in Hove for £1.50 per share. This purchase was made with a view to further purchases in future. The

Brighton directors are not able to exercise any influence over the operating and financial policies of Hove. The shares are currently in the statement of financial position as at 31 December 20X9 at cost and the fair value of a share was £1.70.

Accounting treatment at the year-end

Brighton owns 15% of the Hove issued shares. As the directors are not able to exercise any influence, the investment is dealt with under IAS 39 *Financial Instruments: Recognition and Measurement* and under its provisions the investment is an available-for-sale financial asset. This means that it is to be valued at fair value, with gains or losses taken to equity.

In this case the investment is valued at £25.5 million (15 million × £1.70) and the gain of £3 million (15 million × (£1.70 – £1.50)) is taken to equity through other comprehensive income.

Headings under which reported

Assets are reported as appropriate in the statement of financial position under Other non-current assets, Trade and other receivables, Interest-bearing receivables, or Cash and cash equivalents. For example, financial liabilities measured at amortised cost comprise financial liabilities such as borrowings, trade payables, accrued expenses for services and goods, and certain provisions settled in cash. These are reported in the statement of financial position under Long-term and short-term borrowings, Other provisions, Other long-term liabilities, Trade payables and Other current liabilities.

Impact of classification on the financial statements

The impact of the classification of financial instruments on the financial statements is important as it affects the value of assets and liabilities and also the income recognised. For example, assume that Henry plc had the following financial assets and liabilities at its year-end. All the instruments had been taken out at the start of the current year:

1 A forward exchange contract. At the period-end date the contract was an asset with a fair value of £100,000.

2 An investment of £1,000,000 in a 6% corporate bond. At the period-end date the market rate of interest increased and the bond fair value fell to £960,000.

3 An equity investment of £500,000. This investment was worth £550,000 at the period-end.

The classification of these instruments is important and choices are available as to how they are accounted for. For example, the investment in the corporate bond above could be accounted for as a held-to-maturity investment (if Henry plc had the intent and ability to hold it to maturity), or it could be an available-for-sale investment if so chosen by Henry. The bond and the equity investment could even be recognised as fair value through profit or loss if they met the criteria to be designated as such on initial recognition.

To highlight the impact on the financial statements, the tables below show the accounting positions for the investments on different assumptions. Not all possible classifications are shown in the tables.

Option 1

Instrument	Classification	Statement of financial position	Profit or loss	Other comprehensive income
Forward contract	FV-P&L	£100,000	£100,000	—
Corporate bond	Held-to-maturity	£1,000,000	*(£60,000)	—
Equity investment	Available-for-sale	£550,000	—	£50,000

* *Interest on the bond of £1,000,000 × 6%*

The bond is not revalued because held-to-maturity investments are recognised at amortised cost.

Option 2

Instrument	*Classification*	*Statement of financial position*	*Profit or loss*	*Other comprehensive income*
Forward contract	FV–P&L	£100,000	£100,000	—
Corporate bond	Available-for-sale	£960,000	(£60,000)	(£40,000)
Equity investment	Available-for-sale	£550,000	—	£50,000

Interest is still recognised on the bond but at the year-end it is revalued through equity to its fair value of £960,000.

Option 3

Instrument	*Classification*	*Statement of financial position*	*Profit or loss*	*Other comprehensive income*
Forward contract	FV–P&L	£100,000	£100,000	—
Corporate bond	Held-to-maturity	£1,000,000	(£60,000)	—
Equity investment	FV–P&L	£550,000	£50,000	—

The equity investment is revalued through profit and loss as opposed to through other comprehensive income as it would be if classified as available-for-sale.

14.4.3 Recognition of financial instruments

Initial recognition

A financial asset or liability should be recognised when an entity becomes party to the contractual provisions of the instrument. This means that derivative instruments must be recognised if a contractual right or obligation exists.

Derecognition

Derecognition of financial assets is a complex area which has extensive rules in IAS 39. These rules have been included within IFRS 9 *Financial Instruments* and therefore this is not an area that is expected to change going forward. The main principle is that financial assets should only be derecognised when the entity transfers the risks and rewards that make up the asset. This could be because the benefits are realised, the rights expire or the enterprise surrenders the benefits. The flowchart in Figure 14.3 summarises the approach in the standard.

If it is not clear whether the risks and rewards have been transferred, the entity considers whether control has passed. If control has passed the entity should derecognise the asset, whereas if control is retained the asset is recognised to the extent of the entity's continuing involvement in the asset.

On derecognition any gain or loss should be recorded in profit or loss. Also any gains or losses previously recognised in reserves relating to the asset should be transferred to the profit or loss on sale.

Financial liabilities should only be derecognised when the obligation specified in the contract is discharged, is cancelled or expires.

The requirements for the derecognition of liabilities mean that it is not possible to write off liabilities unless they are discharged, cancelled or expired. In some industries this will

Figure 14.3 Derecognition of financial assets

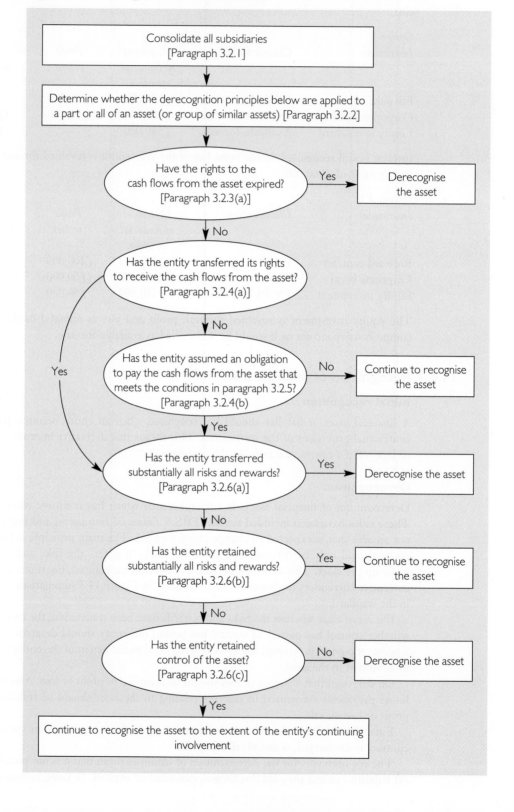

lead to a change in business practice. For example, banks are not allowed to remove dormant accounts from their statements of financial position unless the liability has been legally extinguished.

14.4.4 Embedded derivatives

Sometimes an entity will enter into a contract that includes both a derivative and a host contract, with the effect that some of the cash flows of the combined contract vary in a similar way to a stand-alone derivative.

For example, Cheng Ltd issued a £1m bond 20X1 with an interest rate of 5% redeemable in 20X6. This is referred to as the 'host contract'. If there is a condition which means that the 5% rate is related to a benchmark such as movements in the Consumer Price Index, then this is referred to as an 'embedded derivative'.

Other examples of embedded derivatives that affect the host contract cash flows include:

- contracts for the supply of commodities such as the supply of coal to coal-fired power stations where the price is related to a commodity index;
- contracts where the price of the commodity is fixed payment in a foreign currency such as the supply to the UK from South Africa where the payment is denominated in South African rand;
- leases where the rental might be related to a benchmark such as the lessee's turnover; and
- put options on an equity instrument held by an enterprise, or an equity conversion feature embedded in a debt instrument.

An embedded instrument should be separated from the host contract and accounted for as a derivative under IAS 39 if all of the following conditions are met:

(a) the economic characteristics and risks of the embedded derivative are not closely related to the economic characteristics and risks of the host contract;

(b) a separate instrument with the same terms as the embedded derivative would meet the definition of a derivative; and

(c) the hybrid instrument is not measured at fair value with changes in fair value reported in profit or loss.

If an entity is required to separate the embedded derivative from its host contract but is unable to measure the embedded derivative separately, the entire hybrid instrument should be treated as a financial instrument held at fair value through profit or loss and as a result changes in fair value should be reported through profit or loss.

14.4.5 Measurement of financial instruments

Initial measurement

Financial assets and liabilities (other than those at fair value through profit or loss) should be initially measured at fair value plus transaction costs. In almost all cases this would be at cost. For instruments at fair value through profit and loss, transaction costs are not included and instead are expensed as incurred.

Subsequent measurement

Figure 14.4 summarises the way that financial assets and liabilities are to be subsequently measured after initial recognition.

Figure 14.4 Subsequent measurement

Category	Measurement
Financial assets at fair value through profit or loss*	Fair value without any deduction for transaction costs on sale or disposal
Held-to-maturity investments	Amortised cost using the effective interest method
Loans and receivables	Amortised cost using the effective interest method
Available-for-sale financial assets*	Fair value without any deduction for transaction costs on sale or disposal
Financial liabilities at fair value through profit or loss	Fair value
Other financial liabilities	Amortised cost using the effective interest method

* If these categories include unquoted equity instruments (or derivative liabilities that are settled in unquoted equity instruments) where fair value cannot be measured reliably then they are measured at cost. This, however, should be very rare.

The measurement after initial recognition is at either fair value or amortised cost. The only financial instruments that can be recognised at cost (not amortised) are unquoted equity investments for which there is no measurable fair value. These should be very rare.

The fair value is the amount for which an asset could be exchanged, or a liability settled, between knowledgeable, willing parties in an arm's-length transaction.

The methods for fair value measurement allow a number of different bases to be used for the assessment of fair value. These include published market prices; transactions in similar instruments; discounted future cash flows; or valuation models with market or non-market inputs.

From January 2013 fair value has been assessed under IFRS 13 *Fair Value Measurement*. The method used will be the one which is most reliable for the particular instrument.

In the 2008 financial crisis there were calls on the IASB to either abolish or suspend the fair value measurement basis in IAS 39 as it has been perceived as requiring companies to recognise losses greater than their true value. The reason for this was a concern that the market value was being distorted by a lack of liquidity, and that markets were not functioning efficiently with willing buyers and sellers. The IASB resisted the calls but issued guidance on valuation in illiquid markets that emphasises the different ways that fair value can be determined. For instruments that operate in illiquid markets there is sometimes a need to value the instruments based on valuation models and discounted cash flows; however, these models take into account factors that a market participant would consider in the current circumstances.

Amortised cost – the effective yield calculation

Amortised cost is calculated using the effective interest method on assets and liabilities. For the definition of effective interest it is necessary to look at IAS 39, paragraph 9. The effective rate is defined as:

the rate that exactly discounts estimated future cash receipts or payments through the expected life of the financial instrument.

The definition then goes on to require that the entity shall:

- estimate cash flows considering all contractual terms of the financial instrument (for example, prepayment, call and similar options), but not future credit losses;
- include all necessary fees and points paid or received that are an integral part of the effective yield calculation (IAS 18); and
- make a presumption that the cash flows and expected life of a group of similar financial instruments can be estimated reliably.

Illustration of the effective yield method

George plc lends £10,000 to a customer for fixed interest based on the customer paying 5% interest per annum (annually in arrears) for two years, and then 6% fixed for the remaining three years with the full £10,000 repayable at the end of the five-year term.

The tables below show the interest income over the loan period assuming:

- it is not expected that the customer will repay early (effective rate is 5.55% per annum derived from an internal rate of return calculation); and
- it is expected the customer will repay at the end of year 3 but there are no repayment penalties (effective rate is 5.3% per annum derived from an internal rate of return calculation).

The loan balance will alter as follows:

No early repayment

Period	B/F	Interest income (5.55%)	Cash received	C/F
Year 1	10,000	555	(500)	10,055
Year 2	10,055	558	(500)	10,113
Year 3	10,113	561	(600)	10,074
Year 4	10,074	559	(600)	10,033
Year 5	10,033	557	(10,600)	(10)*

* Difference due to rounding

Early repayment

Period	B/F	Interest income (5.3%)	Cash received	C/F
Year 1	10,000	530	(500)	10,030
Year 2	10,030	532	(500)	10,062
Year 3	10,062	533	(10,600)	(5)*

* Difference due to rounding

Gains or losses on subsequent measurement

When financial instruments are remeasured to fair value the rules for the treatment of the subsequent gain or loss are as shown in Figure 14.5. Gains or losses arising on financial instruments that have not been remeasured to fair value will arise when either the assets are impaired or the instruments are derecognised. These gains and losses are recognised in profit or loss for the period.

14.4.6 Hedging

If a financial instrument has been taken out to act as a hedge, and this position is clearly identified and expected to be highly effective, and is subsequently proved to be highly effective, hedge accounting rules can be followed.

Figure 14.5 Gains or losses on subsequent measurements

Instrument	Gain or loss
Instruments at fair value through profit or loss	Profit or loss
Available-for-sale	Other comprehensive income (except for impairments and foreign exchange gains and losses) until derecognition, at which time the cumulative gain/loss in equity is recognised in profit or loss. Dividend income is recognised in profit or loss when the right to receive payment is established.

There are three types of hedging relationship: fair value hedge, cash flow hedge and net investment hedge.

I Fair value hedge

A fair value hedge arises when a hedging instrument is taken out as a hedge of the exposure to changes in fair value of a recognised asset or liability or an unrecognised firm commitment that will affect reported net income.

The hedge accounting requirements change the way that gains and losses on a hedged item are recognised. Any gain or loss arising on remeasuring the hedging instrument *and* the hedged item should be recognised in profit or loss in the period. Without hedge accounting the gain or loss on the hedged item would not typically be recognised in profit or loss.

An example of a fair value hedge could be protecting against interest rate changes on a fixed rate investment with a matching fixed to floating interest rate swap. This is a fair value hedge because if the interest rate changes, the fair value of the investment will be affected. The fair value change in the swap offsets this. The accounting for this arrangement as a fair value hedge would be:

(i) the interest rate swap would be recognised at fair value with gains or losses in the income statement; and

(ii) the investment would be recognised at fair value with respect to interest rate movements, and the gain or loss will be recognised in the income statement.

The gains and losses on the swap and the investment would offset in the income statement.

2 Cash flow hedge

A cash flow hedge arises when a financial instrument, typically a derivative, is taken out as a hedge of the exposure to variability in cash flows that is attributable to a particular risk associated with the recognised asset or liability, and that will affect reported net income. A hedge of foreign exchange risk on a firm commitment may be a cash flow or a fair value hedge.

Cash flow hedge accounting changes the way that gains and losses on the hedging instrument are recognised. Assuming that the hedging instrument is a derivative the treatment without hedge accounting is to recognise the derivative as fair value through profit and loss. With hedge accounting the gain or loss on the hedging instrument is recognised directly in other comprehensive income, and reflected on the balance sheet in a separate hedge reserve. Any gains or losses recognised in other comprehensive income is included in profit or loss in the period that the hedged item affects profit or loss. If the instrument being hedged results in the recognition of a non-financial asset or liability, the gain or loss on the hedging instrument can be recognised as part of the cost of the hedged item.

Cash flow hedge illustrated

Harvey plc directors agreed at their July 20X6 meeting to acquire additional specialist computer equipment in September 20X7 at an estimated cost of $500,000.

The company entered into a forward contract in July 20X6 to purchase $500,000 in September 20X7 and pay £260,000. At the year-end in December 20X6 the $500,000 has appreciated and has a sterling value of £276,000.

At the year-end the increase of £16,000 will be debited to a Forward Contract asset and credited to a hedge reserve.

In September 20X7 when the equipment is purchased the £16,000 will be deducted in its entirety from the Equipment carrying amount or transferred annually as a reduction of the annual depreciation charge.

3 Net investment hedge

A net investment hedge arises when a hedging instrument, which commonly in this case will be a loan as opposed to a derivative, is entered into to hedge an investment in a foreign entity. The gain or loss on the hedging instrument is recognised directly in other comprehensive income to match against the gain or loss on the hedged investment. Without hedge accounting, the gain or loss on the hedging instrument would be recognised in profit or loss.

A common situation for net investment hedging arises when a foreign equity investment is financed by a foreign loan. As the entity has a foreign asset and liability when exchange rates change one makes a gain and one makes a loss. The gain or loss on the investment is recognised in other comprehensive income under IAS 21 *The Effects of Changes in Foreign Exchange Rates*, and with hedging the opposite gain or loss on the loan is also recognised in other comprehensive income.

Conditions for hedge accounting

In order to be able to apply the hedge accounting techniques detailed above, an entity must meet a number of conditions. These conditions are designed to ensure that only genuine hedging instruments can be hedge accounted, and that the hedged positions are clearly identified and documented. There has been criticism that these hedge criteria can be onerous to comply with in practice and in the future with the introduction of IFRS 9 they are expected to become easier to meet.

The conditions are:

- at the inception of the hedge there is formal documentation of the hedge relationship and the enterprise's risk management objective and strategy for undertaking the hedge;
- the hedge is expected to be highly effective at inception and on an ongoing basis in achieving offsetting changes in fair values or cash flows;
- the effectiveness of the hedge can be reliably measured, that is the fair value of the hedged item and the hedging instrument can be measured reliably;
- for cash flow hedges, a forecasted transaction that is the subject of the hedge must be highly probable; and
- the hedge was assessed on an ongoing basis and determined actually to have been highly effective throughout the accounting period (effective between 80% and 125%).

In order to comply with the conditions it is necessary to businesses to set up the appropriate processes for documenting and monitoring their hedge relationships. In addition if a hedge relationship is not fully effective, but still within the highly effective range of 80% to 125%, the ineffective portion of the hedge will usually be required to be recognised in profit and loss.

14.5 IFRS 7 *Financial Instruments: Disclosure*[7]

14.5.1 Introduction

This standard came out of the ongoing project of improvements to the accounting and disclosure requirements relating to financial instruments.

Prior to the introduction of IFRS 7 disclosures in respect of financial instruments were governed by two standards:

1 IAS 30 *Disclosures in the Financial Statements of Banks and Similar Financial Institutions*; and

2 IAS 32 *Financial Instruments: Disclosure and Presentation*.

In drafting IFRS 7, the IASB:

- reviewed existing disclosures in the two standards, and removed duplicative disclosures;
- simplified the disclosure about concentrations of risk, credit risk, liquidity risk and market risk under IAS 32; and
- transferred disclosure requirements from IAS 32.

Since the original issue of the standard there have been multiple changes to refine and improve the disclosures in respect of financial instruments.

14.5.2 Main requirements

The standard applies to all entities, regardless of the quantity of financial instruments held. However, the extent of the disclosures required will depend on the extent of the entity's use of financial instruments and its exposure to risk.

The standard requires disclosure of the following. These disclosure standards effectively provide the principles that underpin the financial instrument disclosure requirements:

- the significance of financial instruments for the entity's financial position and performance (many of these disclosures were previously in IAS 32); and
- the nature and extent of risks arising from financial instruments to which the entity is exposed during the period and at the end of the reporting period, and how the entity manages those risks.

In addition to the principles above there is extensive detailed guidance in the standard of the disclosures necessary. The disclosures are a mixture of both qualitative and quantitative disclosures. The qualitative disclosures describe management's objectives, policies and processes for managing those risks. The quantitative disclosures provide information about the extent to which the entity is exposed to risk, based on the information provided internally to the entity's key management personnel.

For the disclosure of the significance of financial instruments for the entity's financial position and performance, a key aspect will be to clearly link the statement of financial position and the statement of comprehensive income to the classifications in IAS 39. The detailed requirements from IFRS 7 in this respect are as follows:

8 The carrying amounts of each of the following categories, as defined in IAS 39, shall be disclosed either on the face of the statement of financial position or in the notes:

(a) financial assets at fair value through profit or loss, showing separately (i) those designated as such upon initial recognition and (ii) those classified as held for trading in accordance with IAS 39;

    (b) held-to-maturity investments;

    (c) loans and receivables;

    (d) available-for-sale financial assets;

    (e) financial liabilities at fair value through profit or loss, showing separately (i) those designated as such upon initial recognition and (ii) those classified as held for trading in accordance with IAS 39; and

    (f) financial liabilities measured at amortised cost.

20 An entity shall disclose the following items of income, expense, gains or losses either on the face of the financial statements or in the notes:

    (a) net gains or net losses on:

        (i) financial assets or financial liabilities at fair value through profit or loss, showing separately those on financial assets or financial liabilities designated as such upon initial recognition, and those on financial assets or financial liabilities that are classified as held for trading in accordance with IAS 39;

        (ii) available-for-sale financial assets, showing separately the amount of gain or loss recognised directly in equity during the period and the amount removed from equity and recognised in profit or loss for the period;

        (iii) held-to-maturity investments;

        (iv) loans and receivables; and

        (v) financial liabilities measured at amortised cost;

    (b) total interest income and total interest expense (calculated using the effective interest method) for financial assets or financial liabilities that are not at fair value through profit or loss;

    (c) fee income and expense (other than amounts included in determining the effective interest rate) arising from:

        (i) financial assets or financial liabilities that are not at fair value through profit or loss; and

        (ii) trust and other fiduciary activities that result in the holding or investing of assets on behalf of individuals, trusts, retirement benefit plans, and other institutions;

    (d) interest income on impaired financial assets accrued in accordance with paragraph AG93 of IAS 39; and

    (e) the amount of any impairment loss for each class of financial asset.

The disclosures above, together with further detailed disclosures in the standard provide the explanation of the position and performance of financial instruments. IFRS 7 also requires disclosure of the risks that financial instruments give to entities.

The requirement for qualitative disclosure about risk is as follows:

33 For each type of risk arising from financial instruments, an entity shall disclose:

    (a) the exposures to risk and how they arise;

    (b) its objectives, policies and processes for managing the risk and the methods used to measure the risk; and

    (c) any changes in (a) or (b) from the previous period.

The requirement for quantitative disclosure is as follows:

34 For each type of risk arising from financial instruments, an entity shall disclose:

(a) summary quantitative data about its exposure to that risk at the end of the reporting period. This disclosure shall be based on the information provided internally to key management personnel of the entity (as defined in IAS 24 *Related Party Disclosures*), for example the entity's board of directors or chief executive officer.

(b) the disclosures required by paragraphs 36–42, to the extent not provided in accordance with (a).

(c) concentrations of risk if not apparent from the disclosures made in accordance with (a) and (b).

EXAMPLE ● Extract from the disclosures given by Findel plc in 2013 compliant with IFRS 7:

FINANCIAL INSTRUMENTS

Capital risk management

The group manages its capital to ensure that the group will be able to continue as a going concern while maximising the return to stakeholders through the optimisation of the net debt and equity balance. The board of directors reviews the capital structure of the group regularly considering both the costs and risks associated with each class of capital. The capital structure of the group consists of:

	2013 £000	2012 £000 (Restated)
Net debt		
Borrowings (note 21)	259,176	263,758
Cash at bank and in hand (note 19)	(27,965)	(33,099)
Cash classified as assets held for sale (note 7)	(6,058)	
	225,153	230,659
Equity		
Share capital (note 26)	125,942	125,942
Capital reserves (note 27)	93,857	93,857
Translation reserve (note 32)	756	606
Hedging reserve (note 29)	(89)	(215)
Accumulated losses (note 30)	(119,995)	(116,053)
	100,471	104,137
Gearing (being net debt divided by equity above)	2.24	2.21

Significant accounting policies

Details of the significant accounting policies and methods adopted, including the criteria for recognition, the basis of measurement and the basis on which income and expenses are recognised, in respect of each class of financial asset, financial liability and equity instrument are disclosed in note 1 to the financial statements.

Financial risk management objectives

The group's financial risks include market risk (including currency risk and interest risk), credit risk, liquidity risk and cash flow interest rate risk. The group seeks to minimise the effects of these risks by using derivative financial instruments to manage its exposure. The use of financial derivatives is governed by the group's policies approved by the board of directors. The group does not enter into or trade financial instruments, including derivative financial instruments, for speculative purposes.

Market risk

The group's activities expose it primarily to the financial risks of changes in foreign currency exchange rates and interest rates. The group enters into a variety of derivative financial instruments to manage its exposure to interest rate and foreign currency risk, including:

- forward foreign exchange contracts to hedge the exchange rate risk arising on the purchase of inventory in US dollars; and
- interest rate swaps to mitigate the risk of rising interest rates.

Foreign currency risk management

The group undertakes certain transactions denominated in foreign currencies. Hence, exposures to exchange rate fluctuations arise. Exchange rate exposures are managed utilising forward foreign exchange contracts.

Foreign currency sensitivity analysis

A significant proportion of products sold through the group's Home Shopping and Educational Supplies divisions are procured through the group's Far East buying office. The currency of purchase for these goods is principally the US dollar, with a proportion being in Hong Kong dollars . . . details the group's sensitivity to a 10% increase and decrease in the Sterling against the relevant foreign currencies. 10% represents management's assessment of the reasonably possible change in foreign exchange rates.

Interest rate risk management

The group is exposed to interest rate risk as the group borrows funds at floating interest rates. The risk is managed by the group by the use of interest rate swap contracts and cap contracts when considered necessary. Hedging activities are evaluated regularly to align with interest rate views and defined risk appetite, ensuring hedging strategies are applied by either positioning the balance sheet or protecting interest expense through different interest rate cycles.

Credit risk management

Credit risk refers to the risk that a counterparty will default on its contractual obligations resulting in financial loss to the group. The group's credit risk is primarily attributable to its trade receivables. The amounts presented in the balance sheet are net of allowances for doubtful receivables. An allowance for impairment is made when there is an identified loss event which, based on previous experience, is evidence of a reduction in the recoverability of the cash flows.

Liquidity risk management

Ultimate responsibility for liquidity risk management rests with the board of directors, which has built an appropriate liquidity risk management framework for the management of the group's short-, medium- and long-term funding and liquidity management requirements. The group manages liquidity risk by maintaining adequate reserves, banking facilities and reserve borrowing facilities by continuously monitoring forecast and actual cash flows and matching the maturity profiles of financial assets and liabilities. Included in note 25 is a description of additional undrawn facilities that the group has at its disposal to further reduce liquidity risk.

The group has access to financing and securitisation facilities, the total unused amount of which is £15,566,000 (2012 35,934,000) at the balance sheet date. The group expects to meet its other obligations from operating cash flows and proceeds of maturing financial assets.

Fair value of financial instruments

The directors consider that the carrying amounts of financial assets and financial liabilities recorded at amortised cost in the financial statements approximate their fair value. The group is required to analyse financial instruments that are measured subsequent to initial recognition at fair value, grouped into levels 1 to 3 based on the degree to which the fair value is observable.

- Level 1 fair value measurements are those derived from quoted prices (unadjusted) in active markets for identical assets or liabilities;
- Level 2 fair value measurements are those derived from inputs other than quoted prices included within level 1 that are observable for the asset or liability, either directly (i.e. as prices) or indirectly (i.e. derived from prices); and
- Level 3 fair value measurements are those derived from valuation techniques that include inputs for the asset or liability that are not based on observable market data (unobservable inputs).

The above financial assets and liabilities were measured at fair value on level 2 fair value measurement bases.

[These are extracts from the disclosures; full disclosures can be seen in Findel plc 2013 Annual Report.]

14.6 Financial instruments developments

As a result of the 2008 financial crisis and the subsequent criticism of the accounting standards on financial instruments, the IASB committed to revising IAS 39 and replacing it with a simpler standard that was easier to apply. In order to be able to progress this project quickly, the IASB split the project into a number of areas and IFRS 9 *Financial Instruments* is the outcome of the first part of the project. The areas to be considered are:

(i)   recognition and measurement (IFRS 9);

(ii) impairment and the effective yield model;

(iii) hedge accounting;

(iv) derecognition of financial assets and liabilities (no longer an active project);

(v)   financial liability measurement.

In 2014 the IASB issued the final version of IFRS 9 including guidance on recognition and measurement of financial assets and financial liabilities, impairment and hedge accounting. After considering the requirements for derecognition in some detail the IASB opted not to change the requirements in that area. Therefore the derecognition rules from IAS 39 have been included within IFRS 9. Impairment has proved a complex and difficult area on which to get agreement with users and preparers.

The IASB has also amended the proposed effective date a number of times. As at early 2014, IFRS 9 is expected to be mandatory from accounting periods beginning on or after

1 January 2018, but earlier adoption is permitted. As yet (2014) the European Union has not endorsed IFRS 9 for use in Europe and they have indicated that they will not consider endorsement until the full replacement of IAS 39 has been issued.

14.6.1 IFRS 9 – the replacement of IAS 39

Over the period 2009 to 2014 a number of sections of IFRS 9 have been issued. These sections have to date covered classification and measurement, dercognition, measurement of financial liabilities and hedge accounting. The final version issued in 2014 dealt with impairment and hedge accounting.

IFRS 9 attempts to improve the rules on financial instruments, making them more principles-based and easier to apply. For example, IAS 39 has four different potential classifications of financial assets (held-to-maturity, loans and receivables, available-for-sale and fair value through profit or loss), each with its own measurement requirements. These classifications can be difficult to apply and also can give inconsistencies between entities and between the accounting and the commercial intentions of some instruments (highlighted in the changes made to IAS 39 to allow reclassification in 2008). IFRS 9 simplifies these categories and is clearer in how to determine which instruments are recognised in each category.

Classification and measurement of financial assets and financial liabilities

IFRS 9 has only two measurement bases for financial assets, fair value and amortised cost. It also only allows gains and losses on equity instruments to be presented in other comprehensive income; fair value gains and losses on other instruments are recognised in profit or loss. The diagram in Figure 14.6 summarises the classification approach to financial assets.

The two key factors in determining the accounting treatment are the business model adopted by an entity for the instrument and the nature of the cash flows. The alternative

Figure 14.6 The classification approach of IFRS 9

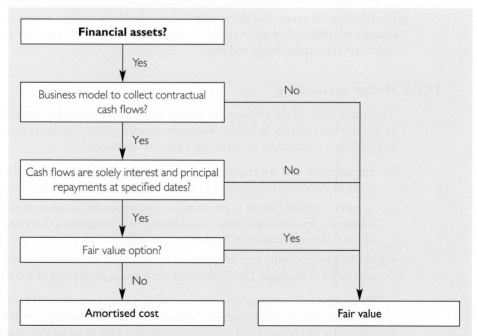

business models could be to collect principal and interest on financial assets, or alternatively to trade the instruments to exploit changes in market price. The contractual cash flows requirement ensures that an instrument held at amortised cost only exhibits the basic loan features of repayment of interest and capital. IFRS 9 does, however, retain the fair value option in IAS 39 although it is not expected to be as significant a choice, as the first two criteria will generally determine the treatment. Reclassification between the categories is acceptable only if an entity changes its business model, and only applies prospectively.

For financial liabilities there are also two measurement models, amortised cost and fair value. Measurement of financial liabilities at fair value under IFRS 9 is unusual although the standard has retained the fair value option in IAS 39 to allow this to occur.

Presentation of gains and losses

Once the measurement of financial assets at fair value or amortised cost is determined, the standard gives a choice of the presentation of fair value gains and losses only for equity instruments. Any debt instruments or derivatives are measured at fair value with gains and losses in profit or loss. However, for equity instruments which are not trading instruments there is a choice for entities to present the gains and losses from movements in fair value in other comprehensive income. This choice is irrevocable and therefore subsequent reclassification is not appropriate.

For financial liabilities the presentation of gains and losses can be more complex. A major problem with allowing financial liabilities to be recognised at fair value with gains and losses in profit and loss is that it can give potentially misleading results. One of the main drivers of changes in fair value of financial liabilities is the issuing entity's own credit risk. For example, if an entity performs poorly and, as a result, its credit rating falls the fair value of its financial liabilities will reduce. This is because the entity is perceived as less able to pay its own debts. If the liabilities are being measured at fair value with gains and losses in profit and loss this will result in gains being recognised. In the most extreme case an entity in significant financial distress could be very profitable.

To respond to this concern the IASB requires, in IFRS 9, that if financial liabilities are fair valued to the extent that the valuation change is due to changes in own credit risk, those changes are reflected in other comprehensive income. Only changes due to other market factors are reflected in profit and loss.

14.6.2 Hedge accounting

The lastest phase of the replacement of IAS 39 to be completed was on hedge accounting. In this area the guidance in IFRS 9 was finally issued in 2014. The hedge accounting models in IAS 39 had come under criticism for a number of reasons:

(i) entities claim that the requirements are complex and difficult to apply. For example, it can be difficult to distinguish a cash flow hedge from a fair value hedge;

(ii) investors regularly resort to pro forma or underlying information produced by entities outside the financial statements to understand risk management (for example, in the case of Rolls-Royce discussed in Section 14.2.1); and

(iii) the hedge accounting requirements can be incompatible with the commercial practices adopted. For example, IAS 39 does not allow hedge accounting of a net position.

The new guidance attempts to address some of the key criticisms of IAS 39. In particular it proposes to follow an approach that attempts to align accounting with risk management activities. The aim is that if risk management activities are to hedge this should be reflected in the financial statements, thereby making the financial statements more meaningful to investors. In addition the IASB is proposing to address specific criticisms, for example by allowing net positions to be designated as hedged items.

In particular there have been changes to the conditions required for hedge accounting. The new criteria are as follows:

A hedging relationship qualifies for hedge accounting only if all of the following criteria are met:

(a) the hedging relationship consists only of eligible hedging instruments and eligible hedged items.

(b) at the inception of the hedging relationship there is formal designation and documentation of the hedging relationship and the entity's risk management objective and strategy for undertaking the hedge. That documentation shall include identification of the hedging instrument, the hedged item, the nature of the risk being hedged and how the entity will assess whether the hedging relationship meets the hedge effectiveness requirements (including its analysis of the sources of hedge ineffectiveness and how it determines the *hedge ratio*).

(c) the hedging relationship meets all of the following hedge effectiveness requirements:

   (i)   there is an economic relationship between the hedged item and the hedging instrument (see paragraphs B6.4.4–B6.4.6);

   (ii) the effect of credit risk does not dominate the value changes that result from that economic relationship (see paragraphs B6.4.7–B6.4.8); and

   (iii) the *hedge ratio* of the hedging relationship is the same as that resulting from the quantity of the hedged item that the entity actually hedges and the quantity of the hedging instrument that the entity actually uses to hedge that quantity of hedged item. However, that designation shall not reflect an imbalance between the weightings of the hedged item and the hedging instrument that would create hedge ineffectiveness (irrespective of whether recognised or not) that could result in an accounting outcome that would be inconsistent with the purpose of hedge accounting (see paragraphs B6.4.9–B6.4.11).

Assuming that hedging instruments and hedged items meet the criteria the mechanics of hedge accounting are generally consistent with IAS 39. There are still three types of hedge relationship: fair value hedges, cash flow hedges and hedges of net investments. The accounting entries required for each type of hedge have largely been retained from IAS 39. However, for a fair value hedge if the hedged item is an equity investment with gains and losses recognised in other comprehensive income, the fair value movements on the hedging instrument are also recognised in other comprehensive income.

14.6.3 Impairment of financial assets

The issues surrounding impairment have proved difficult for the IASB and they have faced significant pressure to change the current impairment models in IAS 39, in particular for instruments measured at amortised cost. To the date of writing this text the IASB has issued two exposure drafts on amortised cost and impairment but final guidance has not been issued. A new standard is expected in 2014. Below we discuss the major concern that the IASB has been asked to address. A revised exposure draft, ED/2013/3 *Financial Instruments: Expected Credit Losses* was issued in early 2013.

Incurred versus expected losses

The debate on impairment largely revolves around whether financial asset impairment should be calculated following an incurred or expected loss model. IAS 39 uses an incurred loss model; however, in the 2008 financial crisis it was suggested that this model delayed the recognition of losses on loans, resulting in misleading results for financial institutions. The key difference between the two approaches is that an incurred loss model provides for impairments only when an event has occurred which causes that impairment. An expected loss model provides for impairment if there is reason to expect that it will arise at some point over the life of the loan (even if it has not arisen at the balance sheet date). For example, if a bank makes a loan to a customer and the customer becomes unemployed and therefore defaults on the loan, under the incurred loss model an impairment would only be recognised when the customer loses their job. Under the expected loss model, the bank would have made an estimate of the likelihood of the customer losing their job from the inception of the loan and provide based on that probability. The provision on the expected loss model is therefore recognised earlier but it does depend much more on the estimation and judgement of management of a company.

ED/2009/12 was issued to address impairment and the way that the amortised cost method of accounting is applied. The ED proposed an expected loss model by proposing changes in the way that the amortised cost model is applied. The amortised cost model determines an effective interest rate by determining the rate at which the initial loan and the cash flows over its life are discounted to zero, effectively the internal rate of return on the loan. IAS 39 requires the rate to be determined on cash flows before future credit losses, whereas the exposure draft requires the calculation on cash flows including expected future credit losses. In every period the expected cash flows would need to be adjusted and discounted back at the original effective rate; any difference in the loan value is then adjusted against profit or loss. The impact of this new approach is that losses would tend to be recognised earlier and no separate impairment model is required; if impairment is expected, the cash flow estimates will automatically adjust for that. It is still to be seen how straightforward the approach will be in practice and whether financial institutions can adapt their systems and processes easily to the revised approach.

A further ED was issued in 2013 as many financial institutions had expressed concerns about the proposed model. The revised approach retains the expected loss model but uses a calculation approach that varies with changes in the credit position since the origination or acquisition of the financial asset.

Summary

This chapter has given some insight into the difficulties and complexities of accounting for financial instruments and the ongoing debate on this topic, highlighted by the financial crisis that began in 2008. The approach of the IASB is to adhere to the principles contained in the *Framework* but to also issue guidance that is robust enough to prevent manipulation and abuse. Whether the IASB has achieved this is open to debate. Some might view the detailed requirements of the standards, particularly IAS 39, to be so onerous that companies will not be able to show their real intentions in the financial statements. This is particularly true, for instance, with the detailed criteria on hedging. These criteria have led to many businesses not hedge accounting even though they are hedging commercially to manage their risks. The hedge accounting criteria do not fit with the way they run or manage their risk profiles. This may change with future developments currently under discussion by the IASB.

The IASB is coming towards the end of their work in replacing IAS 39 with a revised standard, IFRS 9. This revised standard removes some of the complexity in IAS 39 and should make it easier for entities to reflect their true intentions for using financial instruments. However, the revised standard is still likely to be complex, in particular the new requirements for impairment recognition may be difficult and judgemental to apply.

As can be seen, there is much to criticise in these requirements, but it should be borne in mind that the IASB has grasped this issue better than many other standard setters. Financial instruments may be complex and subject to debate but guidance is required in this area, and the IASB has given guidance where many others have not.

In addition to giving an insight into the development of standards, our aim has been that you should be able to calculate the debt/equity split on compound instruments and the finance cost on liability instruments and classify and account for the four categories of financial instrument.

REVIEW QUESTIONS

1 Explain what is meant by the term 'split accounting' when applied to convertible debt or convertible preference shares and the rationale for splitting.

2 Discuss the implications for a business if a substance approach is used for the reporting of convertible loans.

3 Explain how a gain or loss on a forward contract is dealt with in the accounts if the contract is entered into before the period-end but does not close out until post year-end.

4 Explain how redeemable preference shares, perpetual debt, loans and equity investments are reported in the financial statements.

5 The authors[8] contend that the use of current valuations can present an inaccurate view of a firm's true financial status. When assets are illiquid, current value represents only a guess. When assets participate in an economic 'bubble', current value is invariably unsustainable. Accounting standards, the authors conclude, should be flexible enough to fairly assess value in these circumstances. Discuss the alternatives that standard setters could permit in order to fairly assess values in an illiquid market.

6 Explain the difference between a cash flow hedge and a fair value hedge. Does the nature of the hedging instrument (e.g. forward contract, interest rate swap, option) influence the hedging model being used?

7 'Disclosure of the estimated fair values of financial instruments is better than adjusting the values in the financial statements with the resulting volatility that affects earnings and gearing ratios.' Discuss.

8 Companies were permitted in 2008 to reclassify financial instruments that were initially designated as at fair value through profit. Critically discuss the reasons for the standard setters changing the existing standard.

9 'If financial liabilities can be recognised at fair value all gains and losses arising should be reflected in profit or loss.' Discuss.

10 'The only true way to simplify IAS 39 would be for all financial assets and liabilities to be measured at fair value with gains and losses recognised in profit or loss.' Discuss.

EXERCISES

* Question 1

On 1 April year 1, a deep discount bond was issued by DDB AG. It had a face value of £2.5 million and covered a five-year term. The lenders were granted a discount of 5%. The coupon rate was 10% on the principal sum of £2.5 million, payable annually in arrears. The principal sum was repayable in cash on 31 March Year 5. Issuing costs amounted to £150,000.

Required:
Compute the finance charge per annum and the carrying value of the loan to be reported in each year's profit or loss and statement of financial position respectively.

* Question 2

Fairclough plc borrowed €10 million from a bank on 1 January 2011. Fees of €100,000 were charged by the bank which were paid by Fairclough plc at inception of the loan. The terms of the loan are:

Interest

● Interest of 6% until 31 December 2013

● Interest dropping to 5% from 31 December 2013 to 31 December 2015

Repayment schedule

● Repayment of €5 million on 31 December 2013

● Repayment of €5 million on 31 December 2015

Interest is paid annually in arrears.

The effective yield on the loan is 6.07%.

Required:
(a) What is the total finance cost on the loan over the five-year period?
(b) What will be reflected as a liability in the financial statements for each 31 December year-end and what interest costs will be recognised in the statement of comprehensive income?

* Question 3

Isabelle Limited borrows £100,000 from a bank on the following terms:

(i) arrangement fees of £2,000 are charged by the bank and deducted from the initial proceeds on the loan;

(ii) interest is payable at 5% for the first three years of the loan and then increases to 7% for the remaining two years of the loan;

(iii) the full balance of £100,000 is repaid at the end of year 5.

Required:
(a) What interest should be recognised in the statement of comprehensive income for each year of the loan?
(b) If Isabelle Limited repaid the loan after three years for £100,000, what gain or loss would be recognised in the statement of comprehensive income?

* Question 4

On 1 January 2009 Henry Ltd issued a convertible debenture for €200 million carrying a coupon interest rate of 5%. The debenture is convertible at the option of the holders into 10 ordinary shares for each €100 of debenture stock on 31 December 2013. Henry Ltd considered borrowing the €200 million through a conventional debenture that repaid in cash; however, the interest rate that could be obtained was estimated at 7%, therefore Henry Ltd decided on the issue of the convertible.

Required:
Show how the convertible bond issue will be recognised on 1 January 2009 and determine the interest charges that are expected in the statement of comprehensive income over the life of the convertible bond.

Question 5

On 1 October year 1, RPS plc issued one million £1 5% redeemable preference shares. The shares were issued at a discount of £50,000 and are due to be redeemed on 30 September Year 5. Dividends are paid on 30 September each year.

Required:
Show the accounting treatment of the preference shares throughout the lifespan of the instrument calculating the finance cost to be charged to profit or loss in each period.

* Question 6

Milner Ltd issues a 6% cumulative preference share for €1 million that is repayable in cash at par 10 years after issue. The only condition on the dividends is that if the directors declare an ordinary dividend the preference dividend (and any arrears of preference dividend) must be paid first. Arrears of dividend do not need to be paid on redemption of the instrument.

Required:
Explain how this preference share should be accounted for over its life.

Question 7

Creasy plc needs to raise €20 million and is considering two different instruments that could be issued:

(i) A 7% debenture with a par value of €20 million, repayable at par in five years. Interest is paid annually in arrears.

(ii) A 5% convertible debenture with a par value of €20 million, repayable at par in five years or convertible into 5 million €1 shares. Interest is paid annually in arrears.

Required:
Comment on the effect on the statement of comprehensive income and the statement of financial position of issuing these different instruments.

* Question 8

On 1 October 20X1, Little Raven plc issued 50,000 debentures, with a par value of £100 each, to investors at £80 each. The debentures are redeemable at par on 30 September 20X6 and have a coupon rate of 6%, which was significantly below the market rate of interest for such debentures issued at par. In accounting for these debentures to date, Little Raven plc has simply accounted for the cash flows involved, namely:

- On issue: debenture 'liability' included in the statement of financial position at £4,000,000.

- Statements of comprehensive income: interest charged in years ended 30 September 20X2, 20X3 and 20X4 (published accounts) and 30 September 20X5 (draft accounts) – £300,000 each year (being 6% of £5,000,000).

The new finance director, who sees the likelihood that further similar debenture issues will be made, considers that the accounting policy adopted to date is not appropriate. He has asked you to suggest a more appropriate treatment.

Little Raven plc intends to acquire subsidiaries in 20X6.

Statements of comprehensive income for the years ended 30 September 20X4 and 20X5 are as follows:

	Y/e 30 Sept 20X5	Y/e 30 Sept 20X4
	(Draft)	(Actual)
	£000	£000
Turnover	6,700	6,300
Cost of sales	(3,025)	(2,900)
Gross profit	3,675	3,400
Overheads	(600)	(550)
Interest payable – debenture	(300)	(300)
– others	(75)	(50)
Profit for the financial year	2,700	2,500
Retained earnings brought forward	4,300	1,800
Retained earnings carried forward	7,000	4,300

Extracts from the statement of financial position are:

	At 30 Sept 20X5	At 30 Sept 20X4
	(Draft)	(Actual)
	£000	£000
Share capital	2,250	2,250
Share premium	550	550
Retained earnings	7,000	4,300
	9,800	7,100
6% debentures	4,000	4,000
	13,800	11,100

Required:

(a) Outline the considerations involved in deciding how to account for the issue, the interest cost and the carrying value in respect of debenture issues such as that made by Little Raven plc. Consider the alternative treatments in respect of the statement of comprehensive income and refer briefly to the appropriate statement of financial position disclosures for the debentures. Conclude in terms of the requirements of IAS 32 (on accounting for financial instruments) in this regard.

(b) Detail an alternative set of entries in the books of Little Raven plc for the issue of the debentures and subsequently; under this alternative the discount on the issue should be dealt with under the requirements of IAS 32. The constant rate of interest for the allocation of interest cost is given to you as 11.476%. Draw up a revised statement of comprehensive income for the year ended 30 September 20X5, together with comparatives, taking account of the alternative accounting treatment.

* Question 9

George plc adopted IFRS for the first time on 1 January 2008 and has three different instruments whose accounting George is concerned will change as a result of the adoption of the standard. The three instruments are:

1 An investment in 15% of the ordinary shares of Joshua Ltd, a private company. This investment cost €50,000, but had a fair value of €60,000 on 1 January 2008, €70,000 on 31 December 2008 and €65,000 on 31 December 2009.

2 An investment of €40,000 in 6% debentures. The debentures were acquired at their face value of €40,000 on 1 July 2007 and pay interest half-yearly in arrears on 31 December and 30 June each year. The bonds had a fair value of €41,000 at 1 January 2008, €43,000 at 31 December 2008 and €38,000 at 31 December 2009.

3 An interest rate swap taken out to swap floating-rate interest on an outstanding loan to fixed-rate interest. Since taking out the swap the loan has been repaid; however, George plc decided to retain the swap as it was 'in the money' at 1 January 2008. The fair value of the swap was a €10,000 asset on 1 January 2008; however, it became a liability of €5,000 by 31 December 2008 and the liability increased to €20,000 by 31 December 2009. In 2008 George paid €1,000 to the counterparty to the swap and in 2009 paid €5,000 to the counterparty.

Required:

Show the amount that would be recognised for all three instruments in the statement of financial position, in profit and loss and in other comprehensive income on the following assumptions:

(i) Equity and debt investments are available for sale.

(ii) Where possible, investments are treated as held to maturity.

(iii) Where equity investments are treated as fair value through profit and loss, and debt investments are treated as loans and receivables.

Question 10

On 1 January 2009 Hazell plc borrows €5 million on terms with interest of 3% fixed for the period to 31 December 2009, going to variable rate thereafter (at inception the variable rate is 6%). The loan is repayable at Hazell plc's option between 31 December 2011 and 31 December 2013.

Initially Hazell plc estimates that the loan will be repaid on 31 December 2011; however, at 31 December 2010 Hazell plc revises this estimate and assumes the loan will only be repaid on 31 December 2013.

Assume that the variable rate remains at 6% throughout the period and that interest is paid annually in arrears.

Required:
(i) Determine the total expected finance costs and effective yield on the loan at 1 January 2009.
(ii) Show the impact of the loan on the statement of comprehensive income and statement of financial position for periods ended 31 December 2009 and 31 December 2010.

Question 11

Baudvin Ltd has an equity investment that cost €1 million on 1 January 2008. The investment is classified as an available-for-sale investment. The value of the investment at each period-end is:

31 December 2008	€950,000
31 December 2009	€1,030,000
31 December 2010	€1,080,000

Baudvin Ltd sold the investment for €1,100,000 on 31 March 2011.

Required:
Show what should be reflected for the investment in Baudvin Ltd's financial statements for each period from 31 December 2008 to 2011.

* Question 12

A company borrows on a floating-rate loan, but wishes to hedge against interest variations so swaps the interest for fixed rate. The swap should be perfectly effective and has zero fair value at inception. Interest rates increase and therefore the swap becomes a financial asset to the company at fair value of £5 million.

Required:
Describe the impact on the financial statements for the following situations:
(a) The swap is accounted for under IAS 39, but is not designated as a hedge.
(b) The swap is accounted for under IAS 39, and is designated as a hedge.

Question 13

Charles plc is applying IAS 32 and IAS 39 for the first time this year and is uncertain about the application of the standard. Charles plc's balance sheet is as follows:

	£000	Financial asset/liability	IAS 32/39?	Category	Measurement
Non-current assets					
Goodwill	2,000				
Intangible	3,000				
Tangible	6,000				
Investments					
Corporate bond	1,500				
Equity trade investments	900				
	13,400				
Current assets					
Inventory	800				
Receivables	700				
Prepayments	300				
Forward contracts (note 1)	250				
Equity investments held for future sale	1,200				
	3,250				
Current liabilities					
Trade creditors	(3,500)				
Lease creditor	(800)				
Income tax	(1,000)				
Forward contracts (note 1)	(500)				
	(5,800)				
Non-current liabilities					
Bank loan	(5,000)				
Convertible debt	(1,800)				
Deferred tax	(500)				
Pension liability	(900)				
	(8,200)				
Net assets	2,650				

Note

The forward contracts have been revalued to fair value in the balance sheet. They do not qualify as hedging instruments.

Required:

Complete the above balance sheet and consider under IAS 39:

(a) Which items on the balance sheet are financial assets/liabilities?

(b) Are the balances within the scope of IAS 39?

(c) How they should be classified under IAS 39:

HTM	Held-to-maturity
LR	Loans and receivables
FVPL	Fair value through profit and loss
AFS	Available-for-sale
FL	Financial liabilities?

(d) How they should be measured under IAS 39:

FV   Fair value
C   Amortised cost?

Assume that the company includes items in 'fair value through profit and loss' only when required to do so, and also chooses where possible to include items in 'loans and receivables'.

* Question 14

Tan plc owns an available-for-sale equity investment that cost £2 million. At 31 December 2008 the investment was recognised on the balance sheet at £2.1 million.

At 31 December 2009 the investment had declined in value to £1.5 million; however, Tan plc did not assume that the investment had impaired. By 31 December 2010 the investment had not recovered in value (it was still worth £1.5 million) and Tan plc concluded that its decline in value was permanent and it had impaired.

In June 2011 (after the 2010 financial statements had been issued) circumstances changed and the investment recovered in value to £2.2 million.

Required:
Show the accounting entries you would expect Tan plc to make for each December year-end from 2009 to 2011.

Question 15

At the start of the year Cornish plc entered into a number of financial instruments and is considering how to classify these instruments under IAS 39. The instruments are as follows:

(a) Investment in listed 3% government bonds for €2 million. Cornish acquired the bonds when they were issued at their nominal value of €2 million. By the year-end, 31 December 2010, interest rates had fallen and the bonds had a market value of €2,025,000.

(b) Investment in shares in Schaenzler plc, a listed company, for €1,300,000. At 31 December 2010 the investment had fallen in value and was estimated to be worth only €1,200,000.

(c) Cornish plc borrowed €5 million at floating rate in the year and to hedge the interest rate took out an interest rate swap (floating to fixed) on the loan. The swap cost nothing to enter into but by 31 December 2010 because interest rates had fallen it had a fair value (liability) of €50,000. Cornish plc does not use hedge accounting.

Required:
Discuss how the investments could be classified and measured under IAS 39 and the implications of the choices for the financial statements of Cornish plc.

16 Measurement of financial assets and liabilities

Procter Limited, a UK private company has the following financial assets and liabilities in the accounts:

(i)   An equity investment in Milner plc, a UK listed company. Procter recognises the investment as an 'available for sale' investment under IAS 39 *Financial Instruments; Recognition and Measurement*.

(ii)   An investment in government bonds that are classified as 'held to maturity'. The bonds only pay interest and principal and Procter Ltd expects to hold the bonds until maturity.

(iii) A financial liability that Procter Ltd unusually measures at fair value through profit and loss. The liability is measured at fair value because Procter has a documented management policy to manage the liability at fair value.

Required

Discuss how the measurement of the above instruments may change under IFRS 9 *Financial Instruments.*

17 Impairment of Financial Assets

The approach in IAS 39 to the impairment of financial assets was flawed because it did not allow financial institutions to recognise the true losses they expected on loans at the time they had made the loans. How does the final version of IFRS 9 address this?

Discuss.

References

1 IAS 32 *Financial Instruments: Disclosure and Presentation*, IASC, revised 1998.
2 IAS 32 *Financial Instruments: Presentation*, IASC, 2005, para. 4.
3 Ibid., para. 11.
4 Ibid., para. 16.
5 Ibid., para. 28.
6 Ibid., para. 42.
7 IFRS 7 *Financial Instruments: Disclosure*, IASB, 2005.
8 S. Fearnley and S. Sunder, 'Bring back prudence', *Accountancy*, vol. 140 (1370), 2007, pp. 76–7.

CHAPTER 15

Employee benefits

15.1 Introduction

In this chapter we consider the application of IAS 19 *Employee Benefits*.[1] IAS 19 is concerned with the determination of the cost of retirement benefits in the financial statements of **employers** having retirement benefit plans (sometimes referred to as 'pension schemes', 'superannuation schemes' or 'retirement benefit schemes'). The requirements of IFRS 2 *Share-based Payment* will also be considered here. Even though IFRS 2 covers share-based payments for almost any good or service a company can receive, in practice it is employee service that is most commonly rewarded with share-based payments. We also consider the disclosure requirements of IAS 26 *Accounting and Reporting by Retirement Benefit Plans*.[2]

Objectives

By the end of this chapter, you should be able to:

- critically comment on the approaches to pension accounting used under international accounting standards;
- understand the nature of different types of pension plan and account for the different types of pension plan that companies may have;
- explain the accounting treatment for other long-term and short-term employee benefit costs;
- understand and account for share-based payments that are made by companies to their employees;
- outline the required approach of pension schemes to presenting their financial position and performance.

15.2 Greater employee interest in pensions

The percentages of pensioners and public pension expenditure are increasing.

| | % of population over 60 | | Public pensions as % of GDP |
| | 2000 | 2040 | 2040 |
	%	% (projected)	% (projected)
Germany	24	33	18
Italy	24	37	21
Japan	23	34	15
UK	21	30	5
US	17	29	7

This has led to gloomy projections that countries could even be bankrupted by the increasing demand for state pensions. In an attempt to avert what governments see as a national disaster, there have been increasing efforts to encourage private funding of pensions.

As people become more and more aware of the possible failure of governments to provide adequate basic state pensions, they recognise the advisability of making their own provision for their old age. This has raised their expectation that their employers should offer a pension scheme and other post-retirement benefits. These have increased, particularly in Ireland, the UK and the USA, and what used to be a 'fringe benefit' for only certain categories of staff has been broadened across the workforce. This growth in private pension schemes has been encouraged by various governments with favourable tax treatment of both employers' and employees' contributions to pension schemes, and requirements for companies to contribute to pension funds of employees. In the UK, for example, the government now requires employers to make contributions into private pension plans for employees through a process of auto-enrolment.

15.3 Financial reporting implications

The provision of pensions for employees as part of an overall remuneration package has led to the related costs being a material part of the accounts. The very nature of such arrangements means that the commitment is a long-term one that may well involve estimates. The way the related costs are allocated between accounting periods and are reported in the financial statements needs careful consideration to ensure that a fair view of the position is shown.

Over time there has been a shift of view on the way that pension costs should be accounted for, and the principle on which the accounting is based. The first accounting standards on pensions (for example IAS 19 prior to its revision in 1998) required that pension costs should be matched against the period of the employee's service so as to create an even charge for pensions against profit. However, this approach could result in the statement of financial position being misleading. The more recent approach is to make the statement of financial position more sensible, but perhaps accept greater variation in the pension cost in the statement of comprehensive income. The new view is the one endorsed by the current IAS 19 (issued in June 2011) and was the method used in the previous versions of IAS 19 issued since 1998.

Before examining the detail of how IAS 19 (2011) requires pensions and other long-term benefits to be accounted for, we need to consider the types of pension scheme that are commonly used.

15.4 Types of scheme

15.4.1 *Ex gratia* arrangements

These are not schemes at all but are circumstances where an employer agrees to grant a pension to be paid for out of the resources of the firm. Consequently these are arrangements where pensions have not been funded but decisions are made on an *ad hoc* or case-by-case basis, sometimes arising out of custom or practice. No contractual obligation to grant or pay a pension exists, although a constructive obligation may exist which would need to be provided for in accordance with IAS 37 *Provisions, Contingent Liabilities and Contingent Assets*.

15.5 Defined contribution pension schemes

Defined contribution schemes (otherwise known as money purchase schemes) have not presented any major accounting problems. The cost of providing the pension, usually a percentage of salary, is recorded as a remuneration expense in the statement of comprehensive income in the period in which it is due. Assets or liabilities may exist for the pension contributions if the company has not paid the amount due for the period. If a contribution was payable more than 12 months after the reporting date for services rendered in the current period, the liability should be recorded at its discounted amount (using a discount rate based on the market rate for high-quality corporate bonds).

Disclosure is required of the pension contribution charged to the statement of comprehensive income for the period.

Illustration of Andrew plc defined contribution pension scheme costs

Andrew plc has payroll costs of £2.7 million for the year ended 30 June 2013. Andrew plc pays pension contributions of 5% of salary, but for convenience paid £10,000 per month standard contribution with any shortfall to be made up in the July 2013 contribution.

Statement of comprehensive income charge
The pension cost is £2,700,000 × 5% = £135,000.

Statement of financial position
The amount paid over the period is £120,000 and therefore an accrual of £15,000 will be made in the statement of financial position at 30 June 2013.

15.6 Defined benefit pension schemes

15.6.1 The fundamental accounting issue

A problem that arises in accounting for defined benefit schemes that does not arise when accounting for defined contribution schemes is the much greater uncertainty of required contributions to meet the actual benefit payable. There is no guarantee, even after the contributing company has contributed the expected required amount, that the assets of the scheme will be sufficient to settle the likely future liabilities. Therefore the contributing company may have to provide additional contributions to finance a shortfall and these need to be provided for.

15.6.2 Efforts to arrive at a solution

In order to assess the likely level of exposure for contributing companies, it is necessary to look at the work of the actuaries – specialists who advise on the funding of the scheme and its overall financial position. The actuaries will assess the financial position of the scheme on a regular basis and identify whether the liabilities of the scheme (to pay future benefits already earned out of past service) are covered by the market value of the assets into which contributions are invested until required to pay benefits. These actuarial assessments resulted in the identification of deficits or (occasionally) surpluses for such schemes.

Prior to 1998 the international accounting standard in this area focus on the charge to profit and loss for retirement benefits. The charge was typically split into two elements:

- The regular cost – a long-term estimate largely consistent with the contributions made.
- Variations from the regular cost – recognising the actuarial deficit or surplus over the period to the next actuarial assessment.

Differences between the charge to profit and loss and the contributions paid in the period were recognised as liabilities or assets on the statement of financial position. This approach led to a number of problems. Two key problems were:

- The figure in the statement of financial position was difficult to explain.
- The charge to profit and loss depended, *inter alia*, on the frequency of the actuarial assessments.

It was in order to address these issues that a fundamentally different approach was advocated in IAS 19 (revised), first published in 1998 and last revised in 2011.

15.7 IAS 19 (revised 2011) *Employee Benefits*

After a relatively long discussion and exposure period IAS 19 (revised 1998) was issued in 1998 and redefined how all employee benefits were to be accounted for.

IAS 19 has chosen to follow an 'asset or liability' approach to accounting for the pension scheme contributions by the employer and, therefore, it defines how the statement of financial position asset or liability should be built up. The statement of comprehensive income charge is effectively the movement in the asset or liability. The pension fund must be valued sufficiently regularly so that the statement of financial position asset or liability is kept up to date. The valuation would normally be done by a qualified actuary and is based on actuarial assumptions.

In June 2011 the IASB issued an amended IAS 19 which makes changes to the recognition, measurement and presentation of pension liabilities for periods beginning on or after 1 January 2013. The changes did not alter the fundamental principle that pension accounting should reflect a supportable asset or liability on the statement of financial position.

15.8 The asset or liability for pension and other post-retirement costs

The asset or liability for pension costs is made up from the following amounts:

(a) the present value of the defined benefit obligation at the period-end date;

(b) minus the fair value at the period-end date of plan assets (if any) out of which the obligations are to be settled directly.

If this calculation comes out with a negative amount, the company should recognise a pension asset in the statement of financial position. There is a limit on the amount of the asset, known as the 'asset ceiling', if the asset calculated above is greater than the total of:

(i) any unrecognised actuarial losses and past service cost; plus

(ii) the present value of any future refunds from the scheme or reductions in future contributions.

The two elements making up the defined benefit pension asset or liability can now be considered.

15.8.1 Obligations of the fund

The pension fund obligation must be calculated using the 'projected unit credit method'. This method of allocating pension costs builds up the pension liability each year for an extra year of service and a reversal of discounting. Discounting of the liability is done using the market yields on high-quality corporate bonds with similar currency and duration.

The Grado illustration below shows how the obligation to pay pension accumulates over the working life of an employee.

Grado illustration

A lump sum benefit is payable on termination of service and equal to 1% of final salary for each year of service. The salary in year 1 is £10,000 and is assumed to increase at 7% (compound) each year. The discount rate used is 10%. The following table shows how an obligation (in £) builds up for an employee who is expected to leave at the end of year 5. For simplicity, this example ignores the additional adjustment needed to reflect the probability that the employee may leave service at an earlier or a later date.

Year	1	2	3	4	5
Benefit attributed to prior years	0	131	262	393	524
Benefit attributed to current year (1% of final salary)*	131	131	131	131	131
Benefit attributed to current and prior years	131	262	393	524	655
Opening obligation (present value of benefit attributed to prior years)	—	89	196	324	476
Interest at 10%	—	9	20	33	48
Current service cost (present value of benefit attributed to current year)	89	98	108	119	131
Closing obligation (present value of benefit attributed to current and prior years)**	89	196	324	476	655

* Final salary is £10,000 × (1.07)^4 = £13,100.
** Discounting the benefit attributable to current and prior years at 10%.

15.8.2 Fair value of plan assets

This is usually the market value of the assets of the plan (or the estimated value if no immediate market value exists). The plan assets exclude unpaid contributions due from the reporting enterprise to the fund.

15.9 Changes in the pension asset or liability position

The pension asset or liability changes from one period to the next for a number of possible reasons. IAS 19 requires that interest and service costs are recognised in the income statement, whereas remeasurement gains or losses are recognised in other comprehensive income. The possible items to be recognised are as follows:

Income statement recognition

(a) current service cost;

(b) interest cost;

(c) past service cost and the effect of any curtailments or settlements.

Remeasurements

(d) actuarial gains and losses;

(e) the difference in the actual and expected return on plan assets.

Each of the items above is discussed in more detail below.

The items above are all the things that cause the statement of financial position asset or liability for pensions to alter, and the statement of comprehensive income is consequently based on the movement in the liability. Because of the inclusion of actuarial gains and losses and past service costs in comprehensive income, the total comprehensive income is liable to fluctuate much more than the charge made under the original IAS 19.

15.9.1 Current service cost

The current service cost represents the cost of providing pension benefits to employees for the current period. This cost would not be expected to fluctuate significantly from period to period but is influenced by actuarial assumptions.

15.9.2 Interest cost

The interest cost reflects the unwinding of the effect of discounting on the pension asset or liability. The interest cost is calculated by applying the discount rate on high-quality corporate bonds to the pension asset or liability. The effect of this is that if a pension is in an asset position interest income is recognised, and if it is a liability interest cost is recognised.

This area was one of the key changes to IAS 19 (2011) compared to the previous versions. Prior to 2011 an interest cost was calculated on the pension obligations and an expected return was calculated on the pension assets. These were based on different rates. The two amounts were then offset and presented as a net finance cost or income in the income statement. The impact of this approach could be that net interest income was recognised when the pension was in a liability position due to the assumption on the expected return on plan assets being greater than the discount rate on pension obligations.

15.9.3 Past service costs and the effect of curtailments and settlements

Past service costs are costs that arise for a pension scheme as a result of improving the scheme or when a business introduces a plan. They are the extra liability in respect of previous years' service by employees. Note, however, that past service costs can only arise if actuarial assumptions did not take into account the reason why they occurred. Typically they would include:

● estimates of benefit improvements as a result of actuarial gains (if the company proposes to give the gains to the employees);

● the effect of plan amendments that increase or reduce benefits for past service.

A curtailment of a pension scheme occurs when a company is committed to making a material reduction in the number of employees in a scheme or when the employees will receive no benefit for a substantial part of their future service. A settlement occurs when an enterprise enters into a transaction that eliminates any further liability from arising under the fund.

Accounting treatment

Past service costs and the effect of curtailments and settlements should only be recognised after the pension asset or liability has been recognised at up-to-date fair value for the assets and current actuarial assumptions. Following the revaluation the past service cost or gain or loss on settlement or curtailment is recognised at the earlier of the following dates:

(a) when the plan amendment or curtailment occurs; and

(b) when the entity recognises related restructuring costs or termination benefits.

The gain or loss is recognised in profit and loss.

15.9.4 Actuarial gains and losses

Actuarial gains or losses result from either changes in the present value of the defined benefit obligation or changes in the market value of the plan assets. They arise from experience adjustments which are differences between actuarial assumptions and actual experience, or due to the effect of changes in actuarial assumptions. Typical reasons for the gains or losses would be:

● unexpectedly low or high rates of employee turnover;

● the effect of changes in the discount rate;

● changes in life expectancy assumptions.

Accounting treatment

In the version of IAS 19 issued in 2004 there was a significant choice of accounting treatment for actuarial gains and losses. One approach followed a '10% corridor' and required recognition of gains and losses in the profit or loss, whereas other alternatives made no use of the corridor and required gains and losses to be recognised immediately, either in profit and loss or in other comprehensive income.

With the issue of IAS 19 (2011) there was a significant change in the accounting for actuarial gains and losses. The new standard defines them as 'remeasurements' and requires them to be recognised in full in other comprehensive income. The recognition in the income statement and the use of the 10% corridor have been abolished. This change in accounting could impact significantly on the reported pension position of some businesses. For example, British Airways plc had, at 31 December 2012, prior to the new standard taking effect, £2.188bn of unrecognised actuarial losses following the 10% corridor approach. Under the new standard these would be recognised within other comprehensive income. IAS 19 (2011) has transition rules that allow the effect of the change to be spread over a maximum of five years.

15.9.5 Differences in actual and expected return on plan assets

The effect of remeasuring the pension assets is recognised in other comprehensive income together with the actuarial gains and losses. Prior to the issue of IAS 19 (2011) actuaries would estimate the expected return on plan assets and this would be included within net interest income or expense. Any differences between the actual and expected return were included within actuarial gains and losses.

With the introduction of the new standard the overall pension asset or liability generates interest income or expense using a rate on high-quality corporate bonds. To the extent that the actual return on plan assets differs from this the effect is treated as a remeasurement. In effect it is still very similar to an actuarial gain or loss and has the same accounting impact.

15.10 Comprehensive illustration

The following comprehensive illustration demonstrates how a pension liability and statement of comprehensive income statement charge is calculated under IAS 19 (2011). The example does not include the effect of curtailments or settlements.

Illustration

The following information is given about a funded defined benefit plan. To keep the computations simple, all transactions are assumed to occur at the year-end. The present value of the obligation and the market value of the plan assets were both 1,000 at 1 January 20X1.

	20X1	20X2	20X3
Rate on high-quality corporate bonds at start of year	10%	9%	8%
Current service cost	160	140	150
Benefits paid	150	180	190
Contributions paid	90	100	110
Present value of obligations at 31 December	1,100	1,380	1,455
Market value of plan assets at 31 December	1,190	1,372	1,188

In 20X2 the plan was amended to provide additional benefits with effect from 1 January 20X2. The present value as at 1 January 20X2 of additional benefits for employee service before 1 January 20X2 was 50.

Required:
Show how the pension scheme would be shown in the accounts for 20X1, 20X2 and 20X3.

Solution

Step 1 Change in the net pension obligation

	20X1	20X2	20X3
Present value of obligation, 1 January	—	(90)	8
Net interest cost at 10%, 9%, 8%	—	(8)	1
Current service cost	160	140	150
Past service cost	—	50	—
Contributions paid	(90)	(100)	(110)
Actuarial (gain) loss on obligation (balancing figure) *	*(160)*	*16*	*218*
Present value of the obligation (asset), 31 December	(90)	8	267

* In this example we do not distinguish between actuarial gains and losses and differences in the actual and expected return on plan assets.

Step 2 Calculate the impact on the statement of comprehensive income

	20X1	20X2	20X3
Operating costs:			
Current service cost	160	140	150
Past service cost		50	
Net interest cost	—	(8)	1
Profit and loss charge	160	182	151
Other comprehensive income:			
Actuarial gains (losses)	160	(16)	(218)

Step 3 Calculate the statement of financial position

	20X1	20X2	20X3
Present value of pension obligation, 31 December	1,100	1,380	1,455
Fair value of plan assets, 31 December	(1,190)	(1,372)	(1,188)
Liability (asset) recognised	(90)	8	267

15.11 Multi-employer plans

A multi-employer plan is a defined contribution or defined benefit plan that:

(a) pools the assets contributed by various enterprises that are not under common control; and

(b) uses those assets to provide benefits to employees of more than one enterprise, on the basis that contribution and benefit levels are determined without regard to the identity of the enterprise that employs the employees concerned.

An enterprise should account for a multi-employer defined benefit plan as follows:

- it should account for its share of the defined benefit obligation, plan assets and costs associated with the plan in the same way as for any defined benefit plan; or

- if insufficient information is available to use defined benefit accounting, it should:

  - account for the plan as if it were a defined contribution plan; and

  - give extra disclosures.

A group pension plan is not a multi-employer plan and therefore the treatment above is not available. However, a similar exemption exists for group schemes whereby it is not necessary to split the pension liability between the contributing entities. Instead, the contributing individual entities can account for the scheme by recognising only their contributions to the plan. This means that the defined benefit accounting is only necessary in the consolidated accounts and not in the individual company accounts of all companies in the group. It is necessary, however, to recognise the full group defined benefit accounting in at least one entity in the group. The requirements for full defined benefit accounting are included in the individual sponsor company financial statements.

The requirement to reflect the group pension position in the individual sponsor entity can be a concern, particularly for example in the UK, where this treatment for a significant pension liability could impact on the ability of the individual entity to pay dividends.

15.12 Disclosures

The major disclosure requirements of the standard require entities to present information that:

(a) explains the characteristics of its defined benefit plans and risks associated with them;

(b) identifies and explains the amounts in its financial statements rising from its defined benefit plans; and

(c) describes how its defined benefit plans may affect the amount, timing and uncertainty of the entity's future cash flows.

15.13 Other long-service benefits

So far in this chapter we have considered the accounting for post-retirement costs for both defined contribution and defined benefit pension schemes. As well as pensions, IAS 19 (2011) considers other forms of long-service benefit paid to employees. These other forms of long-service benefit include:

(a) long-term compensated absences such as long-service or sabbatical leave;

(b) jubilee or other long-service benefits;

(c) long-term disability benefits;

(d) profit-sharing and bonuses payable 12 months or more after the end of the period in which the employees render the related service; and

(e) deferred compensation paid 12 months or more after the end of the period in which it is earned.

The measurement of these other long-service benefits is not usually as complex or uncertain as it is for post-retirement benefits and therefore a more simplified method of accounting is used for them. For other long-service benefits any remeasurement gains and losses are recognised immediately in profit or loss.

This means that the statement of financial position liability for other long-service benefits is just the present value of the future benefit obligation less the fair value of any assets that the benefit will be settled from directly.

The profit or loss charge for these benefits is therefore the total of:

(a) current and past service cost;

(b) net interest income or cost;

(c) remeasurement gains and losses.

15.14 Short-term benefits

In addition to pension and other long-term benefits considered earlier, IAS 19 gives accounting rules for short-term employee benefits.

Short-term employee benefits include items such as:

● wages, salaries and social security contributions;

● short-term compensated absences (such as paid annual leave and paid sick leave) where the absences are expected to occur within 12 months after the end of the period in which the employees render the related employee service;

- profit-sharing and bonuses payable within 12 months after the end of the period in which the employees render the related service; and

- non-monetary benefits (such as medical care, housing or cars) for current employees.

All short-term employee benefits should be recognised at an undiscounted amount:

- as a liability (after deducting any payments already made); and

- as an expense (unless another international standard allows capitalisation as an asset).

If the payments already made exceed the undiscounted amount of the benefits, an asset should be recognised only if it will lead to a future reduction in payments or a cash refund.

Compensated absences

The expected cost of short-term compensated absences should be recognised:

(a) in the case of accumulating absences, when the employees render service that increases their entitlement to future compensated absences; and

(b) in the case of non-accumulating compensated absences, when the absences occur.

Accumulating absences occur when the employees can carry forward unused absence from one period to the next. They are recognised when the employee renders services regardless of whether the benefit is vesting (the employee would get a cash alternative if they left employment) or non-vesting. The measurement of the obligation reflects the likelihood of employees leaving in a non-vesting scheme.

It is common practice for leave entitlement to be an accumulating absence (perhaps restricted to a certain number of days) but for sick pay entitlement to be non-accumulating.

Profit-sharing and bonus plans

The expected cost of a profit-sharing or bonus plan should be recognised only when:

(a) the enterprise has a present legal or constructive obligation to make such payments as a result of past events; and

(b) a reliable estimate of the obligation can be made.

15.15 Termination benefits

These benefits are treated separately from other employee benefits in IAS 19 (revised) because the event that gives rise to the obligation to pay is the termination of employment as opposed to the service of the employee.

The accounting treatment for termination benefits is consistent with the requirements of IAS 37 and the rules concern when the obligation should be provided for and the measurement of the obligation.

Recognition

Termination benefits can be recognised as a liability at the earlier of:

(a) when the entity can no longer withdraw the offer of those benefits; and

(b) when the entity recognises costs for a restructuring that is within the scope of IAS 37 and involves the payment of termination benefits.

Measurement

If the termination benefits are to be paid more than 12 months after the period-end date, they should be discounted, at a discount rate using the market yield on good-quality corporate bonds. Prudence should also be exercised in the case of an offer made to encourage voluntary redundancy, as provision should only be based on the number of employees expected to accept the offer.

15.16 IFRS 2 *Share-based Payment*[3]

Share awards, either directly through shares or through options, are very common ways of rewarding employee performance. These awards align the interests of the employees with those of the shareholders and, as such, are aimed at motivating the employees to perform in the way that benefits the shareholders. In particular, there is a belief that they will motivate the employees towards looking at the long-term success of the business as opposed to focusing solely on short-term profits. They have additional benefits also to the company and employees, for example in relation to cash and tax. If employees are rewarded in shares or options, the company will not need to pay out cash to reward the employees, and in a start-up situation where cash flow is very limited this can be very beneficial. Many 'dot-com' companies initially rewarded their staff in shares for this reason. There can also be tax benefits to employees with shares in some tax regimes which give an incentive to employees to accept share awards.

Whilst commercially share-based payments have many benefits, the accounting world has struggled in finding a suitable way to account for them. IAS 19 only covered disclosure requirements for share-based payments and had no requirements for the recognition and measurement of the payments when it was issued. The result of this was that companies could give very valuable rewards to their employees in the form of shares or options that did not result in the recognition of any charge against profit. The IASB addressed this by issuing, in February 2004, IFRS 2 *Share-based Payment*, which is designed to cover all aspects of accounting for share-based payments.

15.16.1 Should an expense be recognised?

Historically there has been some debate about whether a charge should be recognised in the statement of comprehensive income for share-based payments. One view is that the reward is given to employees in their capacity as shareholders and, as a result, it is not an employee benefit cost. Also supporters of the 'no-charge' view claimed that to make a charge would be a double hit to earnings per share in that it would reduce profits and increase the number of shares, which they felt was unreasonable.

Supporters of a charge pointed to opposite arguments that claimed having no charge underestimated the reward given to employees and therefore overstated profit. The impact of this was to give a misleading view of the profitability of the company. Also, making a charge gave comparability between companies that rewarded their staff in different ways. Comparability is one of the key principles of financial reporting.

For many years these arguments were not resolved and no standard was in issue, but the IASB eventually decided that a charge is appropriate and issued IFRS 2. In drawing up IFRS 2 a number of obstacles had to be overcome and decisions had to be made, for example:

- What should the value of the charge be – fair value or intrinsic value?
- At what point should the charge be measured – grant date, vesting date or exercise date?
- How should the charge be spread over a number of periods?
- If the charge is made to the statement of comprehensive income, where is the opposite entry to be made?
- What exemptions should be given from the standard?

IFRS 2 has answered these questions, and when introduced it made substantial changes to the profit recognised by many companies. In the UK, for example, the share-based payments charge for many businesses was one of their largest changes to profit on adopting IFRS.

15.17 Scope of IFRS 2

IFRS 2 is a comprehensive standard that covers all aspects of share-based payments. Specifically IFRS 2 covers:

- equity-settled share-based payment transactions, in which the entity receives goods or services as consideration for equity instruments issued;
- cash-settled share-based payment transactions, in which the entity receives goods or services by incurring liabilities to the supplier of those goods or services for amounts that are based on the price of the entity's shares or other equity instruments; and
- transactions in which the entity receives goods or services and either the entity or the supplier of those goods or services may choose whether the transaction is settled in cash (based on the price of the entity's shares or other equity instruments) or by issuing equity instruments.

There are no exemptions from the provisions of the IFRS except for:

(a) acquisitions of goods or other non-financial assets as part of a business combination; and

(b) acquisitions of goods or services under derivative contracts where the contract is expected to be settled by delivery as opposed to being settled net in cash.

15.18 Recognition and measurement

The general principles of recognition and measurement of share-based payment charges are as follows:

- Entities should recognise the goods or services acquired in a share-based payment transaction over the period the goods or services are received.
- The entity should recognise an increase in equity if the share-based payment is equity-settled and a liability if the payment is a cash-settled payment transaction.
- The share-based payment should be measured at fair value.

15.19 Equity-settled share-based payments

For equity-settled share-based payment transactions, the entity shall measure the goods and services received, and the corresponding increase in equity:

- **directly** at the fair value of the goods and services received, unless that fair value cannot be estimated reliably;
- **indirectly**, by reference to the fair value of the equity instruments granted, if the entity cannot estimate reliably the fair value of the goods and services received.

For transactions with employees, the entity shall measure the fair value of services received by reference to the fair value of the equity instruments granted, because typically it is not possible to estimate reliably the fair value of the services received.

In transactions with employees the IASB has decided that it is appropriate to value the benefit at the fair value of the instruments granted at their *grant date*. The IASB could have picked a number of different dates at which the options could have been valued:

- grant date – the date on which the options are given to the employees;
- vesting date – the date on which the options become unconditional to the employees; or
- exercise date – the date on which the employees exercise their options.

The IASB opted for the grant date as it was felt that the grant of options was the reward to the employees, and not the exercise of the options. This means that after the grant date any movements in the share price, whether upwards or downwards, do not influence the charge to the financial statements.

Employee options

In order to establish the fair value of an option at grant date the market price could be used (if the option is traded on a market), but it is much more likely that an option pricing model will need to be used. Examples of option pricing models that are possible include:

- *Black–Scholes.* An option pricing model used for options with a fixed exercise date that does not require adjustment for the inability of employees to exercise options during the vesting period.
- *Binomial model.* An option pricing model used for options with a variable exercise date that will need adjustment for the inability of employees to exercise options during the vesting period.

Disclosures are required of the principal assumptions used in applying the option pricing model.

IFRS 2 does not recommend any one pricing model but insists that whichever model is chosen a number of factors affecting the fair value of the option such as exercise price, market price, time to maturity and volatility of the share price must be taken into account. In practice the Black–Scholes model is probably most commonly used; however, many companies vary the model to some extent to ensure it fits with the precise terms of their options.

Once the fair value of the option has been established at the grant date it is charged to profit or loss over the vesting period. The vesting period is the period in which the employees are required to satisfy conditions, for example service conditions, that allow them to exercise their options. The vesting period might be within the current financial accounting period and all options exercised.

EXAMPLE ● Employees were granted options to acquire 100,000 shares at $20 per share if still in employment at the end of the financial year. The market value of an option was $1.50 per share. All employees exercised their option at the year-end and the company received $2,000,000. There will be a charge in the income statement of $150,000. Although the company has not transferred cash, it has transferred value to the employees. IFRS 2 requires the charge to be measured as the market value of the option, i.e. $1.50 per share.

However, it is more usual for options to be exercised over longer periods. In this case, the charge is spread over the vesting period by calculating a revised cumulative charge each year, and then apportioning that over the vesting period with catch-up adjustments made to amend previous under- or over-charges to profit or loss. The illustration below shows how this approach works.

When calculating the charge in profit or loss the likelihood of options being forfeited due to non-market price conditions (e.g. because the employees leave in the conditional period) should be adjusted for. For non-market conditions the charge is amended each year to reflect any changes in estimates of the numbers expected to vest.

The charge cannot be adjusted, however, for market price conditions. If, for example, the share price falls and therefore the options will not be exercised due to the exercise price being higher than the market price, no adjustment can be made. This means that if options are 'underwater' the statement of comprehensive income will still be recognising a charge for those options.

The charge is made to the statement of comprehensive income but there was some debate about how the credit entry should be made. The credit entry must be made either as a liability or as an entry to equity, and the IASB has decided that it should be an entry to equity. The logic for not including a liability is that the future issue of shares is not an 'obligation to transfer economic benefits' and therefore does not meet the definition of a liability. When the shares are issued it will increase the equity of the company and be a contribution from an owner.

Even though the standard specifies that the credit entry is to equity, it does not specify which item in equity is to be used. In practice it seems acceptable either to use a separate reserve or to make the entry to retained earnings. If a separate reserve is used at the exercise date the reserve will commonly be transferred to retained earnings. Whether the reserve is included within the proceeds of a share issue is not covered under IFRS and would be ultimately a matter for company law in the entities legal jurisdiction.

Illustration of option accounting

Alpha Ltd issued share options to staff on 1 January 20X0, details of which are as follows:

Number of staff	1,000
Number of options to each staff member	500
Vesting period	3 years
Fair value at grant date (per option)	£3
Expected employee turnover (per annum)	5%

In the 31 December 20X1 financial statements, the company revised its estimate of employee turnover to 8% per annum for the three-year vesting period.

In the 31 December 20X2 financial statements, the actual employee turnover had averaged 6% per annum for the three-year vesting period.

Options vest as long as the staff remain with the company for the three-year period.

The charge for share-based payments under IFRS 2 would be as follows:

Year ended 31 December 20X0

In this period the charge would be based on the original terms of the share option issue. The total value of the option award at fair value at the grant date is:

	£000
1,000 staff × 500 options × £3 × (0.95 × 0.95 × 0.95)	1,286

The charge to the statement of comprehensive income for the period is therefore:

£1,286 ÷ 3   429

Year ended 31 December 20X1

In this year the expected employee turnover has risen to 8% per annum. The estimate of the effect of the increase is taken into account. Amended total expected share option award at grant date:

	£000	£000
1,000 staff × 500 options × £3 × (0.92 × 0.92 × 0.92)		1,168

The charge to the statement of comprehensive income is therefore

	£000	£000
£1,168 × $^2/_3$	779	
Less: recognised to date	(429)	
		350

Year ended 31 December 20X2

The actual number of options that vest is now known. The actual value of the option award that vests at the grant date is:

	£000	£000
1,000 staff × 500 options × £3 × (0.94 × 0.94 × 0.94)		1,246

The charge to the statement of comprehensive income is therefore:

	£000	£000
Total value over the vesting period	1,246	
Less: recognised to date	(779)	
		467

Re-priced options

If an entity re-prices its options, for instance in the event of a falling share price, the incremental fair value should be spread over the remaining vesting period. The incremental fair value per option is the difference between the fair value of the option immediately before re-pricing and the fair value of the re-priced option.

Market-related vesting conditions

We have already seen that most vesting conditions (e.g. the requirement for employees to remain employed over the vesting period) are allowed for in estimating the number of options that are likely to vest. Where the vesting condition is a market-related condition, e.g. the share price must exceed a target amount or perform to a specified standard relative to other listed securities, the condition is allowed for in computing the fair value of the option at grant date using an appropriate model. Given that the condition is taken into account in this way, it is ignored when considering the likely vesting of the option to avoid double-counting. This means that, where such conditions exist, it is at least theoretically possible for there to be a charge to profit and loss for an equity-settled share-based payment where the options do not actually vest!

15.20 Cash-settled share-based payments

Cash-settled share-based payments result in the recognition of a liability. The entity measures the goods or services acquired and the liability incurred at fair value. Until the liability is settled, the entity remeasures the fair value of the liability at each reporting date, with any changes in fair value recognised in profit or loss.

For example, an entity might grant share appreciation rights to employees as part of their pay package, whereby the employees will become entitled to a future cash payment (rather than an equity instrument), based on the increase in the entity's share price from a specified level over a specified period.

The entity recognises the services received, and a liability to pay for those services, as the employees render service. For example, some share appreciation rights vest immediately, and the employees are therefore not required to complete a specified period of service to become entitled to the cash payment. In the absence of evidence to the contrary, the entity presumes that the services rendered by the employees in exchange for the share appreciation rights have been received. Thus, the entity recognises immediately the services received and a liability to pay for them. If the share appreciation rights do not vest until the employees have completed a specified period of service, the entity recognises the services received, and a liability to pay for them, as the employees render service during that period.

The liability is measured, initially and at each reporting date until settled, at the fair value of the share appreciation rights, by applying an option pricing model, taking into account the terms and conditions on which the share appreciation rights were granted, and the extent to which the employees have rendered service to date. The entity remeasures the fair value of the liability at each reporting date until settled.

Disclosure is required of the difference between the amount that would be charged to the statement of comprehensive income if the share appreciation rights are paid out in cash as opposed to being paid out with shares.

15.21 Transactions which may be settled in cash or shares

Some share-based payment transactions can be settled in either cash or shares with the settlement option being with the supplier of the goods or services and/or with the entity.

The accounting treatment is dependent upon which counterparty has the choice of settlement.

Supplier choice

If the supplier of the goods or services has the choice over settlement method, the entity has issued a compound instrument. The entity has an obligation to pay out cash (as the supplier can take this choice), but also has issued an equity option, as the supplier may decide to take equity to settle the transaction. The entity therefore recognises both a liability and an equity component.

The fair value of the equity option is the difference between the fair value of the offer of the cash alternative and the fair value of the offer of the equity payment. In many cases these are the same value, in which case the equity option has no value.

Once the split has been determined, each part is accounted for in the same way as other cash-settled or equity-settled transactions.

If cash is paid in settlement, any equity option recognised may be transferred to a different category in equity. If equity is issued, the liability is transferred to equity as the consideration for the equity instruments issued.

Entity choice

For a share-based payment transaction in which an entity may choose whether to settle in cash or by issuing equity instruments, the entity determines whether it has a present obligation to settle in cash and account for the share-based payment transaction accordingly. The entity has a present obligation to settle in cash if the choice of settlement in equity instruments is not substantive, or if the entity has a past practice or a stated policy of settling in cash.

If such an obligation exists, the entity accounts for the transaction in accordance with the requirements applying to cash-settled share-based payment transactions.

If no such obligation exists, the entity accounts for the transaction in accordance with the requirements applying to equity-settled transactions.

15.22 IAS 26 *Accounting and Reporting by Retirement Benefit Plans*[4]

This standard provides complementary guidance in addition to IAS 19 (2011) regarding the way that the pension fund should account and report on the contributions it receives and the obligations it has to pay pensions. The standard mainly contains the presentation and disclosure requirements of the schemes as opposed to the accounting methods that they should adopt.

15.22.1 Defined contribution plans

The report prepared by a defined contribution plan should contain a statement of net assets available for benefits and a description of the funding policy.

With a defined contribution plan it is not normally necessary to involve an actuary, since the pension paid at the end is purely dependent on the amount of fund built up for the employee. The obligation of the employer is usually discharged by the employer paying the agreed contributions into the plan. The main purpose of the report of the plan is to provide information on the performance of the investments, and this is normally achieved by including the following statements:

(a) a description of the significant activities for the period and the effect of any changes relating to the plan, its membership and its terms and conditions;

(b) statements reporting on the transactions and investment performance for the period and the financial position of the plan at the end of the period; and

(c) a description of the investment policies.

15.22.2 Defined benefit plans

Under a defined benefit plan (as opposed to a defined contribution plan) there is a need to provide more information, as the plan must be sufficiently funded to provide the agreed pension benefits at the retirement of the employees. The objective of reporting by the defined benefit plan is to periodically present information about the accumulation of resources and plan benefits over time that will highlight an excess or shortfall in assets.

The report that is required should contain[4] either:

(a) a statement that shows:

   (i) the net assets available for benefits;

   (ii) the actuarial present value of promised retirement benefits, distinguishing between vested benefits and non-vested benefits; and

   (iii) the resulting excess or deficit; or

(b) a statement of net assets available for benefits, including either:

   (i) a note disclosing the actuarial present value of promised retirement benefits, distinguishing between vested benefits and non-vested benefits; or

   (ii) a reference to this information in an accompanying report.

The most recent actuarial valuation report should be used as a basis for the above disclosures and the date of the valuation should be disclosed. IAS 26 does not specify how often actuarial valuations should be done but suggests that most countries require a triennial valuation.

When the fund is preparing the report and using the actuarial present value of the future obligations, the present value could be based on either projected salary levels or current salary levels. Whichever basis has been used should be disclosed. The effect of any significant changes in actuarial assumptions should also be disclosed.

Report format

IAS 26 proposes three different report formats that will fulfil the content requirements detailed above. These formats are:

(a) A report that includes a statement that shows the net assets available for benefits, the actuarial present value of promised retirement benefits, and the resulting excess or deficit. The report of the plan also contains statements of changes in net assets available for benefits and changes in the actuarial present value of promised retirement benefits. The report may include a separate actuary's report supporting the actuarial present value of promised retirement benefits.

(b) A report that includes a statement of net assets available for benefits and a statement of changes in net assets available for benefits. The actuarial present value of the promised retirement benefits is disclosed in a note to the statements. The report may also include a report from an actuary supporting the actuarial value of the promised retirement benefits.

(c) A report that includes a statement of net assets available for benefits and a statement of changes in net assets available for benefits with the actuarial present value of promised retirement benefits contained in a separate actuarial report.

In each format a trustees' report in the nature of a management or directors' report and an investment report may also accompany the statements.

15.22.3 All plans – disclosure requirements

For all plans, whether defined contribution or defined benefit, some common valuation and disclosure requirements exist.

Valuation

The investments held by retirement benefit plans should be carried at fair value. In most cases the investments will be marketable securities and the fair value is the market value. If it is impossible to determine the fair value of an investment, disclosure should be made of the reason why fair value is not used.

Market values are used for the investments because the market value is felt to be the most appropriate value at the report date and the best indication of the performance of the investments over the period.

Disclosure

In addition to the specific reports detailed above for defined contribution and defined benefit plans, the report should also contain:

(a) a statement of net assets available for benefits disclosing:

- assets at the end of the period suitably classified;
- the basis of valuation of assets;
- details of any single investment exceeding either 5% of the net assets available for benefits or 5% of any class or type of security;
- details of any investment in the employer;
- liabilities other than the actuarial present value of promised retirement benefits;

(b) a statement of changes in net assets for benefits showing the following:

- employer contributions;
- employee contributions;
- investment income such as interest or dividends;
- other income;
- benefits paid or payable;
- administrative expenses;
- other expenses;
- taxes on income;
- profits or losses on disposal of investment and changes in value of investments;
- transfers from and to other plans;

(c) a summary of significant accounting policies; and

(d) a description of the plan and the effect of any changes in the plan during the period.

Summary

Accounting for employee benefits has always been a difficult problem with different views as to the appropriate methods.

The different types of pension scheme and the associated risks add to the difficulties in terms of accounting. The accounting treatment for these benefits has recently changed, the current view being that the asset or liability position takes priority over the profit or loss charge. However, one consequence of giving the statement of financial position priority is that this change to the statement of comprehensive income can be much more volatile and this is considered by some to be undesirable.

An interesting more recent development is the option to use 'other comprehensive income' to record remeasurements, i.e. for actuarial gains and losses, rather than taking them to profit or loss.

IFRS 2 is the first serious attempt of the IASB to deal with accounting for share-based payments. It requires companies to recognise that a charge should be made for share-based payments and, in line with other recent standards such as financial instruments, it requires that charge to be recognised at fair value. There has been criticism of the standard in that it brings significant estimation into assessing the amount of charges to profit; however, overall the standard has been relatively well received with companies and with users of the financial statements.

REVIEW QUESTIONS

 1 Outline the differences between a defined benefit and a defined contribution pension scheme.

 2 If a defined contribution pension scheme provided a pension that was 6% of salary each year, the company had a payroll cost of €5 million, and the company paid €200,000 in the year, what would be the statement of comprehensive income charge and the statement of financial position liability at the year-end?

 3 'The approach taken in IAS 19 before its 1998 revision was to match an even pension cost against the period the employees provided service. This follows the accruals principle and is therefore fundamentally correct.' Discuss.

 4 Under the revised IAS 19 (2011) what amount of actuarial gains and losses should be recognised in profit or loss?

 5 In a dissenting view on the issue of IAS 19 (2011) Mr Yamada states that 'the return on high quality corporate bonds would be arbitrary and would not be a faithful representation of the return that investors require or expect from each type of asset'. Discuss.

 6 What is the required accounting treatment for a past service cost in a defined benefit pension scheme?

 7 What distinguishes a termination benefit from the other benefits considered in IAS 19 (2011)?

 8 'The issue of shares by companies, even to employees, should not result in a charge against profits. The contribution in terms of service that employees give to earn their rewards are contributions as owners and not as employees and when owners buy shares for cash there is no charge to profit.' Discuss.

 9 Briefly summarise the required accounting if a company gives its staff a cash bonus directly linked to the share price.

10 Explain what distinguishes the different types of share-based payment: equity-settled, cash-settled and equity with a cash alternative.

EXERCISES

* Question 1

Donna, Inc. operates a defined benefit pension scheme for staff. The pension scheme has been operating for a number of years but not following IAS 19. The finance director is unsure of which accounting policy to adopt under IAS 19 because he has heard very conflicting stories. He went to one presentation in 2010 that referred to a '10% corridor' approach to actuarial gains and losses, recognising them in profit or loss, but went to another presentation in 2012 that said actuarial gains and losses should be recognised in other comprehensive income.

The pension scheme had market value of assets of £3.2 million and a present value of obligations of £3.5 million on 1 January 2013. There were no actuarial gains and losses brought forward into 2013.

The details relevant to the pension are as follows (in £000):

	2013	2014	2015
Discount rate at start of year	6%	5%	4%
Expected rate of return on plan assets at start of year	10%	9%	8%
Current service cost	150	160	170
Benefits paid	140	150	130
Contributions paid	120	120	130
Present value of obligations at 31 December	3,600	3,500	3,200
Market value of plan assets at 31 December	3,400	3,600	3,600

Required:
(a) Advise the finance director of why the presentations from 2010 and 2012 gave different treatments of actuarial gains and losses
(b) Show how the pension scheme would be accounted for for the period 2013–2015 under IAS 19 (2011).

* Question 2

The following information (in £m) relates to the defined benefit scheme of Basil plc for the year ended 31 December 20X7:

Fair value of plan assets at 1 January 20X7 £3,150 and at 31 December 20X7 £3,386; contributions £26; current service cost £80; benefits paid £85; past service cost £150; present value of the obligation at 1 January 20X7 £3,750 and at 31 December 20X7 £4,192.

The discount rate was 7% at 31 December 20X6 and 8% at 31 December 20X7.

Required:
Show the amounts that will be recognised in the statement of comprehensive income and statement of financial position for Basil plc for the year ended 31 December 20X7 under IAS 19 (2011) and the movement in the net liability.

* Question 3

The following information is available for the year ended 31 March 20X6 (values in $m):

Present value of scheme liabilities at 1 April 20X5 $1,007; fair value of plan assets at 1 April 20X5 $844; benefits paid $44; contributions paid by employers $16; current service costs $28; past service costs $1; actuarial gains on assets $31; actuarial losses on liabilities $10; net interest cost $15.

Required:
(a) Calculate the net liability to be recognised in the statement of financial position.
(b) Show the amounts recognised in the statement of comprehensive income.

* Question 4

On 1 October 2005 Omega granted 50 employees options to purchase 500 shares in the entity. The options vest on 1 October 2007 for those employees who remain employed by the entity until that date. The options allow the employees to purchase the shares for $10 per share. The market price of the shares was $10 on 1 October 2005 and $10.50 on 1 October 2006. The market value of the options was $2 on 1 October 2005 and $2.60 on 1 October 2006. On 1 October 2005 the directors estimated that 5% of the relevant employees would leave in each of the years ended 30 September

2006 and 2007 respectively. It turned out that 4% of the relevant employees left in the year ended 30 September 2006 and the directors now believe that a further 4% will leave in the year ended 30 September 2007.

Required:
Show the amounts that will appear in the balance sheet of Omega as at 30 September 2006 in respect of the share options, and the amounts that will appear in the income statement for the year ended 30 September 2006.

You should state where in the balance sheet and where in the income statement the relevant amounts will be presented. Where necessary you should justify your treatment with reference to appropriate international financial reporting standards.

(Dip IFR December 2006)

* Question 5

On 1 January 20X1 a company obtained a contract in order to keep its factory in work but had obtained it on a very tight profit margin. Liquidity was a problem and there was no prospect of offering staff a cash bonus. Instead, the company granted its 80 production employees share options for 1,000 shares each at £10 per share. There was a condition that they would only vest if they still remained in employment at 31 December 20X2. The options were then exercisable during the year ended 31 December 20X3. Each option had an estimated fair value of £6.50 at the grant date.

At 31 December 20X1:

- The fair value of each option was £7.50.
- Four employees had left.
- It was estimated that 16 of the staff would have left by 31 December 20X2.
- The share price had increased from £9 on 1 January 20X1 to £9.90.

Required:
Calculate the charge to the income statement for the year ended 31 December 20X1.

* Question 6

C plc wants to reward its directors for their service to the company and has designed a bonus package with two different elements as follows. The directors are informed of the scheme and granted any options on 1 January 20X7.

1 Share options over 300,000 shares that can be exercised on 31 December 20Y0. These options are granted at an exercise price of €4 each, the share price of C plc on 1 January 20X7. Conditions of the options are that the directors remain with the company, and the company must achieve an average increase in profit of at least 10% per year, for the years ending 31 December 20X7 to 31 December 20X9. C plc obtained a valuation on 1 January 20X7 of the options which gave them a fair value of €3.

No directors were expected to leave the company but, surprisingly, on 30 November 20X9 a director with 30,000 options did leave the company and therefore forfeited his options. At the 31 December 20X7 and 20X8 year-ends C plc estimated that they would achieve the profit targets (they said 80% sure) and by 31 December 20X9 the profit target had been achieved.

By 31 December 20Y0 the share price had risen to €12, giving the directors who exercised their options an €8 profit per share on exercise.

2 The directors were offered a cash bonus payable on 31 December 20X8 based on the share price of the company. Each of the five directors was granted a €5,000 bonus for each €1 rise in the share price or proportion thereof by 31 December 20X8.

On 1 January 20X7 the estimated fair value of the bonus was €75,000; this had increased to €85,000 by 31 December 20X7, and the share price on 31 December 20X8 was €8 per share.

Required:
Show the accounting entries required in the years ending 31 December 20X7, 20X8 and 20X9 for the directors' options and bonus above.

Question 7

Kathryn plc, a listed company, provides a defined benefit pension for its staff, the details of which are given below.

As at 30 April 2013, actuaries valued the company's pension scheme and estimated that the scheme had assets of £10.5 million and obligations of £10.2 million (using the valuation methods prescribed in IAS 19).

The actuaries made assumptions in their valuation that the obligations were discounted using an appropriate corporate bond rate of 10%. The actuaries estimated the current service cost at £600,000. The actuaries informed the company that pensions to retired directors would be £800,000 during the year, and the company should contribute £700,000 to the scheme.

At 30 April 2014 the actuaries again valued the pension fund and estimated the assets to be worth £10.7 million, and the obligations of the fund to be £10.9 million.

Assume that contributions and benefits are paid on the last day of each year.

Required:
(a) Explain the reasons why IAS 19 was revised in 1998, moving from an actuarial income-driven approach to a market-based asset- and liability-driven approach. Support your answer by referring to the *Framework* principles.
(b) Show the extracts from the statement of comprehensive income and statement of financial position of Kathryn plc in respect of the information above for the year ended 30 April 2014. You do not need to show notes to the accounts.

* Question 8

Oberon prepares financial statements to 31 March each year. Oberon makes contributions to a defined benefit post-employment benefit plan for its employees. Relevant data are as follows:

(a) At 1 April 20X0 the plan obligation was €35 million and the fair value of the plan assets was €30 million.

(b) The actuary advised that the current service cost for the year ended 31 March 20X1 was €4 million. Oberon paid contributions of €3.6 million to the plan on 31 March 20X1. These were the only contributions paid in the year.

(c) The appropriate annual interest rate was 6% on 1 April 20X0 and 5.5% on 31 March 20X1.

(d) The plan paid out benefits totalling €2 million to retired members on 31 March 20X1.

(e) At 31 March 20X1 the plan obligation was €41.5 million and the fair value of the plan assets was €32.5 million.

Required:
Compute the amounts that will appear in the statement of comprehensive income of Oberon for the year ended 31 March 20X1 and the statement of financial position at 31 March 20X1 in respect of the post-employment benefit plan. You should indicate where in each statement the relevant amounts will be presented.

* Question 9

On 1 April 20W9 Oliver granted share options to 20 senior executives. The options are due to vest on 31 March 20X2 provided the senior executives remain with the company for the period between 1 April 20W9 and 31 March 20X2. The number of options vesting to each director depends on the cumulative profits over the three-year period from 1 April 20W9 to 31 March 20X2:

● 10,000 options per director if the cumulative profits are between €5 million and €10 million;

● 15,000 options per director if the cumulative profits are more than €10 million.

On 1 April 20W9 and 31 March 20X0 the best estimate of the cumulative profits for the three-year period ending on 31 March 20X2 was €8 million. However, following very successful results in the year ended 31 March 20X1 the latest estimate of the cumulative profits in the relevant three-year period is €14 million.

On 1 April 20W9 it was estimated that all 20 senior executives would remain with Oliver for the three-year period but on 31 December 20W9 one senior executive left unexpectedly. None of the other executives have since left and none are expected to leave before 31 March 20X2.

A further condition for vesting of the options is that the share price of Oliver should be at least €12 on 31 March 20X2. The share price of Oliver over the last two years has changed as follows:

● €10 on 1 April 20W9;

● €11.75 on 31 March 20X0;

● €11.25 on 31 March 20X1.

On 1 April 20W9 the fair value of the share options granted by Oliver was €4.80 per option. This had increased to €5.50 by 31 March 20X0 and €6.50 by 31 March 20X1.

Required:
Produce extracts, with supporting explanations, from the statements of financial position at 31 March 20X0 and 20X1 and from the statements of comprehensive income for the years ended 31 March 20X0 and 20X1 that show how the granting of the share options will be reflected in the financial statements of Oliver. Ignore deferred tax.

Question 10

A plc issues 50,000 share options to its employees on 1 January 2008 which the employees can only exercise if they remain with the company until 31 December 2010. The options have a fair value of £5 each on 1 January 2008.

It is expected that the holders of options over 8,000 shares will leave A plc before 31 December 2010.

In March 2008 adverse press comments regarding A plc's environmental policies and a downturn in the stock market cause the share price to fall significantly to below the exercise price on the options. The share price is not expected to recover in the foreseeable future.

Required:

What charge should A plc recognise for share options in the financial statements for the year ended 31 December 2008?

References

1 IAS 19 *Employee Benefits*, IASB, amended 2011.
2 IAS 26 *Accounting and Reporting by Retirement Benefit Plans*, IASC, reformatted 1994.
3 IFRS 2 *Share-based Payment*, IASB, amended 2013.
4 IAS 26 *Accounting and Reporting by Retirement Benefit Plans*, IASC, reformatted 1994.

CHAPTER 16

Taxation in company accounts

16.1 Introduction

The main purpose of this chapter is to explain the corporation tax system and the accounting treatment of deferred tax.

Objectives

By the end of this chapter, you should be able to:

- discuss the theoretical background to corporation tax systems;
- critically discuss tax avoidance and tax evasion;
- prepare deferred tax calculations;
- critically discuss deferred tax provisions.

16.2 Corporation tax

Limited companies, and indeed all corporate bodies, are treated for tax purposes as being legally separate from their proprietors. Thus, a limited company is itself liable to pay tax on its profits. This tax is known as **corporation tax**. The shareholders are accountable for tax only on the income they receive by way of any dividends distributed by the company. If the shareholder is an individual, then **income tax** becomes due on their dividend income received.

This is in contrast to the position in a partnership, where each partner is individually liable for the tax on their share of the pre-tax profit that has been allocated. A partner is taxed on the profit and not simply on drawings. Note that this is different from the treatment of an employee who is charged tax on the amount of salary that is paid.

In this chapter we consider the different types of company taxation and their accounting treatment. The International Accounting Standard that applies specifically to taxation is IAS 12 *Income Taxes*. The standard was last revised by the IASB in 2012. Those UK unquoted companies that choose not to follow international standards will follow FRSs 100–102 from 2015 (with earlier adoption allowed).

Corporation tax is calculated under rules set by Parliament each year in the Finance Act. The Finance Act may alter the existing rules; it also sets the rate of tax payable. Because of this annual review of the rules, circumstances may change year by year, which makes comparability difficult and forecasting uncertain.

The reason for the need to adjust accounting profits for tax purposes is that although the tax payable is based on the accounting profits as disclosed in the statement of income, the tax rules may differ from the accounting rules which apply prudence to income recognition. For example, the tax rules may not accept that all the expenses which are recognised by the accountant under the IASB's *Framework for the Preparation and Presentation of Financial Statements* and the IAS 1 *Presentation of Financial Statements* accrual concept are deductible when arriving at the taxable profit. An example of this might be a bonus, payable to an employee (based on profits), which is payable in arrears but which is deducted from account-ing profit as an accrual under IAS 1. This expense is allowed in calculating taxable profit on a cash basis only when it is paid in order to ensure that one taxpayer does not reduce his potential tax liability before another becomes liable to tax on the income received.

The accounting profit may therefore be lower or higher than the taxable profit. For example, the formation expenses of a company, which are the costs of establishing it on incorporation, must be written off in its first accounting period; the rules of corporation tax, however, state that these are a capital expense and cannot be deducted from the profit for tax purposes. This means that more tax will be assessed as payable than one would assume from an inspection of the published statement of income.

Similarly, although most businesses would consider that entertaining customers and other business associates was a normal commercial trading expense, it is not allowed as a deduction for tax purposes.

A more complicated situation arises in the case of depreciation. Because the directors have the choice of method of depreciation to use, the legislators have decided to require all companies to use the same method when calculating taxable profits. If one thinks about this, then it would seem to be the equitable practice. Each company is allowed to deduct a uniform percentage from its profits in respect of the depreciation that has arisen from the wear and tear and diminution in value of non-current assets.

The substituted depreciation that the tax rules allow is known as a **capital allowance**. The capital allowance is calculated in the same way as depreciation; the only difference is that the rates are those set out in the Finance Acts.

16.3 Corporation tax systems – the theoretical background

It might be useful to explain that there are three possible systems of company taxation: classical, imputation and partial imputation.[1] These systems differ solely in their tax treat-ment of the relationship between the limited company and those shareholders who have invested in it.

16.3.1 The classical system

In the classical system, a company pays tax on its profits, and then the shareholders suffer a second and separate tax liability when their share of the profits is distributed to them. In effect, the dividend income of the shareholder is regarded as a second and separate source of income from that of the profits of the company. The payment of a dividend creates an additional tax liability which falls directly on the shareholders. It could be argued that this double taxation is inequitable when compared to the taxation system on unincorporated bodies where the rate of taxation suffered overall remains the same whether or not profits are withdrawn from the business. It is suggested that this classical system discourages the distribution of profits to shareholders, since the second tranche of taxation (the tax on dividend income of the shareholders) only becomes payable on payment of the dividend, although some argue that the effect of the burden of double taxation on the economy is less serious than it might seem.[2]

It is also suggested that under this system companies have an incentive to avoid tax as any savings achieved in this way increases the after-tax profits available for dividends.

It is in some countries where the company income tax rates are low such as Ireland and Switzerland that there are classical company income tax systems where dividends are taxed at shareholder's marginal tax rate without credit for company income tax paid.

16.3.2 The imputation system

In an imputation system, the dividend is regarded merely as a flow of the profits on each sale to the individual shareholders, as there is considered to be merely one source of income which could be either retained in the company or distributed to the shareholders. It is certainly correct that the payment of a dividend results from the flow of moneys into the company from trading profits, and that the choice between retaining profits to fund future growth and the payment of a dividend to investing shareholders is merely a strategic choice unrelated to a view as to the nature of taxable profits.

In an imputation system the total of the tax paid by the company and by the shareholder is unaffected by the payment of dividends, and the tax paid by the company is treated as if it were also a payment of the individual shareholders' liabilities on dividends received. It is this principle of the flow of net profits from particular sales to individual shareholders that has justified the repayment of tax to shareholders with low incomes or to non-taxable shareholders of tax paid by the limited company, even though that tax credit has represented a reduction in the overall tax revenue of the state because the tax credit repaid also represented a payment of the company's own corporation tax liability. If the dividend had not been distributed to such a low-income or non-taxable shareholder who was entitled to repayment, the tax revenue collected would have been higher overall.

Australia and New Zealand alone of the OECD countries have a full imputation system where a tax credit is given to shareholders for the full corporate tax. One of the benefits seen is that Australian companies with largely resident shareholders have less incentive to avoid or defer company income tax, so reducing the need for anti-avoidance rules.

16.3.3 The partial imputation system

In a partial imputation system only part of the underlying corporation tax paid is treated as a tax credit. Canada and the United Kingdom operate a partial imputation system where shareholders receive a tax credit for only a portion of the corporate tax. The UK modified its imputation system in 1999, so that a low-income or non-taxable shareholder (such as a charity) could no longer recover any tax credit. Other countries such as France and Germany have moved away from the imputation system.

16.3.4 Common basis

All three systems are based on the taxation of profits earned as shown under the same basic principles used in the preparation of financial statements.

16.4 Corporation tax and dividends

A company pays corporation tax on its income. When that company pays a dividend to its shareholders it is distributing some of its taxed income among the proprietors. In an imputation system the tax paid by the company is 'imputed' to the shareholders who therefore receive a dividend which has already been taxed.

This means that, from the paying company's point of view, the concept of gross dividends does not exist. From the paying company's point of view, the amount of dividend paid shown in the statement of income will equal the cash that the company will have paid.

However, from the shareholder's point of view, the cash received from the company is treated as a net payment after deduction of tax. The shareholders will have received, with the cash dividend, a note of a tax credit, which is regarded as equal to basic rate income tax on the total of the dividend plus the tax credit. For example:

Dividend being the cash paid by the company to the shareholders and disclosed in the company's statement of income	360
Imputed tax credit of 1/9 of dividend paid (being the rate from 6 April 1999)	40
Gross dividend received by the shareholder	400

The imputed tax credit calculation (as shown above) has been based on a basic tax rate of 10% for dividends paid, being the basic rate of income tax on dividend income from 6 April 1999. This means that an individual shareholder who only pays basic rate income tax has no further liability in that the assumption is that the basic rate tax has been paid by the company. A non-taxpayer (including charities and pension funds) cannot obtain a repayment of tax.

Although a company pays corporation tax on its income, when that company pays a dividend to its shareholders it is still considered to be distributing some of its taxed income among the proprietors. In this system the tax payable by the company is 'imputed' to the shareholders who therefore receive a dividend which has already been taxed.

The essential point is that the dividend-paying company makes absolutely no deduction from the dividend, **nor is any payment made by the company to HM Revenue and Customs**. The addition of 1/9 of the dividend paid as an imputed tax credit is purely nominal. A tax credit of 1/9 of the dividend will be deemed to be attached to that dividend (in effect an income tax rate of 10%). That credit is notional in that no payment of the 10% will be made to HM Revenue and Customs. The payment of taxation is not associated with dividends.

Large companies (those with taxable profits of over £1,500,000) pay their corporation tax liability in quarterly instalments starting within the year of account, rather than paying their corporation tax liability nine months thereafter. The payment of taxation is not associated with the payment of dividends. Smaller companies pay their corporation tax nine months after the year-end.

16.5 Corporation tax systems – avoidance and evasion

'Avoidance' means reducing your tax liability legally. In the UK and other countries there has been much discussion and public comment on companies and wealthy individuals minimising the tax they pay, sometimes by artificial tax avoidance schemes. There has been public criticism of the very small UK Corporation Tax paid by some large international companies with significant activities in the UK.

'Evasion' means avoiding tax illegally, i.e. breaking the law.

In attempting to combat evasion and artificial avoidance schemes tax legislation has become increasingly complex – its very complexity providing the opportunity to design yet more ingenious mechanisms to reduce the tax liability.

In the UK the government has passed the Targeted Anti Avoidance Rules (TAARs) and a General Anti Abuse Rule (GAAR). TAARs aim to prevent tax avoidance in specific areas of the legislation whereas the GAAR is targeted at flagrant abusive and artificial schemes. A number of other jurisdictions such as Australia and Canada also have a general anti-avoidance rule.

In 2010 the UK Government asked Graham Aaronson QC to set up a committee to seek ways of preventing major tax avoidance. This committee is called the GAAR Study and it reported in 2011. Its conclusions[3] were:

- there should not be a general anti-avoidance rule; but
- it would be beneficial to introduce a rule targeted at **abusive arrangements**.

Following this, the UK Government introduced legislation into the Finance Act 2013 which aimed to cover abusive arrangements.

Governments have to follow the same basic principles of management as individuals. To spend money, there has to be a source of funds. The sources of funds are borrowing and income. With governments, the source of income is taxation. As with individuals, there is a practical limit as to how much they can borrow; to spend for the benefit of the populace, taxation has to be collected. In a democracy, the tax system is set up to ensure that the more prosperous tend to pay a greater proportion of their income in order to fund the needs of the poorer; this is called a progressive system. As Franklin Roosevelt, the American politician, stated, 'taxes, after all, are the dues that we pay for the privileges of membership in an organized society'.[4] Corporation tax on company profits represents 10% of the taxation collected by HM Revenue and Customs in the UK from taxes on income and wages.

It appears to be a general rule that taxpayers do not enjoy paying taxation (despite the fact that they may well understand the theory underpinning the collection of taxation). This fact of human nature applies just as much to company directors handling company resources as it does to individuals. Every extra pound paid in taxation by a company reduces the resources available for retention for funding future growth or paying dividends.

16.5.1 Tax evasion

Politicians often complain about tax evasion. Evasion is the illegal (and immoral) manipulation of business affairs to escape taxation. An example could be the directors of a family-owned company taking cash sales for their own expenditure. Another example might be the payment of a low salary (below the threshold of income tax) to a family member not working in the company, thus reducing profits in an attempt to reduce corporation tax. It is easy to understand the illegality and immorality of such practices. Increasingly the distinction between tax avoidance and tax evasion has been blurred.[5] When politicians complain of tax evasion, they tend not to distinguish between evasion and avoidance.

16.5.2 Tax avoidance

Tax avoidance could initially be defined as a manipulation of one's affairs, within the law, so as to reduce liability; indeed, as it is legal, it can be argued that it is not immoral. There is a well-established tradition within the UK that 'every man is entitled if he can to order his affairs so that the tax attaching under the appropriate Acts is less than it otherwise would be'.[6]

Indeed the government deliberately sets up special provisions to reduce taxes in order to encourage certain behaviours. The more that employers and employees save for employee retirement, the less social security benefits will be paid out in the future. Thus both companies and individuals obtain full relief against taxation for pension contributions. Another example might be increased tax depreciation (capital allowances) on capital investment, in order to increase industrial investment and improve productivity within the UK economy.

The use of such provisions, as intended by the legislators, is not criticised by anyone, and might better be termed 'tax planning'. The problem area lies between the proper use of such tax planning, and illegal activities. This 'grey area' could best be called 'tax avoidance'.

The Institute for Fiscal Studies has stated:

> We think it is impossible to define the expression 'tax avoidance' in any truly satisfactory manner. People routinely alter their behaviour to reduce or defer their taxation liabilities. In doing so, commentators regard some actions as legitimate tax planning and others as tax avoidance. We have regarded tax avoidance (in contra-indication to legitimate . . . tax planning) as action taken to reduce or defer tax liabilities in a way Parliament plainly did not intend.[7]

The law tends to define tax avoidance as an artificial element in the manipulation of one's affairs, within the law, so as to reduce liability.[8]

16.5.3 The problem of distinguishing between avoidance and evasion

The problem lies in distinguishing clearly between legal avoidance and illegal evasion. It can be difficult for accountants to walk the careful line between helping clients (in tax avoidance) and colluding with them against HM Revenue and Customs.[9]

When clients seek advice, accountants have to be careful to ensure that they have integrity in all professional and business relationships. Integrity implies not merely honesty but fair dealing and truthfulness. 'In all dealings relating to the tax authorities, a member must act honestly and do nothing that might mislead the authorities.'[10]

As an example to illustrate the problems that could arise, a client company has carried out a transaction to avoid taxation, but failed to minute the details as discussed at a directors' meeting. If the accountant were to correct this act of omission in arrears, this would be a move from tax avoidance towards tax evasion. Another example of such a move from tax avoidance to tax evasion might be where an accountant in informing HM Revenue and Customs of a tax-avoiding transaction fails to detail aspects of the transaction which might show it in a disadvantageous light.

Companies can move profit centres from high-taxation countries to low-taxation countries by setting up subsidiaries therein. These areas, known in extreme cases as 'tax havens', are disliked by governments.

Tax havens are countries with very low or zero tax rates on some or all forms of income. They could be classified into two groups:

1 the zero-rate and low-tax havens;
2 the tax havens that impose tax at normal rates but grant preferential treatment to certain activities.

The use of zero-rate and low-tax havens could be considered a form of tax avoidance, although sometimes they are used by tax evaders for their lack of regulation.

A similar problem has arisen in the use of charitable donations where tax relief is allowed to the donor and is a legitimate avoidance to encourage donations, except that the system has been manipulated as a form of tax evasion. This is an international problem and an OECD *Report on Abuse of Charities for Money-laundering and Tax Evasion* issued in 2009 stated that 'Tax evasion and tax fraud through the abuse of charities is a serious and increasing risk in many countries although its impact is variable. Some countries estimate that the abuse of charities costs their treasury many hundreds of millions of dollars and is becoming more prevalent'.[11]

16.5.4 Countering tax avoidance

An interesting discussion paper, *Countering Tax Avoidance in the UK: Which Way Forward?*, was published by the IFS in 2009.[12] It recognises that there is a difficulty in defining what constitutes avoidance.

It is a grey area and possibly not capable of a precise definition. Revenue authorities may often appear to consider tax avoidance to occur where it is sought to reduce the tax burden of individuals, businesses and other entities below the level envisaged by the government; the problem is, however, that the envisaged level is usually unclear.

In the public eye there is a view that certain types of avoidance are unacceptable. This leaves open the question, however, as to what is acceptable. What is acceptable to a taxpayer might be unacceptable to the tax collector. In principle, what is acceptable should be clear from the government as the body responsible for the raising of taxes. In practice it is incredibly difficult to cover all schemes through legislation. Detailed legislation, for instance, to indicate what is acceptable risks becoming more and more complex. This leads to the possibility for schemes to be designed which reflect the legal position but not the commercial substance. It is a similar problem to that faced by standard setters who have adopted the substance over form approach in areas such as accounting for leases.

Adopting a fuzzy approach rather than detailed legislation might appear to give the tax collector greater ability to counteract avoidance, but there is a downside – multinational companies might decide that there is too much uncertainty and base themselves in another jurisdiction.

16.5.5 International approaches

The discussion paper considers the approaches taken to counteracting avoidance in countries such as the UK and the Netherlands. It sets out that the Netherlands, for example, has both a case-based and a practical approach to avoidance.

The **case-based** concept is known as '*fraus legis*'. *Fraus legis* means that the person has acted contrary to the intention of the law even though they have complied with the letter of the law. In order for it to apply, the avoidance of tax must be the only or paramount motive for the transaction and there must be a conflict with the intention and purpose of the law. Once applied, the judges may decide to ignore the tax avoidance transaction or replace it with other transactions if that would better fit with the purpose of the law.

The **practical approach** is to manage the taxpayer relationship. Since 2005, the Dutch tax authorities have entered into 'enforcement covenants' with certain multinationals. Currently, more than 40 have concluded these agreements. The Dutch tax authorities agree to reduce their supervision of the taxpayer's affairs and in return the taxpayer agrees to report tax risks. The taxpayer must be recognised as compliant for this option to be offered to them. The taxpayer effectively agrees to abide by not only the letter of the law but also its spirit and has to be seen to be paying a fair share of tax. A multinational with its tax burden reduced to nil would not be viewed as suitable for this approach.

16.6 IAS 12 – accounting for current taxation

The essence of IAS 12 is that it requires an enterprise to account for the tax consequences of transactions and other events in the same way that it accounts for the transactions and other events themselves. Thus, for transactions and other events recognised in the statement of comprehensive income, any related tax effects are also recognised in the statement of comprehensive income.

The details of how IAS 12 requires an enterprise to account for the tax consequences of transactions and other events follow below.

Statement of comprehensive income disclosure

The standard (paragraph 77) states that the tax expense related to profit or loss from ordinary activities should be presented on the face of the statement of comprehensive income. It also provides that the major components of the tax expense should be disclosed separately. These separate components of the tax expense may include (paragraph 80):

(a) current tax expense for the period of account;

(b) any adjustments recognised in the current period of account for prior periods (such as where the charge in a past year was underprovided);

(c) the amount of any benefit arising from a previously unrecognised tax loss, tax credit or temporary difference of a prior period that is used to reduce the current tax expense; and

(d) the amount of tax expense (income) relating to those changes in accounting policies and fundamental errors which are included in the determination of net profit or loss for the period in accordance with the allowed alternative treatment in IAS 8 *Net Profit or Loss for the Period, Fundamental Errors and Changes in Accounting Policies.*

Statement of financial position disclosure

The standard states that current tax for current and prior periods should, to the extent unpaid, be recognised as a liability. If the amount already paid in respect of current and prior periods exceeds the amount due for those periods, the excess should be recognised as an asset.

The treatment of tax losses

As regards losses for tax purposes, the standard states that the benefit relating to a tax loss that can be carried back to recover current tax of a previous period should be recognised as an asset. Tax assets and tax liabilities should be presented separately from other assets and liabilities in the statement of financial position. An enterprise should offset (paragraph 71) current tax assets and current tax liabilities if, and only if, the enterprise:

(a) has a legally enforceable right to set off the recognised amounts; and

(b) intends either to settle on a net basis, or to realise the asset and settle the liability simultaneously.

The standard provides (paragraph 81) that the following should also be disclosed separately:

(a) tax expense (income) relating to extraordinary items recognised during the period;

(b) an explanation of the relationship between tax expense (income) and accounting profit in either or both of the following forms:

    (i) a numerical reconciliation between tax expense (income) and the product of account-ing profit multiplied by the applicable tax rate(s), disclosing also the basis on which the applicable tax rate(s) is/are computed; or

    (ii) a numerical reconciliation between the average effective tax rate and the applicable tax rate, disclosing also the basis on which the applicable tax rate is computed; and

(c) an explanation of changes in the applicable tax rate(s) compared to the previous account-ing period.

The relationship between tax expense and accounting profit

The following example is an explanation of the relationship between tax expense (income) and accounting profit:

Current Tax Expense

	X5	X6
Accounting profit	8,775	8,740
Add		
Depreciation for accounting purposes	4,800	8,250
Charitable donations	500	350
Fine for environmental pollution	700	—
Product development costs	250	250
Health care benefits payable	2,000	1,000
	17,025	18,590
Deduct		
Depreciation for tax purposes	(8,100)	(11,850)
Taxable profit	8,925	6,740
Current tax expense at 40%	3,570	
Current tax expense at 35%		2,359

16.7 Deferred tax

16.7.1 IAS 12 – background to deferred taxation[13]

The profit on which tax is paid may differ from that shown in the published statement of income. This is caused by two separate factors.

Permanent differences

One factor that we looked at above is that certain items of expenditure may not be legitimate deductions from profit for tax purposes under the tax legislation. These differences are referred to as **permanent** differences because they will not be allowed at a different time and will be permanently disallowed, even in future accounting periods.

Timing differences

Another factor is that there are some other expenses that are legitimate deductions in arriving at the taxable profit which are allowed as a deduction for tax purposes at a later date. These might be simply **timing** differences in that tax relief and charges to the statement of income occur in different accounting periods. The accounting profit is prepared on an accruals basis but the taxable profit might require certain of the items to be dealt with on a cash basis. Examples of this might include bonuses payable to senior management, properly included in the financial statements under the accruals concept but not eligible for tax relief until actually paid some considerable time later, thus giving tax relief in a later period.

Temporary differences[13]

The original IAS 12 allowed an enterprise to account for deferred tax using the statement of comprehensive income liability method which focused on timing differences. IAS 12 (revised) requires the statement of financial position liability method, which focuses on temporary differences, to be used. Timing differences are differences between taxable profit and accounting profit that originate in one period and reverse in one or more subsequent periods. Temporary differences are differences between the tax base of an asset or liability and its carrying amount in the statement of financial position. The tax base of an asset or liability is the amount attributed to that asset or liability for tax purposes. All timing differences are temporary differences.

The most significant temporary difference is depreciation. The depreciation charge made in the financial statements must be added back in the tax calculations and replaced by the official tax allowance for such an expense. The substituted expense calculated in accordance with the tax rules is rarely the same amount as the depreciation charge computed in accordance with IAS 16 *Property, Plant and Equipment*.

Capital investment incentive effect

It is common for legislation to provide for higher rates of tax depreciation than are used for accounting purposes, for it is believed that the consequent deferral of taxation liabilities serves as an incentive to capital investment (this incentive is not forbidden by European Union law or the OECD rules). The classic effect of this is for tax to be payable on a lower figure than the accounting profit in the earlier years of an asset's life because the tax allowances usually exceed depreciation in those years. In later accounting periods, the tax allowances will be lower than the depreciation charges and the taxable profit will then be higher than the accounting profit that appears in the published statement of income.

Deferred tax provisions

The process whereby the company pays tax on a profit that is lower than the reported profit in the early years and on a profit that is higher than reported profit in later years is known as **reversal**. Given the knowledge that, ultimately, these timing differences will reverse, the accruals concept requires that consideration be given to making provision for the future liability in those early years in which the tax payable is calculated on a lower figure. The provision that is made is known as a **deferred tax provision**.

Alternative methods for calculating deferred tax provisions

As you might expect, there has been a history of disagreement within the accounting profession over the method to use to calculate the provision. There have been, historically, two methods of calculating the provision for this future liability – the **deferral** method and the **liability** method.

The deferral method

The deferral method, which used to be favoured in the USA, involves the calculation each year of the tax effects of the timing differences that have arisen in that year. The tax effect is then debited or credited to the statement of income as part of the tax charge; the double entry is effected by making an entry to the deferred tax account. This deferral method of calculating the tax effect ignores the effect of changing tax rates on the timing differences that arose in earlier periods. This means that the total provision may consist of differences calculated at the rate of tax in force in the year when the entry was made to the provision.

The liability method

The liability method requires the calculation of the total amount of potential liability each year at current rates of tax, increasing or reducing the provision accordingly. This means that the company keeps a record of the timing differences and then recalculates at the end of each new accounting period using the rate of corporation tax in force as at the date of the current statement of financial position.

To illustrate the two methods we will take the example of a single asset, costing £10,000, depreciated at 10% using the straight-line method, but subject to a tax allowance of 25% on the reducing balance method. The workings are shown in Figure 16.1. This shows that, if there were no other adjustments, for the first four years the profits subject to tax would be lower than those shown in the accounts, but afterwards the situation would reverse.

Charge to statement of comprehensive income under the deferral method

The deferral method would charge to the statement of income each year the variation multiplied by the current tax rate, e.g. 20X5 at 25% on £1,500 giving £375.00, and 20X8 at 24% on £55 giving £13.20. This is in accordance with the accruals concept which matches the tax expense against the income that gave rise to it. Under this method the deferred tax provision will be credited with £375 in 20X6 and this amount will not be altered in 20X8 when the tax rate changes to 24%. In the example, the calculation for the five years would be as in Figure 16.2.

Charge to statement of comprehensive income under the liability method

The liability method would make a charge so that the total balance on deferred tax equalled the cumulative variation multiplied by the current tax rate. The intention is that the statement of financial position liability should be stated at a figure which represents the tax effect as at the end of each new accounting period. This means that there would be an adjustment made in 20X8 to recalculate the tax effect of the timing difference that was provided for in

Figure 16.1 Deferred tax provision using deferral method

		Accounts (depreciation) £	Tax (allowances) £	Difference (temporary) £	Tax (rate)
01.01.20X5	Cost of asset	10,000	10,000		
31.12.20X5	Depn/tax allowance	1,000	2,500	1,500	25%
		9,000	7,500	1,500	
31.12.20X6	Depn/tax allowance	1,000	1,875	875	25%
		8,000	5,625	2,375	
31.12.20X7	Depn/tax allowance	1,000	1,406	406	25%
		7,000	4,219	2,781	
31.12.20X8	Depn/tax allowance	1,000	1,055	55	24%
		6,000	3,164	2,836	
31.12.20X9	Depn/tax allowance	1,000	791	(209)	24%
		5,000	2,373	2,627	

Figure 16.2 Summary of deferred tax provision using the deferral method

Year ended	Timing difference £	Basic rate %	Deferred tax charge in year £	Deferred tax provision (deferral method) £
31.12.20X5	1,500	25%	375.00	375.00
31.12.20X6	875	25%	218.75	593.75
31.12.20X7	406	25%	101.50	695.25
31.12.20X8	55	24%	13.20	708.45
31.12.20X9	(209)	24%	(50.16)	658.29

earlier years. For example, the provision for 20X6 would be recalculated at 24%, giving a figure of £360 instead of the £375 that was calculated and charged in 20X6. The decrease in the expected liability will be reflected in the amount charged against the statement of income in 20X6. The £15 will in effect be credited to the 20X6 profit statement.

The effect on the charge to the 20X9 profit statement (Figures 16.2 and 16.3) is that there will be a charge of £13.20 using the deferral method and a **credit** of £14.61 using the liability method. The £14.61 is the reduction in the amount provided from £695.25 at the end of 20X8 to the £680.64 that is required at the end of 20X9.

World trend towards the liability method

There has been a move in national standards away from the deferral method towards the liability method, which is a change of emphasis from the statement of comprehensive income to the statement of financial position because the deferred tax liability is shown at current rates of tax in the liability method. This is in accordance with the IASB's conceptual framework which requires that all items in the statement of financial position, other than shareholders' equity, must be either assets or liabilities as defined in the framework. Deferred tax as it is calculated under the traditional deferral method is not in fact a calculation of a liability, but is better characterised as deferred income or expenditure. This is illustrated by the fact that the sum calculated under the deferral method is not recalculated to take account of changes in the rate of tax charged, whereas it is recalculated under the liability method.

The world trend towards using the liability method also results in a change from accounting only for timing differences to accounting for temporary differences.

Figure 16.3 Deferral tax provision using the liability method

Year ended	Temporary difference £	Basic rate	Deferred tax charge in year £	Deferred tax provision (deferral method) £	Rate in 20X9	Deferred tax provision (liability method) £
31.12.20X6	1,500	25%	375.00	375.00	24%	360.00
31.12.20X7	875	25%	218.75	593.75	24%	210.00
31.12.20X8	406	25%	101.50	695.25	24%	97.44
31.12.20X9	55	24%	13.20	708.45	24%	13.20
				708.45		680.64

Temporary versus timing: conceptual difference

These temporary differences are defined in the IASB standard as 'differences between the carrying amount of an asset or liability in the statement of financial position and its tax base'.[13] The conceptual difference between these two views is that under the liability method provision is made for only the future reversal of these timing differences, whereas the temporary difference approach provides for the tax that would be payable if the company were to be liquidated at statement of financial position values (i.e. if the company were to sell all assets at statement of financial position values).

The US standard SFAS 109 argues the theoretical basis for these temporary differences to be accounted for on the following grounds:

> A government levies taxes on net taxable income. Temporary differences will become taxable amounts in future years, thereby increasing taxable income and taxes payable, upon recovery or settlement of the recognized and reported amounts of an enterprise's assets or liabilities . . . A contention that those temporary differences will never result in taxable amounts . . . would contradict the accounting assumption inherent in the statement of financial position that the reported amounts of assets and liabilities will be recovered and settled, respectively; thereby making that statement internally inconsistent.[14]

A consequence of accepting this conceptual argument in IAS 12 is that provision must also be made for the potential taxation effects of asset revaluations.

16.7.2 IAS 12 – deferred taxation

The standard requires that the financial statements are prepared using the liability method described above (which is sometimes known as the statement of financial position liability method).

An example of how deferred taxation operates follows.

EXAMPLE ● An asset which cost £150 has a carrying amount of £100. Cumulative depreciation for tax purposes is £90 and the tax rate is 25% as shown in Figure 16.4.

The tax base of the asset is £60 (cost of £150 less cumulative tax depreciation of £90). To recover the carrying amount of £100, the enterprise must earn taxable income of £100, but will only be able to deduct tax depreciation of £60. Consequently, the enterprise will pay taxes of £10 (£40 at 25%) when it recovers the carrying amount of the asset. The difference between the carrying amount of £100 and the tax base of £60 is a taxable temporary difference of £40. Therefore, the enterprise recognises a deferred tax liability of £10 (£40 at 25%) representing the income taxes that it will pay when it recovers the carrying amount of the asset as shown in Figure 16.5.

The accounting treatment over the life of an asset

The following example illustrates the accounting treatment over the life of an asset.

Figure 16.4 Cumulative depreciation

	In accounts	For tax
Cost	150	150
Depreciation	50	90
Carrying amount	100	60

Figure 16.5 Deferred tax liability

Income to recover	
Carrying amount	£100
Carrying amount for tax	£60
Temporary difference	£40
Tax rate	25%
Deferred tax	£10

EXAMPLE ● An enterprise buys equipment for £10,000 and depreciates it on a straight-line basis over its expected useful life of five years. For tax purposes, the equipment is depreciated at 25% per annum on a straight-line basis. Tax losses may be carried back against taxable profit of the previous five years. In year 0, the enterprise's taxable profit was £0. The tax rate is 40%. The enterprise will recover the carrying amount of the equipment by using it to manufacture goods for resale. Therefore, the enterprise's current tax computation is as follows:

Year	1	2	3	4	5
Taxable income (£)	2,000	2,000	2,000	2,000	2,000
Depreciation for tax purposes	2,500	2,500	2,500	2,500	0
Tax profit (loss)	(500)	(500)	(500)	(500)	2,000
Current tax expense (income) at 40%	(200)	(200)	(200)	(200)	800

The enterprise recognises a current tax asset at the end of years 1 to 4 because it recovers the benefit of the tax loss against the taxable profit of year 0.

The temporary differences associated with the equipment and the resulting deferred tax asset and liability and deferred tax expense and income are as follows:

Year	1	2	3	4	5
Carrying amount (£)	8,000	6,000	4,000	2,000	0
Tax base	7,500	5,000	2,500	0	0
Taxable temporary difference	500	1,000	1,500	2,000	0
Opening deferred tax liability	0	200	400	600	800
Deferred tax expense (income)	200	200	200	200	(800)
Closing deferred tax liability	200	400	600	800	0

The enterprise recognises the deferred tax liability in years 1 to 4 because the reversal of the taxable temporary difference will create taxable income in subsequent years. The enterprise's statement of comprehensive income is as follows:

Year	1	2	3	4	5
Income (£)	2,000	2,000	2,000	2,000	2,000
Depreciation	2,000	2,000	2,000	2,000	2,000
Profit before tax	0	0	0	0	0
Current tax expense (income)	(200)	(200)	(200)	(200)	800
Deferred tax expense (income)	200	200	200	200	(800)
Total tax expense (income)	0	0	0	0	0
Net profit for the period	0	0	0	0	0

Further examples of items that could give rise to temporary differences are:

● Retirement benefit costs may be deducted in determining accounting profit as service is provided by the employee, but deducted in determining taxable profit either when contributions are paid to a fund by the enterprise or when retirement benefits are paid by the enterprise. A temporary difference exists between the carrying amount of the liability (in the financial statements) and its tax base (the carrying amount of the liability for tax purposes); the tax base of the liability is usually nil.

● Research costs are recognised as an expense in determining accounting profit in the period in which they are incurred but may not be permitted as a deduction in determining taxable profit (tax loss) until a later period. The difference between the tax base (the carrying amount of the liability for tax purposes) of the research costs, being the amount the taxation authorities will permit as a deduction in future periods, and the carrying amount of nil is a deductible temporary difference that results in a deferred tax asset.

Treatment of asset revaluations

The original IAS 12 permitted, but did not require, an enterprise to recognise a deferred tax liability in respect of asset revaluations. If such assets were sold at the revalued sum then a profit would arise that could be subject to tax. IAS 12 as currently written requires an enterprise to recognise a deferred tax liability in respect of asset revaluations.

REVALUATION EXAMPLE ● At 31.12.20X1 the company had reported its land and buildings within non-current assets at the following values:

	€000
Land	500
Buildings	1,200

On 1 January 20X2, the land was revalued to €700,000 and its buidings to €1,800,000. At that date the building had a remaining life of 25 years.

Required: Assuming that the Corporation Tax on capital gains is 30%, state the balances on the respective accounts at 1 January 20X2 following the revaluation.

Answer:
The land will be revalued to €700,000 with a capital gain of €200,000.
    This gain will be credited €140,000 to a revaluation reserve (70% of €200,000) and €60,000 (30%) to deferred tax.
    The building will be revalued to €1,800,000 with a capital gain of €600,000.
    This gain will be credited €420,000 to a revaluation reserve (70%) and €180,000 to deferred tax (30%).
    So, deferred tax will increase by €240,000 and the revaluation reserve by €560,000.
    The revaluation of €800,000 and the transfer to deferred tax of €240,000 will appear in 'Other Comprehensive Income'. In the statement of financial position the revaluation reserve will be included under 'other components of equity' and deferred tax will be under 'non-current liabilities'.

Such a deferred tax liability on a revalued asset might not arise for many years, for there might be no intention to sell the asset. Many would argue that IAS 12 should allow for such timing differences by discounting the deferred liability (for a sum due many years in advance is certainly recognised in the business community as a lesser liability than the sum due immediately, for the sum could be invested and produce income until the liability would become due; this is termed the time value of money). The standard does not allow such discounting.[15]

Indeed, it could be argued that in reality most businesses tend to have a policy of continuous asset replacement, with the effect that any deferred liability will be further deferred by these future acquisitions, so that the deferred tax liability would only become payable on a future cessation of trade. Not only does the standard preclude discounting, it also does not permit any account being made for future acquisitions by making a partial provision for the deferred tax.

Deferred tax asset

Except for deferred tax assets arising from taxation, deferred tax assets normally arise where:

(i)   certain cash income is received in the accounting period, but not credited to the income statement until a future period such as lease premiums and rent received in advance; and

(ii) certain expenses are incurred in the current period but not paid until a future accounting period such as where a director's or employee's bonus is charged in the income statement before the year-end but not paid until after the year end.

Unused tax losses

A deferred tax asset should be recognised for the carry-forward of unused tax losses and unused tax credits to the extent that it is probable that future taxable profit will be available against which the unused tax losses and unused tax credits can be utilised. This is reported as in Figure 16.6 with an extract from the Bayer Group 2013 Annual Report.

At each statement of financial position date, an enterprise should reassess unrecognised deferred tax assets. The enterprise recognises a previously unrecognised deferred tax asset to the extent that it has become probable that future taxable profit will allow the deferred tax asset to be recovered. For example, an improvement in trading conditions may make

Figure 16.6 Extract from Bayer Group 2013 Annual Report

14. taxes

The breakdown of tax expenses by origin was as follows:

	2012		2013	
		Of which income taxes		Of which income taxes
	€ million	€ million	€ million	€ million
Taxes paid or accrued				
Income taxes				
Germany	(534)		(795)	
other countries	(1,026)		(849)	
Other taxes				
Germany	(28)		(43)	
other countries	(235)		(188)	
	(1,823)	**(1,560)**	**(1,875)**	**(1,644)**
Deferred taxes				
from temporary differences	782		569	
from tax loss carryforwards and tax credits	55		54	
	837	**837**	**623**	**623**
Total	**(986)**	**(723)**	**(1,252)**	**(1,021)**

it more probable that the enterprise will be able to generate sufficient taxable profit in the future for the deferred tax asset to be recovered.

The Financial Reporting Review Panel in its 2012 Annual Report stated that:

> As reported last year, the Panel continued to have to remind a number of companies with a record of losses of the need to recognise a deferred tax asset for the carry forward of unused tax losses and credits only to the extent that it is probable that future taxable profit will be available against which the temporary differences can be utilised. When a company has a history of losses, in the absence of sufficient taxable temporary differences 'convincing other evidence' is required to support the company's judgement that it is probable that future taxable profits will be available against which the tax losses can be utilised. The Panel sought undertakings that, in future, as required by the standard the deferred tax asset should be quantified and the nature of the evidence supporting its recognition disclosed.

The carrying amount of a deferred tax asset has to be reviewed at the end of each reporting period and reduced to the extent that it is no longer probable that sufficient taxable profit will be available to allow the benefit of the asset to be utilised. However, any such reduction can be reversed later if it becomes probable that sufficient taxable profit will be available.

At the October 2009 joint meeting of the IASB and the FASB, both boards indicated that they would consider undertaking a fundamental review of accounting for income taxes at some time in the future. In the meantime, the IASB issued an amendment in 2012 *Deferred Tax: Recovery of Underlying Assets* relating to the treatment of investment properties.

16.8 A critique of deferred taxation

It could be argued that deferred tax is not a legal liability until it accrues. The consequence of this argument would be that deferred tax should not appear in the financial statements, and financial statements should:

- present the tax expense for the year equal to the amount of income taxes that has been levied based on the income tax return for the year;
- accrue as a receivable any income refunds that are due from taxing authorities or as a payable any unpaid current or past income taxes;
- disclose in the notes to the financial statements differences between the income tax bases of assets and liabilities and the amounts at which they appear in the statement of financial position.

The argument is that the process of accounting for deferred tax is confusing what **did** happen to a company, i.e. the agreed tax payable for the year, and what **did not** happen to the company, which is the tax that would have been payable if the adjustments required by the tax law for timing differences had not occurred. It is felt that the investor should be provided with details of the tax charge levied on the profits for the year and an explanation of factors that might lead to a different rate of tax charge appearing in future financial statements.

The argument against adjusting the tax charge for deferred tax and the creation of a deferred tax provision holds that shareholders are accustomed to giving consideration to many other imponderables concerning the amount, timing and uncertainty of future cash receipts and payments, and the treatment of tax should be considered in the same way. This view has received support from others,[16] who have held that tax attaches to taxable income and not to the reported accounting income and that there is no legal requirement for the tax to bear any relationship to the reported accounting income. Indeed it has been argued that 'deferred tax means income smoothing'.[17]

The creation of a charge in the statement of income for a deferred tax liability has an impact on the EPS in the year in which it arises and when it reverses. However, it is suggested that the arguments for and against deferred taxation accounting must be based solely on the theory underpinning accounting, and be unaffected by commercial considerations.

Accrual accounting assumption

It is also suggested that the above arguments against the use of deferred tax accounting are unconvincing if one considers the IASB's underlying assumption about accrual accounting, as stated in the *Framework*:

> In order to meet their objectives, financial statements are prepared on the accrual basis of accounting . . . Financial statements prepared on the accrual basis inform users not only of past transactions involving the payment and receipt of cash but also of obligations to pay cash in the future and of resources that represent cash to be received in the future.[18]

This underlying assumption confirms that deferred tax accounting makes the fullest possible use of accrual accounting.

Pursuing this argument further, the *Framework* states:

> The future economic benefit embodied in an asset is the potential to contribute, directly or indirectly, to the flow of cash and cash equivalents to the enterprise. The potential may be a productive one that is part of the operating activities of the enterprise.[19]

If a statement of financial position includes current market valuations based on this view of an asset, it is difficult to argue logically that the implicit taxation arising on this future economic benefit should not be provided for at the same time. The previous argument for excluding the deferred tax liability cannot therefore be considered persuasive on this basis.

On the other hand, it is stated in the *Framework* that 'An essential characteristic of a liability is that the enterprise has a present obligation'.[20] One could argue solely from these words that deferred tax is not a liability, but this conflicts with the argument based on the definition of an asset; consequently when considered in context this does not provide a sustainable argument against a deferred tax provision. The fact is that accounting practice has moved definitively towards making such a provision for deferred taxation.

Substance over form assumption

The legal argument that deferred tax is not a legal liability until it accrues runs counter to the criterion of substance over form which gives weight to the economic aspects of the event rather than the strict legal aspects. The *Framework* states:

> **Substance Over Form**
> If information is to represent faithfully the transactions and other events that it purports to represent, it is necessary that they are accounted for and presented in accordance with their substance and economic reality and not merely their legal form. The substance of transactions or other events is not always consistent with that which is apparent from their legal or contrived form.[21]

It is an interesting fact that substance over form has achieved a growing importance since the 1980s and the legal arguments are receiving less recognition. Investments are made on economic criteria, investors make their choices on the basis of anticipated cash flows, and such flows would be subject to the effects of deferred taxation.

16.9 Value added tax (VAT)

VAT is one other tax that affects most companies and for which there is an accounting standard (SSAP 5 *Accounting for Value Added Tax*), which was established on its introduction. This standard was issued in 1974 when the introduction of value added tax was imminent and there was considerable worry within the business community on its accounting treatment. We can now look back, having lived with VAT for well over three decades, and wonder, perhaps, why an SSAP was needed. VAT is essentially a tax on consumers collected by traders and is accounted for in a similar way to PAYE income tax, which is a tax on employees collected by employers.

IAS 18 (paragraph 8) makes clear that the same principles are followed:

> Revenue includes only the gross inflows of economic benefits received and receivable by the enterprise on its own account. Amounts collected on behalf of third parties such as sales taxes, goods and services taxes and value added taxes are not economic benefits which flow to the enterprise and do not result in increases in equity. Therefore, they are excluded from revenue.[22]

16.9.1 The effects of the standard

The effects of the standard vary depending on the status of the accounting entity under the VAT legislation. The term 'trader' appears in the legislation and is the terminology for a business entity. The 'traders' or companies, as we would normally refer to them, are classified under the following headings:

(a) Registered trader

For a registered trader, accounts should only include figures net of VAT. This means that the VAT on the sales will be deducted from the invoice amount. The VAT will be payable to the government and the net amount of the sales invoice will appear in the statement of income in arriving at the sales turnover figure. The VAT on purchases will be deducted from the purchase invoice. The VAT will then be reclaimed from the government and the net amount of the purchases invoice will appear in the statement of income in arriving at the purchases figure.

The only exception to the use of amounts net of VAT is when the input tax is not recoverable, e.g. on entertaining and on 'private' motor cars.

(b) Non-registered or exempt trader

For a company that is classified as non-registered or exempt, the VAT that it has to pay on its purchases and expenses is not reclaimable from the government. Because the company cannot recover the VAT, it means that the expense that appears in the statement of income must be inclusive of VAT. It is treated as part of each item of expenditure and the costs treated accordingly. It will be included, where relevant, with each item of expense (including capital expenditure) rather than being shown as a separate item.

(c) Partially exempt trader

An entity which is partially exempt can only recover a proportion of input VAT, and the proportion of non-recoverable VAT should be treated as part of the costs on the same lines as with an exempt trader. The VAT rules are complex but, for the purpose of understanding the figures that appear in published accounts of public companies, treatment as a registered trader would normally apply.

Summary

Corporation tax is charged on the taxable profit of a company after adjusting the accounting profit for non-allowable deductions and temporary differences.

The imputation system means that dividends are reported at the amount of cash paid out by the company and a credit is allowed on the dividend received by the shareholder.

Deferred tax is provided for under IAS 12 reflecting the amount that is expected to be settled as a liability. The requirement to make such a provision is supported by the *Framework for the Preparation and Presentation of Financial Statements*.

Tax avoidance and tax evasion have been perceived by the public as being unfair and governments have internationally attempted to combat the problem through legislation, case law and encouraging positive consumer reaction to put pressure on companies not appearing to pay a fair amount of tax nationally.

REVIEW QUESTIONS

1 Why does the charge to taxation in a company's accounts not equal the profit multiplied by the current rate of corporation tax?

2 Deferred tax accounting may be seen as an income-smoothing device which distorts the true and fair view. Explain the impact of deferred tax on reported income and justify its continued use.

3 Distinguish between (a) the deferral and (b) the liability methods of company deferred tax.

4 'If a deferred liability or asset is not expected to crystallise they should at least be discounted.' Discuss.

5 'The effective tax rate of all companies should be published and any with a rate below the average for the sector should be subjected to consumer or government commercial pressure to make additional payments.' Discuss.

6 Discuss the problems in distinguishing tax evasion from tax avoidance.

7 Discuss whether there is a socially responsible right amount of tax for a company to pay and who is to determine what is socially responsible.

8 'A tax adviser has a duty of care to a client to legally minimise a company's tax bill and would be professionally negligent not to do so.' Discuss.

9 'A company justified paying little tax on the grounds that it invested funds more effectively than government by creating employment. It further argued that this view was supported when it appears that governments lack the technical skills to control expenditure effectively.' Discuss.

10 The Financial Reporting Review Panel (FRRP) in its 2012 Annual Report stated that: 'Several companies had to be reminded that current and deferred tax liabilities and assets are to be measured using the tax rates that have been enacted or substantively enacted by the end of the reporting period'. Discuss why this is necessary.

11 The following is an extract from the Tesco plc 2013 Annual Report

Unrecognised deferred tax assets

Deferred tax assets in relation to continuing operations have not been recognised in respect of the following items:

	2013 £m	2012 £m
Deductible temporary differences	11	29
Tax losses	170	141
	181	170

Discuss reasons for not recognising the £181m as an asset.

12 Reconciliation of effective tax charge

Tesco plc reported a tax rate on its accounting profit of 24.2% and an effective tax rate of 29.3% in its 2013 Annual Report. Discuss three possible reasons for this difference.

13 A judge ruled in 1929 that 'No man in this country is under the smallest obligation, moral or other, so as to arrange his legal relations to his business or to his property as to enable the Inland Revenue to put the largest possible shovel into his stores'. Discuss the extent to which this approach is permitted in the UK and any impact on an accountant advising a client.

EXERCISES

Question 1

In your capacity as chief assistant to the financial controller, your managing director has asked you to explain to him the differences between tax planning, tax avoidance and tax evasion.

He has also asked you to explain to him your feelings as a professional accountant about these topics.

Required:
Write some notes to assist you in answering these questions.

* Question 2

A non-current asset (a machine) was purchased by Adjourn plc on 1 July 20X2 at a cost of £25,000.

The company prepares its annual accounts to 31 March in each year. The policy of the company is to depreciate such assets at the rate of 15% straight line (with depreciation being charged *pro rata* on a time-apportionment basis in the year of purchase). The company was granted capital allowances at 25% per annum on the reducing balance method (such capital allowances are apportioned *pro rata* on a time-apportionment basis in the year of purchase).

The rate of corporation tax has been as follows:

Year ended		
	31 Mar 20X3	20%
	31 Mar 20X4	30%
	31 Mar 20X5	20%
	31 Mar 20X6	19%
	31 Mar 20X7	19%

Required:
(a) Calculate the deferred tax provision using both the deferred method and the liability method.
(b) Explain why the liability method is considered by commentators to place the emphasis on the statement of financial position, whereas the deferred method is considered to place the emphasis on the statement of income.

* Question 3

The following information is given in respect of Unambitious plc:

(a) Non-current assets consist entirely of plant and machinery. The net book value of these assets as at 30 June 2010 is £100,000 in excess of their tax written-down value.

(b) The provision for deferred tax (all of which relates to fixed asset timing differences) as at 30 June 2010 was £21,000.

(c) The company's capital expenditure forecasts indicate that capital allowances and depreciation in future years will be:

Year ended 30 June	Depreciation charge for year	Capital allowances for year
£	£	£
2011	12,000	53,000
2012	14,000	49,000
2013	20,000	36,000
2014	40,000	32,000
2015	44,000	32,000
2016	46,000	36,000

For the following years, capital allowances are likely to continue to be in excess of depreciation for the foreseeable future.

(d) Corporation tax is to be taken at 21%.

Required:
Calculate the deferred tax charges or credits for the next six years, commencing with the year ended 30 June 2011, in accordance with the provisions of IAS 12.

Question 4

The move from the preparation of accounts under UK GAAP to the users of IFRS by United Kingdom quoted companies for years beginning 1 January 2005 had an effect on the level of profits reported. How will those profits arising from the change in accounting standards be treated for taxation purposes?

Question 5

Discuss the arguments for and against discounting the deferred tax charge.

Question 6

Austin Mitchell MP proposed an Early Day Motion in the House of Commons on 17 May 2005 as follows:

That this House urges the Government to clamp down on artificial tax avoidance schemes and end the . . . tax avoidance loop-holes that enable millionaires and numerous companies trading in the UK to avoid UK taxes; and further urges the Government to . . . so that transactions lacking normal commercial substance and solely entered into for the purpose of tax avoidance are ignored for tax purposes, thereby providing certainty, fairness and clarity, which the UK's taxation system requires to prevent abusive tax avoidance, to protect the interests of ordinary citizens who are committed to making their contribution to society, to avoid an unnecessary burden of tax on individual taxpayers and to ensure that companies pay fair taxes on profits generated in this country.

Required:

(a) The Motion refers to tax avoidance. In your opinion, does the Early Day Motion tend to confuse the boundaries between tax avoidance and tax evasion?

(b) The Motion refers to nullifying the effects of tax avoidance to protect the interests of ordinary citizens who are making their contribution to society, to avoid an unnecessary burden of tax on individual taxpayers. If ordinary citizens require such protection, would it be possible to argue that even if tax avoidance were legal, it might well be immoral?

Question 7

Hanson Products Ltd is a newly formed company. The company commenced trading on 1 January 20X1 when it purchased an item of plant and equipment for $240,000. The plant and equipment has an expected life of five years with zero residual value, and will be depreciated on a straight-line basis on cost over that period. The company's profits before depreciation (of the plant) are expected to be $1 million each year.

Tax allowances for plant are a 40% initial allowance with an annual 25% writing-down allowance on tax written-down value in subsequent years. The company will have a life of five years and, on closure, any unused tax allowances will be allowed as a deduction from the final year's taxable profit.

The rate of corporation tax is 20%. The company does not provide for deferred taxation.

Required:

(a) For each of the years from 20X1 to 20X5, calculate:
   (i)   the capital allowances,
   (ii) the taxable profit,
   (iii) the tax payable on the year's profit.

(b) Discuss the advantages and disadvantages of not providing for deferred taxation.

Question 8

The accountant of Hanson Products Ltd has asked you how your answer to Question 8 above would be affected using the following two methods of calculating deferred taxation.

Required:

(a) For each of the years from 20X1 to 20X5, calculate the deferred tax balance if:
   (i) full provision is made for deferred tax in accordance with IAS 12 *Income Taxes*,
   (ii) the company decided to calculate the deferred tax balance using a discount rate of 5%.

(b) Discuss the advantages and disadvantages of discounting deferred tax balances. Use the following table of discount factors:

Year	Discount factor
1	0.9524
2	0.9070
3	0.8638
4	0.8227
5	0.7835

Question 9

The following information relates to Deferred plc:

- EBITDA (earnings before interest, tax, depreciation and amortisation) for year ended 31.12.20X1 is £300,000

- No interest payable in 20X1

- No amortisation

- Equipment cost £100,000 at 1.1.20X1

  - Depreciation rate is 10% straight line

  - Nil scrap value

- Tax rate is 20%

- Capital allowance is 25% on reducing balance basis.

Required:
Calculate:
(a) deferred tax;
(b) statement of income entries;
(c) statement of financial position entries.

References

1 OECD, *Theoretical and Empirical Aspects of Corporate Taxation*, Paris, 1974; van den Temple, *Corporation Tax and Individual Income Tax in the EEC*, EEC Commission, Brussels, 1974.
2 G.H. Partington and R.H. Chenhall, *Dividends, Distortion and Double Taxation*, Abacus, June 1983.
3 G. Aaronson 'GAAR Study', Chatered Instiutue of Taxation, November 2011.
4 Franklin D. Roosevelt, 1936 Speech at Worcester, Mass., 1936. Roosevelt Museum.
5 *Countering Tax Avoidance in the UK: Which Way Forward?*, A Report for the Tax Law Review Committee, The Institute for Fiscal Studies, 2009, para. 4.2.
6 L.J. Tomlin, in *Duke of Westminster* v *CIR*, HL 1935, 19 TC 490.
7 *Tax Avoidance*, A Report for the Tax Law Review Committee, The Institute for Fiscal Studies, 1997, para. 7.
8 *WT Ramsay Ltd* v *CIR*, HL 1981, 54 TC 101; [1981] STC 174; [1981] 2 WLR 449; [1981] 1 All ER 865.
9 Robert Maas, *Beware Tax Avoidance Drifting into Evasion*, Taxline, Tax Planning 2003–2004, Institute of Chartered Accountants in England & Wales.
10 *Professional Conduct in Relation to Taxation*, Ethical Statement 1.308, Institute of Chartered Accountants in England & Wales, para. 2.13 (this is similar to the statements issued by the other accounting bodies).
11 http://www.oecd.org/tax/exchangeofinformation/42232037.pdf
12 http://www.ifs.org.uk/comms/dp7.pdf
13 IAS 12 *Income Taxes*, IASB, revised 2000, para. 5.
14 SFAS 109, *Accounting for Income Taxes*, FASB, 1992, extracts therefrom.
15 IAS 12 *Income Taxes*, IASB, revised 2000, para. 54.
16 R.J. Chambers, *Tax Allocation and Financial Reporting*, Abacus, 1968.
17 Prof. D.R. Middleton, letter to the Editor, *The Financial Times*, 29 September 1994.
18 *Framework for the Preparation and Presentation of Financial Statements*, IASB, 2001, para. 22.
19 Ibid., para. 53.
20 Ibid., para. 60.
21 Ibid., para. 35.
22 IAS 18 *Revenue*, IASB, 2001, para. 8.

CHAPTER 17

Property, plant and equipment (PPE)

17.1 Introduction

The main purpose of this chapter is to explain how to determine the initial carrying value of PPE and to explain and account for the normal movements in PPE that occur during an accounting period.

Objectives

By the end of this chapter, you should be able to:

- explain the meaning of PPE and determine its initial carrying value;
- account for subsequent expenditure on PPE that has already been recognised;
- explain the meaning of depreciation and compute the depreciation charge for a period;
- account for PPE measured under the revaluation model;
- explain the meaning of impairment;
- compute and account for an impairment loss;
- explain the criteria that must be satisfied before an asset is classified as held for sale and account for such assets;
- explain the accounting treatment of government grants for the purchase of PPE;
- identify an investment property and explain the alternative accounting treatment of such properties;
- explain the impact of alternative methods of accounting for PPE on key accounting ratios.

17.2 PPE – concepts and the relevant IASs and IFRSs

For PPE the accounting treatment is based on the accruals or matching concepts, under which expenditure is capitalised until it is charged as depreciation against revenue in the periods in which benefit is gained from its use. Thus, if an item is purchased that has an economic life of two years, so that it will be used over two accounting periods to help earn profit for the entity, then the cost of that asset should be apportioned in some way between the two accounting periods.

However, this does not take into account the problems surrounding PPE accounting and depreciation, which have so far given rise to six relevant international accounting standards. We will consider these problems in this chapter and cover the following questions.

IAS 16 and IAS 23

- What is PPE (IAS 16)?
- How is the cost of PPE determined (IAS 16 and IAS 23)?
- How is depreciation of PPE computed (IAS 16)?
- What are the regulations regarding carrying PPE at revalued amounts (IAS 16)?

Other relevant international accounting standards and pronouncements

- How should grants receivable towards the purchase of PPE be dealt with (IAS 20)?
- Are there ever circumstances in which PPE should not be depreciated (IAS 40)?
- What is impairment and how does this affect the carrying value of PPE (IAS 36)?
- What are the key changes made by the IASB concerning the disposal of non-current assets (IFRS 5)?

17.3 What is PPE?

IAS 16 *Property, Plant and Equipment*[1] defines PPE as tangible assets that are:

(a) held by an entity for use in the production or supply of goods and services, for rental to others, or for administrative purposes; and

(b) expected to be used during more than one period.

It is clear from the definition that PPE will normally be included in the non-current assets section of the statement of financial position.

17.3.1 Problems that may arise

Problems may arise in relation to the interpretation of the definition and in relation to the application of the materiality concept.

The definitions give rise to some areas of practical difficulty. For example, an asset that has previously been held for use in the production or supply of goods or services but is now going to be sold should, under the provisions of IFRS 5, be classified separately on the statement of financial position as an asset 'held for sale'.

Differing accounting treatments arise if there are different assessments of materiality. This may result in the same expenditure being reported as an asset in the statement of financial position of one company and as an expense in the statement of comprehensive income of another company. In the accounts of a self-employed carpenter, a kit of hand tools that, with careful maintenance, will last many years will, quite rightly, be shown as PPE. Similar assets used by the maintenance department in a large factory will, in all probability, be treated as 'loose tools' and written off as acquired.

Many entities have *de minimis* policies, whereby only items exceeding a certain value are treated as PPE; items below the cut-off amount will be expensed through the statement of comprehensive income.

For example, the MAN 2003 Annual Report stated in its accounting policies:

Tangible assets are depreciated according to the straight-line method over their estimated useful lives. Low-value items (defined as assets at cost of €410 or less) are fully written off in the year of purchase.

17.4 How is the cost of PPE determined?

17.4.1 Components of cost[2]

According to IAS 16, the cost of an item of PPE comprises its purchase price, including import duties and non-refundable purchase taxes, plus any directly attributable costs of bringing the asset to working condition for its intended use. Examples of such directly attributable costs include:

(a) the costs of site preparation;

(b) initial delivery and handling costs;

(c) installation costs;

(d) professional fees such as for architects and engineers;

(e) the estimated cost of dismantling and removing the asset and restoring the site, to the extent that it is recognised as a provision under IAS 37 *Provisions, Contingent Liabilities and Contingent Assets*.

Administration and other general overhead costs are not a component of the cost of PPE unless they can be directly attributed to the acquisition of the asset or bringing it to its working condition. Similarly, start-up and similar pre-production costs do not form part of the cost of an asset unless they are necessary to bring the asset to its working condition.

17.4.2 Self-constructed assets[3]

The cost of a self-constructed asset is determined using the same principles as for an acquired asset. If the asset is made available for sale by the entity in the normal course of business then the cost of the asset is usually the same as the cost of producing the asset for sale. This cost would usually be determined under the principles set out in IAS 2 *Inventories*.

The normal profit that an enterprise would make if selling the self-constructed asset would not be recognised in 'cost' if the asset were retained within the entity. Following similar principles, where one group company constructs an asset that is used as PPE by another group company, any profit on sale is eliminated in determining the initial carrying value of the asset in the consolidated accounts (this will also clearly affect the calculation of depreciation).

If an item of PPE is exchanged in whole or in part for a dissimilar item of PPE then the cost of such an item is the fair value of the asset received. This is equivalent to the fair value of the asset given up, adjusted for any cash or cash equivalents transferred or received.

17.4.3 Capitalisation of borrowing costs

Where an asset takes a substantial period of time to get ready for its intended use or sale then the entity may incur significant borrowing costs in the preparation period. Under the accruals basis of accounting there is an argument that such costs should be included as a directly attributable cost of construction. IAS 23 *Borrowing Costs* was issued to deal with this issue.

IAS 23 states that borrowing costs that are directly attributable to the acquisition, construction or production of a 'qualifying asset' should be included in the cost of that asset.[4] A 'qualifying asset' is one that necessarily takes a substantial period of time to get ready for its intended use or sale.

Borrowing costs that would have been avoided if the expenditure on the qualifying asset had not been undertaken are eligible for capitalisation under IAS 23. Where the funds are borrowed specifically for the purpose of obtaining a qualifying asset, the borrowing costs that are eligible for capitalisation are those incurred on the borrowing during the period less any investment income on the temporary investment of those borrowings. Where the funds are borrowed generally and used for the purpose of obtaining a qualifying asset, the entity should use a capitalisation rate to determine the borrowing costs that may be capitalised. This rate should be the weighted average of the borrowing costs applicable to the entity, other than borrowings made specifically for the purpose of obtaining a qualifying asset. Capitalisation should commence when:

- expenditures for the asset are being incurred;
- borrowing costs are being incurred;
- activities that are necessary to prepare the asset for its intended use or sale are in progress.

When substantially all the activities necessary to prepare the qualifying asset for its intended use or sale are complete, capitalisation should cease.

Borrowing costs treatment in the UK

The UK standard that deals with this issue is FRS 15 *Tangible Fixed Assets*. FRS 15 makes the capitalisation of borrowing costs optional, rather than compulsory. FRS 15 requires that the policy be applied consistently, however. This used to be the treatment under IAS 23 before that standard was revised in 2007.

Borrowing costs for SMEs

IAS 23 *Borrowing Costs* requires borrowing costs directly attributable to the acquisition, construction or production of a qualifying asset (including some inventories) to be capitalised as part of the cost of the asset. For cost–benefit reasons, the IFRS for SMEs requires such costs to be charged to expense.

IFRS for SMEs

All borrowing costs are charged to expense when incurred. Borrowing costs are not capitalised.

17.4.4 Subsequent expenditure

Subsequent expenditure relating to an item of PPE that has already been recognised should normally be recognised as an expense in the period in which it is incurred. The exception to this general rule is where it is probable that future economic benefits in excess of the originally assessed standard of performance of the existing asset will, as a result of the expenditure, flow to the entity. In these circumstances, the expenditure should be added to the carrying value of the existing asset. Examples of expenditure that might fall to be treated in this way include:

- modification of an item of plant to extend its useful life, including an increase in its capacity;
- upgrading machine parts to achieve a substantial improvement in the quality of output;
- adoption of new production processes enabling a substantial reduction in previously assessed operating costs.

Conversely, expenditure that restores, rather than increases, the originally assessed standard of performance of an asset is written off as an expense in the period incurred.

Some assets have components that require replacement at regular intervals. Two examples of such components would be the lining of a furnace and the roof of a building. IAS 16 states[5] that, provided such components have readily ascertainable costs, they should be accounted

Mineral and surface rights

Mineral and surface rights are recorded at cost of acquisition. When there is little likelihood of a mineral right being exploited, or the value of mineral rights have diminished below cost, a write-down is effected against income in the period that such determination is made.

Few jurisdictions have comprehensive accounting standards for extractive activities. IFRS 6 *Exploration for and Evaluation of Mineral Resources* is an interim measure pending a more comprehensive view by the ASB in future. IFRS 6 allows an entity to develop an accounting policy for exploration and evaluation assets without considering the consistency of the policy with the IASB framework. This may mean that for an interim period accounting policies might permit the recognition of both current and non-current assets that do not meet the criteria laid down in the IASB *Framework*. This is considered by some commentators to be unduly permissive. Indeed, about the only firm requirement IFRS 6 can be said to contain is the requirement to test exploration and evaluation assets for impairment whenever a change in facts and circumstances suggests that impairment exists.

17.6 What are the constituents in the depreciation formula?

In order to calculate depreciation it is necessary to determine three factors:

1　Cost (or revalued amount if the company is following a revaluation policy)
2　Economic life
3　Residual value.

A simple example is the calculation of the depreciation charge for a company that has acquired an asset on 1 January 20X1 for £1,000 with an estimated economic life of four years and an estimated residual value of £200. Applying a straight-line depreciation policy, the charge would be £200 per year using the formula of:

$$\frac{\text{Cost} - \text{estimated residual value}}{\text{Estimated economic life}} = \frac{£1,000 - £200}{4} = £200 \text{ per annum}$$

We can see that the charge of £200 is influenced in all cases by the definition of cost, the estimate of the residual value, the estimate of the economic life, and the management decision on depreciation policy.

In addition, if the asset were to be revalued at the end of the second year to £900, then the depreciation for 20X3 and 20X4 would be recalculated using the revised valuation figure. Assuming that the residual value remained unchanged, the depreciation for 20X3 would be:

$$\frac{\text{Revalued asset} - \text{estimated residual value}}{\text{Estimated economic life}} = \frac{£900 - £200}{2} = £350 \text{ per annum}$$

17.6.1 How is the useful life of an asset determined?

The IAS 16 definition of useful life is given in Section 17.5.2 above. This is not necessarily the total life expectancy of the asset. Most assets become less economically and technologically efficient as they grow older. For this reason, assets may well cease to have an economic life long before their working life is over. It is the responsibility of the preparers of accounts to estimate the economic life of all assets.

It is conventional for entities to consider the economic lives of assets by class or category, e.g. buildings, plant, office equipment, or motor vehicles. However, this is not necessarily appropriate, since the level of activity demanded by different users may differ. For example, compare two motor cars owned by a business: one is used by the national sales manager, covering 100,000 miles per annum visiting clients; the other is used by the accountant to drive from home to work and occasionally the bank, covering perhaps one-tenth of the mileage.

In practice, the useful economic life would be determined by reference to factors such as repair costs, the cost and availability of replacements, and the comparative cash flows of existing and alternative assets. The problem of optimal replacement lives is a normal financial management problem; its significance in financial reporting is that the assumptions used within the financial management decision may provide evidence of the expected economic life.

17.6.2 Other factors affecting the useful life figure

We can see that there are technical factors affecting the estimated economic life figure. In addition, other factors have prompted companies to set estimated lives that have no relationship to the active productive life of the asset. One such factor is the wish of management to take into account the effect of inflation. This led some companies to reduce the estimated economic life, so that a higher charge was made against profits during the early period of the asset's life to compensate for the inflationary effect on the cost of replacement. The total charge will be the same, but the timing is advanced. This does not result in the retention of funds necessary to replace; but it does reflect the fact that there is at present no coherent policy for dealing with inflation in the published accounts – consequently, companies resort to *ad hoc* measures that frustrate efforts to make accounts uniform and comparable. *Ad hoc* measures such as these have prompted changes in the standards.

17.6.3 Residual value

IAS 16 defines residual value as the net amount which an entity expects to obtain for an asset at the end of its useful life after deducting the expected costs of disposal. Where PPE is carried at cost, the residual value is initially estimated at the date of acquisition. In subsequent periods the estimate of residual value is revised, the revision being based on conditions prevailing at each statement of financial position date. Such revisions have an effect on future depreciation charges.

Besides inflation, residual values can be affected by changes in technology and market conditions. For example, during the period 1980–90 the cost of small business computers fell dramatically in both real and monetary terms, with a considerable impact on the residual (or second-hand) value of existing equipment.

17.7 Calculation of depreciation

Having determined the key factors in the computation, we are left with the problem of how to allocate that cost between accounting periods. For example, with an asset having an economic life of five years:

	£
Asset cost	11,000
Estimated residual value (no significant change anticipated over useful economic life)	1,000
Depreciable amount	10,000

Figure 17.1 Effect of different depreciation methods

	Straight-line (£2,000) £	Diminishing balance (38%) £	Difference £
Cost	11,000	11,000	
Depreciation for year 1	2,000	4,180	2,180
Net book value (NBV)	9,000	6,820	
Depreciation for year 2	2,000	2,592	592
NBV	7,000	4,228	
Depreciation for year 3	2,000	1,606	(394)
NBV	5,000	2,622	
Depreciation for year 4	2,000	996	(1,004)
NBV	3,000	1,626	
Depreciation for year 5	2,000	618	(1,382)
Residual value	1,000	1,008	

The diminishing balance formula was $1 - \sqrt[n]{(\text{Residual value/Cost})}$

How should the depreciable amount be charged to the statement of comprehensive income over the five years? IAS 16 tells us that it should be allocated on a systematic basis and the depreciation method used should reflect as fairly as possible the pattern in which the asset's economic benefits are consumed. The two most popular methods are **straight-line**, in which the depreciation is charged evenly over the useful life, and **diminishing balance**, where depreciation is calculated annually on the net written-down amount. In the case above, the calculations would be as in Figure 17.1.

Note that, although the diminishing balance is generally expressed in terms of a percentage, this percentage is arrived at by inserting the economic life into the formula as n; the 38% reflects the expected economic life of five years. As we change the life, so we change the percentage that is applied. The normal rate applied to vehicles is 25% diminishing balance; if we apply that to the cost and residual value in our example, we can see that we would be assuming an economic life of eight years. It is a useful test when using reducing balance percentages to refer back to the underlying assumptions.

We can see that the end result is the same. Thus, £10,000 has been charged against income, but with a dramatically different pattern of statement of comprehensive income charges. The charge for straight-line depreciation in the first year is less than half that for reducing balance.

17.7.1 Arguments in favour of the straight-line method

The method is simple to calculate. However, in these days of calculators and computers this seems a particularly facile argument, particularly when one considers the materiality of the figures.

17.7.2 Arguments in favour of the diminishing balance method

First, the charge reflects the efficiency and maintenance costs of the asset. When new, an asset is operating at its maximum efficiency, which falls as it nears the end of its life.

This may be countered by the comment that in year 1 there may be 'teething troubles' with new equipment, which, while probably covered by a supplier's guarantee, will hamper efficiency.

Secondly, the pattern of diminishing balance depreciation gives a net book amount that approximates to second-hand values. For example, with motor cars the initial fall in value is very high.

17.7.3 Other methods of depreciating

Besides straight-line and diminishing balance, there are a number of other methods of depreciating, such as the sum of the units method, the machine-hour method and the annuity method. We will consider these briefly.

Sum of the units method

A compromise between straight-line and reducing balance that is popular in the USA is the sum of the units method. The calculation based on the information in Figure 17.1 is now shown in Figure 17.2. This has the advantage that, unlike diminishing balance, it is simple to obtain the exact residual amount (zero if appropriate), while giving the pattern of high initial charge shown by the diminishing balance approach.

Machine-hour method

The machine-hour system is based on an estimate of the asset's service potential. The economic life is measured not in accounting periods but in working hours, and the depreciation is allocated in the proportion of the actual hours worked to the potential total hours available. This method is commonly employed in aviation, where aircraft are depreciated on the basis of flying hours.

Annuity method

With the annuity method, the asset, or rather the amount of capital representing the asset, is regarded as being capable of earning a fixed rate of interest. The sacrifice incurred in using the asset within the business is therefore twofold: the loss arising from the exhaustion of

Figure 17.2 Sum of the units method

		£
Cost		11,000
Depreciation for year 1	£10,000 × 5/15	3,333
Net book value (NBV)		7,667
Depreciation for year 2	£10,000 × 4/15	2,667
NBV		5,000
Depreciation for year 3	£10,000 × 3/15	2,000
NBV		3,000
Depreciation for year 4	£10,000 × 2/15	1,333
NBV		1,667
Depreciation for year 5	£10,000 × 1/15	667
Residual value		1,000

Figure 17.3 Annuity method

Year	Opening written-down value £	Notional interest (10%) £	Annual payment £	Net movement £	Closing written-down value £
1	10,000	1,000	(2,638)	(1,638)	8,362
2	8,362	836	(2,638)	(1,802)	6,560
3	6,560	656	(2,638)	(1,982)	4,578
4	4,578	458	(2,638)	(2,180)	2,398
5	2,398	240	(2,638)	(2,398)	Nil

the service potential of the asset; and the interest forgone by using the funds invested in the business to purchase the non–current asset. With the help of annuity tables, a calculation shows what equal amounts of depreciation, written off over the estimated life of the asset, will reduce the book value to nil, after debiting interest to the asset account on the diminishing amount of funds that are assumed to be invested in the business at that time, as represented by the value of the asset.

Figure 17.3 contains an illustration based on the treatment of a five-year lease which cost the company a premium of £10,000 on 1 January year 1. It shows how the total depreciation charge is computed. Each year the charge for depreciation in the statement of comprehensive income is the equivalent annual amount that is required to repay the investment over the five-year period at a rate of interest of 10% less the notional interest available on the remainder of the invested funds.

An extract from the annuity tables to obtain the annual equivalent factor for year 5 and assuming a rate of interest of 10% would show:

Year	Annuity $A_{\overline{n}|}^{-1}$
1	1.1000
2	0.5762
3	0.4021
4	0.3155
5	0.2638

Therefore, at a rate of interest of 10% five annual payments to repay an investor of £10,000 would each be £2,638.

A variation of this system involves the investment of a sum equal to the net charge in fixed interest securities or an endowment policy, so as to build up a fund that will generate cash to replace the asset at the end of its life.

This last system has significant weaknesses. It is based on the misconception that depreciation is 'saving up for a new one', whereas in reality depreciation is charging against profits funds already expended. It is also dangerous in a time of inflation, since it may lead management not to maintain the capital of the entity adequately, in which case they may not be able to replace the assets at their new (inflated) prices.

The annuity method, with its increasing net charge to income, does tend to take inflationary factors into account, but it must be noted that the *total* net profit and loss charge only adds up to the cost of the asset.

17.7.4 Which method should be used?

The answer to this seemingly simple question is 'it depends'. On the matter of depreciation IAS 16 is designed primarily to force a fair charge for the use of assets into the statement of comprehensive income each year, so that the earnings reflect a true and fair view.

Straight-line is most suitable for assets such as leases which have a definite fixed life. It is also considered most appropriate for assets with a short working life, although with motor cars the diminishing balance method is sometimes employed to match second-hand values. Extraction industries (mining, oil wells, quarries, etc.) sometimes employ a variation on the machine-hour system, where depreciation is based on the amount extracted as a proportion of the estimated reserves.

Despite the theoretical attractiveness of other methods the straight-line method is, by a long way, the one in most common use by entities that prepare financial statements in accordance with IFRSs. Reasons for this are essentially pragmatic:

- It is the most straightforward to compute.

- In the light of the three additional subjective factors – cost (or revalued amount), residual value and useful life – that need to be estimated, any imperfections in the charge for depreciation caused by the choice of the straight-line method are not likely to be significant.

- It conforms to the accounting treatment adopted by peers. For example, one group reported that it currently used the reducing balance method but, as peer companies used the straight-line method, it decided to change and adopt that policy.

17.8 Measurement subsequent to initial recognition

17.8.1 Choice of models

An entity needs to choose either the cost or the revaluation model as its accounting policy for an entire class of PPE. The cost model (definitely the most common) results in an asset being carried at cost less accumulated depreciation and any accumulated impairment losses.

17.8.2 The revaluation model

Under the revaluation model the asset is carried at revalued amount, being its fair value at the date of the revaluation less any subsequent accumulated depreciation and subsequent accumulated impairment losses. The fair value of an asset is defined in IAS 16 as 'the amount for which an asset could be exchanged between knowledgeable and willing parties in an arm's length transaction'. Thus fair value is basically market value. If a market value is not available, perhaps in the case of partly used specialised plant and equipment that is rarely bought and sold other than as new, then IAS 16 requires that revaluation be based on depreciated replacement cost. Note that the fair value falls within the scope of IFRS 13.

EXAMPLE ● An entity purchased an item of plant for £12,000 on 1 January 20X1. The plant was depreciated on a straight-line basis over its useful economic life, which was estimated at six years. On 1 January 20X3 the entity decided to revalue its plant. No fair value was available for the item of plant that had been purchased for £12,000 on 1 January 20X1 but the replacement cost of the plant at 1 January 20X3 was £21,000.

The carrying value of the plant immediately before the revaluation would have been:

- Cost £12,000
- Accumulated deprecation £4,000 [(£12,000/6) × 2]
- Written-down value £8,000.

Under the principles of IAS 16 the revalued amount would be £14,000 (£21,000 × 4/6). This amount would be reflected in the financial statements by either:

- showing a revised gross figure of £14,000 and reversing out all the accumulated depreciation charged to date so as to give a carrying value of £14,000; or

- restating both the gross figure and the accumulated depreciation by the proportionate change in replacement cost. This would give a gross figure of £21,000, with accumulated depreciation restated at £7,000 to once again give a net carrying value of £14,000.

17.8.3 Detailed requirements regarding revaluations

The frequency of revaluations depends upon the movements in the fair values of those items of PPE being revalued. In jurisdictions where the rate of price changes is very significant revaluations may be necessary on an annual basis. In other jurisdictions revaluations every three or five years may well be sufficient.

Where an item of PPE is revalued, the entire class of PPE to which that asset belongs should be revalued.[6] A class of PPE is a grouping of assets of a similar nature and use in an entity's operations. Examples would include:

- land;
- land and buildings;
- machinery.

This is an important provision because without it entities would be able to select which assets they revalued on the basis of best advantage to the financial statements. Revaluations will usually increase the carrying values of assets and equity and leave borrowings unchanged. Therefore gearing (or leverage) ratios will be reduced. It is important that, if the revaluation route is chosen, assets are revalued on a rational basis.

The following is an extract from the financial statements of Coil SA, a company incorporated in Belgium that prepares financial statements in euros in accordance with international accounting standards: 'Items of PPE are stated at historical cost modified by revaluation and are depreciated using the straight-line method over their estimated useful lives.'

17.8.4 Accounting for revaluations

When the carrying amount of an asset is increased as a result of a revaluation, the increase should be credited directly to other comprehensive income, being shown in equity under the heading of revaluation surplus. The only exception is where the gain reverses a revaluation decrease previously recognised as an expense **relating to the same asset**.

This means that, in the example we considered under Section 17.8.2 above, the revaluation would lead to a credit of £6,000 (£14,000 – £8,000) to other comprehensive income.

If, however, the carrying amount of an asset is decreased as a result of a revaluation, the decrease should be recognised as an expense. The only exception is where that asset had previously been revalued. In those circumstances the loss on revaluation is charged against the revaluation surplus to the extent that the revaluation surplus contains an amount **relating to the same asset**.

EXAMPLE I ● REVALUED BUT NOT SOLD An entity buys freehold land for £100,000 in year 1. The land is revalued to £150,000 in year 3 and £90,000 in year 5. The land is not depreciated.

- In year 3 a surplus of £50,000 (£150,000 – £100,000) is reported as other comprehensive income and included in equity under the heading 'revaluation surplus'.
- In year 5 a deficit of £60,000 (£90,000 – £150,000) arises on the second revaluation. £50,000 of this deficit is deducted from the revaluation surplus and £10,000 is charged as an expense.
- It is worth noting that £10,000 is the amount by which the year 5 carrying amount is lower than the original cost of the land.

EXAMPLE 2 ● REVALUED AND THEN SOLD WITH THE REVALUATION SURPLUS REALISED AT TIME OF SALE Where an asset that has been revalued is sold, the revaluation surplus becomes realised.[7] It may be transferred to retained earnings when this happens but this transfer is not made through the statement of comprehensive income.

Continuing with our example in Section 17.8.2, let us assume that:

- the plant was sold on 1 January 20X5 for £5,000; and
- the carrying amount of the asset in the financial statements immediately before the sale was £7,000 [£14,000 – (2 × £3,500)].

This means that a loss on sale of £2,000 would be taken to the statement of comprehensive income, and the revaluation surplus of £6,000 would be transferred to retained earnings.

EXAMPLE 3 ● REVALUED AND THEN SOLD WITH THE EXCESS DEPRECIATION RECOGNISED EACH YEAR IAS 16 allows for the possibility that the revaluation surplus is transferred to retained earnings as the asset is depreciated. To turn once again to our example, we see that:

- the revaluation on 1 January 20X3 increased the annual depreciation charge from £2,000 (£12,000/6) to £3,500 (£21,000/6);
- following revaluation an amount equivalent to the 'excess depreciation' may be transferred from the revaluation surplus to retained earnings as the asset is depreciated. This would lead in our example to a transfer of £1,500 each year; and
- if this occurs then the revaluation surplus that is transferred to retained earnings on sale is £3,000 [£6,000 – (2 × £1,500)].

17.8.5 IFRS for SMEs

The IFRS for SMEs does not permit the use of the revaluation model and only requires a review if there is an indication that there has been a significant change since the last annual reporting date.

17.9 IAS 36 *Impairment of Assets*

17.9.1 IAS 36 approach

IAS 36 sets out the principles and methodology for accounting for impairments of non-current assets and goodwill. Where possible, individual non-current assets should be individually tested for impairment. However, where cash flows do not arise from the use of a single non-current asset, impairment is measured for the smallest group of assets which generates income that is largely independent of the company's other income streams. This smallest group is referred to as a cash-generating unit (CGU).

Impairment of an asset, or CGU (if assets are grouped), occurs when the carrying amount of an asset or CGU is greater than its recoverable amount, where:

- the carrying amount is the depreciated historical cost (or depreciated revalued amount);
- the recoverable amount is the higher of the net selling price and the value in use, where:
  - the net selling price is the amount at which an asset could be disposed of, less any direct selling costs; and
  - the value in use is the present value of the future cash flows obtainable as a result of an asset's continued use, including those resulting from its ultimate disposal.

When impairment occurs, a **revised carrying amount** is calculated for the statement of financial position as follows:

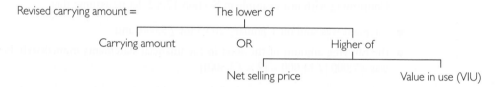

It is not always necessary to go through the potentially time-consuming process of computing the value in use of an asset. If the net selling price can be shown to be higher than the existing carrying value then the asset cannot possibly be impaired and no further action is necessary. However, this is not always the case for non-current assets and a number of assets (e.g. goodwill) cannot be sold, so several value in use computations are inevitable.

The revised carrying amount is then depreciated over the remaining useful economic life.

17.9.2 Dividing activities into CGUs

In order to carry out an impairment review it is necessary to decide how to divide activities into CGUs. There is no single answer to this – it is extremely judgemental, e.g. if the company has multi-retail sites, the cost of preparing detailed cash flow forecasts for each site could favour grouping.

The risk of grouping is that poorly performing operations might be concealed within a CGU and it would be necessary to consider whether there were any commercial reasons for breaking a CGU into smaller constituents, e.g. if a location was experiencing its own unique difficulties such as local competition or inability to obtain planning permission to expand to a more profitable size.

17.9.3 Indications of impairment

A review for impairment is required when there is an indication that an impairment has actually occurred. The following are indicators of impairment:

- External indicators:
  - a fall in the market value of the asset;
  - material adverse changes in regulatory environment;

- material adverse changes in markets;

- material long-term increases in market rates of return used for discounting.

- Internal indicators:

  - material changes in operations;

  - major reorganisation;

  - loss of key personnel;

  - loss or net cash outflow from operating activities if this is expected to continue or is a continuation of a loss-making situation.

If there is such an indication, it is necessary to determine the depreciated historical cost of a single asset, or the net assets employed if a CGU, and compare this with the net realisable value and value in use.

AkzoNobel stated in its 2013 Annual Report:

> We assess the carrying value of intangible assets and property, plant and equipment whenever events or changes in circumstances indicate that the carrying amount of an asset may not be recoverable. In addition, for goodwill and other intangible assets with an indefinite useful life, we review the carrying value annually in the fourth quarter.
>
> The recoverable amount of an asset or its cash-generating unit is the greater of its value in use and its fair value less costs to sell, whereby estimated future cash flows are discounted to their present value. The discount rate used reflects current market assessments of the time value of money and, if appropriate, the risks specific to the assets. If the carrying value of an asset or its cash-generating unit exceeds its estimated recoverable amount, an impairment loss is recognized in the statement of income. The assessment for impairment is performed at the lowest level of assets generating largely independent cash inflows, which we have determined to be at business unit level (one level below segment). We allocate impairment losses in respect of cash-generating units first to goodwill and then to the carrying amount of the other assets on a pro rata basis.

17.9.4 Value in use calculation

Value in use is arrived at by estimating and discounting the income stream. The **income streams**:

- are likely to follow the way in which management monitors and makes decisions about continuing or closing the different lines of business;

- may often be identified by reference to major products or services;

- should be based on reasonable and supportable assumptions;

- should be consistent with the most up-to-date budgets and plans that have been formally approved by management, or if they are for a period beyond that covered by formal budgets and plans should, unless there are exceptional circumstances, assume a steady or declining growth rate;[8]

- should be projected cash flows unadjusted for risk, discounted at a rate of return expected from a similarly risky investment, or should be projected risk-adjusted pre-tax cash flows discounted at a risk-free rate.

The **discount rate** should be:

- calculated on a pre-tax basis;
- an estimate of the rate that the market would expect on an equally risky investment excluding the effects of any risk for which the cash flows have been adjusted:[9]
  - increased to reflect the way the market would assess the specific risks associated with the projected cash flows;
  - reduced to a risk-free rate if the cash flows have been adjusted for risk.

The following illustration is from the Roche Holdings, Inc. 2014 Annual Report:

> When the recoverable amount of an asset, being the higher of its net selling price and its value in use, is less than the carrying amount, then the carrying amount is reduced to its recoverable amount. This reduction is reported in the income statement as an impairment loss. Value in use is calculated using estimated cash flows, generally over a five-year period, with extrapolating projections for subsequent years. These are discounted using an appropriate long-term pre-tax interest rate. When an impairment arises, the useful life of the asset in question is reviewed and, if necessary, the future depreciation/amortisation charge is accelerated.

17.9.5 Treatment of impairment losses

If the carrying value exceeds the higher of net selling price and value in use, then an impairment loss has occurred. The accounting treatment of such a loss is as follows.

Asset not previously revalued

An impairment loss should be recognised in the statement of comprehensive income in the year in which the impairment arises.

Asset previously revalued

An impairment loss on a revalued asset is effectively treated as a revaluation deficit. As we have already seen, this means that the decrease should be recognised as an expense. The only exception is where that asset had previously been revalued. In those circumstances the loss on revaluation is charged against the revaluation surplus to the extent that the revaluation surplus contains an amount **relating to the same asset**.

Allocation of impairment losses

Where an impairment loss arises, the loss should ideally be set against the specific asset to which it relates. Where the loss cannot be identified as relating to a specific asset, it should be apportioned within the CGU to reduce the most subjective values first, as follows:

- first, to reduce any goodwill within the CGU;
- then to the unit's other assets, allocated on a *pro rata* basis;
- with the proviso that no individual asset should be reduced below the higher of:
  - its net selling price (if determinable);
  - its value in use (if determinable);
  - zero.

The following is an example showing the allocation of an impairment loss.

EXAMPLE ● A cash-generating unit contains the following assets:

	£
Goodwill	70,000
Intangible assets	10,000
PPE	100,000
Inventory	40,000
Receivables	30,000
	250,000

The unit is reviewed for impairment due to the existence of indicators and the recoverable amount is estimated at £150,000. The PPE includes a property with a carrying amount of £60,000 and a market value of £75,000. The net realisable value of the inventory is greater than its carrying values and none of the receivables is considered doubtful.

The table below shows the allocation of the impairment loss:

	Pre-impairment £	Impairment £	Post-impairment £
Goodwill	70,000	(70,000)	Nil
Intangible assets	10,000	(6,000)	4,000
PPE	100,000	(24,000)	76,000
Inventory	40,000	Nil	40,000
Receivables	30,000	Nil	30,000
	250,000	(100,000)	150,000

Notes to table:

1 The impairment loss is first allocated against goodwill. After this has been done £30,000 (£100,000 – £70,000) remains to be allocated.

2 No impairment loss can be allocated to the property, inventory or receivables because these assets have a recoverable amount that is higher than their carrying value.

3 The remaining impairment loss is allocated pro rata to the intangible assets (carrying amount £10,000) and the plant (carrying amount £40,000 (£100,000 – £60,000)).

Restoration of past impairment losses

Past impairment losses in respect of an asset other than goodwill may be restored where the recoverable amount increases due to an improvement in economic conditions or a change in use of the asset. Such a restoration should be reflected in the statement of comprehensive income to the extent of the original impairment previously charged to the statement of comprehensive income, adjusting for depreciation which would have been charged otherwise in the intervening period.

17.9.6 Illustration of data required for an impairment review

Pronto SA has a product line producing wooden models of athletes for export. The carrying amount of the net assets employed on the line as at 31 December 20X3 was £114,500. The scrap value of the net assets at 31 December 20X6 is estimated to be £5,000.

There is an indication that the export market will be adversely affected in 20X6 by competition from plastic toy manufacturers. This means that the net assets employed to produce this product might have been impaired.

The finance director estimated the net realisable value of the net assets at 31 December 20X3 to be £70,000. The value in use is now calculated to check if it is higher or lower than £70,000. If it is higher it will be compared with the carrying amount to see if impairment has occurred; if it is lower the net realisable value will be compared with the carrying amount.

Pronto SA has prepared budgets for the years ended 31 December 20X4, 20X5 and 20X6. The assumptions underlying the budgets are as follows:

Unit costs and revenue:

	£
Selling price	10.00
Buying-in cost	(4.00)
Production cost: material, labour, overhead	(0.75)
Head office overheads apportioned	(0.25)
Cash inflow per model	5.00

Estimated sales volumes:

	20X3	20X4	20X5	20X6
Estimated at 31 December 20X2	6,000	8,000	11,000	14,000
Revised estimate at 31 December 20X3	—	8,000	11,000	4,000

Determining the discount rate to be used:

	20X4	20X5	20X6
Rate obtainable elsewhere at same level of risk	10%	10%	10%

The discount factors to be applied to each year are then calculated using cost of capital discount rates as follows:

20X4	$1/1.1$	$= 0.909$
20X5	$1/(1.1)^2$	$= 0.826$
20X6	$1/(1.1)^3$	$= 0.751$

17.9.7 Illustrating calculation of value in use

Before calculating value in use, it is necessary to ensure that the assumptions underlying the budgets are reasonable, e.g. is the selling price likely to be affected by competition in 20X6 in addition to loss of market? Is the selling price in 20X5 likely to be affected? Is the estimate of scrap value reasonably accurate? How sensitive is value in use to the scrap value? Is it valid to assume that the cash flows will occur at year-ends? How accurate is the cost of capital? Will components making up the income stream, e.g. sales, materials, labour, be subject to different rates of inflation?

Assuming that no adjustment is required to the budgeted figures provided above, the estimated income streams are discounted using the normal DCF approach as follows:

	20X4	20X5	20X6
Sales (models)	8,000	11,000	4,000
Income per model	£5	£5	£5
Income stream (£)	40,000	55,000	20,000
Estimated scrap proceeds			5,000
Cash flows to be discounted	40,000	55,000	25,000
Discounted (using cost of capital factors)	0.909	0.826	0.751
Present value	36,360	45,430	18,775

Value in use = £100,565

17.9.8 Illustration determining the *revised* carrying amount

If the carrying amount at the statement of financial position date exceeds net realisable value and value in use, it is revised to an amount which is the higher of net realisable value and value in use. For Pronto SA:

	£
Carrying amount as at 31 December 20X3	114,500
Net realisable value	70,000
Value in use	100,565
Revised carrying amount	**100,565**

17.10 IFRS 5 *Non-current Assets Held for Sale and Discontinued Operations*

IFRS 5 sets out requirements for the classification, measurement and presentation of non-current assets held for sale. The requirements which replaced IAS 35 *Discontinuing Operations* were discussed in Chapter 4. The IFRS is the result of the joint short-term project to resolve differences between IFRSs and US GAAP.

Classification as 'held for sale'

The IFRS (paragraph 6) classifies a non-current asset as 'held for sale' if its carrying amount will be recovered principally through a sale transaction rather than through continuing use. The criteria for classification as 'held for sale' are:

● the asset must be available for immediate sale in its present condition; and

● its sale must be *highly probable*.

The criteria for a sale to be highly probable are:

● the appropriate level of management must be committed to a plan to sell the asset;

● an active programme to locate a buyer and complete the plan must have been initiated;

● the asset must be actively marketed for sale at a price that is reasonable in relation to its current fair value;

● the sale should be expected to qualify for recognition as a completed sale within one year from the date of classification unless the delay is caused by events or circumstances beyond the entity's control and there is sufficient evidence that the entity remains committed to its plan to sell the asset; and

● actions required to complete the plan should indicate that it is unlikely that significant changes to the plan will be made or that the plan will be withdrawn.

Measurement and presentation of assets held for sale

The IFRS requires that assets 'held for sale' should:

● be measured at the lower of carrying amount and *fair value* less costs to sell;

● not continue to be depreciated; and

● be presented separately on the face of the statement of financial position.

The following additional disclosures are required in the notes in the period in which a non-current asset has been either classified as held for sale or sold:

● a description of the non-current asset;

● a description of the facts and circumstances of the sale;

● the expected manner and timing of that disposal;

- the gain or loss if not separately presented on the face of the statement of comprehensive income; and
- the caption in the statement of comprehensive income that includes that gain or loss.

17.10.1 IFRS for SMEs

The IFRS does not require separate presentation in the statement of financial position of 'non-current assets held for sale'. However, if an entity has plans to discontinue or restructure the operation to which an asset belongs and has plans to dispose of an asset before the previously expected date, then this is to be treated as an indication that an asset may be impaired and in such a case an impairment test is required.

17.11 Disclosure requirements

For each class of PPE the financial statements need to disclose:

- the measurement bases used for determining the gross carrying amount;
- the depreciation methods used;
- the useful lives or the depreciation rates used;
- the gross carrying amount and the accumulated depreciation (aggregated with accumulated impairment losses) at the beginning and end of the period;
- a reconciliation of the carrying amount at the beginning and end of the period.

The style employed by British Sky Broadcasting Group plc in its 2010 accounts is almost universally employed for this:

Tangible fixed assets (or PPE)
The movements in the year were as follows:

	Land and freehold buildings	Leasehold improvements	Equipment, furniture and fixtures	Assets not yet available for use	Total
	£m	£m	£m	£m	£m
Cost					
At 1 July 2009	128	77	931	191	1,327
Foreign exchange movements	—	—	(4)	—	(4)
Additions	58	2	152	64	276
Disposals	—	(6)	(69)	(3)	(78)
Transfers	—	—	30	(31)	(1)
At 30 June 2010	186	73	1,040	221	1,520
Depreciation					
At 1 July 2009	22	29	477	—	528
Foreign exchange movements			(4)		(4)
Depreciation	4	4	160	—	168
Impairments	—	—	2	3	5
Disposals	—	(6)	(67)	(3)	(76)
At 30 June 2010	26	27	568	—	621
Carrying amounts					
At 30 June 2009	106	48	454	191	799
At 30 June 2010	160	46	472	221	899

Additionally the financial statements should disclose:

- the existence and amounts of restrictions on title, and PPE pledged as security for liabilities;
- the accounting policy for the estimated costs of restoring the site of items of PPE;
- the amount of expenditures on account of PPE in the course of construction; and
- the amount of commitments for the acquisition of PPE.

17.12 Government grants towards the cost of PPE

The accounting treatment of government grants is covered by IAS 20. The basis of the standard is the accruals concept, which requires the matching of cost and revenue so as to recognise both in the statements of comprehensive income of the periods to which they relate. This should, of course, be tempered with the prudence concept, which requires that revenue is not anticipated. Therefore, in the light of the complex conditions usually attached to grants, credit should not be taken until receipt is assured.

Similarly, there may be a right to recover the grant wholly or partially in the event of a breach of conditions, and on that basis these conditions should be regularly reviewed and, if necessary, provision made.

Should the tax treatment of a grant differ from the accounting treatment, the effect of this would be accounted for in accordance with IAS 12 *Income Taxes*.

IAS 20

Government grants should be recognised in the statement of comprehensive income so as to match the expenditure towards which they are intended to contribute. If this is retrospective, they should be recognised in the period in which they became receivable.

Grants in respect of PPE should be recognised over the useful economic lives of those assets, thus matching the depreciation or amortisation.

IAS 20 outlines two acceptable methods of presenting grants relating to assets in the statement of financial position:

(a) The first method sets up the grant as deferred income, which is recognised as income on a systematic and rational basis over the useful life of the asset.

> **EXAMPLE** ● An entity purchased a machine for £60,000 and received a grant of £20,000 towards its purchase. The machine is depreciated over four years.
>
> The 'deferred income method' would result in an initial carrying amount for the machine of £60,000 and a deferred income credit of £20,000. In the first year of use of the plant the depreciation charge would be £15,000. £5,000 of the deferred income would be recognised as a credit in the statement of comprehensive income, making the net charge £10,000. At the end of the first year the carrying amount of the plant would be £45,000 and the deferred income included in the statement of financial position would be £15,000.
>
> The following is an extract from the 2013 Go-Ahead Annual Report:
>
> **Government grants**
> Government grants are recognised at their fair value where there is reasonable assurance that the grant will be received and all attaching conditions will be complied with. When the grant relates to an expense item, it is recognised in the income statement over the period necessary to match on a systematic basis to the costs that it is intended to compensate. *Where the grant relates to a non-current asset, value is credited to a deferred income account and is released to the income statement over the expected useful life of the relevant asset.*

(b) The second method deducts the grant in arriving at the carrying amount of the relevant asset. If we were to apply this method to the above example then the initial carrying amount of the asset would be £40,000. The depreciation charged in the first year would be £10,000. This is the same as the net charge to income under the 'deferred credit' method. The closing carrying amount of the plant would be £30,000. This is of course the carrying amount under the 'deferred income method' (£45,000) less the closing deferred income under the 'deferred income method' (£15,000).

The following extract is from the 2013 Annual Report of A & J Muklow plc:

Capital grants

Capital grants received relating to the building or refurbishing of investment prperties are deducted from the cost of the relevant property. Revenue grants are deducted from the related expenditure.

17.12.1 Arguments in favour of each approach

The capital approach

Supporters of the capital approach argue that (a) government grants are a means of financing and should therefore be reported as such in the statement of financial position rather than be recognised in profit or loss to offset the items of expense which they finance, and (b) it is inappropriate to recognise government grants in profit or loss, because they are not earned but represent an incentive provided by government without related costs.

The income approach

Supporters of this approach argue that (a) government grants are receipts from a source other than shareholders which should not, therefore, be recognised directly in equity but should be recognised in profit or loss in appropriate periods, and (b) they are not without cost in that the entity earns them through its compliance with their conditions. Their preferred treatment is, therefore, to recognise in profit or loss over the periods in which the entity recognises as expenses the related costs for which the grant was intended to compensate.

17.12.2 IASB future action

The IASB is currently considering drafting an amended standard on government grants. Among the reasons for the Board amending IAS 20 were the following:

● The recognition requirements of IAS 20 often result in accounting that is inconsistent with the *Framework*, in particular the recognition of a deferred credit when the entity has no liability, e.g. the following is an extract from the Annual Report of SSL International plc (now part of Reckitt Benckiser):

Grant income

Capital grants are shown in other creditors within the statement of financial position and released to match the depreciation charge on associated assets.

● IAS 20 contains numerous options. Apart from reducing the comparability of financial statements, the options in IAS 20 can result in understatement of the assets controlled by the entity and do not provide the most relevant information to users of financial statements.

In due course there is the prospect of the IASB issuing a revised standard which requires entities to recognise grants as income as soon as their receipt becomes unconditional. This is consistent with the specific requirements for the recognition of grants relating to agricultural activity laid down in IAS 41 *Agriculture*. This matter is discussed in more detail in Chapter 20.

IFRS for SMEs

Government grants are measured at the fair value of the asset received or receivable and treated as income when the proceeds are receivable if there are no future performance conditions attached. If there are performance conditions, the grant is recognised in profit or loss when the conditions are satisfied.

17.13 Investment properties

While IAS 16 requires all PPE to be subjected to a systematic depreciation charge, this may be considered inappropriate for properties held as assets but not employed in the normal activities of the entity, rather being held as investments. For such properties a more relevant treatment is to take account of the current market value of the property. The accounting treatment is set out in IAS 40 *Investment Property*.

Such properties may be held either as a main activity (e.g. by a property investment company) or by a company whose main activity is not the holding of such properties. In each case the accounting treatment is similar.

Definition of an investment property[10]

For the purposes of the statement, an investment property is property held (by the owner or by the lessee under a finance lease) to earn rentals or capital appreciation or both.

Investment property does **not** include:

(a) property held for use in the production or supply of goods or services or for administrative purposes (dealt with in IAS 16);

(b) property held for sale in the ordinary course of business (dealt with in IAS 2);

(c) an interest held by a lessee under an operating lease, even if the interest was a long-term interest acquired in exchange for a large upfront payment (dealt with in IAS 17);

(d) forests and similar regenerative natural resources (dealt with in IAS 41 *Agriculture*); and

(e) mineral rights, the exploration for and development of minerals, oil, natural gas and similar non-regenerative natural resources (dealt with in IFRS 6).

Accounting models

Under IAS 40, an entity must choose either:

● a fair value model: investment property should be measured at fair value and changes in fair value should be recognised in the statement of comprehensive income; or

● a cost model (the same as the benchmark treatment in IAS 16 *Property, Plant and Equipment*): investment property should be measured at depreciated cost (less any accumulated impairment losses). An entity that chooses the cost model should disclose the fair value of its investment property.

An entity should apply the model chosen to all its investment property. A change from one model to the other model should be made only if the change will result in a more appropriate presentation. The standard states that this is highly unlikely to be the case for a change from the fair value model to the cost model.

In exceptional cases, there is clear evidence when an entity that has chosen the fair value model first acquires an investment property (or when an existing property first becomes investment property following the completion of construction or development, or after a change in use) that the entity will not be able to determine the fair value of the investment property reliably on a continuing basis. In such cases, the entity measures that investment property using the benchmark treatment in IAS 16 until the disposal of the investment property. The residual value of the investment property should be assumed to be zero. The entity measures all its other investment property at fair value.

IFRS for SME treatment

Under this IFRS the accounting for investment property is driven by circumstances. If an entity knows or can measure the fair value without undue cost or effort on an ongoing basis, it must use the fair value through profit or loss model for that investment property. If not, it must use the cost–depreciation–impairment model but, in that case, it is not required to disclose the fair values.

17.14 Effect of accounting policy for PPE on the interpretation of the financial statements

A number of difficulties exist when we attempt to carry out inter-firm comparisons using the external information that is available to a shareholder.

17.14.1 Effect of inflation on the carrying value of the asset

The most serious difficulty is the effect of inflation, which makes the charges based on historical cost inadequate. Companies have followed various practices to take account of inflation. None of these is as effective as an acceptable surrogate for index adjustment using specific asset indices on a systematic annual basis: this is the only way to ensure uniformity and comparability of the cost/valuation figure upon which the depreciation charge is based.

The method that is currently allowable under IAS 16 is to revalue the assets. This is a partial answer, but it results in lack of comparability of ratios such as gearing or leverage.

17.14.2 Effect of revaluation on ratios

The rules of double entry require that when an asset is revalued the 'profit' (or, exceptionally, 'loss') must be credited somewhere. As it is not a 'realised' profit, it would not be appropriate to credit the statement of comprehensive income, so a 'revaluation reserve' must be created. As the asset is depreciated, this reserve may be realised to income; similarly, when an asset is ultimately disposed of, any residue relevant to that asset may be taken into income.

One significant by-product of revaluing assets is the effect on gearing. The revaluation reserve, while not distributable, forms part of the shareholders' funds and thus improves the debt/equity ratio. Care must therefore be taken in looking at the revaluation policies and reserves when comparing the gearing or leverage of companies.

The problem is compounded because the carrying value may be amended at random periods and on a selective category of asset.

17.14.3 Choice of depreciation method

There are a number of acceptable depreciation methods that may give rise to very different patterns of debits against the profits of individual years.

17.14.4 Inherent imprecision in estimating economic life

One of the greatest difficulties with depreciation is that it is inherently imprecise. The amount of depreciation depends on the estimate of the economic life of assets, which is affected not only by the durability and workload of the asset, but also by external factors beyond the control of management. Such factors may be technological, commercial or economic. Here are some examples:

- the production by a competitor of a new product rendering yours obsolete, e.g. watches with battery-powered movements replacing those with mechanical movements;
- the production by a competitor of a product at a price lower than your production costs, e.g. imported goods from countries where costs are lower;
- changes in the economic climate which reduce demand for your product.

This means that the interpreter of accounts must pay particular attention to depreciation policies, looking closely at the market where the entity's business operates. However, this understanding is not helped by the lack of requirement to disclose specific rates of depreciation and the basis of computation of residual values. Without such information, the potential effects of differences between policies adopted by competing entities cannot be accurately assessed.

17.14.5 Mixed values in the statement of financial position

The effect of depreciation on the statement of financial position is also some cause for concern. The net book amount shown for non-current assets is the result of deducting accumulated depreciation from cost (or valuation); it is not intended to be (although many non-accountants assume it is) an estimate of the value of the underlying assets. The valuation of a business based on the statement of financial position is extremely difficult.

17.14.6 *IFRS for SMEs*

This IFRS differs from IAS 16 in that:

- PPE is reported at historical cost less depreciation and less any impairment of the carrying amount. The revaluation model is not permitted.
- A review of the useful life, residual value or depreciation rate is only carried out if there is a significant change in the asset or how it is used. Any adjustment is a change in estimate.
- Assets held for sale are not reported separately, although the fact that an asset is held for sale might be an indication that there has been an impairment.
- Most investment property is treated in the same way as PPE. However, if the fair value of investment property can be measured reliably without excessive cost then the fair value model applies with changes being through profit or loss.
- Separate significant components should be depreciated separately if there are significantly different patterns of consumption of economic benefits.

17.14.7 Different policies may be applied within the same sector

Inter-company comparisons are even more difficult. Two entities following the historical cost convention may own identical assets, which, as they were purchased at different times, may well appear as dramatically different figures in the accounts. This is particularly true of interests in land and buildings.

17.14.8 Effect on the return on capital employed

There is an effect not only on the net asset value, but also on the return on capital employed. To make a fair assessment of return on capital it is necessary to know the current replacement cost of the underlying assets, but, under present conventions, up-to-date valuations are required only for investment properties.

17.14.9 Effect on EPS

IAS 16 is concerned to ensure that the earnings of an entity reflect a fair charge for the use of the assets by the enterprise. This should ensure an accurate calculation of earnings per share. But there is a weakness here. If assets have increased in value without revaluations, then depreciation will be based on the historical cost.

Summary

Before IAS 16 there were significant problems in relation to the accounting treatment of PPE such as the determination of a cost figure and the adjustment for inflation; companies providing nil depreciation on certain types of asset; and revaluations being made selectively and not kept current.

With IAS 16 the IASB has made the accounts more consistent and comparable. This standard has resolved some of these problems, principally requiring companies to provide for depreciation and if they have a policy of revaluation to keep such valuations reasonably current and applied to all assets within a class, i.e. removing the ability to cherry-pick which assets to revalue.

However, certain difficulties remain for the user of the accounts in that there are different management policies on the method of depreciation, which can have a major impact on the profit for the year; subjective assessments of economic life that may be reviewed each year with an impact on profits; and inconsistencies such as the presence of modified historical costs and historical costs in the same statement of financial position. In addition, with pure historical cost accounting, where non-current asset carrying values are based on original cost, no pretence is made that non-current asset net book amounts have any relevance to current values. The investor is expected to know that the depreciation charge is arithmetical in character and will not wholly provide the finance for tomorrow's assets or ensure maintenance of the business's operational base. To give recognition to these factors requires the investor to grapple with the effects of lost purchasing power through inflation; the effect of changes in supply and demand on replacement prices; technological change and its implication for the company's competitiveness; and external factors such as exchange rates. To calculate the effect of these variables necessitates not only considerable mental agility, but also far more information than is contained in a set of accounts. This is an area that needs to be revisited by the standard setters.

REVIEW QUESTIONS

1 Define PPE and explain how materiality affects the concept of PPE.

2 Define depreciation. Explain what assets need not be depreciated and list the main methods of calculating depreciation.

3 What is meant by the phrases 'useful life' and 'residual value'?

4 Define 'cost' in connection with PPE.

5 What effect does revaluing assets have on gearing (or leverage)?

6 How should grants received towards expenditure on PPE be treated?

7 Define an investment property and explain its treatment in financial statements.

8 'Depreciation should mean that a company has sufficient resources to replace assets at the end of their economic lives.' Discuss.

EXERCISES

* Question 1

Simple SA has just purchased a roasting/salting machine to produce roasted walnuts. The finance director asks for your advice on how the company should calculate the depreciation on this machine. Details are as follows:

Cost of machine	SF800,000
Residual value	SF104,000
Estimated life	4 years
Annual profits	SF2,000,000
Annual turnover from machine	SF850,000

Required:

(a) Calculate the annual depreciation charge using the straight-line method and the reducing balance method. Assume that an annual rate of 40% is applicable for the reducing balance method.

(b) Comment upon the validity of each method, taking into account the type of business and the effect each method has on annual profits. Are there any other methods which would be more applicable?

* Question 2

(a) Discuss why IAS 40 *Investment Property* was produced.

(b) Universal Entrepreneurs plc has the following items on its PPE list:

   (i) £1,000,000 – the right to extract sandstone from a particular quarry. Geologists predict that extraction at the present rate may be continued for 10 years.

   (ii) £5,000,000 – a freehold property, let to a subsidiary on a full repairing lease negotiated on arm's-length terms for 15 years. The building is a new one, erected on a greenfield site at a cost of £4,000,000.

(iii) A fleet of motor cars used by company employees. These have been purchased under a contract which provides a guaranteed part exchange value of 60% of cost after two years' use.

(iv) A company helicopter with an estimated life of 150,000 flying hours.

(v) A 19-year lease on a property let out at arm's-length rent to another company.

Required:

Advise the company on the depreciation policy it ought to adopt for each of the above assets.

(c) The company is considering revaluing its interests in land and buildings, which comprise freehold and leasehold properties, all used by the company or its subsidiaries.

Required:

Discuss the consequences of this on the depreciation policy of the company and any special instructions that need to be given to the valuer.

* Question 3

You have been given the task, by one of the partners of the firm of accountants for which you work, of assisting in the preparation of a trend statement for a client, Mercury.

Mercury has been in existence for four years. Figures for the three preceding years are known but those for the fourth year need to be calculated. Unfortunately, the supporting workings for the preceding years' figures cannot be found and the client's own ledger accounts and workings are not available.

One item in particular, plant, is causing difficulty and the following figures have been given to you:

12 months ended 31 March	20X6	20X7	20X8	20X9
	£	£	£	£
(A) Plant at cost	80,000	80,000	90,000	?
(B) Accumulated depreciation	(16,000)	(28,800)	(28,080)	?
(C) Net (written down) value	64,000	51,200	61,920	?

The only other information available is that disposals have taken place at the beginning of the financial years concerned:

	Date of Disposal	Original acquisition	Original cost	Sales proceeds
		12 months ended 31 March	£	£
First disposal	20X8	20X6	15,000	8,000
Second disposal	20X8	20X6	30,000	21,000

Plant sold was replaced on the same day by new plant. The cost of the plant which replaced the first disposal is not known but the replacement for the second disposal is known to have cost £50,000.

Required:

(a) Identify the method of providing for depreciation on plant employed by the client, stating how you have arrived at your conclusion.

(b) Show how the figures shown at line (B) for each of the years ended 31 March 20X6, 20X7 and 20X8 were calculated. Extend your workings to cover the year ended 31 March 20X9.

(c) Produce the figures that should be included in the blank spaces on the trend statement at lines (A), (B) and (C) for the year ended 31 March 20X9.

(d) Calculate the profit or loss arising on each of the two disposals.

Question 4

In the year to 31 December 20X9, Amy bought a new machine and made the following payments in relation to it:

	£	£
Cost as per supplier's list	12,000	
Less: Agreed discount	1,000	11,000
Delivery charge		100
Erection charge		200
Maintenance charge		300
Additional component to increase capacity		400
Replacement parts		250

Required:
(a) State and justify the cost figure which should be used as the basis for depreciation.
(b) What does depreciation do, and why is it necessary?
(c) Briefly explain, without numerical illustration, how the straight-line and diminishing balance methods of depreciation work. What different assumptions does each method make?
(d) Explain the term 'objectivity' as used by accountants. To what extent is depreciation objective?
(e) It is common practice in published accounts in Germany to use the diminishing balance method for PPE in the early years of an asset's life, and then to change to the straight-line method as soon as this would give a higher annual charge. What do you think of this practice? Refer to relevant accounting conventions in your answer.

(ACCA)

* Question 5

The finance director of Small Machine Parts Ltd is considering the acquisition of a lease of a small workshop in a warehouse complex that is being redeveloped by City Redevelopers Ltd at a steady rate over a number of years. City Redevelopers are granting such leases for five years on payment of a premium of £20,000.

The accountant has obtained estimates of the likely maintenance costs and disposal value of the lease during its five-year life. He has produced the following table and suggested to the finance director that the annual average cost should be used in the financial accounts to represent the depreciation charge in the profit and loss account.

Table prepared to calculate the annual average cost					
Years of life	1	2	3	4	5
	£	£	£	£	£
Purchase price	20,000	20,000	20,000	20,000	20,000
Maintenance/repairs					
Year 2		1,000	1,000	1,000	1,000
3			1,500	1,500	1,500
4				1,850	1,850
5					2,000
	20,000	21,000	22,500	24,350	26,350
Resale value	11,500	10,000	8,010	5,350	350
Net cost	8,500	11,000	14,490	19,000	26,000
Annual average cost	8,500	5,500	4,830	4,750	5,200

The finance director, however, was considering whether to calculate the depreciation chargeable using the annuity method with interest at 15%.

Required:

(a) Calculate the entries that would appear in the statement of comprehensive income of Small Machine Parts Ltd for each of the five years of the life of the lease for the amortisation charge, the interest element in the depreciation charge and the income from secondary assets using the *annuity method*. Calculate the net profit for each of the five years assuming that the operating cash flow is estimated to be £25,000 per year.

(b) Discuss briefly which of the two methods you would recommend.

The present value at 15% of £1 per annum for five years is £3.35214.

The present value at 15% of £1 received at the end of year 5 is £0.49717.

Ignore taxation.

(ACCA)

Question 6

(a) IAS 16 *Property, Plant and Equipment* requires that where there has been a permanent diminution in the value of property, plant and equipment, the carrying amount should be written down to the recoverable amount. The phrase 'recoverable amount' is defined in IAS 16 as 'the amount which the entity expects to recover from the future use of an asset, including its residual value on disposal'. The issues of how one identifies an impaired asset, the measurement of an asset when impairment has occurred and the recognition of impairment losses were not adequately dealt with by the standard. As a result the International Accounting Standards Committee issued IAS 36 *Impairment of Assets* in order to address the above issues.

Required:

(i) Describe the circumstances which indicate that an impairment loss relating to an asset may have occurred.

(ii) Explain how IAS 36 deals with the recognition and measurement of the *impairment of assets*.

(b) AB, a public limited company, has decided to comply with IAS 36 *Impairment of Assets*. The following information is relevant to the impairment review:

(i) Certain items of machinery appeared to have suffered a permanent diminution in value. The inventory produced by the machines was being sold below its cost and this occurrence had affected the value of the productive machinery. The carrying value at historical cost of these machines is $290,000 and their net selling price is estimated at $120,000. The anticipated net cash inflows from the machines are now $100,000 per annum for the next three years. A market discount rate of 10% per annum is to be used in any present value computations.

(ii) AB acquired a car taxi business on 1 January 20X1 for $230,000. The values of the assets of the business at that date based on net selling prices were as follows:

	$000
Vehicles (12 vehicles)	120
Intangible assets (taxi licence)	30
Trade receivables	10
Cash	50
Trade payables	(20)
	190

On 1 February 20X1, the taxi company had three of its vehicles stolen. The net selling value of these vehicles was $30,000 and because of non-disclosure of certain risks to the insurance company, the vehicles were uninsured. As a result of this event, AB wishes to recognise an impairment loss of $45,000 (inclusive of the loss of the stolen vehicles) due to the decline in the value in use of the

cash generating unit, that is the taxi business. On 1 March 20X1 a rival taxi company commenced business in the same area. It is anticipated that the business revenue of AB will be reduced by 25%, leading to a decline in the present value in use of the business, which is calculated at $150,000. The net selling value of the taxi licence has fallen to $25,000 as a result of the rival taxi operator. The net selling values of the other assets have remained the same as at 1 January 20X1 throughout the period.

Required:

Describe how AB should treat the above impairments of assets in its financial statements.

(In part (b) (ii) you should show the treatment of the impairment loss at 1 February 20X1 and 1 March 20X1.)

(ACCA)

* Question 7

Infinite Leisure Group owns and operates a number of pubs and clubs across Europe and South East Asia. Since inception the group has made exclusive use of the cost model for the purpose of its annual financial reporting. This has led to a number of shareholders expressing concern about what they see as a consequent lack of clarity and quality in the group's financial statements.

The CEO does not support use of the alternative to the cost model (the revaluation model), believing it produces volatile information. However, she is open to persuasion and so, as an example of the impact of a revaluation policy, has asked you to carry out an analysis (using data concerning '*Sooz*' – one of the group's nightclubs sold during the year to 31 October 2006) to show the impact the revaluation model would have had on the group's financial statements had the model been adopted from the day the club was acquired.

The following extract has been taken from the company's asset register:

Outlet: '*Sooz*'
Acquisition data

Date acquired	1 November 2001
Total cost	€10.24m
Cost components:	
Plant and equipment	
Cost	€0.24m
Economic life	6 years
Residual value	nil
Property	
Buildings	
Cost	€7.0m
Economic life	50 years
Land	
Cost	€3.0m

Updates

1 November 2003   Replacement cost of plant and equipment €0.42m. No fair value available (mainly specialised audio-visual equipment). No change to economic life. Property revaluation €13m (land €4m, buildings €9m). Future economic life as at 1 November 2003 50 years.

Disposal

Date committed to a plan to sell	January 2006
Date sold	June 2006
Net sale price	€9.1m
Sale price components:	
Plant and equipment	€0.1m
Property	€9.0m

Note: the Group accounts for property and for plant and equipment as separate non-current assets in its statement of financial position using straight-line depreciation.

Required:

Prepare an analysis to show the impact on Infinite Leisure's financial statements for each year the 'Sooz' nightclub was owned had the revaluation model been in place from the day the nightclub was acquired.

(The Association of International Accountants)

Question 8

The Blissopia Leisure Group consists of three divisions: Blissopia 1, which operates mainstream bars; Blissopia 2, which operates large restaurants; and Blissopia 3, which operates one hotel – the Eden.

Divisions 1 and 2 have been trading very successfully and there are no indications of any potential impairment. It is a different matter with the Eden, however. The Eden is a 'boutique' hotel and was acquired on 1 November 2006 for $6.90m. The fair value (using net selling price) of the hotel's net assets at that date and their carrying value at the year-end were as follows:

	Fair value 1.11.06 $m	Carrying value 31.10.07 $m
Land and buildings	3.61	3.18
Plant and equipment	0.90	0.81
Cash	1.40	1.12
Vehicles	0.10	0.09
Trade receivables	0.34	0.37
Trade payables	(0.60)	(0.74)
	5.75	4.83

The following facts were discovered following an impairment review as at 31 October 2007:

(i) During August 2007, a rival hotel commenced trading in the same location as the Eden. The Blissopia Leisure Group expects hotel revenues to be significantly affected and has calculated the value-in-use of the Eden to be $3.52m.

(ii) The company owning the rival hotel has offered to buy the Eden (including all of the above net assets) for $4m. Selling costs would be approximately $50,000.

(iii) One of the hotel vehicles was severely damaged in an accident whilst being used by an employee to carry shopping home from a supermarket. The vehicle's carrying value at 31 October 2007 was $30,000 and insurers have indicated that as it was being used for an uninsured purpose the loss is not covered by insurance. The vehicle was subsequently scrapped.

(iv) A corporate client, owing $40,000, has recently gone into liquidation. Lawyers have estimated that the company will receive only 25% of the amount outstanding.

Required:

Prepare a memo for the directors of the Blissopia Leisure Group explaining how the group should account for the impairment to the Eden Hotel's assets as at 31 October 2007.

(The Association of International Accountants)

Question 9

Cryptic plc extracted its trial balance on 30 June 20X5 as follows:

	£000	£000
Land and buildings at cost	750	—
Plant and machinery at cost	480	—
Accumulated depreciation on plant and machinery at 30 June 20X5	—	400
Depreciation on plant and machinery	80	—
Furniture, tools and equipment at cost	380	—
Accumulated depreciation on furniture, etc. at 30 June 20X4	—	95
Receivables and payables	475	360
Inventory of raw materials at 30 June 20X4	112	—
Work in progress at factory cost at 30 June 20X4	76	—
Finished goods at cost at 30 June 20X4	264	—
Sales including selling taxes	—	2,875
Purchases of raw materials including selling taxes	1,380	—
Share premium account	—	150
Advertising	65	—
Deferred taxation	—	185
Salaries	360	—
Rent	120	—
Retained earnings at 30 June 20X4	—	226
Factory power	48	—
Trade investments at cost	240	—
Overprovision for tax for the year ended 30 June 20X4	—	21
Electricity	36	—
Stationery	12	—
Dividend received (net)	—	24
Dividend paid on 15 April 20X5	60	—
Other administration expenses	468	—
Disposal of furniture	—	64
Selling tax control account	165	—
Ordinary shares of 50p each	—	1,000
12% preference shares of £1 each (IAS 32 liability)	—	200
Cash and bank balance	29	—
	5,600	5,600

The following information is relevant:

(i) The company discontinued a major activity during the year and replaced it with another. All non-current assets involved in the discontinued activity were redeployed for the new one. The following expenses incurred in this respect, however, are included in 'Other administration expenses':

	£000
Cancellation of contracts re terminated activity	165
Fundamental reorganisation arising as a result	145

Cryptic has decided to present its results from discontinued operations as a single line on the face of the statement of comprehensive income with analysis in the notes to the accounts as allowed by IFRS 5.

(ii) On 1 January 20X5 the company acquired new land and buildings for £150,000. The remainder of land and buildings, acquired nine years earlier, have *not* been depreciated until this year. The company has decided to depreciate the buildings, on the straight-line method, assuming that one-third of the cost relates to land and that the buildings have an estimated economic life of 50 years. The company policy is to charge a full year of depreciation in the year of purchase and none in the year of sale.

(iii) Plant and machinery was all acquired on 1 July 20X0 and has been depreciated at 10% per annum on the straight-line method. The estimate of useful economic life had to be revised this year when it was realised that if the market share is to be maintained at current levels, the company has to replace all its machinery by 1 July 20X6. The balance in the 'Accumulated provision for depreciation' account on 1 July 20X4 was amended to reflect the revised estimate of useful economic life and the impact of the revision adjusted against the retained earnings brought forward from prior years.

(iv) Furniture acquired for £80,000 on 1 January 20X3 was disposed of for £64,000 on 1 April 20X5. Furniture, tools and equipment are depreciated at 5% p.a. on cost. Depreciation for the current year has not been provided.

(v) Results of the inventory counting at year-end are as follows:

Inventory of raw materials at cost including selling tax	£197,800
Work in progress at factory cost	£54,000
Finished goods at cost	£364,000

(vi) The company allocates its expenditure as follows:

	Production cost	Factory overhead	Distribution cost	Administrative expenses
Salaries and wages	65%	15%	5%	15%
Rent	—	60%	15%	25%
Electricity	—	10%	20%	70%
Depreciation of building	—	40%	10%	50%

(vii) The directors wish to make an accrual for audit fees of £18,000 and estimate the income tax for the year at £65,000. £11,000 should be transferred from the deferred tax account. The directors have to pay the preference dividend.

(viii) The following analysis has been made:

	New activity	Discontinued activity
Sales excluding selling taxes	£165,000	£215,000
Cost of sales	£98,000	£155,000
Distribution cost	£16,500	£48,500
Administrative expenses	£22,500	£38,500

(ix) Assume that the rate of selling taxes applicable to all purchases and sales is 15%, the basic rate of personal income tax is 25% and the corporate income tax rate is 35%.

Required:

(a) Advise the company on the accounting treatment in respect of information stated in (ii) above.

(b) In respect of the information stated in (iii) above, state whether a company is permitted to revise its estimate of the useful economic life of a non-current asset and comment on the appropriateness of the accounting treatment adopted.

(c) Set out a statement of movement of property, plant and equipment in the year to 30 June 20X5.

(d) Set out for publication the statement of income for the year ended 30 June 20X5, the statement of financial position as at that date and any notes other than that on accounting policy, in accordance with relevant standards.

* Question 10

Omega prepares financial statements under International Financial Reporting Standards. In the year ended 31 March 2007 the following transaction occurred.

On 1 April 2006 Omega began the construction of a new production line. Costs relating to the line are as follows:

Details	Amount
	$000
Costs of the basic materials (list price $12.5 million less a 20% trade discount)	10,000
Recoverable sales taxes incurred, not included in the purchase cost	1,000
Employment costs of the construction staff for the three months to 30 June 2006 (Note 1)	1,200
Other overheads directly related to the construction (Note 2)	900
Payments to external advisors relating to the construction	500
Expected dismantling and restoration costs (Note 3)	2,000

Note 1

The production line took two months to make ready for use and was brought into use on 30 June 2006.

Note 2

The other overheads were incurred in the two months ended 31 May 2006. They included an abnormal cost of $300,000 caused by a major electrical fault.

Note 3

The production line is expected to have a useful economic life of eight years. At the end of that time Omega is legally required to dismantle the plant in a specified manner and restore its location to an acceptable standard. The figure of $2 million included in the cost estimates is the amount that is expected to be incurred at the end of the useful life of the production plant. The appropriate rate to use in any discounting calculations is 5%. The present value of $1 payable in eight years at a discount rate of 5% is approximately $0.68.

Note 4

Four years after being brought into use, the production line will require a major overhaul to ensure that it generates economic benefits for the second half of its useful life. The estimated cost of the overhaul, at current prices, is $3 million.

Note 5

Omega computes its depreciation charge on a monthly basis.

Note 6

No impairment of the plant had occurred by 31 March 2007.

The financial statements for the year ended 31 March 2007 were authorised for issue on 15 May 2007.

Required:
Show the impact of the construction of the production line on the income statement of Omega for the year ended 31 March 2007, and on its balance sheet as at 31 March 2007. You should state where in the income statement and the balance sheet relevant balances will be shown. You should make appropriate references to international financial reporting standards.

(IFRS)

References

1 IAS 16 *Property, Plant and Equipment*, IASB, revised 2004, para. 6.
2 Ibid., para. 16.
3 Ibid., para. 22.
4 IAS 23 *Borrowing Costs*, IASB, revised 2007, para. 8.
5 IAS 16 *Property, Plant and Equipment*, IASB, revised 2004, para. 18.
6 Ibid., para. 29.
7 Ibid., para. 41.
8 IAS 36 *Impairment of Assets*, IASB, 2004, para. 33.
9 Ibid., paras 55–56.
10 IAS 40 *Investment Property*, IASB, 2004.

CHAPTER 18

Leasing

18.1 Introduction

The main purpose of this chapter is to introduce the accounting principles and policies that apply to lease agreements.

Objectives

By the end of this chapter, you should be able to:

- critically discuss the reasons for IAS 17;
- account for leases in the financial statements of the lessee;
- account for leases in the financial statements of the lessor;
- critically discuss the reasons for the proposed revision of IAS 17;
- critically appraise the progress in developing a replacement standard for IAS 17.

18.2 Background to leasing

In this section we consider the nature of a lease; why leasing has become popular; and why it was necessary to introduce IAS 17.

18.2.1 What is a lease?

IAS 17 *Leases* provides the following definition:

> A *lease* is an agreement whereby the lessor (the legal owner of the asset) conveys to the lessee (the user of the asset) the right to use an asset for an agreed period of time in return for a payment or series of payments.

In practice, there might well be more than two parties involved in a lease. For example, on leasing a car the parties involved could be the motor dealer, the finance company (who buys the asset from the dealer and acts as lessor) and the company using the car.

18.2.2 Why did leasing initially become popular?

Prior to the issue of IAS 17, three of the main reasons for the popularity of leasing were the tax advantage to the lessor able to make use of investment allowances, the commercial advantages to the lessee and the potential for the lessee to leave the transaction off-balance-sheet.

Commercial advantages for the lessee

There are a number of advantages associated with leases. These are attributable in part to the ability to spread cash payments over the lease period instead of making a one-off lump sum payment. They include the following:

- **Cash flow management**. Leases can be structured with flexible rental patterns. Also, if cash is used to purchase non-current assets, it is not available for the normal operating activities of a company.

- **Conservation of capital**. Lines of credit may be kept open and may be used for purposes where finance might not be easily available (e.g. financing working capital).

- **Continuity**. The lease agreement is itself a line of credit that cannot easily be withdrawn or terminated due to external factors, in contrast to an overdraft which can be called in by the lender.

- **Flexibility of the asset base**. The asset base can be more easily expanded and contracted. In addition, the lease payments can be structured to match the income pattern of the lessee.

18.2.3 Off-balance-sheet financing

Leasing previously provided the lessee with the possibility of off-balance-sheet financing,[1] whereby a company had the use of an economic resource that did not appear in the statement of financial position, with the corresponding omission of the associated liability.

So, if the lease was capitalised then the total of the reported assets would be increased, which would have a negative impact on some of the ratios used by investors and creditors such as the rate of asset turnover and the rate of return on total assets.

An attraction of off-balance-sheet financing was that the exclusion of the liability meant that the gearing appeared lower in that the gearing ratio was not increased. Changes in these ratios can have an impact on pay where they are performance criteria.

18.2.4 Why was IAS 17 necessary?

As with many of the standards, action was required because there was no uniformity in the treatment and disclosure of leasing transactions. The need became urgent following the consistent growth in the leasing industry. By 2007 the total value of leased assets had grown to US$760 billion worldwide.

Leasing has become a material economic activity but the pre-IAS 17 accounting treatment of lease transactions was seen to distort the financial reports of a company so that they did not represent a true and fair view of its commercial activities.

IAS 17, therefore, requires that obligations under lease agreements that transferred substantially all the risks and rewards to the lessee be reported in the financial statements. The asset and liability were both brought onto the statement of financial position.

There was some concern that this might have undesirable economic consequences[2] by reducing the volume of leasing, and that the inclusion of the lease obligation might affect the lessee company's gearing adversely, possibly causing it to exceed its legal borrowing powers. However, in the event, the commercial reasons for leasing and the capacity of the leasing industry to structure lease agreements to circumvent the standard prevented a reduction in lease activity. Evidence of lessors varying the term of the lease agreements to ensure that they remained off-balance-sheet is supported by Cranfield[3] and by Abdel-Khalik et al.[4]

A standard was necessary to ensure uniform reporting and to prevent the accounting message being manipulated.

18.2.5 The approach taken by IAS 17

The approach taken by the standard is to distinguish between two types of lease – finance and operating – and to require a different accounting treatment for each. In brief, the definitions are as follows:

- **Finance lease**: a lease that transfers substantially all the risks and rewards of ownership of an asset. Title may or may not eventually be transferred.
- **Operating lease**: a lease other than a finance lease.

Assets leased under finance leases are required to be capitalised in the lessee's accounts. This means that the leased item should be recorded as an asset in the statement of financial position, and the obligation for future payments should be recorded as a liability in that statement. It is not permissible for the leased asset and lease obligation to be left out of the statement.

In the case of operating leases, the lessee is required only to expense the annual payments as a rental through the statement of comprehensive income.

18.3 Why was the IAS 17 approach so controversial?

The proposal to classify leases into finance and operating leases, and to capitalise those which are classified as finance leases, appears to be a feasible solution to the accounting problems that surround leasing agreements. So, why did the standard setters encounter so much controversy in their attempt to stop the practice of charging all lease payments to the statement of comprehensive income?

The whole debate centres on the concept of faithful representation. Paragraph 5 of Chapter 3 of the IASB's *Conceptual Framework* identifies this concept as one of the fundamental qualitative characteristics of useful finanical information.

IAS 17 has adopted a 'faithful representation' accounting approach which is completely different to the approach based on legal ownership. The IASB argued that in reality there were two separate transactions taking place. In one transaction, the company was borrowing funds to be repaid over a period. In the other, it was making a payment to the supplier for the use of an asset.

In order to faithfully represent the borrowing transaction, it was necessary to include in the lessee's statement of financial position a liability representing the obligation to meet the lease payments, and the correct accounting treatment for the asset acquisition transaction, based on faithful representation, was to include an asset representing the asset supplied under the lease.

IAS 17, paragraph 10, states categorically that 'whether a lease is a finance lease or an operating lease depends on the substance of the transaction rather than the form of the contract'.

18.3.1 How do the accounting and legal professions differ in their approach to the reporting of lease transactions?

The accounting profession sees itself as a service industry that prepares financial reports in a dynamic environment, in which the user is looking for reports that reflect commercial reality. Consequently, the profession needs to be constantly sensitive and responsive to changes in commercial practice.

There was still some opposition within the accounting profession to the inclusion of a finance lease in the statement of financial position as an 'asset'. The opposition rested on the fact that the item that was the subject of the lease agreement did not satisfy the existing criterion for classification as an asset because it was not 'owned' by the lessee. To accommodate this, the definition of an asset has been modified from 'ownership' to 'control' and 'the ability to contribute to the cash flows of the enterprise'.

The legal profession, on the other hand, concentrates on the strict legal interpretation of a transaction. The whole concept of substance over form is contrary to its normal practice.

It is interesting to reflect that, whereas an equity investor might prefer the economic resources to be included in the statement of financial position under the substance-over-form principle, this is not necessarily true for a loan creditor. The equity shareholder is interested in resources available for creating earnings; the lender is interested in the assets available as security.

Another way to view the asset is to think of it as an asset consisting of the ownership of the right to use the facility as opposed to the ownership of the physical item itself. In a way this is similar to owning accounts receivable or a patent or intellectual property. You do not have a physical object but rather a valuable intangible right.

18.4 IAS 17 – classification of a lease

As discussed earlier in the chapter, IAS 17 provides definitions for classifying leases as finance or operating leases, then prescribes the accounting and disclosure requirements applicable to the lessor and the lessee for each type of lease.

The crucial decision in accounting for leases is whether a transaction represents a finance or an operating lease. We have already defined each type of lease, but we must now consider the risks and rewards of ownership.

IAS 17 provides in paragraph 10 a list of the factors that need to be considered in the decision whether risks and rewards of ownership have passed to the lessee. These factors are considered individually and in combination when making the decision, and if any one of them is met this would normally indicate a finance lease:

(a) the lease transfers ownership of the asset to the lessee by the end of the lease term;

(b) the lessee has the option to purchase the asset at a price that is expected to be sufficiently lower than the fair value at the date the option becomes exercisable for it to be reasonably certain, at the inception of the lease, that the option will be exercised;

(c) the lease term is for the major part of the economic life of the asset even if title is not transferred;

(d) at the inception of the lease the present value of the minimum lease payments amounts to at least substantially all of the fair value of the leased asset; and

(e) the leased assets are of such a specialised nature that only the lessee can use them without major modifications.

Leases of land

If land is leased and legal title is not expected to pass at the end of the lease, the lease will normally be an operating lease. The reason is that the useful economic life of land is indefinite so it would seem that no lease term could ever be for the substantial part of the economic life of the asset (criterion (c) above). This means that if a land and buildings lease is entered into it should be classified as two leases, a land lease which is usually an operating lease, and a buildings lease which could be an operating or a finance lease. The lease payments should be allocated between the land and buildings elements in proportion to the relative fair values of the leasehold interests at the lease at its inception.

Figure 18.1 IAS 17 aid to categorising operating and finance leases

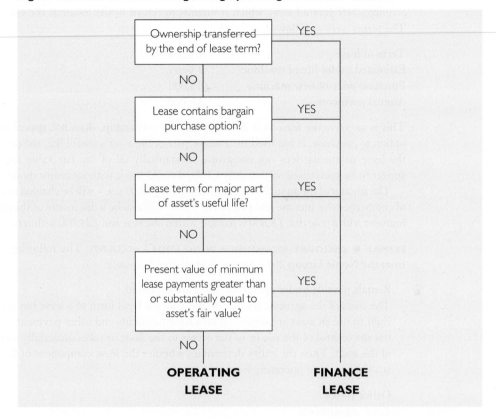

This split is not required by lessees if the land and buildings are an investment property accounted for under IAS 40, and where the fair value model has been adopted.

In its 1997 version , IAS 17 included a helpful flow chart, prepared by the IAS secretariat, which represents examples of some possible positions that would normally be classified as finance leases (Figure 18.1).

18.5 Accounting requirements for operating leases

The treatment of operating leases conforms to the legal interpretation and corresponds to the lease accounting practice that existed before IAS 17. No asset or obligation is shown in the statement of financial position; the operating lease rentals payable are charged to the statement of comprehensive income on a straight-line basis unless another systematic basis is more representative of the time pattern of the user's benefit.

18.5.1 Disclosure requirements for operating leases

IAS 17 requires that the total of operating lease rentals charged as an expense in the statement of comprehensive income should be disclosed, and these rentals should be broken down for minimum lease payments, contingent costs, and sublease payments. Disclosure is required of the payments that a lessee is committed to make during the next year, in the second to fifth years inclusive, and over five years.

EXAMPLE ● OPERATING LEASE Clifford plc negotiates a lease to begin on 1 January 20X1 to acquire a refrigerated lorry which it intends to return to the lessor at the end of the lease. The terms were as follows:

Term of lease	4 years
Estimated useful life of machine	9 years
Purchase price of new machine	£75,000
Annual payments	£8,000

This is an *operating lease* as it does not transfer ownership, does not appear to contain an option to purchase, is not used for a major part of the asset's useful life, the present value of the lease payments does not constitute substantially all of the fair value and it does not appear to be specialised so that only Clifford could use it without major modification.

The amount of the annual rental paid in 20X1 – £8,000 p.a. – will be charged to the statement of comprehensive income and disclosed. There will also be a disclosure of the ongoing commitment with a note that £8,000 is payable within one year and £24,000 within two to five years.

EXAMPLE ● DISCLOSURE REQUIREMENTS IN THE LESSEE'S ACCOUNTS The following is an extract from the Nestlé Group 2013 Annual Report and Accounts:

Rentals payable under operating leases are expensed.
The costs of the agreements that do not take the legal form of a lease but convey the right to use an asset are separated into lease payments and other payments if the entity has the control of the use or of the access to the asset or takes essentially all the output of the asset. Then the entity determines whether the lease component of the agreement is a finance or an operating lease.

Other notes
Lease commitments

The following charges arise from these commitments:

Operating leases
Lease commitments refer mainly to buildings, industrial equipment, vehicles and IT equipment.

In millions of CHF	2013	2012
	Minimum lease payments future value	
Within one year	621	625
In the second year	499	519
In the third to fifth year inclusive	1,042	1,066
After the fifth year	619	657
	2,781	2,867

Operating lease charge for the year 2013 amounts to CHF 734 million (2012: CHF 720 million).

18.6 Accounting requirements for finance leases

We follow a step approach to illustrate the accounting entries in both the statement of financial position and the statement of comprehensive income.

When a lessee enters into a finance lease, both the leased asset and the related lease obligations need to be shown in the statement of financial position.

18.6.1 Statement of financial position step approach to accounting for a finance lease

Step 1 The leased asset should be capitalised in property, plant and equipment (and recorded separately) at the lower of the present value of the minimum lease payments and its fair value at the inception of the lease.

Step 2 The annual depreciation charge for the leased asset should be calculated by depreciating the asset over the shorter of its estimated useful life and the lease period.

Step 3 The net book value of the leased asset should be reduced by the annual depreciation charge.

Step 4 The finance lease obligation is a liability which should be recorded. At the inception of a lease agreement, the value of the leased asset and the leased liability will be the same.

Step 5 (a) The finance charge for the finance lease should be calculated as the difference between the total of the minimum lease payments and the fair value of the asset (or the present value of the minimum lease payments if lower), i.e. it represents the charge made by the lessor for the credit that is being extended to the lessee.

  (b) The finance charge should be allocated to the accounting periods over the term of the lease. Three methods for allocating finance charges are used in practice:

  - **Actuarial method**. This applies a constant periodic rate of charge to the balance of the leasing obligation. The rate to use is the 'rate of interest implicit in the lease'. This is essentially the return earned by the lessor on the funds invested in the leased asset, taking account of any estimated residual value at the end of the lease term. This rate is usually obtainable from the lessor, but if this rate is not available for any reason then IAS 17 states that the lessee's incremental borrowing rate should be used instead.

  - **Sum of digits method**. This method (sometimes known as the 'Rule of 78 Method') is often an acceptable approximation to the actuarial method. The finance charge is apportioned to accounting periods on a reducing scale.

  - **Straight-line method**. This spreads the finance charge equally over the period of the lease (it is only acceptable for immaterial leases).

Step 6 The finance lease obligation should be reduced by the difference between the lease payment and the finance charge. This means that first the lease payment is used to repay the finance charge, and then the balance of the lease payment is used to reduce the book value of the obligation.

18.6.2 Statement of comprehensive income step approach to accounting for a finance lease

Step 1 The annual depreciation charge should be recorded.

Step 2 The finance charge allocated to the current period should be recorded.

18.7 Example allocating the finance charge using the sum of the digits method

EXAMPLE ● FINANCE LEASE Clifford plc negotiates another lease to commence on 1 January 20X1 with the following terms:

Term of lease		3 years
Purchase price of new machine		£16,500
Annual payments (payable in advance)		£6,000

Finance charges are allocated using the sum of the digits method.

18.7.1 Categorise the transaction

First we need to decide whether the lease is an operating or a finance lease. Having established that other factors have not indicated that this is a finance lease we apply the present value criterion.

- Calculate the fair value:
  Fair value of asset = £16,500

- Calculate the present value of minimum lease payments using a discount rate of 10%:
  $$£6,000 + \frac{£6,000}{1.1} + \frac{£6,000}{(1.1)^2} = £16,413$$

- Compare the fair value and the present value. It is a finance lease because the present value of the lease payments is substantially all of the fair value of the asset.

18.7.2 Statement of financial position step approach to accounting for a finance lease

Step 1 Capitalise the lease at fair value (strictly speaking this should be the present value of the minimum lease payments as this is slightly lower, but here the difference is immaterial and we would report at its cost).
Asset value = £16,500

Step 2 Calculate the depreciation (using the straight-line method):
£16,500/3 = £5,500

Step 3 Reduce the asset in the statement of financial position:

Extract as at		*31 Dec 20X1*	*31 Dec 20X2*	*31 Dec 20X3*
Asset	Opening value	16,500	11,000	5,500
(right to	Depreciation	5,500	5,500	5,500
use asset)	Closing value	11,000	5,500	—

Or, if we keep the asset at cost as in published accounts:

Asset	Cost	16,500	16,500	16,500
(right to	Depreciation	5,500	11,000	16,500
use asset)	Closing value	11,000	5,500	—

Step 4 Obligation on inception of finance lease:
Liability = £16,500

Step 5 Finance charge:

Total payments, 3 × £6,000	£18,000
Asset value	£16,500
Finance charge, being the difference	£1,500

Finance charge allocated using sum of digits:

Year 1 = 2/(1 + 2) × £1,500 = £1,000
Year 2 = 1/(1 + 2) × £1,500 = £500

Note that the allocation is over only two periods because the instalments are being made in advance. If the instalments were being made in arrears, the liability would continue over three years and the allocation would be over three years.

Step 6 Reduce the obligation in the statement of financial position:

Extract as at		31 Dec 20X1	31 Dec 20X2	31 Dec 20X3
Liability	Opening value	16,500	11,500	6,000
(obligation	Lease payment	6,000	6,000	6,000
under finance		10,500	5,500	—
lease)	Finance charge	1,000	500	—
	Closing value	11,500	6,000	—

Note that the closing balance on the asset represents unexpired service potential and the closing balance on the liability represents the capital amount outstanding at the period-end date.

18.7.3 Statement of comprehensive income step approach to accounting for a finance lease

Step 1 A depreciation charge is made on the basis of use. The charge would be calculated in accordance with existing company policy relating to the depreciation of that type of asset.

Step 2 A finance charge is levied on the basis of the amount of financing outstanding.

Both then appear in the statement of comprehensive income as expenses of the period:

Extract for year ending	31 Dec 20X1	31 Dec 20X2	31 Dec 20X3
Depreciation	5,500	5,500	5,500
Finance charge	1,000	500	—
Total	6,500	6,000	5,500

18.8 Example allocating the finance charge using the actuarial method

In the Clifford example, we used the sum of the digits method to allocate the finance charge over the period of the repayment. In the following example, we will illustrate the actuarial method of allocating the finance charge.

EXAMPLE ● FINANCE LEASE Sirous Ltd negotiates a four-year finance lease for an item of plant with a cost price of £35,000. The annual lease payments are £10,000 payable in advance. The annual rate of interest implicit in the lease is approximately 9.7%.

First we need to compute the initial carrying value of the asset by comparing the present value of the minimum lease payments with the fair value of the asset. The present value of the minimum lease payments (with payments in advance) is:

$$£10,000 + £10,000/(1.097) + £10,000 (1.097)^2 + £10,000 (10.97)^3 = £35,003$$

In this case the fair value of the asset is slightly lower but (as is often the case with finance leases) the difference is immaterial.

Figure 18.2 shows that the finance charge is levied on the obligation during the period at 9.7.

18.11 Leasing – a form of off-balance-sheet financing

Prior to IAS 17, one of the major attractions of leasing agreements for the lessee was the off-balance-sheet nature of the transaction. However, the introduction of IAS 17 required the capitalisation of finance leases and removed part of the benefit of off-balance-sheet financing.

The capitalisation of finance leases effectively means that all such transactions will affect the lessee's gearing, return on assets and return on investment. Consequently, IAS 17 substantially alters some of the key accounting ratios which are used to analyse a set of financial statements.

Operating leases, on the other hand, are not required to be capitalised. This means that operating leases still act as a form of financing that is off the statement of financial position.[5] Hence, they are extremely attractive to many lessees. Indeed, leasing agreements are increasingly being structured specifically to be classified as operating leases, even though they appear to be more financial in nature.[6]

An important conclusion is that some of the key ratios used in financial analysis become distorted and unreliable in instances where operating leases form a major part of a company's financing.[7]

To illustrate the effect of leasing on the financial structure of a company, we present a buy versus leasing example.

EXAMPLE ● RATIO ANALYSIS OF BUY VERSUS LEASE DECISION Kallend Tiepins plc requires one extra machine for the production of tiepins. The MD of Kallend Tiepins plc is aware that the gearing ratio and the return on capital employed ratio will change depending on whether the company buys or leases (on an operating lease) this machinery. The relevant information is as follows.

The machinery costs £100,000, but it will improve the operating profit by 10% p.a. The current position, the position if the machinery is bought and the position if the machinery is leased are as follows, assuming that lease costs match depreciation charges:

	Current £	Buy £	Lease £
Operating profit	40,000	44,000	44,000
Equity capital	200,000	200,000	200,000
Long-term debt	100,000	200,000	100,000
Total capital employed	300,000	400,000	300,000
Gearing ratio	0.5:1	1:1	0.5:1
ROCE	13.33%	11%	14.66%

It is clear that the impact of a leasing decision on the financial ratios of a company can be substantial.[8] Although this is a very simple illustration, it does show that the buy versus lease decision has far-reaching consequences in the financial analysis of a company.

18.12 Accounting for leases – a new approach

18.12.1 Problems with the current approach

The existing treatment of leased assets in the financial statements of lessees is entirely dependent on the extremely subjective classification of the lease agreement as operating or finance. It is entirely possible that two leasing agreements that are essentially the same in nature are accounted for differently by two entities depending on their respective perception of the risks and rewards inherent in the arrangement.

IAS 17 has also been criticised for possible inconsistencies with the IASB's *Conceptual Framework*. The *Framework*'s definition of a liability is an obligation to transfer economic benefits as a result of past events. Operating leases are often non-cancellable; therefore there is a strong argument that operating lease commitments do meet the definition of a liability. However, no liability is recognised in the lessee's finanical statements.

An implication of the current situation is that arguably the quality of published finanical statements suffers, leaving the users to analyse detailed disclosure notes to get a true perception of the finanical commitments of an entity.

18.12.2 A proposed solution to the perceived problems

For some time the IASB has been working with the FASB – the US standard setter – to produce a converged solution to the problems associated with lease accounting, particularly in the financial statements of lessees. No definitive replacement standard to IAS 17 has yet been issued but in 2013 the IASB published an exposure draft of a revised standard.

The proposals are radical. For all leases other than those with a maximum possible term of 12 months or less the proposal is basically to abolish the distinction between an operating lease and a finance lease in the books of lessees. Lessees will recognise a right of use asset and an associated liability at the inception of the lease. The initial measurement of the asset and liability will be similar to the basis currently used for finance leases – being the present value of the minimum lease payments.

The exposure draft (ED) recognises that, despite the basic need to recognise all significant leasing obligations in the finanical statements, a variety of different lease types exist and it would be misleading to adopt a 'one size fits all' approach. Therefore the ED proposes splitting leases into two types. Type A will be leases of plant and equipment and Type B will be leases of property:

- For a Type A lease the lessee is often using and consuming the asset for a significant part of its useful economic life. The treatment for such leases outlined in the ED is substantially the same as that required for finance leases under the existing IAS 17.

- For a Type B lease the lessee is essentially paying rentals to use the asset for a (much less substantial) portion of the useful life of the asset. For such leases, whilst the statement of finanical position will contain an asset and a liability, the charge to profit or loss will be spread evenly over the lease.

18.12.3 A numerical example of the proposals

A company leases an asset with a useful economic life of 10 years on a four-year lease. Annual lease rentals are £50,000, payable in arrears. The company can borrow at a rate of 8%. Given the facts here the lease would almost certainly be classified as an operating lease under IAS 17. This would mean no asset or liability recognised in the statement of financial position and an annual expense of £50,000 recognised in profit or loss as an operating expense.

Accounting for the lease as a Type A lease

Under the new proposals, if the lease were a Type A lease, an asset and liability of £165,605 (the present value of four payments of £50,000 in arrears discounted at 8%) would be recognised at the inception of the lease.

The right of use asset would be depreciated over the four-year lease term to give an annual charge of £41,401. This charge would be recognised as an operating cost.

The finance cost for the four years would be as shown in the following table:

	Bal b/f	Finance Cost	Rental	Bal c/f
	£	£	£	£
Year 1	165,605	13,249	(50,000)	128,854
Year 2	128,854	10,308	(50,000)	89,162
Year 3	89,162	7,133	(50,000)	46,295
Year 4	46,295	3,705	(50,000)	Nil

Accounting for the lease as a Type B lease

For a Type B lease the initial recognition of the asset and liability would be as for a Type A lease – i.e. as £165,605. However, the pattern of expense recognition is different. For leases of this nature the expense is recognised in profit or loss evenly over the lease term, so a total of £50,000 would be recognised each year in this example as an operating cost. Although the total cost is recognised as an operating cost in profit or loss the other side of the entry needs to be allocated between a reduction in the asset and the liability in the statement of financial position. The proposed solution in the ED is to compute a 'notional' depreciation charge as a balancing figure, after deducting the notional finance cost computed as for a Type A lease. Therefore these notional amounts would be as follows:

Year 1 – £36,751 (£50,000 – £13,249)

Year 2 – £39,692 (£50,000 – £10,308)

Year 3 – £42,867 (£50,000 – £7,133)

Year 4 – £46,295 (£50,000 – £3,705)

It is useful to compare the impacts of the financial statements for year 1 under the three different approaches:

	Existing IAS 17	ED – Type A lease	ED – Type B lease
	£	£	£
Operating cost in profit or loss	50,000	41,401	50,000
Finance cost in profit or loss	Nil	13,249	Nil
Lease liability at year-end	Nil	128,854	128,854
Right of use asset at year-end	Nil	124,204	128,854

18.12.4 An evaluation of the new proposals for lessees

The ED's proposals will have significant impacts on key accounting ratios, whether the leases are Type A or Type B. In both cases the recognition of leased assets and lease liabilities on the statement of financial position will reduce return on capital employed and increase gearing. Measures that focus solely on profits will be less affected, and will only be affected for Type A leases. For Type A leases, reported operating profits are likely to increase, but this increase will be fully compensated by an increase in finance costs.

It is too early to say whether the proposals in the ED will affect the popularity of leasing if they are implemented into a standard. The commercial need for finance will still remain, of course, so in the opinion of this writer there is no reason to suppose that the leasing industry will be adversely affected.

It may well be, though, that short-term leases become more popular, given that they are likely to be treated as currently.

Summary

Off-balance-sheet financing was considered a particular advantage of lease financing. IAS 17 recognised this and attempted to introduce stricter accounting policies and requirements. However, although IAS 17 introduced the concept of 'substance over form', the hazy distinction between finance and operating leases still allows companies to structure lease agreements to achieve either type of lease. This is important because, while stricter accounting requirements apply to finance leases, operating leases can still be used as a form of accounting off the statement of financial position.

We do not know the real extent to which IAS 17 is either observed or ignored. However, it is true to say that creative accountants and finance companies are able to circumvent IAS 17 by using 'structured' leases. Future developments are attempting to reduce the opportunities for such off-balance-sheet financing. It is likely that companies will continue to focus on minimising the amount of the leasing assets and liabilities which they have to include in the statement of financial position.

REVIEW QUESTIONS

1 'Can the legal position on leases be ignored now that substance over form is used for financial reporting?' Discuss.

2 (a) Consider the importance of decisions over the categorisation of lease transactions into operating leases or finance leases when carrying out financial ratio analysis. What ratios might be affected if a finance lease is structured to fit the operating lease classification?

(b) Discuss the effects of renegotiating/reclassifying all operating leases into finance leases. For which industries might this classification have a significant impact on the financial ratios?

3 State the factors that indicate that a lease is a finance lease under IAS 17.

4 The favourite off-balance-sheet financing trick used to be leasing. Use any illustrative numerical examples you may wish to:

(a) Define the term 'off-balance-sheet financing' and state why it is popular with companies.

(b) Illustrate what is meant by the above term in the context of leases and discuss the accounting treatments and disclosures required by IAS 17 which have limited the usefulness of leasing as an off-balance-sheet financing technique.

(c) Suggest two other off-balance-sheet financing techniques and discuss the effect that each technique has on statement of financial position assets and liabilities, and on the income statement.

5 The Tesco 2014 Annual Report included the following accounting policy:

Assets held under finance leases are recognised as assets of the Group at their fair value or, if lower, at the present value of the minimum lease payments, each determined at the inception of the lease. The corresponding liability is included in the Group Balance Sheet as a finance lease obligation. Lease payments are apportioned between finance charges and a reduction of the lease obligations so as to achieve a constant rate of interest on the remaining balance of the liability. Finance charges are charged to the Group Income Statement. Rentals payable under operating leases are charged to the Group Income Statement on a straight-line basis over the term of the lease.

Rental income from operating leases is recognised on a straight-line basis over the term of the lease.

(a) Explain the meaning of 'minimum lease payments and fair value'.

(b) Explain why fair value might be higher than the discounted minimum lease payments.

(c) Explain why the aim is to arrive at a constant rate of interest.

6 Given that the details of operating leases are disclosed in the notes to the accounts, why is it necessary to propose a new standard which incorporates these into the statement of financial position when sophisticated investors already do such adjustments themselves?

7 Under the exposure draft issued in 2013 lessees have to account for all leases other than short-term leases of less than 12 months duration by recognising an asset or liability in the statement of financial position. However, subsequent accounting depends on whether the lease is 'Type A' or 'Type B'. Explain the basis upon which leases are allocated to these two categories and outline the different accounting treatment that results.

8 Companies sometimes get special prices from suppliers if they undertake to purchase specified commodities from the supplier over a designated future period. These supply arrangements do not have to be recorded as assets and liabilities. However, in future leases will give rise to recording of assets and liabilities. Discuss why the transactions are to be treated differently.

EXERCISES

* Question 1

On 1 January 20X8, Grabbit plc entered into an agreement to lease a widgeting machine for general use in the business. The agreement, which may not be terminated by either party to it, runs for six years and provides for Grabbit to make an annual rental payment of £92,500 on 31 December each year. The cost of the machine to the lessor was £350,000, and it has no residual value. The machine has a useful economic life of eight years and Grabbit depreciates its property, plant and equipment using the straight-line method.

Required:

(a) Show how Grabbit plc will account for the above transaction in its statement of financial position at 31 December 20X8, and in its statement of comprehensive income for the year then ended, if it capitalises the leased asset in accordance with the principles laid down in IAS 17. The rate of interest implicit in the lease is 15%.

(b) Explain why the standard setters considered accounting for leases to be an area in need of standardisation and discuss the rationale behind the approach adopted in the standard.

(c) The lessor has suggested that the lease could be drawn up with a minimum payment period of one year and an option to renew. Discuss why this might be attractive to the lessee.

* Question 2

(a) When accounting for finance leases, accountants prefer to overlook legal form in favour of commercial substance.

Required:

Discuss the above statement in the light of the requirements of IAS 17 *Leases*.

(b) State briefly how you would distinguish between a finance lease and an operating lease.

(c) Smarty plc finalises its accounts annually on 31 March. It depreciates its machinery at 20% per annum on cost and adopts the 'Rule of 78' for allocating finance charges among different accounting periods. On 1 August 20X7 it acquired machinery on a finance lease on the following agreement:

  (i) a lease rental of £500 per month is payable for 36 months commencing from the end of the month of acquisition;

  (ii) cost of repairs and insurance are to be met by the lessee;

  (iii) on completion of the primary period the lease may be extended for a further period of two years, at the lessee's option, for a peppercorn rent.

  The cash price of the machine is £15,000.

Required:

(1) Compute the carrying value of the machine in the statement of financial position at 31 March 20X8 and 31 March 20X9.

(2) Compute the finance cost for the years ended 31 March 20X8 and 31 March 20X9.

(3) Compute the lease liability that will be included in the statement of financial position at 31 March 20X8 and 31 March 20X9. In both cases show the split into the current and non-current portions.

* Question 3

The Mission Company Ltd, whose year-end is 31 December, has acquired two items of machinery on leases, the conditions of which are as follows:

Item Y: Ten annual instalments of £20,000 each, the first payable on 1 January 20X0. The machine was completely installed and first operated on 1 January 20X0 and its purchase price on that date was £160,000. The machine has an estimated useful life of 10 years, at the end of which it will be of no value.

Item Z: Ten annual instalments of £30,000 each, the first payable on 1 January 20X2. The machine was completely installed and first operated on 1 January 20X2 and its purchase price on that date was £234,000. The machine has an estimated useful life and is used for 12 years, at the end of which it will be of no value.

The Mission Company Ltd accounts for finance charges on finance leases by allocating them over the period of the lease on the sum of the digits method.

Depreciation is charged on a straight-line basis. Ignore taxation.

Required:

(a) Calculate and state the charges to the statement of comprehensive income for 20X6 and 20X7 if the leases were treated as operating leases.

(b) Calculate and state the charges to the statement of comprehensive income for 20X6 and 20X7 if the leases were treated as finance leases and capitalised using the sum of the digits method for the finance charges.

(c) Show how items Y and Z should be incorporated in the statement of financial position, and notes thereto, at 31 December 20X7, if capitalised.

Question 4

X Ltd entered into a lease agreement on the following terms:

Cost of leased asset	£100,000
Lease term	5 years
Rentals six-monthly in advance	£12,000
Anticipated residual on disposal of the assets at end of lease term	£0
Economic life	5 years
Inception date	1 January 20X4
Lessee's financial year-end	31 December
Implicit rate of interest is applied half-yearly	4.3535%

Required:

Show the statement of comprehensive income entries for the years ended 31 December 20X4 and 20X7 and statement of financial position extracts at those dates in accordance with IAS 17.

Question 5

Alpha entered into an operating lease under which it was committed to five annual payments of £50,000 per year in arrears. Alpha's borrowing rate was 10%.

Required:

Calculate the amounts to be reported in profit and loss and the statement of financial position for each of the five years assuming the lease is accounted for:
(a) Under IAS 17.
(b) Under the 2013 exposure draft as a Type A lease.
(c) Under the 2013 exposure draft as a Type B lease.

Question 6

On 1 January 20X7 Baldwin Brothers entered into a five-year lease of equipment at an annual rental of £39,000 payable in arrears. The interest rate associated with this transaction is 10% and Baldwin Brothers incurred direct costs of £2,700 in setting up the lease. The fair value of the equipment at 1 January 20X7 was £175,000 and its remaining useful economic life at that date was estimated to be eight years.

Required:

Show the relevant amounts that would appear in the financial statements of Baldwin Brothers for the year ended 31 December 20X7 under the following two assumptions:
(1) The lease is accounted for under IAS 17.
(2) The lease is accounted for as a Type A lease under the proposals in the 2013 exposure draft.

Question 7

Magnificant Retailor enters into a five-year lease agreement with Mega Shopping Centres plc to lease a small shop at £50,000 per annum payable at the start of each year. Assume the relevant interest rate is 10% per annum.

Required:

Compute the carrying value of the asset and liability in the financial statements at the end of each year assuming the lease is accounted for as a Type B lease under the 2013 exposure draft.

Question 8

Market Specialists plc leases a motor vehicle for its managing director from K G Financiers plc for a five-year period at £27,618 per annum payable in advance. The expected life of the vehicle is eight years. The relevant interest rate is 10% per annum.

Required:
Compare and contrast the impact on the financial statements of Market Specialists for the first year of the lease of treating the lease as an operating lease under IAS 17 and a 'Type A' lease under the 2013 exposure draft. Your answer should address the impact on key accounting ratios.

Prepare the entries required in the financial statements for the first two years.

References

1 G. Allum et al., 'Fleet focus: to lease or not to lease', *Australian Accountant*, September 1989, pp. 31–58; R.L. Benke and C.P. Baril, 'The lease vs. purchase decision', *Management Accounting*, March 1990, pp. 42–6.
2 B. Underdown and P. Taylor, *Accounting Theory and Policy Making*, Heinemann, 1985, p. 273.
3 Cranfield School of Management, *Financial Leasing Report*, Bedford, 1979.
4 A.R. Abdel-Khalik et al., 'The economic effects on lessees of FASB Statement No. 13', *Accounting for Leases*, FASB, 1981.
5 R.H. Gamble, 'Off-balance-sheet diet: greens on the side', *Corporate Cashflow*, August 1990, pp. 28–32.
6 R.L. Benke and C.P. Baril, 'The lease vs. purchase decision', *Management Accounting*, March 1990, pp. 42–6; N. Woodhams and P. Fletcher, 'Operating lease to take bigger market share with changing standards', *Rydge's (Australia)*, September 1985, pp. 100–10.
7 C.H. Volk. 'The risks of operating lease', *Journal of Commercial Bank Lending*, May 1988, pp. 47–52.
8 Chee-Seong Tah, 'Lease or buy?', *Accountancy*, December 1992, pp. 58–9.

CHAPTER 19

Intangible assets

19.1 Introduction

The statement of financial position or balance sheet has traditionally reported tangible non-current assets and working capital. In order to protect future economic benefits, i.e. profits, companies have paid for legal or contractual intangible assets such as patents and copyright. As these were evidenced by a payment, they satisfied the accounting definition of an asset with a measurable cost and probable future economic benefit.

As business has become more complex, future economic benefits have become more reliant on internally generated intellectual capital that does not satisfy all the criteria for inclusion as an asset in the statement of financial position. For example, expenditure on a skilled workforce, training, research, and the development of a loyal customer base are all charged as an expense in the statement of income.

There are two adverse results that may arise from this: (a) the current year's profits are reduced by the charge and the company may be at risk in the short term from a predatory takeover; and (b) there is a mismatch between the market value of a company and the book value of its net assets.

The main purpose of this chapter is to consider the approach taken by IAS 38 *Intangible Assets*[1] and IFRS 3 *Business Combinations*[2] to the accounting treatment of intangible assets.

Objectives

By the end of this chapter, you should be able to:

- define and explain how to account for:
  - legally enforceable intangibles and internally generated intangibles;
  - research and development (R&D);
  - goodwill;
  - brands; and
  - emissions trading certificates;
- account for development costs;
- comment critically on the IASB requirements in IAS 38 and IFRS 3.

19.2 Intangible assets defined

Intangible assets are identifiable non-monetary assets that cannot be seen, touched or physically measured but are identifiable as a separate asset.

19.2.1 Criteria for recognition as an asset in the statement of financial position

IAS 38 *Intangible Assets* states that an asset is recognised in respect of an intangible item if the asset is characterised by the following properties:

- The asset is identifiable.
  The standard states that for an intangible asset to exist (or be identifiable) it must either be separable or arise from contractual or other legal rights (such as a patent), whether or not the asset can be separately disposed of (such as goodwill).
- The asset is controlled by the entity.
  Control is one of the central features of the *Framework* definition of an asset. Control is said to exist if the entity has the power to obtain the future economic benefits flowing from the underlying resource and to restrict the access of others to those benefits. *It is failing to satisfy the control criterion that prevents the skills of the workforce being recognised as an asset in the statement of financial position.*
- The asset gives future economic benefits.
  Again, it is inherent in the *Framework* definition of an asset that the potential future economic benefits can be identified with reasonable certainty.

If the identifiability and control tests are satisfied then IAS 38 allows recognition of an intangible asset if:

- it is **probable** that the expected future economic benefits that are attributable to the asset will flow to the entity; and
- the cost of the asset can be **measured reliably**.

Application of these criteria means that the costs associated with most internally generated intangible assets are expensed to the statement of income. An exception is development costs, **provided** these meet additional recognition criteria required by the standard.

19.2.2 Examples of intangible assets to be recognised and reported

Examples of intangible assets that should be recognised and reported in the statement of financial position are set out in IAS 38.[3] They include:

- **Marketing-related** intangible assets which are used primarily in the marketing or promotion of products or services such as trademarks, newspaper mastheads, Internet domain names and non-compete agreements.
- **Technology-related** intangible assets which arise from contractual rights to use technology (patented and unpatented), databases, formulae, designs, software, processes and recipes.
- **Customer- or supplier-related** intangible assets which arise from relationships with or knowledge of customers or suppliers such as licensing, royalty and standstill agreements, servicing contracts and use rights such as airport landing slots and customer lists.
- **Artistic-related** intangible assets which arise from the right to benefits such as royalties from artistic works such as plays, books, films and music, and from non-contractual copyright protection.

19.2.3 Recognition criteria illustrated

Devon Cheeses Ltd decided to diversify into the production of vegetarian organic sausages. The project team produced a list of cost headings for the acquisition of:

(a) recipes from an international chef;

(b) a licence to use a specialised computer-controlled oven;

(c) registration of a trade name 'The Organo One'; and

(d) training courses for management in sausage making.

Their auditors were asked for advice on the possibility of capitalising all costs arising in respect of the above. The advice received was that the cost of recipes, the licence and the trade name registration could be capitalised, since:

● they were identifiable arising from contractual rights;

● Devon Cheeses Ltd controlled the future economic benefits;

● the costs could be measured reliably;

● it was probable that there would be future economic benefits; and

● the trade name was a defensive intangible that protected the receipt of the future economic benefits.

The training courses would improve management expertise but failed the control criterion and should be expensed.

19.2.4 Accounting treatment of recognised intangible assets at year-ends

The accounting treatment depends on whether the asset has a finite or an indefinite life.

Recognising intangible assets with a finite life

IAS 38 states that recognised intangible non-current assets should be reported at cost less accumulated amortisation or, as when a parent acquires a subsidiary with intangible assets, fair value less accumulated amortisation.

Amortisation of intangible assets with a finite life

The asset should be amortised on a systematic basis over its estimated useful economic life. This is very similar to the treatment of property, plant and equipment under IAS 16 in that it is frequently on a straight-line basis as for patents with a finite legal life. The following extract is from the Bayer Group 2013 Annual Report:

> Patents are valid for varying periods, depending on the laws of the jurisdiction granting the patent. . . . Intangible assets are recognised at the cost of acquisition or generation. . . . Those with a determinable useful life are amortised accordingly on a straight-line basis over a period of up to 30 years, except where their actual depletion demands a different amortisation pattern.

An acceptable basis for amortisation

Amortisation is to be based on the expected pattern of consumption of the future economic benefits of an asset. A clarification[4] issued by the IASB in 2014 advised that the use of revenue-based methods to calculate the depreciation of an asset is not appropriate because revenue generated by an activity that includes the use of an asset generally reflects factors other than the consumption of the economic benefits embodied in the asset.

Impairment of intangible assets with a finite life

Intangible assets are also tested for impairment where there is a triggering event. The following Accounting Policy extract from the SABMiller 2014 Annual Report explains the amortisation and impairment policy for intangibles with finite lives:

Intangible assets are stated at cost less accumulated amortisation on a straight-line basis (if applicable) and impairment losses . . . Amortisation is included within net operating expenses in the income statement . . . Intangible assets with finite lives are amortised over their estimated useful economic lives, and only tested for impairment where there is a triggering event.

Amortisation of intangible assets with an indefinite life

Where the estimated useful economic life is indefinite there is no amortisation but the asset is subject to annual impairment reviews under IAS 36.

19.2.5 Disclosure of intangible assets under IAS 38

IAS 38 requires the disclosure of the following for each type of intangible asset:[5]

- whether useful lives are indefinite or finite;
- the amortisation methods used for intangible assets with finite useful lives;
- the gross carrying amount and accumulated amortisation at the beginning and end of the period;
- increases or decreases resulting from revaluations and from impairment losses recognised or reversed directly in equity (IAS 36 *Impairment of Assets*); and
- for R&D, disclosure in the financial statements of the charge for research and development in the period.[6]

Where an intangible asset is assessed as having an indefinite useful life, the carrying value of the asset must be stated[7] along with the reasons for supporting the assessment of an indefinite life.

19.3 Accounting treatment for research and development

Under IAS 38 *Intangible Assets*, research expenditure **must be expensed** whereas development expenditure **must be capitalised** provided a strict set of criteria is met. In this section we will consider R&D activities, why research expenditure is written off and the tests for capitalising development expenditure.

19.3.1 Research activities

IAS 38 states[8] 'expenditure on research shall be recognised as an expense when it is incurred'. This means that it cannot be included as an intangible asset in the statement of financial position. The standard gives examples of research activities[9] as:

- activities aimed at obtaining new knowledge;
- the search for, evaluation and final selection of, applications of research findings or other knowledge;
- the search for alternatives for materials, devices, products, processes, systems and services; and
- the formulation, design, evaluation and final selection of possible alternatives for new or improved materials, devices, products, processes, systems or services.

Normally, research expenditure is not related directly to any of the company's products or processes. For instance, development of a high-temperature material which could be used in any aero engine would be 'research', but development of a honeycomb for a particular engine would be 'development'.

Whilst it is in the research phase, the IAS position[10] is that an entity cannot demonstrate that an intangible asset exists that will generate probable future economic benefits. It is this inability that justifies the IAS requirement for research expenditure not to be capitalised but to be charged as an expense when it is incurred.

19.3.2 Development activities

Expenditure on development is recognised[11] as an asset if the entity can demonstrate that the expenditure will generate probable future economic benefits. The standard gives examples of development activities:[12]

(a) the design, construction and testing of pre-production and pre-use prototypes and models;

(b) the design of tools, jigs, moulds and dies involving new technology;

(c) the design, construction and operation of a pilot plant that is not of a scale economically feasible for commercial production; and

(d) the design, construction and testing of a chosen alternative for new or improved materials, devices, products, processes, systems or services.

19.4 Why is research expenditure not capitalised?

Many readers will think of research not as a cost but as a strategic investment which is essential to remain competitive in world markets. Indeed, this was the view[13] taken by the House of Lords Select Committee on Science and Technology, stating that 'R&D has to be regarded as an investment which leads to growth, not a cost'.

It is reported[14] that global R&D spending is in excess of 1.7% of GDP, taking place particularly in the advanced technical industries such as pharmaceuticals, where a sustained high level of R&D investment is required. The regulators, however, do not consider that the expenditure can be classified as an asset for financial reporting purposes.

Why do the regulators not regard research expenditure as an asset?

The IASC in its *Framework for the Preparation and Presentation of Financial Statements*[15] defines an asset as a resource that is controlled by the enterprise, as a result of past events and from which future economic benefits are expected to flow.

Research is controlled by the enterprise and results from past events but there is no reasonable certainty that the intended economic benefits will be achieved. Because of this uncertainty, the accounting profession has traditionally considered it more prudent to write off the investment in research as a cost rather than report it as an asset in the statement of financial position.

The importance to investors of disclosure

It might be thought that this is concealing an asset from investors, but in research on the reactions of both analysts[16] and accountants[17] to R&D expenditure, B. Nixon found that: 'Two important dimensions of the corporate reporting accountants' perspective emerge: first, disclosure is seen as more important than the accounting treatment of R&D expenditure and, second, the financial statements are not viewed as the primary channel of communication for information on R&D.'

This highlights the importance of reading carefully the narrative in financial reports. An interesting study in Singapore[18] examined the impact of annual report disclosures on analysts' forecasts for a sample of firms listed on the Stock Exchange of Singapore (SES) and showed that the level of disclosure affected the accuracy of earnings forecasts among analysts and also led to greater analyst interest in the firm.

Management attitudes to capitalising research expenditure

Management might prefer in general to be able to capitalise research expenditure but there could be circumstances where writing off might be preferred. For example, directors might be pleased to take the expense in a year when they know its impact rather than carry it forward. They are aware of profit levels in the year in which the expenditure arises and could, perhaps, find it embarrassing to take the charge in a subsequent year when profits were lower or the company even reported a trading loss.

Development expenditure, on the other hand, has more probability of achieving future economic benefits and the regulators, therefore, require such expenditure to be capitalised.

19.5 Capitalising development costs

19.5.1 Conditions to be satisfied

The relevant paragraph of IAS 38[19] says an intangible asset for development expenditure must be recognised if and only if an entity can demonstrate **all** of the following:

(a) the technical feasibility of completing the intangible asset so that it will be available for use or sale;

(b) the intention to complete the intangible asset and use or sell it;

(c) its ability to use or sell the intangible asset;

(d) how the intangible asset will generate probable future economic benefits;

(e) the availability of adequate technical, financial and other resources to complete the development and to use or sell the intangible asset;

(f) its ability to measure reliably the expenditure attributable to the intangible asset during its development.

It is important to note that if the answers to all the conditions (a) to (f) above are 'Yes' then the entity *must* capitalise the development expenditure subject to reviewing for impairment.

19.5.2 What costs can be included?

The costs that can be included in development expenditure are similar to those used in determining the cost of inventory (IAS 2 *Inventories*).

It is important to note that only expenditure incurred after the project satisfies the IAS 38 criteria can be capitalised – all expenditure incurred prior to this date must be written off as an expense in the statement of income. Experience tends to indicate that people who develop products are notoriously optimistic. In practice, they encounter many more problems than they imagined and the cost is much greater than estimates. This means that the development project may well be approaching completion before future development costs can be estimated reliably.

At the year-ends development costs are usually amortised over the sales of the product (i.e. the charge in 20X5 would be: 20X5 sales/total estimated sales × capitalised development expenditure) with straight-line as the default.

Whilst development cost can be capitalised there is a requirement in the new EU Accounting Directive that, where the costs of development have not been completely written off, there can be no distribution of profits unless the amount of the reserves available for distribution and profits brought forward is at least equal to that of the costs not written off.

19.6 Disclosure of R&D

R&D is important to many manufacturing companies, such as pharmaceutical companies and car and defence manufacturers. Disclosure is required of the aggregate amount of research and development expenditure recognised as an expense during the period.[4] Normally, this total expenditure will be:

(a) research expenditure;

(b) development expenditure amortised;

(c) development expenditure not capitalised; and

(d) impairment of capitalised development expenditure.

Under IAS 38 more companies may capitalise development expenditure. Management view of the probability of making future profits from the sale of the product is a critical element in making a decision. The following is the R&D policy extract from the Rolls-Royce Annual Report for the year ended 31 December 2013:

Research and development
In accordance with IAS 38 Intangible Assets, expenditure incurred on research and development is distinguished as relating either to a research phase or to a development phase.

All research phase expenditure is charged to the income statement. Development expenditure is capitalised as an internally generated intangible asset only if it meets strict criteria, relating in particular to technical feasibility and generation of future economic benefits. . . . the Group considers that it is not possible to distinguish reliably between research and development activities until relatively late in the programme.

Expenditure capitalised is amortised on a straight-line basis over its useful economic life, up to a maximum of 15 years from the entry into service of the product.

19.7 IFRS for SMEs' treatment of intangible assets

Internally generated intangible assets

The IFRS provides that internally generated intangible assets are not recognised. This means that both research and development costs are expensed.

Separately purchased intangible assets

The IFRS provides the following:

● Such assets should be amortised over the asset's useful life; and if the useful life cannot be estimated, then a 10-year useful life is presumed.

- If there is a significant change in the asset or how it is used, then the useful life, residual value and depreciation rate are reviewed.
- Impairment testing is carried out where there are impairment indications.
- The revaluation of intangible assets is prohibited.

19.8 Internally generated and purchased goodwill

IFRS 3 *Business Combinations* defines goodwill[20] as: 'future economic benefits arising from assets that are not capable of being individually identified and separately recognised'. The definition effectively affirms that the value of a business as a whole is more than the sum of the accountable and identifiable net assets. Goodwill can be internally generated through the normal operations of an existing business or purchased as a result of a business combination.

19.8.1 Internally generated goodwill

Internally generated goodwill falls within the scope of IAS 38 *Intangible Assets* which states that 'Internally Generated Goodwill (or "self-generated goodwill") shall not be recognised as an asset'. If companies were allowed to include internally generated goodwill as an asset in the statement of financial position, it would boost total assets and produce a more favourable view of the statement of financial position, for example by reducing the gearing ratio.

19.8.2 Purchased goodwill

How goodwill is calculated

The key distinction between internally generated goodwill and purchased goodwill is that purchased goodwill has an identifiable 'cost', being the difference between the fair value of the total consideration that was paid to acquire a business and the fair value of the identifiable net assets acquired. This is the initial cost reported in the statement of financial position.

Companies reporting under IFRS are required to disclose the nature of the intangible assets comprising goodwill and explain why they cannot be valued separately.

19.9 The accounting treatment of goodwill

Now that we have a definition of goodwill, we need to consider how to account for it in subsequent years. One might have reasonably thought that a simple requirement to amortise the cost over its estimated useful life would have been sufficient. This has been far from the case. Over the past 40 years, there have been a number of approaches to accounting for purchased goodwill, including:

(a) writing off the cost of the goodwill directly to reserves in the year of acquisition;

(b) reporting goodwill at cost in the statement of financial position (this was attractive to management as there was no charge against profits in any year);

(c) reporting goodwill at cost, amortising over its expected life; and

(d) reporting goodwill at cost, but checking it annually for impairment.

The last (d) is now the treatment required by IFRS 3.

19.9.1 The current IFRS 3 treatment

IFRS 3 prohibits the amortisation of goodwill. It treats goodwill as if it has an indefinite life with the amount reviewed annually for impairment. If the carrying value is greater than the recoverable value of the goodwill, the difference is written off.

Whereas goodwill amortisation gave rise to an annual charge, impairment losses will arise at irregular intervals. This means that the profit for the year will become more volatile. This is why companies and analysts rely more on the EBITDA (earnings before interest, tax, depreciation and amortisation) when assessing a company's performance, assuming that this is a better indication of maintainable profits.

This is illustrated by the following extract from the 2012 Vodafone Annual Report which shows the volatile effect of impairment charges on maintainable profits:

	2012 £m	2011 £m	2010 £m
Revenue	46,417	45,884	44,472
Gross profit	14,871	15,070	15,033
Impairment losses	(4,050)	(6,150)	(2,100)
Operating profit	11,187	5,596	9,480

This illustrates the volatility when impairment charges are included when calculating operating profit or loss with a pre-impairment profit reporting a profit increase in 2012 of 100% instead of 30% and a fall in 2011 instead of an increase.

19.9.2 Identifying intangible assets to reduce the amount of goodwill

Because goodwill is reviewed annually for impairment under IFRS 3 and other intangible assets are mainly amortised annually under IAS 38, standard setters wanted companies to identify any intangible assets that were acquired on an acquisition of another company and not to include them within a global figure of goodwill.

This has two effects: (a) there is greater transparency and control over assets by identifying the asset that the parent acquired; and (b) intangible assets are amortised rather than being reviewed annually for impairment, so reducing the volatility in the reported operating profits.

The following is an extract from the Intel 2013 Annual Report relating to intangible assets:

(In Millions)	Gross Assets	Accumulated Amortization	Net
Acquisition-related developed technology	$ 2,922	$ (1,691)	$ 1,231
Acquisition-related customer relationships	1,760	(828)	932
Acquisition-related trade names	65	(44)	21
Licensed technology and patents	3,093	(974)	2,119
Identified intangible assets subject to amortization	7,840	(3,537)	4,303
Acquisition-related trade names	818	—	818
Other intangible assets	29	—	29
Identified intangible assets not subject to amortization	847	—	847
Total identified intangible assets	$ 8,687	$ (3,537)	$ 5,150

The notes to the accounts relating to amortisation include:

The estimated useful life ranges for substantially all identified intangible assets that are subject to amortization as of December 28, 2013, were as follows:

(In Years)	Estimated Useful Life
Acquisition-related developed technology	4–9
Acquisition-related customer relationships	5–8
Acquisition-related trade names	4–8
Licensed technology and patents	5–17

Greater transparency in relation to the make-up of the goodwill figure should be achieved following the amendment in July 2009 to IFRS 3 which provides that if an intangible asset can be separately identified then it can be measured reliably, as the two conditions are interdependent. This will place further pressure on companies to properly consider the nature and value of any intangible assets they acquire.

However, there is a concern that shareholders might lose sight of the outcome of the acquisition – investors can see if there is a single figure for goodwill which has to be written down if it becomes impaired. This makes it easier to hold management to account for the decision to make the acquisition.

19.10 Critical comment on the various methods that have been used to account for goodwill

Let us consider briefly the alternative accounting treatments.

(a) Reporting goodwill unchanged at cost

It is (probably) wrong to keep goodwill unchanged in the statement of financial position, as its value will decline with time. Its value may be *maintained* by further expenditure, e.g. continued advertising, but this expenditure is essentially creating 'internally generated goodwill' which is not allowed to be capitalised. Sales of most manufactured products often decline during their life and their selling price falls. Eventually, the products are replaced by a technically superior product. An example is computer microprocessors, which initially command a high price and high sales. The selling price and sales quantities decline as faster microprocessors are produced. Much of the goodwill of businesses is represented by the products they sell. Hence, it is wrong to not amortise the goodwill.

(b) Writing off the cost of the goodwill directly to reserves in the year of acquisition

A buyer pays for goodwill on the basis that future profits will be improved. It is wrong, therefore, to write it off in the year of acquisition against previous years in the reserves. The loss in value of the goodwill does not occur at the time of acquisition but occurs over a longer period. The goodwill is losing value over its life, and this loss in value should be charged to the statement of comprehensive income each year. Making the charge direct to reserves stops this charge from appearing in the future income statements.

(c) Amortising the goodwill over its expected useful life

Amortising goodwill over its life could achieve a matching under the accrual concept with a charge in the statement of comprehensive income. However, there are problems

(i) in determining the life of the goodwill and (ii) in choosing an appropriate method for amortising.

(i) What is the life of the goodwill?

Companies wishing to minimise the amortisation charge could make a high estimate of the economic life of the goodwill and auditors have to be vigilant in checking the company's justification. The range of lives can vary widely. For example, goodwill paid to acquire a business in the fashion industry could be quite short compared to that paid to acquire an established business with a loyal customer base.

(ii) The method for amortising

Straight-line amortisation is the simplest method. However, as the benefits are likely to be greater in earlier years than in later ones, amortisation could use the reducing balance method. One might think intuitively that amortisation based on the percentage of actual sales to expected total sales would reasonably match the cost consumed with the revenue – however, it is not an acceptable method under IFRS.

It could be argued that amortising goodwill is equivalent to depreciating tangible fixed assets as prescribed by IAS 16 *Property, Plant and Equipment* and that the amortisation approach appears to be the best way of treating goodwill in the statement of financial position and statement of comprehensive income. This is effectively following a 'statement of comprehensive income' approach to 'expense' (e.g. depreciation) with the expense charged over the life of the asset or in relation to the profits obtained from the acquisition.

There are difficulties but these should not prevent us from using this method. After all, accountants have to make many judgements when valuing items in the statement of financial position, such as assessing the life of property, plant and equipment, the value of inventory and bad debt provisions.

(d) An annual impairment check

IFRS 3 introduced a new treatment for purchased goodwill when it arises from a business combination (i.e. the purchase of a company which becomes a subsidiary). It assumes that goodwill has an indefinite economic life, which means that it is not possible to make a realistic estimate of its economic life and a charge should be made to the statement of income only when it becomes impaired.

This is called a 'statement of financial position' approach to accounting, as the charge is made only when the value (in the statement of financial position) falls below its original cost.

The IFRS 3 treatment is consistent with the *Framework*,[21] which says: 'Expenses are recognised in the statement of comprehensive income when a decrease in future economic benefits related to a decrease in an asset or an increase of a liability has arisen that can be measured reliably.'

Criticism of the 'statement of financial position' approach

However, there has been much criticism of the 'statement of financial position' approach of the *Framework*.

For example, if a company purchased specialised plant which had a resale value of 5% of its cost, then it could be argued that the depreciation charge should be 95% of its cost immediately after it comes into use. This is not sensible, as the purpose of buying the plant is to produce a product, so the depreciation charge should be over the life of the product.

Alternatively, if the 'future economic benefit' approach was used to value the plant, there would be no depreciation until the future economic benefit was less than its original cost.

So, initial sales would incur no depreciation charge, but later sales would have an increased charge.

This example shows the weakness of using impairment and the 'statement of financial position' approach for charging goodwill to the statement of comprehensive income – the charge occurs at the wrong time. The charge should be made earlier when sales, selling prices and profits are high, not when the product becomes out of date and sales and profits are falling.

Why the impairment charge occurs at the wrong time

Although the IFRS 3 treatment of impairment appears to be correct according to the *Framework*, it could be argued that the impairment approach is not correct, as the charge occurs at the wrong time (i.e. when there is a loss in value, rather than when profits are being made), it is very difficult to estimate the future economic benefit of the goodwill and those estimates are likely to be over-optimistic.

In addition, it means that the treatment of goodwill for IFRS 3 transactions is different from the treatment in IAS 38 *Intangible Assets*. This shows the inconsistency of the standards – they should use a single treatment, either IAS 38 amortisation or IFRS 3 impairment.

19.10.1 Why has the IFRS 3 treatment of goodwill differed from the treatment of intangible assets in IAS 38?

The answer is probably related to the convergence of International Accounting Standards to US accounting standards, and pressure from listed companies.

Convergence pressure

In issuing recent International Standards, the IASB has not only aimed to produce 'world-wide' standards but also standards which are acceptable to US standard setters. The IASB wanted their standards to be acceptable for listing on the New York Stock Exchange (NYSE), so there was strong pressure on the IASB to make their standards similar to US standards. The equivalent US standard to IFRS 3 uses impairment of goodwill as the charge against profits (rather than amortisation). Thus, IFRS 3 uses the same method and it prohibits amortisation.

Commercial pressure

A further pressure for impairment rather than amortisation comes from listed companies. Essentially, listed companies want to maximise their reported profit, and amortisation reduces profit. For most of the time, companies can argue that the future economic benefit of the goodwill is greater than its original cost (or carrying value if it has been previously impaired), and thus avoid a charge to the statement of comprehensive income. Also, companies could argue that the 'impairment charge' is an unexpected event and charge it as an exceptional item.

In the UK, many companies publicise their profit before exceptional items and impairment to highlight maintainable profits.

The new EU Accounting Directive provides that in exceptional cases where the useful life of goodwill and development costs cannot be reliably estimated, such assets shall be written off within a maximum period set by the member state – that maximum period to be not shorter than five years and not longer than 10 years. IFRS takes precedence over the Directive.

19.11 Negative goodwill/Badwill

Negative goodwill/badwill arises when the amount paid is less than the fair value of the net assets acquired. IFRS 3 says the acquirer should:

(a) reassess the identification and measurement of the acquiree's identifiable assets, liabilities and contingent liabilities and the measurement of the cost of the combination in case the assets have been undervalued or the liabilities overstated; and

(b) recognise immediately in the statement of comprehensive income any excess remaining after that reassessment.

The immediate crediting of badwill to the statement of comprehensive income seems difficult to justify when, as in many situations, the reason why the consideration is less than the value of the net identifiable assets is that there are expected to be future losses or redundancy payments. Whilst the redundancy payments could be included in the 'contingent liabilities' at the date of acquisition, standard setters are very reluctant to allow a provision to be made for future losses (this has been prohibited in recent accounting standards). This means that the only option is to say the badwill should be credited to the statement of comprehensive income at the date of acquisition. This results in the group profit being inflated when a subsidiary with badwill is acquired.

In some ways, it would be better to credit the badwill to the statement of comprehensive income over the years the losses are expected. However, the 'provision for future losses' (i.e. the badwill) does not fit in very well with the *Framework*'s definition of a liability as being recognised 'when it is probable that an outflow of resources embodying economic benefits will result from the settlement of a present obligation and the amount at which the settlement will take place can be measured reliably'. It is questionable whether future losses are a 'present obligation' and whether they can be 'measured reliably', so it is very unlikely that future losses can be included as a liability in the statement of financial position.

19.12 Brands

We have discussed intangible assets and goodwill above but brands deserve a separate consideration because of their major significance in some companies. For example, the following information appears in the 2013 Diageo annual report:

	£m	£m
Total equity (i.e. net assets)		8,107
Intangible assets:		
Brands	6,244	
Goodwill	1,377	
Other intangible assets	1,200	
Computer software	227	
Total intangible assets		9,048

We can see that brands alone are 77% of total equity. It is interesting to take a look at the global importance of brands within sectors.

19.12.1 The importance of brands to particular sectors

It is interesting to note that certain sectors have high global brand valuations. For example, the Best Global Brands Report 2013[22] showed electronics (Apple $98,316m), Internet services (Google $93,291m), beverages (Coca-Cola $79,213m) and business services (IBM $78,808m). Even the hundredth exceeded $3,000 million (GAP $3,920m).

This indicates the importance of investors having as much information as possible to assess management's stewardship of brands. If this cannot be reported on the face of the statement of financial position then there is an argument for having an additional statement to assist shareholders, including the information that the directors consider when managing brands.

19.12.2 Justifications for reporting all brands as assets

We now consider some other justifications that have been put forward for the inclusion of brands as a separate asset in the statement of financial position.

Reduce equity depletion

For acquisitive companies it could be attributed to the accounting treatment required for measuring and reporting goodwill. The London Business School carried out research into the 'brands phenomenon' and found that 'a major aim of brand valuation has been to repair or pre-empt equity depletion caused by UK goodwill accounting rules'.[23]

Strengthen the statement of financial position

Non-acquisitive companies do not incur costs for acquiring goodwill, so their reserves are not eroded by writing off purchased goodwill. However, these companies may have incurred promotional costs in creating home-grown brands and it would strengthen the statement of financial position if they were permitted to include a valuation of these brands.

Effect on equity shareholders' funds

Immediate goodwill write-off results in a fall in net tangible assets as disclosed by the statement of financial position, even though the market capitalisation of the company increases. One way to maintain the asset base and avoid such a depletion of companies' reserves is to divide the purchased goodwill into two parts: the amount attributable to brands and the remaining amount attributable to pure goodwill.

Effect on borrowing powers

The borrowing powers of public companies may be expressed in terms of multiples of net assets. In Articles of Association there may be strict rules regarding the multiple that a company must not exceed. In addition, borrowing agreements and Stock Exchange listing agreements are generally dependent on net assets.

Effect on ratios

Immediate goodwill write-off distorts the gearing ratios, but the inclusion of brands as intangible assets minimises this distortion by providing a more realistic value for shareholders' funds.

Effect on management decisions

Including brands on the statement of financial position should lead to more informed and improved management decision making. As brands represent one of the most important assets of a company, management should be aware of the success or failure of each individual brand. Knowledge about the performance of brands ensures that management reacts accordingly to maintain or improve competitive advantage.

Effect on management decisions where brands are not capitalised

Whether or not a brand is capitalised, management does take its existence into account when making decisions affecting a company's gearing ratios. For example, in 2007 the Hugo Boss management in explaining its thinking about the advisability of making a Special Dividend payment[24] recognised that one effect was to reduce the book value of equity and increase the gearing ratio, but commented:

> The book value of the equity capital of the HUGO BOSS Group will be reduced by the special dividend. However this perception does not take into consideration that the originally created market value 'HUGO BOSS' is not reflected in the book value of the equity capital. This does not therefore mirror the strong economic position of HUGO BOSS fully.

The implication is that the existence of brand value is recognised by the market and leads to a more sustainable market valuation.

There is also evidence[25] that companies with valuable brand names are not including these in their statements of financial position and are not, therefore, taking account of the assets for insurance purposes.

The above are the justifications for recognising internally generated brands as assets. However, IAS 38 prohibits[26] this by saying: 'Internally generated brands, mastheads, publishing titles, customer lists and items similar in substance shall not be recognised as intangible assets.'

19.13 Accounting for acquired brands

Acquired brands require to be valued. In 2009, the International Valuation Standards Council issued an Exposure Draft, *Valuation of Intangible Assets for IFRS Reporting Purposes*,[27] which considers the need to define more clearly terms used within IFRSs such as 'active' and 'inactive' markets.

A decision is then made in respect of each brand as to whether it should be treated in the financial statements as having a finite or an infinite life. The following is an extract from the accounting policies of WPP in their 2013 Annual Report:

> Corporate brand names . . . acquired as part of acquisitions of business are capitalised separately from goodwill as intangible assets . . . amortisation is provided at rates calculated to write off the cost less estimated residual value of each asset on a straight-line basis over its estimated useful life as follows:
> Brand names (with finite lives) – 10–20 years; Customer-related intangibles – 3–10 years; Other proprietary tools – 3–10 years; Other (including capitalised computer software) – 3–5 years.
> Certain corporate brands of the Group are considered to have an indefinite economic life because of the institutional nature of the corporate brand names, their proven ability to maintain market leadership and profitable operations over long periods of time and the Group's commitment to develop and enhance their value. The carrying value of these intangible assets is reviewed at least annually for impairment and adjusted to the recoverable amount if required.

19.13.1 How effective have IFRS 3 and IAS 38 been?

There is still a temptation for companies to treat the excess paid on acquiring a subsidiary as goodwill. If it is treated as goodwill, then there has been no requirement to make an annual amortisation charge. If any part of the excess is attributed to an intangible, then this has to be amortised.

The position in the UK is that the FRRP will be policing the allocation of any excess on acquisitions to ensure that there is appropriate effort to attribute to intangible asset categories if that is the economic reality.

However, even so, the information is limited in that only acquired brands can be reported on the statement of financial position, which gives an incomplete picture of an entity's value. Even with acquired brands, their value can only remain the same or be revised downward following an impairment review. This means that there is no record of any added value that might have been achieved by the new owners to allow shareholders to assess the current stewardship.

19.14 Emissions trading

Under the European Union Emissions Trading Scheme (EU ETS) governments issue companies with free certificates allowing them to emit a stated amount of CO_2. If a company is not going to emit that quantity of CO_2, it can sell the excess in the market, which companies exceeding the limit can buy. Standard setters have yet to decide (a) how these certificates should be valued in companies' financial statements, and (b) where they should be included in the statement of financial position.

Three possible approaches to valuation could be:

1 If the company receives the certificates free from the government, their value in the financial statements should be zero. It would be unreasonable to put a value on them in the company's financial statement (e.g. number of tonnes of CO_2 × CO_2 emissions value per tonne). This would be 'boosting the statement of financial position'. Presumably their treatment will eventually be addressed by a revision to IAS 20 *Accounting for Government Grants*.

2 If the company is trading in the certificates, they are financial instruments under *FRS 9 Financial Instruments*. They can be valued at cost, with impairment if their value becomes less than cost. However, it is probably more appropriate to treat them as 'fair value through profit or loss', value them at market value, and include profits or losses in the statement of comprehensive income.

3 If a company buys the certificates to use in its business, they could be accounted for like inventory and valued at the lower of original cost and net realisable value. When the CO_2 emission takes place, their cost will be included in cost of sales.

Four possible solutions to the question of where to report could be to include them as:

● an intangible asset subject to the conditions studied in this chapter;
● a financial instrument;
● a prepayment; or
● inventory.

Considering approaches 1 to 3 above in turn:

1 If the certificates have no value, they do not appear in the statement of financial position.
2 If they are classified as a financial instrument, they will be included in current assets if their life is less than one year.
3 This is a problem which will be considered below.

CO_2 emissions certificates have many characteristics of inventory, and the most appropriate accounting treatment is to treat them like inventory. Normally, they will be valued at cost, and they will be charged as cost of sales when the CO_2 emissions take place. Net realisable value (NRV) will apply when the process which produces the CO_2 makes a loss. NRV will be the value which gives a zero profit from the process, but NRV will not be less than zero (negative). The problem with including them as inventory is that inventory is a physical asset, and these emission certificates are not a physical asset; they are an intangible asset.

The certificates could be a financial instrument and valued at cost, or market value or net realisable value. As they are held for use in a production process (which produces CO_2), market value does not seem appropriate. As the CO_2 is emitted, their value will be reduced and the amount charged to cost of sales. It will be like selling part of a holding of shares, but the 'sale' will be a consumption in a production process. Overall, it does not seem appropriate to include the certificates as a financial instrument, as there are more negative factors than when including them as inventory.

The certificates could be included as an intangible asset, like the items considered in this chapter. However, most intangible assets last a number of years, and these certificates will probably be used within a year. The accounting standards prohibit amortisation of certain types of goodwill. This should not apply to emission certificates, as they are being consumed in the production process (i.e. as the CO_2 is being emitted, the units of the emission certificates left diminish).

It is apparent that emission certificates are a current asset, as their life is probably less than a year, and they are consumed in the production process. They come into the category of 'receivables', although they are not an amount owed by a customer. They are more like a prepayment. The company buys the certificates (like buying insurance for the future) and consumes them in the future. Most prepayments relate to payments in advance for a future period (e.g. a year for insurance). Emission certificates are different, as they are consumed in proportion to the amount of CO_2 emitted in the future. However, they are probably more like a prepayment than the other items considered.

This discussion is a view based on various arguments. It is not a definitive answer. You could consider these and other arguments and come to a different conclusion.

There is no standard treatment but there is from 2012 an IASB-only research project which should lead to Board discussions in 2014/15 on:

● an inventory of trading schemes;
● an analysis of common economic characteristics of those schemes;
● an initial assessment of the potential reporting solutions;
● how to account for allowances awarded by a scheme administrator; and
● when, and how, to account for associated liabilities.

19.15 Intellectual capital disclosures (ICDs) in the annual report

The problem of valuing for financial reporting purposes has meant that investors need to look outside the annual report for information which tends to be predominately narrative. This is highlighted in an ICAEW Research Report[28] which comments:

> A wide range of media were used to report ICDs, with the annual report accounting for less than a third of total ICDs across all reporting media . . . examination of ICDs in annual reports was not a good proxy for overall ICD practices in the sample studied . . . disclosures are overwhelmingly narrative. Previous studies have tended to

indicate that monetary expression of IC elements in corporate reports is a relatively rare practice (see, for example, Beattie and Thompson, 2010).[29] This current study of UK ICR practices reinforces this observation.

The report also referred to the fact that preparers of reports did not see that the annual report was the appropriate place to be providing stakeholders with new information on intellectual capital – the annual report being seen as having a confirmatory role in relation to information that was already in the public domain.

19.15.1 The downside of not recognising ICDs in the statement of financial position

There is a common saying 'out of sight – out of mind' which could well be applied to ICDs that are not quantified and reported in the financial statements.

The focus of management's attention may be the physical assets that are reported with a concentration on return on total assets, return on capital employed and return on equity – all of which fail to include the ICDs within the denominator.

The focus of investors' attention may be on assessing the risks attached to achieving maintainable profits. This risk might well be overestimated if there is not an observable asset 'ICD' reported in the financial statements which leads to an increase in the cost of capital.

Investors may be unaware of a failure to achieve the optimum return on ICDs if they are not reported. For example, is the company commercially exploiting its ICDs by, for example, licensing its intellectual property?[30]

Lenders have tended to lend against assets with longer-term debts secured on non-current assets and short-term finance secured by factoring and invoice discounting. There is less confidence and more scepticism in lending against intellectual property. This is seen in SMEs where lenders tend to look for a guarantee supported by a tangible asset in addition to lending on the basis of ICDs. In an increasingly technological world, this is a severe disadvantage to SMEs.

Recognising and reporting ICDs remains an unsolved challenge.

19.15.2 Can internally generated intangibles continue to be unseen?

When a company acquires net assets in another company and pays more than their fair value the difference is treated as goodwill. This is a figure that is evidenced by a payment in the open market, so we know this is an arm's length valuation.

This total figure for goodwill is, for financial report purposes, disaggregated if possible into identifiable intangibles. The IASB view is that if they can be recognised they can be valued as a subset of the total market evidenced figure.

It would seem that companies do not consider that their market value is undervalued by the omission of an 'intellectual property' asset provided they keep investors and analysts up-to-date with developments. A contrary approach could be taken by companies that see an economic value in valuing and reporting in acquisition situations, e.g. payment to acquire customer lists.

19.16 Review of implementation of IFRS 3

In 2014 the IASB issued a Request for Information (RFI). The RFI is part of a post-implementation review and is the second being reviewed under its new review programme (the first reviewed IFRS 8). Its purpose is to seek comments from stakeholders to identify whether IFRS 3 *Business Combinations* provides information that is useful to users of financial statements; whether there are areas of IFRS 3 that are difficult to implement and may prevent the consistent implementation of the standard; and whether unexpected costs have arisen in connection with applying or enforcing the standard.

There are nine questions to which answers are requested – only one relates to this chapter, namely:

> Does the separate recognition of intangible assets and accounting for negative goodwill (badwill) provide useful information?

Responses will be analysed and the IASB will propose such revisions to IFRS 3 as may be appropriate.

19.17 Review of implementation of identified intangibles under IAS 38

We saw in Section 19.9.2 the following in the Intel 2013 Annual Report relating to identified intangible assets subject to amortisation:

Acquisition-related customer relationships	1,760	(828)	932

This item appears in the balance sheet of Intel but not in the balance sheet of the company from which the customer relationships had been purchased. On acquisition intangibles are valued at fair value using one of three IFRS 13 approaches. These are:

- the **market** approach using observable market prices or market transactions – this might be difficult to apply to many of the intangible assets such as brands which are company-specific; or
- the **income** approach which is based on relief from cost (say of a trademark) or the present value of excess earnings over an agreed number of years or the present value of incremental cash flows – this is the most appropriate approach for the majority of intangible assets; or
- the **cost** approach which attaches a value which is no higher than replacement cost.

Treatment of unrealised profits/gains

It is clear that the relationships are capable of being valued and the principal reason for omitting them from the balance sheet is the fact that they have been unrealised.

This is now at odds with the treatment of other assets which have been revalued and the difference between their carrying value and fair value included in Other comprehensive income and carried through into Total equity.

Is it time that the balance sheet reported *Other comprehensive equity* and *Other comprehensive asset* entries with each class of asset disclosed at carrying value and fair value?

Taking human capital as an example, this value would have been built up over time and the expenditure charged against profits – not as an identified charge but subsumed within the cost of recruitment and training.

It would be useful for users to have the asset and equity identified and revalued each year. If human capital improved the asset would increase – if it deteriorated then the asset would be reduced.

19.17.1 How might human resources be valued each year?

One approach might be that taken by Infosys Technologies (www.infosys.com) which became the first software company to value its human resources in India. The company stated in its 2011 Annual Report:

> A fundamental dichotomy in accounting practices is between human and non-human capital. As a standard practice, non-human capital is considered as assets and reported in the financial statements, whereas human capital is mostly ignored by accountants. The definition of wealth as a source of income inevitably leads to the recognition of human capital as one of the several forms of wealth such as money, securities and physical capital.
>
> We have used the Lev & Schwartz model to compute the value of human resources. The evaluation is based on the present value of future earnings of employees and on the following assumptions:
>
> **(a)** Employee compensation includes all direct and indirect benefits earned both in India and overseas
>
> **(b)** The incremental earnings based on group / age have been considered
>
> **(c)** The future earnings have been discounted at the cost of capital of 11.21% (previous year 10.60%).

It produced the following analysis:

	2011	2010
Total income[1]	27,501	22,742
Total employee cost[1]	14,856	12,093
Value-added	25,031	20,935
Net profit[1]	6,823	6,219
Ratios		
Value of human resources per employee	1.03	1.00
Total income / human resources value (ratio)	0.20	0.20
Employee cost / human resources value (%)	11.0	10.7
Value-added / human resources value (ratio)	0.19	0.18
Return on human resources value (%)	5.1	5.5

[1] *As per IFRS (audited) financial statements*

Summary

Intangible assets have grown in importance with the rise of the new economy. This has been principally driven by information and knowledge. It has been identified by the Organisation for Economic Co-operation and Development (OECD) as explaining the increased prominence of intellectual capital as a business and research topic.[31]

Since the industrial revolution, the following chain of events is observable:[32]

(a) Capital and labour were brought together and the factors of production became localised and accessible.

(b) Firms pushed to increase volumes of production to meet the demands of growing markets.

(c) Firms began to build intangibles like brand equity and reputation (goodwill) in order to create a competitive advantage in markets where new entrants limited the profit-making potential of a strategy of mass production.

(d) Firms invested heavily in information technology to increase the quality of products and improve the speed with which those products could be brought to market.

(e) Firms invested heavily in human capital with staff development and customer loyalty creation.

At each stage of this corporate evolution non-current tangible assets became less important, in relative terms, compared with intangible assets in determining a company's success. Accounting and financial reporting practices, however, have remained largely unchanged. Expenditure on intellectual capital (except for development costs) is expensed, net assets are understated and book values of the assets bear little relationship to market values.

This makes it more important for stakeholders to refer to non-financial disclosures in the annual reports and elsewhere. As with all information, more detailed explanations about intellectual capital investment should see a fall in the cost of capital and encourage management to provide more, as found in a research study.[33]

IAS 38 requires development costs to be recognised if they satisfy strict criteria. IFRS 3 requires purchased goodwill to be reviewed annually for impairment and not amortised. There is now a case for recognising internally generated intangible assets such as human capital and customer relationships.

REVIEW QUESTIONS

1 Why do standard setters consider it necessary to distinguish between research and development expenditure, and how does this distinction affect the accounting treatment?

2 Discuss the suggestion that the requirement for companies to write off research investment rather than showing it as an asset exposes companies to short-term pressure from acquisitive companies that are damaging to the country's interest.

3 Discuss why the market value of a business may increase to reflect the analysts' assessment of future growth but the asset(s) responsible for the growth may not appear in the statement of financial position.

4 Discuss the advantages and disadvantages of the proposal that there should be a separate category of asset in the statement of financial position clearly identified as 'research investment – outcome uncertain'.

5 IFRS 3 has introduced a new concept into accounting for purchased goodwill – annual impairment testing, rather than amortisation. Consider the effect of a change from amortisation of goodwill (in IAS 22) to impairment testing and no amortisation in IFRS 3, and in particular:

- the effect on the financial statements;
- the effect on financial performance ratios;
- the effect on the annual impairment or amortisation charge and its timing;
- which method gives the fairest charge over time for the value of the goodwill when a business is acquired;
- whether impairment testing with no amortisation complies with the IASC's *Framework for the Preparation and Presentation of Financial Statements*;
- why there has been a change from amortisation to impairment testing – is this pandering to pressure from the US FASB and/or listed companies?

6 Discuss reasons for the undervaluing of intangibles and subsuming within goodwill.

7 One goodwill impairment indicator is the loss of key personnel. Discuss two further possible indicators.

8 There has been a requirement for companies to disaggregate the amount paid for goodwill into other intangible assets. This has led to the valuation of certain of the relational intellectual capital items such as customer lists. Recent research[34] indicates that there is a variety of structural, human and relational capital components which are considered by a representative cross-section of pre-parers to be significantly more important than others and these key components should be a focus for future research. The researchers raise the need to investigate whether a set of industry-specific standardised metrics can be developed and their disclosure regulated and recommend that IASB include the intangibles project on its active agenda.

Discuss the argument that potentially the future of the accounting profession and its role as the key reporting function could depend on addressing this issue effectively.

9 Critically evaluate the basis of the following assertion: 'I am sceptical that the impairment test will work reliably in practice, given the complexity and subjectivity that lie within the calculations.'[34]

10 Access the annual report of a company (such as WPP plc) in which there is a large amount of goodwill and discuss the effect on earnings if goodwill is required to be amortised over a period of between 5 and 10 years. Discuss how this would affect headline profit.

11 Prior to IFRS 3 some countries permitted goodwill to be written off to equity. Discuss the reason why this was a permitted option and consider whether it is preferable to the estimated amortisation approach.

12 Discuss, after considering the approach taken by Infosys in valuing, whether investors would benefit from having human capital included as an asset in the statement of financial position.

EXERCISES

Question 1

IAS 38 *Intangible Assets* was issued primarily in order to identify the criteria that need to be present before expenditure on intangible items can be recognised as an asset. The standard also prescribes the subsequent accounting treatment of intangible assets that satisfy the recognition criteria and are recognised in the statement of financial position.

Required:
(a) Explain the criteria that need to be satisfied before expenditure on intangible items can be recognised in the statement of financial position as intangible assets.
(b) Explain how the criteria outlined in (a) are applied to the recognition of separately purchased intangible assets, intangible assets acquired in a business combination, and internally generated intangible assets. You should give an example of each category discussed.
(c) Explain the subsequent accounting treatment of intangible assets that satisfy the recognition criteria of IAS 38.

Iota prepares financial statements to 30 September each year. During the year ended 30 September 20X6 Iota (which has a number of subsidiaries) engaged in the following transactions:

1 On 1 April 20X6 Iota purchased all the equity capital of Kappa, and Kappa became a subsidiary from that date. Kappa sells a branded product that has a well-known name and the directors of Iota have obtained evidence that the fair value of this name is $20 million and that it has a useful economic life that is expected to be indefinite. The value of the brand name is not included in the statement of financial position of Kappa, as the directors of Kappa do not consider that it satisfies the recognition criteria of IAS 38 for internally developed intangible assets. However, the directors of Kappa have taken legal steps to ensure that no other entities can use the brand name.

2 On 1 October 20X4 Iota began a project that sought to develop a more efficient method of organising its production. Costs of $10 million were incurred in the year to 30 September 20X5 and debited to the statement of comprehensive income in that year. In the current year the results of the project were extremely encouraging and on 1 April 20X6 the directors of Iota were able to demonstrate that the project would generate substantial economic benefits for the group from 31 March 20X7 onwards as its technical feasibility and commercial viability were clearly evident. Throughout the year to 30 September 20X6 Iota spent $500,000 per month on the project.

Required:
(d) Explain how both of the above transactions should be recognised in the financial statements of Iota for the year ending 30 September 20X6. You should quantify the amounts recognised and make reference to relevant provisions of IAS 38 wherever possible.

Question 2

Environmental Engineering plc is engaged in the development of an environmentally friendly personal transport vehicle. This will run on an electric motor powered by solar cells, supplemented by passenger effort in the form of pedal assistance.

At the end of the current accounting period, the following costs have been attributed to the project:

(a) A grant of £500,000 to the Polytechnic of the South Coast Faculty of Solar Engineering to encourage research.
(b) Costs of £1,200,000 expended on the development of the necessary solar cells prior to the decision to incorporate them in a vehicle.
(c) Costs of £5,000,000 expended on designing the vehicle and its motors, and the planned promotional and advertising campaign for its launch on the market in 12 months' time.

Required:
(i) Explain, with reasons, which of the above items could be considered for treatment as deferred development expenditure, quoting any relevant International Accounting Standard.
(ii) Set out the criteria under which any items can be so treated.

(iii) Advise on the accounting treatment that will be afforded to any such items after the product has been launched.

* Question 3

As chief accountant at Italin NV, you have been given the following information by the director of research:

Project Luca

	€000
Costs to date (pure research 25%, applied research 75%)	200
Costs to develop product (to be incurred in the year to 30 September 20X1)	300
Expected future sales per annum for 20X2–20X7	1,000
Fixed assets purchased in 20X1 for the project:	
Cost	2,500
Estimated useful life	7 years
Residual value	400

(These assets will be disposed of at their residual value at the end of their estimated useful lives.)

The board of directors considers that this project is similar to the other projects that the company undertakes, and is confident of a successful outcome. The company has enough finances to complete the development and enough capacity to produce the new product.

Required:
(a) Prepare a report for the board outlining the principles involved in accounting for research and development and showing what accounting entries will be made in the company's accounts for each of the years ending 30 September 20X1–20X7 inclusive.
(b) Indicate what factors need to be taken into account when assessing each research and development project for accounting purposes, and what disclosure is needed for research and development in the company's published accounts.

* Question 4

Oxlag plc, a manufacturer of pharmaceutical products, has the following research and development projects on hand at 31 January 20X2:

(A) A general survey into the long-term effects of its sleeping pill Chalcedon upon human resistance to infections. At the year-end the research is still at a basic stage and no worthwhile results with any particular applications have been obtained.

(B) A development for Meebach NV in which the company will produce market research data relating to Meebach's range of drugs.

(C) An enhancement of an existing drug, Euboia, which will enable additional uses to be made of the drug and which will consequently boost sales. This project was completed successfully on 30 April 20X2, with the expectation that all future sales of the enhanced drug would greatly exceed the costs of the new development.

(D) A scientific enquiry with the aim of identifying new strains of antibiotics for future use. Several possible substances have been identified, but research is not sufficiently advanced to permit patents and copyrights to be obtained at the present time.

The following costs have been brought forward at 1 February 20X1:

Project	A £000	B £000	C £000	D £000
Specialised laboratory				
Cost	—	—	500	—
Depreciation	—	—	25	—
Specialised equipment				
Cost	—	—	75	50
Depreciation	—	—	15	10
Capitalised development costs	—	—	200	—
Market research costs	—	250	—	—

The following costs were incurred during the year:

Project	A £000	B £000	C £000	D £000
Research costs	25	—	265	78
Market research costs	—	75	—	—
Specialised equipment cost	50	—	—	50

Depreciation on specialised laboratories and special equipment is provided by the straight-line method and the assets have an estimated useful life of 25 and five years respectively. A full year's depreciation is provided on assets purchased during the year.

Required:
(a) Write up the research and development, fixed asset and market research accounts to reflect the above transactions in the year ended 31 January 20X2.
(b) Calculate the amount to be charged as research costs in the statement of comprehensive income of Oxlag plc for the year ended 31 January 20X2.
(c) State on what basis the company should amortise any capitalised development costs and what disclosures the company should make in respect of amounts written off in the year to 31 January 20X3.
(d) Calculate the amounts to be disclosed in the statement of financial position in respect of fixed assets, deferred development costs and work in progress.
(e) State what disclosures you would make in the accounts for the year ended 31 January 20X2 in respect of the new improved drug developed under project C, assuming sales begin on 1 May 20X2, and show strong growth to the date of signing the accounts, 14 July 20X2, with the expectation that the new drug will provide 25% of the company's pre-tax profits in the year to 31 January 20X3.

Question 5

Ross Neale is the divisional accountant for the Research and Development division of Critical Pharmaceuticals PLC. He is discussing the third-quarter results with Tina Snedden who is the manager of the division. The conversation focuses on the fact that whilst they have already fully committed the development capital expenditure budget for the year, the annual expense budget for research is well underspent because of the staff shortages which occurred in the last quarter. Tina mentions that she is under pressure to meet or exceed her expense budgets this year as the industry is renegotiating prescription costs this year and doesn't want to be seen to be too profitable.

Ross suggests that there are several strategies they could employ, namely:

(a) Several of the subcontractors have us as their largest customer and so we could ask them to describe the services in the fourth quarter, which are essentially development cost, as research costs.

(b) We could ask them to charge us in advance for research work that will be required in the first quarter of next year without mentioning that it is an advance in documentation. That would be good for them as it would improve their cash flow and it would guarantee that they would get the work next year.

(c) We could ask some of the subcontractors on development projects to charge us in the first quarter of next year and we could hold out to them that we would give them some better-priced projects next year to compensate them for the interest incurred as a result of the delayed payment.

Required:
Discuss the advantages and disadvantages of adopting these strategies.

Question 6

The brands debate

Under IAS 22, the depletion of equity reserves caused by the accounting treatment for purchased goodwill resulted in some companies capitalising brands on their statements of financial position. This practice was started by Rank Hovis McDougall (RHM) – a company which has since been taken over. Martin Moorhouse, the group chief accountant at RHM, claimed that putting brands on the statement of financial position forced a company to look to their value as well as to profits. It served as a reminder to management of the value of the assets for which they were responsible and that at the end of the day those companies which were prepared to recognise brands on the statement of financial position could be better and stronger for it.[35]

There were many opponents to the capitalisation of brands. A London Business School research study found that brand accounting involves too many risks and uncertainties and too much subjective judgement. In short, the conclusion was that 'the present flexible position, far from being neutral, is potentially corrosive to the whole basis of financial reporting and that to allow brands – whether acquired or homegrown – to continue to be included in the statement of financial position would be highly unwise'.[23]

Required:
Consider the arguments for and against brand accounting. In particular, consider the issues of brand valuation; the separability of brands; purchased versus home-grown brands; and the maintenance/ substitution argument.

Question 7

Brands plc is preparing its accounts for the year ended 31 October 20X8 and the following information is available relating to various intangible assets acquired on the acquisition of Countrywide plc:

(a) A milk quota of 2,000,000 litres at 30p per litre. There is an active market trading in milk and other quotas.

(b) A government licence to experiment with the use of hormones to increase the cream content of milk had been granted to Countrywide shortly before the acquisition by Brands plc. No fee had been required. This is the first licence to be granted by the government and was one of the reasons why Brands acquired Countrywide. The licence is not transferable but the directors estimate that it has a value to the company based on discounted cash flows for a five-year period of £1 million.

(c) A full-cream yoghurt sold under the brand name 'Naughty but Nice' was valued by the directors at £2 million. Further enquiry established that a similar brand name had been recently sold for £1.5 million.

Required:
Explain how each of the above items would be treated in the consolidated financial statements using IAS 38.

Question 8

James Bright has just taken up the position of managing director following the unsatisfactory achievements of the previous incumbent. James arrives as the accounts for the previous year are being finalised.
James wants the previous performance to look poor so that whatever he achieves will look good in
comparison. He knows that if he can write off more expenses in the previous year, he will have lower
expenses in his first year and possibly a lower asset base. He gives directions to the accountants to write
off as many bad debts as possible and to make sure accruals can be as high as they can get past the
auditors. Further, he wants all brand name assets reviewed using assumptions that the sales levels
achieved during the economic downturn are only going to improve slightly over the foreseeable future.
Also he mentions that the cost of capital has risen over the period of the financial crisis so the projected
benefits are to be discounted at a higher rate, preferably at a much higher rate than that used in the
previous reviews!

Required:
Discuss the accountant's professional responsibility and any ethical questions arising in this case.

* Question 9

International Accounting Standards IFRS 3 and IAS 38 address the accounting for goodwill and intangible
assets.

Required:
(a) Describe the requirements of IFRS 3 regarding the initial recognition and measurement of
goodwill and intangible assets.
(b) Explain the proposed approach set out by IFRS 3 for the treatment of positive goodwill in
subsequent years.
(c) Territory plc acquired 80% of the ordinary share capital of Yukon Ltd on 31 May 20X6. The
statement of financial position of Yukon Ltd at 31 May 20X6 was as follows:

	£000
Non-current assets	
Intangible assets	6,020
Tangible assets	38,300
	44,320
Current assets	
Inventory	21,600
Receivables	23,200
Cash	8,800
	53,600
Current liabilities	24,000
Net current assets	29,600
Total assets less current liabilities	73,920
Non-current liabilities	12,100
Provision for liabilities and charges	3,586
	58,234
Capital reserves	
Called-up share capital	10,000
(ordinary shares of £1)	
Share premium account	5,570
Retained earnings	42,664
	58,234

Additional information relating to the above statement of financial position:

(i) The intangible assets of Yukon Ltd were brand names currently utilised by the company. The directors felt that they were worth £7 million but there was no readily ascertainable market value at the statement of financial position date, nor any information to verify the directors' estimated value.

(ii) The provisional market value of the land and buildings was £20 million at 31 May 20X6. This valuation had again been determined by the directors. A valuers' report received on 30 November 20X6 stated the market value of land and buildings to be £23 million as at 31 May 20X6. The depreciated replacement cost of the remainder of the tangible fixed assets was £18 million at 31 May 20X6.

(iii) The replacement cost of inventories was estimated at £25 million and its net realisable value was deemed to be £20 million. Trade receivables and trade payables due within one year are stated at the amounts expected to be received and paid.

(iv) The non-current liability was a long-term loan with a bank. The initial loan on 1 June 20X5 was for £11 million at a fixed interest rate of 10% per annum. The total amount of the interest is to be paid at the end of the loan period on 31 May 20X9. The current bank lending rate is 7% per annum.

(v) The provision for liabilities and charges relates to costs of reorganisation of Yukon Ltd. This provision had been set up by the directors of Yukon Ltd prior to the offer by Territory plc and the reorganisation would have taken place even if Territory plc had not purchased the shares of Yukon Ltd. Additionally Territory plc wishes to set up a provision for future losses of £10 million which it feels will be incurred by rationalising the group.

(vi) The offer made to all of the shareholders of Yukon Ltd was 2.5 £1 ordinary shares of Territory plc at the market price of £2.25 per share plus £1 cash, per Yukon Ltd ordinary share.

(vii) The directors of Yukon Ltd informed Territory plc that as at 31 May 20X7, the brand names were worthless as the products to which they related had recently been withdrawn from sale because they were deemed to be a health hazard.

(viii) In view of the adverse events since acquisition, the directors of Territory plc have impairment-tested the goodwill relating to Yukon Ltd, and they estimate its current value is £1 million.

Calculate the charge for impairment of goodwill in the Group Statement of Comprehensive Income of Territory plc for the accounting period ending on 31 May 20X7.

References

1 IAS 38 *Intangible Assets*, IASC, revised March 2004.
2 IFRS 3 *Business Combinations*, IASB, revised 2008.
3 IAS 38 *Intangible Assets*, IASC, revised March 2004, para. 119.
4 Clarification of Acceptable Methods of Depreciation and Amortisation (Amendments to IAS 16 and IAS 38), IASB, 2014.
5 IAS 38 *Intangible Assets*, IASC, revised March 2004, para. 118.
6 Ibid., para. 126.
7 Ibid., para. 122.
8 Ibid., para. 54.
9 Ibid., para. 56.

10 Ibid., para. 55.

11 Ibid., para. 58.

12 Ibid., para. 59.

13 B. Nixon and A. Lonie, 'Accounting for R&D: the need for change', *Accountancy*, February 1990, p. 91; B. Nixon, 'R&D disclosure: SSAP 13 and after', *Accountancy*, February 1991, pp. 72–3.

14 http://www.rdmag.com/Featured-Articles/2011/12/2012-Global-RD-Funding-Forecast-RD-Spending-Growth-Continues-While-Globalization-Accelerates/

15 IASC, *Framework for the Preparation and Presentation of Financial Statements*, IASB, April 2001, para. 49.

16 A. Goodacre and J. McGrath, 'An experimental study of analysts' reactions to corporate R&D expenditure', *British Accounting Review*, vol. 29, 1997, pp. 155–79.

17 B. Nixon, 'The accounting treatment of research and development expenditure: views of UK company accountants', *European Accounting Review*, vol. 6(2), 1997, pp. 265–77.

18 Li Li Eng and Hong Kiat Teo, 'The relation between annual report disclosures, analysts' earnings forecast and analysts following: evidence from Singapore', *Pacific Accounting Review*, vol. 11 (1/2), 1999, pp. 219–39.

19 IAS 38 *Intangible Assets*, IASC, revised March 2004, para. 57.

20 IFRS 3 *Business Combinations*, IASB, 2004, para. 51.

21 IASC, *Framework for the Preparation and Presentation of Financial Statements*, IASB, April 2001, para. 94.

22 www.interbrand.com/best_global_brands.aspx

23 P. Barwise, C. Higson, A. Likierman and P. Marsh, *Accounting for Brands*, ICAEW, June 1989; M. Cooper and A. Carey, 'Brand valuation in the balance', *Accountancy*, June 1989.

24 http://group.hugoboss.com/en/faq_special_dividend.htm

25 M. Gerry, 'Companies ignore value of brands', *Accountancy Age*, March 2000, p. 4.

26 IAS 38 *Intangible Assets*, IASC, revised March 2004, para. 63.

27 www.ivsc.org

28 J. Unerman, J. Guthrie and M. Striukova, *UK Reporting of Intellectual Capital*, ICAEW, 2007, www.icaew.co.uk

29 V. Beattie and S.J. Thompson, *Intellectual Capital Reporting: Academic Utopia or Corporate Reality in a Brave New World?*, 2010, www.icas.org.uk

30 http://www.ipo.gov.uk/ipresearch-bankingip.pdf

31 OECD, *Final Report: Measuring and Reporting Intellectual Capital: Experience, Issues and Prospects*, Paris: OECD, 2000.

32 J. Guthrie and R. Petty, 'Knowledge management: the information revolution has created the need for a codified system of gathering and controlling knowledge', *Company Secretary*, vol. 9(1), January 1999, pp. 38–41; R. Tissen et al., *Value-Based Knowledge Management*, Longman Nederland BV, 1998, pp. 25–44.

33 M. Mangena, R. Pike and J. Li, *Intellectual Capital Disclosure Practices and Effects on the Cost of Equity Capital: UK Evidence*, ICAS, 2010.

34 V. Beattie and S.J. Thomson, *Intellectual Capital Reporting: Academic Utopia or Corporate Reality in a Brave New World?*, ICAS, 2010, http://www.icas.org.uk/site/cms/contentviewarticle.asp?article=6837

35 M. Moorhouse, 'Brands debate: wake up to the real world', *Accountancy*, July 1990, p. 30.

CHAPTER 20

Inventories

20.1 Introduction

The main purpose of this chapter is to explain the accounting principles involved in the valuation of inventory and biological assets.

> ## Objectives
>
> By the end of this chapter, you should be able to:
>
> - define inventory in accordance with IAS 2;
> - explain why valuation has been controversial;
> - explain the impact of inventory valuation on profits;
> - describe acceptable valuation methods;
> - describe procedure for ascertaining cost;
> - calculate inventory value;
> - explain how inventory could be used for creative accounting;
> - explain IAS 41 provisions relating to agricultural activity;
> - calculate biological value.

20.2 Inventory defined

IAS 2 *Inventories* defines inventories as assets:

(a) held for sale in the ordinary course of business;

(b) in the process of production for such sale;

(c) in the form of materials or supplies to be consumed in the production process or in the rendering of services.[1]

The valuation of inventory involves:

(a) the establishment of physical existence and ownership;

(b) the determination of unit costs;

(c) the calculation of provisions to reduce cost to net realisable value, if necessary.[2]

The resulting evaluation is then disclosed in the financial statements.

These definitions appear to be very precise. We shall see, however, that although IAS 2 was introduced to bring some uniformity into financial statements, there are many areas where professional judgement must be exercised. Sometimes this may distort the financial statements to such an extent that we must question whether they do represent a 'true and fair' view.

20.3 The impact of inventory valuation on profits

The valuation of inventory has impacts on both:

- the statement of income as the earnings per share (EPS) figure is based on the profit after tax. The EPS is then used to calculate the price earnings (PE) ratio; and
- the statement of financial position as the net asset backing for shares and the current ratio are affected.

A small change in inventory valuation may have a material impact on profit

Figure 20.1 presents information relating to Nissan Motor Co. Ltd. It shows that the inventory is material in relation to pre-tax profits. In relation to the profits we can see that an error of 5% in the 2010 inventory values would potentially cause the pre-tax profit to fall by 28%. As inventory is usually a multiple rather than a fraction of profit, inventory errors may have a disproportionate effect on the accounts.

Income smoothing

This is achieved by adjusting the inventory valuation of the closing inventory to increase or decrease the profit for the year. There will be an opposite impact on the profits of the following year – the current year's increase in profit will be the following year's decrease.

Figure 20.2 illustrates the point. Simply by increasing the value of inventory in year 1 by £10,000, profit (and current assets) is increased by a similar amount. Even if the closing inventory value is the same (£15,000) in year 2, such manipulation allows profit to be 'smoothed' and £10,000 profit switched from year 2 to year 1.

Management attitude when estimating need to reduce inventory to net realisable value

According to normal accrual accounting principles, profit is determined by matching costs with related revenues. If it is unlikely that the revenue will in fact be received, prudence dictates that the irrecoverable amount should be written off immediately against current revenue and the inventory stated at net realisable value. If profits are falling or less than forecast it could be a temptation for management to estimate a lower figure for inventory deterioration or obsolescence. Likewise, if profits are rising and there is an expected fall in the following year, management might be inclined to overestimate the provision required.

Such circumstances tend to come to light with a change of management and it was considered important that a definitive statement of accounting practice should be be issued in an attempt to standardise treatment. This is now set out in IAS 2 with the aim of making

Figure 20.1 Nissan Motor Co. Ltd

	2010	Inventory reduced by 5%	% change
Pre-tax profits (million yen)	141,680	101,566	28%
Inventories	802,278	762,164	5%

Figure 20.2 Inventory values manipulated to smooth income

		Year 1		Year 1
				With inventory inflated
Sales		100,000		100,000
Opening inventory	—			—
Purchases	65,000		65,000	
Less: Closing inventory	5,000		15,000	
COST OF SALES		60,000		50,000
PROFIT		40,000		50,000

		Year 2		Year 2
				With inventory inflated
Sales		150,000		150,000
Opening inventory	5,000		15,000	
Purchases	100,000		100,000	
	105,000		115,000	
Less: Closing inventory	15,000		15,000	
COST OF SALES		90,000		100,000
PROFIT		60,000		50,000

inventory valuation consistent, comparable between periods and comparable between companies in the same industry.

20.4 IAS 2 *Inventories*

No area of accounting has produced wider differences in practice than the computation of the amount at which inventory is stated in financial accounts. An accounting standard on the subject needs to define the practices, to narrow the differences and variations in those practices and to ensure adequate disclosure in the accounts.

IAS 2 requires that the amount at which inventory is stated in periodic financial statements should be the total of the lower of cost and net realisable value of the separate items of inventory or of groups of similar items. The standard also emphasises the need to match costs against revenue, and it aims, like other standards, to achieve greater uniformity in the measurement of income as well as improving the disclosure of inventory valuation methods. To an extent, IAS 2 relies on management to choose the most appropriate method of inventory valuation for the production processes used and the company's environment.

Various methods of valuation are theoretically available, including FIFO, LIFO and weighted average or any similar method (see below). In selecting the most suitable method, management must exercise judgement to ensure that the methods chosen provide the fairest practical approximation to cost. IAS 2 does not allow the use of LIFO because it often results in inventory being stated in the statement of financial position at amounts that bear little relation to recent cost levels.

At the end of the day, even though there is an International Accounting Standard in existence, the valuation of inventory can provide areas of subjectivity and choice to management. We will return to this theme many times in the following sections of this chapter.

20.5 Inventory valuation

The valuation rule outlined in IAS 2 is difficult to apply because of uncertainties about what is meant by cost (with some methods approved by IAS 2 and others not) and what is meant by net realisable value.

20.5.1 Methods acceptable under IAS 2

The acceptable methods of inventory valuation include FIFO, AVCO and standard cost.

First-in-first-out (FIFO)

Inventory is valued at the most recent 'cost', since the cost of oldest inventory is charged out first, whether or not this accords with the actual physical flow. FIFO is illustrated in Figure 20.3.

Average cost (AVCO)

Inventory is valued at a 'weighted average cost', i.e. the unit cost is weighted by the number of items carried at each 'cost', as shown in Figure 20.4. This is popular in organisations holding a large volume of inventory at fluctuating 'costs'. The practical problem of actually recording and calculating the weighted average cost has been overcome by the use of sophisticated computer software.

Figure 20.3 First-in-first-out method (FIFO)

	Receipts			Issues			Balance		
Date	Quantity	Rate	£	Quantity	Rate	£	Quantity	Rate	£
January	10	15	150				10		150
February				8	15	120	2		30
March	10	17	170				12		200
April	20	20	400				32		600
May				2	15	30			
				10	17	170			
				12	20	240			
				Cost of goods sold		560			
				Inventory			8	20	160

Figure 20.4 Average cost method (AVCO)

	Receipts			Issues			Balance		
Date	Quantity	Rate	£	Quantity	Rate	£	Quantity	Rate	£
January	10	15	150				10		150
February				8	15	120	2		30
March	10	17	170				12		200
April	20	20	400				32		600
May				24	18.75	450			600
				Cost of goods sold		570			
				Inventory			8	18.75	150

The following is an extract from the J Sainsbury plc 2013 Annual Report:

Inventories

Inventories . . . are valued on a weighted average cost basis and carried at the lower of cost and net realisable value. Net realisable value represents the estimated selling price less all estimated costs of completion and costs to be incurred in marketing, selling and distribution. Cost includes all direct expenditure and other appropriate attributable costs incurred in bringing inventories to their present location and condition.

Standard cost

In many cases this is the only way to value manufactured goods in a high-volume/high-turnover environment, as seen in the following extract from the Next 2013 Annual Report:

Inventories

Inventories (stocks) are valued at the lower of standard cost or net realisable value.

However, the standard is acceptable only if it approximates to actual cost. This means that variances need to be reviewed to see if they affect the standard cost and for inventory evaluation as seen in the following extract from the Noranda Aluminium Holding Corp 2013 Annual Report:

Our Bauxite segment and our Alumina segment are valued using a standard costing system, which gives rise to cost variances. Variances are capitalized to inventory in proportion to the quantity of inventory remaining at period end to quantities produced during the period. Variances are recorded such that ending inventory reflects actual costs on a year-to-date basis.

Retail method

IAS 2 recognises that an acceptable method of arriving at cost is the use of selling price, less an estimated profit margin. This method is only acceptable if it can be demonstrated that it gives a reasonable approximation of the actual cost.

The following is an extract from the Shoprite Holdings Ltd 2013 Annual Financial Statements:

Valuation of inventory: Trading inventories are valued by use of the retail inventory method as an approximation of weighted average cost.

Significant judgment is required in the application thereof, specifically as far as it relates to gross margin percentages, accrual rates for rebates and settlement discounts and shrinkage rates applied.

The retail method approximates the weighted average cost and is determined by reducing the sales value of the inventory by the appropriate percentage gross margin. The percentage used takes into account inventory that has been marked down below original selling price. An average percentage per retail department is used.

Care has to be taken in determining the appropriate percentage write-down. For example, one company, Stein Mart Inc, had to restate its accounts in 2012, having identified that a practice that had been in place for more than ten years was incorrect. It had been accounting for certain markdowns as promotional (temporary) rather than permanent – as a result its inventories were overstated by approximately $3 million.

IAS 2 does not recommend any specific method. This is a decision for each organisation based upon sound professional advice and the organisation's unique operating conditions.

Figure 20.5 Last-in-first-out method (LIFO)

Date	Receipts			Issues			Balance		
	Quantity	Rate	£	Quantity	Rate	£	Quantity	Rate	£
January	10	15	150				10		150
February				8	15	120	2		30
March	10	17	170				12		200
April	20	20	400				32		600
May				20	20	400			
				4	17	68			
				Cost of goods sold		588			
				Inventory			8		132

May closing balance = [(2 × 15) + (6 × 17)]

20.5.2 Methods rejected by IAS 2

Methods rejected by IAS 2 include LIFO and (by implication) replacement cost.

Last-in-first-out (LIFO)

The cost of the inventory most recently received is charged out first at the most recent 'cost'. The practical upshot is that the inventory value is based upon an 'old cost', which may bear little relationship to the current 'cost'. Where LIFO is used companies reconcile the LIFO valuation to the FIFO valuation. LIFO is illustrated in Figure 20.5.

US companies commonly use the LIFO method, as illustrated by this extract from the Deere and Company 2013 Annual Report:

INVENTORIES
Most inventories owned by Deere & Company and its U.S. equipment subsidiaries are valued at cost, on the 'last-in, first-out' (LIFO) basis. . . . If all inventories had been valued on a FIFO basis, estimated inventories . . . would have been in millions of dollars as follows:

	2013	2012
Total FIFO value	6,464	6,591
Less adjustment to LIFO value	1,529	1,421
Inventories	$4,935	$5,170

Although LIFO does not have IAS 2 approval, it is still used in practice. For example, LIFO is commonly used by UK companies with US subsidiaries, since LIFO is the main method of inventory valuation in the USA.

Replacement cost

The inventory is valued at the current cost of the individual item (i.e. the cost to the organisation of replacing the item) rather than the actual cost at the time of manufacture or purchase. The value may be reported as additional information as in the following extract from the A.M. Castle & Co 2013 Annual Report:

Inventories, principally on last-in first-out basis (replacement cost higher by $130,854 and $139,940)

The use of replacement cost is not specifically prohibited by IAS 2 but is out of line with the basic principle underpinning the standard, which is to value inventory at the actual costs incurred in its purchase or production. The IASC *Framework for the Preparation and Presentation of Financial Statements* describes historical cost and current cost as two distinct measurement bases, and where a historical cost measurement base is used for assets and liabilities the use of replacement cost is inconsistent.

20.5.3 Procedure to ascertain cost

Having decided upon the accounting policy of the company, there remains the problem of ascertaining the cost. In a retail environment, the 'cost' is the price the organisation had to pay to acquire the goods, and it is readily established by reference to the purchase invoice from the supplier. However, in a manufacturing organisation the concept of cost is not as simple. Should we use prime cost, or production cost, or total cost? IAS 2 attempts to help by defining cost as 'all costs of purchase, costs of conversion and other costs incurred in bringing the inventories to their present location and condition'.

In a manufacturing organisation each expenditure is taken to include three constituents: direct materials, direct labour and appropriate overhead.

Direct materials

These include not only the costs of raw materials and component parts, but also the costs of insurance, handling (special packaging) and any import duties. An additional problem is waste and scrap. For instance, if a process inputs 100 tonnes at £45 per tonne, yet outputs only 90 tonnes, the output's inventory value **must** be £4,500 (£45 × 100) and not £4,050 (90 × £45). (This assumes the 10 tonnes loss is a normal, regular part of the process.) An adjustment may be made for the residual value of the scrap/waste material, if any. The treatment of component parts will be the same, provided they form part of the finished product.

Direct labour

This is the cost of the actual production in the form of gross pay and those incidental costs of employing the direct workers (employer's national insurance contributions, additional pension contributions, etc.). The labour costs will be spread over the goods' production.

Appropriate overhead

It is here that the major difficulties arise in calculating the true cost of the product for inventory valuation purposes. Normal practice is to classify overheads into five types and decide whether to include them in inventory. The five types are as follows:

- Direct overheads – subcontract work, royalties.
- Indirect overheads – the cost of running the factory and supporting the direct workers, and the depreciation of capital items used in production.
- Administration overheads – the office costs and salaries of senior management.
- Selling and distribution overheads – advertising, delivery costs, packaging, salaries of sales personnel, and depreciation of capital items used in the sales function.
- Finance overheads – the cost of borrowing and servicing debt.

We will look at each of these in turn, to demonstrate the difficulties that the accountant experiences.

Regular, routine direct overhead will be included in the inventory valuation. Special subcontract work would form part of the inventory value where it is readily identifiable to individual units of inventory such as in a customised car manufacturer making 20 cars a month.

Indirect overheads. These always form part of the inventory valuation, as such expenses are incurred in support of production. They include factory rent and rates, factory power and depreciation of plant and machinery; in fact, any indirect factory-related cost, including the warehouse costs of storing completed goods, will be included in the value of inventory.

Administration overheads. This overhead is in respect of the whole business, so only that portion easily identifiable to production should form part of the inventory valuation. For instance, the costs of the personnel or wages department could be apportioned to production on a head-count basis and that element would be included in the inventory valuation. Any production-specific administration costs (welfare costs, canteen costs, etc.) would also be included in the inventory valuation. If the expense cannot be identified as forming part of the production function, it will not form part of the inventory valuation.

Selling and distribution overheads. These costs will not normally be included in the inventory valuation as they are incurred after production has taken place. However, if the goods are on a 'sale or return' basis and are on the premises of the customer but remain the supplier's property, the delivery and packing costs will be included in the inventory value of goods held on a customer's premises.

Finance overheads. Normally these overheads would never be included within the inventory valuation because they are not normally identifiable with production. In a job-costing context, however, it might be possible to use some of this overhead in inventory valuation. Let us take the case of an engineering firm being requested to produce a turbine engine, which requires parts/components to be imported. It is logical for the financial charges for these imports (e.g. exchange fees or fees for letters of credit) to be included in the inventory valuation.

Thus it can be seen that the identification of the overheads to be included in inventory valuation is far from straightforward. In many cases it depends upon the judgement of the accountant and the unique operating conditions of the organisation.

In addition to the problem of deciding **whether** the five types of overhead should be included, there is the problem of deciding **how much** of the total overhead to include in the inventory valuation at the year-end. IAS 2 stipulates the use of 'normal activity' when making this decision on overheads. The vast majority of overheads are 'fixed', i.e. do not vary with activity, and it is customary to share these out over a normal or expected output.

The following is an extract from the Agrana Group 2011/12 Annual Report:

Inventories

Inventories are measured at the lower of cost of purchase and/or conversion and net selling price. The weighted average formula is used. In accordance with IAS 2, the conversion costs of unfinished and finished products include – in addition to directly attributable unit costs – reasonable proportions of the necessary material costs and production overheads inclusive of depreciation of manufacturing plant (*based on the assumption of normal capacity utilisation*) *as well as production-related administrative costs* [our italics]. . . . If this expected output [based on normal capacity utilisation] is not reached, it is not acceptable to allow the actual production to bear the full overhead for inventory purposes.

A numerical example will illustrate this:

Overhead for the year	£200,000
Planned activity	10,000 units
Closing inventory	3,000 units
Direct costs	£2 per unit
Actual activity	6,000 units

Inventory value based on actual activity

Direct costs	$3,000 \times £2$	£6,000
Overhead	$\dfrac{3,000 \times £200,000}{6,000}$	£100,000
Closing inventory value		£106,000

Inventory value based on planned or normal activity

Direct cost	$3,000 \times £2$	£6,000
Overhead	$\dfrac{3,000 \times £200,000}{10,000}$	£60,000
Closing inventory value		£66,000

Comparing the value of inventory based upon actual activity with the value based upon planned or normal activity, we have a £40,000 difference. This could be regarded as increasing the current year's profit by carrying forward expenditure of £40,000 to set against the following year's profit.

The problem occurs because of the organisation's failure to meet expected output level (6,000 actual versus 10,000 planned). By adopting the **actual activity basis**, the organisation makes a profit out of failure. This cannot be an acceptable position when evaluating performance. Therefore, IAS 2 stipulates **the planned or normal activity model** for inventory valuation. The failure to meet planned output could be due to a variety of sources (e.g. strikes, poor weather, industrial conditions); the cause, however, is classed as abnormal or non-routine, and all such costs should be excluded from the valuation of inventory.

20.5.4 What is meant by net realisable value?

IAS 2 requires inventory to be stated at the lower of cost and net realisable value. In arriving at net realisable value a deduction is made from realisable value for any additional expense expected for repackaging, advertising, delivery and, where necessary, repairing damaged inventory prior to sale.

Prudence dictates that net realisable value will be used if it is lower than the 'cost' of the inventory (however that may be calculated). These occasions will vary among organisations, but can be summarised as follows:

- There is a permanent fall in the market price of inventory. Short-term fluctuations should not cause net realisable value to be implemented.

- The organisation is attempting to dispose of high inventory levels or excessively priced inventory to improve its liquidity position (acid test ratio) or reduce its inventory holding costs. Such high inventory volumes or values are primarily a result of poor management decision making.

- The inventory is physically deteriorating or is of an age where the market is reluctant to accept it. This is a common feature of the food industry, especially with the use of 'sell by' dates in the retail environment.

- Inventory suffers obsolescence through some unplanned development. (Good management should never be surprised by obsolescence.) This development could be technical in nature, or due to the development of different marketing concepts within the organisation or a change in market needs.

- The management could decide to sell the goods at 'below cost' for sound marketing reasons. The concept of a 'loss leader' is well known in supermarkets, but organisations also sell below cost when trying to penetrate a new market or as a defence mechanism when attacked.

Such decisions are important and the change to net realisable value should not be undertaken without considerable forethought and planning. Obsolescence should be a decision based upon sound market intelligence and not a managerial 'whim'. The auditors of companies always examine such decisions to ensure they were made for sound business reasons. The opportunities for fraud in such 'price-cutting' operations validate this level of external control.

For example, goods costing £1,000 had been flood damaged and were not covered by insurance. It was estimated that if £200 were spent on cleaning the goods could be sold for £550 giving a NRV for inventory of £350.

A numerical example will demonstrate this concept:

Item	Cost (£)	Net realisable value (£)	Inventory value (£)
1 No. 876	7,000	9,000	7,000
2 No. 997	12,000	12,500	12,000
3 No. 1822	8,000	4,000	4,000
4 No. 2076	14,000	8,000	8,000
5 No. 4732	27,000	33,000	27,000
	(a) 68,000	(b) 66,500	(c) 58,000

The inventory value chosen for the accounts is (c) £58,000, although each item is assessed individually.

20.6 Work in progress

Inventory classified as work in progress (WIP) is mainly found in manufacturing organisations and is simply the production that has not been completed by the end of the accounting period.

The valuation of WIP must follow the same IAS 2 rules and be the lower of cost and net realisable value. We again face the difficulty of deciding what to include in cost. The three basic classes of cost – direct materials, direct labour and appropriate overhead – will still form the basis of ascertaining cost.

20.6.1 Direct materials

It is necessary to decide what proportion of the total materials have been used in WIP. The proportion will vary with different types of organisation, as the following two examples illustrate:

- If the item is complex or materially significant (e.g. a custom-made car or a piece of specialised machinery), the WIP calculation will be based on actual recorded materials and components used to date.

- If, however, we are dealing with mass production, it may not be possible to identify each individual item within WIP. In such cases, the accountant will make a judgement and define the WIP as being $x\%$ complete in regard to raw materials and components. For example, a drill manufacturer with 1 million tools per week in WIP may decide that in respect of raw materials they are 100% complete; WIP then gets the full materials cost of one million tools.

In both cases **consistency** is vital so that, however WIP is valued, the same method will always be used.

20.6.2 Direct labour

Again, it is necessary to decide how much direct labour the items in WIP have actually used. As with direct materials, there are two broad approaches:

- Where the item of WIP is complex or materially significant, the actual time 'booked' or recorded will form part of the WIP valuation.
- In a mass production situation, such precision may not be possible and an accounting judgement may have to be made as to the average percentage completion in respect of direct labour. In the example of the drill manufacturer, it could be that, on average, WIP is 80% complete in respect of direct labour.

20.6.3 Appropriate overhead

The same two approaches as for direct labour can be adopted:

- With a complex or materially significant item, it should be possible to allocate the overhead actually incurred. This could be an actual charge (e.g. subcontract work) or an application of the appropriate overhead recovery rate (ORR). For example, if we use a direct labour hour recovery rate and we have an ORR of £10 per direct labour hour and the recorded labour time on the WIP item is 12 hours, then the overhead charge for WIP purposes is £120.

EXAMPLE ● A custom-car company making sports cars has the following costs in respect of No. 821/C, an unfinished car, at the end of the month:

Materials charged to job 821/C	£2,100
Labour 120 hours @ £4	£480
Overhead £22/DLH × 120 hours	£2,640
WIP value of 821/C	£5,220

This is an accurate WIP value provided *all* the costs have been accurately recorded and charged. The amount of accounting work involved is not great as the information is required by a normal job cost system. An added advantage is that the figure can be formally audited and proven.

- With mass production items, the accountant must either use a budgeted overhead recovery rate approach or simply decide that, in respect of overheads, WIP is $y\%$ complete.

EXAMPLE ● A company produces drills. The costs of a completed drill are:

	£	
Direct materials	2.00	
Direct labour	6.00	
Appropriate overhead	10.00	
Total cost	18.00	(for finished goods inventory value purposes)

Assuming that the company accountant takes the view that for WIP purposes the following applies:

Direct material	100% complete
Direct labour	80% complete
Appropriate overhead	30% complete

then, for one WIP drill:

Direct material	$£2.00 \times 100\% = £2.00$
Direct labour	$£6.00 \times 80\% = £4.80$
Appropriate overhead	$£10.00 \times 30\% = £3.00$
WIP value	$£9.80$

If the company has 100,000 drills in WIP, the value is:

$$100,000 \times £9.80 = £980,000$$

This is a very simplistic view, but the principle can be adapted to cover more complex issues. For instance, there could be 200 different types of drill, but the same calculation can be done on each. Of course, sophisticated software makes the accountant's job mechanically easier.

This technique is particularly useful in processing industries, such as petroleum, brewing, dairy products or paint manufacture, where it might be impossible to identify WIP items precisely. The approach must be consistent and the role of the auditor in validating such practices is paramount.

20.7 Inventory control

Inventory shrinkage can occur for a number of reasons ranging from criminal activity to internal administrative errors. For example, the following statistics were collected in the US in 2012:[3]

Sources of Inventory Shrinkage in Retail	Percent
Employee Theft	40.9%
Shoplifting	33.1%
Administrative Error	15.3%
Vendor Fraud	5.9%
Unknown	7.4%

Management are responsible for internal control but auditors become involved in reporting and advising on internal control procedures – in the case of employee theft, for example, advising on the installation of sales audit and loss prevention software and procedures for authorisation of employee purchases and all documents that allow goods to leave the premises. The final decision on action, however, lies with the management.

Errors can occur at the year-end inventory count when physical inventory is checked against book inventory. A major cause of discrepancy between physical and book inventory is the 'cut-off' date. In matching sales with cost of sales, it may be difficult to identify exactly into which period of account certain inventory movements should be placed, especially when the annual inventory count lasts many days or occurs at a date other than the last day of the financial year. It is customary to make an adjustment to the inventory figure. In many cases the auditor will be present at the inventory count to observe that there are effective systems being followed.

In practice, errors may continue unidentified for a number of years,[4] particularly if there is a paper-based system in operation. This was evident when T.J. Hughes reduced its profit for the year ended 31 January 2001 by £2.5–3 million from a forecast £8 million when stock discrepancies came to light following the implementation of a new stock management system.

20.8 Creative accounting

No area of accounting provides more opportunities for subjectivity and creative accounting than the valuation of inventory. This is illustrated by the report *Fraudulent Financial Reporting: 1987–1997 – An Analysis of U.S. Public Companies* prepared by the Committee of Sponsoring Organizations of the Treadway Commission.[5] This report, which was based on the detailed analysis of approximately 200 cases of fraudulent financial reporting, identified that the fraud often involved the overstatement of revenues and assets with inventory fraud featuring frequently – assets were overstated by understating allowances for receivables, overstating the value of inventory and other tangible assets, and recording assets that did not exist.

This section summarises some of the major methods employed.

20.8.1 Year-end manipulations

There are a number of stratagems companies have followed to reduce the cost of goods sold by inflating the inventory figure. These include the following.

Manipulating cut-off procedures

A major cause of discrepancy between physical and book inventory is the 'cut-off' date. In matching sales with cost of sales, it may be difficult to identify exactly into which period of account certain inventory movements should be placed, especially when the annual inventory count lasts many days or occurs at a date other than the last day of the financial year. It is customary to make an adjustment to the inventory figure, as shown in Figure 20.6.

Goods may be taken into inventory but the purchase invoices are not recorded or sales recorded and goods are still in the warehouse. An accurate record is required of movements between the inventory count date and the financial year-end.

The authors of *Fraudulent Financial Reporting: 1987–1997 – An Analysis of U.S. Public Companies* found that over half the frauds involved overstating revenues by recording revenues prematurely or fictitiously and that such overstatement tended to occur right at the end of the year – hence the need for adequate cut-off procedures. This was illustrated by Ahold's experience in the USA where subsidiary companies took credit for bulk discounts allowed by suppliers before inventory was actually received.

Figure 20.6 Adjusted inventory figure

	£
Inventory on 7 January 20X1	XXX
Less: Purchases	(XXX)
Add: Cost of sales	XXX
Inventory at 31 December 20X0	XXX

Fictitious transfers to overseas locations

Year-end inventory is inflated by recording fictitious transfers of non-existent inventory, e.g. it was alleged by the SEC that certain officers of the Miniscribe Corporation had increased the company's inventory by recording fictitious transfers of non-existent inventory from a Colorado location to overseas locations where physical inventory counting would be more difficult for the auditors to verify or the goods are described as being 'in transit'.[6]

Inaccurate inventory records

Where inventory records are poorly maintained it has been possible for senior management to fail to record material shrinkage due to loss and theft, as in the matter of Rite Aid Corporation.[7]

Journal adjustments

In addition to suppressing purchase invoices, making fictitious transfers and failing to write off obsolete inventory or recognise inventory losses, the senior management may simply reduce the cost of goods sold by adjusting journal entries. Auditors pay particular attention to journal adjustments, questioning whether there have been significant adjusting entries that have increased the inventory balance and whether there have been material reversing entries made to the inventory account after the close of an accounting period.

20.8.2 Net realisable value (NRV)

Although the determination of net realisable value is dealt with extensively in the appendix to IAS 2, the extent to which provisions can be made to reduce cost to NRV is highly subjective and open to manipulation. A provision is an effective smoothing device and allows overcautious write-downs to be made in profitable years and consequent write-backs in unprofitable ones.

20.8.3 Overheads

The treatment of overheads has been dealt with extensively above and is probably the area that gives the greatest scope for manipulation. Including overhead in the inventory valuation has the effect of deferring the overhead's impact and so boosting profits. IAS 2 allows expenses incidental to the acquisition or production cost of an asset to be included in its cost. We have seen that this includes not only directly attributable production overheads, but also those which are indirectly attributable to production and interest on borrowed capital. IAS 2 provides guidelines on the classification of overheads to achieve an appropriate allocation, but in practice it is difficult to make these distinctions and auditors may find it difficult to challenge management on such matters.

The statement suggests that the allocation of overheads included in the valuation needs to be based on the company's normal level of activity. The cost of unused capacity should be written off in the current year. The auditor will insist that allocation should be based on normal activity levels, because if the company underproduces, the overhead per unit increases and can therefore lead to higher year-end values. The creative accountant will be looking for ways to manipulate these year-end values, so that in bad times costs are carried forward to more profitable accounting periods.

20.8.4 Other methods of creative accounting

Over- or understated quantities

A simple manipulation is to show more or less inventory than actually exists. If the commodity is messy and indistinguishable, the auditor may not have either the expertise or the will to verify measurements taken by the client's own employees. This lack of auditor measuring knowledge and involvement allowed one of the biggest frauds ever, which became known as 'the great salad oil swindle'.[8]

Understated obsolete inventory

Another obvious ploy is to include, in the inventory valuation, obsolete or 'dead' inventory. Of course, such inventory should be written off. However, management may be 'optimistic' that it can be sold, particularly in times of economic recession. In high-tech industries, unrealistic values may be placed on inventory that in times of rapid development becomes obsolete quickly.

Lack of marketability

This is a problem that investors need to be constantly aware of, particularly when a company experiences a downturn in demand but a pressure to maintain the semblance of growth. An example is provided by Lexmark[9] which was alleged to have made highly positive statements regarding strong sales and growth for its printers although there was intense competition in the industry – the company reported quarter after quarter of strong financial growth, whereas the actual position appeared to be very different with unmarketable inventory in excess of $25 million to be written down in the fourth quarter of fiscal year 2001. The share price of a company that conceals this type of information is maintained and allows insiders to offload their shareholding on an unsuspecting investing public.

20.9 Audit of the year-end physical inventory count

The problems of accounting for inventory are highlighted at the company's year-end. This is when the closing inventory figure to be shown in both the statement of comprehensive income and the statement of financial position is calculated. In practice, the company will assess the final inventory figure by physically counting all inventory held by the company for trade. The year-end inventory count is therefore an important accounting procedure, one in which the auditors are especially interested.

The auditor generally attends the inventory count to verify both the physical quantities and the procedure of collating those quantities. At the inventory count, values are rarely assigned to inventory items, so the problems facing the auditor relate to the identification of inventory items, their ownership, and their physical condition.

20.9.1 Identification of inventory items

The auditor will visit many companies in the course of a year and will spend a considerable time looking at accounting records. However, it is important for the auditor also to become familiar with each company's products by visiting the shop floor or production facilities during the audit. This makes identification of individual inventory items easier at the year-end. Distinguishing between two similar items can be crucial where there are large differences in value. For example, steel-coated brass rods look identical to steel rods, but their value to the company will be very different. It is important that they are not confused at inventory count because, once recorded on the inventory sheets, values are assigned, production carries on, and the error cannot be traced.

20.9.2 Physical condition of inventory items

Inventory in premium condition has a higher value than damaged inventory. The auditor must ensure that the condition of inventory is recorded at inventory count, so that the correct value is assigned to it. Items that are damaged or have been in inventory for a long period will be written down to their net realisable value (which may be nil) as long as adequate details are given by the inventory counter. Once again, this is a problem of identification, so the auditor must be able to distinguish between, for instance, rolls of first quality and faulty fabric. Similarly, items that have been in inventory for several inventory counts may have little value, and further enquiries about their status should be made at the time of inventory count.

20.9.3 Adjustment if inventory is taken after the year-end date

If inventory is counted after the year-end then an adjustment will need to be made to add back the cost of items sold and deduct the cost of purchases made after the year-end that have been taken into stock.

For example, assume that after the year-end, sales of £100,000 at cost plus 25% were made and dispatched and purchases of £45,000 were made and received. Inventory would be increased by £100,000 × 20/100 = £80,000 and reduced by £45,000.

20.9.4 Adjustment if errors are discovered

Typical errors could include:

- Sales invoices raised and posted but goods are awaiting dispatch – these should be excluded from the year-end inventory.
- Purchase invoices received and posted without waiting for the goods received note – the purchases figure should be reduced.
- Errors on pricing items or casting inventory sheets – these should be corrected when identified.
- Consumable stock might have been included – this should be taken out of inventory. The cost of sales will be higher, gross profit lower and the consumables expense reduced with no effect on the net profit.
- Omitting stock held by third parties on approval or consignment – these would need to be taken into closing inventory at cost.

20.10 Published accounts

Disclosure requirements in IAS 2 have already been indicated. The standard requires the accounting policies that have been applied to be stated and applied consistently from year to year. Inventory should be sub-classified in the statement of financial position or in the notes to the financial statements so as to indicate the amounts held in each of the main categories in the standard statement of financial position formats. But will the ultimate user of those financial statements be confident that the information disclosed is reliable, relevant and useful?

Figure 20.7 Impact of a 5% change in closing inventory

Company:	1	2	3	4	5	6	7	8
	£m	£m	£m	£m	£m	£m	£m	£m
Actual inventory	390.0	428.0	1,154.0	509.0	509.0	280.0	360.0	232.0
Actual pre-tax profit	80.1	105.6	479.0	252.5	358.4	186.3	518.2	436.2
Change in pre-tax profit	19.5	21.4	57.7	25.2	25.5	14.0	18.0	11.6
Impact of a 5% change in closing inventory (%)								
(i) Pre-tax profit	24.3	20.3	12.0	10.0	7.1	7.5	3.5	2.7
(ii) Earnings per share	27.0	25.0	12.0	9.3	8.4	6.9	3.4	3.4

Key to companies:
1 Electrical retailer
2 Textile, etc., manufacturer
3 Brewing, public houses, etc.
4 Retailer – diversified

5 Pharmaceutical and retail chemist
6 Industrial paints and fibres
7 Food retailer
8 Food retailer

We have already indicated many areas of subjectivity and creative accounting, but are such possibilities material?

In 1982 Westwick and Shaw examined the accounts of 125 companies with respect to inventory valuation and its likely impact on reported profit.[10] The results showed that the effect on profit before tax of a 1% error in closing inventory valuation ranged from a low of 0.18% to a high of 25.9% (in one case) with a median of 2.26%. The industries most vulnerable to such errors were household goods, textiles, mechanical engineering, contracting and construction.

Clearly, the existence of such variations has repercussions for such measures as ROCE, EPS and the current ratio. The research also showed that, in a sample of audit managers, 85% were of the opinion that the difference between a pessimistic and an optimistic valuation of the same inventory could be more than 6%.

IAS 2 has since been strengthened and these results may not be so indicative of the present situation. However, using the same principle, let us take a random selection of eight companies' recent annual accounts, apply a 5% increase in the closing inventory valuation and calculate the effect on EPS (taxation is simply taken at 35% on the change in inventory).

Figure 20.7 shows that, in absolute terms, the difference in pre-tax profits could be as much as £57.7 million and the percentage change ranges from 2.7% to 24.3%. Of particular note is the change in EPS, which tends to be the major market indicator of performance. In the case of the electrical retailer (company 1), a 5% error in inventory valuation could affect EPS by as much as 27%. The inventory of such a company could well be vulnerable to such factors as changes in fashion, technology and economic recession.

20.11 Agricultural activity

20.11.1 The overall problem

Agricultural activity is subject to special considerations and so is governed by a separate IFRS, namely IAS 41. IAS 41 defines agricultural activity as 'the management by an entity of the biological transformation of biological assets for sale, into agricultural produce or into additional biological assets'. A biological asset is a living animal or plant.

The basic problem is that biological assets, and the produce derived from them (referred to in IAS 41 as 'agricultural produce'), cannot be measured using the cost-based concepts that form the bedrock of IAS 2 and IAS 16. This is because biological assets, such as cattle, for example, are not usually purchased; they are born and develop into their current state. Therefore different accounting methods are necessary.

20.11.2 The recognition and measurement of biological assets and agricultural produce

IAS 41 states that an entity should recognise a biological asset or agricultural produce when:

- the entity controls the asset as a result of a past event;
- it is probable that future economic benefits associated with the asset will flow to the entity;
- the fair value or cost of the asset can be measured reliably.

Rather than the usual cost-based concepts of measurement that are used for assets, IAS 41 states that assets of this type should be measured at their fair value less estimated costs of sale. The only (fairly rare) exception to this general measurement principle is if the asset's fair value cannot be estimated reliably. In such circumstances a biological asset is measured at cost (if available). Research[11] indicates that the adoption of fair value is avoided in countries such as France where there is a culture of conservatism, which means that they rebut the presumption that fair values can be determined with reliability to justify the use of historical cost. It also means that they are able to avoid the onerous valuation requirements of the standard.

The following is an extract from the 2013 Holmen AB annual report:

Biological assets
The Group divides all its forest assets for accounting purposes into growing forests, which are recognised as biological assets at fair value, and land, which is stated at acquisition cost. Any changes in the fair value of the growing forests are recognised in the income statement. Holmen's assessment is that there are no relevant market prices available that can be used to value forest holdings as extensive as Holmen's. Valuation is therefore carried out by estimating the present value of expected future cash flows (after deduction of selling costs) from the growing forests.

20.11.3 An illustrative example

A farmer owned a dairy herd. At the start of the period the herd contained 100 animals that were two years old and 50 newly born calves. At the end of the period a further 30 calves had been born. None of the herd died during the period. Relevant fair value details were as follows:

	Start of period $	End of period $
Newly born calves	50	55
One-year-old animals	60	65
Two-year-old animals	70	75
Three-year-old animals	75	80

The change in the fair value of the herd is $3,400, made up as follows:

Fair value at end of the year = $(100 \times \$80) = (50 \times \$65) = (30 \times \$55)$ = $12,900
Fair value at start of the year = $(100 \times \$70) = (50 \times \$50)$ = $9,500

IAS 41 requires that the change in the fair value of the herd be reconciled as follows:

	$
Price change – opening newly born calves: 50($55 – $50)	250
Physical change of opening newly born calves: 50($65 – $55)	500
Price change of opening two-year-old animals: 100($75 – $70)	500
Physical change of opening two-year-old animals: 100($80 – $75)	500
Due to birth of new calves: 30 × $55	1,650
Total change	3,400

The costs incurred in maintaining the herd would all be charged in the statement of comprehensive income in the relevant period.

20.11.4 Agricultural produce

Examples of agricultural produce would be milk from a dairy herd or crops from a cornfield. Such produce is sold by a farmer in the ordinary course of business and is inventory. The initial carrying value of the inventory at the point of 'harvest' is its fair value less costs to sell at that date. Agricultural entities then apply IAS 2 to the inventory using the initial carrying value as 'cost'.

20.11.5 Land

Despite its importance in agricultural activity, IAS 41 does not apply to agricultural land, which is accounted for in accordance with IAS 16. Where biological assets are physically attached to land (e.g. crops in a field) then it is often possible to compute the fair value of the biological assets by computing the fair value of the combined asset and deducting the fair value of the land alone.

20.11.6 Minerals

The standard does not apply to the measurement of inventories of producers of agricultural and forest products, agricultural produce after harvest, and minerals and mineral products, to the extent that they are measured at net realisable value in accordance with well-established industry practices.

20.11.7 Government grants relating to biological assets

As mentioned in Chapter 17 such grants are not subject to IAS 20 – the general standard on this subject. Under IAS 41 the IASB view is more consistent with the principles of the *Framework* than the provisions of IAS 20. Under IAS 41 grants are recognised as income when the entity becomes entitled to receive it. This removes the fairly dubious credit balance 'Deferred income' that arises under the IAS 20 approach and does not appear to satisfy the *Framework* definition of a liability.

20.11.8 Fair value or historic cost option?

An interesting research project[10] carrying out an empirical investigation of the implications of IAS 41 for the harmonisation of farm accounting practices in Australia, France and the

UK found that agricultural entities in all three countries are using a variety of valuation methods under IAS 41 and that there is a lack of comparability of disclosure practices. It was their view that IAS 41 has failed to enhance the international comparability of accounting practices in the agricultural sector. The following problems have been identified.

Valuation method

The researchers found that although historical cost is the most common valuation basis for biological assets, a variety of proxies for fair value are used, such as net present value, independent/external valuation, net realisable value and market price, both within and across countries.

National characteristics impact on choice of method

Some countries may be more conservative and private than others. These characteristics and attitudes existed pre-IFRS[12] and do not change merely because the IASB has produced IFRSs.

Fair national or fair global value?

In the European Union IAS 41 requires biological assets to be valued by reference to artificial and highly subsidised or politically mediated market prices. This allows European farmers to export to developing countries at prices which are substantially below production costs.

Cost–benefit considerations

Small and medium-sized companies consider that the cost of compliance is too high and this has been recognised by the IASB which provide that, for biological assets, the fair value through profit or loss model is required only when fair value is readily determinable without undue cost or effort. If fair value is not used SMEs follow the cost–depreciation–impairment model.

Summary

IAS 2 defines inventory and the methods of arriving at cost that are acceptable.

Valuation methods used must result in a reasonable approximation to actual cost.

Inventory manipulation can have a material impact on reported profits and balance sheet ratios to achieve a higher or lower profit in the current reporting period.

Auditors have an involvement in advising on internal controls to protect physical inventory, ensure that proper physical counts are made at the year-end and appropriate methods and procedures are in place to determine cost and net realisable value.

Although legal requirements and IAS 2 have improved the reporting requirements, many areas of subjective judgement can have substantial effects on the reporting of financial information.

REVIEW QUESTIONS

1 Discuss the extent to which individual judgements might affect inventory valuation, e.g. changing the basis of overhead absorption.

2 Discuss the acceptability of the LIFO and replacement cost methods of inventory valuation and why the IASB has not permitted all methods to be used.

3 Explain the criteria to be applied when selecting the method to be used for allocating administrative costs.

4 Discuss the effect on work in progress and finished goods valuation if the net realisable value of the raw material is lower than cost at the statement of financial position date.

5 Discuss why the accurate valuation of inventory is so crucial if the financial statements are to show a true and fair view.

6 The following is an extract from the Anheuser-Busch InBev 2013 Annual Report:

Inventories

Inventories are valued at the lower of cost and net realizable value.

The cost of finished products and work in progress comprises raw materials, other production materials, direct labor, other direct cost and an allocation of fixed and variable overhead based on normal operating capacity.

Discuss the possible effects on profits if the company did not use normal operating activity.

7 It has been suggested that 'Given national characteristics it will be impossible to ensure that financial statements that comply with IFRSs will ever be comparable.' Discuss whether auditors can make this change.

8 The following is an extract from the 2013 Annual Report of SIPEF NV:

Because of the inherent uncertainty associated with the valuation at fair value of the biological assets due to the volatility of the prices of the agricultural produce and the absence of a liquid market, their carrying value may differ from their realisable value.

Given the inherent uncertainty in applying IAS 41, discuss whether the pre-IAS 41 practice of value at historical cost is preferable for the statement of financial position.

EXERCISES

Question 1

Sunhats Ltd manufactures patent hats. It carries inventory of these and sells to wholesalers and retailers via a number of salespeople. The following expenses are charged in the profit and loss account:

Wages of: Storemen and factory foremen

Salaries of: Production manager, personnel officer, buyer, salespeople, sales manager, accountant, company secretary

Other: Directors' fees, rent and rates, electric power, repairs, depreciation, carriage outwards, advertising, bad debts, interest on bank overdraft, development expenditure for new types of hat.

Required:
Which of these expenses can reasonably be included in the valuation of inventory?

* Question 2

Purchases of a certain product during July were:

July	1	100 units @ £10.00
	12	100 units @ £9.80
	15	50 units @ £9.60
	20	100 units @ £9.40

Units sold during the month were:

July	10	80 units
	14	100 units
	30	90 units

Required:
Assuming no opening inventories:
(a) Determine the cost of goods sold for July under three different valuation methods.
(b) Discuss the advantages and/or disadvantages of each of these methods.
(c) A physical inventory count revealed a shortage of five units. Show how you would bring this into account.

* Question 3

Alpha Ltd makes one standard article. You have been given the following information:

1 The inventory sheets at the year-end show the following items:

Raw materials:
100 tons of steel:
Cost £140 per ton
Present price £130 per ton

Finished goods:
100 finished units:
Cost of materials £50 per unit
Labour cost £150 per unit
Selling price £500 per unit

40 semi-finished units
Cost of materials £50 per unit
Labour cost to date £100 per unit
Selling price £500 per unit (completed)

10 damaged finished units:
Cost to rectify the damage £200 per unit
Selling price £500 per unit (when rectified)

2 Manufacturing overheads are 100% of labour cost.
Selling and distribution expenses are £60 per unit (mainly salespeople's commission and freight charges).

Required:
From the information in notes 1 and 2, state the amounts to be included in the statement of financial position of Alpha Ltd in respect of inventory. State also the principles you have applied.

* Question 4

Beta Ltd commenced business on 1 January and is making up its first year's accounts. The company uses standard costs. The company owns a variety of raw materials and components for use in its manufacturing business. The accounting records show the following:

	Standard cost of purchases	Adverse (favourable) variances	
		Price variance	Usage variance
	£	£	£
July	10,000	800	(400)
August	12,000	1,100	100
September	9,000	700	(300)
October	8,000	900	200
November	12,000	1,000	300
December	10,000	800	(200)
Cumulative figures for whole year	110,000	8,700	(600)

Raw materials control account balance at year-end is £30,000 (at standard cost).

Required:
The company's draft statement of financial position includes 'Inventories, at the lower of cost and net realisable value £80,000'. This includes raw materials £30,000: do you consider this to be acceptable? If so, why? If not, state what you consider to be an acceptable figure.

(*Note:* for the purpose of this exercise, you may assume that the raw materials will realise more than cost.)

* Question 5

Uptodate plc's financial year ended on 31 March 20X8. Inventory taken on 7 April 20X8 amounted to £200,000. The following information needs to be taken into account:

(i) Purchases made during the seven days to 7 April amounted to £40,000. Invoices had not been received and only 20% had been delivered by 7 April. These had been taken into inventory.

(ii) Purchases of £10,000, which had been ordered but not paid for before the year-end, had been received before 31 March. However, as the invoices had not been received by 31 March they have not been included in the inventory.

(iii) Purchases of £5,000, which had been ordered and paid for before the year-end, had not been received by 31 March.

(iv) Purchases of £12,000, ordered and paid for by the year-end, were in a bonded warehouse awaiting customs clearance at 31 March. These were eventually delivered to the company on 9 April.

Required:
Calculate the revised year-end inventory as at 31 March 20X8.

Question 6

Hasty plc's financial year ended on 31 March 20X8. Inventory taken on 7 April 20X8 amounted to £100,000. The following information needs to be taken into account:

(i) Sales invoices totalling £9,000 were raised during the seven days after the year-end. £1,500 of this had not been dispatched by 7 April. The company policy was to add 20% to cost.

(ii) Sales returns received on 6 April totalled £600.

(iii) Goods with an invoice value of £6,000 had been sent to customers on approval in February 20X8. £3,600 had been returned in March 20X8. The company policy was to add 20% to cost and not to process the invoice until customers gave notice of purchasing.

(iv) Goods bought in to satisfy a one-off customer order at £575 had been sent on approval in November 20X7 on a pro forma invoice for £850. These had been taken into inventory at the pro forma price.

Required:
Calculate revised inventory as at 31 March 20X8.

Question 7

The statement of income of Bottom, a manufacturing company, for the year ending 31 January 20X2 is as follows:

	$000
Revenue	75,000
Cost of sales	(38,000)
Gross profit	37,000
Other operating expenses	(9,000)
Profit from operations	28,000
Investment income	
Finance cost	(4,000)
Profit before tax	24,000
Income tax expense	(7,000)
Net profit for the period	17,000

Note – accounting policies

Bottom has used the LIFO method of inventory valuation but the directors wish to assess the implications of using the FIFO method. Relevant details of the inventories of Bottom are as follows:

Date	Inventory valuation under:	
	FIFO	LIFO
	$000	$000
1 February 20X1	9,500	9,000
31 January 20X2	10,200	9,300

Required:
Redraft the statement of income of Bottom using the FIFO method of inventory valuation and explain how the change would need to be recognised in the published financial statements, if implemented.

* Question 8

Agriculture is a key business activity in many parts of the world, particularly in developing countries. Following extensive discussions with, and funding from, the World Bank, the International Accounting Standards Committee (IASC) developed an accounting standard relating to agricultural activity. IAS 41 *Agriculture* was published in 2001 to apply to accounting periods beginning on or after 1 January 2003.

Sigma prepares financial statements to 30 September each year. On 1 October 2003 Sigma carried out the following transactions:

● Purchased a large piece of land for $20 million.

● Purchased 10,000 dairy cows (average age at 1 October 2003, two years) for $1 million.

● Received a grant of $400,000 towards the acquisition of the cows. This grant was non-returnable.

During the year ending 30 September 2004 Sigma incurred the following costs:

● $500,000 to maintain the condition of the animals (food and protection).

● $300,000 in breeding fees to a local farmer.

On 1 April 2004, 5,000 calves were born. There were no other changes in the number of animals during the year ended 30 September 2004. At 30 September 2004, Sigma had 10,000 litres of unsold milk in inventory. The milk was sold shortly after the year-end at market prices.

Information regarding fair values is as follows:

Item	Fair value less point-of-sale costs		
	1 October 2003	1 April 2004	30 September 2004
	$ m	$ m	$ m
Land	20	22	24
New-born calves (per calf)	20	21	22
Six-month-old calves (per calf)	23	24	25
Two-year-old cows (per cow)	90	92	94
Three-year-old cows (per cow)	93	95	97
Milk (per litre)	0.6	0.55	0.55

Required:
(a) Discuss how the IAS 41 requirements regarding the recognition and measurement of biological assets and agricultural produce are consistent with the IASC *Framework for the Preparation and Presentation of Financial Statements*.
(b) Prepare extracts from the statement of comprehensive income and the statement of financial position that show how the transactions entered into by Sigma in respect of the purchase and maintenance of the dairy herd would be reflected in the financial statements of the entity for the year ended 30 September 2004. You do not need to prepare a reconciliation of changes in the carrying amount of biological assets.

(ACCA DipIFR 2004)

References

1 IAS 2 *Inventories*, IASB, revised 2004.
2 'A guide to accounting standards – valuation of inventory and work-in-progress', *Accountants Digest*, Summer 1984.
3 http://soccrim.clas.ufl.edu/files/nrssfinalreport2012.pdf
4 M. Perry, 'Valuation problems force FD to quit', *Accountancy Age*, 15 March 2001, p. 2.
5 The report appears on www.coso.org/index.htm
6 See www.sec.gov/litigation/admin/34-41729.htm
7 See www.sec.gov/litigation/admin/34-46099.htm
8 E. Woolf, 'Auditing the stocks – part II', *Accountancy*, May 1976, pp. 108–110.
9 See http://securities.stanford.edu/1022/LXK01-01/
10 C. Westwick and D. Shaw, 'Subjectivity and reported profit', *Accountancy*, June 1982, pp. 129–131.
11 C. Elad and K. Herbohn, *Implementing Fair Value Accounting in the Agricultural Sector*, ICAS Research Report, 2011, http://www.icas.org.uk/site/cms/download/res/elad_Exec_Summary_Feb_2011.pdf
12 C. Nobes, 'Different versions of IFRS practice', in C. Nobes and R. Parker (eds), *Comparative International Accounting* (10th edition), FT Prentice Hall, 2008, pp. 145–156.

CHAPTER 21

Construction contracts

21.1 Introduction

Construction contracts have been given special attention because of the size, duration and special challenges which arise in accounting for them. In this chapter we will be addressing the issues involved in supplying services which are of long duration and thus raise issues of whether revenue should be recognised continuously or whether the contract should be viewed as a series of smaller service provisions (such as a multi-level building which could be viewed as supplying one completed level after another) or whether the completion of the total contract is the delivery of the contracted service. In addition to the accounting for the revenue recognition there are a number of issues relating to the valuation of the work in progress.

The basic principles which apply to revenue recognition both currently and in the future, are applied in construction contracts. So in essence this chapter can be seen as illustrating the application of the revenue standards in a complex business situation together with the application of impairment accounting (see Chapter 17) in arriving at the valuation of the resulting assets. Both these issues will be addressed in this chapter.

We also explain the basic accounting for contracts of public–private partnerships which have become increasingly popular for undertaking major infrastructure construction and operations.

Objectives

By the end of this chapter, you should be:

- aware of some of the historical developments in the accounting for construction contracts such that you understand how to read accounts involving construction activities following IAS 11;
- able to prepare construction accounts in accordance with IFRS 15 revenue recognition rules and to record assets arising from construction contracts; and
- understanding accounting for public–private partnerships.

21.2 The need to replace IAS 11 *Construction Contracts*[1]

Pre-2011 it had been considered that construction contracts were of such complexity that they warranted a separate standard (IAS 11 *Construction Contracts*) and the general rules for revenue recognition set out in IAS 18 *Revenue*[2] were specially excluded. In their place

IAS 11 established a separate set of rules under which revenue is recognised progressively as the item is built, matching expenses, writing off unrecoverable costs and identifying those costs to be carried forward as assets which are expected to be recovered in the future.

However, the presence of two different sets of rules in relation to revenue recognition has not sat comfortably with the idea of having a coherent set of standards. Also IAS 11 was thought to contain insufficient guidance where the construction contract was complex.

Accordingly the IASB and the FASB have produced a joint revenue recognition standard IFRS 15 which includes the rules which will in the future govern accounting for construction contracts. IFRS 15 will be effective for annual periods beginning on or after 1 January 2017 with the option for earlier adoption. Once a company adopts IFRS 15 the old IAS 11 *Construction Costs* will be superseded and will no longer be applicable. Until then IAS 11 is still operational.

21.2.1 IAS 11 *Construction Contracts*

IAS 11 *Construction Contracts* defines a construction contract as:

> A contract specifically negotiated for the construction of an asset or a combination of assets that are closely inter-related or inter-dependent in terms of their design, technology and function or their ultimate purpose or use.

Some construction contracts are **fixed-price contracts**, where the contractor agrees to a fixed contract price. However, where the contract extends over a longer period it is quite normal for such fixed-price contracts to include escalation clauses. An escalation clause essentially means that when specified events beyond the control of the contractor (such as union wage rates or prices of specified material, such as iron reinforcing used in the construction) increase then the price of the contract is amended according to a previously agreed formula to allow the contractor to recover all or part of the cost increases. Thus escalation clauses are a device for sharing or transferring specified risks associated with the contract.

Other construction contracts are **cost-plus contracts**, where the contractor is reimbursed for allowable costs, plus a percentage of these costs or alternatively a fixed fee is added to the allowable costs. This type of contract would be appropriate where the amount of materials or labour needed is unclear as may be the case in an innovative project.

Some examples of construction contracts would involve building ships, aeroplanes, buildings, dams, highways and bridges.

Construction contracts are normally assessed and accounted for individually. However, in certain circumstances construction contracts may be combined or segmented. Combination or segmentation is appropriate when:

● a group of contracts is negotiated as a single package and the contracts are performed together in a continuous sequence (combination); and

● separate proposals have been submitted for each asset and the costs and revenues of each asset can readily be identified (segmentation).

A key accounting issue is when the revenues and costs (and therefore net income) under a construction contract should be recognised. There are two major possibilities:

● Only recognise net income when the contract is complete – the *completed contracts method*.

● Recognise a proportion of net income over the period of the contract – this is currently achieved using the *percentage of completion method*.

IAS 11 requires the latter approach, provided the overall contract result can be predicted with reasonable certainty. If that is not the case then the completed contract method is used.

21.3 Identification of contract revenue under IAS 11

Contract revenue should comprise:

(a) the initial amount of revenue agreed in the contract; and
(b) variations in contract work, claims and incentives payments, to the extent that:
   (i) it is probable that they will result in revenue;
   (ii) they are capable of being reliably measured.

Variations to the initially agreed contract price occur due to events such as:

- cost escalation clauses;
- claims for additional revenue by the contractor due to customer-caused delays or changes in the specification or design;
- incentive payments when specified performance standards are met or exceeded;
- penalty clauses representing agreed damages caused by failure to complete by the contracted date.

Incentive payments might apply to a toll road where early completion would allow additional revenue to be collected by the owner of the road. Penalty clauses might apply to a construction contract because the client would incur additional costs as a result of delays such as temporary storage expenses if they cannot move into the new factory at the agreed handover date. Penalties are a way of ensuring the client can plan ahead for the transfer of their business to the new premises at an agreed date, confident that the contractor will do everything in their power to complete on time.

The same recognition rules apply to variations as apply to the original recognition, namely the probability of occurrence is high and the amount can be predicted with reasonable certainty.

21.4 Identification of contract costs under IAS 11

IAS 11 classifies costs that can be identified with contracts under four headings:

- Costs that directly relate to the specific contract, such as:
  - site labour;
  - cost of materials;
  - depreciation of plant and equipment used on the contract;
  - costs of moving plant and materials to and from the contract site;
  - costs of hiring plant and equipment;
  - costs of design and technical assistance that are directly related to the contract;
  - the estimated costs of rectification and guarantee work;[3]
  - claims from third parties.

- Costs that are attributable to contract activity in general and can be allocated to specific contracts, such as:
  - insurance;
  - costs of design and technical assistance that are not directly related to a specific contract;
  - construction overheads.

  Costs of this nature need to be allocated on a systematic and rational basis, based on the normal level of construction activity.

- The construction contract itself may specify costs which can be recovered under the contract and those of course can be charged to the contract.

- Incremental costs of obtaining a contract.

An interesting area is the cost of tendering. This may seem an insignificant cost but in relation to major complex contracts these costs may amount to many millions of pounds. *The standard says these may be charged to the contract if at the time it is probable that the tender will be successful.* That is a difficult criterion to satisfy. Generally if a contractor is bidding for a contract against other bidders the costs of bidding would only be included in contract costs once the bidder had been selected or had received preferred bidder status. If as a result of the uncertainties involved the cost of tendering is written off in one financial year, and if the contract is awarded in the next financial year, it is **not** possible to write back the tendering expenses of the previous period as an asset. If an asset is established it has to be amortised over the period of the contract.

21.4.1 Examples

Johnson Matthey in their 2013 accounts, which follow IAS 11, state that their accounting policy is:

Long term contracts
Where the outcome of a long term contract can be estimated reliably, revenue and costs are recognized by reference to the stage of completion. This is measured by the proportion that contract costs incurred to date bear to the estimated total contract costs.

Where the outcome of a long term contract cannot be estimated reliably, contract revenue is recognized to the extent of contract costs incurred that it is probable will be recoverable. Contract costs are recognized as expenses in the period in which they are incurred.

When it is probable that the total contract costs will exceed total contract revenue, the expected loss is recognized as an expense immediately.

Note the following interesting issues:

(a) Revenue and expenses are recognised on the basis of stage of completion.

(b) When there is uncertainty regarding the ability of the contract to make a profit overall then the costs incurred that probably won't be recovered are immediately written off as an expense.

Balfour Beatty in their 2013 annual accounts report their accounting policy as:

2.6 Construction and service contracts
When the outcome of individual contracts can be estimated reliably, contract revenue and contract costs are recognised as revenue and expenses respectively by reference to the stage of completion at the reporting date.

Costs are recognised as incurred and revenue is recognised on the basis of the proportion of total costs at the reporting date to the estimated total costs of the contract.

Provision is made for all known or expected losses on individual contracts once such losses are foreseen.

Revenue in respect of variations to contracts, claims and incentive payments is recognised when it is probable it will be agreed by the client. Revenue in respect of claims is recognised when negotiations have reached an advanced stage such that it is probable that the client will accept the claim and the probable amount can be measured reliably.

Profit for the year includes the benefit of claims settled on contracts completed in previous years

2.8 Pre-contract bid costs and recovery

Pre-contract costs are expensed as incurred until it is virtually certain that a contract will be awarded, from which time further pre-contract costs are recognised as an asset and charged as an expense over the period of the contract. Amounts recovered in respect of pre-contract costs that have been written-off are deferred and amortised over the life of the contract.

For construction and services projects, the relevant contract is the construction or services contract respectively. With respect to PPP projects, there are potentially three contracts over which the recovered costs could be amortised, the concession contract itself, the construction contract or the services contract. An assessment is made as to which contractual element the pre-contract costs relate to, in order to determine which is the relevant period for amortisation. The relevant contract is either the construction contract that ultimately gives rise to a financial or intangible asset; or to the services contract where there is no initial construction.

This policy is also in accordance with IAS 11. The difficulties in finalising the contract revenue, given the variations for escalation allowances and modifications to contracts and rectification of claimed deficiencies, are outlined. The treatment of pre-contract costs is also stated. The reference to the PPP projects refers to public–private partnerships. For example, there may be a partnership for a private company to construct a toll road, and then when it is operational to manage the road including collection of revenue and the maintenance of the road. At the end of the contract period the road reverts to government control. The contracts will be designed to make the private company bear some risks such as, say, over expenditure on construction of the road, and to share some risks, say, the government and the private company sharing the risk associated with projections of vehicle usage. Other choices about how risks will be allocated between the parties are possible.

The statement of financial position presentation for construction contracts should show as an asset – *Gross amounts due from customers* – the following net amount:

- total costs incurred to date;
- plus attributable profits (or less foreseeable losses);
- less any progress billings to the customer.

Where for any contract the above amount is negative, it should be shown as a liability – *Gross amounts due to customers*.

Advances – amounts received by the contractor before the related work is performed – should be shown as a liability – effectively a payment on account by the customer.

An extract from the Lend Lease Group 2013 accounts illustrates the above:

b. Construction Work in Progress	2013	2012
Construction work in progress comprises:	A$m	A$m
Contract costs incurred to date	66,411.3	64,388.9
Profit recognised to date	3,143.2	3,095.6
	69,554.5	67,484.5
Less: Progress billings received and receivable on contracts	(69,726.3)	(67,780.4)
Net construction work in progress	(171.8)	(295.9)
Costs in excess of billings – inventories	612.0	505.8
Billings in excess of costs – trade payables	(783.8)	(801.7)
	(171.8)	(295.9)

21.5 Public–private partnerships (PPPs)

PPPs have become a common government policy whereby public bodies enter into contracts with private companies which have included contracts for the building and management of transport infrastructure, prisons, schools and hospitals.

There are inherent risks in any project and the intention is that the government, through a PPP arrangement, should transfer some or all of such risks to private contractors. For this to work equitably there needs to be an incentive for the private contractors to be able to make a reasonable profit provided they are efficient whilst ensuring that the providers, users of the service, taxpayers and employees also receive a fair share of the benefits of the PPP.

The European PPP Expertise Centre (epec) states that: 'In 2013, **the aggregate value of PPP transactions which reached financial close in the European market**[4] totaled **EUR 16.3 billion**, a 27% increase over 2012 (EUR 12.8 billion).' They further indicate that two major UK projects were included in those figures, namely the Thameslink rolling stock (€1.9 billion) and the Royal Liverpool Hospital (€509 million).

Improved public services

It has been recognised that where such contracts satisfy a value for money test it makes economic sense to transfer some or all of the risks to a private contractor. In this way it has been possible to deliver significantly improved public services with:

- increases in the quality and quantity of investment, e.g. by the private contractor raising equity and loan capital in the market rather than relying simply on government funding;

- tighter control of contracts during the construction stage to avoid cost and time overruns, e.g. completing construction contracts within budget and within the agreed time – this is evidenced in a report from the National Audit Office[5] which indicates that the majority are completed on time and within budget; and

- more efficient management of the facilities after construction, e.g. maintaining the buildings, security, catering and cleaning of an approved standard for a specified number of years.

PPP defined

There is no clear definition of a PPP. It can take a number of forms, e.g. in the form of the improved use of existing public assets under the Wider Markets Initiative (WMI) or contracts for the construction of new infrastructure projects and services provided under a Private Finance Initiative (PFI).

The Wider Markets Initiative (WMI)[6]

The WMI encourages public-sector bodies to become more entrepreneurial and to undertake commercial services based on the physical assets and knowledge assets (e.g. patents, databases) they own in order to make the most effective use of public assets. WMI does not relate to the use of surplus assets – the intention would be to dispose of these. However, wanting to become more entrepreneurial leads to the need to collaborate with private enterprises which have the necessary expertise.

Private Finance Initiative (PFI)

The PFI has been described[7] as a form of public–private partnership (PPP) that 'differs from privatisation in that the public sector retains a substantial role in PFI projects, either as the main purchaser of services or as an essential enabler of the project . . . differs from contracting out in that the private sector provides the capital asset as well as the services . . . differs from other PPPs in that the private sector contractor also arranges finance for the project'.

In its 2004 Government Review the HM Treasury stated[8] that:

The Private Finance Initiative is a small but important part of the Government's strategy for delivering high quality public services. In assessing where PFI is appropriate, the Government's approach is based on its commitment to efficiency, equity and accountability and on the Prime Minister's principles of public sector reform. PFI is only used where it can meet these requirements and deliver clear value for money without sacrificing the terms and conditions of staff. Where these conditions are met, PFI delivers a number of important benefits.

By requiring the private sector to put its own capital at risk and to deliver clear levels of service to the public over the long term, PFI helps to deliver high quality public services and ensure that public assets are delivered on time and to budget.

The PFI has meant that more capital projects have or will be undertaken for a given level of public expenditure, and public-service capital projects have been brought on stream earlier. However, it has to be recognised that this increased level of activity must be paid for by higher public expenditure in the future or by additional fees for services paid by the public. The aim is to offset some of those costs by additional income or better efficiency.

Thus the stream of contracted revenue payments to the private sector restricts the options which the current and future governments will have. PFI projects have committed governments to payments to private-sector contractors between 2000/01 and 2025/26 of more than £100 billion. Some of these contracts may be for long periods of time such as 30 years.

Briefly, then, PFI allows the public sector to enter into a contract (known as a concession) with the private sector to provide quality services on a long-term basis, typically twenty-five to thirty years, so as to take advantage of private-sector management skills working under contracts where private-sector finance is at risk. The private sector has the incentive to operate efficiently and effectively if the service requirements are comprehensive and reflect public needs appropriately, and the future risks associated with the project are fairly shared by the two parties.

How does PFI operate?

In principle, private-sector companies accept the responsibility for the design; raise the finance; undertake the construction, maintenance and possibly the operation of assets for the delivery of public services. In return for this the public sector pays for the project by making annual payments that cover all the costs plus a return on the investment through performance payments which include incentives for being efficient.

In practice the construction company and other parties such as the maintenance companies become shareholders in a project company set up specifically to tender for a concession.

- The project company enters into the contract (the 'concession') with the public sector; then enters into two principal subcontracts with;
  - a construction company to build the project assets; and
  - a facilities management company to maintain the asset – this is normally for a period of 5 or so years, after which time it is re-negotiated.

Note: the project company will pass down to the constructor and maintenance subcontractors any penalties or income deductions that arise as a result of their mismanagement.

- The project company raises a mixture of:
  - equity and subordinated debt from the principal private promoters, i.e. the construction company and the maintenance company; and
  - long-term debt.

Note: the long-term debt may be up to 90% of the finance required on the basis that it is cheaper to use debt rather than equity. The loan would typically be obtained from banks and would be without recourse to the shareholders of the project company. As there is no recourse to the shareholders, lenders need to be satisfied that there is a reliable income stream coming to the project company from the public sector, i.e. the lender needs to be confident that the project company can satisfy the contractual terms agreed with the public sector.

The subordinated debt made available to the project company by the promoters will be subordinated to the claims of the long-term lenders in that they will only be repaid after the long-term lenders.

- The project company receives regular payments, usually over a twenty-five- to thirty-year period, from the public sector once the construction has been completed to cover the interest and construction, operating and maintenance costs.

Note: such payments may be conditional on a specified level of performance and the private-sector partners need to have carried out detailed investigation of past practice for accommodation-type projects and/or detailed economic forecasting for throughput projects.

If, for example, it is an accommodation-type project (e.g. prisons, hospitals and schools) then payment is subject to the buildings being available in an appropriate clean and decorated condition – if not, income deductions can result.

If it is a throughput project (e.g. roads, water) with payment made on the basis of throughput such as number of vehicles or litres of water, then payment would be at a fixed rate per unit of throughput and the accuracy of the forecast usage has a significant impact on future income.

- The project company makes interest and dividend payments to the principal promoters.
- Finally, the project company returns the infrastructure assets in agreed condition to the public sector at the end of the twenty-five- to thirty-year contractual period.

This can be shown graphically as in Figure 21.1.

Profit and cash flow profile for the shareholders

Over a typical thirty-year contract the profit and cash flow profiles would follow different growth patterns.

Figure 21.1 The operation of PFI

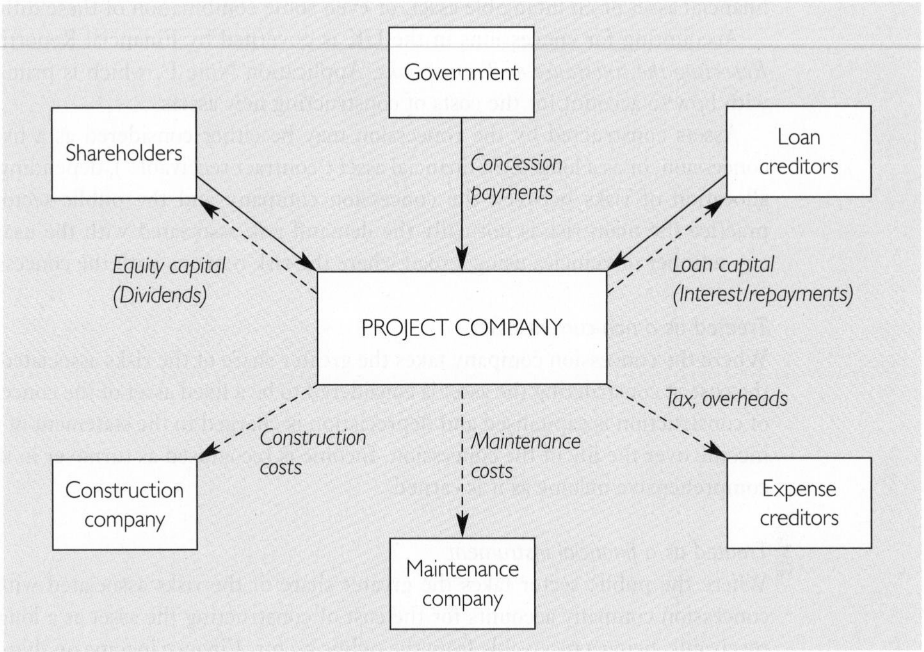

Profit profile

No profits are received as dividends during construction. Before completion the depreciation and loan interest charges can result in losses in the early years. As the loans are reduced the interest charge falls and profits then grow steadily to the end of the concession.

Cash flow

As far as the shareholders are concerned, cash flow is negative in the early years with the introduction of equity finance and subordinated loans. Cash begins to flow in when receipts commence from the public sector and interest payments commence to be made on the subordinated loans, say from year 5, and dividend payments start to be made to the equity shareholders, say from year 15.

How is a concession dealt with in the annual accounts of a construction company?

Statement of comprehensive income entries

The accounting treatment will depend on the nature of the construction company's shareholding in the project company. If it has control, then it would consolidate. Frequently, however, it has significant influence without control and therefore accounts for its investment in concessions by taking to the statement of comprehensive income its share of the net income or expense of each concession, in line with IAS 28 *Investments in Associates*.

How is a concession dealt with in the annual accounts of a concession or project company?

The accounting for service concessions has been a difficult problem for accounting standard setters around the world and different models exist. The main difficulties are in determining the nature of the asset that should be recognised, whether that is a tangible fixed asset, a financial asset or an intangible asset, or even some combination of these different options.

Accounting for concessions in the UK is governed by Financial Reporting Standard 5, *Reporting the Substance of Transactions*, Application Note F, which is primarily concerned with how to account for the costs of constructing new assets.

Assets constructed by the concession may be either considered as a fixed asset of the concession, or as a long-term financial asset ('contract receivable'), depending on the specific allocation of risks between the concession company and the public-sector authority. In practice the main risk is normally the demand risk associated with the usage of the asset, e.g. number of vehicles using a road where the risk remains with the concession company.

Treated as a non-current asset

Where the concession company takes the greater share of the risks associated with the asset, the cost of constructing the asset is considered to be a fixed asset of the concession. The cost of construction is capitalised and depreciation is charged to the statement of comprehensive income over the life of the concession. Income is recognised as turnover in the statement of comprehensive income as it is earned.

Treated as a financial instrument

Where the public sector takes the greater share of the risks associated with the asset, the concession company accounts for the cost of constructing the asset as a long-term contract receivable, being a receivable from the public sector. Finance income on this contract receivable is recorded using a notional rate of return which is specific to the underlying asset, and included as part of non-operating financial income in the statement of comprehensive income.

Under the contract receivable treatment, the revenue received from the public sector is split. The element relating to the provision of services that are considered a separate transaction from the provision of the asset is recognised as turnover in the statement of comprehensive income. The element relating to the contract debtor is split between finance income and repayment of the outstanding principal.

21.6 IFRS 15 treatment of construction contracts[9]

When companies adopt IFRS 15 for revenue recognition then automatically they follow that standard for construction contracts and hence IAS 11 is no longer applicable. Some of the important provisions of the proposed new revenue rules include the requirement to:

- recognise revenue when control passes;
- account for onerous performance obligations as soon as they become apparent so as to be consistent with rules relating to asset recognition and impairment; and
- disclose information to allow report readers to assess the risks and rewards likely to be associated with ongoing contracts.

The main differences relate to the timing of revenue recognition. In many other respects the accounting is the same as under IAS 11.

Recognise revenue when control passes

We discussed in Chapter 8 on revenue recognition the fact that recognition is dependent on the transfer of control rather than transfer of legal title. This will also apply to construction contracts. However, the fact that construction contracts often extend over several years and are not easily subdivided into parts makes the issue of importance to companies involved in substantial contracts. If the *revenue recognition* rules change as proposed, then it is likely that *construction* contracts will be altered in the future to more clearly specify when 'control' of components of the construction contracts pass to the clients. This illustrates that accounting standards are not neutral but are likely to alter how business is done.

Construction over a long time

With construction contracts which take a long time to complete, the identification of when control passes can be complex. Generally these contracts fall under what the standard calls 'performance obligations satisfied over time'. Two relevant types of contracts are provided in paragraph 35:

- the entity's performance *creates* or enhances an asset (for example work in progress) that the customer controls as the asset is created or enhanced . . . ; and
- the entity's performance *does not create* an asset with an alternative use to the entity . . . and the entity has an enforceable right to payment for performance completed to date.

Whilst these conditions depend on the factual and legal situations of each case a typical example of the former would be the construction of houses in a property development where the developer owns the land and sells the houses as they are completed. An example of the latter could be the construction of a motorway which is on government land and hence cannot be diverted to other customers.

An example of an alternative use asset might be the production of a boat which is of a general category such that it could be readily sold in the open market (e.g. a leisure boat). However, if the boat was designed to very specific needs of a particular client and therefore would not be ideal for other potential customers then such a boat would not have an alternative use.

As stated before, it is difficult to generalise and each case has to be assessed on its own facts when the contract commences. One of those facts is the contractual document. If in doubt one has to go back to the basic principal – when does control pass? Additional guidance appears in Appendix B to the standard.

Determining the amount of the contract earned

For items where control passes over time the producer has to determine what part of the contract amount has been earned for the accounting period and the amount to show as assets at the closing date. There are two ways of measuring the amount of revenue earned, namely the output approach and the input approach.

The output method

The output method in many ways appears to be a good way to measure progress, provided reliable information is available at a reasonable cost. In some types of work the contract provides for an independent expert such as an architect to periodically certify the amount of work which has been completed. In such circumstances the certification will either state the contract amount which has been earned or the percentage of the contract which has been completed. Another variant of the output method might be that when certain milestones have been met (say the concrete has been laid in a contract to build a road) an engineer may certify that the work is of the appropriate quality and that the milestone has been met. This is often linked to progress payments with the expert certifying the amount to be paid or allowing that amount to be determined such as when the payment is the work to the milestone less a retention amount to cover future contingencies or future remedial work.

The input method

When the conditions for use of the output method are not met then the input method can be used. The input method measures the rate of progress in terms of the costs incurred to date as a proportion of the total expected costs to complete the contract. As some costs such as inefficiencies or costs to correct errors do not generate revenue or convey value to the customer they should be written off immediately as an expense and are not considered in measuring the rate of progress. The percentage of necessary costs compared to the most recent assessment of the total cost to complete the project approximates the proportion of the revenue which will be deemed earned.

The accounting does not appear to substantially alter previous requirements. The main difference is when to recognise revenue and that includes the requirement to only recognise revenue when it is probable that the entity will be entitled to the consideration.

Onerous contracts

Consistent with that realistic but cautious approach the standard accounts for onerous contracts as soon as it becomes apparent that the contract will not be profitable. In other words, when it becomes apparent that the contract will *not* be profitable the first step will be to recognise that some, or all, of the previously recognised contract work in progress has been impaired and needs to be written off. The impairment is recorded as an expense. When that asset has been extinguished, it is necessary to create a liability to reflect the present value of obligations which are still to be incurred but that will not be recovered through the contract price. This is not a new provision.

Disclosure

The disclosure requirement is explained in the proposed guidance:

> 110 The objective of the disclosure requirements is for an entity to disclose sufficient information to enable users of financial statements to understand the nature, amount, timing, and uncertainty of revenue and cash flows arising from contracts with customers. To achieve that objective, an entity shall disclose qualitative and quantitative information about all of the following:
>
> (a) its contracts with customers . . . ;
>
> (b) the significant judgments, and changes in judgments, made in applying this Standard to those contacts . . . ; and
>
> (c) any assets recognized from the costs to obtain or fulfil a contract with a customer . . .'

21.7 An approach when a contract can be separated into components

To keep track of the construction contracts the accounting has many similarities with job costing where costs are accumulated by job and, if there are distinct components of the job, then separate records may be required for each component as a basis for invoicing.

For example, if you are constructing a shopping centre and you want to complete it in stages so that the landlord can have the first stage operational whilst the second stage is still under construction, you may need two 'jobs' in the books. Care will need to be taken to ensure the expenses are charged to the right stage with subjective decisions such as how to allocate the costs for preparing the land and installing the services. Then when the first stage is complete you may finalise the profit calculation for the first stage. You will have to keep track of the revenue earned to that stage being the contract price for stage one adjusted for contract variations. You will have in the job costing the expenses which relate to stage one. That will enable you to calculate the profit to that stage.

If the contract is not going according to plan you will have to recognise the losses you expect to make on the total contract by forecasting further costs to complete the contract and comparing it with the expected revenue. The total revenue would be the contract price plus allowances which can be invoiced under the escalation clauses plus any revenue arising from agreed variations to the original contract.

So from the above we have to keep track of:

(a) total costs incurred on the contract to date;

(b) the amount of revenue recognised in the accounts to date;

(c) the costs incurred in relation to the revenue which has been recognised;

(d) the amount of the profit or loss recorded on the contract so far;

(e) the amount invoiced to the customer so far; and

(f) the amount unpaid by the customer.

21.8 Accounting for a contract – an example

First year of contract

ABC has two construction contracts (Contract A and Contract B) outstanding at the end of its financial year, 30 June 20X0. Details for Contract A are as follows:

	Contract A £000
Total contract price	25,000
Cost incurred to date	5,500
Anticipated future costs	14,500
Progress billing	—
Agreed price for the component completed	7,000

Step one: Review the anticipated overall position for the contract

	Contract A £000
Total expected cost to complete the contract:	
Costs to date	5,500
Anticipated future costs	14,500
Expected total cost	20,000
Contract price	25,000
Forecast profit on the contract	5,000

Since the contract is expected to be profitable overall, the profit on the component completed to date of £7,000,000 – 5,500,000 or £1,500,000 can be fully realised. If on the other hand the forecast total cost was greater than the revenue, then the anticipated cost overrun would need to be recorded as an expense/loss in the current period.

Step two: The statement of comprehensive income

	Contract A	
	£000	£000
Revenue		7,000
Less:		
Cost incurred to date	5,500	
Allowance for future losses	—	
Total expenses		5,500
Net income		1,500

Step three: The statement of financial position entries

As the statement of financial position is a cumulative statement all figures have to be prepared on that basis.

	Contract A
	£000
Costs incurred to date	5,500
Add: profits to date	1,500
Less: recognised losses to date	—
Gross work done for the customer	7,000
Less: amount billed to the customer	—
Gross amount due from customers	7,000

Note: it is assumed that an identifiable phase of the contract had been completed for which the contract specifies the appropriate revenue is £7,000,000.

If the amount of revenue had not been given in the question then the amount of the revenue has to calculated. Suppose that the conditions have been satisfied for the output method to be used and that revenue is to be based on an independent architect's estimate of the proportion of the contract that has been completed. For example, assuming that the architect said the contract was 28% complete then the revenue to be recognised would be that proportion of the total contract price of £25,000,000 or 25,000,000 × 0.28 equals £7,000,000.

If there was no independent certified percentage of completion then the input method has to be used. In that event the completion percentage would be estimated by dividing the expenses incurred to date (£5,500,000) by the anticipated total cost to complete the project (£20,000,000) to give 27.5% and the revenue recognised at 0.275 × £25,000,000 (the total contract price) i.e. £6,875,000.

The new standard stresses satisfaction of performance obligations. One way in which this can be satisfied is for the company to create or enhance an asset that the customer controls as the creation or enhancement takes place. Alternatively completion of part of the work is associated with an enforceable right to payment, and there is no reason to expect the whole project is not going to be completed in compliance with the contract terms, then the service obligation will be seen to be accomplished. Under this alternative the approach the treatment may be similar to the method applied under IAS 11 but the conditions for its application are different.

Year two of the contract

Details of the transactions for Contract A for the year ended 30 June 20X1 are outlined below.

	Contract A £000
Total contract price	25,000
Costs incurred to date	14,000
Anticipated future costs to complete the contract	6,000
Certified revenue for work completed to date	15,000
Progress billings	12,000
Cash received for items invoiced	12,000
Advance payments	4,000

Step one: Review the anticipated overall result for the contract

	Contract A £000
Costs incurred to date	14,000
Anticipated future costs to complete	6,000
Total expected costs	20,000
Contract price	25,000
Anticipated profit	5,000

Since the project is expected to be profitable there is no requirement to make accruals for losses.

Step two: Prepare the relevant part of the comprehensive income statement

	£000	£000
Total revenue for stages one and two		15,000
Less revenue already recognised		7,000
Revenue for the period		8,000
Less expenses:		
Additional expenses incurred in the period (14,000 – 5,500)	8,500	
Additional anticipated loss accrual	—	
Total expenses		8,500
Loss for the year		(500)

Step three: Prepare the statement of financial position entries

	Contract A £000
Costs incurred to date	14,000
Add: recognised profits	1,500
Less: recognised losses	(500)
Work performed for the customer	15,000
Less: progress billings	12,000
Gross amount due from customers and included in work in progress	3,000

Also there would be a liability for £4,000 revenue paid in advance.

21.9 Illustration – loss-making contract using the step approach

First year of contract

The terms of Contract B specify that 50% of the contract price is due on completion of stage one and 50% on completion of stage two. The details as at 30 June 20X1 are as follows:

	Contract B £000
Total contract price	20,000
Costs incurred to date	13,000
Anticipated future costs	11,000
Progress billings	10,000
Advance payments	nil

At 30 June 20X0 the project had been signed but no work had been done, so no revenue or profit would be recorded. Thus it would only appear in the accounts in the notes when the company discloses the amount of future work contracted for.

Step one: Review the expected overall position of the contract as at 30 June 20X1

	Contract B £000
Costs incurred to date	13,000
Anticipated future costs	11,000
Forecast contract cost	24,000
Contract price (revenue)	20,000
Anticipated Loss on contract	(4,000)

This loss has to be recognised at 30 June 20X1.

Step two: Prepare the relevant part of the comprehensive income statement

	£000	Contract B £000
Revenue earned to date (£20,000,000 × 13,000/24,000)		10,833
Less: revenue previously recognised		nil
Revenue earned in the period		10,833
Costs incurred to date	13,000	
Less: costs previously recorded as an expense	nil	
	(13,000)	
(Anticipated loss in future periods: Rev 9,167 – 11,000 expense)	(1,833)	
Total expenses for the period		(14,833)
Loss on Contract B for the period		(4,000)

The total loss has to be recorded as soon as it is recognised, so the loss of 4,000 for the period is consistent with that. The anticipated revenue in the future was the total of 20,000 for the contract less 10,833 recognised in this period. 11,000 was the forecast additional expenses to complete, giving a difference of negative 1,833.

Step three: Prepare the statement of financial position entries
The entries in the statement of financial position represent the cumulative position.

	Contract B
	£000
Costs incurred to date	13,000
Add: profits recognised to date	nil
Less: losses recognised to date	(4,000)
Work performed to date at cost or net realisable value	9,000
Less: billings to date	(10,000)
Liability for anticipated expenses in excess of future billings in the next period	(1,000)

Recap:

1 First calculate whether the contract is anticipated to be profitable overall. This is done by subtracting from the total revenue under the contract the anticipated total expenses for the contract (that is the expenses already incurred combined with the additional expenses needed to complete the contract). If there is a loss the total of that loss has to be recognised immediately.

2 Do the profit and loss calculation. Use of the appropriate method to calculate the revenue for the period. Deduct the expenses related to the project for the period. If losses are anticipated in future periods the expenses have to be increased by that amount.

3 Prepare the work in progress account which is like a debtor's account for work which has not been billed to a customer yet. It is shown as total expenses to date plus profits less losses which gives a work in progress valued at selling price. Then you deduct from that any billings made to date, leaving the balance of work done valued at selling price which has not been billed (invoiced) to customers.

4 The accounts receivable balance is calculated by taking the opening balance adding any additional amounts billed to the customers during the year less any cash received.

Summary

Long-term contracts are those that cannot be completed within the current financial year. This means that a decision has to be made as to whether or not to include any profit before the contract is actually completed. The view taken by the standard setters pre-2011 is that contract revenue and costs should be recognised under IAS 11 using the percentage of completion method.

There is a proviso that revenue and costs can only be recognised when the amounts are capable of independent verification and the contract has reached a reasonable stage of completion. Although profits are primarily attributed to the financial periods in which the work is carried out, there is a requirement that any foreseeable losses should be recognised immediately in the statement of comprehensive income of the current financial period and not apportioned over the life of the contract.

IAS 11 is to be superseded by IFRS 15 and application of the new standard is required for annual periods beginning on or after 1 January 2017, but early adoption is permitted under IFRS. Under IFRS 15 revenue is recognised when the service has been rendered, control has passed and the results are measurable.

REVIEW QUESTIONS

1 Discuss the relative merits of recognising revenue under the percentage of completion method and the passing of control as major thresholds are met.

2 'Profit on a contract is not realised until completion of the contract.' Discuss.

3 'The use of the passing of control criteria for recognition of revenue will result in less comparability as companies can then manage their earnings.' Discuss.

4 'Profit on a contract that is not complete should be treated as an unrealised holding gain.' Discuss.

5 Discuss what information should be disclosed in the annual report in relation to construction costs in order for it to be useful to report users.

6 The Treasury states that 'Talk of PFI liabilities with a present value of £110 million is wrong. Adding up PFI unitary payments and pretending they present a threat to the public finances is like adding up electricity, gas, cleaning and food bills for the next 30 years.' Discuss.

7 'The operator of an asset in a PFI contract should recognise the tangible assets on its balance sheet.' Discuss.

8 'If the current financial reporting of transactions is well understood by users it is confusing to require a change when none is sought by the users.' Discuss.

EXERCISES

Solutions for exercises marked with an asterisk (*) are accessible in MyAccountingLab.

* Question 1

MACTAR has a series of contracts to resurface sections of motorways. The scale of the contract means several years' work and each motorway section is regarded as a separate contract.

M1	€m
Contract	3.0
Costs to date	2.1
Estimated cost to complete	0.3
Progress billings applied for to date	1.75
Payments received to date	1.5

M6	€m
Contract sum	2.0
Costs to date	0.3
Estimated cost to complete	1.1
Progress billings applied for to date	0.1
Payments received to date	—

M62	€m
Contract sum	2.5
Costs to date	2.3
Estimated costs to complete	0.8
Certified value of sections completed to date	50% of contract
Progress billings applied for to date	1.0
Payments received to date	0.75

The M62 contract has had major difficulties due to difficult terrain, and the contract only allows for a 10% increase in contract sum for such events.

Required:

From the information above, calculate for each contract the amount of profit (or loss) you would show for the year and show how these contracts would appear in the statement of financial position with all appropriate notes.

* Question 2

At 31 October 20X9, Lytax Ltd was engaged in the following five long-term contracts. In each contract Lytax was building cold storage warehouses on five sites where the land was owned by the customer. Details are given below:

	1	2	3	4	5
	£000	£000	£000	£000	£000
Contract price	1,100	950	1,400	1,300	1,200
At 31 October					
Cumulative costs incurred	664	535	810	640	1,070
Estimated further costs to completion	106	75	680	800	165
Estimated cost of post-completion guarantee rectification work	30	10	45	20	5
Cumulative costs incurred transferred to cost of sales	580	470	646	525	900
Progress billings:					
Cumulative receipts	615	680	615	385	722
Invoiced					
– awaiting receipt	60	40	25	200	34
– retained by customer	75	80	60	65	84

It is not expected that any customers will default on their payments.

Up to 31 October 20X8, the following amounts have been included in the revenue and cost of sales figures:

	1	2	3	4	5
	£000	£000	£000	£000	£000
Cumulative revenue	560	340	517	400	610
Cumulative costs incurred transferred to cost of sales	460	245	517	400	610
Foreseeable loss transferred to cost of sales	—	—	—	70	—

It is the accounting policy of Lytax Ltd to arrive at contract revenue by adjusting contract cost of sales (including foreseeable losses) by the amount of contract profit or loss to be regarded as recognised, separately for each contract.

Required:

Show how these items will appear in the statement of financial position of Lytax Ltd with all appropriate notes. Show all workings in tabular form.

* Question 3

During its financial year ended 30 June 20X7 Beavers, an engineering company, has worked on several contracts. Information relating to one for Dam Ltd which is being constructed to a specific customer design is given below:

Contract X201

Date commenced	1 July 20X6
Original estimate of completion date	30 September 20X7
Contract price	£240,000
Proportion of work certified as satisfactorily completed (and invoiced) up to 30 June 20X7	£180,000
Progress payments from Dam Ltd	£150,000
Costs up to 30 June 20X7	
Wages	£91,000
Materials sent to site	£36,000
Other contract costs	£18,000
Proportion of Head Office costs	£6,000
Plant and equipment transferred to the site (at book value on 1 July 20X6)	£9,000

The plant and equipment is expected to have a book value of about £1,000 when the contract is completed.

Inventory of materials at site 30 June 20X7	£3,000
Expected additional costs to complete the contract:	
Wages	£10,000
Materials (including stock at 30 June 20X7)	£12,000
Other (including Head Office costs)	£8,000

At 30 June 20X7 it is estimated that work to a cost value of £19,000 has been completed, but not included in the certifications as the customer did not have control of that work which was performed off-site.

If the contract is completed one month earlier than originally scheduled, an extra £10,000 will be paid to the contractors. At the end of June 20X7 there seemed to be a 'good chance' that this would happen. Assume the output method is appropriate.

Required:

(a) Show the account for the contract in the books of Beavers up to 30 June 20X7 (including any transfer to the statement of comprehensive income which you think is appropriate).

(b) Show the statement of financial position entries.

(c) Calculate the profit (or loss) to be recognised in the 20X6–X7 accounts.

Question 4

Newbild SA commenced work on the construction of a block of flats on 1 July 20X0.

During the period ended 31 March 20X1 contract expenditure was as follows:

	€
Materials issued from stores	13,407
Materials delivered direct to site	73,078
Wages	39,498
Administration expenses	3,742
Site expenses	4,693

On 31 March 20X1 there were outstanding amounts for wages €396 and site expenses €122, and the stock of materials on site amounted to €5,467.

The following information is also relevant:

1 On 1 July 20X0 plant was purchased for exclusive use on site at a cost of €15,320. It was estimated that it would be used for two years after which it would have a residual value of €5,000.

2 By 31 March 20X1 Newbild SA had received €114,580, being the amount of work certified by the architects at 31 March 20X1 on completion and handover of the show flat, less a 15% retention.

3 The total contract price is €780,000. The company estimates that additional costs to complete the project will be €490,000. From costing records it is estimated that the costs of rectification and guarantee work will be 2.5% of the contract price.

Required:

(a) Prepare the contract account for the period, together with a statement showing your calculation of the net income to be taken to the company's statement of comprehensive income on 31 March 20X1. Assume for the purpose of the question that the contract is sufficiently advanced to allow for the taking of profit.

(b) Give the values which you think should be included in the figures of revenue and cost of sales, in the statement of comprehensive income, and those to be included in net amounts due to or from the customer in the statement of financial position in respect of this contract.

Question 5

(a) A concession company, WaterAway, has completed the construction of a wastewater plant. The plant will be transferred to the public sector unconditionally after 25 years. The public sector (the grantor) makes payments related to the volume of wastewater processed.

Required:
Discuss how this will be dealt with in the statement of comprehensive income and statement of financial position of the concession company.

(b) A concession company, LearnAhead, has built a school and receives income from the public sector (the grantor) based on the availability of the school for teaching.

Required:
Discuss how this will be dealt with in the statement of comprehensive income and statement of financial position of the concession company, under IFRIC 12 Service Concessions Arrangements.

* Question 6

Quickbuild Ltd entered into a two-year contract on 1 January 20X7 at a contract price of £250,000. The estimated cost of the contract was £150,000. At the end of the first year the following information was available:

● contract costs incurred totalled £70,000;

● inventories still unused at the contract site totalled £10,000;

● progress payments received totalled £60,000;

● other non-contract inventories totalled £185,000.

Invoices issued £65,000.

The problem is to be considered under two different scenarios:

Case A where the accounting is to be performed under the percentage of completion method as required under IAS 11.

Case B where the accounting is done under the milestone method and the customer gains control of 40% of the facility at the end of the first year.

Required:

(a) Calculate the statement of comprehensive income entries for the contract revenue and the contract costs for each case.

(b) Calculate entries in the statement of financial position for the amounts due from construction contracts and inventories for each case.

Question 7

During 2006, Jack Matelot set up a company, JTM, to construct and refurbish marinas in various ports around Europe. The company's first accounting period ended on 31 October 2006 and during that period JTM won a contract to refurbish a small marina in St Malo, France. During the year ended 31 October 2007, the company won a further two contracts in Barcelona, Spain and Faro, Portugal. The following extract has been taken from the company's contract notes as at 31 October 2007:

Contract:	Barcelona	Faro	St Malo
	€m	€m	€m
Contract value	12.24	10.00	15.00
Work certified as available to clients:			
To 31 October 2006	—	—	6.00
Year to 31 October 2007	6.50	0.50	3.00
To date	6.50	0.50	9.00
Payments received:			
To 31 October 2006	—	—	5.75
Year to 31 October 2007	3.76	—	1.75
To date	3.76	—	7.50
Invoices sent to client:			
To 31 October 2006	—	—	6.00
Year to 31 October 2007	5.00	0.50	2.76
To date	5.00	0.50	8.76
Costs incurred:			
To 31 October 2006	—	—	6.56
Year to 31 October 2007	11.50	1.50	3.94
To date	11.50	1.50	10.50
Estimated costs to complete:			
As at 31 October 2006			5.44
As at 31 October 2007	4.00	5.50	1.50

Notes

Barcelona: Experiencing difficulties. Although JTM does not anticipate any cost increases, the client has offered to increase contract value by €0.76m as compensation.

Faro: No problems.

St Malo: Work has slowed down during 2007. However, company feels it can continue profitably.

The company uses the value of work certified to estimate the percentage completion of each contract.

Required:
(a) For each contract, calculate the profit or loss attributable to the year ended 31 October 2007 and show how it would be recognised in the company's balance sheet at that date. (Show your workings clearly.)
(b) As JTM's 2007 accounts were being prepared, it became evident that the St Malo contract had slowed down due to a dispute with a neighbouring marina which claimed that the JTM refurbishment had damaged part of its quayside. The company has been told that the cost of repairing the damage would be €150,000. Jack Matelot believes it is a fair estimate and, in the interests of completing the contract on time, has decided to settle the claim. He is not unduly concerned about the amount involved as such eventualities are adequately covered by insurance.

Required:
How should this event be dealt with in the 2007 accounts?
(c) During 2007, Jack Matelot had two major worries: (i) the operating performance of JTM had not been as good as expected; and (ii) the planned disposal of surplus property (to finance the agreed acquisition of a competitor, MoriceMarinas, and the payment of a dividend) had not been successful. As a result of these circumstances, Jack had been warning shareholders not to expect a dividend for 2007. However, during November 2007, the property was unexpectedly disposed of for €5m; which enabled the payment of a 2007 dividend of €1m and the acquisition of MoriceMarinas for €4m.

Required:
How should the above events be dealt with in the 2007 accounts?

(The Association of International Accountants)

Question 8

Backwater Construction Company is reviewing a major contract which is in serious difficulty. The contract price is €10,000,000. The project involves the construction of four buildings of equal size and complexity. The first building has been completed and costs to date are €3,000,000. The second building is expected to be completed after one more year, the third building after two years and the final building after three years. The relevant discount rate is 10% p.a.

Required:
Prepare the entries to record the revenue for the year just completed, to record the expenses incurred, and adjust assets (if any) and liabilities (if any) at the balance date.

Question 9

Norwik Construction plc is a large construction company involved in multiple large contracts around the world. One contract to build three stadiums is being undertaken by the Australasian division. Each stadium has an individual contract price. Jim Norwik who is the great-grandson of the founder is in charge of that division. He is concerned that the first stadium is costing more than is included in the tender estimates. Rather than recognise an immediate loss on the contract, he orders his subordinates to charge some of the materials for stadium one to stadium two which seems to be on target.

Discuss the consequences of Jim Norwik's actions for Norwik Construction plc and the likely impact on the behaviour of Jim Norwik's subordinates.

Question 10

Boldwin Construction has entered into a contract with Spears Retailers to construct a new department store on Spears land. The contract sum is £45 million. At 30 June 20X1 the situation is as follows:

(a) the contract is 30% complete;

(b) expenses to date on the contract are £15 million;

(c) additional expenses to complete the contract are estimated at £25 million;

(d) billings during the year £5 million;

(e) payments received from Spears Retailers £4 million.

Required:

(a) Show the profit entries and the resulting assets in the statement of financial position.

(b) Repeat the work in (a) assuming the cost to complete is estimated at £32 million.

(c) If the project is finished by June 20X2 and the additional cost to complete was £23 million, billings for the year £26 million, and cash received £25 million, show the relevant entries in the comprehensive income statement and financial position statement, assuming first scenario (a) occurred, and then repeat the exercise under scenario (b).

References

1 IAS 11 *Construction Contracts*, IASB, 1995.

2 IAS 18 *Revenue*, IASB, revised 2005.

3 It is common in construction work for architects to certify the amount which the contractor is entitled to as various milestones are reached. The amount payable is often the amount earned less a percentage withheld until the contract is completed and all problems resolved. If resolution does not occur the customer can use the money withheld to get a third party to rectify the remaining mistakes.

4 European PPP Expertise Centre, **Market Update**, Review of the European PPP Market in 2013. Europe for this purpose is defined as EU-28, the countries of the Western Balkans and Turkey.

5 National Audit Office, PFI: Construction Performance Feb. 2003, www.nao.org.uk/publications/nao_reports/02-03/0203371.pdf.

6 *Selling government services into wider markets, Policy and Guidance Notes*, Enterprise and Growth Unit, HM Treasury July 1998, www.hmtreasury.gov.uk/mediastore/otherfiles/agswm.pdf.

7 Research Paper 01/0117 *Private Finance Initiative*, G. Allen, Economic Policy and Statistics Section, House of Commons, December 2001.

8 See www.hm-treasury.gov.uk/documents/public_private_partnerships/ppp_index.cfm?ptr=29.

9 IFRS 15 *Revenue from Contracts with Customers*, IASB, 2014.

Consolidated accounts

Accounting for groups at the date of acquisition

22.1 Introduction

The main purpose of this chapter is to explain how to prepare consolidated financial statements at the date of acquisition and the IFRS 10 and 13 requirements.

Objectives

By the end of this chapter, you should be able to:

- prepare consolidated accounts at the date of acquisition:
  - for a wholly owned subsidiary;
  - for a partly owned subsidiary with non-controlling interests, calculating goodwill under the two options available in IFRS 3;
  - where the fair value of a subsidiary's net assets are more or less than their book values;
- explain IFRS 10, IFRS 3 and IFRS 13 provisions;
- discuss the usefulness of group accounts to stakeholders.

22.2 Preparing consolidated accounts for a wholly owned subsidiary

When a company acquires the shares of another company it records the cost as an Investment. If the shares acquired give it control over the acquired company, then the acquirer is referred to as a parent or holding company and the acquired company as a subsidiary.

The shareholders of the parent company want to know how well the directors of their company have managed all of the net assets which they control. This information is provided by the preparation of consolidated accounts which aggregates the assets and liabilities of both companies. In doing this it replaces the Investment in subsidiary in the parent's accounts with the fair value of the assets and liabilities of the subsidiary.

The parent may well have had to pay a premium over and above the fair value of the net assets in order to obtain control – this is referred to as Goodwill.

22.3 IFRS 10 *Consolidated Financial Statements*

IFRS 10 *Consolidated Financial Statements* which defines a group and how to determine control and also requires the use of fair values for a subsidiary.

22.3.1 IFRS 10 definition of a group

One of IASB's main objectives had been to develop a consistent basis for determining when a company consolidates the financial statements of another company to prepare group accounts. For this, it has stated that control should be the determining factor.

Under IFRS 10 *Consolidated Financial Statements*, a group exists where one enterprise (the parent) controls, either directly or indirectly, another enterprise (the subsidiary). A group consists of a parent and its subsidiaries.

22.3.2 IFRS 10 definition of control

Under IFRS 10 an investor is a parent[1] when it is exposed, or has rights, to variable returns from its involvement with the investee and has the ability to affect those returns through its power over the investee.

An investor controls[2] an investee if and only if the investor satisfies all of the following requirements:

- exposure, or rights, to variable returns whether positive or negative from its involvement with the investee;
- power over the investee whereby the investor has existing rights that give it the ability to direct those activities that significantly affect the investee's returns;
- the ability to use its power over the investee to affect the amount of the investor's returns.

The following is an extract from the 2013 Linde AG annual report:

> Scope of consolidation
> The Group financial statements comprise Linde AG and all the companies over which Linde AG is able to exercise control as defined by IFRS 10.

What if the shares acquired are less than 50%?

Even in this situation, it may still be possible to identify an acquirer when one of the combining enterprises, as a result of the business combination, acquires:

(a) power over more than one-half of the voting rights of the other enterprise by virtue of an agreement with other investors;

(b) power to govern the financial and operating policies of the other enterprise under a statute or an agreement;

(c) power to appoint or remove the majority of the members of the board of directors; or

(d) power to cast the majority of votes at a meeting of the board of directors.

What if the parent holds options or potential voting rights?

IFRS 10 provides that if those options give the entity control then this could result in an entity being consolidated. For example, if the investee's management always followed the wishes of the option holder, this may be viewed as having control. The following is an extract from the Reunert Limited 2013 Annual Report:

> **SUBSIDIARIES**
> A subsidiary is an entity over which the group has control. Control exists where the company has the power, directly or indirectly, to govern the financial and operating policies of an entity so as to obtain benefits from its activities. In assessing control, potential voting rights that are currently exercisable or convertible are taken into account.

What if a company has significant voting rights in comparison to other shareholders?

If an investor is so powerful through their voting rights compared to others, for example one investor has 40% while the other 60% is widely dispersed between unconnected investors, this can also give control and result in the investee being consolidated. Both of these areas will require directors to exercise judgement in determining whether control exists.

22.3.3 Requirement to use fair values

When one company acquires a controlling interest in another and the combination is treated as an acquisition, the assets and liabilities of the subsidiary are recorded in the acquirer's consolidated statement of financial position at their fair value.

On consolidation, if the acquirer has acquired less than 100% of the ordinary shares, any differences (positive or negative) between the fair values of the net assets and their book value are recognised in full and the parent and non-controlling interests are credited or debited with their respective percentage interests.

22.4 Fair values

The **fair value** of the consideration paid to acquire an investment in a subsidiary is set against the **fair value** of the identifiable net assets in the subsidiary at the date of acquisition.

Fair value of the consideration

Consideration may be in the form of shares in the acquiring company. If these are quoted then the fair value is the market price of the shares. If the shares are not quoted on an exchange, then they would need to be valued – for the purposes of this chapter the value is given (how to value unquoted shares is discussed in Chapter 29).

Consideration may also be in cash or a combination of shares and cash. If the cash element is deferred for more than a year the consideration is discounted to its present value. The difference is reported as an accrued finance charge.

Fair value of the net assets

The starting point is the book values in the subsidiary's statement of financial position. These are required to be re-stated at fair values when incorporating into the consolidated accounts. First, each asset and liability is reviewed. For example, land may be revalued to market value, raw materials to replacement price, finished goods to selling price less estimated profit, loans may be revalued if there has been a change in interest rates that impacts on their value. In addition to the assets and liabilities in the accounts, it is also necessary to estimate a fair value for any contingent liabilities – if unable to value then they are disclosed.

If the investment is greater than the share of net assets then the difference is regarded as the purchase of goodwill – see the Rose Group example below

22.5 Ilustration where there is a wholly owned subsidiary

The **fair value** of the parent company's investment in a subsidiary is set against the **fair value** of the identifiable net assets in the subsidiary at the date of acquisition. If the investment is greater than the share of net assets then the difference is regarded as the purchase of goodwill – see the Rose Group example below.

The consolidation schedule is as follows:

	Rose	Tulip	Adjustment Dr	Cr	Group	
	£	£	£	£	£	
Non-current assets	20,000	11,000			31,000	Step 3
Investment in Tulip	12,000	—		8,000 (a)	—	
				3,200 (b)		
Goodwill	—	—			800	Step 1
Net current assets	**11,000**	**3,000**			**14,000**	Step 3
Net assets	43,000	14,000			45,800	
Share capital	16,000	10,000	8,000 (a)			Step 4
			2,000 (c)		16,000	
Retained earnings	**27,000**	**4,000**	3,200 (b)			Step 4
			800 (d)		27,000	
	43,000	14,000			43,000	
Non-controlling interest	—	—		**2,000 (c)**		Step 2
				800 (d)	**2,800**	
	43,000	14,000	14,000	14,000	45,800	

(a), (b), (c) and (d) identify the entries in the calculations below.

Step 1. Calculate goodwill

		£	£
The parent company's investment in Tulip			12,000
Less: (a) parent's share of Tulip's share capital	(80% × 10,000)	8,000	
(b) parent's share of the retained earnings	(80% × 4,000)	**3,200**	
Goodwill			11,200
			800

Step 2. Calculate the non-controlling interest in Tulip

(c) Non-controlling interest in the share capital	(20% × 10,000)	2,000
(d) Non-controlling interest in the retained earnings	(20% × 4,000)	800
Representing the non-controlling interest in Tulip's net assets		2,800

In the published consolidated accounts the non-controlling interest will be shown as a separate item in the equity of the group as follows:

Share capital	16,000
Retained earnings	27,000
Rose shareholders' share of equity	43,000
Non-controlling interest	2,800
Total equity	45,800

This recognises that the non-controlling shareholders are part of the ownership of the group rather than a liability.

Step 3. Aggregate the assets and liabilities of the parent and subsidiary

		£
Non-current assets other than goodwill	(20,000 + 11,000)	31,000
Goodwill (as calculated in Step 1)		800
Net current assets	(11,000 + 3,000)	14,000
		45,800

Step 4. Calculate the consolidated share capital and reserves

	£
Share capital *(parent company only)*	16,000
Retained earnings *(parent company only)*	**27,000**
	43,000

Note it is only the parent's share capital that is **ever** reported in the group accounts. As for the retained earnings, it is only the earnings that arise **after** the date when the parent obtains control that are reported as part of the group retained earnings – this is dealt with further in the next chapter.

22.6.2 Illustration where there is a wholly owned subsidiary using Method 2

Let us now consider the impact on the previous example of using Method 2 to measure the non-controlling interest. In order to use this method, we need to know the fair value of the non-controlling interest in the subsidiary at the date of acquisition. Let us assume in this case that the fair value of a share in Tulip is £1.45, giving a value for the 2,000 shares of £2,900.

Two figures are different in the consolidated accounts if this method is used. The use of Method 2 affects two figures – goodwill and the non-controlling interest. Whereas under Method 1 the goodwill represented the cost of Rose obtaining control, under Method 2 we also credit the non-controlling interest with its own goodwill. It is computed as follows:

		£
Fair value of non-controlling interest at date of acquisition		2,900
20% of the net assets at the date of acquisition	(£14,000)	**(2,800)**
Attributable goodwill		100

The consolidated statement of financial position would now be as follows:

		£
Non-current assets other than goodwill		31,000
Goodwill	(£800 + **£100**)	900
Net current assets		**14,000**
		45,900
Share capital		16,000
Retained earnings		27,000
Non-controlling interest	(£2,800 + **£100**)	**2,900**
		45,900

How to determine the value of a share not acquired by the parent

Note that we assumed that the fair value of the non-controlling interest at the date of acquisition was £2,900. If Tulip's shares are quoted then the fair value estimate would be based on the share price prior to a bid. This price could be different from that paid by Rose on the assumption that in seeking to obtain control it would probably have paid more than the current share price. In exercises or exam questions the total figure might be given (as in this example) or a price per share might be given.

In the Rose example the goodwill relating to the parent (Rose's) shareholding of 80% is £800, i.e. 10p per share. The goodwill relating to the non-controlling interest in 2,000 shares, however, based on a £2,900 valuation is £100, i.e. 5p per share.

22.7 The treatment of differences between a subsidiary's fair value and book value

In our examples so far we have assumed that the book value of the net assets in the subsidiary is equal to their fair value. In practice, book value rarely equals fair value and it is necessary to revalue the group's share of the assets and liabilities of the subsidiary prior to consolidation.

The following is an extract from the 2013 EnBW Annual Report:

Basis of consolidation
Business combinations are accounted for in accordance with the acquisition method. The cost of a business combination is measured based on the fair value of the assets acquired and liabilities assumed or entered into at the acquisition date. Non-controlling interests are measured at the proportionate share of fair value of assets identified and liabilities assumed.

Note that, when consolidating, the **parent** company's assets and liabilities remain **unchanged** at book value – it is only the subsidiary's that are adjusted for the purpose of the consolidated accounts.

For example, let us assume that the fair value of Tulip's non-current assets was £600 above their book value at £11,600. If Rose owned 100% of Tulip, then the Rose shareholders would have the benefit of the £600 and the goodwill would be reduced from £800 to £200. However, as Tulip is part-financed by non-controlling shareholders, they are entitled to their 20% share of the £600 as seen in the following schedule:

	Rose £	Tulip £	Group £	Dr	Cr	Group fair value £	
Non-current assets	20,000	11,000	31,000	600		31,600	Step 3
Goodwill	—	—	800		480	320	Step 1
Investment in Tulip	12,000	—	—			—	
Net current assets	11,000	3,000	14,000			14,000	
Net assets	43,000	14,000	45,800			45,920	
Share capital	16,000	10,000	16,000			16,000	
Retained earnings	27,000	4,000	27,000			27,000	
	43,000	14,000	43,000			43,000	
Non-controlling interest	—	—	2,800		120	2,920	Step 2
	43,000	14,000	45,800			45,920	

Step 1. Goodwill is adjusted when fair value exceeds book value

As goodwill is the difference between the consideration and the net assets acquired, any increase in the net assets will mean that the difference is lower.

			£
The parent company's investment in Tulip			12,000
Less: The parent's share of the subsidiary's share capital	(80% × 10,000)	8,000	
The parent's share of retained earnings	(80% × 4,000)	3,200	
The parent's share of the revaluation	(80% × 600)	480	11,680
Goodwill			320

* This is equivalent to the share of net assets, 80% × (11,000 + 3,000 + 600).

Step 2. Non-controlling interest adjusted for fair value in excess of book value

Non-controlling interest in share capital of Tulip	(20% × 10,000)	2,000
Non-controlling interest in retained earnings of Tulip	(20% × 4,000)	800
Revaluation to fair value of the subsidiary's assets	(20% × 600)	120
		2,920

Step 3. Aggregate the parent's non-current assets which remain at book value and the subsidiary's which have been restated to fair value

The non-current assets would be reported as £31,600 (20,000 + 11,000 + 600). Remember that the revaluation of the subsidiary's assets is only necessary for the consolidated accounts. No entries need be made in the individual accounts of the subsidiary or its books of account. The preparation of consolidated accounts is *a separate exercise* that in no way affects the records of the individual companies.

22.8 The parent issues shares to acquire shares in a subsidiary

Shares in another company can be purchased with cash or through an exchange of shares. In the former case, the cash will be reduced and exchanged for another asset called 'investment in the subsidiary company'. If there is an exchange of shares, there will be an increase in the parents' share capital and often in the share premium.

Let us assume that Rose issued its own shares to 80% of the Tulip shareholders who wanted £1.50 for each share, totalling £12,000. Rose would in this case have to set a value of its own shares that was acceptable to the Tulip shareholders.

For illustration purposes, let us assume that the Rose shares were valued at £2.50 each and 4,800 were issued (£12,000/£2.50). The consolidation schedule would show that Rose's cash had not been reduced but the share capital and share premium had increased as follows:

		Rose £	Tulip £	Group £
Non-current assets		20,000	11,600	31,600
Investment in Tulip		12,000	—	—
Goodwill		—	—	320
Net current assets	11,000 + 12,000*	23,000	3,000	26,000
Net assets		55,000	14,000	57,920
Share capital	16,000 + 4,800 at par	20,800	10,000	20,800
Share premium	4,800 × £1.50	7,200		7,200
Retained earnings		27,000	4,000	27,000
Parent company's equity		55,000	14,000	55,000
Non-controlling interest		—	—	2,920
		55,000	14,000	57,920

* This is showing cash at £23,000 which was the position before we assumed that the shares in Tulip had been acquired for cash.

Note that there is no effect on the accounts of the acquired company as the payment of cash or exchange of shares is with the subsidiary company's individual shareholders, not the company itself.

22.9 IFRS 3 *Business Combinations* treatment of goodwill at the date of acquisition

Any differences between the fair values of the net assets and the consideration paid to acquire them is treated as positive goodwill or a bargain purchase (also referred to as badwill or negative goodwill) and dealt with in accordance with IFRS 3 *Business Combinations*.

The treatment of positive goodwill

Positive purchased goodwill, where the investment exceeds the total of the net assets acquired, should be recognised as an asset with no amortisation. In subsequent years goodwill must be subject to impairment tests in accordance with IAS 36 *Impairment of Assets*. These tests will be annual, or more frequently if circumstances indicate that the goodwill might be impaired.[4] Once recognised, an impairment loss for goodwill may not be reversed in a subsequent period, which helps in preventing the manipulation of period profits.

The treatment of a bargain purchase

The acquiring company does not always pay more than the fair value of the identifiable net assets. Paying less (sometimes referred to as negative goodwill) can arise[5] when:

(a) there have been errors measuring the fair value of either the cost of the combination or the acquiree's identifiable assets, liabilities or contingent liabilities; or

(b) future costs such as losses have been taken into account; or

(c) there has been a bargain purchase.

Where a parent pays less than the fair value of the net assets, IFRS 3 requires it to review the fair value exercise to ensure that no asset has been overstated or liability understated. Assuming this review reveals no errors, then the resulting difference is recognised immediately in the statement of income.

22.10 When may a parent company not be required to prepare consolidated accounts?

It may not be necessary for a parent company to prepare consolidated accounts if the parent is itself a wholly owned subsidiary and the ultimate parent produces consolidated financial statements available for public use that comply with International Financial Reporting Standards (IFRSs).[6]

If the parent company is a partially owned subsidiary of another entity, then, if its other owners have been informed and do not object, the parent company need not present consolidated financial statements; nor if its debt or equity instruments are not traded in a public foreign or domestic market.

22.11 When may a parent company exclude or not exclude a subsidiary from a consolidation?

22.11.1 Exclusion permitted

Subsidiaries may be excluded if they are immaterial or there are substantial rights exercisable by non-controlling interests.

Materiality

Exclusion is permissible on grounds of non-materiality[6] as the International Accounting Standards are not intended to apply to immaterial items.

For example, Linde AG states in its 2013 Annual Report:

> Non-consolidated companies are immaterial in aggregate from the Group's point of view in terms of total assets, revenue and net income or net loss for the year and do not have a significant impact on the net assets, financial position and results of operations of the Group. For that reason they are not included in the consolidated financial statements.

Substantial rights exercisable by the non-controlling interest

Exclusion might also be appropriate where there are substantial rights exercisable by a non-controlling interest as seen in the following extract from the Mitsubishi Logistics Corporation 2013 Annual Report:

> The company holds 51% of the voting rights in MICLTL Logistics Company Ltd, however, the other shareholder's agreement is necessary to decide important policies of finance and trade. Therefore, the Company does not treat MICLTL as a subsidiary.

22.11.2 Exclusion not permitted

Exclusion on the grounds that a subsidiary's activities are dissimilar from those of the others within a group is not permitted.[7] This is because information is required under IFRS 8 *Operating Segments* on the different activities of subsidiaries, and users of accounts can, therefore, make appropriate adjustments for their own purposes if required.

22.12 IFRS 13 *Fair Value Measurement*

IFRS 13 *Fair Value Measurement*[8] defines fair value as the price that would be received to sell an asset or paid to transfer a liability in an orderly transaction between market participants at the measurement date. The detailed guidance for determining fair value is also set out in IFRS 3.

The main provisions are that as from the date of acquisition, an acquirer should:

(a) incorporate into the statement of income the results of operations of the acquiree; and

(b) recognise in the statement of financial position the identifiable assets, liabilities and contingent liabilities of the acquiree and any goodwill or negative goodwill arising on the acquisition.

The identifiable assets, liabilities and contingent liabilities acquired that are recognised should be those of the acquiree that existed at the date of acquisition.

Treatment of future liabilities

Liabilities should not be recognised at the date of acquisition if they result from the acquirer's intentions or actions. Therefore liabilities for terminating or reducing the activities of the acquiree should **only** be recognised where the acquiree has, at the acquisition date, an existing liability for restructuring recognised in accordance with IAS 37 *Provisions, Contingent Liabilities and Contingent Assets*.

Treatment of future losses

Liabilities should also not be recognised for future losses[9] or other costs expected to be incurred as a result of the acquisition, whether they relate to the acquirer or the acquiree.

Treatment of contingent liabilities

Under IFRS 3 only those contingent liabilities assumed in a business combination that are a **present** obligation and can be measured reliably are recognised. If not recognised then they are disclosed in the same way as other contingent liabilities.

Treatment of intangible assets

There is a requirement to identify both tangible and intangible assets that are acquired. For example, fair values would be attached to intangibles such as brands and customer lists if these can be measured reliably. If it is not possible to measure them reliably, then the goodwill would be reported at a higher figure, as in the following extract from the 2013 AstraZeneca Annual Report:

> **Business Combinations and Goodwill**
> On the acquisition of a business, fair values are attributed to the identifiable assets and liabilities and contingent liabilities unless the fair value cannot be measured reliably in which case the value is subsumed into goodwill.

Why revalue net assets?

The reason why all the net assets, including the intangible assets that did not appear in the subsidiary's statement of financial position must be identified and fair-valued at the date of acquisition is to prevent distortion of EPS in periods following the acquisition. For example, we have seen in Chapter 19 that intangible assets are required to be amortised with an annual charge against profits, whereas goodwill is not subject to an annual amortisation charge but is reviewed for impairment. Subsuming intangible assets into the goodwill figure means that a regular amortisation charge is avoided. This reason for valuing the intangible assets would not apply if goodwill were to be amortised as it had been in the UK prior to 2008.

22.13 What advantages are there for stakeholders from requiring groups to prepare consolidated accounts?

Advantages include investor protection, help in predicting future earnings per share and means to assess management performance.

(a) **Investor protection**: Consolidation prevents the publication of misleading accounts by such means as inflating the sales through selling to another member of a group.

(b) **Prediction**: Consolidation provides a more meaningful EPS figure. Consolidated accounts show the full earnings on a parent company's investment while the parent's individual accounts only show the dividend received from the subsidiaries.

(c) **Accountability**: Consolidation provides a better measurement of the performance of a parent company's directors as the total earnings of a group can be compared with its total assets in arriving at a group's return on capital employed (ROCE).

It is important to remember that the ROCE prepared from the consolidated financial statements is regarded by management as a ratio that is an important measure of performance and one to be maximised. For example, Northgate plc reported in its 2013 Annual Report:

> Return on capital employed (ROCE) – In a capital intensive business, ROCE is a more important measure of performance than profitability alone, as low margin business returns low value to shareholders . . . ROCE is maximised through a combination of managing utilisation, hire rates, vehicle holding costs and improvements in operational efficiency.

Summary

When one company acquires a controlling interest in another and the combination is treated as an acquisition, the investment in the subsidiary is recorded in the acquirer's statement of financial position at the fair value of the investment.

On consolidation, if the consideration exceeds the fair value of the net assets it is referred to as goodwill and appears as an asset in the statement of financial position. If the consideration is less than the fair value of the net assets it is regarded as a bargain purchase and it will be taken to profit in accordance with IFRS 3.

On consolidation, if the acquirer has acquired less than 100% of the equity shares, any differences between the fair values of the assets or liabilities and their face value are recognised in full and the parent and non-controlling interests credited or debited with their respective percentage interests.

REVIEW QUESTIONS

1 Explain how negative goodwill (bargain purchase) may arise and its accounting treatment.

2 Explain how the fair value is calculated for:
- tangible non-current assets
- inventories
- monetary assets.

3 Explain why only the net assets of the subsidiary and not those of the parent are adjusted to fair value at the date of acquisition for the purpose of consolidated accounts.

4 The 2013 Annual Report of Bayer AG states:

> Subsidiaries that do not have a material impact on the Group's net worth, financial position or earnings, either individually or in aggregate, are accounted for at cost of acquisition less any impairment losses.

Discuss what criteria might have applied in determining that a subsidiary does not have a material impact.

5 Parent plc acquired Son plc at the beginning of the year. At the end of the year there were intangible assets reported in the consolidated accounts for the value of a domain name and customer lists. These assets did not appear in either Parent or Son's statements of financial position.

Discuss why these assets only appeared in the consolidated accounts.

6 In each of the following cases you are required to give your opinion, with reasons, on whether or not there is a parent/subsidiary under IFRS 3. Suggest other information, if any, that might be helpful in making a decision.

(a) Tin acquired 15% of the equity voting shares and 90% of the non-voting preferred shares of Copper. Copper has no other category of shares. The directors of Tin are also the directors of Copper, there is a common head office with shared administration departments and the functions of Copper are mainly the provision of marketing and transport facilities for Tin. Another company, Iron, holds 55% of the equity voting shares of Copper but has never used its voting power to interfere with the decisions of the directors.

(b) Hat plc owns 60% of the voting equity shares in Glove plc and 25% of the voting equity shares in Shoe plc. Glove owns 30% of the voting equity shares in Shoe plc and has the right to appoint a majority of the directors.

(c) Morton plc has 30% of the voting equity shares of Berry plc and also has a verbal agreement with other shareholders, who own 40% of the shares, that those shareholders will vote according to the wishes of Morton.

(d) Bean plc acquired 30% of the shares of Pea plc several years ago with the intention of acquiring influence over the operating and financial policies of that company. Pea sells 80% of its output to Bean. While Bean has a veto over the operating and financial decisions of Pea's board of directors it has only used this veto on one occasion, four years ago, to prevent that company from supplying one of Bean's competitors.

EXERCISES

Questions 1–5

Required in each case:

Prepare the statements of financial position of Parent Ltd and the consolidated statement of financial position as at 1 January 20X7 after each transaction, using for each question the statements of financial position of Parent Ltd and Daughter Ltd as at 1 January 20X7 which were as follows:

	Parent Ltd	Daughter Ltd
	£	£
Ordinary shares of £1 each	40,500	9,000
Retained earnings	4,500	1,800
	45,000	10,800
Cash	20,000	2,000
Other net assets	**25,000**	**8,800**
	45,000	10,800

* Question 1

(a) Assume that on 1 January 20X7 Parent Ltd acquired all the ordinary shares in Daughter Ltd for £10,800 cash. The fair value of the net assets in Daughter Ltd was their book value.

(b) The purchase consideration was satisfied by the issue of 5,400 new ordinary shares in Parent Ltd. The fair value of a £1 ordinary share in Parent Ltd was £2. The fair value of the net assets in Daughter Ltd was their book value.

Required: see above.

* Question 2

(a) On 1 January 20X7 Parent Ltd acquired all the ordinary shares in Daughter Ltd for £16,200 cash. The fair value of the net assets in Daughter Ltd was their book value.

(b) The purchase consideration was satisfied by the issue of 5,400 new ordinary shares in Parent Ltd. The fair value of a £1 ordinary share in Parent Ltd was £3. The fair value of the net assets in Daughter Ltd was their book value.

Required: see above.

* Question 3

(a) On 1 January 20X7 Parent Ltd acquired all the ordinary shares in Daughter Ltd for £16,200 cash. The fair value of the net assets in Daughter Ltd was £12,000.

(b) The purchase consideration was satisfied by the issue of 5,400 new ordinary shares in Parent Ltd. The fair value of a £1 ordinary share in Parent Ltd was £3. The fair value of the net assets in Daughter Ltd was £12,000.

Required: see above.

* Question 4

On 1 January 20X7 Parent Ltd acquired all the ordinary shares in Daughter Ltd for £6,000 cash. The fair value of the net assets in Daughter Ltd was their book value.

Required: see above.

Question 5

On 1 January 20X7 Parent Ltd acquired 75% of the ordinary shares in Daughter Ltd for £9,000 cash. The fair value of the net assets in Daughter Ltd was their book value. Assume in each case that the non-controlling interest is measured using Method 1.

Required: see above.

* Question 6

Rouge plc acquired 100% of the common shares of Noir plc on 1 January 20X0 and gained control. At that date the statements of financial position of the two companies were as follows:

	Rouge € million	Noir € million
ASSETS		
Non-current assets		
Property, plant and equipment	100	60
Investment in Noir	132	
Current assets	**80**	**70**
Total assets	312	130
EQUITY AND LIABILITIES		
Ordinary €1 shares	200	60
Retained earnings	**52**	**40**
	252	100
Current liabilities	**60**	**30**
Total equity and liabilities	312	130

Note: The fair values are the same as the book values.

Required:
Prepare a consolidated statement of financial position for Rouge plc as at 1 January 20X0.

Question 7

Ham plc acquired 100% of the common shares of Burg plc on 1 January 20X0 and gained control. At that date the statements of financial position of the two companies were as follows:

	Ham €000	Burg €000
ASSETS		
Non-current assets		
Property, plant and equipment	250	100
Investment in Burg	90	
Current assets	100	70
Total assets	440	170
EQUITY AND LIABILITIES		
Capital and reserves		
€1 shares	200	100
Retained earnings	**160**	**10**
	360	110
Current liabilities	80	60
Total equity and liabilities	440	170

Notes:
1 The fair value is the same as the book value.
2 €15,000 of the negative goodwill (badwill) arises because the net assets have been acquired at below their fair value and the remainder covers expected losses of €3,000 in the year ended 31/12/20X0 and €2,000 in the following year.

Required:

(a) Prepare a consolidated statement of financial position for Ham plc as at 1 January 20X0.

(b) Explain how the negative goodwill (badwill) will be treated.

Question 8

Set out below is the summarised statement of financial position of Berlin plc at 1 January 20X0.

	£000
ASSETS	
Non-current assets	
Property, plant and equipment	250
Current assets	150
Total assets	400
EQUITY AND LIABILITIES	
Capital and reserves	
Share capital (£5 shares)	200
Retained earnings	80
	280
Current liabilities	120
Total equity and liabilities	400

On 1/1/20X0 Berlin acquired 100% of the shares of Hanover for £100,000 and gained control.

Required:

Prepare the statement of financial position of Berlin immediately after the acquisition if:

(a) Berlin acquired the shares for cash.

(b) Berlin issued 10,000 shares of £5 (market value £10).

Question 9

Bleu plc acquired 80% of the shares of Verte plc on 1 January 20X0 and gained control. At that date the statements of financial position of the two companies were as follows:

	Bleu £m	Verte £m
ASSETS		
Non-current assets		
Property, plant and equipment	150	120
Investment in Verte	210	
Current assets	108	105
Total assets	468	225
EQUITY AND LIABILITIES		
Capital and reserves		
Share capital (£1 shares)	300	120
Retained earnings	78	60
	378	180
Current liabilities	90	45
Total equity and liabilities	468	225

Note: The fair values are the same as the book values.

Required:
Prepare a consolidated statement of financial position for Bleu plc as at 1 January 20X0. Non-controlling interests are measured using Method 1.

Question 10

Base plc acquired 60% of the common shares of Ball plc on 1 January 20X0 and gained control. At that date the statements of financial position of the two companies were as follows:

	Base £000	Ball £000
ASSETS		
Non-current assets		
Property, plant and equipment	250	100
Investment in Ball	90	
Current assets	**100**	**70**
Total assets	440	170
EQUITY AND LIABILITIES		
Capital and reserves		
Share capital	200	80
Share premium		20
Retained earnings	160	10
	360	110
Current liabilities	80	60
Total equity and liabilities	440	170

Note: The fair value of the property, plant and equipment in Ball at 1/1/20X0 was £120,000. The fair value of the non-controlling interest in Ball at 1/1/20X0 was £55,000. The 'fair value method' should be used to measure the non-controlling interest.

Required:
Prepare a consolidated statement of financial position for Base as at 1 January 20X0.

Question 11

Applying the principles of control in IFRS 10 *Consolidated Financial Statements*, as described in Section 22.3.2 of this chapter, you are required to consider whether certain investments of Austin plc are subsidiaries.

Austin plc has investments in a number of companies, and the company's accountant has asked your advice on whether certain of these companies should be treated as subsidiaries under IFRS 10 *Consolidated Financial Statements*.

(a) Austin plc owns 45% of the voting shares of Bond Ltd.

(b) Austin plc owns 60% of the voting shares of Bradford Ltd and Bradford Ltd owns 30% of the voting shares of Derby Ltd. Recently, Austin plc purchased 70% of the voting shares of Coventry Ltd. Coventry Ltd owns 30% of the voting shares of Derby Ltd. The accountant believes Derby Ltd is not a subsidiary of Austin, as Austin effectively owns only 39% of the shares of Derby – 60% × 30% = 18% through Bradford and 70% × 30% = 21% through Coventry.

(c) Recently, Austin plc purchased 60% of the ordinary shares of Norwich plc.

Prior to the purchase, Norwich plc had in issue 6,000,000 'A' shares of £1 each. Each 'A' share carries a single vote. These shares were owned equally by each of the directors of Norwich plc. For the purchase, the directors of Norwich plc sold 2,000,000 'A' shares to Austin plc, and Norwich plc issued 4,000,000 'B' shares of £1 each to Austin plc. 'B' shares do not carry a vote.

Required:
Consider and, where appropriate, discuss whether the following companies are subsidiaries of Austin plc:
(a) Bond Ltd
(b) Derby Ltd
(c) Norwich plc.

References

1 IFRS 10 *Consolidated Financial Statements*, IASB, 2011, B 92–93.
2 Ibid., para. 7.
3 IFRS 3 *Business Combinations*, 2008, B 44.
4 IAS 36 *Impairment of Assets*, IASB, revised 2004, para. 34.
5 IFRS 3 *Business Combinations*, 2008, B 34.
6 IFRS 10 *Consolidated Financial Statements*, IASB, 2011.
7 Ibid., paras 5, 6 and 8.
8 IFRS 13 *Fair Value Measurement*, IASB, 2011, Appendix A.
9 IFRS 10 *Consolidated Financial Settlements*, IASB, 2011, B 92–93.

Step 1. Goodwill calculated as at 1 January 20X1

	£	£
Goodwill on Bend's 80% shareholding in Stretch		
The cost of the parent company's investment in Stretch		12,000
Less:		
(a) Bend's share of Stretch share capital:		
80% × share capital of Stretch (80% × 10,000)	8,000	
(b) Pre-acquisition profit		
Bend's share of Stretch's retained earnings:		
80% × retained earnings as at 1 January 20X1 (80% × 4,000)	3,200	
(c) Fair value adjustment		
Bend's share of any change in the book values:		
80% × revaluation of fixed assets at 1 January 20X1 (80% × 600)	480	
		11,680
Goodwill attributable to the parent company shareholders		320
Goodwill on non-controlling interest's 20% shareholding in Stretch		
Fair value of non-controlling interest at date of acquisition		2,950
20% of net assets at date of acquisition (10,000 + 4,000 + 600)		(2,920)
Goodwill attributable to the non-controlling interest		30
Total goodwill of parent and non-controlling interest (£320 + £30)		£350

Step 2. Non-controlling interest in the net assets of subsidiary calculated as at 31.12.20X1

	£
(a) Subsidiary share capital	
Non-controlling interest in the share capital of Stretch (20% × 10,000)	2,000
(b) Total retained earnings as at 31.12.20X1	
Non-controlling interest in retained earnings of Stretch (20% × 6,000)	1,200
(c) Fair value adjustment of subsidiary's non-current assets	
Non-controlling interest in fair value increase (20% × 600)	120
Non-controlling interest in the net assets of Stretch as at 31.12.20X1	3,320
Non-controlling interest in goodwill	30
Reported in the statement of financial position as at 31 December 20X1	3,350

Step 3. Add together the assets and liabilities of the parent and subsidiary for the group

	Parent		*Subsidiary*	*Group*
	£		£	£
Non-current tangible assets	26,000	+	(12,000 + revaluation 600)	38,600
Goodwill as calculated in Step 1				350
Net current assets	13,000	+	4,000	17,000
Total				55,950

Step 4. Calculate the consolidated share capital and reserves for the group accounts

			£	£
Share capital	Parent only			16,000
Retained earnings	Parent		35,000	
	Bend's share of post-acquisition retained earnings	(80% of (6,000 − 4,000))	1,600	36,600
Total				52,600

Notes:

1 The separation of the retained earnings into pre- and post-acquisition is only of relevance to the parent with the pre-acquisition (£4,000) used when calculating the goodwill and the post-acquisition (£2,000) reported as part of the group earnings.

2 The non-controlling shareholders are entitled to their percentage share of the closing net assets. The pre-acquisition and post-acquisition division is irrelevant to the non-controlling interests – they are entitled to their percentage share of the **total** retained earnings at the date the consolidated statement of financial position is prepared.

23.5 Inter-company transactions

In the Bend example we assumed that there had been no inter-company transactions. In most groups, however, there are inter-company transactions that take place. On consolidation IFRS 10 requires[4] all inter-company transactions to be eliminated. So, if goods have been sold by Many plc, the parent, for £1,500 to Few plc, a subsidiary, the sales that had been reported in Many's statement of income and the cost of sales reported in Few's statement of income would both be eliminated. This is accomplished by a consolidation journal entry:

	Dr	Cr
Sales	1,500	
Cost of sales		1,500
Eliminating intra-group sales		

Note that no entries are made in the individual company's accounts and the elimination is simply to ensure that the consolidated sales and cost of sales only include transactions with non-group parties.

23.5.1 Adjustment when inter-company sales include a profit loading

Where sales have been made between two companies within the group, it is only necessary to provide for an unrealised profit from intra-group sales to the extent that the goods are still in the inventories of the group at the date of the statement of financial position.

We will illustrate the accounting treatment where there is unrealised profit with the Many Group example.

Let us assume that Many plc has bought £1,000 worth of goods for resale and sold them to Few plc for £1,500, making a profit of £500 in Many's own accounts.

We have already seen that one of the consolidated journal entries would be to debit sales £1,500 and credit cost of sales £1,500, whether or not the goods had been sold on to a third party.

If at the year-end Few plc still has these goods in inventory, the group has not yet made a sale to a third party and the £500 profit is therefore 'unrealised'. It must be removed from the consolidated statement of financial position by:

● reducing the retained earnings of Many by £500; and

● reducing the inventories of Few by £500.

The £500 is called a 'provision for unrealised profit'.

If the sale is made by a subsidiary to the parent and there are non-controlling interests, these will be debited with their 'proportion of the unrealised profit'.

23.5.2 Eliminating inter-company current account balances

If the sale were made for cash, then the cash in the seller would have increased and the cash in the buyer would have decreased. On consolidation no adjustment is, therefore, required. However, if the invoice has not been settled, there would be an account receivable in the seller's and an account payable in the buyer's statement of financial position. These must be eliminated by cancelling each.

In our Many Group example, the £1,500 would be cancelled.

Reconciling inter-company balances

In practice, temporary differences may arise for such items as cash or inventory in transit that are recorded in one company's books but of which the other company is not yet aware. If so, this is reconciled on consolidation for cash in transit, debit Cash and credit Accounts receivable and for inventory in transit, debit Inventory and credit Accounts payable.

If Few had sent cash of £400 to Many in part settlement of the £1,500 owing but it had not been recorded in the books of Many at the period-end date, the Many accounts would show £1,500 owing, whereas the Few accounts would show £1,100 owing and the cash balance reduced by the £500. On consolidation the Accounts receivable would be reduced to £1,100 and the cash increased by £400. The Account receivable and payable balances are both £1,100 and would be cancelled.

23.5.3 Inter-company dividends payable/receivable

If the subsidiary company has declared a dividend before the year-end, it will appear in the current liabilities of the subsidiary company and in the current assets of the parent company. It needs to be cancelled by set-off.

If the subsidiary is wholly owned by the parent the whole amount will be cancelled. If, however, there is a non-controlling interest in the subsidiary, the non-cancelled amount of the dividend payable in the subsidiary's statement of financial position will be the amount payable to the non-controlling interest and will be reported as part of the non-controlling interest in the consolidated statement of financial position.

Where a final dividend has not been declared by the year-end date there is no liability under IAS 10 *Events after the Reporting Period Date* and no liability will be reported.

Companies include a reference to these adjustments in their accounting policies, as seen in the following extract from the 2013 Annual Report of Munksjo AB:

> **Transaction eliminated on consolidation**
> Intra-Group receivables and liabilities, income or expenses and unrealised gains or losses arising from intra-Group transactions between Group companies are eliminated in full when preparing the consolidated accounts.

The Prose Group example that follows incorporates the main points dealt with so far on the preparation of a consolidated statement of financial position.

23.6 The Prose Group – assuming there have been inter-group transactions

On 1 January 20X1 Prose plc acquired 80% of the equity shares in Verse plc for £21,100 to gain control and 10% of the 5% loans for £900. The retained earnings as at 1 January 20X1 were £4,000. The fair value of the land in Verse was £1,000 above book value.

During the year Prose sold some of its inventory to Verse for £3,000, which represented cost plus a markup of 25%. Half of these goods are still in the inventory of Verse at 31/12/20X1.

The consolidated statement of financial position as at 31 December 20X1 is shown below with supporting notes. Note that depreciation is not charged on land and Method 1 is used to compute the non-controlling interest.

	Prose £	Verse £	Adjustments Dr	Group Cr	£	
ASSETS						
Non-current assets	25,920	33,400	1,000		60,320	
Investment in Verse/goodwill	22,000	—		21,500	500	Step 1
Current assets						
Inventories	9,600	4,000		300	13,300	Step 3
Verse current account	8,000			8,000	—	Step 2
Loan interest receivable	35			35	—	Step 2
Other current assets	3,965	13,350			17,315	
Total assets	69,520	50,750			91,435	
EQUITY and LIABILITIES						
Equity share capital	24,000	21,000	21,000		24,000	
Retained earnings	30,000	8,500	5,200		33,300	Step 5
Non-controlling interest	—	—		6,100	6,100	Step 4
Non-current liabilities						
5% loan 2017/18	5,000	7,000	700		11,300	Step 2
Current liabilities						
Prose current account		8,000	8,000		—	
Loan interest payable		350	35		315	
Other current liabilities	10,520	5,900			16,420	
	69,520	50,750	35,935	35,935	91,435	

Step 1. Calculation of goodwill (note that this calculation will be the same as when calculated at the date of acquisition)

		£	£
Cost of investment in shares and loan			22,000
Less:			
1 80% × equity shares of Verse	(80% × 21,000)	16,800	
2 80% × retained earnings balance at 1.1.20X1	(80% × 4,000)	3,200	
3 80% × fair value increase at 1.1.20X1	(80% × 1,000)	800	
4 10% × loans of Verse	(10% × 7,000)	700	21,500
5 Goodwill in statement of financial position			500

Step 2. Inter-company elimination by set-off of inter-company balances

1 The current accounts of £8,000 between the two companies are cancelled. Note that the accounts are equal, which indicates that there are no items such as goods in transit or cash in transit which would have required a reconciliation.

2 The loan interest receivable by Prose is cancelled with £35 (10% of £350) of the loan interest payable by Verse, leaving £315 (90% of £350) payable to outsiders. This is not part of the non-controlling interest as loan holders have no ownership rights in the company.

3 The loan of £700 in Prose's accounts is set off against the £7,000 in Verse's accounts, leaving 6,300 owing to non-group members.

Step 3. Unrealised profit in inventory

Markup on the inter-company sales (£3,000 × 20%)	£600
Half the goods are still in inventories at the year-end.	
Unrealised profit	£300

Step 4. Calculation of non-controlling interest as at 31/12/20X1

Note that the non-controlling interest is calculated as at the year-end while goodwill is calculated at the date of acquisition.

		£
Non-controlling interest in the equity shares of Verse	(20% × 21,000)	4,200
Non-controlling interest in the retained earnings of Verse	(20% × 8,500)	1,700
Non-controlling interest in the fair value increase	(20% × 1,000)	200
Statement of financial position figure		6,100

Step 5. Calculation of consolidated share capital and reserves for the group accounts

	£	£
Share capital:		
Equity share capital (parent company's only)		24,000
Retained earnings (parent company's)	30,000	
Less: Provision for unrealised profit	(300)	29,700
Parent's share of the post-acquisition profit of the subsidiary		
(80% × 8,500)	6,800	
Less: 80% of pre-acquisition profits (80% × 4,000)	(3,200)	3,600
Retained earnings in the consolidated statement of financial position		33,300

Summary

When consolidated accounts are prepared after the subsidiary has traded with other members of the group, the goodwill calculation remains as at the date of the acquisition but all inter-company transactions and unrealised profits arising from inter-company transactions must be eliminated.

REVIEW QUESTIONS

1 An accounting policy states that all inter-company transactions, receivables, liabilities and unrealised profits, as well as intra-group profit distributions, are eliminated.

   (a) Discuss three examples of inter-company (also referred to as intra-group) transactions.

   (b) Explain what is meant by 'are eliminated'.

   (c) Explain what effect there could be on the reported group profit if inter-company transactions were not eliminated.

2 Explain why the non-controlling interest is not affected by the pre- and post-acquisition division.

3 Explain why pre-acquisition profits of a subsidiary are treated differently from post-acquisition profits when consolidating.

4 Explain the effect of a provision for unrealised profit on a non-controlling interest:

   (a) where the sale was made by the parent to the subsidiary; and

   (b) where the sale was made by the subsidiary to the parent.

5 A consolidated journal adjustment set off the dividend receivable reported in the parent's statement of financial position against the dividend declared by the subsidiary. Explain why this may not fully eliminate the dividend that is reported in the group statement of financial position.

6 Explain reasons why the current accounts in the parent and subsidiary may not agree. If not, how could the two accounts be set off?

EXERCISES

* Question 1

Sweden acquired 100% of the equity shares of Oslo on 1 March 20X1 and gained control. At that date the balances on the reserves of Oslo were as follows:

Revaluation reserve	Kr10 million
Retained earnings	Kr70 million

The statements of financial position of the two companies at 31/12/20X1 were as follows:

	Sweden Krm	Oslo Krm
ASSETS		
Non-current assets		
Property, plant and equipment	264	120
Investment in Oslo	200	
Current assets	160	140
Total assets	624	260
EQUITY AND LIABILITIES		
Kr10 shares	400	110
Retained earnings	104	80
Revaluation reserve	20	10
	524	200
Current liabilities	100	60
Total equity and liabilities	624	260

Notes:
1 The fair values were the same as the book values on 1/3/20X1.
2 There have been no movements on share capital since 1/3/20X1.
3 20% of the goodwill is to be written off as an impairment loss.
4 Method 1 is to be used to compute the non-controlling interest.

Required:
Prepare a consolidated statement of financial position for Sweden as at 31 December 20X1.

* Question 2

Summer plc acquired 60% of the equity shares of Winter Ltd on 30 September 20X1 and gained control. At the date of acquisition, the balance of retained earnings of Winter was €35,000.

At 31 December 20X1 the statements of financial position of the two companies were as follows:

	Summer €000	Winter €000
ASSETS		
Non-current assets		
Property, plant and equipment	200	200
Investment in Winter	141	
Current assets	100	140
Total assets	441	340
EQUITY AND LIABILITIES		
Equity shares	200	180
Retained earnings	161	40
	361	220
Current liabilities	80	120
Total equity and liabilities	441	340

Notes:

1 The fair value of the non-controlling interest at the date of acquisition was £92,000. The non-controlling interest is to be measured using Method 2. The fair values of the identifiable net assets of Winter at the date of acquisition were the same as their book values.

2 There have been no movements on share capital since 30/9/20X1.

Required:

Prepare a consolidated statement of financial position for Summer plc as at 31 December 20X1.

Question 3

On 30 September 20X0 Gold plc acquired 75% of the equity shares, 30% of the preferred shares and 20% of the bonds in Silver plc and gained control. The balance of retained earnings on 30 September 20X0 was £16,000. The fair value of the land owned by Silver was £3,000 above book value. No adjustment has so far been made for this revaluation.

The statements of financial position of Gold and Silver at 31 December 20X1 were as follows:

	Gold £	Silver £
ASSETS		
Property, plant and equipment (including land)	82,300	108,550
Investment in Silver	46,000	—
Current assets:		
Inventory	23,200	10,000
Silver current account	20,000	
Bond interest receivable	175	
Other current assets	5,000	7,500
Total assets	176,675	126,050
EQUITY AND LIABILITIES		
Equity share capital	60,000	27,600
Preferred shares	10,000	20,000
Retained earnings	75,000	21,200
	145,000	68,800
Non-current liabilities – bonds	12,500	17,500
Current liabilities		
Gold current account		20,000
Bond interest payable	625	875
Other current liabilities	18,550	18,875
Total equity and liabilities	176,675	126,050

Notes:

1 The recoverable amount for purposes of calculating the impairment of goodwill is £50,040.
2 During the year Gold sold some of its inventory to Silver for £3,000, which represented cost plus a markup of 25%. Half of these goods are still in the inventory of Silver at 31.12.20X1.
3 There is no depreciation of land.
4 There has been no movement on share capital since the acquisition.
5 Method 1 is to be used to compute the non-controlling interest.

Required:
Prepare a consolidated statement of financial position as at 31 December 20X1.

Question 4

Prop and Flap have produced the following statements of financial position as at 31 October 2008:

	Prop		Flap	
	$m	$m	$m	$m
ASSETS				
Non-current assets				
Plant and equipment		2,100		480
Investments		800		
Current assets				
Inventories	880		280	
Receivables	580		420	
Cash and cash equivalents	400		8	
		1,860		708
Total assets		4,760		1,188
EQUITY and LIABILITIES				
Equity share capital		2,400		680
Retained earnings		860		200
		3,260		880
Non-current liabilities				
Long-term borrowing		400		
Current liabilities				
Payables	1,100		228	
Bank overdraft	—		80	
		1,100		308
Total equity and liabilities		4,760		1,188

The following information is relevant to the preparation of the financial statements of the Prop Group:

1 Prop acquired 80% of the issued ordinary share capital of Flap many years ago when the retained earnings of Flap were $72 million. Consideration transferred was $800 million. Flap has performed well since acqusition and so far there has been no impairment to goodwill.

2 At the date of acquisition the plant and equipment of Flap was revalued upwards by $40 million, although this revaluation was not recorded in the accounts of Flap. Depreciation would have been $32 million greater had it been based on the revalued figure.

3 Flap buys goods from Prop upon which Prop earns a margin of 20%. At 31 October 2008 Flap's inventories include $180 million goods purchased from Prop.

4 At 31 October 2008 Prop has receivables of $140 million owed by Flap and payables of $60 million owed to Flap.

5 The market price of the non-controlling interest shares just before Flap's acquisition by Prop was $1.30. It is the group's policy to value the non-controlling interest at fair value.

Required:
Prepare the Prop Group consolidated statement of financial position as at 31 October 2008.

(Association of International Accountants)

Question 5

On 1 January 20X0 Hill plc purchased 70% of the ordinary shares of Valley plc for £1.3 million. The fair value of the non-controlling interest at that date was £0.5 million. At the date of acquisition, Valley's retained earnings were £0.4 million.

The statements of financial position of Hill and Valley at 31 December 20X0 were:

	Hill	Valley
Capital and reserves	*£000*	*£000*
Share capital	5,000	1,000
Retained earnings	3,500	200
	8,500	1,200
Net assets	8,500	1,200

Because of Valley's loss in 20X0, the directors of Hill decided to write down the value of goodwill by £0.3 million. The directors of Hill propose to use Method 2 to calculate goodwill in the consolidated statement of financial position. The goodwill is to be written down in proportion to the respective holdings of Valley's shares by Hill and the non-controlling interest.

Required:
(a) Calculate the goodwill of Valley relating to Hill plc and the non-controlling interest.
(b) Show how the goodwill will be written down at 31 December 20X0, for both Hill plc and the non-controlling interest.
(c) Comment on your answer to part (b).

Question 6

The following accounts are the consolidated statement of financial position and parent company statement of financial position for Alpha Ltd as at 30 June 20X2.

	Consolidated statement of financial position		Parent company statement of financial position	
	£	£	£	£
Ordinary shares		140,000		140,000
Capital reserve		92,400		92,400
Retained earnings		79,884		35,280
Non-controlling interest		12,329		—
		324,613		267,680
Non-current assets				
Property		127,400		84,000
Plant and equipment		62,720		50,400
Goodwill		85,680		
Investment in subsidiary (50,400 shares)				151,200
Current assets				
Inventory	121,604		71,120	
Trade receivables	70,429		51,800	
Cash at bank	24,360		—	
	216,393		122,920	
Current liabilities				
Trade payables	140,420		80,920	
Income tax	27,160		20,720	
Bank overdraft	—		39,200	
	167,580		140,840	
Working capital		48,813		(17,920)
		324,613		267,680

Notes:

1 There was only one subsidiary, called Beta Ltd.
2 There were no capital reserves in the subsidiary.
3 Alpha produced inventory for sale to the subsidiary at a cost of £3,360 in May 20X2. The inventory was invoiced to the subsidiary at £4,200 and was still on hand at the subsidiary's warehouse on 30 June 20X2. The invoice had not been settled at 30 June 20X2.
4 The retained earnings of the subsidiary had a credit balance of £16,800 at the date of acquisition. No fair value adjustments were necessary.
5 There was a right of set-off between overdrafts and bank balances.
6 The parent owns 90% of the subsidiary.

Required:
Prepare the statement of financial position as at 30 June 20X2 of the subsidiary company from the information given above. The non-controlling interest is measured using Method 1.

References

1 IFRS 10 *Consolidated Financial Statements*, IASB, 2011, para. 19.
2 Ibid., B 92–93.
3 IAS 16 *Property, Plant and Equipment*, IASB, revised 2003, para. 31.
4 IFRS 10 *Consolidated Financial Statements*, IASB, 2011, B 86.

CHAPTER 24

Preparation of consolidated statements of income, changes in equity and cash flows

24.1 Introduction

The main purpose of this chapter is to explain how to prepare a consolidated statement of income.

Objectives

By the end of this chapter, you should be able to:

- eliminate inter-company transactions;
- prepare a consolidated statement of income;
- attribute income to the non-controlling shareholders;
- prepare a consolidated statement of changes in equity;
- prepare a consolidated statement of income when a subsidiary is acquired partway through a year;
- prepare a consolidated statement of cash flows.

24.2 Eliminate inter-company transactions

Many business combinations occur because the acquirer seeks closer links with the acquired company. There are many examples of this, such as a clothing manufacturer in Europe acquiring a denim supplier in Hong Kong with inter-company purchases and sales following the acquisition.

Inter-company sales

When the consolidated statement of income is prepared the inter-company sales are eliminated. This avoids the possibility that the group could inflate its revenue merely by group companies selling to each other. The sales and purchases both need to be reduced by the invoiced amount of the inter-company sales. This is achieved in the consolidation process by reducing the aggregate sales and aggregate cost of sales figures.

Unrealised profit

In the previous chapter we treated any unrealised profit by reducing the inventory figure and reducing the retained earnings figure. The retained earnings figure would have incorporated the retained earnings balance from the statement of income, i.e. the adjustment for the unrealised profit would have already been reported in the statement of income.

In the consolidation process the unrealised profit is added to the cost of sales to achieve the reduction in group gross profit.

Dividends and interest

Having set off the sales and cost of sales and adjusted for any unrealised profit, further adjustments may be required[1] to establish the profit before tax earned by the group as a whole. This requires us to eliminate any dividends (and interest if any) that have been credited in the parent's statement of income for amounts paid or payable to the parent by the subsidiaries.

If this were not done, there would be double-counting because we would be including in the consolidated statement of income the subsidiary's profit from operations and again as dividends and interest received/receivable by the parent.

Group profits before tax

We can see, therefore, that group profit before tax is arrived at after setting-off inter-company sales against the cost of sales, adding the unrealised profit to the cost of sales figure, and eliminating any dividends or interest received or receivable from a subsidiary.

We will illustrate this in the following Ante Group example.

24.3 Preparation of a consolidated statement of income – the Ante Group

The following information is available:

At the date of acquisition on 1 January 20X3:
Ante plc acquired 75% of the ordinary shares in Post plc. (*This shows that Ante had control.*) At that date the retained earnings of Post were £30,000. (*These are pre-acquisition profits and should not be included in the group profit for the year.*)

At the end of 20X3:
The retained earnings of Ante were £69,336 and the retained earnings of Post were £54,000.

During the year ended 31 December 20X4:
Ante had sold Post goods at their cost price of £9,000 plus a markup of one-third. These were the only inter-company sales. (*This indicates that the group sales and cost of sales require reducing.*)

At the end of the financial year on 31 December 20X4:
Half of these goods were still in the inventory at the end of the year. (*There is unrealised profit to be removed from the group gross profit by adding the unrealised amount to the cost of sales figure.*)

Dividends paid in 20X4 by group companies were as follows:

	Ante	Post
On ordinary shares	£40,000	£5,000

Set out below are the individual statements of income of Ante and Post together with the consolidated statement of income for the year ended 31 December 20X4 with explanatory notes.

Statements of comprehensive income for the year ended 31 December 20X4

	Ante £	Post £	Consolidated £	
Sales	200,000	120,000	308,000	Note 1
Cost of sales	60,000	60,000	109,500	Notes 1 and 2
Gross profit	140,000	60,000	198,500	
Expenses	59,082	40,000	99,082	Note 3
Profit from operations	80,918	20,000	99,418	
Dividends received – ordinary shares	3,750	—	—	
Profit before tax	84,668	20,000	99,418	Note 4
Income tax expense	14,004	6,000	20,004	Note 5
Profit for the period	70,664	14,000	79,414	
Attributable to:				
Ordinary shareholders of Ante (balance)			75,914	
Non-controlling shareholders in Post			3,500	Note 6
			79,414	

Notes:

1 Eliminate inter-company sales on consolidation

Cancel the inter-company sales of £12,000 (£9,000 × 1⅓) by

(i) reducing the sales of Ante from £200,000 to £188,000; and

(ii) reducing the cost of sales of Post by the same amount from £60,000 to £48,000.

(Remember that the same amount is deducted from both sales and cost of sales – a sale to one party is the amount of the purchase by the other party.)

(iii) Group sales are £188,000 + £120,000 = £308,000.

(iv) Group cost of sales (before any adjustment for unrealised profit) is £60,000 + £48,000 = £108,000.

2 Eliminate unrealised profit on inter-company goods still in closing inventory

(i) Ante had sold the goods to Post at a markup of £3,000.

(ii) Half of the goods remain in the inventory of Post at the year-end.

(iii) From the group's view there is an unrealised profit of half of the markup, i.e. £1,500. Therefore:

- deduct £1,500 from the gross profit of Ante by adding this amount to the cost of sales;

- reduce the inventories in the consolidated statement of financial position by the amount of the provision (as explained in the previous chapter).

(iv) Cost of sales has been increased from £108,000 to £109,500.

3 Aggregate expenses

In this example we do not have any inter-company transactions such as Head Office management fees that need to be set off. No adjustment is, therefore, required to the parent or subsidiary total figures.

4 Profit before tax, accounting for the inter-company dividends
The ordinary dividend of £3,750 received by Ante is an inter-company item that does not appear in the group profit before tax.

5 Aggregate the taxation figures
No adjustment is required to the parent or subsidiary total figures.

6 Allocation of profit to equity holders and non-controlling interest
Adjustment is required[2] to establish how much of the profit after tax is attributable to equity holders of the parent. The amount is that remaining after deducting the non-controlling interest's percentage of the subsidiary's after-tax figure, i.e. 25% of £14,000 = £3,500.

24.4 The statement of changes in equity (SOCE)[3]

In practice the opening figures for the SOCE would be available from the 20X3 group accounts. It is not uncommon in an examination context to require you to calculate the opening figure for the group SOCE. The calculation is as follows:

Opening balance for the Ante group

	£
Ante's retained earnings at the start of the year	69,336
Group share of Post's post-acquisition earnings (75% × (54,000 − 30,000))	18,000
	87,336

Opening balance for the non-controlling interest

Total retained earnings as at 31.12.20X3	25% of 54,000	13,500

We can then complete the group SOCE as follows:

	Ante	Non-controlling interest	Total
	£	£	£
Opening balance	87,336	13,500	100,836
Income for the period	75,914	3,500	79,414
Dividends paid	(40,000)	(1,250)	(41,250)
Closing balance	123,250	15,750	139,000

Dividends paid

In the Ante column the dividends paid are those of the parent only. The parent company's share of Post's dividend cancels out with the parent company's investment income. The non-controlling share is £5,000 minus the £3,750 paid to the parent. This is the amount dealt with in their column.

24.5 Other consolidation adjustments

In the above example we dealt with adjustments for intra-group sale of goods, unrealised profit on inventories and dividends received from a subsidiary. There are other adjustments that often appear in examinations relating to depreciation and dividends paid by a subsidiary out of pre-acquisition profits.

24.5.1 Depreciation adjustment when fair value is higher than book value

If the fair value of depreciable non-current assets is different from their book value, it is necessary to adjust the depreciation that has been charged in the subsidiary's books.

For example, assume that the parent acquired a non-current asset from a subsidiary which had a book value of £100,000 that was being depreciated by the subsidiary on a straight-line basis over five years and the scrap value was nil. The annual charge in the subsidiary's statement of income would be £20,000.

If the fair value on acquisition was £150,000, the charge in the consolidated statement of income should be based on the £150,000, i.e. £30,000 (£150,000/5) with the depreciation increased by £10,000. If there is no information as to the type of non-current asset, the £10,000 would be added to the cost of sales figure. If the type of asset is identified, for example as delivery vehicles, then the adjustment would be made to the appropriate expense, e.g. distribution costs.

24.5.2 Depreciation adjustment when transfer has been at cost plus a profit loading

Let us consider Digdeep plc, a civil engineering company that has a subsidiary, Heavylift plc, that manufactures digging equipment. Assume that at the beginning of the financial year Heavylift sold equipment costing £80,000 to Digdeep for £100,000. It is Digdeep's depreciation policy to depreciate at 5% using the straight-line method with nil scrap value.

On consolidation, the following adjustments are required:

(i) Revenue is reduced by £20,000 and the asset is reduced by £20,000 to bring the asset back to its cost of £80,000.

Dr: Revenue	£20,000	
Cr: Asset		£20,000

(ii) Revenue is then reduced by £80,000 and cost of sales reduced by £80,000 to eliminate the inter-company sale.

Dr: Revenue	£80,000	
Cr: Cost of sales		£80,000

(iii) Depreciation needs to be based on the cost of £80,000. The depreciation charge was £5,000 (5% of £100,000); it should be £4,000 (5% of £80,000) so the adjustment is:

Dr: Accumulated depreciation	£1,000	
Cr: Depreciation in the statement of income		£1,000

24.5.3 Dividends or interest paid by the subsidiary out of pre-acquisition profits

When a parent acquires the net assets of a subsidiary it is paying for all of the assets including the cash. If the subsidiary then pays part of this to the parent as a dividend it is in effect transferring an asset that the parent had already paid for. The dividend received by the parent is not, therefore, income but a return of part of the purchase price. It is credited by the parent to the investment in subsidiary account. This is illustrated in the Bow plc example below.

Illustration of a dividend paid out of pre-acquisition profits

Bow plc acquired 75% of the shares in Tie plc on 1 January 20X4 for £80,000 when the balance of the retained earnings of Tie was £40,000. On 10 January 20X4 Bow received a dividend of £3,000 from Tie out of the profits for the year ended 31/12/20X3. The draft summarised statements of income for the year ended 31/12/20X4 were as follows:

	Bow £	Tie £	Consolidated £
Gross profit	130,000	70,000	200,000
Expenses	50,000	40,000	90,000
Profit from operations	80,000	30,000	110,000
Dividends received from Tie (see note)	3,000	—	—
Profit before tax	83,000	30,000	110,000
Income tax expense	24,000	6,000	30,000
Profit for the period	59,000	24,000	80,000

Note:

The treatment is incorrect. The £3,000 dividend received from Tie is not income and must not therefore appear in Bow's statement of income. The correct treatment is to deduct it from the investment in Tie, which will then become £77,000 (80,000 – 3,000) with a debit to dividends received and a credit to the Investment in Tie.

24.5.4 Goodwill

We know that there is no amortisation charge for goodwill. However, if there has been any impairment then this would appear as an expense in the group column of the consolidated statement of income.

For example, if in our Ante example above you were informed that the goodwill on acquisition was £10,000 and that it had been impaired by £2,000, the consolidated statement of income would have an entry in the group column and appear as follows:

	Ante £	Post £	Consolidated £	
Sales	200,000	120,000	308,000	Note 1
Cost of sales	60,000	60,000	109,500	Notes 1/2
Gross profit	140,000	60,000	198,500	
Expenses	59,082	40,000	99,082	Note 3
Goodwill impairment			2,000	
Profit from operations	80,918	20,000	97,418	

24.6 A subsidiary acquired part-way through the year

It would be attractive for a company whose results had not been as good as expected to acquire a profitable subsidiary at the end of the year and take its current year's profit into the group accounts. However, this is window dressing and it is not permitted. The group can only bring in a subsidiary's profits from the date of the acquisition when it assumed control. The Tight plc example below illustrates the approach.

24.6.1 Illustration of a subsidiary acquired part-way through the year – Tight plc

The following information is available:

At the date of acquisition on 30 September 20X1
Tight acquired 75% of the shares and 20% of the 5% long-term loans in Loose. The book value and fair value were the same amount.

During the year
There have been no inter-company sales. If there had been then normal set-off would apply. All income and expenses are deemed to accrue evenly through the year and the dividend received may be apportioned to pre- and post-acquisition on a time basis.

At the end of the financial year
The Tight Group prepares its accounts as at 31 December each year.

Set out below are the individual statements of income of Tight and Loose together with the consolidated statement of income for the year ended 31 December 20X1.

	Tight	Loose	Time-apportion		Consolidated
	£	£		£	£
Revenue	200,000	120,000	3/12	30,000	230,000
Cost of sales	60,000	60,000	3/12	15,000	75,000
Gross profit	140,000	60,000	3/12	15,000	155,000
Expenses	59,082	30,000	3/12	7,500	66,582
Interest paid on 5% loans		10,000		2,500	2,000
Interest received on Loose loans	2,000		Set off		NIL
	82,918	20,000			86,418
Dividends received	3,600	NIL	Set off		NIL
Profit before tax	86,518	20,000			86,418
Income tax expense	14,004	6,000	3/12	1,500	15,504
Profit for the period after tax	72,514	14,000		3,500	70,914
Attributable to:					
Ordinary shareholders of Tight (balance)					70,039
Non-controlling shareholders in Loose					875
					70,914

Notes:

1 Time-apportion and aggregate the revenue, cost of sales, expenses and income tax
   Group items include a full year for the parent company and three months for the subsidiary (1 October to 31 December).

2 Account for inter-company interest
   Inter-company expense items need to be eliminated or cancelled by set-off against the interest paid by Loose. Interest is an expense which is normally deemed to accrue evenly over the year and is to be apportioned on a time basis.

(i)   It has been assumed that interest is paid annually in arrears. This means that the interest received by Tight has to be apportioned on a time basis: $^{9}/_{12} \times £2,000 = £1,500$ is treated as being pre-acquisition. It is therefore deducted from the cost of the investment in Loose.

(ii) The remainder ($£500$) is cancelled with $£500$ of the post-acquisition element of the interest paid by Loose. The interest paid figure in the consolidated financial statements will be the post-acquisition interest less the inter-company elimination, which represents the amount payable to the holders of 80% of the loan capital.

(iii) The interest of $£10,000$ paid by Loose to its loan creditors is time-apportioned with $£7,500$ being pre-acquisition. The post-acquisition amount of $£2,500$ includes $£500$ that was included in the $£2,000$ reported by Tight in its statement of income. This is cancelled, leaving $£2,000$ which was paid to the 80% non-group loan creditors.

3   Account for inter-company dividends

	£
Amount received by Tight =	3,600
The dividend received by Tight is apportioned on a time basis, and the pre-acquisition element is credited to the cost of investment in Tight, i.e. $^{9}/_{12} \times £3,600 =$	(2,700)
The post-acquisition element is cancelled	(900)
Amount credited to consolidated statement of income	NIL

4   Calculate the share of post-acquisition consolidated profits belonging to the non-controlling interest

As only the post-acquisition proportion of the subsidiary's profit after tax has been included in the consolidated statement of income, the amount deducted as the non-controlling interest in the profit after tax is also time-apportioned, i.e. 25% of $£3,500 = £875$.

24.7 Published format statement of income

The statement of comprehensive income follows the classification of expenses by function as illustrated in IAS 1:

	£
Revenue	230,000
Cost of sales	75,000
Gross profit	155,000
Distribution costs	42,562
Administrative expense	24,020
	66,582
	88,418
Finance cost	2,000
	86,418
Income tax expense	15,504
Profit for the period	70,914
Attributable to:	
Equity holders of the parent	70,039
Non-controlling interest	875
	70,914

24.8 Consolidated statements of cash flows

Statements of cash flows are explained in Chapter 5 for a single company. A consolidated statement of cash flows differs from that for a single company in two respects:

(a) there are additional items such as dividends paid to non-controlling interests; and

(b) adjustments may be required to the actual amounts to reflect the assets and liabilities brought in by the subsidiary which did not arise from cash movements.

24.8.1 Adjustments to changes between opening and closing statements of financial position

Adjustments are required if the closing statement of financial position items have been increased or reduced as a result of non-cash movements. Such movements occur if there has been a purchase of a subsidiary to reflect the fact that the assets and liabilities from the new subsidiary have not necessarily resulted from cash flows. The following illustrates such adjustments in relation to a subsidiary acquired at the end of the financial year where the net assets of the subsidiary were as follows:

Net assets acquired	£000	Effect in consolidated statement of cash flows
Working capital:		
Inventory	10	Reduce inventory increase
Trade payables	(12)	Reduce trade payables increase
Non-current assets:		
Vehicles	20	Reduce capital expenditure
Cash/bank:		
Cash	5	Reduce amount paid to acquire subsidiary in investing section
Net assets acquired	23	

Let us assume that the consideration for the acquisition was as follows:

Shares	10	Reduce share cash inflow
Share premium	10	Reduce share cash inflow
Cash	3	Payment to acquire subsidiary in investing section
	23	

The consolidated statement of cash flows can then be prepared using the indirect method.

Statement of cash flows using the indirect method

		£000	£000
Cash flows from operating activities			
Net profit before tax		500	
Adjustments for:			
Depreciation		102	
Operating profit before working capital changes		602	
Increase in inventories	(400)		
Less: **Inventory brought in on acquisition**	10	(390)	
Decrease in trade payables	(40)		
Add: **Trade payables brought in on acquisition**	(12)	(52)	
Cash generated from operations		160	
Income taxes paid (200 + 190 − 170)		(220)	
Net cash from operating activities			(60)
Cash flows from investing activities			
Purchase of property, plant and equipment	(563)		
Less: **Vehicles brought in on acquisition**	20	(543)	
Payment to acquire subsidiary		(3)	
Cash acquired with subsidiary		5	
Net cash used in investing activities			(541)
Cash flows from financing activities			
Proceeds from issuance of share capital	300		
Less: **Shares issued on acquisition not for cash**	(20)	280	
Dividends paid (from statement of income)		(120)	
Net cash from financing activities			160
Net decrease in cash and cash equivalents			(441)
Cash and cash equivalents at the beginning of the period			72
Cash and cash equivalents at the end of the period			(369)

Supplemental disclosure of acquisition

	£
Total purchase consideration	23,000
Portion of purchase consideration discharged by means of cash or cash equivalents	3,000
Amount of cash and cash equivalents in the subsidiary acquired	5,000

Summary

The retained earnings of the subsidiary brought forward are divided into pre-acquisition profits and post-acquisition profits – the group share of the former are used in the good-will calculation, and the share of the latter are brought into the consolidated shareholders' equity.

Revenue and cost of sales are adjusted in order to eliminate intra-group sales and unrealised profits.

Finance expenses and income are adjusted to eliminate inter-company payments of interest and dividends.

The non-controlling interest in the profit after tax of the subsidiary is deducted to arrive at the profit for the year attributable to the equity holders of the parent.

If a subsidiary is acquired during a financial year, the items in its statement of income require apportioning. In the illustration in the text we assumed that trading was evenly spread throughout the year – in practice you would need to consider any seasonal patterns that would make this assumption unrealistic, remembering that the important consider-ation is that the group accounts should only be credited with profits arising whilst the subsidiary was under the parent's control.

REVIEW QUESTIONS

1 Explain why the dividends deducted from the group in the statement of changes in equity are only those of the parent company.

2 Explain two ways in which unrealised profits might arise from transactions between companies in a group and why it is important to remove them.

3 Explain why it is necessary to apportion a subsidiary's profit or loss if acquired part-way through a financial year.

4 Explain why dividends paid by a subsidiary to a parent company are eliminated on consolidation.

5 Give four examples of inter-company income and expense transactions that will need to be eliminated on consolidation and explain why each is necessary.

6 A shareholder was concerned that following an acquisition the profit from operations of the parent and subsidiary were less than the aggregate of the individual profit from operations figures. She was concerned that the acquisition, which the directors had supported as improving earnings per share, appeared to have reduced the combined profits. She wanted to know where the profits had gone. Give an explanation to the shareholder.

7 Explain how a management charge made by a parent company would be dealt with on consolidation.

8 Explain how the impairment of goodwill is dealt with on consolidation.

9 Explain why unrealised profits on inventory purchased from another member of the group is added to the cost of sales when it is not a cost.

10 Explain why differences between the opening and closing statements of financial position are adjusted when preparing a consolidated statement of cash flows when a subsidiary is acquired.

EXERCISES

* Question 1

Hyson plc acquired 75% of the shares in Green plc on 1 January 20X0 for £6 million when Green plc's accumulated profits were £4.5 million. At acquisition, the fair value of Green's non-current assets were £1.2 million in excess of their carrying value. The remaining life of these non-current assets is six years.

The summarised statements of comprehensive income for the year ended 31.12.20X0 were as follows:

	Hyson	Green
	£000	£000
Revenue	23,500	6,400
Cost of sales	16,400	4,700
Gross profit	7,100	1,700
Expenses	4,650	1,240
Profit before tax	2,450	460
Income tax expense	740	140
Profit for the period	1,710	320

There were no inter-company transactions. Depreciation of non-current assets is charged to cost of sales.

Required:
Prepare a consolidated statement of comprehensive income for the year ended 31 December 20X0.

* Question 2

Forest plc acquired 80% of the ordinary shares of Bulwell plc some years ago. At acquisition, the fair values of the assets of Bulwell plc were the same as their carrying value. Bulwell plc manufacture plant and equipment.

On 1 January 20X3, Bulwell sold an item of plant and equipment to Forest plc for $2 million. Forest plc depreciate plant and equipment at 10% per annum on cost, and charge this expense to cost of sales. Bulwell plc made a gross profit of 30% on the sale of the plant and equipment to Forest plc.

The income statements of Forest and Bulwell for the year ended 31 December 20X3 are:

	Forest	Bulwell
	$000	$000
Revenue	21,300	8,600
Cost of sales	14,900	6,020
Gross profit	6,400	2,580
Other operating expenses	3,700	1,750
Profit before tax	2,700	830
Taxation	820	250
Profit after tax	1,880	580

Required:
Prepare an income statement for the Forest plc group for the year ended 31 December 20X3.

* Question 3

Bill plc acquired 80% of the common shares and 10% of the preferred shares in Ben plc on 31 December three years ago when Ben's retained profits were €45,000. During the year Bill sold Ben goods for €8,000 plus a markup of 50%. Half of these goods were still in stock at the end of the year. There was goodwill impairment loss of €3,000. Non-controlling interests are measured using Method 1.

The statements of comprehensive income of the two companies for the year ended 31 December 20X1 were as follows:

	Bill	Ben
	€	€
Revenue	300,000	180,000
Cost of sales	90,000	90,000
Gross profit	210,000	90,000
Expenses	88,623	60,000
	121,377	30,000
Dividends received – common shares	6,000	—
Dividends received – preferred shares	450	—
Profit before tax	127,827	30,000
Income tax expense	21,006	9,000
Profit for the period	106,821	21,000

Required:
Prepare a consolidated statement of comprehensive income for the year ended 31 December 20X1.

* Question 4

Morn Ltd acquired 90% of the shares in Eve Ltd on 1 January 20X1 for £90,000 when Eve Ltd's accumulated profits were £50,000. On 10 January 20X1 Morn Ltd received a dividend of £10,800 from Eve Ltd out of the profits for the year ended 31/12/20X0. On 31/12/20X1 Morn increased its non-current assets by £30,000 on revaluation. The summarised statements of comprehensive income for the year ended 31/12/20X1 were as follows:

	Morn	Eve
	£	£
Gross profit	360,000	180,000
Expenses	120,000	110,000
	240,000	70,000
Dividends received from Eve Ltd	10,800	—
Profit before tax	250,800	70,000
Income tax expense	69,000	18,000
Profit for the period	181,800	52,000

There were no inter-company transactions, other than the dividend. There was no goodwill.

Required:
Prepare a consolidated statement of comprehensive income for the year ended 31 December 20X1.

* Question 5

River plc acquired 90% of the common shares and 10% of the 5% bonds in Pool Ltd on 31 March 20X1. All income and expenses are deemed to accrue evenly through the year. On 31 January 20X1 River sold Pool goods for £6,000 plus a markup of one-third. 75% of these goods were still in stock at the end of the year. There was a goodwill impairment loss of £4,000. On 31/12/20X1 River increased its non-current assets by £15,000 on revaluation. Non-controlling interests are measured using Method 1. Set out below are the individual statements of comprehensive income of River and Pool:

Statements of comprehensive income for the year ended 31 December 20X1

	River £	Pool £
Net turnover	100,000	60,000
Cost of sales	30,000	30,000
Gross profit	70,000	30,000
Expenses	20,541	15,000
Interest payable on 5% bonds		5,000
Interest receivable on Pool Ltd bonds	500	
	49,959	10,000
Dividends received	2,160	NIL
Profit before tax	52,119	10,000
Income tax expense	7,002	3,000
Profit for the period	45,117	7,000

Required:
Prepare a consolidated statement of comprehensive income for the year ended 31 December 20X1.

Question 6

The statements of financial position of Mars plc and Jupiter plc at 31 December 20X2 are as follows:

	Mars £	Jupiter £
ASSETS		
Non-current assets at cost	550,000	225,000
Depreciation	220,000	67,500
	330,000	157,500
Investment in Jupiter	187,500	
Current assets		
Inventories	225,000	67,500
Trade receivables	180,000	90,000
Current account – Jupiter	22,500	
Bank	36,000	18,000
	463,500	175,500
Total assets	**981,000**	**333,000**
EQUITY AND LIABILITIES		
Capital and reserves		
£1 common shares	196,000	90,000
General reserve	245,000	31,500
Retained earnings	225,000	135,000
	666,000	256,500
Current liabilities		
Trade payables	283,500	40,500
Taxation	31,500	13,500
Current account – Mars		22,500
	315,000	76,500
Total equity and liabilities	**981,000**	**333,000**

Statements of comprehensive income for the year ended 31 December 20X2

	£	£
Sales	1,440,000	270,000
Cost of sales	1,045,000	135,000
Gross profit	395,000	135,000
Expenses	123,500	90,000
Dividends received from Jupiter	9,000	NIL
Profit before tax	280,500	45,000
Income tax expense	31,500	13,500
Profit for the period	249,000	31,500
Dividends paid	180,000	11,250
	69,000	20,250
Retained earnings brought forward from previous years	156,000	114,750
	225,000	135,000

Mars acquired 80% of the shares in Jupiter on 1 January 20X0 when Jupiter's retained earnings were £80,000 and the balance on Jupiter's general reserve was £18,000. Non-controlling interests are measured using Method 1. During the year Mars sold Jupiter goods for £18,000 which represented cost plus 50%. Half of these goods were still in stock at the end of the year.

During the year Mars and Jupiter paid dividends of £180,000 and £11,250 respectively. The opening balances of retained earnings for the two companies were £156,000 and £114,750 respectively.

Required:
Prepare a consolidated statement of income for the year ended 31/12/20X2, a statement of financial position as at that date, and a consolidated statement of changes in equity. Also prepare the retained earnings columns of the consolidated statement of changes in equity for the year.

Question 7

The statements of financial position of Red Ltd and Pink Ltd at 31 December 20X2 are as follows:

	Red	Pink
	$	$
ASSETS		
Non-current assets	225,000	100,000
Depreciation	80,000	30,000
	145,000	70,000
Investment in Pink Ltd	110,000	
Current assets		
Inventories	100,000	30,000
Trade receivables	80,000	40,000
Current account – Pink Ltd	10,000	
Bank	16,000	8,000
	206,000	78,000
Total assets	**461,000**	**148,000**
EQUITY AND LIABILITIES		
Capital and reserves		
$1 common shares	176,000	40,000
General reserve	20,000	14,000
Revaluation reserve	25,000	
Retained earnings	100,000	60,000
	321,000	114,000
Current liabilities		
Trade payables	125,996	18,000
Taxation payable	14,004	6,000
Current account – Red Ltd		10,000
	140,000	34,000
Total equity and liabilities	**461,000**	**148,000**

Statements of comprehensive income for the year ended 31 December 20X2

	$	$
Sales	200,000	120,000
Cost of sales	60,000	60,000
Gross profit	140,000	60,000
Expenses	59,082	40,000
Dividends received	3,750	NIL
Profit before tax	84,668	20,000
Income tax expense	14,004	6,000
	70,664	14,000
Surplus on revaluation	25,000	—
Total comprehensive income	**95,664**	**14,000**

Red Ltd acquired 75% of the shares in Pink Ltd on 1 January 20X0 when Pink Ltd's retained earnings were $30,000 and the balance on Pink's general reserve was $8,000. The fair value of the non-controlling interest at the date was £32,000. Non-controlling interests are to be measured using Method 2.

On 31 December 20X2 Red revalued its non-current assets. The revaluation surplus of £25,000 was credited to the revaluation reserve.

During the year Pink sold Red goods for $9,000 plus a markup of one-third. Half of these goods were still in inventory at the end of the year. Goodwill suffered an impairment loss of 20%.

Required:
Prepare a consolidated statement of comprehensive income for the year ended 31/12/20X2 and a statement of financial position as at that date.

Question 8

H Ltd has one subsidiary, S Ltd. The company has held a controlling interest for several years. The latest financial statements for the two companies and the consolidated financial statements for the H Group are as shown below:

Statements of comprehensive income for the year ended 30 September 20X4

	H Ltd €000	S Ltd €000	H Group €000
Turnover	4,000	2,200	5,700
Cost of sales	(1,100)	(960)	(1,605)
	2,900	1,240	4,095
Administration	(420)	(130)	(550)
Distribution	(170)	(95)	(265)
Dividends received	180	—	—
Profit before tax	2,490	1,015	3,280
Income tax	(620)	(335)	(955)
Profit after tax	1,870	680	2,325
Attributable to:			
Equity shareholders of H Ltd			2,155
Non-controlling shareholders in S Ltd			170
			2,325

Statements of financial position at 30 September 20X4

	H Ltd €000	€000	S Ltd €000	€000	H Group €000	€000
Non-current assets:						
Tangible	7,053		2,196		9,249	
Investment in S Ltd	1,700	8,753	—	2,196	—	9,249
Current assets:						
Inventory	410		420		785	
Receivables	535		220		595	
Bank	27	972	19	659	46	1,426
Current liabilities:						
Payables	(300)		(260)		(355)	
Dividend to non-controlling interest	—		—		(45)	
Taxation	(605)	(905)	(375)	(635)	(980)	(1,380)
		8,820		2,220		9,295

	H Ltd £000	S Ltd £000	H Group £000
Share capital	4,500	760	4,500
Retained earnings	4,320	1,460	4,240
	8,820	2,220	8,740
Non-controlling interest	—	—	555
	8,820	2,220	9,295

Goodwill of €410,000 was written off at the date of acquisition following an impairment review.

Required:

(a) Calculate the percentage of S Ltd which is owned by H Ltd.
(b) Calculate the value of sales made between the two companies during the year.
(c) Calculate the amount of unrealised profit which had been included in the inventory figure as a result of inter-company trading and which had to be cancelled on consolidation.
(d) Calculate the value of inter-company receivables and payables cancelled on consolidation.
(e) Calculate the balance on S Ltd's retained earnings when H Ltd acquired its stake in the company. Non-controlling interests are measured using Method 1.

(CIMA)

Question 9

The following are the financial statements of White and its subsidiary Brown as at 30 September 20X9:

Statement of income for the year ended 30 September 20X9	White £000	Brown £000
Sales revenue	245,000	95,000
Cost of sales	(140,000)	(52,000)
Gross profit	105,000	43,000
Distribution costs	(12,000)	(10,000)
Admin expenses	(55,000)	(13,000)
Profit from operations	38,000	20,000
Dividend from Brown	7,000	—
Profit before tax	45,000	20,000
Tax	(13,250)	(5,000)
Net profit for the year	31,750	15,000

Statements of financial position as at 30 September 20X9	White £000	Brown £000
Non-current assets:		
Property, plant & equipment	110,000	40,000
Investments – 21 million shares in Brown	24,000	—
Current assets:		
Inventory	13,360	3,890
Trade receivables & dividend receivable	14,640	6,280
Bank	3,500	2,570
	165,500	52,740
Equity and reserves:		
Ordinary shares of £1 each	100,000	30,000
Reserves	9,200	1,000
Retained earnings	27,300	9,280
	136,500	40,280
Current liabilities:		
Trade payables	9,000	2,460
Dividend declared	20,000	10,000
	165,500	52,740

The following information is also available:

(i) White purchased its ordinary shares in Brown on 1 September 20X4 when Brown had credit balances on reserves of £0.5 million and on retained earnings of £1.5 million.

(ii) At 1 September 20X8 goodwill on the acquisition of Brown was £960,000. The impairment review at 30 September 20X9 reduced this to £800,000.

(iii) During the year ended 30 September 20X9 White sold goods which originally cost £12 million to Brown and were invoiced to Brown at cost plus 40%. Brown still had 30% of these goods in inventory as at 30 September 20X9.

(iv) Brown owed White £1.5 million at 30 September 20X9 for goods supplied during the year.

Required:
(a) Calculate the goodwill arising at the date of acquisition.
(b) Prepare the consolidated statement of income for the year ended 30 September 20X9.
(c) Prepare the consolidated statement of financial position at 30 September 20X9.

* Question 10

Alpha has owned 80% of the equity shares of Beta since the incorporation of Beta. On 1 July 20X6. Alpha purchased 60% of the equity shares of Gamma. The statements of comprehensive income and summarised statements of changes in equity of the three entities for the year ended 31 March 20X7 are given below:

Statement of comprehensive income

	Alpha	Beta	Gamma
	$000	$000	$000
Revenue (Note 1)	180,000	120,000	106,000
Cost of sales	(90,000)	(60,000)	(54,000)
Gross profit	90,000	60,000	52,000
Distribution costs	(9,000)	(8,000)	(8,000)
Administrative expenses	(10,000)	(9,000)	(8,000)
Investment income (Note 2)	26,450	NIL	NIL
Finance cost	(10,000)	(8,000)	(5,000)
Profit before tax	87,450	35,000	31,000
Income tax expense	(21,800)	(8,800)	(7,800)
Net profit for the period	65,650	26,200	23,200

Summarised statements of changes in equity

	Alpha	Beta	Gamma
Balance at 1 April 20X6	152,000	111,000	102,000
Net profit for the period	65,650	26,200	23,200
Dividends paid on 31 January 20X7	(30,000)	(13,000)	(15,000)
Revaluation of non-current assets	—	20,000	—
Balance at 31 March 20X7	187,650	144,200	110,200

Notes to the financial statements

Note 1 – Inter-company sales

Alpha sells products to Beta and Gamma, making a profit of 30% on the cost of the products sold. All the sales to Gamma took place in the post-acquisition period. Details of the purchases of the products by Beta and Gamma, together with the amounts included in opening and closing inventories in respect of the products, are given below:

	Purchased in year	Included in opening inventory	Included in closing inventory
	$000	$000	$000
Beta	20,000	2,600	3,640
Gamma	10,000	Nil	1,950

Note 2 – Investment income

Alpha's investment income includes dividends received from Beta and Gamma and interest receivable from Beta. The dividend received from Gamma has been credited to the statement of comprehensive income of Alpha without time-apportionment. The interest receivable is in respect of a loan of $60 million to Beta at a fixed rate of interest of 6% per annum. The loan has been outstanding for the whole of the year ended 31 March 20X7.

Note 3 – Details of acquisition of shares in Gamma

On 1 July 20X6 Alpha purchased 15 million of Gamma's issued equity shares by a share exchange. Alpha issued four new equity shares for every three shares acquired in Gamma. The market value of the shares in Alpha and Gamma at 1 July 20X6 was $5 and $5.50 respectively. The non-controlling interest in Gamma is measured using Method 1.

The fair values of the net assets of Gamma closely approximated to their carrying values in Gamma's financial statements with the exception of the following items:

(i)   A property that had a carrying value of $20 million at the date of acquisition had a market value of $30 million. $16 million of this amount was attributable to the building, which had an estimated useful future economic life of 40 years at 1 July 20X6. In the year ended 31 March 20X7 Gamma had charged depreciation of $200,000 in its own financial statements in respect of this property.

(ii) Plant and equipment that had a carrying value of $6 million at the date of acquisition had a market value of $8 million. The estimated useful future economic life of the plant at 1 July 20X6 was four years. None of this plant and equipment had been sold or scrapped prior to 31 March 20X7.

(iii) Inventory that had a carrying value of $3 million at the date of acquisition had a fair value of $3.5 million. This entire inventory had been sold by Gamma prior to 31 March 20X7.

Note 4 – Other information

(i)   Gamma charges depreciation and impairment of assets to cost of sales.

(ii) On 31 March 20X7 the directors of Alpha computed the recoverable amount of Gamma as a single cash-generating unit. They concluded that the recoverable amount was $150 million.

(iii) When the directors of Beta and Gamma prepared the individual financial statements of these companies no impairment of any assets of either company was found to be necessary.

(iv) On 31 March 20X7 Beta revalued its non-current assets. This resulted in a surplus of £20,000 which was credited to Beta's revaluation reserve.

Required:

Prepare the consolidated statement of comprehensive income and consolidated statement of changes in equity of Alpha for the year ended 31 March 20X7. Notes to the consolidated statement of comprehensive income are not required. Ignore deferred tax.

Question 11

Splash plc has a number of subsidiaries, one of which, Muck Ltd, was acquired during the year ended 31 December 2009.

The draft consolidated financial statements for the year ended 31 December 2009 are as follows:

Consolidated statement of comprehensive income of Splash plc for the year ended 31 December 2009

	€000
Profit from operations	1,210
Interest	(100)
	1,110
Share of profits of associates	240
Profit before taxation	1,350
Taxation	(482)
	868
Non-controlling interest	(104)
Group profit	764

Statements of financial position

	Splash plc consolidated		Muck Ltd
	at 31/12/2009	at 31/12/2008	at acquisition
	€000	€000	€000
Assets			
Non-current assets			
Property, plant and equipment	4,730	2,610	610
Intangibles	350	310	—
Investment in associates	520	500	—
	5,600	3,420	610
Current assets			
Inventories	740	610	150
Trade and other receivables	390	350	85
Cash and cash equivalents	40	85	20
Total assets	6,770	4,465	865
Equity and liabilities			
€1 ordinary shares	1,400	1,000	500
Share premium	300	200	100
Retained earnings	1,615	865	80
	3,315	2,065	680
Non-controlling interest	580	610	—
	3,895	2,675	680
Non-current liabilities			
Long-term loans	1,900	1,100	—
Current liabilities			
Trade payables	520	480	75
Taxation	455	210	110
Total equity and liabilities	6,770	4,465	865

Additional information:

1   Splash plc issued 400,000 €1 ordinary shares at a premium of 25 cents and paid a cash consideration of €197,500 to acquire 75% of Muck Ltd. At the date of acquisition, Muck Ltd's assets and liabilities were recorded at their fair value with the exception of some plant which had a fair value of €90,000 in excess of its carrying value. Goodwill on acquisition was €120,000.

2   The property, plant and equipment during the year to 31 December 2009 shows plant with a carrying value of €800,000 which was sold for €680,000. Total depreciation for the year was €782,000.

Required:

(a) Prepare a consolidated statement of cash flows in accordance with IAS 7 *Statement of Cash Flows* for the year ended 31 December 2009.

(b) The Managing Director of Splash plc has asked you to draft a memorandum, briefly explaining the following:

(i)   Why is it important to remove unrealised profits arising from transactions between companies in a group?

(ii) Is it possible for a business to make losses year after year but still increase its bank balance?

(iii) Explain the difference between the direct method and indirect methods of calculating the net cash flow from operating activities.

(Institute of Certified Public Accountants (ICPA), Professional I Stage I Corporate Reporting Examination, August 2010)

References

1   IFRS 10 *Consolidated Financial Statements*, IASB, 2011, B 86.
2   *Ibid.*, B 94, B 89.
3   IAS 1 *Presentation of Financial Statements*, IASB, revised 2007, Implementation Guidance.

CHAPTER 25

Accounting for associates and joint arrangements

25.1 Introduction

The previous three chapters have focused on the need for consolidated financial statements where an investor has control over an entity. In those circumstances line-by-line consolidation is appropriate. Where the size of an investment is not sufficient to give sole control, but where the investment gives the investor significant influence or joint control, then a modified form of accounting is appropriate. We will consider this issue further in this chapter.

Objectives

By the end of this chapter, you should be able to:

- define an associate;
- incorporate a profit-making associate into the consolidated financial statements using the equity method;
- incorporate a loss-making associate into the consolidated financial statements using the equity method;
- define and describe a joint operation and a joint venture and prepare financial statements incorporating interests in joint ventures;
- explain disclosure requirements.

25.2 Definitions of associates and of significant influence

An associate is an entity over which the investor has significant influence and which is neither a subsidiary nor a joint venture of the investor.[1] **Significant influence** is the power to participate in the financial and operating policy decisions of the investee but is not control over these policies.[2]

Significant influence will be assumed in situations where one company has 20% or more of the voting power in another company, unless it can be shown that there is no such influence. Unless it can be shown to the contrary, a holding of less than 20% will be assumed insufficient for associate status. The circumstances of each case must be considered.[2]

IAS 28 *Investments in Associates and Joint Ventures* suggests that one or more of the following might be evidence of an associate:

(a) representation on the board of directors or equivalent governing body of the investee;

(b) participation in policy-making processes;

(c) material transactions between the investor and the investee;

(d) interchange of managerial personnel; or

(e) provision of essential technical information.[3]

25.3 The treatment of associated companies in consolidated accounts

Associated companies will be shown in consolidated accounts under the equity method, unless the investment meets the criteria of a disposal group held for sale under IFRS 5 *Non-current Assets Held for Sale and Discontinued Operations*. If this is the case it will be accounted for under IFRS 5 at the lower of carrying value and fair value less costs to sell.

The equity method is a method of accounting whereby:

● The investment is reported in the consolidated statement of financial position in the non-current asset section.[4] It is reported initially at cost adjusted, at the end of each financial year, for the post-acquisition change in the investor's share of the net assets of the investee.[5]

● In the consolidated statement of comprehensive income, income from associates is reported after profit from operations together with finance costs and finance expenses.[6] The income reflects the investor's share of the post-tax results of operations of the investee.[5]

25.4 The Brill Group – group accounts with a profit-making associate

Brill plc was the parent of the Brill Group which consisted of Brill and a single subsidiary, Bream plc. On 1 January 20X0 Brill acquired 20% of the ordinary shares in Cod Ltd for £20,000. At that date the retained earnings of Cod were £22,500 and the general reserve was £6,000.

Set out below are the consolidated accounts of Brill and its subsidiary Bream and the individual accounts of the associated company, Cod, together with the consolidated group accounts.

25.4.1 Consolidated statement of financial position

Statements of financial position of the Brill Group (parent plus subsidiary already consolidated) and Cod (an associate company) as at 31 December 20X2 are as follows:

	Brill and subsidiary £	Cod £	Group £	
Non-current assets				
Property, plant and equipment	172,500	59,250	172,500	
Goodwill on consolidation	13,400		13,400	
Investment in Cod	20,000		23,600	Note 1
Current assets				
Inventories	132,440	27,000	132,440	
Trade receivables	151,050	27,000	151,050	
Current account – Cod	2,250		2,250	Note 2
Bank	36,200	4,500	36,200	
Total assets	527,840	117,750	531,440	
Current liabilities				
Trade payables	110,250	25,500	110,250	
Taxation	27,750	6,000	27,750	
Current account – Brill		2,250		
	138,000	33,750	138,000	
Total net assets	389,840	84,000	393,440	
EQUITY				
£1 ordinary shares	187,500	37,500	187,500	
General reserve	24,900	9,000	25,500	Note 3
Retained earnings	145,940	37,500	148,940	Note 4
	358,340	84,000	361,940	
Non-controlling interest	31,500	—	31,500	Note 5
	389,840	84,000	393,440	

Notes:

1 **Investment in associate**

	£	£
Initial cost of the 20% holding		20,000
Share of post-acquisition reserves of Cod:		
Retained earnings 20% × (37,500 – 22,500)	3,000	
General reserve 20% × (9,000 – 6,000)	600	3,600
		23,600

Note that (a) unlike subsidiaries the assets and liabilities are not joined line-by-line with those of the companies in the group; (b) where necessary the investment in the associate is tested for impairment under IAS 28;[7] and (c) goodwill is not reported separately and is only calculated initially to establish a figure when considering possible impairment.

2 **The Cod current account** is received from outside the group and must therefore continue to be shown as receivable by the group. *It is not cancelled.*

3 General reserve consists of:

	£
Parent's general reserve	24,900
General reserve of Cod:	
The group share of the post-acquisition general reserve,	
i.e. 20% × (9,000 − 6,000)	600
Consolidated general reserve	**25,500**

4 Retained earnings consist of:

Brill group's retained earnings	145,940
Retained earnings of Cod:	
The group share of the post-acquisition retained profits,	
i.e. 20% × (37,500 − 22,500)	3,000
Consolidated retained earnings	**148,940**

5 Non-controlling interest

Note that there is no non-controlling interest in Cod. Only the group share of Cod's net assets has been brought into the total net assets above (see Note 1). This is unlike the consolidation of a subsidiary when all of the subsidiary's assets and liabilities are aggregated into the consolidation.

25.4.2 Consolidated statement of income

Statements of income for the year ended 31 December 20X2 are as follows:

	Brill and subsidiary	Cod	Group	
	£	£	£	
Sales	329,000	75,000	329,000	
Cost of sales	114,060	30,000	114,060	
Gross profit	214,940	45,000	214,940	
Expenses	107,700	22,500	107,700	
Profit from operations	107,240	22,500	107,240	
Dividends received	1,200	—	NIL	Note 1
Share of associate's **post-tax** profit	—	—	3,300	Note 2
Profit before tax	108,440	22,500	110,540	
Income tax expense	27,750	6,000	27,750	
Profit for the period	80,690	16,500	82,790	

Notes:

1 **Dividend received from Cod** is not shown because the share of Cod's profits (before dividend) has been included in the group account (see Note 2). To include the dividend as well would be double-counting.

2 **Share of Cod's profit after tax = 20% × £16,500 = £3,300**

As in the statement of financial position, there is no need to account for a non-controlling interest in Cod. This is because the consolidated statement of income only included the group share of Cod's profits.

There are no additional complications in the statement of changes in equity. The group retained earnings column will include the group share of Cod's post-acquisition retained earnings. There will be no additional column for a non-controlling interest in Cod.

25.4.3 The treatment of unrealised profits

It is never appropriate in the case of associated companies to remove 100% of any unrealised profit on inter-company transactions because only the group's share of the associate's profit and net assets are shown in the group accounts. For example, let us assume that Brill had purchased goods from Cod during the year at an agreed markup of £10,000, and a quarter of the goods were held by Brill in inventory at the year-end.

The Brill Group will provide for 20% of £2,500 (i.e. £500) by reducing the group share of the associate's profit in the statement of income and reducing the investment in the associate reported in the statement of financial position.

If the sale had been made by Brill, the cost of sales would be increased by £500 and the investment in the associate would be reduced by £500.

25.5 The Brill Group – group accounts with a loss-making associate

The treatment of losses in and impairment of an associate are described below.

Losses

Losses in an associate are normally treated the same way as profits. The group statement of income will show a loss after tax of the associate, and the statement of financial position will continue to show the associate at cost plus its share of post-acquisition profits or less its share of post-acquisition losses.

If the losses were such that they exceeded the carrying amount of the investment in the associate, the investment would be reduced to zero. After that point, additional losses are recognised by a provision (liability) only to the extent that the investor has incurred legal or constructive obligations or made payments on behalf of the associate.

If the associate subsequently reports profits, the investor resumes recognising its share of those profits only after its share of the profits equals the share of losses not recognised.[8]

Impairment

IAS 36 *Impairment of Assets* says (paragraph 9): 'An entity shall assess at the end of each reporting period whether there is any indication that an asset may be impaired. If any such indication exists [*such as making losses or small profits*], the entity shall estimate the recoverable amount of the asset.'

Brill and its subsidiary have a loss-making associate, Herring, which is 20% owned by Brill. On 1 January 20X0 Brill acquired 20% of the ordinary shares in Herring for £20,000. At that date the retained earnings of Herring were £22,500 and the general reserve was £6,000.

Because of the losses incurred by Herring, Brill has carried out an impairment test on the value of the investment in the associate. The recoverable amount of a 20% shareholding in Herring at 31 December 20X2 is £10,000.

Statements of financial position of the Brill Group and Herring as at 31 December 20X2

	Brill group £	Herring £	Group £	
Non-current assets				
Property, plant and equipment	172,500	59,250	172,500	
Goodwill on consolidation	13,400		13,400	
Investment in Herring	20,000		10,000	Note 1
Current assets				
Inventories	132,440	10,500	132,440	
Trade receivables	151,050	12,000	151,050	
Current account – Herring	2,250		2,250	Note 2
Bank	36,200	500	36,200	
	527,840	82,250	517,840	
Current liabilities				
Trade payables	110,250	25,500	110,250	
Taxation	27,750	—	27,750	
Current account – Herring		2,250		
	138,000	27,750	138,000	
Total net assets	389,840	54,500	379,840	
EQUITY				
£1 ordinary shares	187,500	37,500	187,500	
General reserve	24,900	7,000	25,100	Note 3
Retained earnings	145,940	10,000	135,740	Note 4
	358,340	54,500	348,340	
Non-controlling interest	31,500	—	31,500	
	389,840	54,500	379,840	

Notes:

1 Investment in associate

	£	£
Initial cost of 20% holding		20,000
Share of post-acquisition reserves of Herring		
20% × (10,000 – 22,500) (retained earnings)	(2,500)	
20% × (7,000 – 6,000) (general reserves)	200	(2,300)
Carrying value (before impairment)		17,700
Impairment (write down to recoverable amount)		(7,700)
Value in statement of financial position		10,000

The post-acquisition loss of £12,500 gives a loss of £2,500 in the group financial statements. As the carrying value (before impairment) is higher than the recoverable amount of £10,000, the value of the associate in Brill's statement of financial position is reduced to £10,000.

2 The Herring current account remains at £2,250.

3 General reserve consists of:

	£
Parent's general reserve	24,900
General reserve of Herring:	
The group share of the post-acquisition general reserve,	
i.e. 20% × (7,000 – 6,000)	200
Consolidated general reserve	25,100

4 **Retained earnings consist of:**

	£
Parent's retained earnings	145,940

Retained earnings of Herring:

The group share of the post-acquisition retained earnings, i.e. 20% × (10,000 – 22,500)	(2,500)
Impairment of investment in associate (see Note 1)	(7,700)
Consolidated general reserve	135,740

Statements of income for the year ended 31 December 20X2

	Brill group £	Herring £	Group £	
Sales	329,000	75,000	329,000	
Cost of sales	114,060	66,000	114,060	
Gross profit	214,940	9,000	214,940	
Expenses	107,700	27,500	107,700	
Profit/(loss) from operations	107,240	(18,500)	107,240	
Share of associate's after-tax loss			(3,700)	Note 1
Impairment of investment in associate			(7,700)	Note 2
Profit before tax	107,240	(18,500)	95,840	
Income tax expense	27,750	—	27,750	
Profit/(loss) for the period	79,490	(18,500)	68,090	

Notes:

1 Share of associate's loss after tax = 20% × (18,500) = (3,700).

2 Impairment of investment in associate: this figure comes from the investment in associate in the statement of financial position. It reduces the carrying value of £17,700 to its recoverable amount of £10,000.

25.6 The acquisition of an associate part-way through the year

In order to match the cost (the investment) with the benefit (share of the associate's net assets), the associate's profit will only be taken into account from the date of acquiring the holding in the associate. The associate's profit at the date of acquisition represents part of the net assets that are being acquired at that date. The Puff example below is an illustration of the accounting treatment. The adjustment for unrealised profit is made against the group's share of the associate's profit and investment in the associate.

25.6.1 The Puff Group

At date of acquisition on 31 March 20X4 of shares in the associate:

● Puff plc acquired 30% of the shares in Blow plc.

● At that date the retained earnings of Blow were £61,500.

During the year:

● On 1/10/20X4 Blow sold Puff goods for £15,000 which was cost plus 25%.

● All income and expenditure for the year in Blow's statement of comprehensive income accrued evenly throughout the year.

At end of financial year on 31 December 20X4:

- 75% of the goods sold to Puff by Blow were still in inventory.

Set out below are the consolidated statement of income of Puff and its subsidiaries and the individual statement of income of an associated company, Blow, together with the consolidated group statement of income.

	Puff and subsidiaries	Blow	Group accounts	
	£	£	£	
Revenue	225,000	112,500	225,000	Note 1
Cost of sales	75,000	56,250	75,000	Note 2
Gross profit	150,000	56,250	150,000	
Expenses	89,850	30,000	89,850	
	60,150	26,250	60,150	
Dividends received from associate	1,350	NIL	NIL	Note 3
Share of associate's profit	—	—	3,713	Note 4
Profit before taxation	61,500	26,250	63,863	
Income tax for the period	15,000	6,750	15,000	
Profit for the period	46,500	19,500	48,863	

Notes:

1 The revenue, cost of sales and all other income and expenses of the associated company are not added on a line-by-line basis with those of the parent company and its subsidiaries. The group's share of the profit after taxation of the associate is shown as one figure (see Note 4) and added to the remainder of the group's profit before taxation.

2 The group accounts 'cost of sales' figure has not been adjusted for unrealised profit, as this has been deducted from the share of the associate's profit.

3 The dividend received of £1,350 is eliminated, being replaced by the group share of its underlying profits.

4 Share of profits after tax of the associate:

	£
Profit after tax	19,500
Apportion for 9 months ($^9/_{12} \times 19,500$)	14,625
Less: Unrealised profit ($^{25}/_{125} \times 15,000$) $\times 75\%$	2,250
	12,375
Group share (30% $\times$ 12,375)	3,713

5 There is no share of the associated company's retained earnings brought forward because the shares in the associate were purchased during the year.

25.7 Joint arrangements

IFRS 11 *Joint Arrangements* was issued by the IASB in 2011. Under this standard, joint arrangements are classified as either *joint operations* or *joint ventures* depending upon the parties' rights and obligations.

Joint control[9]

Notice that both joint operations and joint ventures require that there should be joint control. Joint control exists where there is a contractually agreed sharing of control of an arrangement under which decisions require the *unanimous* consent of the parties sharing control.

This may be by implicit agreement such as when two parties establish an arrangement in which each has 50% of the voting rights and the contractual arrangement between them specifies that at least 51% of the voting rights are required to make decisions, which results in joint control.

This does not mean the unanimous consent of *all* parties but of those who *collectively control* an arrangement, as illustrated in the following example.[10]

EXAMPLE ● Assume that three parties establish an arrangement: A has 50% of the voting rights in the arrangement, B has 30% and C has 20%. The contractual arrangement between A, B and C specifies that at least 75% of the voting rights are required to make decisions. Even though A can block any decision, it does not control the arrangement because it needs the agreement of B. The terms of their contractual arrangement requiring at least 75% of the voting rights to make decisions about the relevant activities imply that A and B have joint control of the arrangement, because decisions about the relevant activities of the arrangement cannot be made without both A and B agreeing.

Joint operations[11]

This is where the parties, called joint operators, have joint control of the arrangement which gives rights to the assets and obligations for the liabilities. It is the existence of rights and obligations that is critical to determining whether a joint operation exists as opposed to the legal structure of the joint venture. In the predecessor standard, IAS 31, a joint operation could only exist where no new entity was formed.

There may be situations where the ownership rights have been varied by contract. For example, the contractual arrangement might provide for the allocation of revenues and expenses on the basis of the relative performance of each party to the joint arrangement, such as when companies control and finance an oil pipeline equally but pay according to the amount of their throughput. In other instances, the parties might have agreed to share the profit or loss on the basis of a specified proportion such as the parties' ownership interest in the arrangement. These contractual arrangements would not prevent the arrangement from being a joint operation so long as the parties have rights to the assets and obligations for the liabilities.

Joint ventures[12]

This is where the parties, called joint venturers, have joint control of the arrangement which gives rights to the *net* assets of the arrangement. Typically in a joint venture the venturers take a share of the overall profit or loss earned by the joint venture as opposed to taking a share of the output of the venture.

25.7.1 Consolidated financial statements

Joint operations

IFRS 11 says:[13]

A joint operator shall recognise in relation to its interest in a joint operation:

(a) its assets, including its share of any assets held jointly;

(b) its liabilities, including its share of any liabilities incurred jointly;

(c) its revenue from the sale of its share of the output arising from the joint operation;

(d) its share of the revenue from the sale of the output by the joint operation; and

(e) its expenses, including its share of any expenses incurred jointly.

As there are no numerical examples in the standard, it is not clear how the assets, liabilities, revenue and expenses of the joint operation will be shown in the financial statements of each contributor to the joint venture.

The following example suggests how the joint operation would be shown in the financial statements of Sherwood plc:

EXAMPLE ● The joint operators are Sherwood plc and Arnold plc. Sherwood provides the land and buildings for the joint operation, and Sherwood and Arnold have provided equal cash sums to set up the joint venture. The profit is allocated equally between Sherwood and Arnold, after a payment to Sherwood of 5% of the carrying value of the land and buildings. On liquidation of the joint operation, the land and buildings will be returned to Sherwood, and the remaining assets and liabilities split equally between Sherwood and Arnold.

Sherwood's statement of financial position will include all the value of the land and buildings, and half the value of all the other assets and liabilities. It appears that these figures will be combined with the other assets and liabilities of Sherwood and not shown separately.

On the income statement, the joint operation's revenue will be included with other revenue. It would be helpful to the users of the financial statements if the revenue of the joint operation was shown separately.

On expenses, the standard is not clear whether they will be shown separately (in total) or combined with the other individual expense items of Sherwood. It may be shown as a separate figure (in total) as this would be helpful to the users of the financial statements. The rent on the land and buildings (of 5% of their carrying value) is likely to be shown separately in the income statement.

Joint operations can be very complex and require detailed analysis to identify the specific rights and obligations. Already audit firms are considering the practical implication of applying the standard and the possible restatement of prior years' financial statements if accounting policy and treatment changes.

Joint ventures

A joint venturer recognises[14] its interest in a joint venture as an investment which is accounted for using the equity method in accordance with IAS 28 *Investments in Associates and Joint Ventures*.

What if a party participates but does not have joint control?

If it is a joint operation the party would include its interest in the assets and liabilities. If a joint venture, the treatment then depends on the extent of influence that can be exerted. If it is significant then it is accounted for as an associate in accordance with IAS 28. If it is not significant then it is accounted for accordance with IFRS 9 *Financial Instruments*.

25.7.2 Determining whether we are dealing with a joint operation or a joint venture

The following is a helpful extract from www.kpmg.com:

An entity determines the type of joint arrangement by considering the structure, the legal form, the contractual arrangement and other facts and circumstances.

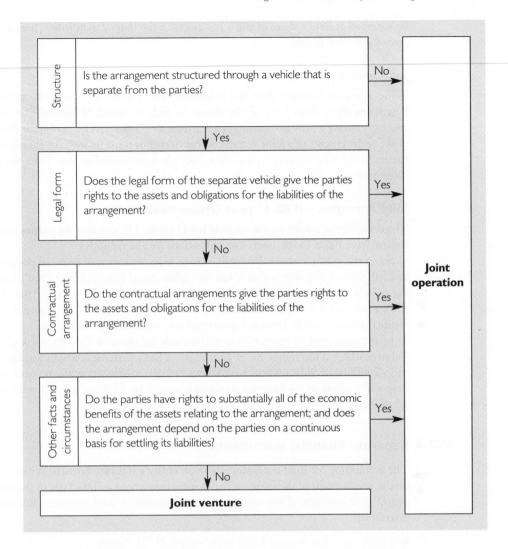

25.7.3 The accounting treatment required where the investment is a subsidiary, associate or joint operation

To illustrate the accounting treatments, we will take a parent company, Pete, which has an investment in another company, Sid.

Subsidiary – IFRS 10 Consolidated Financial Statements applies

If Pete owns more than 50% of the shares of Sid then Pete is presumed to have control of Sid. Sid is, therefore, classified as a subsidiary and the Pete group accounts will include all Sid's assets and liabilities as Pete has *control* over *all* of Sid's assets and liabilities.

Associate – IAS 28 Investments in Associates and Joint Ventures applies

If Pete owns between 20% and 50% of Sid's shares, then it is presumed that Pete is able to exercise significant influence and Sid will be classified as an associated company. In the Pete group accounts the investment in Sid is shown as a single figure in non–current assets. It will be reported at Pete's share of Sid's net assets at the date of acquisition plus goodwill plus post-acquisition profits.

Joint venture

If Sid is a joint venture, then Sid is treated like an associated company of Pete. Even if Pete owns more than 50% of the shares in Sid, it would be treated like an associated company (using equity accounting) rather than a subsidiary. The reason for this is that one of the requirements of a joint venture is that decisions must be with the agreement of all the parties to the joint venture so that Pete does not have control of Sid. Thus Sid cannot be a subsidiary.

Joint operation – IFRS 11 *Joint Arrangements* applies

The definition of a joint operation in IFRS 11 (para. 15) says that the parties that have joint control have rights to the assets and obligations for the liabilities.

So, in the example above, Pete will include in its statement of financial position the assets and liabilities of Sid over which it has the rights. So, if Pete purchased Sid's building, then the building will be included in Pete's statement of financial position. After allocating all Sid's assets and liabilities attributable to Pete in Pete's statement of financial position, any residual amount will be included in current assets and current liabilities.

In the statement of income Pete will include its share of the revenue and expenses of Sid over which it has rights. So, if £1 million of Sid's revenue is wholly attributable to Pete and £2 million of Sid's revenue over which Pete has a 40% share, the revenue of Sid which Pete will include in its financial statements would be £1.8 million (£1 million + 40% of £2 million).

25.7.4 Separate financial statements

The accounting for joint arrangements in an entity's separate financial statements depends on the involvement of the entity in that joint arrangement and the type of the joint arrangement.[15] For example, if the entity is a joint operator or joint venturer, it accounts for its interest in

- a joint operation in accordance with Section 25.7.1 above;
- a joint venture in accordance with paragraph 10 of IAS 27 *Separate Financial Statements*.

25.8 Disclosure in the financial statements

IFRS 12 *Disclosure of Interests in Other Entities* was issued by the IASB in 2011 to bring 'off-balance-sheet finance' onto the financial statements and to enable[18] users of financial statements to evaluate:

(a) the nature of, and risks associated with, its interests in other entities; and

(b) the effects of those interests on its financial position, financial performance and cash flows.

In terms of this chapter, there is a general requirement to disclose information about significant judgements and assumptions made when determining control, joint control, significant influence and classification of joint arrangements. In relation to interest in subsidiaries and joint arrangements, there are specific disclosure requirements.

Interests in subsidiaries disclosures

An entity shall disclose information that enables[17] users of its consolidated financial statements to:

(a) understand the composition of the group and the interest that non-controlling interests have in the group's activities and cash flows; and

(b) evaluate the nature and extent of significant restrictions on its ability to access or use assets, and settle liabilities, of the group; the nature of, and changes in, the risks associated with its interests in consolidated structured entities; the consequences of changes in its ownership interest in a subsidiary that do not result in a loss of control; and the consequences of losing control of a subsidiary during the reporting period.

Interests in joint arrangements and associates disclosures

An entity shall disclose information that enables[18] users of its financial statements to evaluate:

(a) the nature, extent and financial effects of its interests in joint arrangements and associates; and

(b) the nature of, and changes in, the risks associated with its interests in joint ventures and associates.

25.9 Parent company use of the equity method in its separate financial statements

In December 2013, IASB issued Exposure Draft ED/2013/10 which proposes to amend IAS 27 *Separate Financial Statements* so that an entity can use the equity method to account for investments in subsidiaries, joint ventures and associates in its separate financial statements.

IAS 27 (2011) required entities to account for investments in subsidiaries, joint ventures and associates either:

(a) at cost, or

(b) in accordance with IFRS 9 *Financial Instruments*.

This proposed amendment would allow the entity to use equity accounting to account for such investments.

Essentially, both IAS 28 *Investments in Associates* and, for joint ventures, IFRS 11 *Joint Arrangements* require equity accounting to be used for these arrangements.

Allowing subsidiaries to use equity accounting is a new alternative. Using equity accounting will reduce disclosure of the subsidiary:

(a) in the statement of financial position, a single figure in non-current assets of the cost plus share of profit since acquisition less any impairment (or share of net assets plus goodwill less impairment) (note: the two methods produce the same figure);

(b) in the statement of income, the share of the profit of the subsidiary after tax.

This means that there is far less disclosure in the parent company's financial statements. This reduces the work of the preparer of the financial statements but provides less information to the user.

Where the parent company elects not to prepare consolidated financial statements and instead prepares separate financial statements, it must disclose:

(a) the fact the financial statements are separate financial statements and that the exemption from consolidation has been used;

(b) a list of significant investments in subsidiaries, joint ventures and associates, including:

   (i) the name of those investees;

   (ii) the principal place of business of those investees;

   (iii) the proportion of ownership interest held in those investees;

(c) a description of the method used to account for the investments.

The option to use equity accounting (rather than cost or treating it as an investment) provides more useful information to shareholders as it gives the share of profit for the year and a measure of the value of the subsidiary/associated company/joint venture.

Comments on the Exposure Draft ED/2013/10 were favourable and the IASB published 'Equity method in Separate Financial Statements' amending IAS 27.

25.9.1 Critique of Equity Method in Separate Financial Statements

For associates and joint ventures

The permission to use equity accounting will produce the same figures and disclosure as are required by IAS 28 *Investments in Associates* and, for joint ventures, IFRS 11 *Joint Arrangements*. This is certainly much better than valuing them at cost, and, where these investments are to be held for the long term, it is more relevant information (and less subject to fluctuation) than using IFRS 7 *Financial Instruments*.

For subsidiaries

It is much easier for the preparer of the financial statements to include the subsidiary using the equity method, as all that is required is:

(i) for the income statement, the share of the profit after tax;

(ii) for the statement of financial position, the cost plus the share of profit since acquisition.

However, for the user/investor significantly less information is provided than a full consolidation. This would be acceptable if the subsidiary is small compared with the parent company (or the parent company and other subsidiaries). Compared with full consolidation, the reduced information will give the same profit (after tax) in the income statement and the same value of equity in the statement of financial position.

This is ignoring, however, the definition of a subsidiary which is that the parent company has **control** – so the subsidiary's results should be combined with those of the parent company. This is not what happens when the proposed amendment to IAS 27 is applied (essentially, only the parent company's detailed results are included in the separate financial statements).

Thus, applying the equity method to the parent's financial statements could well produce misleading financial statements, particularly when the subsidiary is a significant part of the group's activities (separate financial statements would suggest the group is much smaller than it is, and it would hide any risks associated with the subsidiary).

In conclusion, unless the subsidiary is immaterial, it is apparent that the user's needs are not satisfied by the reduced disclosure of IAS 27, and these reduced financial statements could be misleading. The separate financial statements may reduce the work of the preparer of the financial statements, but there is a greater loss to the user of the financial statement. Thus, the parent company should continue to consolidate the subsidiary, and not be permitted to reduce the subsidiary's disclosure solely to that required by equity accounting.

Summary

Associates and joint ventures are accounted for under IAS 28 (revised 2011) using the equity method whereby there is a single-line entry in the statement of financial position carried initially at cost and the balance adjusted annually for the investor's share of the associate's current year's profit or loss. For joint venture entities, IAS 31 (now superseded by IFRS 11) permitted alternative treatments with investors able to adopt the equity accounting method or proportionate consolidation. Proportionate consolidation is no longer permitted. Joint operations are accounted for in accordance with IFRS 11.

REVIEW QUESTIONS

1 Why are associated companies accounted for under the equity method rather than consolidated?

2 IAS 28, paragraph 17, states:

> The recognition of income on the basis of distributions received may not be an adequate measure of the income earned by an investor on an investment in an associate.

Explain why this may be so.

3 How does the treatment of inter-company unrealised profit differ between subsidiaries and associated companies?

4 The result of including goodwill by valuing the non-controlling shares at their market price using Method 2 is to value the non-controlling shares on a different basis to valuing an equity investment in an associate. Discuss whether there should be a uniform approach to both.

5 Where an associate has made losses, IAS 28, paragraph 30, states:

> After the investor's interest is reduced to zero, additional losses are provided for, and a liability is recognised, only to the extent that the investor has incurred legal or constructive obligations or made payments on behalf of the associate. If the associate subsequently reports profits, the investor resumes recognising its share of those profits only after its share of the profits equals the share of losses not recognised.

Explain why profits are recognised only after its share of the profits equals the share of losses not recognised.

6 The following is an extract from the notes to the 2013 consolidated financial statements of the Chugoku Electric Power Company, Incorporated:

> For the year ended March 31, 2013, 10 affiliated companies were stated at cost without applying the equity method.

Discuss:

(a) why these affiliated companies are being reported at cost;

(b) the effect on the consolidated retained earnings if they had been reported using the equity method.

7 Explain the difference between a joint operation and a joint venture.

8 Explain the approach to determining whether an arrangement is a joint operation or a joint venture.

EXERCISES

* Question 1

The statements of income for Continent plc, Island Ltd and River Ltd for the year ended 31 December 20X9 were as follows:

	Continent plc €	Island Ltd €	River Ltd €
Revenue	825,000	220,000	82,500
Cost of sales	(616,000)	(55,000)	(8,250)
Gross profit	209,000	165,000	74,250
Administration costs	(33,495)	(18,700)	(3,850)
Distribution costs	(11,000)	(14,300)	(2,750)
Dividends receivable from Island and River	4,620	—	—
Profit before tax	169,125	132,000)	67,650
Income tax	(55,000)	(33,000)	(11,000)
Profit after tax	114,125	99,000	56,650

Continent plc acquired 80% of Island Ltd for €27,500 on 1 January 20X3, when Island Ltd's retained earnings were €22,000 and share capital was €5,500. During the year, Island Ltd sold goods costing €2,750 to Continent plc for €3,850. At the year end, 10% of these goods were still in Continent plc's inventory.

Continent plc acquired 40% of River Ltd for €100,000 on 1 January 20X5, when River Ltd's share capital and reserves totalled €41,250 (share capital consisted of 11,000 50c shares). During the year River Ltd sold goods costing €1,650 to Continent plc for €2,200. At the year-end, 50% of these goods were still in Continent plc's inventory.

Goodwill in Island Ltd had suffered impairment charges in previous years totalling €2,200 and goodwill in River Ltd impairment charges totalling €7,700. Impairment has continued during 2009, reducing the goodwill in Island by €550 and the goodwill in River by €3,850.

Continent plc includes in its revenue management fees of €5,500 charged to Island Ltd and €2,750 charged to River Ltd. Both companies treat the charge as an administration cost.

Non-controlling interests are measured using Method 1.

Required:
Prepare Continent plc's consolidated statement of income for the year ended 31 December 20X9.

Question 2

The statements of comprehensive income for Highway plc, Road Ltd and Lane Ltd for the year ended 31 December 20X9 were as follows:

	Highway plc $	Road Ltd $	Lane Ltd $
Revenue	184,000	152,000	80,000
Cost of sales	(48,000)	(24,000)	(16,000)
Gross profit	136,000	128,000	64,000
Administration costs	(13,680)	(11,200)	(20,800)
Distribution costs	(11,200)	(17,600)	(8,000)
Dividends receivable from Road	2,480		
Profit before tax	113,600	99,200	35,200
Income tax	(32,000)	(8,000)	(4,800)
Profit for the period	81,600	91,200	30,400

Highway plc acquired 80% of Road Ltd for $160,000 on 1.1.20X6 when Road Ltd's share capital was $64,000 and reserves were $16,000.

Highway plc acquired 30% of Lane Ltd for $40,000 on 1.1.20X7 when Lane Ltd's share capital was $8,000 and reserves were $8,000.

Goodwill of Road Ltd had suffered impairment charges of $14,400 in previous years and $4,800 was to be charged in the current year. Goodwill of Lane Ltd had suffered impairment charges of $3,520 in previous years and $1,760 was to be charged in the current year.

During the year Road Ltd sold goods to Highway plc for $8,000. These goods had cost Road Ltd $1,600. 50% were still in Highway's inventory at the year-end.

During the year Lane Ltd sold goods to Highway plc for $6,400. These goods had cost Lane Ltd $3,200. 50% were still in Highway's inventory at the year-end.

Highway's revenue included management fees of 5% of Road and Lane's turnover. Both of those companies have treated the charge as an administration cost.

Non-controlling interests are measured using Method 1.

Required:
Prepare Highway's consolidated statement of comprehensive income for the year ended 31.12.20X9.

Question 3

Alpha has owned 75% of the equity shares of Beta since the incorporation of Beta. Therefore, Alpha has prepared consolidated financial statements for some years. On 1 July 20X6 Alpha purchased 40% of the equity shares of Gamma. The statements of comprehensive income and summarised statements of changes in equity of the three entities for the year ended 30 September 20X6 are given below:

Statements of income

	Alpha	Beta	Gamma
	$000	$000	$000
Revenue (Note 1)	150,000	100,000	96,000
Cost of sales	(110,000)	(78,000)	(66,000)
Gross profit	40,000	22,000	30,000
Distribution costs	(7,000)	(6,000)	(6,000)
Administrative expenses	(8,000)	(7,000)	(7,200)
Profit from operations	25,000	9,000	16,800
Investment income (Note 2)	6,450	NIL	NIL
Finance cost	(5,000)	(3,000)	(4,200)
Profit before tax	26,450	6,000	12,600
Income tax expense	(7,000)	(1,800)	(3,600)
Net profit for the period	19,450	4,200	9,000
Summarised statements of changes in equity			
Balance at 1 October 20X5	122,000	91,000	82,000
Net profit for the period	19,450	4,200	9,000
Dividends paid on 31 July 20X6	(6,500)	(3,000)	(5,000)
Balance at 30 September 20X6	134,950	92,200	86,000

Notes to the financial statements

Note 1 – Inter-company sales

Alpha sells products to Beta and Gamma, making a profit of 25% on the cost of the products sold. All the sales to Gamma took place in the post-acquisition period. Details of the purchases of the products by Beta and Gamma, together with the amounts included in opening and closing inventories in respect of the products, are given below:

	Purchased in year	Included in opening inventory	Included in closing inventory
	$000	$000	$000
Beta	20,000	2,000	3,000
Gamma	10,000	NIL	1,500

There were no other inter-company sales between Alpha, Beta or Gamma during the period.

Note 2 – Investment income

Alpha's investment income includes dividends received from Beta and Gamma and interest receivable from Beta. The dividend received from Gamma has been credited to the statement of comprehensive income of Alpha without time-apportionment. The interest receivable is in respect of a loan of $20 million to Beta at a fixed rate of interest of 6% per annum. The loan has been outstanding for the whole of the year ended 30 September 20X6.

Note 3 – Details of acquisitions by Alpha

Entity	Date of acquisition	Fair value adjustment at date of acquisition $000
Beta	1 July 20X5	NIL
Gamma	1 June 20X6	6,400

There has been no impairment of the goodwill arising on the acquisition of Beta or of the investment in Gamma since the dates of acquisition of either entity.

The fair value adjustment has the effect of increasing the fair value of property, plant and equipment above the carrying value in the individual financial statements of Gamma. Group policy is to depreciate property, plant and equipment on a monthly basis over its estimated useful economic life. The estimated life of the property, plant and equipment of Gamma that was subject to the fair value adjustment is five years, with depreciation charged against cost of sales.

Note 4 – Other information

- The purchase of shares in Gamma entitled Alpha to appoint a representative to the board of directors of Gamma. This meant that Alpha was potentially able to participate in, and significantly influence, the policy decisions of Gamma.

- No other investor is able to control the operating and financial policies of Gamma, but on one occasion since 1 July 20X6 Gamma made a policy decision with which Alpha did not fully agree.

- Alpha has not entered into a contractual relationship with any other investor to exercise joint control over the operating and financial policies of Gamma.

- All equity shares in Beta carry one vote at general meetings.

- The policy of Alpha regarding the treatment of equity investments in its consolidated financial statements is as follows:

  - subsidiaries are fully consolidated;

  - joint ventures are proportionally consolidated;

  - associates are equity accounted; and

  - other investments are treated as available for sale financial assets.

Your assistant has been reading the working papers for the consolidated financial statements of Alpha for previous years. He has noticed that Beta has been consolidated as a subsidiary and has expressed the view that this must be because Alpha owns more than 50% of its shares. He has further stated that Gamma should be treated as an available-for-sale financial asset since Alpha is unable to control its operating and financial policies.

Required:

(a) Prepare the consolidated statement of income and consolidated statement of changes in equity of Alpha for the year ended 30 September 20X6. Notes to the consolidated statement of comprehensive income are not required. Ignore deferred tax.

(b) Assess the observations of your assistant regarding the appropriate method of consolidating Beta and Gamma. Your assessment need *not* include an explanation of the detailed mechanics of consolidation. You should refer to the provisions of international financial reporting standards where you consider they will assist your explanation.

* Question 4

The following are the financial statements of the parent company Alpha plc, a subsidiary company Beta and an associate company Gamma.

Statements of financial position as at 31 December 20X9

	Alpha £	Beta £	Gamma £
ASSETS			
Non-current assets			
Land at cost	540,000	256,500	202,500
Investment in Beta	216,000		
Investment in Gamma	156,600		
Current assets			
Inventories	162,000	54,000	135,000
Trade receivables	108,000	72,900	91,800
Dividend receivable from Beta	12,420		
Current account – Beta	10,800		
Current account – Gamma	13,500		
Cash	237,600	62,100	67,500
Total current assets	544,320	189,000	294,300
Total assets	1,456,920	445,500	496,800
EQUITY AND LIABILITIES			
£1 shares	540,000	67,500	27,000
Retained earnings	769,500	329,400	391,500
	1,309,500	396,900	418,500
Current liabilities			
Trade payables	93,420	24,300	59,400
Dividends payable	54,000	13,500	5,400
Current account – Alpha	—	10,800	13,500
Total equity and liabilities	1,456,920	445,500	496,800

On 1 January 20X5 Alpha plc acquired 80% of Beta plc for £216,000 when Beta plc's share capital and reserves were £81,000, and 30% of Gamma Ltd for £156,600 when Gamma Ltd's share capital and reserves were £40,500. The fair value of the land at the date of acquisition was £337,500 in Beta plc and £270,000 in Gamma Ltd. Both companies have kept land at cost in their statement of financial position. All other assets are recorded at fair value. There have been no further share issues or purchases of land since the date of acquisition.

At the year-end, Alpha plc has inventory acquired from Beta plc and Gamma Ltd. Beta plc had invoiced the inventory to Alpha plc for £54,000 – the cost to Beta plc had been £40,500. Gamma Ltd had invoiced Alpha plc for £13,500 – the cost to Gamma Ltd had been £8,100. Goodwill has been impaired by £52,650. The whole of the impairment relates to Beta.

Non-controlling interests are measured using Method 1.

Required:
Prepare Alpha plc's consolidated statement of financial position as at 31.12.20X9.

Question 5

The following are the statements of financial position of Garden plc, its subsidiary Rose Ltd and its associate Petal Ltd:

Statements of financial position as at 31 December 20X9

	Garden £	Rose £	Petal £
ASSETS			
Non-current assets			
Land at cost	240,000		84,000
Land at valuation		180,000	
Investment in Rose	300,000		
Investment in Petal	72,000		
Investments	18,000		
Current assets			
Inventories	15,000	99,000	5,400
Trade receivables	33,000	98,400	1,200
Current account – Rose	18,000		
Current account – Petal	2,400		
Cash	6,600	67,200	300
Total current assets	75,000	264,600	6,900
Total assets	705,000	444,600	90,900
EQUITY AND LIABILITIES			
£1 shares	300,000	120,000	30,000
Revaluation reserve		90,000	
Retained earnings	270,000	216,000	57,600
	570,000	426,000	87,600
Current liabilities			
Trade payables	135,000	3,600	900
Current account – Garden	—	15,000	2,400
Total equity and liabilities	705,000	444,600	90,900

On 1 January 20X3 Garden plc acquired 75% of Rose Ltd for £300,000 when Rose's share capital and reserves were £252,000. Prior to the acquisition, the net book value of Rose's non-current assets was £90,000. Rose revalued its non-current assets immediately prior to the acquisition to fair value and included the revaluation in its statement of financial position.

On 1 January 20X5 Garden acquired 20% of Petal Ltd for £72,000 when the fair value of Petal's net assets were £42,000.

Goodwill has been impaired in Rose by £77,700 and in Petal by £31,800.

At the year-end, Garden plc has inventory acquired from Rose and Petal. Rose had invoiced the inventory to Garden for £6,000 – the cost to Rose had been £1,200. Petal had invoiced Garden for £3,000 – the cost to Petal had been £1,800.

Non-controlling interests are measured using Method 1.

Required:
Prepare Garden plc's consolidated statement of financial position as at 31.12.20X9.

* Question 6

The following are the financial statements of the parent company Swish plc, a subsidiary company Broom and an associate company Handle.

Statements of financial position as at 31 December 20X3

	Swish £	Broom £	Handle £
ASSETS			
Non-current assets			
Property, plant and equipment at cost	320,000	180,000	100,000
Depreciation	200,000	70,000	21,000
	120,000	110,000	79,000
Investment in Broom	140,000		
Investment in Handle	40,000		
Current assets			
Inventories	120,000	60,000	36,000
Trade receivables	130,000	70,000	36,000
Current account – Broom	15,000		
Current account – Handle	3,000		
Bank	24,000	7,000	6,000
Total current assets	292,000	137,000	78,000
Total assets	592,000	247,000	157,000
EQUITY AND LIABILITIES			
£1 ordinary shares	250,000	60,000	50,000
General reserve	30,000	20,000	12,000
Retained earnings	150,000	120,000	50,000
	430,000	200,000	112,000
Current liabilities			
Trade payables	132,000	25,000	34,000
Taxation payable	30,000	7,000	8,000
Current account – Swish		15,000	3,000
Total equity and liabilities	592,000	247,000	157,000

Statement of income for the year ended 31 December 20X3

	£	£	£
Sales	300,000	160,000	100,000
Cost of sales	90,000	80,000	40,000
Gross profit	210,000	80,000	60,000
Expenses	95,000	50,000	30,000
Dividends received from Broom and Handle	11,000	NIL	NIL
Profit before tax	126,000	30,000	30,000
Income tax expense	30,000	7,000	8,000
Profit for the period	96,000	23,000	22,000
Dividend paid (shown in equity)	40,000	10,000	8,000

Swish acquired 90% of the shares in Broom on 1 January 20X1 when the balance on the retained earnings of Broom was £60,000 and the balance on the general reserve of Broom was £16,000. Swish also acquired 25% of the shares in Handle on 1 January 20X2 when the balance on Handle's accumulated retained profits was £30,000 and the general reserve £8,000.

During the year Swish sold Broom goods for £16,000, which included a markup of one-third. 80% of these goods were still in inventory at the end of the year.

Required:
(a) Prepare a consolidated statement of income, including the associated company Handle's results, for the year ended 31 December 20X3.
(b) Prepare a consolidated statement of financial position as at 31 December 20X3. The group policy is to measure non-controlling interests using Method 1.

Question 7

Set out below are the financial statements of Ant Co., its subsidiary Bug Co. and an associated company Nit Co. for the accounting year-end 31 December 20X9.

Statements of financial position as at 31 December 20X9

	Ant $	Bug $	Nit $
ASSETS			
Non-current assets			
Property, plant and equipment at cost	240,000	135,000	75,000
Depreciation	150,000	52,500	15,750
	90,000	82,500	59,250
Investment in Bug	90,000		
Investment in Nit	30,000		
Current assets			
Inventories	105,000	45,000	27,000
Trade receivables	98,250	52,500	27,000
Current account – Bug	11,250		
Current account – Nit	2,250		
Bank	17,250	5,250	4,500
Total current assets	234,000	102,750	58,500
Total assets	444,000	185,250	117,750
EQUITY AND LIABILITIES			
$1 ordinary shares	187,500	45,000	37,500
General reserve	22,500	15,000	9,000
Retained earnings	112,500	90,000	37,500
	322,500	150,000	84,000
Current liabilities			
Trade payables	99,000	18,750	25,500
Taxation payable	22,500	5,250	6,000
Current account – Ant		11,250	2,250
Total equity and liabilities	444,000	185,250	117,750

Statements of comprehensive income for the year ended 31 December 20X9

	Ant	Bug	Nit
	$	$	$
Sales	225,000	120,000	75,000
Cost of sales	67,500	60,000	30,000
Gross profit	157,500	60,000	45,000
Expenses	70,500	37,500	30,000
Dividends received	7,500	NIL	7,500
Profit before tax	94,500	22,500	22,500
Taxation	22,500	5,250	6,000
Profit for the year	72,000	17,250	16,500
Dividends paid in year	30,000	7,500	6,000

Ant Co. acquired 80% of the shares in Bug Co. on 1 January 20X7 when the balance on the retained earnings of Bug Co. was $45,000 and the balance on the general reserve of Bug Co. was $12,000. The fair value of the non-controlling interest in Bug on 1 January 20X7 was $21,000. Group policy is to measure non-controlling interests using Method 2. Ant Co. also acquired 25% of the shares in Nit Co. on 1 January 20X8 when the balance on Nit's retained earnings was $22,500 and the general reserve $6,000.

During the year Ant Co. sold Bug Co. goods for $12,000, which included a markup of one-third. 90% of these goods were still in inventory at the end of the year.

Required:
(a) Prepare a consolidated statement of income for the year ending 31/12/20X9, including the associated company Nit's results.
(b) Prepare a consolidated statement of financial position at 31/12/20X9, including the associated company.

* Question 8

Epsilon acquired 40% of Zeta when Zeta's retained earnings were $50,000, 25% of Kappa when Kappa's retained earnings were $40,000, and 25% of Lambda when Lambda's retained earnings were $50,000.

The four companies' statements of financial position as at 31 October 2011 were as follows:

	Epsilon	Zeta	Kappa	Lambda
	$000	$000	$000	$000
ASSETS				
Non-current assets	1,900	170	140	160
Investment in Zeta	100			
Investment in Kappa	55			
Investment in Lambda	60			
	2,115	170	140	160
Current assets:				
Inventory	8	6	12	11
Trade receivables	12	5	4	7
Bank	5	4	3	2
	25	15	19	20
Total assets	2,140	185	159	180
LIABILITIES				
Equity:				
Share capital	500	50	60	70
Reserves	1,563	124	91	98
	2,063	174	151	168
Non-current liabilities	50			
Current liabilities:				
Trade payables	27	11	8	12
	2,140	185	159	180

Epsilon is entitled to appoint three members of Zeta's board. Zeta's articles state that the board of directors is restricted to five members and that board decisions are binding whenever a simple majority of the directors agree.

Epsilon used its voting rights to secure a place on Kappa's board for one of its own directors. This director has access to internal management reports and can exert some influence on decision making within the company.

Epsilon does not have a representative on the board of Lambda. The directors of Epsilon attempted to secure a place on the board, but were rebuffed by Ms Strong, who owns 75% of the shares. Ms Strong takes a very direct role in the management of Lambda.

Required:
(a) Discuss how each of Epsilon shareholdings should be accounted for in the Epsilon group's consolidated financial statements.
(b) Prepare a consolidated statement of financial position for the Epsilon Group as at 31 October 2011.

(The Association of International Accountants)

Question 9

This question concerns an associated company making a loss and possible impairment of goodwill.

Hyson plc acquired a 30% interest in the ordinary shares of Green plc on 1 January 20X3 when Green's general reserve was £25,000 and its retained earnings were £40,000.

In the year ended 31 December 20X8 Green made a loss after tax of £65,000 because of a recession in its principal sales market.

The statements of financial position of Hyson plc and Green plc at 31 December 20X8 are as follows:

	Hyson £	Green £
ASSETS		
Non-current assets:		
Property, plant and equipment	650,000	230,000
Depreciation	(310,000)	(105,000)
	340,000	125,000
Investment in Green	90,000	
Current assets:		
Inventories	145,000	64,000
Trade receivables	180,000	85,000
Current account – Green	5,000	
Bank	25,000	3,000
Total current assets	355,000	152,000
Total assets	785,000	277,000
EQUITY AND LIABILITIES		
£1 ordinary shares	300,000	200,000
General reserve	60,000	30,000
Retained earnings	225,000	(57,000)
	585,000	173,000
Current liabilities:		
Trade payables	163,000	99,000
Taxation payable	37,000	—
Current account – Hyson	—	5,000
Total equity and liabilities	785,000	277,000

The statements of income of Hyson and Green for the year ended 31 December 20X8 are:

	£	£
Sales	1,045,000	350,000
Cost of sales	683,000	320,000
Gross profit	362,000	30,000
Distribution expenses	42,000	20,000
Administration expenses	152,000	75,000
Profit/(loss) before tax	168,000	(65,000)
Income tax expense	33,000	—
Profit for the period	135,000	(65,000)
Dividend paid (shown in equity)	40,000	—

Because of the losses of Green in 20X8, the recoverable amount of a 30% interest in Green is £40,000 at 31 December 20X8.

Required:

(a) Prepare a consolidated statement of financial position of Hyson plc as at 31 December 20X8.

(b) Prepare a consolidated statement of income of Hyson plc, including the associated company Green, for the year ended 31 December 20X8.

(c) State the changes to your answers in (a) and (b) above if the recoverable amount of a 30% interest in Green was £65,000.

References

1 IAS 28 *Investments in Associates and Joint Ventures*, IASB, revised 2011, para. 3.
2 Ibid., para. 5.
3 Ibid., para. 8.
4 Ibid., para. 15.
5 Ibid., para. 10.
6 IAS 1 *Presentation of Financial Statements*, IASB, revised 2003, Implementation Guidance.
7 IAS 28 *Investments in Associates and Joint Ventures*, IASB, revised 2011, para. 40.
8 Ibid., para. 30.
9 Ibid., para. 7.
10 IFRS 11 *Joint Arrangements*, IASB, 2011, B8.
11 Ibid., para. 15.
12 Ibid., para. 16.
13 Ibid., para. 20.
14 Ibid., paras. 24–5.
15 Ibid., para. 26.
16 IFRS 12 *Disclosure of Interest in Other Entities*, IASB, 2011, para. 1.
17 Ibid., para. 10.
18 Ibid., para. 20.

CHAPTER 26

Introduction to accounting for exchange differences

26.1 Introduction

The increasing globalisation of business means that it is becoming more and more common for companies to enter into transactions that have to be paid for in a foreign currency.

When currency fluctuations occur the exchange rate will have changed between the date the goods or services have been invoiced and the date that payment is made. The difference impacts on cash flows and will be reported as a realised exchange gain or loss in the statement of income.

In this chapter we also consider how to prepare consolidated accounts when there is a foreign subsidiary that maintains its own accounts in the local currency which is different from that of its parent. IAS 21 refers to the local currency as the **functional** currency and the parent's currency as the **presentation** currency. The restatement of the functional currency into the presentation is referred to as translation. Any difference on exchange arising on translation has not been realised and is reported as other comprehensive income.

Objectives

By the end of this chapter, you should be able to:

- account for foreign transactions where differences arise on actual cash inflows and outflows resulting in realised gains or losses;
- translate the financial statements of foreign subsidiaries into the parent company's currency and report any exchange differences under other comprehensive income;
- explain the criteria when determining 'functional' and 'presentation' currency;
- prepare consolidated financial statements to include subsidiaries whose financial statements prepared using the local functional currency have to be translated into a different presentation currency on consolidation; and
- explain the characteristics of a hyperinflationary economy and restatement of the functional currency financial statements.

26.2 How to record foreign currency transactions in a company's own books

We will comment briefly on the IAS 21 provisions relating to (i) how a foreign currency transaction is defined, (ii) the amount entered into the company's accounting records on entering into a transaction, (iii) the accounting treatment of exchange differences when the transaction is settled within the current accounting period, (iv) the accounting treatment when settlement occurs in the next accounting period, (v) the accounting treatment when settlement occurs in an accounting period beyond the next, and (vi) hedging the amount payable.

26.2.1 Defining foreign transactions

IAS 21 *The Effects of Changes in Foreign Exchange Rates* defines foreign transactions as follows:[1]

A foreign transaction is a transaction which is denominated in or requires settlement in a foreign currency, including transactions arising when an entity:

(a) buys or sells goods or services whose price is denominated in a foreign currency;

(b) borrows or lends funds when the amounts payable or receivable are denominated in a foreign currency;

(c) otherwise acquires or disposes of assets, or incurs or settles liabilities, denominated in a foreign currency.

26.2.2 The amount recorded on entering into a transaction

On initial recognition,[2] transactions are entered in the books at the spot currency exchange rate at the transaction date.

For example, let us assume that Brie SA buys vintage cheese from a UK company, Cheddar Ltd, on 1 October 20X1 for £100,000 when the exchange rate was £1 = €1.20. This will be recorded by Brie as Purchases €120,000 and Trade payable (Cheddar) at €120,000.

Where it is more practical an average rate may be used for a period to translate the month's purchases (it will be inappropriate where exchange rates fluctuate significantly).

26.2.3 The accounting treatment of exchange differences when the transaction is settled within the current accounting period

Amounts paid or received in settlement of foreign currency monetary items during an accounting period are translated at the date of settlement, and any exchange difference is taken to the statement of income as a realised gain or loss.

For example, if the rate at the date of payment on 31 October 20X1 was £1 = €1.22, then Brie would pay €122,000 to obtain the sterling amount of £100,000. The exchange difference of €2,000 (€122,000 − €120,000) is debited to the statement of income as an operating expense. If the rate had changed to £1 = €1.18 then there would have been an operating income of €2,000.

The following is an extract from the accounting policies in Nemetschek's 2011 report:

Currency translation
Exchange rate differences arising on the settlement of monetary items at rates different from those at which they were initially recorded during the period, are recognized as other operating income or other operating expenses in the period in which they arise.

26.2.4 The accounting treatment of exchange differences at the year-end when settlement is to occur in the next accounting period

The treatment depends on the ledger balances outstanding. For instance:

- Monetary balances are retranslated at the closing rate as at the date of the statement of financial position.
- Non-monetary items such as property, plant, equipment and inventory reported at historical cost remain translated at their original transaction rate.
- Non-monetary items at fair value are translated at the rate on the date the fair value was determined.[3]

Continuing with our Brie example on accounting for inventory, there are the following possibilities.

Inventory has been sold but the account payable is still outstanding

Assuming that all of the cheese had been sold but Cheddar had still not been paid, then there is no inventory to consider, only the amount payable to Cheddar. This balance is required to be translated at the closing rate. If the closing rate is £1 = €1.24 then the liability to Cheddar would be restated at €124,000 and there would be a resulting exchange loss of €4,000 which is reported in the statement of income.

Inventory has still not been sold and the account payable is still outstanding

If the cheese had not been sold and was still held as inventory, it is required to be reported at the rate as at the date of the initial transaction and not the closing rate, i.e. reported in the statement of financial position as €120,000 – the cost as at 1 October 20X1, the date of purchase. The account payable would still be reported at €124,000.

Inventory has still not been sold but net realisable value is lower than cost

If enquiry established that the cheese had deteriorated and the net realisable value was 50% of cost, then this would be translated at the closing rate as €62,000 (£50,000 × 1.24) and a loss reported of €58,000.

26.2.5 The accounting treatment of exchange differences when settlement occurs in a yet later accounting period

If a monetary item remains unpaid beyond the next accounting period then it will need to be retranslated at the closing rate as at the end of that period.

Let us assume the following:

- Brie has translated the €120,000 due to Cheddar as €124,000 and recognised an operating loss of €4,000 in the year ended 31 December 20X1.
- Brie has reached an agreement with Cheddar that the cheese needs a further period to mature and settlement in full is to be on 1 January 20X3.
- The exchange rate is £1 = €1.23 on 31 December 20X2.

At 31 December 20X2 Brie would report that there was €123,000 owing and there would be an operating gain reported in the 20X2 statement of income of €1,000 (€124,000 – €123,000). If the exchange rate had weakened to a rate higher than £1 = €1.24 there would have been a further operating loss reported in 20X2.

26.2.6 Hedging a foreign currency transaction to crystallise the amount of any exchange difference

A company might enter into a hedging transaction under IAS 39 or IFRS 9 *Financial Instruments*. The intention is to neutralise the exchange risk so that the company knows exactly how much a transaction will cost when settlement is required at a later date. This can be achieved in a number of ways such as entering into a forward contract or an options contract.

For example, let us continue with our Brie example and assume that the euro is weakening and Brie's finance director wants to fix the exact amount it is required to pay in euros on 31 October 20X1 to settle the debt currently recorded as €120,000. His worry is that the end-of-month rate might be £1 = €1.30 which would result in an operating loss of €10,000. In order to take away the uncertainty, Brie enters into a forward contract to buy £100,000 at the end of the month at a rate of say €1.25. This means that there is a known loss of €5,000 as opposed to the risk of a potential loss of up to €10,000.

26.3 Boil plc – a more detailed illustration

Let us assume the following transactions were entered into by Boil plc, a UK company that buys and sells catering equipment in New Zealand, during the year ended 31 December 20X4:

1/11   Buys goods for $30,000 on credit from Napier Ltd
15/11   Sells goods for $40,000 on credit to Wellington Ltd
15/11   Pays Napier Ltd $20,000 on account for the goods purchased
10/12   Receives $25,000 on account from Wellington Ltd in payment for the goods sold
10/12   Buys machinery for $80,000 from Auckland Ltd on credit
22/12   Pays Auckland Ltd $80,000 for the machinery

Boil's functional currency is sterling and the New Zealand companies' functional currency is NZ$.

The exchange rates at the relevant dates were:

1/11   £1 = $2.00   15/11   £1 = $2.20   10/12   £1 = $2.40
22/12   £1 = $2.50   31/12   £1 = $2.60

(Assume that Boil plc buys foreign currency to pay for goods and non-current assets on the day of settlement and immediately converts into sterling any currency received from sales.)

Translating monetary accounts

We need to calculate any exchange differences on monetary accounts that are to be reported in the statement of income which arise on changes between the date of the initial transaction and the rate on the date of its settlement or the statement of financial position date, whichever is the earlier. Profits or losses on exchange differences will arise on the following monetary accounts:

Napier Ltd   Trade payables
Wellington Ltd   Trade receivable
Auckland Ltd   Payable for machinery

The profit or loss on foreign exchange in these cases will be as follows:

| | Napier Payable | | | Wellington Receivable | | | Auckland Payable | | |
| | NZ$ | Rate | £ | NZ$ | Rate | £ | NZ$ | Rate | £ |

(i) Record using the exchange rate on the date of transaction:
    30,000 @ 2.00 = 15,000   40,000 @ 2.20 = (18,182)   80,000 @ 2.40 = 33,333

(ii) Record using the exchange rate at the settlement date:
    20,000 @ 2.20 = (9,091)   25,000 @ 2.40 = 10,417   80,000 @ 2.50 = (32,000)

(iii) Retranslate and record using the closing exchange rate as at the year-end:
    10,000 @ 2.60 = (3,846)   15,000 @ 2.60 = 5,769

(iv) Calculate any gain (loss) on exchange:
                    2,063   (1,996)   1,333

The exchange gains of £2,063 and £1,333 and exchange loss of £1,996 have been realised and are reported in the statement of income as operating income and operating expense.

Accounting treatment of other balances

All other balances, i.e. purchases and sales in the statement of income and machinery (non-monetary), will be translated on the day of the initial transaction and no profit or loss on foreign exchange will arise. These balances will therefore appear in the financial statements as follows:

Purchases	$30,000/2.00 = £15,000
Sales	$40,000/2.20 = £18,182
Machinery	$80,000/2.40 = £33,333

26.4 IAS 21 *Concept of Functional and Presentation Currencies*

All companies have a functional and a presentation currency. In a group with foreign subsidiaries these currencies often differ.

Many groups consist of a parent with a number of foreign subsidiaries that prepare their accounts in the local currency, their functional currency. At the year-end each set of foreign subsidiary accounts is translated into the currency of the parent, the presentation currency or presentational currency.

26.4.1 The functional currency

The functional currency is the currency of the primary economic environment in which the entity operates. For example, the following extract is from the Rio Tinto 2013 financial statements:

> The functional currency for each entity in the Group . . . is the currency of the primary economic environment in which that entity operates. For many entities, this is the currency of the country in which they are located.

Factors to consider when determining the functional currency for an individual company

IAS 21 sets out the factors which a reporting entity (a company preparing financial statements) will consider in determining its functional currency.[1] These are:

- the currency that mainly influences sales prices for goods and services;
- the currency that mainly influences labour, materials and other costs of providing goods and services; and
- the currency in which funds from financing activities are generated and the currency in which the receipts from operating activities are usually retained, which also provide evidence of an entity's functional currency.[4]

If the functional currency is not obvious from the above, then managers have to make a judgement as to which currency most represents the economic effects of its transactions.

Factors a parent considers when deciding with a subsidiary on the subsidiary's functional currency

In making its decision the following factors will be considered:[5]

(a) Whether the activities of the foreign operation are carried out as an extension of the reporting entity (the parent), rather than being carried out with a significant degree of autonomy. An example of the former is when the foreign operation only sells goods imported from the parent and remits the proceeds to it. An example of the latter is when the operation accumulates cash and other monetary items, incurs expenses, generates income and arranges borrowings, all substantially in its local currency.

(b) Whether transactions with the parent are a high or low proportion of the foreign operation's activities.

(c) Whether cash flows from the activities of the foreign operation directly affect the cash flows of the parent and are readily available for remittance to it.

(d) Whether cash flows from the activities of the foreign operation are sufficient to service existing and normally expected debt obligations without funds being made available by the parent.

If the functional currency of the foreign operation is the same as that of the parent, there will of course be no need for translation and the consolidation will be just as for any other subsidiary.

26.4.2 The presentation currency[6]

The **presentation currency** is the currency a parent chooses for its financial statements. The parent is entitled to present its group accounts in any currency, so that in some cases the parent's presentation currency may differ from its own functional currency. There are various reasons for this, such as the principal or potential investors tending to function in a country with a different currency. For example, a parent whose functional currency is the euro might decide to raise finance in the US and so translates its euro financial statements into US$.

The following is an extract from a Press Announcement in 2010 by Tullow Oil plc:

Change in presentation currency
Tullow Oil plc ('the Company', together with its subsidiaries, 'the Group') will present its results in US dollars with effect from 1 January 2010. The Group has decided it is appropriate to change the presentational currency from Sterling as the majority of the Group's activities are in Africa where oil revenues and costs are dollar denominated.

26.5 Translating the functional currency into the presentation currency

Whenever the presentational currency is different from the functional currency, it is necessary to translate the financial statements into the presentational currency. In this situation there is no impact on cash flows and so there is no realised exchange gain or loss to be reported in the statement of income.

Any gain or loss will, therefore, be reported as other comprehensive income. The translation rules used in this situation are set out in paragraph 39 of IAS 21 as follows:

(a) assets and liabilities . . . shall be translated at the closing rate at the date of the statement of financial position;

(b) income and expenses . . . shall be translated at exchange rates at the dates of the transactions [or average rate if this is a reasonable approximation]; and

(c) all resulting exchange differences shall be recognised as a separate component of equity.

The following is an extract from Nemetschek AG's 2011 annual report:

Currency translation
The group's consolidated financial statements are presented in Euros, which is the group's presentation currency.

Functional currency policy
Each entity in the group determines its own functional currency. That is the currency of the primarily economic environment in which the company operates. Items included in the financial statements of each entity are measured using the functional currency. Transactions in foreign currencies are initially recorded at the functional currency rate ruling on the date of the transaction.

Monetary assets and liabilities denominated in foreign currencies are retranslated at the functional currency spot rate of exchange ruling at the balance sheet date. Foreign exchange differences are recorded in profit or loss . . .

Non-monetary items that are measured in terms of historical cost in a foreign currency are translated using the exchange rate as of the date of the initial transaction. Non-monetary items measured at fair value in a foreign currency are translated using the exchange rates at the date when the fair value is determined.

Group policy re subsidiaries
Assets and liabilities of foreign companies are translated to the Euro at the closing rate (incl. goodwill). Income and expenses are translated at the average exchange rate. Any resulting exchange differences are recognized separately in equity.

26.6 Preparation of consolidated accounts

The consolidated accounts are prepared for the Pau Group from the following data.

On 1 January 20X1 Pau Inc. acquired 80% of the ordinary shares of a Brazilian company Briona for $18m when Briona's retained earnings were R$2m and the share premium was R$7m. Briona's financial statements have been audited in their functional currency of Brazilian reals and comply with IAS 21. The summarised statements of income and financial position as at 31 December 20X1 were as follows:

Statements of income for the year ended 31 December 20X1

		Pau		Briona
	US$000	US$000	R$000	R$000
Sales		200,000		80,000
Opening inventories	20,000		10,000	
Purchases	130,000		60,000	
Closing inventories	(40,000)		(30,000)	
Cost of sales		110,000		40,000
Gross profit		90,000		40,000
Other expenses		(15,000)		(14,000)
Interest paid				(1,000)
Total expenses		(15,000)		(15,000)
Profit before taxation		75,000		25,000
Taxation		(15,000)		(5,000)
Profit after taxation		60,000		20,000

Statement of financial position as at 31 December 20X1

	US$000	R$000
Non-current assets	70,000	30,000
Investment in Briona	18,000	
Current assets		
Inventories	40,000	30,000
Trade receivables	27,000	25,000
Cash	2,000	1,000
Total current assets	69,000	56,000
Current liabilities		
Trade payables	35,000	12,000
Taxation	15,000	5,000
Total current liabilities	50,000	17,000
Debentures		6,000
Total assets less liabilities	107,000	63,000
Share capital	20,000	34,000
Share premium		7,000
Retained earnings	87,000	22,000
	107,000	63,000

The following information is also available:

(i) The opening inventory was acquired when the exchange rate was US$1 = R$2.0 and the closing inventory when the rate was US$1 = R$2.4.

(ii) Exchange rates were as follows:

At 1 January 20X1	US$1 = R$2.0
Average for the year ending 31 December 20X4	US$1 = R$2.25
At 31 December 20X1	US$1 = R$2.5

Required:

(a) Prepare a consolidated statement of income.

(b) Prepare a consolidated statement of financial position:

    (i) Show the goodwill calculation.

    (ii) Show the non-controlling interest calculation.

    (iii) Complete with retained earnings as a balancing figure.

(c) Reconcile the retained earnings figure showing exchange gains and losses.

26.6.1 Pau Group draft consolidated accounts

(a) Statement of income

	Pau US$000	Briona R$000	Rate	Briona US$000	Pau Group US$000
Sales	200,000	80,000	2.25	35,555.6	235,555.6
Opening inventories	20,000	10,000	2.0	5,000.0	25,000.0
Purchases	130,000	60,000	2.25	26,666.7	156,666.7
Closing inventories	−40,000	−30,000	2.4	−12,500.0	−52,500.0
Cost of sales	110,000	40,000		19,166.7	129,166.7
Gross profit	90,000	40,000		16,388.9	106,388.9
Other expenses	−15,000	−14,000	2.25	−6,222.2	−21,222.2
Interest paid		−1,000	2.25	−444.4	−444.4
Total expenses	15,000	15,000		6,666.7	21,666.7
Profit before tax	75,000	25,000		9,722.2	84,722.2
Income tax	−15,000	−5,000	2.5	−2,000.0	−17,000.0
Profit after tax	60,000	20,000		7,722.2	67,722.2

(b) Statement of financial position

	Pau US$000	Briona R$000	Rate	Briona US$000		Pau Group US$000
Non-current assets	70,000	30,000	2.5	12,000		82,000
Investment in Briona	18,000				Goodwill (b1)	640
Current assets						
Inventories	40,000	30,000	2.5	12,000		52,000
Trade receivables	27,000	25,000	2.5	10,000		37,000
Cash	2,000	1,000	2.5	400		2,400
	69,000	56,000		22,400		91,400
Current liabilities						
Trade payables	35,000	12,000	2.5	4,800		39,800
Taxation	15,000	5,000	2.5	2,000		17,000
Total current liabilities	50,000	17,000		6,800		56,800
Debentures		6,000	2.5	2,400		2,400
Total assets less liabilities	107,000	63,000		25,200		114,840
Share capital	20,000	34,000	2.5	13,600		20,000
Share premium		7,000	2.5	2,800		
Retained earnings	87,000	22,000	2.5	8,800	(b3)	89,800
	107,000	63,000		25,200		109,800
Non-controlling interest					(b2)	5,040
	107,000	63,000		25,200		114,840

(b1) Goodwill

	R$000		R$000	Rate at 1.1.20X1	US$000
Cost			36,000		
Share capital	34,000				
Share premium	7,000				
Retained earnings	2,000				
	43,000 × 80%		34,400		
Goodwill			1,600	2.0	800
Required to restate at year-end:					
Goodwill			1,600	2.5	640

(b2) Non-controlling interest (NCI) at 31.12.20X1

	R$000	Rate	US$000	US$000
Share capital	34,000	2.5	13,600	
Share premium	7,000	2.5	2,800	
Retained earnings	22,000	2.5	8,800	
			25,200 × 20%	5,040

(b3) The consolidated statement of financial position could be completed by inserting a balancing figure of US$89,800 for the retained earnings made up of Pau's retained earnings of £87,000 and Briona's post-acquisition profit of £2,800. This can be proved as follows:

(c) Subsidiary post-acquisition profit included in the $89,800 group retained earnings

	R$000	Rate	US$000	US$000 Parent	US$000 NCI
Retained profit per Income Statement			7,722.2	6,177.8	1,544.4
At closing rate	20,000	2.5	8,000		
Gain on exchange			277.8	222.2	55.6
Loss on opening shareholders' funds					
Share capital	34,000				
Share premium	7,000				
Retained earnings	2,000				
		Opening rate			
	43,000	2.0	21,500		
		Closing rate			
		2.5	17,200		
Loss			−4,300	(3,440)	(860)
Loss on goodwill		Opening rate			
Goodwill	1,600	2.0	800		
		Closing rate			
		2.5	640		
Loss			−160	(160)	
Post-acquisition profit of Briona attributable to Pau				2,800	
Post-acquisition profit attributable to NCI					740
Opening NCI (43,000 × 20%/2.0)					4,300
Closing NCI (63,000 × 20%/2.5)					5,040
Group retained profit at 31.12.20X1		US$000			
Pau		87,000			
Briona post-acquisition (above)		2,800			
Group retained profit		89,800			

Note that there is no post-acquisition share premium, as the subsidiary's balance at acquisition and at 31.12.20X1 is the same at R$7m.

26.7 How to reduce the risk of translation differences

We have seen that when a parent invests in a foreign subsidiary it is required at each year-end to translate the assets and liabilities from the subsidiary's functional currency into that of the parent.

For example, let us assume that a UK parent has spent $10m on acquiring a US trading subsidiary and at the year-end the net assets of the US subsidiary are also $10m.

On consolidation by the UK parent, the $10m net assets of the US subsidiary are translated into sterling for inclusion in the consolidated statement of financial position. At each year-end the sterling value of any foreign exchange differences are taken to reserves. This means that the group's consolidated shareholders' funds will fluctuate up and down as exchange rates move.

The parent is able to reduce the extent of such fluctuations by hedging the translation risk. It normally does so by acquiring a matching foreign exchange liability. One way is to take on a debt such as a $10m loan.

Assuming that the opening exchange rate is £1 = $2 and the closing rate is £1 = $2.5, without hedging there would be an exchange loss on holding the net assets of £1,000. If the same amount is borrowed there would be an exchange gain on holding the debt of £1,000 ($10m at 2.0 less $10m at 2.5).

In practice, it may be difficult for the parent to borrow as much as $10m unless it gave a guarantee to the lender. By borrowing all its investment in its subsidiary in dollars, the parent would minimise its exchange gains and losses in its subsidiary. In this example, the dollar depreciates against the pound and there is a loss (without hedging). If the dollar appreciated against the pound, there would be a gain (without hedging).

26.8 Critique of use of presentational currency

Multinational companies may have subsidiaries in many different countries, each of which may report by choice or legal requirement internally in their local currency. With globalisation, reporting the group in a presentation currency assists the efficiency of international capital markets, particularly where a group raises funds in more than one market. Although each subsidiary might be controlled through financial statements prepared in the local currency, realism requires the use of a single presentation currency.

26.9 IAS 29 *Financial Reporting in Hyperinflationary Economies*[7]

IAS 29 applies where an entity's functional currency is that of a hyperinflationary economy. Its objective is to give guidance on (a) determining when an economy is hyperinflationary, and (b) restating financial statements to make them meaningful.

26.9.1 Determining when an economy is hyperinflationary

There is no precise criterion, although there is a view that there is hyperinflation when the cumulative inflation rate over three years exceeds 100%. The IAS 29 approach is to leave it as a matter of judgement, based on indicators (IAS 29.3) such as:

- the general population preferring to keep its wealth in non-monetary assets or in a relatively stable foreign currency, with amounts of local currency held being immediately invested to maintain purchasing power;
- the general population regarding monetary amounts not in terms of the local currency but in terms of a relatively stable foreign currency in which prices may be quoted;
- sales and purchases on credit taking place at prices that compensate for the expected loss of purchasing power during the credit period, even if the period is short;
- interest rates, wages and prices being linked to a price index; and
- the cumulative inflation rate over three years approaching, or exceeding, 100%.

26.9.2 How to restate financial statements

Having decided that hyperinflation has occurred, the standard requires the financial statements (and corresponding figures for previous periods) of an entity with a functional currency that is hyperinflationary to be restated for the changes in the general pricing power of the functional currency using the measuring unit current at the year-end date.

26.9.3 Restatement treatment of statements of income and financial position

The statement of comprehensive income

All items in the statement of comprehensive income are expressed in terms of the measuring unit current at the end of the reporting period. All amounts need to be restated by applying the change in the general price index from the dates when the items of income and expenses were initially recorded in the financial statements.

The statement of financial position

Amounts not already expressed in terms of the measuring unit current at the end of the reporting period are restated by applying a general price index.

Monetary items

- **Monetary items are not restated because they are already expressed in terms of the monetary unit.**
- Inventory also as it has been written down to net realisable value under IAS 2.

Non-monetary items

Non-monetary items are restated from the date of acquisition if they are historical cost financial statements or from date of valuation if any asset has been reported at valuation.

If they are current cost financial statements then there is no restatement because they are already expressed in the unit of measurement current at the end of the reporting period.

26.9.4 Disclosures

The following disclosures are required:

(a) the fact that the financial statements and the corresponding figures for previous periods have been restated in terms of the measuring unit current at the end of the reporting period;

(b) whether the financial statements are based on a historical cost approach or a current cost approach; and

(c) the price index that has been used and the level of the price index at the end of the reporting period and the movement in the index during the current and the previous reporting periods.

Summary

When accounts are prepared in the functional currency, exchange differences arising on the settlement of monetary items or on translating monetary items at rates that are different from those which applied on initial recognition (i.e. settled in a later accounting period) are recognised in profit or loss in the period in which they arise.

When functional currency financial statements are translated into a different presentation currency, assets and liabilities are translated at closing rate, income and expenses are translated at the rate as at the date of the transactions (an average may be practical if appropriate), and resulting exchange gains are recognised in other comprehensive income.

REVIEW QUESTIONS

1 Discuss the desirability or otherwise of isolating profits or losses caused by exchange differences from other profit or losses in financial statements.

2 Explain the term functional currency and describe the factors an entity should take into account when determining which is the functional currency.

3 Explain why exchange differences are treated differently in financial statements prepared in a functional currency and those prepared in a presentation currency.

4 Discuss why a company that is not part of a group might decide to translate its financial statements into a presentation currency.

5 Explain why exchange differences might appear in other comprehensive income.

6 How does the treatment of changes in foreign exchange rates relate to the prudence and accruals concepts?

7 It was reported[8] that 'Belarus' cumulative inflation index will exceed 100%, which means that IAS 29 is likely to be applicable to Belarus up to 2014 . . . does not expect any significant microeconomic consequences of Belarus' qualifying as a country with hyperinflationary economy, except for the significant deterioration in financial performance indicators of banks and enterprises applying IFRS.'

Discuss what financial performance indicators might be adversely affected by applying IAS 29.

EXERCISES

Question 1

Fry Ltd has the following foreign currency transactions in the year to 31/12/20X0:

15/11	Buys goods for $40,000 on credit from Texas Inc
15/11	Sells goods for $60,000 on credit to Alamos Inc
20/11	Pays Texas Inc $40,000 for the goods purchased
20/11	Receives $30,000 on account from Alamos Inc in payment for the goods sold
20/11	Buys machinery for $100,000 from Chicago Inc on credit
20/11	Borrows $90,000 from an American bank
21/12	Pays Chicago Inc $80,000 for the machinery

The exchange rates at the relevant dates were:

15/11	£1 = $2.60
20/11	£1 = $2.40
21/12	£1 = $2.30
31/12	£1 = $2.10

Required:
Calculate the profit or loss to be reported in the financial statements of Fry Ltd at 31/12/20X0.

* Question 2

On 1 January 20X1 Fibre plc acquired 80% of the ordinary shares of a Singaporean company, Fastlink Ltd, for £6m when Fastlink's retained earnings were $15.5m and the share premium was $0.8m. Fastlink's financial statements have been prepared in their functional currency of Singapore dollars and comply with IAS 21. The summarised statements of income and financial position as at 31 December 20X1 were as follows:

Statements of income for the year ended 31 December 20X1

	Fibre £000	Fastlink $000
Sales	200,000	50,000
Opening inventories	20,000	8,000
Purchases	130,000	30,000
Closing inventories	(40,000)	(6,000)
Cost of sales	110,000	32,000
Gross profit	90,000	18,000
Expenses	(15,000)	(6,500)
Profit before taxation	75,000	11,500
Taxation	(15,000)	(3,000)
Profit after taxation	60,000	8,500

Statement of financial position as at 31 December 20X1

	£000	$000
Non-current assets	90,000	25,000
Investment in Fastlink	6,000	
Current assets:		
Inventories	40,000	6,000
Trade receivables	27,000	5,000
Cash	2,000	4,000
Total current assets	69,000	15,000
Current liabilities:		
Trade payables	35,000	11,000
Taxation	15,000	3,000
Total current liabilities	50,000	14,000
Total assets less liabilities	115,000	26,000
Share capital	20,000	1,200
Share premium		800
Retained earnings	95,000	24,000
	115,000	26,000

The following information is also available:

(i) The opening inventory was acquired when the exchange rate was £1 = $2.6 and the closing inventory when the rate was £1 = $2.2.

(ii) Exchange rates were as follows:

At 1 January 20X1   £1 = $2.5
Average for the year ending 31 December 20X4   £1 = $2.25
At 31 December 20X1   £1 = $2.0

Required:
(a) Prepare a consolidated statement of income.
(b) Prepare a consolidated statement of financial position:
 (i) Show the goodwill calculation.
 (ii) Show the non-controlling interest calculation.
 (iii) Complete with retained earnings as a balancing figure.
(c) Reconcile the retained earnings figure showing exchange gains and losses.

Question 3

On 1 January 20X0 Walpole Ltd acquired 90% of the ordinary shares of a French subsidiary Paris SA. At that date the balance on the retained earnings of Paris SA was €10,000. The non-controlling interest in Paris was measured as a percentage of identifiable net assets. No shares have been issued by Paris since acquisition. Paris SA's dividend was paid on 31 December 20X2. The summarised statements of comprehensive income and statements of financial position of Walpole Ltd and Paris SA at 31 December 20X2 were as follows:

Statements of comprehensive income for the year ended 31 December 20X2

	Walpole Ltd £000	Paris SA £000
Sales	317,200	200,000
Opening inventories	50,000	22,000
Purchases	180,000	90,000
Closing inventories	60,000	12,000
Cost of sales	170,000	100,000
Gross profit	147,200	100,000
Dividend received from Paris SA	1,800	NIL
Depreciation	30,000	30,000
Other expenses	15,000	7,000
Interest paid	6,000	3,000
Total expenses	51,000	40,000
Profit before taxation	98,000	60,000
Taxation	21,000	15,000
Profit after taxation	77,000	45,000
Dividend paid	20,000	10,000

Statement of financial position as at 31 December 20X2

	Walpole Ltd £000	Paris SA £000
Non-current assets	94,950	150,000
Investment in Paris SA	41,050	
Current assets:		
Inventories	60,000	12,000
Trade receivables	59,600	40,000
Paris SA	2,400	
Cash	11,000	11,000
Total current assets	133,000	63,000
Current liabilities:		
Trade payables	45,000	18,000
Walpole Ltd		12,000
Taxation	21,000	15,000
Total current liabilities	66,000	45,000
Debentures	40,000	10,000
Total assets less liabilities	163,000	158,000
Share capital	80,000	60,000
Share premium	6,000	20,000
Revaluation reserve	10,000	12,000
Retained earnings	67,000	66,000
	163,000	158,000

The following information is also available:

(i) The revaluation reserve in Paris SA arose from the revaluation of non-current assets on 1/1/20X2.

(ii) No impairment of goodwill has occurred since acquisition.

(iii) Exchange rates were as follows:

At 1 January 20X0	£1 = €2
Average for the year ending 31 December 20X2	£1 = €4
At 31 December 20X1/1 January 20X2	£1 = €3
At 31 December 20X2	£1 = €5

Required:
Assuming that the functional currency of Paris SA is the euro, prepare the consolidated accounts for the Walpole group at 31 December 20X2.

* Question 4

(a) According to IAS 21 *The Effects of Changes in Foreign Exchange Rates*, how should a company decide what its functional currency is?

(b) Until recently Eufonion, a UK limited liability company, reported using the euro (€) as its functional currency. However, on 1 November 2007 the company decided that its functional currency should now be the dollar ($).

The summarised balance sheet of Eufonion as at 31 October 2008 in € million was as follows:

ASSETS		€m
Non-current assets		420
Current assets		
Inventories	26	
Trade and other receivables	42	
Cash and cash equivalents	8	
		76
Total assets		496
EQUITY AND LIABILITIES		
Equity		
Share capital		200
Retained earnings		107
		307
Non-current liabilities	85	
Current liabilities		
Trade and other payables	63	
Current taxation	41	
	104	
Total liabilities		189
Total equity and liabilities		496

Non-current liabilities includes a loan of $70 million which was raised in dollars ($) and translated at the closing rate of $1 = €0.72425.

Trade receivables include an amount of $20 million invoiced in dollars ($) to an American customer which has been translated at the closing rate of $1 = €0.72425.

All items of property, plant and equipment were purchased in euros (€) except for plant which was purchased in British pounds (£) in 2007 and which cost £150 million. This was translated at the exchange rate of £1 = €1.46015 as at the date of purchase. The carrying value of the equipment was £90 million as at 31 October 2008.

Required:

Translate the balance sheet of Eufonion as at 31 October 2008 into dollars ($m), the company's new functional currency.

(c) The directors of Eufonion (as in (b) above) are now considering using the British pound (£) as the company's presentation currency for the financial statements for the year ended 31 October 2009.

Required:

Advise the directors how they should translate the company's income statement for the year ended 31 October 2009 and its balance sheet as at 31 October 2009 into the new presentation currency.

(d) Discuss whether or not a reporting entity should be allowed to present its financial statements in a currency which is different from its functional currency.

(The Association of International Accountants)

Question 5

Helvatia GmbH is a Swiss company which is a wholly owned subsidiary of Corolli, a UK company. Helvatia GmbH was formed on 1 November 2005 to purchase and manage a property in Zürich in Switzerland. The reporting and functional currency of Helvatia GmbH is the Swiss franc (CHF).

As a financial accountant in Corolli you are converting the financial statements of Helvatia GmbH into £ sterling in order to be consolidated with the results of Corolli which reports in £s.

The following are the summarised income statements and balance sheet (in thousands of Swiss francs) of Helvatia GmbH:

Helvatia GmbH income statement and retained earnings for the year ended 31 October 2007

	CHF (000)
Revenue	8,800
Depreciation	(1,370)
Other operating expenses	(1,900)
Net income	5,530
Retained earnings at 1 November 2006	3,760
	9,290
Dividends paid	(1,000)
Retained earnings at 31 October 2007	8,290

Helvatia GmbH balance sheet as at 31 October

		2007 CHF (000)		2006 CHF (000)
ASSETS				
Non-current assets				
Land		6,300		3,300
Buildings		12,330		13,700
		18,630		17,000
Current assets				
Receivables	550		1,550	
Cash	5,610		610	
		6,160		2,160
		24,790		19,160
LIABILITIES AND EQUITY				
Non-current liabilities				
Mortgage loan		10,800		10,000
Current liabilities				
Payables		700		400
Equity				
Issued share capital	5,000		5,000	
Retained earnings	8,290	13,290	3,760	8,760
		24,790		19,160

The following exchange rates are available:

	1 Swiss franc = £
At 1 November 2005	0.40
At 1 November 2006	0.55
At 30 November 2006	0.53
At 31 January 2007	0.53
At 31 October 2007	0.45
Weighted average for the year ended 31 October 2007	0.50

The non-current assets and mortgage loan of Helvatia GmbH as at 31 October 2006 all date from 1 November 2005. Helvatia GmbH purchased additional land and increased the mortgage loan on 31 January 2007. There were no other purchases of non-current assets. Land is not depreciated but the building is depreciated at 10% a year using the reducing balance method. Helvatia GmbH's dividends were paid on 31 January 2007.

The sterling equivalent of Helvatia GmbH's retained earnings as at 31 October 2006 was £1,222,000.

Required:
Prepare the following statements for Helvatia GmbH in £000 sterling:
(a) A summarised income statement for the year ended 31 October 2007.
(b) A summarised balance sheet as at 31 October 2007.
(c) A statement of cash flows for the year ended 31 October 2007 using the indirect method. Additional notes are not required.

(The Association of International Accountants)

References

1 IAS 21 *The Effects of Changes in Foreign Exchange Rates*, IASB, revised 2003, para. 20.
2 Ibid., para. 21.
3 Ibid., para. 23.
4 Ibid., para. 10.
5 Ibid., para. 11.
6 Ibid., para. 38.
7 IAS 29 *Financial Reporting in Hyperinflationary Economies*, IASB, 1989.
8 http://www.prime-tass.by/english/News/show.asp?id=96939

PART 6

Interpretation

CHAPTER 27

Earnings per share

27.1 Introduction

The main purpose of this chapter is to undertand the importance of earnings per share (EPS) and the PE ratio as a measure of the financial performance of a company (or 'an enterprise'). This chapter will enable you to calculate the EPS according to IAS 33 for both the current year and prior years, when there is an issue of shares in the year. Also, it will enable you to understand and calculate the diluted earnings per share, for future changes in share capital arising from exercising of share options and conversion of other financial instruments into shares.

Objectives

By the end of this chapter, you should be able to:

● define earnings per share and the PE ratio;
● comment critically on alternative EPS figures;
● calculate the basic earnings per share;
● calculate the diluted earnings per share.

27.2 Why is the earnings per share figure important?

One of the most widely publicised ratios for a public company is the price/earnings or PE ratio. The PE ratio is significant because, by combining it with a forecast of company earnings, analysts can decide whether the shares are currently over- or undervalued.[1]

The ratio is published daily in the financial press and is widely employed by those making investment decisions. The following is a typical extract:

Breweries, Pubs and Restaurants

Company	Price 31/10/12	PE ratio
Company A	283	8.9
Company B	471	11.0
Company C	705	17.0

The PE ratio is calculated by dividing the market price of a share by the earnings that the company generated for that share. Alternatively, the PE figure may be seen as a multiple of the earnings per share, where the multiple represents the number of years' earnings required to recoup the price paid for the share. For example, it would take a shareholder in Company B 11 years to recoup her outlay if all earnings were to be distributed, whereas it would take a shareholder in Company A almost nine years to recoup his outlay, and one in Company C 17 years.

27.2.1 What factors affect the PE ratio?

The PE ratio for a company will reflect investors' confidence and hopes about the international scene, the national economy and the industry sector, as well as about the current year's performance of the company as disclosed in its financial report. It is difficult to interpret a PE ratio in isolation without a certain amount of information about the company, its competitors and the industry within which it operates.

For example, a **high PE ratio** might reflect investor confidence in the existing management team: people are willing to pay a high multiple for expected earnings because of the underlying strength of the company. Conversely, it might also reflect lack of investor confidence in the existing management, but an anticipation of a takeover bid which will result in transfer of the company assets to another company with better prospects of achieving growth in earnings than has the existing team.

A **low PE ratio** might indicate a lack of confidence in the current management or a feeling that even a new management might find problems that are not easily surmounted. For example, there might be extremely high gearing, with little prospect of organic growth in earnings or new capital inputs from rights issues to reduce it.

These reasons for a difference in the PE ratios of companies, even though they are in the same industry, are market-based and not simply a function of earnings. However, both the current earnings per share figure and the individual shareholder's expectation of future growth relative to that of other companies also have an impact on the share price.

27.3 How is the EPS figure calculated?

Because of the importance attached to the PE ratio, it is essential that there be a consistent approach to the calculation of the EPS figure. IAS 33 *Earnings per Share*[2] was issued in 1998 for this purpose. A revised version of the standard was issued in 2003.

The EPS figure is of major interest to shareholders not only because of its use in the PE ratio calculation, but also because it is used in the earnings yield percentage calculation. It is a more acceptable basis for comparing performance than figures such as dividend yield percentage because it is not affected by the distribution policy of the directors. The formula is:

$$\text{EPS} = \frac{\text{Earnings}}{\text{Weighted number of ordinary shares}}$$

The standard defines two EPS figures for disclosure, namely,

- **basic** EPS based on ordinary shares currently in issue; and
- **diluted** EPS based on ordinary shares currently in issue *plus* potential ordinary shares.

27.3.1 Basic EPS

Basic EPS (BEPS) is defined in IAS 33 as follows:[3]

- Basic earnings per share is calculated by dividing the net profit or loss for the period attributable to ordinary shareholders by the weighted average number of ordinary shares outstanding during the period.

For the purpose of the BEPS definition:

- **Net profit** is the profit for the period attributable to the parent entity after deduction of preference dividends (assuming preference shares are equity instruments).[4]
- The **weighted average number of ordinary shares** should be adjusted for events, other than the conversion of potential ordinary shares, that have changed the number of ordinary shares outstanding, without a corresponding change in resources.[5]
- An **ordinary share** is an equity instrument that is subordinate to all other classes of equity instruments.[6]

Earnings per share is calculated on the overall profit attributable to ordinary shareholders but also on the profit from continuing operations if this is different from the overall profit for the period.

27.3.2 Diluted EPS

Diluted EPS is defined as follows:

- For the purpose of calculating diluted earnings per share, the net profit attributable to ordinary shareholders and the weighted average number of shares outstanding should be *adjusted for the effects of all dilutive potential ordinary* shares.[7]

This means that *both* the earnings *and* the number of shares used *may* need to be adjusted from the amounts that appear in the profit and loss account and statement of financial position.

- **Dilutive** means that earnings in the future may be spread over a larger number of ordinary shares.
- **Potential ordinary shares** are financial instruments that may entitle the holders to ordinary shares.

27.4 The use to shareholders of the EPS

Shareholders use the reported EPS to estimate future growth which will affect the future share price. It is an important measure of growth over time. There are, however, limitations in its use as a performance measure and for inter-company comparison.

27.4.1 How does a shareholder estimate future growth in the EPS?

The current EPS figure allows a shareholder to assess the wealth-creating abilities of a company. It recognises that the effect of earnings is to add to the individual wealth of shareholders in two ways: first, by the payment of a dividend which transfers cash from the company's control to the shareholder; and, secondly, by retaining earnings in the company for reinvestment, so that there may be increased earnings in the future.

The important thing when attempting to arrive at an estimate is to review the statement of comprehensive income of the current period and identify the earnings that can reasonably be expected to continue. In accounting terminology, you should identify the **maintainable post-tax earnings** that arise in the **ordinary course of business**.

Companies are required to make this easy for the shareholder by disclosing separately, by way of note, any unusual items and by analysing the profit and loss on trading between discontinuing and continuing activities.

Shareholders can use this information to estimate for themselves the maintainable post-tax earnings, assuming that there is no change in the company's trading activities. Clearly, in a dynamic business environment it is extremely unlikely that there will be no change in the current business activities. The shareholder needs to refer to any information on capital commitments which appear as a note to the accounts and also to the chairman's statement and any coverage in the financial press. This additional information is used to adjust the existing maintainable earnings figure.

27.4.2 Limitations of EPS as a performance measure

EPS is thought to have a significant impact on the market share price. However, there are limitations to its use as a performance measure.

The limitations affecting the use of EPS as an inter-period performance measure include the following:

- It is based on historical earnings. Management might have made decisions in the past to encourage current earnings growth at the expense of future growth, e.g. by reducing the amount spent on capital investment and research and development. Growth in the EPS cannot be relied on as a predictor of the rate of growth in the future.

- EPS does not take inflation into account. Real growth might be materially different from the apparent growth.

The limitations affecting inter-company comparisons include the following:

- The earnings are affected by management's choice of accounting policies, e.g. whether non-current assets have been revalued or interest has been capitalised.

- EPS is affected by the capital structure, e.g. changes in number of shares by making bonus issues.

However, the **rate of growth** of EPS is important and this may be compared between different companies and over time within the same company.

27.5 Illustration of the basic EPS calculation

Assume that Watts plc had post-tax profits for 20X1 of £1,250,000 and an issued share capital of £1,500,000 comprising 1,000,000 ordinary shares of 50p each and 1,000,000 £1 10% preference shares that are classified as equity. The basic EPS (BEPS) for 20X1 is calculated at £1.15 as follows:

	£000
Profit on ordinary activities after tax	1,250
Less preference dividend	(100)
Profit for the period attributable to ordinary shareholders	1,150

BEPS = £1,150,000/1,000,000 shares = **£1.15**

Note that it is the *number* of issued shares that is used in the calculation and *not the nominal value* of the shares. The market value of a share is not required for the BEPS calculation.

27.6 Adjusting the number of shares used in the basic EPS calculation

The earnings per share is frequently used by shareholders and directors to demonstrate the growth in a company's performance over time. Care is required to ensure that the number of shares is stated consistently to avoid distortions arising from changes in the capital structure that have changed the number of shares outstanding without a corresponding change in resources during the whole or part of a year. Such changes occur with (a) bonus issues and share splits; (b) new issues and buybacks at full market price during the year; and (c) the bonus element of a rights issue.

We will consider the appropriate treatment for each of these capital structure changes in order to ensure that EPS is comparable between accounting periods.

27.6.1 Bonus issues

A bonus issue, or capitalisation issue as it is also called, arises when a company capitalises reserves to give existing shareholders more shares. In effect, a simple transfer is made from reserves to issued share capital. In real terms, neither the shareholder nor the company is giving or receiving any immediate financial benefit. The process indicates that the reserves will not be available for distribution, but will remain invested in the physical assets of the company. There are, however, more shares.

Treatment in current year

In the Watts plc example, assume that the company increased its shares in issue in 20X1 by the issue of another 1 million shares and achieved identical earnings in 20X1 as in 20X0. The EPS reported for 20X1 would be immediately halved from £1.15 to £0.575. Clearly, this does not provide a useful comparison of performance between the two years.

Restatement of previous year's BEPS

The solution is to restate the EPS for 20X0 that appears in the 20X1 accounts, using the number of shares in issue at 31.12.20X1, i.e. £1,150,000/2,000,000 shares = BEPS of £0.575.

27.6.2 Share splits

When the market value of a share becomes high some companies decide to increase the number of shares held by each shareholder by changing the nominal value of each share. The effect is to reduce the market price per share but for each shareholder to hold the same total value. A share split would be treated in the same way as a bonus issue.

For example, if Watts plc split the 1,000,000 shares of 50p each into 2,000,000 shares of 25p each, the 20X1 BEPS would be calculated using 2,000,000 shares. It would seem that the BEPS had halved in 20X1. This is misleading and the 20X0 BEPS is therefore restated using 2,000,000 shares. The total market capitalisation of Watts plc would remain unchanged. For example, if, prior to the split, each share had a market value of £4 and the company had a total market capitalisation of £4,000,000, after the split each share would have a market price of £2 and the company market capitalisation would remain unchanged at £4,000,000.

A split is frequently taken as a sign that the board is confident of improved future performance, plus the fact that the fall in the share price makes the shares become more attractive to smaller investors means that the share price might rise above £4 due to the increased demand.

The following is an extract relating to The Coca-Cola Company:

April 25, 2012 – The Board of Directors of The Coca-Cola Company today voted to recommend a two-for-one stock split to shareowners. The split would be the 11th in the stock's 92-year history and the first in 16 years.

'Our recommended two-for-one stock split reflects the Board of Directors' continued confidence in the long-term growth and financial performance of our Company . . . 'Our system's 2020 Vision to double our revenues over this decade provides a clear roadmap for creating value for our consumers, customers, bottling partners and shareowners. A stock split reflects our desire to share value with an ever-growing number of people and organizations around the world.'

Effect on ratios of a share split

Those ratios that are expressed as 'per share' are restated in proportion to the split as with the earnings, dividends and asset per share ratios. For example, if the earnings per share are 50c before a two-for-one split, this will be restated as 25c per share.

Reverse share split

The board might decide to recommend a reverse if the share price is considered too low. There are different reasons for a company making this decision. For example:

● it might be to avoid the shares appearing to be low-quality following poor results, or

● it might be to satisfy listing requirements for a minimum share price in order to avoid being delisted, or

● it might be to make the shares more attractive to institutional investors who might avoid low price shares.

Just as with a share split, the market capitalisation is unchanged. The ratio tends to be significantly higher than for a share split with 1 for 10 or higher being the norm. For example, if there were 5 million shares with a market value of 80c and there was a reverse split of one share for twenty currently held, the share price would increase to €16; however, the market capitalisation would remain at €4m.

27.6.3 New issue at full market value

Selling more shares to raise additional capital should generate additional earnings. In this situation we have a real change in the company's capital and there is no need to adjust any comparative figures. However, a problem arises in the year in which the issue took place. Unless the issue occurred on the first day of the financial year, the new funds would have been *available to generate profits* for only a part of the year. It would therefore be misleading to calculate the EPS figure by dividing the earnings generated during the year by the number of shares in issue at the end of the year. The method adopted to counter this is to use a time-weighted average for the number of shares.

For example, let us assume in the Watts example that the following information is available:

	No. of shares
Shares (nominal value 50p) in issue at 1 January 20X1	1,000,000
Shares issued for cash at market price on 30 September 20X1	500,000

The time-weighted number of shares for EPS calculation at 31 December 20X1 will be:

	No. of shares
Shares in issue for 9 months to date of issue ($1,000,000 \times 9/12$)	750,000
Shares in issue for 3 months from date of issue ($1,500,000 \times 3/12$)	375,000
Time-weighted shares for use in BEPS calculation EPS for 20X1	1,125,000

will be £1,150,000/1,125,000 shares = **£1.02**

27.6.4 Buybacks at market value

Companies are prompted to buy back their own shares when there is a fall in the stock market. The main arguments that companies advance for purchasing their own shares are:

- to reduce the cost of capital when equity costs more than debt;
- the shares are undervalued;
- to return surplus cash to shareholders; and
- to increase the apparent rate of growth in BEPS.

The following is an extract from the 2012 Vodafone Group plc Annual Report:

Our business is highly cash generative and in the last four years we have returned over 30% of our market capitalisation to shareholders in the form of dividends and share buybacks, while still investing around £6 billion a year in our networks and infrastructure.

Earnings per share
Adjusted earnings per share was 14.91 pence, a decline of 11.0% year-on-year, reflecting the loss of our 44% interest in SFR and Polkomtel's profits, the loss of interest income from investment disposals and mark-to-market items charged through finance costs, *partially offset by a reduction in shares arising from the Group's share buyback programme.*

Shares bought back by the company are included in the basic EPS calculation time-apportioned from the beginning of the year to the date of buyback.

For example, let us assume in the Watts example that the following information is available:

	No. of shares
Shares (50p nominal value) in issue at 1 January 20X1	1,000,000
Shares bought back on 31 May 20X1	240,000
Profit attributable to ordinary shares	£1,150,000

The time-weighted number of shares for EPS calculation at 31 December 20X1 will be:

1.1.20X1	Shares in issue for 5 months to date of buyback	$(1,000,000 \times 5/12)$	416,667
31.5.20X1	Number of shares bought back by company	(240,000)	
31.12.20X1	Opening capital less shares bought back	$(760,000 \times 7/12)$	443,333
Time-weighted shares for use in BEPS calculation			860,000

BEPS for 20X1 will be £1,150,000/860,000 shares = **£1.34**

Note that the effect of this buyback has been to increase the BEPS for 20X1 from £1.15 as calculated in Section 27.5 above. This is a mechanism for management to lift the BEPS and achieve EPS growth.

27.7 Rights issues

A rights issue involves giving existing shareholders 'the right' to buy a set number of additional shares at a price below the fair value which is normally the current market price. A rights issue has two characteristics, being both an issue for cash and, because the price is below fair value, a bonus issue. Consequently the rules for *both* a cash issue *and* a bonus issue need to be applied in calculating the weighted average number of shares for the basic EPS calculation.

This is an area where students frequently find difficulty with Step 1 and we will illustrate the rationale without accounting terminology.

The following four steps are required:

Step 1: Calculate the average price of shares before and after a rights issue to identify the amount of the bonus the company has granted.

Step 2: Calculate the weighted average number of shares for the current year.

Step 3: Calculate the BEPS for the current year.

Step 4: Adjust the previous year's BEPS for the bonus element of the rights issue.

Step 1: Calculate the average price of shares before and after a rights issue to identify the amount of the bonus the company has granted

Assume that Mr Radmand purchased two 50p shares at a market price of £4 each in Watts plc on 1 January 20X1 and that on 2 January 20X1 the company offered a 1:2 rights issue (i.e. one new share for every two shares held) at £3.25 per share.

If Mr Radmand had bought at the market price, the position would simply have been:

		£
Two shares at market price of £4 each on 1 January 20X1	=	8.00
One share at market price of £4 on 2 January 20X1	=	4.00
Total cost of three shares as at 2 January 20X1		12.00
Average cost per share unchanged at		4.00

However, this did not happen. Mr Radmand paid only £3.25 for the new share. This meant that the total cost of three shares to him was:

		£
Two shares at market price of £4 each on 1 January 20X1	=	8.00
One share at discounted price of £3.25 on 2 January 20X1	=	3.25
Total cost of three shares	=	11.25
Average cost per share (£11.25/3 shares)	=	3.75

The rights issue has had the effect of reducing the cost per share of each of the three shares held by Mr Radmand on 2 January 20X1 by £0.25 per share.

The accounting terms applied are:

● The average cost per share after the rights issue (£3.75) is *the theoretical ex-rights value.*

● The amount by which the average cost of each share is reduced (£0.25) is *the bonus element.*

In accounting terminology, Step 1 is described as follows:

Step 1: *Theoretical ex-rights calculation.* The bonus element is ascertained by calculating the theoretical ex-rights value, i.e. the £0.25 is ascertained by calculating the £3.75 and deducting it from £4 pre-rights market price.

In accounting terminology, this means that existing shareholders get an element of bonus per share (£0.25) at the same time as the company receives additional capital (£3.25 per new share). The bonus element may be quantified by the calculation of a **theoretical ex-rights price (£3.75)**, which is compared with the last market price (£4.00) prior to the issue; the difference is a bonus. The theoretical ex-rights price is calculated as follows:

	£
Two shares at fair value of £4 each prior to rights issue =	8.00
One share at discounted rights issue price of £3.25 each =	3.25
Three shares at fair value after issue (i.e. ex-rights) =	11.25
Theoretical ex-rights price (£11.25/3 shares) =	3.75
Bonus element (fair value £4 less £3.75) =	0.25

Note that for the calculation of the number of shares and the time-weighted number of shares for a bonus issue, share split and issue at full market price per share, the market price per share is not relevant. The position for a rights issue is different and the market price becomes a relevant factor in calculating the number of bonus shares.

Step 2: Calculate the weighted average number of shares for the current year

Assume that Watts plc made a rights issue of one share for every two shares held on 1 January 20X1. There would be no need to calculate a weighted average number of shares. The total used in the BEPS calculation would be as follows:

	No. of shares
Shares to date of rights issue:	
1,000,000 shares held for a full year =	1,000,000
Shares from date of rights issue:	
500,000 shares held for a full year =	500,000
Total shares for BEPS calculation	1,500,000

However, if a rights issue is made part-way through the year, a time-apportionment is required. For example, if we assume that a rights issue is made on 30 September 20X1, the time-weighted number of shares is calculated as follows:

	No. of shares
Shares to date of rights issue:	
1,000,000 shares held for a full year =	1,000,000
Shares from date of rights issue:	
500,000 shares held for 3 months (500,000 × 3/12) =	125,000
Weighted average number of shares	1,125,000

Note, however, that the 1,125,000 has not taken account of the fact that the new shares had been issued at less than market price and that the company had effectively granted the existing shareholders a bonus. We saw above that when there has been a bonus issue the number of shares used in the BEPS is increased. We need, therefore, to calculate the number of bonus shares that would have been issued to achieve the reduction in market price from £4.00 to £3.75 per share. This is calculated as follows:

Total market capitalisation 1,000,000 shares @ £4.00 per share	= £4,000,000
Number of shares that would reduce the market price to £3.75	= £4,000,000/£3.75
	= 1,066,667 shares
Number of shares prior to issue	= 1,000,000
Bonus shares deemed to be issued to existing shareholders	= 66,667
Bonus shares for period of 9 months to date of issue (66,667 × 9/12) =	**50,000**

The bonus shares for the nine months are added to the existing shares and the time-apportioned new shares as follows:

Figure 27.1 Formula approach to calculating weighted average number of shares

						No. of shares
Shares to date of rights issue:						
No. of shares	×	Increase by bonus fraction		×	Time adjustment	
1,000,000				×	9/12	=	750,000
Bonus:		((1,000,000 × 4/3.75) − 1,000,000)		×	9/12	=	50,000
Shares from date of issue:						
1,500,000	×			×	3/12	=	375,000
Weighted average number of shares						1,175,000

	No. of shares
Shares to date of rights issue:	
1,000,000 shares held for a full year =	1,000,000
Shares from date of rights issue:	
500,000 shares held for 3 months (500,000 × 3/12) =	125,000
Weighted average number of shares	1,125,000
Bonus shares:	
66,667 shares held for 9 months (66,667 × 9/12) =	50,000
	1,175,000

The same figure of 1,175,000 can be derived from the following approach using the relationship between the market price of £4.00 and the theoretical ex-rights price of £3.75 to calculate the number of bonus shares.

The relationship between the actual cum-rights price and theoretical ex-rights price is shown by the bonus fraction:

$$\frac{\text{Actual cum-rights share price}}{\text{Theoretical ex-rights share price}}$$

This fraction is applied to the number of shares before the rights issue to adjust them for the impact of the bonus element of the rights issue. This is shown in Figure 27.1.

Step 3: Calculate the BEPS for the current year

The BEPS for 20X1 is then calculated as £1,150,000/1,175,000 shares = **£0.979**.

Step 4: Adjust the previous year's BEPS for the bonus element of the rights issue

The 20X0 BEPS of £1.15 needs to be restated, i.e. reduced to ensure comparability with 20X1.

In Step 2 above we calculated that the company had made a bonus issue of 66,667 shares to existing shareholders. In recalculating the BEPS for 20X0 the shares should be increased by 66,667 to 1,066,667. The restated BEPS for 20X0 is as follows:

Earnings/restated number of shares
£1,150,000/1,066,667 = £1.078125

Assuming that the earnings for 20X0 and 20X1 were £1,150,000 in each year, the 20X0 BEPS figures will be reported as follows:

As reported in the 20X0 accounts as at 31.12.20X0 = £1,150,000/1,000,000 = £1.15
As restated in the 20X1 accounts as at 31.12.20X1 = £1,150,000/1,066,667 = £1.08

The same result is obtained using the bonus element approach by reducing the 20X0 BEPS as follows by multiplying it by the reciprocal of the bonus fraction:

$$\frac{\text{Theoretical ex-rights fair value per share}}{\text{Fair value per share immediately before the exercise of rights}} = \frac{£3.75}{£4.00}$$

As restated in the 20X1 accounts as at 31.12.20X1 = £1.15 × (3.75/4.00) = **£1.08**.

27.7.1 Would the BEPS for the current and previous years be the same if the company had made a separate full market price issue and a separate bonus issue?

This section is included to demonstrate that the BEPS is the same, i.e. £1.08, if we approach the calculation on the assumption that there was a full price issue followed by a bonus issue. This will demonstrate that the BEPS is the same as that calculated using theoretical ex-rights. There are five steps, as follows.

Step 1: Calculate the number of full value and bonus shares in the company's share capital

	No. of shares
Shares in issue *before* bonus	1,000,000
Rights issue at full market price	
(500,000 shares × £3.25 issue price/£4 full market price)	406,250
	1,406,250
Total number of bonus shares	93,750
Total shares	1,500,000

Step 2: Allocate the total bonus shares to the 1,000,000 original shares

(Note that the previous year will be restated using the proportion of original shares: original shares + bonus shares allocated to these original 1,000,000 shares.)

		No. of shares
Shares in issue before bonus		1,000,000
Bonus issue applicable to pre-rights:		
93,750 bonus shares × (1,000,000/1,406,250) = 66,667 × 9/12	= 50,000	
Bonus issue applicable to post-rights:		
93,750 bonus shares × (1,000,000/1,406,250) = 66,667 × 3/12	= 16,667	
Total bonus shares allocated to existing 1,000,000 shares		66,667
Total original holding plus bonus shares allocated to that holding		1,066,667

Step 3: Time-weight the rights issue and allocate bonus shares to rights shares

Rights issue at full market price:

500,000 shares × (£3.25 issue price/£4 full market price) = 406,250 × 3/12 =   101,563

Bonus issue applicable to rights issue:

93,750 bonus shares × (406,250/1,406,250) = 27,083 × 3/12   =   6,770

Weighted average ordinary shares (includes shares from Steps 2 and 3)   1,175,000

Step 4: BEPS calculation for 20X1

Calculate the BEPS using the post-tax profit and weighted average ordinary shares, as follows:

$$20\text{X1 BEPS} = \frac{£1,150,000}{1,175,000} = £0.979$$

Step 5: BEPS restated for 20X0

There were 93,750 bonus shares issued in 20X1. The 20X0 BEPS needs to be reduced, therefore, by the same proportion as applied to the 1,000,000 ordinary shares in 20X1, i.e. 1,000,000:1,066,667:

20X0 BEPS × bonus adjustment = restated 20X0 BPES

$$= £1.15 \times (1,000,000/1,066,667) = £1.08.$$

This approach illustrates the rationale for the time-weighted average and the restatement of the previous year's BEPS. The adjustment using the theoretical ex-rights approach produces the same result and is simpler to apply but the rationale is not obvious.

27.8 Adjusting the earnings and number of shares used in the diluted EPS calculation

We will consider briefly what dilution means and the circumstances which require the weighted average number of shares and the net profit attributable to ordinary shareholders used to calculate BEPS to be adjusted.

27.8.1 What is dilution?

In a modern corporate structure, a number of classes of person such as the holders of convertible bonds, the holders of convertible preference shares, members of share option schemes and share warrant holders may be entitled as at the date of the statement of financial position to become equity shareholders at a future date.

If these people exercise their entitlements at a future date, the EPS would be reduced. In accounting terminology, the EPS will have been *diluted*. The effect on future share price could be significant. Assuming that the share price is a multiple of the EPS figure, any reduction in the figure could have serious implications for the existing shareholders; they need to be aware of the potential effect on the EPS figure of any changes in the way the capital of the company is or will be constituted. This is shown by calculating and disclosing both the basic and 'diluted EPS' figures.

IAS 33 therefore requires a diluted EPS figure to be reported using as the denominator potential ordinary shares that are dilutive, i.e. would decrease net profit per share or increase net loss from continuing operations.[7]

27.8.2 Circumstances in which the number of shares used for BEPS is increased

The holders of convertible bonds, the holders of convertible preference shares, members of share option schemes and the holders of share warrants will each be entitled to receive ordinary shares from the company at some future date. Such additional shares, referred to as potential ordinary shares, *may* need to be added to the basic weighted average number *if they are dilutive*. It is important to note that if a company has potential ordinary shares they are not automatically included in the fully diluted EPS calculation. There is a test to apply to see if such shares actually are dilutive – this is discussed further in Section 27.9 below.

27.8.3 Circumstances in which the earnings used for BEPS are increased

The earnings are increased to take account of the post-tax effects of amounts recognised in the period relating to dilutive potential ordinary shares that will no longer be incurred on their conversion to ordinary shares, e.g. the loan interest payable on convertible loans will no longer be a charge after conversion and earnings will be increased by the post-tax amount of such interest.

27.8.4 Procedure where there are share warrants and options

Where options, warrants or other arrangements exist which involve the issue of shares below their fair value (i.e. at a price lower than the average for the period) then the impact is calculated by notionally splitting the potential issue into shares issued at fair value and shares issued at no value for no consideration.[8] Since shares issued at fair value are not dilutive, that number is ignored, but the number of shares at no value is employed to calculate the dilution. The calculation is illustrated here for Watts plc.

Assume that Watts plc had at 31 December 20X1:

- an issued capital of 1,000,000 ordinary shares of 50p each nominal value;
- profit attributable to shareholders of £1,150,000;
- an average market price per share of £4; and
- share options in existence 500,000 shares issuable in 20X2 at £3.25 per share.

The computation of basic and diluted EPS is as follows:

	Per share	*Earnings*	*Shares*
Profit attributable to shareholders		£1,150,000	
Weighted average shares during 20X1			1,000,000
Basic EPS (£1,150,000/1,000,000)	1.15		
Number of shares under option			500,000
Number that would have been issued at fair value (500,000 × £3.25/£4)			(406,250)
Adjusted earnings and number of shares		£1,150,000	1,093,750
Diluted EPS (£1,150,000/1,093,750)	1.05		

27.8.5 Procedure where there are convertible bonds or convertible preference shares

The post-tax profit should be adjusted[9] for:

- any dividends on dilutive potential ordinary shares that have been deducted in arriving at the net profit attributable to ordinary shareholders;
- interest recognised in the period for the dilutive potential ordinary shares; and
- any other changes in income or expense that would result from the conversion of the dilutive potential ordinary shares, e.g. the reduction of interest expense related to convertible bonds results in a higher post-tax profit but this could lead to a consequential increase in expense if there were a non-discretionary employee profit-sharing plan.

27.8.6 Convertible preference shares calculation

Assume that Watts plc had at 31 December 20X1:

- an issued capital of 1,000,000 ordinary shares of 50p each nominal value;
- profit attributable to ordinary shareholders of £1,150,000;
- convertible 8% preference shares of £1 each totalling £1,000,000, convertible at one ordinary share for every five convertible preference shares.

The computation of basic and diluted EPS for convertible bonds is as follows:

	Per share	Earnings	Shares
Post-tax net profit for 20X1 (after interest)		£1,150,000	
Weighted average shares during 20X1			1,000,000
Basic EPS (£1,150,000/1,000,000)	£1.15		
Number of shares resulting from conversion			200,000
Add back the preference dividend paid in 20X1		80,000	
Adjusted earnings and number of shares		1,230,000	1,200,000
Diluted EPS (£1,230,000/1,200,000)	£1.025		

27.8.7 Convertible bonds calculation

Assume that Watts plc had at 31 December 20X1:

- an issued capital of 1,000,000 ordinary shares of 50p each nominal value;
- profit attributable to ordinary shareholders of £1,150,000;
- convertible 10% loan of £1,000,000;
- an average market price per share of £4;

and the convertible loan is convertible into 250,000 ordinary shares of 50p each.
The computation of basic and diluted EPS for convertible bonds is as follows:

	Per share	Earnings	Shares
Post-tax net profit for 20X1 (after interest)		£1,150,000	
Weighted average shares during 20X1			1,000,000
Basic EPS (£1,150,000/1,000,000)	£1.15		
Number of shares resulting from conversion			250,000
Interest expense on convertible loan		100,000	
Tax liability relating to interest expense, assuming the firm's marginal tax rate is 40%		(20,000)	
Adjusted earnings and number of shares		1,230,000	1,250,000
Diluted EPS (£1,230,000/1,250,000)	£0.98		

27.9 Procedure where there are several potential dilutions

Where there are several potential dilutions the calculation must be done in progressive stages starting with the most dilutive and ending with the least.[10] Any potential 'antidilutive' issues (i.e. potential issues that would increase earnings per share) are ignored.
Assume that Watts plc had at 31 December 20X1:

- an issued capital of 1,000,000 ordinary shares of 50p each nominal value;
- profit attributable to ordinary shareholders of £1,150,000;

- an average market price per share of £4;
- share options in existence of 500,000 shares exercisable in year 20X2 at £3.25 per share;
- a convertible 10% loan of £1,000,000 convertible in year 20X2 into 250,000 ordinary shares of 50p each; and
- convertible 8% preference shares of £1 each totalling £1,000,000 convertible in year 20X4 at one ordinary share for every 40 preference shares.

There are two steps in arriving at the diluted EPS, namely:

Step 1: Determine the increase in earnings attributable to ordinary shareholders on conversion of potential ordinary shares.

Step 2: Determine the potential ordinary shares to include in the computation of diluted earnings per share.

Step 1: Determine the increase in earnings attributable to ordinary shareholders on conversion of potential ordinary shares

	Increase in earnings	Increase in number of ordinary shares	Earnings per incremental share
Options			
Increase in earnings			
Incremental shares issued for no consideration			
500,000 × (£4 − 3.25)/£4	NIL	93,750	NIL
Convertible preference shares			
Increase in net profit 8% of £1,000,000	80,000		
Incremental shares 1,000,000/40		25,000	3.20
10% convertible bond			
Increase in net profit £1,000,000 × 0.10 × (60%)	60,000		
(assuming a marginal tax rate of 40%)			
Incremental shares 1,000,000/4		250,000	0.24

Step 2: Determine the potential ordinary shares to include in the computation of diluted earnings per share

	Net profit attributable to continuing operations	Ordinary shares	Per share
As reported for BEPS	1,150,000	1,000,000	1.15
Options	—	93,750	
	1,150,000	1,093,750	1.05 dilutive
10% convertible bonds	60,000	250,000	
	1,210,000	1,343,750	0.90 dilutive
Convertible preference shares	80,000	25,000	
	1,290,000	1,368,750	0.94 antidilutive

Since the diluted earnings per share is increased when taking the convertible preference shares into account (from 90p to 94p), the convertible preference shares are antidilutive and are ignored in the calculation of diluted earnings per share. The lowest figure is selected and the diluted EPS will, therefore, be disclosed as 90p.

27.10 Exercise of conversion rights during the financial year

Shares actually issued will be in accordance with the terms of conversion and will be included in the BEPS calculation on a time-apportioned basis from the date of conversion to the end of the financial year.

27.10.1 Calculation of BEPS assuming that convertible loan has been converted and options exercised during the financial year

This is illustrated for the calculation for the year 20X2 accounts of Watts plc as follows. Assume that Watts plc had at 31 December 20X2:

● an issued capital of 1,000,000 ordinary shares of 50p each as at 1 January 20X2;

● a convertible 10% loan of £1,000,000 **converted** on 1 January 20X2 into 250,000 ordinary shares of 50p each; and

● share options for 500,000 ordinary shares of 50p each **exercised** on 1 January 20X2.

The weighted average number of shares for BEPS is calculated as follows:

	Net profit attributable to continuing operations	Ordinary shares	Per share
As reported for BEPS	1,150,000	1,000,000	1.15
Options	—	93,750	
	1,150,000	1,093,750	1.05
10% convertible bonds	60,000	250,000	
	1,210,000	1,343,750	0.90
Convertible preference shares	80,000	25,000	
	1,290,000	1,368,750	0.94

27.11 Disclosure requirements of IAS 33

The standard[11] requires the following disclosures. For the current year:

● Companies should disclose the basic and diluted EPS figures for profit or loss from continuing operations and for profit or loss with equal prominence, whether positive or negative, on the face of the statement of comprehensive income for each class of ordinary share that has a different right to share in the profit for the period.

● The amounts used as the numerators in calculating basic and diluted earnings per share, and a reconciliation of those amounts to the net profit or loss for the period.

● The weighted average number of shares used as the denominator in calculating the basic and diluted earnings per share and a reconciliation of these denominators to each other.

For the previous year (if there has been a bonus issue, rights issue or share split):

● BEPS and diluted EPS should be adjusted retrospectively.

27.11.1 Alternative EPS figures

In the UK the Institute of Investment Management and Research (IIMR) published Statement of Investment Practice No. 1, entitled *The Definition of Headline Earnings*,[12] in which it identified two purposes for producing an EPS figure:

- as a measure of the company's **maintainable earnings** capacity, suitable in particular for forecasts and for inter-year comparisons, and for use on a per-share basis in the calculation of the price/earnings ratio;
- as a factual headline figure for historical earnings, which can be a benchmark figure for the **trading outcome for the year**.

The Institute recognised that the maintainable earnings figure required exceptional or non-continuing items to be eliminated, which meant that, in view of the judgement involved in adjusting the historical figures, the calculation of maintainable earnings figures could not be put on a standardised basis. It took the view that there was a need for an earnings figure, calculated on a standard basis, which could be used as an unambiguous reference point among users. The Institute accordingly defined a **headline earnings** figure for that purpose.

An interesting recent research study[13] supports the finding that the additional EPS figures provide a better indication of future operating earnings one year ahead.

27.11.2 Definition of IIMR headline figure

The Institute criteria for the headline figure are that it should be:

- **A measure of the trading performance.** This means that it will:
  - (a) *exclude* capital items such as profits/losses arising on the sale or revaluation of non-current assets, profits/losses arising on the sale or termination of a discontinued operation and amortisation charges for goodwill, because these are likely to have a different volatility from trading outcomes;

    The following is an extract from the Ricardo 2013 Annual Report:

    Underlying earnings per share is shown because the directors consider that this provides a more useful indication of underlying performance and trends over time.

	2013 £m	2012 £m
Earnings	17.0	15.1
Add back amortization of acquired intangible assets (net of tax)	0.5	–
Add back acquisition costs (net of tax)	1.0	–
Underlying earnings	18.5	15.1

  - (b) *exclude* provisions created for capital items such as profits/losses arising on the sale of non-current assets or on the sale or termination of a discontinued operation; and
  - (c) *include* abnormal items with a clear note and profits/losses arising on operations discontinued during the year.
- **Robust,** in that the result could be arrived at by anyone using the financial report produced in accordance with IAS 1 and IFRS 5.
- **Factual,** in that it will not have been adjusted on the basis of subjective opinions as to whether a cost is likely to continue in the future.

The strength of the Institute's approach is that, by defining a headline figure, it is producing a core definition. Additional earnings, earnings per share and price/earnings ratio figures can be produced by individual analysts, refining the headline figure in the light of their own evaluation of the quality of earnings.

27.11.3 IAS 33 disclosure requirements

If an enterprise discloses an additional EPS figure using a reported component of net profit other than net profit for the period attributable to ordinary shareholders, IAS 33 requires that:

- it must still use the weighted average number of shares determined in accordance with IAS 33;
- if the net profit figure used is not a line item in the statement of comprehensive income, then a reconciliation should be provided between the figure and a line item which is reported in the statement of comprehensive income; and
- the additional EPS figures cannot be disclosed on the face of the statement of comprehensive income.

The following is an extract from the 2012 Annual Report of Vodacom (Pty) Ltd:

	2012 Cents per share	2011 Cents per share
As calculated from IAS 33	694	561
Impairment losses	(15)	(95)
Headline earnings per share as defined by IIMR	709	656

27.12 The Improvement Project

IAS 33 was one of the IASs revised by the IASB as part of its Improvement Project. The objective of the revised standard was to continue to prescribe the principles for the determination and presentation of earnings per share so as to improve comparisons between different entities and different reporting periods. The Board's main objective when revising was to provide additional guidance on selected complex issues such as the effects of contingently issuable shares and purchased put and call options. However, the Board did not reconsider the fundamental approach to the determination and presentation of earnings per share contained in the original IAS 33.

27.13 The Convergence Project

The earnings used as the numerator and the number of shares used as the denominator are both calculated differently under IAS 33 and the US SFAS 128 *Earnings per Share* and so produce different EPS figures.

In 2008, as part of the Convergence Project, the IASB and FASB issued an Exposure Draft which aimed to achieve some convergence in the calculation of the denominator of earnings per share. They are, in the meanwhile, conducting a joint project on financial statement presentation. When they have completed that project and their joint project on liabilities and equity, they may consider whether to conduct a more fundamental review of the method for determining EPS which would look at an agreed approach to determining earnings and number of shares to be used in both the basic and diluted EPS calculation.

Summary

The increased globalisation of stock market transactions places an increasing level of importance on international comparisons. The EPS figure is regarded as a key figure with a widely held belief that management performance could be assessed by the comparative growth rate in this figure. This has meant that the earnings available for distribution, which was the base for calculating EPS, became significant. Management action has been directed towards increasing this figure: sometimes by healthy organic growth; sometimes by buying in earnings by acquisition; sometimes by cosmetic manipulation, e.g. structuring transactions so that all or part of the cost bypassed the statement of comprehensive income; and at other times by the selective exercise of judgement, e.g. underestimating provisions. Regulation by the IASB has been necessary.

IAS 33 permits the inclusion of an EPS figure calculated in a different way, provided that there is a reconciliation of the two figures. Analysts have expressed the view that EPS should be calculated to show the future maintainable earnings and in the UK have arrived at a formula designed to exclude the effects of unusual events and of activities discontinued during the period.

REVIEW QUESTIONS

1 Explain: (i) basic earnings per share; (ii) diluted earnings per share; (iii) potential ordinary shares; and (iv) limitation of EPS as a performance measure.

2 In connection with IAS 33 *Earnings per Share*:

(a) Define the profit used to calculate basic and diluted EPS.

(b) Explain the relationship between EPS and the price/earnings (PE) ratio. Why may the PE ratio be considered important as a stock market indicator?

3 Would the following items justify the calculation of a separate EPS figure under IAS 33?

(a) A charge of £1,500 million that appeared in the accounts, described as additional provisions relating to exposure to countries experiencing payment difficulties.

(b) Costs of £14 million that appeared in the accounts, described as redundancy and other non-recurring costs.

(c) Costs of £62.1 million that appeared in the accounts, described as cost of rationalisation and withdrawal from business activities.

(d) The following items that appeared in the accounts:

(i) Profit on sale of property £80m

(ii) Reorganisation costs £35m

(iii) Disposal and discontinuance of hotels £659m.

4 Explain the adjustments made to earnings when reporting an underlying or headline EPS figure and discuss why this is more relevant than an EPS calculated in accordance with IAS 33.

5 The following note appeared in the 2013 Annual Report of Mercer International Inc.:

Net income (loss) per share attributable to common shareholders:

	2013	2012	2011
Basic	$(0.47)	$(0.28)	$1.39
Diluted	$(0.47)	$(0.28)	$1.24

The calculation of diluted net income (loss) per share attributable to common shareholders does not assume the exercise of any instruments that would have an anti-dilutive effect on net income (loss) per share.

Explain what is meant by antidilutive.

6 Why are issues at full market value treated differently from rights issues?

7 Explain why companies buy back shares and the effect that this has on the earnings per share figure.

8 Explain reverse share splits and the effect that this has on a company's market capitalisation.

9 Discuss the limitations of an IAS 33 calculated EPS figure for performance reporting.

10 Discuss the limitations of EPS as a criterion for setting executive remuneration targets.

EXERCISES

Question 1

Alpha plc had an issued share capital of 2,000,000 ordinary shares at 1 January 20X1. The nominal value was 25p and the market value £1 per share. On 30 September 20X1 the company made a rights issue of 1 for 4 at a price of 80p per share. The post-tax earnings were £4.5m and £5m for 20X0 and 20X1 respectively.

Required:
(a) Calculate the basic earnings per share.
(b) Restate the basic earnings per share for 20X0.

* Question 2

Beta Ltd had the following changes during 20X1:

1 January	1,000,000 shares of 50c each
31 March	500,000 shares of 50c each issued at full market price of $5 per share
30 April	Bonus issue made of 1 for 2
31 August	1,000,000 shares of 50c each issued at full market price of $5.50 per share
31 October	Rights issue of 1 for 3. Rights price was $2.40 and market value was $5.60 per share.

Required:
Calculate the time-weighted average number of shares for the basic earnings per share denominator. Note that adjustments will be required for time, the bonus issue and the bonus element of the rights issue.

* Question 3

The computation and publication of earnings per share (EPS) figures by listed companies are governed by IAS 33 *Earnings per Share*.

Nottingham Industries plc
Statement of comprehensive income for the year ended 31 March 20X6
(extract from draft unaudited accounts)

		£000
Profit on ordinary activities before taxation	(Note 2)	1,000
Tax on profit on ordinary activities	(Note 3)	(420)
Profit on ordinary activities after taxation		580

Notes:

1 Called-up share capital of Nottingham Industries plc:
   In issue at 1 April 20X5:
   16,000,000 ordinary shares of 25p each
   1,000,000 10% cumulative preference shares of £1 each classified as equity
   1 July 20X5: Bonus issue of ordinary shares, 1 for 5.
   1 October 20X5: Market purchase of 500,000 of own ordinary shares at a price of £1.00 per share.

2 In the draft accounts for the year ended 31 March 20X6, 'profit on ordinary activities before taxation' is arrived at after charging or crediting the following items:

   (i)   accelerated depreciation on fixed assets, £80,000;

   (ii) book gain on disposal of a major operation, £120,000.

3 Profit after tax included a write-back of deferred taxation (accounted for by the liability method) in consequence of a reduction in the rate of corporation tax from 45% in the financial year 20X4 to 40% in the financial year 20X5.

4 The following were charged:

   (i)   Provision for bad debts arising on the failure of a major customer, £150,000. Other bad debts have been written off or provided for in the ordinary way.

   (ii) Provision for loss through expropriation of the business of an overseas subsidiary by a foreign government, £400,000.

5 In the published accounts for the year ended 31 March 20X5, basic EPS was shown as 2.2p; fully diluted EPS was the same figure.

6 Dividends paid totalled £479,000.

Required:

(a) On the basis of the facts given, compute the basic EPS figures for 20X6 and restate the basic EPS figure for 20X5, stating your reasons for your treatment of items that may affect the amount of EPS in the current year.

(b) Compute the diluted earnings per share for 20X6 assuming that on 1 January 20X6 executives of Nottingham plc were granted options to take up a total of 200,000 unissued ordinary shares at a price of £1.00 per share: no options had been exercised at 31 March 20X6. The average fair value of the shares during the year was £1.10.

(c) Give your opinion as to the usefulness (to the user of financial statements) of the EPS figures that you have computed.

* Question 4

The following information relates to Simrin plc for the year ended 31 December 20X0:

	£
Turnover	700,000
Operating costs	476,000
Trading profit	224,000
Net interest payable	2,000
	222,000
Exceptional charges	77,000
	145,000
Tax on ordinary activities	66,000
Profit after tax	79,000

Simrin plc had 100,000 ordinary shares of £1 each in issue throughout the year. Simrin plc has in issue warrants entitling the holders to subscribe for a total of 50,000 shares in the company. The warrants may be exercised after 31 December 20X5 at a price of £1.10 per share. The average fair value of shares was £1.28. The company had paid an ordinary dividend of £15,000 and a preference dividend of £9,000 on preference shares classified as equity.

Required:
(a) Calculate the basic EPS for Simrin plc for the year ended 31 December 20X0, in accordance with best accounting practice.
(b) Calculate the diluted EPS figure, to be disclosed in the statutory accounts of Simrin plc in respect of the year ended 31 December 20X0.
(c) Briefly comment on the need to disclose a diluted EPS figure and on the relevance of this figure to the shareholders.
(d) In the past, the single most important indicator of financial performance has been earnings per share. In what way has the profession attempted to destroy any reliance on a single figure to measure and predict a company's earnings, and how successful has this attempt been?

* Question 5

Gamma plc had an issued share capital at 1 April 20X0 of:

● £200,000 made up of 20p shares; and
● 50,000 £1 convertible preference shares classified as equity receiving a dividend of £2.50 per share. These shares were convertible in 20X6 on the basis of one ordinary share for one preference share.

There was also loan capital of:

● £250,000 10% convertible loans. The loan was convertible in 20X9 on the basis of 500 shares for each £1,000 of loan, and the tax rate was 40%.

Earnings for the year ended 31 March 20X1 were £5,000,000 after tax.

Required:
(a) Calculate the diluted EPS for 20X1.
(b) Calculate the diluted EPS assuming that the convertible preference shares were receiving a dividend of £6 per share instead of £2.50.

Question 6

Delta NV has share capital of €1m in shares of €0.25 each. At 31 May 20X9 shares had a market value of €1.1 each. On 1 June 20X9 the company makes a rights issue of one share for every four held at €0.6 per share. Its profits were €500,000 in 20X9 and €440,000 in 20X8. The year-end is 30 November.

Required:

Calculate

(a) the theoretical ex-rights price;

(b) the bonus issue factor;

(c) the basic earnings per share for 20X8;

(d) the basic earnings per share for 20X9.

Question 7

The following information is available for X Ltd for the year ended 31 May 20X1:

Net profit after tax and minority interest	£18,160,000
Ordinary shares of £1 (fully paid)	£40,000,000
Average fair value for year of ordinary shares	£1.50

Notes:

1 Share options have been granted to directors giving them the right to subscribe for ordinary shares between 20X1 and 20X3 at £1.20 per share. The options outstanding at 31 May 20X1 were 2,000,000 in number.

2 The company has £20 million of 6% convertible loan stock in issue. The terms of conversion of the loan stock per £200 nominal value of loan stock at the date of issue were:

Conversion date	*No. of shares*
31 May 20X0	24
31 May 20X1	23
31 May 20X2	22

No loan stock has as yet been converted. The loan stock had been issued at a discount of 1%.

3 There are 1,600,000 convertible preference shares in issue classified as equity. The cumulative dividend is 10p per share and each preference share can convert into two ordinary shares. The preference shares can be converted in 20X2.

4 Assume a corporation tax rate of 33% when calculating the effect on income of converting the convertible loan stock.

Required:

(a) Calculate the diluted EPS according to IAS 33.

(b) Discuss why there is a need to disclose diluted earnings per share.

Question 8

(a) The issued share capital of Manfred, a quoted company, on 1 November 2004 consisted of 36,000,000 ordinary shares of 75 cents each. On 1 May 2005 the company made a rights issue of 1 for 6 at $1.46 per share. The market value of Manfred's ordinary shares was $1.66 before announcing the rights issue. Tax is charged at 30% of profits.

Manfred reported a profit after taxation of $4.2 million for the year ended 31 October 2005 and $3.6 million for the year ended 31 October 2004. The published figure for earnings per share for the year ended 31 October 2004 was 10 cents per share.

Required:
Calculate Manfred's earnings per share for the year ended 31 October 2005 and the comparative figure for the year ended 31 October 2004.

(b) Brachly, a publicly quoted company, has 15,000,000 ordinary shares of 40 cents each in issue throughout its financial year ended 31 October 2005. There are also:

● 1,000,000 8.5% convertible preference shares of $1 each in issue classified as equity. Each preference share is convertible into 1.5 ordinary shares.

● $2,000,000 12.5% convertible loan notes. Each $1 loan note is convertible into two ordinary shares.

● Options granted to the company's senior management giving them the right to subscribe for 600,000 ordinary shares at a cost of 75 cents each.

The statement of comprehensive income of Brachly for the year ended 31 October 2005 reports a net profit after tax of $9,285,000 and preference dividends paid of $85,000. Tax on profits is 30%. The average market price of Brachly's ordinary shares was 84 cents for the year ended 31 October 2005.

Required:
Calculate Brachly's basic and diluted earnings per share figures for the year ended 31 October 2005.

(The Association of International Accountants)

* Question 9

(a) The Dent group earned profits from continuing operations attributable to the parent company for the year ended 31 October 2011 of $13.6 million. Losses from discontinued operations attributable to the parent company were $4 million. The group has a complex capital structure. The following transactions and events relate to changes in Dent's capital structure during the year ended 31 October 2011:

● *Ordinary shares.* The number of ordinary shares outstanding at 1 November 2010 was 6 million. On 1 February 2011 1 million ordinary shares were issued for cash.

● *Convertible bonds.* In 2009 10,000 $1,000 4% convertible bonds were issued for cash at par value. Each $1,000 bond is convertible into 35 ordinary shares. The entire issue was converted on 1 April 2011.

● *Preference shares.* On 1 November 2010, 100,000 non-convertible, non-redeemable cumulative preference shares classified as equity each with a par value of $100 were issued at $89 each. The shares are entitled to a cumulative annual dividend starting in two years' time and equivalent to the market rate dividend yield at the time of issue of 6%. In 2008 Dent had issued 1 million 5% convertible preference shares. Each share is convertible into one ordinary share. 75% of these were converted into ordinary shares on 1 May 2011. Preference dividends are paid half-yearly in arrears.

● *Warrants.* On 1 November 2010 Dent issued warrants to purchase 2 million ordinary shares at $6 per share for an exercise period of three years. All warrants were exercised on 1 October 2011. The average market price of each warrant during the period to 1 October 2011 was $7.50.

Required:

Calculate the basic earnings per share figures for Dent for the year ended 31 October 2011 and show how the information would be presented in Dent's financial statements.

(b) On 1 January 2011 Ram issued 100,000 ordinary shares for cash at $10 each. On 18 July 2011, Ram reacquired 10% of these shares for cash at a cost of $15 each. One month later the reacquired shares were all reissued for cash of $17 each. This type of transaction is new to the directors of Ram and they are unsure how they should be accounted for in the company's financial statements.

Required:

Advise Ram on the appropriate accounting treatment of these transactions in the company's financial statements.

(The Association of International Accountants)

References

1 J. Day, 'The use of annual reports by UK investment analysts', *Accounting Business Research*, Autumn 1986, pp. 295–307.
2 IAS 33 *Earnings per Share*, IASB, 2003.
3 Ibid., para. 10.
4 Ibid., para. 12.
5 Ibid., para. 26.
6 Ibid., para. 5.
7 Ibid., para. 31.
8 Ibid., para. 45.
9 Ibid., para. 33.
10 Ibid., para. 44.
11 Ibid., paras. 66 and 70.
12 Statement of Investment Practice No. 1, *The Definition of Headline Earnings*, IIMR, 1993.
13 Young-soo Choi, M. Walker and S. Young, 'Bridging the earnings GAAP', *Accounting*, February 2005, pp. 77–8.

CHAPTER 28

Review of financial statements for management purposes

28.1 Introduction

The key objective of financial statements is to provide useful financial information to the stakeholders, or 'users' – those with legitimate rights to such information. Different users have different information needs, for example:

- Existing and potential equity investors will be primarily interested in the profitability of an entity but will also require reassurance that the entity's liquidity (ability to generate cash) is such that it can continue in operational existence for the foreseeable future as a going concern.
- Lenders (both short- and long-term) will be primarily interested in the ability of the entity to generate the cash that is required to repay them and will focus on liquidity issues.
- Management will be concerned with both profitability (to satisfy the legitimate needs of the investors to whom they are accountable) and liquidity (to satisfy the legitimate needs of the lenders and suppliers to receive repayment of the amounts owed to them).

A financial analyst needs to be able to extract useful information from financial data, whether this is produced internally as detailed statements for the benefit of management or published externally for the benefit of external stakeholders, primarily the equity investors. The purpose of this chapter is to provide a framework for the analysis of financial data in order to write a report.

Objectives

By the end of this chapter, you should be able to:

- appreciate the potential of ratio analysis as an analytical tool;
- carry out an initial overview of financial statements;
- discuss the relationship between the return on capital employed and supporting accounting ratios through the 'pyramid of ratios';
- analyse the financial statements of a single entity;
- draft a report based on an inter-period and inter-firm comparison;
- explain the limitations of comparisons based on ratios.

28.2 Overview of techniques for the analysis of financial data

28.2.1 The 'golden rule of analysis'

This might be described as 'identify your yardstick of comparison'. Analysis without comparison is meaningless. For example, if you were simply told that an entity generated revenue of £10 million and made a profit of £900,000 it would be difficult or impossible to assess whether that was 'good' or 'bad' without reference to factors such as:

- the previous year's revenues and profits;
- the budgeted revenues and profits;
- the revenues and profits of competitors in the same industry; and
- the underlying expectations of the analyst based on their knowledge of relevant internal and external factors.

If we are making an inter-firm comparison for management purposes care has to be taken to select a company that is in the same industry.

Whilst it is possible to compare the return on investment that is obtainable in different industries when deciding whether to invest, it would be extremely difficult to compare management ratios in different industries looking at, say, how well managers are controlling the cash cycle. For example, the cash cycle of a retail company where customers generally pay on receipt of goods is completely different from that of a construction company. We need to be sure, as far as possible, that we are making a valid comparison.

28.2.2 The benefits of ratio analysis

The use of accounting ratios for analysis purposes has a number of important benefits for analysts:

- Ratios allow comparison with peers through inter-firm comparison schemes and comparison with industry averages so that possible strengths and weaknesses can be identified.
- Through the pyramid approach it is possible to carry out a structured analysis of financial performance and financial position by drilling down to identify ratios in ever greater detail, building up to the return on capital employed.
- It enables, for certain ratios, the comparison of entities of different sizes. For example, it is very difficult to compare the absolute profits of two entities without an appreciation of how 'large' one entity is relative to another. However, it might be perfectly legitimate to compare the ratio of profit to revenue of two entities of very different sizes in the same industry.

28.2.3 Ratio analysis – some notes of caution

In order to evaluate a ratio, it is customary to make a comparison with that of the previous year or with the industry average. However, remember to check if:

- The same accounting policies have been applied; for example, have non-current assets been reported using the same measurement bases (i.e. at depreciated cost or revalued amounts in both cases)?
  Inter-firm comparison schemes overcome this problem by requiring all member companies to report using uniform defined ratios.

- Note has been taken of different commercial practices. For example, some retail entities lease their properties on operating leases whilst others purchase them. *Accounting ratios that use assets as their denominator will be affected.*

- The ratios have been defined in the same way. This is important when comparing ratios from different companies' Annual Reports – check to see if the company has defined its ratios.

28.3 Ratio analysis – a case study

Vertigo plc is a family company which deals in building materials and garden supplies. It has been managed by non-family members since the principal shareholder/managing director retired from active management at the end of 20X6 on health grounds. Let us assume that you are a trainee in an accounting firm that has been approached by a client who is a family member for a report on the company's financial position and financial performance following a fall in profit available for dividend and a request by the management for an injection of more capital.

We will use the financial statements of Vertigo (see below) to illustrate the technique.

28.3.1 Financial statements for the case study

Vertigo plc: statement of income for year ended 31 December

	20X9		20X8	
	£000	£000	£000	£000
Revenues		3,461		3,296
Opening inventory	398		253	
Purchases	2,623		2,385	
Closing inventory	(563)		(398)	
Cost of goods sold		(2,458)		(2,240)
Gross profit		1,003		1,056
Distribution costs:				
Depreciation	187		239	
Irrecoverable debts	17		32	
Advertising	24		94	
		(228)		(365)
Administrative expenses:				
Rent	60		60	
Salaries and wages	362		316	
Miscellaneous expenses	177		159	
		(599)		(535)
Operating profit		176		156
Dividend received		—		51
Finance costs		(60)		(53)
Profit before tax		116		154
Income tax expense		(25)		(39)
Profit after taxation		91		115

Vertigo plc: statement of financial position at 31 December

	20X9	20X8
	£000	£000
ASSETS		
Non-current assets:		
Machinery	2,100	2,240
Motor vehicles	394	441
Investments	340	340
	2,834	3,021
Current assets:		
Inventory	563	398
Trade receivables	1,181	912
Cash and cash equivalents	9	11
	1,753	1,321
	4,587	4,342
EQUITY AND LIABILITIES		
Equity:		
Ordinary shares of 50p each	3,000	3,000
Retained earnings	353	262
	3,353	3,262
Non-current liabilities:		
Long-term borrowings (repayable in 8 years)	600	600
Current liabilities:		
Trade payables	498	398
Accrued expenses	15	12
Taxation	24	29
Short-term borrowings	97	41
	634	480
	4,587	4,342

28.4 Introductory review

Before embarking on detailed ratio analysis, an analyst (whether an internal or an external user of the financial statements) would carry out a review to gain an overall impression of:

(a) the external trading conditions for the building materials sector, for example, refer to subscription sources such as the Markit/CIPS Purchasing Managers' Index (PMI) indices; and

(b) the financial statements as a whole.

We will illustrate one approach using common-sized statements for Vertigo before proceeding to consider more detailed ratios and the preparation of a report.

Overall impressions from initial review

Common-sized statements are a useful aid when making an initial review of a company's financial structure, such as seeing the percentage of cash to current assets, and cost structures such as the percentage of sales revenue that goes on administration.

28.4.1 The company's financial structure

Our first thought might be to gain an impression of the financial structure of a company.

Vertical analysis – common-sized statement

The vertical analysis approach highlights the structure of the statement of financial position by presenting non-current assets, working capital, debt and equity as a percentage of debt plus equity. It allows us to form a view on the financing of the business, in particular the extent to which a business is reliant on debt to finance its non-current assets. In times of recession this is of particular interest and is described as indicating the strength of the financial position.

	20X8 £000	20X8 %	20X9 £000	20X9 %
Non-current assets	3,021	69.6	2,834	61.8
Current assets	1,321	30.4	1,753	38.2
Total	4,342	100	4,587	100
Equity	3,262	75.1	3,353	73.1
Debt	600	13.8	600	13.1
Current liabilities	480	11.1	634	13.8
Total	4,342	100	4,587	100

This indicates that the financial strength is maintained in terms of the amount of debt compared to the amount of capital put in by the shareholders.

However, the non-current assets have fallen and the fall appears to be due to the depreciation charge. We need to assess whether this lack of investment in non-current assets is likely to be a concern for the future and to check if the management has identified and quantified future capital expenditure commitments.

Horizontal analysis – common-sized statement

A horizontal analysis looks at the percentage change that has occurred. We could calculate the percentage change for every asset and liability, but it is more helpful in Vertigo to concentrate on the area that seems to require closer investigation, i.e. current assets and liabilities. The analysis is as follows:

	20X8 £000	20X9 £000	Percentage change
Current assets:			
Inventory	398	563	+41.5
Trade receivables	912	1,181	+29.5
Cash and cash equivalents	11	9	−18.2
Trade payables	398	498	+25.1
Accrued expenses	12	15	+25.0
Taxation	29	24	−17.2
Bank overdraft	41	97	+136.6

Inventories and (to a lesser extent) trade receivables have risen significantly when we consider that sales have increased by only 5%.

This raises questions in our mind. For example, is it possibly because greater quantities of inventory are expected to be required in anticipation of growth in future sales? Alternatively, is the inventory slow-moving with the possibility that net realisable value is lower than cost?

Trade payables have increased significantly. This could be due to poor cash flow putting pressure on liquidity (short-term borrowings have increased by around £50,000 and there has been no additional long-term equity or loan finance).

The common-sized analysis of the financial position has given us questions to have in our minds when carrying out a more detailed analysis. The next step would be to extract detailed turnover ratios for inventory, trade receivables and payables and ascertain the terms and limit of the overdraft. Before doing that we carry out a similar common-sized exercise to form a view of a company's cost structure.

28.4.2 The company's cost structure

Again, both a vertical and horizontal analysis is helpful.

Vertical analysis – common-sized statement

An overview is obtained by restating by function into a vertical common-sized statement format as follows:

	20X8 £000	20X8 %	20X9 £000	20X9 %
Sales	3,296	100.0	3,461	100.0
Cost of sales	2,240	68.0	2,458	71.0
Total gross profit	1,056	32.0	1,003	29.0
Distribution costs	365	11.1	228	6.6
Administration expenses	588	17.8	659	19.0
Net profit before tax	103	3.1	116	3.4

We can see that there has been a change in the cost structure with a fall in the gross profit from 32% to 29% compensated for by a significant fall in the distribution costs.

Horizontal analysis – common-sized statement

An overview is obtained by calculating the percentage change as follows:

	20X8 £000	20X9 £000	Percentage change
Sales	3,296	3,461	+5.0
Cost of sales	2,240	2,458	+9.7
Total gross profit	1,056	1,003	−5.0
Distribution costs	365	228	−37.5
Administration expenses	588	659	+12.1
Net profit before tax and dividend income	103	116	+12.6

Our initial observations are as follows:

● Revenues have risen slightly but gross profits have fallen. We need to establish the reasons for this.

● Other operating expenses (distribution costs and administrative expenses) have fallen significantly. This appears in the main to be caused by the reduction in depreciation charges and advertising expenditure.

● No income has been received from the financial asset in the period. This may be due to timing issues (given that dividend income is basically recognised only when received). However, we would need to carry out further investigations here.

We can now go on to a more detailed analysis.

28.5 Financial statement analysis, part I – financial performance

Return on investment

If the analysis is being performed exclusively for the shareholders then an appropriate ROI measure might be 'Return on Equity (ROE)'. This ratio would be calculated as:

$$\frac{\text{Profit attributable to the shareholders}}{\text{Equity}}$$

For Vertigo, ROE would be

	20X9	20X8
Profit after tax	91	115
Equity	3,353	3,262
So ROE equals	2.7%	3.5%

This shows a fall of more than 20%.

Return on capital employed

If the analysis is of the overall performance of the entity (however it is financed) then the appropriate ratio is 'Return on Capital Employed (ROCE)'. Management would be likely to consider this to be the best measure of ROI, as it shows the return on the assets under their control without any effect from the rates of tax and interest which operational management might regard as outside their control.

This ratio would be calculated as:

$$\frac{\text{Profit before interest and tax (PBIT)}}{\text{Capital employed (CE) (equity + borrowings)}}$$

Definitions of ratios vary

It should be remembered that there is no 'accounting standard' that governs the exact composition of this ratio and care needs to be taken when making inter-firm comparisons. For example, capital employed might be defined as:

(a) total assets, also expressed as equity plus long-term loans plus current liabilities; or

(b) net assets, also expressed as equity plus long-term loans. Even here, though, care is needed – if a company maintains a high level of relatively permanent overdraft it might be added to the long-term loans.

For the purposes of this analysis, we will take 'borrowings' to be long-term borrowings only. Therefore our ROCE would be as follows:

	20X9	20X8
Profit before interest and tax	116 + 60	154 + 53
Capital employed	3,353 + 600	3,262 + 600
So ROCE equals	4.4%	5.4%

Our initial conclusion would be that Vertigo is less profitable in 20X9 than it was in 20X8. We would need to investigate further to establish the reasons for this. In looking for a reason for the fall from 5.4% to 4.4% we propose to follow the pyramid approach.

Figure 28.1 Pyramid for return on capital employed

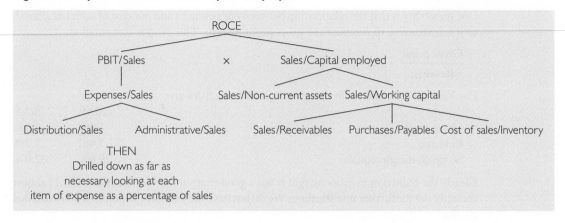

28.5.1 The pyramid approach

In this approach we start at the top of the pyramid with the return on capital employed and systematically analyse those ratios that impact on the profit and those that impact on the assets employed in the business. This approach is also the basis for a number of inter-firm comparison schemes.

Diagrammatically the pyramid is shown in Figure 28.1. We can see the pyramid starts with the following relationship:

$$\frac{\text{Profit before interest and tax}}{\text{Capital employed}} = \frac{\text{Profit before interest and tax}}{\text{Revenue}} \times \frac{\text{Revenue}}{\text{Capital employed}}$$

This is often expressed as:

ROCE = Profit margin × Asset turnover

This shows us that the two key components of the return on capital employed are **'margin'** (PBIT/Revenue) and **'volume'** (Revenue/Capital employed). It is to these two aspects that we now turn.

28.5.2 Margin and expense analysis

The first 'margin ratio' we compute is the 'net profit margin'. This is simply:

$$\frac{\text{'Profit' (PBIT as used in the ROCE ratio)}}{\text{Revenue}}$$

For Vertigo, the profit margins for 20X9 and 20X8 are:

	20X9	20X8
Profit	176	207
Revenue	3,461	3,296
So profit margin equals	5.1%	6.3%

This shows us that one of the reasons for the decline in ROCE is a decline in the profit margin. For our report we 'drill down' into the detail and investigate further why the margin has reduced. Is it because the gross profit has fallen or is it due to expenses?

Gross profit

One possibility is that the relationship between our revenues and our cost of sales has altered, so it is instructive to compute the gross profit margin. This ratio is computed as:

$$\frac{\text{Gross profit}}{\text{Revenue}}$$

For Vertigo, the gross profit margins for 20X9 and 20X8 are:

	20X9	20X8
Gross profit	1,003	1,056
Revenue	3,461	3,296
So profit margin equals	29.0%	32.0%

Clearly the reduction in gross margin is not a good thing and internal analysts would almost certainly call for further investigation. We do not have the data here to perform more detailed checks.

Remember, however, that in answering any interpretation question it is always important to identify the further questions you would ask and the further information you would request, giving your reasons. For example, questions would be asked as to whether there has been:

- a change in the sales mix, with a greater proportion of lower-margin items being sold this year than last year;
- a change to maintain sales volume at the expense of the profit margin;
- discounting or longer-running sales;
- a rise in raw material costs that could not be passed on to customers in the form of increased sales prices; or
- a rise in the employment costs of production workers that could not be passed on to customers in the form of increased sales prices. This is unlikely to be the reason for the change in the gross margin here, given that cost of sales appears to include purchases, rather than production costs. Apparently Vertigo is a retail organisation rather than a manufacturing organisation – unless, perhaps, it is involved in also constructing any of the building products such as conservatories and garden studios.

Operating expenses – administrative expenses

We could also compute:

	20X9	20X8
Administrative expenses	599	535
Revenues	3,461	3,296
Ratio	17.3%	16.2%

This ratio reveals a slightly less satisfactory position in 20X9 compared with 20X8. The information we have shows us that a key factor behind the increase is the rise in salary costs of approximately 15%. This seems excessive given that revenues have grown by only 5%.

Operating expenses – distribution costs

A further part of the analysis of the profit margin is to investigate the relationship between other operating expenses and revenues. For Vertigo, this would involve computing:

	20X9	20X8
Distribution costs	228	365
Revenues	3,461	3,296
Ratio	6.6%	11.1%

Clearly, for Vertigo, the adverse movement in the gross profit margin is at least partly mitigated by a reduction of the percentage of distribution costs to revenues. Given the information we have for Vertigo (not necessarily available to an external analyst), we can see that there has been a significant reduction in depreciation (all of which has been charged to this expense heading) and advertising costs.

We would at this stage drill down further, in the same way as when designing audit tests, to target areas of significant change.

	20X8 £000	20X9 £000	Percentage change
Sales revenue	3,296	3,461	+5.0
Inventory – opening	253	398	
Purchases	2,385	2,623	+10.0
Inventory – closing	(398)	(563)	+41.5
Cost of goods sold	(2,240)	(2,458)	+9.7
Gross profit	1,056	1,003	−5.0
Distribution costs:			
Depreciation	239	187	−21.8
Bad debts	32	17	−46.9
Advertising	94	24	−74.5
Administrative expenses:			
Rent	60	60	—
Salaries and wages	316	362	+14.6
Miscellaneous expenses	159	177	+11.3
Operating profit	156	176	+12.8

It is interesting to see that discretionary costs in the form of advertising have been reduced by 74.5%. However, if the advertising had been maintained at 20X8 levels the operating profit would be reduced by £70,000 to £106,000, which would have shown a fall from the previous year of 32% rather than an increase of 12.8%.

This is where it is important to look at trends, in particular from 1 January 20X6 which was the last year when the previous Managing Director had been in control. There should be further enquiry to establish (a) the normal level over the previous three years – whether there was heavier advertising in 20X8 to achieve the 5% increase in sales in the light of the company's intention to attempt to obtain further investment in 20X9; (b) whether the reduction is likely to have an adverse effect on future sales; (c) what the company's reason was for reduced spending; and (d) the necessity or otherwise to return to a higher level in future years. This is more of commercial relevance to the client who is already concerned about the fall in profits than audit relevance.

28.5.3 Volume analysis – asset turnover

The basic 'volume ratio' is:

$$\frac{\text{Revenue}}{\text{Capital employed}}$$

This ratio is commonly referred to as the *asset turnover ratio*. For Vertigo, this ratio is:

	20X9	20X8
Revenue	3,461	3,296
Capital employed	3,953	3,862
Asset turnover	87.6%	85.3%
Turnover expressed as a multiple	0.876 ×	0.853 ×

This shows us that the asset turnover has in fact slightly improved in 20X9 compared with 20X8. Therefore an overall conclusion we can make is that the decline in ROCE is due to a declining margin rather than a decline in the utilisation of assets. Using our formula we can now see that:

ROCE (4.4%) = profit margin (5.1%) × asset turnover (0.876)

Although the asset turnover rate has improved, we still need to analyse the reasons for the change, because the change can have resulted from changes in sales or any of the non-current and current assets.

Non-current asset turnover

The non-current asset turnover is:

$$\frac{\text{Revenue}}{\text{Non-current assets}}$$

For Vertigo, this ratio is:

	20X9	20X8
Revenue	3,461	3,296
Non-current assets	2,834	3,021
Non-current asset turnover	122.1%	109.1%
Turnover expressed as a multiple	1.22 ×	1.09 ×

From a profitability point of view, this is an improvement. However, we should remember that there has been no investment in non-current assets this year and, after depreciation, the asset turnover would appear to have improved simply because the written-down value of the non-current assets is lower.

An increasing ratio is not always an improving ratio and might not always be good for the long-term health of the business. For example, if we had made an investment in non-current assets this year we would have quite possibly replaced older, fully depreciated, assets with newer assets that have higher net book values. This might be good for the long term but in the short term the fall in the rate of turnover of non-current assets would have a negative impact on the ROCE.

New, growth companies are likely to have a fall in the rate of non-current asset turnover as they expand. We must take care that our use of ratios does not take us into 'short-term thinking'.

Asset turnover – working capital

When we analyse net current assets (or 'working capital') we generally do this by an individual focus on the three key components of inventory, trade receivables and trade payables.

Inventory turnover

The ratio we use to assess the effectiveness of our inventory management is the 'inventory days ratio'. This would normally be computed as:

$$\frac{\text{Closing inventory} \times 365}{\text{Cost of sales}}$$

The rationale behind the ratio is that we are effectively dividing closing inventory by 'one day's usage' to give us a hypothetical period for how long it will take us to sell the inventory. Whilst this analysis can be useful, we need to sound two notes of caution:

- We are relating the closing inventory to the average 'usage' in the previous year. The closing inventory will of course be used next year and so a more 'realistic' figure would be to base it on next year's projected usage, but of course this often is not available to the analyst.

- With this (and other) ratios we are comparing a 'point of time' figure (closing inventory) with a 'period' figure (cost of sales).

To a certain extent, both of the above factors are at least partly mitigated by the fact that, when using ratio analysis, we are comparing one ratio with another, and if the above factors apply to both the ratio and its comparative, to a certain extent the above 'defects' can cancel each other out.

That said, our inventory days ratio will be:

	20X9	20X8
Closing inventory × 365	$\dfrac{563 \times 365}{2,458}$	$\dfrac{398 \times 365}{2,240}$
Cost of sales		
Inventory days	84 days	65 days
Turnover expressed as a multiple	4.4 ×	5.6 ×

Inventory is not being turned over as quickly in 20X9. This is not a positive sign. Not only does it affect the profitability of Vertigo but it also affects its liquidity, as we will see in the next section.

Trade receivables

The second key component of working capital is trade receivables. The equivalent ratio for trade receivables is:

$$\frac{\text{Trade receivables} \times 365}{\text{Revenue}}$$

For Vertigo, this ratio would be

	20X9	20X8
Trade receivables × 365	$\dfrac{1,181 \times 365}{3,461}$	$\dfrac{912 \times 365}{3,296}$
Sales		
Trade receivables days	125 days	101 days
Turnover expressed as a multiple	2.9 ×	3.6 ×

It appears that Vertigo is collecting its cash from its customers less quickly in 20X9 than was the case in 20X8. This has a negative impact on profitability as the working capital cycle is lengthened when customers take longer to pay. This in turn has a negative impact on liquidity.

Late payment is a serious problem and a study in 2012 by the Clydesdale Bank and Yorkshire Bank in the UK reported that 10% of businesses say closing or seriously scaling back operations would have to be looked at if customers took more than 90 days to pay invoices. This poses a problem for management who need to tighten up their systems and controls and introduce procedures such as agreeing payment terms and conditions upfront or using incentives for early payment. Vertigo's management need to review their current procedures.

Trade payables

The third key component of working capital is trade payables. The equivalent ratio for trade payables is:

$$\frac{\text{Trade payables} \times 365}{\text{Credit purchases}}$$

For Vertigo, this ratio would be

	20X9	20X8
Trade payables × 365	498 × 365	398 × 365
Credit purchases	2,623	2,385
Trade payables days	69 days	61 days
Turnover expressed as a multiple	5.3 ×	6.0 ×

It appears that Vertigo is taking slightly longer to pay its suppliers in 20X9 than in 20X8. Given the way we have computed the profitability ratios (capital employed is total assets less current liabilities) this will actually improve the asset turnover and hence the ROCE. Given that our suppliers effectively provide us with interest-free finance there is, in a sense, a liquidity benefit in extending the credit we take from our suppliers.

However, this can also be indicative of liquidity problems that make it difficult for us to settle our debts as they fall due and, if we allow the level of our trade payables to get too high, it could lead to problems with future supplies and ultimately could lead to the entity being wound up. Overall the 'real' level of trade payables of Vertigo is probably not a major concern but management will need to monitor this going forward.

It should be noted that, whilst the trade payables ratio can be calculated from the accounts of Vertigo, those accounts are more detailed than the information available in the published financial statements. Credit purchases would not normally be available from the published financial statements. In practice external analysts would use cost of sales as a 'proxy' for credit purchases. As stated before, whilst this practice clearly isn't strictly correct, the fact that interpretation involves a comparison of ratios means that, if used consistently, this slightly contrived ratio can be used as a means of comparing the payment policies of a single entity over time or two comparable entities over a corresponding period.

The cash cycle

The cash cycle, also referred to as the cash conversion cycle, measures the number of days it takes to acquire and sell inventory and convert sales into cash. It measures how effective managers are in managing this process.

For Vertigo the cash cycle is:

Accounts Receivable days + Inventory days − Accounts Payable days = Cash Cycle
125 + 84 − 69 = 140 days

This means it takes Vertigo 140 days from the time the company acquires inventory from its suppliers, completes the sale of the inventory to its customers and collects the cash from accounts receivable.

The 140 days can be regarded as the length of time the company needs to have cash to cover the cash cycle or, thinking defensively, to cover its operating expenses. The means that the management of the cash cycle is critical to the cash flow and profitability of the company.

High working capital turnover rate

As with all ratios, a high rate does not always indicate that it is acceptable. For example, a high turnover rate can indicate overtrading, i.e. the sales volume is excessive in relation to the equity investment in the business. A high turnover might be an indication that the business relies too much on credit granted by suppliers or the bank instead of providing an adequate margin of operating funds.

28.6 Financial statement analysis, part 2 – liquidity

Liquidity is the lifeblood of any business. The ultimate price for poor liquidity is insolvency and therefore internal managers cannot ignore it. External users who have lent or who are thinking about lending money to the entity, whether on a short-term or a long-term basis, will almost certainly be more concerned with liquidity than with profitability.

Analysts can consider the liquidity of an entity in two ways. The first is through ratio analysis. We discussed in the previous section the fact that investment in working capital, as revealed when calculating changes in inventory days, trade receivables days and trade payables days, had an impact on liquidity.

However, there are also other ratios that are commonly used to assess liquidity. These include the current ratio, the quick ratio and cash flow ratios.

28.6.1 The current ratio

This ratio is simply the ratio of current assets to current liabilities. In the case of Vertigo, this ratio would be:

	20X9	20X8
Current assets	1,753	1,321
Current liabilities	634	480
So current ratio equals	2.76	2.75

The rationale behind the ratio is that the current assets are a short-term source of cash for the entity, whilst the current liabilities are the amounts that need settling reasonably quickly.

It is very difficult to give a general level for this ratio which analysts would regard as 'satisfactory' because different entities vary so much in their working capital cycles. In most cases you would expect this ratio to be well in excess of 1 for analysts to feel comfortable. However, entities that can generate cash easily are often able to operate with current ratios well below 1.

Consider a food retailer: food retailers have little if any trade receivables, since they sell to their customers for cash. Their inventory levels necessarily have to be quite low, since their products are often perishable. However, their trade payables days would be just as large as for any manufacturing entity and, if they reinvest the cash they generate quickly, their current ratios are often less than 1/2:1. This does not mean they have liquidity problems, however!

Rather, therefore, than identifying an absolute level at which the current ratio should be, it is probably better to monitor whether or not there has been a significant change from one period to another and compare with the industry average or peer group. Comparing it with the previous year we can see that it is virtually unchanged – this does not mean, however, that it is acceptable. We would need to look further at the trend over the past four years and also at competitors' current ratios.

An increase in the current ratio beyond the company's own normal range may arise for a number of reasons, some beneficial, others unwelcome.

Beneficial reasons

These include:

- A build-up of inventory in order to support increased sales following an advertising campaign or increasing popular demand as for, say, a PlayStation. Management action will be to establish from a cash budget that the company will not experience liquidity problems from holding such inventory, e.g. there may be sufficient cash in hand or from operations, short-term loans, extended credit or bank overdraft facilities.

- A permanent expansion of the business which will require continuing higher levels of inventory. Management action will be to consider existing cash resources or future cash flows from operations or arrange additional long-term finance, e.g. equity or long-term borrowings to finance the increased working capital.

Unwelcome reasons

These include:

- Operating losses may have eroded the working capital base. Management action will vary according to the underlying problem, e.g. disposing of underperforming segments, arranging a sale of non-current assets or inviting a takeover.
- Inefficient control over working capital, e.g. poor inventory or accounts receivable control allowing a build-up of slow-moving inventories or doubtful trade receivables.
- Adverse trading conditions, e.g. inventory becoming obsolete or introduction of new models by competitors.

28.6.2 The quick ratio

Another ratio that is used for liquidity assessment purposes is the quick ratio (also known as the acid test ratio). This ratio is:

$$\frac{\text{Current assets} - \text{inventory}}{\text{Current liabilities}}$$

The rationale for using the quick ratio is that entities cannot regard their inventory as a short-term source of cash because of the time it takes to realise cash through its sale. Whether this is true depends on the nature of the entity. This would certainly be true for entities in the construction sector, but for many entities in the retail sector, particularly those entities that sell their goods directly to the general public for cash, the current ratio would be a better measure of liquidity.

For Vertigo, the quick ratio would be:

	20X9	20X8
Current assets – inventory	1,753 – 563	1,321 – 398
Current liabilities	634	480
So quick ratio equals	1.88	1.92

There has been a small decline in this ratio given the higher trade payables levels but the decline is not significant.

28.6.3 Cash flow ratios

Even if a statement of cash flows is not provided in a question it is worth preparing and analysing one. If we prepared such a statement for Vertigo for the year ended 31 December 20X9 we would get the following:

	£000	£000
Profit before tax	116	
Finance costs	60	
Depreciation	187	
Increase in inventory	(165)	
Increase in trade receivables	(269)	
Increase in trade payables	100	
Increase in accrued expenses	3	
Cash generated from operations		32
Interest paid		(60)
Tax paid		(30)
Reduction in cash and cash equivalents		(58)
Cash and cash equivalents, 1 January 20X9 (11 – 41)		(30)
Cash and cash equivalents, 31 December 20X9 (9 – 97)		(88)

This statement shows that the entity is struggling to generate cash from its operations. This is mainly due to the increased levels of working capital; all three components have increased in real terms as we have already seen.

This increase has absorbed significant amounts of cash such that cash from operating activities is negative. There has been no investment in non-current assets or additional equity or loan capital raised, and cash flow fails to cover the current year's interest and any dividend payments.

Interest cover

The lenders would be interested in their interest cover, i.e. the number of times that their interest could be paid out of cash generated by the operations. In this case, the interest cover is $32/60 = 0.53$ times. Notice that this is far worse than the interest cover based on the statement of income which indicates that there is adequate cover at 2.93 times (176/60).

Servicing future debt

As we have already seen, the entity has not purchased any non-current assets this year but sooner or later they may have to. Their borrowing levels are not currently excessive (see Section 28.7 below) but their ability to service additional debt is questionable. Based on this, it would appear that consideration may need to be given to raising further long-term finance if future expansion of the business is envisaged.

28.6.4 The cash ratio

This is a more conservative ratio than the quick ratio as it shows the ratio of cash and cash equivalents to current liabilities. Suppliers are able to see whether these are enough to settle the amount owed to them. In this case, of course, it is a negative figure.

It is certainly not a problem that faces Vertigo but there are companies sitting on hoards of cash to meet cyclical demands or because they are nervous about investing in the uncertain economic climate or they are unable to find investment opportunities. We see major companies like Apple, therefore, setting aside US$10 billion for stock buybacks.

28.7 Financial statement analysis, part 3 – financing

One of the key issues for analysts is the way a business is financed. Of particular concern is the relationship between borrowings (debt finance) and equity finance. Because most equity investors are risk-averse, the return required by the providers of debt finance is lower than that required by equity investors as they would normally have fixed or floating security. However, management must balance the benefit of 'cheaper' debt finance against the fact that the greater the proportion of finance provided through borrowings the greater the risk for both as measured by the gearing ratio.

28.7.1 The gearing ratio

There are a number of ways in which the gearing ratio can be computed but the two most common are:

$$\frac{\text{Debt finance}}{\text{Debt finance} + \text{equity finance}}$$

and

$$\frac{\text{Debt finance}}{\text{Equity finance}}$$

Both these ratios will increase as the proportion of debt finance gets greater. We will use the former ratio to illustrate the gearing of Vertigo:

	20X9	20X8
Debt finance (long-term only)	600	600
Debt finance + equity finance	600 + 3,353	600 + 3,262
So gearing ratio equals	15.2%	15.5%

Gearing is relatively stable, the only fluctuation being caused by the retention of 20X9 profits increasing equity whilst long-term borrowings stay static. It is difficult to generalise, but this is a relatively low gearing ratio – ratios of less than one-third would normally be regarded as 'low' and gearing would normally only be regarded as 'high' when it exceeded 50%. There would appear to be plenty of scope for Vertigo to obtain more debt finance subject to being able to produce forecasts showing its ability to service the debt.

28.7.2 How should a potential investor decide on an acceptable level of gearing?

This is initially influenced by the political and economic climate of the time. We have seen that prior to the credit crisis arising in 2007 high gearing was not seen by many as risky and there was a general feeling that borrowing was good, leverage was respectable, and capital gains were inevitable. This might have reduced the importance of questions that would normally have been asked, such as the following.

Asset values

- Are the values in the statement of financial position reasonably current? If much lower than current then the gearing ratio may be significantly overstated.

Gearing ratios

- Is the gearing ratio constant or has it increased over time with heavier borrowing? If higher:
  - further borrowing might be difficult;
  - it might indicate that there has been investment that will lead to higher profits, so details are needed as to how the funds borrowed have been used.
- What covenants are in place and what is the risk that they might be breached? A breach could lead to a company having to renegotiate finance at a higher interest rate or even go into administration or liquidation.
- How does the gearing compare to other companies in the same sector?

Use of funds

- If gearing has increased, what were the funds used for? Was it to:
  - restructure debt following inability to meet current repayment terms?
  - finance new maintenance/expansion capital expenditure?
  - improve liquid ratios?

Interest commitment

- How variable is the rate of interest that is being charged on the borrowings? If rates are falling then equity shareholders benefit, but if rates rise then expenses are higher.
- How many times does the earnings before tax cover the interest? A highly geared company is more at risk if the business cycle moves into recession because the company has to continue to service the debts even if sales fall substantially.
- How many times does the cash flow from operations currently cover the interest? This is a useful ratio if profits are not converted into cash, e.g. they might be reinvested in working capital.

Cash flows

- How variable is the company's cash flow from operations? A company with a stable cash flow is less at risk, so the trend is important.
- What is the likely effect of contingent liabilities if they crystallise on the cash flows and debt ratio? Could it have a significant adverse impact?

A company's attitude to leverage may vary over time

This is often dependent on the availability of finance and the possibility of profitable capital investment. If there is uncertainty about either then there will an unwillingness to lend and an unwillingness to borrow.

28.8 Peer comparison

We have so far prepared internal ratios for two years making our comparison with 20X8. We have now selected comparative ratios from a competitor and set out some comparative ratios where Vertigo's ratios seem too high or too low:

	20X9	20X8	20X7	20X6	20X5
Asset turnover ratio:					
Vertigo	0.88	0.85			
Competitor*	2.77	2.10	1.96	1.59	1.43
Inventory turnover:					
Vertigo	84 days	65 days			
Competitor	58 days	70 days	62 days	80 days	87 days

* For illustration, the competitor comparisons were ratios reported in the Everest Industries 2012 Annual Report.

	20X9	20X8	20X7	20X6	20X5
Profit before interest and tax margin:					
Vertigo	5.1%	6.3%			
Competitor	7.9%	6.41%	7.2%	6.9%	4.38%
Debt / equity ratio:					
Vertigo	0.15	0.15			
Competitor	0.28	0.53	0.69	1.13	0.94
Current ratio:					
Vertigo	2.76	2.75			
Competitor	0.86	1.33	1.07	0.90	0.89
Quick ratio:					
Vertigo	1.88	1.92			
Competitor	0.66	0.63	0.70	0.67	0.70

Looking at the profit before interest and tax, it is interesting to see that the competitor has had a rising trend over the five years with alternating positive and negative changes but the overall trend is up. An examination of the past five years' figures for Vertigo would be helpful in identifying its trend.

The inventory turnover has risen in 20X9 for Vertigo but it is interesting to see that again the trend with the competitor is falling with uneven positive and negative changes over the five years. The competitor has clearly addressed the level of inventory held in the last year. This could well indicate that a target of 70 days for Vertigo should be achievable.

The debt/equity ratio is steady at 0.15 in Vertigo. This is almost half of the gearing in the competitor where the gearing has fallen year on year to less threatening levels.

The asset turnover, however, paints a different picture with the competitor turning over its assets three times faster than Vertigo. This would seem to indicate that Vertigo needs to work its assets more effectively and aim at increasing its sales. A 5% increase in sales compares with a 22% increase in the competitor's sales.

The current ratio and quick ratio are more than double those of the competitor whose trend figures show that it is operating on levels of less than 1:1 for both ratios.

Note that it is important to obtain a comparator from the same industry and size, as far as possible.

28.9 Report based on the analysis

A report based on the above analysis might read as follows:

Report:
From:
To:
Date:

Subject: Financial Performance of Vertigo Ltd

Profitability

Vertigo's profitability has declined compared with 20X8, with the ROCE declining from 5.4% in 20X8 to 4.4% in 20X9. This decline is mainly due to a reduction in the profit margin (see Appendix). The reduction is a combination of three factors:

- A reduction in the gross margin. Reasons for this need to be investigated further.

- An increase in administrative expenses. This is mainly caused by a 15% rise in salary costs which is a little surprising given the rise in revenue is only 5%.

- The reduction in the profit margin is slightly mitigated by a fall in distribution costs. The key reason for this is a significant reduction (almost 75%) in advertising expenditure. This reduction might be beneficial for profitability in the short term, but as a long-term measure this may be unwise.

Liquidity

Liquidity ratios are conservative but seem excessive when compared to the current and quick ratios of the competitor (see Appendix).

The cash generated from operations is very low given the level of profits and this amount does not cover the interest and tax payments made in the year. During 20X9 the cash balances declined by £88,000. A key reason for the disappointing cash flow is the significant increase in working capital, particularly inventory and trade receivables.

The reason for the increase in the inventory turnover needs to be discussed further with management. As far as the impact on cash is concerned, if inventory is brought back to the 65 days turnover level, the increase of £165,000 would be reduced by more than £120,000 – more than enough to pay off the existing short-term borrowings. An improvement to 70 days would be sufficient to clear all short-term borrowings.

Further investigation of the management of receivables is required, particularly in the present credit climate.

The overall rise in working capital is mitigated to a certain extent by a rise in trade payables. This needs to be carefully monitored to ensure that the credit status of Vertigo is not compromised.

Financial position

As stated above, overall liquidity ratios are unchanged in both years but the management of the working capital needs to be addressed. There has been no investment in non-current assets during 20X9 and the shareholders have not received a dividend. Both these factors may be due to a cash shortage and Vertigo would appear to require additional long-term finance. Compared to the competitor the gearing is low which, on the basis of the current level of borrowing, would allow Vertigo to seek additional debt finance.

Conclusion

Profits are under pressure. Although revenues are continuing to rise there appears to be a decline in the gross margin which needs investigating further.

As far as the possibility of an improvement in profitability is concerned, there is concern that the asset turnover is low and sales are increasing but at a slower rate than the competitor's. There has at the same time been a significant reduction in advertising spend, which seems strange in a competitive environment and with the slow rate of sales growth.

It is noted that there has been no investment in non-current assets in 20X9. It is not clear without further enquiry whether the current level of non-current assets can sustain an increase in sales. If not, the need for further capital expenditure could not be provided by the current level of operating cash-flow.

Further attention urgently needs to be paid to working capital management.

As far as obtaining additional loan or equity capital, profitability needs to be addressed. The ROE is low at 2.7% and operating cash is insufficient to fully cover interest payments. The more positive aspect is that, given an improvement in profitability, it would appear possible to obtain this through issuing more debt as gearing levels are fairly low.

To support a request for additional funding a feasible three-year forecast would be required and we would be pleased to assist with this if so instructed.

Appendix – detailed ratios (not reproduced here as computed earlier)

Subscription sources are available for inter-firm ratios such as *RMA Annual Statement Studies* (Risk Management Association).[1]

28.10 Caution when using ratios for prediction

At the beginning of the chapter we mentioned the importance of taking an overview which influenced your expectations as to, say, the level of sales or profits that could be expected.

The same approach has to be taken when interpreting the ratios. This involves considering external and internal factors that could help explain current ratios and what might be predicted from them.

28.10.1 External factors

There are a number of external factors that need to be considered when interpreting ratios bearing in mind the economic context within which a business has been and will be operating. Consider, for example, assuming that Vertigo is a retail company:

- Have the retail sales been adversely or positively affected by growth of Internet sales?
- Has there been a change in fashion or downturn in the market?
- Will this mean inventory write-downs? Discounted sales?
- Have wage costs gone up (or will they be going up) following legislation for equal pay for women, fairer pay for part-time staff and legislation for maternity and paternity leave?
- Have credit sales been affected by less being spent on non-essential items?
- Has the company had to respond to pressure to pay small suppliers on time?
- Has there been a change in the sales mix that has impacted (or will impact) on sales or profits?
- Is property leased and, if so, are any rent reviews due? Are there any onerous covenants on the leases?

28.10.2 Internal factors

There are internal factors to consider:

- Ratios need to be interpreted in conjunction with reading the narrative and notes in the annual reports. The narrative could be helpful in explaining changes in the ratios, e.g. whether an inventory build-up is in anticipation of sales or a fall in demand. The notes could be helpful in corroborating the narrative, e.g. if the narrative explains that the increase in inventory is due to anticipated further production and sales, check whether the non-current assets have increased or whether there is a note about future capital expenditure.
- Ratios might be distorted because they are based on period-end figures. The end-of-year figures are static and might not be a fair reflection of normal relationships such as when a business is seasonal, e.g. an arable farm might have no inventory until the harvest and a toy manufacturer might have little inventory after supplying wholesalers in the lead-up to Christmas. Any ratios based on the inventory figure such as inventory turnover could be misleading if calculated at, say, a 31 December year-end.
- The use of norms can be misleading, e.g. the current ratio of 2:1 might be totally inappropriate for an entity like Asda which does not have long inventory turnover periods and, as its sales are for cash, it would not produce trade receivable collection period ratios.
- Factors that could invalidate inter-firm comparisons, such as:
  - use of different measurement bases with non-current assets reported at historical cost or revaluation and revaluations carried out at different dates;
  - use of different commercial practices, e.g. factoring trade receivables so that cash is increased – a perfectly normal transaction but one that could cause the comparative ratio of days' credit allowed to be significantly reduced;
  - applying different accounting practice, e.g. adopting different depreciation methods such as straight-line and reducing balance; adopting different inventory valuation methods such as FIFO and weighted average; or assuming different degrees of optimism or pessimism when making judgement-based adjustments to non-current and current assets;

- having different definitions for ratios, e.g. the numerator for ROCE could be operating profit, profit before interest, profit before interest and tax (PBIT), earnings before interest, tax, depreciation and amortisation (EBITDA), profit after tax, etc.; the denominator for ROCE could be total assets, total assets less intangibles, net assets, average total assets, etc.

28.10.3 Degree of scepticism

This depends on the role of the person using the ratios. For example, a financial controller/ FD preparing a report to the Board would have local knowledge of the company's business activities. In the Vertigo circumstances a reporting accountant, and to a lesser extent an external auditor, might not have this local knowledge and their starting point would be to form an overall impression followed by a more detailed analysis.

In expressing an opinion they might need to be more investigative and consider:

- Whether there is a risk of window dressing to improve sales, e.g. dispatching goods at the end of the period knowing them to be defective so that they appear in the current year's sales and accepting that they will be returned later in the next period.

- Whether liabilities have been omitted to improve the quick ratio, e.g. simply by suppressing purchase invoices at the year-end.

- Whether liabilities have been omitted to improve gearing, e.g. by the use of off-balance-sheet finance such as structuring the terms of a lease to ensure that it is treated as an operating lease and not a finance lease and special-purpose enterprises to keep debts off the statement of financial position.

- Whether there has been full disclosure in the notes of, say, contingent liabilities, which could result in ratios not being accurate predictors of future earnings and solvency.

Summary

Ratios are an aid in interpreting financial performance and liquidity. Comparison with prior periods and competitor/industry averages can provide a business with an indication of its relative performance – has it improved and how does it compare to its competitors? In this chapter we have followed a common-sized approach to the initial overview and the pyramid approach to calculating the ratios for two years to provide a basis for a report.

A comparison was made with a competitor's ratios for those areas that required further investigation. In practice it would be helpful to have data for 3–5 years in order to review trends. Reference was then made to the need to be cautious when using the ratios for prediction – remembering that at all times there needs to be a degree of scepticism when interpreting the ratios.

REVIEW QUESTIONS

1 State and express two ratios that can be used to analyse each of the following:

   (i)   profitability;

   (ii)   liquidity;

   (iii) management control.

2 Discuss the importance of the disclosure of exceptional items to the users of the annual report in addition to the operating profit.

3 Explain how a reader of the accounts might be able to assess whether the non-current asset base is being maintained.

4 Explain in what circumstances an increase in the revenue to current assets might be an indication of a possible problem.

5 Explain in what circumstances a decrease in the rate of non-current asset turnover might be a positive indicator.

6 Discuss why an increasing current ratio might not be an indicator of better working capital management.

7 The management of Alpha Ltd calculates ROCE using profit before interest and tax as a percentage of net closing assets. Discuss how this definition might be improved.

8 The asset turnover rate has increased by 50% over the previous year. Explain the questions you would have in mind and what other ratios you would review.

9 The current ratio has doubled since the previous year. Explain the questions that you would have in mind when reviewing the accounts.

10 Explain the problems a creditor might have when assessing the creditworthiness of a subsidiary entity.

11 You ascertain that inventories and (to a lesser extent) trade receivables have risen significantly when you consider that sales have increased by only 5%. Discuss the questions that you ask and the possible impact of each answer on the ratios.

12 Access the annual reports of two companies in the same industry and identify (a) the ratios that they report in common, (b) how these have been defined, and (c) why some ratios are not common to both.

13 A company has a very high rate of inventory turnover. Discuss circumstances when this might be of concern to management.

14 The ratio of current liabilities to net worth (equity + retained earnings) was 75%. Discuss how this would be viewed by suppliers and management.

15 The ratio of non-current assets to net worth was 75%. Discuss the risk that this poses for a company.

EXERCISES

Question 1

Flash Fashions plc has had a difficult nine months and the management team is discussing strategy for the final quarter.

In the last nine months the company has survived by cutting production, reducing staff and reducing overheads wherever possible. However, the share market, whilst recognising that sales across the industry have been poor, has worried about the financial strength of the business and as a result the share price has fallen 40%.

The company is desperate to increase sales. It has been recognised that the high fixed costs of the factory are not being fully absorbed by the lower volumes which are costed at standard cost. If sales and production can be increased then more factory costs will be absorbed and increased sales volume will raise staff morale and make analysts think the firm is entering a turnaround phase.

The company decides to drop prices by 15% for the next two months and to change the terms of sale so that property does not pass until the clothes are paid for. This is purely a reflection of the tough economic conditions and the need to protect the firm against customer insolvency. Further, it is decided that if sales have not increased enough by the end of the two months, the company representatives will be advised to ship goods to customers on the understanding that they will be invoiced but if they don't sell the goods in two months they can return them. Volume discounts will be stressed to keep the stock moving.

These actions are intended to increase sales, increase profitability, justify higher stocks, and ensure that more overheads are transferred out of the profit statement into stocks.

For the purposes of annual reporting it was decided not to spell out sales growth in financial figure terms in the managing director's report but rather to focus on units shipped in graphs using scales (possibly log scales) designed to make the fall look less dramatic. Also comparisons will be made against industry volumes as the fashion industry has been more affected by economic conditions than the economy as a whole.

To make the ratios look better, the company will enter into an agreement on the last week of the year with a so-called 'two-dollar company' called Upstart Ltd owned by Colleen Livingston, friend of the managing director of Flash Fashions, Sue Cotton. Upstart Ltd will sign a contract to buy a property for £30 million from Flash Fashions and will also sign promissory notes payable over the next three quarters for £10 million each. The auditors will not be told, but Flash Fashions will enter into an agreement to buy back the property for £31 million any time after the start of the third month in the new financial year.

Required:
Critically discuss each of the proposed strategies.

* Question 2

Relationships plc

You are informed that the non-current assets totalled €350,000, current liabilities €156,000, the opening retained earnings totalled €103,000, the administration expenses totalled €92,680 and that the available ratios were the current ratio 1.5, the acid test ratio 0.75, the trade receivables collection period was six weeks, the gross profit was 20% and the net assets turned over 1.4 times.

Required:
Prepare the Relationships plc statement of financial position from the above information.

* Question 3

The major shareholder/director of Esrever Ltd has obtained average data for the industry as a whole. He wishes to see what the forecast results and position of Esrever Ltd would be if in the ensuing year its performance were to match the industry averages.

At 1 July 20X0, actual figures for Esrever Ltd included:

	£
Land and buildings (at written-down value)	132,000
Fixtures, fittings and equipment (at written-down value)	96,750
Inventory	22,040
12% loan (repayable in 20X5)	50,000
Ordinary share capital (50p shares)	100,000

For the year ended 30 June 20X1 the following forecast information is available:

1 Depreciation of non-current assets (on reducing balance)

Land and buildings	2%
Fixtures, fittings and equipment	20%

2 Net current assets will be financed by a bank overdraft to the extent necessary.

3 At 30 June 20X0 total assets minus current liabilities will be £231,808.

4 Profit after tax for the year will be 23.32% of gross profit and 11.16% of total assets minus all external liabilities, both long-term and short-term.

5 Tax will be at an effective rate of 20% of profit before tax.

6 Cost of sales will be 68% of turnover (excluding VAT).

7 Closing inventory will represent 61.9 days' average cost of sales (excluding VAT).

8 Any difference between total expenses and the aggregate of expenses ascertained from this given information will represent credit purchases and other credit expenses, in each case excluding VAT input tax.

9 A dividend of 2.5p per share will be proposed.

10 The collection period for the VAT-exclusive amount of trade receivables will be an average of 42.6 days of the annual turnover. All the company's supplies are subject to VAT output tax at 15%.

11 The payment period for the VAT-exclusive amount of trade payables (purchases and other credit expenses) will be an average of 29.7 days. All these items are subject to (reclaimable) VAT input tax at 15%. This VAT rate has been increased to 17.5% and may be subject to future changes, but for the purpose of this question the theory and workings remain the same irrespective of the rate.

12 Payables, other than trade payables, will comprise tax due, proposed dividends and VAT payable equal to one-quarter of the net amount due for the year.

13 Calculations are based on a year of 365 days.

Required:
Construct a forecast statement of comprehensive income for Esrever Ltd for the year ended 30 June 20X1 and a forecast statement of financial position at that date in as much detail as possible. (All calculations should be made to the nearest £1.)

* Question 4

Saddam Ltd is considering the possibility of diversifying its operations and has identified three firms in the same industrial sector as potential takeover targets. The following information in respect of the companies has been extracted from their most recent financial statements.

	Ali Ltd	Baba Ltd	Camel Ltd
ROCE before tax %	22.1	23.7	25.0
Net profit %	12.0	12.5	3.75
Asset turnover ratio	1.45	1.16	3.73
Gross profit %	20.0	25.0	10.0
Sales/non-current assets	4.8	2.2	11.6
Sales/current assets	2.1	5.2	5.5
Current ratio	3.75	1.4	1.5
Acid test ratio	2.25	0.4	0.9
Average number of weeks' receivables outstanding	5.6	6.0	4.8
Average number of weeks' inventory held	12.0	19.2	4.0
Ordinary dividend %	10.0	15.0	30.0
Dividend cover	4.3	5.0	1.0

Required:
(a) **Prepare a report for the directors of Saddam Ltd, assessing the performance of the three companies from the information provided and identifying areas which you consider require further investigation before a final decision is made.**
(b) **Discuss briefly why a firm's statement of financial position is unlikely to show the true market value of the business.**

Question 5

You work for Euroc, a limited liability company, which seeks growth through acquisitions. You are a member of a team that is investigating the possible purchase of Choggerell, a limited liability company that manufactures a product complementary to the products currently being sold by Euroc.

Your team leader wants you to prepare a report for the team evaluating the recent performance of Choggerell and the quality of its management, and has given you the following financial information which has been derived from the financial statements of Choggerell for the three years ended 31 March 2006, 2007 and 2008.

Financial year ended 31 March	2006	2007	2008
Revenue (€ million)	2,243	2,355	2,237
Cash and cash equivalents (€ million)	−50	81	−97
Return on equity	13%	22%	19%
Sales revenue to total assets	2.66	2.66	2.01
Cost of sales to sales revenue	85%	82%	79%
Operating expenses to sales revenue	11%	12%	15%
Net income to sales revenue	2.6%	4.3%	4.2%
Current/Working capital ratio (to 1)	1.12	1.44	1.06
Acid test ratio (to 1)	0.80	1.03	0.74
Inventory turnover (months)	0.6	0.7	1.0
Credit to customers (months)	1.3	1.5	1.7
Credit from suppliers (months)	1.5	1.5	2.0
Net assets per share (cents per share)	0.86	0.2	0.97
Dividend per share (cents per share)	10.0	14.0	14.0
Earnings per share (cents per share)	11.5	20.1	18.7

Required:

Use the above information to prepare a report for your team leader which:

(a) reviews the performance of Choggerell as evidenced by the above ratios;

(b) makes recommendations as to how the overall performance of Choggerell could be improved; and

(c) indicates any limitations in your analysis.

(The Association of International Accountants)

Question 6

Liz Collier runs a small delicatessen. Her profits in recent years have remained steady at around £21,000 per annum. This type of business generally earns a uniform rate of net profit on sales of 20%.

Recently, Liz has found that this level of profitability is insufficient to enable her to maintain her desired lifestyle. She is considering three options to improve her profitability.

Option 1 Liz will borrow £10,000 from her bank at an interest rate of 10% per annum, payable at the end of each financial year. The whole capital sum will be repaid to the bank at the end of the second year. The money will be used to hire the services of a marketing agency for two years. It is anticipated that turnover will increase by 40% as a result of the additional advertising.

Option 2 Liz will form a partnership with Joan Mercer, who also runs a local delicatessen. Joan's net profits have remained at £12,000 per annum since she started in business five years ago. The sales of each shop in the combined business are expected to increase by 20% in the first year and then remain steady. The costs of the amalgamation will amount to £6,870, which will be written off in the first year. The partnership agreement will allow each partner a partnership salary of 2% of the revised turnover of their own shop. Remaining profits will be shared in the ratio of Liz 3/5, Joan 2/5.

Option 3 Liz will reduce her present sales by 80% and take up a franchise to sell Nickson's Munchy Sausage. The franchise will cost £80,000. This amount will be borrowed from her bank. The annual interest rate will be 10% flat rate based on the amount borrowed. Sales of Munchy Sausage yield a net profit to sales percentage of 30%. Sales are expected to be £50,000 in the first year, but should increase annually at a rate of 15% for the following three years then remain constant.

Required:
(a) Prepare a financial statement for Liz comparing the results of each option for each of the next two years.
(b) Advise Liz which option may be the best to choose.
(c) Discuss any other factors that Liz should consider under each of the options.

Question 7

Chelsea plc has embarked on a programme of growth through acquisitions and has identified Kensington Ltd and Wimbledon Ltd as companies in the same industrial sector, as potential targets.

Using recent financial statements of both Kensington and Wimbledon and further information obtained from a trade association, Chelsea plc has managed to build up the following comparability table:

	Kensington	Wimbledon	Industrial average
Profitability ratios			
ROCE before tax %	22	28	20
Return on equity %	18	22	15
Net profit margin %	11	5	7
Gross profit ratio %	25	12	20
Activity ratios			
Total assets turnover = times	1.5	4.0	2.5
Non-current asset turnover = times	2.3	12.0	5.1
Receivables collection period in weeks	8.0	5.1	6.5
Inventory holding period in weeks	21.0	4.0	13.0
Liquidity ratios			
Current ratio	1.8	1.7	2.8
Acid test	0.5	0.9	1.3
Debt–equity ratio %	80.0	20.0	65.0

Required:
(a) Prepare a performance report for the two companies for consideration by the directors of Chelsea plc indicating which of the two companies you consider to be a better acquisition.
(b) Indicate what further information is needed before a final decision can be made.

Question 8

The Housing Department of Chaldon District Council has invited tenders for re-roofing 80 houses on an estate. Chaldon Direct Services (CDS) is one of the Council's direct services organisations and it has submitted a tender for this contract, as have several contractors from the private sector.

The Council has been able to narrow the choice of contractor to the four tenderers who have submitted the lowest bids, as follows:

	£
Nutfield & Sons	398,600
Chaldon Direct Services	401,850
Tandridge Tilers Ltd	402,300
Redhill Roofing Contractors plc	406,500

The tender evaluation process requires that the three private tenderers be appraised on the basis of financial soundness and quality of work. These tenderers were required to provide their latest final accounts (year ended 31 March 20X4) for this appraisal; details are as follows:

	Nutfield & Sons	Tandridge Tilers Ltd	Redhill Roofing Contractors plc
Profit and loss account for year ended 31 March 20X4			
	£	£	£
Revenue	611,600	1,741,200	3,080,400
Direct costs	(410,000)	(1,190,600)	(1,734,800)
Other operating costs	(165,000)	(211,800)	(811,200)
Interest	—	(85,000)	(96,000)
Net profit before taxation	36,600	253,800	438,400
Statement of financial position as at 31 March 20X4			
	£	£	£
Non-current assets (net book value)	55,400	1,542,400	2,906,800
Inventories and work in progress	26,700	149,000	449,200
Receivables	69,300	130,800	240,600
Bank	(11,000)	10,400	(6,200)
Payables	(92,600)	(140,600)	(279,600)
Dividend declared	—	(91,800)	(70,000)
Loan	—	(800,000)	(1,200,000)
	47,800	800,200	2,040,800
Capital	47,800	—	—
Ordinary shares @ £1 each	—	250,000	1,000,000
Reserves	—	550,200	1,040,800
	47,800	800,200	2,040,800

Nutfield & Sons employ a workforce of six operatives and have been used by the Council for four small maintenance contracts worth between £60,000 and £75,000 which they have completed to an appropriate standard. Tandridge Tilers Ltd have been employed by the Council on a contract for the replacement of flat roofs on a block of flats, but there have been numerous complaints about the standard of the work. Redhill Roofing Contractors plc is a company which has not been employed by the Council in the past and, as much of its work has been carried out elsewhere, its quality of work is not known.

CDS has been suffering from the effects of increasing competition in recent years and achieved a return on capital employed of only 3.5% in the previous financial year. CDS's manager has successfully renegotiated more beneficial service-level agreements with the Council's central support departments with effect from 1 April 20X4. CDS has also reviewed its non-current asset base which has resulted in the disposal of a depot which was surplus to requirements and in the rationalisation of vehicles and plant. The consequence of this is that CDS's average capital employed for 20X4/X5 is likely to be some 15% lower than in 20X3/X4.

A further analysis of the tender bids is provided below:

	Nutfield & Sons	Chaldon Direct Services	Tandridge Tilers Ltd
	£	£	£
Labour		234,000	251,400
Materials	140,000	100,000	80,000
Overheads (including profit)	24,600	50,450	18,700

The Council's Client Services Committee can reject tenders on financial and/or quality grounds. However, each tender has to be appraised on these criteria and reasons for acceptance or rejection must be justified in the appraisal process.

Required:
In your capacity as accountant responsible for reporting to the Client Services Committee, draft a report to the Committee evaluating the tender bids and recommending to whom the contract should be awarded.

Question 9

The statements of financial position, cash flows, income and movements of non-current assets of Dragon plc for the year ended 30 September 20X6 are set out below:

(i) *Statement of financial position*

	20X5		20X6	
	£000	*£000*	*£000*	*£000*
Tangible non-current assets		1,200		1,160
Freehold land and buildings, at cost		700		1,700
Plant and equipment, at net book value		1,900		2,860
Current assets:				
Inventory	715		1,020	
Trade receivables	590		826	
Short-term investments	52		—	
Cash at bank and in hand	15		47	
	1,372		1,893	
Current liabilities:				
Trade payables	520		940	
Taxation payable	130		45	
Dividends payable	90		105	
	740		1,090	
Net current assets		632		803
		2,532		3,663
Long-term liability and provisions:				
8% debentures, 20X9		500		1,500
Provisions for deferred tax		100		180
		1,932		1,983
Capital and reserves		1,400		1,400
Ordinary shares of £1 each				
Share premium account		250		250
Retained earnings		282		333
		1,932		1,983

(ii) *Statement of income (extract) for the year ended 30 September 20X6*

EBITDA		1,161
Depreciation		660
Operating profit		501
Interest payable: debentures		150
Profit before taxation		351
Income tax		125
Profit attributable to shareholders		226
Dividends: paid	70	
: proposed	105	175
Retained earnings for year		51
Retained earnings brought forward		282
Retained earnings carried forward		333

(iii) *Statement of cash flows*

Net cash flow from operating activities		1,033
Interest paid	(150)	
Income taxes paid	(130)	(280)
Net cash from operating activities:		753
Cash flows from investing activities		
Purchase of property, plant and equipment	(1,620)	
Net cash used in investing activities:		(1,620)
Cash flows from financing activities		
Proceeds from sale of short-term investments	59	
Proceeds from long-term borrowings	1,000	
Dividends paid	(160)	
Net cash from financing activities:		899
Net increase in cash and cash equivalents		32
Cash and cash equivalents at the beginning of the period		15
Cash and cash equivalents at the end of the period		47

(iv) *Tangible non-current assets (or PPE)*

The movements in the year were as follows:

	Freehold land and buildings £000	Plant and machinery £000	Total £000
Cost:			
At 1 October 20X5	2,000	1,600	3,600
Additions	—	1,620	1,620
At 30 September 20X6	2,000	3,220	5,220
Depreciation:			
At 1 October 20X5	800	900	1,700
Charge during the year	40	620	660
At 30 September 20X6	840	1,520	2,360
Net book value:			
Beginning of year	1,200	700	1,900
End of year	1,160	1,700	2,860

You are also provided with the following information:

(i) There was a debenture issue on 1 October 20X5 with interest payable on 30 September each year.

(ii) An interim dividend of £70,000 was paid on 1 July 20X6.

(iii) The short-term investment was sold for £59,000 on 1 October 20X5.

(iv) Business activity increased significantly to meet increased consumer demand.

Required:
(a) Prepare a reconciliation of operating profit to net cash inflow from operating activities.
(b) Discuss the financial developments at Dragon plc during the financial year ended 30 September 20X6 with particular regard to its financial position at the year-end and prospects for the following financial year, supported by appropriate financial ratios.

* Question 10

Amalgamated Engineering plc makes specialised machinery for several industries. In recent years, the company has faced severe competition from overseas businesses, and its sales volume has hardly changed. The company has recently applied for an increase in its bank overdraft limit from £750,000 to £1,500,000. The bank manager has asked you, as the bank's credit analyst, to look at the company's application.

You have the following information:

(i) *Statements of financial position as at 31 December 20X5 and 20X6*

| | 20X5 | | 20X6 | |
	£000	£000	£000	£000
Tangible non-current assets:				
Freehold land and buildings, at cost		1,800		1,800
Plant and equipment, at net book value		3,150		3,300
		4,950		5,100
Current assets:				
Inventory	1,125		1,500	
Trade receivables	825		1,125	
Short-term investments	300		—	
	2,250		2,625	
Current liabilities:				
Bank overdraft	225		675	
Trade payables	300		375	
Taxation payable	375		300	
Dividends payable	225		225	
	1,125		1,575	
Net current assets		1,125		1,050
		6,075		6,150
Long-term liability				
8% debentures, 20X9		1,500		1,500
		4,575		4,650
Capital and reserves:				
Ordinary shares of £1 each		2,250		2,250
Share premium account		750		750
Retained earnings		1,575		1,650
		4,575		4,650

(ii) *Statements of comprehensive income for the years ended 31 December 20X5 and 20X6*

	20X5		20X6	
	£000	£000	£000	£000
Revenue		6,300		6,600
Cost of sales: materials	1,500		1,575	
: labour	2,160		2,280	
: production: overheads	750		825	
		4,410		4,680
		1,890		1,920
Administrative expenses		1,020		1,125
Operating profit		870		795
Investment income		15		—
		885		795
Interest payable: debentures	120		120	
: bank overdraft	15		75	
		135		195
Profit before taxation		750		600
Taxation		375		300
Profit attributable to shareholders		375		300
Dividends		225		225
Retained earnings for year		150		75

(iii) The general price level rose on average by 10% between 20X5 and 20X6. Average wages also rose by 10% during this period.

(iv) The debenture stock is secured by a fixed charge over the freehold land and buildings, which have recently been valued at £3,000,000. The bank overdraft is unsecured.

(v) Additions to plant and equipment in 20X6 amounted to £450,000: depreciation provided in that year was £300,000.

Required:

(a) Prepare a statement of cash flows for the year ended 31 December 20X6.

(b) Calculate appropriate ratios to use as a basis for a report to the bank manager.

(c) Draft the outline of a report for the bank manager, highlighting key areas you feel should be the subject of further investigation. Mention any additional information you need, and where appropriate refer to the limitations of conventional historical cost accounts.

(d) On receiving the draft report the bank manager advised that he also required the following three cash-based ratios:

(i) Debt service coverage ratio defined as EBITDA/annual debt repayments and interest.

(ii) Cash flow from operations to current liabilities.

(iii) Cash recovery rate defined as ((cash flow from operations proceeds from sale of non-current assets)/average gross assets) × 100.

The director has asked you to explain why the bank manager has requested this additional information given that he has already been supplied with profit-based ratios.

Reference

1 http://www.rmahq.org/

CHAPTER 29

Analysis of published financial statements

29.1 Introduction

In Chapter 28 we considered the way in which we could 'make the numbers talk' from a set of published financial statements. We explained the importance of taking a 'helicopter perspective' initially and identifying key issues before focusing on specific areas of detail. We showed how powerful ratio analysis could be as an analytical tool provided the ratios were interpreted appropriately. A particularly important issue was the need to differentiate between changes to ratios that were caused by operational and business factors and changes caused by accounting policies and accounting estimates.

When we are interpreting the financial statements of an entity a key issue is the amount of financial information actually available. If we are performing an analysis on behalf of management, or a controlling shareholder, then the amount of financial information available to us is likely to be sufficient to perform any analysis we consider appropriate. On the other hand, where we are performing an analysis from a purely external perspective there will be a limit to the amount of information available to us, because published financial statements generally contain only the information that is required by the appropriate regulatory framework.

Objectives

By the end of this chapter, you should be able to:

- discuss steps taken to improve information for shareholders;
- critically discuss the limitations of published financial data as a source of useful information for interpretation purposes;
- discuss additional entity-wide cash-based performance measures;
- explain the use of ratios in determining whether a company is shariah-compliant;
- explain the use of ratios in debt covenants;
- critically discuss various scoring systems for predicting corporate failure;
- critically discuss remuneration performance criteria;
- critically discuss the role of credit rating agencies;
- calculate the value of unquoted investments.

29.2 Improvement of information for shareholders

There have been a number of discussion papers, reports and voluntary code provisions from professional firms and regulators making recommendations on how to provide additional information to allow investors to form a view as to the business's future prospects by (a) making financial information more understandable and easier to analyse and (b) improving the reliability of the historical financial data. This would help ensure the equal treatment of all investors and improve accountability for stewardship.

29.2.1 Making financial information more understandable

There has been a view that users should bring a reasonable level of understanding when reading an annual report. This view could be supported when transactions were relatively simple. It no longer applies when even professional accountants comment that the only people who understand some of the disclosures are the technical staff of the regulator and the professional accounting firms.

Statutory measures

Users need the financial information to be made more accessible. This is being achieved in part by initiatives such as the Strategic Report in the UK with the requirement to publish information on the past year, including a fair review of the company's business, a description of the principal risks and uncertainties facing the company, and a balanced and comprehensive analysis of the performance of the company's business during the financial year.

As regards the future, a description of the company's strategy, a description of the company's business model and the main trends and factors likely to affect the future development, performance and position of the company's business are also required.

Need to understand volatility

There is a need on the part of investors to understand the volatility that can arise as a result of a company's strategy, such as recognising the short- and medium-term impact on earnings of R&D investment. There has been a view that investors are unhappy with an uneven profit trend and that companies have responded by smoothing earnings from year to year to maintain investor confidence.

An ICAEW report[1] produced in 1999 *No Surprises: The Case for Better Risk Reporting* recognised the need for management to disclose their strategies and how they managed risk whilst stating that the intention was not to encourage profit-smoothing but rather a better management of risk and a better understanding by investors of volatility.

29.2.2 Improving the reliability of financial information

Investors rely on the fact that annual reports are audited and so present a fair view of a company's financial performance and position. However, accounting scandals, such as in Enron, Satyam and the SEC probe in 2012 into the auditing of Chinese companies, have led to a feeling that auditors are not protecting their interests. The profession is aware of this view and of the existence of an expectation gap between what investors expect from an audit and what can reasonably be delivered. This is discussed further in Chapter 31.

Reliability of narrative information in the Annual Report

The following is an indication of the work carried out by an auditor.

- Other information contained in the Annual Report is read and considered as to whether it is consistent with the audited financial statements.

- The other information comprises only the Directors' Report, the unaudited part of the Directors' Remuneration Report, the Chairman's Statement, the Operating and Financial Review, the Strategic Report and the Corporate Governance Statement.

- The implications for the audit report are considered if there is an awareness of any apparent misstatements or material inconsistencies with the financial statements.

- The responsibilities of the auditor do not extend to any other information.

29.3 Published financial statements – their limitations for interpretation purposes

Assuming that the financial statements have been audited and present a fair view, there remain limitations such as lack of detail and the impact of unaudited information when attempting to analyse the statements.

Limitation 1 – Lack of detail

This limitation is due to the amount that corporate entities are required to disclose by the appropriate regulatory framework. Only that information that is required to be disclosed would be subject to objective external scrutiny through audit and that information is strictly limited. For example:

- When analysing the profitability of a corporate entity, whether gross or net profit, the extent to which expenses can be broken down into categories is strictly limited. Most current frameworks require the disclosure of cost of sales and other operating expenses but do not require further analysis. Therefore, when, say, the gross margin shows a variation (either from one period to another for single-entity comparison or between entities) we cannot further investigate the components of gross margin because the published financial statements do not provide the required detail.

- Most frameworks require analysis of expenses into a number of headings but do not prescribe exactly where certain expenses (e.g. advertising) would fit. This means that when we compare the gross margin of one corporate entity with that of another we may not be comparing like with like, because one may have treated advertising as part of cost of sales and another may have treated equivalent costs as other operating expenses and the amount could be significant.

- Lack of detailed information prevents the computation of certain useful ratios in their 'purest' from. For example, one of the ratios we discussed in Chapter 28 was 'payables days' – trade payables as a number of days' credit purchases. If we tried to compute this ratio from the published financial statements we would have a problem – credit purchases are not required to be disclosed in the published financial statements of corporate entities in most regulatory frameworks. It is possible to use cost of sales as a proxy for credit purchases. However, this 'contrived' ratio is not as useful as the ratio would be were credit purchases to be available.

Limitation 2 – The impact of unaudited information

There is a varying amount of information relating to areas such as strategy, risk and KPIs and an ongoing move for improvement. For example, an interesting report issued by the FRC in 1999, *Rising to the Challenge: A Review of Narrative Reporting by UK Listed Companies*,[2] found the following:

- For KPIs, the best companies linked KPIs to strategy and provided an explanation of each measure along with some targets, reconciliations, graphical illustrations of year-on-year comparatives and tables to link KPIs to strategy and targets or future intentions. However, many reports still featured an isolated KPI table with no accompanying discussion or link to the remainder of the document.

- For principal risks, best-practice reports provided some context for the risk, indicating whether it was increasing or decreasing, and provided some idea of the impact of a risk crystallising, supported by numbers. However, users would find it difficult to assess risk where there was too little detail or too many risks identified that obscured which were important.

29.4 Published financial statements – additional entity-wide cash-based performance measures

When making inter-firm comparisons there is the problem that accrual accounting requires a number of subjective judgements to be made such as the non-cash adjustments for depreciation, amortisation and impairment. Inter-firm comparison schemes overcome this by requiring member companies to restate their results using uniform policies such as restating non-current assets at current values and applying uniform depreciation policies.

External analysts are unable to achieve this and have, therefore, developed additional performance measures which are becoming more frequently met in published financial statements. However, there are concerns that they are not mandatory or uniformly defined. This is being addressed by a number of bodies including the International Federation of Accountants (IFAC) with its exposure draft for an International Good Practice Guidance on *Developing and Reporting Supplementary Financial Measures – Definition, Principles, and Disclosures*, the European Securities and Markets Authority (ESMA) with its draft *Guidelines on Alternative Performance Measures* and the IASB which has indicated an intention to research the presentation and disclosure of non-IFRS financial information as part of its *Disclosure Initiative project*.

The position then at present is that management defines the additional performance measures that they report which they consider best assist users to understand how these are used by management in making business decisions.

We discuss some of these measures below.

29.4.1 EBITDA

EBITDA is fairly widely used by external analysts. It stands for 'earnings before interest, tax, depreciation and amortisation'.

EBITDA more closely reflects the cash effect of earnings by adding back depreciation and amortisation charges to the operating profit. The figure can be derived by adding back the depreciation and amortisation that is disclosed in the statement of cash flows.

By taking earnings before depreciation and amortisation we eliminate differences due to different ages of plant and equipment when making inter-period comparisons of performance and also differences arising from the use of different depreciation methods when making inter-firm comparisons. By taking earnings before interest it shows how much is available to pay interest.

Note that there is no standard definition – for example some companies define it as earnings before interest, depreciation, tax, amortisation, *impairment* and *exceptional items*.

EBITDA shows an approximation to the cash impact of earnings. It differs from the cash flow from operations reported in the statement of cash flows in that it is before adjustment for working capital changes.

Comparing segment performance

EBITDA information is useful where an entity has a number of segments. It allows performance to be compared by calculating the EBITDA for each segment which provides a figure that is independent of the age structure of the non-current assets.

For example, the following is an extract from the Vodafone 2011 Annual Report:

	EBITDA £m	EBITDA margin %
31 March 2011		
Germany	2,952	37.4
Italy	2,643	46.2
Spain	1,562	30.4
UK	1,233	23.4
Other Europe	2,433	
Europe	**10,823**	

Interestingly, the company states that it uses EBITDA as an operating performance measure which is reviewed by the Chief Executive to assess internal performance in conjunction with EBITDA margin, which is an alternative sales margin figure.

29.4.2 Other 'EBITDA-based' ratios commonly produced

These include the following.

EV (Enterprise value)/EBITDA

EV is the value of the whole business calculated as the market capitalisation of equity plus debt, non-controlling interest and preference shares less total cash and cash equivalents.

Assuming for the current year an EV of $199,283m and EBITDA of $29,806m, the ratio EV/EBITDA is 6.69. This is compared to the industry average which is, say, 5.99 which indicates that the company is valued above the industry average. If the company ratio were significantly below the 5.99 it could invite the interest of a takeover. It would be normal to calculate the ratio for a period of say five years to note the trend.

Net debt/EBITDA

This ratio shows the number of years that it would take to 'pay off' the net debt. For example, the following is an extract from the AMEC 2011 Annual Report:

> The group is currently in a net cash position. If debt is subsequently required, the long-term net debt is expected to be no more than two times EBITDA. The group may exceed this operating parameter should the business profile require it. However, it is expected that any increases would be temporary given the net operational cash flows of the group.

Debt service coverage ratio

This is defined as EBITDA/annual debt repayments and interest. This ratio is often used in setting debt covenants and by banks assessing a company's ability to repay debt on the terms being sought by the borrower.

EBITDA/interest

This shows the number of times interest is covered. This is also a ratio that banks set as covenant thresholds when agreeing bank credit limits. The following is an extract from the Wienerberger 2011 Annual Report:

	2010	2011	Threshold
Net debt/EBITDA	1.8	1.7	<3.50
Operating EBITDA/interest	4.9	6.8	>3.75

EBITDAR

EBITDAR is a variant of EBITDA that has become popular with analysts in recent times. It stands for 'earnings before interest, tax, depreciation, amortisation and rental expense'.

Adding this rental expense back allegedly makes performance comparisons between entities with different proportions of assets leased under operating leases more valid. It also removes the subjectivity introduced by lease classification as operating or finance.

The following is an extract from the J Sainsbury plc 2012 Annual Report:

Key financial ratios		
Adjusted net debt to EBITDAR[1]	**4.1 times**	4.1 times
Interest cover[2]	**7.5 times**	7.9 times
Fixed charge cover[3]	**3.1 times**	3.1 times
Gearing[4]	**35.2%**	33.4%

1 Net debt plus capitalised lease obligations (5.5% NPV) divided by EBITDAR.
2 Underlying profit before interest and tax divided by underlying net finance costs.
3 EBITDAR divided by net rent and underlying net finance costs.
4 Net debt divided by net assets.

EBITDAR is used as a comparator between companies. For example, Tesco plc in their 2012 Annual Report state that their fixed charge cover remained broadly flat due to increased rent offsetting their reduced interest and increase in operating cash flow. Their target was stated to be a level of cover in the band of 4 to 4.5 times. In its 2011 Annual Report Tesco plc had charted their EBITDAR against Sainsbury's and Morrisons.

EBITDARM

EBITDARM stands for 'earnings before interest, tax, depreciation, amortisation, rental expense and management fees'. The rationale behind this measure is that management fees are extracted from different entities in different proportions. The following is an example from the healthcare sector:

	Care homes for the elderly	*Mental health services*
	£	£
Fee income	457m	76m
EBITDARM	132m	15m
% margin	28.8%	19.0%

Management charges may not always be totally representative of the services provided. Therefore management fees might sometimes be a form of profit extraction rather than a genuine expense and adding them back once again facilitates inter-entity comparison.

29.4.3 Evaluating the use of EBITDA

EBITDA is often used when valuing a company. It helps when comparing the performance of companies which may have differently geared capital structures, depreciation policies and tax rates.

However, it is not a substitute for cash flow in that it does not take into account changes in working capital that may be significant in a fast-growing company, material finance charges that may exist in a highly leveraged company and potential cash required by a capital-intensive company.

It needs to be used in conjunction with other ratios. For example, in reviewing a highly leveraged company the debt service coverage ratio and EBITDA/Interest would be considered. In reviewing a capital-intensive company reference would be made to Free cash flow discussed in Chapter 5. In assessing dividend potential the Free cash flow to Equity would be considered by calculating the cash after interest, taxes and reinvestment have been paid.

29.5 Ratio thresholds to satisfy shariah compliance

In addition to considering the range of cash-based earnings ratios, investors might also require a company to satisfy certain threshold ratios *before* making an investment. An example is seen with the ratios relevant for shariah compliance.

Shariah law is a regulatory system that is derived from the Islamic religion. Islam commands followers to avoid consumption of alcohol and pork and so adherents do not condone investments in those industries. There is screening to check that (a) business activities are not prohibited and (b) certain of the financial ratios do not exceed specified limits.

This use of ratios is included because of the growing importance of investment in shariah-compliant companies. Islamic banking is gaining popularity all over the world. Global Islamic finance assets reached a record $1.46 trillion in 2012, with industry to top $2 trillion by 2014.

There are a number of shariah indices including the Dow Jones Islamic Indexes, the FTSE Global Islamic Index Series, the FTSE SGX Shariah Index Series, the FTSE DIFX Shariah Index Series and the FTSE Bursa Malaysia Index Series. The indices include companies such as Google Inc., TOTAL SA, BP plc, Exxon Mobil Corporation, Petroleo Brasileiro, Novartis AG, Roche Holding, GlaxoSmithKline plc, BHP Billiton Ltd, Siemens AG, Samsung Electronics, International Business Machines Corporation, Nestlé SA, and Coca-Cola.

Investors interested in establishing whether an entity is shariah-compliant are assisted by the service provided by various Islamic indices where the constituent companies have been screened to confirm that they are shariah-compliant with reference to the nature of the business and debt ratios.

The indices are compiled after:

- screening companies to confirm that their business activities are not prohibited (or fall within the 5% permitted threshold);
- calculating three financial ratios based on total assets; and
- calculating a dividend adjustment factor which results in more relevant benchmarks, as they reflect the total return to an Islamic portfolio net of dividend purification.

Details are provided below.

Screening

Shariah investment principles do not allow investment in entities which are directly active in, or derive more than 5% of their revenue (cumulatively) from, the following activities ('prohibited activities'):

- Alcohol: distillers, vintners and producers of alcoholic beverages, including producers of beer and malt liquors, owners and operators of bars and pubs.
- Tobacco: cigarettes and other tobacco products manufacturers and retailers.
- Pork-related products: companies involved in the manufacture and retail of pork products.
- Conventional financial services: an extensive range including commercial banks, investment banks, insurance companies, consumer finance such as credit cards, and leasing.
- Defence/weapons: manufacturers of military aerospace and defence equipment, parts or products, including defence electronics and space equipment.
- Gambling/casinos: owners and operators of casinos and gaming facilities, including companies providing lottery and betting services.
- Music: producers and distributors of music, owners and operators of radio broadcasting systems.
- Hotels: owners and operators of hotels.

Key ratios

Shariah investment principles do not allow investment in companies deriving significant income from interest or companies that have excessive leverage. MSCI Barra uses the following three financial ratios to screen for these companies:

- total debt over total assets;
- sum of an entity's cash and interest-bearing securities over total assets;
- sum of an entity's accounts receivables and cash over total assets.

None of the financial ratios may exceed 33.33%.

Dividend adjustment (or 'purification')

If an entity does derive part of its total income from interest income and/or from prohibited activities, shariah investment principles state that this proportion must be deducted from the dividend paid out to shareholders and given to charity.

Dividend purification may be calculated by dividing prohibited income (including interest income) by total income and multiplying by the dividend received. An alternative is to divide total prohibited income (including interest income) by the number of shares issued at the end of the period and multiply by the number of shares held. MSCI Barra applies a 'dividend adjustment factor' to all reinvested dividends.

The 'dividend adjustment factor' is defined as:

$$\frac{\text{Total earnings} - (\text{Income from prohibited activities} + \text{Interest income})}{\text{Total earnings}}$$

In this formula, total earnings are defined as gross income, and interest income is defined as operating and non-operating interest.

29.6 Use of ratios in restrictive loan covenants

Whereas the shariah compliance criteria apply *before* making a financial commitment, lenders might set specific threshold ratios that a company must comply with *after* making a loan in order to limit the lender's risk – these are described as affirmative or negative debt covenants.

When a corporate entity borrows, the borrowing agreement often includes a provision which requires that specified accounting ratios such as gearing (relationship between debt and equity) of the entity be kept below a certain level. The loan agreement would of course have to specify exactly how any ratio is computed for this purpose.

The existence of a debt covenant or covenants has a number of potential implications for an entity and for analysts:

- An entity with a debt covenant that is close to its limit will be unable to raise funds by borrowing, so it will need to raise any required funds by an equity issue. Given the attitude of investors to risk, the return required by equity shareholders in a highly geared entity will be higher than that of an entity in which the gearing is lower. This will affect the overall amount of funding an entity can raise.

- Where a ratio of an entity subject to a debt covenant approaches the limit set out in the covenant, there is an inevitable temptation for the preparers to ensure the ratio is kept within the limit, leading to a potential temptation to misstate the financial statements.

The potential existence of a debt covenant is a factor that should be borne in mind by external analysts. The problem is that the existence of such debt covenants is not normally a required disclosure by relevant regulatory frameworks. Therefore a concerned analyst would need to attempt to obtain this information from the management of the entity. The success or otherwise of this attempt will depend on the bargaining power of the analyst.

29.6.1 Affirmative and negative covenants

Lenders may require borrowers to do certain things by affirmative covenants or refrain from doing certain things by negative covenants.

Affirmative covenants may, for example, include requiring the borrower to:

- provide quarterly and annual financial statements;
- remain within certain ratios whilst ensuring that each agreed ratio is not so restrictive that it impairs normal operations:
  - maintain a current ratio of not less than an agreed ratio – say 1.6 to 1;
  - maintain a ratio of total liabilities to tangible net worth at an agreed rate – say no greater than 2.5 to 1;
  - maintain tangible net worth in excess of an agreed amount – say £1 million;
- maintain adequate insurance.

Negative covenants may, for example, include requiring the borrower *not* to:

- grant any other charges over the company's assets;
- repay loans from related parties without prior approval;
- change the group structure by acquisitions, mergers or divestment without prior agreement.

29.6.2 What happens if a company is in breach of its debt covenants?

Borrowers will normally have prepared forecasts to assure themselves and the lenders that compliance is reasonably feasible. Such forecasts will also normally include the worst-case scenario, e.g. taking account of seasonal fluctuations that may trigger temporary violations with higher borrowing required to cover higher levels of inventory and trade receivables.

If any violation has occurred, the lender has a range of options, such as:

● amending the covenant, e.g. accepting a lower current ratio; or

● granting a waiver period when the terms of the covenant are not applied; or

● renegotiating the credit facility and restructuring the finance, as in the following extract from the 2009 Annual Report of Sunshine Holdings 3 Ltd:

> The Group faces more restrictive financial covenants . . . the directors believe it is likely that the Group will not meet the financial covenants required under the first lien credit agreement.
>
> **Directors' report**
> While the directors fully expect to resolve the covenant issues with a restructuring and/or amendment to the facility agreements, these circumstances represent a material uncertainty regarding the Group's going concern status. . . . the directors have a reasonable expectation that the Group will satisfactorily conclude its covenant issues and will have adequate resources to continue in operational existence for the foreseeable future. Therefore the accounts have been prepared on a going concern basis.

In addition, companies may increase their equity capital, possibly by a rights issue as the current shareholders have a greater incentive to provide additional capital than new investors.

For example, it was reported in 2012 that Lonmin planned a $800m rights issue to avoid possibly breaching its covenants. A rights issue is often made in these circumstances as existing shareholders have a greater incentive than new shareholders to inject further equity capital.

In times of recession a typical reaction is for companies to also take steps to reduce their operating costs, align production with reduced demand, tightly control their working capital and reduce discretionary capital expenditure.

29.6.3 Risk of aggressive earnings management

In 2001, before the collapse of Enron, there was a consensus amongst respondents to the UK Auditing Practices Board Consultation Paper *Aggressive Earnings Management* that aggressive earnings management was a significant threat and actions should be taken to diminish it.

Reasons for earnings management

It was considered that aggressive earnings management could occur to increase earnings in order to avoid losses, to meet profit forecasts, to ensure compliance with loan covenants and when directors' and managements' remuneration were linked to earnings. It could also occur to reduce earnings to reduce tax liabilities or to allow profits to be smoothed.

In 2004, as a part of the *Information for Better Markets* initiative, the Audit and Assurance Faculty commissioned a survey.[3] This showed that the vulnerability of corporate reporting to manipulation is perceived as being always with us but at a lower level following the greater awareness and scrutiny by non-executive directors and audit committees.

Sector variations

The analysts interviewed in the survey believed the potential for aggressive earnings management varied from sector to sector, e.g. in the older, more established sectors followed by the same analysts for a number of years, they believed that company management would find it hard to disguise anything aggressive even if they wanted to. However, this was not true of newer sectors (e.g. IT) where the business models may be loss-making initially and imperfectly understood.

Levels of confidence

Whilst analysts and journalists tend to have low confidence in the reported earnings where there are pressures to manipulate, there is a research report[4] which paints a rather more optimistic picture. This report aimed to assess the level of confidence investors had in different sources of company information, including audited financial information, when making investment decisions. As far as audited financial information was concerned, the levels of confidence in UK audited financial information amongst UK and US investors remained very high, with 87% of UK respondents having either a 'great deal' or a 'fair amount' of confidence in UK audited financial information.

Need for scepticism

The auditing profession continues to respond to the need to contain aggressive earnings management. This is not easy because it requires a detailed understanding not only of the business but also of the process management follows when making its estimates. ISA 540 Revised, *Auditing Accounting Estimates, including Fair Value Accounting Estimates, and Related Disclosures*, requires auditors to exercise greater rigour and scepticism and to be particularly aware of the cumulative effect of estimates which in themselves fall within a normal range but which, taken together, are misleading.

29.6.4 Audit implications when there is a breach of a debt covenant

Auditors are required to bring a healthy scepticism to their work. This applies particularly at times such as when there is a potential debt covenant breach. There may then well be a temptation to manipulate to avoid reporting a breach. This will depend on the specific covenant. For example, if the current ratio is likely to fall below the agreed figure, management might be more optimistic when setting inventory obsolescence and accounts receivable provisions and assessing the probability of contingent liabilities crystallising.

29.6.5 Impact on share price

If there is a risk of bank covenants being breached, there can be a significant adverse effect on the share price. For example, it was reported in 2012 that Lonmin's share price dropped sharply by 4.6%, a new 52-week low for the company, following the announcement that it may be in the breach of its covenants with its financial lenders.

However, both the company and the lender might prefer to keep potential breaches private unless there is a risk that enforced disclosure is imminent.

29.7 Investor-specific ratios

The analysis we carried out in Chapter 28 (and the additional performance measures we discussed in Section 29.4 above) was done from the perspective of the performance and position of the entity. In this section we will focus on additional ratios and measures that have as their focus the position of the shareholders of the entity. Some of these measures are 'financial statement measures' and others are 'market-based measures'.

29.7.1 Return on equity

We discussed ROE in Chapter 28 so this section is included as a brief reminder. In Chapter 28 we stated that a primary entity profitability measure is 'Return on equity', i.e. 'Profit'/ Capital employed. Where the focus is on the equity shareholders the applicable ratio is ROE where the numerator is the post-tax profit.

If capital employed is funded by sources other than equity, then there is a financial leverage impact on the ROCE when calculating the ROE to reflect the potential benefit to equity shareholders of the company borrowing and investing at a higher rate.

As far as the equity shareholders are concerned, it might appear that the higher the financial leverage the better. However:

- If borrowings are high, it might be difficult to obtain additional loans to take advantage of new opportunities. For example, HSBC raised £12.5 billion in 2009 by a rights issue on the basis that this would give the bank a competitive advantage over its rivals by restoring its position as having the strongest statement of financial position, i.e. high borrowings limit a company's flexibility.

- Interest has to be paid even in bad years with the risk that loan creditors could put the company into administration if interest is not paid.

The relationship is illustrated using data from the financial statements of Vertigo plc for the year ended 31 December 20X9 presented in Section 28.3.1:

	£000
Total assets	4,587
Equity	3,353
Pre-tax profit	116
Tax	25
Sales	3,461

The effect of leverage on ROE is:

Pre-tax margin (3.35%) × Asset turnover (0.755) = Return on assets (2.53%)
    (116/3,461)   (3,461/4,587)

Return on assets (2.53%) × Leverage (1.37) × (1 − tax rate) (0.785) = ROE (2.72)
                            (4,587/3,353)   (1 − 0.215)

The effect of leverage on EPS (assuming 3 million shares in issue) is:

ROE (2.72) × Book value (1.12) = EPS (3.05)
            (3,353/3,000)

29.7.2 Price/earnings (PE) ratio

The PE ratio is computed as:

$$\frac{\text{Market value of a share}}{\text{Earnings per share}}$$

The PE ratio is a market-based measure and a high ratio indicates that investors are relatively confident in the maintainability and quality of the earnings of the entity. Entities in certain sectors (e.g. the retail sector) tend to have higher PE ratios than those in other sectors (e.g. the construction sector). Higher PE ratios imply a greater level of market confidence, which usually means that (given the attitude an average investor takes to risk) the entity with a higher PE ratio operates in a sector which is less cyclical.

We will see in Section 29.11 that competitor or industry PE ratios (or their reciprocal, the earnings yield) are used as a base for valuing shares in unquoted companies – comparators being obtained from trade association schemes or sites such as http://biz.yahoo.com/p/industries.html.

Earnings yield

This is the reciprocal of the PE ratio. For example, a PE ratio of 10 becomes an earnings yield of 10% ($1/10 \times 100$).

29.7.3 Earnings per share (EPS)

EPS is computed as:

$$\frac{\text{Profit attributable to the ordinary (equity) shareholders}}{\text{Weighted average number of ordinary shares in issue during the period}}$$

The detailed calculation of basic and diluted EPS was dealt with in Chapter 27.

EPS could be said to be a more reliable indicator of the true trend in profitability than the actual profit numbers because the denominator of the fraction factors in any change in the issued capital during the period. The fact that the weighted average number is used removes the potential inconsistency that arises when dividing a 'period' number like profit by a 'point of time' number like the number of shares.

How EPS might be manipulated

The appropriateness of EPS as a performance measure can be influenced by the subjectivity of the directors when preparing the financial statements. For example, their remuneration may be based on the growth in EPS. Looking at calculation of the EPS of 3.05 (rounded) we can see that it is affected by:

- the number of shares in issue which can be changed by issuing bonus shares, share splits and reverse share splits;
- the profit which can be manipulated by adjusting accrued liabilities, depreciation, amortisation and impairment charges.

Simply buying back one-sixth of the shares can lift the EPS by more than 10%.

29.7.4 Dividend cover

Dividend cover is computed as:

$$\frac{\text{Profit for the period}}{\text{Dividends paid}} \quad \text{or} \quad \frac{\text{EPS}}{\text{Dividend per share}}$$

Dividend cover is a measure of the vulnerability of the dividend to a fall in profits. The legality of a dividend payment is normally based on cumulative profits rather than the profits for a single period, but in practice an entity would wish the dividend declared for a particular period to be 'covered' by profits made in that period. Therefore this ratio is seen as a measure of the 'security' of the dividend.

An issue with this ratio is whether a high dividend cover is good or bad. In one sense, a shareholder might be content with a high dividend cover, because this would mean that profits could potentially fall quite significantly without the dividend necessarily falling, and retained earnings are being employed profitably within the company. Alternatively, a shareholder might feel disgruntled that the dividend itself is not higher. Therefore conclusions about whether a change in dividend cover is 'good' or 'bad' need to be made with caution – the trend and inter-firm comparators from the same industry need to be looked at. For example, some companies may target the rate of dividend cover as a key performance indicator as shown in the following extract from the Morrisons 2011 Annual Report:

> Our aim is that dividend cover will be the same as the average for the European food retail sector. Our dividend cover is 2.4 times, in line with the European food retail sector average. This has resulted in dividend growth of 17%.

29.7.5 Dividend yield

Dividend yield is computed as:

$$\frac{\text{Dividend per share}}{\text{Market value of a share}} \text{ (expressed as a percentage)}$$

This ratio measures the 'effective' current investment by the shareholder in the entity, because by deciding to keep the share rather than dispose of it the shareholder is forgoing an amount that would be available were the shareholder to make a disposal decision.

This ratio is a 'market-based' ratio, because it is influenced by the share price of the entity. We need to interpret any 'market-based' ratio with caution. In this case a high dividend yield could mean that the shareholder is receiving a very healthy dividend (which would be very positive) or that the share price was very low (which would clearly not be a desirable position either for the entity or for the shareholder).

Indeed, in times of disappointing prices on securities markets, dividend yields often tend to be very high because entities are reluctant to cut their dividends for fear the share price will fall even further. A combination of a static dividend and falling share prices leads inevitably to a rise in dividend yields. This would become more apparent if dividend growth were considered in addition to dividend yield.

29.8 Determining value

There are three aspects to consider. One is to assess from an entity viewpoint whether adequate returns (EVA) are being generated, the other is to assess from a shareholder's viewpoint the total shareholder return (TSR), and the third is how either is used in setting directors' remuneration.

29.8.1 Economic value added (EVA)

Value is created when the return on a company's economic capital employed is greater than the cost of that capital. This is expressed as EVA which is the profit that a company earns less the cost of capital.

Companies are increasingly becoming aware that investors need to be confident that the company can achieve growth, and that communication of a positive EVA is the key. This is why companies are using the annual report to provide shareholders and potential shareholders with a measure of the company's performance that will give them confidence to maintain or make an investment in the company. This is the view expressed in the 2009 Annual Report of Geveke NV Amsterdam:

> A positive EVA indicates that over a specific period economic value has been created. Net operating profit after tax is then greater than the cost of finance (i.e. the company's weighted average cost of capital). Research has shown that a substantial part of the long-term movement in share price is explained by the development of EVA. The concept of EVA can be a very good method of performance measurement and monitoring of decisions. It is for this reason that it is being incorporated into corporate strategy worldwide.

29.8.2 Formula for calculating economic value added

The formula applied by Geveke is as follows:

> EVA measures economic value achieved over a specific period. It is equal to net operating profit after tax (NOPAT), corrected for the cost of capital employed (the sum of interest bearing liabilities and shareholders' equity). The cost of capital employed is the required yield (R) times capital employed (CE).
>   In the form of a formula: $NOPAT - (R \times CE) = EVA$

We will illustrate the formula for Alpha NV, which has the following data (in euros):

	31 March 20X3	31 March 20X4	31 March 20X5
NOPAT	10m	11m	12.5m
Weighted average cost of capital (WACC)	12%	11.5%	11%
Capital employed	70m	77m	96m

The EVA is:

	Percentage change
31 March 20X3: EVA = 10m − (12% of 70m) = 1.6m	—
31 March 20X4: EVA = 11m − (11.5% of 77m) = 2.145m	34%
31 March 20X5: EVA = 12.5m − (11% of 96m) = 1.94m	(10%)

The formula allows weight to be given to the capital employed to generate operating profit. The percentage change is an important management tool in that the annual increase is seen as the created value rather than the absolute level, i.e. the 34% is the key figure rather than the 2.145 million. Further enquiry is necessary to assess how well Alpha NV will employ the increase in capital employed in future periods.

It is useful to calculate rate of change over time. However, as for all inter-company comparisons of ratios, it is necessary to identify how the WACC and capital employed have been defined. This may vary from company to company.

WACC calculation

This figure depends on the capital structure and risk in each country in which a company has a significant business interest. For example, the following is an extract from the 2003 Annual Report of the Orkla Group:

Capital structure and cost of capital

The Group's average cost of capital is calculated as a weighted average of the costs of borrowed capital and equity. The calculations are based on an equity-to-total-assets ratio of 60%. The cost of equity is calculated with the help of the Capital Asset Pricing Model. The cost of borrowed capital is based on a long-term, weighted interest rate for relevant countries in which Orkla operates . . .

A table included in the company's report showed:

Description	Rates	Relative %	Weighted cost
Weighted average beta	1.0		
× Market risk premium	4.0%		
= Risk premium for equity	4.0%		
+ Risk free long-term interest rate	4.9%		
= Cost of equity	8.9%	60%	5.3%
Imputed borrowing rate before tax	5.9%		
Imputed tax charge	28%		
= Imputed borrowing rate after tax	4.2%	40%	1.7%
WACC after tax			7.0%

In 2012 Orkla's WACC was 10%.

Capital employed definition

The norm is to exclude non-interest-bearing liabilities including current liabilities when determining net total assets. However, there are variations in the treatment of intangible assets, e.g. goodwill may be excluded from the net assets.

29.8.3 Achieving increases in EVA

EVA can be improved in three ways: by increasing NOPAT, reducing WACC and/or improving the utilisation of capital employed.

- Increasing NOPAT: this is achieved by optimising strategic choices by comparing the cash flows arising from different strategic opportunities, e.g. appraising geographic and product segmental information, cost reduction programmes, appraising acquisitions and divestments.

- Reducing WACC: this is achieved by reviewing the manner in which a company is financed, e.g. determining a favourable gearing ratio and reducing the perceived risk factor by a favourable spread of products and markets.

- Improving the utilisation of capital employed: this is achieved by consideration of activity ratios, e.g. non-current asset turnover, working capital ratio.

29.8.4 Management attitude to use of EVA

One study[5] identified a number of companies that used value-based measures at head office level, but retained traditional profit measures in their divisions. KPMG, in a 1995 survey of value-based management, described this type of company as 'light users', who report overall results in value-based terms but retain traditional measures within their performance measurement systems.

Turning EVA into a comparative ratio

EVA momentum[6] relates the change in the EVA £ value to the previous period's sales. The formula is:

$$EVA\ momentum = \frac{EVA^{Period\ 2} - EVA^{Period\ 1}}{Revenue^{Period\ 1}}$$

Companies are now being ranked by EVA momentum and it is reported[7] that because it is based on the change in EVA rather than the level, EVA momentum captures profitability performance where it matters most – at the margin. Companies that are losing money but cut their losses dramatically score well on EVA momentum. In contrast, even extremely profitable companies can score poorly on this performance measure if their economic profits are static or declining. As a result, EVA momentum is a great measure for spotting turning points in performance.

29.8.5 Total shareholder return approach

Shareholder value (SV)

It has been a long-standing practice for analysts to arrive at shareholder value of a share by calculating the internal rate of return (IRR %) on an investment from the dividend stream and realisable value of the investment at date of disposal, i.e. taking account of dividends received and capital gains. However, it is not a generic measure in that the calculation is specific to each shareholder. The reason for this is that the dividends received will depend on the length of period the shares are held and the capital gain achieved will depend on the share price at the date of disposal – and, as we know, the share price can move significantly even over a week.

For example, consider the SV for each of the three shareholders, Miss Rapid, Mr Medium and Miss Undecided, who each invested £10,000 on 1 January 20X6 in Spacemobile Ltd which pays a dividend of £500 on these shares on 31 December each year. Miss Rapid sold her shares on 31 December 20X7. Mr Medium sold his on 31 December 20X9, whereas Miss Undecided could not decide what to do with her shares. The SV for each shareholder is as follows:

Shareholder	Date acquired	Investment at cost	Dividends amount (total)	Date of disposal	Sale proceeds	IRR %
Miss Rapid	1.1.20X6	10,000	1,000	31.12.20X7	11,000	10%*
Mr Medium	1.1.20X6	10,000	2,000	31.12.20X9	15,000	15%
Miss Undecided	1.1.20X6	10,000	2,000	Undecided		

* $(500 \times .9091) + (11,500 \times .8265) - 10,000 = 0$

We can see that Miss Rapid achieved a shareholder value of 10% on her shares and Mr Medium, by holding until 31.12.20X9, achieved an increased capital gain raising the SV to 15%. We do not have the information as to how Miss Rapid invested from 1.1.20X8 and so we cannot evaluate her decision – it depends on the subsequent investment and the economic value added by that new company.

29.8.6 Total shareholder return

Miss Undecided has a notional SV at 31.12.20X9 of 15% as calculated for Mr Medium. However, this has not been realised and, if the share price changed the following day, the SV would be different. The notional 15% calculated for Miss Undecided is referred to as the total shareholder return (TSR) – it takes into account market expectation on the assumption that share prices reflect all available information but it is dependent on the assumption made about the length of the period the shares are held.

29.8.7 Performance-based remuneration using EVA and TSR

EVA and managers' performance

In some organisations EVA has been used as a basis for determining bonus payments made to managers. There is some evidence that managers rewarded under such a scheme do perform better than those operating under more traditional schemes. However, research[8] indicated that this occurs when managers understand the concept of EVA and that it is not universally appropriate as other factors need to be taken into account such as the area of the firm in which a manager is employed. The following is an extract from the ThyssenKrupp 2009 Annual Report:

> This management and controlling system is linked to the bonus system in such a way that the amount of the performance-related remuneration is determined by the achieved EVA.

However, there is a risk that this approach can encourage short-termism by focusing on annual targets.

TSR and managers' performance

TSR has been used for performance monitoring, as a criterion for performance-based remuneration and to satisfy statutory requirements.

Performance monitoring

TSR has been used by companies to monitor their performance by comparing their own TSR with that of comparator companies. It is also used to set strategic targets. For example, Unilever set itself a TSR target in the top third of a reference group of 21 international consumer goods companies. Unilever calculates the TSR over a three-year rolling period which it considers 'sensitive enough to reflect changes but long enough to smooth out short-term volatility'.

Statutory requirement

The Directors' Report Regulations 2002 now require a line graph to be prepared showing such a comparison. Marks & Spencer Group's 2012 Annual Report contained the following:

Total shareholder return performance graph
The graph illustrates the performance of the Company against the FTSE 100 over the past five years. The FTSE 100 has been chosen as it is a recognised broad equity market index of which the Company has been a member throughout the period.

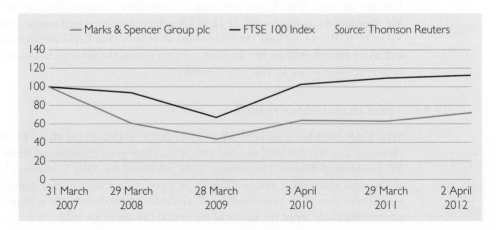

Management and investors assess a company's performance based on the use of ratios described in Chapter 28 and earlier in the present chapter. This follows the pyramid approach of starting with the ROE and drilling down to identify possible causes of change.

The models that attempt to predict corporate failure combine selected ratios to produce a single-figure score. There are a number of such models and we will discuss a selection.

29.9 Predicting corporate failure

In the preceding chapter we extolled the virtues of ratio analysis for the interpretation of financial statements. However, ratio analysis is an excellent indicator only when applied properly. Unfortunately, a number of limitations impede its proper application. How do we know which ratios to select for the analysis of company accounts? Which ratios can be combined to produce an informative end-result? How should individual ratios be ranked to give the user an overall picture of company performance? How reliable are all the ratios – can users place more reliance on some ratios than others? We will consider which ratios have been selected to produce Z-scores and H-scores.

Z-score analysis can be employed to overcome some of the limitations of traditional ratio analysis. It evaluates corporate stability and, more importantly, predicts potential instances of corporate failure. All the forecasts and predictions are based on publicly available financial statements.[9] The aim is to identify potential failures so that 'the appropriate action to reverse the process [of failure] can be taken before it is too late'.[10]

29.9.1 What are Z-scores?

Inman[11] describes what Z-scores are designed for:

Z-scores attempt to replace various independent and often unreliable and misleading historical ratios and subjective rule-of-thumb tests with scientifically analysed ratios which can reliably predict future events by identifying benchmarks above which 'all's well' and below which there is imminent danger.

Z-scores provide a single-value score to describe the combination of a number of key characteristics of a company. Some of the most important predictive ratios are weighted according to perceived importance and then summed to give the single Z-score. This is then evaluated against the identified benchmark.

The two best-known Z-scores are Altman's Z-score and Taffler's Z-score.

Altman's Z-score

The original Z-score equation was devised by Professor E. Altman in 1968 and developed further in 1977.[12] The original equation was:

$$Z = 0.012X_1 + 0.014X_2 + 0.033X_3 + 0.006X_4 + 0.999X_5$$

where:

X_1 = Working capital/Total assets
(Liquid assets are being measured in relation to the business's size and this may be seen as a better predictor than the current and acid test ratios which measure the interrelationships within working capital. For X_1 the more relative Working capital, the more liquidity.)

X_2 = Retained earnings/Total assets
(In early years the proportion of retained earnings used to finance the total asset base may be quite low and the length of time the business has been in existence has been seen as a factor in insolvency. In later years the more earnings that are retained the more funds that could be available to pay creditors. X_2 also acts as an indication of a company's dividend policy – a high dividend payout reduces the retained earnings with impact on solvency and creditors' position.)

X_3 = Earnings before interest and tax (EBIT)/Total assets
(Adequate operating profit is fundamental to the survival of a business.)

X_4 = Market capitalisation/Book value of debt
(This is an attempt to include market expectations which may be an early warning as to possible future problems. Solvency is less likely to be threatened if shareholders' interest is relatively high in relation to the total debt.)

X_5 = Sales/Total assets
(This indicates how assets are being used. If efficient, then profits available to meet interest payments are more likely. It is a measure that might have been more appropriate when Altman was researching companies within the manufacturing sector. It is a relationship that varies widely between manufacturing sectors and even more so within knowledge-based companies.)

Altman identified two benchmarks. Companies scoring over 3.0 are unlikely to fail and should be considered safe, while companies scoring under 1.8 are very likely to fail. The value of 3.0 has since been revised down to 2.7.[13] Z-scores between 2.7 and 1.8 fall into the grey area. The 1968 work is claimed to be able to distinguish between successes and failures up to two or three years before the event. The 1977 work claims an improved prediction period of up to five years before the event.

The Zeta model

This was a model developed by Altman and Zeta Services, Inc. in 1977. It is the same as the Z-score for identifying corporate failure one year ahead but it is more accurate in identifying potential failure in the period two to five years ahead. The model is based on the following variables:

X_1 return on assets: earnings before interest and tax/total assets;
X_2 stability of earnings: normalised return on assets around a five- to ten-year trend;
X_3 interest cover: earnings before interest and tax/total interest;
X_4 cumulative profitability: retained earnings/total assets;
X_5 liquidity: the current ratio;
X_6 capitalisation: equity/total market value;
X_7 size: total tangible assets.

Zeta is available as a subscription service and the coefficients have not been published.

Taffler's Z-score

The exact definition of Taffler's Z-score[10,14] is unpublished, but the following components form the equation:

$$Z = c_0 + c_1 X_1 + c_2 X_2 + c_3 X_3 + c_4 X_4$$

where

X_1 = Profit before tax/Current assets (53%)
X_2 = Current assets/Current liabilities (13%)
X_3 = Current liabilities/Total assets (18%)
X_4 = No credit interval = Length of time which the company can continue to finance its
operations using its own assets with no revenue inflow (16%)

In the equation, c_0 to c_4 are the coefficients, and the percentages in brackets represent the ratios' contributions to the power of the model.

The benchmark used to detect success or failure is 0.2. Companies scoring above 0.2 are unlikely to fail, while companies scoring less than 0.2 demonstrate the same symptoms as companies that have failed in the past.

PAS-score: performance analysis score

Taffler adapted the Z-score technique to develop the PAS-score. The PAS-score evaluates company performance relative to other companies in the industry and incorporates changes in the economy.

The PAS-score ranks all company Z-scores in percentile terms, measuring relative performance on a scale of 0 to 100. A PAS-score of X means that $100 - X\%$ of the companies have scored higher Z-scores. So, a PAS-score of 80 means that only 20% of the companies in the comparison have achieved higher Z-scores.

The PAS-score details the relative performance trend of a company over time. Any downward trends should be investigated immediately and the management should take appropriate action.

SMEs and failure prediction

The effectiveness of applying a failure prediction model is not restricted to large companies. This is illustrated by research[15] conducted in New Zealand where such a model was applied to 185 SMEs and found to be useful. As with all models, it is also helpful to refer to other supplementary information that may be available, e.g. other credit reports, credit managers' assessments and trade magazines.

29.9.2 H-scores

An H-score is produced by Company Watch to determine overall financial health. The H-score is an enhancement of the Z-score technique in giving more emphasis to the strength of the statement of financial position. The Company Watch system calculates a score ranging from 0 to 100 with below 25 being in the danger zone. It takes into account profit management, asset management and funding management using seven factors: profit from the statement of income; three factors from the asset side of the statement of financial position, namely current asset cover, inventory and trade receivables management and liquidity; and three factors from the liability side of the statement of financial position, namely equity base, debt dependence and current funding.

The factors are taken from published financial statements, which makes the approach taken by the IASB to bring off-balance-sheet transactions onto the statement of financial position particularly important.

The ability to chart each factor against the sector average and to 25 level criteria over a five-year period means that it is valuable for a range of user needs, from trade creditors considering extending or continuing to allow credit to potential lenders and equity investors and the big four accounting firms in reviewing audit risk. The model also has the ability to process 'what-ifs'.

It appears to be a robust, useful and exciting tool for all user groups. It is not simply a tool for measuring risk. It can also be used by investors to identify companies whose share price might have fallen but which might be financially strong with the possibility of the share price recovering – it can indicate 'buy' situations. It is also used by leading firms of accountants for the purpose of targeting companies in need of turnaround. Further information appears on the company's website at www.companywatch.net which includes additional examples.

29.9.3 A-scores

A-scores concentrate on non-financial signs of failure.[16] This method sets out to quantify different judgemental factors.

Management defects and strategic mistakes

The whole basis of the analysis is that financial difficulties are the direct result of management defects and strategic mistakes which can be evidenced by symptoms. A weighting is then attached to individual defects and mistakes.

For example, in looking at management defects a weighting system might be applied such as:

	Weight
Defects in operational management:	
The chief executive is an autocrat	8
The chief executive is also the chairman	4
The board is unbalanced, e.g. too few with finance experience	2
Defects in financial management:	
There are no budgets for budgetary control	3
Weak finance director	3
There is a poor response to change, e.g. out-of-date plant, old-fashioned products, poor marketing	15

To calculate a company A-score, different scores are allocated to each defect, mistake and symptom according to their importance. Then this score is compared with the benchmark values. If companies achieve an overall score of over 25, or a defect score of over 10, or a mistakes score of over 15, then the company is demonstrating typical signs leading up to failure. Generally, companies not at risk will score below 18, and companies which are at risk will score well over 25.

Symptoms

With an adverse A-score, symptoms of failure will start to arise. These are directly attributable to preceding management mistakes. Typical symptoms are financial signs (e.g. poor ratios, poor Z-scores); creative accounting (management might attempt to 'disguise' signs of failure in the accounts); non-financial signs (e.g. investment decisions delayed; market share drops); and terminal signs (when the financial collapse of the company is imminent).

It is interesting to see the weighting given to the chief executive being an autocrat, which is supported by the experience in failures such as Worldcom in 2002 with the following comment:[17]

> **'Autocratic style'**
> Worldcom pursued an aggressive strategy under Ebbers . . . In 1998, Ebbers cemented his reputation when Worldcom purchased MCI for $40bn – the largest acquisition in corporate history at that time . . . But according to one journalist in Mississippi who followed Worldcom from its inception, the seeds of the disaster were sown from the start by Ebbers' aggressive autocratic management style.

However, there are also limitations to participative management which could lead to slow reaction to change in a fast-moving environment.[18]

29.9.4 Failure prediction combining cash flow and accrual data

There is a continuing interest in identifying variables which have the ability to predict the likelihood of corporate failure – particularly if this only requires a small number of variables. One study[19] indicated that a parsimonious model that included only three financial variables, namely a cash flow, a profitability and a financial leverage variable, was accurate in 83% of the cases in predicting corporate failure one year ahead.

29.10 Professional risk assessors

Credit agencies such as Standard & Poor and Moody's Investor Services assist investors, lenders and trade creditors by providing a credit rating service. Companies are given a rating that can range from AAA for companies with a strong capacity to meet their financial commitments down to D for companies that have been unable to make contractual payments or have filed for bankruptcy, with more than 10 ratings in between, e.g. BBB for companies that have adequate capacity but which are vulnerable to internal or external economic changes.

29.10.1 How are ratings set?

The credit agencies take a broad range of internal company and external factors into account. Internal company factors may include:

- an appraisal of the financial reports to determine:
  - trading performance, e.g. return on equity (ROE) and return on assets (ROA); earnings volatility; how well a company has coped with business cycles and severe competition;
  - cash flow adequacy, e.g. EBITDA interest cover; EBIT interest cover; free operating cash flow;
  - capital structure, e.g. gearing ratio; any off-balance-sheet financing;
- a consideration of the notes to the accounts to determine possible adverse implications, e.g. contingent liabilities, whether the company is fixed-capital- or working-capital-intensive or has heavy capital investment commitments;
- meetings and discussions with management;
- monitoring expectation, e.g. against quarterly reports, company press releases, profit warnings;
- monitoring changes in company strategy, e.g. changes to funding structure with company buyback of shares, new divestment or acquisition plans and implications for any debt covenants.

However, experience with companies such as Enron makes it clear that off-balance-sheet transactions can make appraisal difficult even for professional agencies if companies continue to avoid transparency in their reporting.

External factors may include:

- growth prospects, e.g. trends in industry sector; technology possible changes; peer comparison;

- competitors, e.g. the major domestic and foreign competitors; product differentiation; barriers to entry;

- keeping a watching brief on macroeconomic factors, e.g. environmental statutory levies, tax changes, political changes such as restrictions on the supply of oil, foreign currency risks.

29.10.2 Regulation of credit rating agencies

Since the credit crisis there has been severe criticism that credit rating agencies had not been independent when rating financial products. The agencies have been self-regulated but this has been totally inadequate in curtailing conflicts of interest. The conflicts have arisen because they were actively involved in the design of products (collateralised debt obligations) to which they then gave an 'objective' credit rating which did not clearly reflect the true risks associated with investing in them. This conflict of interest was compounded by the fact that (a) agency staff were free to join a company after rating its products, and (b) the companies issuing the products paid their fees.

The following swingeing comments were made by the ACCA:[17]

Regulation of credit agencies
It's a joke that an industry with such influence, particularly during the current volatile economic climate, is self-regulated and only subject to a toothless voluntary code of conduct.

   The mere fact that credit rating agencies are paid by the companies they rate puts their independence in jeopardy . . . greater transparency is required . . . We have to strike the right balance when regulating the market between protecting and over-burdening. A range of measures is necessary to bring about transparency in the ratings process . . . Regulation would be part of the solution, but it can't be used in isolation . . . This is a perfect example for when an international set of regulations and other measures are imperative to regain trust in financial markets and avoid further credit crunched victims.

This has led to a call for both Europe and the US to regulate the agencies.

European Commission Agency Regulation[20]

In November 2008, the European Commission adopted a proposal for a Regulation on Credit Rating Agencies, which would require agencies to have procedures in place to ensure that:

- ratings are not affected by conflicts of interest;
- credit rating agencies have a high standard for the quality of the rating methodology and the ratings; and
- credit rating agencies act in a transparent manner.

The intention is that the agencies would remain responsible for the content of the ratings.

   In 2014 the Commission adopted a report on the feasibility of a network of smaller credit rating agencies to facilitate their growth to become more competitive market players.

29.11 Valuing shares of an unquoted company – quantitative process

The valuation of shares brings together a number of different financial accounting procedures that we have covered in previous chapters. The assumptions may be highly subjective, but there is a standard approach. This involves the following:

- Estimate the maintainable income flow based on earnings defined in accordance with the IIMR guidelines, as described in Chapter 27. Normally the profits of the past five years are used, adjusted for any known or expected future changes.

- Estimate an appropriate dividend yield, as described in Section 29.7.5, if valuing a non-controlling holding.

- Estimate an appropriate PE or earnings yield if valuing a majority holding. In the UK there is now a Valuation Index[21] focused on SMEs which is the result of UK200's Corporate Finance members providing key data on actual transactions involving the purchase or sale of real businesses (in the form of asset or share deals) over the past five years. The average PE ratio at November 2011 stood at 6.0 and the ratio of deal value to EBITDA had increased from 4.6 to 4.9 times. Average deal size in the last two years continued to be just under £3m.

- Make a decision on any adjustment to the required yields. For example, the shares in the unquoted company might not be as marketable as those in the comparative quoted companies and the required yield would therefore be increased to reflect this lack of marketability; or the statement of financial position might not be as strong with lower current/acid test ratios or higher gearing, which would also lead to an increase in the required yield.

- Calculate the economic capital value, as described in Chapter 6, by applying the required yield to the income flow.

- Compare the resulting value with the net realisable value (NRV), as described in Chapter 7, when deciding what action to take based on the economic value.

EXAMPLE • The Doughnut Ltd is an unlisted company engaged in the baking of doughnuts. The statement of financial position of the Doughnut Ltd as at 31 December 20X9 showed:

	£000	£000
Freehold land		100
Non-current assets at cost	240	
Accumulated depreciation	40	
		200
Current assets	80	
Current liabilities	(60)	
		20
		320
Share capital in £1 shares		300
Retained earnings		20
		320
Estimated net realisable values:		
Freehold land		180
Plant and equipment		120
Current assets		70

The company achieved the following profit after tax (adjusted to reflect maintainable earnings) for the past five years ended 31 December:

	20X5	20X6	20X7	20X8	20X9
Maintainable earnings (£000)	36	40	44	38	42
Dividend payout history: Dividends	10%	10%	12%	12%	12%

Current yields for comparative quoted companies as at 31 December 20X9:

	Earnings yield %	Dividend yield %
Ace Bakers plc	14	8
Busi-Bake plc	10	8
Hard-to-beat plc	13	8

Acquiring a majority holding

You are required to value a holding of 250,000 shares for a shareholder, Mr Quick, who makes a practice of buying shares for sale within three years.

Now, the 250,000 shares represent an 83% holding. This is a majority holding and the steps to value it are as follows:

1 Calculate average maintainable earnings (in £000):

$$\frac{36,000 + 40,000 + 44,000 + 38,000 + 42,000}{5} = £40,000$$

2 Estimate an appropriate earnings yield:

$$\frac{14\% + 10\% + 13\%}{3} = 12.3\%$$

3 Adjust the rate for lack of marketability by, say, 3% and for the lower current ratio (of 1.3:1) by, say, 2%. Both these adjustments are subjective and would be a matter of negotiation between the parties.

Required yield	= 12.3
Lack of marketability weighting	= 3.0
Statement of financial position weakness =	2.0
Required earnings yield	17.3

The adjustments depend on the actual circumstances. For instance, there might be negotiation over the use of the average of £40,000 with differing views on growth and, if Mr Quick were intending to hold the shares as a long-term investment, there might be less need to increase the required return for lack of marketability.

4 Calculate share value:

$$(£40,000 \times 100/17.3)/300,000 = 77p$$

5 Compare with the net realisable values on the basis that the company was to be liquidated:

	£
Net realisable values = 70,000 + 120,000 + 180,000 =	370,000
Less: Current liabilities	60,000
	310,000
Net asset value per share = £310,000/300,000 =	£1.03

The comparison indicates that, on the information we have been given, Mr Quick is paying less than the net realisable value, but the difference may not be enough to justify acquiring the shares in order to asset strip and liquidate the company to make an immediate capital gain.

Acquiring a minority holding

Let us extend our illustration by assuming that, if Mr Quick acquires control, it is intended to replace the non-current assets at a cost of £20,000 per year out of retained earnings. One of the remaining minority shareholders, Ms Croissant, wishes to dispose of shares and is in discussion with Mr Small who has £10,000 to invest. You are required to calculate for Mr Small how many shares he should aim to acquire from Ms Croissant.

There are two significant changes: the cash available for distribution as dividends will be reduced by £20,000 per year, which is used to replace non-current assets; and Mr Small is acquiring only a minority holding, which means that the appropriate valuation method is the **dividend yield** rather than the **earnings yield**.

The share value will be calculated as follows:

1 Estimate income flow:

	£
Maintainable earnings	40,000
Less: CAPEX	20,000
Cash available for distribution	20,000

Note that we are here calculating not distributable profits, but the available cash flow.

2 Required dividend yield:

	%
Average dividend yield	8.0
Lack of negotiability, say	2.0
Financial risk, say	1.5
	11.5

3 Share value:

$$\frac{£20,000}{300,000} \times \frac{100}{11.5} = 58p$$

At this price it would be possible for Mr Small to acquire (£10,000/58p) = 17,241 shares.

29.12 Valuing shares of an unquoted company – qualitative process

In the section above we illustrated how to value shares using the capitalisation of earnings and capitalisation of dividends methods. However, share valuation is an extremely subjective exercise.

A company's future cash flows may be affected by a number of factors. These may occur as a result of a change of control, action within the company (e.g. management change, revenue investment) or external events (e.g. change in the rate of inflation, change in competitive pressures).

- **Change of control:**
  - Aer Lingus said the offer in 2012 of €1.30 (£1.02) per share by Ryanair was 31% below the €1.87 cash per share based on the company's cash balance of €1bn.
  - Ryanair in its offer document said it would grow jobs at Aer Lingus and raise the flag carrier's passenger numbers from 9.5 million a year to 14 million, by cutting Aer Lingus ticket prices and improving the productivity of Aer Lingus staff in order to hold down costs and maintain profit margins.

- **Management change** often heralds a significant change in a company's share price. For example, car and bike parts retailer Halfords' share price jumped after the company appointed a new Chief Executive Officer in October 2012 following the abrupt departure of David Wild in the summer, as it revealed that full-year profits would be at the top end of guidance after a strong second quarter.

- **Revenue investment** refers to discretionary revenue expenditure, such as charges to the income statement for research and development, training and advertising. It also relates to expenditure on costs such as amount of office space provided and travel expense allowed. Where in the recession there had been a reduction in face-to-face meetings and an increase in video- and web-conferencing, there is ongoing pressure to maintain this process into the future.

- **Changes in the rate of inflation** can affect the required yield. If, for example, it is expected that inflation will fall, this might mean that past percentage yields will be higher than the percentage yield that is likely to be available in the future.

- **Change in competitive pressures** can affect future sales. For example, increased foreign competition could mean that past maintainable earnings are not achievable in the future and the historic average level might need to be reduced.

These are a few of the internal and external factors that can affect the valuation of a share. The factors that are relevant to a particular company may be industry-wide (e.g. change in rate of inflation), sector-wide (e.g. change in competitive pressure) or company-specific (e.g. loss of key managers or employees).

If the company supports the acquisition of the shares, the valuer will be able to gain access to relevant internal information. For example, details of research and development expenditure may be available analysed by type of technology involved, by product line, by project and by location, and distinguishing internal from externally acquired R&D.

If the acquisition is being considered without the company's knowledge or support, the valuer will rely more heavily on information gained from public sources, e.g. statutory and voluntary disclosures in the annual accounts and industry information such as trade journals. Information on areas such as R&D may be provided in the OFR (Operating and Financial Review or Strategic Report), but probably in an aggregated form, constrained by management concerns about use by potential competitors.[22]

There is an increasing wealth of financial and narrative disclosures to assist investors in making their investment decisions. There are external data such as the various multivariate Z-scores and H-scores and professional credit agency ratings; and there is greater internal disclosure of financial data such as TSR and EVA data indicating how well companies have managed value in comparison with a peer group and of narrative information such as the IFRS Practice Statement *Management Commentary*. It is also easier to access companies' financial data through the Web.

Literature search of qualitative factors which can lead to improved or reduced valuations

There is an interesting research report[23] investigating the nature of SME intangible assets in which the researchers have reported the following:

- **Factors identified in the literature as enhancing achieved price**: transportable business with a transferable customer base; non-cancellable service agreements and beneficial contractual arrangements; unexploited property situations; synergistic and cost-saving benefits; under-exploited brands and products; customer base providing cross-selling opportunities; competitor elimination, increased market share; complementary product or service range; market entry – a quick way of overcoming entry barriers; buy into new technology; access to distribution channels; and non-competition agreements.

- **Factors identified in the literature as diminishing achieved price**: confused accounts; poor housekeeping, doubtful debts, under-utilised equipment, outstanding litigation, etc.; over-dependence upon owner and key individuals; over-dependence on a small number of customers; unrelated side activities; poor or out-of-date company image; long-term contracts about to finish; poor liquidity; poor performance; minority and 'messy' ownership structures; inability to substantiate ownership of assets; and uncertainties surrounding liabilities.

Not all of these satisfy the criteria for recognition in annual financial statements.

Summary

This chapter has introduced a number of additional analytical techniques to complement the pyramid approach to ratio analysis discussed in the previous chapter.

The increasing use of 'non-GAAP' cash-based ratios was discussed to reduce the effect of subjective judgements. The use of ratio thresholds was discussed in determining shariah compliance and in setting debt covenants.

The calculation of EVA and TSR was explained, statuary disclosures in the UK were illustrated and their use in the context of performance-related remuneration was discussed. In addition, this chapter has described the use of ratios in the valuation of unquoted shares.

All users of financial statements (both internal and external users) should be prepared to utilise any or all of the interpretive techniques suggested in this chapter and the preceding one. These techniques help to evaluate the financial health and performance of a company. Users should approach these financial indicators with real curiosity – any unexplained or unanswered questions arising from this analysis should form the basis of a more detailed examination of the company accounts.

REVIEW QUESTIONS

1 It has been suggested that the growth in profits can be achieved by accounting sleight of hand rather than genuine economic growth. Consider how 'accounting sleight of hand' can be used to report increased profits and discuss what measures can be taken to mitigate against the possibility of this happening.

2 Explain how the use of debt can improve returns to equity shareholders in good years and increase their losses in poor years.

3 Telecomsabroad plc has a dividend payout ratio of 95%. Discuss why using the ratio of free cash flow to dividend might influence your assessment of dividend growth.

4 Discuss the difficulties when attempting to identify comparator companies for benchmarking as, for example, when selecting a TSR peer group.

5 The Unilever annual review stated:

> Total Shareholder Return (TSR) is a concept used to compare the performance of different companies' stocks and shares over time. It combines share price appreciation and dividends paid

to show the total return to the shareholder. The absolute size of the TSR will vary with stock markets, but the relative position is a reflection of the market perception of overall performance relative to a reference group. The Company calculates the TSR over a three-year rolling period . . . Unilever has set itself a TSR target in the top third of a reference group of 21 . . . companies.

Discuss (a) why a three-year rolling period has been chosen, and (b) the criteria you consider appropriate for selecting the reference group of companies.

6 Discuss Z-score analysis with particular reference to Altman's Z-score and Taffler's Z-score. In particular:

(i)   What are the benefits of Z-score analysis?

(ii) What criticisms can be levelled at Z-score analysis?

7 Identify the two most significant variables in the Altman's and Taffler's Z-scores and discuss why each variable might have been selected.

8 Discuss three situations when management might be under pressure to adopt an aggressive earnings management approach.

9 Explain how and why EVA is calculated.

10 Discuss the advantages and disadvantages of all companies adopting the ratio criteria required to be shariah-compliant.

11 Describe the measures taken to reduce the risk that credit rating agencies can mislead investors.

12 The following is an extract from the Bayer AG 2012 Annual Report:

The value-based indicators aid management's decision-making, especially regarding strategic portfolio optimization and the allocation of resources for acquisitions and capital expenditures. The focus at the operational level is on the key drivers of enterprise value: growth (sales), cost efficiency (EBITDA) and capital efficiency (working capital, capital expenditures), since these directly affect value creation.

Discuss (a) why and how EBITDA is used as a driver for cost efficiency, and (b) how capital efficiency is determined in relation to working capital and capital expenditures.

13 Discuss how the following might be used by a shareholder and by the management:

(i)   The ratio of dividends plus share price movement to the opening share price.

(ii) Accounting profit less an additional charge for the use of equity capital.

14 The finance director was investigating a potential acquisition. As part of the exercise she gave your colleague the current value of total assets, the post-tax operating income, the economic life of the assets and the scrap value of the assets with a request to calculate the cash flow return on investment (CFROI) for the company. Your colleague has asked you to explain to him (a) how this is done or where he could find further information about this on the Web, and (b) how the CFROI will be used.

15 Hard Times Ltd has been just about breaking even. It has recently identified a new project which will improve its ROC. Discuss whether entering into the project will always be to the advantage of the shareholders considering WACC.

16 There are differences of opinion as to whether alternative performance measures (APMs) should be prescribed by the IASB or whether they should remain as defined by management. Discuss arguments for and against prescription.

EXERCISES

* Question 1

Belt plc and Braces plc were in the same industry. The following information appeared in their 20X9 accounts:

	Belt	Braces
	€m	€m
Revenue	200	300
Total operating expenses	180	275
Average total assets during 20X9	150	125

Required:

(a) Calculate the following ratios for each company and show the numerical relationship between them:

    (i) Their rate of return on the average total assets.

    (ii) The net profit percentages.

    (iii) The ratio of revenue to average total assets.

(b) Comment on the relative performance of the two companies.

(c) State any additional information you would require as:

    (i) A potential shareholder.

    (ii) A potential loan creditor.

* Question 2

Quickserve plc is a food wholesale company. Its financial statements for the years ended 31 December 20X8 and 20X9 are as follows:

Statements of income

	20X9	20X8
	£000	£000
Sales revenue	12,000	15,000
Gross profit	3,000	3,900
Distribution costs	500	600
Administrative expenses	1,500	1,000
Operating profit	1,000	2,300
Interest receivable	80	100
Interest payable	(400)	(350)
Profit before taxation	680	2,050
Income taxation	240	720
Profit after taxation	440	1,330
Dividends in SOCE	800	600

Statements of financial position

	20X9	20X8
	£000	£000
Non-current assets:		
Intangible assets	200	—
Tangible assets	4,000	7,000
Investments	600	800
	4,800	7,800
Current assets:		
Inventory	250	300
Trade receivables	1,750	2,500
Cash & bank	1,500	200
	3,500	3,000
Total assets	8,300	10,800
Equity and reserves:		
Ordinary shares of 10p each	1,000	1,000
Share premium account	1,000	1,000
Revaluation reserve	1,110	1,750
Retained earnings	3,190	3,550
	6,300	7,300
Debentures	1,000	2,000
Current liabilities	1,000	1,500
	8,300	10,800

Required:

(a) Describe the concerns of the following users and how reading an annual report might help satisfy these concerns:
   (i)   employees;
   (ii) bankers;
   (iii) shareholders.
(b) Calculate relevant ratios for Quickserve and suggest how each of the above user groups might react to these.

* Question 3

The following are the accounts of Bouncy plc, a company that manufactures playground equipment, for the year ended 30 November 20X6.

Statements of comprehensive income for years ended 30 November

	20X6	20X5
	£000	£000
Profit before interest and tax	2,200	1,570
Interest expense	170	150
Profit before tax	2,030	1,420
Taxation	730	520
Profit after tax	1,300	900
Dividends paid in SOCE	250	250

Statements of financial position as at 30 November 20X6

	20X6 £000	20X5 £000
Non-current assets (written-down value)	6,350	5,600
Current assets		
Inventories	2,100	2,070
Receivables	1,710	1,540
Total assets	10,160	9,210
Creditors: amounts due within one year		
Trade payables	1,040	1,130
Taxation	550	450
Bank overdraft	370	480
Total assets less current liabilities	8,200	7,150
Creditors: amounts due after more than one year		
10% debentures 20X7/20X8	1,500	1,500
	6,700	5,650
Capital and reserves		
Share capital: ordinary shares of 50p fully paid up	3,000	3,000
Share premium	750	750
Retained earnings	2,950	1,900
	6,700	5,650

The directors are considering two schemes to raise £6,000,000 in order to repay the debentures and finance expansion estimated to increase profit before interest and tax by £900,000. It is proposed to make a dividend of 6p per share whether funds are raised by equity or loan. The two schemes are:

1 an issue of 13% debentures redeemable in 30 years;

2 a rights issue at £1.50 per share. The current market price is £1.80 per share (20X5: £1.50; 20X4: £1.20).

Required:
(a) Calculate the return on equity and any three investment ratios of interest to a potential investor.
(b) Calculate three ratios of interest to a potential long-term lender.
(c) Report briefly on the performance and state of the business from the viewpoint of a potential shareholder and lender using the ratios calculated above and explain any weaknesses in these ratios.
(d) Advise management which scheme they should adopt on the basis of your analysis above and explain what other information may need to be considered when making the decision.

Question 4

Sally Gorden seeks your assistance to decide whether she should invest in Ruby plc or Sapphire plc. Both companies are quoted on the London Stock Exchange. Their shares were listed on 20 June 20X4 as Ruby 110p and Sapphire 120p.

The performance of these two companies during the year ended 30 June 20X4 is summarised as follows:

	Ruby plc	Sapphire plc
	£000	£000
Operating profit	588	445
Interest and similar charges	(144)	(60)
	444	385
Taxation	(164)	(145)
Profit after taxation	280	240
Interim dividend paid	(30)	—
Preference dividend paid	(90)	—
Ordinary dividend paid	(60)	(160)

The companies have been financed on 30 June 20X4 as follows:

	Ruby plc	Sapphire plc
	£000	£000
Ordinary shares of 50p each	1,000	1,500
15% preference shares of £1 each	600	—
Share premium account	60	—
Retained earnings	250	450
17% debentures	800	—
12% debentures	—	500
	2,710	2,450

On 1 October 20X3 Ruby plc issued 500,000 ordinary shares of 50p each at a premium of 20%. On 1 April 20X4 Sapphire plc made a 1 for 2 bonus issue. Apart from these, there has been no change in the issued capital of either company during the year.

Required:
(a) Calculate the earnings per share (EPS) of each company.
(b) Determine the price/earnings ratio (PE) of each company.
(c) Based on the PE ratio alone, which company's shares would you recommend to Sally?
(d) On the basis of appropriate accounting ratios (which should be calculated), identify three other matters Sally should take account of before she makes her choice.
(e) Describe the advantages and disadvantages of gearing.

* Question 5

Growth plc made a cash offer for all of the ordinary shares of Beta Ltd on 30 September 20X9 at £2.75 per share. Beta's accounts for the year ended 31 March 20X9 showed:

	£000
Profit for the year after tax	750
Dividends paid	250

Statement of financial position as at 31 March 20X9

	£000
Buildings	1,600
Other tangible non-current assets	1,400
	3,000

		£000
Current assets	2,000	
Current liabilities	1,400	
		600
		3,600
£1 ordinary shares		2,500
Retained earnings		1,100
		3,600

Additional information:

(i) The half yearly profits to 30 September 20X9 show an increase of 25% over those of the corresponding period in 20X8. The directors are confident that this pattern will continue, or increase even further.

(ii) The Beta directors hold 90% of the ordinary shares.

(iii) The following valuations are available:

Realisable values

	£000
Buildings	2,500
Other non-current assets	700
Current assets	2,500

Net replacement values

Buildings	2,600
Other non-current assets	1,800
Current assets	2,200

(iv) Shares in quoted companies in the same sector have a PE ratio of 10. Beta Ltd is an unquoted company.

(v) One of the shareholders is a bank manager who advises the directors to press for a better price.

(vi) The extra risk for unquoted companies is 25% in this sector.

Required:
(a) Calculate valuations for the Beta ordinary shares using four different bases of valuation.
(b) Draft a report highlighting the limitations of each basis and advise the directors whether the offer is reasonable.

Question 6

R. Johnson inherited 810,000 £1 ordinary shares in Johnson Products Ltd on the death of his uncle in 20X5. His uncle had been the founder of the company and managing director until his death. The remainder of the issued shares were held in small lots by employees and friends, with no one holding more than 4%.

R. Johnson is planning to emigrate and is considering disposing of his shareholding. He has had approaches from three parties, who are:

1 A competitor – Sonar Products Ltd. Sonar Products Ltd considers that Johnson Products Ltd would complement its own business and is interested in acquiring all of the 810,000 shares. Sonar Products Ltd currently achieves a post-tax return of 12.5% on capital employed.

2 Senior employees. Twenty employees are interested in making a management buyout with each acquiring 40,500 shares from R. Johnson. They have obtained financial backing, in principle, from the company's bankers.

3 A financial conglomerate – Divest plc. Divest plc is a company that has extensive experience of acquiring control of a company and breaking it up to show a profit on the transaction. It is its policy to seek a pre-tax return of 20% from such an exercise.

The company has prepared draft accounts for the year ended 30 April 20X9. The following information is available.

(a) Past earnings and distributions:

Year ended 30 April	Profit/(Loss) after tax	Gross dividends declared
£	%	
20X5	79,400	6
20X6	(27,600)	—
20X7	56,500	4
20X8	88,300	5
20X9	97,200	6

(b) Statement of financial position of Johnson Products Ltd as at 30 April 20X9:

	£000	£000
Non-current assets		
Land at cost		376
Premises at cost	724	
Aggregate depreciation	216	
		508
Equipment at cost	649	
Aggregate depreciation	353	
		296
Current assets		
Inventories	141	
Receivables	278	
Cash at bank	70	
	489	
Payables due within one year	(335)	
Net current assets		154
Non-current liabilities		(158)
		1,176
Represented by:		
£1 ordinary shares		1,080
Retained earnings		96
		1,176

(c) Information on the nearest comparable listed companies in the same industry:

Company	Profit after tax for 20X9 £000	Retention %	Gross dividend yield %
Eastron plc	280	25	15
Westron plc	168	16	10.5
Northron plc	243	20	13.4

Profit after tax in each of the companies has been growing by approximately 8% per annum for the past five years.

(d) The following is an estimate of the net realisable values of Johnson Products Ltd's assets as at 30 April 20X9:

	£000
Land	480
Premises	630
Equipment	150
Receivables	168
Inventories	98

Required:

(a) As accountant for R. Johnson, advise him of the amount that could be offered for his shareholding with a reasonable chance of being acceptable to the seller, based on the information given in the question, by each of the following:
(i)   Sonar Products Ltd;
(ii) the 20 employees;
(iii) Divest plc.

(b) As accountant for Sonar Products Ltd, estimate the maximum amount that could be offered by Sonar Products Ltd for the shares held by R. Johnson.

(c) As accountant for Sonar Products Ltd, state the principal matters you would consider in determining the future maintainable earnings of Johnson Products Ltd and explain their relevance.

(ACCA)

Question 7

Harry is about to start negotiations to purchase a controlling interest in NX, an unquoted limited liability company. The following is the statement of financial position of NX as at 30 June 2006, the end of the company's most recent financial year.

NX
Statement of financial position as at 30 June 2006

ASSETS	$
Non-current assets	3,369,520
Current assets	
Inventories, at cost	476,000
Trade and other receivables	642,970
Cash and cash equivalents	132,800
	1,251,770
Total assets	4,621,290
LIABILITIES AND EQUITY	
Non-current liabilities	
8% loan note	260,000
	260,000
Current liabilities	
Trade and other payables	467,700
Current tax payable	414,700
	882,400
Equity	
Ordinary shares, 40 cent shares	2,000,000
5% preferred shares of $1	200,000
Retained profits	1,278,890
	3,478,890
Total liabilities	1,142,400
Total liabilities and equity	4,621,290

The non-current assets of NX comprise:

	Cost	Depreciation	Net
	$	$	$
Property	2,137,500	262,500	1,875,000
Equipment	1,611,855	515,355	1,096,500
Motor vehicles	696,535	298,515	398,020
	4,445,890	1,076,370	3,369,520

NX has grown rapidly since its formation in 2000 by Albert Bell and Candy Dale who are currently directors of the company and who each own half of the company's issued share capital. The company was formed to exploit knowledge developed by Albert Bell. This knowledge is protected by a number of patents and trademarks owned by the company. Candy Dale's expertise was in marketing and she was largely responsible for developing the company's customer base. Figures for turnover and profit after tax taken from the statements of comprehensive income of the company for the past three years are:

	Turnover	Profit after tax
	$	$
Profit for 2004	8,218,500	1,031,000
Profit for 2005	10,273,100	1,288,720
Profit for 2006	11,414,600	991,320

NX's property has recently been valued at $3,000,000 and it is estimated that the equipment and motor vehicles could be sold for a total of $1,568,426. The net realisable values of inventory and receivables are estimated at $400,000 and $580,000 respectively. It is estimated that the costs of selling off the company's assets would be $101,000.

The 8% loan note is repayable at a premium of 30% on 31 December 2006 and is secured on the company's property. It is anticipated that it will be possible to repay the loan note by issuing a new loan note bearing interest at 11% repayable in 2012.

As directors of the company, Albert Bell and Candy Dale receive annual remuneration of $99,000 and £74,000 respectively. Both would cease their relationship with NX because they wish to set up another company together. Harry would appoint a general manager at an annual salary of $120,000 to replace Albert Bell and Candy Dale.

Investors in quoted companies similar to NX are currently earning a dividend yield of 6% and the average PE ratio for the sector is currently 11. NX has been paying a dividend of 7% on its common stock for the past two years.

Ownership of the issued common stock and preferred shares is shared equally between Albert Bell and Candy Dale.

Harry wishes to purchase a controlling interest in NX.

Required
(a) On the basis of the information given, prepare calculations of the values of a preferred share and an ordinary share in NX on each of the following bases:
  (i)   net realisable values;
  (ii) future maintainable earnings.
(b) Advise Harry on other factors which he should be considering in calculating the total amount he may have to pay to acquire a controlling interest in NX.

(The Association of International Accountants)

Question 8

The directors of Chekani plc, a large listed company, are engaged in a policy of expansion. Accordingly, they have approached the directors of Meela Ltd, an unlisted company of substantial size, in connection with a proposed purchase of Meela Ltd.

The directors of Meela Ltd have indicated that the shareholders of Meela Ltd would prefer the form of consideration for the purchase of their shares to be in cash and you are informed that this is acceptable to the prospective purchasing company, Chekani plc.

The directors of Meela Ltd have now been asked to state the price at which the shareholders of Meela Ltd would be prepared to sell their shares to Chekani plc. As a member of a firm of independent accountants, you have been engaged as a consultant to advise the directors of Meela Ltd in this regard.

In order that you may be able to do so, the following details, extracted from the most recent financial statements of Meela, have been made available to you.

Meela Ltd accounts for year ended 30 June 20X4
Statement of financial position extracts as at 30 June 20X4:

	£000
Purchased goodwill unamortised	15,000
Freehold property	30,000
Plant and machinery	60,000
Investments	15,000
Net current assets	12,000
10% debentures 20X9	(30,000)
Ordinary shares of £1 each	(40,000)
7% preference shares of £1 each (cumulative)	(12,000)
Share premium account	(20,000)
Retained earnings	(30,000)

Meela Ltd disclosed a contingent liability of £3.0m in the notes to the statement of financial position.
(Amounts in brackets indicate credit balances.)

Statement of comprehensive income extracts for the year ended 30 June 20X4:

	£000
Profit before interest payments and taxation and exceptional items	21,000
Exceptional items	1,500
Interest	(3,000)
Taxation	(6,000)
Dividends paid – Preference	(840)
– Ordinary	(3,000)
Retained profit for the year	9,660

(Amounts in brackets indicate a charge or appropriation to profits.)

The following information is also supplied:

(i) Profit before interest and tax for the year ended 30 June 20X3 was £24.2 million and for the year ended 30 June 20X2 it was £30.3 million.

(ii) Assume tax at 30%.

(iii) Exceptional items in 20X4 relate to the profit on disposal of an investment in a related company. The related company contributed to profit before interest as follows:

To 30 June 20X4	£0
To 30 June 20X3	£200,000
To 30 June 20X2	£300,000

(iv) The preference share capital can be sold independently, and a buyer has already been found. The agreed purchase price is 90p per share.

(v) Chekani plc has agreed to purchase the debentures of Meela Ltd at a price of £110 for each £100 debenture.

(vi) The current rental value of the freehold property is £4.5 million per annum and a buyer is available on the basis of achieving an 8% return on their investment.

(vii) The investments of Meela Ltd have a current market value of £22.5 million.

(viii) Meela Ltd is engaged in operations substantially different from those of Chekani plc. The most recent financial data relating to two listed companies that are engaged in operations similar to those of Meela Ltd are:

	NV per share	Market price per share	P/E	Net dividend per share	Cover	Yield
Ranpar plc	£1	£3.06	11.3	12 pence	2.6	4.9
Menner plc	50p	£1.22	8.2	4 pence	3.8	4.1

Required:

Write a report, of approximately 2,000 words, to the directors of Meela Ltd, covering the following:

(a) Advise them of the alternative methods used for valuing unquoted shares and explain some of the issues involved in the choice of method.

(b) Explain the alternative valuations that could be placed on the ordinary shares of Meela Ltd.

(c) Recommend an appropriate strategy for the board of Meela Ltd to adopt in its negotiations with Chekani plc.

Include, as appendices to your report, supporting schedules showing how the valuations were calculated.

Question 9

Briefly state:

(i) the case for segmental reporting;

(ii) the case against segmental reporting.

Question 10

Discuss the following issues with regard to financial reporting for risk:

(a) How can a company identify and prioritise its key risks?

(b) What actions can a company take to manage the risks identified in (a)?

(c) How can a company measure risk?

References

1 *No Surprises: The Case for Better Risk Reporting*, ICAEW, 1999.
2 www.frc.org.uk/Our-Work/Publications/ASB/Rising-to-the-Challenge/Full-results-of-a-Review-of-Narrative-Reporting-by.aspx
3 J. Collier, *Aggressive Earnings Management: Is It Still a Significant Threat?*, ICAEW, October 2004.
4 Alpa A. Virdi, *Investors' Confidence in Audited Financial Information*, Research Report, ICAEW, December 2004.
5 C. Minchington and G.Francis, 'Shareholder value', *Management Quarterly*, Part 6, January 2000.
6 www.evadimensions.com
7 http://www3.cfo.com/article/2011/11/benchmarking_top-and-bottom-25-eva-momentum-ranking-of-large-companies

8 J. Stern, 'Management: its mission and its measure', *Director*, October 1994, pp. 42–44.

9 C. Pratten, *Company Failure*, Financial Reporting and Auditing Group, ICAEW, 1991, pp. 43–45.

10 R.J. Taffler, 'Forecasting company failure in the UK using discriminant analysis and financial ratio data', *Journal of the Royal Statistical Society*, Series A, vol. 145, part 3, 1982, pp. 342–358.

11 M.L. Inman, 'Altman's Z-formula prediction', *Management Accounting*, November 1982, pp. 37–39.

12 E.I. Altman, 'Financial ratios, discriminant analysis and the prediction of corporate bankruptcy', *Journal of Finance*, vol. 23(4), 1968, pp. 589–609.

13 M.L. Inman, 'Z-scores and the going concern review', *ACCA Students' Newsletter*, August 1991, pp. 8–13.

14 R.J. Taffler, 'Z-scores: an approach to the recession', *Accountancy*, July 1991, pp. 95–97.

15 K. Van Peursem and M. Pratt, 'Failure prediction in New Zealand SMEs: measuring signs of trouble', *International Journal of Business Performance Management (IJBPM)*, vol. 8 (2/3), 2006.

16 J. Argenti, 'Predicting corporate failure', *Accountants Digest*, no. 138, Summer 1983, pp. 18–21.

17 http://news.bbc.co.uk/1/hi/business/4352553.stm

18 www.managementstudyguide.com/limitations-of-participitative-management.htm

19 www.accaglobal.com/databases/pressandpolicy/unitedkingdom/3107831

20 http://ec.europa.eu/internal_market/consultations/docs/securities_agencies/consultation-cra-framework_en.pdf

21 http://www.uk200group.co.uk/Members/SpecialistGroups/SpecialistPanels/CorporateFinance/Sp_Valuations/Sp_Valuations_Home.aspx

22 W.A. Nixon and C.J. McNair, 'A measure of R&D', *Accountancy*, October 1994, p. 138.

23 C. Martin and J. Hartley, *SME Intangible Assets*, Certified Accountants Research Report 93, London, 2006.

CHAPTER 30

An introduction to financial reporting on the Internet

30.1 Introduction

The main objective of this chapter is to explain the developments in the way that investors and analysts obtain published financial reports and analyse the data.

Objectives

By the end of this chapter, you should be able to:

- discuss the comparative requirements for financial statements that are both useful for assessing stewardship and decision-useful;
- explain the facilities that assist users to analyse the financial data;
- understand the reason for the development of a business reporting language;
- explain the benefits of tagging in XML and XBRL code data for financial reporting;
- understand why companies should adopt XBRL;
- list the processes a company needs to take to adopt XBRL.

30.2 The objectives of financial reporting

There are two objectives of financial reporting. These are that the statements should (a) provide the means for investors to assess the management's stewardship of the company's resources and (b) provide data that is decision-useful.

30.2.1 Stewardship

Stewardship is backward-looking. It relies on historical information. It requires this information to be reliable and complete. This historical information has been traditionally in the form of hard copy and this is now supplemented or replaced by reporting on the Internet.

Hard copy

This is provided for:

- annual reports which have been:
  - prepared in accordance with IFRSs, EU *Revised Transparency Directive 2013*[1] or national accounting standards;

- audited; and
- published within 4 months of the period end;
- half-yearly condensed financial statements:
  - prepared in accordance with IAS 34 *Interim Financial Reports* or national stanards such as the *Disclosure and Transparency Rules* of the UK's Financial Conduct Authority;
  - not audited but reviewed by an auditor in accordance with the *International Standard on Review Engagements*; and
  - published within 2 months of the period end;
- quarterly reports or interim management statements:
  - not audited and published with a caveat such as;[2]

This announcement contains forward looking statements which are made in good faith based on the information available at the time of its approval. It is believed that the expectations reflected in these statements are reasonable but they may be affected by a number of risks and uncertainties that are inherent in any forward looking statement which could cause actual results to differ materially from those currently anticipated. Nothing in this document should be regarded as a profits forecast.

PDF files

At an individual company level we find that most companies have a website to communicate all types of information to interested parties including financial information. Stakeholders or other interested parties can then download this information for their own particular use. Most of the financial information is in the format of PDF files created by a software program called Adobe® Acrobat®. This program is used for the conversion of all their documents, which make up the financial information contained within the annual general reports, into one document, a PDF file, for publication on the Internet. This PDF file can be formatted to include encryption and digital signatures to ensure that the document cannot be changed.

In order for the user to be able to read the PDF files, a special software program called Adobe Reader® needs to be downloaded from the Adobe website www.adobe.com.

30.2.2 Decision-usefulness

Investors have been assisted in their decision making by access to:

- commercial databases where selected financial reports have been formatted by each database into a standardised format;
- individual company sites where the company provides data in downloadable format. For example, an increasing number of companies such as BP, BMW, Colgate, Dell, Lloyds TSB and Vodacom, have been providing their annual report in a multi-year downloadable Excel format.

  However, because companies have different items in their financial reports, it is not possible to make a line-by-line comparison. The user needs to synchronise items and to do this needs to be conversant with the accounting definition of each individual item; and
- individual company reports where the information has been described uniformly and tagged as achieved with eXtensible Business Reporting Language (XBRL).

Commercial databases

Various financial databases have been developed to assist in ratio analysis and the appraisal of inter-period and inter-company performance. This allows subscribers to select peer groups and search across a variety of variables. Students having access to such databases at their own institution may carry out a range of assignments and projects such as selecting companies suitable for takeover based on stated criteria such as ROCE, % sales and % earnings growth.

Financial databases that might be available to students include:

- Fame
  This database provides detailed, financial, descriptive and ownership information on over 2.8 million public and private companies in the UK and Ireland.

- Amadeus
  This database provides company information for both Western and Eastern Europe, with a focus on private company information and company financials in a standard format so that you can compare companies across borders.

- Compustat
  These bases provide academic researchers with historical fundamental and market data in a standardised format that allows comparison across companies, industries and business cycles.

- Datastream
  This database providing current/historical financial data for international companies/indices and bond data.

- PI NAVIGATOR
  This database provides the ability to download non-rekeyed financial statements from original PDFs into Excel and locate comparable reports from multiple companies.

XBRL

We will now discuss reports and the flow of information possible with the use of XBRL. We will see that information in this format could put the investor in the same position as management itself. This moves on from quarterly reporting towards continuous reporting. Whilst this might be technologically possible an investor would not have the same contextual understanding of the information with an awareness of the probability of change.

Its current strength lies in the possibility it provides of peer review across any of the elements in the statement of income, financial position and cash flows.

30.3 Reports and the flow of information pre-XBRL

The information flow from an organisation reporting to stakeholders and regulatory bodies and banks is considerable. The information required is not the same for each of the external parties and so one report is not appropriate.

A typical flow is set out in Figure 30.1 demonstrating how information is collated from Operational Data Stores and coded to the General Ledger (GL) using the chart of accounts (C of A). Once the data have been captured in the GL, statements of comprehensive income, financial position and cash flows can be produced for shareholders and for statutory filing. In addition, separate reports are produced for a variety of other stakeholders such as the tax authorities, stock exchanges, banks and creditors.

The reports can be in different formats such as printed statements for internal management and audit use, hard-copy annual reports for investors, and summary or full reports on a company's home web page in PDF or HTML format, now that this is becoming mandatory or encouraged. This is a very costly process which has led to the development of a special

Figure 30.1 Today: a convoluted information supply chain

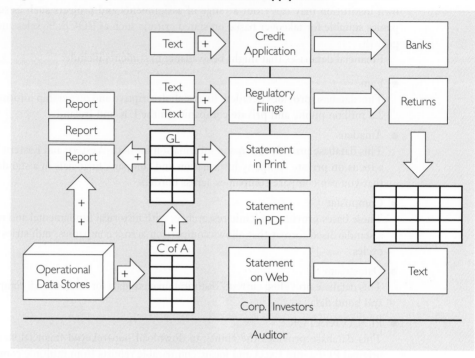

Source: www.xbrl.org.au/training/NSWWorkshop.pdf

business reporting language called e**X**tensible **B**usiness **R**eporting **L**anguage or **XBRL** which is based on XML.

30.4 What are HTML, XML and XBRL?

XBRL is based upon the eXtensible Markup Language or XML. XML itself is an extension of the Hyper Text Markup Language (HTML) which controls the format and display of web pages. We will briefly comment on each.

HTML

HTML is extensively used in website creation for the purposes of display. For example, the following text using HTML would have tags that describe the format and placement of the text:

Assets $50,000
Liabilities $25,000

`<p><b>Assets    $50,000</b></p>`
`<p><b>Liabilities    $25,000</b></p>`

where <p> instructs the item to be printed on the screen (and also where on the screen or in what format) and <b> instructs the item to be displayed in bold print. The </p> denotes the end of the commands and instructs the data to be 'printed' on the computer screen.

XML

XML is a language developed by the World Wide Web Consortium.[3] It goes one step further by allowing for 'tags' to be created which convey identification and meaning of the data within the tags. Thus instead of looking simply at format and presentation, the XML code looks for the text displayed within the code. For example, the user can design the tags used in XML as follows:

Assets $50,000 in this example of XML would be written as:

<Assets>$50,000</Assets>

and similarly for **Liabilities $25,000** the XML code would be:

<Liabilities>$25,000</Liabilities>

The computer program reading the XML code would thus know that the value found of $50,000 within the tags relates to Assets.

XBRL

XBRL has taken XML one step further and designed 'tags' based upon the common financial language used. For example, the terms ASSETS and LIABILITIES are common terms used in financial reports even though the calculations or valuations and the definitions used in different accounting standards may be dependent on those accounting standards applicable to the company.

30.4.1 Advantages of XBRL

Using XBRL means that it is easier for direct system-to-system information-sharing between a company and its stakeholders and allows for improved analytical capacity. The numeric data in the financial statements of all companies filing their annual reports will be uniformly defined and presented and available for analysis, e.g. downloaded into Excel and other analytical software. The advantage of using XBRL according to XBRL International[4] is that:

> Computers can treat XBRL data 'intelligently': they can recognise the information in an XBRL document, select it, analyse it, store it, exchange it with other computers and present it automatically in a variety of ways for users. XBRL greatly increases the speed of handling of financial data, reduces the chance of error and permits automatic checking of information.

30.5 Reports and the flow of information post-XBRL

When XBRL is used, (a) information flows from an organisation to stakeholders are much simpler as seen in Figure 30.2, and (b) it is possible for stakeholders to receive information that can be understood by computer software and allow them to analyse the data obtained, as seen in Figure 30.3.

Figure 30.2 With XBRL: multiple outlets from a single specification

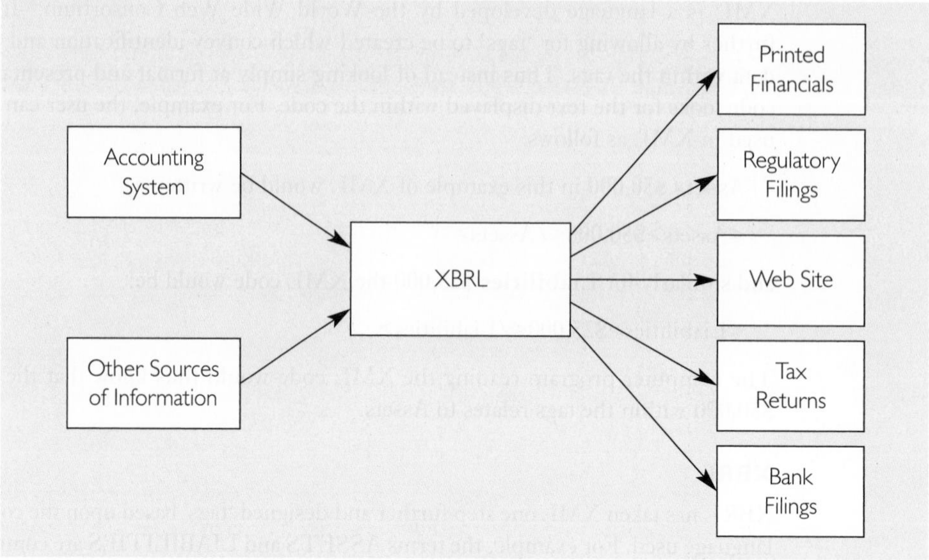

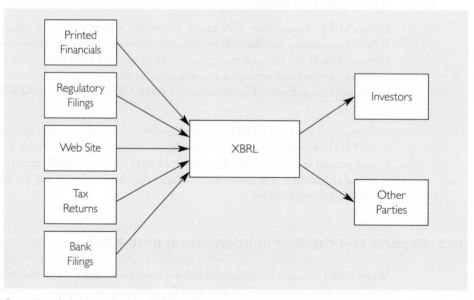

Source: http://xbrl.org.au/training/NSWWorkshop.pdf

Figure 30.3 XBRL: information flow to stakeholders

Source: http://xbrl.org.au/training/NSWWorkshop.pdf

30.6 Why are companies adopting XBRL?

We will consider briefly the influence of regulatory requirements, commercial benefits and the views of stakeholders.

30.6.1 Regulatory requirements

One of the driving forces has been the pressure from national regulatory bodies for companies to file corporate tax returns, stock exchange and corporate statutory financial statements in XBRL format. In some countries there are specific requirements for financial statements filing.

US regulatory requirements

The US Securities and Exchange Commission (SEC) has required[5] public and foreign companies with a float over $5 billion that prepare financial statements based on US GAAP to lodge their reports in XBRL since 2009. Foreign companies using IFRS have been required to lodge their financial reports since 2011. After 2014 XBRL-based statements will have the same legal status as any other financial report. This will have implications for auditors and preparers of the financial reports.

UK regulatory requirements

UK companies filing accounts online at Companies House have been required since 2011 to use Inline XBRL (iXBRL). iXBRL is a specific form of XBRL that focuses on the human-readable format.

HM Revenue and Customs (HMRC) have required companies with a turnover of more than £100,000 to lodge online since 2010.

Singaporean regulatory requirements

Singapore's Accounting Corporate Regulatory Authority (ACRA),[6] the regulating authority for businesses incorporated in Singapore, has been receiving company reports for most incorporated commercial companies in XBRL format since 2007.

30.6.2 Commercial benefit

The main benefit is that companies can easily generate tailored reports from a single data set and the data can be readily accessed at a lower cost by regulators, auditors, credit rating agencies, investors and research institutions.

30.6.3 The views of stakeholders

A report in 2009 *XBRL: The Views of Stakeholders*[7] concluded that overall, there is considerable lack of knowledge of XBRL within UK business. Some of its policy recommendations were that HMRC, Companies House, professional bodies such as ACCA, and IT specialists should publicise the business case for XBRL more widely with the provision of 'hands-on', user-focused sessions that highlight the interoperability and flexibility of XBRL.

30.7 What are the processes followed to adopt XBRL for outputting information?

There are four processes, supported by the appropriate software, to be completed to adopt XBRL. The processes are (a) taxonomy design, (b) mapping, (c) creating an instance document, and (d) selecting and applying a stylesheet.

(a) The taxonomy needs to be designed

Taxonomy has two functions. It establishes relationships and defines elements acting like a dictionary. For example, the taxonomy for assets in the statement of financial position would be to show how total assets are derived by aggregating each asset and defining each asset as follows:

	Relationship	Definitions
Non-current assets	a	Not expected to be converted into cash within one year
Current assets		Expected to be turned into cash in less than one year
Inventory	v	Finished goods ready for sale, goods in course of production and raw materials
Trade receivables	w	Amounts owed by customers
Cash	x	Cash and cash equivalents
Subtotal	$v + w + x$	
Total assets	$a + v + w + x$	

The IFRS Taxonomy 2014 edition translates International Financial Reporting Standards (IFRSs) into eXtensible Business Reporting Language (XBRL) with separate modules now for full Standards, IFRS for SMEs and IFRS Practice Management Commentary.

The taxonomy also contains **linkbases** which provide additional information. For example:

- a means to cross-reference with the paragraph in the relevant IFRS;
- an indication of the language used in the financial report, e.g. English, French;
- prompts when a note to the accounts is required for a particular element.

(b) Mapping

The term 'mapping' relates to equating the terminology used in the financial statements to 'names' used in the taxonomy. For example, if the taxonomy refers to 'Inventory' as being products held for sale, but the organisation refers to this as 'Stock in Trade' in the financial statements, then this needs to be 'mapped' to the taxonomy. All the names used in the financial statements, or any other reports, need thus to be compared and mapped to (identified with) the taxonomy. This 'mapping' is done the first time the taxonomy is used.

(c) Instance documents

The instance document holds the data which are to be reported. For example, if preparing the statement of financial position at 30 September 2010, entries of individual asset values would be made in this document. This data would then be input to a stylesheet to produce the required report.

	Values	Date
Non-current assets	1,250	30.9.2010
Inventory	650	30.9.2010
Trade receivables	310	30.9.2010
Cash	129	30.9.2010

(d) Stylesheets

The format of a required report is specified in a template referred to as a 'stylesheet' where the display is pre-designed. A stylesheet can be used repeatedly as, for example, for an annual report, or new stylesheets can be designed if reports are more variable as in interim reports. The annual report would be displayed in the correct format with appropriate headings, currency and scale. For example:

Statement of financial position as at 30 September 2015

	000	000
Non-current assets		1,250
Current assets		
Inventory	650	
Trade receivables	310	
Cash	129	
		1,089
Total assets		2,339

The taxonomy and stylesheets do not need to be changed every time a report is produced. The only changes that are made are those in the instance documents regarding data entries.

Summary of the four processes

A summary is set out in Figure 30.4.

30.7.1 XBRL certification

The XBRL Foundation Certificate Program[8] is an intensive training and examination process that will prepare stakeholders at all levels to understand the ramifications of using XBRL to meet their business objectives. It includes coverage of the relation between taxonomies and instance documents, linkbases and the content of XBRL instance documents.

Figure 30.4 Summary of the four processes

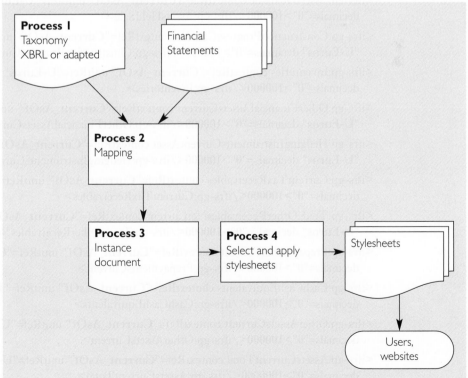

In the US the AICPA and XBRL US have developed for CPAs a comprehensive interactive online learning programme, the *XBRL U.S. GAAP Certificate Program*, to provide a sound understanding of XBRL financials.

Internationally companies are accessing training from the accounting profession, as in India[9] following the requirement in 2011 to file financial reports in XBRL.

30.8 What is needed when receiving XBRL output information?

Institutional users

Institutions which receive XBRL formatted financial information from companies, such as revenue authorities, stock exchanges, banks and insurance companies, normally require the information to be lodged according to a predetermined format and their software is specifically designed to be able to extract and display the XBRL data.

Non-institutional users

For other interested parties, specific software is needed to make the XBRL format data readable. In order for the text to be understood by a human in a way that indicates that we are looking at a financial report, it needs to be 'translated', a process known as **rendering**, by computer. 'Rendering' the items contained within XBRL is the current challenge.

Example of rendering

The text below represents the code for XBRL formatted data in an instance document.

Instance document in XBRL

```
<ifrs-gp:AssetsHeldSale contextRef=vCurrent_AsOf" unitRef="U-Euros"
    decimals="0">100000</ifrs-gp:AssetsHeldSale>
<ifrs-gp:ConstructionProgressCurrent contextRef="Current_AsOf" unitRBf=
    "U-Euros" decimals="0">100000</ifrs-gp:ConstructionProgressCurrent>
<ifrs-gp:Inventories contextRef="Current_AsOf" unitRef="U-Euros"
    decimals="0">100000</ifrs-gp:Inventories>
<ifrs-gp:OtherFinancialAssetsCurrent contextRef="Current_AsOf" unitRef=
    "U-Euros" decimals="0">100000</ifrs-gp:OtherFinancialAssetsCurrent>
<ifrs-gp:HedgingInstrumentsCurrentAsset contextRef="Current_AsOf" unitRef=
    "U-Euros" decimals="0">100000</ifrs-gp:HedgingInstrumentsCurrentAsset>
<ifrs-gp:CurrentTaxReceivables contextRef="Current_AsOf" unitRef="U-Euros"
    decimals="0">100000</ifrs-gp:CurrentTaxReceivables>
<ifrs-gp:TradeOtherReceivablesNetCurrent contextRef="Current_AsOf" unitRef=
    "U-Euros" decimals="0">100000</ifrs-gp:TradeOtherReceivablesNetCurrent>
<ifrs-gp:PrepaymentsCurrent contextRef="Current_AsOf" unitRef="U-Euros"
    decimals="0">100000</ifrs-gp:PrepaymentsCurrent>
<ifrs-gp:CashCashEquivalents contextRef="Current_AsOf" unitRef="U-Euros"
    decimals="0">100000</ifrs-gp:CashCashEquivalents>
<ifrs-gp:OtherAssetsCurrent contextRef="Current_AsOf" unitRef="U-Euros"
    decimals="0">100000</ifrs-gp:OtherAssetsCurrent>
<ifrs-gp:AssetsCurrentTotal contextRef="Current_AsOf" unitRef="U-Euros"
    decimals="0">1000000</ifrs-gp:AssetsCurrentTotal>
```

Looking at the first two lines of code, it is possible to see that the data contain financial information about assets held for sale, that these are 'Current' and that the unit of measurement is in euros and has zero decimals with a value of 100,000. This is possible for a few lines but it would not be feasible to do this for a complex financial statement. Rendering translates the code into readable format as follows.

Rendered XBRL data

CURRENT ASSETS	€
Assets held for sale	100,000
Construction in progress, current	100,000
Inventories	100,000
Other financial assets, current	100,000
Hedging instruments, current [asset]	100,000
Current tax receivables	100,000
Trade and other receivables, net, current	100,000
Prepayments, current	100,000
Cash and cash equivalents	100,000
Other assets, current	100,000
Current assets, total	1,000,000

The data can now be recognised as belonging to that part of the financial statement where the current assets are listed. This example can be found at http://www.xbrl.org/Example1/.

The rendering process is of particular interest to investors and other third parties who may want to access financial data in XBRL format for evaluation purposes and who may not have software capable of rendering the instance document into human readable format. The ability to render an XBRL document becomes even more important for an investor or analyst seeking to carry out trend or inter-firm comparison analysis. For a more in-depth discussion of the processes involved in rendering visit www.xbrl.org/uk/Rendering/.

30.8.1 How might **XBRL** assist the user?

If we take the revenue authorities as an example, they have had their own in-house-developed software for carrying out a risk analysis in an attempt to identify those that look as though they should be investigated. Such risk analysis was routine before XBRL but XBRL has allowed the existing analysis software to be refined – this allows obviously compliant companies to be identified and investigation to be targeted where there is possible or probable non-compliance. However, it was still not possible to present the data in human-readable form.

Auditors also gain advantages such as there being reduced risk of manual re-entry errors and, with XBRL GL, they can extract information in a single standardised form and review more data in greater detail, potentially in real time.

Society might benefit if it becomes possible to evaluate a company holistically by tagging, for instance, environmental and social information such as carbon disclosures. (see http://www.fm-magazine.com/feature/depth/xbrl-era#).

30.8.2 Development of **iXBRL**

Inline XBRL (known as iXBRL) has been developed so that the XBRL data are capable of being read by the user. It achieves this by embedding the XBRL coding in an HTML document so that it is similar to reading a web page. iXBRL takes a report, say a company's published accounts, in Excel, MS Word or PDF and then 'translates' this to iXBRL. It is then still able to be viewed in human-readable format. This would be an advantage for smaller businesses where there may not be accountants with XBRL skills, or where the cost would be prohibitive and they also do not need more advanced software. Corefiling's software[10] would be a good example of iXBRL.

This is newly developed software and, if you want to explore it a little further, there are helpful websites – one that offers a software company's view[11] and one that offers another perspective.[12]

How has iXBRL assisted the user?

If we consider the position of the regulators we can see that there is an impact on narrative in reports. For example, there is the requirement of the SEC to include the notes to the business reports in a certain format in the near future. For larger companies[13] this means that tags have to be developed for many more items than before. The inclusion of the notes to the accounts is not new, but the new requirements from the SEC involve different levels of disclosure within the notes and this will require further development of either a linkbase or a standard approach to the application of stylesheets.

If we continue with the revenue authorities example, the availability of the data in readable format has meant that, once a high-risk case has been identified, it is possible to drill down into the data and highlight relationships that would not have been possible before iXBRL.

We have discussed the use of XBRL and iXBRL for submitting reports to statutory authorities. There have also been interesting developments in their use for internal accounting.

30.9 Progress of XBRL development for internal accounting

Development in the general ledger area is continuing and will probably be one of the most important developments for companies with consolidation requirements when multiple general ledgers are involved. The general ledger specification has the advantage that organisational data are classified at source and the classification decision with respect to XBRL names will have been made at the Chart of Accounts level.

This is quite a task as the financial statements usually report aggregated data. For example, the total for administration expenses in the Income Statement is usually made up by aggregating a number of different account classifications in the General Ledger. A further consideration is the effect of IFRSs when aggregating expense accounts. For example, the Chart of Account structures for the disclosure of segmentation by product class and by geographical areas are distinctly different. The XBRL code also needs to reflect this.

The XBRL for the General Ledger may also bring great cost savings as data collection at source is automated and the extraction and processing of data into reports can be achieved in a much shorter time. A company such as General Electric has more than 150 general ledgers which are not compatible in use. XBRL has the potential to streamline consolidation processes considerably.

The XBRL Global General Ledger Working Group (XBRL GL WG) within XBRL has released an updated GL module[14] to include the SRCD (Summary Reporting Contextual Data):

> SRCD is a module of the XBRL Global Ledger Framework (XBRL GL) designed to facilitate the link between detailed data represented with XBRL GL and end reporting represented with XBRL for financial reporting (XBRL FR) or other XML schemas.

This module should enable a streamlined preparation of business reports from the general ledger. In February 2010 the GL framework also released the GL module with Japanese labels and this is now awaiting feedback and then a final recommendation.

In the US the American Institute of Certified Public Accountants has developed a data standard for the general ledger and accounts receivable ledger which can assist the analytical work of both the internal and external audit.

30.10 Real-time reporting

We know that investors are not a homogeneous group. It will include investors who have a long-term interest in a company and investors who have a short-term interest.

This influences their view on the timeliness of financial reporting – its reliability and relevance.

Long-term view

For those with a long-term view the Annual Report is valued because it has been audited and is regarded as reliable as a means of assessing management's stewardship. Regulators have been taking steps to require companies to reduce the time between the end of the financial period and the publication of the report and companies themselves see that there could be a competitive advantage in doing this, with some taking only 10 days after the period end. There are indications that investors associate the speed of reporting with robust corporate governance and enhanced investment returns. The speed is also affected, of course, by the complexity of a company's operations and structure.

Regulators have also been addressing the need for investors to be aware of a company's business model and its procedures for dealing with risk through Management Commentaries and Strategic Reports.

Whilst there is an interest in receiving information during the year, it is not clear that investors would be happy for the company to incur the extra cost of auditing interim data – it is of interest but perhaps not of major significance.

Short-term view

Investors looking to buy and sell shares during the financial year would welcome real-time reporting. There is a preparedness to accept that the data will be less reliable and might result in greater volatility in share price movements, which for some investors provides in itself an opportunity. The short-term investors are also not a homogeneous group. Some will have access to company briefings, others will need to rely on company reports.

The form of interim reporting

There are practical problems involved with real-time reporting. Data may be less reliable and there is a problem auditing moving data. At present there is a requirement by some regulators to produce quarterly reports, but even with these reports there are critics who question their value.

A practical solution at present would seem to be for a company to provide warnings if there is an unexpected change in performance such as with profit warnings and qualitative interim reports with assurance rather than audit.

Survey results

A survey carried out in 2013 on behalf of the ACCA[15] found that there was a genuine demand for 'real-time' reporting among investors and this would increase investor returns and enhance the level of confidence in corporate reporting. For example, 85% said that real-time data would improve their ability to react quickly and 71% said it would increase their understanding of corporate performance.

However, almost two-thirds believed real-time reporting would create further financial instability and lead to an increased tendency to short-termism in financial markets, which raised the question as to whether what investors were asking for would be positive for the market as a whole.

30.11 Further developments

The XBRL International website (www.xbrl.org) has an extensive listing of companies and authorities currently using XBRL. The reader is encouraged to investigate further any of the resources available on the XBRL and other websites such as learn.vubiz.com/ChAccess/XBRL/ where there is an introductory course 'Advances in Business and Financial Reporting'. A number of the links provided will lead to good discussions of the projects and demonstrate how XBRL is applied. Some of the links will also bring the reader to websites in languages other than English (Google translation toolbar may be helpful) and may be of particular interest to readers of this text living in non-English-speaking countries.

In the UK the FRC announced an 18-month project starting in 2014, *Corporate Reporting in a Digital World*, conducted through the Financial Reporting Lab to investigate how companies do, and might in the future, use digital media in their corporate reporting to improve investors' access to information. The Lab will initially review how companies currently use a wide range of digital media, including websites, videos, apps, social media platforms and blogs in their external communications to investors, and how investors use what is produced. It will then progress to considering barriers to the use of digital media in reporting and how companies might make the most of technological opportunities. The Lab is inviting listed companies, investors and analysts to express their interest in taking part in the project.

Summary

XBRL is still a developing area relating to organisational reporting. In the coming years this will continue and extend beyond the current focus on published financial statements. The general ledger area is developing and this will benefit the organisational information supply chain. Accounting software suppliers are also adopting XBRL in their developments and this will increase accessibility to XBRL. Accounting software companies are also using XBRL in their new developments aimed at smaller organisations. Future software development may also make it easier for accountants to use XBRL, especially when a country's taxonomies are in the 'final' approved stage.

Financial statements presented in XBRL format are capable of being downloaded into an analyst's/investor's own spreadsheet (such as Microsoft Excel). The advantage of this is that the analyst/investor does not need to retype the information. The commercial databases which compile specific information for analysts/investors are usually only concerned with public companies listed on the Stock Exchange. XBRL allows any type of financial information to be transferred to a statistical package without having to retype the information.

XBRL could thus also benefit not-for-profit organisations and trusts. Professional accounting consultants would also be able to use XBRL in transferring information from a client's accounting package into an analytical tool to prepare information to evaluate business efficiency. This information is often more extensive than the end-of-year financial information.

The large software developer SAP announced in February 2009 that 'SAP® BusinessObjects™' is now available for financial publications in XBRL. This conforms to Security and Exchange Commission (SEC) requirements to lodge specific financial information from June 2009. SAP also stated that its software can be used to lodge financial information using XBRL with HM Revenue and Customs in the UK. The software allows for automatic and easy tagging of the information (see www.xbrlspy.org/sap_announces_xbrl_publishing_support).

Accountants and students wishing to keep up to date with these developments are gaining a competitive advantage by creating and developing a 'niche' skill which can only add value to an organisation employing these professionals.

REVIEW QUESTIONS

1 Discuss how an investor might benefit from annual reports being made available in XBRL.

2 Explain how a body such as a tax authority might benefit from XBRL.

3 Explain what you understand by taxonomy and mapping.

4 Explain the use of instance documents.

5 Explain the use of stylesheets.

6 Explain iXBRL and where it is used.

7 XBRL will make it easier to prepare quicker quarterly reports. Discuss the suggestion that this encourages short-termism and that companies should not therefore be required to produce on a quarterly basis.

8 'Interim assessment of performance is important but should only be discursive.' Discuss.

9 Investor demand for assurance takes precedence over their demand for speed when it comes to general financial information and liquidity. Access www.accaglobal.com/reporting and identify from its report *Understanding investors: the road to real time reporting* three situations where a company seeks real-time reporting and whether this should be made available to investors at the same time.

EXERCISES

Question 1

Visit www.us.kpmg.com/microsite/xbrl/kkb.asp to attempt the XBRL tutorial and write a brief note on how you think it will affect the work of a financial accountant.

Question 2

Find the financial reports for a company of your own choice. List the company and describe the format of the Annual Report. See if you can also find information on the *company's own* website about its use of XBRL.

Question 3

The following is an extract from *Digital reporting: a progress report*, an initiative from the Institute of Chartered Accountants in England & Wales commenting in 2004 on possible barriers to the development of increased digital reporting:

> Technology is unlikely to be the barrier ... The barrier, if any, will arise from organisations being unwilling to provide the necessary level of access or information to external users. In addition, issues relating to systems security, particularly control of access to programs and data, would need to be further addressed, as would questions of liability and assurance. The latter two are potentially very big issues indeed.

> Whilst there has been major progress by regulators in the US, Australia and the Netherlands, an ICAEW and University of Birmingham (ISARG) Workshop on 25th January 2011 titled 'The future of XBRL in Europe: Impetus, institutions and interrelationships' http://www.icaew.com/~/media/ Files/About-ICAEW/What-we-do/thought-leadership/the-future-of-xbrl-in-europe-final-summary-for-release.pdf explored the future of XBRL in Europe and contrasted the US and Pan European drivers:

> **Pan European development**
> The U.S. SEC project provides a model of the effectiveness of visionary leadership from the top of a regulatory body to achieve a relatively rapid mandatory implementation. As a model for pan European developments it is limited because of the significantly different institutional environment, goals and processes. It does however raise important questions for consideration. The SEC's stated target market was retail (non-professional) investors in line with its mandate. What could be the focus of the driver for developments in Europe? The SEC was also under pressure to review a significant portion of filings under the Sarbanes Oxley Act that provided a strong internal driver for more efficient processing internally.

Required:
Review the two sources and discuss what equivalent pressures member states are experiencing.

Question 4

Find out more about any of the following topics and write a one-page summary on:

(a) the XBRL general ledger work;

(b) use of XBRL by stock exchanges;

(c) the commitment by the IFRS to the XBRL project;

(d) accounting software companies involved in providing XBRL capabilities;

(e) public utilities that are using XBRL;

(f) government involvement in XBRL.

References

1 http://ec.europa.eu/internal_market/securities/transparency/index_en.htm
2 http://www.keller.co.uk/N/media/Files/K/Keller/storage/pdfs/2014/interim-report-2014.pdf
3 www.w3.org/Consortium
4 www.xbrl.org
5 xbrl.sec.gov
6 www.acra.gov.sg/
7 T. Dunne, C. Helliar, A. Lymer and R. Mousa, *XBRL: The Views of Stakeholders*, ACCA Research report 111, 2009
8 http://www.xbrl.org/xbrl-foundation-certificate
9 http://www.kpmg.com/in/en/services/advisory/advisorytrainings/pages/xbrltraining.aspx
10 www.corefiling.com/products/seahorse.html
11 www.tcsl.co.uk/
12 www.xbrlspy.org/to_render_or_not_to_render_xbrl
13 www.claritysystems.com/ap/events/webcasts/Pages/XBRL4Tagging.aspx
14 www.xbrl.org/GLFiles/
15 http://www.accaglobal.com/content/dam/acca/global/PDF-technical/financial-reporting/pol-afb-ui03.pdf

Bibliography

H. Ashbaugh, K.M. Johnstone and T.D. Warfield, 'Corporate reporting on the Internet', *Accounting Horizons*, vol. 13(3), 1999, pp. 241–57.

R. Debreceny and G. Gray, 'Financial reporting on the Internet and the external audit', *European Accounting Review*, vol. 8(2), 1999, pp. 335–50.

R. Debreceny and G. Gray, 'The production and use of semantically rich accounting reports on the Internet: XML and XBRL', *International Journal of Accounting Information Systems*, vol. 1(3), 2000.

D. Deller, M. Stubenrath and C. Weber, 'A survey of the use of the Internet for investor relations in the USA, UK and Germany', *European Accounting Review*, vol. 8(2), pp. 351–64.

Ernst & Young, *Web Enabled Business Reporting. De invloed van XBRL op het verslaggevingsprocess*, Kluwer, 2004.

M. Ettredge, V.J. Richardson and S. Scholz, *Accounting Information at Corporate Web Sites: Does the Auditor's Opinion Matter?*, University of Kansas, February 1999.

M. Ettredge, V.J. Richardson and S. Scholz, 'Going concern auditor reports at corporate web sites', *Research in Accounting Regulation*, vol. 14, 2000, pp. 3–21.

Neil Hannon, 'XBRL grows fast in Europe', *Strategic Finance*, October 2004, pp. 55–6.

Mark Huckelsby and Josef Macdonald, 'The three tenets of XBRL – adoption, adoption, adoption!', *Chartered Accountants Journal of New Zealand*, March 2004, pp. 46–7.

V. Richardson and S. Scholz, 'Corporate reporting and the Internet: vision reality and intervening obstacles', *Pacific Accounting Review*, vol. 11(2), 2000, pp. 153–60.

Mike Rondel, 'XBRL – do I need to know more?', *Chartered Accountants Journal of New Zealand*, vol. 83(5), June 2004, pp. 37–40.

G. Trites, *The Impact of Technology on Financial and Business Reporting*, Toronto: Canadian Institute of Chartered Accountants, 1999.

The following websites were accessed:
www.adobe.com
www.xbrl.org/FRTaxonomies/
www.xbrl.org
www.ubmatrix.com/home/
www.semansys.com
www.edgar.com

PART **7**

Accountability

CHAPTER 31

Corporate governance

31.1 Introduction

The main aim of this chapter is to create an awareness of what constitutes good corporate governance – how to achieve it, the threats to achieving it and the role of accountants and auditors.

Objectives

By the end of this chapter, you should be able to:

- understand the concept of corporate governance;
- have an awareness of how and why governance mechanisms may differ from jurisdiction to jurisdiction;
- have an appreciation of the role which accounting and auditing play in the governance process;
- have a greater sensitivity to areas of potential conflicts of interest.

31.2 A systems perspective

Corporations do not act in a vacuum. They are corporate citizens of society with rights and responsibilities. The way in which they exercise these rights and responsibilities is influenced by the history, institutions and cultural expectations of society. A systems perspective recognises that an entity is not independent but is interdependent with its environment. This has given rise to the need for corporate governance.

Corporate governance is defined by C. Oman[1] as:

> private and public institutions, including laws, regulations and accepted business practices, which together govern the relationship, in a market economy, between corporate managers and entrepreneurs ('corporate insiders') on the one hand, and those who invest resources in corporations on the other.

31.2.1 Good corporate governance – investor perspective

When we pause to contemplate the contribution of corporations to our standard of living, we are reminded how important their contribution is to most aspects of our existence. It is therefore vital that they operate as good citizens in their treatment of the investors who provide their funds and of other stakeholders. This includes actions by management when dealing with investors such as:

- complying with the laws and norms of society;
- striving to achieve the company objectives in a manner which does not involve taking risks which are greater than expected or acceptable to investors;
- balancing short- and long-term performance;
- establishing mechanisms to ensure that managers are acting in the interests of shareholders and are not directly or indirectly using their knowledge or positions to gain inappropriate benefits at the expense of shareholders;
- providing investors with relevant, reliable and timely information that allows them to assess the performance, solvency and financial stability of the business; and
- providing investors with an independent opinion that the financial statements are a fair representation.

This list does not cover all eventualities but is intended to indicate what could be expected from corporate governance – good being determined by the degree that the actions and information flows achieve fair outcomes.

31.2.2 Good corporate governance – other stakeholder perspective

A stakeholder perspective addresses all the other parties whose continued support is necessary to ensure the satisfactory performance of the business. The parties are normally seen as belonging to one of the following categories: loan creditors, employees, trade unions representing employees, customers, governments and suppliers.

Good corporate governance might include actions by management such as:

- fair treatment of employees, avoiding discrimination;
- establishing mechanisms for resolving conflicts of interests;
- establishing mechanisms for whistle-blowing so that if inappropriate behaviour is taking place it is highlighted as quickly as possible so as to minimise the cost to the organisation and society;
- paying suppliers, particularly small businesses, promptly within the agreed credit period; and
- providing suppliers with relevant, reliable and timely information that allows them to assess the solvency of the business.

31.2.3 Good corporate governance – stakeholder pressure

As well as there being conflicting interests, there are also differences in the influence that a stakeholder can exert. For example, dominant shareholders, institutional investors and major customers have a greater ability to hold management to account and achieve good corporate governance outcomes. The existence of the ability does not necessarily mean that it is put into effect, since the individual stakeholder's private interest might not be advanced by taking action – it might, for example, divert their management's attention away from their own business.

31.2.4 Good corporate governance – all sectors

The objective is to influence behaviour so that all parties act within the spirit of good governance. The actions and information flows above have been oriented towards business entities but we should expect all organisations to behave in the same way. For example, in the case of a not-for-profit enterprise such as a charity it is important that the money raised be used in a manner consistent with the uses envisaged by the donors, and that an appropriate balance be achieved between administrative costs and the money devoted to assisting the beneficiaries of the charity. The approach to enforcement of good corporate governance by charities varies internationally.

31.3 Different jurisdictions have different governance priorities

The predominant conflicts of interest will vary from country to country depending on each country's history, economic and legal developments, norms and religion.

In the UK and the United States, with their similar considerable reliance on stock exchanges for the financing of public companies, there is a need for an active, efficient capital market. This leads to their focus being on potential conflicts between management and shareholders.

In Germany, where companies have a board of directors made up of investors as well as an advisory board representing both management and employees, there is a recognition that there is a need to reconcile both management and employee long-term interests and to ensure that both groups are motivated to achieve the organisation's long-term goals.

In south-east Asia, with many of the large corporations having substantial shareholdings owned by members of a single family, the emphasis has been on avoiding conflicts between family and minority shareholders.

In Muslim countries companies should not be involved in activities related to alcohol and gambling; they cannot pay or charge interest and they have religious obligations to make a minimum level of donations. This means there is a need for corporate governance mechanisms to ensure that there is no conflict between commercial activity and religious obligations.

From the above we can see how the governance priorities differ from country to country. They result from the role of the political institutions, the stage of economic development, the diversity of stakeholder perspectives and a country's heritage in so far as it shapes the law, the religion and the social norms.

The large number of multinational companies means that these companies have to be sensitive to the approaches taken in all countries in which they have subsidiary companies and joint ventures. They also have to be aware of the provisions of the US Foreign Corrupt Practices Act 1977 and the UK Bribery Act 2010, particularly as the Bribery Act creates a corporate offence of failing to prevent bribery by persons associated with a corporation.

Companies identifying bribery and corruption risk management in their Annual Reports

The following is an extract from the Centrica 2012 Annual Report:

> Anti-bribery and corruption is a business priority. Centrica shall regularly and systematically identify bribery and corruption risks in its business and implement adequate risk-based procedures aimed at preventing bribery and corruption occurring including:
>
> ● Audit – Our internal control systems will be subject to regular internal and independent audit to provide assurance that they are effective in countering bribery and corruption . . .

- Business relationships – We will ensure that our business partners – including contractors, suppliers, agents, brokers and joint venture partners – are fit to do business with . . .

- Conflicts of interest – Gifts and hospitality – We will address conflicts of interest and the risks created by gifts and hospitality through the implementation of our internal policies.

- Government officials – We will implement procedures applicable to our (or our agents', or those suppliers in our supply chains') dealings with government officials, political parties and related persons or organisations.

Just as governance priorities differ, so do the institutions and methods for controlling corporate governance. The institutions include statutory bodies enforcing detailed prescriptive requirements and statutory bodies that encourage voluntary adoption of good practices with disclosure, through voluntary organisations such as Transparency International UK to professional accounting bodies that have built the awareness of good corporate governance into their examination syllabi.

31.3.1 Corporate governance culture

In China, Russia and the former communist countries in Eastern Europe, the economies are being changed from state-controlled businesses to privately owned companies. The 'model' of these companies is similar to those in the US and UK. So, the trend is towards the US and UK model of companies' shares being listed on their national stock exchange. This trend to wider share ownership will encourage the development of corporate governance criteria similar to those in the US and the UK. For some countries this is a real cultural shift and it will take time for the concept of good corporate governance to be applied. The following is an extract from an OECD Note of a meeting on Corporate Governance Development in State-owned Enterprises in Russia:[2]

> Finally, as stressed by investors, the OECD, and government officials at this expert's meeting, the emergence of a true corporate governance culture is vital. Such a culture-based approach should involve the understanding of the principles and values behind corporate governance, and replace the 'box-ticking' mechanistic approach in which superficial institutions fulfill certain criteria but do not bring real benefits in terms of effective achievement of corporate goals. This would complement the creation of specific incentives intended to guide the behaviour of economic actors.

31.4 Pressures on good governance behaviour vary over time

History shows that business behaviour is influenced by where we are in the economic cycle, whether it's a time of boom or bust.

31.4.1 Behaviour in boom times

During the booms there has always been a tendency to be over-optimistic and to expect the good times to continue indefinitely. In such periods there is a tendency for everyone to focus on making profits. The safeguards that are in the system to prevent conflicts of interest and to limit undesirable behaviour are seen as slowing down the business and causing genuine opportunities to be missed. Over-optimism leads to a business taking risks that the shareholders had not sanctioned and is, to that extent, excessive.

This is accompanied by a tendency to water down the controls or to simply ignore them. When that happens there will always be some unethical individuals who will exploit some of the opportunities for themselves rather than for the business.

31.4.2 Behaviour in bust times

We see a repetitive reaction from bust to bust. When it occurs some of the malpractices will come to light, there will be a public outcry and governance procedures will be tightened up. Although controls are weakly enforced during boom times, it is a fact of life that vigilance is required at all times. Fraud, misrepresentation, misappropriation and anti-social behaviour will be constantly with us and robust corporate governance systems need to be in place and monitored.

The ideal would be that the controls in place develop a culture that makes individuals constrain their own behaviour to that which is ethical, having previously sensitised themselves to recognise the potential conflicts of interest. It is interesting to see the approach taken by the professional accounting bodies which are concentrating on sensitising students and members to ethical issues.

31.5 Types of past unethical behaviour

Some of the unethical behaviour which has been identified in earlier periods and which our governance systems should attempt to prevent are listed below:

- Looting is a term applied to executives who strip corporations of money for their own use, i.e. misappropriation of funds. The misappropriation is often concealed by normal corporate activities such as entering into transactions with associates of the management or dominant shareholders at inflated prices or by falsifying the accounting records and financial statements.

  For example, in the US the SEC filed a complaint[3] against Richard E. McDonald, former CEO and chairman of World Health Alternatives, Inc. ('World Health'):

  > The Commission's complaint alleges that McDonald was the principal architect of a wide-ranging financial fraud at World Health by which McDonald misappropriated approximately $6.4 million for his personal benefit. Also named as defendants are Deanna Seruga of Pittsburgh, the company's former controller and a CPA, . . .
  >
  > A key aspect of the fraud involved the manipulation of World Health's accounting entries . . . repeatedly falsified accounting entries in World Health's financial books and records, understating expenses and liabilities. This made the Company appear more financially sound, and masked McDonald's misappropriation of funds.

- Insider trading, particularly around major events such as a forthcoming company buyout, takeover or development of a new product. Detection is actively pursued by the SEC in the US and penalties are exacted. For example, the SEC charged a former major league baseball player and three others with insider trading ahead of a company buyout and obtaining more than $1.7 million in illegal profits. $2.5 million was paid to settle the SEC's charges.

- Excessive remuneration so that the rewards flow disproportionately to management compared to other stakeholders and often with the major risks being borne by the other stakeholders.

- Excessive risk taking which is hidden from shareholders and stakeholders until after the catastrophe has struck.

- Unsuccessful managers being given 'golden handshakes' to leave and thus being rewarded for poor performance. For example, in Denmark it was reported that 'Banks are facing criticism for giving their CEOs million-kroner "golden handshakes", despite poor performances'.[4]

- Auditors, bankers, lawyers, credit rating agencies, and stock analysts, who might put their fees before the interests of the public for honest reporting.

- Directors who do not stand up to authoritarian managing directors or seriously question their ill-advised plans. For example, it was reported in 2010 that 'A dominant CEO and a weak board of directors was a recipe for disaster at Orion Bank of Naples [Florida]. Orion failed last November because Chief Executive Jerry Williams and his inexperienced board could not handle the bank's overly aggressive growth strategy, according to a new report by the Federal Reserve's Office of Inspector General'.[5]

- Management setting incentives for employees which encourage action that is not in the firm's interests.

31.6 The effect on capital markets of good corporate governance

Good governance is important to facilitate large-scale commerce. The mechanism of legal structures such as limited liability of companies exists because it allows the capital of many investors to be combined in the pursuit of economic activities which need large quantities of capital to be economically viable. There are also statutory provisions relating to directors' duties and shareholders' rights. This is a good backcloth which is necessary but not sufficient to ensure the effective working of the capital market.

In addition there has to be a high level of trust by shareholders in their relationship with management. Firstly, they need to believe the company will deal with them in an honest and prudent manner and act diligently. This means that shareholders need to be confident that:

- their money will be invested in ventures of an appropriate degree of risk;
- efforts will be made to achieve a competitive return on equity;
- management will not take personal advantage of their greater knowledge of events in the business; and
- the company will provide a flow of information that will contribute to the market fairly valuing shares at the times of purchase and sale.

Failure to achieve appropriate levels of trust will lead to the risk of the loss of potential investors or the provision of lesser amounts of funds at higher costs. Similarly if other stake-holders, such as the bank, do not trust the management, there will be fewer participants and the terms will be less favourable. Another way of addressing this is to say that people have a strong sense of what is or is not fair. Whilst economic necessity may lead to participation, the level of commitment is influenced by the perceived fairness of the transaction.

Also from a macro perspective, the more efficient and effective the individual firms, the better allocation of resources and the higher the average standard of living. If management as a group is not diligent in its activities and fair in its treatment of stakeholders, there will be lower standards of living both economically and socially.

In addition the current focus on corporate social responsibility could be seen as a response to governance failures by some companies. For example, some managers ignored externalities such as the costs to society of rectifying pollution because management was only judged on the financial results of the firm, and not the net benefit to society.

31.7 Risk management

We have seen with the issue by the IASB of its *Practice Statement* and the UK with its *Strategic Report* that there is a growing pressure internationally for a company to disclose its risk management policy. In any company there is a range of risks that have to be managed. It is not a matter of just avoiding risks but rather of systematically analysing the risks and then deciding how to decide what risks should be borne, which to avoid, and how to minimise the possible adverse consequences of those which it is not economic to shift. A good governance system will ensure that (a) comprehensive risk management occurs as a normal course of events and (b) there is transparent disclosure to shareholders and regulators of the nature, extent and management of these risks.

There is a variety of approaches which could be adopted to the process of identifying the types of risks associated with a company. In this chapter we will discuss briefly strategic, operational and legal/regulatory risks.

31.7.1 Strategic risks

Strategic risk is associated with maintaining the attractiveness and economic viability of the product and service offerings. In other words, current product decisions have to be made with a strong sense of their probable future consequences. To do that the business has to be constantly monitoring trends in the current markets, potential merging of markets,[6] shifting demographics and consumer tastes, technological developments, political developments and regulations so as to capitalise on opportunities and to counter threats. It must be remembered that to do nothing may involve as much or more risk as entering into new ventures. When entering into new projects there needs to be a thorough risk analysis to ensure that there are no false assumptions in the projections, there has been pilot testing, and the question of the exit strategy if the project fails has been seriously considered and costed.

31.7.2 Operational risks

Operational risks include (a) insurable risks, (b) transferable risks and (c) potential hazards.

Insurable risks

These include such risks as physical damage from fire, flood or accident and reputational damage from quality and public liability issues. The question then is 'If this event should happen could we comfortably bear the cost?' If the answer is no, then we should insure at least for the amount we couldn't afford to bear.

Transferable risks

These include such risks as difficulty recruiting skilled staff to meet orders or dependence on a key supplier.

On staffing, the question may be 'Could work be outsourced?' However, that in itself creates risks such as dependence, quality control, reliability of delivery, lack of involvement in technological developments, and financial risks associated with the subcontractor.

On supply policy, the question may be 'Should the company opt for multiple suppliers?' This would protect against normal hazards such as strikes at the supplier, adverse weather conditions blocking supply, or threats to supply caused by political factors but at a probable increase in cost.

Hazards

In relation to risks like occupational health and safety, the steps involve identification of potential hazards, identifying the best physical process for handling them, and developing standard ways of operating, then training personnel in those standard operating procedures, and regularly checking to ensure those procedures are being followed.

31.7.3 Legal and regulatory risks

This refers to the possibility that the firm will breach its legal or regulatory requirements and thus expose the company to fines and injury to its reputation. This involves being aware of the requirements of each country in which it operates or in which its products and services are used. Further, the staff of the company need to know of the relevant requirements which apply to their activities. They should also have access to advice in order to avoid problems or to address issues that do arise. Once again, standard operating procedures and standard documentation can help reduce the risks. Many businesses now have a Compliance Officer responsible for making sure outside regulatory requirements are being followed. Many of these now have a degree in accountancy, business or finance. In some companies the protection of intellectual property should be of considerable relevance.

31.8 The role of internal control and internal audit in corporate governance

Good governance is supported by (a) adequate internal controls, (b) effective internal audit and (c) full disclosure in the financial statements.

31.8.1 Adequate internal control

In some jurisdictions the company and the external auditors have to explicitly state that the company has adequate internal controls and the accounts present a fair view. In other jurisdictions it is implied that if the company receives a clean audit report then the internal controls are adequate.

In the US when the explicit requirement was introduced many companies spent considerable sums after the introduction of the Sarbanes–Oxley Act 2002 in upgrading their systems, particularly as the CEO and CFO were made personally liable for the effectiveness of the internal controls.

There is an opportunity cost in CEOs focusing on compliance issues rather than on strategic issues and an actual cost in upgrading systems. This led to some arguing that the costs were unjustified.

Whilst the need to consider cost–benefit considerations in relation to all corporate governance measures is a valid concern, it is also important to remember the costs of bad corporate governance. Good governance will not stop all fraud and excessive risk taking but it will stop then from being so widespread. The internal control systems should limit the ability of management to misdirect resources to their personal use or to publish financial statements with material misrepresentation.

31.8.2 Effective internal audit

Sound internal controls combined with an effective internal audit unit should make it more difficult for senior managers to misappropriate resources or misrepresent the financial position. Naturally we know that the more senior the managers the more likely it is that they can override the internal controls or pressure others to do so. It can be argued that such a situation justifies the requirement that the internal audit unit (if one exists) should report direct to the Audit Committee.

Role of the Audit Committee in relation to internal auditors

The Institute of Internal Auditors Model Audit Committee Charter[7] states that the Audit Committee should:

- Approve:
  - the internal audit charter and the risk-based internal audit annual plan.
- Review:
  - with management and the chief audit executive the activities, staffing, and organisational structure of the internal audit function;
  - at least once per year, the performance of the chief audit executive and concur with the annual compensation and salary adjustment;
  - the effectiveness of the internal audit function, including compliance with The Institute of Internal Auditors' International Professional Practices Framework for Internal Auditing consisting of the Definition of Internal Auditing, the Code of Ethics and the Standards;
  - significant accounting and reporting issues, including complex or unusual transactions and highly judgmental areas, and recent professional and regulatory pronouncements, and understand their impact on the financial statements.
- Ensure:
  - there are no unjustified restrictions or limitations; and
  - review and concur in the appointment, replacement, or dismissal of the chief audit executive.

The following sets out some of the activities of the Rank Group Audit Committee in 2013 in relation to internal control. The committee examined the effectiveness of the Group's approach to internal control by approving a new internal audit charter, reviewing and approving an audit universe; approving the 2013/14 internal operational audit plan and a recommended plan and approach for a corporate and systems audit cycle; reviewing internal audit's findings; reviewing the effectiveness of the Group's internal audit function; and action plans to address any failings or weaknesses of internal control.

31.8.3 Full disclosure

There are two aspects to full disclosure, namely the financial data and the narrative.

Financial data

There has been a serious problem with the use by companies of off-balance-sheet finance and special-purpose entities (SPEs) which conceal certain of the company's activities.

For example, in the case of Enron, assets whose values were expected to have to be written down or were vulnerable to substantial market fluctuations were sold to special-purpose entities. Under US rules at that time if there was a 3% outside interest then the special-purpose entity's financial affairs did not have to be consolidated. The exclusion of such items led to the group accounts being misleading.

Good governance requires full disclosure in group accounts and it could be argued that the legislators were at fault in creating the possibility for such off-balance-sheet opportunities and the accounting profession at fault for not making representation against their use.

In the US steps have been taken to address three issues: (i) improving disclosure to investors in the securities about the nature of the underlying financial assets; (ii) limiting conflicts of interest between originators of those financial assets and investors in securities issued by corporate SPEs purchasing those assets; and (iii) increasing rating agency scrutiny of securitisation transactions.[8]

Narrative information

The financial data are backward-looking. Comprehensive information would also include items which are likely to be very important in the future even though they are not currently required to be reported. This could in many jurisdictions include matters relating to future sustainability and comprehensive assessments of environmental impacts. Other future-oriented information would have to relate to the company's strategic drivers and opportunities.

Annual reports can also be used as devices to disclose information which is solely oriented to corporate governance. For example, the disclosure of related-party transactions is intended to make it more difficult for a major shareholder to exploit the company for their own benefit. Whilst the disclosure does not prevent that, it allows shareholders to view the level of activity and if they find the level of activity a matter of concern they can raise the issue at an Annual General Meeting.

We have discussed the more general issue of the need for the board of directors to identify and report the potential risks which could have a significant impact on the organisation, such as over-reliance on a supplier or customer and exposure to environmental or product liabilities.

Adequate disclosure of risk is now being addressed by the IASB *Practice Statement Management Commentary* and in the UK by the *Strategic Report*.

31.9 External audits in corporate governance

External audits are intended to increase participation in financial investing and to lower the cost of funds. They may be *ad hoc* reports or audit reports giving an opinion on the fair view of annual financial statements.

Ad hoc reports

In the case of lending to companies it is not uncommon for lenders to impose restrictions to protect the interests of the lenders. Such restrictions or covenants include compliance with certain ratios such as liquidity and leverage or gearing ratios. Auditors then report to lenders or trustees for groups of lenders on the level of compliance. In this way auditors facilitate the flows of funds at good rates.

Statutory audit reports

Similarly for shareholders the audit report is intended to create confidence that the financial statements are presenting a fair view of financial performance and position. If that confidence is undermined by examples of auditors failing to detect misrepresentation or material misstatement, the public becomes wary of holding shares, share prices in the market tend to fall and the availability of new funds shrinks. Confidence depends on shareholders accepting that:

(a) the auditors:

- are independent;
- approach the audit with a degree of scepticism;
- are professionally competent;
- have industry knowledge;
- carry out a quality audit;
- report the results of the audit in a clear manner; and

(b) the profession enforces audit standards.

31.9.1 Auditor independence

The external auditors should keep in mind that their main responsibility is to shareholders. However, there is a potential governance conflict in that for all practical purposes they are appointed by the board, their remuneration is agreed with the board and their day-to-day dealings are with the management. Appointments and remuneration have to be approved by the shareholders but this is normally a rubber-stamping exercise.

It is not uncommon for auditors to talk of the management as the customer, which is of course the wrong mindset. To reduce the identification with management and loss of independence arising from a personal interest in the financial performance of the client, a number of controls are often put in place, for example:

- Financial threats to independence:
  - Auditors and close relatives should not have shares or options in the company, particularly if the value of their financial interest could be directly affected by their decisions.
  - Auditors must not accept contingency fees or gifts, nor should relatives or close associates receive benefits.
  - Undertaking non-audit work the loss of which, if a significant amount, might be perceived as affecting the auditor's independence. This is a contentious issue with some advocating that auditors should not undertake non-audit work, whereas the client might consider it to be cost-effective. All the indications are that current practice will continue with disclosure of the amounts involved.

- Familiarity threats:
  - Appointments, terminations and the remuneration of auditors should be handled by the audit committee.
  - Auditors should not have worked for the company or its associates.
  - Audit partners should be rotated periodically so the audit is looked at with fresh eyes.
  - Audit tests should vary so that employees cannot anticipate what will be audited.
  - It is not desirable that audit staff be transferred to senior positions in a client company. This happens but it does mean that they will continue to have close relations with the auditors and knowledge of their audit procedures. Clients might regard this as a benefit.

However, the above are indications and not to be seen as taking a rule-based approach. Good governance is not just a matter of compliance with rules as ways can always be found to comply with rules whilst not complying with their spirit. It is really a question of behaviour – good governance depends on the auditors behaving independently, with professional competence, and identifying with shareholders and other stakeholders whose interests they are supposed to be protecting. Failing to do this leads to what is described as the expectation gap.

31.9.2 Lack of independence – Enron

The following is an extract from the United Nations Conference on Trade and Development G-24 Discussion Paper Series illustrating the dangers when there is a lack of independence:[9]

> Regarding auditing good corporate governance requires high-quality standards for preparation and disclosure, and independence for the external auditor. Enron's external auditor was Arthur Andersen, which also provided the firm with extensive internal auditing and consulting services. In 2000 consultancy fees (at $27 million) accounted for more than 50 per cent of the approximately $52 million earned by Andersen for work on Enron . . . the following assessment by the Powers Committee: The evidence available to us suggests that Andersen did not fulfil its professional responsibilities in connection with its audits of Enron's financial statements, . . . lack of independence linked to its multiple consultancy roles was a crucial factor in Andersen's failure to fulfill its obligations as Enron's external auditor.

31.9.3 Professional scepticism

Professional scepticism is defined[10] as 'an attitude that includes a questioning mind, being alert to conditions which may indicate possible misstatement due to error or fraud, and a critical assessment of audit evidence'. The auditor is explicitly required to plan and perform an audit with professional scepticism recognising that circumstances may exist that cause the financial statements to be materially misstated.

Warning flags

The individual circumstances will vary but there are indicators such as the following that should be considered:

- Internal conditions:
  - Lack of personnel with appropriate accounting and financial reporting skills.
  - Changes in key personnel including departure of key executives.
  - Deficiencies in internal control, especially those not addressed by management.
  - Changes in the IT environment.
- Trading conditions:
  - Operations in regions that are economically unstable, for example countries with significant currency devaluation or highly inflationary economies.
  - Changes in the industry in which the entity operates.
  - Developing or offering new products or services, or moving into new lines of business.
  - Changes in the supply chain.
- Scale of operations:
  - Expanding into new locations.
  - Changes in the entity such as large acquisitions or reorganisations or other unusual events.
  - Entities or business segments likely to be sold.
  - The existence of complex alliances and joint ventures.

- Financial conditions:
  - Going concern and liquidity issues including loss of significant customers.
  - Constraints on the availability of capital and credit.
  - Use of off-balance-sheet finance, special-purpose entities, and other complex financing arrangements.
  - Significant transactions with related parties.
  - Excessive reliance on management representations

31.9.4 Example of where there was excessive reliance on management representations

In the normal course of an audit it is usual to obtain a letter of representation from management, for example providing information regarding a subsequent event occurring after year-end and the existence of off-balance-sheet contingencies. It is confirmation to the auditor that management has made full disclosure of all material activities and transactions in its financial records and statements.

However, the representations do not absolve the auditor from obtaining sufficient and appropriate audit evidence. The following is an extract[11] from an SEC finding relating to two Certified Public Accountants who were auditing a company (Structural Dynamics Research Corporation) which had improperly recorded sales and then written them off in the following accounting period:

> Despite the fact that the language in purchase orders clearly stated the orders were conditional and subject to cancellation, the auditors accepted the controller's explanation and did not take exception to the recognition of revenue on these orders. This undue reliance on management's representations constitutes insufficient professional skepticism by Present [the engagement partner].
>
> Moreover, Present failed to corroborate management's representations regarding conditional purchase orders with sufficient additional evidence that these sales were properly recorded . . . Overall, Present failed to exercise due professional care in the performance of the audit.

31.9.5 Developing and enforcing audit standards

There are international audit standards set by the International Auditing and Assurance Standards Board (IAASB). The IAASB in developing standards has to have regard to developments in financial reporting which has grown more complex with a greater variety of disclosures than those traditionally disclosed and the need for greater transparency in the audit work that has been carried out. As with the development of IFRSs, the IAASB follows a process of Discussion Papers, Exposure Drafts and IASs.

It has responded by taking steps such as publishing a Discussion Paper in 2011 *The Evolving Nature of Financial Reporting: Disclosure and Its Audit Implications* in which it addressed areas such as the increasing disclosure of judgements made in applying accounting policies and disclosure of material uncertainties. In 2014 it published a *Framework for Audit Quality: Key Elements that Create an Environment for Audit Quality*.

Whilst the standards are international, the enforcement of the standards is carried out nationally. National practice varies. In the UK, the Financial Reporting Council (FRC) is the independent regulator for corporate reporting and corporate governance. Through its Codes and Standards Committee the FRC has primary responsibility for setting, monitoring and enforcement of auditing standards in the UK.

31.9.6 Governance within audit firms

Within the audit practices there is also the need to apply systems to ensure that there are adequate reviews of the performance of individual auditors and that the individual partners do not take advantage of their positions of trust.

The greatest control mechanism within an audit firm is the culture of the firm. Arthur Wyatt made the following observation:[12]

> The leadership of the various firms needs to understand that the internal culture of firms needs a substantial amount of attention if the reputation of the firms is to be restored. No piece of legislation is likely to solve the behavioural changes that have evolved within the past thirty years.

Impact of consultancy on audit attitudes

Wyatt, drawing on his experience in Arthur Andersen and his observation of competitors, indicated that in earlier times there was a culture of placing the maintenance of standards ahead of retention of clients; the smaller size of firms meant there was more informal monitoring of compliance with firm rules and ethical standards. Promotion was more likely to flow to those with the greatest technical expertise and compliance with ethical standards, rather than an ability to bring in more fees. The values of conservative accountants predominated over the risk-taking orientation of consultants.

Wyatt's view was that the growth and risk orientations of consulting are incompatible with the values needed to perform auditing in a manner which is independent in attitude.

31.9.7 The expectation gap

Another area of corporate governance and auditing relates to the expectation gap. The gap is between the stakeholders' expectation of the outcomes that can be expected from the auditors' performance and the outcomes that could reasonably be expected given the audit work that should have been performed.

The stakeholders' expectation is that the auditor guarantees that the financial statements are accurate, that every transaction has been 100% checked and any fraud would have been detected. The auditors' expectation is that the audit work carried out should identify material errors and misstatements based on a judgemental or statistical sampling approach.

Loss of confidence following corporate scandals

There have been a number of high-profile corporate failures and irregularities; for example, in the US, Enron failed, having inflated its earnings and hidden liabilities in SPEs (special-purpose entities). In 2008 the same problem of hiding liabilities appears to have occurred with Lehman Brothers where according to the Examiner's report[13] Lehman used what amounted to financial engineering to temporarily shuffle $50 billion of troubled assets off its books in the months before its collapse in September 2008 to conceal its dependence on borrowed money, and senior Lehman executives as well as the bank's accountants at Ernst & Young were aware of the moves. In Italy, Parmalat created a false paper trail and created assets where none existed; and in the US, the senior management of Tyco looted the company.

This raises questions such as (a) Were the auditors independent? (b) Did they carry out the work with due professional competence? and (c) Did they rely unduly on management representations?

31.9.8 Action by auditors to limit liability

In each of the above there is good reason for the expectation gap in that the audit had not been conducted in accordance with generally accepted audit standards and there was a lack of due professional care. If the auditors have been negligent then they are liable to be sued in a civil action. In the UK the profession has sought to obtain a statutory limit on their liability and, failing that, some have registered as limited liability partnerships – the path taken by Ernst & Young in 1996 and KPMG in 2002. In Australia some accountants operate under a statutory limit on their liability and in return ensure they have a minimum level of professional indemnity insurance.

31.9.9 Detection of fraud

An audit is designed to obtain evidence that the financial statements present a fair view and do not contain material misstatements. It is not a forensic investigation commissioned to detect fraud. Such an investigation would be expensive and in the majority of cases not be cost-effective. It has been argued that auditors should be required to carry out a fraud and detection role to avoid public concerns that arise when hearing about the high-profile corporate failures. However, it would appear that it is not so much a question of making every audit a forensic investigation to detect fraud but rather enforcing the exercise of due professional care in the conduct of all audits. The audit standards reinforce this when they emphasise the importance of scepticism.

31.9.10 Educating users

Many surveys have shown that there has been a considerable difference between auditors and audit report users regarding auditors' responsibilities for discovering fraud and predicting failure. Users of published financial statements need to be made aware that auditors rely on systems reviews and *sample testing* to evaluate the company's annual report. Based on those evaluations they form an opinion on the *likelihood* that the accounts provide a true and fair view or fairly present the accounts. However, they cannot guarantee the accounts are 100% accurate.

A number of major companies have collapsed without warning signs and the public have criticised the auditors. It is difficult when there are such high-profile corporate failures to persuade the public that lack of due professional care is not endemic. In response to these pressures auditors have modified their audit standards to place more emphasis on scepticism.

31.10 Executive remuneration in the UK

It is worth reinforcing the fact that the objective of corporate governance is to focus management on achieving the objectives of the company whilst keeping risks to appropriate levels and positioning the firm for a prosperous future. At the same time, sufficient safeguards must be in place to reduce the risks of resources being inappropriately diverted to any group at the expense of other groups involved.

31.10.1 The problem

The following is an extract[14] from a speech by Vince Cable, UK Secretary of State for business and industry, in 2012:

> The issue is **partly** about 'rewards for failure'. But it is not just that.
> There is also a ratchet in executive pay with everyone believing that they should be paid well above average and that they should be benchmarked against US peers when

they live and work in the UK. It is of course a logical absurdity for everyone to be paid above the average, let alone in the top quartile. Imagine if this happened with workers' pay awards. There would be galloping wage inflation and loud business objections about our loss of competitiveness.

While it is true that rising executive pay is a global phenomenon, trends in inequality at the very top are very divergent between countries, despite them all operating in the same global economy. There are world class companies in the Nordic countries, Japan, Holland and Germany who take a very different approach to the UK and US.

But let me be clear. There is a legitimate role for high pay for exceptional talent and performance – quite apart from high returns to successful entrepreneurs – and I will defend that.

Since management is the group with the most discretion and power, it is important to ensure they do not obtain excessive remuneration or perks, or be allowed to shirk, or to gamble with company resources by taking excessive risks.

31.10.2 UK government response

In the UK there are new requirements[15] for directors remuneration reports from 2014. The report is split into three parts – a statement from the Chairman of the Remuneration Committee, a Policy Report and an Annual Report on remuneration. The Policy Report contains details of the performance measures (not the actual targets which are commercially sensitive) and is subject to a shareholder binding vote. The Annual Report includes a single figure for each director and the link between pay and performance.

31.10.3 What is fair?

The statistics show[16] that in recent years the remuneration of executives relative to the average employee has been considerably higher than it was 20 years ago, and the remuneration of the top executive compared to the average of the next four executives is also higher than in the past. That seems intuitively unfair – but is it a valid comparison to be referring back to a relationship that existed 20 years ago? Perhaps the question should be whether this higher relative remuneration reflects a greater contribution to performance, whether it reflects that as businesses increase in size the remuneration of the chief executive tends to increase to reflect the higher responsibilities, or whether it has been achieved simply because directors have been effectively able to set their own remuneration.

31.10.4 How to set criteria – in principle

There are a number of issues that will require a judgement to be made:

- What is the right balance between short-term performance and long-term performance?
- What if there are revenues and costs that are beyond the control or influence of management? Should these be excluded from the measure?
- Also, to the extent that performance may be influenced by general economic conditions, should managers be assessed on absolute performance or relative performance?

Relative performance means that if the performance fell from 10% to minus 3% during an economic downturn, and competitors' performance fell to minus 5%, managers would qualify for a bonus recognising that their performance had been relatively better. This may be resented by shareholders who have seen the share price fall.

Often companies resort to outside consultants, but the observation has been made that one doesn't hear of outside consultants recommending a pay cut and they are in part responsible for ratcheting up the levels of remuneration.

31.10.5 Where do accountants feature in setting directors' remuneration?

The equity of the remuneration is not normally seen as an accounting matter, but accountants should ensure transparent disclosure of the performance criteria and of the payments. In some jurisdictions there is legislation setting out in some detail what has to be disclosed.

For example, in the UK The Large and Medium-sized Companies and Groups (Accounts and Reports) (Amendment) Regulations 2013 requires the annual report to contain a single total figure table comprising six columns, reporting for each director (with certain conditions):

● the total amount of salary and fees;

● all taxable benefits;

● money or other assets received or receivable for the relevant financial year as a result of the achievement of performance measures and targets relating to a period ending in that financial year;

● money or other assets received or receivable for periods of more than one financial year where final vesting is determined as a result of the achievement of performance measures or targets relating to a period ending in the relevant financial year;

● all pension-related benefits including payments (whether in cash or otherwise) in lieu of retirement benefits and all benefits in year from participating in pension schemes; and

● the total amount of the sums set out in the previous five columns.

31.10.6 Performance criteria

Directors are expected to produce increases in the share price and dividends. Traditional measures have been largely based on growth in earnings per share (EPS), which has encouraged companies to seek to increase short-term earnings at the expense of long-term earnings, e.g. by cutting back capital programmes. Even worse, concentrating on growth in earnings per share can result in a reduction in shareholder value, e.g. by companies borrowing and invest-ing in projects that produce a return in excess of the interest charge, but less than the return expected by equity investors.

31.10.7 Institutional investor guidelines

One of the problems is the innovative nature of the remuneration packages that companies might adopt and the fact that there is no uniquely correct scheme. The following are examples of the various criteria which have evolved and which have been adopted:

Absolute Measures or Targets
Normalised earnings per share measured by reference to a percentage margin, for example 2% per annum growth, in excess of inflation over a 3 year period. It is important that the figures for earnings be smoothed where appropriate to avoid distortions arising from one-off extraordinary or exceptional items included within the FRS 3 definition of earnings per share.

Comparative Measures

Outperformance of an index or of the median or weighted average of a pre-defined peer group in the case of basic options: or the achievement of top quartile performance in the case of super-options:

(i) *Normalised earnings per share*
Outperformance of the median or weighted average rate of increase in normalised earnings of a peer group.

(ii) *Net Asset Value per Share*
Net asset value per share measured, for example against a predefined peer group or index.

(iii) *Total Shareholder Return (ie share price performance plus gross dividend per share)*
Where total shareholder return is used this should be based on exceeding the relevant benchmark within a predefined peer group but, as this formula relies substantially on share price, attainment of the criterion should also be supported by a defined secondary criterion validating sustained and significant improvement in the underlying financial performance.

(iv) *Comparative Share Price*
Comparative share price relative to a peer group would be an acceptable alternative to total shareholder return, conditional in the same way on a secondary performance criterion validating sustained and significant improvement in underlying financial performance over the same period.

31.10.8 Institutional investors' statements of principles

In the UK, in response to The Large and Medium-sized Companies and Groups (Accounts and Reports) (Amendment) Regulations 2013, the Association of British Insurers (ABI)[17] and the National Association of Pension Funds (NAPF)[18] have issued statements of principles that they expect companies to consider when setting remuneration policies.

These include proactive proposals that schemes should ensure that executive rewards reflect long-term returns to shareholders by expecting executive management to make a material long-term investment in shares of the businesses they manage. There are also proposals to address the criticisms that have been made that poor performance has still been rewarded by proposing that there should be provisions that allow a company to forfeit all or part of a bonus or long-term incentive award before it has vested and been paid and claw back moneys already paid.

31.11 Corporate governance, legislation and codes

Investors looking to the safety and adequacy of the return on their investment are influenced by their level of confidence in the ability of the directors to achieve this. Good governance has not been fully defined and various reports have attempted to set out principles and practices which they perceive to be helpful in making directors accountable. These principles and practices are set out in a variety of Acts, e.g. Sarbanes–Oxley in the US and the Companies Act and regulations in the UK, and codes such as the Singapore Code of Corporate Governance 2012 and the UK Corporate Governance Code (formerly the Combined Code).

The various laws and codes that have been published set out principles and recommended best practice relating to the board of directors, directors' remuneration, relations with shareholders, accountability and audit.

The European Corporate Governance Institute[19]

This is an excellent resource that covers pretty well all the corporate governance codes in the world. It is interesting to refer to the Institute's website to observe the number of new and amended codes since 2010 which reflects the growing importance attached to corporate governance in terms of investor confidence.

31.11.1 Codes as a partial solution

As the nature of business and expectations of society change, the governance requirements evolve to reflect the new laws and regulations. By anticipating changing requirements, companies can prepare for the future. At the same time they should identify the special areas of potential conflict in their own operations and develop policies to manage those relationships.

Good governance is a question of having the right attitudes. All the corporate governance codes will not achieve much if they focus on form rather than substance. Codes work because people want to achieve good governance. People can always find ways around rules.

The FRC has taken the view in 2014 that more effective application of, and reporting on, existing code principles may often have a greater impact on actual standards of governance and stewardship than managing further change.

Furthermore, rules cannot cover all cases, so good governance needs a commitment to the fundamental idea of fairness.

The research on whether good governance leads to lower cost of capital is very mixed, reflecting both the difficulty of identifying the impact of good governance and the fact that some engage with the spirit of the concept and some do not. There are those who question the impact of good corporate governance, and supporting the case of those who doubt that there is a positive impact on performance is an Australian research project[20] looking at companies in the S&P/ASX 200 index which found that companies which the researcher classified as having poor corporate governance outperformed companies classified as having good corporate governance over a range of measures including EBITDA growth and return on assets. There is an ongoing need for further research, particularly as to the effect on smaller listed companies, and it will be interesting to await the outcome.

31.12 Corporate governance – the UK experience

In the UK there have been a number of initiatives in attempting to achieve good corporate governance through (a) legislation, (b) the UK Corporate Governance Code, (c) non-executive directors (NEDs), (d) shareholder activism and (e) audit. We discuss each of these briefly below.

31.12.1 Legislation

Legislation is in place that attempts to ensure that investors receive sufficient information to make informed judgements. For example, there are requirements for the audit of financial statements, majority voting on directors' remuneration policy and disclosure of directors' remuneration. There could be a case for increased statutory involvement in the affairs of a company by, for example, putting a limit on benefits and specifying how share options should be structured. However, the government has gone down the road of disclosure and transparency to encourage and empower shareholder activism.

31.12.2 The UK Corporate Governance Code[21]

The Code is routinely reviewed every two years. The current code was published by the Financial Reporting Council (FRC) in 2012. It sets out standards of good practice in relation to board leadership and effectiveness, remuneration, accountability and relations with shareholders. It is not a rule book but is principles-based and sets out best practice. It relies for its effectiveness on disclosure by requiring companies listed on a stock exchange to explain if they do not comply with its provisions.

The original code made an interesting development by separating its proposals into two parts:

- Part 1 containing Principles of Good Governance (Main and Supplementary) relating to:
  - A: directors;
  - B: directors' remuneration;
  - C: relations with shareholders;
  - D: accountability and audit.
- Part 2 containing Codes of Best Practice with procedures to make the Principles operational.

The current updated Code continues to set out broad principles from which companies are largely free to choose their own method of implemention. The detailed code provisions are those which companies are required to say whether they have complied with and, where they have not complied, to explain why not. The intention is to combine flexibility over detailed implementation with clarity where there was non-compliance.

The principles and code provisions relating to the board of directors are set out below to illustrate the code's approach. The six principles that relate to directors cover:

- **A1** the board
- **A2** chairman and chief executive
- **A3** board balance and independence
- **A4** appointments to the board
- **A5** information and professional development
- **A6** performance evaluation.

As an illustration of the level of detail, the Principles (A1) and Provisions (A1.3 and A1.4) relating to the board are set out below.

A1 The Board
Main Principle
Every company should be headed by an effective board, which is collectively responsible for the success of the company.

Supporting Principles include:

- The board should
  - set the company's values and standards; and
  - ensure that its obligations to its shareholders and others are understood and met.
- As part of their role as members of a unitary board, non-executive directors should
  - constructively challenge and help develop proposals on strategy;
  - scrutinise the performance of management in meeting agreed goals and objectives and monitor the reporting of performance;

- satisfy themselves on the integrity of financial information and that financial controls and systems of risk management are robust and defensible.
- As non-executive directors they
  - are responsible for determining appropriate levels of remuneration of executive directors; and
  - have a prime role in appointing, and where necessary removing, executive directors, and in succession planning.

Code Provisions (*relating to NEDs*)

A.1.3 The chairman should hold meetings with the non-executive directors without the executives present. Led by the senior independent director, the non-executive directors should meet without the chairman present at least annually to appraise the chairman's performance (as described in A.6.1) and on such other occasions as are deemed appropriate.

A.1.4 Where directors have concerns which cannot be resolved about the running of the company or a proposed action, they should ensure that their concerns are recorded in the board minutes. On resignation, a non-executive director should provide a written statement to the chairman, for circulation to the board, if they have any such concerns.

Revisions to the code in 2012

The revisions included:

- Boards will be expected to confirm that the report and accounts, taken as a whole, is fair, balanced and understandable and provides the information needed for shareholders to assess the company's performance, business model and strategy.
- A description of the board's policy on diversity, including gender, any measurable objectives that it has set for implementing the policy, and progress on achieving the objectives.
- Evaluation of the board should consider the balance of skills, experience, independence and knowledge of the company on the board, its diversity, including gender, how the board works together as a unit, and other factors relevant to its effectiveness.

31.12.3 Non-executive directors (NEDs)

The main function of non-executive directors is to ensure that the executive directors are pursuing policies consistent with shareholders' interests.[22]

Review of their contribution

Considering the qualities that are required, the Cadbury Report recommended that the board should include non-executive directors of sufficient calibre and number for their views to carry significant weight in the board's decisions. Research[23] indicated that they are concerned to maintain their reputation in the external market in order to maintain their marketability.

NEDs on many boards bring added or essential commercial and financial expertise, for example on a routine basis as members of the audit committee, or on an *ad hoc* basis providing experience when a company is preparing to float or having specific industry knowledge. They are also valued as having a role in questioning investment decisions and entering into unduly risky projects.

Limitations

However, NEDs are not and never can be a universal panacea. It has to be recognised that there may be constraints such as:

- They might have divided loyalties, having been nominated by the chairman, the CEO or another board member.
- This has been addressed by the Code which states 'An explanation should be given if neither an external search consultancy nor open advertising has been used in the appointment of a nonexecutive director. Where an external search consultancy has been used, it should be identified in the report and a statement should be made as to whether it has any other connection with the company.'
- They might have other NED appointments and/or executive appointments which limit the time they can give to the company's affairs.
- This is addressed by some companies such as BUPA[24] which requires non-executive directors to disclose their other significant commitments to the board before appointment, with a broad indication of the time involved; and inform the board of any subsequent changes.
- They might not be able to restrain an overbearing CEO, particularly if the CEO is also the chairman.
- A 2009 survey[25] indicated that a third of non-executive directors feel they are unable to control their chairmen and chief executives, and almost 40% feel they would be unable to sack underperforming board colleagues.

With so many caveats, it would be reasonable to assume that NEDs could not easily divert a dominant CEO or executive directors from a planned course of action. In such cases, their influence on good corporate governance is reduced unless the interest of directors and shareholders already happen to coincide. However, if the issue is serious enough for one or more independent director to resign it is likely that the market will certainly take note.

Independent NEDs and risk – a negative view

Research[26] commented that the view that outside directors brought experience and strategic expertise, together with vigilance in monitoring management decisions, to prevent strategic mistakes and/or opportunistic behaviour by management was not supported by much evidence that governance reduces risks. The researchers found little evidence that governance was effective in reducing the volatility of share prices or the chance of large adverse share price movements. As with the financial sector in the credit crunch, independent directors seem not to be a protection against companies adopting risky strategies.

Independent NEDs and risk – a positive view

However, on a more positive note, the presence of NEDs is perceived to be indicative of good corporate governance, and a research report[27] indicated that good governance has a positive impact on investor confidence. The research examined 654 UK FTSE All-Share companies from 2003 to 2007 using unique governance data from the ABI's Institutional Voting and Information Service (IVIS). An extract from the ABI research is as follows:

> New research from the ABI (Association of British Insurers) shows that companies with the best corporate governance records have produced returns 18% higher than those with poor governance. It was also revealed that a breach of governance best practice (known as a red top in the ABI's guidance) reduces a company's industry-adjusted return on assets (ROA) by an average of 1 percentage point a year. For even the best

performing companies (those within the top quartile of ROA performance), that equates to an actual fall of 8.6% in returns per year.

The research also shows that shareholders investing in a poorly governed company suffer from low returns. £100 invested in a company with no corporate governance problems leads to an average return of £120 but if invested in the worst governed companies the return would have been just £102.

There are many highly talented, well-experienced NEDs but their ability to influence good governance should not be overestimated. Their effectiveness might be reduced if they have limited time, limited access to documents, limited respect from full-time executive directors and limited expertise within the remuneration and/or audit committees. When a company is prospering their influence could be extremely beneficial; when there are problems they may not have the authority to ensure good governance.

31.12.4 Shareholder activism

In the UK the need for good corporate governance is affected by how widely shares are held.

In the US and the UK, a large number of financial institutions and individuals hold shares in listed companies, so there is a greater need for corporate governance requirements. In Japan and most European countries (except the UK) shares in listed companies tend to be held by a small number of banks, financial institutions and individuals. Where there are few shareholders in a company, they can question the directors directly, so there is less need for corporate governance requirements.

The following table is an extract from the UK Office for National Statistics[28] showing the holdings in UK shares:

Beneficial ownership of UK shares in 2010

	Pounds (bn) *2010*	*Percentages* *2010*
Rest of the world	732.6	41.2
Insurance companies	153.6	8.6
Pensions funds	91.3	5.1
Individuals	204.5	11.5
Unit trusts	118.8	6.7
Investment trusts	37.2	2.1
Other financial institutions	284.5	16.0
Charities	15.1	0.9
Private non-financial companies	40.3	2.3
Public sector	54.4	3.1
Banks	45.0	2.5
Total	1,777.5	100.0

Figures show that at the end of 2010 the UK stock market was valued at £1,777.5 billion. At that date:

- Rest-of-the-world investors owned 41.2% of the value of the UK stock market, up from 30.7% in 1998. This figure had increased to 53.2% in 2012.

- Other financial institutions held 16.0%, up from 2.7% in 1998.

- Insurance companies held 8.6% and pension funds held 5.1% by value. These are the lowest percentages since the share ownership survey began in 1963.

- UK individuals owned 11.5%, down from 16.7% in 1998. This figure had fallen to 10.7% in 2012.

Individual shareholder influence on corporate governance

With the rest of the world holding 53.2% and individual shareholders holding only 10.7%, it is difficult for the latter group to exercise any significant group influence on management behaviour. In passing legislation, there is an implicit view that individual shareholders have a responsibility to achieve good corporate governance. Statutes can provide for disclosure and be fine-tuned in response to changing needs but they are not intended to replace shareholder activism. When the economy is booming there is a temptation to sit back, collect the dividends and capital gains, bin the annual report and post in proxy forms.

Shareholder influence has to rely on that exercised by the institutional investors.

Large-block investors' influence on corporate governance

There is mixed evidence about the influence of large-block shareholders. The following is an extract from a Department of Trade and Industry report:[29]

> The report observed from a review of economics, corporate finance and 'law and economics' research literature that there was no unambiguous evidence that presence of large-block and institutional investors among the firm's shareholders performed monitoring and resource functions of 'good' corporate governance. However, management and business strategy research suggests that it does have a significant effect on *critical* organisational decisions, such as executive turnover, value-enhancing business strategy, and limitations on anti-takeover defences.

Feedback from the experts' evaluation of the governance roles of various types of shareholders provided the following pattern:[29]

	Mean	Standard deviation
Pension funds, mutual funds, foundations	4.58	1.50
Private equity funds	4.52	1.76
Individual (non-family) blockholders	4.36	1.70
Family blockholders	4.20	1.63
Corporate pension funds	3.85	1.55
Insurance companies	3.69	1.69
Banks	3.31	1.49
Dispersed individual shareholders	2.18	1.41

The highest scores were assigned to the governance roles of pension funds, mutual funds, foundations and private equity investors:

> Some respondents also suggested that various associations of institutional investors such as NAPF, ABI, etc., play strong governance roles, as do individual blockholders and family owners. At the other end of the spectrum are dispersed individual shareholders whose governance roles received the lowest score. However, it must be kept in mind that none of the individual scores is above 5 indicating that, on average, our experts were rather sceptical about the effectiveness of large blockholders from the 'good' governance perspective.[29]

A further related factor is that US and UK companies have tended to have a low gearing with most of the finance provided by shareholders. However, in other countries the gearing of companies is much higher, which indicates that most finance for companies comes from banks. If the majority of the finance is provided by shareholders, then there is a greater need for corporate governance requirements than if finance is in the form of loans where the lenders are able to stipulate conditions and loan covenants, e.g. the maximum level of gearing and action available to them if interest payments or capital repayments are missed.

However, institutional investors do not represent a majority in any company. Their role is to achieve the best return on the funds under their management consistent with their attitude to environmental and social issues. Their expertise has been largely directed towards the strategic management and performance of the company with, perhaps, an excessive concern with short-term gains. Issues such as directors' remuneration might well be of far less significance than the return on their investment.

The Walker Review of Corporate Governance of the UK Banking Industry[30]

This reviewed corporate governance in the UK banking industry and financial institutions and made recommendations on the effectiveness of risk management at board level, including the incentives in remuneration policy to manage risk effectively; the balance of skills, experience and independence required on the boards; the effectiveness of board practices and the performance of audit, risk, remuneration and nomination committees; and the role of institutional shareholders in engaging effectively with companies and monitoring of boards. The role of the institutional investors is important and a Stewardship Code has been proposed.

The Stewardship Code[31]

The **Stewardship Code** is a set of principles or guidelines issued by the FRC in 2010. Its principal aim is to make institutional investors take an active role to protect the interests of the people who have placed their money with them to invest.

The code consists of seven principles which, if followed, should benefit corporate governance. The principles are:

1 **Institutional investors should publicly disclose their policy on how they will discharge their stewardship responsibilities.** Such a policy should include how investee companies will be monitored with an active dialogue on the board and its policy on voting and the use made of proxy voting.

2 **Institutional investors should have a robust policy on managing conflicts of interest in relation to stewardship and this policy should be publicly disclosed.** Such a policy should include how to manage conflicts of interest when, for example, voting on matters affecting a parent company or client.

3 **Institutional investors should monitor when it is necessary to enter into an active dialogue with their boards.** Such monitoring should include satisfying themselves that the board and sub-committee structures are effective, and that independent directors provide adequate oversight and maintain a clear audit trail of the institution's decisions. The objective is to identify problems at an early stage to minimise any loss of shareholder value.

4 **Institutional investors should establish clear guidelines on when and how they will escalate their activities as a method of protecting and enhancing shareholder value.** Such guidelines should say the circumstances when they will actively intervene. Instances when institutional investors may want to intervene include when they have concerns about the company's strategy and performance, its governance or its approach to the risks arising from social and environmental matters. If there are concerns, then any action could escalate from meetings with management specifically to discuss the concerns through to requisitioning an EGM, possibly to change the board.

5 **Institutional investors should be willing to act collectively with other investors where appropriate.** Such action is proposed in extreme cases when the risks posed threaten the ability of the company to continue.

6 **Institutional investors should have a clear policy on voting and disclosure of voting activity.** Institutional investors should seek to vote all shares held. They should not automatically support the board and if they have been unable to reach a satisfactory outcome through active dialogue then they should register an abstention or vote against the resolution.

7 **Institutional investors should report periodically on their stewardship and voting activities.** Such reports should be made regularly and explain how they have discharged their responsibilities. However, it is recognised that confidentiality in specific situations may well be crucial to achieving a positive outcome.

The FRC proposes to extend these principles to all listed companies.

The Kay Review[32]

The Kay Review of UK Equity Markets and Long-term Decision Making was published in 2012. It recommended that the Stewardship Code should be developed to incorporate a more expansive form of stewardship, focusing on strategic issues as well as questions of corporate governance. This is in keeping with the increasing pressure for companies to report risks and how they are addressing them. It also recommended that an investors' forum should be established to facilitate collective engagement by investors in UK companies. This recommendation has been acted on[33] and the 'Investors Forum' has been launched by the Collective Engagement Working Group.

- There has been recognition that a cultural change is needed with investors and companies developing a shared sense of partnership to promote long-term strategies that can generate sustainable wealth creation for all stakeholders. One way forward is for all major listed companies to hold an annual strategy meeting for institutional investors, outside the results cycle, where investors and company executives can link governance to the company's long-term strategy without the focus on short-term results. Where there are shared concerns about a particular company it is proposed that an Engagement Action Group should operate.

- An investors' forum should be established to facilitate collective engagement by investors in UK companies.

- Companies should consult their major long-term investors over major board appointments.

- High-quality, succinct narrative reporting should be strongly encouraged.

- Companies should structure directors' remuneration to relate incentives to sustainable long-term business performance. Long-term performance incentives should be provided only in the form of company shares to be held at least until after the executive has retired from the business.

- Asset management firms should similarly structure managers' remuneration so as to align the interests of asset managers with the interests and timescales of their clients. Pay should therefore not be related to short-term performance of the investment fund or asset management firm. Rather a long-term performance incentive should be provided in the form of an interest in the fund (either directly or via the firm) to be held at least until the manager is no longer responsible for that fund.

Legal safeguards

Corporate governance has to react to changing circumstances and threats. It evolves and will continue to need to be revised and updated. The law provides minimum safeguards but in the ever-changing complexities of global trade and finance, good governance is dependent on the behaviour of directors and their commitment to principles and values. The UK system is heavily dependent on codes which set out principles and the requirement for directors to explain if they fail to comply. The UK Corporate Governance Code and Stewardship Code will rely for their effectiveness on investor engagement. This recognises that investors cannot delegate all responsibility to their agents, the directors, accept their dividends and be dormant principals.

UK experience and international initiatives

It is interesting to note that what constitutes good corporate governance is evolving with new initiatives being taken globally. For example, the OECD is responding to weaknesses in corporate governance that became apparent in the financial crisis by developing recommendations for improvements in board practices, the remuneration process and how shareholders should actively exercise their rights. It is also reviewing governance in relation to risk management which featured as such a threat in the way financial institutions conducted their business. However, there is some concern that measures that might be essential for the control of the financial sector should not be imposed arbitrarily on non-financial-sector organisations.

31.12.5 Audit

There has been audit reform in the EU[34] with a Directive setting out new rules which addresses independence, the expectation gap and competition.

Independence
These include the mandatory rotation of auditors every 10 years (or 20 years subject to retendering), the prohibition on the provision of certain non-audit services to audit and the introduction of a cap on the fees that can be earned from the provision of permitted non-audit services to PIEs.

The expectation gap
This is to be improved by ensuring increased audit quality and more detailed and informative audit reports with meaningful data for investors and better accountability with a provision for 5% of the shareholders of a company to initiate actions to dismiss the auditors.

Competition
Medium sized audit practices have been precluded from obtaining audit work with some major companies that have had a 'Big Four only' policy. This restrictive practice is now prohibited. In addition there will be the impact of the prohibition of certain non-audit services to audit clients.

Auditors are subject to professional oversight to ensure that they are independent, up-to-date and competent. However, where there is a determined effort to mislead the auditors, for example by creating false paper trails and misstatement at the highest level, then there is the risk that fraud will be missed.

There have been allegations of audit negligence in some high-profile corporate failures, in some of which auditors have been found liable. In part, the financial crisis arising from issues such as the use of special-purpose entities and complex financial instruments has made life more difficult for auditors. This is because pressure groups, such as the investment banks, have influenced the regulators to allow, or not question, practices that have since been found to be highly risky and undisclosed in group accounts.

In general, whilst the audit appears to be a reliable mechanism for ensuring that the financial statements give a true and fair view, there is a need for audit staff to acquire a detailed understanding of the industry being audited and the risks attaching to financial instruments.

Summary

Good governance is achieved when all parties feel that they have been fairly treated. It is achieved when behaviour is prompted by the idea of fairness to all parties. Independent behaviour is expected of the NEDs and auditors and they are expected to have the strength of character to act professionally with proper regard for the interest of the shareholders. The shareholders in turn should be exercising their rights and not be inert. They have a role to play and it is not fair of them to sit on their hands and complain.

Good corporate governance cannot be achieved by rules alone. The principle-based approach such as that of the FRC with the UK Corporate Governance Code recognises that it is behaviour that is the key – it sets out broad principles and a recommended set of provisions/rules which are indicative of good practice, and disclosure is required if there is a reason why they are not appropriate in a specific situation.

Good corporate governance depends on directors behaving in the best interest of shareholders. Corporate governance mechanisms to achieve this include legislation, corporate governance codes, appointment of NEDs, shareholder activism and audit. Such mechanisms are necessary when companies are financed largely by equity capital. It is noticeable that they are being developed in many countries in response to wider share ownership.

Corporate governance best practice is being regularly reviewed and improved internationally.

REVIEW QUESTIONS

1 Explain in your own words what you understand corporate governance to mean.

2 Explain why governance procedures may vary from country to country.

3 What are the implications of governance for audit practices?

4 Auditors should take a more combative position and start with presumptive doubt and a more sceptical frame of mind, even though past experience of the FD and client staff has never revealed any cause for suspicion. Discuss the extent to which the requirement to adopt a different approach will increase the auditor's responsibility for detecting fraud.

5 The Association of British Insurers held the view that options should be exercised only if the company's earnings per share growth exceeded that of the Retail Price Index. The National Association of Pension Funds preferred the criterion to be a company's outperformance of the FTA All-Share Index.

(a) Discuss the reasons for the differences in approach.

(b) Discuss the implication of each approach to the financial reporting regulators and the auditors.

6 Research[35] suggests that companies whose managers own a significant proportion of the voting share capital tend to violate the UK Corporate Governance Code recommendations on board composition far more frequently than other companies. Discuss the advantages and disadvantages of enforcing greater compliance.

7 'Good corporate governance is a myth – just look at these frauds and irregularities:

- Enron www.sec.gov/litigation/litreleases/lr18582.htm
- WorldCom www.sec.gov/litigation/litreleases/lr17588.htm
- Xerox Corporation www.sec.gov/litigation/complaints/complr17465.htm
- Dell www.sec.gov/news/press/2010/2010-131.htm
- Lehman http://lehmanreport.jenner.com/VOLUME%201.pdf'

How realistic is it to expect good governance to combat similar future behaviour?

8 'Stronger corporate governance legislation is emerging globally but true success will only come from self-regulation, increased internal controls and the strong ethical corporate culture that organisations create.' Discuss.

9 In the modern commercial world, auditors provide numerous other services to complement their audit work. These services include the following:

(a) Accountancy and book-keeping assistance, e.g. in the maintenance of ledgers and in the preparation of monthly and annual accounts.

(b) Consultancy services, e.g. advice on the design of information systems and organisational structures, advice on the choice of computer equipment and software packages, and advice on the recruitment of new executives.

(c) Investigation work, e.g. appraisals of companies that might be taken over.

(d) Taxation work, e.g. tax planning advice and preparation of tax returns to HM Revenue and Customs for both the company and the company's senior management.

Discuss:

(i) Whether any of these activities is unacceptable as a separate activity because it might weaken an auditor's independence.

(ii) The advantages and disadvantages to the shareholders of the audit firm providing this range of service.

10 The following is an extract from the *Sunday Times* of 8 March 2009:

> Marc Jobling, the ABI's assistant director of investment affairs, said: 'Pay consultants are a big contributor to the problems around executive pay. We have heard of some who admit that they work for both management and independent directors – which is a clear conflict of interest and not acceptable. We believe that remuneration consultants, whose livelihood appears to depend on pushing an ever-upward spiral in executive pay, should be obliged to develop a code of ethics.'

Discuss the types of issues which should be included in such a code of ethics and how effective they would be in achieving good corporate governance.

11 There has been much criticism of the effectiveness of non-executive directors following failures such as Enron. Some consider that their interests are too close to those of the executive directors and they have neither the time nor the professional support to allow them to be effective monitors of the executive directors. Draft a job specification and personal criteria that you think would allay these criticisms.

12 In 2000, the chairman of the US Securities and Exchange Commission (SEC), Arthur Levitt, proposed that other services provided by audit firms to their audit clients should be severely restricted, probably solely to audit and tax work.[36] Discuss why this has still not happened.

13 Discuss how remuneration policies may adversely affect good corporate governance and how these effects may be reduced or prevented.

14 Discuss the major risks which will need to be managed by a pharmaceutical company and the extent to which these should be disclosed.

15 Egypt is a country in which many of the public companies have substantial shareholders in the form of founding families or government shareholders. How do you think that would affect corporate governance?

16 'Management will become accountable only when shareholders receive information on corporate strategy, future-based plans and budgets, and actual results with explanations of variances.' Discuss whether this is necessary, feasible and in the company's interest.

17 The Chartered Institute of Management Accountants (CIMA) has warned that linking directors' pay to EPS or return on assets is open to abuse, since these are not the objective measures they might appear.

    (a) Identify four ways in which the directors might manipulate the EPS and return on assets without breaching existing standards.

    (b) Suggest two alternative bases for setting criteria for bonuses.

18 Review reporting requirements in relation to disclosure of related party transactions and discuss their adequacy in relation to the avoidance of conflicts of interest.

19 Discuss in what situations audit independence could be compromised.

20 In 2012 the Chancellor unveiled a £100m 'employee–owner' scheme that will allow shares worth £2,000 to £50,000 to be exempt from tax if employees give up certain work rights, such as the right to claim unfair dismissal. He said the measure was aimed at the tens of thousands of small and medium-sized firms.

Discuss the effect of this proposal on corporate governance. Would your view change if the company should be badly managed and trading poorly?

21 It has been suggested that an Investors Forum will make mangament more accountable to the shareholders. Discuss how this might be achieved when shareholdings are so widely held.

22 It has been suggested that there would be less of an expectation gap if there were to be a note to the accounts giving in relation to those assets and liabilities which involved estimates the range of values and confidence level in the reported figure – for example, land, inventory, trade receivables.

Discuss the pros and cons of this suggestion from the viewpoint of the shareholder and the auditor.

23 Access the FRC Guidance on the Strategic Report issued in June 2014 and refer to section 7 'The strategic report: content elements'. Then select a set of published accounts and review the extent to which the company has satisfied the guidance. https://www.frc.org.uk/Our-Work/Publications/Accounting-and-Reporting-Policy/Guidance-on-the-Strategic-Report.pdf

EXERCISES

Question 1

Manufacturing Co. has been negotiating with Fred Paris regarding the sale of some property that represented an old manufacturing site which is now surplus to requirements. Because part of the site was used for manufacturing, it has to be decontaminated before it can be subdivided as a new housing development. This has complicated negotiations. Fred is a property developer and has a private company (Paris Property Development Pty Ltd) and is also a major (15%) shareholder of FP Development of which he is chairman. The negotiators for Manufacturing Co. note that the documents keep switching between Paris Property Development and FP Development and they use that as feedback as to how well they are negotiating.

Required:
Is there a corporate governance failure? Discuss.

Question 2

Harvey Storm is chief executive of West Wing Savings and Loans. Harvey authorises a loan to Middleman Properties secured on the land it is about to purchase. Middleman Properties has little money of its own. Middleman Properties subdivides the land and builds houses on them. It offers buyers a house and finance package under which West Wing provides the house loans up to 97% of the house price even to couples with poor credit ratings. This allows Middleman Properties to ask for higher prices for the houses.

Middleman Properties appoints Frontman Homes as the selling agent who kindly provides buyers with the free services of a solicitor to handle all the legal aspects including the conveyancing. Most of the profits from the developments are paid to Frontman Homes as commissions. Harvey Storm's wife has a 20% interest in Frontman Homes.

Required:
Are these corporate governance failures? Discuss.

Question 3

Conglomerate plc was a family company which was so successful that the founding Alexander family could not fully finance its expansion. So the company was floated on the Stock Exchange with the Alexander family holding 'A' class shares and the public holding 'B' class shares. 'A' class shares held the right to appoint six of the eleven directors. 'B' class shares could appoint five directors and had the same dividend rights as the 'A' class shares. The company could not be wound up unless a resolution was passed by 75% or more of 'A' class shareholders.

Required:
Is there any risk of a governance failure? Discuss.

Question 4

The board of White plc is discussing the filling of a vacant position arising from the death of Lord White. A list of possible candidates is as follows:

(a) Lord Sperring, who is a well-known company director and who was the managing director of Sperring Manufacturers before he switched to being a professional director.

(b) John Spate, B.Eng., PhD, who is managing director of a successful, innovative high-technology company and will be taking retirement in four months' time.

(c) Gerald Stewart, B. Com, who is the retired managing director of Spry and Montgomery advertising agency which operates in six countries, being the UK and five other Commonwealth countries.

The managing director leads the discussion and focuses on the likelihood of the three candidates being able to work in harmony with other members of the board. He suggests that John Spate is too radical to be a member of the board of White plc. The other members of the board agree that he has a history of looking at things differently and would tend to distract the board.

The chairman of the board suggests that Lord Sperring is very well connected in the business community and would be able to open many doors for the managing director. It was unanimously agreed that the chairman should approach Lord Sperring to see if he would be willing to join the board.

Required:
Critically discuss the appointment process.

Question 5

(a) Describe the value to the audit client of the audit firm providing consultancy services.

(b) Why is it undesirable for audit firms to provide consultancy services to audit clients?

(c) Why do audit firms want to continue to provide consultancy services to audit clients?

Question 6

How is the relationship between the audit firm and the audit client different for:

(a) the provision of statutory audit when the auditor reports to the shareholders;

(b) the provision of consultancy services by audit firms?

Question 7

Why is there a prohibition of auditors owning shares in client companies? Is this prohibition reasonable? Discuss.

References

1 C. Oman (ed.), *Corporate Governance in Development: The Experiences of Brazil, Chile, India and South Africa*, OECD Development Centre and Center for International Private Enterprise, Paris and Washington, DC, 2003, cited in N. Meisel, *Governance Culture and Development*, Development Centre, OECD, Paris, 2004, p. 16.
2 www.oecd.org/dataoecd/28/62/38699164.pdf
3 www.sec.gov/litigation/litreleases/2009/lr21350.htm
4 http://cphpost.dk/business/ruin-bank-and-earn-ten-million-kroner
5 http://www.heraldtribune.com/article/20100719/COLUMNIST/7191018
6 An example of this is the convergence of computing, telephone, television and entertainment markets as new devices impinge on all fields compared to ten years ago when they were quite distinct fields.
7 https://na.theiia.org/standards-guidance/Public%20Documents/MODEL_AUDIT_COMMITTEE_CHARTER.pdf
8 http://scholarship.law.duke.edu/cgi/viewcontent.cgi?article=3074&context=faculty_scholarship

9 A. Cornford, 'Enron and internationally agreed principles for corporate governance and the financial sector', G-24 Discussion Paper Series, United Nations.

10 ISA 200, *Overall Objectives of the Independent Auditor and the Conduct of an Audit in Accordance with International Standards on Auditing*, paragraph 13(l).

11 www.sec.gov/litigation/admin/3438494.txt

12 A.R. Wyatt (2003), 'Accounting professionalism – they just don't get it!', http://aaahq.orgAM2003/WyattSpeech.pdf

13 http://lehmanreport.jenner.com/

14 http://www.bis.gov.uk/news/speeches/vince-cable-executive-pay-remuneration-2012

15 The Large and Medium-sized Companies and Groups (Accounts and Reports) (Amendment) Regulations 2013 http://www.legislation.gov.uk/uksi/2013/1981/schedule/made

16 L. Bebchuk, M. Cremers and U. Peyer, 'Higher CEO salaries don't always pay off', *The Australian Financial Review*, 12 February 2010, p. 59.

17 https://www.ivis.co.uk/guidelines

18 http://www.napf.co.uk/PolicyandResearch/DocumentLibrary/~/media/Policy/Documents/0351_3_remuneration_principles_for_building_and_reinforcing%20_longterm_business_success_nov2013.pdf

19 www.ecgi.org

20 M. Gold, 'Corporate governance reform in Australia: the intersection of investment fiduciaries and issuers', in P. Ali and G. Gregoriou (eds), *International Corporate Governance after Sarbanes–Oxley*, John Wiley and Sons, New York, 2006.

21 http://www.frc.org.uk/Our-Work/Codes-Standards/Corporate-governance/UK-Corporate-Governance-Code.aspx

22 E. Fama, 'Agency problems and the theory of the firms', *Journal of Political Economy*, vol. 88, 1980, pp. 288–307.

23 E. Fama and M. Jensen, 'Separation of ownership and control', *Journal of Law and Economics*, vol. 26, 1983, pp. 301–325.

24 http://www.bupa.com/investor-relations/our-status-and-governance/our-corporate-governance/role-of-the-sid

25 http://www.guardian.co.uk/business/2009/feb/01/ftse-royal-bank-scotland-group

26 A. Abdullah and M. Page, *Corporate Governance and Corporate Performance: UK FTSE 350 Companies*, The Institute of Chartered Accountants of Scotland, Edinburgh, 2009.

27 ABI Research, *Corporate Governance 'Pays' for Shareholders and Company Performance*, ABI, 27 February 2008, Ref: 12/08.

28 http://www.ons.gov.uk/ons/rel/pnfc1/share-ownership---share-register-survey-report/2010/stb-share-ownership-2010.html

29 G. Igor Filatochev, H.G. Jackson and D. Allcock, *Key Drivers to Good Corporate Governance and Appropriateness of UK Policy Responses*, DTI, 2007, www.berr.gov.uk/files/file36671.pdf

30 www.hm-treasury.gov.uk/walker_review_information.htm

31 http://www/frc.org.uk/Our-Work/Codes-Standards/Corporate-governance/UK-Corporate-Governance-Code.aspx

32 http://www.bis.gov.uk/assets/biscore/business-law/docs/k/12-917-kay-review-of-equity-markets-final-report.pdf

33 file:///C:/Users/Barry/Downloads/20131203-cewginvestorforum%20(1).pdf

34 http://europa.eu/rapid/press-release_STATEMENT-14-104_en.htm

35 K. Peasnell, P. Pope and S. Young, 'A new model board', *Accountancy*, July 1998, p. 115.

36 'PwC and E&Y in favour of rules to restrict services', *Accountancy*, October 2000, p. 7.

CHAPTER 32

Sustainability – environmental and social reporting

32.1 Introduction

In previous chapters we have discussed the constituent parts of an Annual Report which included:

- highlights such as an introduction to the business model, KPIs and brief comment on performance;
- Chairman's Statement explaining how effective the business has been in achieving the strategy;
- Strategic Report setting out the business model, strategy, significant business risks, review of the year's performance and comments on the future;
- corporate governance explaining how the various boards and committees are run;
- Directors' Report to satisfy statutory requirements;
- Director's Remuneration Report to satisfy statutory requirements;
- Auditor's Report on the financial statements;
- the financial statements.

We have also discussed the steps taken to achieve improvements in each of these areas with, for example:

- increased mandatory requirements with improvements by the IAASB in audit reporting to address the expectations gap;
- improved corporate governance disclosures with the various Corporate Governance and Stewardship Codes;
- development by the IASB of a Conceptual Framework to provide a conceptual basis in the future for IFRSs affecting the financial statements;
- increased statutory requirements addressing concerns about excessive directors' rewards with a more transparent Director's Remuneration Report;
- increased statutory requirements for a Strategic Report.

The Strategic Report

This explains the business model, the objectives and strategies, the principal risks and how these are addressed and future prospects. The business model employs various capitals for its inputs and converts these into outputs of goods and services. A business has to be aware of the threats and opportunities in relation to both its inputs and its output. These may be industry- and/or company-specific. The proposal is that there should be a Framework within which the inputs or capitals are considered by management. This is provided by the Integrated Reporting Framework.

The Integrated Reporting Framework[1]

This describes the following capitals that a business might employ in achieving outputs from its business model. These are:

- financial capital, i.e. the resources that it controls which is represented by the net assets reported in a statement of financial position;

- manufactured capital, i.e. the non-current assets and finished goods in its inventory reported within the resources;

- intellectual capital, i.e. intangible assets which are reported in the statement of financial position and also organisational intellectual capital in the form of systems and procedures that do not satisfy the Conceptual Framework definition of an asset;

- human capital, which is dependent on the way in which they carry out their work and the degree to which they identify with culture and support the business model. Their contribution does not fall within the Conceptual Framework definition of an asset;

- social and relationship capital, i.e. customer-based intangibles such as brands that may be reported in the statement of financial position and key stakeholder relationships with shareholders, employees and society;

- natural capital, i.e. the renewable (forests) and non-renewable (gas, oil, mineral) environmental resources that a business uses and which are reported as expenses in the Statement of income.

It aims to explain how the company interacts with the external environment and the capitals to create value over the short term, medium term and long term.

We have discussed financial, manufactured and intellectual capital in previous chapters. In this chapter we will look at developments in environmental and social reporting.

Early stages of environmental and social reporting

In the early stages of environmental, social and sustainability reporting the emphasis was on reporting to *other stakeholders*. It is interesting to see that sustainability and corporate governance are beginning to merge as companies see that the two do not have separate audiences. We are at the beginning of a move towards integrated financial reporting.

Reports that started out as PR exercises are now gradually becoming valued by investors and analysts who are requiring a greater degree of detail. It has become important for companies to be transparent about their corporate responsibility and sustainability efforts if they are to remain competitive in a world with a growing number of informed, concerned consumers.

In Part 6 on Interpretation we saw that there is a well-understood process of presenting ratios with the ability to benchmark by inter-period, inter-company and industry comparators. We have not yet achieved this with regard to benchmarking a company's environmental performance. It is still not possible to make comparisons between companies, and attention has concentrated on the comparative year-on-year percentage changes in annual published reports.

Objectives

By the end of this chapter, you should be able to:

- discuss development of Integrated Financial Reporting;
- discuss an overview of the development of corporate social reporting;
- discuss the evolution of sustainability reporting including:
  - the Global Reporting Initiative (GRI);
  - the Eco-Management and Audit Scheme (EMAS);
  - the International Organization for Standardization (ISO);
- discuss the accountant's role in a capitalist industrial society;
- discuss the accountant's role in sustainability reporting;
- discuss the evolution of social accounting in the annual report;
- discuss corporate social responsibility reporting;
- comment critically on all of the above areas.

32.2 An overview – stakeholders' growing interest in corporate social responsibility (CSR)

32.2.1 Primary stakeholders

When corporate bodies were first created the primary stakeholders were the shareholders who had invested the capital and it was seen as the directors' responsibility to maximise their return by way of dividends and capital growth. This view was promoted by Milton Friedman writing that:

> few trends would so thoroughly undermine the very foundations of our free society as the acceptance by corporate officials of a social responsibility other than to make as much money for their shareholders as they possibly can.

It follows from this that directors were accountable to the shareholders who in turn should hold them to account. The Friedman approach offered protection for shareholders provided they actually did exercise their ability to hold directors accountable. However, it did not have regard to the interests of any other group affected by a company's decisions, such as employees, suppliers, consumers or the community for the environmental and social impact of a company's operations unless there was a direct financial benefit or risk to the company.

32.2.2 Other stakeholders

Since Friedman's writing in the 1960s companies have been under pressure to be accountable to a growing number of stakeholders. The pressure can be seen to come from various quarters, such as:

- European Union Directives, e.g. the Landfill Directive whose aim was to prevent or reduce negative effects on the environment;
- national legislation affecting financial reports, e.g. in the UK the Companies Act 2006 has a requirement that the business/strategic review in the Annual Report must include information about:
  - environmental matters (including the impact of the company's business on the environment);

- the company's employees; and
- social and community issues;
- investor pressure, e.g. see the European Sustainable Investment Forum[2] which is a pan-European network and think-tank whose mission is to develop sustainability through European financial markets; and
- consumers – although consumers are currently largely unaware that negative production externalities are not internalised and that this is why products are cheaper than they should otherwise be; basically, pricing today does not reflect the true environmental cost of production. In the future, an increased ability for consumers to understand the negative environmental impacts associated with production may well influence their purchasing choices. Responsible consumerism will impact on business and will be both a threat and an opportunity.

32.3 An overview – business's growing interest in corporate social responsibility

Companies are now under pressure to act responsibly in their relationships with other stakeholders who have a legitimate interest in the business. Although there was a fear within companies that their financial performance would be damaged if public costs and other stakeholder interests were taken into account, societal pressure has grown since the 1990s.

Management response has polarised between those companies that see environmental expenditure as a means of improving financial performance by gaining a competitive edge and those that act with non-profit motives. Whatever their view, CSR is on the international agenda. For example, the following is a quote from the World Business Council for Sustainable Development:

> CSR is the continuing commitment by business to behave ethically and contribute to economic development while improving the quality of life of the workforce and their families as well as of the local community and society at large.

There are three interesting points to highlight in this quotation. The first is the reference to behaving ethically, the second is the acknowledgement that a company has an economic objective, and the third is the extension to improve the quality of life of other stakeholders which includes environmental and social impacts.

Need to trade profitably

Whilst accepting these three objectives there is an underlying need for a company to trade profitably and make an economic return if it is to be in a position to satisfy the environmental and social benefits. This is reflected in the following extract from the 2013 Annual Report of Imperial Holdings Limited, a company listed on the Johannesburg stock exchange:

Sustainability, business integrity and ethics
The board has adopted a written code of ethics for the group, to which all operations are required to adhere.

Without satisfactory profits and a strong financial foundation, it would not be possible to fulfill our responsibilities to shareholders, employees, society and those with whom we do business.

However, our corporate actions are not governed solely by economic criteria and take into account social, environmental and political consideration.

32.3.1 The executives' commitment to sustainability

It is interesting to note that a survey, *A New Era of Sustainability: UN Global Compact-Accenture CEO Study 2010*,[3] reported that 93% of CEOs surveyed say that sustainability will be critical to the future success of their companies and will be fully integrated within a decade.

32.3.2 The board's commitment to sustainability

As far back as 1975 there have been various initiatives in the UK such as the Corporate Report proposing the disclosure of additional information, such as employment and value added reports.

External corporate reporting has been evolving from the simple financial reporting of profits and losses, assets and liabilities to, for example, the *ad hoc* inclusion of information on governance (e.g. disclosure of directors' remuneration), as well as non-financial information such as environmental and social policies.

For the board it is a balance between improving the total shareholder return (TSR) and acting as good corporate citizens. However, adopting environmental and social policies need not mean a reduction in TSR.

Adopting environmental and social policies improves stock market performance

This reflects an understanding as evidenced by research[4] that companies that voluntarily adopted environmental and social policies by 1993 significantly outperform their counterparts over the long term, in terms of both stock market and accounting performance. The evidence indicated that the outperformance is stronger in sectors where the customers are individual consumers, where companies compete on the basis of brands and reputation, and in sectors where companies' products significantly depend upon extracting large amounts of natural resources.

Board composition affects environmental policies

We talk about the board as a unified body. It is, of course, made up of a group of individuals who might well hold different views on committing resources for what they might perceive to be simply a public good. Indeed, with regard to the composition of the board, research[5] indicates that having a higher proportion of outside board directors and having three or more female directors seem to have some positive effect on environmental performance.

Another interesting research finding[6] was that there was a negative relation between board members who had shown past unethical risk behaviour and the environmental performance of the firm. Boards with a higher proportion of risk-prone, unethical members seem to focus less on the environmental concerns of their businesses, which is probably what one would intuitively feel.

If the share capital is provided by a few, the impact of the expenditure becomes far more personal and the board could be under pressure with a reluctance to regard sustainability policies as a source of cost-efficiencies and revenue growth, rather seeing them as an avoidable cost.

32.3.3 Reporting environmental performance and impact in the annual report

Narrative references to environmental issues have appeared on an *ad hoc* basis in annual reports. In some jurisdictions there may be, as in the UK, a statutory requirement for a director to act in the way most likely to promote the success of the company for the benefit of its members as a whole, and in doing so to have regard to the interests of the company's employees, the need to foster the company's business relationships with suppliers, customers and others, and the impact of the company's operations on the community and the environment.

Danger of information overload

When considering inclusion of CSR reporting in the annual report, one of the problems has been the volume of data. Companies are overcoming this by issuing summary CSR reports in hard copy and uploading the full CSR report onto their websites. The following is an extract[7] relating to experience in the Netherlands:

> In the Netherlands, various companies are aiming for further integration of the CSR information into their annual reports. Companies are increasingly using the Internet in order to reduce the size of their CSR reports. They publish hard copy summary reports, with the full versions available on the Internet.

A survey[8] commissioned by the Global Reporting Initiative (GRI) and The Prince's Accounting for Sustainability Project (A4S) commented that the future steps to be taken should include developing a considered and targeted approach to online investor and analyst communication, focusing on how users interact with online content and the usability of this source of information.

Evolution of stand-alone environmental reports

Companies might find that the environmental data are already so integrated that a single combined report is advisable, or decide that a separate report has the advantage of raising more awareness both in the company and with other stakeholders.

Jurisdictions where there is a legislative requirement

There is legislation in some jurisdictions, such as Denmark, the Netherlands, Norway and Sweden, requiring environmental statements from environmentally sensitive industries either in their financial statements or in a stand-alone report; in other countries, voluntary disclosures are proposed.

The following is an extract relating to a Danish experience[9] where most companies now produce separate Corporate Social Responsibility reports:

- Demonstrate upside or business potential stemming from the sustainability agenda, and how the enterprises respond to this;
- Demonstrate how CSR is integrated into the way the companies run their businesses.

As a result, these companies have fully integrated non-financial data with financial data in the report. In doing this, the companies clearly demonstrate their full understanding of the value of reporting on non-financial data. They demonstrate that the data are being used as a serious management and communication tool and that they are able to link CSR to business strategy.

Publishing an Environmental Profit and Loss Account

This is a step forward. Developments in financial accounting and reporting remain dynamic and it is fascinating to see the wealth of academic research and business innovation. Many companies are actively engaging with ways in which to report the environmental impact of their operations. One such is the publication[10] by Puma in 2011 of an Environmental Profit and Loss Account which identifies the reasons and the cost of £145m as the environmental impact. It is well worth accessing the corporate site for a more detailed look at its approach. It was not surprising that Puma was also finalist in the 2012 Finance for the Future Awards.

32.4 Companies' voluntary adoption of guidelines and certification

Although there are no mandatory reporting standards, there are a number of schemes such as the Connected Reporting Framework, the Global Reporting Initiative (GRI), the Eco-Management and Audit Scheme (EMAS) and the International Organization for Standardization (ISO).

32.4.1 The Connected Reporting Framework[11]

The Connected Reporting Framework has been adopted by a range of organisations including Aviva, BT, EDF Energy and HSBC. It explains how all areas of organisational performance can be presented in a connected way, reflecting the organisation's strategy and the way it is managed.

It proposes five key environmental indicators, which all organisations should consider reporting: polluting emissions, energy use, water use, waste and significant use of other finite resources, and the inclusion of industry benchmarks, when available, for key perform-ance indicators, to aid performance appraisal. It is seeking consistency in presentation to aid comparability between years and organisations.

The following is an extract from the 2012 Aviva plc Annual Report:

Accounting for Sustainability
We also report our performance using Accounting for Sustainability's connected reporting framework, which integrates financial and non-financial data to provide a comprehensive picture of our impacts.

We were one of the first companies to help develop the framework and have used this approach for environmental reporting in our Annual Report and Accounts since 2007. We continue to explore ways to extend this framework and have included customer and community indicators since 2009. We have reported the following indicators for 2012: Greenhouse gas emissions, Waste, Resource usage, Customer advocacy and Community development

32.4.2 The Global Reporting Initiative (GRI)[12]

The GRI has a mission to develop global sustainability reporting guidelines for voluntary use by organisations reporting on the three linked elements of sustainability, namely the economic, environmental and social dimensions of their activities, products and services.

- The *economic dimension* includes financial and non-financial information on R&D expenditure, investment in the workforce, current staff expenditure and outputs in terms of labour productivity.

- The *environmental dimension* includes any adverse impact on air, water, land, biodiversity and human health by an organisation's production processes, products and services.

- The *social dimension* includes information on health and safety and recognition of rights, e.g. human rights for both employees and outsourced employees.

The Global Reporting Initiative means that parties contemplating a relationship such as assessing investment risk and obtaining goods or services will have available to them a clear picture of the human and ecological impact of the business. Its influence has been growing and there are jurisdictions, such as that of Sweden, that require a GRI report.

GRI reports have assurance, having been verified by independent, competent and impartial external assurance providers. The assurance providers themselves now have a standard – the AA1000 Assurance Standard.[13] As an example, in the 2011 Aviva Annual Report, Ernst &

Young, who were the assurance providers, stated that they were forming a conclusion on matters such as

- inclusivity – whether Aviva had engaged with stakeholders across the business to further develop its approach to corporate responsibility;

- materiality – whether Aviva had provided a balanced representation of material issues concerning Aviva's corporate responsibility performance;

- completeness – whether Aviva had complete information on which to base a judgement of what was material for inclusion in the Report; and

- responsiveness – whether Aviva had responded to stakeholder concerns.

32.4.3 The Eco-Management and Audit Scheme (EMAS)[14]

The Eco-Management and Audit Scheme (EMAS) was adopted by the European Council in 1993, allowing voluntary participation in an environmental management scheme. Its aim is to promote continuous environmental performance improvements of activities by committing organisations to evaluating and improving their own environmental performance.

Just as with GRI reports, verification is seen as an important element and environmental audits, covering all activities at the organisation concerned, must be conducted within an audit cycle of no longer than three years.

In addition to helping internal management, the adoption of such a scheme also assists external auditors when reviewing compliance by the company with legislation.

32.4.4 The International Organization for Standardization (ISO)[15]

The ISO is a non-governmental organisation whose aim is to establish international standards to reduce barriers to international trade. Its standards, including environmental standards, are voluntary and companies may elect to join in order to obtain ISO certification.

One group of standards, the ISO 14000 series, is intended to encourage organisations to systematically assess the environmental impacts of their activities through a common approach to environmental management systems. Within the group, the ISO 14001 standard states the requirements for establishing an EMS and companies must satisfy its requirements in order to qualify for ISO certification.

Benefits from ISO certification

These include:

- *Top-level management become involved.* They are required to define an overall policy and, in addition, they recognise significant financial considerations from certification, e.g. customers might in the future prefer to deal with ISO-compliant companies, insurance premiums might be lower and there is the potential to reduce costs by greater production efficiency.

- *Environmental management.* ISO 14001 establishes a framework for a systematic approach to environmental management which can identify inefficiencies that were not apparent beforehand, resulting in operational cost savings and reduced environmental liabilities. Aviva, for example, which we mentioned above, reduced its water consumption by over 25% in 2011.

What criticisms are there of a compliance approach?

Compliance approaches which set out criteria such as a commitment to minimise environmental impact can allow companies to set low objectives for improvement and report these as achievements with little confidence that there has been significant environmental benefit.

This is gradually being addressed by industry comparison and by governments also seeking to produce targets and benchmarks.

32.5 The accountant's role in a capitalist industrial society

Shareholder interest initially dominant

In a capitalist, industrial society, production requires the raising and efficient use of capital largely through joint stock companies. These operate within a legal framework which grants them limited liability subject to certain obligations. The obligations include capital maintenance provisions to protect creditors (e.g. restriction on distributable profits) and disclosure provisions to protect shareholders (e.g. the publication of annual reports).

Accountants issue standards to ensure there is reliable and relevant information to the owners to support an orderly capital market. This has influenced the nature of accounting standards, with their concentration on earnings and monetary values.

Other stakeholder interests gradually recognised

However, production and distribution involve complex social relationships between private ownership of property and wage labour[16] and other stakeholders. This raises the question of the role of accountants. Should their primary concern be to serve the interests of the shareholders, or the interests of management, or to focus on equity issues and social welfare?[17]

Pressure for reports to reflect substance

Prior to the formation in the UK of the ASB, the profession identified with management and it was not unusual to allow information to be reported to suit management. If managers were unhappy with a standard, they were able to frustrate or delay its implementation. This meant that the reported results might bear little resemblance to the commercial substance of the underlying transactions.

Creating standards relating to environmental liabilities and assets

Reporting environmental liabilities – IFRSs

Possible environmental liabilities that give rise to a provision include waste disposal, pollution, decommissioning and restoration expenses. There may also be liabilities arising from participation in a specific market, such as vehicle production or the manufacture of electrical and electronic equipment.

We have seen in Chapter 13 that a provision is recognised when an entity has a present obligation as a result of a past event, it is probable that a transfer of economic benefits will be required to settle the obligation, and a reliable estimate can be made of the amount of the obligation (IAS 37).

This information can be commercially important to potential corporate investors as, for example, when deciding on the price to pay to acquire shares on a takeover. For example, acquisitive companies needed to be aware of contingent environmental liabilities,[18] which can be enormous. In the USA the potential cost of clearing up past industrially hazardous sites has been estimated at $675 billion.

Even in relation to individual companies the scale of the contingency can be large, as in the Love Canal case. In this case a housing project was built at Love Canal in upper New York State on a site that until the 1950s had been used by the Hooker Chemicals Corporation for dumping chemical waste containing dioxin. Occidental, which had acquired Hooker Chemicals, was judged liable for the costs of clean-up of more than $260 million.[19] Existing shareholders and the share price would also be affected by these increased costs.

Reporting environmental assets – IFRSs

There was a question as to whether physical assets acquired to comply with environmental legislation could be classified as an asset for reporting purposes as they were not in isolation creating revenue or reducing costs. This was addressed by IAS 16 *Property, Plant and Equipment* which provides that some elements of tangible non-current assets can be acquired for reasons of safety or are environmental in nature. Although the acquisition of such property, plant and equipment does not in itself qualify them for recognition, they qualify as assets because they enable the institution to obtain additional economic benefits from the rest of its assets which would not have been earned otherwise.

32.6 The nature of the accountant's involvement

32.6.1 Finance directors

FDs have become increasingly involved internally in dealing with issues such as recommending which voluntary code to adopt. Voluntary codes have developed in a largely unstructured way and accountants may be involved in identifying those codes which are appropriate to the business, such as GRI and EMAS.

32.6.2 Accounting staff and systems design

Responding to public interest in the ethical sourcing of raw materials and products, FDs have also become responsible for the design and monitoring of purchasing policies to establish an audit trail. This will become more important as more attention is focused on sourcing. For example, the UN Global Compact[20] encourages signatories to engage with their suppliers around the Ten Principles, and thereby to develop more sustainable supply chain practices. However, many companies lack the knowledge or capacity to effectively integrate the principles into their existing supply chain programmes and operations. In particular, a challenge remains to ensure that sustainability considerations are embedded within all sourcing processes.

32.6.3 Accounting staff and routine collection of data

In addition, there will be other typical inputs from accountants in each of the three elements, often requiring a greater degree of quantification than at present for the economic and environmental dimensions. For example:

- The economic dimension may require economic indicators such as:
  - profit: segmental gross margin, EBITDA, EBIT, return on average capital employed;
  - intangible assets: ratio of market valuation to book value;
  - investments: human capital, R&D, debt/equity ratio;
  - wages and benefits: totals by country;
  - labour productivity: levels and changes by job category;
  - community development: jobs by type and country showing absolute figures and net change;
  - suppliers: value of goods and services outsourced, performance in meeting credit terms.
- The environmental dimension may require environmental indicators such as:
  - products and services: major issues, e.g. disposal of waste, paper usage, packaging practices, percentage of product reclaimed after use;

- suppliers: supplier issues identified through stakeholder consultation, e.g. forest stewardship, sustainable logging;

- travel: objectives and targets, e.g. product distribution, fleet operation, quantitative estimates of miles travelled by transport type.

- Social dimensions may require social indicators such as:

  - quality of management: employee retention rates, ratio of jobs offered to jobs accepted, ranking as an employer in surveys;

  - health and safety: reportable cases, lost days, absentee rate, investment per worker in injury prevention;

  - wages and benefits: ratio of lowest wage to local cost of living, health and pension benefits provided;

  - training and education: ratio of training budget to annual operating cost, programmes to encourage worker participation in decision making;

  - freedom of association: grievance procedures in place, number and types of legal action concerning anti-union practices.

32.6.4 Auditors

Auditors will be increasingly required to provide assurance on the application of standards in the supply chain.[21]

There are mixed feelings towards the need for assurance. Companies that are committed to the voluntary codes (GRI, EMAS, ISO 14001) are more supportive of the need than other companies that are not as familiar with assurance reports. There has, however, been a growing pressure for CSR information to be subject to assurance reports to give stakeholders the same confidence as they have obtained from the audit reports on financial statements. Challenges for UK companies in applying the AA1000 Assurance Standard (2008) include:

- providing more detailed commentary on methodology and recommendations in the statement; and

- focusing more on the materiality, completeness and responsiveness principles rather than just simply checking accuracy of information.[22]

32.6.5 Environmental auditing – international initiatives

Europe

Since 1999 the European Federation of Accountants (Fédération des Experts Comptables Européans – FEE) Sustainability Working Party (formerly Environmental) has been active in the project Providing Assurance on Environental Reports[23] and is actively participating with other organisations and collaborating on projects such as with GRI Sustainability Guidelines.

Canada

It is interesting to see the multidisciplinary approach that is now being taken. For example, the Canadian Environmental Auditing Association (CEAA)[24] is a multidisciplinary organisation whose international membership base now includes environmental managers, ISO 14001 registration auditors, EMS (Eco-Management Systems) consultants, **corporate environmental auditors**, engineers, chemists, government employees, **accountants** and lawyers. The CEAA is now accredited by the Standards Council of Canada as a certifying body for EMS Auditors.

32.6.6 The profession

All of the professional accounting bodies are active in various ways in promoting developments in reporting on environmental and sustainability issues. We briefly comment on a couple of the ICAEW and ACCA activities.

ICAEW – Environmental Issues and Annual Financial Reporting

In 2009 the report *Turning Questions into Answers: Environmental Issues and Annual Financial Reporting*[25] was issued jointly by the ICAEW and the Environment Agency. It identified the main concerns of the users regarding environmental information currently provided in annual reports to be in the areas of:

- Consistency and comparability – in the absence of well-defined disclosure requirements, there is a lack of consistency and comparability.
- Relevance and usefulness – the lack of focus on environmental issues that are of critical importance to specific industry sectors. Disclosure of data does not necessarily reveal the effectiveness of a policy.
- Reliability and assurance – this is regarded as of secondary importance to the above qualities. There are also concerns as to the skills and experience required by assurance providers.
- Materiality – the business review requirement for disclosure of any material environmental issues is unsatisfactory without further guidance.
- Presentation – integration of material financial and non-financial information within annual reports is generally preferred. It will also assist international comparability.

These are all indicators of the need for further work by the accounting profession.

The ICAEW together with NatWest and The Prince's Accounting for Sustainability Project were founder members of the Finance for the Future Awards. These awards look at long-term sustainability from a finance perspective or celebrate accounting for sustainability as the key to creating an innovative business model. They focus in particular on the role that the finance team has to play in reaching these outcomes. One of the award's objectives is to share best practice through development of case studies and educational resources based on the experiences of the nominees and winners.

The ACCA Award schemes

One of the earliest of the award schemes was that of the ACCA. Its 'Awards for Sustainability Reporting' scheme has three award categories for Environmental Reporting, Social Reporting and Sustainability Reporting. Details of the UK and European Sustainability Reporting Award can be found on the ACCA website. Other schemes have been set up using the ACCA criteria,[26] such as the ACCA Malaysia Sustainability Reporting Awards (MaSRA) 2013 which was launched with the theme 'Diversity and Inclusion'. The ACCA MaSRA programme[26] recognises the best practices in sustainability reporting within corporate Malaysia. Since its inception in 2002, these Awards have served as an important platform to create awareness on corporate environment and social responsibility and benefits of reporting to business.

32.7 Summary on environmental reporting

Environmental reporting is in a state of evolution ranging from *ad hoc* comments in the annual report to a more systematic approach in the annual report to stand-alone environmental reports.

Environmental investment is no longer seen as an additional cost but as an essential part of being a good corporate citizen, and environmental reports are seen as necessary in communicating with stakeholders to address their environmental concerns.

Companies are realising that it is their corporate responsibility to achieve sustainable development whereby they meet the needs of the present without compromising the ability of future generations to meet their own needs. Economic growth is important for shareholders and other stakeholders alike in that it provides the conditions in which protection of the environment can best be achieved; and environmental protection, in balance with other human goals, is necessary to achieve growth that is sustainable.

However, there is still a long way to go and the EU's Sixth Action Programme 'Environment 2010: Our Future, Our Choice'[27] recognises that effective steps have not been taken by all member states to implement EC environmental directives and there is weak ownership of environmental objectives by stakeholders. In 2014 it issued its Seventh Action Plan which will guide European policy until 2020. This identifies three key objectives:

- to protect, conserve and enhance the Union's natural capital;
- to turn the Union into a resource-efficient, green, and competitive low-carbon economy;
- to safeguard the Union's citizens from environment-related pressures and risks to health and wellbeing.

32.8 Concept of social accounting

This is difficult because there are so many definitions of social accounting.[26] The main points are that it includes non-financial as well as financial information and addresses the needs of stakeholders other than the shareholders. Stakeholders can be broken down into three categories:

- internal stakeholders – managers and workers;
- external stakeholders – shareholders, creditors, banks and debtors; and
- related stakeholders – society as represented by national and local government and the increasing role of pressure groups such as Amnesty International and Greenpeace.

32.8.1 Reporting at corporate level

Prior to 1975, social accounting was viewed as being in the domain of the economist and concerned with national income and related issues. In 1975, *The Corporate Report* gave a different definition:

the reporting of those costs and benefits, which may or may not be quantifiable in money terms, arising from economic activities and subsequently borne or received by the community at large or particular groups not holding a direct relationship with the reporting entity.[28]

This is probably the best working definition of the topic and it establishes the first element of the social accounting concept, namely **reporting at a corporate level** and interpreting corporate in its widest sense as including all organisations of economic significance regardless of the type of organisation or the nature of ownership.

32.8.2 Accountability

The effect of the redefinition by *The Corporate Report* was to introduce the second element of our social accounting concept: accountability. The national income view was only of interest to economists and could not be related to individual company performance – *The Corporate Report* changed that. Social accounting moved into the accountants' domain and it should be the aim of accountants to learn how accountability might be achieved and to define a model against which to judge their own efforts and the efforts of others.

32.8.3 Comprehensive coverage

The annual report is concerned mainly with monetary amounts or clarifying monetary issues. Despite the ASB identifying employees and the public within the user groups,[29] no standards have been issued that deal specifically with reporting to employees or the public.

Instead, the ASB prefers to assume that financial statements that meet the needs of investors will meet most of the needs of other users.[30] For all practical purposes, it disassociates itself from the needs of non-investor users by assuming that there will be more specific information that they may obtain in their dealings with the enterprise.[31]

The information needs of different categories, e.g. employees and the public, need not be identical. The provision of information of particular interest to the public has been referred to as **public interest accounting**,[32] but there is a danger that, whilst valid as an approach, it could act as a constraint on matters that might be of legitimate interest to the employee user group. For example, safety issues at a particular location might be of little interest to the public at large but of immense concern to an employee exposed to work-related radiation or asbestos. The term 'social accounting' as defined by *The Corporate Report* is seen as embracing all interests, even those of a small group.

Equally, the information needs within a category – say, employees – can differ according to the level of the employees. One study identified that different levels of employee ranked the information provided about the employer differently, e.g. lower-level employees rated safety information highest, whereas higher-level employees rated organisation information highest.[33] There were also differences in opinion about the need for additional information, with the majority of lower-level but a minority of higher-level employees agreeing that the social report should also contain information on corporate environmental effects.[34]

The need for social accounting to cope with both inter-group and intra-group differences was also identified in a Swedish study.[33]

32.8.4 Independent review

The degree of credibility accorded a particular piece of information is influenced by factors such as whether it is historical or deals with the future; whether appropriate techniques exist for obtaining it; whether its source causes particular concern about deliberate or unintentional bias towards a company view; whether past experience has been that the information was reasonably complete and balanced; and, finally, the extent of independent verification.[35]

Given that social accounting is complex and technically underdeveloped, that it deals with subjective areas or future events, and that it is reported on a selective basis within a report prepared by the management, it is understandable that its credibility will be called into question. Questions will be raised as to why particular items were included or omitted – after all, it is not that unusual for companies to want to hide unfavourable developments.

32.9 Background to social accounting

A brief consideration of the history of social accounting in the UK could be helpful in putting the subject into context. *The Corporate Report* (1975) was the starting point for the whole issue.

It was a discussion paper which represented the first UK conceptual framework. Its approach was to identify users and their information needs. It identified seven groups of user, which included employees and the public, and their information needs. However, although it identified that there were common areas of interest among the seven groups, such as assessing liquidity and evaluating management performance, it concluded that a single set of general-purpose accounts would not satisfy each group – a different conclusion from that stated by the ASB in 1991, as discussed above.[29] The conclusions reached in *The Corporate Report* were influenced by the findings of a survey of the chairmen of the 300 largest UK listed companies. They indicated a trend towards acceptance of multiple responsibilities towards groups affected by corporate decision-making and their interest as stakeholders.[36]

It was proposed in *The Corporate Report* that there should be additional reports to satisfy the needs of the other stakeholders. These included a Statement of corporate objectives, an Employment report, a Statement of future prospects and a Value added statement.

Statement of corporate objectives

Would this be the place for social accounting to start? Would this be the place for vested interests to be represented so that agreed objectives take account of the views of all stakeholders and not merely the management and, indirectly, the shareholders? At present, social accounting appears as a series of add-ons, e.g. a little on charity donations, a little on disabled recruitment policy. Corporate objectives or the mission statement are often seen as something to be handed down; could they assume a different role?

Employment report

The need for an employment report was founded on the belief that there is a trust relationship between employers and employees and an economic relationship between employment prospects and the welfare of the community. The intention was that such a report should contain statistical information relating to such matters as numbers, reasons for change, training time and costs, age and sex distribution, and health and safety.

Statement of future prospects

There has always been resistance to publishing information focusing on the future. The arguments raised against it have included competitive disadvantage and the possibility of misinterpretation because the data relate to the future and are therefore uncertain. The writers of *The Corporate Report* nevertheless considered it appropriate to publish information on future employment and capital investment levels that could have a direct impact on employees and the local community.

Value added statement

A value added report was intended to give a different focus from the profit and loss account with its emphasis on the bottom-line earnings figure. It was intended to demonstrate the interdependence of profits and payments to employees, shareholders, the government and the company via inward investment. It reflected the mood picked up from the survey of chairmen that distributable profit could no longer be regarded as the sole or prime indicator of company performance.[36]

The value added statement became a well-known reporting mechanism to measure how effectively an organisation utilised its resources and added value to its raw materials to turn them into saleable goods. Figure 32.1 is an example of a value added statement.

Several advantages have been claimed for these reports, including improving employee attitudes by reflecting a broader view of companies' objectives and responsibilities.[37]

Figure 32.1 Imperial Holdings Ltd value added statement

for the year ended 30 June 2012

	2012 Rm	%	2011 Rm	%
Revenue	80,830		64,667	
Paid to suppliers for materials and services	62,699		49,933	
Total wealth created	**18,131**		14,734	
Wealth distribution				
Salaries, wages and other benefits (note 1)	10,703	59	8,713	59
Providers of capital	1,772	10	1,547	11
– Net financing costs	681	4	554	4
– Dividends, share buybacks and cancellations	1,091	6	993	7
Government (note 2)	1,572	9	1,543	10
Reinvested in the group to maintain and develop operations	4,084	22	2,931	20
– Depreciation, amortisation and recoupments	1,822		1,488	
– Future expansion	2,262		1,443	
	18,131	**100**	14,734	100
Value-added ratios				
– Number of employees (continuing operations)	47,699		40,898	
– Revenue per employee (000)	1,695		1,581	
– Wealth created per employee (000)	380		360	
Notes				
1. Salaries, wages and other benefits				
Salaries, wages, overtime, commissions, bonuses, allowances	9,959		8,070	
Employer contributions	744		643	
	10,703		8,713	
2. Central and local governments				
SA normal taxation	1,102		1,131	
Secondary tax on companies	90		108	
Foreign taxation	192		151	
Rates and taxes	72		69	
Skills development levy	41		43	
Unemployment Insurance Fund	48		41	
Carbon emissions tax	27			
	1,572		1,543	

Value added 2012 (%)

Employees 59%
Providers of capital 10%
Government 9%
Reinvested in the group 22%

Value added 2011 (%)

Employees 59%
Providers of capital 11%
Government 10%
Reinvested in the group 20%

There have also been criticisms, e.g. they are merely a restatement of information that appears in the annual report; they only report data capable of being reported in monetary terms; and the individual elements of societal benefit are limited to the traditional ones of shareholders, employees and the government, with others such as society and the consumers ignored.

There was also criticism that there was no standard so that expenditures could be aggregated or calculated to disclose a misleading picture, e.g. the inclusion of PAYE tax and welfare payments made to the government in the employee classification so that wages were shown gross, whereas distributions to shareholders were shown net of tax. The effect of both was to overstate the apparent employee share and understate the government and shareholders' share.[38]

In the years immediately following the publication of *The Corporate Report*, companies published value added statements on a voluntary basis but their importance has declined. There was a move away from industrial democracy and the standard-setting regulators did not make the publication of value added statements mandatory.

32.10 Corporate social responsibility reporting

There is recognition that CSR has moved from a PR exercise to being a core business value. This is evidenced by a 2010 survey[39] of CSR trends which reported that 81% of surveyed companies had CSR information on their website and 31% had their reports assured. Companies are also concentrating on those aspects that are of significance to their business. For example, this change in emphasis is illustrated by the following extract from the Ford Motor Company 2009/10 Sustainability Report:

> A key part of our reporting strategy has been the development of a materiality analysis process, which has been a critical tool in helping shape the content of this report. We used the analysis to focus our reporting on those issues determined to be most material to the company.

This has progressed with the approach to CSR becoming increasingly formalised with the setting up of committees reporting to the board and more comprehensive, targeted CSR reports.

There is now a wealth of opportunities for further study of the topic as we now see CSR modules in many university undergraduate and Business School programmes and postgraduate degrees such as the MA in Corporate Social Responsibility at London Metropolitan University.

32.11 Need for comparative data

There is evidence[40] that environmental performance could be given a higher priority when analysts assess a company if there were comparable data by sector on a company's level of corporate responsibility.

We will consider one approach that has taken place to satisfy this need for comparable data using benchmarking.

32.11.1 Benchmarking

There are a number of benchmarking schemes and we will consider one by way of illustration – the London Benchmarking Group,[41] established in 1994. The Group consists of companies which join in order to measure and report their involvement in the community, which is a key part of any corporate social responsibility programme, and which have a tool to assist them effectively to assess and target their community programmes. Organisations such as British Airways, Deloitte & Touche, Lloyds TSB and Pearson are members.

The scheme is concerned with corporate community involvement. It identifies three categories into which different forms of community involvement can be classified, namely charity donations, social or community investment and commercial initiatives, and includes only contributions made over and above those that result from the basic business operations.

It uses an input/output model, putting a monetary value on the 'input' costs which include contributions made in cash, in time or in kind, together with full cost of staff involved; and collecting 'output' data on the community benefit, e.g. number who benefited, leveraged resources and benefit accruing to the business.

32.12 Investors

Investors are gradually beginning to require information on a company's policy and programmes for environmental compliance and performance in order to assess the risk to earnings and financial position. One would expect that the more transparent these are the less volatile the share prices will be, which could be beneficial for both the investor and the company. This will be a fruitful field for research as environmental reporting evolves with more consistent, comparable, relevant and reliable numbers and narrative disclosures.

32.12.1 Socially Responsible Investing (SRI)

This has also given rise to Socially Responsible Investing (SRI) which considers both the investor's financial needs and the investee company's impact on society to an extent that in 2010 over £900 billion in assets were invested in 'ethical' investment funds.

In the UK there is pressure from bodies such as the Association of British Insurers for institutional investors to take SRI principles into account. A number of rating and bench-marking systems are used on behalf of investors and others to grade organisations through the use of ratings and benchmarks based on environmental and other sustainability criteria such as the Dow Jones Sustainability Indices and the FTSE4Good Index.

32.12.2 ABI Socially Responsible Investment Guidelines

In 2007, the ABI issued its updated guidelines on responsible investment disclosure. These are a modification of the Socially Responsible Investment Guidelines launched by the ABI in 2001.

They take account of the EU Accounts Modernisation Directive and the UK Companies Act, as well as recent experience of narrative reporting and the clarification by the UK government of directors' liability for narrative statements. They do not involve substantial change but aim to highlight aspects of responsibility reporting on which shareholders place particular value. This is narrative reporting which:

- sets environmental, social and governance (ESG) risks in the context of the whole range of risks and opportunities facing the company;
- contains a forward-looking perspective; and
- describes the actions of the board in mitigating these risks.

Institutional investors support the revised guidelines, which encourage listed companies to include narrative discussion of the environmental, social and governance (ESG) risks they face. The guidelines also encourage companies to explain what steps they are taking to mitigate and address those risks.

32.12.3 Dow Jones Sustainability Indices

The main impetus for rating and benchmarking systems comes from the growth of interest in SRI. The Dow Jones Sustainability Indices were begun in 1999 and were the first global indices tracking the financial performance of the leading sustainability-driven companies worldwide covering 58 sectors.

Companies are selected for the indices based on a comprehensive assessment of long-term economic, environmental and social criteria that account for general as well as industry-specific sustainability trends. Only firms that lead their industries based on this assessment are included in the indices. There are sub-indices excluding alcohol, gambling, tobacco, armaments and adult entertainment, and global and regional blue-chip indices.

SAM (an investment boutique focused exclusively on Sustainability Investing) annually identifies[42] the top company in each of the 19 supersectors derived from the 58 sectors. The 2012–2013 Supersector Leaders include companies in Australia (2) (Banks and Real Estate); Belgium (1) (Media); Brazil (1) (Financial Services); Finland (1) (Basic Resources); France (2) (Technology and Travel & Leisure); Germany (2) (Automobiles and Industrial Goods); Netherlands (3) (Chemicals, Food & Beverage and Personal & Household); South Korea (3) (Construction, Retail and Telecommunications); Spain (Oil & Gas and Utilities); and Switzerland (2) (Health Care and Insurance). The UK is noticeable for its absence.

32.12.4 FTSE4Good Index Series

The FTSE4Good Index Series provides potential investors with a measure of the performance of companies that meet globally recognised corporate responsibility standards. FTSE4Good is helpful as a basis for socially responsible investment and as a benchmark for tracking the performance of socially responsible investment portfolios.

There is still some way to go, however, with research[43] carried out by Trucost and commissioned by the Environment Agency into quantitative disclosures finding that direct links between management of environmental risks and shareholder value are almost non-existent, with only 11% of FTSE 350 companies making a link between the environment and some aspect of their financial performance and only 5% explicitly linking it to shareholder value.

32.13 The accountant's changing role

Professional accountants have provided the expertise for the design and operation of sound systems of accounting and internal control. They have had a central role in external reporting of financial data including, for example, identifying and measuring environmental costs to demonstrate efficiency in the use of energy and measuring social costs such as the cost of staff turnover and absenteeism.

They are becoming increasingly involved with the assurance of the additional narrative disclosures required by national statute and EU Directives.

National statutes

In the UK quoted companies are from 2013 disclosing in their Strategic Report information on the company's strategy, business model, human rights and gender diversity and in their

Directors' Report information on greenhouse gas emissions. The new Directive will extend the disclosures required on diversity with their policies on age, gender, educational and professional background and will for the first time require reporting on bribery and corruption.

EU Directives

In 2014 the EU issued the *Directive on Non-financial Disclosure* requiring large companies to report from 2017 their policies and how they manage risk on environmental, social, employee, human rights, anti-corruption and bribery matters. It does not require an integrated report but disclosure on a comply or explain basis. It accepts that its requirements might be satisfied by disclosures made under GRI and national disclosure requirements.

32.13.1 Sustainability and the profession

The professional accounting bodies are responding in a number of ways. For example, they are inviting advice through forums and are building sustainability into their examination syllabi and continuing professional development (CPD) programmes.

Global forums

The ACCA has created global forums to bring together experts from around the world to provide an international perspective when advising when the ACCA develops its policies on key issues, including consultation exercises from national and international standard-setting and regulatory bodies, and to advise on proactive initiatives such as events, publications and research projects. Its global forums include those on Corporate Reporting, Environmental Accountability and Governance, Risk and Performance.

Exam syllabus

The ACCA states that its members need to ensure that they are aware of the changing face of reporting, accounting and assurance and are advising their organisations and clients on the latest developments. It has embedded sustainable development (SD) and corporate social responsibility (CSR) issues in its exam syllabus and from December 2014 will examine students on integrated reporting.

32.13.2 Academic programmes

There are numerous undergraduate CSR modules and postgraduate Masters degrees such as at Erasmus University Rotterdam, University of Birmingham, Nottingham University and London Metropolitan University.

There is also active research such as at the Centre for Social and Environmental Accounting Research (CSEAR),[44] the International Centre for Corporate Social Responsibility[45] at Nottingham University Business School, and the Centre for the Study of Global Ethics[46] at the University of Birmingham.

Research interests include environmental and social issues. For example, the Centre for the Study of Global Ethics at the University of Birmingham was a partner in a project aimed at the promotion of corporate social responsibility by way of a methodology for monitoring corporate observance of work-related human rights, core labour and social standards as set out in the Decent Work Agenda and the Social Agenda of the European Union.

The International Centre for Corporate Social Responsibility at Nottingham University has undertaken a number of research projects on CSR and on corporate reporting on gender equality in the workplace.

There is progress being made at varying rates around the world. One of the stimuli has been the environmental, social and sustainability award schemes.

There is a growing interest and experimenting in many companies. As with all initiatives there will be varying views from those opposed to the idea of sustainability reporting, to those who think that the existing initiatives such as GRI are already too complex, to the discussions that will be needed to try to arrive at a uniform model.

Accounting and finance students will find that they will become increasingly involved within companies, not-for-profit organisations, and the profession – working within their accounting speciality and also in multidisciplinary teams. Accounting qualifications will give access to more innovative and rewarding scenarios.

Given the time taken to arrive at a Conceptual Framework for Financial Reporting, it might be quite a while before an Integrated Reporting Framework is adopted by all companies. However, given the pressure from stakeholders for improvements in narrative reporting, its influence is already being widely felt. In the meantime there will be exciting ongoing academic and professional research and empirical experimenting by companies.

Summary

Sustainability is now recognised as having three elements. These are the economic, environmental and social. It is recognised that advances in environmental and social improvement are dependent on the existence of an economically viable organisation.

As environmental and social reporting evolves, proposals are being made to harmonise the content and disclosure. This can be seen with the publications such as those of the Connected Framework, the GRI and the International Integrated Reporting Council all working towards companies adopting an integrated reporting framework with the long-term objective of perhaps achieving a global standard – although this might be a number of years off. Each of these bodies has been a catalyst for change – now, however, it is the time for there to be a coordinated development overseen perhaps by the IASB.

In addition there are benchmark schemes which allow stakeholders to compare corporate social reports and evaluate an individual company's performance. The management systems that are being developed within companies should result in data that are consistent and reliable and capable of external verification. The benchmarking systems should assist in both identifying best practice and establishing relevant perform-ance indicators.

Corporate social reporting is coming of age. Initially there were fears that it would add to costs and there are present concerns that it is diverting too much of a finance director's attention away from commercial and stragetic planning. However, it is becoming generally recognised that a company's reputation and its attractiveness to potential investors are influenced by a company's behaviour and attitude to corporate governance, social reporting and sustainability.

Companies are reacting positively to the need to be good corporate citizens and it is interesting to see the developments around the world where sustainability, good corporate governance and strategic planning are merging into an integrated system.

This will take time but companies are taking up the challenge to be transparent and innovative in their financial reporting. Companies are integrating their non-financial narrative and using the Internet to get their message out to a wider public.

Time has passed since corporate governance, sustainability, environmental and social reporting were seen purely as a PR exercise.

REVIEW QUESTIONS

1 Obtain a copy of the environmental report of a company that has taken part in the ACCA Awards for Sustainability Reporting and critically discuss from an investor's and public interest viewpoint.

2 'Charters and guidelines help make reports reliable but inhibit innovation and reduce their relevance.' Discuss how the Integrated Reporting Framework attempts to avoid this.

3 Discuss the implications of the Global Reporting Initiative for the accountancy profession.

4 Discuss the value added concept, giving examples and ways to improve the statement.

5 Outline the arguments for and against a greater role for the audit function in corporate social reporting.

6 Discuss the challenges that accountant and auditor will face with the increasing demand for environmental impact to be reported.

7 'Human capital is always incapable of being valued for inclusion in the financial statements.' Discuss.

8 The following is an extract from the 2011 Tottenham Hotspur Annual Report when considering carrying out an impairment test:

> The Group does not consider that it is possible to determine the value in use of an individual football player in isolation as that player (unless via a sale or insurance recovery) cannot generate cash flows on his own. Furthermore, the Group also considers that all of the players are unable to generate cash flows even when considered together.

Discuss why the players taken separately or together do not have a value in use.

9 Discuss the impact of the following groups on the accounting profession:

  (a) environmental groups;

  (b) ethical investors.

10 Nissan, the Japanese car company, decided that 'any environmentalism should pay for itself and for every penny you spend you must save a penny. You can spend as many pennies as you like as long as other environmental actions save an equal number.'[47] Discuss the significance of this for each of the stakeholders.

11 (a) Discuss the significant direct KPIs relating to the air transport industry. (b) Identify which industries you consider to be significant supplier industries.

  (Access to the following is helpful: http://archive.defra.gov.uk/environment/business/reporting/pdf/envkpi-guidelines.pdf)

12 'Consumer-oriented models are more likely to be influenced by ethical principles.' Discuss.

13 A chemical entity installed new manufacturing equipment at a cost of €1m to comply with environmental regulations concerning the production and storage of chemicals.

Discuss how this should be dealt with in the Annual Report.

14 Discuss how it would be possible to apply sustainability criteria in determining executive remuneration.

15 A research report[48] found that comparability of extra-financial information between companies is an issue: 61% said they find social information difficult to compare and 41% said they find environmental information difficult to compare, while only 3% said they find it difficult to compare financial information between companies.

Discuss whether comparability is achievable and, if so, the measures that could be taken to make it possible.

16 It has been suggested that there should be either a single integrated report or a set of reports containing at the discretion of the management integrated information such as the annual financial statements plus any one or more of the following reports on: ethics, governance, sustainability and social and environmental impacts. Discuss the pros and cons of allowing a company to have such a choice.

EXERCISES

Question 1

(a) You are required to prepare a value added statement to be included in the corporate report of Hythe plc for the year ended 31 December 20X6, including the comparatives for 20X5, using the information given below:

	20X6 £000	20X5 £000
Non-current assets (net book value)	3,725	3,594
Trade receivables	870	769
Trade payables	530	448
14% debentures	1,200	1,080
6% preference shares	400	400
Ordinary shares (£1 each)	3,200	3,200
Sales	5,124	4,604
Materials consumed	2,934	2,482
Wages	607	598
Depreciation	155	144
Fuel consumed	290	242
Hire of plant and machinery	41	38
Salaries	203	198
Auditors' remuneration	10	8
Corporation tax provision	402	393
Ordinary share dividend	9p	8p
Number of employees	40	42

(b) Although value added statements were recommended by *The Corporate Report*, as yet there is no accounting standard related to them. Explain what a value added statement is and provide reasons as to why you think it has not yet become mandatory to produce such a statement as a component of current financial statements through either a Financial Reporting Standard or company law.

Question 2

The following items have been extracted from the accounts:

	2005 (€m)	2004 (€m)
Other income	844	980
Cost of materials	25,694	24,467
Financial income	−188	54
Depreciation/amortisation	4,207	3,589
Providers of finance	1,351	1,059
Retained	1,815	1,823
Revenues	46,656	44,335
Government	1,590	1,794
Other expenses	4,925	5,093
Shareholders	424	419
Employees	7,306	7,125

Required:
(a) **Prepare a value added statement showing % for each year and % change.**
(b) **Draft a note for inclusion in the annual report commenting on the statement you have prepared.**

Question 3

David Mark is a sole trader who owns and operates supermarkets in each of three villages near Ousby. He has drafted his own accounts for the year ended 31 May 20X4 for each of the branches. They are as follows:

	Arton		Blendale		Clifearn	
	£	£	£	£	£	£
Sales		910,800		673,200		382,800
Cost of sales		633,100		504,900		287,100
Gross profit		277,700		168,300		95,700
Less: Expenses:						
David Mark's salary	10,560		10,560		10,560	
Other salaries and wages	143,220		97,020		78,540	
Rent			19,800			
Rates	8,920		5,780		2,865	
Advertising	2,640		2,640		2,640	
Delivery van expenses	5,280		5,280		5,280	
General expenses	11,220		3,300		1,188	
Telephone	2,640		1,980		1,584	
Wrapping materials	7,920		3,960		2,640	
Depreciation:						
Fixtures	8,220		4,260		2,940	
Vehicle	3,000	203,620	3,000	157,580	3,000	111,237
Net profit/(loss)		74,080		10,720		(15,537)

The figures for the year ended 31 May 20X4 follow the pattern of recent years. Because of this, David Mark is proposing to close the Clifearn supermarket immediately.

David Mark employs 12 full-time and 20 part-time staff. His recruitment policy is based on employing one extra part-time assistant for every £30,000 increase in branch sales. His staff deployment at the moment is as follows:

	Arton	Blendale	Clifearn
Full-time staff (including managers)	6	4	2
Part-time staff	8	6	6

Peter Gaskin, the manager of the Clifearn supermarket, asks David to give him another year to make the supermarket profitable. Peter has calculated that he must cover £125,500 expenses out of his gross profit in the year ended 31 May 20X5 in order to move into profitability. His calculations include extra staff costs and all other extra costs.

Additional information:

1 General advertising for the business as a whole is controlled by David Mark. This costs £3,960 per annum. Each manager spends a further £1,320 advertising his own supermarket locally.

2 The delivery vehicle is used for deliveries from the Arton supermarket only.

3 David Mark has a central telephone switchboard which costs £1,584 rental per annum. Each supermarket is charged for all calls actually made. For the year ended 31 May 20X4 these amounted to:

Arton	£2,112
Blendale	£1,452
Clifearn	£1,056

Required:

(a) A report addressed to David Mark advising him whether to close the Clifearn supermarket. Your report should include a detailed financial statement based on the results for the year ended 31 May 20X4 relating to the Clifearn branch.

(b) Calculate the increased turnover and extra staff needed if Peter's suggestion is implemented.

(c) Comment on the social implications for the residents of Clifearn if (i) David Mark closes the supermarket, (ii) Peter Gaskin's recommendation is undertaken.

* Question 4

Gettry Doffit plc is an international company with worldwide turnover of £26 million. The activities of the company include the breaking down and disposal of noxious chemicals at a specialised plant in the remote Scottish countryside. During the preparation of the financial statements for the year ended 31 March 20X5, it was discovered that:

1 Quantities of chemicals for disposals on site at the year-end included:

(A)	Axylotl peroxide	40,000 gallons
(B)	Pterodactyl chlorate	35 tons

Chemical A is disposed of for a South Korean company, which was invoiced for 170 million won on 30 January 20X5, for payment in 120 days. It is estimated that the costs of disposal will not exceed £75,000. £60,000 of costs have been incurred at the year-end.

Chemical B is disposed of for a British company on a standard contract for 'cost of disposal plus 35%', one month after processing. At the year-end the chemical has been broken down into harmless by-products at a cost of £77,000. The by-products, which belong to Gettry Doffit plc, are worth £2,500.

2 To cover against exchange risks, the company entered into two forward contracts on 30 January 20X5:

No. 03067   Sell 170 million won at 1,950 won = £1:   31 May 20X5
No. 03068   Buy $70,000 at $1.60   = £1:   31 May 20X5

Actual sterling exchange rates were:

	won	$
30 January 20X5	1,900	1.70
31 March 20X5	2,000	1.38
30 April 20X5 (today)	2,100	1.80

The company often purchases a standard chemical used in processing from a North American company, and the dollars will be applied towards this purpose.

3 The company entered into a contract to import a specialised chemical used in the breaking down of magnesium perambulate from a Nigerian company which demanded the raising of an irrevocable letter of credit for £65,000 to cover 130 tons of the chemical. By 31 March 20X5 bills of lading for 60 tons had been received and paid for under the letter of credit. It now appears that the total needed for the requirements of Gettry Doffit plc for the foreseeable future is only 90 tons.

4 On 16 October 20X4 Gettry Doffit plc entered into a joint venture as partners with Dumpet Andrunn plc to process perfidious recalcitrant (PR) at the Gettry Doffit plc site using Dumpet Andrunn plc's technology. Unfortunately, a spillage at the site on 15 April 20X5 has led to claims being filed against the two companies for £12 million. A public inquiry has been set up, to assess the cause of the accident and to determine liability, which the finance director of Gettry Doffit plc fears will be, at the very least, £3 million.

Required:
Discuss how these matters should be reflected in the financial statements of Gettry Doffit plc as on and for the year ended 31 March 20X5.

Question 5

In 2010 there was an explosion, fire, sinking and loss of life on board the mobile offshore drilling unit, the Deepwater Horizon, in the Gulf of Mexico. There has since been a final report issued by the Bureau of Ocean Energy Management, Regulation and Enforcement (BOEMRE)/US Coast Guard Joint Investigation Team (JIT).

Required:
Examine the BP 2011 Sustainability Report[49] and prepare a brief presentation to the group explaining the company's Operating Management System in relation to environmental and social issues and commenting on the extent to which such reports can provide investors with the possible social costs of environmental disasters.

References

1 http://www.theiirc.org/
2 http://www.eurosif.org/
3 http://www.unglobalcompact.org/news/42-06-22-2010
4 R.G. Eccles, I. Ioannou and G. Serafeim, 'The impact of a corporate culture of sustainability on corporate behavior and performance', NBER Working Paper No. 17950, March 2012, http://www.nber.org/papers/w17950

5 C. Post, N. Rahman and E. Rubow, 'Green governance: boards of directors' composition and environmental corporate social responsibility', *Business & Society*, March 2011, vol. 50, pp. 189–223.

6 L.G. Hassel, J.-P. Kallunki and H. Nilsson, 'Implications of past unethical risk behaviour of board members and CEOs on the environmental performance and reporting quality of firms', www.sirp.se/getfile.ashx?cid=48377&cc=3&refid=35file:///C:/Users/Barry/Downloads/PU

7 www.sustainabilityreporting.eu/netherlands/index.htm

8 'The value of extra-financial disclosure: what investors and analysts said', https://www.globalreporting.org/resourcelibrary/The-value-of-extra-financial-disclosure.pdf

9 www.sustainablereporting.eu/denmark/index.htm

10 http://about.puma.com/wp-content/themes/aboutPUMA_theme/media/pdf/2011/en/PRESS_KIT_E_P&L.pdf

11 http://www.accountingforsustainability.org/wp-content/uploads/2011/10/Connected-Reporting.pdf

12 www.globalreporting.org

13 www.accountability.org.uk

14 See http://ec.europa.eu/environment/emas/index_en.htm

15 www.iso.org

16 C. Lehman, *Accounting's Changing Role in Social Conflict*, Markus Weiner Publishing, 1992, p. 64.

17 Ibid., p. 17.

18 KPMG Peat Marwick McLintock, 'Environmental considerations in acquiring', *Corporate Finance Briefing*, 17 May 1991.

19 M. Jones, 'The cost of cleaning up', *Certified Accountant*, May 1995, p. 47.

20 http://www.unglobalcompact.org/issues/supply_chain

21 Information for Better Markets, 'Sustainability: the role of accountants', ICAEW, 2004.

22 www.sustainablereporting.eu/uk/index.htm

23 See www.fee.be/issues/other.htm#Sustainability

24 http://www.qualidade.eng.br/ambiente/conheca_canadian_auditing.htm

25 http://www.icaew.com/~/media/Files/Technical/Sustainability/environmental-issues-and-annual-financial-reporting-2009.pdf

26 http://www2.accaglobal.com/allnews/national/malaysia/3538611

27 See http://ec.europa.eu/environment/newprg/

28 R. Gray, D. Owen and K. Maunders, *Corporate Social Reporting*, Prentice Hall, 1987, p. 75.

29 ASB, *Statement of Principles: The Objective of Financial Statements*, 1991, para. 9.

30 Ibid., para. 10.

31 Ibid., para. 11.

32 F. Okcabol and A. Tinker, 'The market for positive theory: deconstructing the theory for excuses', *Advances in Public Interest Accounting*, vol. 3, 1990.

33 H. Sebreuder, 'Employees and the corporate social report: the Dutch case', in S.J. Gray (ed.), *International Accounting and Transnational Decisions*, Butterworth, 1983, p. 287.

34 Ibid., p. 289.

35 AICPA, *The Measurement of Corporate Social Performance*, 1977, p. 243.

36 R. Gray, D. Owen and K. Maunders, *Corporate Social Reporting*, Prentice Hall, 1987, p. 44.

37 S.J. Gray and K.T. Maunders, 'Value added reporting: uses and measurement', ACCA, 1980; B. Underwood and P.J. Taylor, *Accounting Theory and Policy Making*, Heinemann, 1985, p. 298.

38 Ibid., p. 74.

39 http://admin.csrwire.com/system/report_pdfs/1189/original/CSR_TRENDS_2010.pdf

40 Business in the Environment, *Investing in the Future*, May 2001.

41 See www.lbg-online.net/

42 http://www.sustainability-indexes.com/review/supersector-leaders-2012.jsp

43 http://www.trucost.com/news-2006/57/trucost-and-the-environment-agency-report-on-the-environmental-disclosures-of-ftse-all-share-under-the-requirements-of-the-european-union-accounts-modernisation-directive

44 http://www.st-andrews.ac.uk/~csearweb/aboutcsear/

45 http://www.nottingham.ac.uk
46 www.business.bham.ac.uk
47 M. Brown, 'Greening the bottom line', *Management Today*, July 1995, p. 73.
48 http://www.totalecomanagement.co.uk/2/post/2012/09/what-investors-and-analysts-said-the-value-of-extra-financial-disclosure.html
49 http://www.bp.com/assets/bp_internet/globalbp/STAGING/global_assets/e_s_assets/e_s_assets_2010/downloads_pdfs/bp_sustainability_review_2011.pdf

Bibliography

R.B. Adams and D. Ferreira, 'Women in the boardroom and their impact on governance and performance', *Journal of Financial Economics*, vol. 94, 2009, pp. 291–309.

T. Artiach, D. Lee, D. Nelson and J. Walker, 'The determinants of corporate sustainability performance', *Accounting and Finance*, vol. 50, 2009, pp. 31–51.

J.A. Batten and T.A. Fetherston (eds), *Social Responsibility: Corporate Governance Issues* (Research in International Business and Finance, Volume 17), JAI Press, 2003.

S. Bear, N. Rahman and C. Post, 'The impact of board diversity and gender composition on corporate social responsibility and firm reputation', *Journal of Business Ethics*, vol. 97, 2011, pp. 207–222.

S. Bhagat and B. Bolton, 'Corporate governance and firm performance', *Journal of Corporate Finance*, vol. 14, 2008, pp. 257–273.

K. Bondy, D. Matten and J. Moon, 'Codes of conduct as a tool for sustainable governance in MNCs', in S. Benn and D. Dunphy (eds), *Corporate Governance and Sustainability: Challenges for Theory and Practice*, Routledge, 2006.

S. Brammer and S. Pavelin, 'Voluntary environmental disclosures by large UK companies', *Journal of Business Finance and Accounting*, vol. 33, 2006, pp. 1168–1188.

M.C. Branco and L.L. Rodrigues, 'Issues in corporate social and environmental reporting research: an overview', *Issues in Social and Environmental Accounting*, vol. 1(1), 2007, pp. 72–90.

D. Campbell, 'A longitudinal and cross-sectional analysis of environmental disclosure in UK companies – a research note', *British Accounting Review*, vol. 36(1), 2004, pp. 107–17.

W. Chapple and J. Moon, 'Corporate social responsibility (CSR) in Asia: a seven country study of CSR website reporting', *Business and Society*, vol. 44(4), 2005, pp. 115–36.

P.M. Clarkson, M.B. Overell and L. Chapple, 'Environmental reporting and its relation to corporate environmental performance', *Abacus*, vol. 47, 2011, pp. 27–60.

S.M. Cooper and D.I. Owen, 'Corporate social reporting and stakeholder accountability: the missing link', *Accounting, Organizations and Society*, vol. 32(7–8), October–November 2007, pp. 649–67.

P.K. Cornelius and B. Kogul, *Corporate Governance and Capital Flows in the Global Economy*, Oxford University Press, 2003.

B. Coyle, *Risk Awareness and Corporate Governance*, Financial World Publishing, 2004.

A. Crane, D. Matten and J. Moon, *Corporations and Citizenship*, Cambridge University Press, 2006.

D. Crowther and K.T. Caliyurt, *Stakeholders and Social Responsibility*, Ansted University Press, 2004.

D. Crowther and L. Rayman-Bacchus (eds), *Perspectives on Corporate Social Responsibility*, Ashgate, 2004.

P. De Moor and I. De Beelde, 'Environmental auditing and the role of the accountancy profession: a literature review', *Environmental Management*, vol. 36(2), 2005, pp. 205–19.

J. Derwall, K. Koedijk and J.T. Horst, 'A tale of values-driven and profit-seeking social investors', *Journal of Banking & Finance*, vol. 35, 2011, pp. 2137–47.

R. Gray, J. Bebbington and M. Houldin, *Accounting for the Environment* (2nd edition), Sage Publications, 2001.

R. Gray, *Social and Environmental Accounting* (Sage Library in Accounting and Finance) (with J. Bebbington and S. Gray), London: Sage, Volumes I–IV, 1664, p. 2010.

K. Grosser and J. Moon, 'Developments in company reporting on workplace gender equality: a Corporate Social Responsibility perspective', *Accounting Forum*, vol. 32, 2008, pp. 179–98.

N. Guenster, R. Bzauer, J. Derwall and K. Koedijk, 'The economic value of corporate eco-efficiency', *European Financial Management*, vol. 17, 2011, pp. 679–704.

D. Hawkins, *Corporate Social Responsibility: Balancing Tomorrow's Sustainability and Today's Profitability*, Palgrave Macmillan, 2006.

G. Heal, 'Corporate social responsibility: an economic and financial framework', *Geneva Papers*, vol. 30, 2005, pp. 387–409. http://www.genevaassociation.org/PDF/Geneva_papers_on_Risk_and_Insurance/GA2005_GP30(3)_Heal.pdf

L. Holland and Y.B. Foo, 'Differences in environmental reporting practices in the UK and the US: the legal and regulatory context', *British Accounting Review*, vol. 35(1), 2003, pp. 1–18.

A.W. Savitz and K. Weber, *The Triple Bottom Line: How Today's Best-Run Companies Are Achieving Economic, Social and Environmental Success – and How You Can Too*, Jossey-Bass, San Francisco, 2006.

J.L. Walls, P. Berrone and P.H. Phan, 'Corporate governance and environmental performance: is there really a link?', *Strategic Management Journal*, vol. 33(8), 2012, pp. 885–913.

R. Wearing, *Cases in Corporate Governance*, Sage, 2005.

Index

A & J Muklow plc 434
A M Castle & Co 502
A-scores 734–5
A4S (Accounting for Sustainability) 813, 819
AA1000 Assurance Standard 814, 818
Aaronson, Graham 392
ABI *see* Association of British Insurers
absences, compensated 372
academic programmes on environmental, social and sustainability issues 827–8
ACCA *see* Association of Chartered Certified Accountants
accountability 560, 820–1
accountants
  in business 252–3
  commercial awareness 4
  environmental reporting 817–20, 826–7
  ethical behaviour *see* ethics
  executive remuneration and 791
  external reporting 3–4
  internal reporting 3, 4–8
  money laundering implications for 265
  in practice 251–2
  problems, ethics and 251–3
  reporting roles 5
  sustainability roles 816–18, 826–7
  technical expertise 4
  view of income, capital and value 132–6
  work environments, ethics and 256–8
  see also accounting profession; auditors
accounting capital 133–4
Accounting Corporate Regulatory Authority (ACRA), Singapore 761
Accounting Council, FRC 216
Accounting Directive 219–21, 225
Accounting for Sustainability (A4S) 813, 819
accounting income 133
accounting numbers, role in defining contractual entitlements 211–12
accounting policies 44–5, 70, 190–1, 436–8
accounting profession
  credibility 214
  cultural differences and 256

environmental issues and 819
ethical behaviour and 250–3
public view of 213
status of 228–9
sustainability and 819, 827–8
accounting ratios *see* ratios
accounting standards
  arguments against 214–15
  arguments for 214
  conceptual framework 230, 234–45
  consensus-seeking 215
  enforcement
    Europe 221
    UK 215–17
    USA 222
  ethics and 253–6
  global, advantages and disadvantages for publicly accountable entities 223–4
  mandatory 212–14
  need for 211–12
  overload 215
  setting 215–17
    European Union 219–21
    process, ethics and 249–50
    USA 221–3
  see also Financial Reporting Standards; International Accounting Standards; International Financial Reporting Standards; Statements of Financial Accounting Standards; Statements of Standard Accounting Practice
Accounting Standards Board (ASB)
  accrual accounting 27
  Considering the Effects of Accounting Standards (Discussion Paper 2011) 250
  inflation accounting approach 169–71
  Statement of Principles (1999) 170–1, 230
  see also Financial Reporting Standards; Statements of Standard Accounting Practice
Accounting Standards Committee (ASC) 50, 230
accounting system differences 49
accounting theory 229, 235–7
Accounts Modernisation Directive 825
accrual accounting 16, 21–2

basis 22–3
cash flow statements and 27–8
cash payments adjustments 23–4
cash receipts adjustments 23–4
deferred taxation 405
historical cost convention 22
non-current assets 24–7
statements of financial position 23–4
summary 28
views on 27
accrual data, failure predictions and 735
accruals 242
acid test ratios 694
acquisitions
  cash 557
  fair value 551–2, 555–7, 559–60
  foreign currencies and 639–43
  goodwill 475
  share issues 278–80
  standards 212–13
  see also business combinations; consolidated accounts; groups
ACRA (Accounting Corporate Regulatory Authority), Singapore 761
actual performance 130
Actuarial Council, FRC 216
actuarial gains and losses, pension schemes 368
actuarial method, finance lease accounting 455, 457–8
adjusting events 68
adjustments 36, 159–61
administration overheads 503, 504
administrative expenses 36–7
administrative expenses ratios 688
Adobe® 756
advocacy threats 253
AEI, takeover by GEC 212–13
Aer Lingus 41, 739
affirmative covenants 721
agency costs 8
aggressive earnings management 722–3
Agrana Group 504
agricultural activities 513–16
Ahold 213–14
AIA (Association of International Accountants) 266
AICPA (American Institute of Certified Public Accountants) 764, 767

AkzoNobel 417, 427
Altman's Z-score 731–2
Amadeus 757
AMEC 717
American Institute of Certified Public
    Accountants (AICPA) 764, 767
AML (anti-money laundering) standard
    265
amortisation 36, 212, 417–18, 470, 471,
    476–8
amortised costs 322, 338–9, 347–8, 349
analysis
    cash flow statements 107–12
    financial statements *see* financial
        statements
    techniques
        A-scores 734–5
        golden rule 681
        H-scores 733–4, 740
        overview 681–2
        predicting corporate failure 731–5
        Z-scores 731–3, 740
    see also ratios
annual refunds, revenue recognition 191
annual reports 67
    directors' reports 54
    discontinued operations 79–80
    environmental reporting 812–13,
        819–20
    events after reporting periods 67–9
    held for sale 80–2
    IAS 8: 69–71
    intellectual capital disclosures in
        484–5
    related party disclosures 82–6
    segment reporting 71–8
    Strategic Reports 53–4, 109
    see also financial statements; published
        accounts
annuity method, depreciation 421–2
Anti-Abuse Rule, General (GAAR) 392
anti-avoidance rules, targeted (TAAR)
    392
anti-money laundering (AML) standard
    265
anti-social behaviour 779
APB *see* Auditing Practices Board
Apple 480
appointments, ethics and 251
appropriate overheads 503–5, 507–8
arm's length transactions 82
Arthur Andersen 260, 264, 266, 786
artistic-related intangible assets 469
ASB *see* Accounting Standards Board
ASC (Accounting Standards Committee)
    50, 230
Aspen Pharmacare Holdings Ltd 327
assets
    bases 450
    biological 513–16
    capital maintenance and 281–2
    consolidated accounts 570
    contingent 306–7, 310–11
    current values 155
    deferred tax 403–4
    definitions 24, 242

disaggregation 200
environmental, reporting 817
held for sale 80–2
inflation accounting 154, 155, 156
jointly controlled 613
mining 417–18
net 134
non-monetary 133–4, 161–2, 634, 643,
    644
pension schemes *see* pension schemes
recognition 243
related parties 86
return on (ROA) 796–7
revaluation 133, 402–3, 436
self-constructed 414
turnover ratios 689–92, 697
see also current assets; financial assets;
        intangible assets; non-current
        assets; property, plant and
        equipment
associates
    acquired during the year 611–12
    consolidated accounts 605–12, 617
    definition 605–6
    disclosures 617
    equity method 606
    unrealised profits 609
Association of British Insurers (ABI)
    corporate governance research 796–7
    Principles of Executive Remuneration
        792
    Socially Responsible Investment
        Guidelines 825–6
Association of Chartered Certified
    Accountants (ACCA)
    Awards for Sustainability Reporting
        819
    credit agencies regulation 736
    ethics
        codes, assistance in designing
            262–3
        examinations 266
    Global Forums 827
    narrative reporting survey 54–5
    real-time reporting survey 767–8
    sustainable development 827
Association of International Accountants
    (AIA) 266
assurance 819
AstraZeneca 48, 278, 560
Audit and Assurance Council, FRC 216
Audit and Assurance Faculty 722
audit committees 783
auditing, environmental 818
Auditing Practices Board (APB)
    Aggressive Earnings Management
        (Consultation Paper 2001) 722
    Practice Note 12 *Money Laundering*
        265
auditors
    competition 801
    confidentiality breaches 265
    environmental assurance 818
    ethics and 260
    expectation gap 801
    governance within audit firms 788

independence 785–6, 801
liabilities 788–9
limited liability partnerships 789
negligence 801
professional oversight 801
professional scepticism 786–7
scepticism 723
unethical behaviour and 259, 780
work carried out by 714–15
XBRL and 765
audits 782–9, 801–2
Australia
    accounting profession 229
    accounting theory 229
    agricultural accounting 515–16
    auditor liability 789
    corporate governance 793
    legal system 226
    sustainable investing companies 826
    tax imputation system 390
available-for-sale financial assets 332–4,
    338
average cost (AVCO) 47–8, 500–1
Aviva 814–15
avoidance of tax 391–4

badwill 480
BAE Systems 262
balance sheets *see* statements of financial
        position
Balfour Beatty plc 326, 526–7
bankers 780
banks 228, 323, 798, 799
basic earnings per share (BEPS) 656–7,
    661–70
Baumol, William 255
Bayer Group 403, 470
Beder, S. 261
behaviour
    corporate 778–80
    ethical *see* ethics
    professional 251
    unethical 258–63, 779–80
behavioural entity, economy as 138
Belgium, sustainable investing companies
    826
benchmarking 824–5
BEPS (basic earnings per share) 656–7,
    661–70
BFCF (business free cash flows) 111
BHP Billiton 719
binomial option pricing model 375
biological assets 513–16
Black–Scholes option pricing model 375
BMW 756
BNP Paribas Group 331
boards of directors *see* directors
bonds 667, 668
bonus issues 659
bonus plans, employees 372
bonuses
    directors 211–12
    for performance, revenue recognition
        192
boom times, corporate behaviour in
    778–9

borrowings
  costs, capitalisation 414–15
  inflation accounting 164–5
  powers 481
  see also loans
BP 260–1, 290, 719, 756
brands 480–3
Brazil, sustainable investing companies
  826
bribery 262, 265, 777–8
British Airways 264–5, 368, 825
British Sky Broadcasting plc 432
British Telecommunications plc (BT)
  130, 257, 278
buildings, leases 452–3, 459
BUPA 796
business combinations
  brands 482–3
  fair values 173, 559–60
  goodwill 475–80, 482–3
  non-controlling interests 553
  review of implementation of 485–6
  see also acquisitions; consolidated
    accounts
business free cash flows (BFCF) 111
business reviews 819
bust times, corporate behaviour in 779
buyback of shares 290–1, 661

Cable, Vince 789–90
Cadbury Report 795
Canada
  accounting profession 229
  accounting theory 229
  environmental auditing 818
  legal system 226
capacity to adapt concept 153
CAPEX (capital expenditure) 110, 111
capital
  accountants' views 132–6
  conservation 450
  costs of 258, 728
  definitions 142–3
  economists' view 136–42
  financial 809
  human 809
  intellectual 484–5, 809
  invested 130
  manufactured 809
  money 142, 143
  natural 809
  operating capacity 142
  potential consumption 142, 143
  reduction 284–91
  relationship 809
  restructuring 276
  social 809
  structures 728
  weighted average cost (WACC)
    727–8
  see also share capital; working capital
capital allowances 389
capital asset pricing model (CAPM) 728
capital employed 728
capital expenditure (CAPEX) 110, 111
capital grants 433–5

capital investment incentive effect,
  deferred taxation 397
capital maintenance 142–3, 276, 816
  creditor protection 280–2
  financial 27, 151–3, 167–8
  operating 153, 157–68
capital markets 184, 780
capital profits 135
capital reconstructions 287–9
capital transactions with owners 43
capitalisation 472–4, 659
CAPM (capital asset pricing model) 728
carbon dioxide allowances emissions
  trading 483–4
carrying amounts 426, 431
case-based approach to tax avoidance 394
Case Management Committee, FRC 216,
  217
cash conversion cycle 692
cash conversion ratios (CCR) 111
cash cycle 692
cash dividend coverage ratio (CDCR)
  109, 111
cash equivalents 103, 106, 107
cash flow accounting 3–17, 23–4, 117
cash flow hedges 340–1
cash flow ratios 694–5
cash flow statements 99
  accrual accounting 27–8
  analysing 107–12
  consolidated accounts 590–1
  development of 99–100
  direct method 100–1, 102
  disclosures 112
  examination question answering with
    time constraints 112–15
  financing activities 101, 106, 111
  IAS 7: 99–117
  indirect method 100, 102, 103–6, 591
  information needs 239
  investing activities 101, 103, 106,
    109–10
  notes to 106–7, 112
  operating activities 100–3, 106
  operations 101, 106, 108–9
  preparation when statements of
    income not available 115–16
  summary 117
cash flows 9
  failure predictions and 735
  free (FCF) 110–11, 719
  future 46, 109, 130
  management 450
  private finance initiative 530–1
cash generating units (CGU) 426–9
cash ratios 695
cash-settled share-based payments 377–9
CCA (current cost accounting) 157–70
CCR (cash conversion ratios) 111
CDCR (cash dividend coverage ratio)
  109, 111
Centre for Social and Environmental
  Accounting Research (CSEAR)
  827
Centre for the Study of Global Ethics
  827

Centrica 777–8
Certified Public Accountants (CPA),
  USA 266
CGU (cash generating units) 426–9
chairmen's statements 54, 109
changes in equity statements (SOCE)
  42–3, 585
charities 393–4
chief operating decision makers (CODM)
  72, 74, 78
China
  corporate governance 778
classical corporation tax system 389–90
classification
  current assets 43–4, 212
  errors 70
  financial instruments 334–5, 347–8
  held for sale 80–2
  income 36–9
  leases 452–3, 456, 459
  non-current assets 43–4, 212, 431
  operating expenses 36–40
CO_2 emissions trading 483–4
Coca-Cola 481, 719
CoCoA (Continuous Contemporary
  Accounting) 150
Codes and Standards Committee, FRC
  216
codes of corporate governance 792–802
CODM (chief operating decision makers)
  72, 74, 78
Coil SA 424
Colgate 756
Collective Engagement Working Group
  800
Colt SA 110
commercial awareness 4
commercial databases, ratio analysis
  756–7
commissions, ethics and 252
Committee of Sponsoring Organizations
  of the Treadway Commission
  (COSO)
  *Fraudulent Financial Reporting:
    1987–1997 — An Analysis of
    U.S. Public Companies* 509
commodity contracts 325
common law legal systems 226
common size statements 684–5
community issues
  key performance indicators 54
companies *see* limited liability companies;
  private companies; public
  companies; unquoted companies
Companies Act 1929: 129
Companies Act 1981: 226
Companies Act 2006: 53, 219, 276, 281,
  283–4, 291, 792, 825
Companies House, XBRL use 761
Company Watch 733
comparability 214, 241, 819
comparative data 824–5
comparative share prices 792
compensated absences 372
compensation of key management
  personnel 85

competence 250
competition, auditors 801
competitive pressures, unquoted
        companies, share valuations 740
completed contracts method 524
completeness 15, 815
complexity 223
compliance
    creative 226
    with fundamental principles, threats
        to 253
    IFRS 221
components, construction contracts
        534–5
compound instruments 326–7
compound interest 137
comprehensive income 39–40, 43
    see also statements of comprehensive
        income
Compustat 757
conceptual framework 234
    approach 236–7
    financial accounting theory historical
        overview 235–7
    IASB Conceptual Framework for
        Financial Reporting 2010 see
        International Accounting
        Standards Board
    IASC Framework for the Presentation
        and Preparation of Financial
        Statements see International
        Accounting Standards Committee
    move towards 230
    national differences 234–5
concessions, construction contracts
        529–32
Conduct Committee, FRC 216, 217
confidentiality 251, 265
confirmatory values 240
Connected Reporting Framework 814
consistency 42, 242, 819
consolidated accounts
    accounting for groups at date of
        acquisition 549–6
    associates 605–12
    cash flow statements 590–1
    depreciation 586
    excluded subsidiaries 558–9
    foreign currency transactions 639–42
    IFRS 10: 549–51
    inter-company transactions 571–4,
        582–3
    joint ventures 613–18
    non-current assets acquired from
        subsidiaries 586
    partly owned subsidiaries 553–5
    pre- and post-acquisition profits and
        losses 568–9
    preparation not required 558
    profits 568–9, 583, 586–7
    statements of changes in equity 585
    statements of comprehensive
        income 582–92, 608, 611,
        639, 640
    statements of financial position
        568–74, 607–8, 610

subsidiaries acquired during the year
        587–9, 590–1
    uniform policies 568
    wholly owned subsidiaries 549, 551–3
constant purchasing power measurement
        of financial capital maintenance
        27
construction contracts 523–39
constructive obligations 311
consultancies, auditors 788
consumer price index 150
contingent assets 306–7, 310–11
contingent liabilities 306, 307, 309–10,
        311, 560
Continuous Contemporary Accounting
        (CoCoA) 150
contracts
    construction 523–39
    equally unperformed, revenue
        recognition 199
    modification of terms of, revenue
        recognition 199–200
    onerous see onerous contracts
    revenue recognition 190–1, 194–6
control
    changes, unquoted companies, share
        valuations 739
    construction contracts 533
    definition 550–1
    internal 782–4
    inventories 508–9
convertible bonds 667, 668
convertible loans 326
convertible preference shares 279,
        667–8
Corefiling 766
corporate failure, prediction 731–5
corporate behaviour 778–80
corporate governance 775
    audits 801–2
    boards of directors 792, 794–7
    capital markets 780
    codes 792–802
    corporate behaviour and 778–80
    culture 778
    definition 775
    executive remuneration 789–92
    external audits 784–9
    good 775–7
    internal controls and internal audits
        782–4
    investor perspective 775–6
    legal safeguards 800–1
    legislation 792, 793
    national differences 777–8
    risk management 781–2
    shareholder activism 797–801
    stakeholder perspective 776
    systems perspective 775–7
    UK experience 793–802
corporate objectives, statements of 822
The Corporate Report 820–4
corporate social responsibility (CSR) 250,
        810–13, 818, 824
Corporate Social Responsibility,
        International Centre for 827

corporation tax 388–94
corruption 777–8
COSA (cost of sales adjustment) 159–60
COSO see Committee of Sponsoring
        Organizations of the Treadway
        Commission
cost approach to valuation of intangibles
        486
cost-plus contracts 524
costs
    agency costs 8
    amortised see amortised costs
    of assets, unconsumed 134
    of borrowing, capitalisation 414–15
    of capital 258, 728
    construction contracts 525–8
    current service 367
    decommissioning 305
    development 473–4
    distribution 36
    finance 37, 327–9
    interest see interest
    opportunity 156
    overheads 503–5
    past service, pension schemes 367–8
    pension schemes 365–9
    post retirement 365–9
    promotion 36
    property, plant and equipment
        414–16
    redundancy 312
    research see research costs
    of sales 36, 159–60
    selling 36
    structures 685
    termination 312
    transport 36
    warehousing 36
    see also expenses
counter-terrorism financing (CTF)
        standard 265
covenants, loans 721–3
CPA (Certified Public Accountants),
        USA 266
CPPA (current purchasing power
        accounting) 149–50, 151–5
creative accounting 217, 509–11
creative compliance 226
credibility 214
credit ratings 735–6, 780
creditors 275
    protection of 280–2
    share capital reduction 286–7
    see also loan creditors
CSEAR (Centre for Social and
        Environmental Accounting
        Research) 827
CSR see corporate social responsibility
CTF (counter-terrorism financing)
        standard 265
culture
    corporate governance 778
    differences, ethics and 255–6
    ethical influences of 249
cumulative preference shares 279
currency see foreign currency

current assets
  classification 43–4, 212
  emissions trading certificates 484
  held for sale 80–2
current cost accounting (CCA) 157–70
current cost reserves 159–67
current entry cost accounting *see*
  replacement cost accounting
current exit cost accounting *see* net
  realisable value accounting
current purchasing power accounting
  (CPPA) 149–50, 151–5
current ratios 693–4, 697
current service costs 367
current taxation 37–8, 395–6
current value accounting (CVA) 149
current values 169–71
curtailments, pension schemes 367–8
customer-related intangible assets 469
customers
  key performance indicators 54
CVA (current value accounting) 149

damaged inventory items 512
Datastream 757
DCF (discounted cash flow) 139–40, 142
debentures 289
debt
  covenants 721–3
  coverage 108
  debt/equity ratios 695–6, 697
  future, servicing 695
  perpetual 327
  service coverage ratios 717, 719
decision makers, chief operating
  (CODM) 72, 74, 78
decision-usefulness, financial reporting
  756–7
decommissioning costs 305
deductive approach, financial reporting
  236–7
Deere and Company 502
deferral method, deferred tax provisions
  397–9
deferred income, grants as 433–5
deferred taxation 396–404
defined benefit schemes *see* pension
  schemes
defined contribution schemes *see* pension
  schemes
Dell 756
Deloitte & Touche 825
Denmark
  environmental reporting 813
  unethical behaviour 780
depreciation 36
  accrual accounting 25
  annuity method 421–2
  calculations 419–23
  choice of method 49, 423, 436–7
  consolidated accounts 586
  corporation tax 389
  current cost accounting 160
  deferred tax and 397, 400–1
  definition 416–18
  diminishing balance method 420–1, 423

  formula 418
  machine-hour method 421, 423
  net realisable value accounting 156
  residual values 418, 419
  straight-line method 420, 423
  sum of the units method 421
deprival value 170
derecognition
  financial instruments 335–7
  non-financial liabilities 311
derivatives 330, 337
detail, lack of, published account
  limitations 715
development
  costs 473–4
  key performance indicators 53
Diageo 480
diluted earnings per share 657, 666–70
diminishing balance method,
  depreciation 420–1, 423
direct costs 36
direct labour 503, 507
direct materials 503, 506–7
direct method, cash flow statements
  100–1, 102
direct overheads 503, 504
directors
  boards
    corporate governance 792, 794–7
    sustainability commitment 812
  bonuses 211–12
  discipline 214
  independent 795, 796–7
  non-executive (NED) 794–7
  performance 791
  remuneration 789–92
  reports 54
disaggregation 200
discipline 214
disclosures
  associates 617
  cash flow statements 112
  compensation of key management
    personnel 85
  construction contracts 534
  controlling relationships 84–5
  discontinued operations 79–80
  earnings per share (EPS) 670–2
  financial instruments 342–6
  full 783–4
  government-related entities,
    exemptions 84–5
  hyperinflation 644
  intangible assets 471
  intellectual capital 484–5
  inventories 512–13
  joint ventures and joint operations
    617
  leases *see* leases
  mandatory 51–2
  non-financial liabilities 312–13
  operating leases 453–4
  pension schemes 371, 380–1
  property, plant and equipment (PPE)
    432–3
  provisions 305–6

related parties 82–6
research and development 472–3, 474
revenue recognition 200–1
of risks 342–6
segment information 74–6
shareholder protection 816
subsidiaries 617
tax expense 395
discontinued operations 79–80
discount rates 428
discounted cash flow (DCF) 139–40, 142
discounting 136, 302
discretionary expenditure, deferring 212
disposal groups 80–2
distributable profits 282–4
distributable reserves 276, 277, 281
distribution costs 36
distribution costs ratios 688–9
distribution overheads 503, 504
dividends
  adjustment factors, Shariah principles
    720
  cash dividend coverage ratio (CDCR)
    109, 111
  consolidated accounts 583, 586–7
  cover 109, 725–6
  events after reporting dates 68–9
  inflation accounting and 154
  inter-company 572
  policies 130
  purification, Shariah principles 720
  revenue recognition 186
  taxation 389–91
  yield 726
donations, charitable 393–4
Dow Jones Islamic Market Indices 719
Dow Jones Sustainability Indices 825,
  826
due care 250

earnings
  aggressive management 722–3
  before interest, tax, depreciation and
    amortisation (EBITDA) 111, 476,
    716–19, 793
  before interest, tax, depreciation,
    amortisation and rental expense
    (EBITDAR) 718
  before interest, tax, depreciation,
    amortisation, rental expense and
    management fees (EBITDARM)
    718
  construction contracts 533–4
  headline 671–2
  maintainable 671
  normalised 254
  per share (EPS) 655, 725
    basic (BEPS) 656–7, 661–70
    calculations 656–7, 658–61
    consolidated accounts 560
    conversion rights exercised 670
    diluted 657, 666–70
    disclosures 670–2
    importance 655–6
    income measurement and 130
    inventories and 513

earnings (*continued*)
    non-current assets and 438
    normalised 791–2
    rights issues 661–6
    share prices 277
    shareholders and 657–8
    quality of 41
    sustainable 254
    yields 725
Eastern Europe
    accounting profession 229
    corporate governance 778
Ebbers, Bernard 734
EBITDA (earnings before interest, tax,
    depreciation and amortisation)
    111, 476, 716–19, 793
EBITDA-based ratios 717–18
EBITDA/interest ratios 718, 719
EBITDAR (earnings before interest, tax,
    depreciation, amortisation and
    rental expense) 718
EBITDARM (earnings before interest,
    tax, depreciation, amortisation,
    rental expense and management
    fees) 718
ECGI (European Corporate Governance
    Institute) 793
Eco-Management and Audit Scheme
    (EMAS) 815
economic benefits 301
    see also future economic benefits
economic dimension 814, 817
economic income 136–42
economic lives 416–17, 418–19, 437, 470–1
economic measures of ability to create
    value 53
economic value added (EVA) 726–9, 730–1
economic value of businesses 131
economists, view of income measurement
    136–42
economy as behavioural entity 138
EEA (European Economic Area) 221
EECS (European Enforcers
    Co-ordination Sessions) 221
EFRAG *see* European Financial
    Reporting Advisory Group
Eighth Directive 219
elements of financial statements 242–3
EMAS (Eco-Management and Audit
    Scheme) 815
embedded derivatives 337
emissions trading 483–4
empirical inductive approach, financial
    reporting 236
employees
    benefits 360–81
    key performance indicators 54
    revenue restatement effects 184
employment reports 822
EnBW 556
enjoyment 138
Enron 213, 221, 222, 264, 714, 722, 783,
    786, 788
enterprise value/EBITDA ratios 717
entity-wide cash-based performance
    measures 716–19

Environment Agency
    Turning Questions into Answers:
    Environmental Issues and Annual
    Financial Reporting (joint 2009
    report with ICAEW) 819
environmental assets reporting 817
environmental auditing 818
environmental dimension 814, 817–18
environmental liabilities 305, 816
environmental issues
    key performance indicators 54
environmental profit and loss accounts 813
environmental reporting 809, 812–13,
    819–20
environmental, social and governance
    (ESG) risks 825
EPEC (European PPP Expertise Centre)
    528
EPS (earnings per share) *see* earnings
equally unperformed contracts, revenue
    recognition 199
equipment *see* property, plant and
    equipment
equitable exchanges 253–4
equity 276
    definition 243
    depletion 481
    impact on IFRS adoption 223–4
    investors 275–6
    liabilities and 325–6
    statements of changes in (SOCE)
    42–3, 585
equity accounting method 606, 614,
    617–18
equity compensation plans 363
equity instruments 324
equity market capitalisation 227–8
equity-settled share-based payments
    374–7, 378–9
Ernst & Young 788, 789
errors
    adjusting events 68
    freedom from 241
    inventories, adjustments for 512
    prior-period 70–1
    strategic 734
ESG (environmental, social and
    governance) risks 825
ESMA *see* European Securities and
    Markets Authority
estimates 71, 133, 138, 174, 300, 437, 737
ethics 248
    in accountants' work environments,
    research report 256–8
    accounting standards setting process
    and 249–50
    advice 257–8
    Centre for the Study of Global Ethics
    827
    collapse in organisations, effects of 259
    company codes 260–1
    fundamental principles 251, 253
    guidelines 248–9
    IFAC *Code of Ethics for Professional*
    Accountants 250–3
    individual guidelines 248–9

    meaning of ethical behaviour 248–9
    neutrality and 250
    personal 248–9
    policies 257–8
    professional guidelines 249
    seniors' perspectives 257
    students and 257, 266–7
    trainees' perspectives 257
    unethical behaviour 258–63, 779–80
    values, implications for principles
    versus rules-based approaches to
    accounting standards 253–6
    whistleblowing 263–5
Europe
    awards for sustainability reporting 819
    see also European Union
European Corporate Governance
    Institute (ECGI) 793
European Economic Area (EEA) 221
European Enforcers Co-ordination
    Sessions (EECS) 221
European Federation of Accountants
    (FEE) 221, 818
European Financial Reporting Advisory
    Group (EFRAG)
    Considering the Effects of Accounting
    Standards (Discussion Paper
    2011) 250
    Towards a Disclosure Framework for the
    Notes (Discussion Paper 2012) 215
European Parliament
    IFRS 8 reservations 77
European PPP Expertise Centre (EPEC)
    528
European Securities and Markets
    Authority (ESMA)
    Considerations of materiality in financial
    reporting (consultation paper
    2011) 241
European Union
    Accounting Directive 219–21, 225
    Accounts Modernisation Directive 825
    credit rating agency regulation 736
    Eco-Management and Audit Scheme
    (EMAS) 815
    Eighth Directive 219
    Emissions Trading Scheme 483–4
    environmental auditing 818
    environmental reporting 818
    financial instruments concerns 323
    Fourth Directive 219, 226, 229
    IFRS adoption 221
    Member State Options (MSO) 220–1
    regulatory burden on SMEs and
    micro-undertakings 220–2
    Seventh Directive 219
    Sixth Action Programme
    'Environment 2010: Our Future,
    Our Choice' 820
    standard setting and enforcement 219–21
EV (enterprise value)/EBITDA ratios 717
EVA (economic value added) 726–9, 730–1
EVA Momentum 729
evasion of tax 391–4
events after reporting periods 67–9
exceptional items 40–2, 52

excluded subsidiaries 558–9
Executive Committee, FRC 216, 217
executive remuneration 789–92
exercise dates, share-based payments 375
expectation gap, audits 788, 801
expected income 139–41
expected losses model 350
expenditure, capital (CAPEX) 110
expenses
  analyses by function 37
  definition 243
  recognition 243
  share-based payments 373–4
  see also costs; operating expenses
explanatory notes to financial statements
  44–7
exposure drafts see International
  Accounting Standards Board
eXtensible Business Reporting Language
  (XBRL) 756, 757–68
eXtensible Mark-up Language (XML)
  758–9
external audits 784–9
Exxon Mobil 719

failure prediction 731–5
fair exchanges 253–4
fair override 50–1
fair presentation 50–1
  see also true and fair view
fair values
  acquisitions 551–2, 555–7, 559–60
  agricultural produce 514–16
  biological assets 514–16
  consolidation depreciation adjustments
    586
  definition 173
  distribution decisions and 284
  emissions trading certificates 483
  estimates 174
  financial assets 322–3, 330–1, 338
  financial instruments 284, 322–3,
    337–9, 347–8
  financial liabilities 322–3, 330–1, 338
  financial statements, effects on 174–5
  hedging 340
  hierarchy 173–4
  increasing use 173
  intangible assets 486
  judgement 174
  leases 454, 456, 457
  measurement, IFRS 13 see
    International Financial Reporting
    Standards
  pension schemes 366, 380
  share-based payments 375–7
fair view 44
  see also true and fair view
fairness 249–50
faithful representation 16, 241
Fame 757
familiarity threats 253
families, ethical influences of 249
farming 513–16
FASB see Financial Accounting
  Standards Board

FATF (Financial Action Task Force)
  265
FCF (free cash flow) 110–11, 719
FD (finance directors) 817
FEB see future economic benefits
FEE (European Federation of
  Accountants) 221, 818
fictitious transfers 510
FIFO (first-in-first-out) 47–8, 499–500,
  502
Finance Acts 388, 391
finance costs 37, 327–9
finance directors 817
Finance for the Future Awards 819
finance overheads 503, 504
finance leases see leases
Financial Accounting Standards Board
  (FASB)
  accounting standards 223
  Accounting Standards Codification 221
  accrual accounting 27
  Convergence Project 672
  fair values 173
  Norwalk Agreement 223
  revenue recognition 185, 187
  standard setting 221–2
  see also Statements of Financial
    Accounting Standards
financial accounting theory historical
  overview 235–7
Financial Action Task Force (FATF) 265
financial assets 322–3, 324–5, 330–1,
  332–4, 338, 349–50
financial capital 143, 809
financial capital maintenance 27, 151–3,
  167–8
financial conditions 787
financial crisis (2008) 322–4
financial databases, ratio analysis 756–7
financial decision model 5–6
financial information
  qualitative characteristics 240–1
  reliability improvement 714–15
  useful 240–1
financial instruments 321
  amortised costs 338–9, 347–8
  classification 334–5, 347–8
  definitions 324–5, 330–5
  derecognition 335–7
  developments 346–50
  disclosures 342–6
  ED/2013/3: 349
  emissions trading certificates 483
  EU concerns 323
  fair values 284, 322–3, 337–9, 347–8
  financial crisis (2008) 322–4
  gains presentation 348
  hedging 339–41
  IAS 32: 322, 324–9, 342
  IAS 39: 322–5, 329–41, 342, 346, 347–
    50
  IASB problems 321–4
  IFRS 7: 321, 342–6
  IFRS 9: 321, 346–8
  losses presentation 348
  measurement 337–9, 347–8

offsetting 329
presentation 325–7
private finance initiative 532
recognition 335–7, 347–8
rules versus principles 322
US GAAP 322–3
financial liabilities 322–3, 324–5, 330–1,
  335–7, 338
financial performance 686–92
financial ratios see ratios
financial reporting 713
  analysis of published financial
    statements 713–41
  conceptual framework 230, 234–45
  decision-usefulness 756–7
  deductive approach 236–7
  empirical inductive approach 236
  global standards evolution 211–30
  hard copy 755–6
  history of 235–7
  information flows post-XBRL 759–60
  information flows pre-XBRL 757–8
  interim 767
  international standards 217–30
  internet 755–69
  materiality in 241
  national differences 226–30
  objectives 755–8
  pension schemes 361
  shareholder information 714–15
  real-time 767–8
  small and medium-sized enterprises
    225
  stewardship 755–6
  see also published accounts
Financial Reporting Council (FRC)
  Accounting Council 216
  Actuarial Council 216
  Audit and Assurance Council 216
  Board 216–17, 234
  Case Management Committee 216,
    217
  Codes and Standards Committee 216,
    787
  Conduct Committee 216, 217
  corporate governance 793
  Corporate Reporting in a Digital World
    (project) 768
  exceptional items, note on 41
  Executive Committee 216, 217
  Financial Reporting Review Panel see
    Financial Reporting Review Panel
  financial statements, consistency in
    preparation 42
  function 215–16
  going concern 16
  Monitoring Committee 216, 217
  narrative reporting review 715–16
  Stewardship Code 799–800
  Strategic Reports, guidance 54
  structure 216
  true and fair view 50
  UK Corporate Governance Code 794
Financial Reporting Lab (FRL)
  Corporate Reporting in a Digital World
    (project) 768

Financial Reporting Review Panel
(FRRP) 42, 51, 77–8, 215, 217, 483
Financial Reporting Standard for Smaller
Entities (FRSSE) 225
Financial Reporting Standards (FRS)
5 *Reporting the Substance of
Transactions* 532
15 *Tangible Fixed Assets* 415
100 *Application of financial reporting
requirements* 388
101 *Reduced disclosure framework* 388
102 *The financial reporting standard
applicable in the UK and Republic
of Ireland* 388
financial reports
ethics and 254–5
unethical behaviour, implications for
stakeholders using 258–63
financial risks 781
financial securities, revenue recognition
198
financial statements
analysis
financial performance 686–92
financing 695–7
liquidity 692–5
peer comparisons 697–8
published financial statements 713–41
reports based on 698–9
comparability 214
Conceptual Framework 239–40
consistency in preparation 42
current tax accounting 37–8
elements 242–3
explanatory notes 44–7
for external users, contents 22
fair presentation 50–1
fair values effects on 174–5
financial instruments classification and
334–5
formats of statements for publication
35–40, 47–9
hyperinflation 643
IAS 1 alternative method 39–40
joint ventures 613–18
internal statements of income from
trial balances 32–5
investors' needs other than 51–5
misstatements 70
narrative reporting survey 54–5
notes to 41–2
objectives 21–2, 239–40, 680
omissions 70–1
other comprehensive income 39–40
preparation
under cash flow concept 10–12
published statements 32–55
property, plant and equipment 436–8
prospective changes 69, 71
retrospective changes 69, 70
reviews for management purposes
680–701
separate 616, 617–8
statements of changes in equity 42–3
statements of financial position 43–7
XBRL 756, 757–68

see also annual reports; cash flow
statements; financial reporting;
published accounts; statements of
comprehensive income;
statements of financial position
financial structures of companies 684–5
financing
activities, cash flows 101, 106, 111
financial statement analysis, ratios
695–7
Findel plc 344–6
finite lives, intangible assets 470–1
Finland, sustainable investing companies
826
first-in-first-out (FIFO) 47–8, 499–500,
502
Fiscal Studies, Institute for *see* Institute
for Fiscal Studies
Fisher, Irving 138–9
fixed assets *see* non-current assets
fixed-price contracts 524
football clubs, money laundering and 265
Ford 824
Foreign Corrupt Practices Act 1977,
USA 777
foreign currency
definitions 633
exchange differences, accounting for
632–45
formats
financial statements for publication
35–40, 47–9
Fourth Directive 219, 226, 229
France
agricultural accounting 515–16
legal system 226
sustainable investing companies 826
tax and reporting systems 228
fraud 68, 70–1, 222, 259, 509, 779, 789
see also creative accounting
fraus legis 394
FRC *see* Financial Reporting Council
free cash flow (FCF) 110–11, 719
freehold land 417–18
Friedman, Daniel 255
Friedman, Milton 253, 810
FRL *see* Financial Reporting Lab
FRRP *see* Financial Reporting Review
Panel
FRS *see* Financial Reporting Standards
FRSSE (Financial Reporting Standard
for Smaller Entities) 225
FTSE Shariah indices 719
FTSE4Good Index Series 825, 826
full disclosures 783–4
functional currencies 632, 636–8
fundamental principles, threats to
compliance with 253
future cash flows *see* cash flows
future economic benefits (FEB) 468, 469,
470, 472, 478–9
future liabilities 559
future losses 560
future performance 3, 130
future prospects, statements of 822
future values 137

GAAP, US *see* United States
GAAR (General Anti-Abuse Rule) 392
gains
actuarial, pension schemes 368
financial instruments 339–40, 348
holding 155, 161–2, 167–8
inflation accounting 155
unrealised 486
see also income; profits
GAP 481
GDP (gross domestic product) 227
gearing 164, 436
ratios 481, 695–7
General Anti-Abuse Rule (GAAR) 392
General Electric 212–13, 766
general price indices (GPI) 150, 153, 154,
167
general purchasing power (GPP) *see*
current purchasing power
accounting
Generally Accepted Accounting
Principles (GAAP)
United States *see* United States
Germany
corporate governance 777
finance of industry 227
legal system 226
sustainable investing companies 826
tax and reporting systems 228
Getronics 279
Geveke nv Amsterdam 727
GlaxoSmithKline 719
global accounting standards 223–4
Global Ethics, Centre for the Study of
827
Global Forums, ACCA 827
Global Reporting Initiative (GRI) 813,
814–15, 818, 828
Go-Ahead 433
going concern 16, 24–6, 69, 135–6, 156,
242
golden handshakes 780
Goldfields 417–18
good corporate governance 775–7
goodwill
accounting treatment 475–9
amortisation 475–8
consolidated accounts 556, 558, 570,
573, 746
definition 475
IAS 38: 476, 479
IFRS 3: 475–80, 482–3
impairment 475–6, 478–9
intangible assets identifying to reduce
amount of 476–7
internally generated 475
keeping in statements of financial
position unchanged 475, 477
negative 480, 558
purchased 475, 481, 551–2
review of implementation of IFRS 3:
485–6
statements of financial position
approach 478–9
writing off to reserves 475, 477, 481
Google 481, 719

governance
  in audit firms 788
  corporate *see* corporate governance
  environmental, social and governance
    (ESG) risks 825
government-related entities 84–5
governments
  ethics actions 260, 262
  grants 433–5, 515
  whistleblowing and 264
GPI *see* general price indices
GPP (general purchasing power) *see* current
    purchasing power accounting
grant dates, share-based payments 375
grants 433–5, 515
GRI *see* Global Reporting Initiative
Grontmij N.V. 136
gross domestic product (GDP) 227
gross profit margin ratios 688
groups
  accounting at the date of acquisition
    549–61
  definition 550
  see also acquisitions; business
    combinations; consolidated
    accounts

H-scores 733–4, 740
Halfords 740
hard copy financial reporting 755–6
hazards 782
HCA (historical cost accounting) *see*
    historical cost
headline earnings 671–2
hedge funds 228
hedging 322, 339–41, 348–9, 635
held for sale 80–2
held-for-trading investments 330
held-to-maturity investments 331–2, 338
Hewlett-Packard Company 190–1
Hicks income model 139–41
hidden liabilities, unethical behaviour
    and 258–9
historical cost
  accounting (HCA) 132–3
    acceptance 236
    illustrations 151–60
    problems 148–9
    profits 168–9
    unsuccessful attempts to replace
      170–1
  convention 22
  principles 135
  valuations, agricultural activities
    515–16
HM Revenue and Customs 130–2, 761
HMV 136
holding gains 155, 161–2, 167–8
Holmen AB 514
Hooker Chemicals Corporation 816
horizontal analysis 684–5
HSBC 724
HTML (Hyper Text Mark-up
    Language) 757, 758
Hugo Boss 482
human capital 809

human resources valuation 487
Hyper Text Mark-up Language
    (HTML) 757, 758
hyperinflation 172, 643–4

IAASB (International Auditing and
    Assurance Standards Board) 787
IAS *see* International Accounting
    Standards
IASB *see* International Accounting
    Standards Board
IASC *see* International Accounting
    Standards Committee
IBM 481, 719
ICAEW *see* Institute of Chartered
    Accountants in England and
    Wales
ICAS *see* Institute of Chartered
    Accountants of Scotland
ICD (intellectual capital disclosures)
    484–5
ICMA (International Capital Market
    Association) 259
IFAC *see* International Federation of
    Accountants
IFAD (International Forum on
    Accountancy Development) 235
IFRIC (International Financial Reporting
    Interpretations Committee) 215
IFRS *see* International Financial
    Reporting Standards
IFS *see* Institute for Fiscal Studies
IIA *see* Institute of Internal Auditors
IIMR (Institute of Investment
    Management and Research)
    671–2
impairment 82, 349–50, 425–31, 470–1,
    476, 478–9, 609–11
Imperial Holdings Limited 811, 823
Improvement Project, IASB 672
imputation corporation tax system 390–1
incentives 780
inclusivity 815
income
  accountants' view 132–6
  classification 36–9
  comprehensive 39–40, 43
  concept of 132
  deferred, grants as 433–5
  definition 243
  economic 136–42
  economists' view 136–42
  expected 139–41
  grants as 433–5
  information 130
  as means of control 130–1
  as means of prediction 130
  measurement, role and objective 129–32
  operating 40–2, 155
  recognition 243
  replacement cost 142–3
  smoothing, inventories 498
  statements *see* statements of
    comprehensive income
  streams 427
  see also gains; profits

income approach to valuation of
    intangibles 486
income tax 388
incurred losses model 350
indefinite lives, intangible assets 471
independence
  auditors 801
  ethics and 252
independent directors 795, 796–7
independent reviews 821
indirect method *see* cash flow statements
indirect overheads 503, 504
individual ethical guidelines 248–9
inflation 143, 172, 436, 643–4, 740
inflation accounting 148–75
influence, significant 605–6
information
  flows
    financial reporting post-XBRL
      759–60
    financial reporting pre-XBRL 757–8
  needs
    financial statements 239–40
    of internal users 3, 5–8
    of managers 3, 4–5
    of shareholders 3–4, 714–15
  overload 813
  unaudited, published account
    limitations 715–16
  see also financial information
Infosys Technologies 487
Inline XBRL (iXBRL) 761, 765–6
Inman, M.L. 731
input method, construction contract
    earnings 534
insider trading 779
instance documents, XBRL 762, 763, 764
Institute for Fiscal Studies (IFS) 393
  *Countering Tax Avoidance in the UK:
    Which Way Forward?* (discussion
    paper 2009) 394
Institute of Chartered Accountants in
    England and Wales (ICAEW)
  Ethics Advisory Service 259
  The Future Shape of Financial Reports 99
  risk reporting 714
  Support Members Scheme 263
  Technical Release 02/10 *Determination
    of realised profits and losses in the
    context of distributions under the
    Companies Act 2006*: 284
  *Turning Questions into Answers:
    Environmental Issues and Annual
    Financial Reporting* (joint 2009
    report with Environment Agency)
    819
Institute of Chartered Accountants of
    Scotland (ICAS)
  ethics 266
  Taking Ethics to Heart (discussion
    paper report) 256–8, 259
  The Future Shape of Financial Reports 99
  *The implementation of IFRS in the UK,
    Italy and Ireland* (research report
    2008) 224
  principles-based approach 244

Institute of Internal Auditors (IIA)
    Model Audit Committee Charter 783
Institute of Investment Management and
    Research (IIMR) 671–2
institutional investors 228, 791–2, 798–800
institutional uses of XBRL 764
Institutional Voting and Information
    Service (IVIS) 796
insurable risks 781
insurance companies 228
intangible assets 468
    accounting treatment 470–1
    acquisitions 560
    amortisation 470, 471
    brands 480–3
    definition 468
    disclosures 471
    emissions trading certificates 483
    examples of 469
    fair values 486
    finite lives 470–1
    goodwill *see* goodwill
    human resources valuation 487
    identifying to reduce amount of
        goodwill 476–7
    IFRS for SMEs 474–5
    impairment 470–1
    indefinite lives 471
    intellectual capital 484–5
    measurement 469
    recognition criteria 469–71
    research and development 471–4
    review of implementation of identified
        intangibles under IAS 38: 486–7
    useful economic lives 470–1
    valuation 486, 487
Integrated Reporting Framework 809,
    828
integrity 250
Intel 476–7, 486
intellectual capital 809
intellectual capital disclosures (ICD)
    484–5
intellectual property
    transfers, revenue recognition 198–9
inter-company performance 756–7
inter-company transactions 571–4, 582–3
inter-period performance 756–7
interest
    compound 137
    consolidated accounts 583, 586–7
    costs, pension schemes 367
    cover 108, 695
    leases 457
    present values and 138
    revenue recognition 186
interests
    in associates 617
    non-controlling *see* non-controlling
        interests
    in joint operations and joint ventures
        617
    shareholders 816
    stakeholders 816
    in subsidiaries 617
    vested 212

interim financial reporting 767
internal accounting, XBRL 766–7
internal audits 782–4
internal conditions 786
internal control 782–4
internal rate of return (IRR) 7–8, 729
internal reporting 4–8
internal statements of income from trial
    balances 32–5
internally generated goodwill 475
International Accounting Standards
    (IAS)
    1 *Presentation of Financial Statements*
        alternative method 39–40
        cash and cash equivalents 103
        exceptional items 52
        fair presentation 50
        formats of statements of income
            35–40
        statements of comprehensive
            income 39–40
        statements of financial position 43–6
        tax rules and 389
    2 *Inventories*
        agricultural produce 515
        definition 497–8
        development expenditure 473
        disclosure requirements 512–13
        properties held for sale 435
        self-constructed assets 414
        valuations 499
    7 *Statement of cash flows* 27–8, 99–117
    8 *Accounting Policies, Changes in
        Accounting Estimates and Errors*
        45, 50, 69–71, 240, 395
    10 *Events After the Reporting Period*
        67–9, 572
    11 *Construction Contracts* 523–39
    12 *Income Taxes* 388, 395–404, 433
    14 *Segment Reporting* 77
    15 *Information Reflecting the Effects of
        Changing Prices* 171–2
    16 *Property, Plant and Equipment*
        413–25, 478, 514, 569, 817
    17 *Leases* 435, 449–63
    18 *Revenue* 185–6, 187–8, 406, 523
    19 *Employee Benefits* 312, 360–81, 379
    20 *Accounting for Government Grants
        and Disclosure of Government
        Assistance* 413, 433–4, 515
    21 *The Effects of Changes in Foreign
        Exchange Rates* 632, 633, 636–8
    23 *Borrowing Costs* 413, 414–15
    24 *Related Party Disclosures* 82–6
    26 *Accounting and Reporting by
        Retirement Benefit Plans* 360, 379–
        81
    27 *Separate Financial Statements* 616,
        617–18
    28 *Investments in Associates* 531, 606,
        607, 614, 615–16, 617, 618
    29 *Financial Reporting in
        Hyperinflationary Economies* 172,
        643–4
    30 *Disclosures in the Financial
        Statements of Banks* 342

    31 *Interests in Joint Ventures* 613
    32 *Financial Instruments: Disclosure and
        Presentation* 324–9, 342
    32 *Financial Instruments: Presentation*
        322
    33 *Earnings per Share* 225, 656–7,
        670–2
    34 *Interim financial reporting* 225
    35 *Discontinuing Operations* 431
    36 *Impairment of assets* 82, 413,
        425–31, 471, 558, 609
    37 *Provisions, Contingent Liabilities
        and Contingent Assets* 298
        constructive obligations 311
        contingent assets 307, 310–11
        contingent liabilities 306, 307,
            309–10
        disclosures for non-financial
            liabilities 312–13
        disclosures for provisions 305–6
        ED (exposure draft) *Non-financial
            Liabilities* 300, 307–14
        ED/2010/1 *Measurement of
            Liabilities in IAS 37*: 314
        environmental liabilities 816
        ex gratia pension arrangements 361
        legal obligations 311
        measurement of provisions 301–2
        onerous contracts 199, 312
        present values 311
        property, plant and equipment 414
        recognition of provisions 300–1
        restructuring 312, 559
        subsequent measurement and
            derecognition 312
    38 *Intangible Assets* 310–11, 468–74,
        479, 482–3
        review of implementation of
            identified intangibles under
            486–7
    39 *Financial Instruments: Recognition
        and Measurement* 322–5, 329–41,
        342, 347–50
        fair values 284
        replaced by IFRS 9: 346
    40 *Investment property* 413, 435–6, 453
    41 *Agriculture* 435, 513–16
    harmonisation 184–5
    list 219
    see also International Financial
        Reporting Standards
International Accounting Standards
    Board (IASB)
    accrual accounting 27
    *Conceptual Framework for Financial
        Reporting* 238–9, 242, 828
        accrual accounting 23
        definition of elements 24, 242–3
        goodwill 478–9
        measurement of elements 243
        objective of financial statements
            21–2, 239–40
        recognition of elements 243
        *Review of the Conceptual Framework
            for Financial Reporting* (discussion
            paper DP/2013/1) 243–4

International Accounting Standards
    Board (IASB) (*continued*)
  qualitative characteristics of useful
    financial information 240–1
  taxation 389, 405
  Convergence Project 672
  exposure drafts
    ED *Non-financial Liabilities* 300,
      307–14
    ED/2009/12 *Financial Instruments:
      Amortised Cost and Impairment*
      350
    ED/2010/1 *Measurement of
      Liabilities in IAS 37*: 314
    ED/2010/2 *Conceptual Framework
      for Financial Reporting – The
      Reporting Entity* 240
    ED/2013/3 *Financial Instruments:
      Expected Credit Losses* 349
    ED/2013/10 *Equity Method in
      Separate Financial Statements*
      617–18
  fair values 173
  financial crisis (2008) and 322–4
  financial statements composition 22
  governance 323
  IFRS 8 post-implementation review 78
  Improvement Project 672
  inflation accounting approach 172
  Management Commentary (IFRS
    Practice Statement) 52–3, 109
  Norwalk Agreement 223
  objective of financial statements 21–2
  political pressures 323–4
  revenue recognition 184–5
  small and medium-sized entities 225
  work of 217–19
International Accounting Standards
    Committee (IASC)
  *Framework for the Presentation and
    Preparation of Financial
    Statements* 218, 230, 237–8
  assets 307, 310
  financial capital maintenance 27
  liabilities 307, 307, 310
  measurement bases 503
  research expenditure 472
  revision 238
  underlying assumptions 242
International Auditing and Assurance
    Standards Board (IAASB) 787
International Capital Market Association
    (ICMA) 259
International Centre for Corporate Social
    Responsibility 827
International Federation of Accountants
    (IFAC)
  *Code of Ethics for Professional
    Accountants* 250–3
International Financial Reporting
    Interpretations Committee
    (IFRIC) 215
International Financial Reporting
    Standards (IFRS)
  1 *First-time Adoption of International
    Financial Reporting Standards* 223

  2 *Share-Based Payments* 360, 373–9
  3 *Business Combinations*
    brands 482–3
    fair values 173, 559–60
    goodwill 475–80, 482–3
    non-controlling interests 553
    review of implementation of 485–6
  4 *Insurance contracts* 225
  5 *Non-current Assets Held for Sale and
    Discontinued Operations* 79–80,
    225, 413, 431–2, 606
  6 *Exploration for and Evaluation of
    Mineral Resources* 418
  7 *Financial Instruments: Disclosures*
    321, 342–6, 618
  8 *Operating Segments* 71–8, 107, 225,
    559
  9 *Financial Instruments* 321, 346–8,
    483, 614, 617, 635
  10 *Consolidated Financial Statements*
    549–51, 568, 615
  11 *Joint Arrangements* 612, 613–14,
    616, 617, 618
  12 *Disclosure of Interests in Other
    Entities* 616
  13 *Fair Value Measurement* 173–5, 338,
    423, 486, 559–60
  15 *Revenue from Contracts with
    Customers* 185, 187–201
  compliance 221
  ED/2009/12 *Financial Instruments:
    Amortised Cost and Impairment*
    350
  ED/2010/2 *Conceptual Framework for
    Financial Reporting – The
    Reporting Entity* 240
  ED/2013/3 *Financial Instruments:
    Expected Credit Losses* 349
  environmental assets reporting 817
  environmental liabilities reporting 816
  EU adoption 221
  harmonisation 184–5
  IFRS for SMEs
    government grants 435
    intangible assets 474–5
    investment properties 436
    issue 225
    measurement 425
    non-current assets held for sale 432
    property, plant and equipment 415,
      425, 432, 435, 436, 437
  *The implementation of IFRS in the UK,
    Italy and Ireland* (ICAS research
    report 2008) 224
  list 219
  production of 217–19
  United States and 223
  XBRL and 762
  see also International Accounting
    Standards
International Forum on Accountancy
    Development (IFAD) 235
International Organization for
    Standardization (ISO) 815–16
International Standards on Auditing
    (ISA)

  540 *Auditing Accounting Estimates,
    including Fair Value Accounting
    Estimates, and Related Disclosures*
    723
International Valuation Standards
    Council (IVSC) 482
Internet, financial reporting 755–69
intimidation threats 253
introductory reviews, ratio analysis 683–5
inventories 497
  adjusting events 68
  agricultural activities 513–16
  control 508–9
  creative accounting 509–11
  current cost accounting 159
  definition 497–8
  disclosures 512–13
  emissions trading certificates 484
  errors, adjustments for 512
  foreign exchange differences 634
  identification of items 511–12
  inaccurate records 510
  income smoothing 498
  obsolete 511
  physical condition 512
  profits, valuation impact on 498–9
  turnover ratios 690–1, 697
  unmarketable 511
  valuation 44, 47–8, 132, 497–506
  work-in-progress 506–8
  year-end count 511–12
investing activities, cash flows 101, 103,
    106, 109–10
investment companies, distributable
    profits 283
investment, inward 822
investment property 435–6
investor-specific ratios 723–6
investors
  corporate governance perspective
    775–6
  decision-making needs 51–5
  environmental issues and 825–6
  IFRS 8, views on 78
  institutional 228, 791–2, 798–800
  protection 560
  Sarbanes–Oxley Act 222–3
Investors Forum 800
inward investment 822
Ireland
  *The implementation of IFRS in the UK,
    Italy and Ireland* (ICAS research
    report 2008) 224
  legal system 226
IRR (internal rate of return) 7–8, 729
ISA *see* International Standards on
    Auditing
Islamic Indices 719–20
ISO 14001: 815
Italy
  *The implementation of IFRS in the UK,
    Italy and Ireland* (ICAS research
    report 2008) 224
ITV 41
IVIS (Institutional Voting and
    Information Service) 796

IVSC (International Valuation Standards Council) 482
iXBRL (Inline XBRL) 761, 765–6

James Hardie Group 258–9
Japan
    finance of industry 227
    legal system 226
Jennings, M.M. 259
Johnson Matthey 211–12, 526
joint operations 612–18
joint ventures 612–18
jointly controlled assets 613
journal adjustments 510–30

Kay Review 800
key management personnel compensation 85
key performance indicators (KPI) 53–4
Kingfisher Group 110–11
KPI (key performance indicators) 53–4
KPMG 224, 252, 728, 789

land
    agricultural 515
    freehold 417–18
    leases 452–3, 459
language, and accounting systems 229–30
Large and Medium-sized Companies and Groups (Accounts and Reports) (Amendment) Regulations 2013: 791, 792
large-block investors see institutional investors
last-in-first-out (LIFO) 499, 502
lawyers 780
leases 449
    accounting requirements
        finance leases 454–8
        new approach 460–2
        operating leases 453–4
    background 449–51
    classification 452–3, 456, 459
    commercial advantages 450
    controversy 451–2
    definition 449
    disclosures
        finance leases 458–9
        operating leases 453–4
    finance leases 451, 452–3, 454–9, 461
    IAS 17: 449–63
    land and buildings 452–3, 459
    lessee accounting 461–2
    liabilities associated to right of use 461
    off-balance sheet financing 450, 460
    operating leases 451, 452–4, 459, 461
    plant and equipment 461
    problems perceived, proposed solution 461–2
    property 461
    revenue recognition 197
    right-of-use approach 461–2
    Type A 461–2
    Type B 461, 462
legal obligations 311
legal risks 782

legal systems, national 226–7
Lehman Bros 188, 788
lessees and lessors see leases
Lev & Schwartz model 487
leverage see gearing
Lexmark 511
liabilities 298
    adjusting events 68
    auditors 788–9
    capital maintenance and 281
    consolidated accounts 570
    contingent 306, 307, 309–10, 311
    definitions 24, 242, 307
    environmental, reporting 816
    equity and 325–6
    financial see financial liabilities
    future 559
    hidden, unethical behaviour and 258–9
    leases, associated to right of use 461
    non-financial 307–14
    pension scheme changes 366–7
    recognition 243
liability method, deferred tax provisions 398, 399
LIFO (last-in-first-out) 499, 502
limitations of published accounts see published accounts
limited liability companies
    creditor protection 281
limited liability partnerships (LLP), auditors 789
Linde AG 550, 559
liquidity ratios 692–5
listed companies see public companies
Lloyds TSB 756, 825
LLP (limited liability partnerships), auditors 789
loan creditors 275, 287
loans
    convertible 326
    covenants 721–3
    financial assets 332, 338
    related parties 86
    see also borrowings
local currencies 643
London Benchmarking Group 824–5
long-service benefits 371
    see also pension schemes
long-term contracts, revenue recognition 197
Lonmin 722
looting 779, 788
losses
    actuarial, pension schemes 368
    associates 609–11
    capital reduction 284–9
    consolidated accounts 568–9
    expected losses model 350
    financial instruments 339–40, 350
    future 560
    impairment 428–9
    incurred losses model 350
    inflation accounting 155
    operating 304
    tax losses 395–6

machine-hour method, depreciation 421, 423
macroeconomic concept 130–1
maintainable earnings 671
Malaysia
    awards for sustainability reporting 819
MAN 413
management
    attitudes 49
    changes, unquoted companies, share valuations 740
    commentaries 52–3
    compensation of key management personnel 85
    decisions, brands and 481–2
    defects 734
    EVA use attitudes 728–9
    by objectives 261
    orientation 154
    performance 730–1
    reporting to 3, 4–5
    representations 787
    reviews of financial statements for purposes of 680–701
    USA 222
managers see management
mandatory standards 212–14
manufactured capital 809
mapping, XBRL 762, 763
margin ratios 687–9
market approach to valuation of intangibles 486
market positioning 53
market-related vesting conditions 377
market share, key performance indicators 53
marketing, ethics and 252
marketing-related intangible assets 469
markets
    morals and, realignment 255
Marks and Spencer 730
materiality 70–1, 240–1, 559, 815, 819
McDonald, Richard E. 779
measurement
    agricultural produce 514–16
    biological assets 514–16
    conceptual framework 244
    contingent liabilities 310
    fair values 173–5
    financial capital maintenance 27
    financial instruments 337–9, 347–8
    IASB Conceptual Framework 243
    income measurement 129–44
    intangible assets 469
    non-controlling interests 553
    non-current assets held for sale 431–2
    non-financial liabilities 310, 312, 314
    pension assets or liabilities 367
    property, plant and equipment 423–5
    provisions 301–2
    segment information 74
    share-based payments 374
    termination benefits 373
medium-sized companies see small and medium-sized enterprises
Member State Options (MSO), EU 220–1

Merck 111
mergers *see* acquisitions; business combinations; consolidated accounts
micro-undertakings
  regulatory burden on 220–2
minerals 417–18, 515
minimum share capital 282
mining assets 417–18
Miniscribe Corporation 510
minority interests *see* non-controlling interests
misappropriation 779
misrepresentation 779
misstatements 70
mistakes *see* errors
Mitsubishi Logistics Corporation 559
Model Audit Committee Charter 783
modification of terms of contracts, revenue recognition 199–200
monetary items 155, 634, 644
monetary working capital adjustment (MWCA) 160–1
money capital 142, 143
money laundering 265
Monitoring Committee, FRC 216, 217
Moody's 735
morals 255
  see also ethics
Morrisons 718, 726
MSCI Barra 720
multinational companies, ethics 256, 262
Munksjo AB 568, 572
Muslim countries
  corporate governance 777
  shariah compliance 719–20
MWCA (monetary working capital adjustment) 160–1

NAPF *see* National Association of Pension Funds
narrative reporting 54–5, 784, 812–13, 825–6
National Association of Pension Funds (NAPF)
  remuneration principles 792
national differences
  conceptual framework 234–5
  financial reporting 226–30
  legal systems 226
natural capital 809
NED (non-executive directors) 794–7
negative covenants 721
negative goodwill 480, 558
negligence, auditors 801
Nemetschek AG 633, 638
Nestlé 45, 454, 719
net asset value per share 792
net assets 134, 560, 570
net cash positions 16
net debt/EBITDA ratios 717
net investment hedges 341
net operating profit after tax (NOPAT) 727, 728
net present value (NPV) 7, 9, 328–9
net profits

impact on IFRS adoption 223–4
  margin ratios 687
net realisable value accounting (NRVA) 150, 151–6, 173
net realisable values (NRV) 484, 498, 505–6, 510
Netherlands
  accounting profession 229
  accounting standards 224
  accounting theory 229
  environmental reporting 813
  sustainable investing companies 826
  tax and reporting systems 228
  tax avoidance 394
neutrality 15, 241, 250
New York v *Ernst & Young* (2010) 187
New Zealand
  legal system 226
  tax imputation system 390
Next plc 289–90
Nike 261
Nissan 498
Nixon, B. 472–3
Nokia 362–3
nominal monetary units measurement of financial capital maintenance 27
nominal values, shares 279
non-adjusting events 68–9
non-cash transactions 106–7
non-controlling interests 553–5, 559, 570, 574
non-cumulative preference shares 279
non-current assets
  accrual accounting 24–7
  adjusting events 68
  capital reductions and 286–9
  cash flow model 14–15
  classification 43–4, 212
  consolidated accounts 586
  held for sale 431–2
  impairment 425–31
  private finance initiative 532
  revaluations to fair values on acquisition 586
  turnover ratios 690
  see also intangible assets
non-distributable reserves 276, 277, 279–80, 281
non-executive directors (NED) 794–7
non-financial liabilities 307–14
non-monetary assets 133–4, 161–2, 634, 643, 644
non-monetary items 155, 158
non-recurring items 40–2
NOPAT (net operating profit after tax) 727, 728
normalised earnings 254
normalised earnings per share 791–2
normative statements 236
Norwalk agreement 223
Norway
  accounting system 229
  environmental reporting 813
notes to financial statements 41–2, 215
Nottingham Forest Football Club Ltd 306

Novartis 719
NPV (net present value) 7, 9, 328–9
NRV *see* net realisable values
NRVA (net realisable value accounting) 150, 151–6, 173

objectives
  corporate, statements of 822
  financial reporting 755–8
  financial statements 21–2, 239–40, 680
objectivity 132, 250
obligating events 301, 303
obligations
  contingent liabilities 309, 311
  pension funds 366
  performance, revenue recognition 191, 194–9
  provisions 300–1, 303, 311
Occidental 816
OCI (other comprehensive income) 39–40
OECD *see* Organization for Economic Co-operation and Development
off-balance sheet finance 783–4
  leasing 450, 460
offers for subscription of shares 278
offsetting financial instruments 329
Oman, C. 775
omissions 70–1
onerous contracts 199, 304, 312, 534
operating activities, cash flows 100–3, 106
operating capacity capital 142
operating capital maintenance 153, 157–68
operating expenses
  classification 36–40
  ratios 688–9
operating income 40–2, 155
operating leases *see* leases
operating losses 304
operational risks 781–2
operations
  cash flows 101, 106, 108–9
  scale of 786
opportunity costs 156
options
  share-based payments 373–9
  shares 667, 670, 792
ordinary shares 276, 277–8
Organization for Economic Cooperation and Development (OECD)
  corporate governance 778, 801
  Report on Abuse of Charities for Money-laundering and Tax Evasion 394
Orion Bank of Naples 780
Orkla Group 727–8
other comprehensive income (OCI) 39–40
other operating income or expense 37
output method, construction contract earnings 533–4
over-optimism 778
overheads 36, 503–5, 507–8, 510
overstated quantities 511
own shares, purchase of 290–1

owners
  capital transactions with 43

P/E ratios see price/earnings ratios
par values, shares 279
parent companies
  equity method use in separate financial
    statements 617–18
parents, ethical influences of 248
Parmalat 788
partial imputation corporation tax system
  390
participating preference shares 276, 279
partly owned subsidiaries 553–5
partnerships, capital maintenance 280–1
PAS-score (performance analysis score)
  733
past events 301, 303, 309
past service costs, pension schemes 367–8
payments
  deferred, revenue recognition 192–3
PBIT (profit before interest and tax)
  686–7, 697
PCaW (Public Concern at Work) 263
PDF files 756, 757
PE ratios see price/earnings ratios
Pearson 825
peer comparisons, financial statement
  analysis 697–8
peer groups, ethical influences of 249
pension funds 228
pension schemes
  assets
    actual and expected returns 368–9
    changes 366–7
  current service costs 367
  curtailments 367–8
  deferred tax 402
  defined benefit schemes 362–3, 364–5,
    379–81
  defined contribution schemes 362–3,
    379, 380–1
  disclosures 371, 380–1
  employee interest 360–1
  equity compensation plans 363
  ex gratia arrangements 361
  fund obligations 366
  illustrations 366, 369–70
  interest costs 367
  liability for costs 365–9
  multi-employer plans 370
  settlements 367–8
percentage of completion method 524–5
performance
  actual 130
  bonuses, revenue recognition 192
  directors 791
  entity-wide cash-based measures 716–
    19
  future 3, 130
  inter-company 756–7
  inter-period 756–7
  key performance indicators 53
  monitoring 730
  obligations, revenue recognition 191,
    194–9

poor, rewards for 780
predicted 130
remuneration related to 730–1
stewardship 130
performance analysis score (PAS-score)
  733
Pergamon Press 212, 213
permanent differences, deferred taxation
  396
perpetual debt 327
personal ethics 248–9
Petroleo Brasileiro 719
PFI (private finance initiative) 529–32
Pfizer 278
PI NAVIGATOR 757
placings of shares 278
plant see property, plant and equipment
policies
  accounting 44–5, 70, 436–8
  depreciation 437
  dividends 130
  ethics 257–8
  uniform, consolidated accounts 568
post retirement costs 365–9
poor performance, rewards for 780
position, key performance indicators 53
positive NPV 9
possible obligations 309
potential consumption capital 142, 143
PPE see property, plant and equipment
PPP (public–private partnerships) 528–32
pre-acquisition losses 568–9
pre-acquisition profits 568–9, 586–7
predicted performance 130
prediction
  consolidated accounts 560
  income as means of 130
  orientation 16
  ratios for, caution when using 699–701
predictive values 240
preference shares see shares
Premier Foods 80
premiums, shares 279–80, 557
prepayments, emissions trading
  certificates 484
present obligations 300–1, 303, 309
present values (PV) 136–8, 142, 302, 311,
  456, 457
  see also net present value
presentation
  annual reports 819
  financial instruments 325–7
  non-current assets held for sale 431–2
presentation currencies 632, 636, 637–8,
  643
price/earnings (PE) ratios 130, 277,
  655–6, 724–5
price index systems 149–50, 154
  see also general price indices
price-level changes, accounting for
  148–75
prices
  changing 143, 235
    see also inflation; inflation
    accounting
  shares 723

PricewaterhouseCoopers (PwC) 224
pricing transactions, revenue recognition
  191–9
principles-based approaches 243–4,
  253–6, 322
prior period adjustments 42
prior-period errors 70–1
private companies
  distributable profits 282–3
  share capital reductions 284
private finance initiative (PFI) 529–32
produce, agricultural 515
professional behaviour 251
professional bodies
  ethics 249, 259–60, 262–3
  see also accounting profession
professional competence 250
professional oversight, auditors 801
professional scepticism, auditors 786–7
profit and loss accounts
  environmental 813
  see also statements of comprehensive
    income
profit before interest and tax (PBIT)
  686–7, 697
profit-sharing 372
profitable trading 811
profits
  capital profits 135
  concept of 132
  consolidated accounts 568–9, 586–7,
    583
  current purchasing power 155
  distributable 282–4
  gross, margin ratios 688
  inflation accounting 159–61
  inter-company sales 571
  inventories valuation impact on 498–9
  maximisation 253
  net see net profits
  pre-acquisition 568–9, 586–7
  private finance initiative 530–1
  realisation 135
  unrealised 135, 486, 609
  see also gains; income
projected unit credit method 366
promotion costs 36
property, plant and equipment (PPE)
  412–13
  cost of 414–16
  current cost accounting 158–9
  definition 413
  depreciation see depreciation
  disclosures 432–3
  financial statements interpretation
    436–8
  government grants 433–5
  impairment of assets 425–31
  investment properties 435–6
  measurement subsequent to initial
    recognition 423–5
  revaluations 423–5
  statements of financial position
    explanatory notes 45–6
  subsequent expenditure 415–16
  see also land; leases

proportionate consolidation 619
proportionate share method 553–5
prospective changes to financial
      statements 69, 71
provisions
   creation, decision tree 299–300
   decommissioning costs 305
   decreases 305
   deferred tax 397–9
   definition 300, 308
   ED IAS 37 approach 308
   environmental liabilities 305
   increases 305
   measurement 301–2
   onerous contracts 304
   operating losses 304
   recognition 300–1
   restructuring 304–5
   treatment 300
   unrealised profits 609
prudence 15, 135
public companies
   distributable profits 283
Public Concern at Work (PCaW) 263
public interest 251, 817, 821
Public Interest Disclosure Act 1999:
      263–4
public–private partnerships (PPP)
      528–32
published accounts
   additional entity-wide cash-based
         performance measures 716–19
   analysis 713–41
   formats for 35–40, 47–9
   inventories 512–13
   limitations for interpretation purposes
         715–16
      lack of detail 715
      unaudited information 715–16
   preparation 32–55
   reliability of financial information
         714–15
   see also annual reports; financial
         reporting; financial statements
Puma 813
purchased goodwill 475, 481, 551–2
purchasing power 150, 153
PV see present values
PwC (PricewaterhouseCoopers) 224
pyramid approach, ratios 687

qualitative characteristics 238, 240–1
quality of earnings 41
quick ratios 694, 697

R&D (research and development) 471–4
R&R Ice Cream plc 49
Rank Group 783
ratios
   acid test 694
   administrative expenses 688
   analysis
      benefits 681
      case study 682–3
      cautionary notes
      commercial databases 756–7

cost structures of companies 685
financial databases 756–7
financial structures of companies
      684–5
   horizontal 684–5
   introductory reviews 683–5
   vertical 684, 685
asset turnover 689–92, 697
brand effects on 481
cash 695
cash conversion (CCR) 111
cash dividend coverage (CDCR) 109,
      111
cash flow 694–5
current 693–4, 697
debt covenants 721–3
debt/equity 695–6, 697
debt service coverage 717, 719
distribution costs 688–9
EBITDA-based 717–18
EBITDA/interest 718, 719
ED/2010/1 potential impact on 314
EV (enterprise value)/EBITDA 717
financing 695–7
gearing 695–7
gross profit margin 688
   external factors 700
   financial performance 686–92
   interest cover 695
   internal factors 700–1
   inventory turnover 690–1, 697
   investor-specific 723–6
   liquidity 692–5
   margin 687–9
   net debt/EBITDA 717
   net profit margin 687
   non-current assets
      revaluation effects on 436
      turnover 690
   operating expenses 688–9
   peer comparisons 697–8
   predictive use 699–701
   price/earnings see price/earnings
         ratios
   profit before interest and tax (PBIT)
         686–7, 697
   pyramid approach 687
   quick 694, 697
   reports based on 698–9
   restrictive loan covenants 721–3
   return on capital employed see return
         on capital employed
   return on equity (ROE) 686
   return on investment (ROI) 686
   revaluation effects 436
   scepticism 701
   shariah compliance 719–20
   trade payables 691–2
   trade receivables 691
   volume analysis 689–92
   working capital 690–2
raw materials, inflation accounting 154
RCA (replacement cost accounting) 150,
      151–6, 502–3
real terms system, inflation accounting
      167–8

real-time financial reporting 767–8
receivables
   emissions trading certificates 484
   financial assets 332, 338
   trade 691
recognition
   agricultural produce 514–16
   assets 243
   biological assets 514–16
   expenses 243
   financial instruments 335–7, 347–8
   foreign currency transactions 633
   IASB Conceptual Framework 243
   income 243
   intangible assets 469–71
   liabilities 243
   pension assets or liabilities 367
   provisions 300–1
   share-based payments 373–4
   termination benefits 372
reconstructions 287–9
redeemable preference shares 279
redundancy costs 312
refunds, annual, revenue recognition 191
Registrar of Companies (Companies
      House), XBRL use 761
regulatory risks 782
reimbursements 311
related party disclosures 82–6
relationship capital 809
relevance 240–1, 819
relevant accounts 284
reliability 15–16, 714–15, 819
religions, ethical influences of 249
remuneration
   directors 789–92
   ethics and 252
   excessive 779
   executive 789–92
   performance criteria 730–1
rendering, XBRL 764–5
replacement cost 131
   accounting (RCA) 150, 151–6
   income 142–3
   inventory valuation 502–3
reporting
   accountants' roles 5
   entities, conceptual framework 240
   internal 4–8
   see also annual reports; financial
         reporting; published accounts
research and development (R&D) 471–4
research costs 472–3
   deferred tax 402
reservation of title
   revenue recognition 197–8
reserves
   consolidated accounts 570, 574
   current cost 159–67
   distributable 276, 277, 281
   goodwill, writing off to 475, 477, 481
   non-distributable see non-distributable
         reserves
   revaluation 43, 436
   share premiums 279–80, 557
residual values 418, 419

responsiveness 815
restrictive loan covenants 721–3
restructuring 304–5, 312, 559
retail inventory valuation method 501
retail price index (RPI) 150, 154
retention of title *see* reservation of title
retirement benefits *see* pension schemes
retirement funds 228
retrospective changes to financial
    statements 69, 70
return on assets (ROA) 796–7
return on capital employed (ROCE) 438,
    513, 560–1, 686–7
return on equity (ROE) 686, 724
return on investment (ROI) 686
revaluations
    assets 133, 402–3, 436, 560
    property, plant and equipment 423–5
    reserves 43, 436
revenue
    accrual accounting 22–3
    assets disaggregation 200
    construction contracts 525
    definition 186, 188
    disaggregation 200
    investment, unquoted companies,
        share valuations 740
    recognition 184–5
        amounts identification 189–200
        annual refunds 191
        challenges 187–8
        construction contracts 532–4
        disclosures 200–1
        dividends 186
        equally unperformed contracts 199
        financial securities 198
        IAS 18 *Revenue* 185–6, 187–8
        identifying contracts with customers
            190–1
        IFRS 15 *Revenue from Contracts
            with Customers* 185, 187–201
        intellectual property transfers 198–9
        interest 186
        issues in developing a new standard
            186–7
        judgements 201
        leases 197
        long-term contracts 197
        modification of terms of contracts
            199–200
        onerous contracts 199
        payments deferred 192–3
        performance bonuses 192
        performance obligations 191, 194–9
        pricing transactions 191–9
        reservation of title 197–8
        royalties 186
        sale of goods 185, 196
        sales and repurchases 198
        services 186, 196
        time value of money 192–3
        timing 189–200
        uncertainties 189
        warranties 196–7
    terms and conditions 201
reversal, deferred tax 397

reviews of financial statements for
    management purposes 680–701
Revsine 131, 142
right-of-use approach, leases 461–2
rights issues 278, 661–6
Rio Tinto 636
risks
    analysis using XBRL 765
    assessment 735–6
    disclosures 342–6
    environmental, social and governance
        (ESG) 825
    financial 781
    legal 782
    management, corporate governance
        781–2
    operational 781–2
    regulatory 782
    strategic 781
    taking, excessive 779
ROA (return on assets) 796–7
ROCE *see* return on capital employed
Roche Holdings, Inc. 428, 719
ROE (return on equity) 686, 724
Rohan plc 326
ROI (return on investment) 686
Rolls-Royce 262, 322, 348, 474
Romalpa clauses *see* reservation of title
Roman law legal systems 226
Roosevelt, Franklin D. 392
Royal Liverpool Hospital 528
royalties, revenue recognition 186
RPI (retail price index) 150, 154
rule of 78 method, finance lease
    accounting 455
rules-based approaches 243–4, 253–6,
    322
Russia, corporate governance 778
Ryanair 739

SABMiller 470–1
Sainsbury's 501, 718
salaries *see* remuneration
sales
    cost of 36, 159–60
    of goods, revenue recognition 185, 196
    inter-company 571
    and repurchases, revenue recognition
        198
SAM 826
Samsung Electronics 719
Sarbanes-Oxley Act 2002, USA 222–3,
    260, 792
satisfaction 138
scale of operations 786
scepticism
    auditors 786–7
    ratios 701
screening, Shariah principles 720
second opinions, ethics and 252
Securities and Exchange Commission
    (SEC)
    enforcement by 222
    reason for establishment 229, 234
    XBRL and 761, 766
segment performance 717

segment reporting 71–8, 107
self-interest threats 253
self-review threats 253
selling
    costs 36
    overheads 503, 504
seniors' perspectives of ethics 257
separate financial statements 616, 617–8
Serious Fraud Office (SFO) 262
Seruga, Deanna 779
services, revenue recognition 186, 196
set-off of inter-company balances 573
settlements, pension schemes 367–8
Seventh Directive 219
SFAS *see* Statements of Financial
    Accounting Standards
SFO (Serious Fraud Office) 262
share-based payments 373–9
share capital
    consolidated accounts 570, 574
    minimum 282
    paid-in, repayment of part 289–90
    reduction 284–91
    unpaid, cancellation 289–90
    see also capital maintenance
shareholders
    activism 797–801
    capital reduction and 285–9
    earnings per share and 657–8
    funds 165, 277–9, 481
    information needs 3–4, 714–15
    interests 816
    orientation, ethics and 55
    preference, capital reductions and 287
    private finance initiative 530–1
    returns 729–31
    total funds 277–9
    total return (TSR) *see* total shareholder
        return
    value (SV) 729
shares
    accounting entries 279–80
    beneficial ownership statistics 797
    buybacks 290–1, 661
    exchanges 557
    issuing 43, 276–7, 278–80
    new issues at full market value 660
    nominal values 279
    offers for subscription 278
    options 667, 670, 792
    ordinary 276, 277–8
    own, purchase of 290–1
    par values 279
    placings 278
    preference 276, 277, 279, 287, 290,
        325–7, 667–8
    premiums 279–80, 557
    prices 723, 792
    rights issues 278, 661–6
    splits 659–60
    treasury shares 290–1
    warrants 667
    see also earnings: per share; share
        capital
shariah compliance 719–20
short-term employee benefits 371–2

Siemens 719
significant influence 605–6
Singapore
    annual report disclosures 473
    Code of Corporate Governance 792
    XBRL 761
small and medium-sized enterprises
        (SME)
    borrowing costs 415
    definition 225
    failure prediction and 733
    IFRS for see International Financial
        Reporting Standards: IFRS for
        SMEs
    regulatory burden on 220–2
    reporting requirements 224–5
    Valuation Index 737
small companies see small and
        medium-sized enterprises
SME see small and medium-sized
        enterprises
Smith, Adam 253–4
SOCE (statements of changes in equity)
        42–3, 240, 585
social accounting 820–4
Social and Environmental Accounting
        Research, Centre for (CSEAR)
        827
social capital 809
social dimension 814, 818
social groups, ethical influences of 249
social issues, key performance indicators
        54
social risks
    environmental, social and governance
        (ESG) risks 825
Socially Responsible Investing (SRI) 825
    ABI Guidelines 825–6
sole traders, capital maintenance 280–1
solvency 100
South Korea, sustainable investing
        companies 826
sovereign wealth funds 228
SOX see Sarbanes-Oxley Act 2002
Spacek, Leonard 250
Spain, sustainable investing companies
        826
Special Purpose Entities (SPE) 783–4,
        788
Spirit Pub Company plc 304
splits, shares 659–60
SRCD (Summary Reporting Contextual
        Data) 766–7
SRI see Socially Responsible Investing
SSAP see Statements of Standard
        Accounting Practice
SSL International 434
stakeholders
    company accountability to 810–11
    corporate governance 776
    interests 816
    social accounting 820
    using financial reports, unethical
        behaviour implications for 258–63
    XBRL views 761
Standard & Poor's 735

standard cost, inventory valuation 501
standards see accounting standards;
        Financial Reporting Standards;
        International Accounting
        Standards; International Financial
        Reporting Standards; Statements
        of Financial Accounting
        Standards; Statements of
        Standard Accounting Practice
statements of cash flows see cash flow
        statements
statements of changes in equity (SOCE)
        42–3, 240, 585
statements of comprehensive income
    accrual accounting 23
    cash flow statements preparation when
        not available 115–16
    consolidated accounts 582–92, 608,
        611, 639, 640
    construction contracts 536, 537, 538
    current cost accounting 165, 168
    decision-usefulness 42
    deferred taxation 398
    discontinued operations 79–80
    foreign currencies and 639, 640
    formats 589
    hyperinflation 644
    inflation accounting 159–61
    information needs 239
    internal 32–5
    leases 453–4, 455–6, 457
    as linking statements 134
    pension schemes 367, 370
    private finance initiative 531–2
    real terms system 168
    tax expense 395
statements of corporate objectives 822
Statements of Financial Accounting
        Standards (SFAS) (USA)
    109 Accounting for Income Taxes 400
    128 Earnings per Share 672
    131 Disclosures about Segments of an
        Enterprise and Related Information
        77
statements of financial position 43–7
    accrual accounting 23–6
    brands 481
    cash flow accounting 12–14
    consolidated accounts 568–74, 607–8,
        610
    construction contracts 527–8, 536, 537,
        539
    current cost accounting 157–8, 162–3,
        166–7
    current purchasing power 155
    depreciation effects 437
    emissions trading certificates 483
    foreign currencies and 639, 641
    goodwill, keeping in unchanged 475,
        477
    hyperinflation 644
    inflation accounting 157–8, 162–3,
        166–7
    information needs 239
    leases 453, 454–5, 459
    opening, restating to current cost 157–9

pension schemes 370
    tax expense 395
    see also off-balance sheet finance
statements of future prospects 822
statements of income see statements of
        comprehensive income
Statements of Standard Accounting
        Practice (SSAP)
    5 Accounting for Value Added Tax 406
    7 Accounting for Changes in the
        Purchasing Power of Money 170
    16 Current Cost Accounting 170
stewardship 12–13, 16, 21–2, 130, 174,
        214, 755–6, 799–800, 801
stock see inventories; shares
stock analysts 780
Stohl, C. 261
straight-line methods
    amortisation of goodwill 478
    depreciation 420, 423
    finance lease accounting 455
strategic mistakes 734
Strategic Reports 53–4, 109
strategic risks 781
Structural Dynamics Research
        Corporation 787
students, ethics and 257, 266–7
stylesheets, XBRL 762–3
subsidiaries
    acquired during year 587–9, 590–1
    disclosures 617
    excluded 558–9
    partly owned 553–5
    wholly owned 549, 551–3
    see also acquisitions; business
        combinations; consolidated
        accounts
substance of transactions 254, 816
substance over form 16, 325, 405,
        451–2
sum of digits method, finance lease
        accounting 455
sum of the units method, depreciation
        421
Summary Reporting Contextual Data
        (SRCD) 766–7
Sunshine Holdings 3 Ltd 722
superfairness 255
supplier-related intangible assets 469
suppliers
    key performance indicators 54
sustainability 808–28
Sustainable Development, World
        Business Council for (WBCSD)
        811
sustainable earnings 254
SV (shareholder value) 729
Sweden, environmental reporting 813,
        814
Switzerland, sustainable investing
        companies 826
symptoms of failure 734–5

T J Hughes 509
TAAR (targeted anti-avoidance rules)
        392

Taffler's Z-score 731, 732–3
tags, XBRL 758–60
takeovers *see* acquisitions; business combinations; consolidated accounts
Tanzania 262
targeted anti-avoidance rules (TAAR) 392
targets, ethical codes and, conflicts between 261–2
tax havens 393
taxation
    accounting basis required 16
    avoidance 391–4
    corporation tax 388–94
    current 37–8, 395–6
    deferred 396–404
    evasion 391–4
    income measurement 130–2
    national differences 228
    value added tax 406
taxonomies, XBRL 762, 763
technical circulars 236
technical expertise 4
technology-related intangible assets 469
Technotrans 306
temporary differences, deferred taxation 397–9, 400
tendering, construction contracts 526
termination benefits 372–3
termination costs 312
Tesco 76–7, 718
Thameslink 528
threats to compliance with fundamental principles 253
ThyssenKrupp 730
time value of money 192–3
timeliness 241
timing differences, deferred taxation 396–9, 400
timing of revenue recognition 189–200
title, reservation of *see* reservation of title
total owners' equity 277–8
TOTAL SA 719
total shareholder return (TSR) 726, 729–31, 792, 812
total shareholders' funds 277–9
trade creditors *see* creditors
trade payables 276
    capital reductions and 286–7
    ratios 691–2
trade receivables ratios 691
trading
    conditions 786
    losses, capital reduction 284–9
    profitably 811
    related parties 86
trainees' perspectives on ethics 257
training, ethics and 266–7
transactions
    foreign currencies 632–45
    related parties 86
    substance of 254, 816
transferable risks 781
translation, foreign currencies 632, 633–6, 638, 642–3

transparency
    ED/2010/1 potential impact on 314
transport costs 36
Treadway Commission *see* Committee of Sponsoring Organizations of the Treadway Commission
treasury shares 290–1
trial balances 32–5
true and fair view 50–1, 254, 255
TSR *see* total shareholder return
Tullow Oil 637
turnover
    asset turnover ratios 689–92, 697
    non-current asset turnover ratios 690
Tweedie, Sir David 300
Tyco 788
Type A leases 461–2
Type B leases 461, 462

UK Corporate Governance Code 792, 793–802
unaudited information, published account limitations 715–16
uncertainties
    revenue recognition 189
    see also provisions
understandability 241
understated quantities 511
unethical behaviour 258–63, 779–80
Unilever 86, 730
unincorporated businesses, capital maintenance 280–1
United Kingdom
    accounting profession 229
    accounting theory 229
    agricultural accounting 516
    Bribery Act 2010: 777
    corporate governance 777, 778
    executive remuneration 789–92
    The implementation of IFRS in the UK, Italy and Ireland (ICAS research report 2008) 224
    legal system 226–9
    mandatory standards 212–14
    own shares, buyback 291
    regulatory framework 215–17
    Treasury shares 291
    UK Corporate Governance Code 792, 793–802
    XBRL 761
United States
    accounting profession 229
    accounting standards 229, 479
    accounting theory 229
    American Institute of Certified Public Accountants (AICPA) 764, 767
    Certified Public Accountants (CPA) 266
    corporate governance 777, 778
    Department of Justice 262
    finance of industry 227
    Foreign Corrupt Practices Act 1977: 777
    ·GAAP 222, 223, 256, 322–3, 761, 764
    IFRS and 223

international standards, progress towards adoption 223
    legal system 226–9
    own shares, buyback 291
    standard setting and enforcement 221–3
    tax and reporting systems 228
    XBRL 761
    see also Financial Accounting Standards Board; Securities and Exchange Commission; Statements of Financial Accounting Standards
unquoted companies
    share valuations 736–41
    see also private companies
unrealised gains 486
unrealised profits 135, 486, 609
useful economic lives 416–17, 418–19, 437, 470–1
usefulness 819
user needs, financial statements 240

valuation accounting 172
Valuation Index, SMEs 737
valuations
    emissions trading certificates 483, 484
    errors 71
    historical cost, agricultural activities 515–16
    intangible assets 486, 487
    inventories *see* inventories
    pension schemes 380
    see also revaluations
value added statements 822–4
value added tax (VAT) 406
values
    accountants' view 132–6
    to the business 170
    capital 142–3
    current 169–71
    current value accounting (CVA) 149
    determining value 726–31
    economists' view 136–42
    ethical, implications for principles versus rules-based approaches to accounting standards 253–6
    future 137
    residual 418, 419
    shareholder value (SV) 729
    of unconsumed costs of assets 134
    in use 427–8, 430
    see also fair values; net realisable values; present values
VAT (value added tax) 406
verifiability 241
vertical analysis 684, 685
vested interests 212
vesting dates, share-based payments 375
Virgin Atlantic 265
Vodacom 756
Vodafone 476, 661
volatility 224, 714
volume analysis, ratios 689–92

WACC (weighted average cost of capital) 727–8

Walker Review of Corporate Governance of UK Banking Industry 799
warehousing costs 36
warranties, revenue recognition 196–7
warrants, shares 667
Wassenan 224
WBCSD (World Business Council for Sustainable Development) 811
wealth distribution 212
weighted average cost of capital (WACC) 727–8
whistleblowing 263–5
wholly owned subsidiaries 549, 551–3
wider markets initiative (WMI) 529
Wienerberger 718

Williams, Jerry 780
WIP (work-in-progress) 506–8
WMI (wider markets initiative) 529
work-in-progress (WIP) 506–8
working capital 13–14, 100, 102, 103, 106, 108, 109, 690–2
World Business Council for Sustainable Development (WBCSD) 811
World Health Alternatives 779
world wide web *see* Internet
WorldCom 734
WPP 482
Wyatt, Arthur 788

XBRL (eXtensible Business Reporting Language) 756, 757–68

XBRL Foundation Certificate Program 763
XML (eXtensible Mark-up Language) 758–9

year-end adjustments 34–5
year-end manipulations 509–10
yields
  dividends 726
  earnings 725
  unquoted companies, share valuations 737

Z-scores 731–3, 740
Zeta model 732